Design Challenges

At the beginning of each chapter the reader is challenged by a design problem that can be solved using the material presented in that chapter. A solution to the design problem is presented at the end of the chapter. The **Problem Solving Method** illustrated by the flow chart shown below is used to solve each Design Challenge.

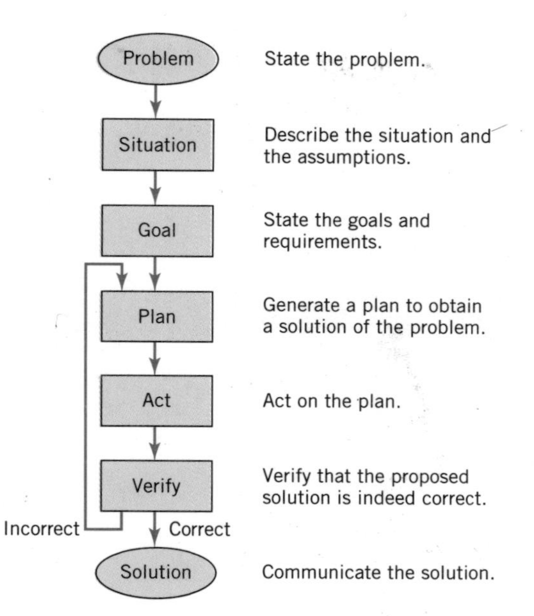
Problem
State the problem.
Situation
Describe the situation and the assumptions.
Goal
State the goals and requirements.
Plan
Generate a plan to obtain a solution of the problem.
Act
Act on the plan.
Verify
Verify that the proposed solution is indeed correct.
Incorrect
Correct
Solution
Communicate the solution.

The Problem Solving Method

Here is a list of the Design Challenges:

Chapter 1 Jet Valve Controller
Chapter 2 Temperature Sensor
Chapter 3 Adjustable Voltage Source
Chapter 4 Potentiometer Angle Display
Chapter 5 Strain Gauge Bridge
Chapter 6 Transducer Interface Circuit
Chapter 7 Integrator and Switch
Chapter 8 Computer and Printer
Chapter 9 Auto Airbag Igniter
Chapter 10 Op Amp Circuit
Chapter 11 Maximum Power Transfer
Chapter 12 Power Factor Correction
Chapter 13 Radio Tuner
Chapter 14 Space Shuttle Cargo Door
Chapter 15 DC Power Supply
Chapter 16 Anti-Aliasing Filter
Chapter 17 Transistor Amplifier

Introduction to
ELECTRIC CIRCUITS

ACQUISITIONS EDITOR: BILL ZOBRIST
MARKETING MANAGER: KATHERINE HEPBURN
SENIOR PRODUCTION EDITOR: PATRICIA MCFADDEN
COVER DESIGN: NORM CHRISTIANSEN
INTERIOR DESIGN: MICHAEL JUNG
COVER PHOTOGRAPH: PROVIDED BY AALOG DEVICES, INC., NORWOOD, MASS.
ILLUSTRATION EDITOR: SIGMUND MALINOWSKI
ELECTRONIC ILLUSTRATIONS: WELLINGTON STUDIO

This book was set in Times Roman by York Graphic Services and printed and bound by Von Hoffmann Press. The cover was printed by Phoenix Color Corp.

This book is printed on acid-free paper.♾

The paper in this book was manufactured by a mill whose forest management programs include sustained yield harvesting of its timberlands. Sustained yield harvesting principles ensure that the numbers of trees cut each year does not exceed the amount of new growth.

Library of Congress Cataloging-in-Publication Data

Dorf, Richard C.
Introduction to electric circuits / Richard C. Dorf, James A. Svoboda.--5th ed.
p. cm.
ISBN 0-471-38689-8 (alk. paper)
1. Electric circuits. I. Svoboda, James A. II. Title.

TK454 .D67 2001
621.319′24--dc21 00-042290

ISBN 0-471-38689-8

Printed in the United States of America

10 9 8 7 6 5

Introduction to

ELECTRIC CIRCUITS

Fifth Edition

Richard C. Dorf
University of California, Davis

James A. Svoboda
Clarkson University

John Wiley & Sons, Inc.
New York • Chichester • Brisbane • Toronto • Singapore

The scientific nature of the ordinary man
Is to go on out and do the best he can.

—John Prine

But Captain, I cannot change the laws of physics.

—Lt. Cmdr. Montgomery Scott (Scotty), USS Enterprise

Dedicated to our grandchildren:

Ian Christopher Boilard, Kyle Everett Schafer, and Graham Henry Schafer

and

Heather Lynn Svoboda, James Hugh Svoboda, and Jacob Arthur Leis

ABOUT THE AUTHORS

Richard C. Dorf, professor of electrical and computer engineering at the University of California, Davis, teaches graduate and undergraduate courses in electrical engineering in the fields of circuits and control systems. He earned a Ph.D. in electrical engineering from the U.S. Naval Postgraduate School, an M.S. from the University of Colorado, and a B.S. from Clarkson University. Highly concerned with the discipline of electrical engineering and its wide value to social and economic needs, he has written and lectured internationally on the contributions and advances in electrical engineering.

Professor Dorf has extensive experience with education and industry and is professionally active in the fields of robotics, automation, electric circuits, and communications. He has served as a visiting professor at the University of Edinburgh, Scotland; The Massachusetts Institute of Technology; Stanford University; and the University of California, Berkeley.

A Fellow of the Institute of Electrical and Electronic Engineers, Dr. Dorf is widely known to the profession for his *Modern Control Systems*, Eighth Edition (Addison-Wesley, 1998) and *The International Encyclopedia of Robotics* (Wiley, 1988). Dr. Dorf is also the coauthor of *Circuits, Devices and Systems* (with Ralph Smith), Fifth Edition (Wiley, 1992). Dr. Dorf edited the widely used *Electrical Engineering Handbook*, Second Edition (CRC Press and IEEE Press) published in 1997.

James A. Svoboda is an associate professor of electrical and computer engineering at Clarkson University where he teaches courses on topics such as circuits, electronics, and computer programming. He earned a Ph.D. in electrical engineering from the University of Wisconsin, Madison, an M.S. from the University of Colorado, and a B.S. from General Motors Institute.

Sophomore Circuits is one of Professor Svoboda's favorite courses. He has taught this course to 2500 undergraduates at Clarkson University over the past 21 years. In 1986, he received Clarkson University's Distinguished Teaching Award.

Professor Svoboda has written several research papers describing the advantages of using nullors to model electric circuits for computer analysis. He is interested in the way technology affects engineering education and has developed several software packages for use in Sophomore Circuits.

Professor Svoboda's email address is svoboda@clarkson.edu. His spot on the internet is located at http://www.clarkson.edu/~svoboda/.

Preface

The central theme of *Introduction to Electric Circuits* is the concept that electric circuits are part of the basic fabric of modern technology. Given this theme, we endeavor to show how the analysis and design of electric circuits are inseparably intertwined with the ability of the engineer to design complex electronic, communication, computer, and control systems as well as consumer products.

APPROACH

This book is designed for a one- to three-semester course in electric circuits or linear circuit analysis. The presentation is geared to readers who are being exposed to the basic concepts of electric circuits for the first time, and the scope of the work is broad. Students should come to the course with a basic knowledge of differential and integral calculus.

The emphasis on circuit design is enhanced in this edition by a design challenge that is introduced at the start of each chapter. The design of circuits is a unique advantage of this book.

The text is designed with maximum flexibility in mind. A flowchart immediately following this preface demonstrates alternative chapter organizations that can accommodate different course outlines without disrupting continuity.

Since this book is designed for introductory courses, a major effort has been made to provide the history as well as the current motivation for each topic. Thus, circuits are shown to be the results of real invention and the answers to real needs in industry, the office, and the home. Although the tools of electric circuit analysis may be partially abstract, electric circuits are the building blocks of modern society. The analysis and design of electric circuits are critical skills for all engineers.

CHANGES IN THE FIFTH EDITION

The fifth edition introduces the "Electronic Teaching Assistant" (ETA), a set of interactive exercises and examples. The ETA consists of the Electric Circuits Workout, Circuit Design Lab, and the Interactive Illustrations.

The Electric Circuits Workout provides an opportunity to develop and practice circuit analysis skills by providing exercises similar to quiz or exam problems. The Electric Circuits Workout poses a problem, then accepts and checks the user's answer. A calculator is provided, as well as help screens that refer to appropriate sections of this text. At any time, the user can ask to see the answer to the problem that has been posed or ask for a new problem. A scorecard measures the student's progress. Figure P-1 shows a screen shot from the Electric Circuits Workout.

Workout

The Circuit Design Lab provides an opportunity for experimentation and design. Each lab exercise presents a circuit together with scrollbars that control the values of important circuit parameters. This provides an opportunity for "What if..." explorations as the effects of changing parameter values are observed. The Circuit Design Lab also provides an opportunity for circuit design exercises. Circuit performance specifications are given, and the student is invited to determine parameter values required to satisfy those specifications. Figure P-2 shows a screen shot from the Circuit Design Lab.

Lab

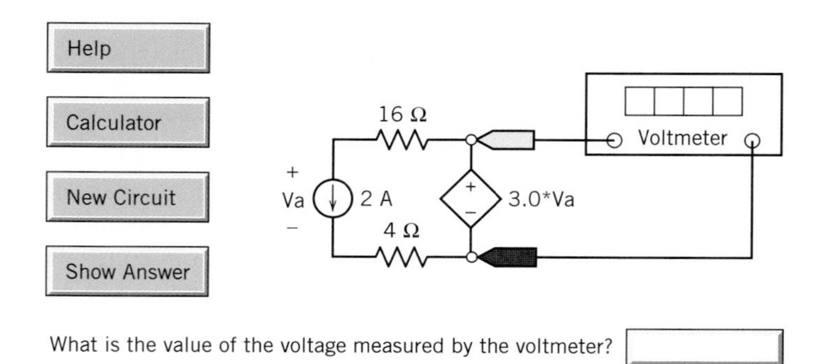

Figure P-1
Screen shot from Electric Circuits Workout.

The Interactive Illustrations provide an opportunity to change some aspect of an illustration and observe the consequence. Such illustrations emphasize the connection between concepts. For example, one of the Interactive Illustrations (Figure P-3) shows a phasor and the corresponding sinusoid (Chapter 10). The student interacts with the illustration by changing the phasor using the computer mouse. The Interactive Illustration responds by changing the sinusoid appropriately, emphasizing the relationship between phasor and sinusoid.

The ETA consists of html pages containing Java applets and so can be viewed using any Java equipped browser, e.g., Netscape 4.0 or Explorer 4.0.

Examples and exercises in the text have been revised to illustrate solution of problems from the ETA.

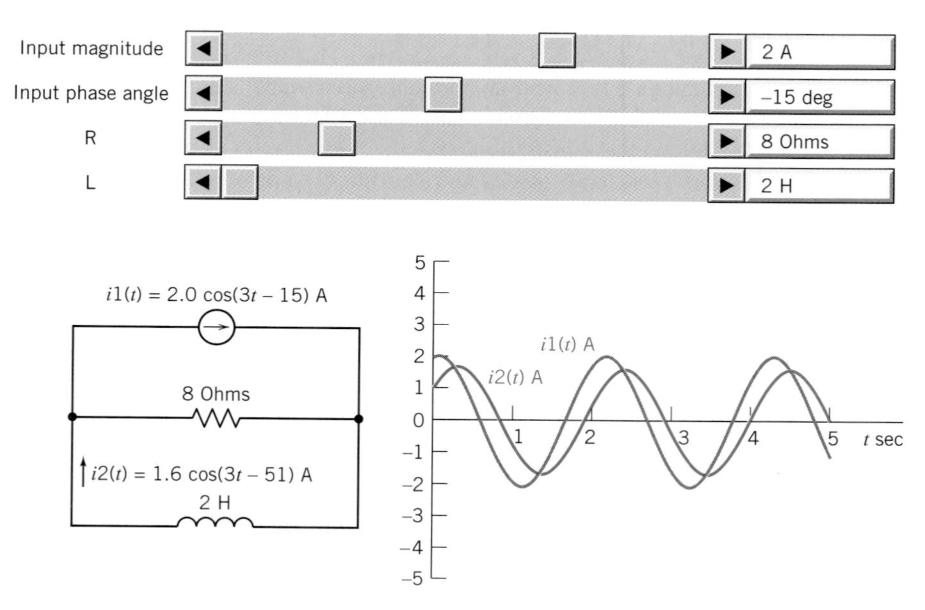

Figure P-2
Screen shot from Circuit Design Lab.

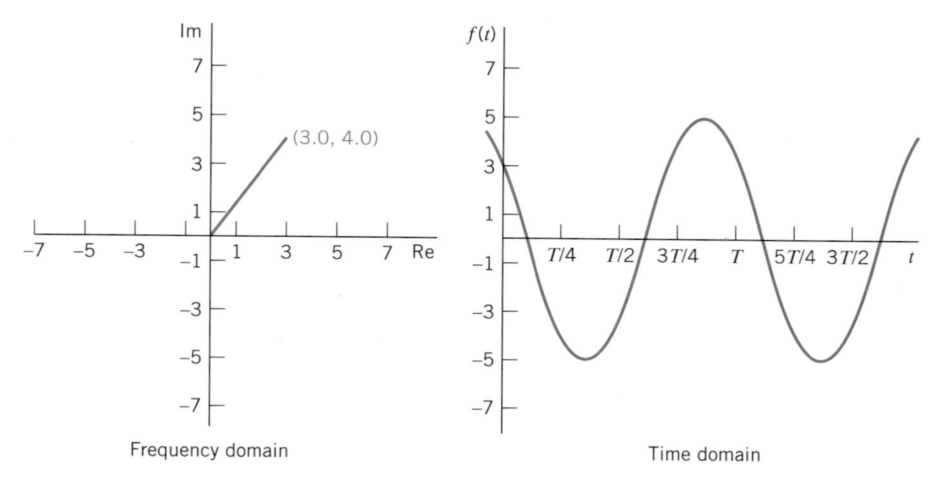

Figure P-3
Screen shot from Interactive Illustrations.

ALTERNATIVE CHAPTER ORGANIZATIONS

The flowchart shown on pages x–xi illustrates the alternate chapter organizations that can accommodate different course outlines without disrupting continuity.

CHAPTER FEATURES

Preview Each chapter opens with a preview outlining the content and objective of the chapter.

Design Challenge Following the preview, the reader is challenged by a design problem that can only be solved by utilizing the material presented in the chapter. The student is asked to consider the challenge and then proceed to read the content of the chapter.

Design Challenge Solution At the end of each chapter, the Design Challenge is solved utilizing the chapter's methods and approach. The design method utilizes a step-by-step methodology for solving design problems. The steps of the methodology are (1) state the problem, (2) describe the situation, (3) state the goal, (4) generate a plan, (5) act on the plan, and (6) verify the proposed solution.

Historical Vignettes A section of each chapter is dedicated to one aspect of electrical engineering history and current practice. This enables the reader to witness past and current motivations for the development of modern electrical engineering devices and methods as well as to grasp the excitement of engineering in the 2000s.

Illustrative Examples Because this book is oriented toward gaining expertise in problem solving, we have included more than 200 illustrative examples. Here, as in the end-of-chapter problems, the derivation of results is precise from a mathematical standpoint, but the solution of practical problems is emphasized.

Summary Each chapter includes a summary of concepts covered in that chapter.

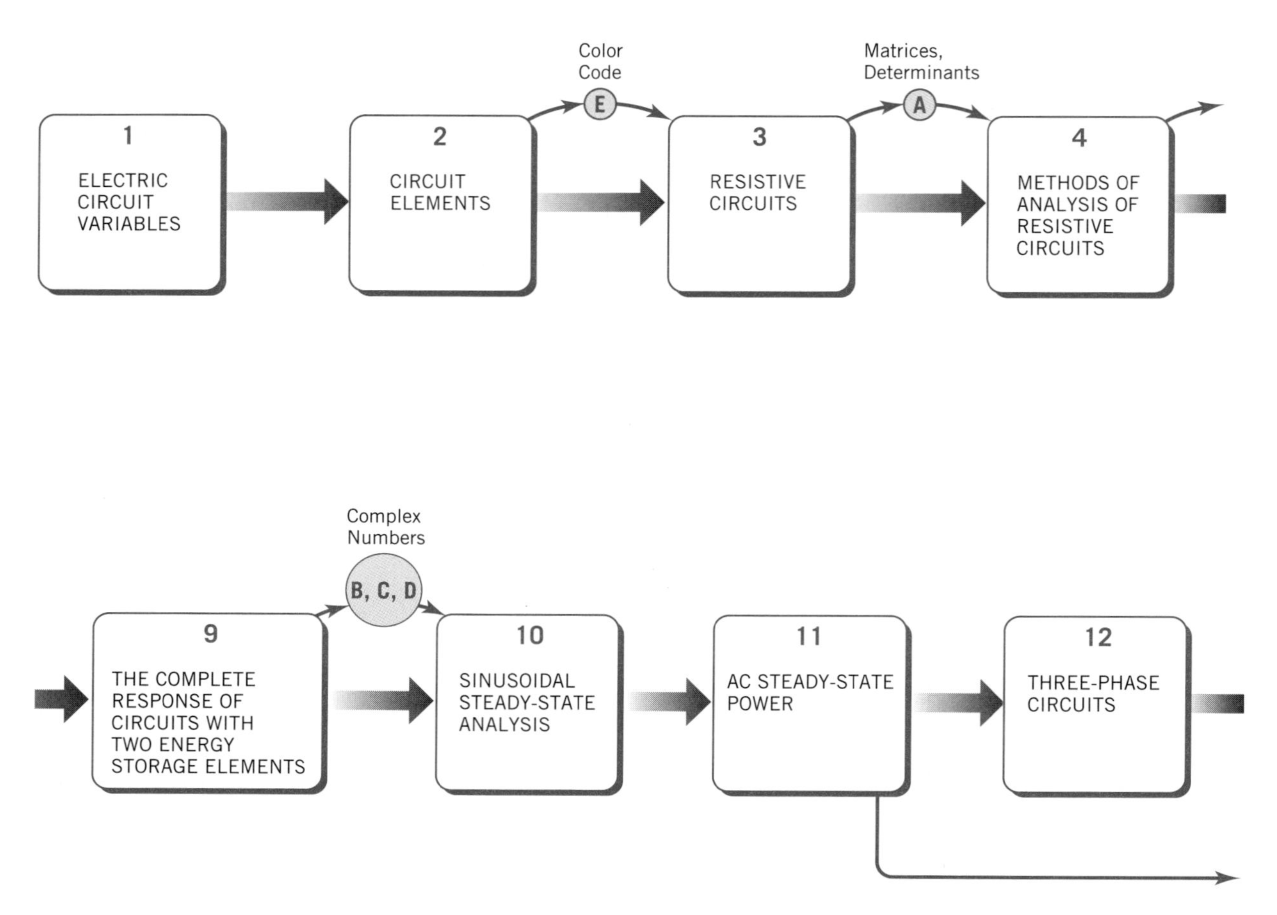

Exercises and Problems The exercises and problems are organized into five types

- Exercises provide the students with factual review; answers are provided immediately following the exercise.
- Problems serve as extensions of the Illustrative Examples within the chapter; answers to selected problems are provided in the text, and other answers can be found in the Instructor's Manual.

- Verification problems challenge the reader to consider a problem with a stated set of results. The reader must then verify the correctness of the results provided in the problem statement.
- Design problems, written with ABET accreditation standards in mind, provide practice in applying the material to interesting design situations. For example, students design a car airbag ignition circuit and an electric fence to deflect sharks from the beach. These open-ended problems vary in level of sophistication.
- Optional PSpice problems encourage computer use.
- Optional sections illustrate the use of the computer program MATLAB for analyzing electric circuits.

There are more than 300 exercises and 800 homework problems. The reader should first consider the exercises and then attempt selected problems. Answers are liberally provided to the readers, enabling a check of the solutions. Answers are available for all exercises.

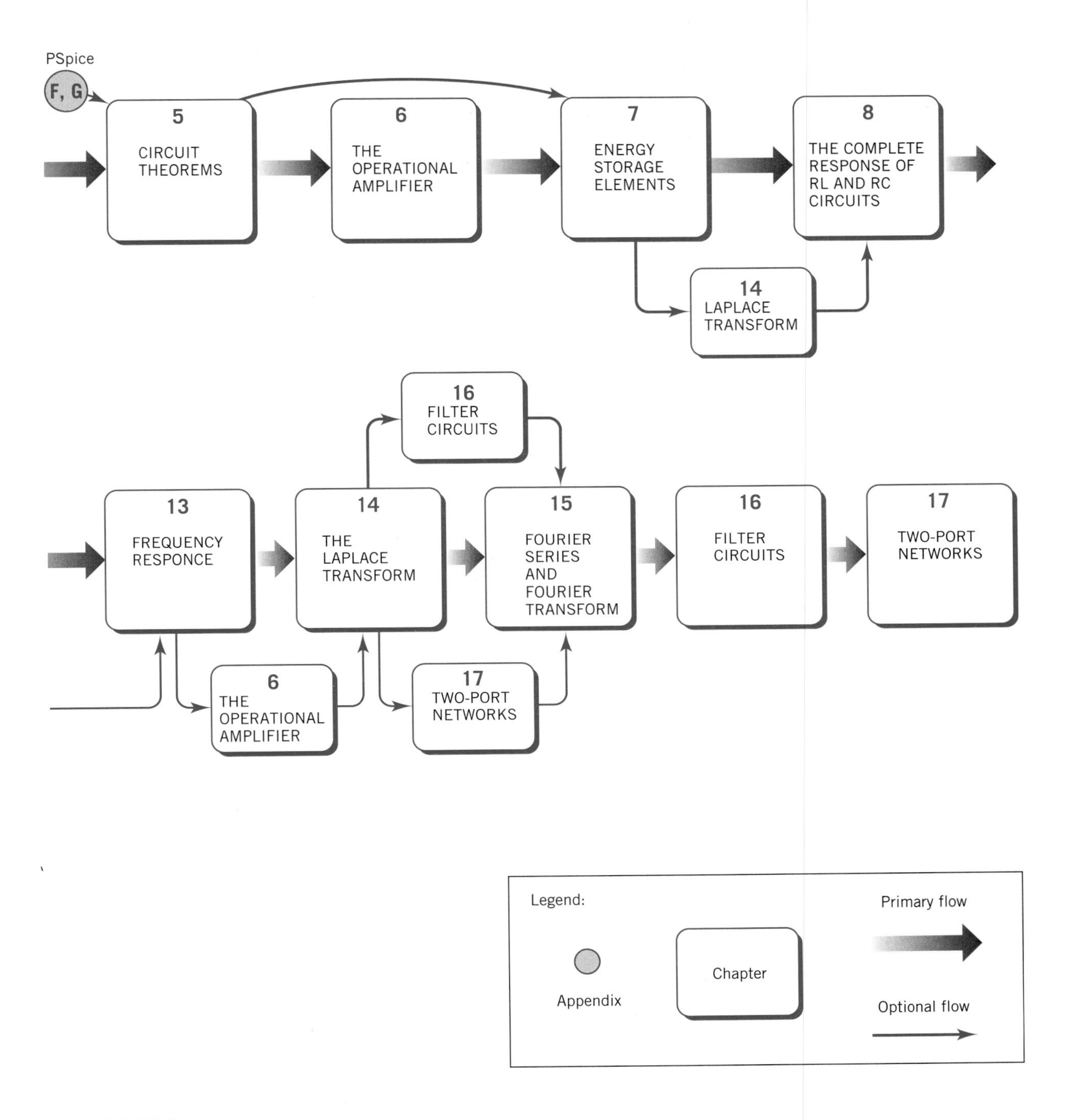

CONTENTS

The book begins with the development of electric circuit elements and the electric variables used to describe them. Then resistive circuits are studied in Chapters 1–5 to provide an in-depth introduction to the concept of a circuit and its analysis. Chapter 6 focuses on the widely used electronic element, the operational amplifier. The large number of useful circuits that can be constructed by using the operational amplifier with resistors and capacitors is demonstrated.

Next the many useful theorems and principles developed for insightful analysis of electric circuits are considered. Then the discussion moves to a description of energy storage elements—the inductor and the capacitor—which are widely used in industrial and office systems and in consumer products, such as the television receiver or the automobile.

Chapter 8 begins the study of the time response of electric circuits that incorporate energy storage elements. We also introduce the concept of switched circuits, which play a very important role in modern circuits. Chapter 9 is concerned with the study of the time response of circuits with two or more energy storage elements, which are described by differential equations.

In Chapter 10, the discussion focuses on the steady-state behavior of circuits with sinusoidal sources and develops the concepts of the phasor transform and impedance.

Chapter 11 is concerned with the study of ac steady-state power. In this chapter we also introduce the electric transformer, another important circuit device. In Chapter 12 three-phase circuits and their general utility in modern power systems are studied.

Chapter 13 provides an introduction to the concept of frequency response of a circuit and the utility of Bode diagrams. Then in Chapter 14 the study of the Laplace transform and its use in the analysis of electric circuits are considered.

Chapter 15 provides an introduction to the concept of describing signals by the Fourier series or Fourier Transform and their use in circuit analysis and design. Chapter 16 provides the background needed to design active filter circuits.

Chapter 17 is concerned with the description of two-port networks, the T-to-TT transformation, and the interconnection of circuits.

The appendices provide a review of matrices, determinants, and complex numbers. Two appendices describe Euler's formula and the standard resistor color code. An extensive appendix describes the analysis and system program, PSpice®, which can be used to obtain the value of many variables in a circuit and to verify a solution obtained analytically. Finally, a new appendix describes the Design Center™, the windows version of PSpice®.

SUPPLEMENTS

The almost ubiquitous use of computers and the web has provided for an exciting opportunity to rethink supplementary material. The supplements available have been greatly enhanced:

Student Resource CD Bound into each text, this CD contains the complete Electronic Teaching Assistant (ETA). The ETA's Electronic Circuits Workout, Circuit Design Lab, and Interactive Illustrations are linked to appropriate topics and questions throughout the fifth edition text. An updated version of the *MATLAB Tutorial* by Gary Ybarra and Michael Gustafson of Duke University is also included. The Tutorial builds upon the MATLAB examples in the text. By providing these additional examples, the authors show how this powerful tool is easily utilized in appropriate areas of introductory circuit analysis. A total of ten example problems are created in HTML.

Websites Additional resources can be found on both the John Wiley & Sons text website at www.wiley.com/college/dorf and the Circuits Extra website at www.wiley.com/college/circuits. Specific to this text, Electronics Workbench® v.6 Student Edition on the Student Companion Site has been integrated to include the Design Challenges within every chapter, and these circuits will be available for download. This leading schematic capture and simulation tool allows students to design and simulate their own circuits—fast! The complete ETAs and *MATLAB Tutorial* will also be available on the Dorf website. Circuits Extra is a general circuits resource for use with any text. It provides a variety of tools and links that enhance a student's understanding of the first circuits course.

The Instructor Companion Site includes the Solutions Manual for the text, PowerPoint slides for all of the in-chapter figures, a sample syllabus, and instructions on how to incorporate ETA Workout problems into exams and quizzes. These resources are available only to instructors who have adopted the text for classroom use. Visit the web site to register for a password.

Matlab Ordering information for the Student Edition of MATLAB® is available from the MathWorks web page located at http://www.mathworks.com/.

PSpice An evaluation version of PSpice® is available, free of charge, from OrCAD MicroSim. The evaluation version of PSpice® can be requested using the OrCAD MicroSim web page located at http://www.microsim.com/.

ACKNOWLEDGMENTS

We are grateful to many people whose efforts have gone into the making of this textbook. We are especially grateful to our editor Bill Zobrist and marketing manager Katherine Hepburn for their support and enthusiasm. We are grateful to Pamela Kennedy, Patricia McFadden, Sigmund Malinowski, and Jeanine Furino for their efforts in producing this textbook. We wish to thank Jennifer Welter, Penny Perrotto, and Ray Paquin for their significant contributions to this project.

We gratefully acknowledge the contribution of Joseph Tront in the development of the original content used in Appendix F on PSpice. We wish to thank reviewers who provided many useful suggestions for improvement. We specifically thank the following individuals:

Jonny Anderson, University of Washington
Daniel Bukofzer, California State University, Fresno
Mauro Caputo, Hofstra University
Yu Chang, Union College
Sherif Embabi, Texas A&M University
Dennis Fitzgerald, California State Polytechnic, Pomona
Gary Ford, University of California, Davis
Prashant Gandhi, Villanova University
Charles Gauder, University of Dayton
Ed Gerber, Drexel University
Victor Gerez, Montana State University
Paul Gordy, Tidewater Community College
Cliff Griggs, Rose Hulman Institute
Nazli Gundes, University of California, Davis
Robert Herrick, Purdue University
DeVerls Humphreys, Brigham Young University
Aziz Inan, University of Portland
Ashok Iyer, University of Nevada, Las Vegas
J. Michael Jacob, Purdue University
Rich Johnson, Lawrence Tech University
Ravindra Joshi, Old Dominion University
Gerald Kane, University of Tulsa
Steven Kaprielian, Lafayette College
Richard Klafter, Temple University
Joseph Kozikowski, Villanova University
Robert Laramore, Purdue University
Haniph Latchman, University of Florida
Gerald Lemay, University of Massachusetts, Dartmouth
Gary Lipton, Lafayette College
Steve McFee, McGill University
Robert Miller, Virginia Tech
David Moffatt, Ohio State University
Paul Murray, Mississippi State University
Burkes Oakley, University of Illinois, Urbana-Champaign
Anthony J. A. Oxtoby, Purdue University

Peng, Central Michigan University
rkins, University of Illinois, Urbana-
ampaign
os Saiedpazouki, Villanova University
ar Sanchez, Texas A&M University
omas Schubert, University of San
Diego
avid Skitek, University of Missouri,
Kansas City
Charles Smith, University of Mississippi
Jerry Suran, University of California,
Davis
Arthur Sutton, California State Polytechnic
University
William Sutton, George Mason University
Xiao-Bang Xu, Clemson University
Mark Yoder, Rose Hulman University
Gary Ybarra, Duke University
Barrett Hazeltine, Brown University
Steven Schennum, Gonzaga University
Glenn Swift, University of Manitoba
Mesut Muslu, University of Wisconsin
Peddapullaiah Sunnuti, Rutgers University
Y. P. Kakad, University of North Carolina
David deWolf, Virginia Polytechnic
Institute
Roobik Gharabagi, St. Louis University—
Parks College
Michael Werter, University of California,
Los Angeles
Said Ahmed-Zaid, Boise State University
David Luneau, University of Arkansas,
Little Rock
Mark Jupina, Villanova University
Pierra Tremblay, University Laval
Wilfrido Moreno, University of South
Florida
Jay Harris, San Diego State University
Sohrab Rabii, University of Pennsylvania
Sung-won Park, Texas A&M University
Naim Kheir, Oakland University
Gene Moriarty, San Jose State University
Glen Hower, Washington State University
Michael Sain, Notre Dame University
Ilya Grinberg, Buffalo State University
Steve McFee, McGill University
R. Mark Nelms, Auburn University
H. Jack Allison, Oklahoma State University
Saroj Biswas, Temple University
Mostafa Chinichian, California Polytechnic
State University, San Luis Obispo
Kathleen Cummings, Georgia Institute of
Technology
Robert Erlandson, Wayne State University
John Fleming, Texas A&M University
Ward Helms, University of Washington
Richard Johnston, Lawrence Technological
University
Sharlene Katz, California State University,
Northridge
Robert Krueger, University of Wisconsin,
Milwaukee
Gary Ybarra, Duke University

Richard C. Dorf
James Svoboda

Contents

CHAPTER 1

CHAPTER 2

TER 3

CHAPTER 4

CHAPTER 5

CHAPTER 8

CHAPTER 9

CHAPTER 10

CHAPTER 11

CHAPTER 14

CHAPTER 15

CHAPTER 16

CHAPTER 17

CHAPTER 1

Electric Circuit Variables

Preview

From the beginning of recorded time, humans have explored the electrical phenomena they have experienced in everyday life. As scientists developed the knowledge of electrical charge, they formulated the laws of electricity as we know them today.

In this chapter we illustrate the thinking process underlying the design of an electric circuit and briefly review the history of electrical science up to the late 1800s. With the knowledge of electricity available by the turn of the century, scientists and engineers analyzed and built electric circuits. Here we explore how electric elements may be described and analyzed in terms of the variables charge, current, voltage, power, and energy.

The design of electric circuits is the process of combining electric elements to provide desired values of these circuit variables. We now illustrate the design process in the context of a jet valve controller circuit.

1.1 Design Challenge

JET VALVE CONTROLLER

Often we need a circuit that provides energy to a device such as a pump or a valve. It is necessary to determine the required current and voltage provided to the device so that it will operate for a desired time period. Here we consider a jet valve controller that requires 40 mJ of energy for one minute of operation. The energy will be supplied to the jet valve controller by another element, that is, a battery. We need to draw the circuit model of this jet valve controller and its energy supply. Let us proceed to consider simple electric circuits and describe the voltage and current in terms of the energy delivered to an element such as a jet valve controller. Then, at the end of the chapter we will determine the current and voltage required to provide 40 mJ of energy for one minute of operation and describe the necessary battery.

We will return to the design challenge for the jet valve controller in Section 1.10 after briefly reviewing the history of electrical science and examining the circuit variables in detail.

1.2 THE EARLY HISTORY OF ELECTRICAL SCIENCE

Electricity is a natural phenomenon controlled for the purposes of humankind. Through this phenomenon we have developed communications, lighting, and computing devices.

Electricity is the physical phenomenon arising from the existence and interaction of electric charge.

Prehistoric people experienced the properties of magnetite—permanently magnetized pieces of ore, often called lodestones. These magnetic stones were strong enough to lift pieces of iron.

The philosopher Thales of Miletus (640–546 B.C.) is thought to have been the first person who observed the electrical properties of amber. He noted that when amber was rubbed, it acquired the ability to pick up light objects such as straw and dry grass. He also experimented with the lodestone and knew of its power to attract iron. By the thirteenth century, floating magnets were used for compasses.

William Gilbert of England published the book *De Magnete* in 1600, and it represented the greatest forward step in the study of electricity and magnetism up to that time. In his studies, he found a long list of materials that could be electrified. Gilbert also presented a procedure for analyzing physical phenomena by a series of experiments, which we now call the scientific method.

Following Gilbert's lead, Robert Boyle published his many experimental results in 1675, as shown in Figure 1.2-1. Boyle was one of the early experimenters with electricity in a vacuum.

Otto von Guericke (1602–1686) built an electrical generator and reported it in his *Experimenta Nova* of 1672. This device, shown in Figure 1.2-2, was a sulfur globe on a shaft that could be turned on its bearings. When the shaft was turned with a dry hand held on the globe, an electrical charge gathered on the globe's surface. Guericke also noted small sparks when the globe was discharged.

A major advance in electrical science was made in Leyden, Holland, in 1746, when Pieter van Musschenbroek introduced a jar that served as a storage apparatus for static

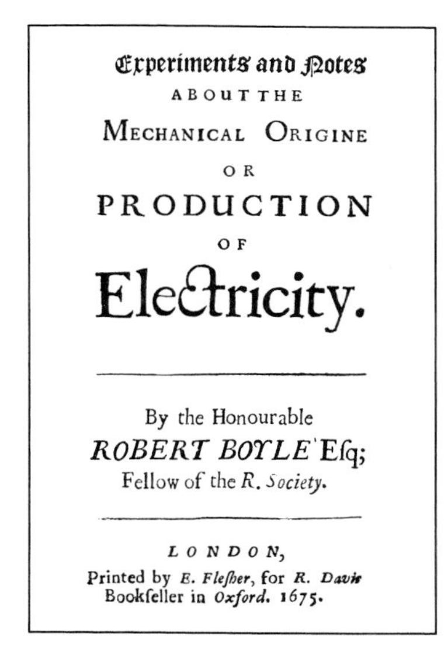
Experiments and Notes
ABOUT THE
MECHANICAL ORIGINE
OR
PRODUCTION
OF
Electricity.

By the Honourable
ROBERT BOYLE Esq;
Fellow of the *R. Society*.

LONDON,
Printed by *E. Flesher*, for *R. Davis*
Bookseller in *Oxford*. 1675.

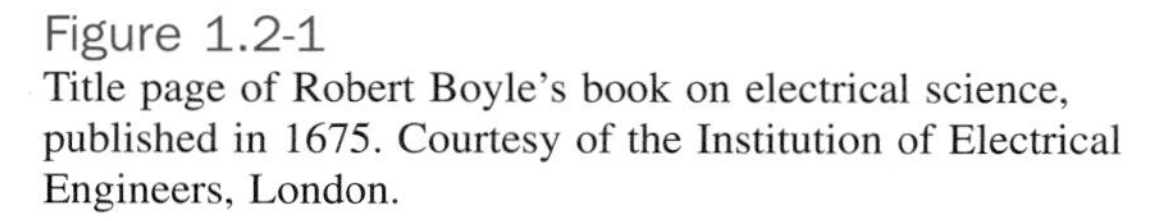
Figure 1.2-1
Title page of Robert Boyle's book on electrical science, published in 1675. Courtesy of the Institution of Electrical Engineers, London.

electricity. The jar was coated inside and out with tinfoil, and a metallic rod was attached to the inner foil lining and passed through the lid. As shown in Figure 1.2-3, Leyden jars were gathered in groups (called batteries) and arranged with multiple connections, thereby further improving the discharge energy.

People have always watched but few have analyzed that great display of power present in the electrical discharge in the sky called lightning and illustrated by Figure 1.2-4. In the late 1740s, Benjamin Franklin developed the theory that there are two kinds of charge, positive and negative. With this concept of charge, Franklin developed his famous kite experiment in June 1752 and his innovation, the lightning rod, for draining the electrical

Figure 1.2-2 In his studies of attraction and gravitation, Guericke devised the first electrical generator. When a hand was held on a sulfur ball revolving in its frame, the ball attracted paper, feathers, chaff, and other light objects. Courtesy of Burndy Library.

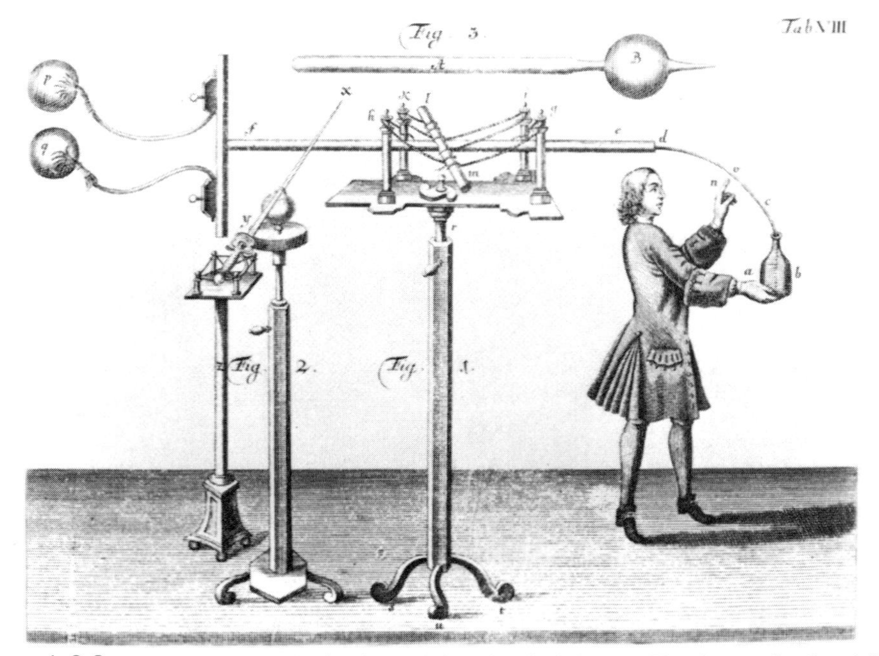

Figure 1.2-3 Illustration of Leyden jar for retaining electrical charges. The charges developed from the rotating glass globes (left) were transmitted through the central conductor and led down a wire to the bottle, which was partly filled with water. With the bottle held in one hand, the other hand completed the circuit—with a resulting shock. Courtesy of Burndy Library.

charge from the clouds. Franklin, shown in Figure 1.2-5, was the first great American electrical scientist.

In 1767 Joseph Priestley published the first book on the history of electricity; the title page is shown in Figure 1.2-6. Twenty years later, Professor Luigi Galvani of Bologna, Italy, carried out a series of experiments with a frog's legs. Galvani noted that the leg of

Figure 1.2-4 A display of lightning. Courtesy of the National Severe Storms Laboratory.

Figure 1.2-5 Benjamin Franklin and a section of the lightning rod erected in his Philadelphia home in September 1752. Divergence of the balls indicated a charged cloud overhead. With this apparatus, Franklin discovered that most clouds were negatively charged and that "'tis the earth that strikes into the clouds, and not the clouds that strike into the earth." Courtesy of Burndy Library.

a dead frog would twitch when dissected with a metal scalpel, and he published his findings in 1791. Galvani found that the frog's legs twitched when subjected to an electrical discharge, as shown in Figure 1.2-7.

Alessandro Volta of Padua, Italy, shown in Figure 1.2-8, recognized that the twitch was caused by two dissimilar metals that were moist and touching at one end and in contact with the frog's leg nerves at the other end. He went on to construct an electrochemical

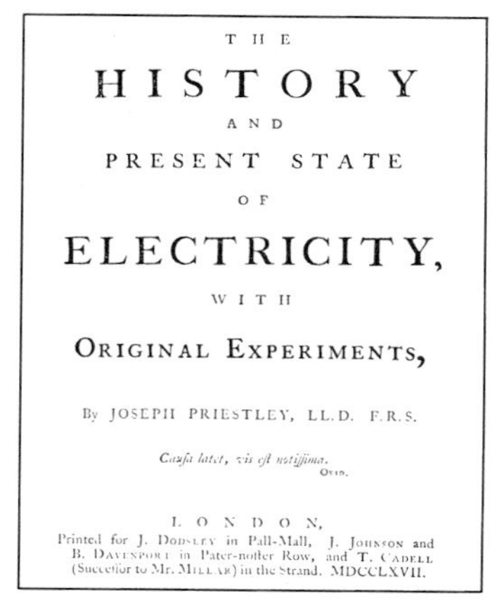

THE

HISTORY

AND

PRESENT STATE

OF

ELECTRICITY,

WITH

ORIGINAL EXPERIMENTS,

By JOSEPH PRIESTLEY, LL.D. F.R.S.

Causa latet, vis est notissima.
OVID.

LONDON,
Printed for J. DODSLEY in Pall-Mall, J. JOHNSON and B. DAVENPORT in Pater-noster Row, and T. CADELL (Successor to Mr. MILLAR) in the Strand. MDCCLXVII.

Figure 1.2-6 Title page of the book on electrical science by Joseph Priestley (1767). Courtesy of the Institution of Electrical Engineers, London.

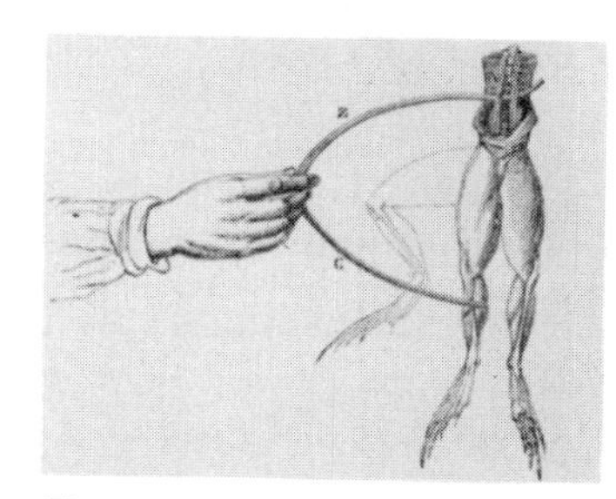

Figure 1.2-7 Luigi Galvani in 1786 noticed that a frog's leg would twitch if touched by two dissimilar metals, copper and zinc. Courtesy of Burndy Library.

Figure 1.2-8 Alessandro Volta. Courtesy of Burndy Library.

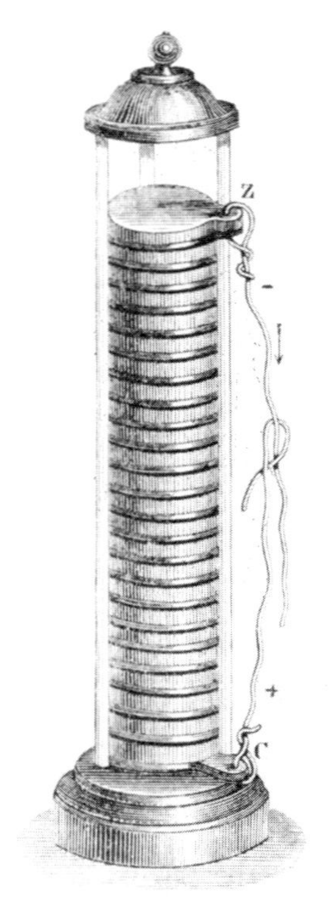

Figure 1.2-9 The voltaic pile, which was a series of successions of three conducting substances—a plate of silver, a plate of zinc, and a similar piece of spongy matter impregnated with a saline solution—repeated 30 or 40 times. Courtesy of Burndy Library.

Figure 1.2-10 André-Marie Ampère. Courtesy of the Library of Congress.

Figure 1.2-11 James Prescott Joule (1818-1889). Courtesy of the Smithsonian Institution.

pile, as illustrated in Figure 1.2-9, which consisted of pairs of zinc and silver disks separated by brine-soaked cloth or paper. This voltaic pile could cause the sensation of current flowing through a person who placed a hand at either end of the pile.

Volta, with the invention of his pile, or electric battery, in 1800, was able to show a steady current in a closed circuit. Volta was remembered 54 years after his death when the unit of electromotive force was officially named the *volt*.

The foundation of electrodynamics was laid by André-Marie Ampère. He called the former studies of electricity electrostatics, in order to highlight the difference. During the 1820s he defined the electric current and developed the means of measuring it. Ampère, shown in Figure 1.2-10, was honored by having the unit of electric current, the ampere, named after him in 1881.

A paper published by James Prescott Joule in 1841 claimed the discovery of the relationship between a current and the heat or energy produced, which today we call Joule's law. Joule is shown in Figure 1.2-11, and the unit of energy is called the joule in his honor.

The theories of electrodynamics were stated in mathematical terms by James Clerk Maxwell, a Scottish mathematical physicist, in papers published between 1855 and 1864. His famous book *Treatise on Electricity and Magnetism* was published in 1873. Maxwell in his student days is shown in Figure 1.2-12.

The major events in electrical science and engineering are summarized in Table 1.2-1.

Table 1.2-1 **Major Events in Electrical Science and Engineering**

1600	William Gilbert published *De Magnete.*
1672	Otto von Guericke published *Experimenta Nova.*
1675	Robert Boyle published *Production of Electricity.*
1746	The Leyden jar was demonstrated in Holland.
1750	Benjamin Franklin invented the lightning conductor.
1767	Joseph Priestley published *The Present State of Electricity.*
1786	Luigi Galvani observed electrical convulsion in the legs of dead frogs.

Table 1.2-1 ***(Continued)***

1800	Alessandro Volta announced the voltaic pile.
1801	Henry Moyes was the first to observe an electric arc between carbon rods.
1820	Hans Oersted discovered the deflection of a magnetic needle by current in a wire.
1821	Michael Faraday produced magnetic rotation of a conductor and magnet—the first electric motor.
1825	André-Marie Ampère defined electrodynamics.
1828	Joseph Henry produced silk-covered wire and more powerful electromagnets.
1831	Michael Faraday discovered electromagnetic induction and carried out experiments with an iron ring and core. He also experimented with a magnet and rotating disk.
1836	Samuel Morse devised a simple relay.
1836	Electric light from batteries was shown at the Paris Opéra.
1841	James Joule stated the relation between current and energy produced.
1843	Morse transmitted telegraph signals from Baltimore to Washington, D.C.
1850	First channel telegraph cable was laid from England to France.
1858	Atlantic telegraph cable was completed and the first message sent.
1861	Western Union established telegraph service from New York to San Francisco.
1863	James Clerk Maxwell determined the ohm.
1873	Maxwell published *Treatise on Electricity and Magnetism.*
1875	Alexander Graham Bell invented the telephone.
1877	Thomas Edison invented the carbon telephone transmitter.
1877	Edison Electric Light Company was formed.
1881	First hydropower station was brought into use at Niagara, New York.
1881	Edison constructed the first electric power station at Pearl Street, New York.
1883	Overhead trolley electric railways were started at Portrush and Richmond, Virginia.
1884	Philadelphia electrical exhibition was held.
1884	American Institute of Electrical Engineers (AIEE) was formed as a professional society for electrical engineers.
1885	The American Telephone and Telegraph Company was organized.
1886	H. Hollerith introduced his tabulating machine.
1897	J.J. Thomson discovered the electron.
1899	Guglielmo Marconi transmitted radio signals from South Foreland to Wimereux, England.
1904	John Ambrose Fleming invented the thermionic diode.
1906	Lee De Forest invented the triode.
1912	The Institute of Radio Engineers (IRE) was formed as a professional society for radio engineers.
1915	Commercial telephone service was established from New York to San Francisco.
1927	Television was established experimentally.
1933	Edwin Armstrong demonstrated FM radio transmission.
1936	Boulder Dam hydroelectric scheme was completed with 115,000-horsepower turbines.
1946	The electronic digital computer ENIAC was introduced.
1948	William Shockley, John Bardeen, and Walter Brattain produced the first practical transistor.
1958	First voice transmission was sent from a satellite and the laser was invented.
1959	The integrated circuit was invented by Jack Kilby and Robert Noyce.
1963	The Institute of Electrical and Electronics Engineers was formed as a merger of AIEE and IRE.
1980	The first fiber optic cable was installed in Chicago.
1987	Superconductivity was demonstrated at 95 K.
1995	Internet established.

Figure 1.2-12
Maxwell in 1855 as a student at Cambridge University, England. Courtesy of Burndy Library.

1.3 ELECTRIC CIRCUITS AND CURRENT FLOW

The outstanding characteristics of electricity when compared with other power sources are its mobility and flexibility. Electrical energy can be moved to any point along a couple of wires and, depending on the user's requirements, converted to light, heat, or motion.

An **electric circuit** or electric network is an interconnection of electrical elements linked together in a closed path so that an electric current may flow continuously.

Consider a simple circuit consisting of two well-known electrical elements, a battery and a resistor, as shown in Figure 1.3-1. Each element is represented by the two-terminal element shown in Figure 1.3-2. Elements are sometimes called devices, and terminals are sometimes called nodes.

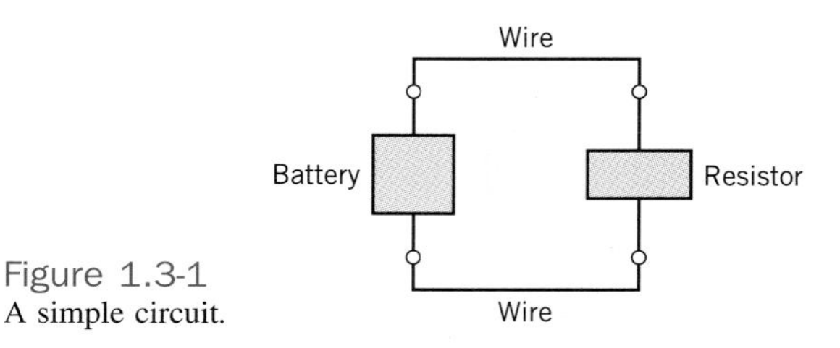

Figure 1.3-1
A simple circuit.

Figure 1.3-2 A general two-terminal electrical element with terminals a and b.

The basic element shown in Figure 1.3-2 has two terminals, cannot be subdivided into other elements, and can be described mathematically in terms of the electrical variables, voltage and current. Wires interconnect the elements to close the circuit so that a continuous current may flow. A more complex set of elements, with a particular terminal interconnection, is shown in Figure 1.3-3.

A *current* may flow in an electric circuit. *Current is the time rate of change of charge past a given point.* Charge is the intrinsic property of matter responsible for electric phenomena. The quantity of charge q can be expressed in terms of the charge on one electron, which is -1.602×10^{-19} coulombs. Thus, -1 coulomb is the charge on 6.24×10^{18} electrons. The current flows through a specified area and is defined by the electric charge passing through the area per unit of time. Thus, q is defined as the charge expressed in coulombs (C).

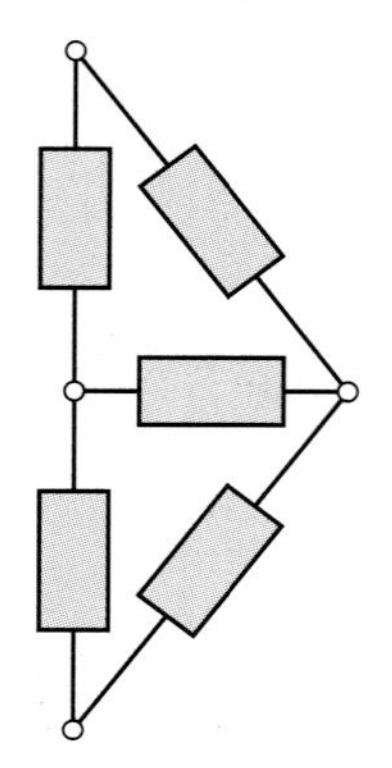

Figure 1.3-3 A circuit with five elements.

Charge is the quantity of electricity responsible for electric phenomena.

Then we can express current as

$$i = \frac{dq}{dt} \tag{1.3-1}$$

The unit of current is the ampere (A); an ampere is 1 coulomb per second.

Current is the time rate of flow of electric charge past a given point.

Note that throughout this chapter we use a lowercase letter, such as q, to denote a variable that is a function of time, $q(t)$. We use an uppercase letter, such as Q, to represent a constant.

The flow of current is conventionally represented as a flow of positive charges. This convention was initiated by Benjamin Franklin. Of course, we now know that charge flow in metal conductors results from electrons with a negative charge. Nevertheless, we will conceive of current as the flow of positive charge, according to accepted convention.

Figure 1.3-4 shows the notation that we use to describe a current. There are two parts to this notation: a value (perhaps represented by a variable name) and an assigned direction. As a matter of vocabulary, we say that a current exists *in* or *through* an element. Figure 1.3-4 shows that there are two ways to assign the direction of the current through an element. The current i_1 is the rate of flow of electric charge from terminal a to terminal b. On the other hand, the current i_2 is the flow of electric charge from terminal b to terminal a. The currents i_1 and i_2 are similar but different. They are the same size but have different directions. Therefore, i_2 is the negative of i_1 and

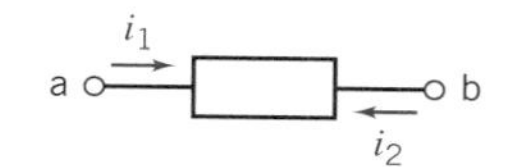

Figure 1.3-4 Current in a circuit element.

$$i_1 = -i_2$$

We always associate an arrow with a current to denote its direction. A complete description of current requires both a value (which can be positive or negative) and a direction (indicated by an arrow).

If the current flowing through an element is constant, we represent it by the constant I, as shown in Figure 1.3-5. A constant current is called a *direct current* (dc).

A **direct current** (dc) is a current of constant magnitude.

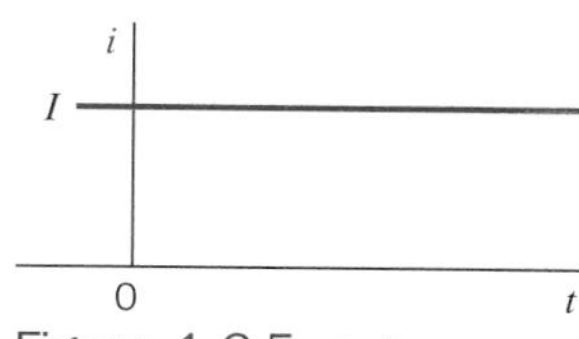

Figure 1.3-5 A direct current of magnitude I.

A time-varying current $i(t)$ can take many forms, such as a ramp, a sinusoid, or an exponential, as shown in Figure 1.3-6. The sinusoidal current is called an alternating current, or ac.

If the charge q is known, the current i is readily found using Eq. 1.3-1. Alternatively, if the current i is known, the charge q is readily calculated. Note that from Eq. 1.3-1 we obtain

$$q = \int_{-\infty}^{t} i\, d\tau = \int_{0}^{t} i\, d\tau + q(0) \tag{1.3-2}$$

where $q(0)$ is the charge at $t = 0$.

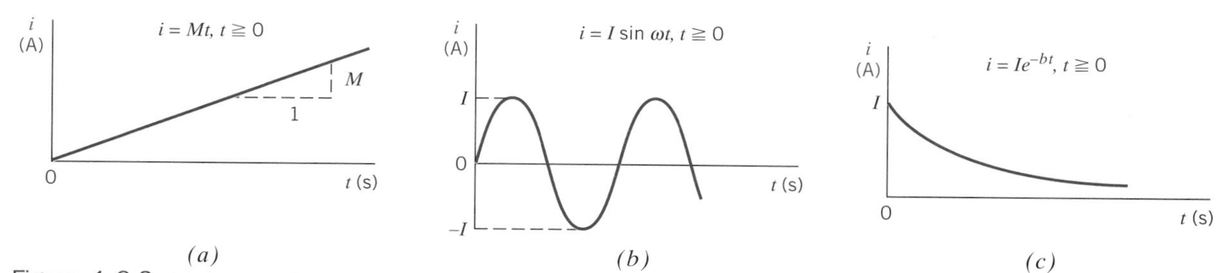

Figure 1.3-6 (*a*) A ramp with a slope M. (*b*) A sinusoid. (*c*) An exponential. I is a constant. The current i is zero for $t < 0$.

Example 1.3-1

Find the current in an element when the charge entering the element is

$$q = 12t \quad \text{C}$$

where t is the time in seconds.

Solution

Recall that the unit of charge is coulombs, C. Then the current, from Eq. 1.3-1, is

$$i = \frac{dq}{dt} = 12 \text{ A}$$

where the unit of current is amperes, A.

Example 1.3-2

Find the charge that has entered the terminal of an element by time t when the current is

$$i = Mt, \qquad t \geq 0$$

as shown in Figure 1.3-6*a,* and M is a constant. Assume that the charge is zero at $t = 0$ $(q(0) = 0)$.

Solution

Using Eq. 1.3-2, we have

$$q = \int_0^t M\tau \, d\tau = M\frac{t^2}{2} \quad \text{C}$$

Example 1.3-3

Find the charge that has entered the terminal of an element from $t = 0$ s to $t = 3$ s when the current is as shown in Figure 1.3-7.

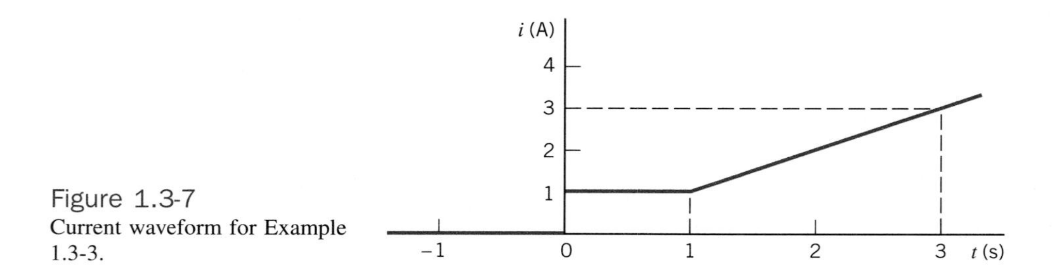

Figure 1.3-7
Current waveform for Example 1.3-3.

Solution

From Figure 1.3-7 we can describe $i(t)$ as

$$i(t) = \begin{cases} 0 & t < 0 \\ 1 & 0 \leq t \leq 1 \\ t & t > 1 \end{cases}$$

Using Eq. 1.3-2, we have

$$q = \int_0^3 i(t)\,dt = \int_0^1 1\,dt + \int_1^3 t\,dt$$

$$= t\Big|_0^1 + \frac{t^2}{2}\Big|_1^3$$

$$= 1 + \frac{1}{2}(9 - 1) = 5\text{ C}$$

Alternatively, we note that integration of $i(t)$ from $t = 0$ to $t = 3\ s$ simply requires the calculation of the area under the curve shown in Figure 1.3-7. Then, we have

$$q = 1 + 2 \times 2 = 5\text{C}$$

Example 1.3-4

Find the charge, $q(t)$, and sketch its waveform when the current entering a terminal of an element is as shown in Figure 1.3-8. Assume that $q(0) = 0$.

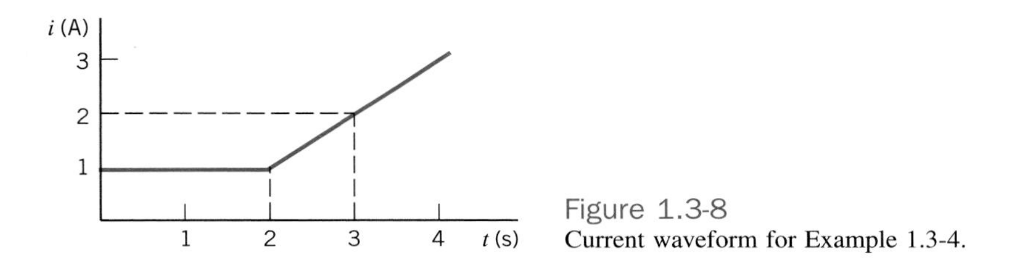

Figure 1.3-8
Current waveform for Example 1.3-4.

Solution

From Figure 1.3-8, we can describe $i(t)$ as

$$i(t) = \begin{cases} 0 & t < 0 \\ 1 & 0 \le t \le 2 \\ t - 1 & t > 2 \end{cases}$$

Using Eq. 1.3-2, we have

$$q(t) = \int_0^t i(\tau)\,d\tau$$

Hence, when $0 \le t \le 2$, we have

$$q = \int_0^t 1\,d\tau = t\text{ C}$$

When $t \ge 2$, we obtain

$$q = \int_0^t i(\tau)\,d\tau = \int_0^2 1\,d\tau + \int_2^t (\tau - 1)\,d\tau$$

$$= \tau\Big|_0^2 + \frac{\tau^2}{2}\Big|_2^t - \tau\Big|_2^t = \frac{t^2}{2} - t + 2\text{ C}$$

The sketch of $q(t)$ is shown in Figure 1.3-9. Note that $q(t)$ is a continuous function of time even though $i(t)$ has a discontinuity at $t = 0$.

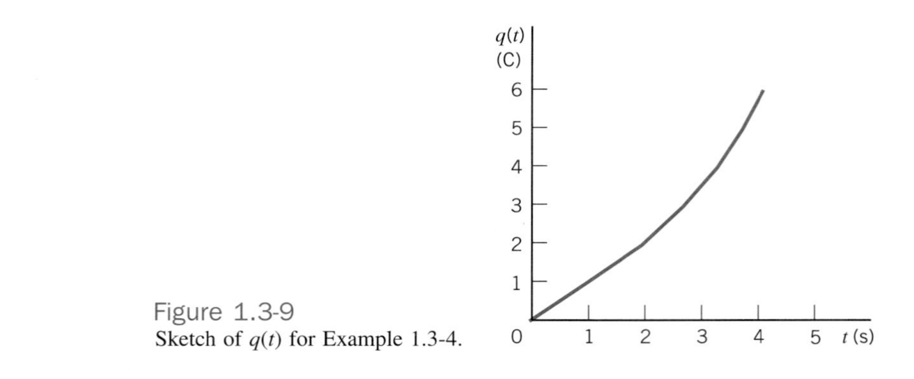

Figure 1.3-9
Sketch of $q(t)$ for Example 1.3-4.

Exercise 1.3-1 Find the charge that has entered an element by time t when $i = 8t^2 - 4t$ A, $t \geq 0$. Assume $q(0) = 0$.

Answer: $q(t) = \frac{8}{3}t^3 - 2t^2 \quad \text{C}$

Exercise 1.3-2 The total charge that has entered a circuit element is $q(t) = 4 \sin 3t$ C when $t \geq 0$ and $q(t) = 0$ when $t < 0$. Determine the current in this circuit element for $t \geq 0$.

Answer: $i(t) = \frac{d}{dt} 4 \sin 3t = 12 \cos 3t \quad \text{A}$

Exercise 1.3-3 The current in a circuit element is $i(t) = 4 \sin 3t$ A when $t \geq 0$ and $i(t) = 0$ when $t < 0$. Determine the total charge that has entered a circuit element for $t \geq 0$.

Hint: $q(0) = \int_{-\infty}^{0} i(\tau)\, d\tau = \int_{-\infty}^{0} 0\, d\tau = 0$

Answer: $q(t) = -\frac{4}{3} \cos 3t + \frac{4}{3}$ C for $t \geq 0$.

Exercise 1.3-4 The total charge $q(t)$, in coulombs, that has entered the terminal of an element is

$$q(t) = \begin{cases} 0 & t < 0 \\ 2t & 0 \leq t \leq 2 \\ 3 + e^{-2(t-2)} & t > 2 \end{cases}$$

Find the current $i(t)$ and sketch its waveform for $t \geq 0$.

Answer:

$$i(t) = \begin{cases} 0 & t < 0 \\ 2 & 0 \leq t \leq 2 \\ -2e^{-2(t-2)} & t > 2 \end{cases}$$

1.4 SYSTEMS OF UNITS

In representing a circuit and its elements, we must define a consistent system of units for the quantities occurring in the circuit. At the 1960 meeting of the General Conference of Weights and Measures, the representatives modernized the metric system and created the Système International d'Unités, commonly called SI units.

SI is *Système International d'Unités* or the International System of Units.

The fundamental, or base, units of SI are shown in Table 1.4-1. Symbols for units that represent proper (persons') names are capitalized; the others are not. Periods are not used after the symbols, and the symbols do not take on plural forms. The derived units for other physical quantities are obtained by combining the fundamental units. Table 1.4-2 shows the more common derived units along with their formulas in terms of the fundamental units or preceding derived units. Symbols are shown for the units that have them.

The basic units such as length in meters (m), time in seconds (s), and current in amperes (A) can be used to obtain the derived units. Then, for example, we have the unit for

Table 1.4-1 **SI Base Units**

	SI Unit	
Quantity	Name	Symbol
Length	meter	m
Mass	kilogram	kg
Time	second	s
Electric current	ampere	A
Thermodynamic temperature	kelvin	K
Amount of substance	mole	mol
Luminous intensity	candela	cd

Table 1.4-2 **Derived Units in SI**

Quantity	Unit Name	Formula	Symbol
Acceleration—linear	meter per second per second	m/s^2	
Velocity—linear	meter per second	m/s	
Frequency	hertz	s^{-1}	Hz
Force	newton	$kg \cdot m/s^2$	N
Pressure or stress	pascal	N/m^2	Pa
Density	kilogram per cubic meter	kg/m^3	
Energy or work	joule	$N \cdot m$	J
Power	watt	J/s	W
Electric charge	coulomb	$A \cdot s$	C
Electric potential	volt	W/A	V
Electric resistance	ohm	V/A	Ω
Electric conductance	siemens	A/V	S
Electric capacitance	farad	C/V	F
Magnetic flux	weber	$V \cdot s$	Wb
Inductance	henry	Wb/A	H

Table 1.4-3 **SI Prefixes**

Multiple	Prefix	Symbol
10^{12}	tera	T
10^{9}	giga	G
10^{6}	mega	M
10^{3}	kilo	k
10^{-2}	centi	c
10^{-3}	milli	m
10^{-6}	micro	μ
10^{-9}	nano	n
10^{-12}	pico	p
10^{-15}	femto	f

charge (C) derived from the product of current and time (A · s). The fundamental unit for energy is the joule (J), which is force times distance or N · m.

The great advantage of the SI system is that it incorporates a decimal system for relating large or smaller quantities to the basic unit. The powers of 10 are represented by standard prefixes given in Table 1.4-3. An example of the common use of a prefix is the centimeter (cm), which is 0.01 meter.

The decimal multiplier must always accompany the appropriate units and is never written by itself. Thus, we may write 2500 W as 2.5 kW. Similarly, we write 0.012 A as 12 mA.

Example 1.4-1

A mass of 150 grams experiences a force of 100 newtons. Find the energy or work expended if the mass moves 10 centimeters. Also, find the power if the mass completes its move in 1 millisecond.

Solution

The energy is found as

$$\begin{aligned} \text{Energy} &= \text{force} \times \text{distance} \\ &= 100 \times 0.1 \\ &= 10 \text{ J} \end{aligned}$$

Note that we used the distance in units of meters. The power is found from

$$\text{Power} = \frac{\text{energy}}{\text{time period in seconds}}$$

where the time period is 10^{-3} s. Thus,

$$\text{Power} = \frac{10}{10^{-3}} = 10^4 \text{ W}$$

Exercise 1.4-1 Which of the three currents, $i_1 = 45\,\mu\text{A}$, $i_2 = 0.03\,\text{mA}$ and $i_3 = 25 \times 10^{-4}$ A, is largest?

Answer: i_3 is largest.

Exercise 1.4-2 A constant current of 4 kA flows through an element. What is the charge that has passed through the element in the first millisecond?

Answer: 4 C

Exercise 1.4-3 A charge of 45 nC passes through a circuit element during a particular interval of time that is 5 ms in duration. Determine the average current in this circuit element during that interval of time.

Answer: $i = 9\,\mu\text{A}$

Exercise 1.4-4 Ten billion electrons per second pass through a particular circuit element. What is the average current in that circuit element?

Answer: $i = 1.602\,\text{nA}$

1.5 VOLTAGE

The basic variables in an electrical circuit are current and voltage. These variables describe the flow of charge through the elements of a circuit and the energy required to cause charge to flow. Figure 1.5-1 shows the notation that we use to describe a voltage. There are two parts to this notation: a value (perhaps represented by a variable name) and an assigned direction. The value of a voltage may be positive or negative. The direction of a voltage is given by its polarities $(+,-)$. As a matter of vocabulary, we say that a voltage exists *across* an element. Figure 1.5-1 shows that there are two ways to label the voltage across an element. The voltage v_{ba} is proportional to the work required to move a positive charge from terminal b to terminal a. On the other hand, the voltage v_{ab} is proportional to the work required to move a positive charge from terminal a to terminal b. We sometimes read v_{ba} as "the voltage at terminal b with respect to terminal a." Similarly, v_{ab} can be read as "the voltage at terminal a with respect to terminal b." Alternatively, we sometimes say that v_{ab} is the voltage drop from terminal a to terminal b. The voltages v_{ab} and v_{ba} are similar but different. They have the same magnitude but different directions. This means that

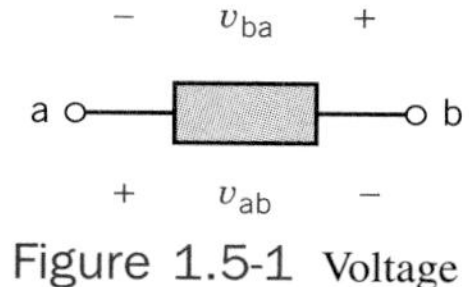

Figure 1.5-1 Voltage across a circuit element.

$$v_{\text{ab}} = -v_{\text{ba}}$$

When considering v_{ba}, terminal b is called the "+ terminal" and terminal a is called the "− terminal." On the other hand, when talking about v_{ab}, terminal a is called the "+ terminal" and terminal b is called the "− terminal."

The **voltage** across an element is the work (energy) required to move a unit positive charge from the − terminal to the + terminal. The unit of voltage is the volt, V.

The equation for the voltage across the element is

$$v = \frac{dw}{dq} \qquad (1.5\text{-}1)$$

where v is voltage, w is energy (or work), and q is charge. A charge of 1 coulomb delivers an energy of 1 joule as it moves through a voltage of 1 volt.

Energy is the capacity to perform work.

1.6 POWER AND ENERGY

The power and energy delivered to an element are of great importance. For example, the useful output of an electric light bulb can be expressed in terms of power. We know that a 300-watt bulb delivers more light than a 100-watt bulb.

Power is the time rate of expending or absorbing energy.

Thus, we have the equation

$$p = \frac{dw}{dt} \tag{1.6-1}$$

where p is power in watts, w is energy in joules, and t is time in seconds. The power associated with the current flow through an element is

$$p = \frac{dw}{dt} = \frac{dw}{dq} \cdot \frac{dq}{dt} = v \cdot i \tag{1.6-2}$$

From Eq. 1.6-2 we see that the power is simply the product of the voltage across an element times the current through the element. The power has units of watts.

Two circuit variables correspond to each element of a circuit: a voltage and a current. Figure 1.6-1 shows that there are two different ways to arrange the directions of this current and voltage. In Figure 1.6-1*a*, the current is directed from the + terminal of the voltage to the − terminal. In contrast, in Figure 1.6-1*b*, the assigned direction of the current is directed from the − terminal of the voltage to the + terminal.

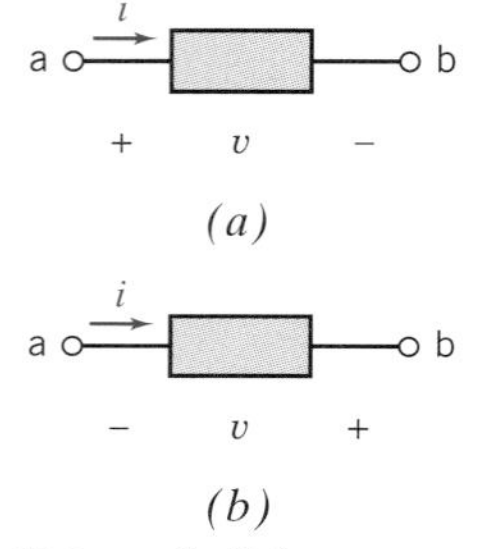

Figure 1.6-1 (*a*) The passive convention is used for element voltage and current. (*b*) The passive convention is not used.

First, consider Figure 1.6-1*a*. This situation, the assigned direction of the current is directed from the + terminal of the voltage direction to the − terminal, is called "the passive convention." In the passive convention, the voltage indicates the work required to move a positive charge in the direction indicated by the current. Accordingly, the power calculated by multiplying the element voltage by the element current

$$p = vi$$

is the power **absorbed** by the element. (This power is also called "the power dissipated by the element" and "the power delivered *to* the element.") The power absorbed by an element can be either positive or negative. This will depend on the values of the element voltage and current.

Next, consider Figure 1.6-1*b*. Here the passive convention has not been used. Instead, the direction of the current is from the − terminal of the voltage to the + terminal. In this case, the voltage indicates the work required to move a positive charge in the direction opposite to the direction indicated by the current. Accordingly, the power calculated by multiplying the element voltage by the element current when the passive convention is not used is

$$p = vi$$

and is the power **supplied** by the element. (This power is also called "the power delivered *by* the element.") The power supplied by an element can be either positive or negative. This will depend on the values of the element voltage and current.

The power absorbed by an element and the power supplied by that same element are related by

$$\textit{power absorbed} = -\textit{power supplied}$$

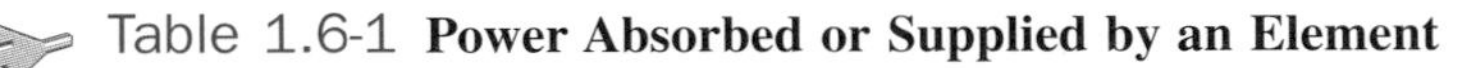

Table 1.6-1 **Power Absorbed or Supplied by an Element**

Power Absorbed by an Element	Power Supplied by an Element
a ○ $\xrightarrow{i}$ [element] ○ b; + v −	a ○ $\xrightarrow{i}$ [element] ○ b; − v +
Since the reference directions of v and i adhere to the passive convention, the power $p = vi$ is the power absorbed by the element.	Since the reference directions of v and i do not adhere to the passive convention, the power $p = vi$ is the power supplied by the element.

The rules for the passive convention are summarized in Table 1.6-1. When the element voltage and current adhere to the passive convention, the energy absorbed by an element can be determined from Eq. 1.6-1 by rewriting it as

$$dw = p\,dt \tag{1.6-3}$$

On integrating, we have

$$w = \int_{-\infty}^{t} p\,d\tau \tag{1.6-4}$$

If the element only receives power for $t \geq t_0$ and we let $t_0 = 0$, then we have

$$w = \int_{0}^{t} p\,d\tau \tag{1.6-5}$$

Example 1.6-1

Let us consider the element shown in Figure 1.6-1*a* when $v = 4$ V and $i = 10$ A. Find the power absorbed by the element and the energy absorbed over a 10 s interval.

Solution

The power absorbed by the element is

$$p = vi = 4 \cdot 10 = 40 \text{ W}$$

The energy absorbed by the element is

$$w = \int_{0}^{10} p\,dt = \int_{0}^{10} 40\,dt = 40 \cdot 10 = 400 \text{ J}$$

Example 1.6-2

Consider the element shown in Figure 1.6-2. The current i and voltage v_{ab} adhere to the passive convention, so the power *absorbed* by this element is

$$\textit{power absorbed} = i \cdot v_{ab} = 2 \cdot (-4) = -8 \text{ W}$$

The current i and voltage v_{ba} do not adhere to the passive convention, so the power *supplied* by this element is

$$\textit{power supplied} = i \cdot v_{ba} = 2 \cdot (4) = 8 \text{ W}$$

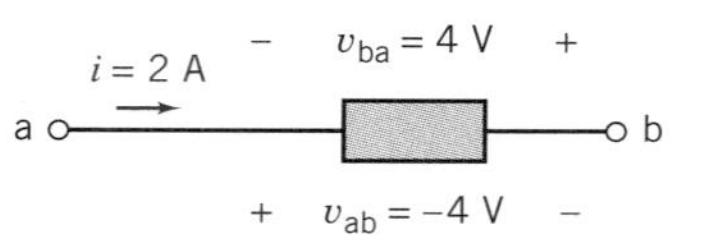

Figure 1.6-2
The element considered in Example 1.6-2.

As expected

$$\textit{power absorbed} = -\textit{power supplied}$$

Now let us consider an example when the passive convention is not used. Then $p = vi$ is the power supplied by the element.

Example 1.6-3

Consider the circuit shown in Figure 1.6-3 with $v = 8e^{-t}$ V and $i = 20e^{-t}$ A for $t \geq 0$. Find the power supplied by this element and the energy supplied by the element over the first second of operation. We assume that v and i are zero for $t < 0$.

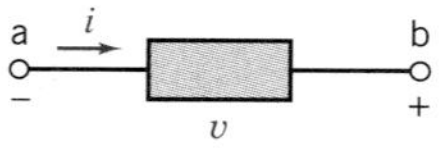

Figure 1.6-3 An element with the current flowing into the terminal with a negative voltage sign.

Solution

The power supplied is

$$p = vi = (8e^{-t})(20e^{-t}) = 160e^{-2t}\text{ W}$$

This element is providing energy to the charge flowing through it.

The energy supplied during the first second is

$$w = \int_0^1 p\,dt = \int_0^1 (160e^{-2t})dt$$

$$= 160\frac{e^{-2t}}{-2}\bigg|_0^1 = \frac{160}{-2}(e^{-2} - 1) = 80(1 - e^{-2}) = 69.2\text{ J}$$

Example 1.6-4

The average current in a typical lightning thunderbolt is 2×10^4 A and its typical duration is 0.1 s (Williams 1988). The voltage between the clouds and the ground is 5×10^8 V. Determine the total charge transmitted to the earth and the energy released.

Solution

The total charge is

$$Q = \int_0^{0.1} i(t)\,dt = \int_0^{0.1} 2 \times 10^4\,dt = 2 \times 10^3\text{ C}$$

The total energy released is

$$w = \int_0^{0.1} i(t) \times v(t)\,dt = \int_0^{0.1} (2 \times 10^4)(5 \times 10^8)\,dt = 10^{12}\text{ J} = 1\text{ TJ}$$

Exercise 1.6-1 Find the power and the energy for the first 10 seconds of operation of the element shown in Figure 1.6-1*a* when $v = 10$ V and $i = 20$ A.

Answer: $p = 200$ W, $w = 2$ kJ

Exercise 1.6-2 Find the power and energy supplied during the first 10 seconds of operation for the element shown in Figure 1.6-3 when $v = 50e^{-10t}$ V and $i = 5e^{-10t}$ A. The circuit begins operation at $t = 0$.

Answer: $p = 250e^{-20t}$ W, $w = 12.5$ J

Exercise 1.6-3 A hydroelectric power plant can deliver power to remote users. A representation of the power plant is shown in Figure 1.6-1*b*. If $v = 100$ kV and $i = 120$ A, find the power supplied by the hydroelectric plant and the energy supplied each day.

Answer: $p = 12$ MW, $w = 1.04$ TJ

1.7 VOLTMETERS AND AMMETERS

Measurements of dc current and voltage are made with direct-reading (analog) or digital meters, as shown in Figure 1.7-1. A direct-reading meter has an indicating pointer whose angular deflection depends on the magnitude of the variable it is measuring. A digital meter displays a set of digits indicating the measured variable value.

An ideal ammeter measures the current flowing through its terminals, as shown in Figure 1.7-2*a*, and has zero voltage, v_{m}, across its terminals. An ideal voltmeter measures the voltage across its terminals, as shown in Figure 1.7-2*b*, and has a terminal current i_{m} equal to zero. Practical measuring instruments only approximate the ideal conditions. For a practical ammeter, the voltage across its terminals is negligibly small. Similarly, the current into the terminal of a voltmeter is usually negligible.

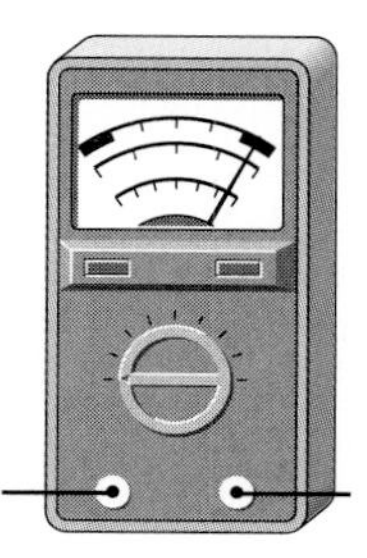

(*a*)

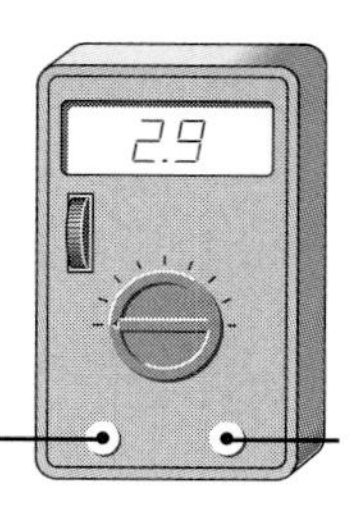

(*b*)

Figure 1.7-1 (*a*) A direct-reading (analog) meter and (*b*) a digital meter.

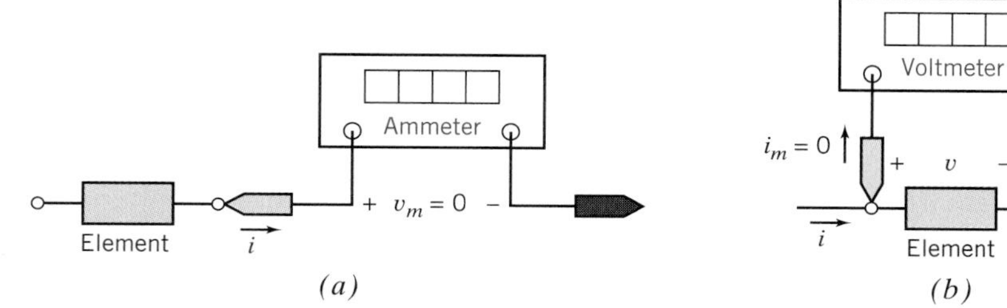

(*a*) (*b*)

Figure 1.7-2 (*a*) Ideal ammeter and (*b*) ideal voltmeter.

1.8 CIRCUIT ANALYSIS AND DESIGN

The analysis and design of electric circuits are the primary activities described in this book and are key skills for an electrical engineer. The *analysis* of a circuit is concerned with the methodical study of a given circuit designed to obtain the magnitude and direction of one or more circuit variables, such as a current or voltage.

The analysis process begins with a statement of the problem and usually includes a given circuit model. The goal is to determine the magnitude and direction of one or more circuit variables, and the final task is to verify that the proposed solution is indeed correct. Usually, the engineer first identifies what is known and the principles that will be used to determine the unknown variable.

The problem-solving method that will be used throughout this book is shown in Figure 1.8-1. Generally, the problem statement is given. The analysis process then moves sequentially through the five steps shown in Figure 1.8-1. First, we describe the situation and the assumptions. We also record or review the circuit model that is provided. Second, we state the goals and requirements, and we normally record the required circuit variable to be determined. The third step is to create a plan that will help obtain the solution of the problem. Typically, we record the principles and techniques that pertain to this problem. The fourth step is to act on the plan and carry out the steps described in the plan. The final step is to verify that the proposed solution is indeed correct. If it is correct, we communicate this solution by recording it in writing or by presenting it verbally. If the verification step indicates that the proposed solution is incorrect or inadequate, then we return to the plan steps, reformulate an improved plan, and go back again to steps 4 and 5.

In order to illustrate this analytical method, we will consider an example. In Example 1.8-1 we use the steps described in the problem-solving method of Figure 1.8-1.

Example 1.8-1

An experimenter in a lab assumes that an element is absorbing power, and uses a voltmeter and ammeter to measure the voltage and current as shown in Figure 1.8-2. The mea-

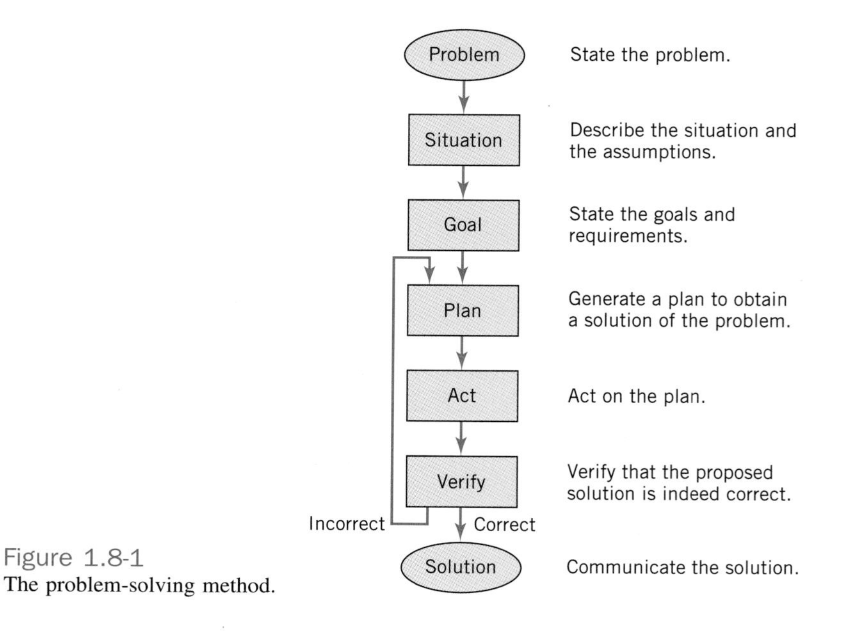
Problem
State the problem.
Situation
Describe the situation and the assumptions.
Goal
State the goals and requirements.
Plan
Generate a plan to obtain a solution of the problem.
Act
Act on the plan.
Verify
Verify that the proposed solution is indeed correct.
Incorrect
Correct
Solution
Communicate the solution.

Figure 1.8-1
The problem-solving method.

Figure 1.8-2 An element with a voltmeter and ammeter.

Figure 1.8-3 The circuit from Figure 1.8-2 with the ammeter probes reversed.

surements indicate that the voltage is $v = +12$ V and the current is $i = -2$A. Determine if the experimenter's assumption is correct.

Describe the Situation and the Assumptions Strictly speaking, the element *is* absorbing power. The value of the power absorbed by the element may be positive or zero or negative. When we say that someone "assumes that an element is absorbing power," we mean that someone assumes that the power absorbed by the element is positive.

The meters are ideal. These meters have been connected to the element in such a way as to measure the voltage labeled v and the current labeled i. The values of the voltage and current are given by the meter readings.

State the Goals Calculate the power absorbed by the element in order to determine if the value of the power absorbed is positive.

Generate a Plan Verify that the element voltage and current adhere to the passive convention. If so, the power absorbed by the device is $p = vi$. If not, the power absorbed by the device is $p = -vi$.

Act on the Plan Referring to Table 1.6-1, we see that the element voltage and current do adhere to the passive convention. Therefore, power absorbed by the element is

$$p = vi = 12 \cdot (-2) = -24\text{W}$$

The value of the power absorbed is not positive.

Verify the Proposed Solution Let's reverse the ammeter probes as shown in Figure 1.8-3. Now the ammeter measures the current i_1 rather than the current i, and so $i_1 = 2$A and $v = 12$ V. Since i_1 and v do not adhere to the passive convention $p = i_1 \cdot v = 24$ W is the power supplied by the element. Supplying 24 W is equivalent to absorbing -24 W, thus verifying the proposed solution.

Design is a purposeful activity in which a designer visualizes a desired outcome. It is the process of originating circuits and predicting how these circuits will fulfill objectives. Engineering design is the process of producing a set of descriptions of a circuit that satisfy a set of performance requirements and constraints.

The design process may incorporate three phases: analysis, synthesis, and evaluation. The first task is to diagnose, define, and prepare—that is, to understand the problem and produce an explicit statement of goals; the second task involves finding plausible solutions; the third concerns judging the validity of solutions relative to the goals and selecting among alternatives. A cycle is implied in which the solution is revised and improved by reexamining the analysis. These three phases are part of a framework for planning, organizing, and evolving design projects.

Design is the process of creating a circuit to satisfy a set of goals.

The problem-solving process shown in Figure 1.8-1 is used to solve the Design Challenges included in each chapter.

1.9 VERIFICATION EXAMPLE

Engineers are frequently called upon to verify that a solution to a problem is indeed correct. For example, proposed solutions to design problems must be checked to confirm that all of the specifications have been satisfied. In addition, computer output must be reviewed to guard against data entry errors, and claims made by vendors must be examined critically.

Engineering students are also asked to verify the correctness of their work. For example, occasionally just a little time remains at the end of an exam. It is useful to be able to identify quickly those solutions that need more work.

This text includes some examples, called verification examples, that illustrate techniques useful for checking the solutions of the particular problems discussed in that chapter. At the end of each chapter are some problems, called verification problems, presented so that the reader will have an opportunity to practice these techniques.

Let's proceed to demonstrate the verification of a calculation provided in a laboratory report.

A laboratory report states that the measured values of v and i for the circuit element shown in Figure 1.9-1 are -5 V and -2 A, respectively. The report also states that the power absorbed by the element is 10 W. Verify this statement.

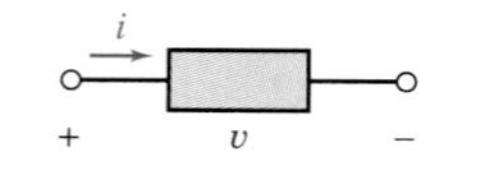

Figure 1.9-1 A circuit element with a measured current and voltage.

Solution

The circuit shown in Figure 1.9-1 adheres to the passive sign convention.

Then, the power absorbed is

$$p = vi$$

Substituting v and i, we have

$$P = (-5)(-2) = 10 \text{ W}$$

Thus, we have verified that the circuit element is absorbing 10 W.

1.10 Design Challenge Solution

JET VALVE CONTROLLER

Problem

A small, experimental space rocket uses a two-element circuit, as shown in Figure 1.10-1, to control a jet valve from point of liftoff at $t = 0$ until expiration of the rocket after one minute. The energy that must be supplied by element 1 for the one-minute period is 40 mJ. Element 1 is a battery to be selected.

It is known that $i(t) = \mathrm{D}e^{-t/60}$ mA for $t \geq 0$, and the voltage across the second element is $v_2(t) = \mathrm{B}e^{-t/60}$ V for $t \geq 0$. The maximum magnitude of the current, D, is lim-

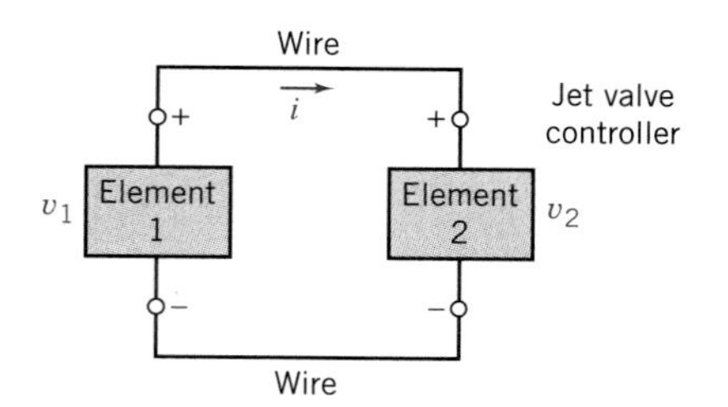

Figure 1.10-1
The circuit to control a jet valve for a space rocket.

ited to 1 mA. Determine the required constants D and B, and describe the required battery.

Describe the Situation, and the Assumptions

1. The current enters the plus terminal of the second element.
2. The current leaves the plus terminal of the first element.
3. The wires are perfect and have no effect on the circuit (they do not absorb energy).
4. The model of the circuit, as shown in Figure 1.10-1, assumes that the voltage across the two elements is equal; that is, $v_1 = v_2$.
5. The battery voltage v_1 is $v_1 = \mathrm{B}e^{-t/60}$ V where B is the initial voltage of the battery that will discharge exponentially as it supplies energy to the valve.
6. The circuit operates from $t = 0$ to $t = 60$ s.
7. The current is limited, so $\mathrm{D} \leq 1\mathrm{mA}$.

State the Goal

Determine the energy supplied by the first element for the one-minute period and then select the constants D and B. Describe the battery selected.

Generate a Plan

First, find $v_1(t)$ and $i(t)$ and then obtain the power, $p_1(t)$, supplied by the first element. Next, using $p_1(t)$, find the energy supplied for the first 60 s.

Goal	Equation	Need	Information
The energy w_1 for the first 60 s	$w_1 = \int_0^{60} p_1(t)\,dt$	$p_1(t)$	v_1 and i known except for constants D and B

Act on the Plan

First, we need $p_1(t)$, so we first calculate

$$\begin{aligned} p_1(t) &= iv_1 = (\mathrm{D}e^{-t/60} \times 10^{-3}\,\mathrm{A})(\mathrm{B}e^{-t/60}\,\mathrm{V}) \\ &= \mathrm{DB}e^{-t/30} \times 10^{-3}\,\mathrm{W} = \mathrm{DB}e^{-t/30}\,\mathrm{mW} \end{aligned}$$

Second, we need to find w_1 for the first 60 s as

$$\begin{aligned} w_1 &= \int_0^{60} (\mathrm{DB}e^{-t/30} \times 10^{-3})\,dt = \left.\frac{\mathrm{DB} \times 10^{-3} e^{-t/30}}{-1/30}\right|_0^{60} \\ &= -30\mathrm{DB} \times 10^{-3}(e^{-2} - 1) = 25.9\mathrm{DB} \times 10^{-3}\,\mathrm{J} \end{aligned}$$

Since we require $w_1 = 40$ mJ,

$$40 = 25.9\mathrm{DB}$$

Next, select the limiting value, D = 1, to get

$$B = \frac{40}{(25.9)(1)} = 1.54 \text{ V}$$

Thus, we select a 2-V battery so that the magnitude of the current is less than 1 mA.

Verify the Proposed Solution

We must verify that at least 40 mJ is supplied using the 2-V battery. Since $i = e^{-t/60}$mA and $v_2 = 2e^{-t/60}$ V, the energy supplied by the battery is

$$w = \int_0^{60} (2e^{-t/60})\,(e^{-t/60} \times 10^{-3})\, dt = \int_0^{60} 2e^{-t/30} \times 10^{-3}\, dt = 51.8 \text{ mJ}$$

Thus, we have verified the solution, and we communicate it by recording the requirement for a 2-V battery.

1.11 SUMMARY

- The uses of electric power are diverse and very important to modern societies. However, electrical science developed slowly over the centuries, with many studies focusing on the nature of charge. As scientists became aware of the ability to store and control charge, they formulated the idea of a circuit.
- A circuit consists of electrical elements linked together in a closed path so that charge may flow.
- Charge is the intrinsic property of matter responsible for electric phenomena. The quantity of charge q can be expressed in terms of the charge on one electron, which is -1.602×10^{-19} coulombs. Current i is the time rate of change of charge past a given point. We can express current as $i = \dfrac{dq}{dt}$. The unit of current is the ampere (A); an ampere is 1 coulomb per second.
- The SI units are used by today's engineers and scientists. Using decimal prefixes, we may simply express electrical quantities with a wide range of magnitudes.
- The voltage across an element is the work required to move a unit of charge through the element. When the passive convention is used to assign the reference directions, the product of the element current and the element voltage gives the power absorbed by the element.

PROBLEMS

Section 1.3 Electric Circuits and Current Flow

P 1.3-1 A wire carries a constant current of 10 mA. How many coulombs pass a cross section of the wire in 20 s?
Answer: $q = 0.2$ C

P 1.3-2 The total charge that has entered a circuit element is $q(t) = 4(1 - e^{-5t})$ when $t \geq 0$ and $q(t) = 0$ when $t < 0$. Determine the current in this circuit element for $t \geq 0$.
Answer: $i(t) = 20e^{-5t}$ A

P 1.3-3 The current in a circuit element is $i(t) = 4(1 - e^{-5t})$ A when $t \geq 0$ and $i(t) = 0$ when $t < 0$. Determine the total charge that has entered a circuit element for $t \geq 0$.

Hint: $q(0) = \int_{-\infty}^{0} i(\tau)d\tau = \int_{-\infty}^{0} 0\, d\tau = 0$

Answer: $q(t) = 4t + 0.8e^{-5t} - 0.8$ C for $t \geq 0$.

P 1.3-4 The charge entering the terminal of a device is given by $q(t) = 2k_1 t + k_2 t^2$ C. If $i(0) = 4$ and $i(3) = -4$, find k_1 and k_2.
Answer: $k_1 = 2$, $k_2 = -4/3$

P 1.3-5 In a closed electric circuit, the number of electrons that pass a given point is known to be 10 billion per second. Find the corresponding current flow in amperes.

P 1.3-6 The current in a circuit element is

$$i(t) = \begin{cases} 0 & t < 2 \\ 2 & 2 < t < 4 \\ -1 & 4 < t < 8 \\ 0 & 8 < t \end{cases}$$

where the units of current are A and the units of time are s. Determine the total charge that has entered a circuit element for $t \geq 0$.

$$\textbf{Answer: } q(t) = \begin{cases} 0 & t < 2 \\ 2t - 4 & 2 < t < 4 \\ 8 - t & 4 < t < 8 \\ 0 & 8 < t \end{cases}$$ where the units of charge are C.

P 1.3-7 An electroplating bath, as shown in Figure P 1.3-7, is used to plate silver uniformly onto objects such as kitchenware and plates. A current of 600 A flows for 20 minutes, and each coulomb transports 1.118 mg of silver. What is the weight of silver deposited in grams?

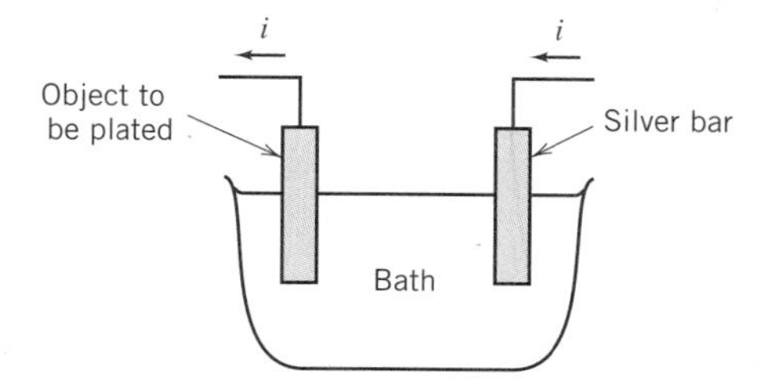

Figure P 1.3-7 An electroplating bath.

Section 1.6 Power and Energy

P 1.6-1 Modern technology has produced a small 1.5-V alkaline battery with a nominal stored energy of 150 joules. For how many days will it power a calculator that draws a 2-mA current? Can you see why automatic shutoff is a good idea?

P 1.6-2 An electric range has a constant current of 10 A entering the positive voltage terminal with a voltage of 110 V. The range is operated for two hours. (a) Find the charge in coulombs that passes through the range. (b) Find the power absorbed by the range. (c) If electric energy costs 6 cents per kilowatt-hour, determine the cost of operating the range for two hours.

P 1.6-3 A walker's cassette tape player uses four AA batteries in series to provide 6 V to the player circuit. The four alkaline battery cells store a total of 200 watt-seconds of energy. If the cassette player is drawing a constant 10 mA from the battery pack, how long will the cassette operate at normal power?

P 1.6-4 A large storage battery is required for fishing and trolling boats that run on electric motors. One such battery, the Die-Hard, provides 675 A at 12 V for 30 s to start a big boat moving. Once the boat is moving, the battery is able to provide 20 A at 11 V for 200 minutes. (a) Calculate the power provided during the starting period and the trolling period. (b) Calculate the energy provided over the total of the startup and trolling periods.

P 1.6-5 The energy w absorbed by a two-terminal device is shown in Figure P 1.6-5 as a function of time. If the voltage across the device is $v(t) = 12 \cos \pi t$ V, where t is in ms, find the current entering the positive terminal at $t = 1, 3, 6$ ms. The element voltage and current adhere to the passive convention.

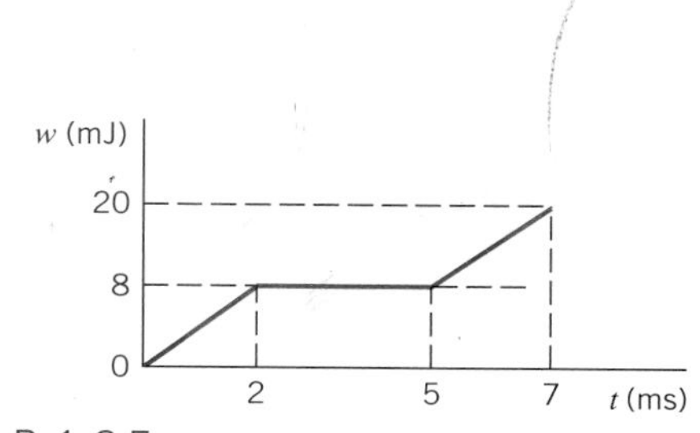

Figure P 1.6-5 Waveform of energy $w(t)$.

P 1.6-6 The current through and voltage across an element vary with time as shown in Figure P 1.6-6. Sketch the power delivered to the element for $t > 0$. What is the total energy delivered to the element between $t = 0$ and $t = 25$ s? The element voltage and current adhere to the passive convention.

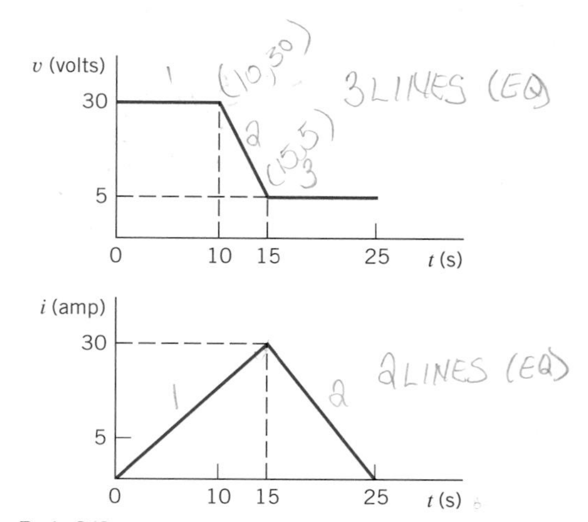

Figure P 1.6-6 (*a*) Voltage $v(t)$ and (*b*) current $i(t)$ for an element.

P 1.6-7 An automobile battery is charged with a constant current of 2 A for five hours. The terminal voltage of the battery is $v = 11 + 0.5t$ V for $t > 0$, where t is in hours. (a) Find the energy delivered to the battery during the five hours. (b) If electric energy costs 10 cents/kWh, find the cost of charging the battery for five hours.

Answer: (b) 1.23 cents

P 1.6-8 The circuit shown in Figure P 1.6-8*a* has a current source i as shown in Figure P 1.6-8*b*. The resulting voltage across the circuit is shown in Figure P 1.6-8*c*. (a) Determine the power $p(t)$ and the energy $w(t)$ absorbed by the element. (b) Sketch $p(t)$ and $w(t)$.

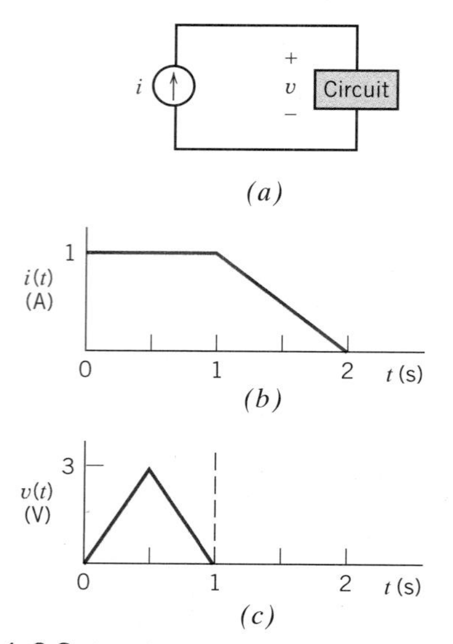

Figure P 1.6-8 (*a*) Element and its (*b*) current and (*c*) voltage.

P 1.6-9 Find the power, $p(t)$, supplied by the element shown in Figure P 1.6-9 when $v(t) = 4 \cos 3t$ V and $i(t) = \dfrac{\sin 3t}{12}$ A. Evaluate $p(t)$ at $t = .5$ s and also at $t = 1$ s. Observe that the power supplied by this element has a positive value at some times and a negative value at other times.

Hint: $(\sin at)(\cos bt) = \dfrac{1}{2}(\sin (a + b)t + \sin (a - b)t)$

Answer: $p(t) = \dfrac{1}{6} \sin 6t$ W, $p(0.5) = 0.0235$ W, $p(1) = -0.0466$ W

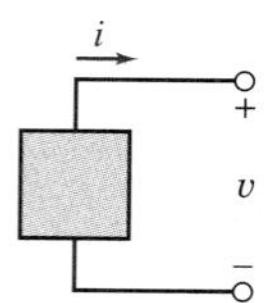

Figure P 1.6-9 An element.

P 1.6-10 Find the power, $p(t)$, supplied by the element shown in Figure P 1.6-9 when $v(t) = 8 \sin 3t$ V and $i(t) = 2 \sin 3t$ A.

Hint: $(\sin at)(\sin bt) = \dfrac{1}{2}(\cos(a - b)t - \cos(a + b)t)$

Answer: $p(t) = 8 - 8 \cos 6t$ W

P 1.6-11 Find the power, $p(t)$, supplied by the element shown in Figure P 1.6-9. The element voltage is represented as $v(t) = 4(1 - e^{-2t})$V when $t \geq 0$ and $v(t) = 0$ when $t < 0$. The element current is represented as $i(t) = 2e^{-2t}$ A when $t \geq 0$ and $i(t) = 0$ when $t < 0$.

Answer: $p(t) = 8(1 - e^{-2t})e^{-2t}$ W

P 1.6-12 A battery is delivering power to an automobile starter. The battery current $i = 10e^{-t}$ A and battery voltage $v = 12e^{-t}$ V do not adhere to the passive convention. (a) Find the power supplied by the source. (b) Find the energy $w(t)$ delivered by the source to the starter.

P 1.6-13 The current i and voltage v of an element adhere to the passive convention. Using measuring devices, we find that $i = 4e^{-50t}$ mA and $v = 10 - 20e^{-50t}$ V for $t \geq 0$. (a) How much power is absorbed by the element at $t = 10$ ms? (b) How much energy is absorbed by the element in the interval $0 \leq t \leq \infty$?

P 1.6-14 A 12-V car battery is connected so that it supplies power to car headlights while the engine is disconnected. (a) Find the power delivered by the car battery when the current is 1 A. (b) Find the power absorbed by the headlights when the current is 1 A. (c) Find the energy absorbed by the headlights over an interval of 10 minutes.

P 1.6-15 Calculate the power absorbed or supplied by each element in Figure P 1.6-15. State in each case whether positive power is absorbed or supplied.

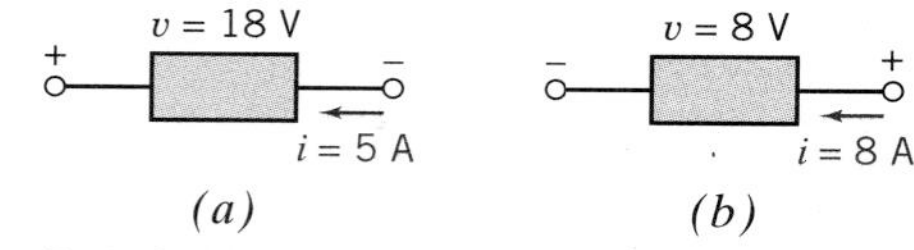

Figure P 1.6-15

P 1.6-16 Neglecting losses, determine the power that can be developed from Niagara Falls, which has an average height of 168 ft and over which the water flows at 500,000 tons per minute.

Answer: 3.8 GW

P 1.6-17 The battery of a flashlight develops 3 V, and the current through the bulb is 200 mA. What power is absorbed by the bulb? Find the energy absorbed by the bulb in a five-minute period.

P 1.6-18 A Die-Hard 12-V automobile battery can deliver 2×10^6 J over a 10-hour interval. What is the current through the battery?

Answer: 4.63 A

P 1.6-19 A large thunderbolt has a current of 160 A and transfers 10^{20} electrons in 0.1 s. Determine the power generated when the voltage from cloud to ground is 100 kV (Williams 1988).

P 1.6-20 Sixty years after it was conceived, the electromagnetic launcher known as the railgun is finally coming of age (Metzger 1989). A railgun launches a projectile with a powerful pulse of electric current. The railgun projectile completes the circuit between two parallel rails of conductive material, creating a magnetic field that pushes against the projectile and slings it forward. It continues to accelerate for its entire ride down the launcher. The greater the current or the longer the rails, the higher the speed the projectile can reach.

One experimental railgun uses 14,000 12-V experimental high-current batteries in series to generate 168 kV. They can produce a 5-s jolt of 2.5 MA. Calculate the energy generated and determine whether a 10-g projectile can be launched to 10 km above the earth.

VERIFICATION PROBLEMS

VP 1-1 Conservation of energy requires that the sum of the power absorbed by all of the elements in a circuit be zero. Figure VP 1.1 shows a circuit. All of the element voltages and currents are specified. Are these voltage and currents correct? Justify your answer.
Hint: Calculate the power absorbed by each element. Add up all of these powers. If the sum is zero, conservation of energy is satisfied and the voltages and currents are probably correct. If the sum is not zero the element voltages and currents cannot be correct.

VP 1-2 Conservation of energy requires that the sum of the power absorbed by all of the elements in a circuit be zero. Figure VP 1.2 shows a circuit. All of the element voltages and currents are specified. Are these voltage and currents correct? Justify your answer.
Hint: Calculate the power absorbed by each element. Add up all of these powers. If the sum is zero, conservation of energy is satisfied and the voltages and currents are probably correct. If the sum is not zero the element voltages and currents cannot be correct.

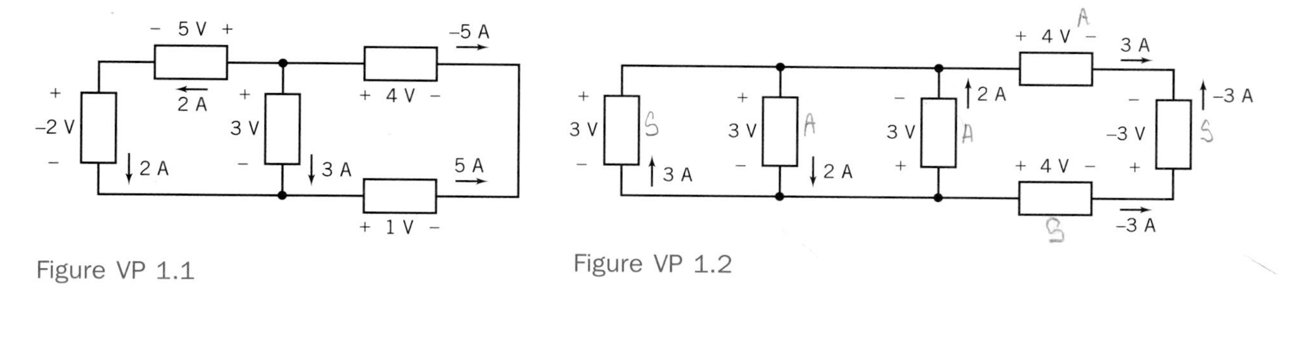

Figure VP 1.1

Figure VP 1.2

DESIGN PROBLEMS

DP 1-1 A particular circuit element is available in three grades. Grade A guarantees that the element can safely absorb 1/2 W continuously. Similarly Grade B guarantees that 1/4 W can be absorbed safely and Grade C guarantees that 1/8 W can be absorbed safely. As a rule, elements that can safely absorb more power are also more expensive and bulkier.

The voltage across an element is expected to be about 20 V and the current in the element is expected to be about 8 mA. Both estimates are accurate to within 25%. The voltage and current reference adhere to the passive convention.

Specify the Grade of this element. Safety is the most important consideration, but don't specify an element that is more expensive than necessary.

DP 1-2 The voltage across a circuit element is $v(t) = 20\,(1 - e^{-8t})$ V when $t \geq 0$ and $v(t) = 0$ when $t < 0$. The current in this element is $i(t) = 30e^{-8t}$ mA when $t \geq 0$ and $i(t) = 0$ when $t < 0$. The element current and voltage adhere to the passive convention. Specify the power that this device must be able to absorb safely.
Hint: Use MATLAB, or a similar program, to plot the power.

CHAPTER 2

Circuit Elements

Preview

Electric circuits consist of interconnections of circuit elements. In this chapter we first consider the great example of Thomas A. Edison, who may be called the first electrical engineer. In the 1880s Edison demonstrated his electric lighting system, and electrical engineering began to emerge as a profession.

Circuit elements, such as Edison's electric lamp, may be represented by a model of their behavior described in terms of the terminal current and voltage of each element. In this chapter we consider how electrical elements may be described in order to interconnect them in a circuit and then describe the circuit's overall behavior.

We describe how the resistance of an element depends on the material used in its construction. Then we consider the linear model for the voltage and current of the resistor as developed by Georg Ohm in 1827. In addition we describe the linear model for voltage and current sources. Finally, we describe the models for switches and transducers.

2.1 Design Challenge

TEMPERATURE SENSOR

Currents can be measured easily using ammeters. A temperature sensor such as Analog Devices' AD590 can be used to measure temperature by converting temperature to current (Dorf 1998). Figure 2.1-1 shows a symbol used to represent a temperature sensor. In order for this sensor to operate properly, the voltage v must satisfy the condition

$$4 \text{ volts} \leq v \leq 30 \text{ volts}$$

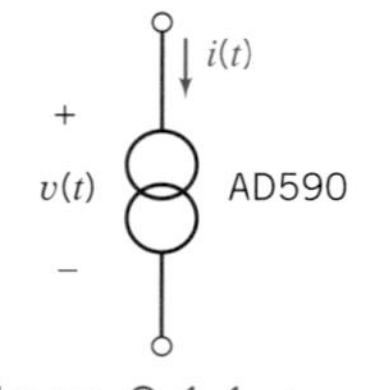

Figure 2.1-1 A temperature sensor.

When this condition is satisfied, the current i, in μA, is numerically equal to the temperature T, in °K. The phrase "numerically equal" indicates that the two variables have the same value but different units.

$$i = k \cdot T \quad \text{where} \quad k = 1 \frac{\mu\text{A}}{°\text{K}}$$

The goal is to design a circuit using the AD590 to measure the temperature of a container of water. In addition to the AD590 and an ammeter, several power supplies and an assortment of standard 2 percent resistors are available. The power supplies are voltage sources. Power supplies having voltages of 10, 12, 15, 18, or 24 volts are available.

The solution to this problem will have to wait until we know more about the devices used to build electric circuits. We will return to this problem at the end of the chapter.

2.2 THOMAS A. EDISON—THE FIRST ELECTRICAL ENGINEER

Electrical energy can be converted into heat, light, and mechanical energy. Nature exhibits the first two conversions in lightning. When the battery was invented in 1800, it was expected that the continuous current available could be used to produce the familiar spark. This continuous current was ultimately used in the arc light for illumination.

During the first two-thirds of the nineteenth century, gas was used to provide illumination and was the marvel of the age. By 1870 there were 390 companies manufacturing gas for lighting in the United States.

In 1878, in Ohio, Charles F. Brush invented an electric arc lighting system. Later, a search was undertaken to find a suitable light that would give less glare than an arc light. Thus, research was centered on finding a material that could be heated to incandescence without burning when an electric current was passed through it.

Thomas A. Edison (1847–1931) realized in 1878 that the development of an electric lighting device was of great commercial importance. His success was both a technical and a commercial achievement. Edison presided over a well-staffed laboratory in Menlo Park, New Jersey, for many years. In an interview, Edison stated:

I have an idea that I can make the electric light available for all common uses, and supply it at a trifling cost, compared with that of gas. There is no difficulty about dividing up the electric currents and using small quantities at different points. The trouble is in finding a candle that will give a pleasant light, not too intense, which can be turned on or off as easily as gas. (McMahon 1984)

Edison was on the right track, seeking a direct replacement for the gas light that would be easily replaceable and easy to control. As we may recall, Edison eventually found the "ideal" filament for his light bulb in the form of a carbonized thread in 1879. Other materials that could be used for a thin high-resistance thread were later substituted. Edison also undertook a program to improve the vacuum in a light bulb, thus reducing the effects of occluding gases and improving the efficiency of the filament and bulb. Edison is shown in Figure 2.2-1 with several of his lamps.

With his early engineering insight, Edison saw that a complete lighting system was required, as he stated:

> The problem then that I undertook to solve was . . . the production of the multifarious apparatus, methods, and devices, each adapted for use with every other, and all forming a comprehensive system. (McMahon 1984)

By 1882 the system had been conceived, designed, patented, and tested at Menlo Park. The equipment for the lighting system was manufactured by the Edison companies. In his prime and during the development of his electric lighting system, Edison depended on investment bankers for funds. By 1882 he had built a system with 12,843 light bulbs within a few blocks of Wall Street, New York. How consistent he was in his approach can be seen by the way he finally marketed the electric light. To emphasize to customers that they were buying an old familiar product—light—and not a new unfamiliar product—electricity—the bills were for light-hours rather than kilowatts. Edison exhibited his lights at the Philadelphia Exhibition in 1884, as shown in Figure 2.2-2. Today, nearly one-quarter of the electricity sold in the United States is devoted to lighting uses.

Figure 2.2-1 Thomas A. Edison in the laboratory at Menlo Park, New Jersey. He is shown with his Edison lamps, discovered in 1879. Courtesy of Edison National Historical Site.

Figure 2.2-2 Edison's tower of lights, shown at the Philadelphia Exhibition, 1884. Courtesy of Edison National Historical Site.

The great Electrical Exhibition of 1884 provided the site for the first annual meeting of a new U.S. professional society, the American Institute of Electrical Engineers. Edison, among others, called for a college course of study in electrical engineering. By the mid-1880s such courses had begun at the Massachusetts Institute of Technology and Cornell University. By 1890 electrical engineering as a profession and an academic discipline had begun with Edison as role model and leader and then progressed to the formation of a professional society and a college course of study.

2.3 ENGINEERING AND LINEAR MODELS

One view of engineering describes it as the activity of problem solving under constraints. A somewhat expanded definition follows:

Engineering combines the study of mathematics and natural and social sciences to direct the forces of nature for the benefit of humankind.

The art of engineering is to take a bright idea and, using money, materials, knowledgeable people, and a regard for the environment, produce something the buyer wants at an affordable price.

The ultimate objective of engineering work is the design and production of specific items, often called hardware. Many such devices incorporate electric circuits. The engineer, in the accomplishment of a task,

1. Analyzes the problem.
2. Synthesizes a solution.
3. Evaluates the results and, possibly, resynthesizes a solution.

Another way of looking at the task is to say that the engineer must

1. Understand the problem.
2. Devise a plan to solve the problem.
3. Carry out the plan.
4. Look back and check the solution obtained.
5. Based on the solution, possibly reformulate the initial problem and solve it again.

Engineers use *models* to represent the elements of an electric circuit. We generate models for manufactured elements and devices in order to manipulate parameters and establish bounds on the devices' operating characteristics. Intuition demands a simple model and ease of solution. This may lead to simplification in the form of aggregations of variables and properties of a device or circuit. Although the model serves to illuminate the real thing, it is *not* the real thing. Thus, a model is constructed to facilitate understanding and enhance prediction. In our work we will construct models of elements and then interconnect them to form a *circuit model*.

A **model** is an object or pattern of objects or an equation that represents an element or circuit.

An automobile ignition circuit is shown in Figure 2.3-1*a*, and the circuit model of the ignition circuit is shown in Figure 2.3-1*b*.

The idealized models of electric devices are precisely defined. It is important to distinguish between actual devices and their idealized models, which we call circuit elements.

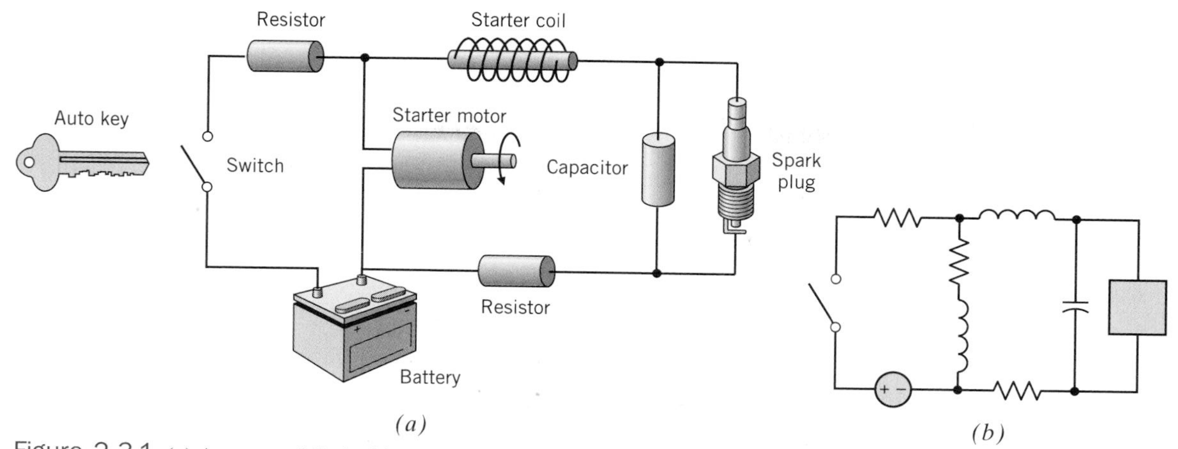

Figure 2.3-1 (*a*) An automobile ignition circuit and (*b*) model of the ignition circuit for starting a car.

The goal of circuit analysis is to predict the quantitative electrical behavior of physical circuits. Its aim is to predict and to explain the terminal voltages and terminal currents of the circuit elements and thus the overall operation of the circuit.

A device or element is *linear* if the element's excitation and response satisfy certain properties. Consider the element shown in Figure 2.3-2. Suppose that the excitation is the current i, and the response is the voltage v. When the element is subjected to a current i_1, it provides a response v_1. Furthermore, when the element is subjected to a current i_2, it provides a response v_2. For a linear circuit, it is necessary that the excitation $i_1 + i_2$ result in a response $v_1 + v_2$. This is usually called the *principle of superposition.*

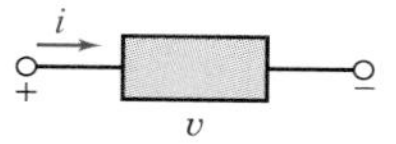

Figure 2.3-2 An element with an excitation current i and a response v.

Furthermore, it is necessary that the magnitude scale factor be preserved for a linear element. If the element is subjected to an excitation of ki, where k is a constant multiplier, then it is necessary that the response of a linear device be equal to kv. This is called the *property of homogeneity*. A circuit is linear if, and only if, the properties of superposition and homogeneity are satisfied for all excitations and responses.

A **linear element** satisfies the properties of superposition and homogeneity.

Let us restate mathematically the two required properties of a linear circuit, using the arrow notation to imply the transition from excitation to response:

$$i \longrightarrow v$$

Then we may state the two properties required as follows.

Superposition:

$$\begin{aligned} i_1 &\longrightarrow v_1 \\ i_2 &\longrightarrow v_2 \\ \text{then} \quad i_1 + i_2 &\longrightarrow v_1 + v_2 \end{aligned} \tag{2.3-1}$$

Homogeneity:

$$\begin{aligned} i &\longrightarrow v \\ \text{then} \quad k\,i &\longrightarrow k\,v \end{aligned} \tag{2.3-2}$$

A device that does not satisfy either the superposition or homogeneity principles is said to be nonlinear.

Example 2.3-1

Consider the element represented by the relationship between current and voltage as

$$v = Ri$$

Determine whether this device is linear.

Solution

The response to a current i_1 is

$$v_1 = Ri_1$$

The response to a current i_2 is

$$v_2 = Ri_2$$

The sum of these responses is

$$v_1 + v_2 = Ri_1 + Ri_2 = R(i_1 + i_2)$$

Since the sum of the responses to i_1 and i_2 is equal to the response to $i_1 + i_2$, the principle of superposition is satisfied. Next, consider the principle of homogeneity. Since

$$v_1 = Ri_1$$

we have for an excitation $i_2 = ki_1$

$$v_2 = Ri_2 = Rki_1$$

Therefore,

$$v_2 = kv_1$$

satisfies the principle of homogeneity. Because the element satisfies both the properties of superposition and homogeneity, it is linear.

Example 2.3-2

Now let us consider an element represented by the relationship between current and voltage:

$$v = i^2$$

Determine whether this device is linear.

Solution

The response to a current i_1 is

$$v_1 = i_1^2$$

The response to a current i_2 is

$$v_2 = i_2^2$$

The sum of these responses is

$$v_1 + v_2 = i_1^2 + i_2^2$$

The response to $i_1 + i_2$ is

$$(i_1 + i_2)^2 = i_1^2 + 2i_1i_2 + i_2^2$$

Since,

$$i_1^2 + i_2^2 \neq (i_1 + i_2)^2$$

the principle of superposition is not satisfied. Therefore, the device is nonlinear.

The distinguishing feature of an electric circuit component is that its behavior is described in terms of a voltage–current relation. The "$v-i$" characteristic may be obtained experimentally or from physical principles. Although no device is exactly linear for all values of current, we often assume a range of linear operation. Edison's first commercially available lamp is shown in Figure 2.3-3*a*. For example, consider the $v-i$ characteristic of the incandescent lamp, shown in Figure 2.3-3*b*. The lamp is essentially linear for the range

$$-i_{\mathrm{m}} < i < i_{\mathrm{m}}$$

That is, i_{m} indicates the range of current for linearity.

In this book we will be concerned only with linear models of circuit components. In other words, only linear circuits will be considered. Of course, we recognize that these linear models are accurate representations for only some limited range of operation.

Exercise 2.3-1 Determine whether the following element is linear.

$$v = \frac{di}{dt}$$

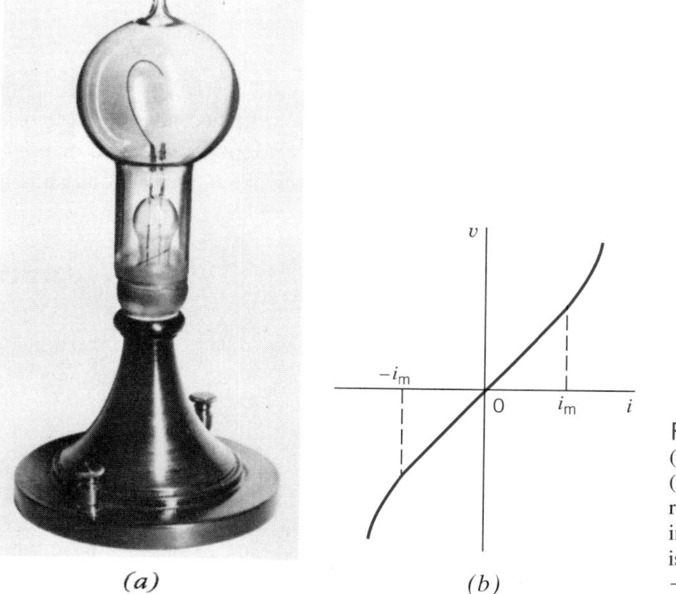

Figure 2.3-3
(*a*) An incandescent lamp.
(*b*) Voltage–current relationship for an incandescent lamp. The lamp is linear within the range $-i_{\mathrm{m}} < i < i_{\mathrm{m}}$.

Exercise 2.3-2 Show that the following element is nonlinear.

$$v = Ri, \qquad i \geq 0$$

$$v = 0, \qquad i < 0$$

Exercise 2.3-3 The voltage–current relationship for two elements is shown in Figure E 2.3-3. Determine a linear model for each element and indicate the range of linearity. Assume that the elements of Figure E 2.3-3 normally operate around $i = 0$.

Answers: (a) $v = 2.5i$ for $-1 < i < 1$

(b) $v = -1.333i$ for $-1.5 < i < 1.5$

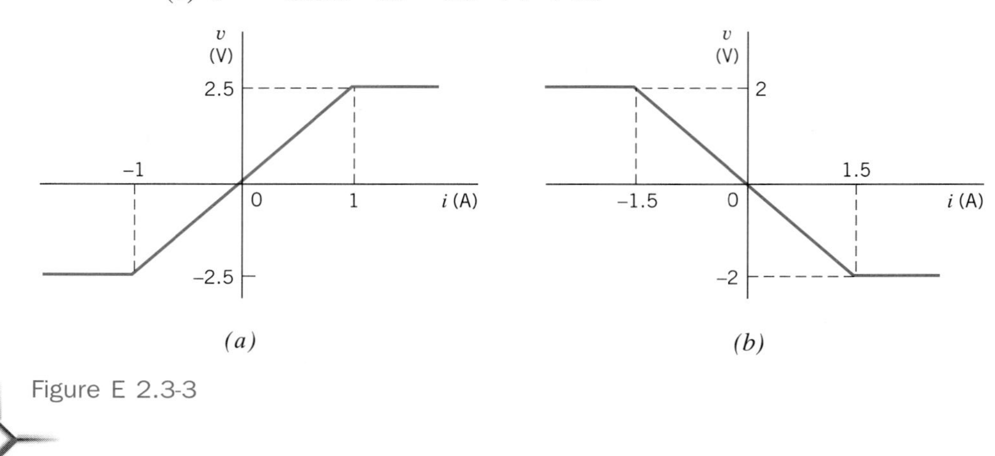

Figure E 2.3-3

2.4 ACTIVE AND PASSIVE CIRCUIT ELEMENTS

Entry node, Exit node, $+$, v, $-$, i, Exit node, Entry node

(a) (b)

Figure 2.4-1 (*a*) The entry node of the current i is the positive node of the voltage v, (*b*) the entry node of the current i is the negative node of the voltage v. The current flows from the entry node to the exit node.

We may classify circuit elements in two categories, *passive* and *active*, by determining whether they absorb energy or supply energy. An element is said to be passive if the total energy delivered to it from the rest of the circuit is always nonnegative (zero or positive). Then for a passive element, with the current flowing into the + terminal as shown in Figure 2.4-1*a*, this means that

$$w = \int_{-\infty}^{t} vi \, d\tau \geq 0 \qquad (2.4\text{-}1)$$

for all values of t.

A **passive element** absorbs energy.

An element is said to be *active* if it is capable of delivering energy. Thus, an active element violates Eq. 2.4-1 when it is represented by Figure 2.4-1*a*. In other words, an active element is one that is capable of generating energy. Active elements are potential sources of energy, whereas passive elements are sinks or absorbers of energy. Examples of active elements include batteries and generators. Consider the element shown in Figure 2.4-1*b*. Note that the current flows into the negative terminal and out of the positive terminal. This element is said to be active if

$$w = \int_{-\infty}^{t} vi \, d\tau \geq 0 \qquad (2.4\text{-}2)$$

for at least one value of t.

An **active element** is capable of supplying energy.

Example 2.4-1

A circuit has an element represented by Figure 2.4-1*b* where the current is a constant 5 A and the voltage is a constant 6 V. Find the energy supplied over the time interval 0 to T.

Solution

Since the current enters the negative terminal, the energy *supplied by* the element is given by

$$w = \int_{0}^{T} (6)\,(5)\, d\tau = 30T \text{ J}$$

Thus, the device is a generator or an active element, in this case a dc battery.

A collection of circuit elements or components is shown in Figure 2.4-2. Useful circuits include both active and passive elements that are assembled into a circuit.

Figure 2.4-2 Collection of active and passive circuit elements used for an electrical circuit. Courtesy of Hewlett-Packard Co.

2.5 RESISTORS

The ability of a material to resist the flow of charge is called its *resistivity,* ρ. Materials that are good electrical insulators have a high value of resistivity. Materials that are good conductors of electric current have low values of resistivity. Resistivity values for selected materials are given in Table 2.5-1. Copper is commonly used for wires since it permits current to flow relatively unimpeded. Silicon is commonly used to provide resistance in semiconductor electric circuits. Polystyrene is used as an insulator.

Table 2.5-1 **Resistivities of Selected Materials**

Material	Resistivity ρ (ohm-cm)
Polystyrene	1×10^{18}
Silicon	2.3×10^{5}
Carbon	4×10^{-3}
Aluminum	2.7×10^{-6}
Copper	1.7×10^{-6}

Resistance is the physical property of an element or device that impedes the flow of current; it is represented by the symbol R.

Georg Simon Ohm was able to show that the current in a circuit composed of a battery and a conducting wire of uniform cross section could be expressed as

Figure 2.5-1 Georg Simon Ohm (1787-1854), who determined Ohm's law in 1827. The ohm was chosen as the unit of electrical resistance in his honor.

$$i = \frac{Av}{\rho L} \tag{2.5-1}$$

where A is the cross-sectional area, ρ the resistivity, L the length, and v the voltage across the wire element. Ohm, who is shown in Figure 2.5-1, defined the constant resistance R as

$$R = \frac{\rho L}{A} \tag{2.5-2}$$

Ohm's law, which related the voltage and current, was published in 1827 as

$$v = Ri \tag{2.5-3}$$

The unit of resistance R was named the ohm in honor of Ohm and is usually abbreviated by the symbol Ω (capital omega), where 1 Ω = 1 V/A. The resistance of a 10-m length of common TV cable is 2 mΩ.

An element that has a resistance R is called a *resistor*. A resistor is represented by the two-terminal symbol shown in Figure 2.5-2. Ohm's law, Eq. 2.5-3, requires that the i-versus-v relationship be linear. As shown in Figure 2.5-3, a resistor may become nonlinear outside its normal rated range of operation. We will assume that a resistor is linear

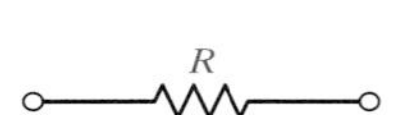

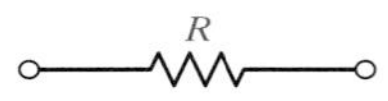

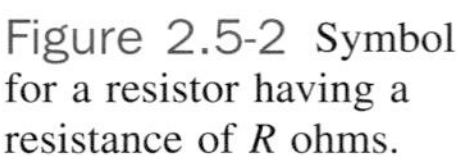

Figure 2.5-2 Symbol for a resistor having a resistance of R ohms.

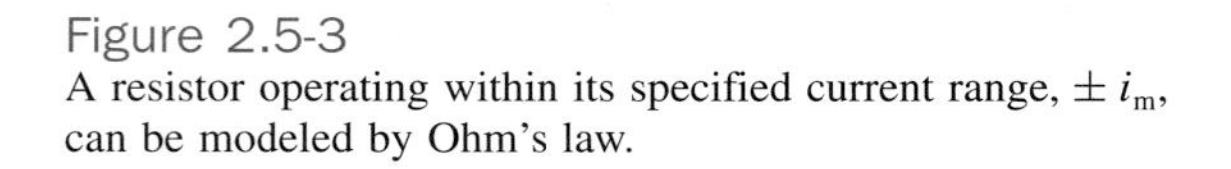

Figure 2.5-3
A resistor operating within its specified current range, $\pm\, i_m$, can be modeled by Ohm's law.

unless stated otherwise. Thus, we will use a linear model of the resistor as represented by Ohm's law.

In Figure 2.5-4 the element current and element voltage of a resistor are labeled. The relationship between the directions of this current and voltage is important. The voltage direction marks one resistor terminal + and the other −. The current flows from the terminal marked + to the terminal marked −. This relationship between the current and voltage reference directions is a convention called the passive convention. Ohm's law states that when the element voltage and the element current adhere to the passive convention, then

Figure 2.5-4 A resistor with element current and element voltage.

$$v = Ri \qquad (2.5\text{-}4)$$

Consider Figure 2.5-4. The element currents i_a and i_b are the same except for the assigned direction, so

$$i_a = -i_b$$

The element current i_a and the element voltage v adhere to the passive convention,

$$v = Ri_a$$

Replacing i_a by $-i_b$ gives

$$v = -Ri_b$$

There is a minus sign in this equation because the element current i_b and the element voltage v do not adhere to the passive convention. We must pay attention to the current direction so that we don't overlook this minus sign.

Ohm's law, Eq. 2.5-4, can also be written as

$$i = Gv \qquad (2.5\text{-}5)$$

where G denotes the *conductance* in siemens (S) and is the reciprocal of R; that is, $G = 1/R$. Many engineers denote the units of conductance as mhos with the symbol ℧, which is an inverted omega (mho is ohm spelled backward). However, we will use SI units and retain siemens as the units for conductance.

Most discrete resistors fall into one of four basic categories: carbon composition, carbon film, metal film, or wirewound. Carbon composition resistors have been in use for nearly 100 years and are still popular. Carbon film resistors have supplanted carbon composition resistors for many general-purpose uses because of their lower cost and better tolerances. Two wirewound resistors are shown in Figure 2.5-5.

Resistors are sensitive to temperature change from an assumed ambient temperature of 27°C. The temperature sensitivity is defined as

$$k = \frac{1}{R}\frac{dR}{dT}$$

where R is the resistance, T is the temperature in degrees Celsius, and the units of k are parts per million per degree Celsius. A typical carbon film resistor has a coefficient of −400 ppm/°C.

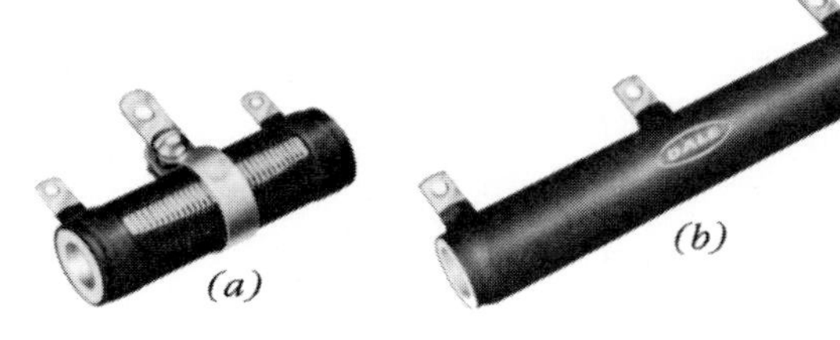

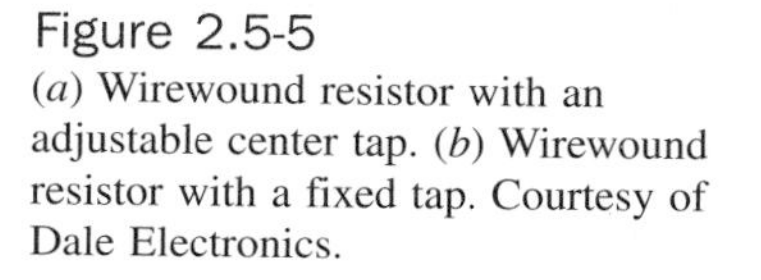

Figure 2.5-5
(*a*) Wirewound resistor with an adjustable center tap. (*b*) Wirewound resistor with a fixed tap. Courtesy of Dale Electronics.

Figure 2.5-6 Small thick-film resistor chips used for miniaturized circuits. Courtesy of Corning Electronics.

Figure 2.5-7 A 1/4-watt metal film resistor. The body of the resistor is 6-mm long. Courtesy of Dale Electronics.

Thick-film resistors, as shown in Figure 2.5-6, are used in circuits because of their low cost and small size. General-purpose resistors are available in standard values for tolerances of 2, 5, 10, and 20 percent. Carbon composition resistors and some wirewounds have a color code with three to five bands. A color code is a system of standard colors adopted for identification of the resistance of resistors. Figure 2.5-7 shows a metal film resistor with its color bands. This is a 1/4-watt resistor, implying that it should be operated at or below 1/4 watt of power delivered to it. The normal range of resistors is from less than 1 ohm to 10 megohms. Typical values readily available are $R = N10^n$ ohms, where $n = 1, 2, \ldots, 8$.[1]

The power delivered to a resistor (when the passive convention is used) is

$$p = vi = v\left(\frac{v}{R}\right) = \frac{v^2}{R} \tag{2.5-6}$$

Alternatively, since $v = iR$, we can write the equation for power as

$$p = vi = (iR)i = i^2R \tag{2.5-7}$$

Thus, the power is expressed as a nonlinear function of the current i through the resistor or of the voltage v across it.

Recall the definition of a passive element as one for which the energy absorbed is always nonnegative. The equation for energy delivered to a resistor is

$$w = \int_{-\infty}^{t} p \, d\tau = \int_{-\infty}^{t} i^2R \, d\tau \tag{2.5-8}$$

Since i^2 is always positive, the energy is always positive and the resistor is a passive element.

Resistance is a measure of an element's ability to dissipate power irreversibly.

[1] See Appendix E for a description of color code and the standard values for N.

Example 2.5-1

Let us devise a model for a car battery when the lights are left on and the engine is off. We have all experienced or seen a car parked with its lights on. If we leave the car for a period, the battery will "run down" or "go dead." An auto battery is a 12-V constant-voltage source, and the light bulb can be modeled by a resistor of 6 ohms. The circuit is shown in Figure 2.5-8. Let us find the current i, the power p, and the energy supplied by the battery for a four-hour period.

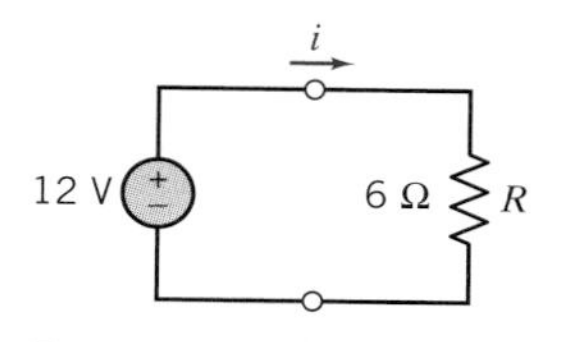

Figure 2.5-8 Model of a car battery and the headlight lamp.

Solution

According to Ohm's law, Eq. 2.5-4, we have

$$v = Ri$$

Since $v = 12$ V and $R = 6\ \Omega$, we have $i = 2$ A.

In order to find the power delivered by the battery, we use

$$p = vi = 12(2) = 24 \text{ W}$$

Finally, the energy delivered in the four-hour period is

$$w = \int_0^t p\, d\tau$$

$$= 24t = 24(60 \times 60 \times 4) = 3.46 \times 10^5 \text{J}$$

Since the battery has a finite amount of stored energy, it will deliver this energy and eventually be unable to deliver further energy without recharging. We then say the battery is run down or dead until recharged. A typical auto battery may store 10^6 J in a fully charged condition.

Exercise 2.5-1 Find the power absorbed by a 100-ohm resistor when it is connected directly across a constant 10-V source.

Answer: 1 W

Exercise 2.5-2 A voltage source $v = 10 \cos t$ V is connected across a resistor of 10 ohms. Find the power delivered to the resistor.

Answer: $10 \cos^2 t$ W

2.6 INDEPENDENT SOURCES

Some devices are intended to supply energy to a circuit. These devices are called sources. Sources are categorized as being one of two types: voltage sources and current sources. Figure 2.6-1a shows the symbol that is used to represent a voltage source. The voltage of a voltage source is specified, but the current is determined by the rest of the circuit. A voltage source is described by specifying the function $v(t)$, for example,

$$v(t) = 12 \cos 1000t \quad \text{or} \quad v(t) = 9 \quad \text{or} \quad v(t) = 12 - 2t$$

An active two-terminal element that supplies energy to a circuit is a *source* of energy. An *independent voltage source* provides a specified voltage independent of the current through it and is independent of any other circuit variable.

A **source** is a voltage or current generator capable of supplying energy to a circuit.

An *independent current source* provides a current independent of the voltage across the source element and is independent of any other circuit variable. Thus, when we say a source is independent, we mean it is independent of any other voltage or current in the circuit.

An **independent source** is a voltage or current generator not dependent on other circuit variables.

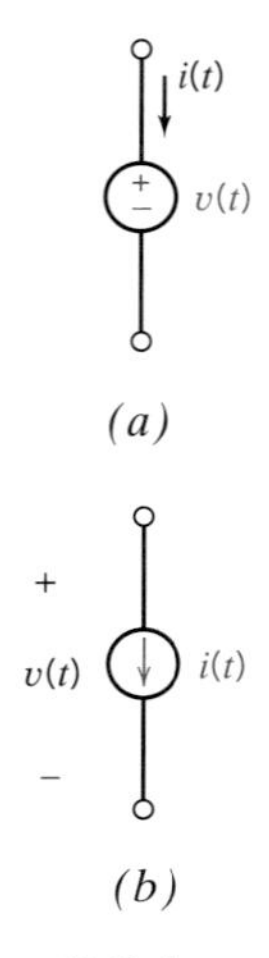

Figure 2.6-1
(*a*) Voltage source.
(*b*) Current source.

Suppose the voltage source is a battery and

$$v(t) = 9 \text{ volts}$$

The voltage of this battery is known to be 9 volts regardless of the circuit in which the battery is used. In contrast, the current of the voltage source is not known and depends on the circuit in which the source is used. The current could be 6 amps when the voltage source is connected to one circuit and 6 milliamps when the voltage source is connected to another circuit.

Figure 2.6-1*b* shows the symbol that is used to represent a current source. The current of a current source is specified, but the voltage is determined by the rest of the circuit. A current source is described by specifying the function $i(t)$, for example,

$$i(t) = 6 \sin 500t \quad \text{or} \quad i(t) = -0.25 \quad \text{or} \quad i(t) = t + 8$$

A current source specified by $i(t) = -0.25$ milliamps will have a current of -0.25 milliamps in any circuit in which it is used. The voltage across this current source will depend on the particular circuit.

The preceding paragraphs have ignored some complexities in order to give a simple description of the way sources work. The voltage across a 9-volt battery may not actually be 9 volts. This voltage depends on the age of the battery, the temperature, variations in manufacturing, and the battery current. It is useful to make a distinction between real sources, such as batteries, and the simple voltage and current sources described in the preceding paragraphs. It would be *ideal* if the real sources worked like these simple sources. Indeed, the word *ideal* is used to make this distinction. The simple sources described in the previous paragraph are called the **ideal voltage source** and the **ideal current source.**

The voltage of an **ideal voltage source** is given to be a specified function, say $v(t)$. The current is determined by the rest of the circuit.

The current of an **ideal current source** is given to be a specified function, say $i(t)$. The voltage is determined by the rest of the circuit.

An **ideal source** is a voltage or a current generator independent of the current through the voltage source or the voltage across the current source.

Example 2.6-1

Consider the plight of the engineer who needs to analyze a circuit containing a 9-volt battery. Is it really necessary for this engineer to include the dependence of battery voltage on the age of the battery, the temperature, variations in manufacturing, and the battery

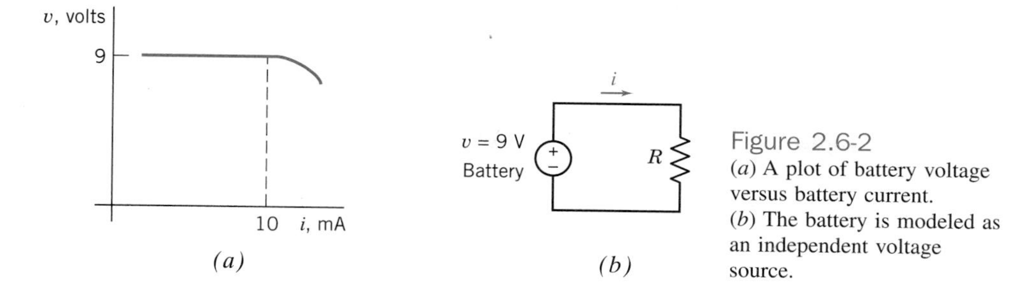

Figure 2.6-2
(*a*) A plot of battery voltage versus battery current.
(*b*) The battery is modeled as an independent voltage source.

current in this analysis? Hopefully not. We expect the battery to act enough like an ideal 9-volt voltage source that the differences can be ignored. In this case it is said that the battery is **modeled** as an ideal voltage source.

To be specific, consider a battery specified by the plot of voltage versus current shown in Figure 2.6-2*a*. This plot indicates that the battery voltage will be $v = 9$ volts when $i \leq 10$ milliamps. As the current increases above 10 milliamps the voltage decreases from 9 volts. When $i \leq 10$ milliamps the dependence of the battery voltage on the battery current can be ignored and the battery can be modeled as an independent voltage source.

Suppose a resistor is connected across the terminals of the battery as shown in Figure 2.6-2*b*. The battery current will be

$$i = \frac{v}{R} \tag{2.6-1}$$

The relationship between v and i shown in Figure 2.6-2*a* complicates this equation. This complication can be safely ignored when $i \leq 10$ milliamps. When the battery is modeled as an ideal 9-volt voltage source, the voltage source current is given by

$$i = \frac{9}{R} \tag{2.6-2}$$

The distinction between these two equations is important. Equation 2.6-1, involving the $v-i$ relationship shown in Figure 2.6-2*a*, is more accurate but also more complicated. Eq. 2.6-2 is simpler but may be inaccurate.

Suppose that $R = 1000$ ohms. Eq. 2.6-2 gives the current of the ideal voltage source:

$$i = \frac{9 \text{ volts}}{1000 \text{ ohms}} = 9 \text{ milliamps} \tag{2.6-3}$$

Since this current is less than 10 milliamps, the ideal voltage source is a good model for the battery and it is reasonable to expect that the battery current is 9 milliamps.

Suppose instead that $R = 600$ ohms. Once again, Eq. 2.6-2 gives the current of the ideal voltage source.

$$i = \frac{9 \text{ volts}}{600 \text{ ohms}} = 15 \text{ milliamps} \tag{2.6-4}$$

Since this current is greater than 10 milliamps, the ideal voltage source is not a good model for the battery. In this case it is reasonable to expect that the battery current is different from the current for the ideal voltage source.

Engineers frequently face a trade-off when selecting a model for a device. Simple models are easy to work with but may not be accurate. Accurate models are usually more complicated and harder to use. The conventional wisdom suggests that simple models be used

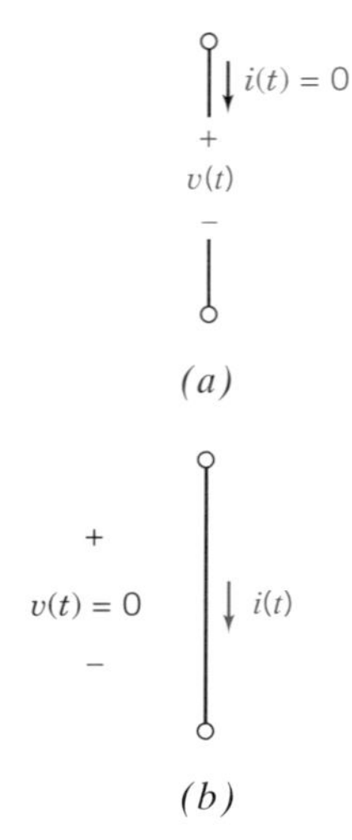

Figure 2.6-3 (*a*) Open circuit. (*b*) Short circuit.

first. The results obtained using the models must be checked to verify that use of these simple models is appropriate. More accurate models are used when necessary.

The **short circuit** and **open circuit** are special cases of ideal sources. A **short circuit** is an ideal voltage source having $v(t) = 0$. The current in a short circuit is determined by the rest of the circuit. An **open circuit** is an ideal current source having $i(t) = 0$. The voltage across an open circuit is determined by the rest of the circuit. Figure 2.6-3 shows the symbols used to represent the short circuit and the open circuit. Notice that the power absorbed by each of these devices is zero.

Open and short circuits can be added to a circuit without disturbing the branch currents and voltages of all the other devices in the circuit. Figure 2.7-3 shows how this can be done. Figure 2.7-3*a* shows an example circuit. In Figure 2.7-3*b* an open circuit and a short circuit have been added to this example circuit. The open circuit was connected between two nodes of the original circuit. In contrast, the short circuit was added by cutting a wire and inserting the short circuit. Adding open circuits and short circuits to a network in this way does not change the network.

Open circuits and short circuits can also be described as special cases of resistors. A resistor with resistance $R = 0\,(G = \infty)$ is a short circuit. A resistor with conductance $G = 0\,(R = \infty)$ is an open circuit.

2.7 VOLTMETERS AND AMMETERS

Measurements of dc current and voltage are made with direct-reading (analog) or digital meters, as shown in Figure 2.7-1. A direct-reading meter has an indicating pointer whose angular deflection depends on the magnitude of the variable it is measuring. A digital meter displays a set of digits indicating the measured variable value.

To measure a voltage or current, a meter is connected to a circuit using terminals called probes. These probes are color coded to indicate the reference direction of the variable being measured. Frequently, meter probes are colored red and black. In this text, the probes will be shown as blue and black. An ideal voltmeter measures the voltage from the red to the black probe. The red terminal is the positive terminal, and the black terminal is the negative terminal (see Figure 2.7-2*b*).

An ideal ammeter measures the current flowing through its terminals, as shown in Figure 2.7-2*a* and has zero voltage, v_m, across its terminals. An ideal voltmeter measures the voltage across its terminals, as shown in Figure 2.7-2*b*, and has terminal current, i_m, equal to zero. Practical measuring instruments only approximate the ideal conditions. For a prac-

(*a*)

2.9

(*b*)

Figure 2.7-1 (*a*) A direct-reading (analog) meter. (*b*) A digital meter.

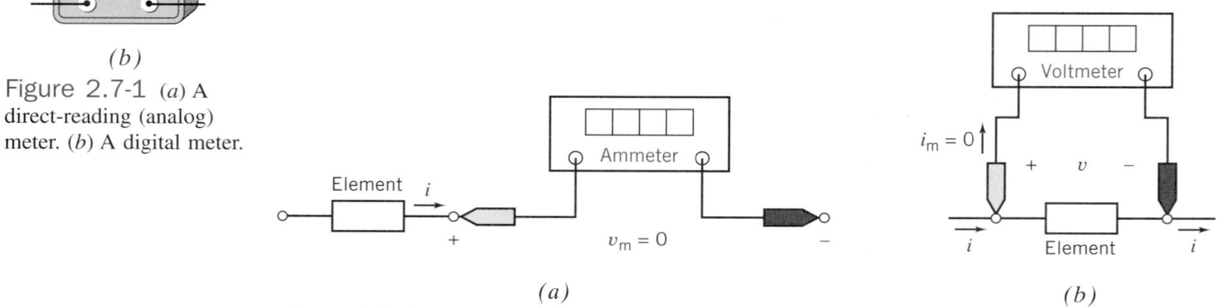

(*a*) (*b*)

Figure 2.7-2 (*a*) Ideal ammeter. (*b*) Ideal voltmeter.

tical ammeter the voltage across its terminals is usually negligibly small. Similarly, the current into a voltmeter is usually negligible.

Ideal voltmeters act like open circuits, and ideal ammeters act like short circuits. In other words, the model of an ideal voltmeter is an open circuit and the model of an ideal ammeter is a short circuit. Consider the circuit of Figure 2.7-3*a* and then add an open circuit with a voltage v and a short circuit with a current i as shown in Figure 2.7-3*b*. In Figure 2.7-3*c* the open circuit has been replaced by a voltmeter, and the short circuit has been replaced by an ammeter. The voltmeter will measure the voltage labeled v in Figure 2.7-3*b* while the ammeter will measure the current labeled i. Notice that Figure 2.7-3*c* could be obtained from Figure 2.7-3*a* by adding a voltmeter and an ammeter. Ideally, adding the voltmeter and ammeter in this way does not disturb the circuit. One more interpretation of Figure 2.7-3 is useful. Figure 2.7-3*b* could be formed from Figure 2.7-3*c* by replacing the voltmeter and the ammeter by their (ideal) models.

The reference direction is an important part of an element voltage or element current. Figures 2.7-4 and 2.7-5 show that attention must be paid to reference directions when measuring an element voltage or element current. Figure 2.7-4*a* shows a voltmeter. Voltmeters have two color-coded probes. This color coding indicates the reference direction of the voltage being measured. In Figures 2.7-4*b* and 2.7-4*c* the voltmeter is used to measure the voltage across the 6-kΩ resistor. When the voltmeter is connected to the circuit as shown in Figure 2.7-4*b*, the voltmeter measures v_a, with + on the left, at the red probe. When the voltmeter probes are interchanged as shown in Figure 2.7-4*c*, the voltmeter measures v_b, with + on the right, again at the red probe. Note $v_b = -v_a$.

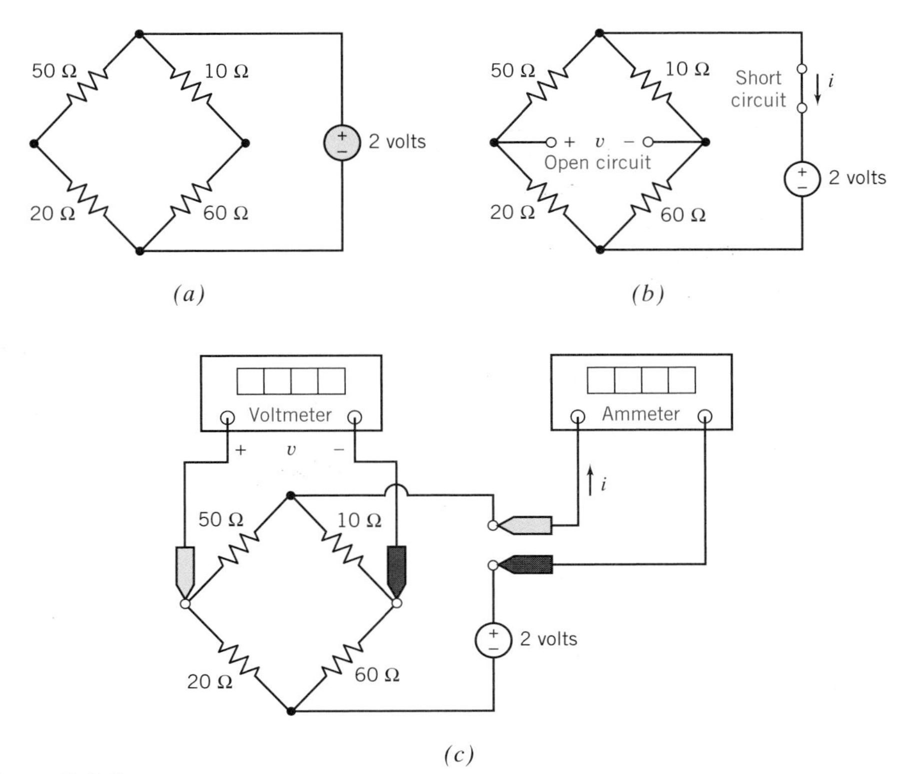

Figure 2.7-3 (*a*) An example circuit, (*b*) plus an open circuit and a short circuit. (*c*) The open circuit is replaced by a voltmeter, and the short circuit is replaced by an ammeter. All resistances are in ohms.

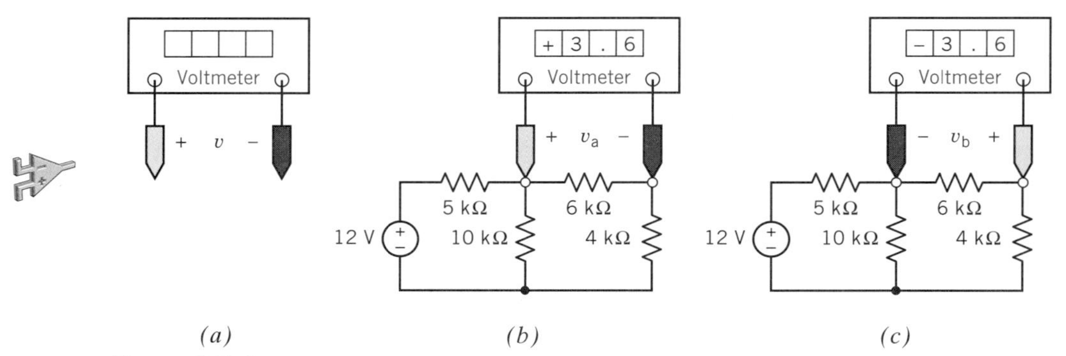

Figure 2.7-4 (*a*) The correspondence between the color-coded probes of the voltmeter and the reference direction of the measured voltage. In (*b*) the + sign of v_a is on the left, while in (*c*) the + sign of v_b is on the right. The colored probe is shown here in blue. In the laboratory this probe will be red. We will refer to the colored probe as the "red probe".

Figure 2.7-5*a* shows an ammeter. Ammeters have two color-coded probes. This color coding indicates the reference direction of the current being measured. In Figures 2.7-5*b,c* the ammeter is used to measure the current in the 6-kΩ resistor. When the ammeter is connected to the circuit as shown in Figure 2.7-5*b*, the ammeter measures i_a, directed from the red probe toward the black probe. When the ammeter probes are interchanged as shown in Figure 2.7-5*c*, the ammeter measures i_b, again directed from the red probe toward the black probe. Note $i_b = -i_a$.

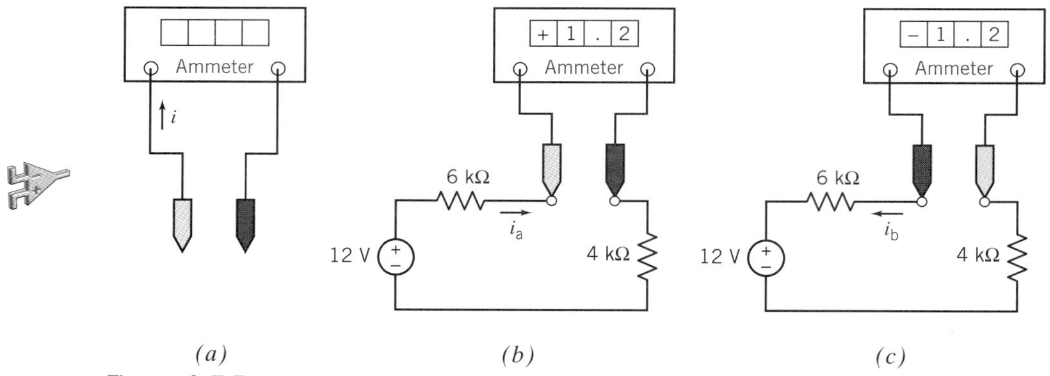

Figure 2.7-5 (*a*) The correspondence between the color-coded probes of the ammeter and the reference direction of the measured current. In (*b*) the current i_a is directed to the right, while in (*c*) the current i_b is directed to the left. The colored probe is shown here in blue. In the laboratory this probe will be red. We will refer to the colored probe as the "red probe".

2.8 DEPENDENT SOURCES

Some devices, such as transistors and amplifiers, act like controlled sources. For example, the output voltage of an amplifier is controlled by the input voltage of that amplifier. Such devices can be modeled using dependent sources. Dependent sources consist of two elements: the controlling element and the controlled element. The controlling element is either an open circuit or a short circuit. The controlled element is either a voltage source or a current source. There are four types of dependent source that correspond to the four ways of choosing a controlling element and a controlled element. These four dependent sources are called the voltage-controlled voltage source (VCVS), current-controlled voltage source (CCVS), voltage-controlled current source (VCCS), and current-controlled current source (CCCS). The symbols that represent dependent sources are shown in Table 2.8-1. Notice that the symbol for the controlled element is different from the symbol used to represent independent voltage and current sources. Indeed, the terms **independent voltage source** and **independent current source** are used to emphasize the distinction between these sources and dependent sources.

A **dependent source** is a voltage or current generator that depends on another circuit variable.

Consider the CCVS shown in Table 2.8-1. The controlling element is a short circuit. The element current and voltage of the controlling element are denoted as i_c and v_c. The voltage across a short circuit is zero, so $v_c = 0$. The short circuit current, i_c, is the controlling signal of this dependent source. The controlled element is a voltage source. The element current and voltage of the controlled element are denoted as i_d and v_d. The voltage v_d is controlled by i_c:

$$v_d = r\, i_c$$

The constant r is called the gain of the CCVS. The current i_d, like the current in any voltage source, is determined by the rest of the circuit.

Table 2.8-1 **Dependent Sources**

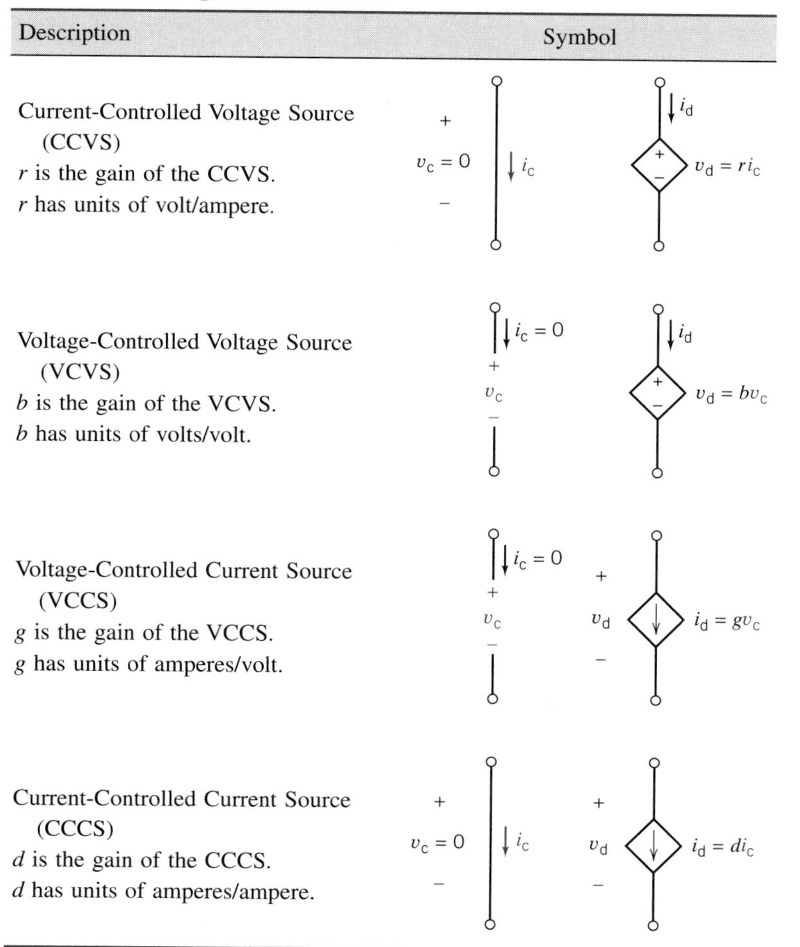

Description	Symbol
Current-Controlled Voltage Source (CCVS) r is the gain of the CCVS. r has units of volt/ampere.	$v_c = 0$, i_c; i_d, $v_d = ri_c$
Voltage-Controlled Voltage Source (VCVS) b is the gain of the VCVS. b has units of volts/volt.	$i_c = 0$, v_c; i_d, $v_d = bv_c$
Voltage-Controlled Current Source (VCCS) g is the gain of the VCCS. g has units of amperes/volt.	$i_c = 0$, v_c; v_d, $i_d = gv_c$
Current-Controlled Current Source (CCCS) d is the gain of the CCCS. d has units of amperes/ampere.	$v_c = 0$, i_c; v_d, $i_d = di_c$

Next consider the VCVS shown in Table 2.8-1. The controlling element is an open circuit. The current in an open circuit is zero, so $i_c = 0$. The open circuit voltage, v_c, is the controlling signal of this dependent source. The controlled element is a voltage source. The voltage v_d is controlled by v_c:

$$v_d = b\, v_c$$

The constant b is called the gain of the VCVS. The current i_d is determined by the rest of the circuit.

The controlling element of the VCCS shown in Table 2.8-1 is an open circuit. The current in this open circuit is $i_c = 0$. The open circuit voltage, v_c, is the controlling signal of this dependent source. The controlled element is a current source. The current i_d is controlled by v_c:

$$i_d = g\, v_c$$

The constant g is called the gain of the VCCS. The voltage v_d, like the voltage across any current source, is determined by the rest of the circuit.

The controlling element of the CCCS shown in Table 2.8-1 is a short circuit. The voltage across this open circuit is $v_c = 0$. The short circuit current, i_c, is the controlling signal of this dependent source. The controlled element is a current source. The current i_d is controlled by i_c:

$$i_d = d\, i_c$$

The constant d is called the gain of the CCCS. The voltage v_d, like the voltage across any current source, is determined by the rest of the circuit.

Figure 2.8-1 illustrates the use of dependent sources to model electronic devices. In certain circumstances, the behavior of the transistor shown in Figure 2.8-1*a* can be represented using the model shown in Figure 2.8-1*b*. This model consists of a dependent source and a resistor. The controlling element of the dependent source is an open circuit connected across the resistor. The controlling voltage is v_{be}. The gain of the dependent source is g_m. The dependent source is used in this model to represent a property of the transistor, namely, that the current i_c is proportional to the voltage v_{be}, that is,

$$i_c = g_m v_{be}$$

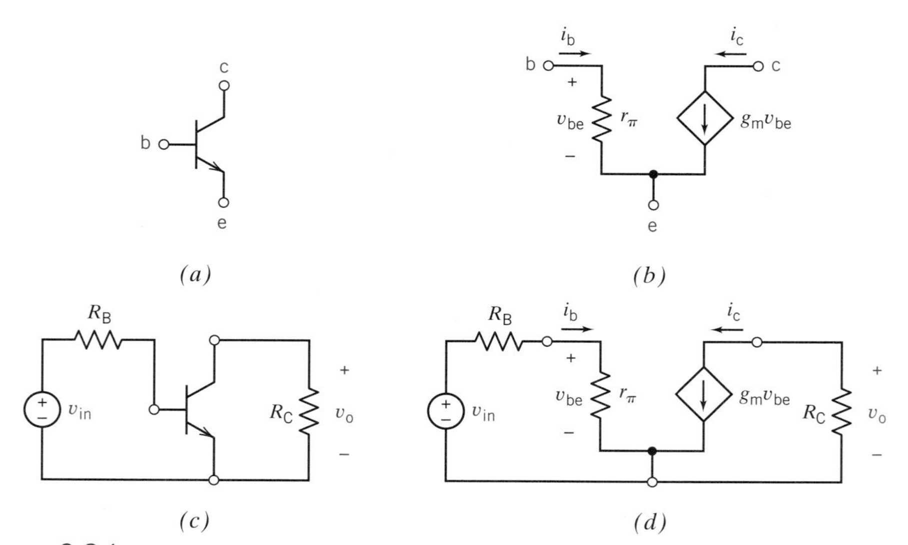

Figure 2.8-1 (*a*) A symbol for a transistor. (*b*) A model of the transistor. (*c*) A transistor amplifier. (*d*) A model of the transistor amplifier.

where g_m has units of amperes/volt. Figures 2.8-1*c,d* illustrate the utility of this model. Figure 2.8-1*d* is obtained from Figure 2.8-1*c* by replacing the transistor by the transistor model. The voltage gain of the transistor amplifier shown in Figure 2.8-1*c* is defined as

$$A = \frac{v_o}{v_{in}}$$

This gain can be calculated by analyzing Figure 2.8-1*d* instead of Figure 2.8-1*c*. Analysis of circuits such as Figure 2.8-1*d* will be discussed in Chapter 3.

Example 2.8-1

Determine the power absorbed by the VCVS in Figure 2.8-2.

Solution

The VCVS consists of an open circuit and a controlled voltage source. There is no current in the open circuit so no power is absorbed by the open circuit.

The voltage, v_c, across the open circuit is the controlling signal of the VCVS. The voltmeter measures v_c to be

$$v_c = 2\text{ V}$$

The voltage of the controlled voltage source is

$$v_d = 2\,v_c = 4\text{ V}$$

The ammeter measures the current in the controlled voltage source to be

$$i_d = 1.5\text{ A}$$

The element current, i_d, and voltage, v_d, adhere to the passive convention. Therefore

$$p = i_d v_d = (1.5)(4) = 6\text{ W}$$

is the power absorbed by the VCVS.

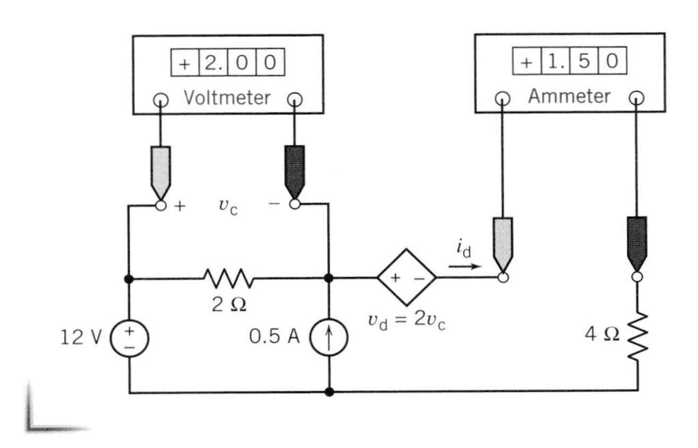

Figure 2.8-2
A circuit containing a VCVS. The meters indicate that the voltage of the controlling element is v_c = 2.0 volts and that the current of the controlled element is i_d = 1.5 amperes.

Exercise 2.8-1 Find the power absorbed by the CCCS in Figure E 2.8-1.

Hint: The controlling element of this dependent source is a short circuit. The voltage across a short circuit is zero. Hence, the power absorbed by the controlling element is zero. How much power is absorbed by the controlled element?

Answer: −115.2 watts are absorbed by the CCCS. (The CCCS delivers +115.2 watts to the rest of the circuit.)

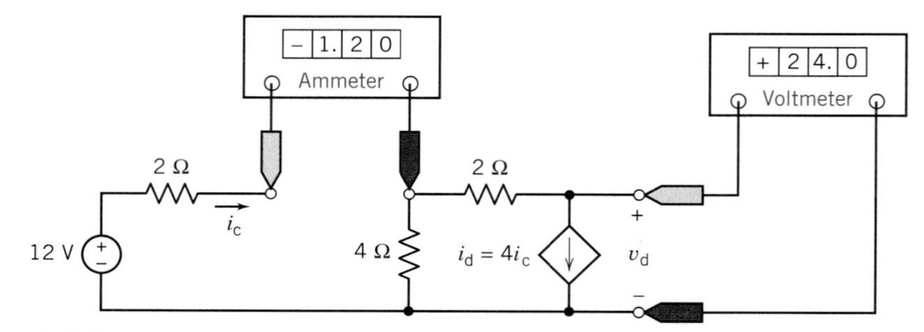

Figure E 2.8-1 A circuit containing a CCCS. The meters indicate that the current of the controlling element is $i_c = -1.2$ amperes and that the voltage of the controlled element is $v_d = 24$ volts.

Exercise 2.8-2 Find the power supplied by the VCCS in Figure E 2.8-2.

Answer: 17.6 watts are supplied by the VCCS. (−17.6 watts are absorbed by the VCCS.)

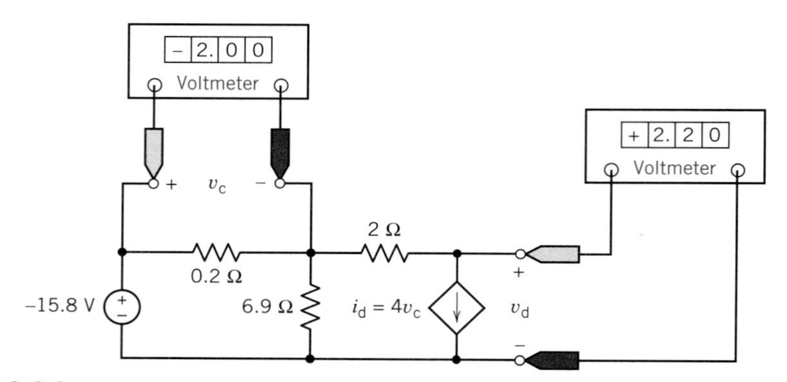

Figure E 2.8-2 A circuit containing a VCCS. The meters indicate that the voltage of the controlling element is $v_c = -2.0$ volts and that the voltage of the controlled element is $v_d = 2.2$ volts.

Exercise 2.8-3 Find the power absorbed by the CCVS in Figure E 2.8-3.

Answer: 4.375 watts are delivered to the CCVS.

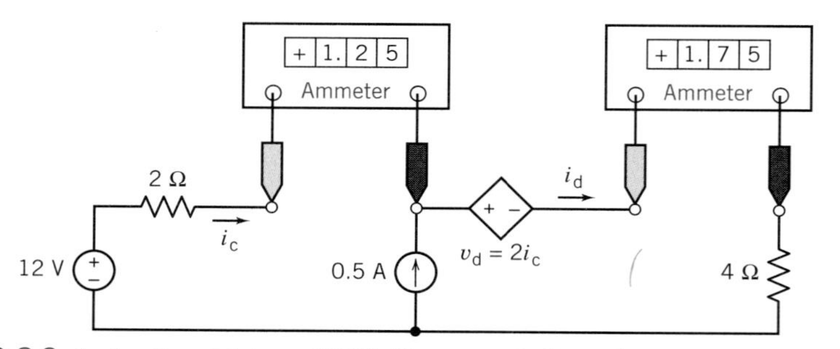

Figure E 2.8-3 A circuit containing a CCVS. The meters indicate that the current of the controlling element is $i_c = 1.25$ amperes and that the current of the controlled element is $i_d = 1.75$ amperes.

2.9 TRANSDUCERS

Transducers are devices that convert physical quantities to electrical quantities. This section describes two transducers: potentiometers and temperature sensors. Potentiometers convert position to resistance, and temperature sensors convert temperature to current.

Figure 2.9-1*a* shows the symbol for the potentiometer. The potentiometer is a resistor having a third contact, called the wiper, that slides along the resistor. Two parameters, R_p and a, are needed to describe the potentiometer. The parameter R_p specifies the potentiometer resistance ($R_p > 0$). The parameter a represents the wiper position and takes values in the range $0 \le a \le 1$. The values $a = 0$ and $a = 1$ correspond to the extreme positions of the wiper.

Figure 2.9-1*b* shows a model for the potentiometer that consists of two resistors. The resistances of these resistors depend on the potentiometer parameters R_p and a.

Frequently, the position of the wiper corresponds to the angular position of a shaft connected to the potentiometer. Suppose θ is the angle in degrees and $0 \le \theta \le 360$. Then

$$a = \frac{\theta}{360}$$

Example 2.9-1

Figure 2.9-2*a* shows a circuit in which the voltage measured by the meter gives an indication of the angular position of the shaft. In Figure 2.9-2*b* the current source, the potentiometer, and the voltmeter have been replaced by models of these devices. Analysis of Figure 2.9-2*b* yields

$$v_m = R_p\, I a = \frac{R_p I}{360}\,\theta$$

Solving for the angle gives

$$\theta = \frac{360}{R_p I}\, v_m$$

Suppose $R_p = 10$ kΩ and $I = 1$ mA. An angle of 163° would cause an output of $v_m = 4.53$ V. A meter reading of 7.83 V would indicate that $\theta = 282°$.

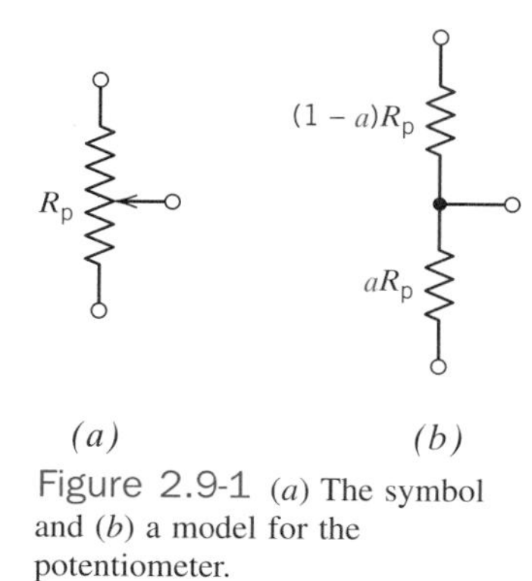

Figure 2.9-1 (*a*) The symbol and (*b*) a model for the potentiometer.

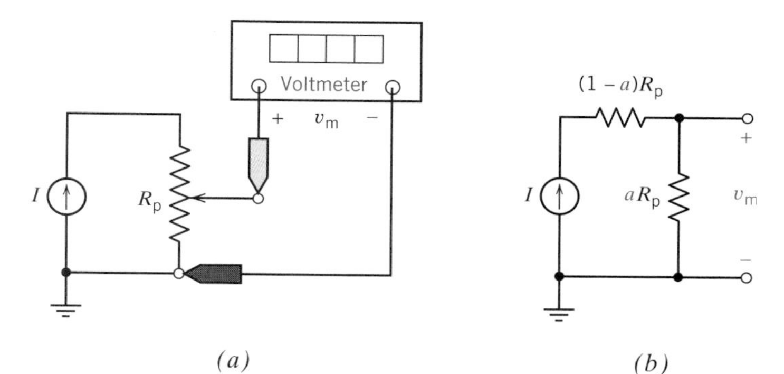

Figure 2.9-2 (*a*) A circuit containing a potentiometer. (*b*) An equivalent circuit containing a model of the potentiometer.

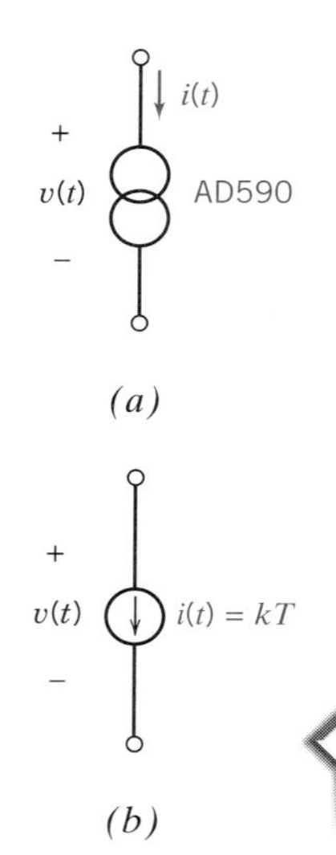

Figure 2.9-3 (*a*) The symbol and (*b*) a model for the temperature sensor.

Temperature sensors, such as the AD590 manufactured by Analog Devices, are current sources having current proportional to absolute temperature. Figure 2.9-3*a* shows the symbol used to represent the temperature sensor. Figure 2.9-3*b* shows the circuit model of the temperature sensor. In order for the temperature sensor to operate properly, the branch voltage v must satisfy the condition

$$4 \text{ volts} \leq v \leq 30 \text{ volts}$$

When this condition is satisfied, the current, i, in microamps, is numerically equal to the temperature T, in degrees Kelvin. The phrase "numerically equal" indicates that the current and temperature have the same value but different units. This relationship can be expressed as

$$i = k \cdot T$$

where $k = 1 \dfrac{\mu\text{A}}{°\text{K}}$, a constant associated with the sensor.

Exercise 2.9-1 For the potentiometer circuit of Figure 2.9-2, calculate the meter voltage, v_m, when $\theta = 45°$, $R_p = 20\,\text{k}\Omega$, and $I = 2\,\text{mA}$.

Answer: $v_m = 5\,\text{V}$.

Exercise 2.9-2 The voltage and current of an AD590 temperature sensor of Figure 2.9-3 are 10 V and 280 μA, respectively. Determine the measured temperature.

Answer: $T = 280°\text{K}$ or approximately 6.8°C.

2.10 SWITCHES

Switches have two distinct states: open and closed. Ideally, a switch acts as a short circuit when it is closed and as an open circuit when it is open. Figures 2.10-1 and 2.10-2 show several types of switches. In each case, the time when the switch changes state is indicated. Consider first the Single-Pole, Single-Throw (SPST) switches shown in Figure 2.10-1. The switch in Figure 2.10-1*a* is initially open. This switch changes state, becoming closed, at time $t = 0$ s. When this switch is modeled as an ideal switch, it is treated like an open circuit when $t < 0$ s and like a short circuit when $t > 0$ s. The ideal switch changes state instantaneously. The switch in Figure 2.10-1*b* is initially closed. This switch changes state, becoming open, at time $t = 0$ s.

Next, consider the Single-Pole, Double-Throw (SPDT) switch shown in Figure 2.10-2*a*. This SPDT switch acts like two SPST switches, one between terminals c and a, another between terminals c and b. Before $t = 0$ s, the switch between c and a is closed and the switch between c and b is open. At t = 0 s both switches change state; that is, the switch between a and c opens, and the switch between c and b closes. Once again, the ideal switches are modeled as open circuits when they are open and as short circuits when they are closed.

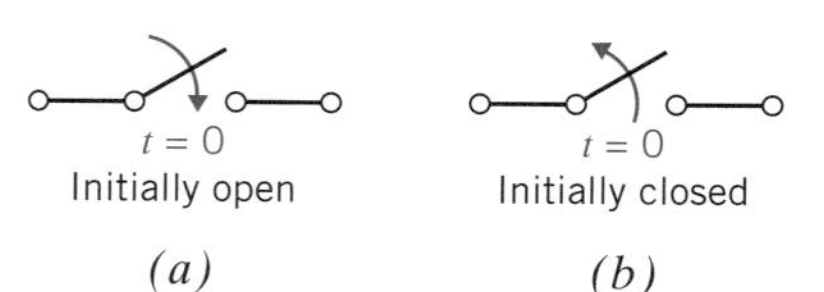

Figure 2.10-1 SPST switches. (*a*) Initially open and (*b*) initially closed.

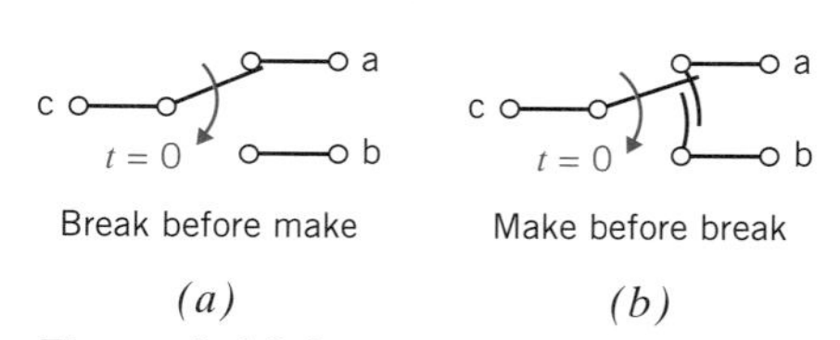

Figure 2.10-2 SPDT switches. (*a*) Break before make and (*b*) make before break.

In some applications, it makes a difference whether the switch between c and b closes before, or after, the switch between c and a opens. Different symbols are used to represent these two types of double-pole single-throw switch. The break-before-make switch is manufactured so that the switch between c and b closes after the switch between c and a opens. The symbol for the break-before-make switch is shown in Figure 2.10-2*a*. The make-before-break switch is manufactured so that the switch between c and b closes before the switch between c and a opens. The symbol for the make-before-break switch is shown in Figure 2.10-2*b*. Remember: the switch transition from terminal a to terminal b is assumed to take place instantaneously. Thus, the switch wiper makes before breaks, but this transition is very fast compared to the circuit time response.

Example 2.10-1

Figure 2.10-3 illustrates the use of open and short circuits for modeling ideal switches. In Figure 2.10-3*a* a circuit containing three switches is shown. In Figure 2.10-3*b* the circuit is shown as it would be modeled before $t = 0$ s. The two single-pole single-throw switches change state at time $t = 0$ s. Figure 2.10-3*c* shows the circuit as it would be modeled when the time is between 0 s and 2 s. The single-pole double-throw switch changes state at time $t = 2$ s. Figure 2.10-3*d* shows the circuit as it would be modeled after 2 s.

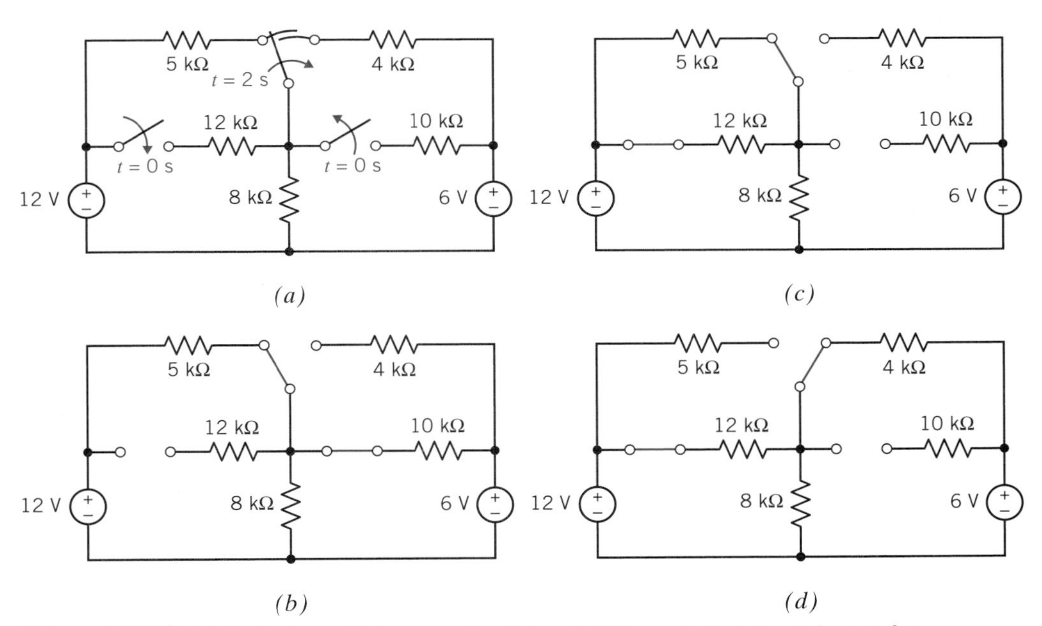

Figure 2.10-3 (*a*) A circuit containing several switches. (*b*) The equivalent circuit for $t \leq 0$ s. (*c*) The equivalent circuit for $0 < t < 2$ s. (*d*) The equivalent circuit for $t > 2$ s.

Exercise 2.10-1 What is the value of the current i in Figure E 2.10-1 at time $t = 1$ s? At $t = 5$ s?

Answers: $i = 4$ milliamps at $t = 1$ s (the switch is closed), and $i = 0$ amperes at $t = 5$ s (the switch is open).

Exercise 2.10-2 What is the value of the current i in Figure E 2.10-2 at time $t = 4$ s?

Answer: $i = 0$ amperes at $t = 4$ s (both switches are open).

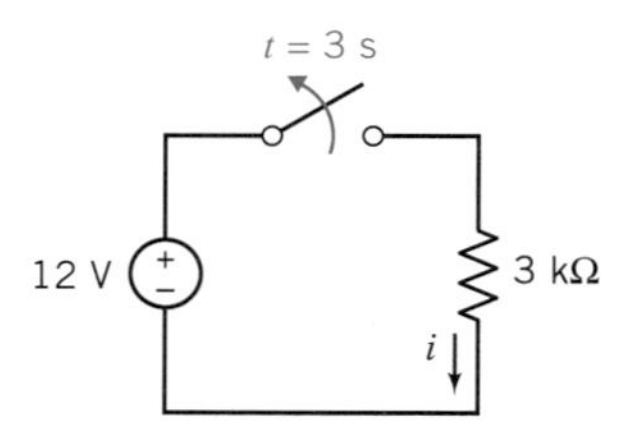

Figure E 2.10-1 A circuit with an SPST switch that opens at $t = 3$ seconds.

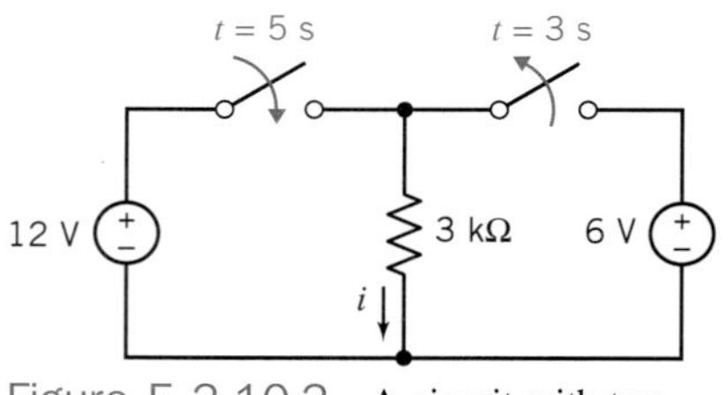

Figure E 2.10-2 A circuit with two SPST switches.

Exercise 2.10-3 What is the value of the voltage v in Figure E 2.10-3 at time $t = 4$ s? At $t = 6$ s?

Answers: $v = 6$ volts at $t = 4$ s, and $v = 0$ volts at $t = 6$ s.

Exercise 2.10-4 What is the value of the current i in Figure E 2.10-4 at time $t = 1$ s? At $t = 3$ s?

Answer: $i = 2$ milliamps at $t = 1$ s, and $i = 4$ milliamps at $t = 3$ s.

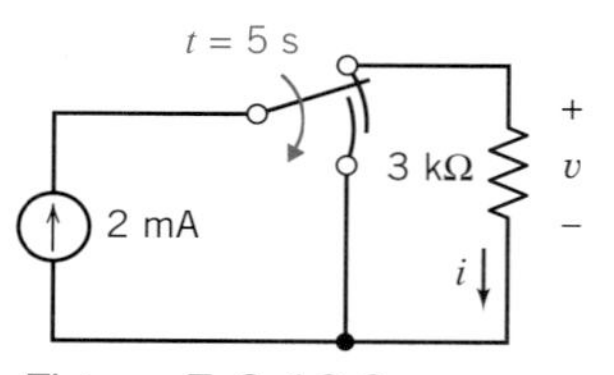

Figure E 2.10-3 A circuit with a make-before-break SPDT switch.

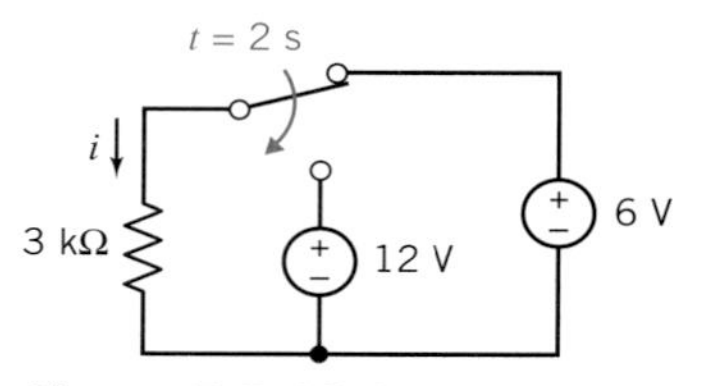

Figure E 2.10-4 A circuit with a break-before-make SPDT switch.

2.11 VERIFICATION EXAMPLE

The meters in the circuit of Figure 2.11-1 indicate that $v_1 = -4$ V, $v_2 = 8$ V and that $i = 1$ A.

(a) Verify that these values satisfy Ohm's law.

(b) Verify that the power supplied by the voltage source is equal to the power absorbed by the resistors.

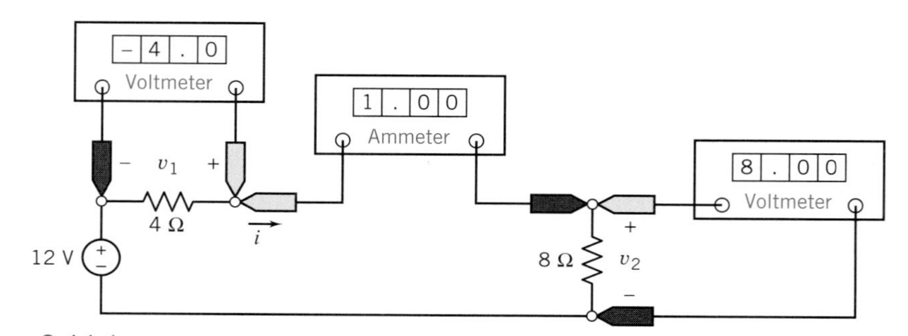

Figure 2.11-1 A circuit with meters.

Solution

(a) Consider the 8-Ω resistor. The current i flows through this resistor from top to bottom. Thus the current i and the voltage v_2 adhere to the passive convention. Therefore, Ohm's law requires that $v_2 = 8i$. The values $v_2 = 8$ V and $i = 1$ A satisfy this equation.

Next, consider the 4-Ω resistor. The current i flows through this resistor from left to right. Thus the current i and the voltage v_1 do not adhere to the passive convention. Therefore, Ohm's law requires that $v_1 = 4(-i)$. The values $v_1 = -4$ V and $i = 1$ A satisfy this equation.

Thus, Ohm's law is satisfied.

(b) The current i flows through the voltage source from bottom to top. Thus the current i and the voltage 12 V do not adhere to the passive convention. Therefore, $12i = 12(1) = 12$ W is the power supplied by the voltage source. The power absorbed by the 4-Ω resistor is $4i^2 = 4(1^2) = 4$ W, and the power absorbed by the 8-Ω resistor is $8i^2 = 8(1^2) = 8$ W. The power supplied by the voltage source is indeed equal to the power absorbed by the resistors.

2.12 Design Challenge Solution

TEMPERATURE SENSOR

Currents can be measured easily using ammeters. A temperature sensor, such as Analog Devices' AD590, can be used to measure temperature by converting temperature to current. Figure 2.12-1 shows a symbol used to represent a temperature sensor. In order for this sensor to operate properly, the voltage v must satisfy the condition

$$4 \text{ volts} \leq v \leq 30 \text{ volts}$$

When this condition is satisfied, the current i, in μA, is numerically equal to the temperature T, in °K. The phrase "numerically equal" indicates that the two variables have the same value but different units.

$$i = k \cdot T \quad \text{where} \quad k = 1\frac{\mu\text{A}}{°\text{K}}$$

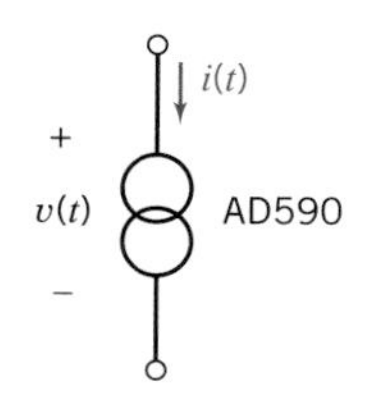

Figure 2.12-1 A temperature sensor.

The goal is to design a circuit using the AD590 to measure the temperature of a container of water. In addition to the AD590 and an ammeter, several power supplies and an assortment of standard 2 percent resistors are available. The power supplies are voltage sources. Power supplies having voltages of 10, 12, 15, 18, or 24 volts are available.

Describe the Situation and the Assumptions

In order for the temperature transducer to operate properly, its element voltage must be between 4 volts and 30 volts. The power supplies and resistors will be used to establish this voltage. An ammeter will be used to measure the current in the temperature transducer.

The circuit must be able to measure temperatures in the range from 0°C to 100°C since water is a liquid at these temperatures. Recall that the temperature in °C is equal to the temperature in °K minus 273°.

State the Goal

Use the power supplies and resistors to cause the voltage, v, of the temperature transducer to be between 4 volts and 30 volts.

Use an ammeter to measure the current, i, in the temperature transducer.

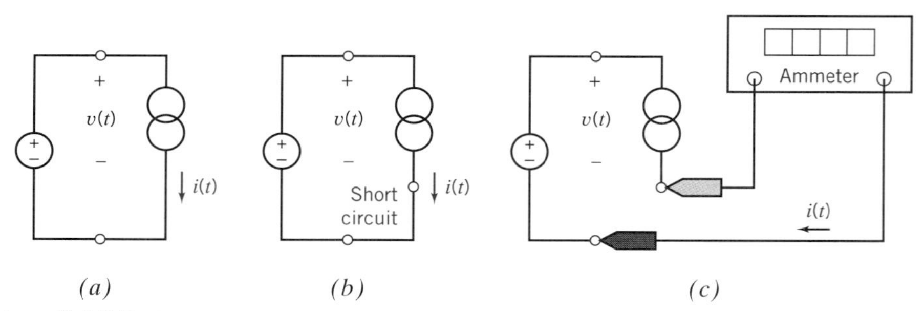

Figure 2.12-2 (*a*) Measuring temperature with a temperature sensor. (*b*) Adding a short circuit. (*c*) Replacing the short circuit by an ammeter.

Generate a Plan

Model the power supply as an ideal voltage source and the temperature transducer as an ideal current source. The circuit shown in Figure 2.12-2*a* causes the voltage across the temperature transducer to be equal to the power supply voltage. Since all of the available power supplies have voltages between 4 volts and 30 volts, any one of the power supplies can be used. Notice that the resistors are not needed.

In Figure 2.12-2*b* a short circuit has been added in a way that does not disturb the network. In Figure 2.12-2*c* this short circuit has been replaced with an (ideal) ammeter. Since the ammeter will measure the current in the temperature transducer, the ammeter reading will be numerically equal to the temperature in °K.

Although any of the available power supplies is adequate to meet the specifications, there may still be an advantage to choosing a particular power supply. For example, it is reasonable to choose the power supply that causes the transducer to absorb as little power as possible.

Act on the Plan

The power absorbed by the transducer is

$$p = v \cdot i$$

where v is the power supply voltage. Choosing v as small as possible, 10 volts in this case, makes the power absorbed by the temperature transducer as small as possible. Figure 2.12-3*a* shows the final design. Figure 2.12-3*b* shows a graph that can be used to find the temperature corresponding to any ammeter current.

Verify the Proposed Solution

Let's try an example. Suppose the temperature of the water is 80.6°F. This temperature is equal to 27°C or 300°K. The current in the temperature sensor will be

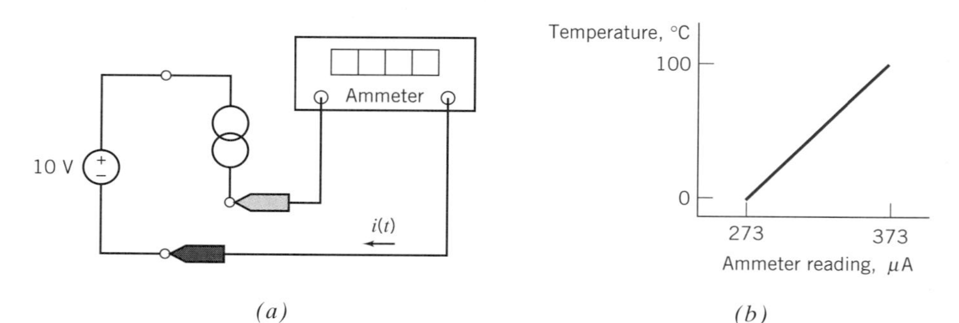

Figure 2.12-3 (*a*) Final design of a circuit that measures temperature with a temperature sensor. (*b*) Graph of temperature versus ammeter current.

$$i = \left(1 \frac{\mu\text{A}}{°\text{K}}\right) 300°\text{K} = 300\,\mu\text{A}$$

Next, suppose that the ammeter in Figure 2.12-3*a* reads 300 μA. A sensor current of 300 μA corresponds to a temperature of

$$T = \frac{300\ \mu\text{A}}{1 \frac{\mu\text{A}}{°\text{K}}} = 300°\text{K} = 27°\text{C} = 80.6°\text{F}$$

The graph in Figure 2.12-3*b* indicates that a sensor current of 300 μA does correspond to a temperature of 27°C.

This example shows that the circuit is working properly.

2.13 SUMMARY

- Thomas A. Edison, the first electrical engineer, invented and developed many electric circuits and elements. By 1882 Edison had conceived, designed, patented, and tested an electric lighting system to replace the prevalent gas lighting of that period. By 1884 he had demonstrated his lighting system in New York and at the Philadelphia Exhibition.
- Engineering is a problem-solving activity. It can also be described as the profession that strives to find ways to utilize economically the materials and forces of nature for the benefit of humankind.
- The engineer uses models, called circuit elements, to represent the devices that make up a circuit. In this book, we consider only linear elements or linear models of devices. A device is linear if it satisfies the properties of both superposition and homogeneity. An example of a physical circuit made up of devices and a circuit model made up of circuit elements is shown in Figure 2.13-1.
- The relationship between the directions of the current and voltage of a circuit element is important. The voltage direction marks one terminal + and the other −. The element voltage and current adhere to the passive convention, if the current flows from the terminal marked + to the terminal marked −.
- Resistors are widely used as circuit elements. When the resistor voltage and current adhere to the passive convention, resistors obey Ohm's law: the voltage across the terminals of the resistor is related to the current into the positive terminal as $v = Ri$. The power delivered to a resistance is $p = i^2R = v^2/R$ watts.
- An independent source provides a current or a voltage independent of other circuit variables. A dependent source provides a current (or a voltage) that is dependent on another variable elsewhere in the circuit.
- The **short circuit** and **open circuit** are special cases of independent sources. A **short circuit** is an ideal voltage source having $v(t) = 0$. The current in a short circuit is determined by the rest of the circuit. An **open circuit** is an ideal current source having $i(t) = 0$. The voltage across an open circuit is determined by the rest of the circuit. Open circuits and short

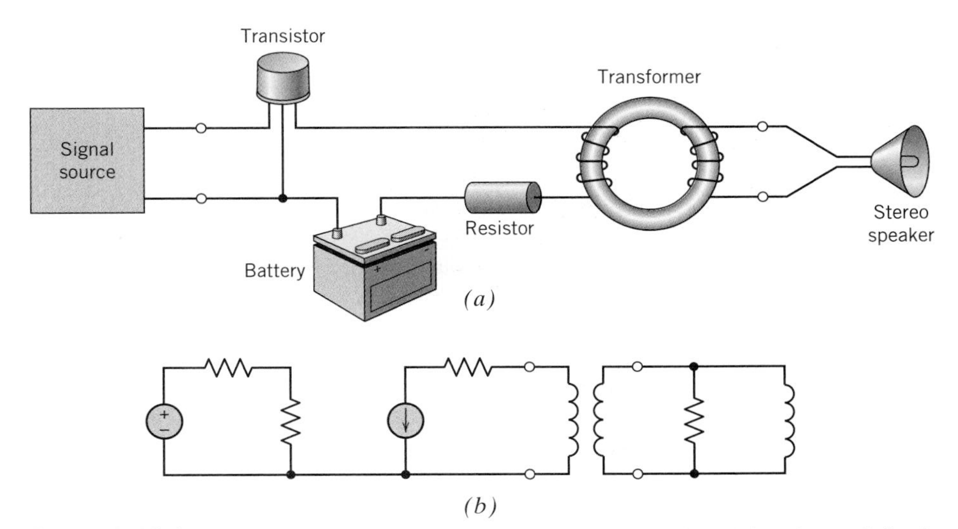

Figure 2.13-1 (*a*) Physical circuit made up of devices. (*b*) Its circuit model made up of circuit elements.

circuits can also be described as special cases of resistors. A resistor with resistance $R = 0 (G = \infty)$ is a short circuit. A resistor with conductance $G = 0 (R = \infty)$ is an open circuit.

• An ideal ammeter measures the current flowing through its terminals and has zero voltage across its terminals. An ideal voltmeter measures the voltage across its terminals and has terminal current equal to zero. Ideal voltmeters act like open circuits, and ideal ammeters act like short circuits.

• Transducers are devices that convert physical quantities, such as rotational position, to an electrical quantity such as voltage. In this chapter we describe two transducers: potentiometers and temperature sensors.

• Switches are widely used in circuits to connect and disconnect elements and circuits. They can also be used to create discontinuous voltages or currents.

PROBLEMS

Section 2.3 Engineering and Linear Models

P 2.3-1 An element has voltage v and current i as shown in Figure P 2.3-1*a*. Values of the current i and corresponding voltage v have been tabulated as shown in Figure P 2.3-1*b*. Determine if the element is linear.

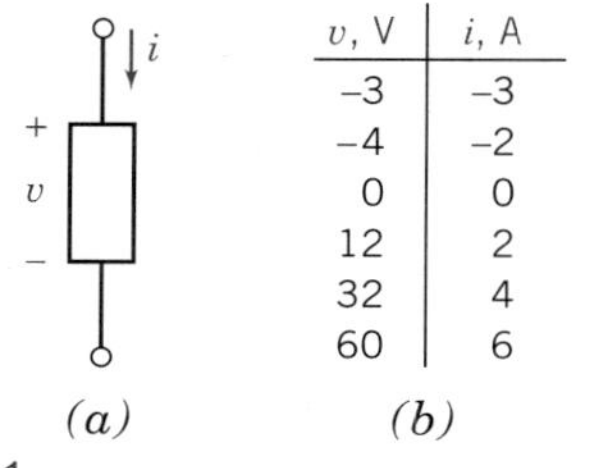

v, V	i, A
–3	–3
–4	–2
0	0
12	2
32	4
60	6

(*b*)

Figure P 2.3-1

P 2.3-2 An element has voltage v and current i as shown in Figure P 2.3-2*a*. Values of the current i and corresponding voltage v have been tabulated as shown in Figure P 2.3-2*b*. Is it possible that the element is linear? Justify your answer.

Hint: Plot the data. Is the plot a straight line? If so, determine the equation of the line.

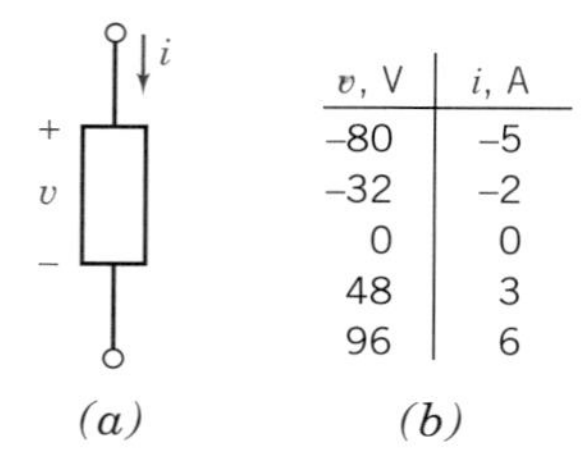

v, V	i, A
–80	–5
–32	–2
0	0
48	3
96	6

(*b*)

Figure P 2.3-2

P 2.3-3 A linear element has voltage v and current i as shown in Figure P 2.3-3*a*. Values of the current i and corresponding voltage v have been tabulated as shown in Figure P 2.3-3*b*. Represent the element by an equation that expresses v as a function of i. This equation is a model of the element. (a) Verify that the model is linear. (b) Use the model to predict the value of v corresponding to a current of $i = 40\,\text{mA}$. (c) Use the model to predict the value of i corresponding to a voltage of $v = 4\,\text{V}$.

Hint: Plot the data. We expect the data points to lie on a straight line. Obtain a linear model of the element by representing that straight line by an equation.

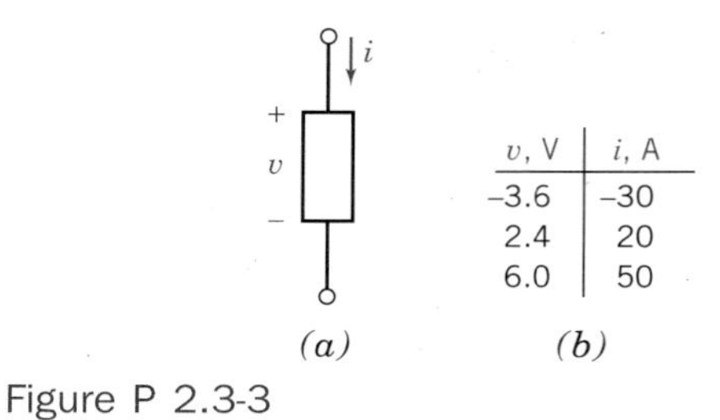

v, V	i, A
-3.6	-30
2.4	20
6.0	50

(*b*)

Figure P 2.3-3

P 2.3-4 A linear element has voltage v and current i as shown in Figure P 2.3-4*a*. Values of the current i and corresponding voltage v have been tabulated as shown in Figure P 2.3-4*b*. Represent the element by an equation that expresses v as a function of i. This equation is a model of the element. (a) Verify that the model is linear. (b) Use the model to predict the value of v corresponding to a current of $i = 4\,\text{mA}$. (c) Use the model to predict the value of i corresponding to a voltage of $v = 12\,\text{V}$.

(*a*)

v, V	i, mA
3.078	12
5.13	20
12.825	50

(*b*)

Figure P 2.3-4

Hint: Plot the data. We expect the data points to lie on a straight line. Obtain a linear model of the element by representing that straight line by an equation.

P 2.3-5 An element is represented by the relation between current and voltage as

$$v = \sqrt{i}$$

Determine if the element is linear.

P 2.3-6 An element is represented by the relation between current and voltage as

$$v = 3i + 5$$

Determine if the element is linear.

P 2.3-7 The Nelson River direct-current power line in Canada delivers 1.2 GW, at 900 kV, over a distance of 800 km. At the other end, at which the power is generated, the voltage is 950 kV. (a) Draw a circuit representation of this problem. (b) What is the efficiency of the line? That is, what is the power delivered divided by the power generated? (c) Where does the energy lost go? (d) How many joules are transmitted in a day if the conditions do not vary throughout the day?

P 2.3-8 A diagram of an auto battery charger is shown in Figure P 2.3-8*a*. The linear model of the charger circuit is shown in Figure P 2.3-8*b*, where element x represents the transformer and diode and element y represents the power losses as heat within the car battery. Determine the power delivered to charge the battery (the 12-V source) and how long it will take to deliver 3360 coulombs of charge. Calculate the power supplied by the charger and verify that it is equal to the power absorbed by the auto battery.

Section 2.4 Active and Passive Circuit Elements

P 2.4-1 An electrical element has a current i entering the positive terminal with a voltage v across the element. For $t > 0$ it is known that

$$v(t) = 10 \sin 100t \text{ V}$$
$$i(t) = 2 \cos 100t \text{ mA}$$

(a) Find the power $p(t)$ and plot $p(t)$ versus t.

(b) Determine intervals of time when the element is supplying or absorbing power.

P 2.4-2 The current entering the positive terminal of an element is $i = 2 \sin t$ A for $t \geq 0$ and $i = 0$ for $t < 0$. The voltage across the element is

$$v = 2\frac{di}{dt}$$

Determine whether the element is active or passive by calculating the energy absorbed by the element.

Section 2.5 Resistors

P 2.5-1 A current source and a resistor are connected in series in the circuit shown in Figure P 2.5-1. Elements connected in series have the same current, so $i = i_s$ in this circuit. Suppose that $i_s = 3$ A and R $= 7\Omega$. Calculate the voltage v across the resistor and the power absorbed by the resistor.

Answer: $v = 21$ V and the resistor absorbs 63 W

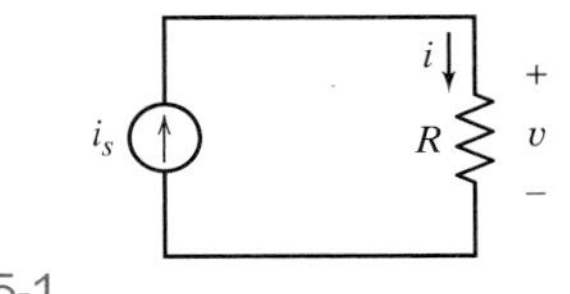

Figure P 2.5-1

P 2.5-2 A current source and a resistor are connected in series in the circuit shown in Figure P 2.5-1. Elements connected in series have the same current, so $i = i_s$ in this circuit. Suppose that $i = 3$ mA and $v = 24$ V. Calculate the resistance R and the power absorbed by the resistor.

Answer: R $= 8$ kΩ and the resistor absorbs 72 mW.

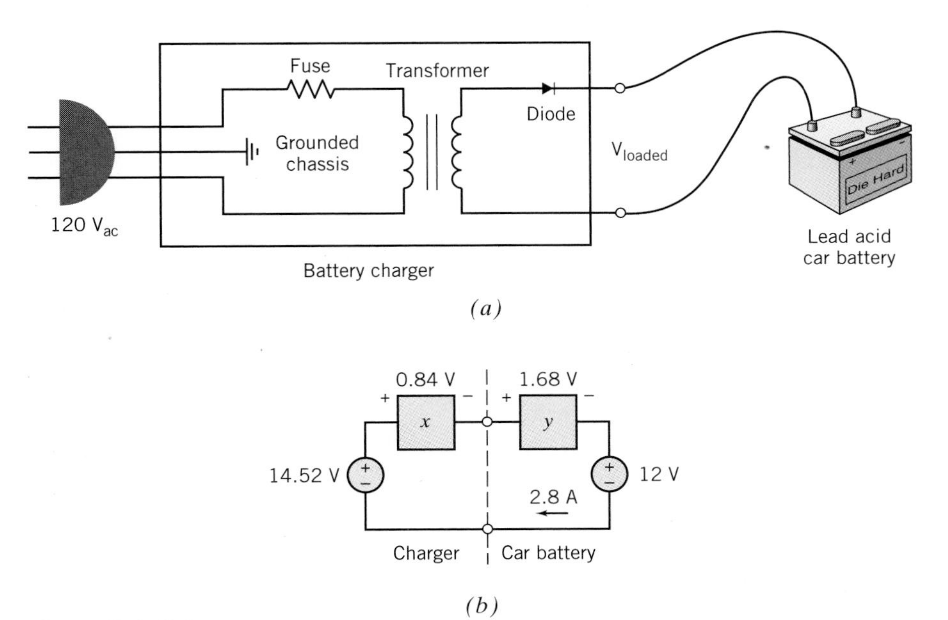

Figure P 2.3-8 (*a*) Auto battery charger. (*b*) Circuit model.

P 2.5-3 A voltage source and a resistor are connected in parallel in the circuit shown in Figure P 2.5-3. Elements connected in parallel have the same voltage, so $v = v_s$ in this circuit. Suppose that $v_s = 10$ V, and $R = 5\Omega$. Calculate the current i in the resistor and the power absorbed by the resistor.
Answer: $i = 2$ A and the resistor absorbs 20 W.

P 2.5-4 A voltage source and a resistor are connected in parallel in the circuit shown in Figure P 2.5-3. Elements connected in parallel have the same voltage, so $v = v_s$ in this circuit. Suppose that $v_s = 24$ V and $i = 2$ A. Calculate the resistance R and the power absorbed by the resistor.
Answer: R $= 12\Omega$ and the resistor absorbs 48 W.

P 2.5-5 A voltage source and two resistors are connected in parallel in the circuit shown in Figure P 2.5-5. Elements connected in parallel have the same voltage, so $v_1 = v_s$ and $v_2 = v_s$ in this circuit. Suppose that $v_s = 150$ V, $R_1 = 50\Omega$, and $R_2 = 25\Omega$. Calculate the current in each resistor and the power absorbed by each resistor.
Hint: Notice the reference directions of the resistor currents.
Answer: $i_1 = 3$ A and $i_2 = -6$ A. R_1 absorbs 450 W and R_2 absorbs 900 W.

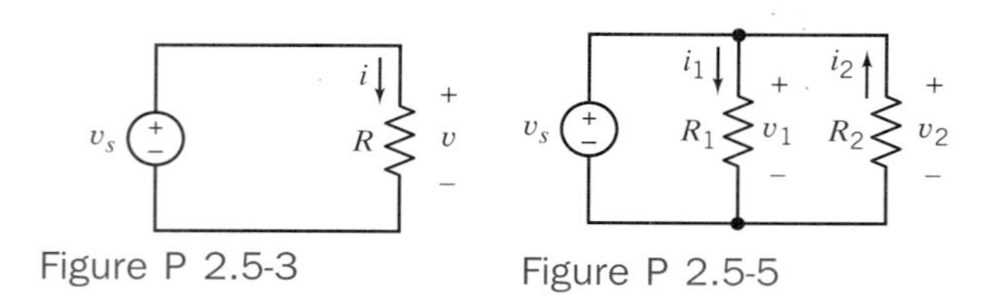

Figure P 2.5-3 Figure P 2.5-5

P 2.5-6 A current source and two resistors are connected in series in the circuit shown in Figure P 2.5-6. Elements connected in series have the same current, so $i_1 = i_s$ and $i_2 = i_s$ in this circuit. Suppose that $i_s = 2$ A, $R_1 = 4\Omega$, and $R_2 = 8\Omega$. Calculate the voltage across each resistor and the power absorbed by each resistor.
Hint: Notice the reference directions of the resistor voltages.
Answer: $v_1 = -8$ V and $v_2 = 16$ V. R_1 absorbs 16 W and R_2 absorbs 32 W.

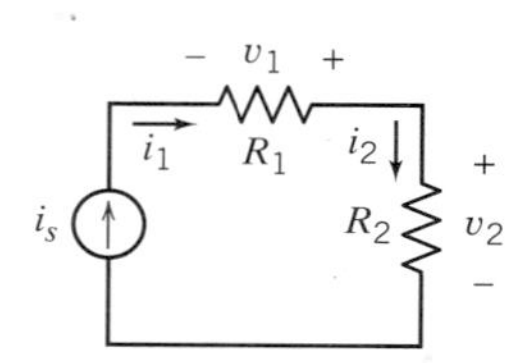

Figure P 2.5-6

P 2.5-7 An electric heater is connected to a constant 250-V source and absorbs 1000 W. Subsequently, this heater is connected to a constant 210-V source. What power does it absorb from the 210-V source? What is the resistance of the heater?
Hint: Model the electric heater as a resistor.

P 2.5-8 The portable lighting equipment for a mine is located 100 meters from its dc supply source. The mine lights use a total of 5 kW and operate at 120 V dc. Determine the required cross-sectional area of the copper wires used to connect the source to the mine lights if we require that the power lost in the copper wires be less than or equal to 5 percent of the power required by the mine lights.
Hint: Model both the lighting equipment and the wire as resistors.

Section 2.6 Independent Sources

P 2.6-1 A current source and a voltage source are connected in parallel with a resistor as shown in Figure P 2.6-1. All of the elements connected in parallel have the same voltage, v_s in this circuit. Suppose that $v_s = 15$ V, $i_s = 3$ A, and $R = 5\Omega$. (a) Calculate the current i in the resistor and the power absorbed by the resistor. (b) Change the current source current to $i_s = 5$ A and recalculate the current, i, in the resistor and the power absorbed by the resistor.
Answer: $i = 3$ A and the resistor absorbs 45 W both when $i_s = 3$ A and when $i_s = 5$ A.

P 2.6-2 A current source and a voltage source are connected in series with a resistor as shown in Figure P 2.6-2. All of the elements connected in series have the same current, i_s, in this circuit. Suppose that $v_s = 10$ V, $i_s = 2$ A, and $R = 5\Omega$. (a) Calculate the voltage v across the resistor and the power absorbed by the resistor. (b) Change the voltage source voltage to $v_s = 5$ V and recalculate the voltage, v, across the resistor and the power absorbed by the resistor.
Answer: $v = 10$ V and the resistor absorbs 20 W both when $v_s = 10$ V and when $v_s = 5$ V.

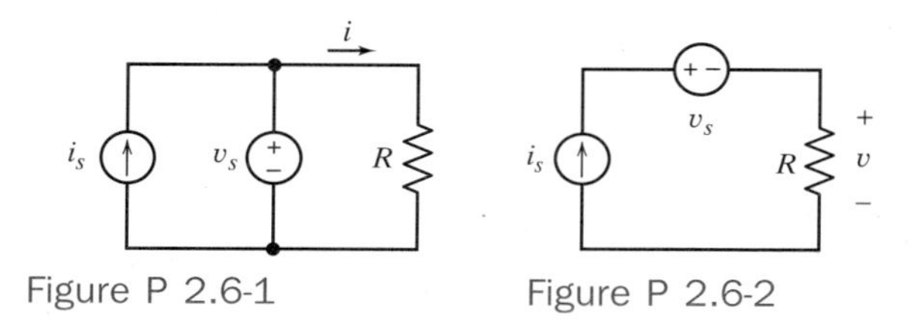

Figure P 2.6-1 Figure P 2.6-2

P 2.6-3 The current source and voltage source in the circuit shown in Figure P 2.6-3 are connected in parallel so that they both have the same voltage, v_s. The current source and voltage source are also connected in series so that they both have the same current, i_s. Suppose that $v_s = 12$ V and $i_s = 3$ A. Calculate the power supplied by each source.
Answer: The voltage source supplies -36 W, and the current source supplies 36 W.

P 2.6-4 The current source and voltage source in the circuit shown in Figure P 2.6-4 are connected in parallel so that they both have the same voltage, v_s. The current source and voltage source are also connected in series so that they both have the same current, i_s. Suppose that $v_s = 12$ V and $i_s = 3$ A. Calculate the power supplied by each source.
Answer: The voltage source supplies 36 W, and the current source supplies -36 W.

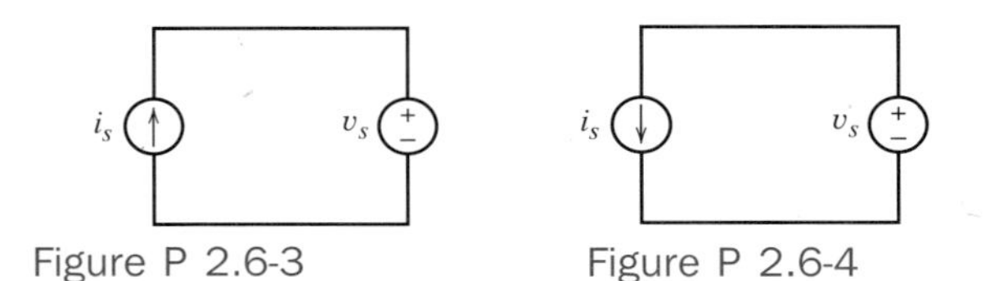

Figure P 2.6-3 Figure P 2.6-4

P 2.6-5 (a) Find the power supplied by the voltage source shown in Figure P 2.6-5 when for $t \geq 0$ we have

$$v = 2 \cos t \text{ V}$$

and

$$i = 10 \cos t \text{ mA}$$

(b) Determine the energy supplied by this voltage source for the period $0 \leq t \leq 1$ s.

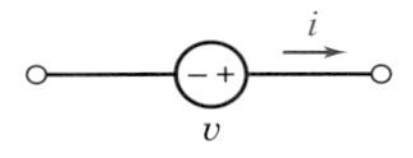

Figure P 2.6-5

Section 2.7 Voltmeters and Ammeters

P 2.7-1 For the circuit of Figure P 2.7-1
(a) What is the value of the resistance R?
(b) How much power is delivered by the voltage source?

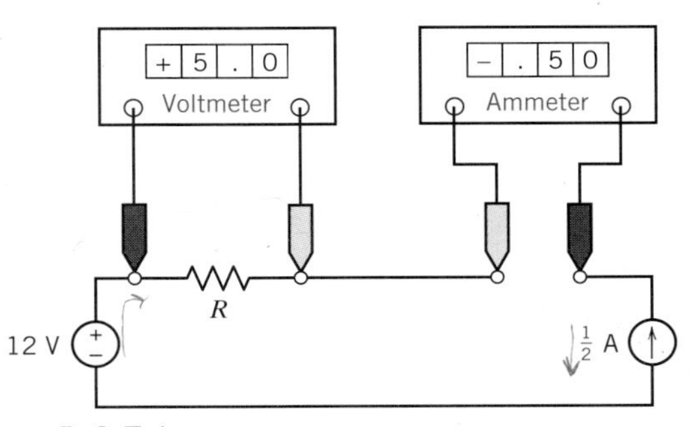

Figure P 2.7-1

P 2.7-2 The current source in Figure 2.7-2 supplies 40W. What values do the meters in Figure P 2.7-2 read?

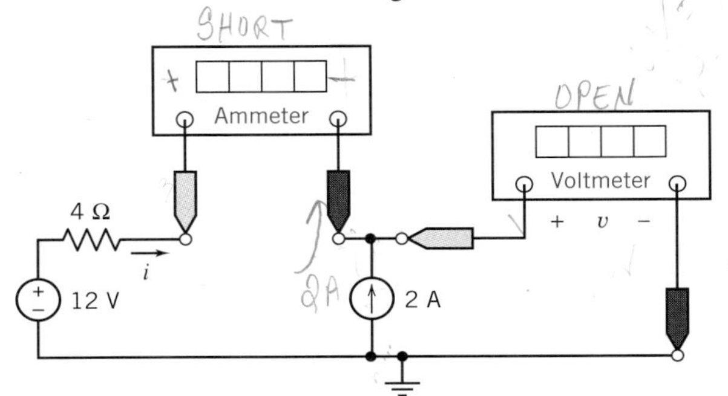

Figure P 2.7-2

Section 2.8 Dependent Sources

P 2.8-1 The ammeter in the circuit shown in Figure P 2.8-1 indicates that $i_a = 2$ A, and the voltmeter indicates that $v_b = 8$ V. Determine the value of r, the gain of the CCVS.
Answer: $r = 4$ V/A

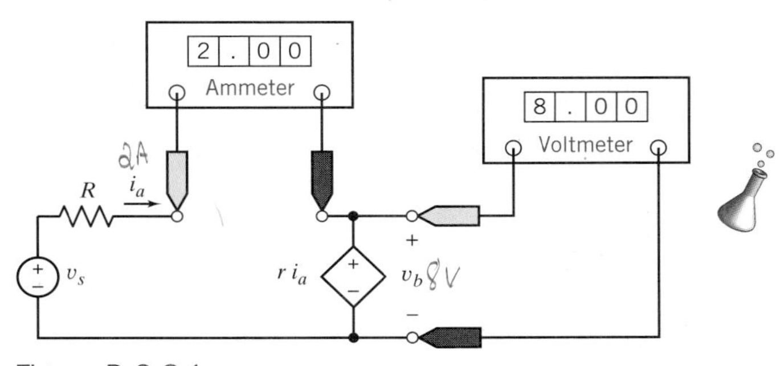

Figure P 2.8-1

P 2.8-2 The ammeter in the circuit shown in Figure P 2.8-2 indicates that $i_a = 2$ A, and the voltmeter indicates that $v_b = 8$ V. Determine the value of g, the gain of the VCCS.
Answer: $g = 0.25$ A/V

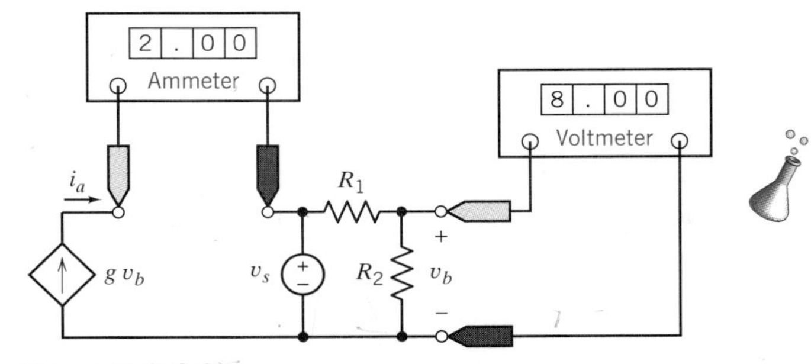

Figure P 2.8-2

P 2.8-3 The ammeters in the circuit shown in Figure P 2.8-3 indicate that $i_a = 32$ A that $i_b = 8$ A. Determine the value of d, the gain of the CCCS.
Answer: $d = 4$ A/A

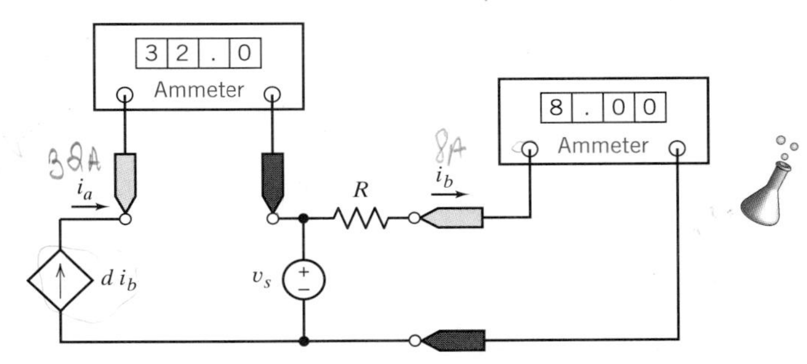

Figure P 2.8-3

P 2.8-4 The voltmeters in the circuit shown in Figure P 2.8-4 indicate that $v_a = 2$ V that $v_b = 8$ V. Determine the value of b, the gain of the VCVS.
Answer: $b = 4$ V/V

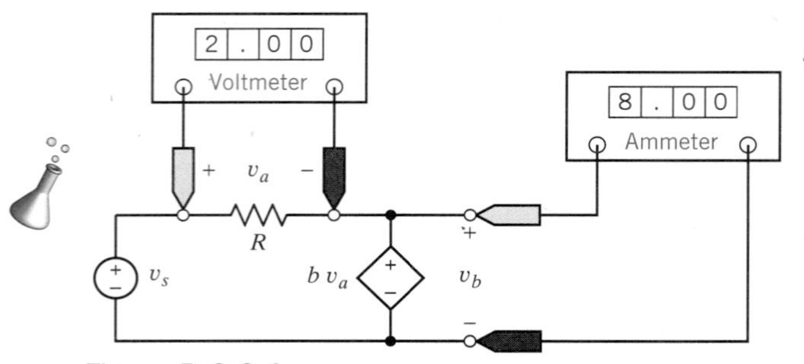

Figure P 2.8-4

Section 2.9 Transducers

P 2.9-1 For the potentiometer circuit of Figure 2.9-2 the current source and potentiometer resistance are 1.1 mA and 100 kΩ, respectively. Calculate the required angle, θ, so that the measured voltage is 23 V.

P 2.9-2 An AD590 sensor has an associated constant $k = 1\dfrac{\mu A}{^\circ K}$. The sensor has a voltage $v = 20$ V; and the measured current, $i(t)$, as shown in Figure 2.9-3 is $4\ \mu A < i < 13\ \mu A$ in a laboratory setting. Find the range of measured temperature.

Section 2.10 Switches

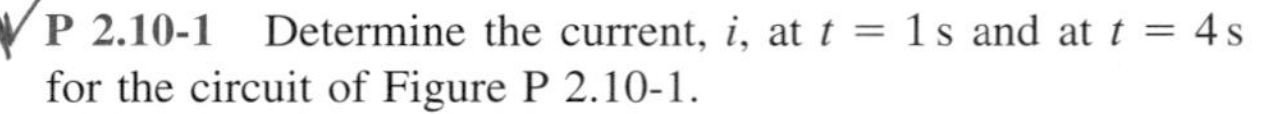

P 2.10-1 Determine the current, i, at $t = 1$ s and at $t = 4$ s for the circuit of Figure P 2.10-1.

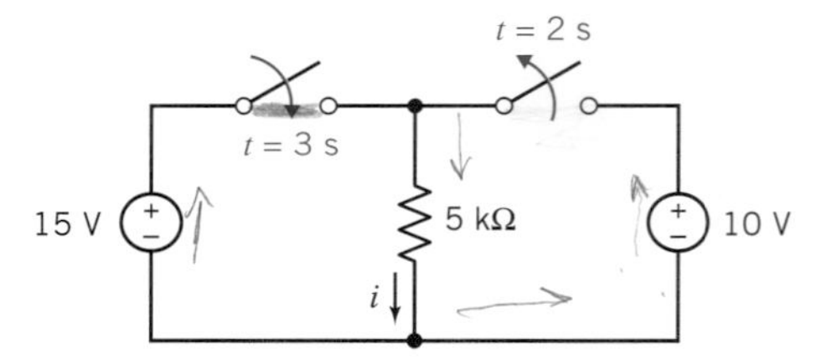

Figure P 2.10-1

P 2.10-2 Determine the voltage, v, at $t = 1$ s and at $t = 4$ s for the circuit shown in Figure P 2.10-2.

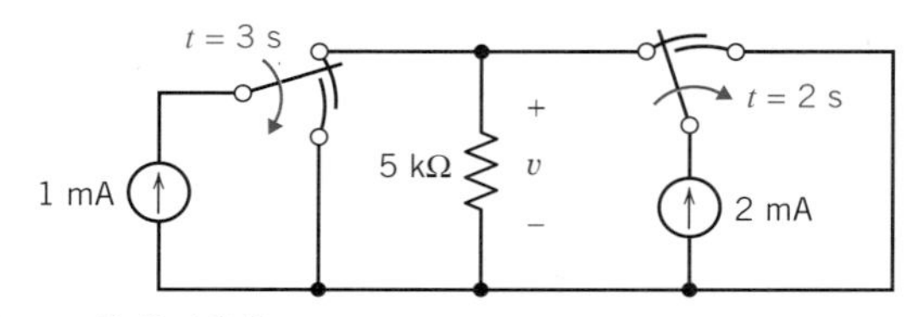

Figure P 2.10-2

VERIFICATION PROBLEMS

VP 2-1 The circuit shown in Figure VP 2.1 is used to test the CCVS. Your lab partner claims that this measurement shows that the gain of the CCVS is −20 V/A instead of +20 V/A. Do you agree? Justify your answer.

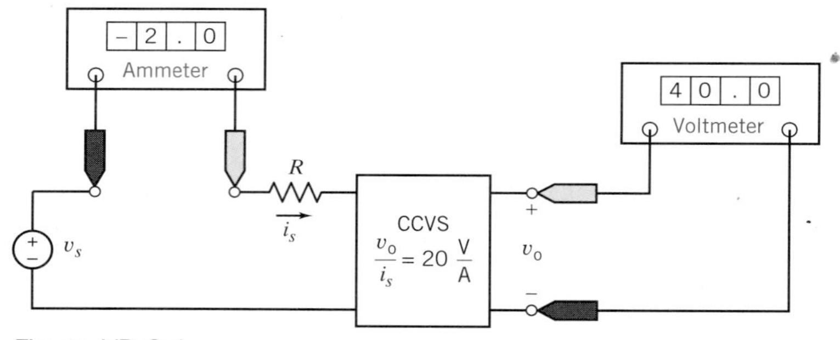

Figure VP 2.1

VP 2-2 The circuit of Figure VP 2.2 is used to measure the current in the resistor. Once this current is known, the resistance can be calculated as $R = \dfrac{v_s}{i}$. The circuit is constructed using a voltage source with $v_s = 12$ V and a 25Ω, 1/2 W resistor. After a puff of smoke and an unpleasant smell, the ammeter indicates that $i = 0$ A. The resistor must be bad. You have more 25Ω, 1/2 W resistors. Should you try another resistor? Justify your answer.

Hint: 1/2 W resistors are able to safely dissipate one 1/2 W of power. These resistors may fail if required to dissipate more than one half-watt of power.

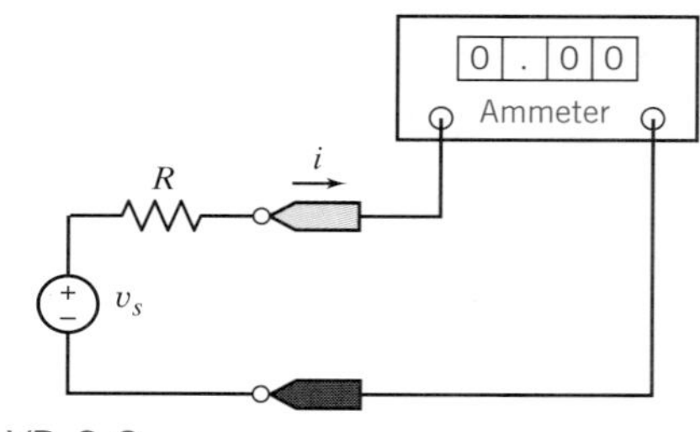

Figure VP 2.2

DESIGN PROBLEMS

√**DP 2-1** Specify the resistance R in Figure DP 2.1 so that both of the following conditions are satisfied:
1. $i > 40$ mA
2. the power absorbed by the resistor is less than 0.5 W

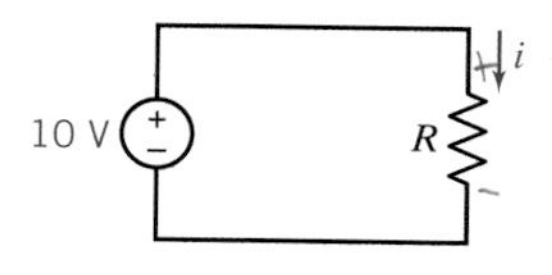

Figure DP 2.1

DP 2-2 Specify the resistance R in Figure DP 2.2 so that both of the following conditions are satisfied:
1. $v > 40$ V
2. the power absorbed by the resistor is less than 15 W

Hint: There is no guarantee that specifications can always be satisfied.

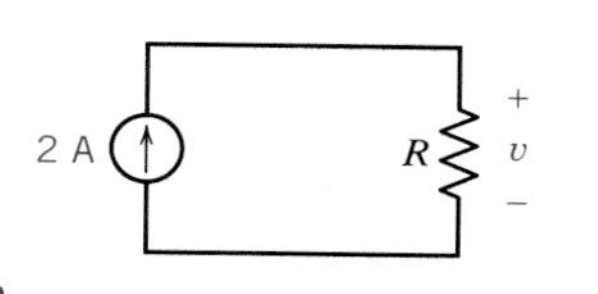

Figure DP 2.2

DP 2-3 Resistors are given a power rating. For example, resistors are available with ratings of 1/8 W, 1/4 W, 1/2 W and 1W. A 1/2 W resistor is able to safely dissipate 1/2 W of power, indefinitely. Resistors with larger power ratings are more expensive and bulkier than resistors with lower power ratings. Good engineering practice requires that resistor power ratings be specified to be large, but not larger than necessary.

Consider the circuit shown in Figure DP 2.3. The values of the resistances are

$$R_1 = 1000\ \Omega,\ R_2 = 2000\ \Omega,\ \text{and}\ R_3 = 4000\ \Omega$$

The value of the current source current is

$$i_s = 30\ \text{mA}$$

Specify the power rating for each resistor.

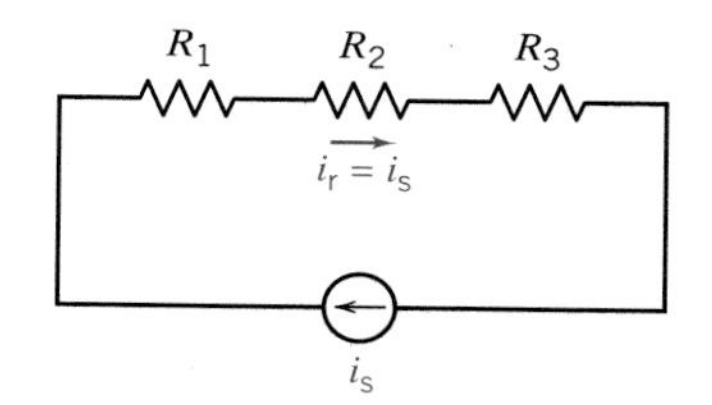

Figure DP 2.3

CHAPTER 3

Resistive Circuits

Preview

The resistor, with resistance R, is an element commonly used in most electric circuits. In this chapter we consider the analysis of circuits consisting of resistors and sources.

In addition to Ohm's law, we need two laws for relating (1) current flow at connected terminals and (2) the sum of voltages around a closed circuit path. These two laws were developed by Gustav Kirchhoff in 1847.

Using Kirchhoff's laws and Ohm's law, we are able to complete the analysis of resistive circuits and determine the currents and voltages at desired points in the circuit. This analysis may be accomplished for circuits with both independent and dependent sources.

3.1 Design Challenge

ADJUSTABLE VOLTAGE SOURCE

A circuit is required to provide an adjustable voltage. The specifications for this circuit are:

1. It should be possible to adjust the voltage to any value between -5 V and $+5\text{ V}$. It should not be possible to accidentally obtain a voltage outside this range.
2. The load current will be negligible.
3. The circuit should use as little power as possible.

The available components are:

1. Potentiometers: resistance values of 10 kΩ, 20 kΩ, and 50 kΩ are in stock.
2. A large assortment of standard 2 percent resistors having values between 10 Ω and 1 MΩ (see Appendix E).
3. Two power supplies (voltage sources): one 12-V supply and one −12-V supply; both rated for a maximum of 100 mA (milliamps).

Describe the Situation and the Assumptions

Figure 3.1-1 shows the situation. The voltage v is the adjustable voltage. The circuit that uses the output of the circuit being designed is frequently called the "load." In this case, the load current is negligible, so $i = 0$.

State the Goal

A circuit providing the adjustable voltage

$$-5\text{ V} \leq v \leq +5\text{ V}$$

must be designed using the available components.

Generate a Plan

Make the following observations.

1. The adjustability of a potentiometer can be used to obtain an adjustable voltage v.
2. Both power supplies must be used so that the adjustable voltage can have both positive and negative values.
3. The terminals of the potentiometer cannot be connected directly to the power supplies because the voltage v is not allowed to be as large as 12 V or −12 V.

These observations suggest the circuit shown in Figure 3.1-2*a*. The circuit in Figure 3.1-2*b* is obtained by using the simplest model for each component in Figure 3.1-2*a*.

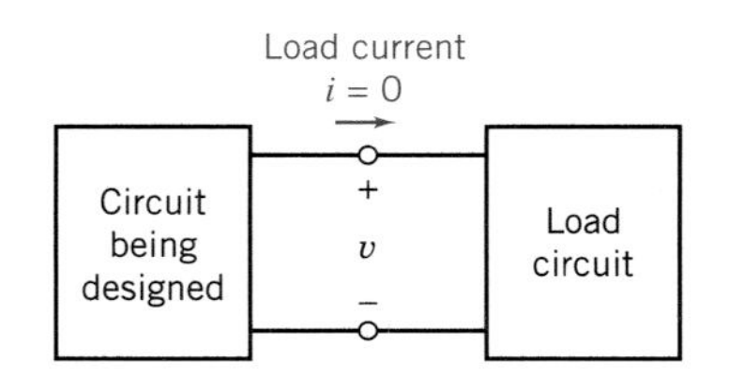

Figure 3.1-1
The circuit being designed provides an adjustable voltage, v, to the load circuit.

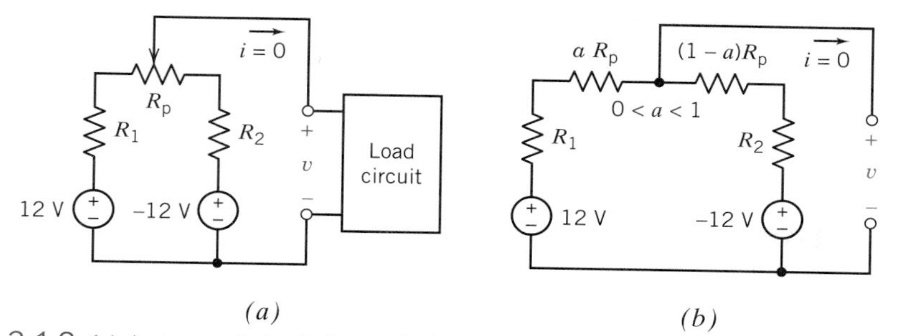

Figure 3.1-2 (*a*) A proposed circuit for producing the variable voltage v, and (*b*) the equivalent circuit after the potentiometer is modeled.

To complete the design, values need to be specified for R_1, R_2, and R_p. Then, several results need to be checked and adjustments made, if necessary.

1. Can the voltage v be adjusted to any value in the range -5 V to 5 V?
2. Are the voltage source currents less than 100 mA? This condition must be satisfied if the power supplies are to be modeled as ideal voltage sources.
3. Is it possible to reduce the power absorbed by R_1, R_2, and R_p?

Act on the Plan

Taking action on this plan requires analyzing the circuit in Figure 3.1-2*b*. Analysis of this type of circuit is discussed in this chapter. We will return to this problem at the end of this chapter.

3.2 ELECTRIC CIRCUIT APPLICATIONS

Overseas communications have always been of great importance to nations. One of the most brilliant chapters of the history of electrical technology was the development of underwater electric cable circuits. Underwater electric cables were used to carry electric telegraph communications. In late 1852 England and Ireland were connected by cable, and a year later there was a cable between Scotland and Ireland. In June 1853 a cable was strung between England and Holland, a distance of 115 miles.

It was Cyrus Field and Samuel Morse who saw the potential for a submarine cable across the Atlantic. By 1857 Field had organized a firm to complete the transatlantic telegraph cable and issued a contract for the production of 2500 miles of cable. The cable laying began in June 1858. After several false starts, a cable was laid across the Atlantic by August 5, 1858. However, this cable failed after only a month of operation.

Another series of cable-laying projects commenced, and by September 1865 a successful Atlantic cable was in place. This cable stretched over 3000 miles, from England to eastern Canada. There followed a flurry of cable laying. Approximately 150,000 km (90,000 mi) were in use by 1870, linking all continents and all major islands. An example of modern undersea cable is shown in Figure 3.2-1.

One of the greatest uses of electricity in the late 1800s was for electric railways. In 1884 the Sprague Electric Railway was incorporated. Sprague built an electric railway for Richmond, Virginia, in 1888. By 1902 the horse-drawn street trolley was obsolete, and there were 22,576 miles of electric railway track in the United States. A 1900 electric railway is shown in Figure 3.2-2.

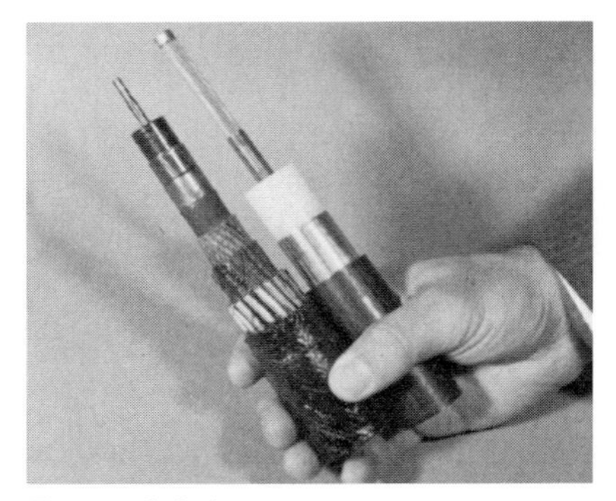

Figure 3.2-1 Examples of undersea cable. Courtesy of Bell Laboratories.

Figure 3.2-2 View of the Sprague electric railway car on the Brookline branch of the Boston system about 1900. This electric railway branch operates as an electric trolley railroad today with modern electric cars. Courtesy of General Electric Company.

3.3 KIRCHHOFF'S LAWS

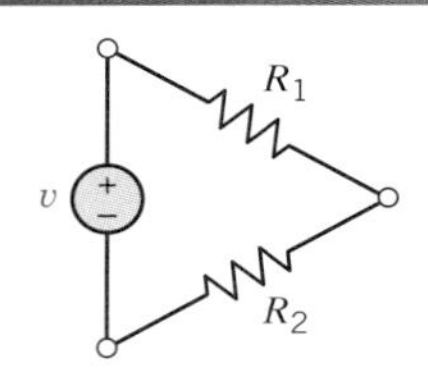

Figure 3.3-1 Simple two-resistor circuit with a voltage source.

It is very important to be able to determine the current and voltage relationships when a circuit consists of two or more circuit elements. In this chapter we consider circuits made up of resistors and we show a simple two-resistor circuit in Figure 3.3-1. This circuit also has a voltage source. Each element is connected at its terminal to another element.

The circuit may conveniently be redrawn as shown in Figure 3.3-2, where a perfectly conducting wire is used to connect terminals d and c. Thus, this model of the wire has zero resistance. A wire has zero voltage across it regardless of the current through it and is sometimes called a short circuit.

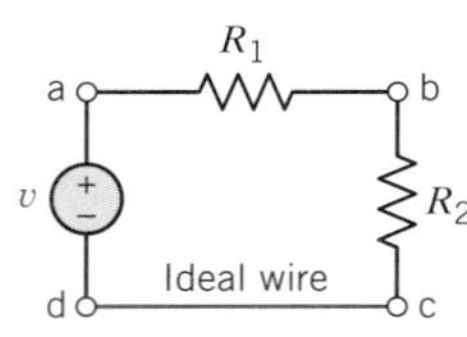

Figure 3.3-2 Alternative form of the circuit shown in Figure 3.3-1. The wire connecting terminals d and c is an ideal, perfectly conducting wire.

When we disconnect an element such as R_2 from the circuit of Figure 3.3-2, we say that we have an *open circuit* at terminals b–c as shown in Figure 3.3-3. The current between terminals c–b is identically zero, regardless of the voltage across terminals c–b, when the resistor is removed (or R_2 is infinite). A junction in which two or more elements have a common connection is called a *node*. More correctly, a node is a junction of conductors composed of ideal wires. If we start at terminal a and traverse around the circuit, passing through each node in turn (that is, a to b to c to d, and back to a), then we have traversed a *closed path.*

A *closed path* in a circuit is a traversal through a series of nodes ending at the starting node without encountering a node more than once. A closed path is often called a *loop*.

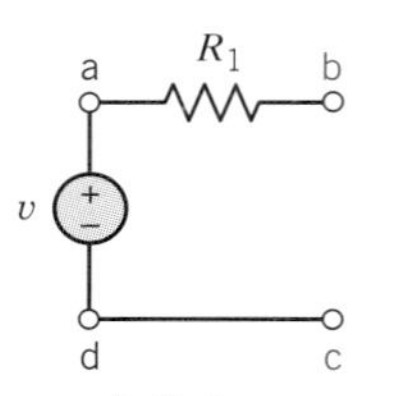

Figure 3.3-3 An open circuit at terminals b–c obtained by disconnecting R_2 from the circuit of Figure 3.3-2.

Example 3.3-1

Identify the closed paths in the circuit of Figure 3.3-4.

Solution

There are three closed paths:

1. a–b–c–d–e–f–a
2. a–b–e–f–a

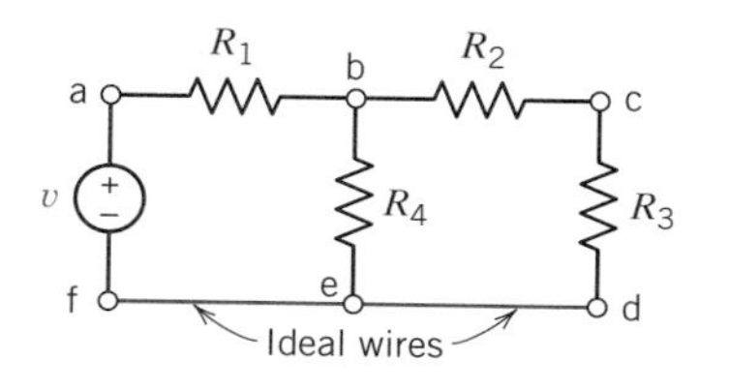

Figure 3.3-4
Circuit with three closed paths or loops.

3. b–c–d–e–b

Note that ideal wires imply that terminals d, e, and f are actually one identical node.

Ohm's law gives the voltage and current relationship for one resistor. However, it remained for Gustav Robert Kirchhoff, a professor at the University of Berlin, to formulate two laws that relate the current and voltage in a circuit with two or more resistors. Kirchhoff, who formulated his laws in 1847, is shown in Figure 3.3-5.

Kirchhoff's current law (KCL) states that the algebraic sum of the currents entering any node is identically zero for all instants of time. This statement is a consequence of the fact that charge cannot accumulate at a node.

Figure 3.3-5 Gustav Robert Kirchhoff (1824–1887). Kirchhoff stated two laws in 1847 regarding the current and voltage in an electrical circuit. Courtesy of the Smithsonian Institution.

By **Kirchhoff's current law,** the algebraic sum of the currents into a node at any instant is zero.

Let us consider the node shown in Figure 3.3-6. The sum of the currents entering the node is

$$-i_a + i_b - i_c + i_d = 0$$

Note that we have $-i_a$ since the current i_a is leaving the node. Another way to state KCL is that the sum of currents entering the node is equal to the sum of currents leaving the node.

If the sum of the currents entering a node were not equal to zero, then charge would accumulate at a node. However, a node is a perfect conductor and cannot accumulate or store charge. Thus, the sum of the currents entering a node is equal to zero.

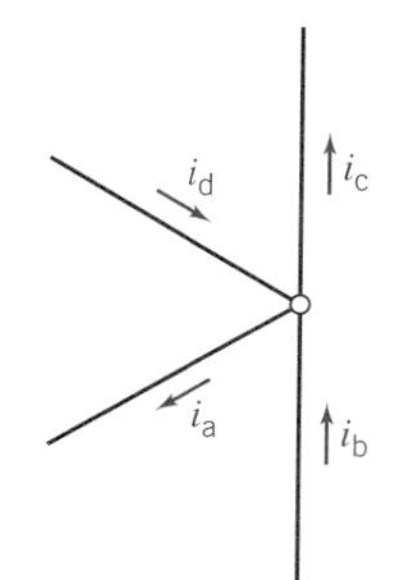

Figure 3.3-6 Currents at a node. The remaining circuit is not shown.

By **Kirchhoff's voltage law** (KVL), the algebraic sum of the voltages around any closed path in a circuit is identically zero for all time.

The term *algebraic* implies the dependency on the voltage polarity encountered as the closed path is traversed.

Let us consider a circuit with two elements and a voltage source connected, as shown in Figure 3.3-7. The voltage across each element is shown with the sign of the voltage displayed. Starting at node c, the sum of the voltage drops around the loop is

$$-v_1 + v_2 - v_3 = 0$$

A common convention is to use the voltage sign on the first terminal of an element encountered as we traverse a path. Therefore, leaving terminal c, we encounter the minus sign on v_1, then the plus sign on v_2, and finally the minus sign on v_3.

Consider the circuit shown in Figure 3.3-8, where the voltage for each element is identified with its sign. The ideal wire has zero resistance, and thus the voltage across it is equal to zero. The sum of the voltages around the loop incorporating v_6, v_3, v_4, and v_5 is

$$-v_6 - v_3 + v_4 + v_5 = 0$$

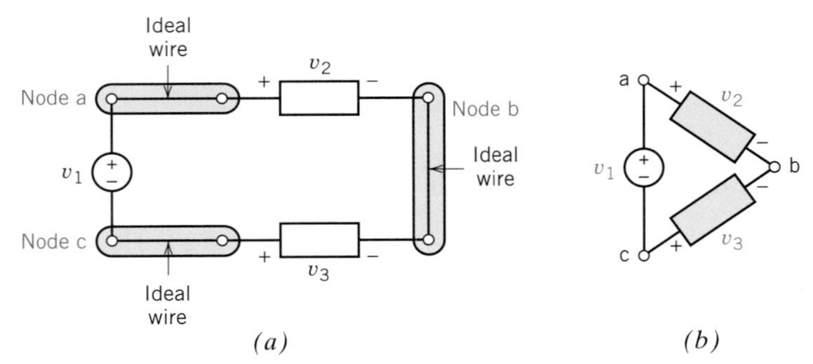

Figure 3.3-7 Circuit with three circuit elements. (*a*) Circuit with ideal wires and nodes identified. (*b*) Circuit with ideal wires removed, displaying nodes.

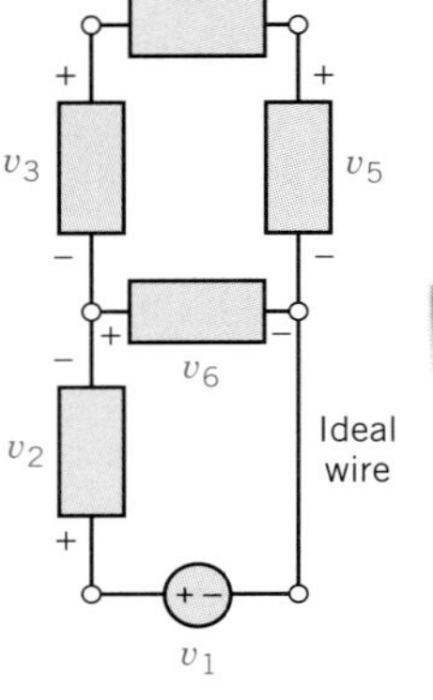

Figure 3.3-8 Circuit with three closed paths. The ideal wire has zero resistance, and thus the voltage across the wire is zero.

The sum of the voltages around a loop is equal to zero. A circuit loop is a conservative system, which means that the energy required to move a charge around a closed path is zero. Thus, since voltage is work per unit charge, the voltage around the loop is zero.

Nevertheless, it is important to note that not all electrical systems are conservative. An example of a nonconservative system is a radio wave broadcasting system.

Example 3.3-2

Consider the circuit shown in Figure 3.3-9. Notice that the passive convention was used to assign reference directions to the resistor voltages and currents. This anticipates using Ohm's law. Find each current and each voltage when $R_1 = 8\ \Omega$, $v_2 = -10$ V, $i_3 = 2$ A, and $R_3 = 1\ \Omega$. Also, determine the resistance R_2.

Solution

The sum of the currents entering node a is

$$i_1 - i_2 - i_3 = 0$$

Using Ohm's law for R_3, we find that

$$v_3 = R_3 i_3 = 1(2) = 2\text{ V}$$

Kirchhoff's voltage law for the bottom loop incorporating v_1, v_3, and the 10-V source is

$$-10 + v_1 + v_3 = 0$$

Therefore,

$$v_1 = 10 - v_3 = 8\text{ V}$$

Ohm's law for the resistor R_1 is

$$v_1 = R_1 i_1$$

or

$$i_1 = v_1/R_1 = 8/8 = 1\text{ A}$$

Since we have now found $i_1 = 1$ A and $i_3 = 2$ A as originally stated, then

$$i_2 = i_1 - i_3 = 1 - 2 = -1\text{ A}$$

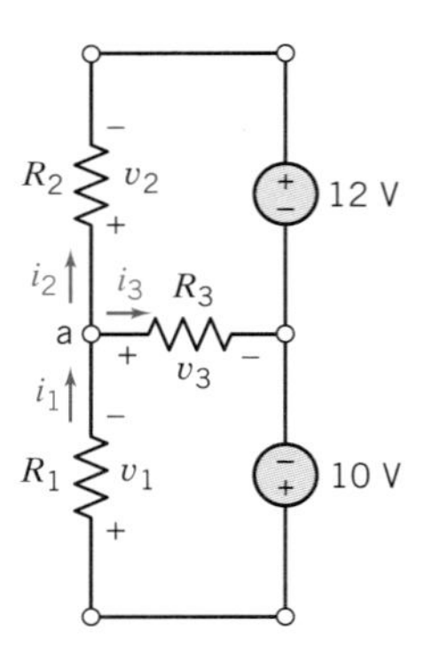

Figure 3.3-9 Circuit with two constant-voltage sources.

We can now find the resistance R_2 from

$$v_2 = R_2 i_2$$

or

$$R_2 = v_2/i_2 = -10/-1 = 10\ \Omega$$

Example 3.3-3

Determine the value of the current, in amps, measured by the ammeter in Figure 3.3-10*a*.

Solution

An ideal ammeter is equivalent to a short circuit. The current measured by the ammeter is the current in the short circuit. Figure 3.3-10*b* shows the circuit after replacing the ammeter by the equivalent short circuit.

The circuit has been redrawn in Figure 3.3-11 to label the nodes of the circuit. This circuit consists of a voltage source, a dependent current source, two resistors and two short circuits. One of the short circuits is the controlling element of the CCCS and the other short circuit is a model of the ammeter.

Applying KCL twice, once at node **d** and again at node **a** shows that the current in the voltage source and the current in the 4 Ω resistor are both equal to i_a. These currents are labeled in Figure 3.3-11. Applying KCL again, at node c, shows that the current in the 2 Ω resistor is equal to i_m. This current is labeled in Figure 3.3-11.

Next, Ohm's Law tells us that the voltage across the 4 Ω resistor is equal to $4\,i_a$ and that the voltage across the 2 Ω resistor is equal to $2\,i_m$. Both of these voltages are labeled in Figure 3.3-11.

Applying KCL at node **b** gives

$$-i_a - 3i_a - i_m = 0$$

Applying KVL to closed path **a - b - c - e - d - a** gives

$$0 = -4\,i_a + 2\,i_m - 12 = -4\left(-\frac{1}{4}i_m\right) + 2\,i_m - 12 = 3\,i_m - 12$$

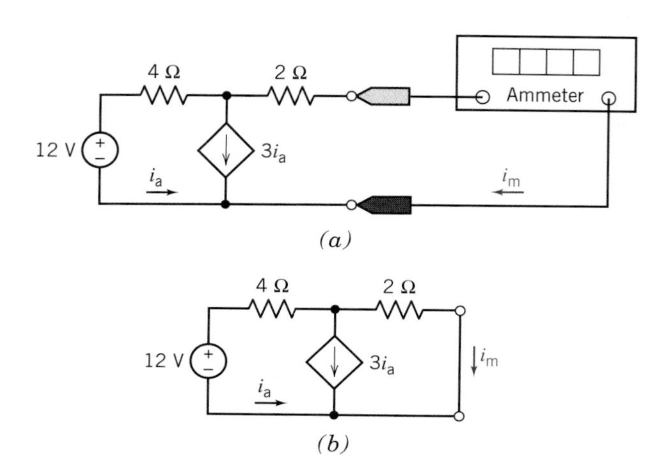

Figure 3.3-10
(*a*) A circuit with dependent source and an ammeter. (*b*) The equivalent circuit after replacing the ammeter by a short circuit.

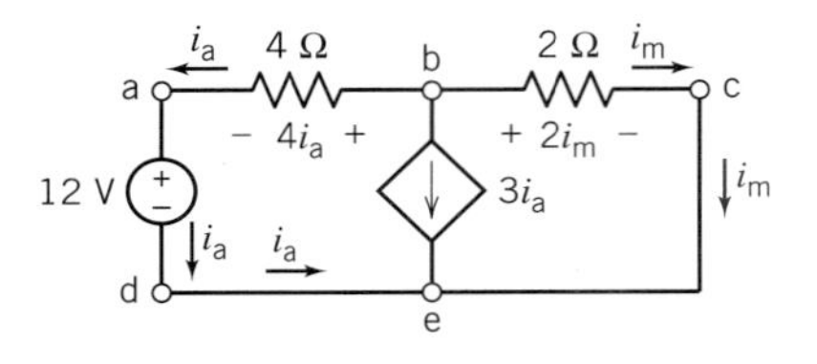

Figure 3.3-11
The circuit of Figure 3.3-10 after labeling the nodes and some element currents and voltages.

Finally, solving this equation gives

$$i_m = 4\,\text{A}$$

Example 3.3-4

Determine the value of the voltage, in volts, measured by the voltmeter in Figure 3.3-12*a*.

Solution

An ideal voltmeter is equivalent to an open circuit. The voltage measured by the voltmeter is the voltage across the open circuit. Figure 3.3-12*b* shows the circuit after replacing the voltmeter by the equivalent open circuit.

The circuit has been redrawn in Figure 3.3-13 to label the nodes of the circuit. This circuit consists of a voltage source, a dependent voltage source, two resistors, a short circuit, and an open circuit. The short circuit is the controlling element of the CCVS and the open circuit is a model of the voltmeter.

Applying KCL twice, once at node **d** and again at node **a,** shows that the current in the voltage source and the current in the 4 Ω resistor are both equal to i_a. These currents are labeled in Figure 3.3-13. Applying KCL again, at node c, shows that the current in the 5 Ω resistor is equal to the current in the open circuit, that is, zero. This current is labeled in Figure 3.3-13. Ohm's Law tells us that the voltage across the 5 Ω resistor is also equal to zero. Next, applying KVL to the closed path **b - c - f - e - b** gives $v_m = 3\,i_a$.

Applying KVL to the closed path **a - b - e - d - a** gives

$$-4\,i_a + 3\,i_a - 12 = 0$$

so

$$i_a = -12\,\text{A}$$

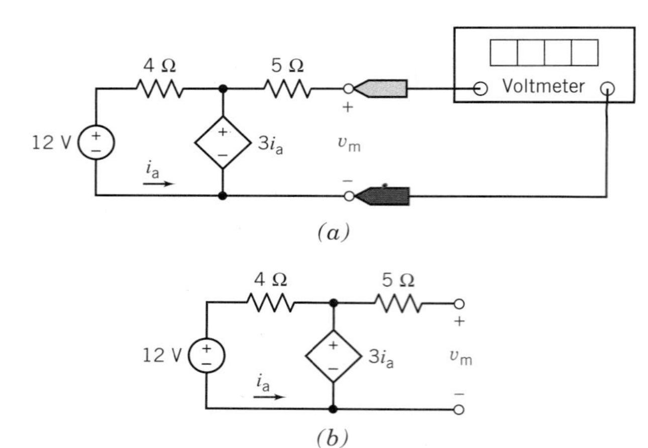

Figure 3.3-12
(*a*) A circuit with dependent source and a voltmeter. (*b*) The equivalent circuit after replacing the voltmeter by a open circuit.

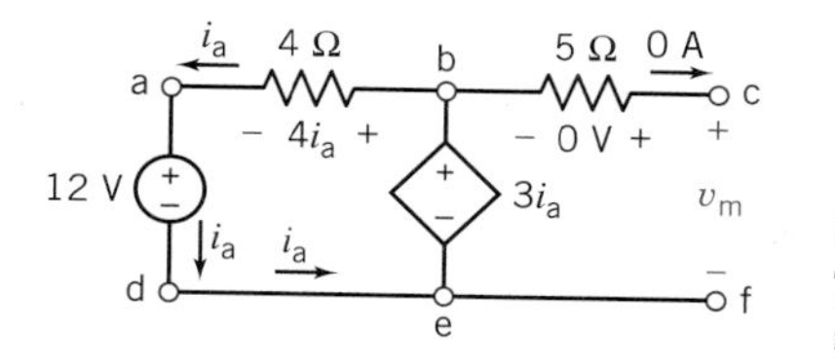

Figure 3.3-13
The circuit of Figure 3.3-12*b* after labeling the nodes and some element currents and voltages.

Finally

$$v_m = 3\,i_a = 3(-12) = -36\,\text{V}$$

Exercise 3.3-1 For the circuit in Figure E 3.3-1, find i_2 and v_2 when $v_3 = 6\,\text{V}$.

Answer: $i_2 = -1\,\text{A}$, $v_2 = 7\,\text{V}$

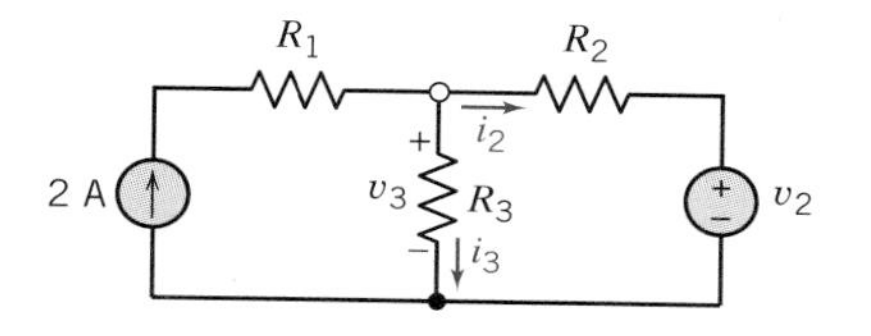

Figure E 3.3-1
Circuit with a current source and an unknown voltage source. $R_1 = R_2 = 1\ \Omega$ and $R_3 = 2\ \Omega$. Also $v_3 = 6\ \text{V}$.

Exercise 3.3-2 Determine the value of the current i_m in Figure E 3.3-2*a*.

Hint: Apply KVL to the closed path **a - b - d - c - a** in Figure E 3.3-2*b* to determine v_a. Then apply KCL at node b to find i_m.

Answer: $i_m = 9\,\text{A}$.

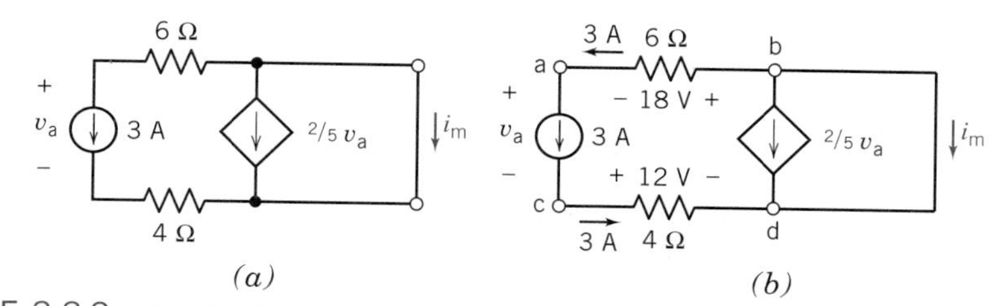

Figure E 3.3-2 (*a*) A circuit containing a VCCS. (*b*) The circuit after labeling the nodes and some element currents and voltages.

Exercise 3.3-3 Determine the value of the voltage v_m in Figure 3.3-3*a*.

Hint: Apply KVL to the closed path **a - b - d - c - a** in Figure E 3.3-3*b* to determine v_a.

Answer: $v_m = 24\,\text{V}$.

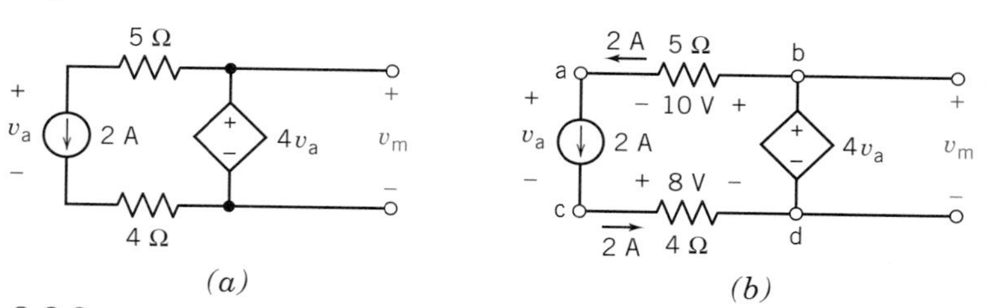

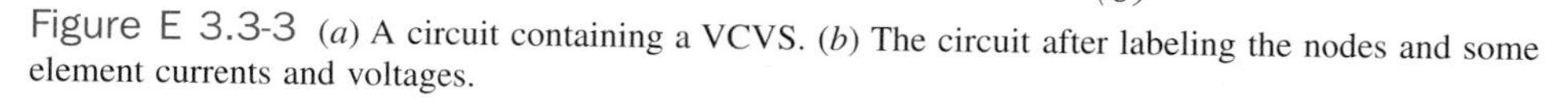

Figure E 3.3-3 (*a*) A circuit containing a VCVS. (*b*) The circuit after labeling the nodes and some element currents and voltages.

3.4 A SINGLE-LOOP CIRCUIT—THE VOLTAGE DIVIDER

Let us consider a single-loop circuit, as shown in Figure 3.4-1. In anticipation of using Ohm's law, the passive convention has been used to assign reference directions to resistor voltages and currents. Using KCL at each node, we obtain

$$\text{a:} \quad i_s - i_1 = 0 \tag{3.4-1}$$

$$\text{b:} \quad i_1 - i_2 = 0 \tag{3.4-2}$$

$$\text{c:} \quad i_2 - i_3 = 0 \tag{3.4-3}$$

$$\text{d:} \quad i_3 - i_s = 0 \tag{3.4-4}$$

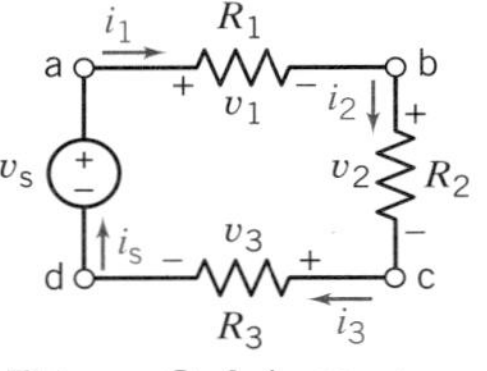

Figure 3.4-1 Single-loop circuit with a voltage source v_s.

We have four equations, but any one of the four can be derived from the other three equations. In any circuit with n nodes, $n - 1$ independent current equations can be derived from Kirchhoff's current law.

Of course, we also note that

$$i_s = i_1 = i_2 = i_3$$

so that the current i_1 can be said to be the loop current and flows continuously around the loop from a to b to c to d and back to a.

The connection of resistors in Figure 3.4-1 is said to be a *series* connection since all the elements carry the same current.

In order to determine i_1, we use KVL around the loop to obtain

$$-v_s + v_1 + v_2 + v_3 = 0 \tag{3.4-5}$$

where v_1 is the voltage across the resistor R_1. Using Ohm's law for each resistor, Eq. 3.4-5 can be written as

$$-v_s + i_1R_1 + i_1R_2 + i_1R_3 = 0$$

Solving for i_1, we have

$$i_1 = \frac{v_s}{R_1 + R_2 + R_3}$$

Thus, the voltage across the nth resistor R_n is v_n and can be obtained as

$$v_n = i_1R_n = \frac{v_sR_n}{R_1 + R_2 + R_3} \tag{3.4-6}$$

For example, the voltage across resistor R_2 is

$$v_2 = \frac{R_2}{R_1 + R_2 + R_3} v_s$$

Thus, the voltage appearing across one of a series connection of resistors connected in series with a voltage source will be the ratio of its resistance to the total resistance times the source voltage. This circuit demonstrates the principle of *voltage division,* and the circuit is called a *voltage divider.*

In general, we may represent the voltage divider principle by the equation

$$v_n = \frac{R_n}{R_1 + R_2 + \cdots + R_N} v_s \tag{3.4-7}$$

where the voltage is across the nth resistor of N resistors connected in series.

Example 3.4-1

Let us consider the circuit shown in Figure 3.4-2 and determine the resistance R_2 required so that the voltage across R_2 will be one-fourth of the source voltage when $R_1 = 9\ \Omega$. Determine the current flowing when $v_s = 12\text{ V}$.

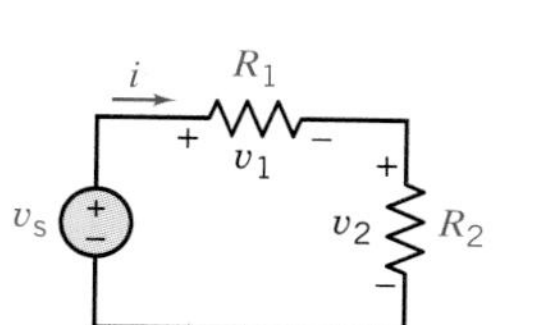

Figure 3.4-2 Voltage divider circuit with $R_1 = 9\ \Omega$.

Solution

The voltage across resistor R_2 will be

$$v_2 = \frac{R_2}{R_1 + R_2} v_s$$

Since we desire $v_2/v_s = 1/4$, we have

$$\frac{R_2}{R_1 + R_2} = \frac{1}{4}$$

or

$$R_1 = 3R_2$$

Since $R_1 = 9\ \Omega$, we require that $R_2 = 3\ \Omega$. Using KVL around the loop, we have

$$-v_s + v_1 + v_2 = 0$$

or

$$v_s = iR_1 + iR_2$$

Therefore,

$$i = \frac{v_s}{R_1 + R_2} \tag{3.4-8}$$

or

$$i = \frac{12}{12} = 1\text{ A}$$

Let us consider the simple circuit of the voltage source connected to a resistance R_s as shown in Figure 3.4-3. For this circuit

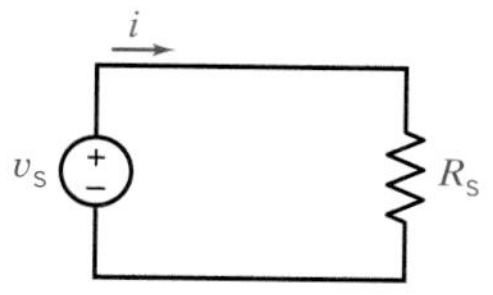

$$i = \frac{v_s}{R_s} \tag{3.4-9}$$

Figure 3.4-3 Equivalent circuit for a series connection of resistors.

Comparing Eqs. 3.4-8 and 3.4-9, we see that the currents are identical when

$$R_s = R_1 + R_2$$

The resistance R_s is said to be an *equivalent resistance* of the series connection of resistors R_1 and R_2. In general, the equivalent resistance of a series of N resistors is

$$R_s = R_1 + R_2 + \cdots + R_N \tag{3.4-10}$$

In this specific case

$$R_s = R_1 + R_2 = 9 + 3 = 12\ \Omega$$

Each resistor in Figure 3.4-2 absorbs power, so the power absorbed by R_1 is

$$p_1 = \frac{v_1^2}{R_1}$$

and the power absorbed by the second resistor is

$$p_2 = \frac{v_2^2}{R_2}$$

The total power absorbed by the two resistors is

$$p = p_1 + p_2 = \frac{v_1^2}{R_1} + \frac{v_2^2}{R_2} \qquad (3.4\text{-}11)$$

However, according to the voltage divider principle,

$$v_n = \frac{R_n}{R_1 + R_2} v_s$$

Then we may rewrite Eq. 3.4-11 as

$$p = \frac{R_1}{(R_1 + R_2)^2} v_s^2 + \frac{R_2}{(R_1 + R_2)^2} v_s^2$$

Since $R_1 + R_2 = R_s$, the equivalent series resistance, we have

$$p = \frac{R_1 + R_2}{R_s^2} v_s^2 = \frac{v_s^2}{R_s}$$

Thus, the total power absorbed by the two series resistors is equal to the power absorbed by the equivalent resistance R_s. The power absorbed by the two resistors is equal to that supplied by the source v_s. We show this by noting that the current is

$$i = \frac{v_s}{R_s}$$

and therefore the power supplied is

$$p = v_s i = \frac{v_s^2}{R_s}$$

Example 3.4-2

For the circuit of Figure 3.4-4*a*, find the current measured by the ammeter. Then show that the power absorbed by the two resistors is equal to that supplied by the source.

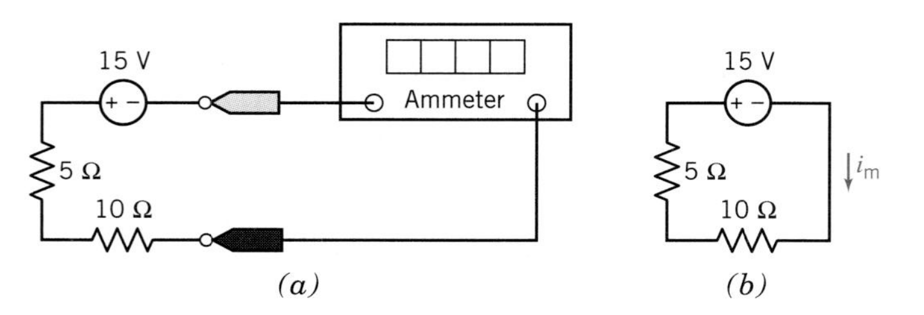

Figure 3.4-4
(*a*) A circuit containing series resistors. (*b*) The circuit after the ideal ammeter has been replaced by the equivalent short circuit and a label has been added to indicate the current measured by the ammeter, i_m.

Solution

Figure 3.4-4*b* shows the circuit after the ideal ammeter has been replaced by the equivalent short circuit and a label has been added to indicate the current measured by the ammeter, i_m. Applying KVL gives

$$15 + 5\,i_m + 10\,i_m = 0$$

The current measured by the ammeter is

$$i_m = -\frac{15}{5+10} = -1\text{ A}$$

(Why is i_m negative? Why can't we just divide the source voltage by the equivalent resistance? Recall that when we use Ohm's Law, the voltage and current must adhere to the passive convention. In this case, the current calculated by dividing the source voltage by the equivalent resistance does not have the same reference direction as i_m, and so we need a minus sign.)

The total power absorbed by the two resistors is

$$p_R = 5i_m^2 + 10i_m^2 = 15(1^2) = 15\text{ W}$$

The power supplied by the source is

$$p_s = v_s\,i_m = 15(1) = 15\text{ W}$$

Thus, the power supplied by the source is equal to that absorbed by the series connection of resistors.

Exercise 3.4-1 For the circuit of Figure E 3.4-1, find the voltage v_3 and the current i and show that the power delivered to the three resistors is equal to that supplied by the source.

Answer: $v_3 = 3\text{ V},\ i = 1\text{ A}$

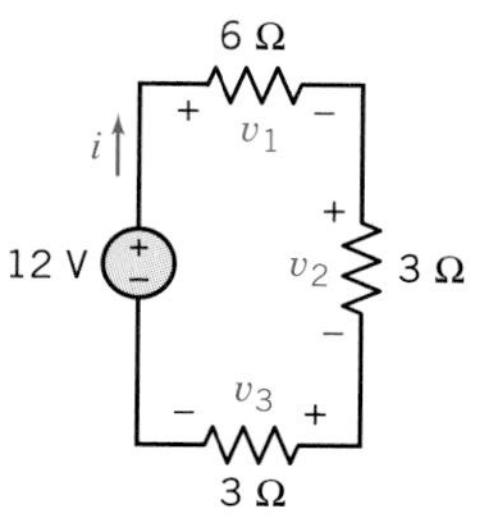

Figure E 3.4-1 Circuit with three series resistors (for Exercise 3.4-1).

Exercise 3.4-2 Consider the voltage divider shown in Figure E 3.4-2 when $R_1 = 6\ \Omega$. It is desired that the output power absorbed by $R_1 = 6\ \Omega$ be 6 W. Find the voltage v_o and the required source v_s.

Answer: $v_s = 14\text{ V},\ v_o = 6\text{ V}$

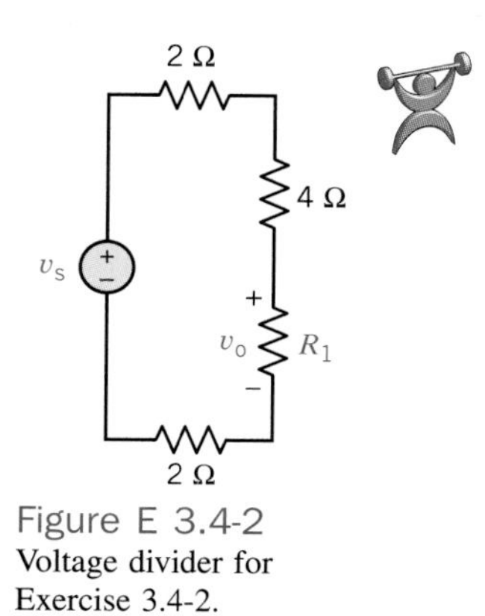

Figure E 3.4-2 Voltage divider for Exercise 3.4-2.

Exercise 3.4-3 Determine the voltage measured by the voltmeter in the circuit shown in Figure E 3.4-3*a*.

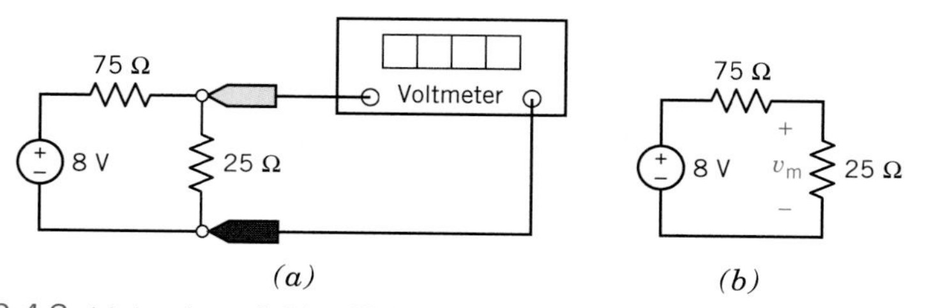

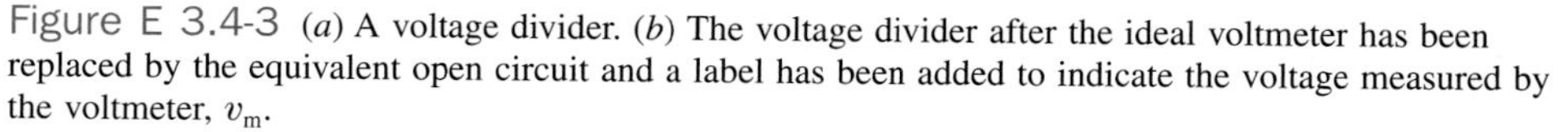

Figure E 3.4-3 (*a*) A voltage divider. (*b*) The voltage divider after the ideal voltmeter has been replaced by the equivalent open circuit and a label has been added to indicate the voltage measured by the voltmeter, v_m.

Hint: Figure E 3.4-3*b* shows the circuit after the ideal voltmeter has been replaced by the equivalent open circuit and a label has been added to indicate the voltage measured by the voltmeter, v_m.

Answer: $v_m = 2$ V.

Exercise 3.4-4 Determine the voltage measured by the voltmeter in the circuit shown in Figure E 3.4-4*a*.

Hint: Figure E 3.4-4*b* shows the circuit after the ideal voltmeter has been replaced by the equivalent open circuit and a label has been added to indicate the voltage measured by the voltmeter, v_m.

Answer: $v_m = -2$ V.

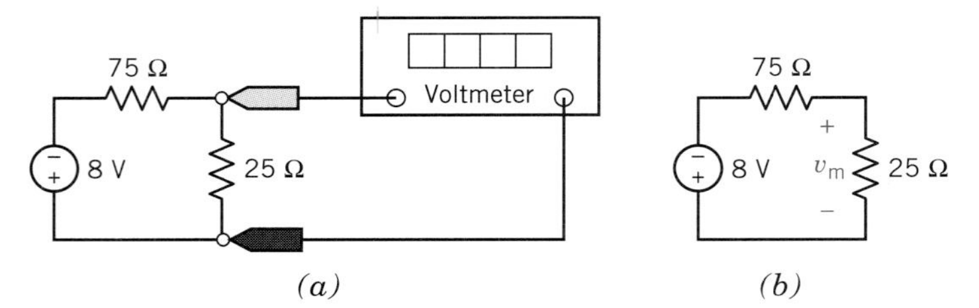

Figure E 3.4-4 (*a*) A voltage divider. (*b*) The voltage divider after the ideal voltmeter has been replaced by the equivalent open circuit and a label has been added to indicate the voltage measured by the voltmeter, v_m.

3.5 PARALLEL RESISTORS AND CURRENT DIVISION

Edison reasoned that an electric light system required high-resistance lamps connected in parallel. A parallel combination in which the lamps are strung between lines is shown in Figure 3.5-1. In this arrangement each lamp is modeled by a resistance R_n and the source voltage v_s appears across each lamp. Each lamp will thus have the same voltage applied, and turning on and off any number of the lamps does not affect the voltage applied to the others. Since all of the lamps have the same voltage across them, the current supplied by the source is the sum of the currents in all of the lamps. (In a series arrangement, on the other hand, the current in the lamps and the lines is the same.) To avoid excessively large and expensive line conductors, the current in each of the lamps in a parallel connection must be kept small. The only answer was to have a lamp whose filament had a high resistance. Edison, therefore, was looking not only for a material that would remain incandescent for a long time when a current was passed through it but also for something that would have a high resistance—a commercial rather than a technical requirement. Edison finally found a thin high-resistance carbonized thread for the filament.

Circuit elements, such as resistors, are connected in *parallel* when the voltage across each element is identical. Elements in parallel are connected at both terminals. The circuit in Figure 3.5-1 is a parallel circuit since each resistor has v_s across its terminals.

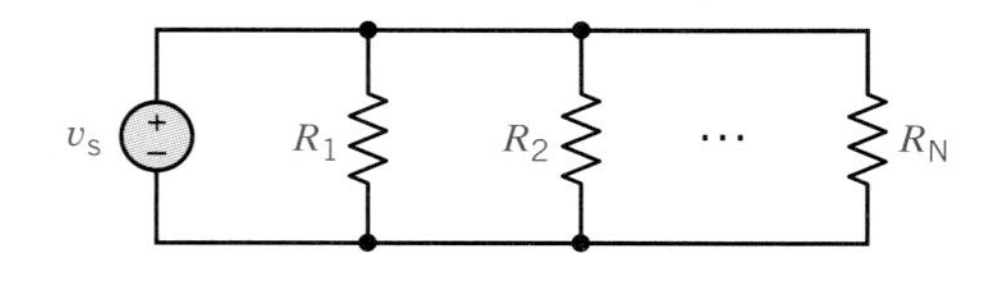

Figure 3.5-1 Edison's parallel lamp scheme with the *n*th lamp represented by its resistance R_n. This circuit has a total of *N* lamps.

Consider the circuit with two resistors and a current source shown in Figure 3.5-2. Note that both resistors are connected to terminals a and b and that the voltage v appears across each parallel element. In anticipation of using Ohm's law, the passive convention is used to assign reference directions to the resistor voltages and currents. We may write KCL at node a (or at node b) to obtain

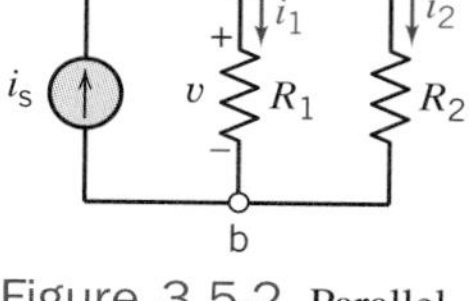

Figure 3.5-2 Parallel circuit with a current source.

$$i_s - i_1 - i_2 = 0$$

or

$$i_s = i_1 + i_2$$

However, from Ohm's law

$$i_1 = \frac{v}{R_1} \quad \text{and} \quad i_2 = \frac{v}{R_2}$$

Then

$$i_s = \frac{v}{R_1} + \frac{v}{R_2} \qquad (3.5\text{-}1)$$

Recall that we defined conductance G as the inverse of resistance R. We may therefore rewrite Eq. 3.5-1 as

$$i_s = G_1 v + G_2 v = (G_1 + G_2)\, v \qquad (3.5\text{-}2)$$

Thus, the equivalent circuit for this parallel circuit is a conductance G_P, as shown in Figure 3.5-3, where

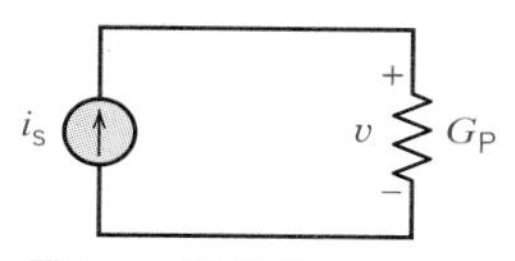

Figure 3.5-3 Equivalent circuit for a parallel circuit.

$$G_P = G_1 + G_2$$

The equivalent resistance for the two-resistor circuit is found from

$$G_P = \frac{1}{R_1} + \frac{1}{R_2}$$

Since $G_P = 1/R_P$, we have

$$\frac{1}{R_P} = \frac{1}{R_1} + \frac{1}{R_2}$$

or

$$R_P = \frac{R_1 R_2}{R_1 + R_2} \qquad (3.5\text{-}3)$$

Note that the total conductance increases as additional parallel elements are added and that the total resistance declines as each resistor is added.

The circuit shown in Figure 3.5-2 is called a *current divider* circuit since it divides the source current. Note that

$$i_1 = G_1 v \qquad (3.5\text{-}4)$$

Also, since $i_s = (G_1 + G_2)v$, we solve for v, obtaining

$$v = \frac{i_s}{G_1 + G_2} \qquad (3.5\text{-}5)$$

Substituting v from Eq. 3.5-5 into Eq. 3.5-4, we obtain

$$i_1 = \frac{G_1 i_s}{G_1 + G_2} \qquad (3.5\text{-}6)$$

Similarly,

$$i_2 = \frac{G_2 i_s}{G_1 + G_2}$$

Note that we may use $G_2 = 1/R_2$ and $G_1 = 1/R_1$ to obtain the current i_2 in terms of two resistances as follows:

$$i_2 = \frac{R_1 i_s}{R_1 + R_2}$$

The current of the source divides between conductances G_1 and G_2 in proportion to their conductance values.

Let us consider the more general case of current division with a set of N parallel conductors as shown in Figure 3.5-4. The KCL gives

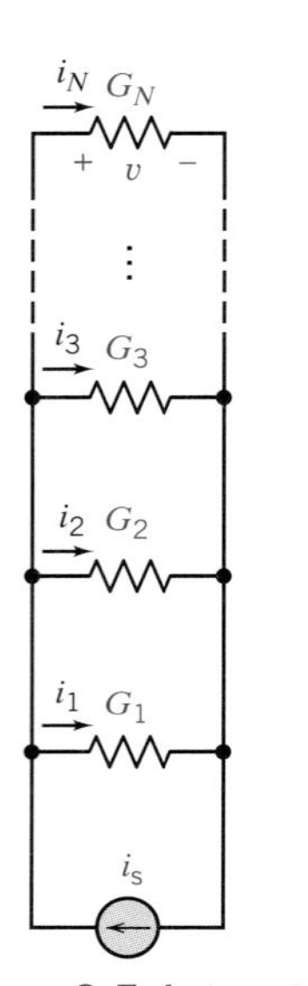

Figure 3.5-4 Set of N parallel conductances with a current source i_s.

$$i_s = i_1 + i_2 + i_3 + \cdots + i_N \tag{3.5-7}$$

for which

$$i_n = G_n v \tag{3.5-8}$$

for $n = 1, \ldots, N$. We may write Eq. 3.5-7 as

$$i_s = (G_1 + G_2 + G_3 + \cdots + G_N)v \tag{3.5-9}$$

Therefore

$$i_s = v \sum_{n=1}^{N} G_n \tag{3.5-10}$$

Since $i_n = G_n v$, we may obtain v from Eq. 3.5-10 and substitute it in Eq. 3.5-8, obtaining

$$i_n = \frac{G_n i_s}{\sum_{n=1}^{N} G_n} \tag{3.5-11}$$

Recall that the equivalent circuit, Figure 3.5-3, has an equivalent conductance G_P such that

$$G_P = \sum_{n=1}^{N} G_n \tag{3.5-12}$$

Therefore

$$i_n = \frac{G_n i_s}{G_P} \tag{3.5-13}$$

which is the basic equation for the current divider with N conductances. Of course, Eq. 3.5-12 can be rewritten as

$$\frac{1}{R_P} = \sum_{n=1}^{N} \frac{1}{R_n} \tag{3.5-14}$$

Example 3.5-1

For the circuit in Figure 3.5-5 find (a) the current in each branch, (b) the equivalent circuit, and (c) the voltage v. The resistors are

$$R_1 = \frac{1}{2}\,\Omega, \quad R_2 = \frac{1}{4}\,\Omega, \quad R_3 = \frac{1}{8}\,\Omega$$

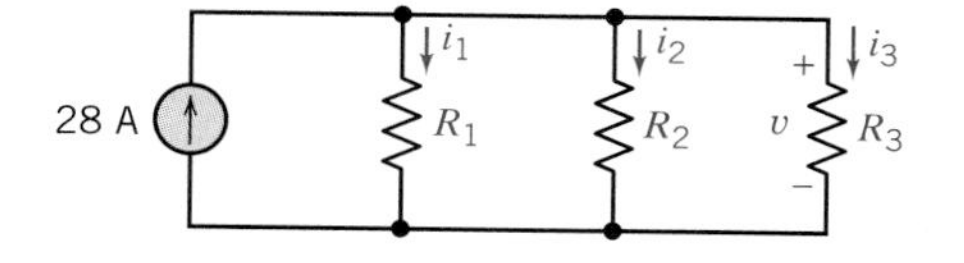

Figure 3.5-5
Parallel circuit for Example 3.5-1.

Solution

The current divider follows the equation

$$i_n = \frac{G_n\, i_s}{G_P}$$

so it is wise to find the equivalent circuit, as shown in Figure 3.5-6, with its equivalent conductance G_P. We have

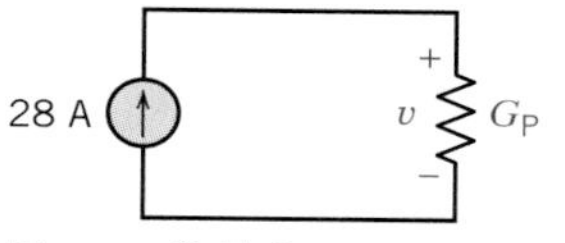

Figure 3.5-6
Equivalent circuit for the parallel circuit of Figure 3.5-5.

$$G_P = \sum_{n=1}^{N} G_n = G_1 + G_2 + G_3 = 2 + 4 + 8 = 14\,\text{S}$$

Recall that the units for conductance are siemens (S). Then

$$i_1 = \frac{G_1\, i_s}{G_P} = \frac{2}{14}(28) = 4\,\text{A}$$

Similarly,

$$i_2 = \frac{G_2\, i_s}{G_P} = \frac{4(28)}{14} = 8\,\text{A}$$

and

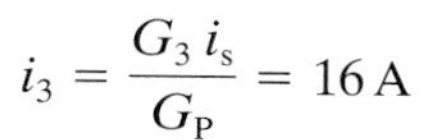

$$i_3 = \frac{G_3\, i_s}{G_P} = 16\,\text{A}$$

Since $i_n = G_n v$, we have

$$v = \frac{i_1}{G_1} = \frac{4}{2} = 2\,\text{V}$$

Example 3.5-2

For the circuit of Figure 3.5-7*a*, find the voltage measured by the voltmeter. Then show that the power absorbed by the two resistors is equal to that supplied by the source.

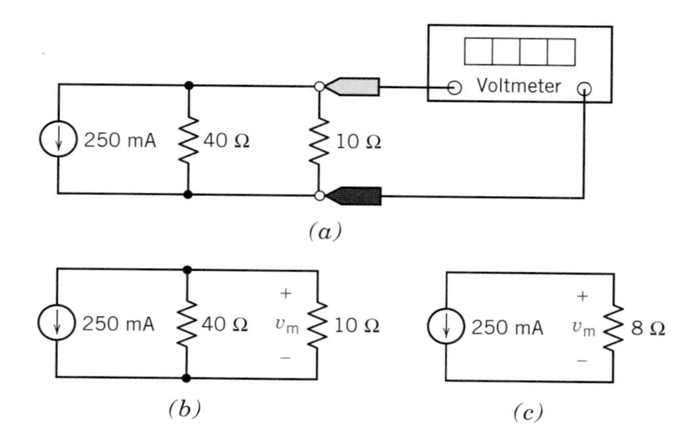

Figure 3.5-7
(*a*) A circuit containing parallel resistors. (*b*) The circuit after the ideal voltmeter has been replaced by the equivalent open circuit and a label has been added to indicate the voltage measured by the voltmeter, v_m. (*c*) The circuit after the parallel resistors have been replaced by an equivalent resistance.

Solution

Figure 3.5-7*b* shows the circuit after the ideal voltmeter has been replaced by the equivalent open circuit and a label has been added to indicate the voltage measured by the voltmeter, v_m. The two resistors are connected in parallel and can be replaced with a single equivalent resistor. The resistance of this equivalent resistor is calculated as

$$\frac{40 \cdot 10}{40 + 10} = 8\ \Omega$$

Figure 3.5-7*c* shows the circuit after the parallel resistors have been replaced by the equivalent resistor. The current in the equivalent resistor is 250 mA, directed upward. This current and the voltage v_m do not adhere to the passive convention. The current in the equivalent resistance can also be expressed as −250 mA, directed downward. This current and the voltage v_m do adhere to the passive convention. Ohm's Law gives

$$v_m = 8\,(-0.25) = -2\text{ V}$$

The voltage v_m in Figure 3.5-7*b* is equal to the voltage v_m in Figure 3.5-7*c*. This is a consequence of the equivalence of the 8 Ω resistor to the parallel combination of the 40 Ω and 10 Ω resistors. Looking at Figure 3.5-7*b*, the power absorbed by the resistors is

$$p_R = \frac{v_m^2}{40} + \frac{v_m^2}{10} = \frac{2^2}{40} + \frac{2^2}{10} = 0.1 + 0.4 = 0.5\text{ W}$$

The power supplied by the current source is

$$p_s = 2(0.25) = 0.5\text{ W}$$

Thus, the power absorbed by the two resistors is equal to that supplied by the source.

Exercise 3.5-1 A resistor network consisting of parallel resistors is shown in a package used for printed circuit board electronics in Figure E 3.5-1*a*. This package is only 2 cm × 0.7 cm, and each resistor is 1 kΩ. The circuit is connected to use four resistors as shown in Figure E 3.5-1*b*. Find the equivalent circuit for this network. Determine the current in each resistor when $i_s = 1$ mA.

Answer: $R_p = 250\ \Omega$.

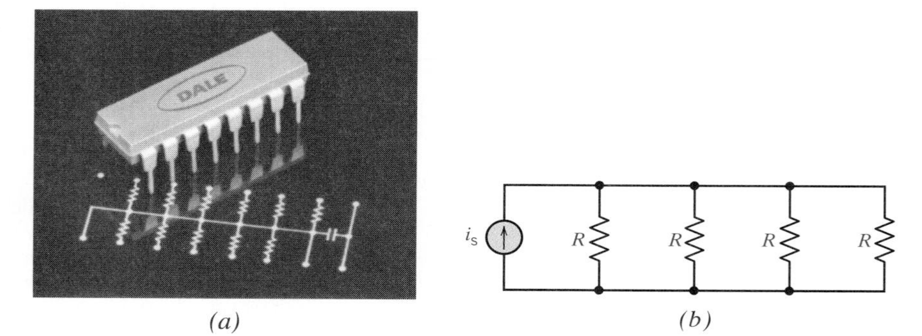

Figure E 3.5-1 (*a*) A parallel resistor network. Courtesy of Dale Electronics. (*b*) The connected circuit uses four resistors where $R = 1\text{ k}\Omega$.

Exercise 3.5-2 Determine the current measured by the ammeter in the circuit shown in Figure E 3.5-2*a*.

Hint: Figure E 3.5-2*b* shows the circuit after the ideal ammeter has been replaced by the equivalent short circuit and a label has been added to indicate the current measured by the ammeter, i_m.

Answer: $i_m = -1$ A.

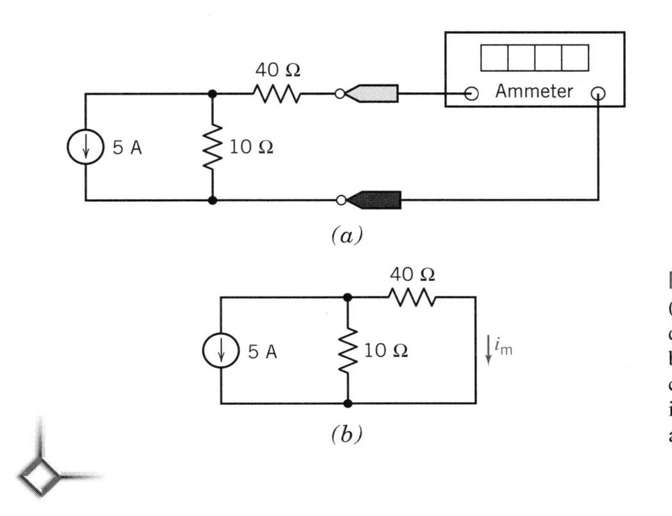

Figure E 3.5-2
(*a*) A current divider. (*b*) The current divider after the ideal ammeter has been replaced by the equivalent short circuit and a label has been added to indicate the current measured by the ammeter, i_m.

3.6 SERIES VOLTAGE SOURCES AND PARALLEL CURRENT SOURCES

Voltage sources connected in series are equivalent to a single voltage source. The voltage of the equivalent voltage source is equal to the sum of voltages of the series voltage sources.

Consider the circuit shown in Figure 3.6-1*a*. Notice that the currents of both voltage sources are equal. Accordingly, define the current, i_s, to be

$$i_s = i_a = i_b \tag{3.6-1}$$

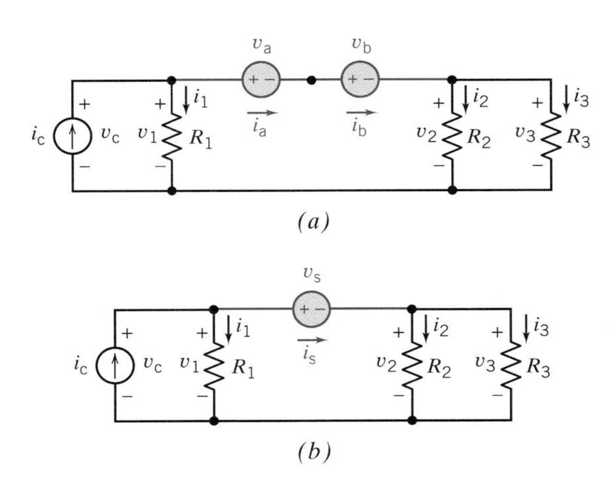

Figure 3.6-1
(*a*) A circuit containing voltage sources connected in series and (*b*) an equivalent circuit.

Next, define the voltage, v_s, to be

$$v_s = v_a + v_b \tag{3.6-2}$$

Using KCL, KVL, and Ohm's law, we can represent the circuit in Figure 3.61*a* by the equations

$$i_c = \frac{v_1}{R_1} + i_s \tag{3.6-3}$$
$$i_s = \frac{v_2}{R_2} + i_3 \tag{3.6-4}$$
$$v_c = v_1 \tag{3.6-5}$$
$$v_1 = v_s + v_2 \tag{3.6-6}$$
$$v_2 = i_3 R_3 \tag{3.6-7}$$

where $i_s = i_a = i_b$ and $v_s = v_a + v_b$. These same equations result from applying KCL, KVL, and Ohm's law to the circuit in Figure 3.6-1*b*. If $i_s = i_a = i_b$ and $v_s = v_a + v_b$, then the circuits shown in Figures 3.6-1*a* and 3.6-1*b* are equivalent because they are both represented by the same equations.

For example, suppose that $i_c = 4$ A, $R_1 = 2\ \Omega$, $R_2 = 6\ \Omega$, $R_3 = 3\ \Omega$, $v_a = 1$ V, and $v_b = 3$ V. The equations describing the circuit in Figure 3.6-1*a* become

$$4 = \frac{v_1}{2} + i_s \tag{3.6-8}$$
$$i_s = \frac{v_2}{6} + i_3 \tag{3.6-9}$$
$$v_c = v_1 \tag{3.6-10}$$
$$v_1 = 4 + v_2 \tag{3.6-11}$$
$$v_2 = 3i_3 \tag{3.6-12}$$

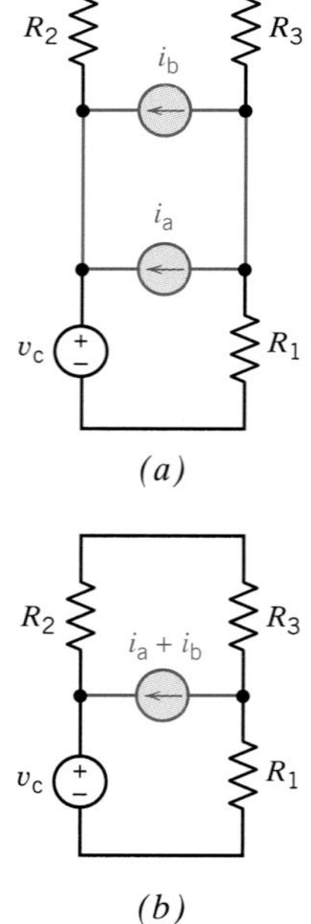

Figure 3.6-2 (*a*) A circuit containing parallel current sources and (*b*) an equivalent circuit.

The solution to this set of equations is $v_1 = 6$ V, $i_s = 1$ A, $i_3 = 0.66$ A, $v_2 = 2$ V, and $v_c = 6$ V. Equations 3.6-8 to 3.6-12 also describe the circuit in Figure 3.6-1*b*. Thus, $v_1 = 6$ V, $i_s = 1$ A, $i_3 = 0.66$ A, $v_2 = 2$ V, and $v_c = 6$ V in both circuits. Replacing series voltage sources by a single, equivalent voltage source does not change the voltage or current of other elements of the circuit.

Figure 3.6-2*a* shows a circuit containing parallel current sources. The circuit in Figure 3.6-2*b* is obtained by replacing these parallel current sources by a single, equivalent current source. The current of the equivalent current source is equal to the sum of the currents of the parallel current sources.

We are not allowed to connect independent current sources in series. Series elements have the same current. This restriction prevents series current sources from being independent. Similarly, we are not allowed to connect independent voltage sources in parallel.

Table 3.6-1 summarizes the parallel and series connections of current and voltage sources.

Table 3.6-1 **Parallel and Series Voltage and Current Sources**

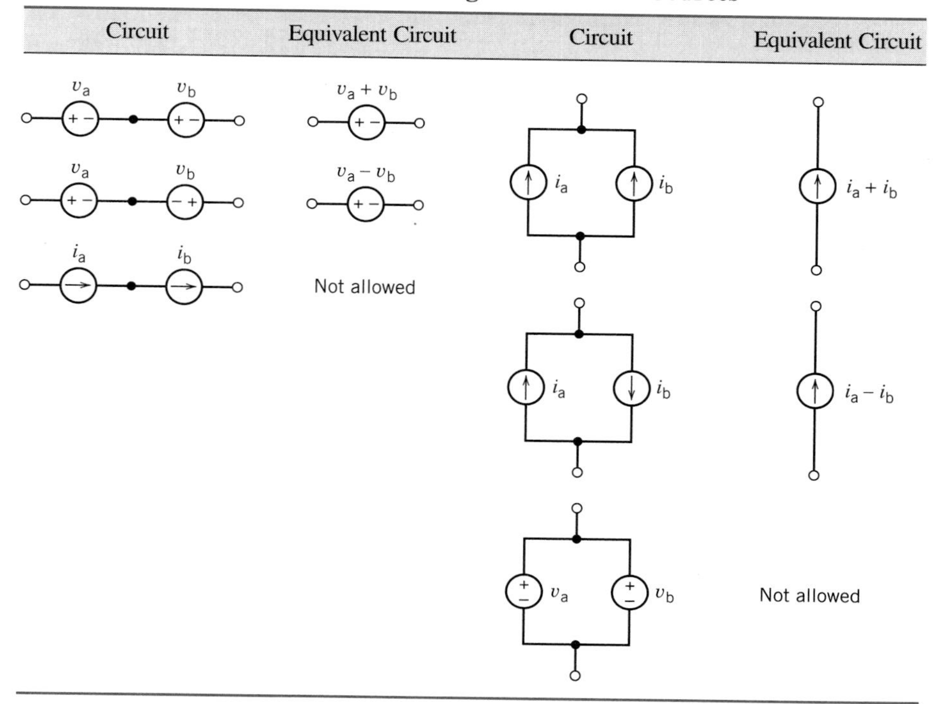

3.7 CIRCUIT ANALYSIS

In this section we consider the analysis of a circuit by replacing a set of resistors with an equivalent resistance, thus reducing the network to a form easily analyzed.

Consider the circuit shown in Figure 3.7-1. Note that it includes a set of resistors that is in series and another set of resistors that is in parallel. It is desired to find the output voltage v_o, so we wish to reduce the circuit to the equivalent circuit shown in Figure 3.7-2.

Two circuits are **equivalent** if they exhibit identical characteristics at the same two terminals.

We note that the equivalent series resistance is

$$R_s = R_1 + R_2 + R_3$$

and the equivalent parallel resistance is

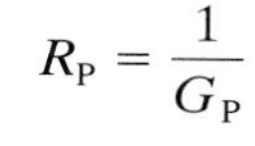

$$R_P = \frac{1}{G_P}$$

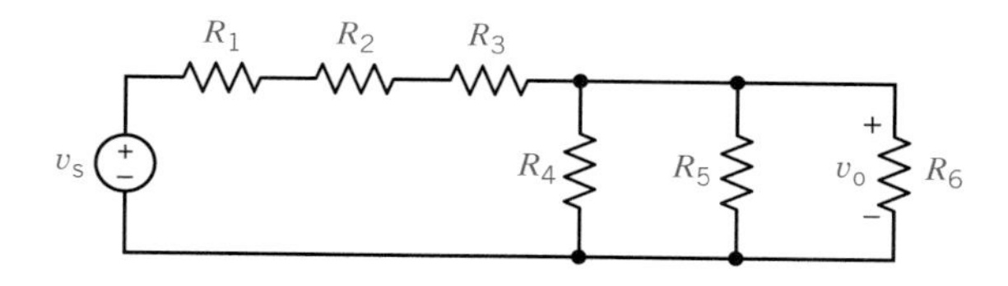

Figure 3.7-1
Circuit with a set of series resistors and a set of parallel resistors.

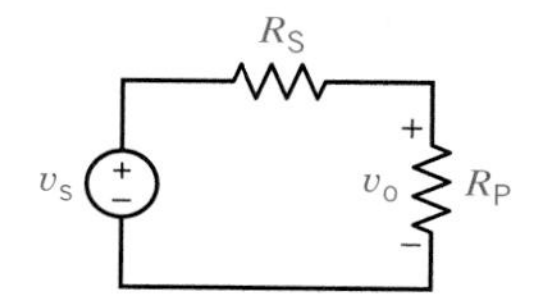

Figure 3.7-2
Equivalent circuit for the circuit of Figure 3.7-1.

where

$$G_{\rm P} = G_4 + G_5 + G_6$$

Then, using the voltage divider principle, with Figure 3.7-2, we have

$$v_{\rm o} = \frac{R_{\rm P}}{R_{\rm s} + R_{\rm P}} v_{\rm s}$$

If a circuit has several combinations of sets of parallel resistors and sets of series resistors, one works through several steps, reducing the network to its simplest form.

Example 3.7-1

Consider the circuit shown in Figure 3.7-3. Find the current i_1 when

$$R_4 = 2\ \Omega \quad \text{and} \quad R_2 = R_3 = 8\ \Omega$$

Solution

Since the objective is to find i_1, we will attempt to reduce the circuit so that the 3-Ω resistor is in parallel with one resistor and the current source i_s. Then we can use the current divider principle to obtain i_1. Since R_2 and R_3 are in parallel, we find an equivalent resistance as

$$R_{\rm P1} = \frac{R_2 R_3}{R_2 + R_3} = 4\ \Omega$$

Then adding $R_{\rm P1}$ to R_4, we have a series equivalent resistor

$$R_{\rm s} = R_4 + R_{\rm P1} = 2 + 4 = 6\ \Omega$$

Now the $R_{\rm s}$ resistor is in parallel with three resistors as shown in Figure 3.7-3*b*. However, we wish to obtain the equivalent circuit as shown in Figure 3.7-4, so that we can find i_1. Therefore, we combine the 9-Ω resistor, the 18-Ω resistor, and $R_{\rm s}$ shown to the right of terminals a–b in Figure 3.7-3*b* into one parallel equivalent conductance $G_{\rm P2}$. Thus, we find

$$G_{\rm P2} = \frac{1}{9} + \frac{1}{18} + \frac{1}{R_{\rm s}} = \frac{1}{9} + \frac{1}{18} + \frac{1}{6} = \frac{1}{3}\,\text{S}$$

Then, using the current divider principle,

$$i_1 = \frac{G_1 i_{\rm s}}{G_{\rm P}}$$

where

$$G_{\rm P} = G_1 + G_{\rm P2} = \frac{1}{3} + \frac{1}{3} = \frac{2}{3}$$

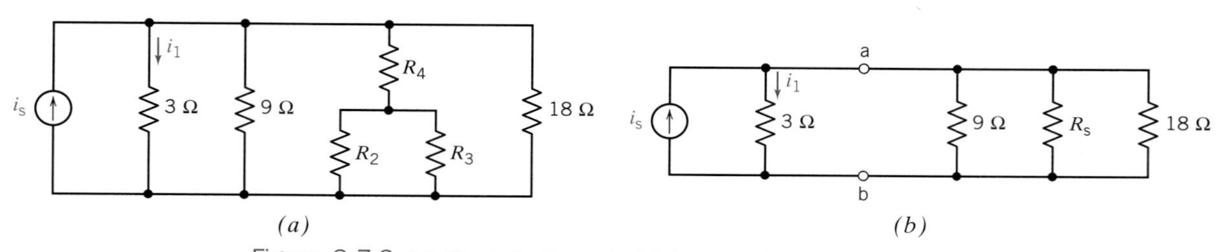

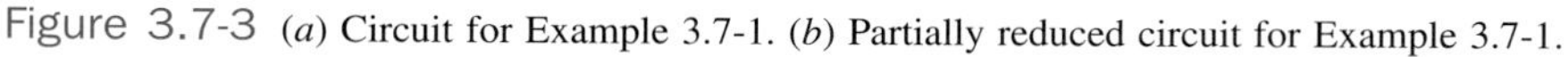

Figure 3.7-3 (*a*) Circuit for Example 3.7-1. (*b*) Partially reduced circuit for Example 3.7-1.

Therefore,

$$i_1 = \frac{1/3}{2/3} i_s = \frac{1}{2} i_s$$

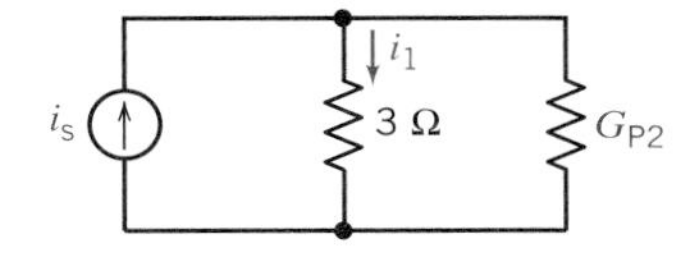

Figure 3.7-4
Equivalent circuit for Figure 3.7-3.

Example 3.7-2

The circuit in Figure 3.7-5*a* contains an ohmmeter. An ohmmeter is an instrument that measures resistance in ohms. The ohmmeter will measure the equivalent resistance of the resistor circuit connected to its terminals. Determine the resistance measured by the ohmmeter in Figure 3.7-5*a*.

Solution

Working from left to right, the 30 Ω resistor is parallel to the 60 Ω resistor. The equivalent resistance is

$$\frac{60 \cdot 30}{60 + 30} = 20\ \Omega$$

In Figure 3.7-5*b* the parallel combination of the 30 Ω and 60 Ω resistors has been replaced with the equivalent 20 Ω resistor. Now the two 20 Ω resistors are in series. The equivalent resistance is

$$20 + 20 = 40\ \Omega$$

In Figure 3.7-5*c* the series combination of the two 20 Ω resistors has been replaced with the equivalent 40 Ω resistor. Now the 40 Ω resistor is parallel to the 10 Ω resistor. The equivalent resistance is

$$\frac{40 \cdot 10}{40 + 10} = 8\ \Omega$$

In Figure 3.7-5*d* the parallel combination of the 40 Ω and 10 Ω resistors has been replaced with the equivalent 8 Ω resistor. Thus, the ohmmeter measures a resistance equal to 8 Ω.

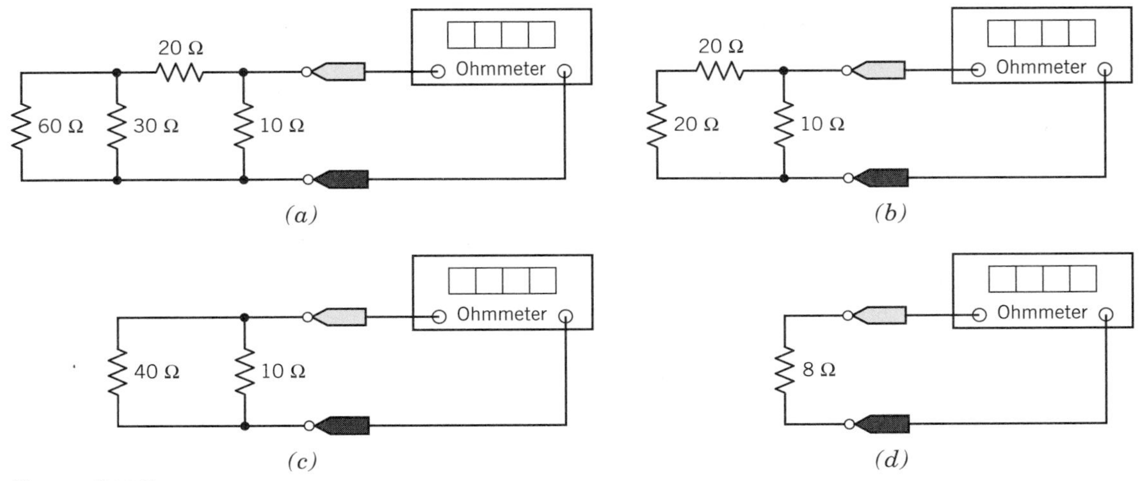

Figure 3.7-5

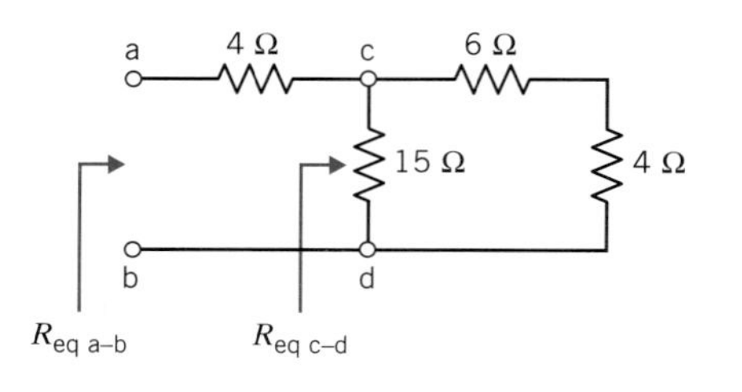

Figure 3.7-6
The equivalent resistance looking into terminals c–d is denoted as $R_{\text{eq c-d}}$.

In general, we may find the equivalent resistance (or conductance) for a portion of a circuit consisting only of resistors and then replace that portion of the circuit with the equivalent resistance. For example, consider the circuit shown in Figure 3.7-6. We use $R_{\text{eq x-y}}$ to denote the equivalent resistance seen looking into terminals x–y. We note that the equivalent resistance to the right of terminals c–d is

$$R_{\text{eq c-d}} = \frac{15(6+4)}{15+(6+4)} = \frac{150}{25} = 6\ \Omega$$

Then the equivalent resistance of the circuit to the right of terminals a–b is

$$R_{\text{eq a-b}} = 4 + R_{\text{eq c-d}} = 4 + 6 = 10\ \Omega$$

Exercise 3.7-1 Determine the resistance measured by the ohmmeter in Figure E 3.7-1.

Answer: $\dfrac{(30+30)\cdot 30}{(30+30)+30} + 30 = 50\ \Omega$

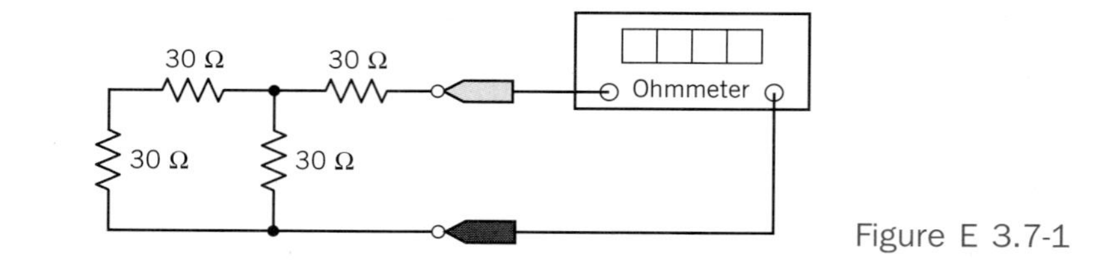

Figure E 3.7-1

Exercise 3.7-2 Determine the resistance measured by the ohmmeter in Figure E 3.7-2.

Answer: $12 + \dfrac{40\cdot 10}{40+10} + 4 = 24\ \Omega$

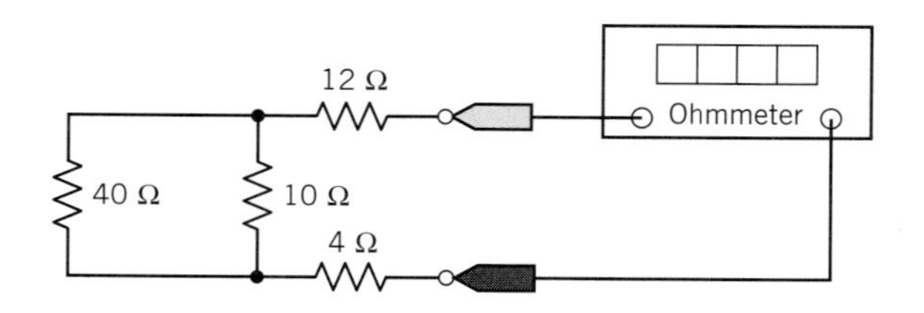

Figure E 3.7-2

Exercise 3.7-3 Determine the resistance measured by the ohmmeter in Figure E 3.7-3.

Answer: $\dfrac{(60 + 60 + 60) \cdot 60}{(60 + 60 + 60) + 60} = 45\ \Omega$

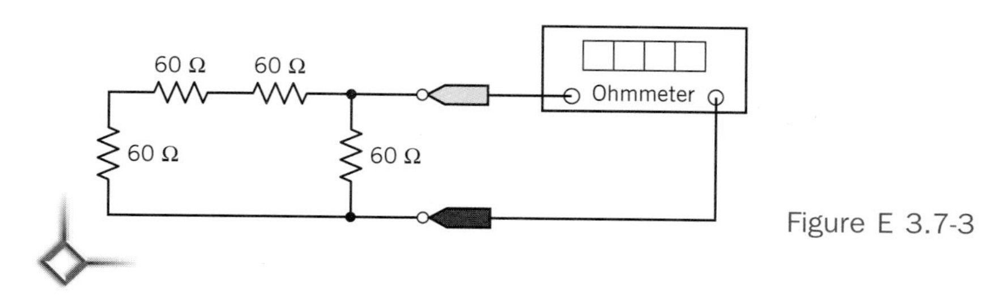

Figure E 3.7-3

3.8 ANALYZING RESISTIVE CIRCUITS USING MATLAB

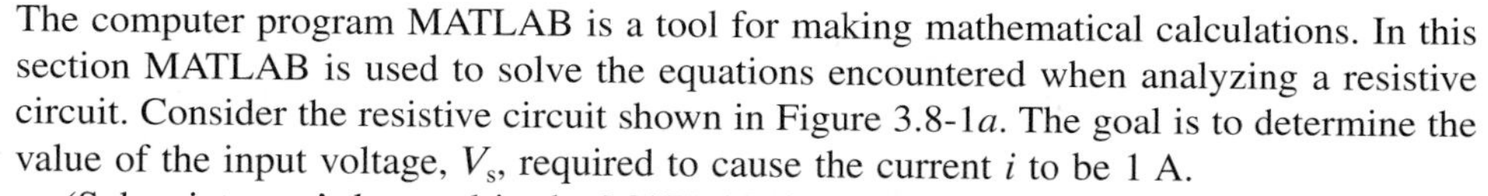

The computer program MATLAB is a tool for making mathematical calculations. In this section MATLAB is used to solve the equations encountered when analyzing a resistive circuit. Consider the resistive circuit shown in Figure 3.8-1*a*. The goal is to determine the value of the input voltage, V_s, required to cause the current i to be 1 A.

(Subscripts can't be used in the MATLAB input file. Thus V_s and R_p in Figure 3.8-1 become Vs and Rp in the MATLAB input file. We have been using lowercase letters to represent element voltages and currents, but in MATLAB examples we will use capital letters to represent currents and voltages to improve the readability of the MATLAB input file. For example, the input and output voltage are denoted as V_s and V_o rather than v_s and v_o.)

Resistors R_1, R_2, and R_3 are connected in series and can be replaced by an equivalent resistor, R_s, given by

$$R_s = R_1 + R_2 + R_3 \tag{3.8-1}$$

Resistors R_4, R_5, and R_6 are connected in parallel and can be replaced by an equivalent resistor, R_P, given by

$$R_P = \frac{1}{\dfrac{1}{R_4} + \dfrac{1}{R_5} + \dfrac{1}{R_6}} \tag{3.8-2}$$

Figure 3.8-1*b* shows the circuits after R_1, R_2, and R_3 are replaced by R_s and R_4, R_5, and R_6 are replaced by R_P. Applying voltage division to the circuit in Figure 3.8-1*b* gives

$$V_o = \frac{R_P}{R_s + R_P} V_s \tag{3.8-3}$$

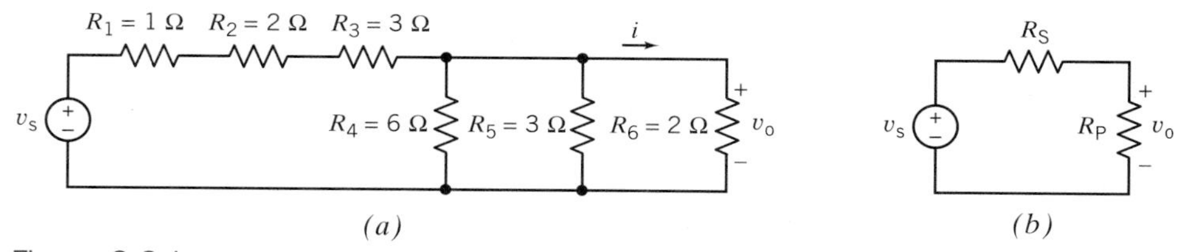

Figure 3.8-1 (*a*) A resistive circuit and (*b*) an equivalent circuit.

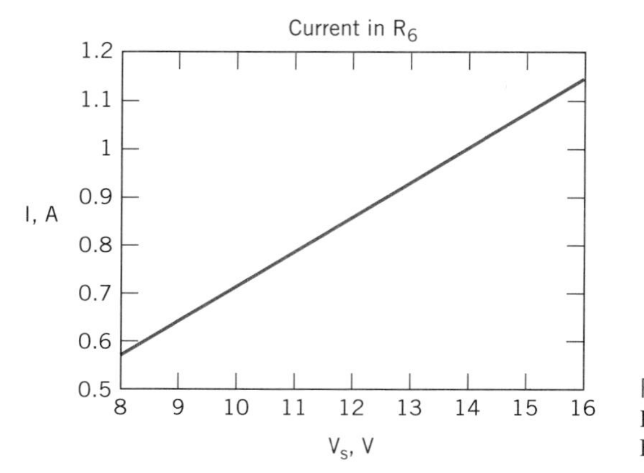

Figure 3.8-2 Plot of I versus V_s for the circuit shown in Figure 3.8-1.

where V_o is the voltage across R_P in Figure 3.8-1*b* and is also the voltage across the parallel resistors in Figure 3.8-1*a*. Ohm's law indicates that the current in R_6 is given by

$$I = \frac{V_o}{R_6} \tag{3.8-4}$$

Figure 3.8-2 shows a plot of the output current I versus the input voltage V_s. This plot shows that I will be 1 A when $V_s = 14$ V. Figure 3.8-3 shows the MATLAB input file that was used to obtain Figure 3.8-2. The MATLAB program first causes V_s to vary over a range of voltages. Next, MATLAB calculates the value of I corresponding to each value of V_s using Eqs. 3.8-1 through 3.8-4. Finally, MATLAB plots the current I versus the voltage V_s.

```
% Analyzing Resistive Circuits Using MATLAB - ch3ex.m
%------------------------------------------------------------
% Vary the input voltage from 8 to 16 volts in 0.1 volt steps.
%------------------------------------------------------------
Vs = 8:0.1:16;
%------------------------------------------------------------
%             Enter values of the resistances.
%------------------------------------------------------------
R1 = 1; R2 = 2; R3 = 3;      % series resistors, ohms
R4 = 6; R5 = 3; R6 = 2;      % parallel resistors, ohms
%------------------------------------------------------------
%    Find the current, I, corresponding to each value of Vs.
%------------------------------------------------------------
Rs = R1 + R2 + R3;                        % Equation 3.8-1
Rp = 1 / (1/R4 +1/R5 +1/R6);              % Equation 3.8-2
for k = 1:length(Vs)
    Vo(k) = Vs(k) * Rp / (Rp + Rs);       % Equation 3.8-3
    I(k)  = Vo(k) / R6;                   % Equation 3.8-4
end
%------------------------------------------------------------
%                      Plot I versus Vs
%------------------------------------------------------------
plot(Vs, I)
grid
xlabel('Vs, V'), ylabel('I,A')
title('Current in R6')
```

Figure 3.8-3 MATLAB input file used to obtain the plot of I versus V_s shown in Figure 3.8-2.

3.9 VERIFICATION EXAMPLE

The circuit shown in Figure 3.9-1*a* was analyzed by writing and solving a set of simultaneous equations:

$$12 = v_2 + 4i_3, \quad i_4 = \frac{v_2}{5} + i_3$$

$$v_5 = 4i_3, \quad \frac{v_5}{2} = i_4 + 5i_4$$

A computer and the program Mathcad (Mathcad User's Guide, 1991) was used to solve the equations as shown in Figure 3.9-1*b*. It was determined that

$$v_2 = -60\,\text{V}, \quad i_3 = 18\,\text{A}, \quad i_4 = 6\,\text{A}, \quad v_5 = 72\,\text{V}$$

Are these currents and voltages correct?

The current i_2 can be calculated from v_2, i_3, i_4, and v_5 in a couple of different ways. First, Ohm's law gives

$$i_2 = \frac{v_2}{5} = \frac{-60}{5} = -12\,\text{A}$$

Next, applying KCL at node b gives

$$i_2 = i_3 + i_4 = 18 + 6 = 24\,\text{A}$$

Clearly, i_2 cannot be both -12 and 24 A, so the values calculated for v_2, i_3, i_4, and v_5 cannot be correct. Checking the equations used to calculate v_2, i_3, i_4, and v_5, we find a sign error in the KCL equation corresponding to node b. This equation should be

$$i_4 = \frac{v_2}{5} - i_3$$

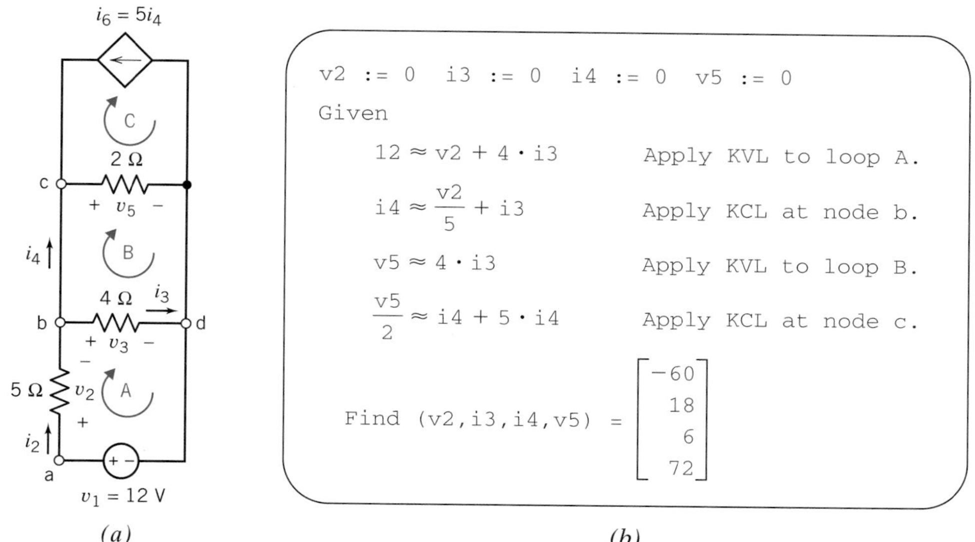

(*a*) (*b*)

Figure 3.9-1 (*a*) An example circuit and (*b*) computer analysis using Mathcad.

After making this correction, v_2, i_3, i_4, and v_5 are calculated to be

$$v_2 = 7.5\,\text{V}, \quad i_3 = 1.125\,\text{A}, \quad i_4 = 0.375\,\text{A}, \quad v_5 = 4.5\,\text{V}$$

Now

$$i_2 = \frac{v_2}{5} = \frac{7.5}{5} = 1.5$$

and

$$i_2 = i_3 + i_4 = 1.125 + 0.375 = 1.5$$

This checks as we expected.

As an additional check, consider v_3. First, Ohm's law gives

$$v_3 = 4\,i_3 = 4(1.125) = 4.5$$

Next, applying KVL to the loop consisting of the voltage source and the 4 Ω and 5 Ω resistors gives

$$v_3 = 12 - v_2 = 12 - 7.5 = 4.5$$

Finally, applying KVL to the loop consisting of the 2 Ω and 4 Ω resistors gives

$$v_3 = v_5 = 4.5$$

The results of these calculations agree with each other, indicating that

$$v_2 = 7.5\,\text{V}, \quad i_3 = 1.125\,\text{A}, \quad i_4 = 0.375\,\text{A}, \quad v_5 = 4.5\,\text{V}$$

are the correct values.

3.10 Design Challenge Solution

ADJUSTABLE VOLTAGE SOURCE

A circuit is required to provide an adjustable voltage. The specifications for this circuit are:

1. It should be possible to adjust the voltage to any value between −5 V and +5 V. It should not be possible to accidentally obtain a voltage outside this range.
2. The load current will be negligible.
3. The circuit should use as little power as possible.

The available components are:

1. Potentiometers: resistance values of 10 kΩ, 20 kΩ, and 50 kΩ are in stock.
2. A large assortment of standard 2 percent resistors having values between 10 Ω and 1 MΩ (see Appendix E).
3. Two power supplies (voltage sources): one 12-V supply and one −12-V supply, both are rated at 100 mA (maximum).

Describe the Situation and the Assumptions

Figure 3.10-1 shows the situation. The voltage v is the adjustable voltage. The circuit that uses the output of the circuit being designed is frequently called the "load." In this case, the load current is negligible, so $i = 0$.

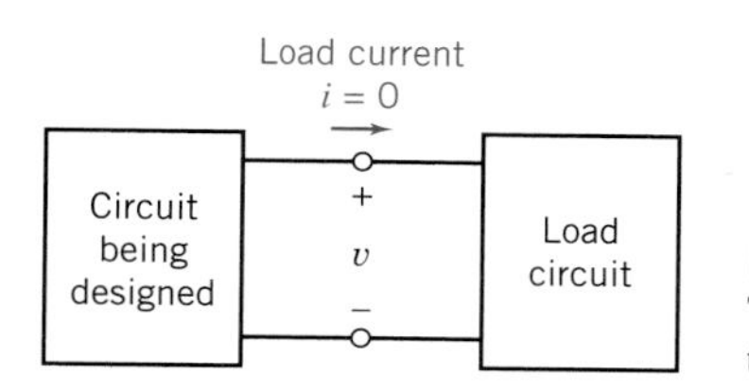

Figure 3.10-1
The circuit being designed provides an adjustable voltage, v, to the load circuit.

State the Goal

A circuit providing the adjustable voltage

$$-5\text{ V} \leq v \leq +5\text{ V}$$

must be designed using the available components.

Generate a Plan

Make the following observations.

1. The adjustability of a potentiometer can be used to obtain an adjustable voltage v.
2. Both power supplies must be used so that the adjustable voltage can have both positive and negative values.
3. The terminals of the potentiometer cannot be connected directly to the power supplies because the voltage v is not allowed to be as large as 12 V or -12 V.

These observations suggest the circuit shown in Figure 3.10-2*a*. The circuit in Figure 3.10-2*b* is obtained by using the simplest model for each component in Figure 3.10-2*a*.

To complete the design, values need to be specified for R_1, R_2, and R_p. Then, several results need to be checked and adjustments made, if necessary.

1. Can the voltage v be adjusted to any value in the range -5 V to $+5$ V?
2. Are the voltage source currents less than 100 mA? This condition must be satisfied if the power supplies are to be modeled as ideal voltage sources.
3. Is it possible to reduce the power absorbed by R_1, R_2, and R_p?

Act on the Plan

It seems likely the R_1 and R_2 will have the same value, so let $R_1 = R_2 = R$. Then it is convenient to redraw Figure 3.10-2*b* as shown in Figure 3.10-3.

Applying KVL to the outside loop yields

$$-12 + Ri_a + a\,R_P\,i_a + (1 - a)\,R_P\,i_a + Ri_a - 12 = 0$$

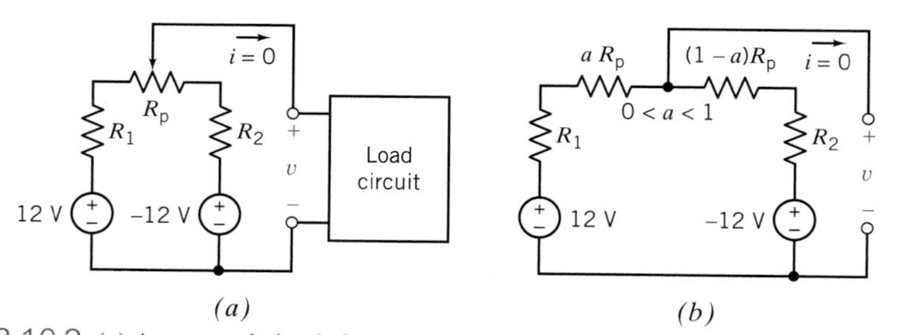

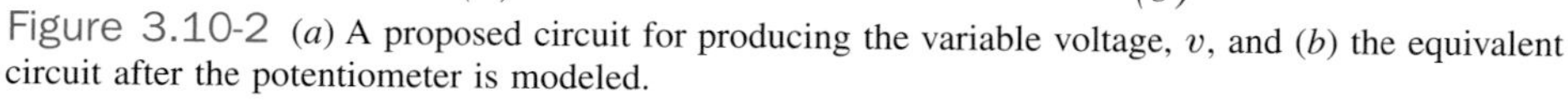
Figure 3.10-2 (*a*) A proposed circuit for producing the variable voltage, v, and (*b*) the equivalent circuit after the potentiometer is modeled.

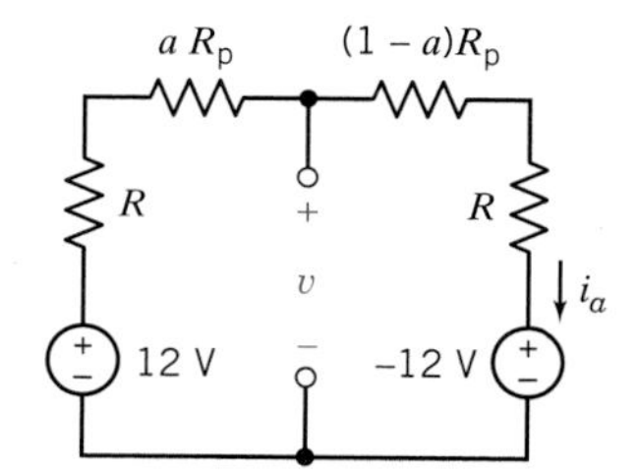

Figure 3.10-3 The circuit after setting $R_1 = R_2 = R$.

so

$$i_a = \frac{24}{2R + R_P}$$

Next applying KVL to the left loop gives

$$v = 12 - (R + a\,R_P)i_a$$

Substituting for i_a gives

$$v = 12 - \frac{24(R + a\,R_P)}{2R + R_P}$$

When a = 0, v must be 5 V, so

$$5 = 12 - \frac{24R}{2R + R_P}$$

Solving for R gives

$$R = 0.7R_P$$

Suppose the potentiometer resistance is selected to be $R_P = 20\,\mathrm{k}\Omega$, the middle of the three available values. Then

$$R = 14\,\mathrm{k}\Omega$$

Verify the Proposed Solution

As a check, notice that when $a = 1$

$$v = 12 - \left(\frac{14\text{ k} + 20\text{ k}}{28\text{ k} + 20\text{ k}}\right)24 = -5$$

as required. The specification that

$$-5\text{ V} \leq v \leq 5\text{ V}$$

has been satisfied. The power absorbed by the three resistances is

$$p = i_a^2(2R + R_P) = \frac{24^2}{2R + R_P}$$

so

$$p = 12\,\mathrm{mW}$$

Notice that this power can be reduced by choosing R_P to be as large as possible, $50\,\mathrm{k}\Omega$ in this case. Changing R_P to $50\,\mathrm{k}\Omega$ requires a new value of R:

$$R = 0.7 \times R_p = 35\,\mathrm{k}\Omega$$

Since

$$-5\text{ V} = 12 - \left(\frac{35\text{ k} + 50\text{ k}}{70\text{ k} + 50\text{ k}}\right)24 \leq v \leq 12 - \left(\frac{35\text{ k}}{70\text{ k} + 50\text{ k}}\right)24 = 5\text{ V}$$

the specification that

$$-5\text{ V} \leq v \leq 5\text{ V}$$

has been satisfied. The power absorbed by the three resistances is now

$$p = \frac{24^2}{50\text{ k} + 70\text{ k}} = 5\,\mathrm{mW}$$

Finally, the power supply current is

$$i_a = \frac{24}{50\text{ k} + 70\text{ k}} = 0.2\text{ mA}$$

which is well below that 100 mA that the voltage sources are able to supply. The design is complete.

3.11 DESIGN EXAMPLE—VOLTAGE DIVIDER

PROBLEM

A voltage divider is connected to a source and a voltmeter as shown in Figure 3.11-1. Ideally, $R_s = 0$ and $R_m = \infty$. However, for one practical circuit, $R_s \leq 125\,\Omega$ and $R_m \geq 10\,\text{k}\Omega$. Select R_1 and R_2 to minimize the error introduced by R_s and R_m when it is desired that $v/v_s = 0.75$ (Svoboda 1992).

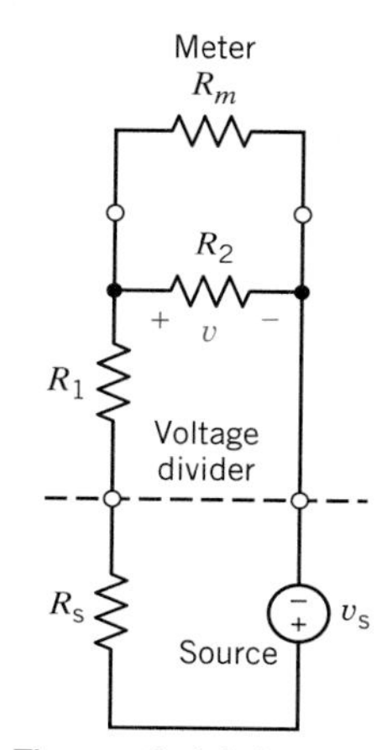

Figure 3.11-1 A voltage divider with a practical source and a meter.

Describe the Situation and the Assumptions

1. A voltage divider is used with a practical source resistance R_s.
2. The resistor R_2 is "loaded" by the meter resistance R_m.
3. The desired ratio is $v/v_s = 0.75$.

State the Goal

Determine R_1 and R_2 to minimize the difference between the values of the voltage v for the ideal and practical cases.

Generate a Plan

1. Determine the voltage v for the ideal case.
2. Determine the voltage v' for the practical case.
3. Define a measure of the error.
4. Minimize the error and then determine R_2 and R_1.

Act on the Plan

The ideal voltage divider is obtained when $R_s = 0$ and $R_m = \infty$. Then, the divider ratio is

$$v = \frac{R_2}{R_1 + R_2} v_s = a\, v_s \tag{3.11-1}$$

where $a = R_2/(R_1 + R_2)$.

For the practical case, the output voltage v' is

$$v' = \frac{R_p}{R_s + R_1 + R_p} v_s$$

Substitute for R_p and $R_1 = R_2(1 - a)/a$ to obtain

$$v' = \frac{a\, R_2 R_m}{(a\, R_s + R_2)(R_m + R_2) - a\, R_2^2} v_s \tag{3.11-2}$$

where $R_p = R_2R_m/(R_2 + R_m)$ and a is as defined in Eq. 3.11-1. We will define the error as

$$e = \frac{v - v'}{v} = \frac{av_s - v'}{av_s} \tag{3.11-3}$$

Substituting Eq. 3.11-2 into Eq. 3.11-3, we have

$$e = 1 - \frac{R_2R_m}{(a\,R_s + R_2)(R_m + R_2) - a\,R_2^2} \tag{3.11-4}$$

The objective is to minimize the error e by selecting R_2. We can use calculus and set $de/dR_2 = 0$ to find the best value of R_2. Alternatively, we can use a spreadsheet computer program and determine the best value of R_2. If we use calculus, we find that we require

$$R_2 = (3\,R_mR_s)^{1/2}$$

when $a = 0.75$. Then, taking the worst case values, $R_m = 10\,\text{k}\Omega$ and $R_s = 125\,\Omega$, we require that

$$R_2 = 1936.5\ \Omega$$

Also, we have from Eq. 3.11-1

$$R_2 = a\,(R_1 + R_2)$$

or

$$R_1 = \frac{R_2(1 - a)}{a} = \frac{1936.5\,(0.25)}{0.75} = 645.5\ \Omega$$

Verify the Proposed Solution

Evaluating Eq. 3.11-4, we find that the minimum error is 9.6 percent when $R_1 = 645.5\ \Omega$ and $R_2 = 1936.5\ \Omega$. The error increases slowly if R_2 is either increased or decreased, verifying that 9.6% is the minimum error.

3.12 SUMMARY

- Gustav Robert Kirchhoff formulated the laws that enable us to study a circuit. Kirchhoff's current law (KCL) states that the algebraic sum of the currents entering a node is zero. Kirchhoff's voltage law (KVL) states that the algebraic sum of the voltage around a closed path (loop) is zero.
- Two special circuits of interest are the series circuit and the parallel circuit. Table 3.12-1 summarizes the results from this chapter regarding series and parallel elements.

⇒ The first row of the table shows series resistors connected to a circuit. These series resistors can be replaced by an equivalent resistor. Doing so does not disturb the circuit connected to the series resistors. The element voltages and currents in this circuit don't change. In particular, the terminal voltage and current (labeled v and i, respectively, in Table 3.12-1) do not change when the series resistors are replaced by the equivalent resistor. Indeed, this is what is meant by "equivalent."

⇒ Replacing parallel resistors by an equivalent resistor does not change any voltage or current in the circuit connected to the parallel resistors.

⇒ Replacing series voltage sources by an equivalent voltage source does not disturb the circuit connected to the series voltage sources.

⇒ Replacing parallel current sources by an equivalent current source does not disturb the circuit connected to the parallel current sources.

- A circuit consisting of series resistors is sometimes called a voltage divider because the voltage across the series resistors divides between the individual resistors. Similarly, a circuit consisting of parallel resistors is called a current divider because the current through the parallel resistors divides between the individual resistors. Table 3.12-1 provides the equations describing voltage division for series resistors and current division for parallel resistors.

Table 3.12-1 **Equivalent Circuits for Series and Parallel Elements**

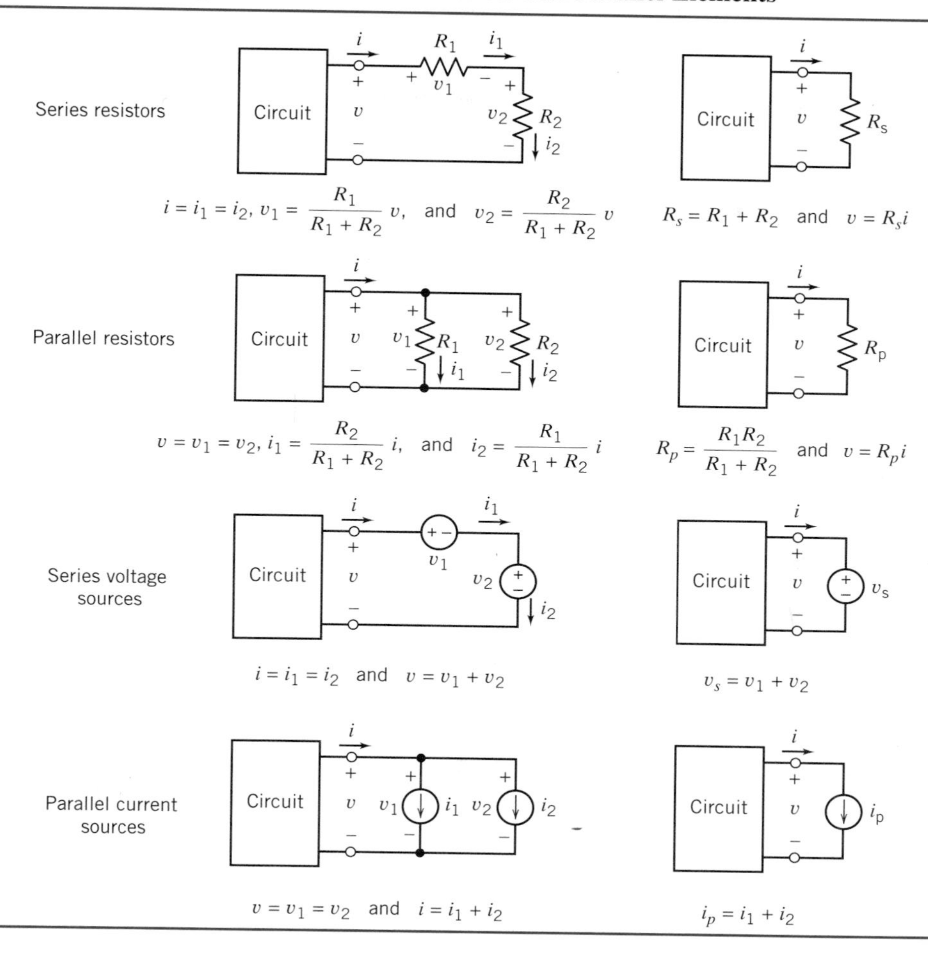

PROBLEMS

Section 3-3 Kirchhoff's Laws

P 3.3-1 Determine the values of v_1, v_2, and i_3, in the circuit shown in Figure P 3.3-1.

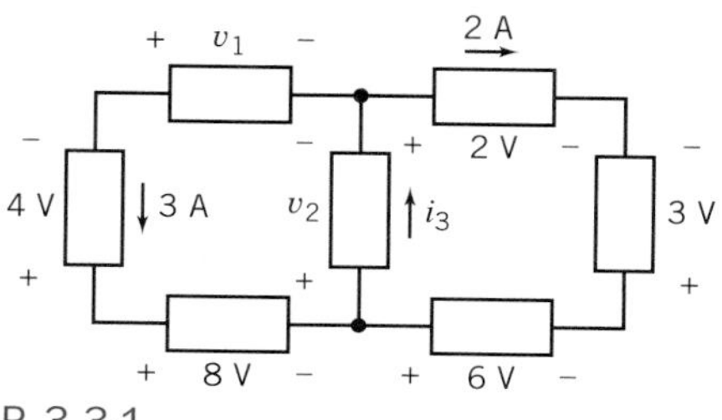

Figure P 3.3-1

P 3.3-2 Determine the values of v_1, i_2, i_3, and i_4 in the circuit shown in Figure P 3.3-2.

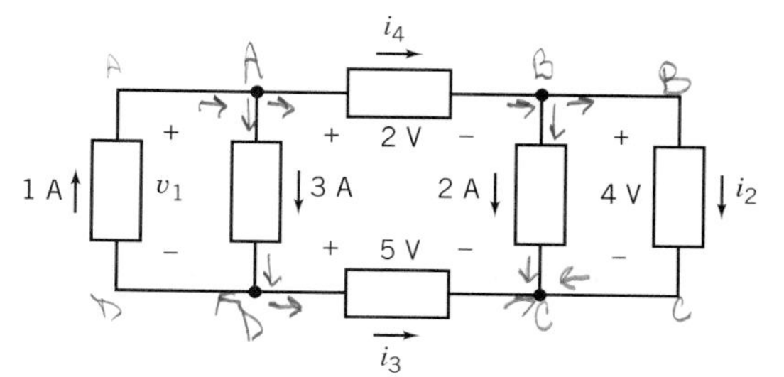

Figure P 3.3-2

P 3.3-3 Consider the circuit shown in Figure P 3.3-3.

(a) Suppose that $R_1 = 6\ \Omega$ and $R_2 = 3\ \Omega$. Find the current i and the voltage v.

(b) Suppose, instead, that $i = 1.5$ A and $v = 2$ V. Determine the resistances R_1 and R_2.

(c) Suppose, instead, that the voltage source supplies 24 W of power and that the current source supplies 9 W of power. Determine the current i, the voltage v, and the resistances R_1 and R_2.

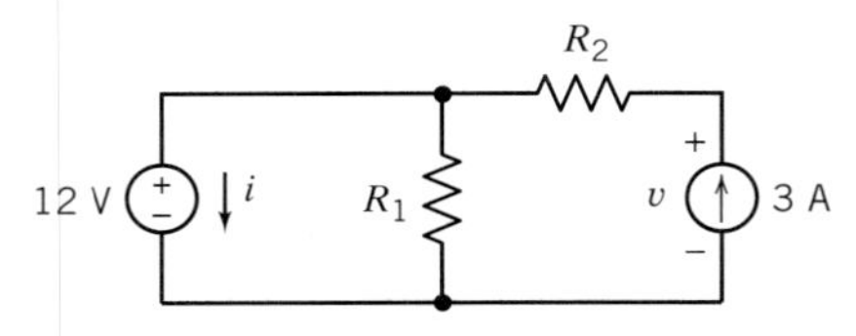

Figure P 3.3-3

P 3.3-4 Determine the power absorbed by each of the resistors in the circuit shown in Figure P 3.3-4.
Answer: The 4-Ω resistor absorbs 100 W, the 6-Ω resistor absorbs 24 W, and the 8-Ω resistor absorbs 72 W.

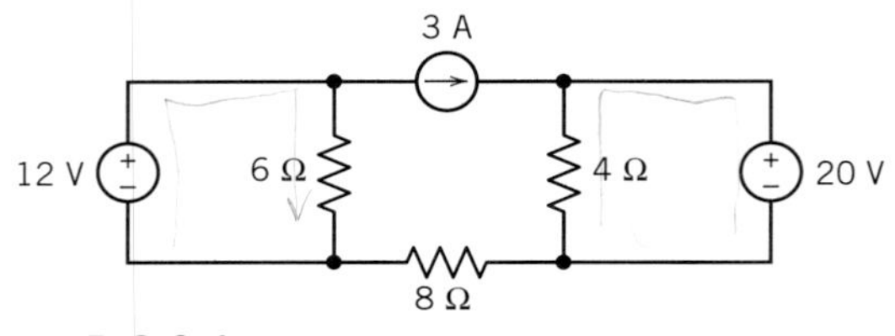

Figure P 3.3-4

P 3.3-5 Determine the power absorbed by each of the resistors in the circuit shown in Figure P 3.3-5.
Answer: The 4-Ω resistor absorbs 16 W, the 6-Ω resistor absorbs 24 W, and the 8-Ω resistor absorbs 8 W.

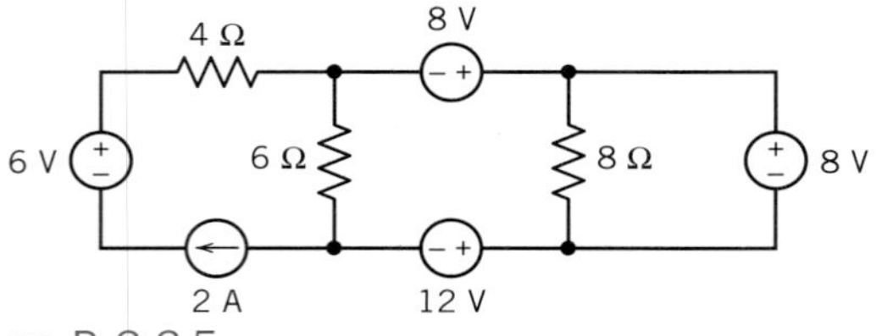

Figure P 3.3-5

P 3.3-6 Determine the power supplied by each current source in the circuit of Figure P 3.3-6.
Answer: The 2 mA current source supplies 6 mW and the 1 mA current source supplies –7 mW.

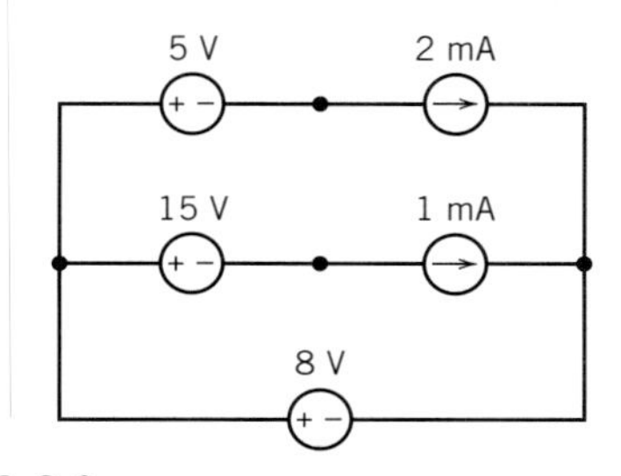

Figure P 3.3-6

P 3.3-7 Determine the power supplied by each voltage source in the circuit of Figure P 3.3-7.
Answer: The 2 V voltage source supplies 2 mW and the 3 V voltage source supplies –6 mW.

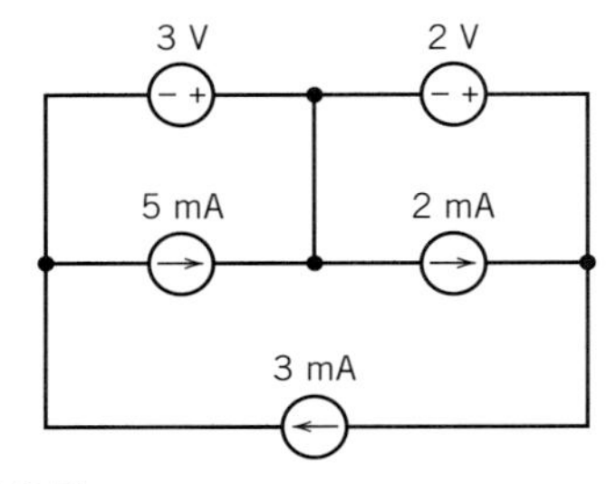

Figure P 3.3-7

P 3.3-8 Consider the circuit shown in Figure P 3.3-8. Some element voltages and element currents are given. What is the power delivered to element a? What is the power delivered to element b?

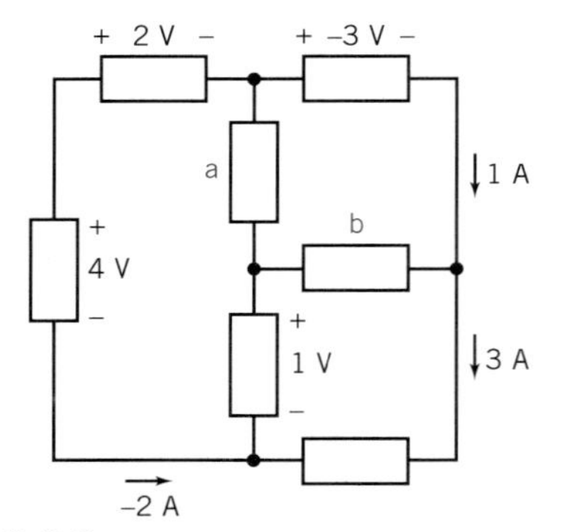

Figure P 3.3-8 Circuit with branch currents in amperes and voltages in volts.

P 3.3-9 For the circuit of Figure P 3.3-9
a. Find the power delivered by each source.
b. Find the power delivered to each resistor.
c. Is energy conserved?

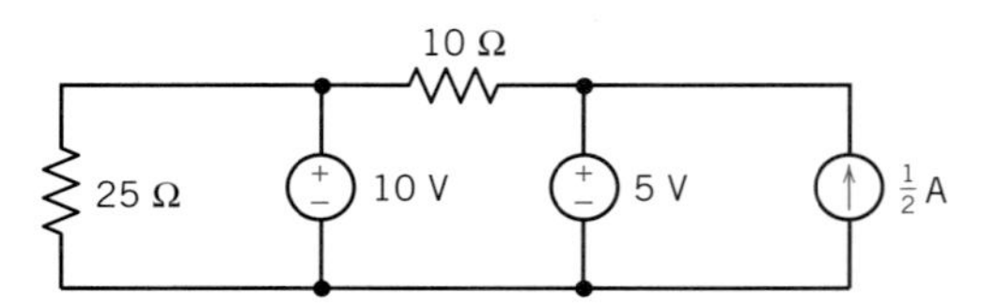

Figure P 3.3-9

P 3.3-10 Determine the currents i_a and i_b for the circuit of Figure P 3.3-10.
Hint: Use KCL to find the current in the 3Ω and the 4Ω resistors. Apply KCL at the center node and KVL to the lower, left-hand mesh to get two equations in i_a and i_b.
Answer: i_a = 1.67 mA and i_b = 3.33 mA.

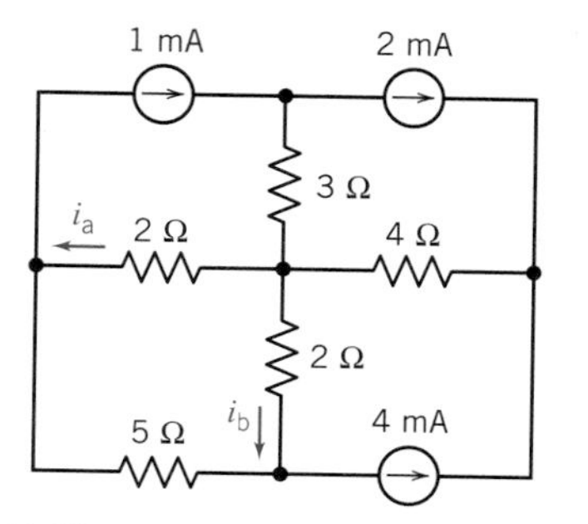

Figure P 3.3-10

Section 3.4 A Single-Loop Circuit—The Voltage Divider

P 3.4-1 Use voltage division to determine the voltages v_1, v_2, v_3, and v_4, in the circuit shown in Figure P 3.4-1.

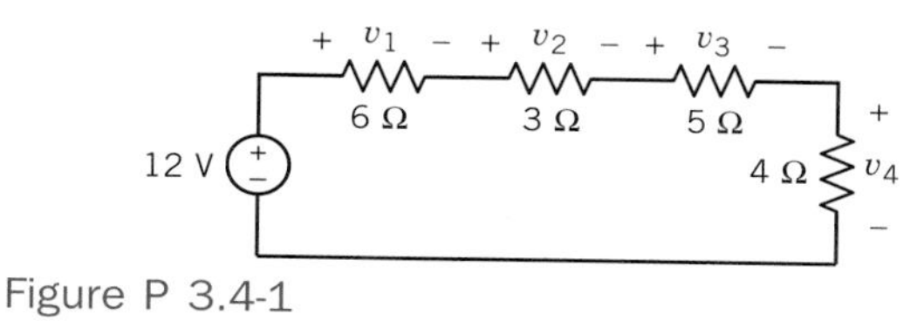

Figure P 3.4-1

P 3.4-2 Consider the circuits shown in Figure P 3.4-2.

(a) Determine the value of the resistance R in Figure P 3.4-2b that makes the circuit in Figure P 3.4-2b equivalent to the circuit in Figure P 3.4-2a.

(b) Determine the current i in Figure P 3.4-2b. Because the circuits are equivalent, the current i in Figure P 3.4-2a is equal to the current i in Figure P 3.4-2b.

(c) Determine the power supplied by the voltage source.

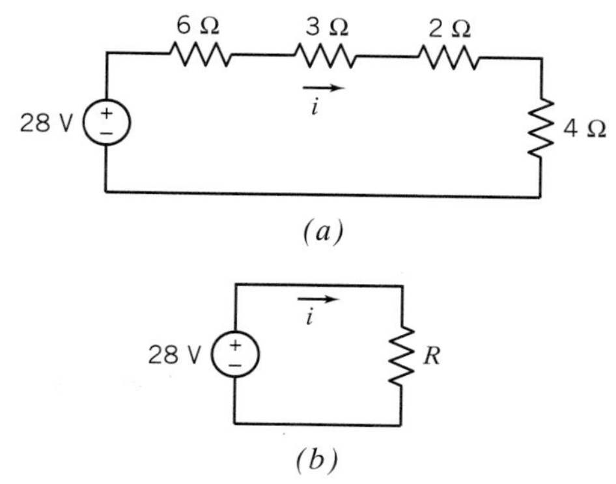

Figure P 3.4-2

P 3.4-3 The ideal voltmeter in the circuit shown in Figure P 3.4-3 measures the voltage v.

(a) Suppose $R_2 = 100\ \Omega$. Determine the value of R_1.

(b) Suppose, instead, $R_1 = 100\ \Omega$. Determine the value of R_2.

(c) Suppose, instead, that the voltage source supplies 1.2 W of power. Determine the values of both R_1 and R_2.

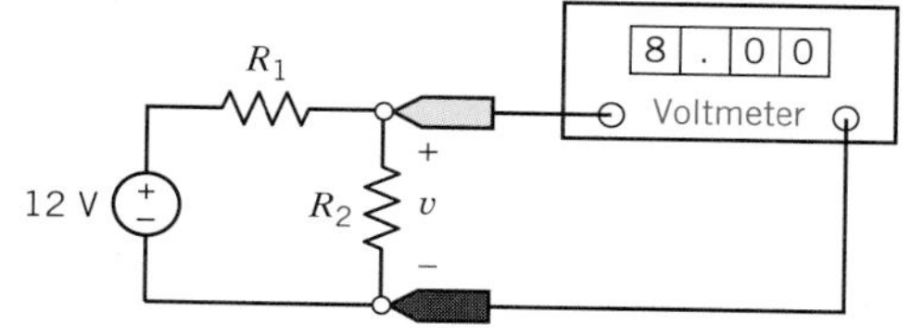

Figure P 3.4-3

P 3.4-4 Determine the voltage v in the circuit shown in Figure P 3.4-4.

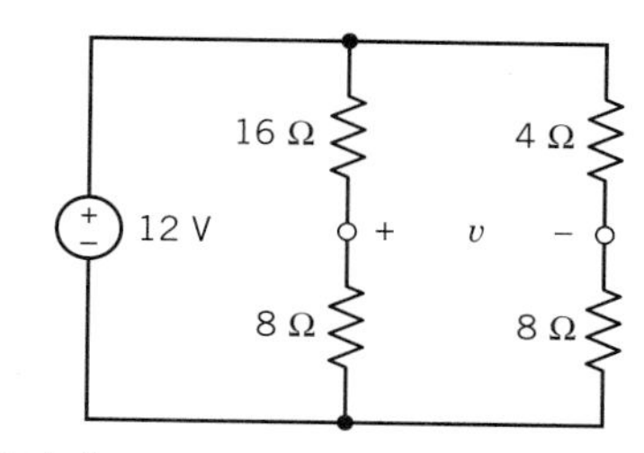

Figure P 3.4-4

P 3.4-5 The model of a cable and load resistor connected to a source is shown in Figure P 3.4-5. Determine the appropriate cable resistance, R, so that the output voltage, v_o, remains between 9 V and 13 V when the source voltage, v_s, varies between 20 V and 28 V. The cable resistance can only assume integer values in the range $20 < R < 100\ \Omega$.

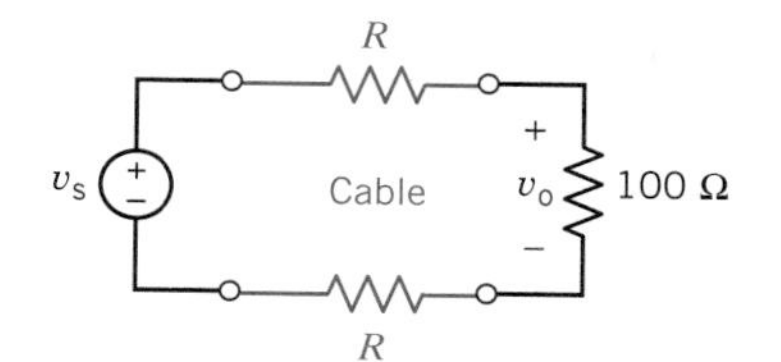

Figure P 3.4-5 Circuit with a cable.

P 3.4-6 The portable lighting equipment for a mine is located 100 meters from its dc supply source. The mine lights use a total of 5 kW and operate at 120 V dc. Determine the required cross-sectional area of the copper wires used to connect the source to the mine lights if we require that the power lost in the copper wires be less than or equal to 5 percent of the power required by the mine lights.

Hint: Model both the lighting equipment and the wire as resistors.

Section 3.5 Parallel Resistors and Current Division

P 3.5-1 Use current division to determine the currents i_1, i_2, i_3, and i_4 in the circuit shown in Figure P 3.5-1.

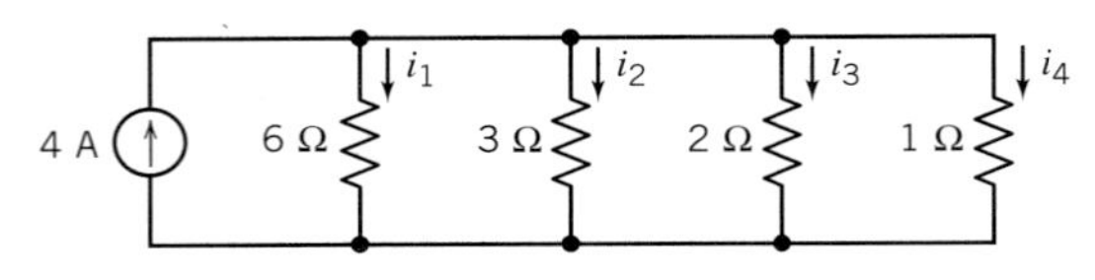

Figure P 3.5-1

P 3.5-2 Consider the circuits shown in Figure P 3.5-2.

(a) Determine the value of the resistance R in Figure P 3.5-2b that makes the circuit in Figure P 3.4-2b equivalent to the circuit in Figure P 3.5-2a.

(b) Determine the voltage v in Figure P 3.5-2b. Because the circuits are equivalent, the voltage v in Figure P 3.5-2a is equal to the voltage v in Figure P 3.5-2b.

(c) Determine the power supplied by the current source.

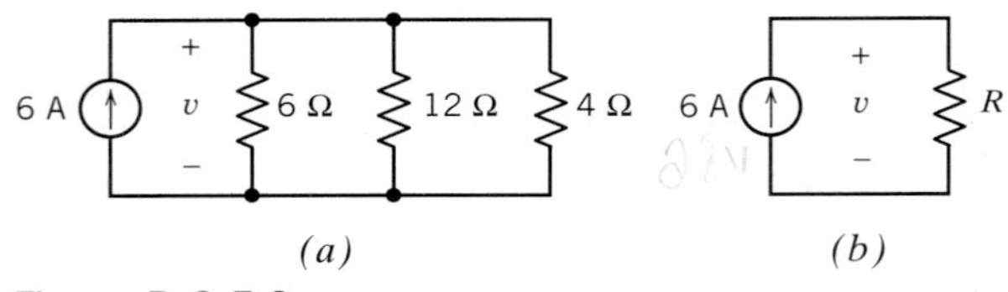

Figure P 3.5-2

P 3.5-3 The ideal voltmeter in the circuit shown in Figure P 3.5-3 measures the voltage v.

(a) Suppose $R_2 = 12\ \Omega$. Determine the value of R_1 and of the current i.

(b) Suppose, instead, $R_1 = 12\ \Omega$. Determine the value of R_2 and of the current i.

(c) Instead, choose R_1 and R_2 to minimize the power absorbed by any one resistor.

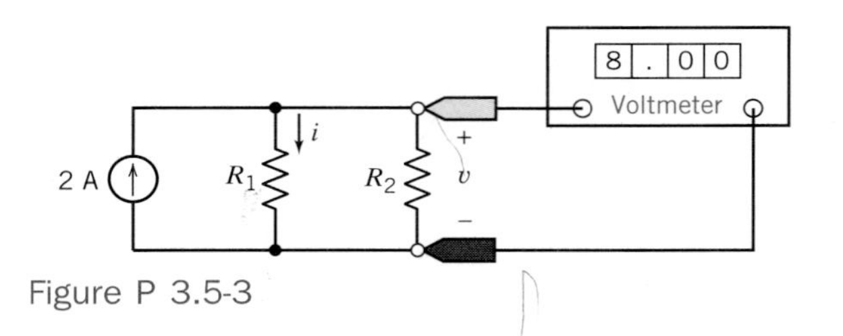

Figure P 3.5-3

P 3.5-4 Determine the current i in the circuit shown in Figure P 3.5-4.

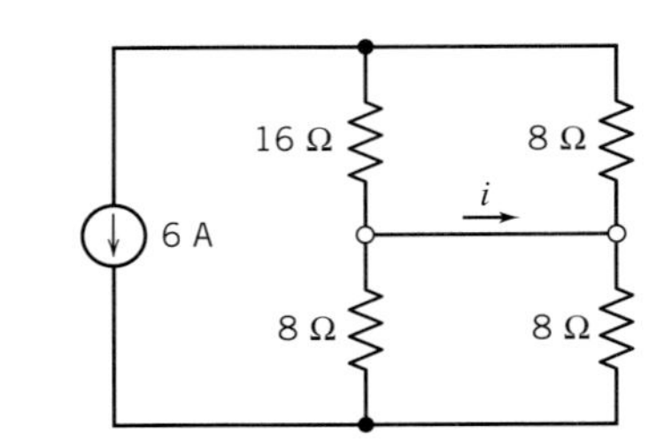

Figure P 3.5-4

P 3.5-5 Consider the circuit shown in Figure P 3.5-5 when $4\ \Omega \le R_1 \le 6\ \Omega$ and $R_2 = 10\ \Omega$. Select the source i_s so that v_o remains between 9 V and 13 V.

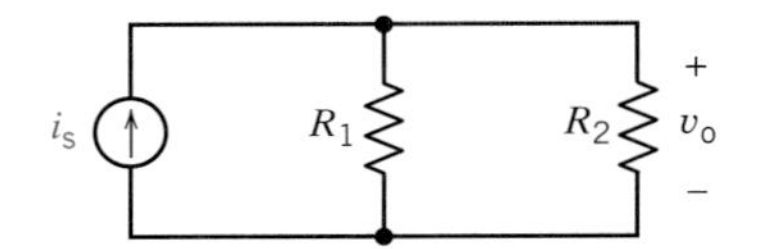

Figure P 3.5-5

P 3.5-6 A solar photovoltaic panel may be represented by the model shown in Figure P 3.5-6, where R_L is the load resistor. The source current is 2 A and $v_{ab} = 40\,\text{V}$. Find R_1 when $R_2 = 10\ \Omega$ and $R_L = 30\ \Omega$.

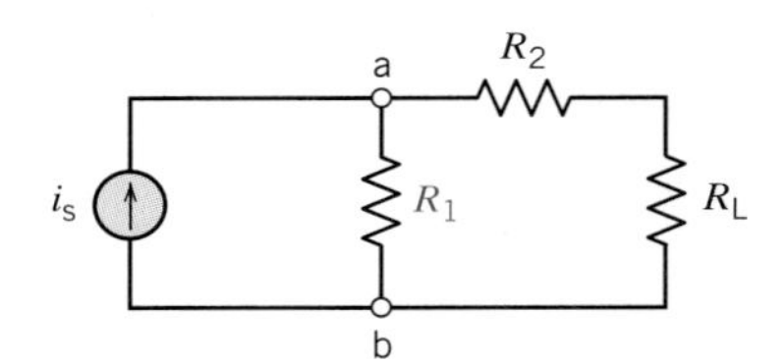

Figure P 3.5-6 Circuit model for solar photovoltaic panel.

Section 3.7 Circuit Analysis

P 3.7-1 The circuit shown in Figure P 3.7-1a has been divided into two parts. In Figure P 3.7-1b, the right-hand part has been replaced with an equivalent circuit. The left-hand part of the circuit has not been changed.

(a) Determine the value of the resistance R in Figure P 3.7-1b that makes the circuit in Figure P 3.7-1b equivalent to the circuit in Figure P 3.7-1a.

(b) Find the current i and the voltage v shown in Figure P 3.7-1b. Because of the equivalence, the current i and the voltage v shown in Figure P 3.7-1a are equal to the current i and the voltage v shown in Figure P 3.7-1b.

(c) Find the current i_2 shown in Figure P 3.7-1a using current division.

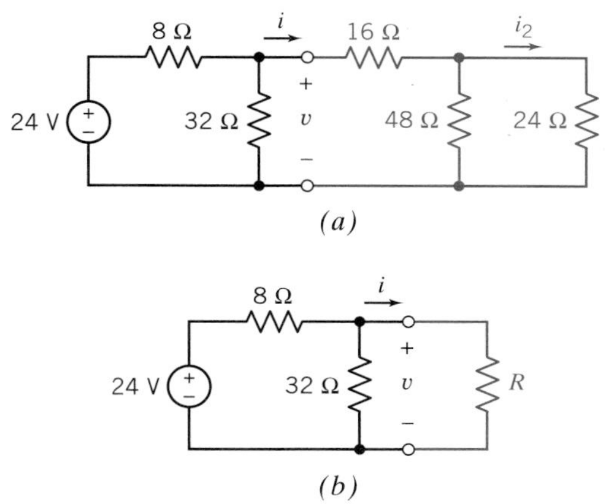

Figure P 3.7-1

P 3.7-2 The circuit shown in Figure P 3.7-2*a* has been divided into three parts. In Figure P 3.7-2*b*, the rightmost part has been replaced with an equivalent circuit. The rest of the circuit has not been changed. The circuit is simplified further in Figure 3.7-2*c*. Now the middle and rightmost parts have been replaced by a single equivalent resistance. The leftmost part of the circuit is still unchanged.

(a) Determine the value of the resistance R_1 in Figure P 3.7-2*b* that makes the circuit in Figure P 3.7-2*b* equivalent to the circuit in Figure P 3.7-2*a*.

(b) Determine the value of the resistance R_2 in Figure P 3.7-2*c* that makes the circuit in Figure P 3.7-2*c* equivalent to the circuit in Figure P 3.7-2*b*.

(c) Find the current i_1 and the voltage v_1 shown in Figure P 3.7-2*c*. Because of the equivalence, the current i_1 and the voltage v_1 shown in Figure P 3.7-2*b* are equal to the current i_1 and the voltage v_1 shown in Figure P 3.7-2*c*.
Hint: $24 = 6(i_1 - 2) + i_1 R_2$

(d) Find the current i_2 and the voltage v_2 shown in Figure P 3.7-2*b*. Because of the equivalence, the current i_2 and the voltage v_2 shown in Figure P 3.7-2*a* are equal to the current i_2 and the voltage v_2 shown in Figure P 3.7-2*b*.
Hint: Use current division to calculate i_2 from i_1.

(e) Determine the power absorbed by the 3-Ω resistance shown at the right of Figure P 3.7-2*a*.

P 3.7-3 Find i using appropriate circuit reductions and the current divider principle for the circuit of Figure P 3.7-3.

Figure P 3.7-3

P 3.7-4 (a) Determine values of R_1 and R_2 in Figure P 3.7-4*b* that make the circuit in Figure P 3.7-4*b* equivalent to the circuit in Figure P 3.7-4*a*.

(b) Analyze the circuit in Figure P 3.7-4*b* to determine the values of the currents i_a and i_b.

(c) Because the circuits are equivalent, the currents i_a and i_b shown in Figure P 3.7-4*b* are equal to the currents i_a and i_b shown in Figure P 3.7-4*a*. Use this fact to determine values of the voltage v_1 and current i_2 shown in Figure P 3.7-4*a*.

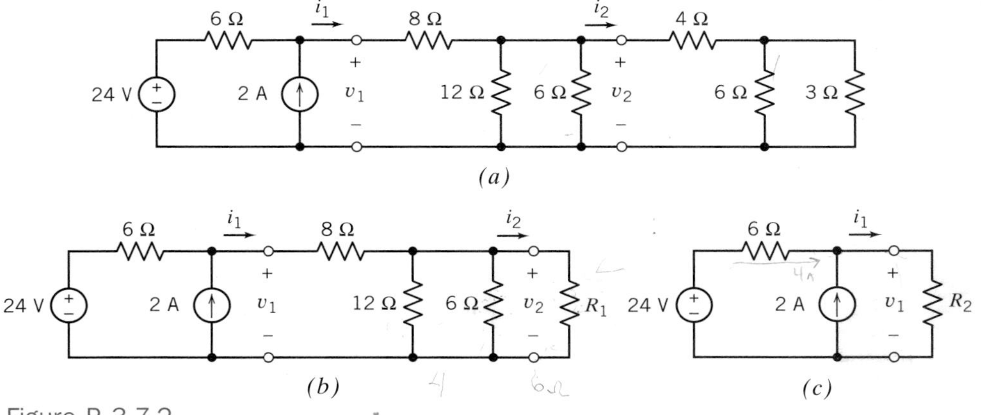

Figure P 3.7-2

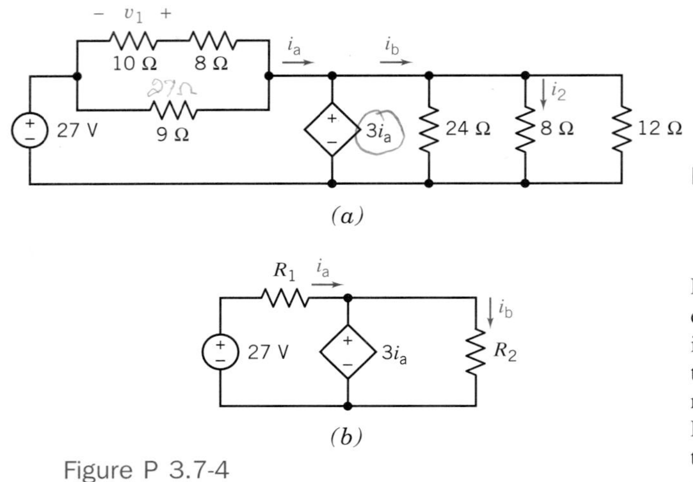

Figure P 3.7-4

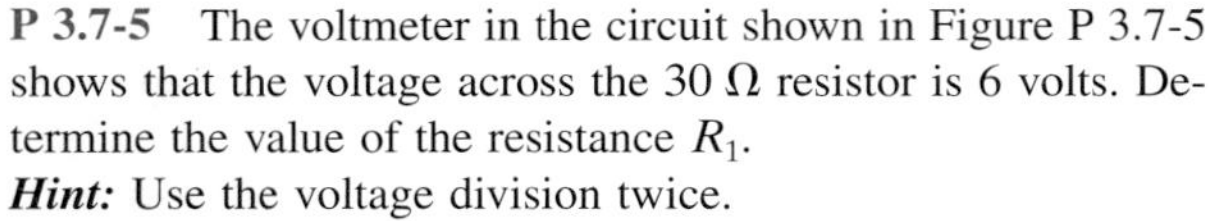

P 3.7-5 The voltmeter in the circuit shown in Figure P 3.7-5 shows that the voltage across the 30 Ω resistor is 6 volts. Determine the value of the resistance R_1.
Hint: Use the voltage division twice.
Answer: $R_1 = 40\ \Omega$.

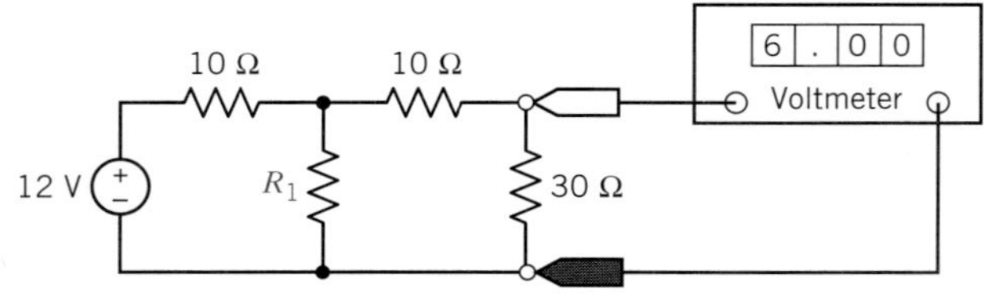

Figure P 3.7-5

P 3.7-6 Determine the voltages v_a and v_c and the currents i_b and i_d for the circuit shown in Figure P 3.7-6.
Answer: $v_a = -2\text{ V}$, $v_c = 6\text{ V}$, $i = -16\text{ mA}$ and $i_d = 2\text{ mA}$.

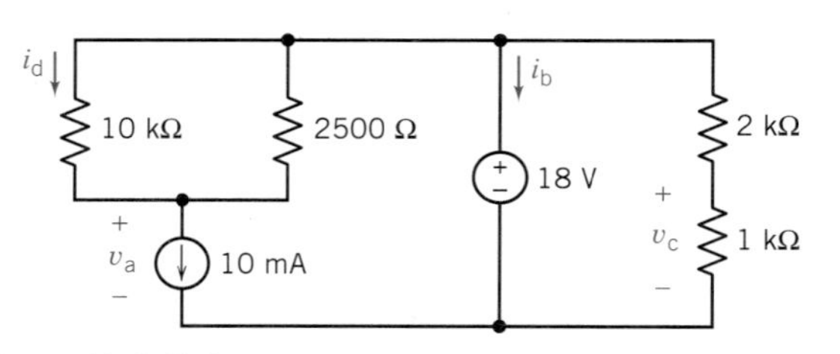

Figure P 3.7-6

P 3.7-7 Determine the value of the resistance R in Figure 3.7-7.
Answer: $R = 28\text{ k}\Omega$.

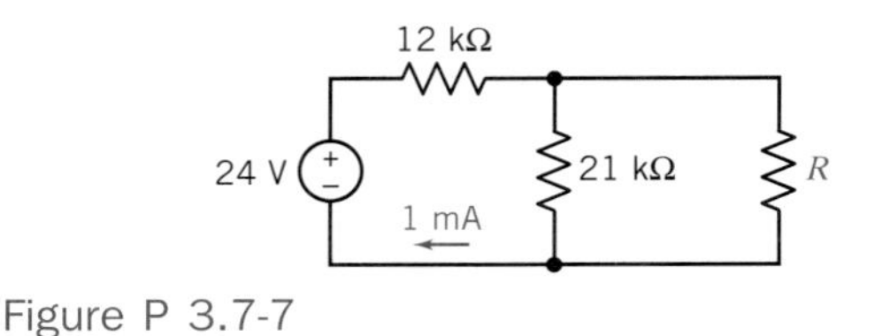

Figure P 3.7-7

P 3.7-8 Most of us are familiar with the effects of a mild electric shock. The effects of a severe shock can be devastating and often fatal. Shock results when current is passed through the body. A person can be modeled as a network of resistances. Consider the model circuit shown in Figure P 3.7-8. Determine the voltage developed across the heart and the current flowing through the heart of the person when he or she firmly grasps one end of a voltage source whose other end is connected to the floor. The heart is represented by R_h. The floor has resistance to current flow equal to R_f, and the person is standing barefoot on the floor. This type of accident might occur at a swimming pool or boat dock. The upper-body resistance R_U and lower-body resistance R_L vary from person to person.

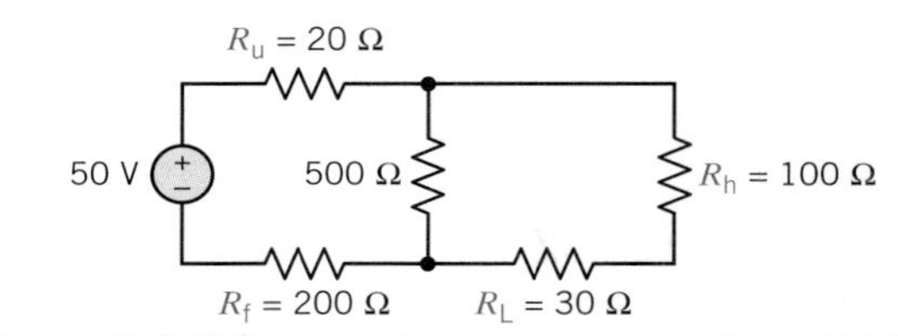

Figure P 3.7-8 The resistance of the heart. $R_h = 100\ \Omega$.

P 3.7-9 Determine the value of the current i in Figure 3.7-9.
Answer: $i = 0.5\text{ mA}$.

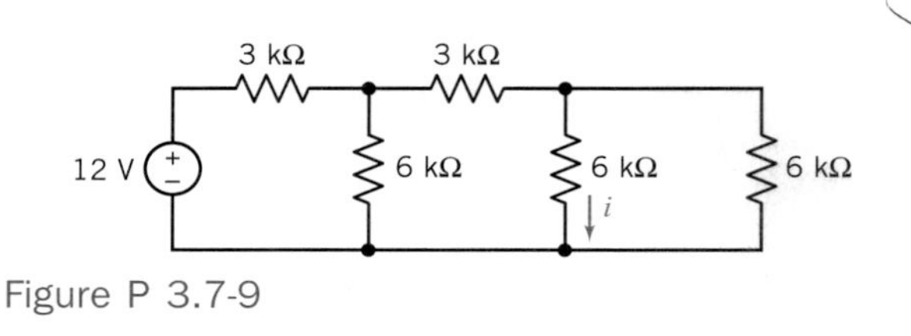

Figure P 3.7-9

P 3.7-10 Electric streetcar railways have been used for about 100 years. An electric railway was established by Werner Siemens from Charlottenbourg to Spandau in Berlin in 1885, as shown in Figure P 3.7-10a. This railway may be represented by the electric circuit shown in Figure P 3.7-10b. Find the power delivered to the motor R_L by the central power source.

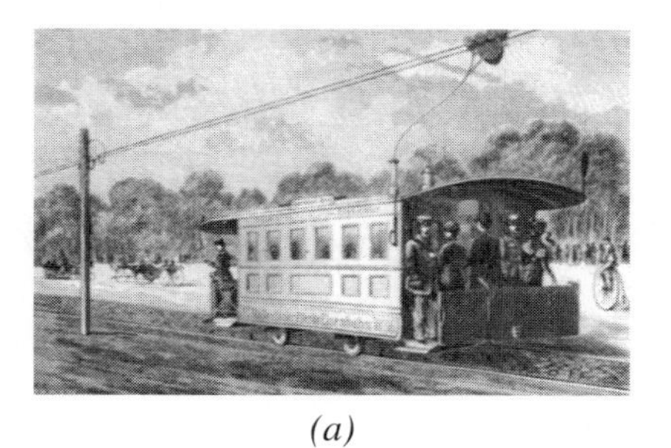

(*a*)

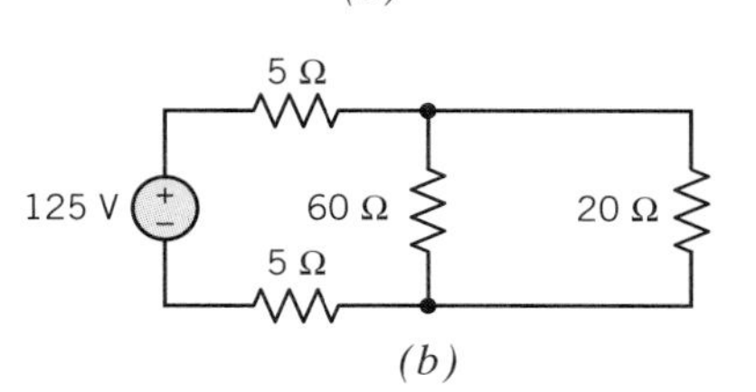

(*b*)

Figure P 3.7-10 (*a*) The electric street railway in Berlin in 1885. Courtesy of Burndy Library. (*b*) Circuit model of the railway, where $R_L = 20\ \Omega$.

P 3.7-11 Find i and $R_{eq\ a\text{–}b}$ if $v_{ab} = 40$ V in the circuit of Figure P 3.7-11.
Answer: $R_{eq\ a\text{–}b} = 8\ \Omega$, $i = 5/6$ A

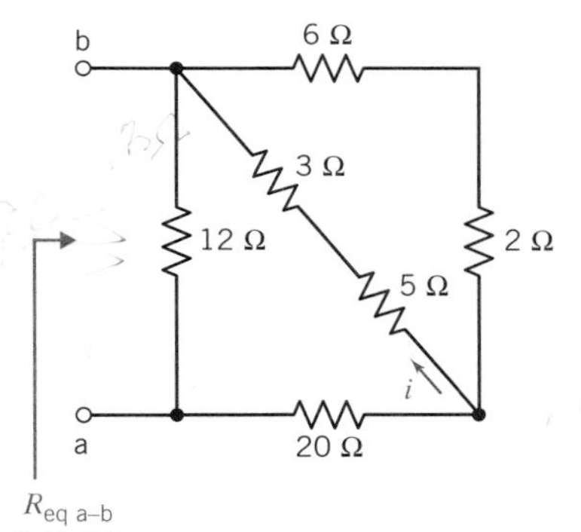

Figure P 3.7-11

P 3.7-12 Find R_{eq}, i, and v if $v_{ab} = 12$ V for the circuit of Figure P 3.7-12.
Answer: $R_{eq} = 12\ \Omega$, $i = 2/3$ A, $v = 16/3$ V

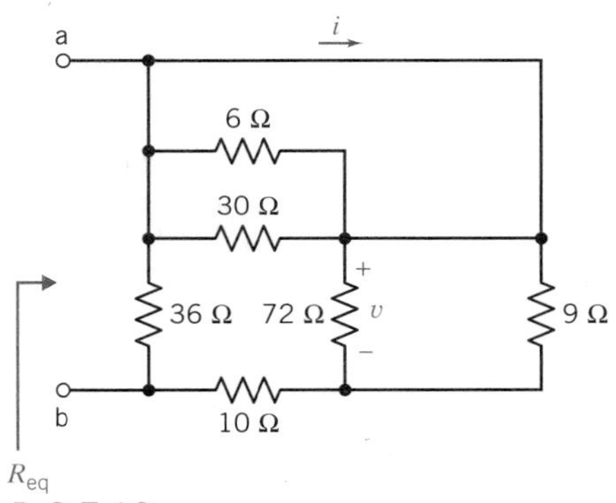

Figure P 3.7-12

P 3.7-13 The source $v_s = 240$ volts is connected to three equal resistors as shown in Figure P 3.7-13. Determine R when the voltage source delivers 1920 W to the resistors.
Answer: $R = 45\ \Omega$

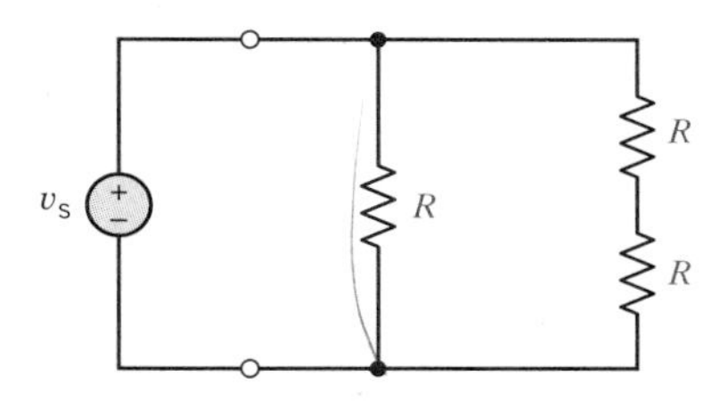

Figure P 3.7-13

P 3.7-14 Find the R_{eq} at terminals a–b in Figure P 3.7-14. Also determine i, i_1, and i_2.
Answer: $R_{eq} = 8\ \Omega$, $i = 5$ A, $i_1 = 5/3$ A, $i_2 = 5/2$ A

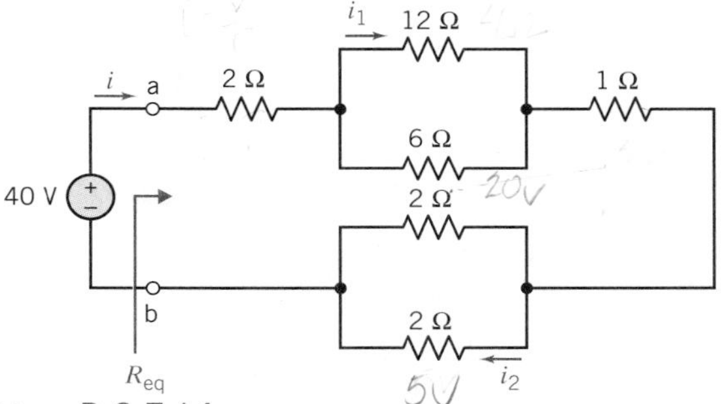

Figure P 3.7-14

P 3.7-15 For the circuit of Figure P 3.7-15, given that $R_{eq} = 9\ \Omega$, find R.
Answer: $R = 15\ \Omega$

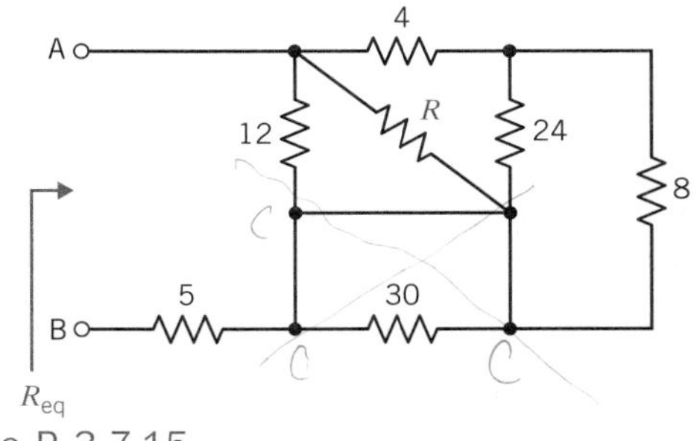

Figure P 3.7-15

P 3.7-16 Determine R when $R_{eq} = 20\ \Omega$ for the circuit shown in Figure P 3.7-16.

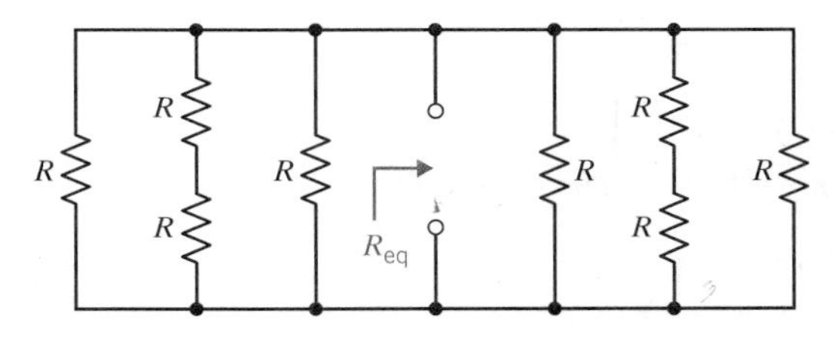

Figure P 3.7-16

VERIFICATION PROBLEMS

VP 3-1 A computer analysis program, used for the circuit of Figure VP 3.1, provides the following branch currents and voltages: $i_1 = -0.833$, $i_2 = -0.333$, $i_3 = -1.167$, and $v = -2.0$. Are these answers correct?
Hint: Verify that KCL is satisfied at the center node and that KVL is satisfied around the outside loop consisting of the two 6 Ω resistors and the voltage source.

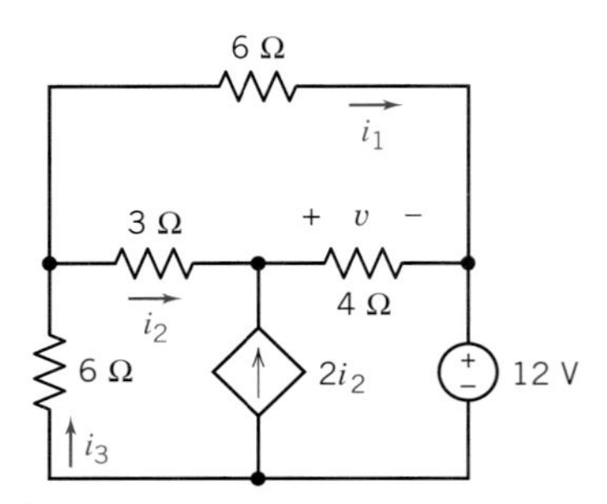

Figure VP 3.1

VP 3-2 The circuit of Figure VP 3.2 was assigned as a homework problem. The answer in the back of the textbook says the current, i, is 1.25 A. Verify this answer using current division.

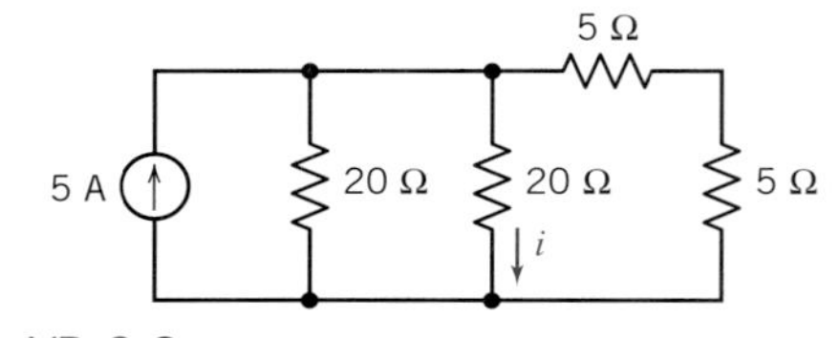

Figure VP 3.2

VP 3-3 The circuit of Figure VP 3.3 was built in the lab and v_o was measured to be 6.25 V. Verify this measurement using the voltage divider principle.

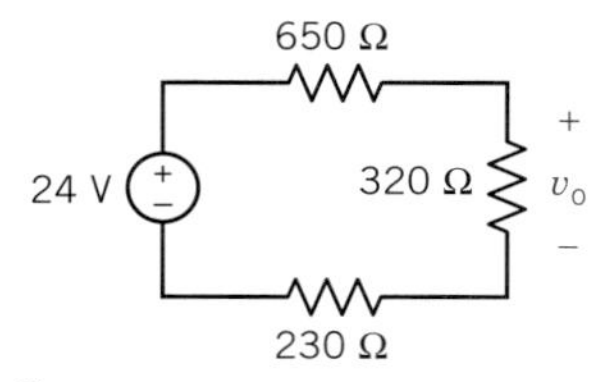

Figure VP 3.3

VP 3-4 The circuit of Figure VP 3.4 represents an auto's electrical system. A report states that $i_H = 9$ A, $i_B = -9$ A, and $i_A = 19.1$ A. Verify that this result is correct.
Hint: Verify that KCL is satisfied at each node and that KVL is satisfied around each loop.

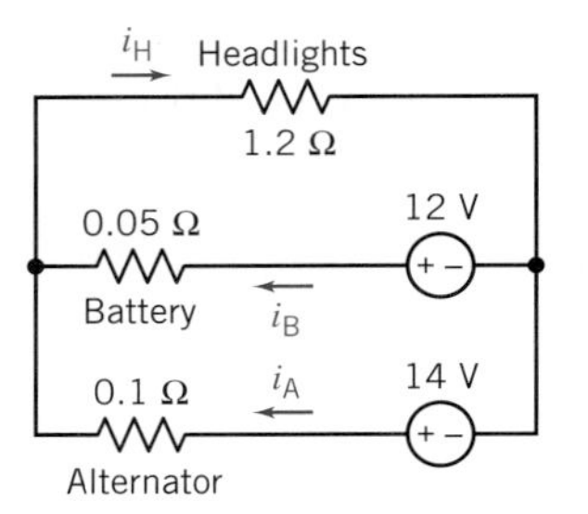

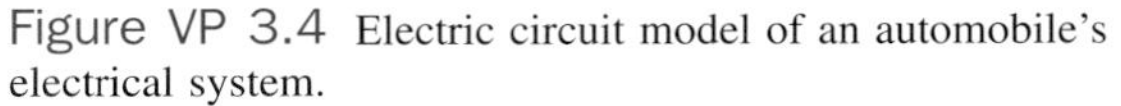

Figure VP 3.4 Electric circuit model of an automobile's electrical system.

VP 3-5 Computer analysis of the circuit in Figure VP 3.5 shows that $i_a = -0.5$ mA and $i_b = -2$ mA. Was the computer analysis done correctly?
Hint: Verify that the KVL equations for all three meshes are satisfied when $i_a = -0.5$ mA and $i_b = -2$ mA.

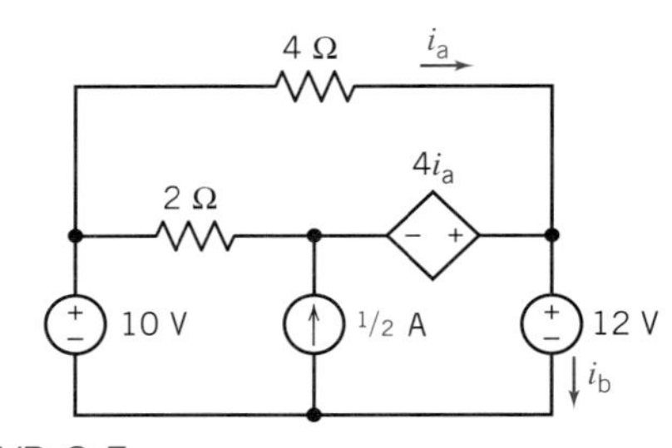

Figure VP 3.5

VP 3-6 Computer analysis of the circuit in Figure VP 3.6 shows that $i_a = 0.5$ mA and $i_b = 4.5$ mA. Was the computer analysis done correctly?
Hint: First, verify that the KCL equations for all five nodes are satisfied when $i_a = 0.5$ mA and $i_b = 4.5$ mA. Next, verify that the KVL equation for the lower left mesh (a-e-d-a) is satisfied. (The KVL equations for the other meshes aren't useful because each involves an unknown voltage.)

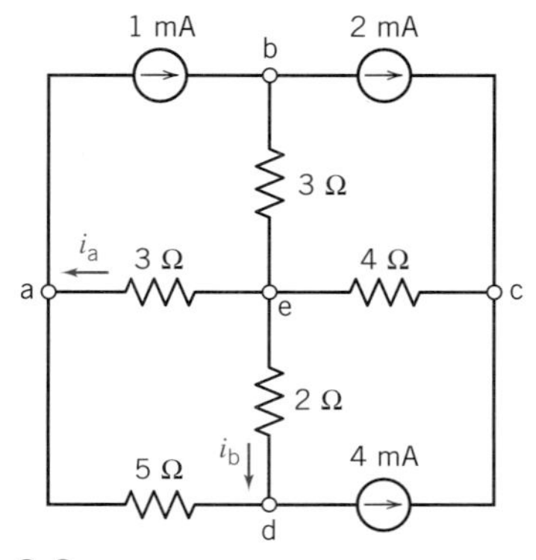

Figure VP 3.6

DESIGN PROBLEMS

DP 3-1 The circuit shown in Figure DP 3.1 uses a potentiometer to produce a variable voltage. The voltage v_m varies as a knob connected to the wiper of the potentiometer is turned. Specify the resistances R_1 and R_2 so that the following three requirements are satisfied:

1. The voltage v_m varies from 8 V to 12 V as the wiper moves from one end of the potentiometer to the other end of the potentiometer.
2. The voltage source supplies less than 0.5 W of power.
3. Each of R_1, R_2, and R_P dissipate less than 0.25 W.

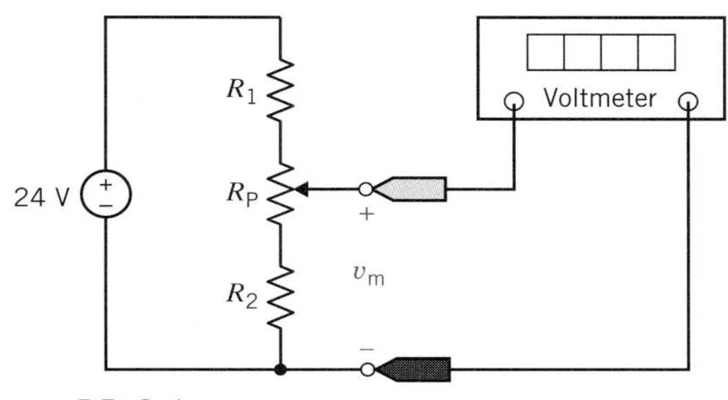

Figure DP 3.1

DP 3-2 The resistance R_L in Figure DP 3.2 is the equivalent resistance of a pressure transducer. This resistance is specified to be 200 Ω ± 5%. That is, 190 Ω ≤ R_L ≤ 210 Ω. The voltage source is a 12 V ±1% source capable of supplying 5 W. Design this circuit, using 5%, 1/8 Watt resistors for R_1 and R_2, so that the voltage across R_L is

$$v_o = 4\text{ V} \pm 10\%.$$

(A 5%, 1/8 Watt 100 Ω resistor has a resistance between 95 and 105 Ω and can safely dissipate 1/8 W continuously.)

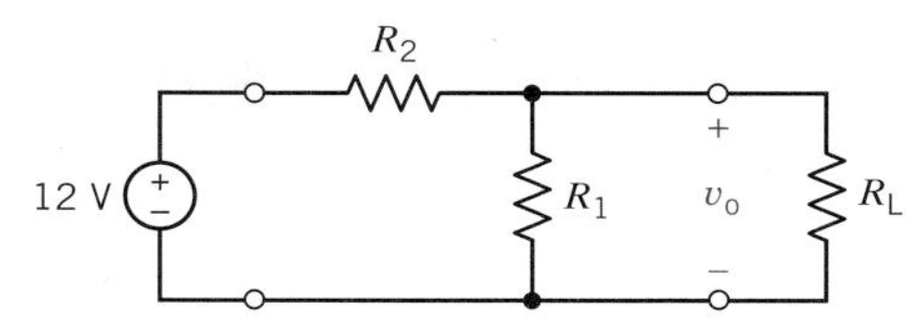

Figure DP 3.2

DP 3-3 A phonograph pickup, stereo amplifier, and speaker are shown in Figure DP 3.3*a* and redrawn as a circuit model as shown in Figure DP 3.3*b*. Determine the resistance R so that the voltage v across the speaker is 16 V. Determine the power delivered to the speaker.

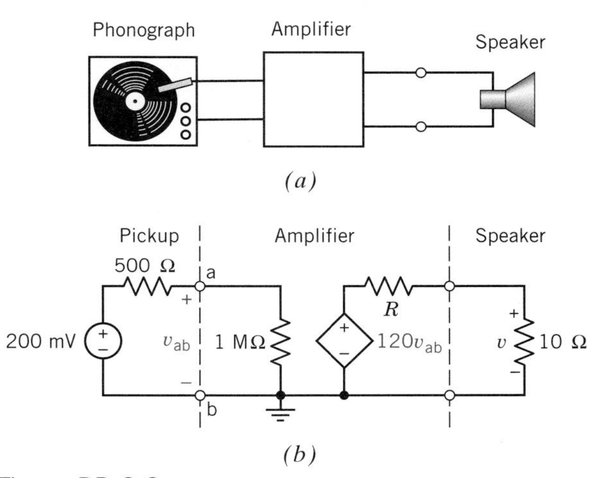

Figure DP 3.3 A phonograph stereo system.

DP 3-4 A Christmas tree light set is required that will operate from a 6-V battery on a tree in a city park. The heavy-duty battery can provide 9 A for the four-hour period of operation each night. Design a parallel set of lights (select the maximum number of lights) when the resistance of each bulb is 12 Ω.

CHAPTER 4

Methods of Analysis of Resistive Circuits

Preview

Increasingly complex circuits are required to meet the needs of ever-growing communication systems. These circuits were effective in transmitting information over great distances. In Chapter 3 we analyzed circuits consisting of resistors as well as voltage and current sources. As the complexity of circuits increased, analysis techniques were developed that incorporated rigorous systematic methods.

In this chapter we define and utilize two widely used methods of analysis: (1) node voltage and (2) mesh current. These very powerful methods are widely used today for the analysis of large complex circuits in communications and electrical systems.

4.1 Design Challenge

POTENTIOMETER ANGLE DISPLAY

A circuit is needed to measure and display the angular position of a potentiometer shaft. A *potentiometer* is a variable resistor with three terminals. Two terminals are connected to the opposite ends of the resistive element and the third terminal connects to the sliding contact. The angular position, θ, will vary from -180 degrees to 180 degrees.

Figure 4.1-1 illustrates a circuit that could do the job. The $+15$-V and -15-V power supplies, the potentiometer, and resistors R_1 and R_2 are used to obtain a voltage, v_i, that is proportional to θ. The amplifier is used to change the constant of proportionality in order to obtain a simple relationship between θ and the voltage, v_o, displayed by the voltmeter. In this example, the amplifier will be used to obtain the relationship

$$v_o = k \cdot \theta \quad \text{where } k = 0.1 \frac{\text{volt}}{\text{degree}} \tag{4.1-1}$$

so that θ can be determined by multiplying the meter reading by 10. For example, a meter reading of -7.32 V indicates that $\theta = -73.2$ degrees.

Describe the Situation and the Assumptions

The circuit diagram in Figure 4.1-2 is obtained by modeling the power supplies as ideal voltage sources, the voltmeter as an open circuit, and the potentiometer by two resistors. The parameter, a, in the model of the potentiometer varies from 0 to 1 as θ varies from -180 degrees to 180 degrees. That means

$$a = \frac{\theta}{360^\circ} + \frac{1}{2} \tag{4.1-2}$$

Solving for θ gives

$$\theta = \left(a - \frac{1}{2}\right) \cdot 360^\circ \tag{4.1-3}$$

State the Goal

Specify values of resistors R_1 and R_2, the potentiometer resistance R_p, and the amplifier gain b that will cause the meter voltage, v_o, to be related to the angle θ by Eq. 4.1-1.

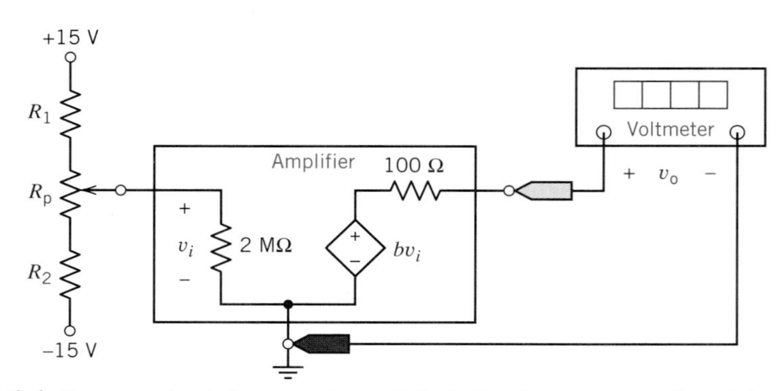

Figure 4.1-1 Proposed circuit for measuring and displaying the angular position of the potentiometer shaft.

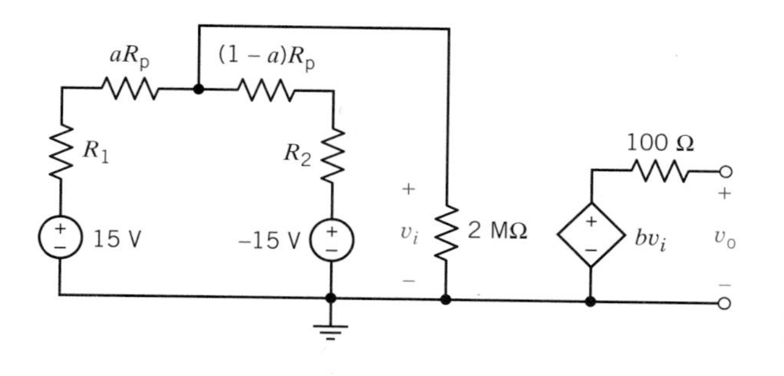

Figure 4.1-2
Circuit diagram containing models of the power supplies, voltmeter, and potentiometer.

Generate a Plan

Analyze the circuit shown in Figure 4.1-2 to determine the relationship between v_i and θ. Select values of R_1, R_2, and R_p. Use these values to simplify the relationship between v_i and θ. If possible, calculate the value of b that will cause the meter voltage, v_o, to be related to the angle θ by Eq. 4.1-1. If this isn't possible, adjust the values of R_1, R_2, and R_p and try again.

Chapter 4 describes two methods, called nodal analysis and mesh analysis, for analyzing circuits like the one shown in Figure 4.1-2. We will return to this problem at the end of the chapter after learning about node and mesh analysis.

4.2 ELECTRIC CIRCUITS FOR COMMUNICATIONS

Samuel F. B. Morse became engaged in the search for an efficient electric telegraph in the 1830s. He worked on the design of a sending device and a code of dots and dashes for letters and numbers. Morse, shown in Figure 4.2-1, publicly demonstrated his telegraph on January 24, 1838. On February 21, 1838, President Van Buren witnessed the transmission of messages over 10 miles of wires. On April 7, 1838, Morse applied for a patent.

By 1843 the U.S. Senate appropriated funds for the construction of a telegraph line between Baltimore and Washington, D.C. On May 24, 1844, the famous first message was sent: "What hath God wrought!"

Figure 4.2-1 Samuel F. B. Morse, the inventor of the electric telegraph.

Figure 4.2-2 The first telephone transmitter with its parchment diaphragm attached to the magnetized metallic reed. This instrument was used to transmit the first speech sounds electrically in 1875. Courtesy of Bell Laboratories.

Figure 4.2-3
Alexander Graham Bell at the New York end of the circuit to Chicago as this line was opened on October 18, 1892, as part of the ceremonies accompanying the Columbian Exposition. Courtesy of Bell Laboratories.

By 1851 there were over 50 telegraph companies in the United States. Telegraphs were widely used by railways and the military, in addition to private individuals. However, the telegraph did not offer the advantages of voice communication and of dialogue, although it did provide an expanding communications network in the United States and Europe. The development of a telephone became a central activity by the 1870s.

Alexander Graham Bell (1847–1922) is commonly singled out as the inventor who effectively produced a practical telephone. Bell produced the first telephone transmitter, shown in Figure 4.2-2, in 1875.

In New Haven, Connecticut, in 1878, twenty-one subscribing parties could use eight lines to the first central switchboard. New York was first connected to Chicago in 1892. Bell is shown in Figure 4.2-3 at the opening of the Chicago line in 1892.

There were twice as many telephones in the United States—over 155,000—as in Europe in 1885. However, by 1898 the number of lines was greater in Europe. After that time, the United States maintained a widening lead in the use of the telephone. The growth in the number of telephones in the United States from 1925 to 1990 is shown in Figure 4.2-4. By 1925 there were 20 million telephones in use in the United States. The telephone has become almost universal.

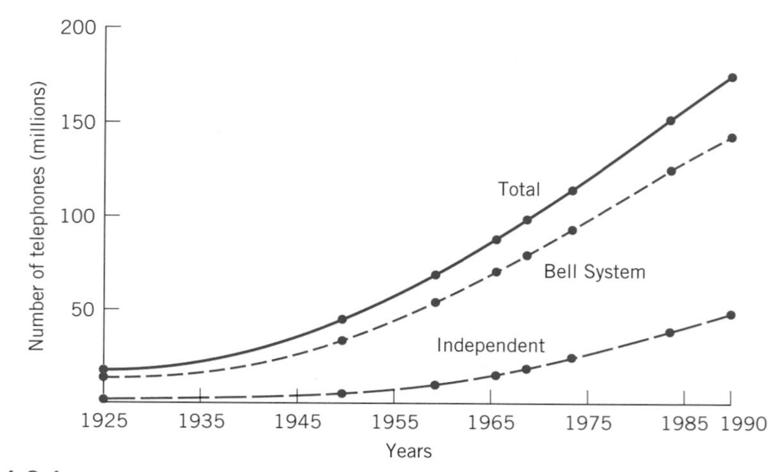

Figure 4.2-4 Growth of telephone service in the United States. *Source:* Bell Laboratories.

With the advent of the telephone system and the electric power system, complex electric circuits became increasingly common. Thus, it became necessary to develop rigorous useful methods of analysis of complex circuits.

4.3 NODE VOLTAGE ANALYSIS OF CIRCUITS WITH CURRENT SOURCES

In the previous chapter we analyzed simple circuits that could be reduced to one containing only two nodes. We then obtained a single equation for the one unknown quantity, typically the voltage between the two nodes. We now wish to consider more complicated circuits and to develop an analysis method for obtaining *node voltages.*

Consider the circuit shown in Figure 4.3-1*a*. This circuit contains four elements: three resistors and a current source. The *nodes* of a circuit are the places where the elements are connected together. The circuit shown in Figure 4.3-1*a* has three nodes. It is customary to draw the elements horizontally or vertically and to connect these elements by horizontal and vertical lines that represent wires. In other words, nodes are drawn as points, or are drawn using horizontal or vertical lines. Figure 4.3-1*b* shows the same circuit, redrawn so that all three nodes are drawn as points rather than lines. In Figure 4.3-1*b* the nodes are labeled as node a, node b and node c.

Analyzing a connected circuit containing n nodes will require $n - 1$ KCL equations. One way to obtain these equations is to apply KCL at each node of the circuit except for one. The node at which KCL is not applied is called the reference node. Any node of the circuit can be selected to be the reference node. We will often choose the node at the bottom of the circuit to be the reference node. (When the circuit contains a grounded power supply, the ground node of the power supply is usually selected as the reference node.) In Figure 4.3-1*b*, node c is selected as the reference node and marked with the symbol used to identify the reference node.

The voltage at any node of the circuit, relative to the reference node, is called a **node voltage.** In Figure 4.3-1*b*, there are two node voltages: the voltage at node a with respect to the reference node, node c, and also the voltage at node b, again with respect to the reference node, node c. In Figure 4.3-1*c*, voltmeters are added to measure the node volt-

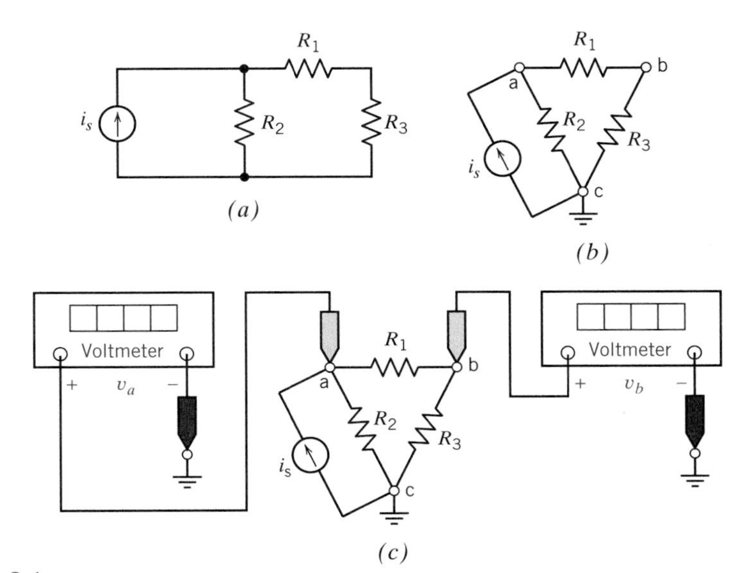

Figure 4.3-1 (*a*) A circuit with three nodes. (*b*) The circuit after the nodes have been labeled and a reference node has been selected and marked. (*c*) Using voltmeters to measure the node voltages.

ages. To measure node voltage at node a, connect the red probe of the voltmeter at node a and connect the black probe at the reference node, node c. To measure node voltage at node b, connect the red probe of the voltmeter at node b and connect the black probe at the reference node, node c.

The node voltages in Figure 4.3-1c can be represented as v_{ac} and v_{bc}, but it is conventional to drop the subscript c and refer to these as v_a and v_b. Notice that the node voltage at the reference node is 0 V, since a voltmeter measuring the node voltage at the reference node would have both probes connected to the same point.

In order to determine the node voltages, we use Kirchhoff's current law at each of the circuit's nodes, except at the reference node. This set of equations, called the node equations, enables us to find the node voltages.

Figure 4.3-2 shows how to express the current in any resistor as a function of the node voltages. Figure 4.3-2a shows the circuit from Figure 4.3-1. The voltage across each resistor has been labeled. Notice that the voltage across R_2 is equal to the node voltage at node a and that the voltage across R_3 is equal to the node voltage at node b. Applying KVL to the right-hand mesh provides the equation

$$-v_a + v_c + v_b = 0$$

Solving for v_c gives

$$v_c = v_a - v_b$$

This equation expresses the voltage across R_1 as a function of the node voltages, v_a and v_b. In Figure 4.3-2b, the label for the voltage across R_1 is $v_a - v_b$ instead of v_c. Now all three resistor voltages are expressed as functions of the node voltages. Applying Ohm's law to each resistor in Figure 4.3-2b expresses the current in each resistor as a function of the node voltages and resistances. These currents are labeled in Figure 4.3-2c.

The node equations representing the circuit in Figure 4.3-2 are obtained by applying Kirchhoff's Current Law (KCL) at nodes a and b. Using KCL at node a gives

$$i_s = \frac{v_a}{R_2} + \frac{v_a - v_b}{R_1} \tag{4.3-1}$$

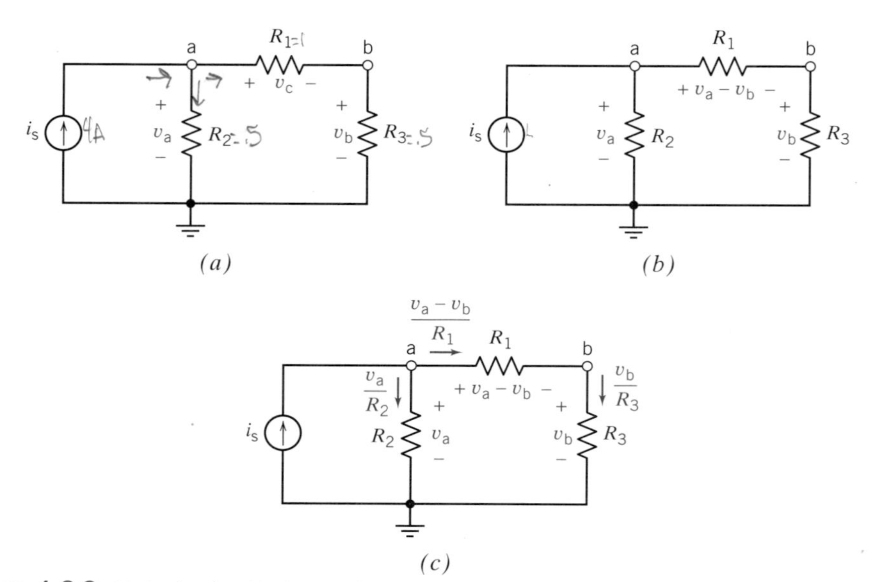

Figure 4.3-2 (a) A circuit with three resistors. (b) The resistor voltages expressed as functions of the node voltages. (c) The resistor currents expressed as functions of the node voltages.

Similarly, the KCL equation at node b is

$$\frac{v_a - v_b}{R_1} = \frac{v_b}{R_3} \tag{4.3-2}$$

If $R_1 = 1\ \Omega$, $R_2 = R_3 = 0.5\ \Omega$, and $i_s = 4$ A, the two node equations (4.3-1) and (4.3-2) may be rewritten as

$$4 = \frac{v_a - v_b}{1} + \frac{v_a}{0.5} \tag{4.3-3}$$

$$\frac{v_a - v_b}{1} = \frac{v_b}{0.5} \tag{4.3-4}$$

Solving Eq. 4.3-4 for v_b gives

$$v_b = \frac{v_a}{3} \tag{4.3-5}$$

Substituting Eq. 4.3-5 into Eq. 4.3-3 gives

$$4 = v_a - \frac{v_a}{3} + 2v_a = \frac{8}{3}v_a \tag{4.3-6}$$

Solving Eq. 4.3-6 for v_a gives

$$v_a = \frac{3}{2}\text{ V}$$

Finally, Eq. 4.3-5 gives

$$v_b = \frac{1}{2}\text{ V}$$

Thus, the node voltages of this circuit are

$$v_a = \frac{3}{2}\text{ V} \quad \text{and} \quad v_b = \frac{1}{2}\text{ V}$$

Example 4.3-1

Determine the value of the resistance R in the circuit shown in Figure 4.3-3*a*.

Solution

Let v_a denote the node voltage at node a and v_b denote the node voltage at node b. The voltmeter in Figure 4.3-3 measures the value of the node voltage at node b, v_b. In Figure 4.3-3*b*, the resistor currents are expressed as functions of the node voltages. Apply KCL at node a to obtain

$$4 + \frac{v_a}{10} + \frac{v_a - v_b}{5} = 0$$

Using $v_b = 5$ V, gives

$$4 + \frac{v_a}{10} + \frac{v_a - 5}{5} = 0$$

Solving for v_a, we get

$$v_a = -10\text{ V}$$

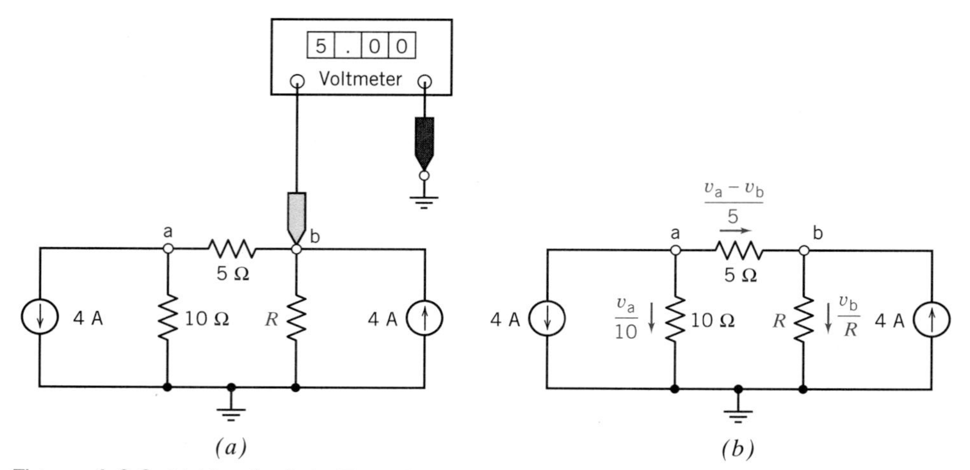

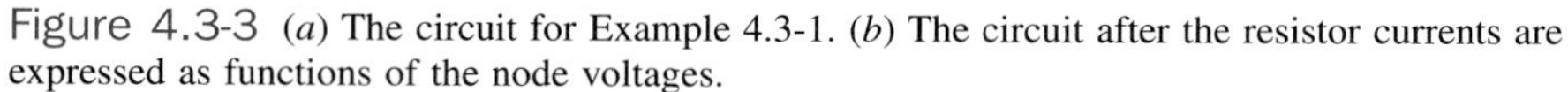

Figure 4.3-3 (*a*) The circuit for Example 4.3-1. (*b*) The circuit after the resistor currents are expressed as functions of the node voltages.

Next, apply KCL at node b to obtain

$$-\left(\frac{v_a - v_b}{5}\right) + \frac{v_b}{R} - 4 = 0$$

Using $v_a = -10$ V and $v_b = 5$V gives

$$-\left(\frac{-10 - 5}{5}\right) + \frac{5}{R} - 4 = 0$$

Finally, solving for R gives

$$R = 5\ \Omega$$

Example 4.3-2

Obtain the node equations for the circuit in Figure 4.3-4.

Solution

Let v_a denote the node voltage at node a, v_b denote the node voltage at node b and v_c denote the node voltage at node c. Apply KCL at node a to obtain

$$-\left(\frac{v_a - v_c}{R_1}\right) + i_1 - \left(\frac{v_a - v_c}{R_2}\right) + i_2 - \left(\frac{v_a - v_b}{R_5}\right) = 0$$

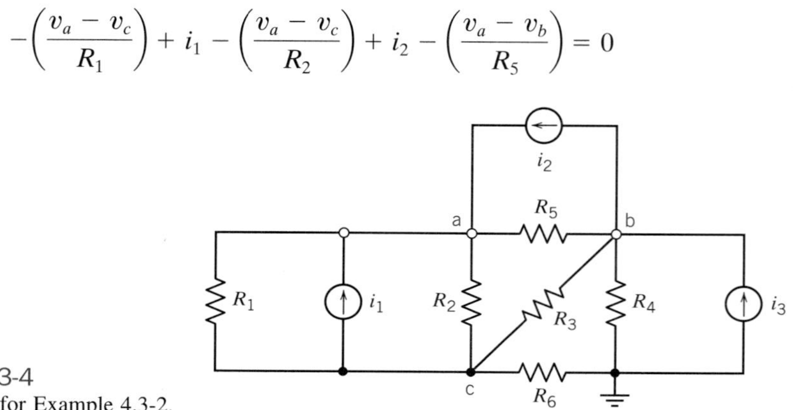

Figure 4.3-4
The circuit for Example 4.3-2.

Separate the terms of this equation that involve v_a from the terms that involve v_b and the terms that involve v_c to obtain

$$\left(\frac{1}{R_1}+\frac{1}{R_2}+\frac{1}{R_5}\right)v_a-\left(\frac{1}{R_5}\right)v_b-\left(\frac{1}{R_1}+\frac{1}{R_2}\right)v_c=i_1+i_2$$

There is a pattern in the node equations of circuits that contain only resistors and current sources. In the node equation at node a, the coefficient of v_a is the sum of the reciprocals of the resistances of all resistors connected to node a. The coefficient of v_b is minus the sum of the reciprocals of the resistances of all resistors connected between node b and node a. The coefficient of v_c is minus the sum of the reciprocals of the resistances of all resistors connected between node c and node a. The right hand side of this equation is the algebraic sum of current source currents directed into node a.

Apply KCL at node b to obtain

$$-i_2+\left(\frac{v_a-v_b}{R_5}\right)-\left(\frac{v_b-v_c}{R_3}\right)-\left(\frac{v_b}{R_4}\right)+i_3=0$$

Separate the terms of this equation that involve v_a from the terms that involve v_b and the terms that involve v_c to obtain

$$-\left(\frac{1}{R_5}\right)v_a+\left(\frac{1}{R_3}+\frac{1}{R_4}+\frac{1}{R_5}\right)v_b-\left(\frac{1}{R_3}\right)v_c=i_3-i_2$$

As expected, this node equation adheres to the pattern for node equations of circuits that contain only resistors and current sources. In the node equation at node b, the coefficient of v_b is the sum of the reciprocals of the resistances of all resistors connected to node b. The coefficient of v_a is minus the sum of the reciprocals of the resistances of all resistors connected between node a and node b. The coefficient of v_c is minus the sum of the reciprocals of the resistances of all resistors connected between node c and node b. The right hand side of this equation is the algebraic sum of current source currents directed into node b.

Finally, use the pattern for the node equations of circuits that contain only resistors and current sources to obtain the node equation at node c

$$\left(\frac{1}{R_1}+\frac{1}{R_2}\right)v_a-\left(\frac{1}{R_3}\right)v_b-\left(\frac{1}{R_1}+\frac{1}{R_2}+\frac{1}{R_3}+\frac{1}{R_6}\right)v_c=-i_1$$

Example 4.3-3

Determine the node voltages for the circuit in Figure 4.3-4 when $i_1 = 1$ A, $i_2 = 2$ A, $i_3 = 3$ A, $R_1 = 5\ \Omega$, $R_2 = 2\ \Omega$, $R_3 = 10\ \Omega$, $R_4 = 4\ \Omega$, $R_5 = 5\ \Omega$, and $R_6 = 2\ \Omega$.

Solution

The node equations are

$$\left(\frac{1}{5}+\frac{1}{2}+\frac{1}{5}\right)v_a-\left(\frac{1}{5}\right)v_b-\left(\frac{1}{5}+\frac{1}{2}\right)v_c=1+2$$

$$-\left(\frac{1}{5}\right)v_a+\left(\frac{1}{10}+\frac{1}{5}+\frac{1}{4}\right)v_b-\left(\frac{1}{10}\right)v_c=-2+3$$

$$-\left(\frac{1}{5}+\frac{1}{2}\right)v_a-\left(\frac{1}{10}\right)v_b+\left(\frac{1}{5}+\frac{1}{2}+\frac{1}{10}+\frac{1}{2}\right)v_c=-1$$

or

$$\begin{aligned} 0.9v_a - 0.2v_b - 0.7v_c &= 3 \\ -0.2v_a + 0.55v_b - 0.1v_c &= 1 \\ -0.7v_a - 0.1v_b + 1.3v_c &= -1 \end{aligned}$$

The node equations can be written using matrices as

$$\begin{bmatrix} 0.9 & -0.2 & -0.7 \\ -0.2 & 0.55 & -0.1 \\ -0.7 & -0.1 & 1.3 \end{bmatrix} \begin{bmatrix} v_a \\ v_b \\ v_c \end{bmatrix} = \begin{bmatrix} 3 \\ 1 \\ -1 \end{bmatrix}$$

This matrix equation can be solved using Cramer's Rule (see Appendix A). First calculate

$$\Delta = \begin{vmatrix} 0.9 & -0.2 & -0.7 \\ -0.2 & 0.55 & -0.1 \\ -0.7 & -0.1 & 1.3 \end{vmatrix} = 0.2850$$

Next, we find

$$v_a = \frac{\begin{vmatrix} 3 & -0.2 & -0.7 \\ 1 & 0.55 & -0.1 \\ -1 & -0.1 & 1.3 \end{vmatrix}}{\Delta} = 7.1579, \quad v_b = \frac{\begin{vmatrix} 0.9 & 3 & -0.7 \\ -0.2 & 1 & -0.1 \\ -0.7 & -1 & 1.3 \end{vmatrix}}{\Delta} = 5.0526 \text{ and}$$

$$v_c = \frac{\begin{vmatrix} 0.9 & -0.2 & 3 \\ -0.2 & 0.55 & 1 \\ -0.7 & -0.1 & -1 \end{vmatrix}}{\Delta} = 3.4737$$

Exercise 4.3-1 Determine the node voltages, v_a and v_b, for the circuit of Figure E 4.3-1.

Answer: $v_a = 3$ V and $v_b = 11$ V.

Exercise 4.3-2 Determine the node voltages, v_a and v_b, for the circuit of Figure E 4.3-2.

Answer: $v_a = -4/3$ V and $v_b = 4$ V.

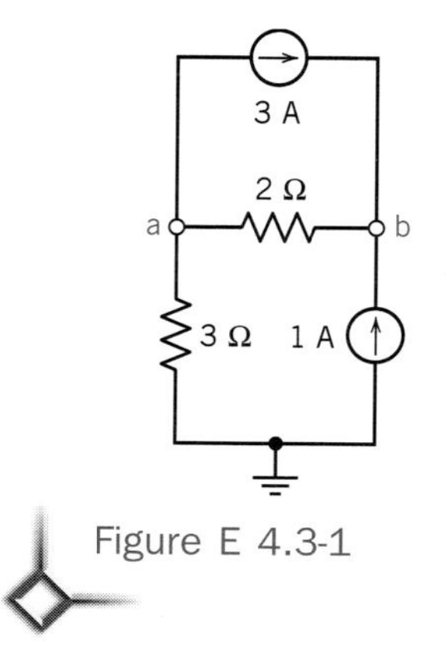

Figure E 4.3-1

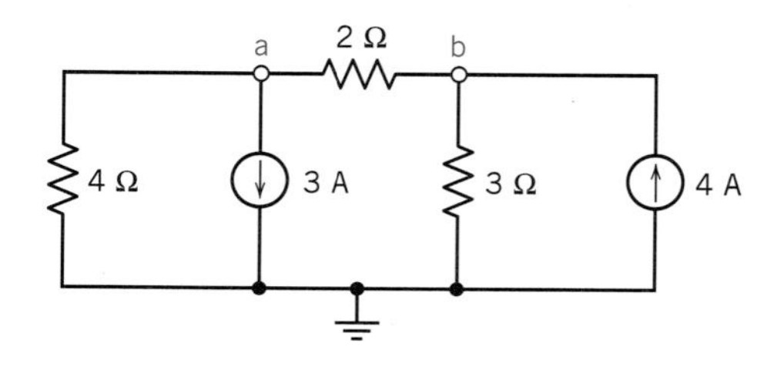

Figure E 4.3-2

4.4 NODE VOLTAGE ANALYSIS OF CIRCUITS WITH CURRENT AND VOLTAGE SOURCES

In the preceding section we determined the node voltages of circuits with independent current sources only. In this section we consider circuits with both independent current and voltage sources.

First we consider the circuit with a voltage source between ground and one of the other nodes. Since we are free to select the reference node, this particular arrangement is easily achieved. Such a circuit is shown in Figure 4.4-1. We immediately note that the source is connected between terminal a and ground, and therefore

$$v_a = v_s$$

Thus, v_a is known and only v_b is unknown. We write the KCL equation at node b to obtain

$$i_s = \frac{v_b}{R_3} + \frac{v_b - v_a}{R_2}$$

However, $v_a = v_s$. Therefore

$$i_s = \frac{v_b}{R_3} + \frac{v_b - v_s}{R_2}$$

Then, solving for the unknown node voltage v_b, we get

$$v_b = \frac{R_2 R_3 i_s + R_3 v_s}{R_2 + R_3}$$

Next, let us consider the circuit of Figure 4.4-2, which includes a voltage source between two nodes. Since the source voltage is known, use KVL to obtain

$$v_a - v_b = v_s$$

or

$$v_a = v_s + v_b$$

To account for the fact that the source voltage is known, we consider both node a and node b as part of one larger node represented by the shaded ellipse shown in Figure 4.4-2. We require a larger node since v_a and v_b are dependent. This larger node is often called a *supernode* or a *generalized node*. KCL says that the algebraic sum of the currents entering a supernode is zero. That means that we apply KCL to a supernode in the same way that we apply KCL to a node.

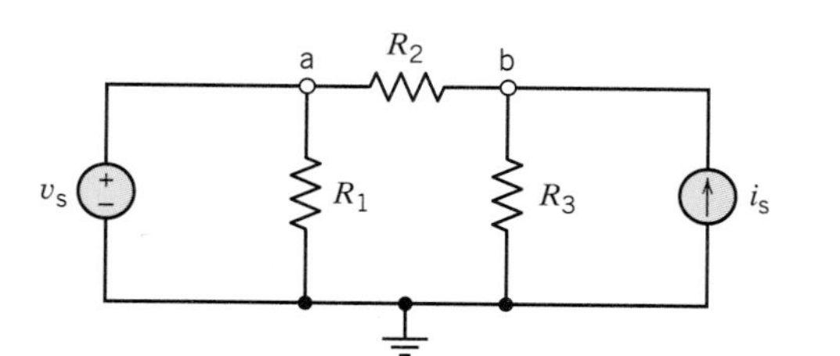

Figure 4.4-1 Circuit with an independent voltage source and an independent current source.

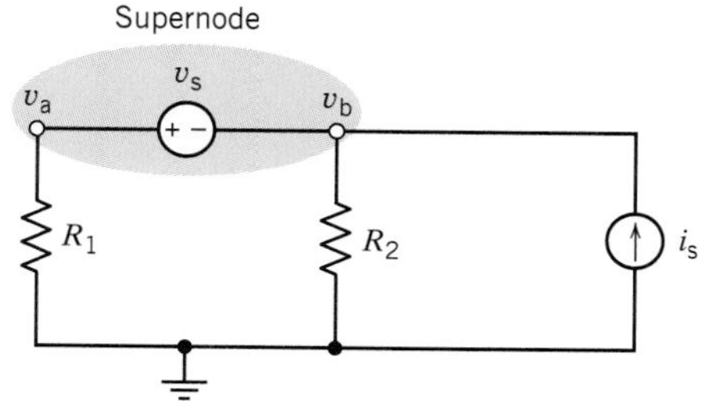

Figure 4.4-2 Circuit with a supernode that incorporates v_a and v_b.

Table 4.4-1 **Node Voltage Analysis Methods with a Voltage Source**

Case	Method
1. The voltage source connects node q and the reference node (ground).	Set v_q equal to the source voltage accounting for the polarities and proceed to write the KCL at the remaining nodes.
2. The voltage source lies between two nodes, a and b.	Create a supernode that incorporates a and b and equate the sum of all the currents into the supernode (both nodes a and b) to zero.

A **supernode** consists of two nodes connected by an independent or a dependent voltage source.

We then can write the KCL equation at the supernode as

$$\frac{v_a}{R_1} + \frac{v_b}{R_2} = i_s$$

However, since $v_a = v_s + v_b$, we have

$$\frac{v_s + v_b}{R_1} + \frac{v_b}{R_2} = i_s$$

Then, solving for the unknown node voltage v_b, we get

$$v_b = \frac{R_1 R_2 i_s - R_2 v_s}{R_1 + R_2}$$

We can now compile a summary of both methods of dealing with independent voltage sources in a circuit we wish to solve by node voltage methods, as recorded in Table 4.4-1.

Example 4.4-1

Determine the node voltages for the circuit shown in Figure 4.4-3.

Solution

The methods summarized in Table 4.4-1 are exemplified in this solution. The 4-V voltage source connected to node a exemplifies method 1. The 8-V source between nodes b and c exemplifies method 2.

Using method 1 for the 4-V source, we note that

$$v_a = -4 \text{ V}$$

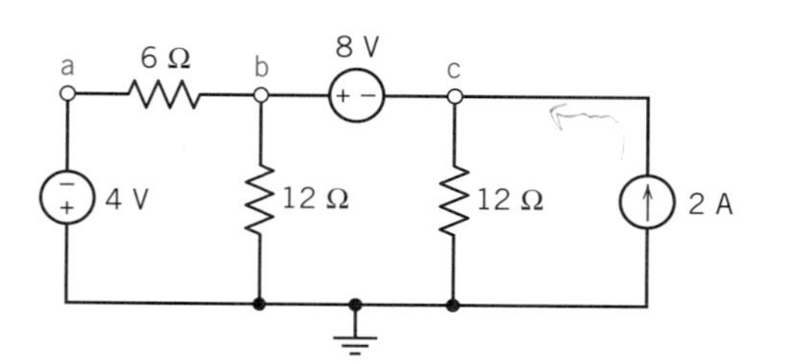

Figure 4.4-3 A circuit containing two voltage sources, only one of which is connected to the reference node.

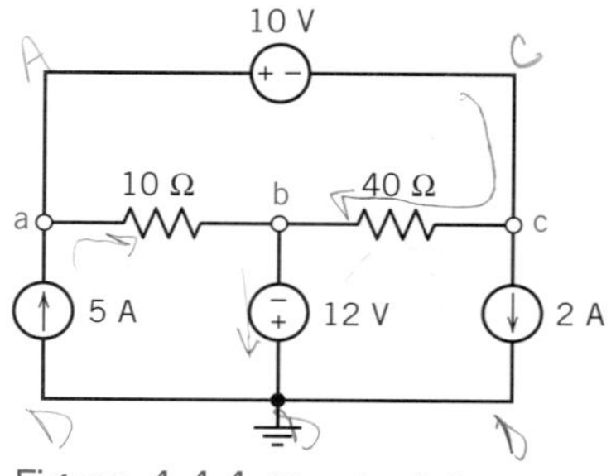

Figure 4.4-4 The circuit for Example 4.4-2.

Using method 2 for the 8-V source, we have a supernode at nodes b and c. The node voltages at nodes b and c are related by

$$v_b = v_c + 8$$

Writing a KCL equation for the supernode, we have

$$\frac{v_b - v_a}{6} + \frac{v_b}{12} + \frac{v_c}{12} = 2$$

or

$$3\,v_b + v_c = 24 + 2\,v_a$$

Using $v_a = -4$ V and $v_b = v_c + 8$ to eliminate v_a and v_b, we have

$$3(v_c + 8) + v_c = 24 + 2(-4)$$

Solving this equation for v_c we get

$$v_c = -2 \text{ V}$$

Now we calculate v_b to be

$$v_b = v_c + 8 = -2 + 8 = 6 \text{ V}$$

Example 4.4-2

Determine the node voltages for the circuit shown in Figure 4.4-4.

Solution

We will calculate the node voltages of this circuit by writing a KCL equation for the supernode corresponding to the 10 V voltage source. First notice that

$$v_b = -12 \text{ V}$$

and that

$$v_a = v_c + 10$$

Writing a KCL equation for the supernode, we have

$$\frac{v_a - v_b}{10} + 2 + \frac{v_c - v_b}{40} = 5$$

or

$$4\,v_a + v_c - 5\,v_b = 120$$

Using $v_a = v_c + 10$ and $v_b = -12$ to eliminate v_c and v_b, we have

$$4(v_c + 10) + v_c - 5(-12) = 120$$

Solving this equation for v_c we get

$$v_c = 4 \text{ V}$$

Exercise 4.4-1 Find the node voltages for the circuit of Figure E 4.4-1.

Hint: Write a KCL equation for the supernode corresponding to the 10 V voltage source.

Answer: $2 + \dfrac{v_b + 10}{20} + \dfrac{v_b}{30} = 5 \quad \Rightarrow \quad v_b = 30 \text{ V and } v_a = 40 \text{ V}$

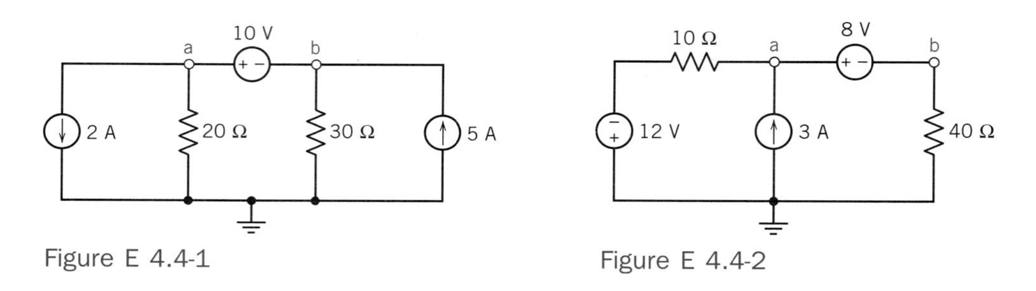

Figure E 4.4-1

Figure E 4.4-2

Exercise 4.4-2 Find the voltages v_a and v_b for the circuit of Figure E 4.4-2.

Answer: $$\frac{(v_b + 8) - (-12)}{10} + \frac{v_b}{40} = 3 \quad \Rightarrow \quad v_b = 8\,\text{V and } v_a = 16\,\text{V}$$

4.5 NODE VOLTAGE ANALYSIS WITH DEPENDENT SOURCES

When a circuit contains a dependent source, the controlling current or voltage of that dependent source must be expressed as a function of the node voltages. It is then a simple matter to express the controlled current or voltage as a function of the node voltages. The node equations are then obtained using the techniques described in the previous two sections.

Example 4.5-1

Determine the node voltages for the circuit shown in Figure 4.5-1.

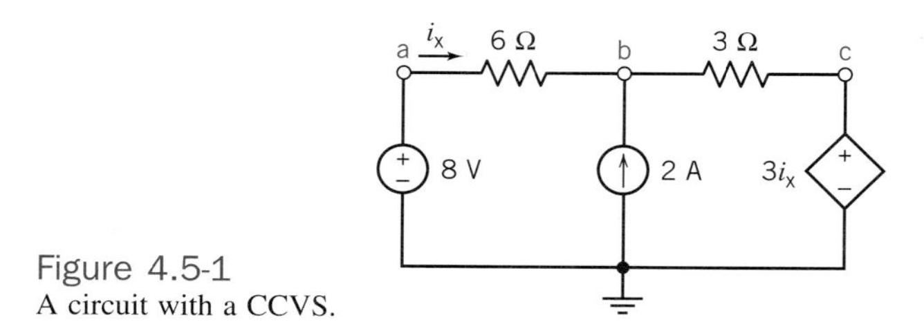

Figure 4.5-1
A circuit with a CCVS.

Solution

The controlling current of the dependent source is i_x. Our first task is to express this current as a function of the node voltages:

$$i_x = \frac{v_a - v_b}{6}$$

The value of the node voltage at node a is set by the 8 V voltage source to be

$$v_a = 8\text{ V}$$

so

$$i_x = \frac{8 - v_b}{6}$$

The node voltage at node c is equal to the voltage of the dependent source, so:

$$v_c = 3\,i_x = 3\left(\frac{8 - v_b}{6}\right) = 4 - \frac{v_b}{2} \tag{4.5-1}$$

Next apply KCL at node b to get

$$\frac{8 - v_b}{6} + 2 = \frac{v_b - v_c}{3} \tag{4.5-2}$$

Using Eq. 4.5-1 to eliminate v_c from Eq. 4.5-2 gives

$$\frac{8 - v_b}{6} + 2 = \frac{v_b - \left(4 - \frac{v_b}{2}\right)}{3} = \frac{v_b}{2} - \frac{4}{3}$$

Solving for v_b gives

$$v_b = 7 \text{ V}$$

Then

$$v_c = 4 - \frac{v_b}{2} = \frac{1}{2} \text{ V}$$

Example 4.5-2

Determine the node voltages for the circuit shown in Figure 4.5-2.

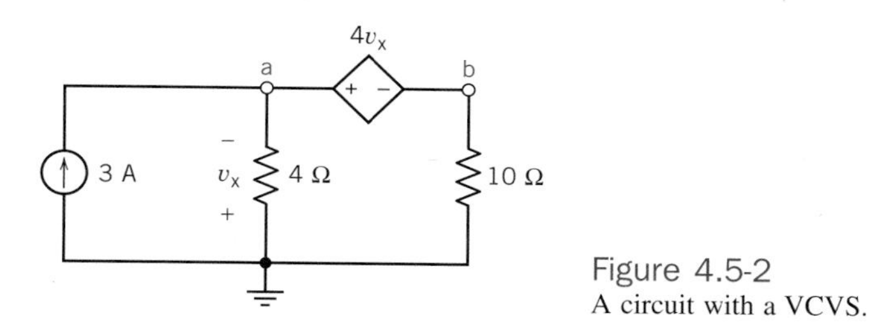

Figure 4.5-2
A circuit with a VCVS.

Solution

The controlling voltage of the dependent source is v_x. Our first task is to express this voltage as a function of the node voltages:

$$v_x = -v_a$$

The difference between the node voltages at nodes a and b is set by voltage of the dependent source:

$$v_a - v_b = 4\,v_x = 4\,(-v_a) = -4\,v_a$$

Simplifying this equation gives

$$v_b = 5\,v_a \tag{4.5-3}$$

Applying KCL to the supernode corresponding to the dependent voltage source gives

$$3 = \frac{v_a}{4} + \frac{v_b}{10} \tag{4.5-4}$$

Using Eq. 4.5-3 to eliminate v_b from Eq. 4.5-4 gives

$$3 = \frac{v_a}{4} + \frac{5\,v_a}{10} = \frac{3}{4}\,v_a$$

Solving for v_a, we get

$$v_a = 4\,\text{V}$$

Finally,

$$v_b = 5\,v_a = 20\,\text{V}$$

Example 4.5-3

Determine the node voltages for the circuit shown in Figure 4.5-3.

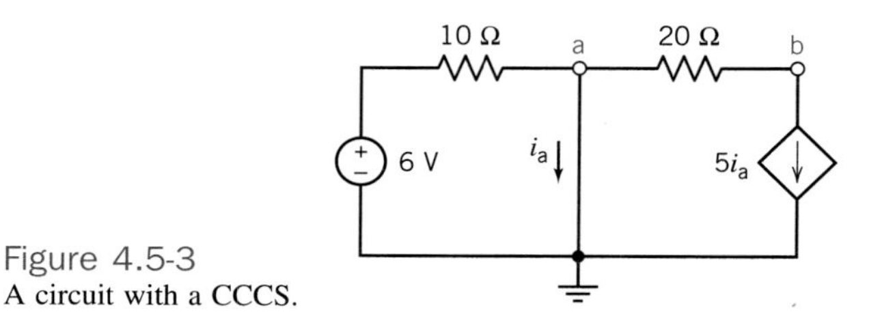

Figure 4.5-3
A circuit with a CCCS.

Solution

The controlling current of the dependent source is i_a. Our first task is to express this current as a function of the node voltages. Apply KCL at node a to get

$$\frac{6 - v_a}{10} = i_a + \frac{v_a - v_b}{20}$$

Node a is connected to the reference node by a short circuit, so $v_a = 0$ V. Substituting this value of v_a into the above equation and simplifying gives

$$i_a = \frac{12 + v_b}{20} \tag{4.5-5}$$

Next, apply KCL at node b to get

$$\frac{0 - v_b}{20} = 5\,i_a \tag{4.5-6}$$

Using Eq. 4.5-5 to eliminate i_a from Eq. 4.5-6 gives

$$\frac{0 - v_b}{20} = 5\left(\frac{12 + v_b}{20}\right)$$

Solving for v_b gives

$$v_b = -10\,\text{V}$$

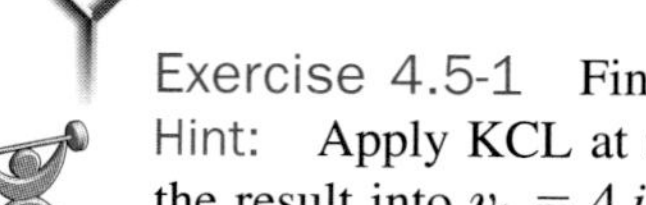

Exercise 4.5-1 Find the node voltage v_b for the circuit shown in Figure E 4.5-1.
Hint: Apply KCL at node a to express i_a as a function of the node voltages. Substitute the result into $v_b = 4\,i_a$ and solve for v_b.

Answer: $v_b = 4\left(\dfrac{9 + v_b}{12}\right) \quad \Rightarrow \quad v_b = 4.5\,\text{V}$

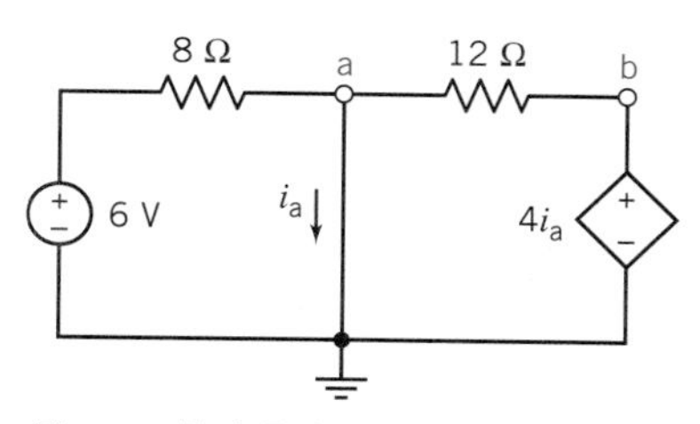

Figure E 4.5-1
A circuit with a CCVS.

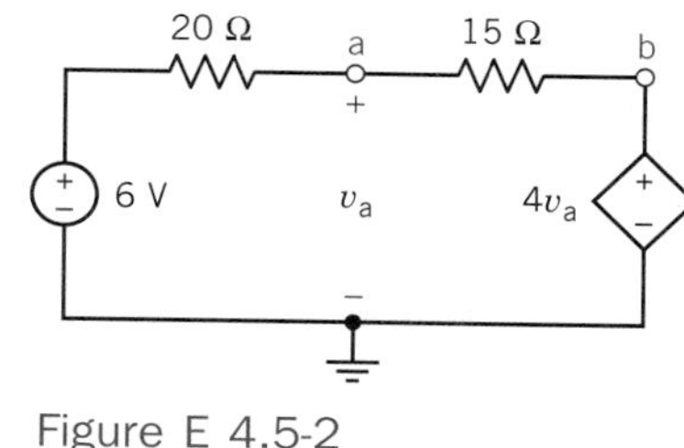

Figure E 4.5-2
A circuit with a VCVS.

Exercise 4.5-2 Find the node voltages for the circuit shown in Figure E 4.5-2.
Hint: The controlling voltage of the dependent source is a node voltage so it is already expressed as a function of the node voltages. Apply KCL at node a.

Answer: $\frac{v_a - 6}{20} + \frac{v_a - 4\,v_a}{15} = 0 \quad \Rightarrow \quad v_a = -2\text{ V}$

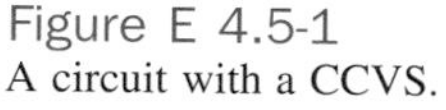

4.6 MESH CURRENT ANALYSIS WITH INDEPENDENT VOLTAGE SOURCES

In this and succeeding sections, we consider the analysis of circuits using Kirchhoff's voltage law (KVL) around a closed path. A *closed path* or a *loop* is drawn by starting at a node and tracing a path such that we return to the original node without passing an intermediate node more than once.

A mesh is a special case of a loop. A *mesh* is a loop that does not contain any other loops within it. Mesh current analysis is applicable only to planar networks. A planar circuit is one that can be drawn on a plane, without crossovers. An example of a nonplanar circuit is shown in Figure 4.6-1, where the crossover is identified and cannot be removed by redrawing the circuit. For planar networks, the meshes in the network look like "windows." There are four meshes in the circuit shown in Figure 4.6-2. They are identified as M_i. Mesh 2 contains the elements R_3, R_4, and R_5. Note that the resistor R_3 is common to both mesh 1 and mesh 2.

We define a *mesh current* as the current that flows through the elements constituting the mesh. We will use the convention of a mesh current flowing clockwise, as shown in Figure 4.6-3. This circuit has three mesh currents. Note that the current in an element common to two meshes is the algebraic sum of the mesh currents. Therefore, for Figure 4.6-3 the current in R_4 flowing downward is

$$i_{R4} = i_1 - i_2$$

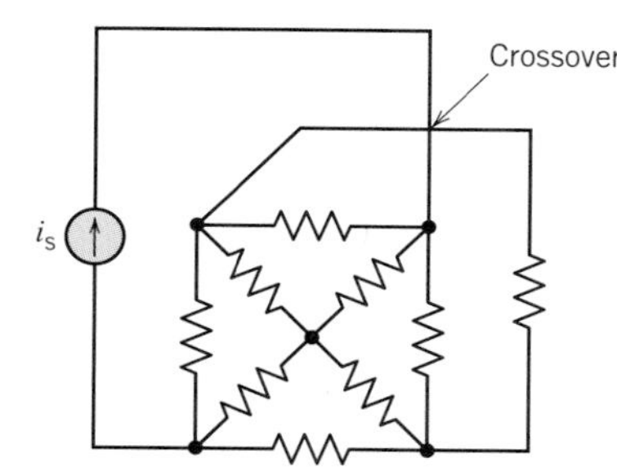

Figure 4.6-1 Nonplanar circuit with a crossover.

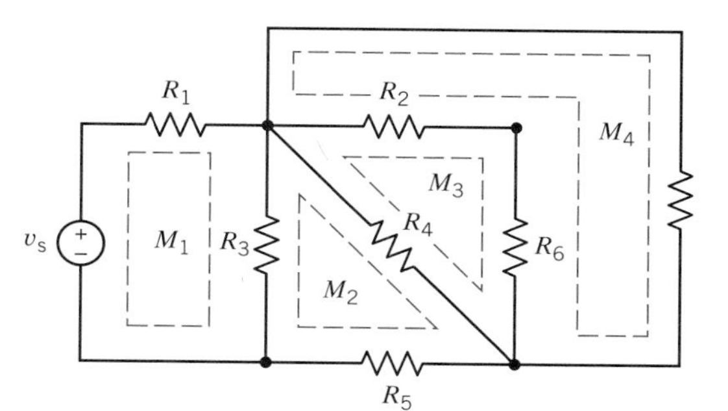

Figure 4.6-2 Circuit with four meshes. Each mesh is identified by dashed lines.

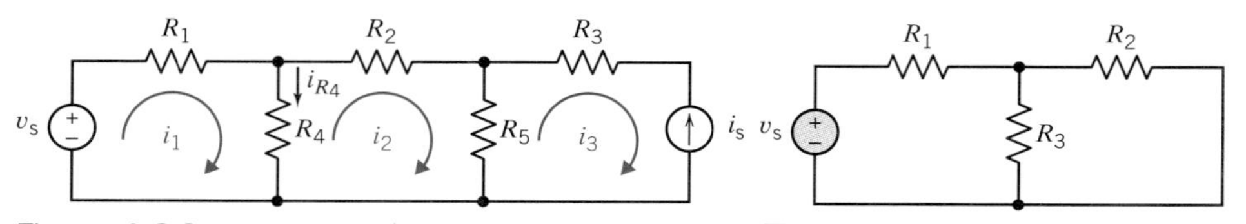

Figure 4.6-3 Circuit with three mesh currents.

Figure 4.6-4 Circuit with two meshes.

Let us consider the two-mesh circuit of Figure 4.6-4. A mesh cannot have other loops within it. Thus we cannot choose the outer loop $v_s \rightarrow R_1 \rightarrow R_2 \rightarrow v_s$ as one mesh, since it would contain the loop $v_s \rightarrow R_1 \rightarrow R_3 \rightarrow v_s$ within it. We are required to choose the two mesh currents as shown in Figure 4.6-5.

We may use Kirchhoff's voltage law around each mesh. We will use the following convention for obtaining the algebraic sum of voltages around a mesh. We will move around the mesh in the clockwise direction. If we encounter the + sign of the voltage reference direction of an element voltage before the − sign, we add that voltage. Conversely, if we encounter the − of the voltage reference direction of an element voltage before the + sign, we subtract that voltage. Thus, for the circuit of Figure 4.6-5, we have

$$\text{mesh 1:} \quad -v_s + R_1 i_1 + R_3(i_1 - i_2) = 0 \tag{4.6-1}$$

$$\text{mesh 2:} \quad R_3(i_2 - i_1) + R_2 i_2 = 0 \tag{4.6-2}$$

Note that the voltage across R_3 in mesh 1 is determined from Ohm's law, where

$$v = R_3 i_a = R_3(i_1 - i_2)$$

where i_a is the actual element current flowing downward through R_3.

The two equations 4.6-1 and 4.6-2 will enable us to determine the two mesh currents i_1 and i_2. Rewriting the two equations, we have

$$i_1(R_1 + R_3) - i_2 R_3 = v_s$$

and

$$-i_1 R_3 + i_2(R_3 + R_2) = 0$$

If $R_1 = R_2 = R_3 = 1\ \Omega$, we have

$$2i_1 - i_2 = v_s$$

and

$$-i_1 + 2i_2 = 0$$

Add twice the first equation to the second equation, obtaining $3i_1 = 2v_s$. Then we have

$$i_1 = \frac{2v_s}{3} \quad \text{and} \quad i_2 = \frac{v_s}{3}$$

Thus, we have obtained two independent mesh current equations that are readily solved for the two unknowns. If we have N meshes and write N mesh equations in terms of N mesh currents, we can obtain N independent mesh equations. This set of N equations is independent, and thus guarantees a solution for the N mesh currents.

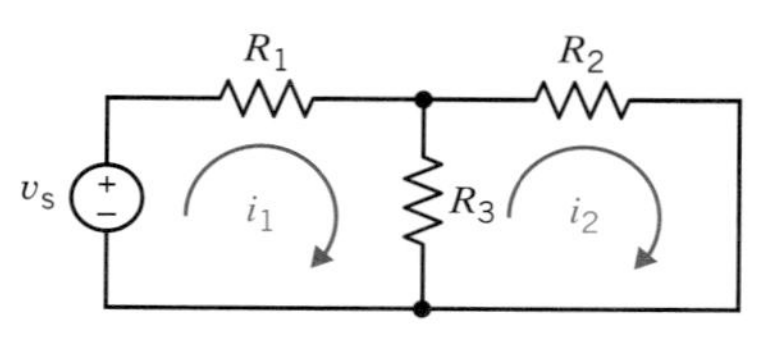

Figure 4.6-5
Mesh currents for the circuit of Figure 4.6-4.

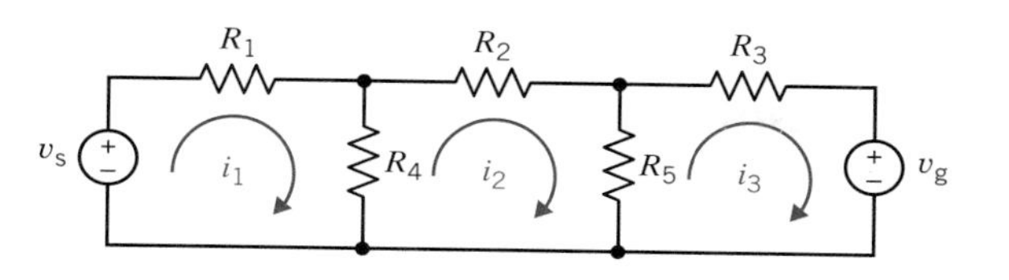

Figure 4.6-6
Circuit with three mesh currents and two voltage sources.

A circuit that contains only independent voltage sources and resistors results in a specific format of equations that can readily be obtained. Consider a circuit with three meshes, as shown in Figure 4.6-6. Assign the clockwise direction to all of the mesh currents. Using KVL, we obtain the three mesh equations

$$\text{mesh 1:} \quad -v_s + R_1 i_1 + R_4(i_1 - i_2) = 0$$

$$\text{mesh 2:} \quad R_2 i_2 + R_5(i_2 - i_3) + R_4(i_2 - i_1) = 0$$

$$\text{mesh 3:} \quad R_5(i_3 - i_2) + R_3 i_3 + v_g = 0$$

These three mesh equations can be rewritten by collecting coefficients for each mesh current as

$$\text{mesh 1:} \quad (R_1 + R_4)\, i_1 - R_4 i_2 = v_s$$

$$\text{mesh 2:} \quad -R_4 i_1 + (R_4 + R_2 + R_5) i_2 - R_5 i_3 = 0$$

$$\text{mesh 3:} \quad -R_5 i_2 + (R_3 + R_5) i_3 = -v_g$$

Hence, we note that the coefficient of the mesh current i_1 for the first mesh is the sum of resistances in the loop and that the coefficient of the second mesh current is the negative of the resistance common to meshes 1 and 2. In general, we state that for mesh current i_n, the equation for the nth mesh with independent voltage sources only is obtained as follows:

$$-\sum_{q=1}^{Q} R_k\, i_q + \sum_{j=1}^{P} R_j\, i_n = \sum_{n=1}^{N} v_{sn} \tag{4.6-3}$$

That is, for mesh n we multiply i_n by the sum of all resistances R_j around the mesh. Then we add the terms due to the resistances in common with another mesh as the negative of the connecting resistance R_k, multiplied by the mesh current in the adjacent mesh i_q for all Q adjacent meshes. Finally, the independent voltage sources around the loop appear on the right side of the equation as the negative of the voltage sources encountered as we traverse the loop in the direction of the mesh current. Remember that the above result is obtained assuming all mesh currents flow clockwise.

The general matrix equation for the mesh current analysis for independent voltage sources present in a circuit is

$$\mathbf{R}\,\mathbf{i} = \mathbf{v}_s \tag{4.6-4}$$

where **R** is a symmetric matrix with a diagonal consisting of the sum of resistances in each mesh and the off-diagonal elements are the negative of the sum of the resistances common to two meshes. The matrix **i** consists of the mesh currents as

$$\mathbf{i} = \begin{bmatrix} i_1 \\ i_2 \\ \vdots \\ i_N \end{bmatrix}$$

For N mesh currents, the source matrix $\mathbf{v}_s$ is

$$\mathbf{v}_s = \begin{bmatrix} v_{s1} \\ v_{s2} \\ \vdots \\ v_{sN} \end{bmatrix}$$

where v_{sj} is the sum of the voltages of the voltage sources in the jth mesh with the appropriate sign assigned to each source voltage.

For the circuit of Figure 4.6-5 and the matrix Eq. 4.6-4, we have

$$\mathbf{R} = \begin{bmatrix} (R_1 + R_4) & -R_4 & 0 \\ -R_4 & (R_2 + R_4 + R_5) & -R_5 \\ 0 & -R_5 & (R_3 + R_5) \end{bmatrix}$$

Note that **R** is a symmetric matrix, as we expected.

Exercise 4.6-1 Determine the value of the voltage measured by the voltmeter in Figure E 4.6-1.

Answer: -1 V

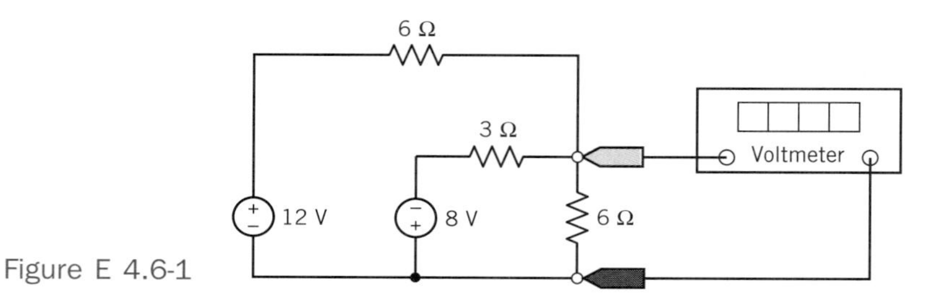

Figure E 4.6-1

4.7 MESH CURRENT ANALYSIS WITH CURRENT SOURCES

Heretofore, we have considered only circuits with independent voltage sources for analysis by the mesh current method. If the circuit has an independent current source, as shown in Figure 4.7-1, we recognize that the second mesh current is equal to the negative of the current source current. We can then write

$$i_2 = -i_s$$

and we need only determine the first mesh current i_1. Writing KVL for the first mesh, we obtain

$$(R_1 + R_2)i_1 - R_2 i_2 = v_s$$

Since $i_2 = -i_s$, we have

$$i_1 = \frac{v_s - R_2 i_s}{R_1 + R_2} \tag{4.7-1}$$

where i_s and v_s are sources of known magnitude.

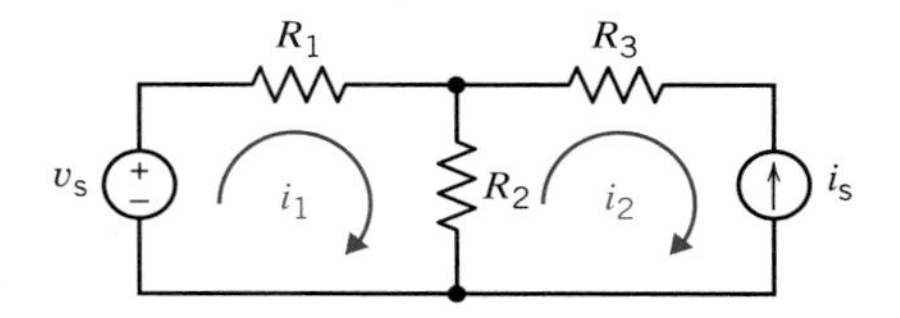

Figure 4.7-1 Circuit with an independent voltage source and an independent current source.

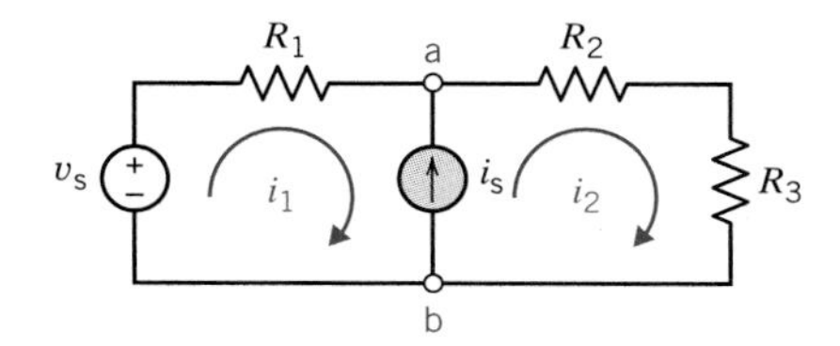

Figure 4.7-2 Circuit with an independent current source common to both meshes.

If we encounter a circuit as shown in Figure 4.7-2, we have a current source i_s that has an unknown voltage v_{ab} across its terminals. We can readily note that

$$i_2 - i_1 = i_s \tag{4.7-2}$$

by writing KCL at node a. The two mesh equations are

$$\text{mesh 1:} \quad R_1 i_1 + v_{ab} = v_s \tag{4.7-3}$$

$$\text{mesh 2:} \quad (R_2 + R_3) i_2 - v_{ab} = 0 \tag{4.7-4}$$

We note that if we add Eqs. 4.7-3 and 4.7-4 we eliminate v_{ab}, obtaining

$$R_1 i_1 + (R_2 + R_3) i_2 = v_s$$

However, since $i_2 = i_s + i_1$, we obtain

$$R_1 i_1 + (R_2 + R_3)(i_s + i_1) = v_s$$

or

$$i_1 = \frac{v_s - (R_2 + R_3)\, i_s}{R_1 + R_2 + R_3} \tag{4.7-5}$$

Thus, we account for independent current sources by recording the relationship between the mesh currents and the current source current. If the current source influences *only one* mesh current, we record that constraining equation and write the KVL equations for the remaining meshes. If the current source influences two mesh currents, we write the KVL equation for both meshes, assuming a voltage v_{ab} across the terminals of the current source. Then, adding these two mesh equations, we obtain an equation independent of v_{ab}.

Example 4.7-1

Consider the circuit of Figure 4.7-3 where $R_1 = R_2 = 1\,\Omega$ and $R_3 = 2\,\Omega$. Find the three mesh currents.

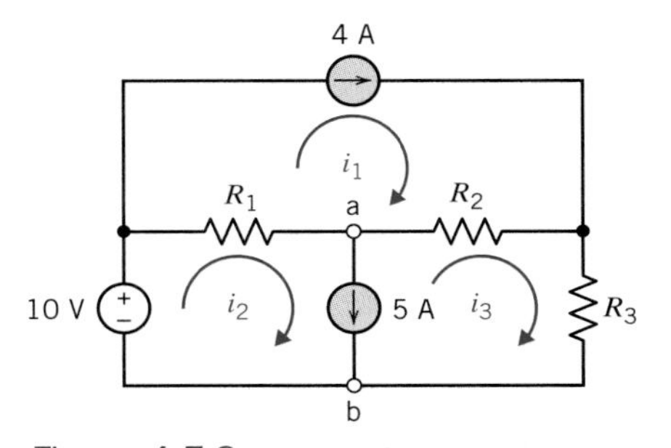

Figure 4.7-3 Circuit with two independent current sources.

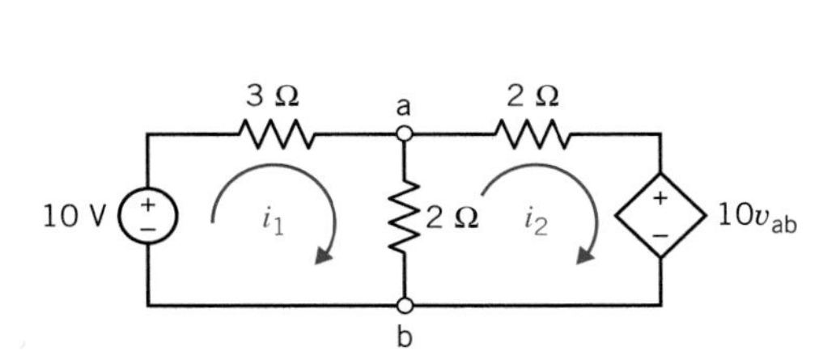

Figure 4.7-4 Circuit with one dependent voltage source.

Solution

Since the 4-A source flows only in mesh 1, we note that

$$i_1 = 4$$

For the 5-A source, we have

$$i_2 - i_3 = 5 \tag{4.7-6}$$

Writing KVL for mesh 2 and mesh 3, we obtain

$$\text{mesh 2:} \quad R_1(i_2 - i_1) + v_{ab} = 10 \tag{4.7-7}$$

$$\text{mesh 3:} \quad R_2(i_3 - i_1) + R_3 i_3 - v_{ab} = 0 \tag{4.7-8}$$

We substitute $i_1 = 4$ and add Eqs. 4.7-7 and 4.7-8 to obtain

$$R_1(i_2 - 4) + R_2(i_3 - 4) + R_3 i_3 = 10 \tag{4.7-9}$$

From Eq. 4.7-6, $i_2 = 5 + i_3$. Substituting into Eq. 4.7-9, we have

$$R_1(5 + i_3 - 4) + R_2(i_3 - 4) + R_3 i_3 = 10$$

Using the values for the resistors, we obtain

$$i_2 = \frac{33}{4}\text{ A} \quad \text{and} \quad i_3 = \frac{13}{4}\text{ A}$$

If the circuit includes dependent sources, we must add the constraining equation imposed by each dependent source. For example, the circuit shown in Figure 4.7-4 contains one dependent voltage source. Writing the KVL equations for the two meshes, we have

$$\text{mesh 1:} \quad 5i_1 - 2i_2 = 10 \tag{4.7-10}$$

$$\text{mesh 2:} \quad -2i_1 + 4i_2 = -10v_{ab} \tag{4.7-11}$$

However, $v_{ab} = 2(i_1 - i_2)$. Therefore, we obtain for mesh 2 (Eq. 4.7-11)

$$-2\,i_1 + 4\,i_2 = -20\,(i_1 - i_2)$$

or

$$18\,i_1 - 16\,i_2 = 0 \tag{4.7-12}$$

Subtracting 8 times Eq. 4.7-10 from Eq. 4.7-12, we obtain $22\,i_1 = 80$. Therefore, we have

$$i_1 = \frac{80}{22}\text{ A} \quad \text{and} \quad i_2 = \frac{90}{22}\text{ A}$$

A more general technique for the mesh analysis method when a current source is common to two meshes involves the concept of a supermesh. A *supermesh* is one mesh created from two meshes that have a current source in common, as shown in Figure 4.7-5. We then reduce the number of meshes by one when we have a common current source between two meshes. This current source is an element common to the two meshes and thus reduces the number of independent mesh equations by one.

A **supermesh** is one larger mesh created from two meshes that have an independent or dependent current source in common.

For example, consider the circuit of Figure 4.7-5. The 5-A current source is common to mesh 1 and mesh 2. The supermesh consists of the interior of mesh 1 and mesh 2.

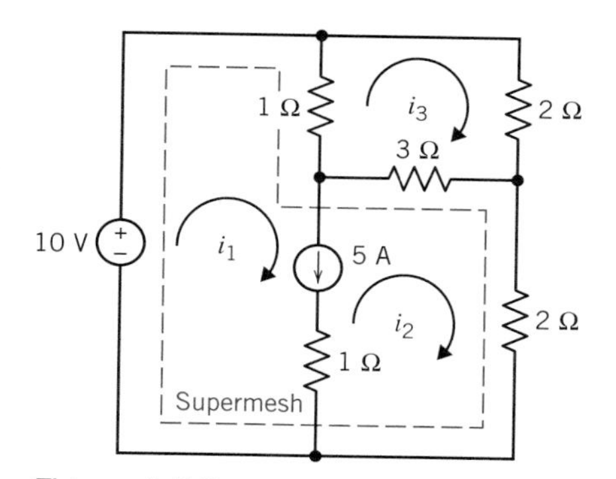

Figure 4.7-5 Circuit with a supermesh that incorporates mesh 1 and mesh 2. The supermesh is indicated by the dashed line.

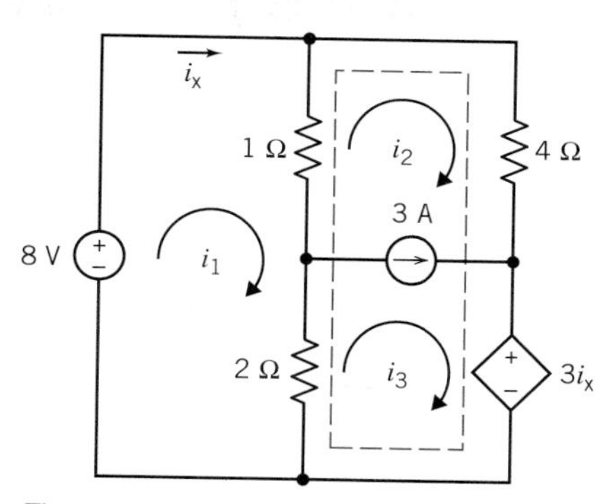

Figure 4.7-6
Circuit with a supermesh and a dependent voltage source. The supermesh is identified by the heavy dashed line.

Writing KVL around the periphery of the supermesh shown by the heavy dashed lines, we obtain

$$-10 + 1(i_1 - i_3) + 3\,(i_2 - i_3) + 2\,i_2 = 0$$

For mesh 3, we have

$$1(i_3 - i_1) + 2\,i_3 + 3\,(i_3 - i_2) = 0$$

Finally, the constraint equation required by the current source common to meshes 1 and 2 is

$$i_1 - i_2 = 5$$

Then the three equations may be reduced to

$$\begin{aligned} &\text{supermesh:} && 1i_1 + 5\,i_2 - 4\,i_3 = 10 \\ &\text{mesh 3:} && -1i_1 - 3\,i_2 + 6\,i_3 = 0 \\ &\text{current source:} && 1i_1 - 1i_2 \qquad = 5 \end{aligned}$$

Therefore, solving the three equations simultaneously, we find that $i_2 = 2.5$ A, $i_1 = 7.5$ A, and $i_3 = 2.5$ A.

The methods of mesh current analysis utilized when a current source is present are summarized in Table 4.7-1.

Table 4.7-1 **Mesh Current Analysis Methods with a Current Source**

Case	Method
1. A current source appears on the periphery of only one mesh, n.	Equate the mesh current i_n to the current source current, accounting for the direction of the current source.
2. A current source is common to two meshes	A. Assume a voltage v_{ab} across the terminals of the current source, write the KVL equations for the two meshes, and add them to eliminate v_{ab} or B. Create a supermesh as the periphery of the two meshes and write one KVL equation around the periphery of the supermesh. In addition, write the constraining equation for the two mesh currents in terms of the current source.

As a final example, let us consider the circuit shown in Figure 4.7-6. This circuit includes a current source common to two meshes and a dependent voltage source. We select a supermesh since mesh 2 and mesh 3 have a current source in common. We obtain a KVL equation for mesh 1 and the supermesh as follows:

$$\text{mesh 1:} \quad 3\,i_1 - i_2 - 2\,i_3 = 8 \tag{4.7-13}$$

$$\text{supermesh:} \quad -3\,i_1 + 5\,i_2 + 2\,i_3 + 3\,i_x = 0 \tag{4.7-14}$$

We note that

$$i_x = i_1$$

Also, the constraint equation for the current source is

$$i_3 - i_2 = 3$$

Substituting $i_x = i_1$ and $i_3 = 3 + i_2$ into Eqs. 4.7-13 and 4.7-14, we obtain

$$3\,i_1 - i_2 - 2\,(3 + i_2) = 8$$

and

$$-3\,i_1 + 5\,i_2 + 2\,(3 + i_2) + 3\,i_1 = 0$$

Then, we rearrange these equations obtaining

$$3\,i_1 - 3\,i_2 = 14 \tag{4.7-15}$$

and

$$7\,i_2 = -6 \tag{4.7-16}$$

From Eq. 4.7-16, we have

$$i_2 = \frac{-6}{7}\text{ A}$$

Then, from Eq. 4.7-15, we obtain

$$i_1 = \frac{80}{21}\text{ A}$$

Exercise 4.7-1

Determine the value of the voltage measured by the voltmeter in Figure E 4.7-1.

Hint: Write and solve a single mesh equation to determine the current in the 3Ω resistor.

Answer: −4 V

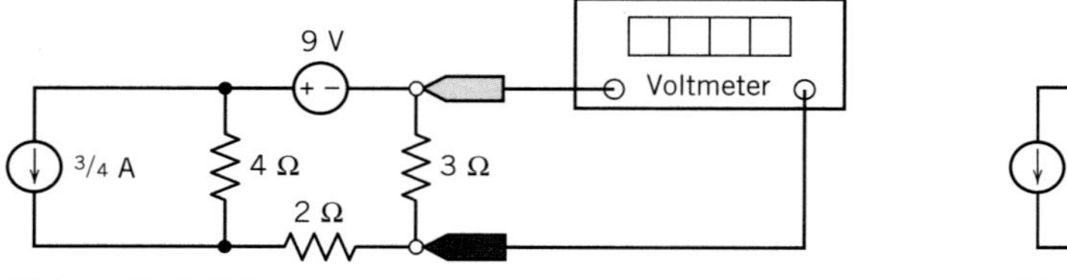

Figure E 4.7-1

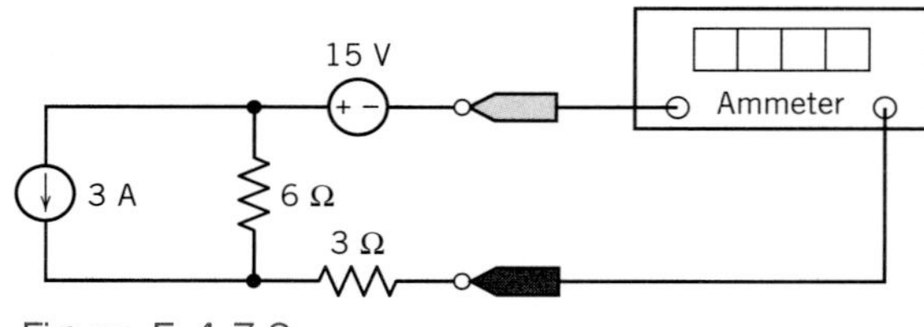

Figure E 4.7-2

Exercise 4.7-2

Determine the value of the current measured by the ammeter in Figure E 4.7-2.

Hint: Write and solve a single mesh equation.

Answer: -3.67 A

Exercise 4.7-3

Determine the value of the voltage measured by the voltmeter in Figure E 4.7-3.

Hint: Apply KVL to a supermesh to determine the current in the 2Ω resistor.

Answer: $4/3$ V

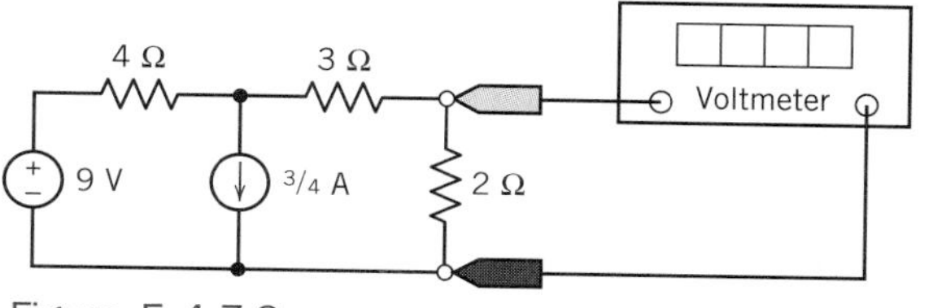

Figure E 4.7-3

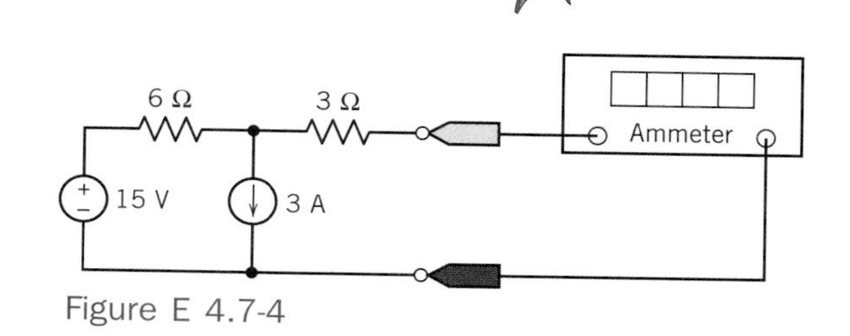

Figure E 4.7-4

Exercise 4.7-4

Determine the value of the current measured by the ammeter in Figure E 4.7-4.

Hint: Apply KVL to a supermesh.

Answer: -0.333 A

Exercise 4.7-5

Modern household electric appliances are commonplace today. However, imagine the pleasure of those who first purchased a vacuum cleaner, as shown in Figure E 4.7-5*a*. During the first half of the twentieth century, the average American household was transformed by the introduction of electric appliances. The circuit model of the cleaner and its power source is shown in Figure E 4.7-5*b*. Find the voltage source v_s required to deliver 150 W to the motor connected between terminals a and b.

Answer: 60 V

(*a*)

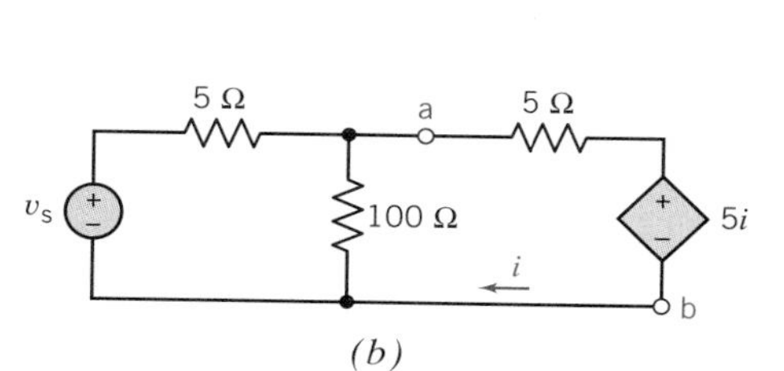

(*b*)

Figure E 4.7-5 (*a*) The modern convenience of the vacuum cleaner is serenely demonstrated in about 1910. Courtesy of Brown Brothers. (*b*) Circuit model of the cleaner and its power source.

Exercise 4.7-6

Determine the current i in the circuit shown in Figure E 4.7-6.

Answer: $i = 3\text{ A}$

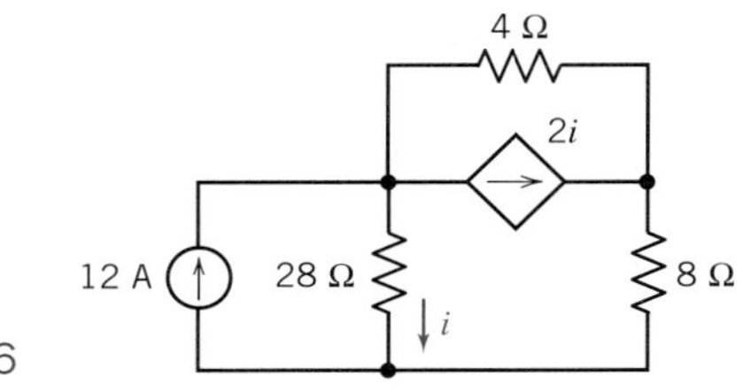

Figure E 4.7-6

4.8 THE NODE VOLTAGE METHOD AND MESH CURRENT METHOD COMPARED

The analysis of a complex circuit can usually be accomplished by either the node voltage or the mesh current method. The advantage of using these methods is the systematic procedures provided for obtaining the simultaneous equations.

In some cases one method is clearly preferred over another. For example, when the circuit contains only voltage sources, it is probably easier to use the mesh current method. When the circuit contains only current sources, it will usually be easier to use the node voltage method.

If a circuit has both current sources and voltage sources, it can be analyzed by either method. One approach is to compare the number of equations required for each method. If the circuit has fewer nodes than meshes, it may be wise to select the node voltage method. If the circuit has fewer meshes than nodes, it may be easier to use the mesh current method.

Another point to consider when choosing between the two methods is what information is required. If you need to know several currents, it may be wise to proceed directly with mesh current analysis. Remember, mesh current analysis only works for planar networks.

It is often helpful to determine which method is more appropriate for the problem requirements and to consider both methods.

Example 4.8-1

Determine the best analysis method for the circuits of Figure 4.8-1 when it is required to determine

(a) The voltage v_{ab} in Figure 4.8-1*a*.
(b) The current through resistor R_2 in Figure 4.8-1*b*.
(c) The current i in Figure 4.8-1*c*.

Solution

(a) The circuit of Figure 4.8-1*a* is most suitable for the node voltage method. Since $v_a = v_s$, we need only write KCL at node b.
(b) The circuit of Figure 4.8-1*b* is most suitable for mesh current analysis. Since the current source defines the right-hand mesh current, we need only write the KVL equation for the left-hand mesh.

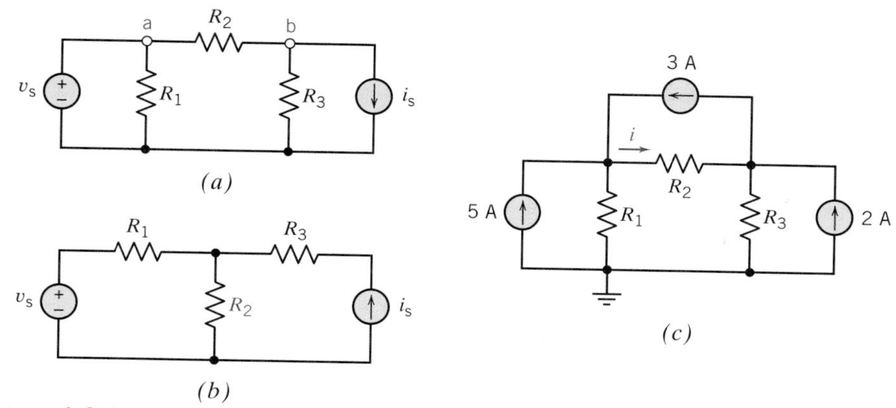

Figure 4.8-1 The circuits of Example 4.8-1.

(c) The circuit has two nodes in addition to the ground node, so two node equations would be required. The circuit has four meshes; however, the currents in three meshes are defined by the three current sources. Therefore, only one mesh current, that of the mesh including R_1, R_2, and R_3, is unknown. It would be easiest, therefore, to find i by using the mesh current method.

Exercise 4.8-1

Determine the best analysis method for the circuits of Figure E 4.8-1*a* and E 4.8-1*b*. For circuit (*a*), it is required to find the voltage v_a. For circuit (*b*), we wish to find the current i.

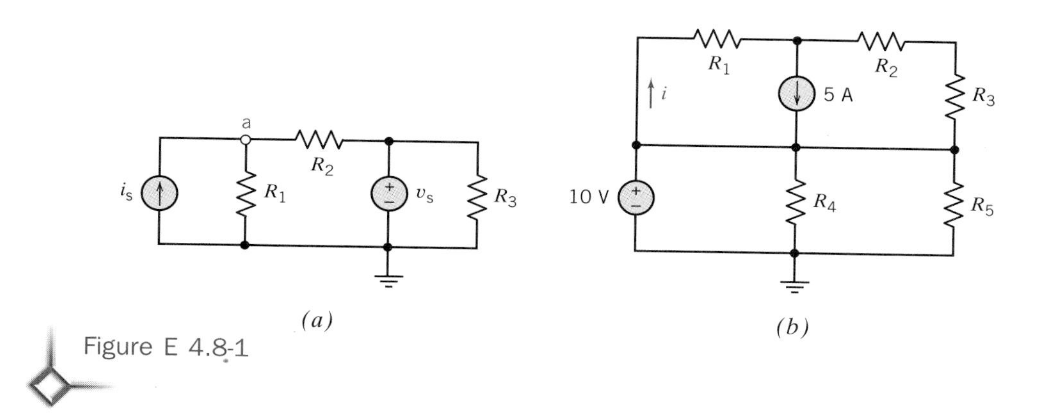

Figure E 4.8-1

4.9 DC ANALYSIS USING PSPICE[1]

The calculation of currents and voltages in a complicated circuit can be facilitated by using PSpice. In this section we will illustrate the utility of PSpice while also using algebraic methods for insight.

[1]This section may be omitted or considered with a later chapter. See Appendix F for an introduction to PSpice.

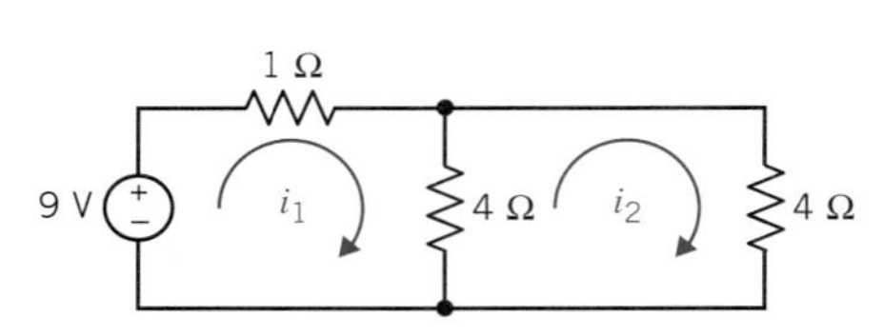

Figure 4.9-1 A simple two-mesh circuit.

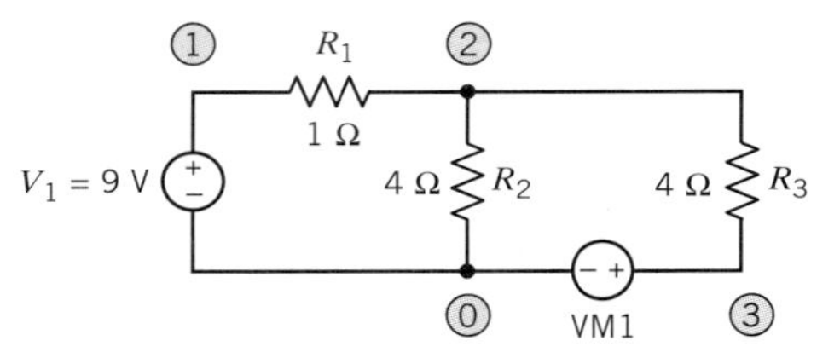

Figure 4.9-2 The circuit of Figure 4.9-1 redrawn with node numbers indicated. The reference node is denoted as zero.

First, let us consider a simple two-mesh circuit as shown in Figure 4.9-1 and determine i_1 and i_2. The two mesh equations are

$$\text{mesh 1:}\quad 5i_1 - 4i_2 = 9 \tag{4.9-1}$$

$$\text{mesh 2:}\quad -4i_1 + 8i_2 = 0 \tag{4.9-2}$$

Multiplying Eq. 4.9-2 by one-half and adding it to Eq. 4.9-1, we obtain

$$3i_1 = 9$$

or

$$i_1 = 3\text{ A}$$

Then, using Eq. 4.9-2, we find that $i_2 = 1.5$ A. This simple circuit is easily analyzed using algebraic methods. We now proceed to use PSpice to determine i_1 and i_2.

First, we redraw the circuit numbering the nodes with the reference node chosen as zero, as shown in Figure 4.9-2. The dummy source VM1 (VM1 = 0) is inserted to measure the current i_2. Recall that the source will measure the current with the convention of current flowing from + to − through the source. The PSpice program is provided in Figure 4.9-3. Note that the title statement is the first line. Also note that the second line beginning with an asterisk is a comment line and allows us to set up column headings. The output is provided in Figure 4.9-4. It is verified that $i_1 = 3$ A and $i_2 = 1.5$ A. Note that the printout gives a current through the source of −3 A because PSpice convention has the current flow from + to − through the source. The printout also provides the voltage at each numbered node with reference to the ground (0) node.

In this section we are using the single-point dc analysis, which is called the "small signal bias solution," in order to calculate the dc currents and node voltages.

Clearly, the utility of PSpice increases as the circuit becomes more complex. Consider the circuit shown in Figure 4.9-5, where we wish to determine i_1 and i_3. The mesh equations are

$$\text{mesh 1:}\quad 11i_1 - 8i_2 = 42$$

$$\text{mesh 2:}\quad -8i_1 + 18i_2 - 6i_3 = 0$$

$$\text{mesh 3:}\quad -6i_2 + 18i_3 = 0$$

Rewriting these equations in matrix form, we obtain

$$\mathbf{Ri} = \mathbf{v}_s \tag{4.9-3}$$

```
*** TWO-MESH ANALYSIS (Title Statement)

* ELEM    NODE    NODE    VALUE
  R1      1       2       1
  R2      2       0       4
  R3      2       3       4
  V1      1       0       DC      9
  VM1     3       0       DC      0
  .END
```

Figure 4.9-3 PSpice program for the circuit of Figure 4.9-2.

```
*TWO-MESH ANALYSIS

****    SMALL SIGNAL BIAS SOLUTION     TEMPERATURE=   27.000 DEG C

NODE  VOLTAGE  NODE  VOLTAGE   NODE   VOLTAGE  NODE   VOLTAGE
( 1)    9.0000  ( 2)    6.0000  ( 3)     0.0000

VOLTAGE  SOURCE  CURRENTS
  NAME            CURRENT
  V1           -3.000E+00
  VM1           1.500E+00
  TOTAL POWER DISSIPATION  2.70E+01  WATTS

 JOB CONCLUDED
 TOTAL JOB TIME            .50
```

Figure 4.9-4 Output of PSpice calculation.

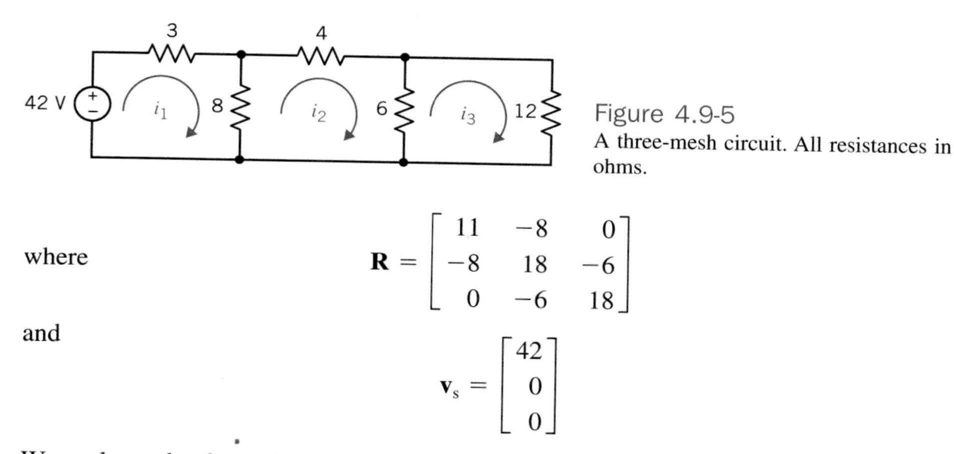

Figure 4.9-5 A three-mesh circuit. All resistances in ohms.

where

$$\mathbf{R} = \begin{bmatrix} 11 & -8 & 0 \\ -8 & 18 & -6 \\ 0 & -6 & 18 \end{bmatrix}$$

and

$$\mathbf{v}_s = \begin{bmatrix} 42 \\ 0 \\ 0 \end{bmatrix}$$

We evaluate the determinant of **R** as $\Delta = 2016$. Then, using Cramer's rule, we find that $i_1 = 6$ A and $i_3 = 1$ A. This relatively simple circuit would require more complicated calculations as controlled, and other independent sources are added so that all the elements of **R** and $\mathbf{v}_s$ become nonzero. This circuit with three meshes is about as far as we can go without introducing calculation tedium. A four-mesh circuit with many sources would be difficult to solve without causing calculation errors. Thus, PSpice is a real aid when the circuit becomes complicated.

The three-mesh circuit of Figure 4.9-5 is redrawn for PSpice analysis as shown in Figure 4.9-6. The PSpice program is provided in Figure 4.9-7. Note that it is verified that $i_1 = 6$ A and $i_3 = 1$ A as shown in the output of Figure 4.9-8.

At the end of Chapter 5 we will demonstrate the analysis of circuits with dependent sources and the use of the .DC command.

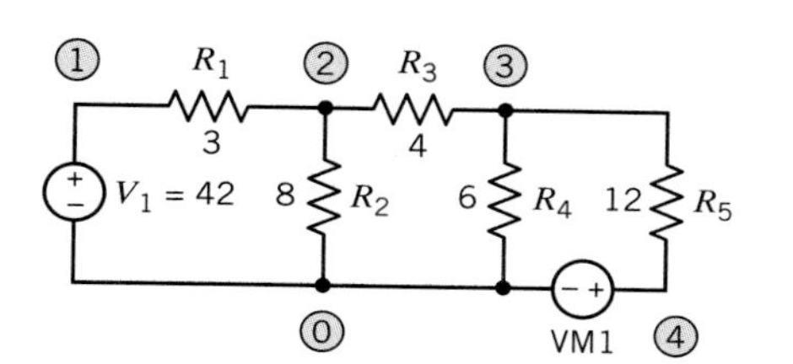

Figure 4.9-6 The three-mesh circuit redrawn for PSpice analysis. All resistances in ohms.

```
*** THREE-MESH ANALYSIS

* ELEM   NODE   NODE   VALUE
R1        1      2       3
R2        2      0       8
R3        2      3       4
R4        3      0       6
R5        3      4      12
V1        1      0      DC     42
VM1       4      0      DC      0
.END
```

Figure 4.9-7 The PSpice program for the three-mesh circuit.

```
 *THREE-MESH ANALYSIS

****   SMALL SIGNAL BIAS SOLUTION   TEMPERATURE=   27.000 DEG C

NODE VOLTAGE NODE VOLTAGE   NODE   VOLTAGE NODE   VOLTAGE
( 1)   42.0000  ( 2)   24.0000 ( 3)   12.0000  ( 4)    0.0000

VOLTAGE SOURCE CURRENTS
  NAME          CURRENT
  V1          -6.000E+00
  VM1           1.000E+00
  TOTAL POWER DISSIPATION 2.52E+02 WATTS
```

Figure 4.9-8 The output of the PSpice program for the three-mesh circuit.

4.10 DC ANALYSIS USING MATLAB

We have seen that circuits that contain resistors and independent or dependent sources can be analyzed by

1. Writing a set of node or mesh equations.
2. Solving those equations.

In this section, we will use the computer program MATLAB to solve the equations.

Consider the circuit shown in Figure 4.10-1*a*. This circuit contains a potentiometer. In Figure 4.10-1*b*, the potentiometer has been replaced by a model of a potentiometer. R_p is the resistance of the potentiometer. The parameter a varies from 0 to 1 as the wiper of the potentiometer is moved from one end of the potentiometer to the other. The resistances R_4 and R_5 are described by the equations

$$R_4 = aR_p \tag{4.10-1}$$

and

$$R_5 = (1 - a)R_p \tag{4.10-2}$$

Our objective is to analyze this circuit to determine how the output voltage changes as the position of the potentiometer wiper is changed.

The circuit in Figure 4.10-1*b* can be represented by mesh equations as

$$R_1 i_1 + R_4 i_1 + R_3(i_1 - i_2) - v_1 = 0$$

$$R_5 i_2 + R_2 i_2 + [v_2 - R_3(i_1 - i_2)] = 0 \tag{4.10-3}$$

These mesh equations can be rearranged as

$$(R_1 + R_4 + R_3)i_1 - R_3 i_2 = v_1$$

$$-R_3 i_1 + (R_5 + R_2 + R_3)i_2 = -v_2 \tag{4.10-4}$$

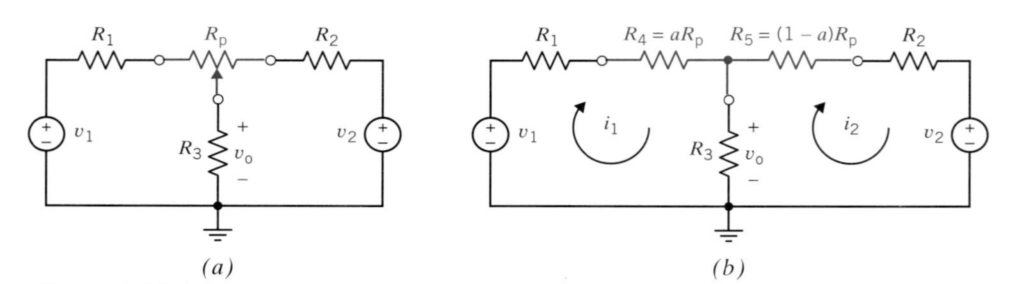

Figure 4.10-1 (*a*) A circuit that contains a potentiometer and (*b*) an equivalent circuit formed by replacing the potentiometer by a model of a potentiometer ($0 \leq a \leq 1$).

Substituting Eqs. 4.10-1 and 4.10-2 into Eq. 4.10-4 gives

$$(R_1 + aR_p + R_3)i_1 - R_3 i_2 = v_1$$
$$-R_3 i_1 + [(1-a)R_p + R_2 + R_3]i_2 = -v_2 \quad (4.10\text{-}5)$$

Equation 4.10-5 can be written using matrices as

$$\begin{bmatrix} R_1 + aR_p + R_3 & -R_3 \\ -R_3 & (1-a)R_p + R_2 + R_3 \end{bmatrix} \begin{bmatrix} i_1 \\ i_2 \end{bmatrix} = \begin{bmatrix} v_1 \\ -v_2 \end{bmatrix} \quad (4.10\text{-}6)$$

Next, i_1 and i_2 are calculated by using MATLAB to solve the mesh equation, Eq. 4.10-6. Then the output voltage is calculated as

$$v_o = R_3(i_1 - i_2) \quad (4.10\text{-}7)$$

Figure 4.10-2 shows the MATLAB input file. The parameter a varies from 0 to 1 in increments of 0.05. At each value of a, MATLAB solves Eq. 4.10-6 and then uses

```
% mesh.m solves mesh equations
%--------------------------------------------------------------
%  Enter values of the parameters that describe the circuit.
%--------------------------------------------------------------
                     % circuit parameters
R1=1000;             % ohms
R2=1000;             % ohms
R3=5000;             % ohms
V1=  15;             % volts
V2=-15;              % volts

                     % potentiometer parameters
Rp=20e3;             % ohms
%--------------------------------------------------------------
%   the parameter a varies from 0 to 1 in 0.05 increments.
%--------------------------------------------------------------

a=0:0.05:1;          % dimensionless

for k=1:length(a)
    %----------------------------------------------------------
    % Here is the mesh equation, RV=I:
    %----------------------------------------------------------

    R = [R1+a(k)*Rp+R3        -R3;                       % ------
              -R3      (1-a(k))*Rp+R2+R3];                % eqn.
    V = [ V1;                                             % 4.10-6
         -V2];                                            % ------

    %----------------------------------------------------------
    % Tell MATLAB to solve the mesh equation:
    %----------------------------------------------------------
    I=V'/R;

    %----------------------------------------------------------
    % Calculate the output voltage from the mesh currents.
    %----------------------------------------------------------

    Vo(k)=R3*(I(1)-I(2)); % eqn. 4.10-7

end
%--------------------------------------------------------------
%   Plot Vo versus a
%--------------------------------------------------------------
plot(a, Vo)
axis([0 1 -15 15])
xlabel('a, dimensionless')
ylabel('Vo, V')
```

Figure 4.10-2 MATLAB input file used to analyze the circuit shown in Figure 4.10-1.

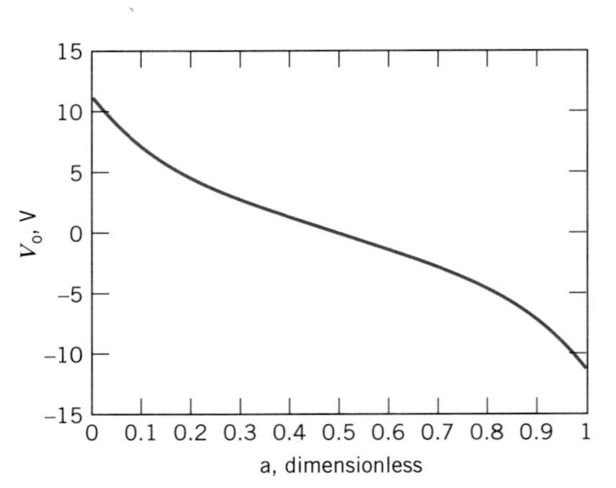

Figure 4.10-3
Plot of v_o versus a for the circuit shown in Figure 4.10-1.

Eq. 4.10-7 to calculate the output voltage. Finally, MATLAB produces the plot of v_o versus a that is shown in Figure 4.10-3.

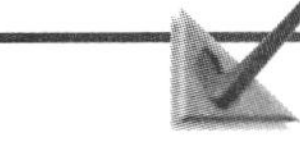

4.11 VERIFICATION EXAMPLES

Problem

The circuit shown in Figure 4.11-1*a* was analyzed using PSpice. The PSpice output file, Figure 4.11-1*b,* includes the node voltages of the circuit. Are these node voltages correct?

Solution

The node equation corresponding to node 2 is

$$\frac{V(2) - V(1)}{100} + \frac{V(2)}{200} + \frac{V(2) - V(3)}{100} = 0$$

where, for example, $V(2)$ is the node voltage at node 2. When the node voltages from Figure 4.11-1*b* are substituted into the left-hand side of this equation, the result is

$$\frac{7.2727 - 12}{100} + \frac{7.2727}{200} + \frac{7.2727 - 5.0909}{100} = 0.011$$

The right-hand side of this equation should be 0 instead of 0.011. It looks like something is wrong. Is a current of only 0.011 negligible? Probably not in this case. If the node voltages were correct, then the currents of the 100 Ω resistors would be 0.047 A and 0.022 A, respectively. The current of 0.011 A does not seem negligible when compared to currents of 0.047 A and 0.022 A.

Is it possible that PSpice would calculate the node voltages incorrectly? Probably not, but the PSpice input file could easily contain errors. In this case, the value of the resistance connected between nodes 2 and 3 has been mistakenly specified to be 200 Ω. After changing this resistance to 100 Ω, PSpice calculates the node voltages to be

$$V(1) = 12.0,\ V(2) = 7.0,\ V(3) = 5.5,\ V(4) = 8.0$$

Substituting these voltages into the node equation gives

$$\frac{7.0 - 12.0}{100} + \frac{7.0}{200} + \frac{7.0 - 5.5}{100} = 0.0$$

so these node voltages do satisfy the node equation corresponding to node 2.

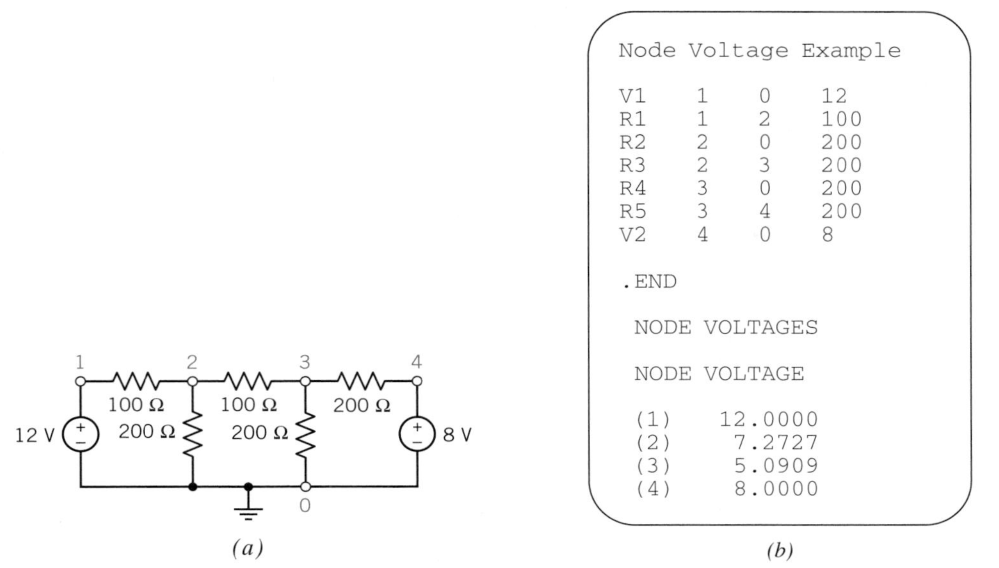


```
Node Voltage Example

V1    1    0    12
R1    1    2    100
R2    2    0    200
R3    2    3    200
R4    3    0    200
R5    3    4    200
V2    4    0    8

.END

 NODE VOLTAGES

 NODE VOLTAGE

 (1)     12.0000
 (2)      7.2727
 (3)      5.0909
 (4)      8.0000
```

(b)

Figure 4.11-1 *(a)* A circuit and *(b)* the node voltages calculated using PSpice. The bottom node has been chosen as the reference node, which is indicated by the ground symbol and the node number 0. The voltages and resistors have units of voltages and ohms, respectively.

Problem

The circuit shown in Figure 4.11-2*a* was analyzed using PSpice. The PSpice output file, Figure 4.11-2*b*, includes the mesh currents of the circuit. Are these mesh currents correct?

The PSpice output file will include the currents through the voltage sources. Recall that PSpice uses the passive convention so that the current in the 8-V source will be $-i_1$ instead of i_1. The two 0-V sources have been added to include mesh currents i_2 and i_3 in the PSpice output file.

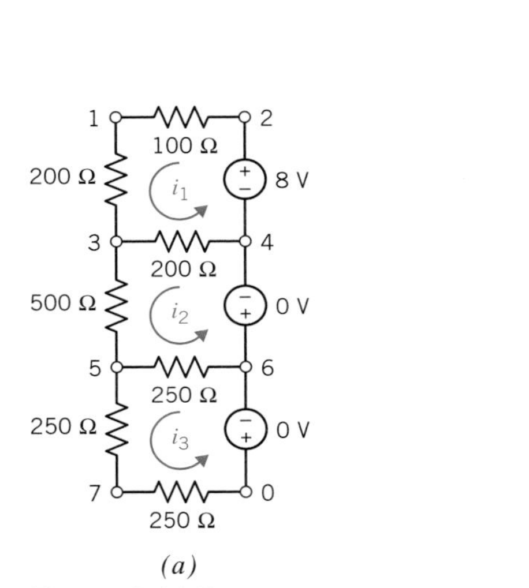

(a)

```
Mesh Current Example

R1    1    2    100
R2    1    3    200
V1    2    4    8
R3    3    4    200
R5    3    5    500
V2    4    6    0
R6    5    6    250
R7    5    7    250
V3    6    0    0
R8    7    0    250

.END

 MESH CURRENTS

 NAME    CURRENT

 I1     1.763E-02
 I2    -4.068E-03
 I3    -1.356E-03
```

(b)

Figure 4.11-2 *(a)* A circuit and *(b)* the mesh currents calculated using PSpice. The voltages and resistances are given in volts and ohms, respectively.

Solution

The mesh equation corresponding to mesh 2 is

$$200(i_2 - i_1) + 500i_2 + 250(i_2 - i_3) = 0$$

When the mesh currents from Figure 4.11-2*b* are substituted into the left-hand side of this equation, the result is

$$200(-0.004068 - 0.01763) + 500(-0.004068) + 250(-0.004068 - (-0.001356)) = 1.629$$

The right-hand side of this equation should be 0 instead of 1.629. It looks like something is wrong. Most likely, the PSpice input file contains an error. This is indeed the case. The nodes of both 0-V voltage sources have been entered in the wrong order. Recall that the first node should be the positive node of the voltage source. After correcting this error, PSpice gives

$$i_1 = 0.01763, \quad i_2 = 0.004068, \quad i_3 = 0.001356$$

Using these values in the mesh equation gives

$$200(0.004068 - 0.01763) + 500(0.004068) + 250(0.004068 - 0.001356) = 0.0$$

These mesh currents do indeed satisfy the mesh equation corresponding to mesh 2.

4.12 Design Challenge Solution

POTENTIOMETER ANGLE DISPLAY

A circuit is needed to measure and display the angular position of a potentiometer shaft. The angular position, θ, will vary from -180 degrees to 180 degrees.

Figure 4.12-1 illustrates a circuit that could do the job. The $+15$-V and -15-V power supplies, the potentiometer, and resistors R_1 and R_2 are used to obtain a voltage, v_i, that is proportional to θ. The amplifier is used to change the constant of proportionality to obtain a simple relationship between θ and the voltage, v_o, displayed by the voltmeter. In this example, the amplifier will be used to obtain the relationship

$$v_o = k \cdot \theta \quad \text{where} \quad k = 0.1 \frac{\text{volt}}{\text{degree}} \tag{4.12-1}$$

so that θ can be determined by multiplying the meter reading by 10. For example, a meter reading of -7.32 V indicates that $\theta = -73.2$ degrees.

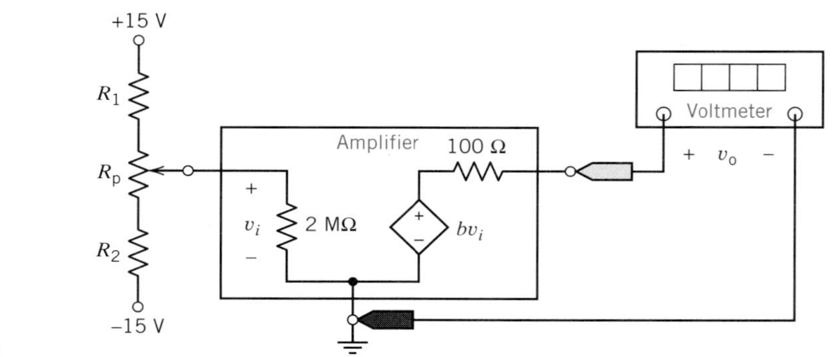

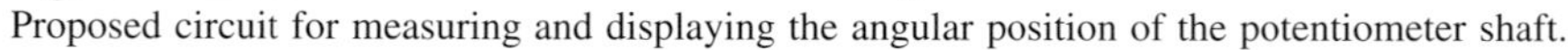

Figure 4.12-1
Proposed circuit for measuring and displaying the angular position of the potentiometer shaft.

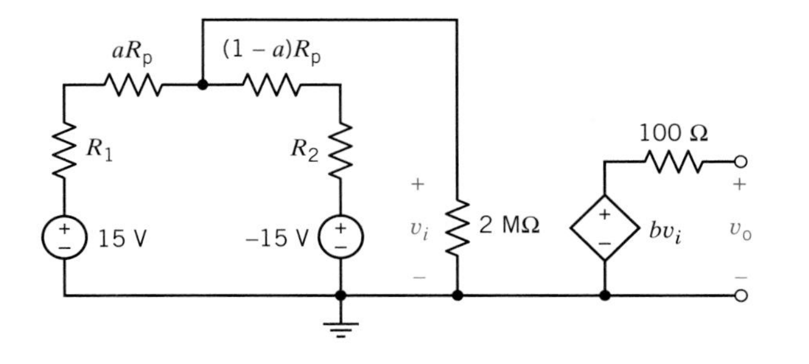

Figure 4.12-2
Circuit diagram containing models of the power supplies, voltmeter, and potentiometer.

Describe the Situation and the Assumptions

The circuit diagram in Figure 4.12-2 is obtained by modeling the power supplies as ideal voltage sources, the voltmeter as an open circuit, and the potentiometer by two resistors. The parameter, a, in the model of the potentiometer varies from 0 to 1 as θ varies from -180 degrees to 180 degrees. That means

$$a = \frac{\theta}{360^\circ} + \frac{1}{2} \tag{4.12-2}$$

Solving for θ gives

$$\theta = \left(a - \frac{1}{2}\right) \cdot 360^\circ \tag{4.12-3}$$

State the Goal

Specify values of resistors R_1 and R_2, the potentiometer resistance R_p, and the amplifier gain b that will cause the meter voltage, v_o, to be related to the angle θ by Eq. 4.12-1.

Generate a Plan

Analyze the circuit shown in Figure 4.12-2 to determine the relationship between v_i and θ. Select values of R_1, R_2, and R_p. Use these values to simplify the relationship between v_i and θ. If possible, calculate the value of b that will cause the meter voltage, v_o, to be related to the angle θ by Eq. 4.12-1. If this isn't possible, adjust the values of R_1, R_2, and R_p and try again.

Act on the Plan

The circuit has been redrawn in Figure 4.12-3. A single-node equation will provide the relationship between v_i and θ:

$$\frac{v_i}{2\ \text{M}\Omega} + \frac{v_i - 15}{R_1 + a\,R_p} + \frac{v_i - (-15)}{R_2 + (1 - a)\,R_p} = 0$$

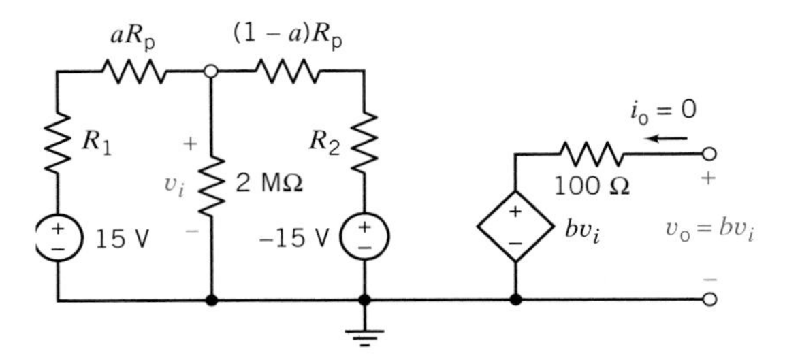
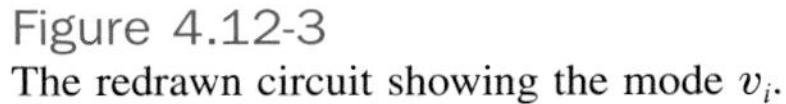

Figure 4.12-3
The redrawn circuit showing the mode v_i.

Solving for v_i gives

$$v_i = \frac{2\,\text{M}\Omega\,(R_P\,(2a - 1) + R_1 - R_2)15}{(R_1 + a\,R_P)(R_2 + (1 - a)\,R_P) + 2\,\text{M}\Omega\,(R_1 + R_2 + R_P)} \tag{4.12-4}$$

This equation is quite complicated. Let's put some restrictions on R_1, R_2, and R_p that will make it possible to simplify this equation. First, let $R_1 = R_2 = R$. Second, require that both R and R_p be much smaller than 2 MΩ (for example, $R < 20$ kΩ). Then

$$(R + a\,R_p)(R + (1 - a)\,R_p) \ll 2\,\text{M}\Omega\,(2R + R_p)$$

That is, the first term in the denominator of the left side of Eq. 4.12-4 is negligible compared to the second term. Eq. 4.12-4 can be simplified to

$$v_i = \frac{R_p\,(2a - 1)15}{2R + R_p}$$

Next, using Eq. 4.12-4

$$v_i = \left(\frac{R_p}{2R + R_p}\right)\left(\frac{15\text{ V}}{180°}\right)\theta$$

It is time to pick values for R and R_p. Let $R = 5$ kΩ and $R_p = 10$ kΩ; then

$$v_i = \left(\frac{7.5\text{ V}}{180°}\right)\theta$$

Referring to Figure 4.12-2, the amplifier output is given by

$$v_o = bv_i \tag{4.12-5}$$

so

$$v_o = b\left(\frac{7.5\text{ V}}{180°}\right)\theta$$

Comparing this equation to Eq. 4.12-1 gives

$$b\left(\frac{7.5\text{ V}}{180°}\right) = 0.1\frac{\text{volt}}{\text{degree}}$$

or

$$b = \frac{180}{7.5}(0.1) = 2.4$$

The final circuit is shown in Figure 4.12-4.

Verify the Proposed Solution

As a check, suppose $\theta = 150°$. From Eq. 4.12-2 we see that

$$a = \frac{150°}{360°} + \frac{1}{2} = 0.9167$$

Using Eq. 4.12-4, we calculate

$$v_i = \frac{2\,\text{M}\Omega\,(10\,\text{k}\Omega(2 \times 0.9167 - 1))15}{(5\,\text{k}\Omega + 0.9167 \times 10\,\text{k}\Omega)(5\,\text{k}\Omega + (1 - 0.9167)10\,\text{k}\Omega) + 2\,\text{M}\Omega(2 \times 5\,\text{k}\Omega + 10\,\text{k}\Omega)} = 6.24$$

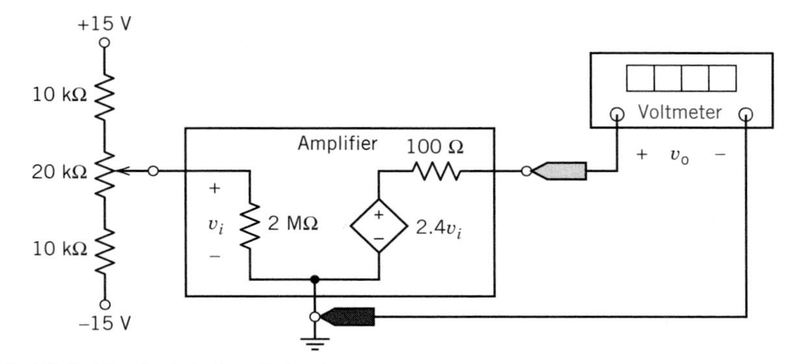

Figure 4.12-4 The final designed circuit.

Finally, Eq. 4.12-5 indicates that the meter voltage will be

$$v_o \times 2.4 \cdot 6.24 = 14.98$$

This voltage will be interpreted to mean that the angle was

$$\theta = 10 \cdot v_o = 149.8°$$

which is correct to three significant digits.

4.13 SUMMARY

- Electric circuits are widely used for communication systems such as the telegraph and telephone. These circuits are increasingly complex and require a disciplined method of analysis.
- The node voltage method of circuit analysis identifies the nodes of a circuit where two or more elements are connected. When the circuit consists of only resistors and current sources, the following procedure is used to obtain the node equations.
  1. We choose one node as to the reference node. Label the node voltages at the other nodes.
  2. Express element currents as functions of the node voltages. Figure 4.13-1*a* illustrates the relationship between the current in a resistor and the voltages at the nodes of the resistor.
  3. Apply KCL at all nodes except for the reference node.

  Solution of the simultaneous equations results in knowledge of the node voltages. All the voltages and currents in the circuit can be determined once the node voltages are known.
- When a circuit has voltage sources as well as current sources, we can still use the node voltage method by utilizing the concept of a supernode. A supernode is a "large node" that includes two nodes connected by a known voltage source. If the voltage source is directly connected between a node q and the reference node, we may set $v_q = v_s$ and write the KCL equations at the remaining nodes.
- If the circuit contains a dependent source, we first express the controlling voltage or current of the dependent source as a function of the node voltages. Next, we express the controlled voltage or current as a function of the node voltages. Finally, we apply KCL to nodes and supernodes.

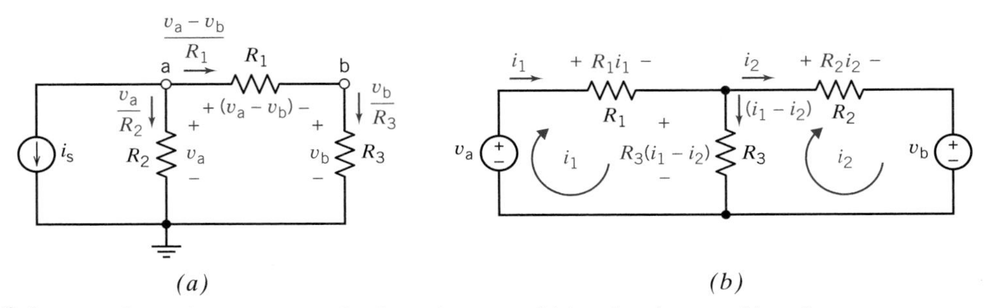

Figure 4.13-1 Expressing resistor currents and voltages in terms of *(a)* node voltage or *(b)* mesh currents.

- Mesh current analysis is accomplished by applying KVL to the meshes of a planar circuit. When the circuit consists of only resistors and voltage sources, the following procedure is used to obtain the mesh equations.
    1. Label the mesh currents.
    2. Express element voltages as functions of the mesh currents. Figure 4.13-1*b* illustrates the relationship between the voltage across a resistor and the currents of the meshes that include the resistor.
    3. Apply KVL to all meshes.

  Solution of the simultaneous equations results in knowledge of the mesh currents. All the voltages and currents in the circuit can be determined once the mesh currents are known.
- If a current source is common to two adjoining meshes, we define the interior of the two meshes as a supermesh. We then write the mesh current equation around the periphery of the supermesh. If a current source appears at the periphery of only one mesh, we may define that mesh current as equal to the current of the source, accounting for the direction of the current source.
- If the circuit contains a dependent source, we first express the controlling voltage or current of the dependent source as a function of the mesh currents. Next, we express the controlled voltage or current as a function of the mesh currents. Finally, we apply KVL to meshes and supermeshes.
- In general, either node voltage or mesh current analysis can be used to obtain the currents or voltages in a circuit. However, a circuit with fewer node equations than mesh current equations may require that we select the node voltage method. Conversely, mesh current analysis is readily applicable for a circuit with fewer mesh current equations than node voltage equations.
- MATLAB greatly reduces the drudgery of solving node or mesh equations.

PROBLEMS

Section 4.3 Node Voltage Analysis of Circuits with Current Sources

P 4.3-1 The node voltages in the circuit of Figure P 4.3-1 are $v_1 = -4$ V and $v_2 = 2$ V. Determine *i*, the current of the current source.

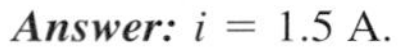

Answer: $i = 1.5$ A.

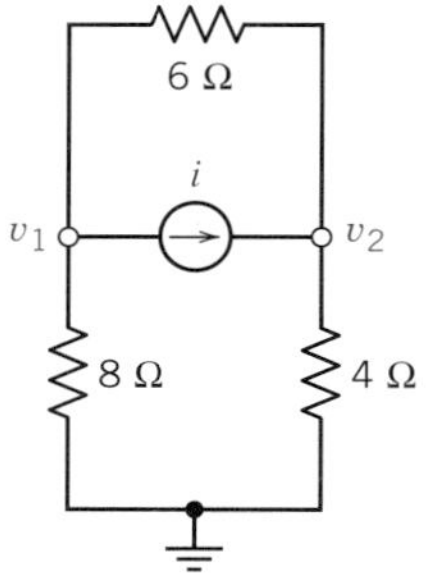

Figure P 4.3-1

P 4.3-2 Determine the node voltages for the circuit of Figure P 4.3-2.

Answer: $v_1 = 2$ V, $v_2 = 30$ V and $v_3 = 24$ V.

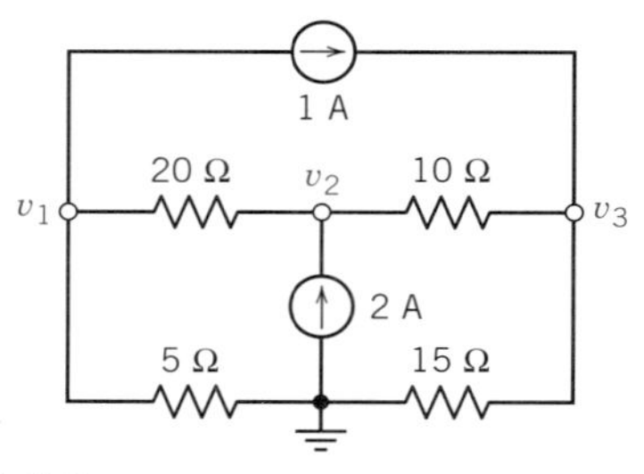

Figure P 4.3-2

P 4.3-3 The node voltages in the circuit of Figure P 4.3-3 are $v_1 = 4$ V, $v_2 = 15$ V and $v_3 = 18$ V. Determine i_1 and i_2, the currents of the current sources.

Answer: $i_1 = -2$ A and $i_2 = 2$ A.

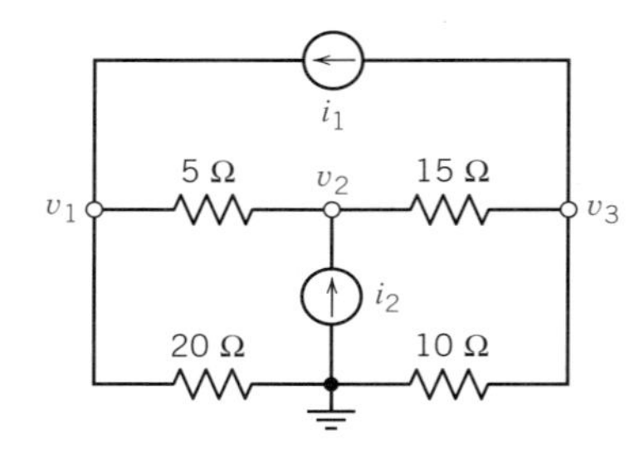

Figure P 4.3-3

P 4.3-4 Consider the circuit shown in Figure P 4.3-4. Find values of the resistances R_1 and R_2 that cause the voltages v_1 and v_2 to be $v_1 = 1$ V and $v_2 = 2$ V.

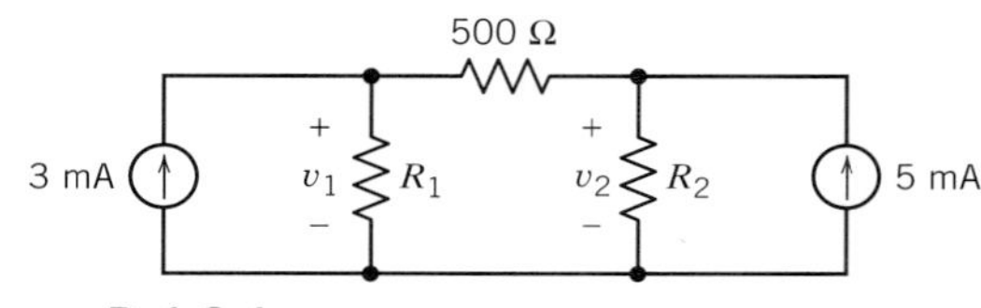

Figure P 4.3-4

P 4.3-5 Find the voltage v for the circuit shown in Figure P 4.3-5.

Answer: $v = 21.7$ mV

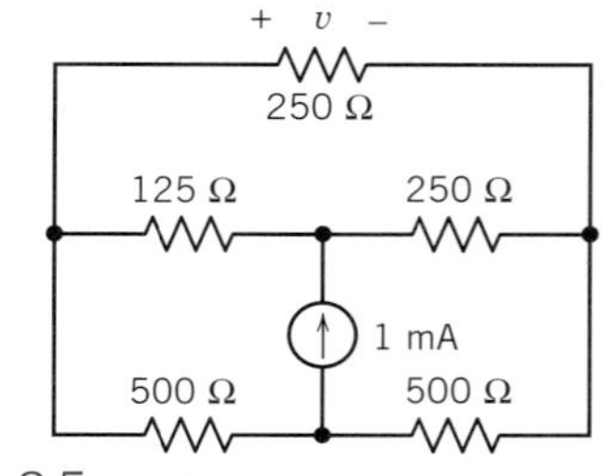

Figure P 4.3-5

Section 4.4 Node Voltage Analysis of Circuits with Current and Voltage Sources

P 4.4-1 Determine the node voltage v_a for the circuit of Figure P 4.4-1.
Answer: $v_a = 4$ V

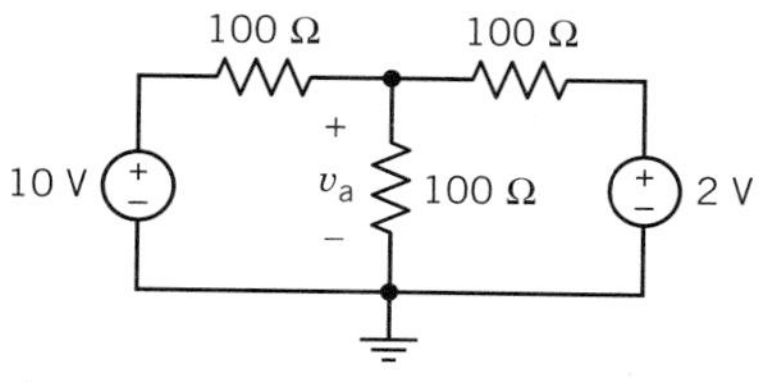

Figure P 4.4-1

P 4.4-2 Determine the node voltage v_a for the circuit of Figure P 4.4-2.
Answer: $v_a = -2$ V

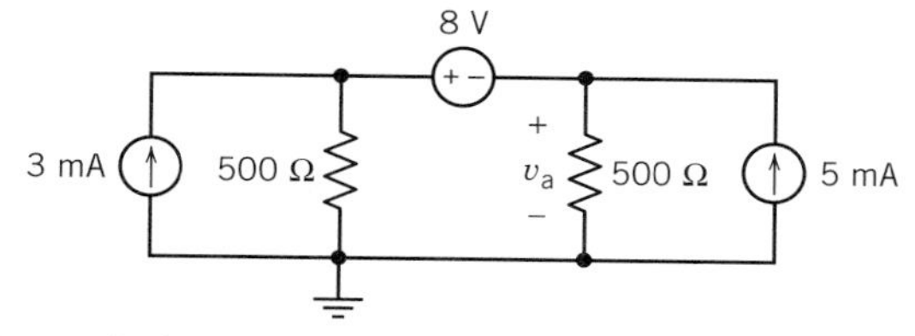

Figure P 4.4-2

P 4.4-3 Determine the node voltage v_a for the circuit of Figure P 4.4-3.
Answer: $v_a = 7$ V

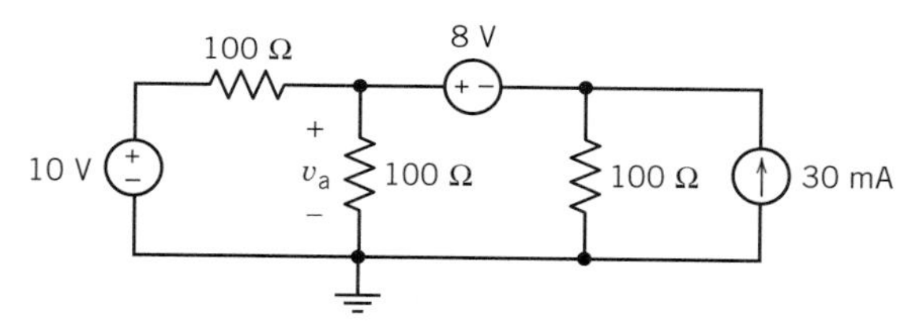

Figure P 4.4-3

P 4.4-4 Determine the node voltage v_a for the circuit of Figure P 4.4-4.
Answer: $v_a = 4$ V

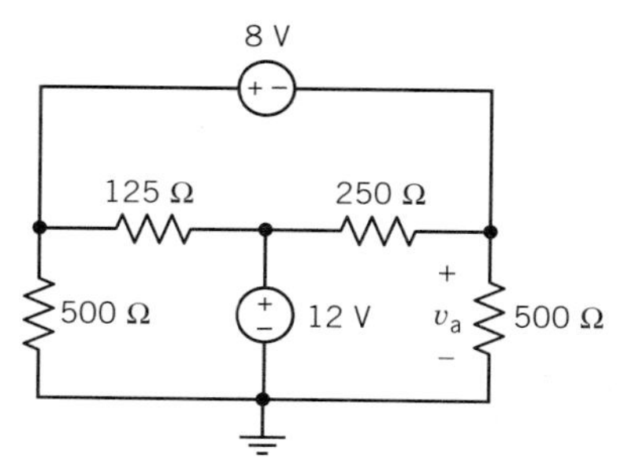

Figure P 4.4-4

P 4.4-5 Consider the circuit shown in Figure P 4.4-5. The voltage across R_3 is 6 V. Find the value of the resistance R_3.

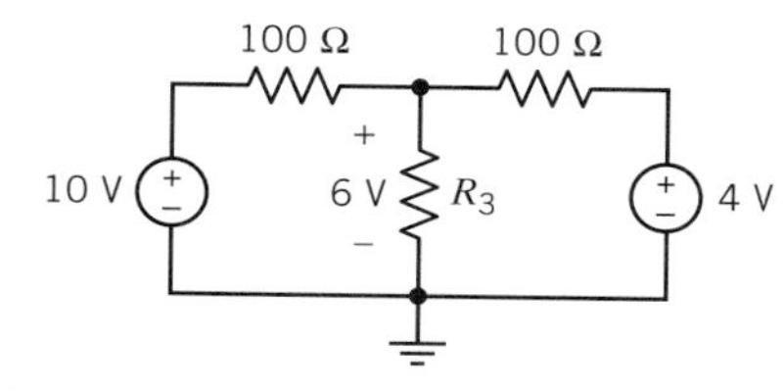

Figure P 4.4-5

P 4.4-6 An electronic instrument incorporates a 15-V power supply. A digital display is added that requires a 5-V power supply. Unfortunately, the project is over budget and you are instructed to use the existing power supply. Using a voltage divider, as shown in Figure P 4.4-6, you are able to obtain 5 V. The specification sheet for the digital display shows that the display will operate properly over a supply voltage range of 4.8 V to 5.4 V. Furthermore, the display will draw 300 mA (I) when the display is active and 100 mA when quiescent (no activity).

(a) Select values of R_1 and R_2 so that the display will be supplied with 4.8 V to 5.4 V under all conditions of current I.

(b) Calculate the maximum power dissipated by each resistor, R_1 and R_2, and the maximum current drawn from the 15-V supply.

(c) Is the use of the voltage divider a good engineering solution? If not, why? What problems might arise?

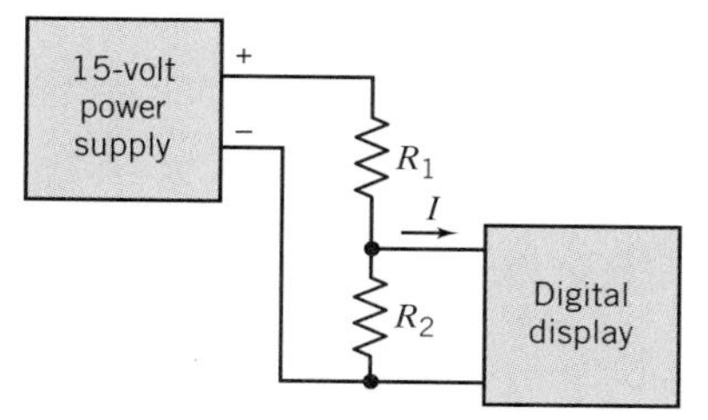

Figure P 4.4-6

P 4.4-7 The voltmeter in the circuit of Figure P 4.4-7 measures a node voltage. The value of that node voltage depends on the value of the resistance R.

(a) Determine the value of the resistance R that will cause the voltage measured by the voltmeter to be 4 V.

(b) Determine the voltage measured by the voltmeter when $R = 1.2\ \text{k}\Omega = 1200\ \Omega$.

Answers: (a) 6 kΩ (b) 2 V

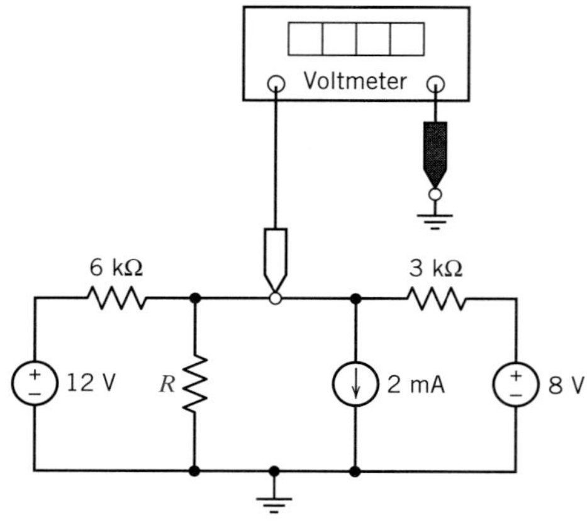

Figure P 4.4-7

P 4.4-8 Find the voltage v for the circuit of Figure P 4.4-8.
Answer: $v = 3.33$ V

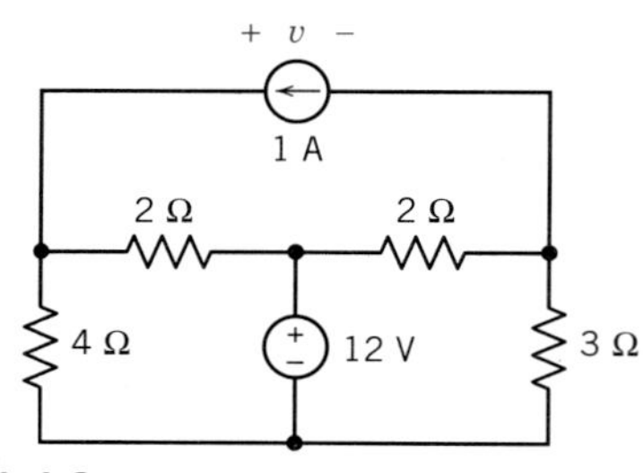

Figure P 4.4-8

Section 4.5 Node Voltage Analysis with Dependent Sources

P 4.5-1 Determine the node voltage v_b for the circuit of Figure P 4.5-1.
Answer: $v_b = 18$ V

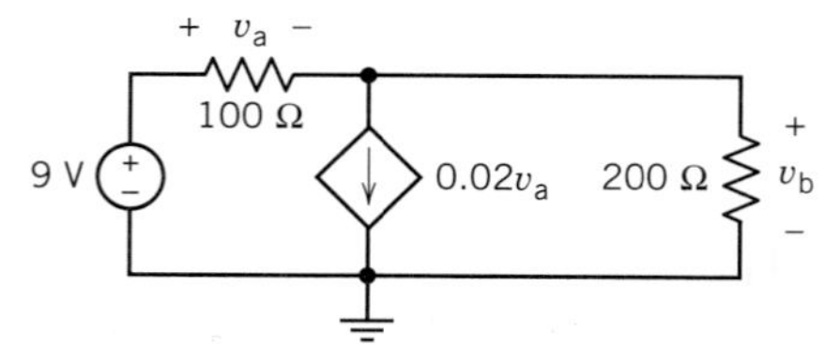

Figure P 4.5-1

P 4.5-2 Using node voltage analysis, find v_a for the circuit of Figure P 4.5-2.
Answer: $v_a = 10$ V

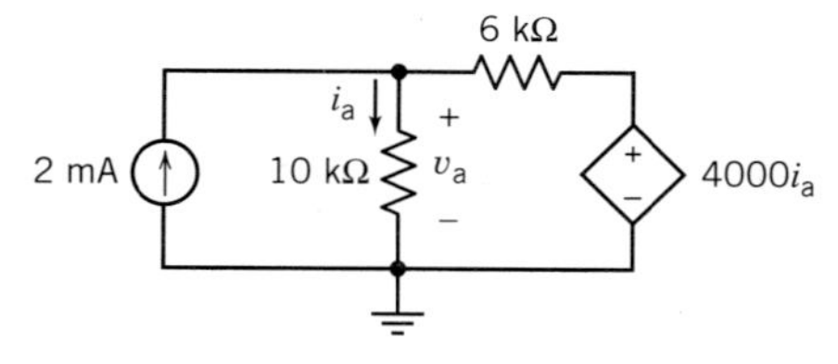

Figure P 4.5-2

P 4.5-3 Find i_b for the circuit shown in Figure P 4.5-3.
Answer: $i_b = -12$ mA

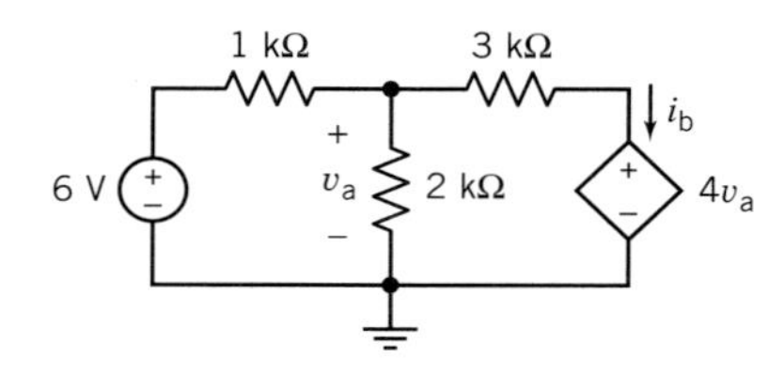

Figure P 4.5-3

P 4.5-4 Determine the node voltage v_b for the circuit of Figure P 4.5-4.
Answer: $v_b = 1.5$ V

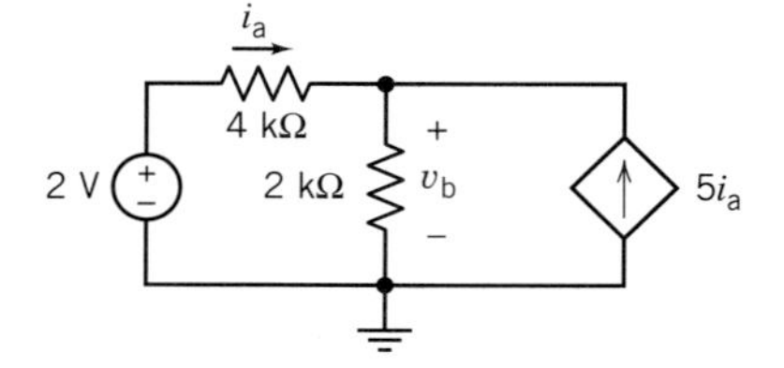

Figure P 4.5-4

P 4.5-5 The encircled numbers in Figure P 4.5-5 are node numbers. The node voltages of this circuit are $v_1 = 10$ V, $v_2 = 14$ V and $v_3 = 12$ V.
(a) Determine the value of the current i_b.
(b) Determine the value of r, the gain of the CCVS.
Answers: (a) −2 A (b) 4 V/A

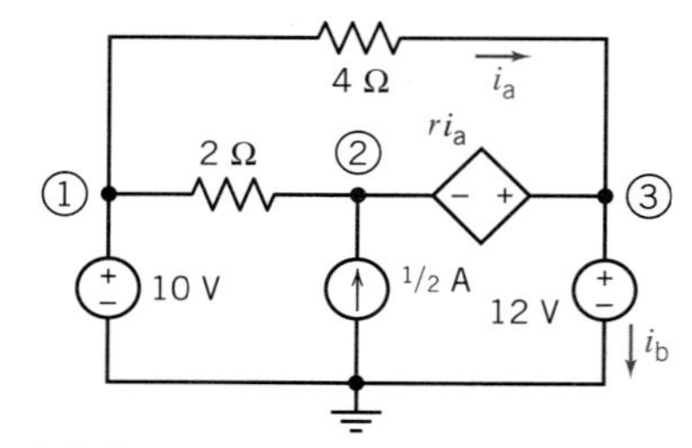

Figure P 4.5-5

P 4.5-6 Determine the value of the current i_x in the circuit of Figure P 4.5-6.
Answer: $i_x = 2.4$ A

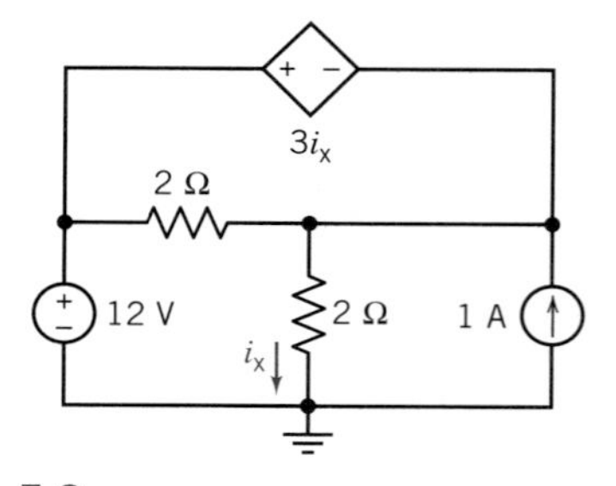

Figure P 4.5-6

P 4.5-7 Determine v_x for the circuit shown in Figure P 4.5-7.

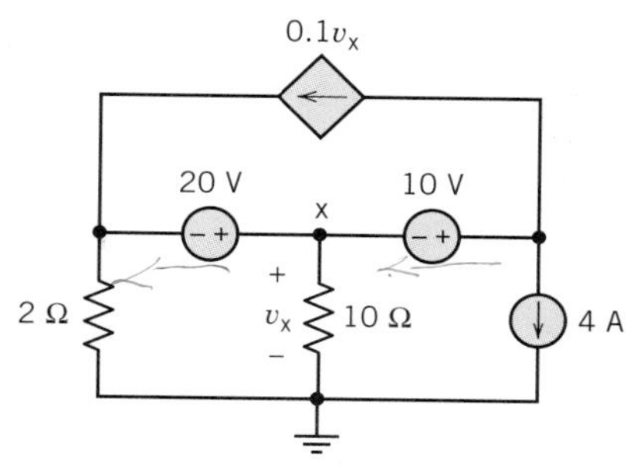

Figure P 4.5-7

Section 4.6 Mesh Current Analysis with Independent Voltage Sources

P 4.6-1 Determine the mesh currents, i_1, i_2 and i_3, for the circuit shown in Figure P 4.6-1.
Answers: $i_1 = 3$ A, $i_2 = 2$ A and $i_3 = 4$ A.

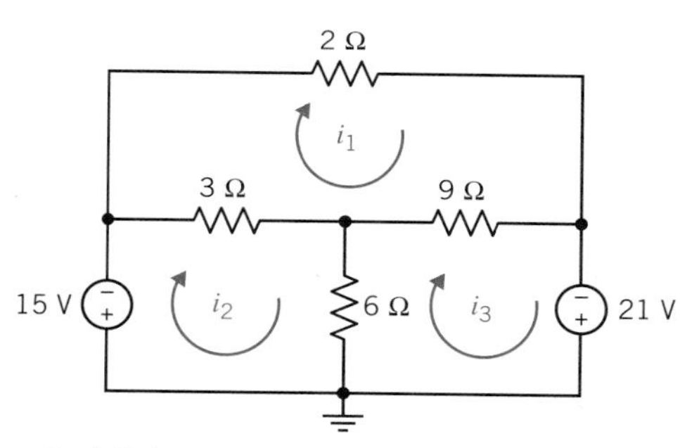

Figure P 4.6-1

P 4.6-2 The values of the mesh currents in the circuit shown in Figure P 4.6-2 are $i_1 = 2$ A, $i_2 = 3$ A and $i_3 = 4$ A. Determine the values of the resistance R and of the voltages v_1 and v_2 of the voltage sources.
Answers: $R = 12\ \Omega$, $v_1 = -4$ V and $v_2 = -28$ V.

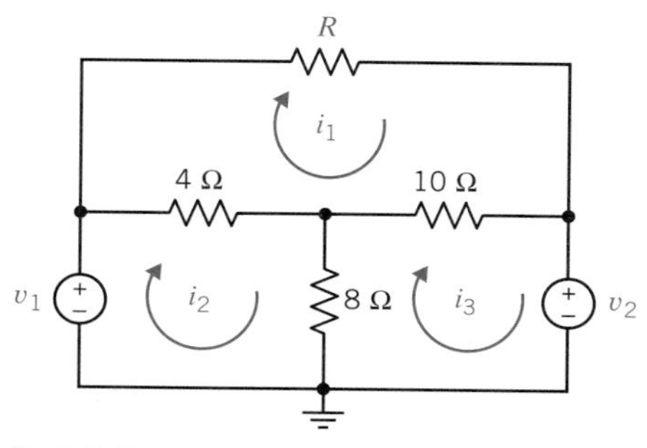

Figure P 4.6-2

P 4.6-3 Find the mesh currents, i_a and i_b, in the circuit shown in Figure P 4.6-3.
Answer: $i_a = -20$ mA and $i_b = -30$mA.

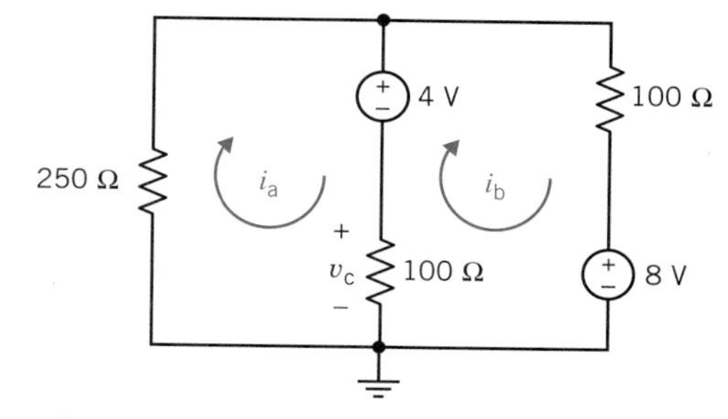

Figure P 4.6-3

P 4.6-4 Determine the mesh currents, i_a and i_b, in the circuit shown in Figure P 4.6-4.

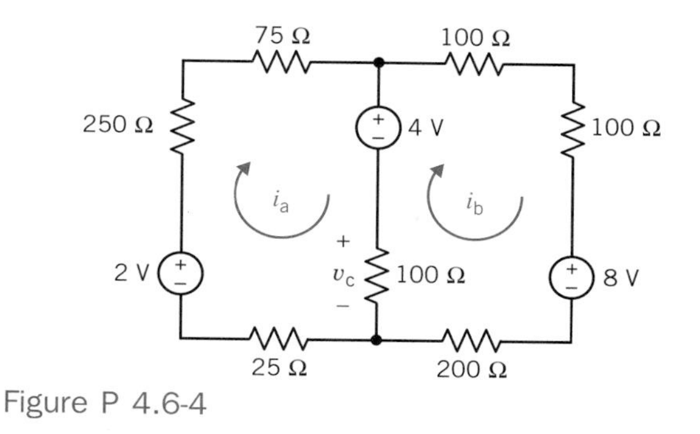

Figure P 4.6-4

P 4.6-5 Find the current i for the circuit of Figure P 4.6-5.
Hint: A short circuit can be treated as a 0-V voltage source.

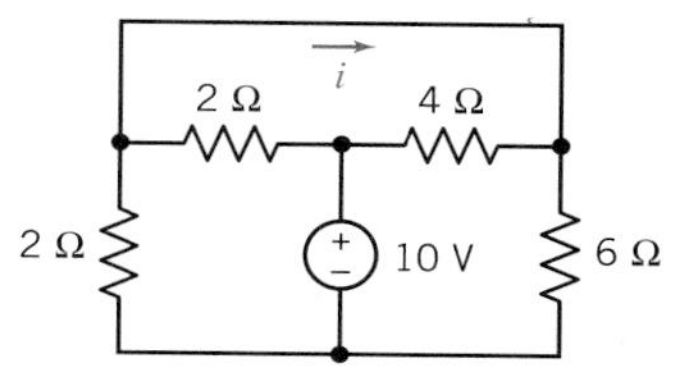

Figure P 4.6-5

Section 4.7 Mesh Current Analysis with Current Sources: (a) Independent Sources Only

P 4.7-1 Find i_b for the circuit shown in Figure P 4.7-1.
Answer: $i_b = 0.6$ A

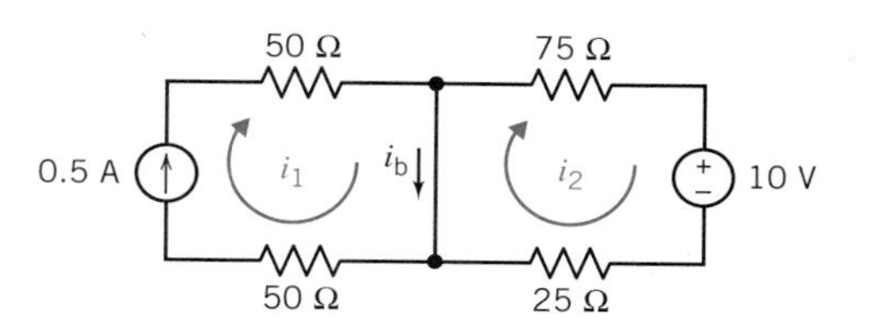

Figure P 4.7-1

P 4.7-2 Find v_c for the circuit shown in Figure P 4.7-2.
Answer: $v_c = 15$ V

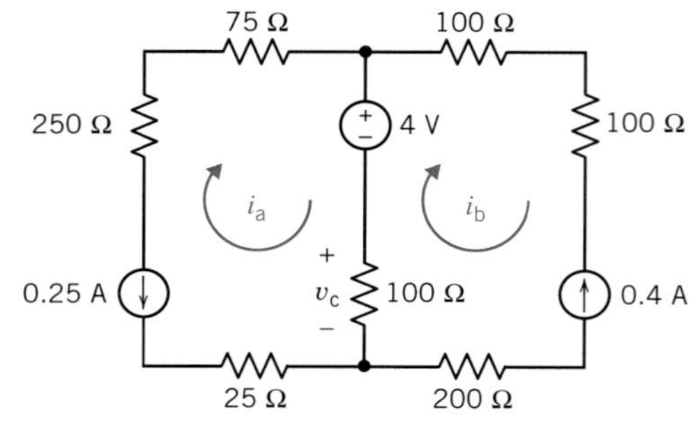

Figure P 4.7-2

P 4.7-3 Find v_2 for the circuit shown in Figure P 4.7-3.
Answer: $v_2 = 2$V

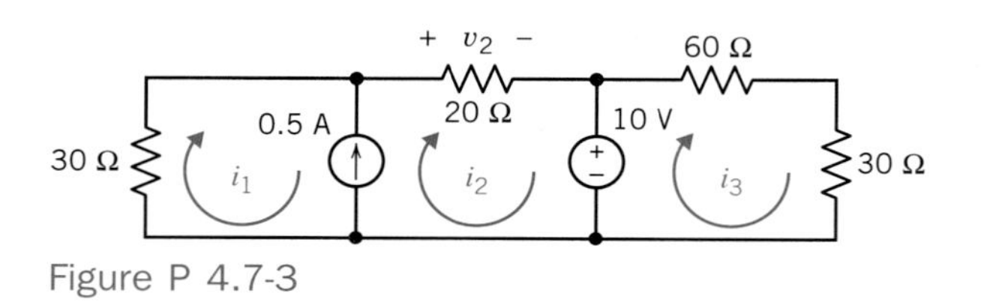

Figure P 4.7-3

P 4.7-4 Find v_c for the circuit shown in Figure P 4.7-4.

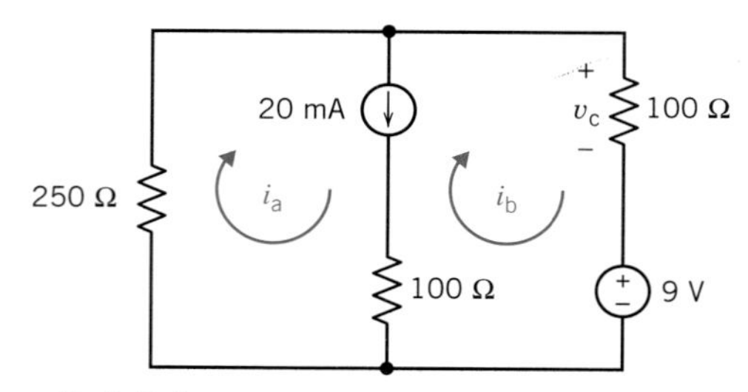

Figure P 4.7-4

P 4.7-5 Find i_1 and v_1 using mesh analysis methods for the circuit of Figure P 4.7-5.

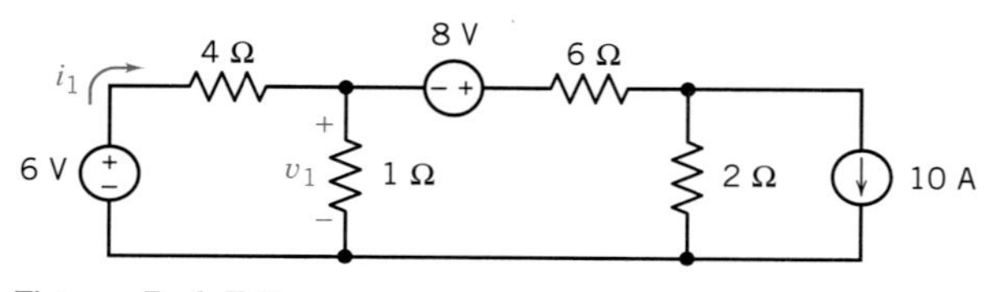

Figure P 4.7-5

P 4.7-6 Determine the value of the voltage measured by the voltmeter in Figure P 4.7-6.
Answer: 8 V

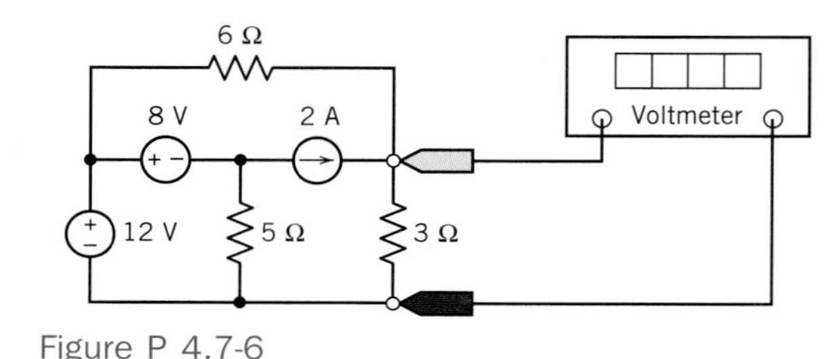

Figure P 4.7-6

P 4.7-7 Determine the value of the current measured by the ammeter in Figure P 4.7-7.
Hint: Write and solve a single mesh equation.
Answer: $-5/6$ A

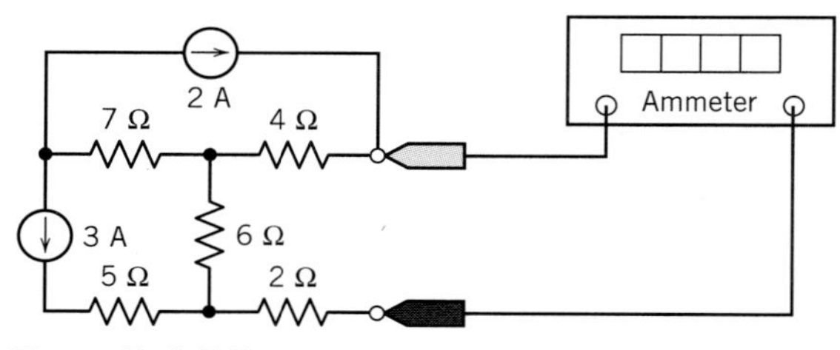

Figure P 4.7-7

P 4.7-8 Using mesh analysis, find i_1 for the circuit shown in Figure P 4.7-8.
Answer: $i_1 = 3$ mA

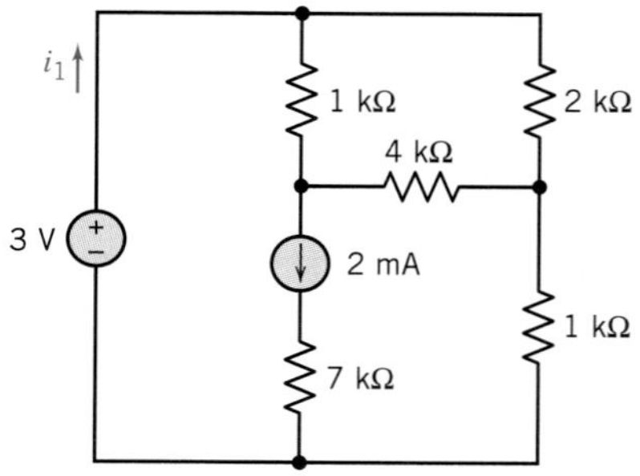

i_1

1 kΩ 2 kΩ 4 kΩ 3 V 2 mA 1 kΩ 7 kΩ

Figure P 4.7-8

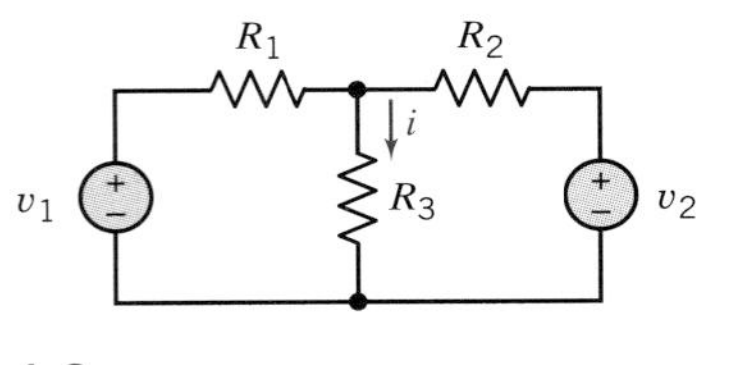

Figure DP 4.6
A circuit for earth fault line stress measurement.

DP 4-7 For the circuit of Figure DP 4.7, determine the required value of g so that $v_a - v_c = 20/3$ V.
Answer: $g = 4$

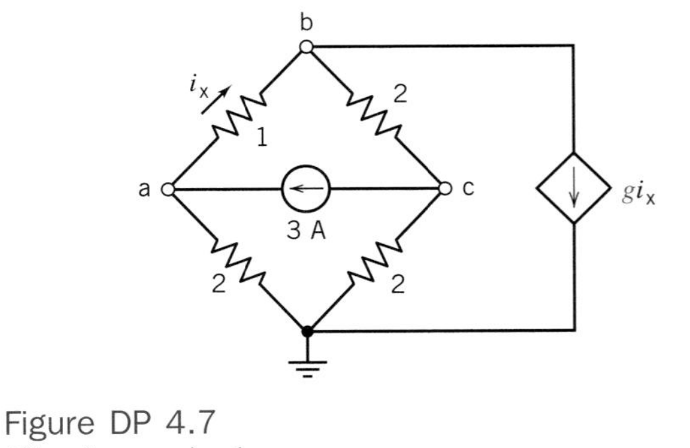

Figure DP 4.7
All resistances in ohms.

VP 4-5 Computer analysis of the circuit shown in Figure VP 4.5 indicates that the mesh currents are $i_1 = 2$ A, $i_2 = 4$ A and $i_3 = 3$ A. Verify that this analysis is correct.
Hint: Use the mesh currents to calculate the element voltages. Verify that KVL is satisfied for each mesh.

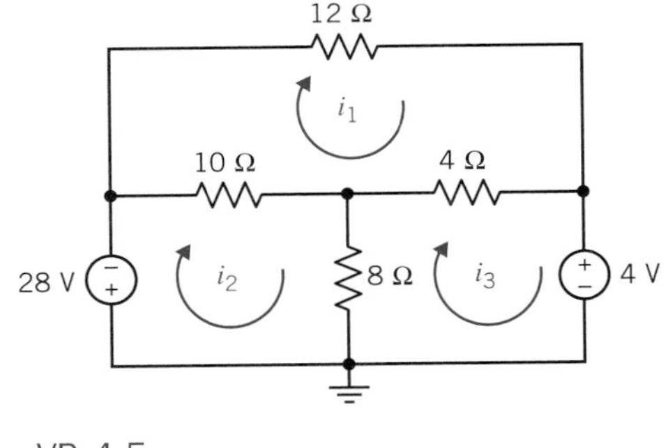

Figure VP 4.5

DESIGN PROBLEMS

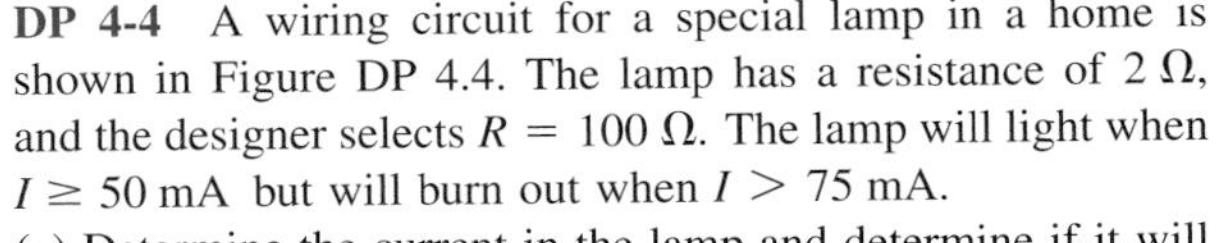

DP 4-1 For the circuit shown in Figure DP 4.1, it is desired that $v_{ba} = 3$ V by adjusting the current source i_s. Select i_s to achieve the desired voltage v_{ba}.

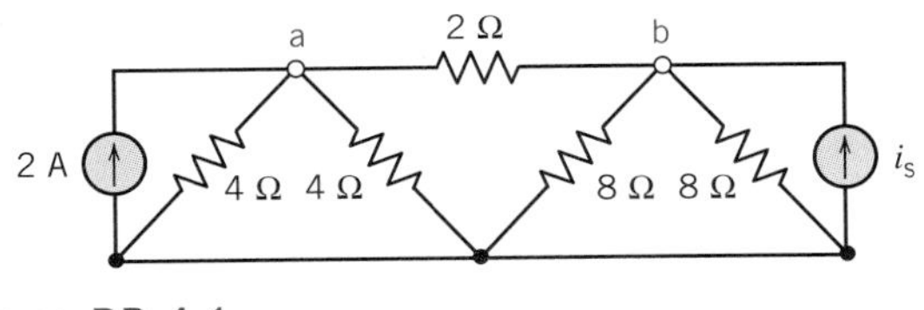

Figure DP 4.1

DP 4-2 For the circuit shown in Figure DP 4.2, determine the required voltage source v_s so that the current i equals 3.0 A.

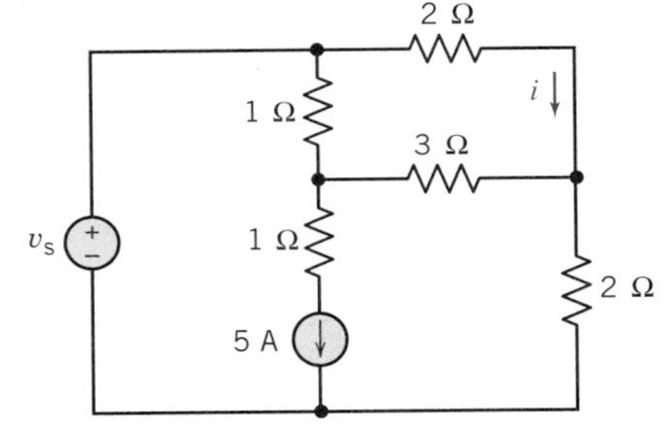

Figure DP 4.2

DP 4-3 For the circuit shown in Figure DP 4.3, it is desired to set the voltage at node a equal to 0 V in order to control an electric motor. Select voltages v_1 and v_2 in order to achieve $v_a = 0$ V when v_1 and v_2 are less than 20 V and greater than zero and $R = 2\ \Omega$.

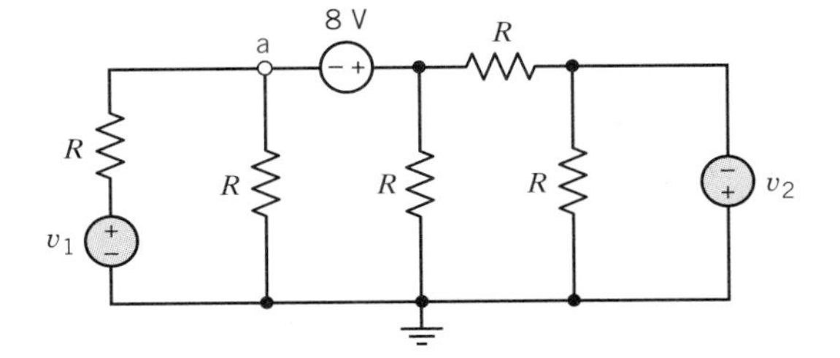

Figure DP 4.3

DP 4-4 A wiring circuit for a special lamp in a home is shown in Figure DP 4.4. The lamp has a resistance of 2 Ω, and the designer selects $R = 100\ \Omega$. The lamp will light when $I \geq 50$ mA but will burn out when $I > 75$ mA.

(a) Determine the current in the lamp and determine if it will light for $R = 100\ \Omega$.

(b) Select R so that the lamp will light but will not burn out if R changes by $\pm 10\%$ because of temperature changes in the home.

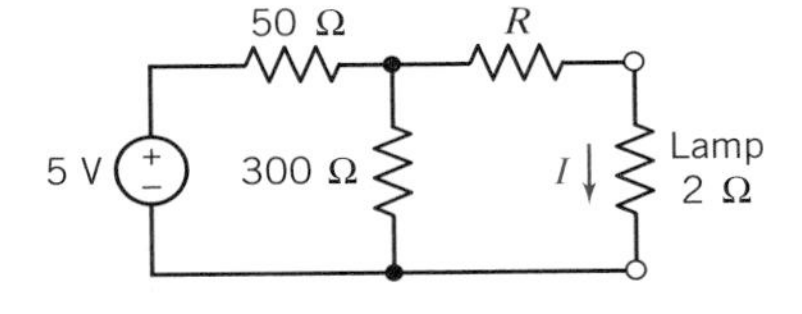

Figure DP 4.4
A lamp circuit.

DP 4-5 In order to control a device using the circuit shown in Figure DP 4.5, it is necessary that $v_{ab} = 10$ V. Select the resistors when it is required that all resistors be greater than 1 Ω and $R_3 + R_4 = 20\ \Omega$.

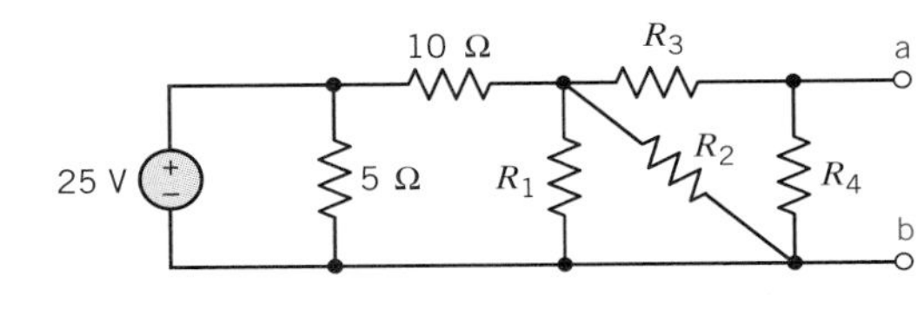

Figure DP 4.5

DP 4-6 The current i shown in the circuit of Figure DP 4.6 is used to measure the stress between two sides of an earth fault line. Voltage v_1 is obtained from one side of the fault, and v_2 is obtained from the other side of the fault. Select the resistances R_1, R_2, and R_3 so that the magnitude of the current i will remain in the range between 0.5 mA and 2 mA when v_1 and v_2 may each vary independently between +1 V and +2 V (1 V $\leq v_n \leq$ 2 V).

SP 4-7 Find $v_a - v_c$ for the circuit of Figure SP 4.7.

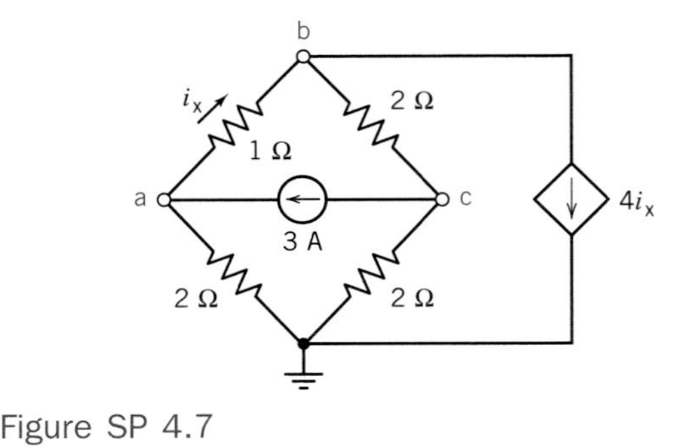

Figure SP 4.7

VERIFICATION PROBLEMS

VP 4-1 Computer analysis of the circuit shown in Figure VP 4.1 indicates that the node voltages are $v_a = 5.2$ V, $v_b = -4.8$ V, and $v_c = 3.0$ V. Is this analysis correct?

Hint: Use the node voltages to calculate all the element currents. Check to see that KCL is satisfied at each node.

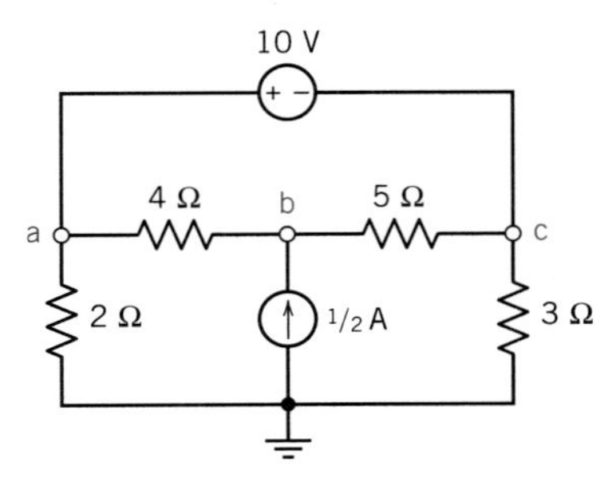

Figure VP 4.1

VP 4-2 An old lab report asserts that the node voltages of the circuit of Figure VP 4.2 are $v_a = 4$ V, $v_b = 20$ V, and $v_c = 12$ V. Are these correct?

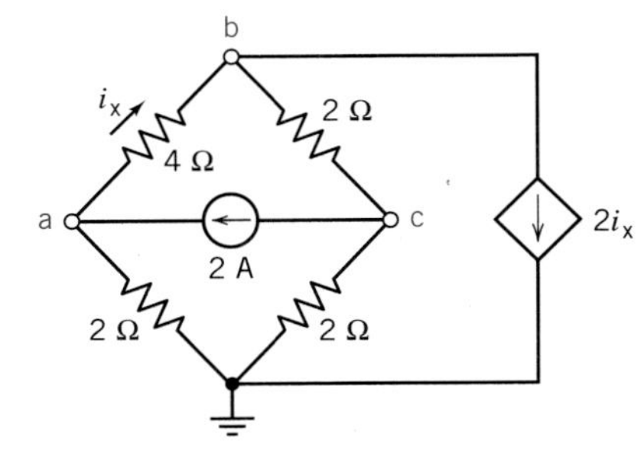

Figure VP 4.2

VP 4-3 Your lab partner forgot to record the values of R_1, R_2, and R_3. He thinks that two of the resistors in Figure VP 4.3 had values of 10 kΩ and that the other had a value of 5 kΩ. Is this possible? Which resistor is the 5 kΩ resistor?

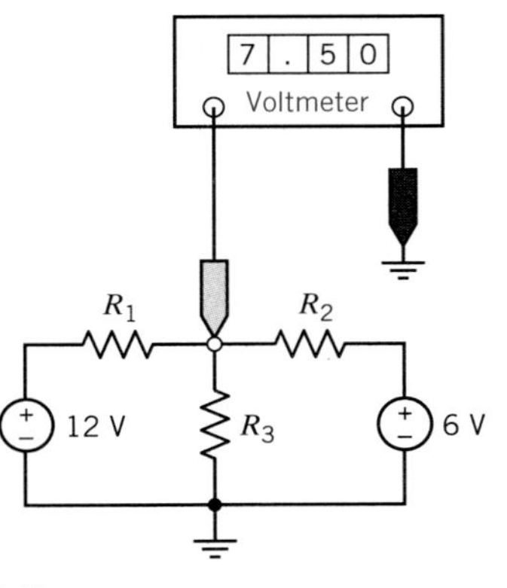

Figure VP 4.3

VP 4-4 Computer analysis of the circuit shown in Figure VP 4.4 indicates that the node voltages are $v_1 = -8$ V, $v_2 = -20$ V and $v_3 = -6$ V. Verify that this analysis is correct.

Hint: Use the node voltages to calculate the element currents. Verify that KCL is satisfied at each node.

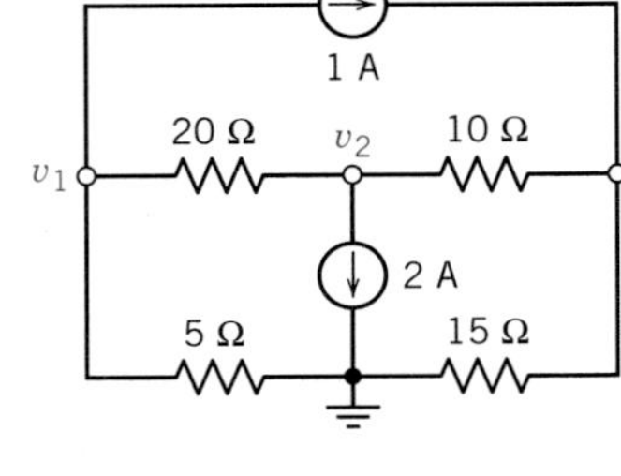
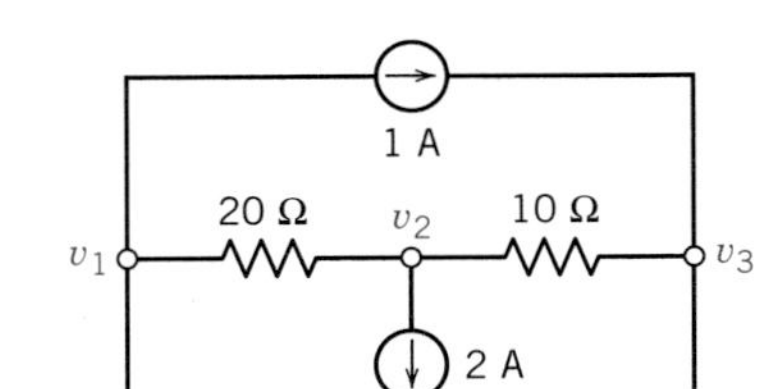

Figure VP 4.4

CHAPTER 5

Circuit Theorems

Preview

Many electric circuits are complex, but it is the engineer's goal to reduce their complexity in order to analyze them readily. In this chapter we consider the reduction of complex circuits to a simpler form. We also show how some simple sources can be transformed from a current source form to a voltage source form, and vice versa.

Furthermore, using the property of superposition for linear circuits, we develop a relatively simple method of analyzing a complex circuit when we wish to determine only one voltage or current within the circuit.

The function of many circuits is to deliver maximum power to a load such as an audio speaker in a stereo system. Here we develop the required relationship between a load resistor and a fixed series resistor that can represent the remaining portion of the circuit.

5.1 Design Challenge

STRAIN GAUGE BRIDGE

Strain gauges are transducers that measure mechanical strain, which is a deformation caused by force. Electrically, the strain gauges are resistors. The strain causes a change in resistance that is proportional to the strain.

Figure 5.1-1 shows four strain gauges connected in a configuration called a bridge. Strain gauge bridges are used to measure force or pressure (Doebelin 1966).

The bridge output is usually a small voltage. In Figure 5.1-1 an amplifier multiplies the bridge output, v_i, by a gain to obtain a larger voltage, v_o, which is displayed by the voltmeter.

Describe the Situation and the Assumptions

A strain gauge bridge is used to measure force. The strain gauges have been positioned so that the force will increase the resistance of two of the strain gauges while, at the same time, decreasing the resistance of the other two strain gauges.

The strain gauges used in the bridge have nominal resistances of $R = 120\ \Omega$. (The nominal resistance is the resistance when the strain is zero.) This resistance is expected to increase or decrease by no more than $2\ \Omega$ due to strain. This means that

$$-2\ \Omega \leq \Delta R \leq 2\ \Omega \tag{5.1-1}$$

The output voltage, v_o, is required to vary from -10 V to $+10$ V as ΔR varies from $-2\ \Omega$ to $2\ \Omega$.

State the Goal

Determine the amplifier gain, b, needed to cause v_o to be related to ΔR by the equation

$$v_o = 5 \frac{\text{volt}}{\text{ohm}} \cdot \Delta R \tag{5.1-2}$$

Generate a Plan

Analyze the circuit shown in Figure 5.1-1 to determine the relationship between and v_i and ΔR. Calculate the amplifier gain needed to satisfy Eq. 5.1-2.

The required analysis could be accomplished using node or mesh equations. Chapter 5 presents Thévenin's theorem. This theorem will significantly simplify the required analysis. We will postpone the analysis until the end of the chapter so that we can make use of Thévenin's Theorem.

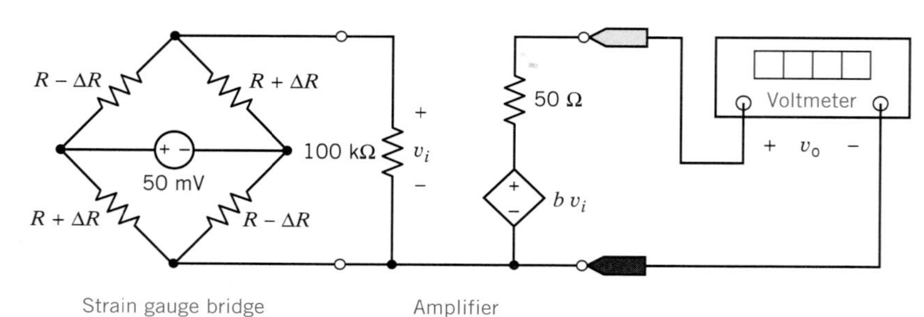

Figure 5.1-1 Design problem involving a strain gauge bridge.

5.2 ELECTRIC POWER FOR CITIES

The outstanding advantage of electricity compared to other power sources is its transmittability and flexibility. Electrical energy can be moved to any point along a pair of wires and, depending on the user's requirements, converted to light, heat, motion, or other forms.

Although the first demand for large-scale generation of electrical power was for lighting, that use was soon matched by other applications such as street railways and industrial processes. In all cases, the basic problems are the same: (1) to generate the electric power from another energy source such as coal or oil, (2) to transmit it to the place of use, and (3) to convert it to the form in which it is to be used. The first two decades of the 20th century brought solutions to these three problems and thus the ascendancy of electric power as the predominant choice.

Perhaps one of the most interesting steps in the beginning of power generation was the opening of the Pearl Street Station, as shown in Figure 5.2-1. With the transmission of power, cities were able to use this power for lighting and electric street railways and could develop the infrastructure necessary for their modern operation.

Although lighting and vehicle traction were the first uses of electric power, they were soon eclipsed by industrial use. By 1920 industry's use of electricity was predominant. Factories used electric power for motors, heating, machines, and electrochemistry, among others. One of the first products to be made on a large scale by electric heating was carborundum (silicon carbide). Later, in 1906, the electrolytic process for producing aluminum was introduced.

By 1910 electric automobiles, as shown in Figure 5.2-2, were commonplace. Nevertheless, they were replaced by gasoline-fueled automobiles by 1920 because electric cars operated at lower top speeds and over shorter ranges without recharging than gasoline cars could achieve. However, the availability of electric motive power remained a critical factor in the development of cities. Electrically powered elevators permitted the construction of high-rise, multistory office and apartment complexes. Also, the modern electric railway and subway permitted cities to spread out into suburbs while accommodating the needs for mass transit. A modern electric railway is shown in Figure 5.2-3. With the construction of dense, modern cities, the demand for electric power became intensive. The skyline of nighttime New York City, shown in Figure 5.2-4, illustrates the dependence on electric power for lighting and the elevator.

Figure 5.2-1 Dynamo Room at the Pearl Street Station. This was Edison's first central station for incandescent electric lighting. It began operation in New York City in 1882. Courtesy of General Electric Company.

Figure 5.2-2 Baker electric car, 1910. Courtesy of Motor Vehicle Manufacturers Assoc.

Figure 5.2-3 Bay Area Rapid Transit (BART) railway. Photograph copyright © by Ron May.

Figure 5.2-4 Skyline of New York City at night. Photograph copyright © by Ron May.

Figure 5.2-5 Original tinfoil phonograph patented by Edison in 1877. Courtesy of Science Museum, London.

Consumption of electric energy in the home increased as electric appliances became widely available. By 1940 about one-half of American homes were using electric vacuum cleaners and clothes washers. By 1980 over 95 percent of American homes had vacuum cleaners and over 40 percent had electric dishwashers.

The early development of the electric phonograph can be traced to Edison's tinfoil phonograph, shown in Figure 5.2-5. With the addition of a battery-driven electric motor in 1893, Edison's phonograph became one of the most widely sought-after entertainment devices. By 1893 Edison had 65 patents on the phonograph and its improvements.

5.3 SOURCE TRANSFORMATIONS

We found in Chapter 4 that it is generally easier to use mesh current analysis when all the sources are voltage sources. Similarly, it is usually easier to use node current analysis when all the sources are current sources. If we have both voltage and current sources in a circuit, it is valuable to make a set of adjustments to the circuit so that all the sources are of one type.

It is possible to transform an independent voltage source in series with a resistor into a current source in parallel with a resistance, or vice versa. Consider the pair of circuits shown in Figures 5.3-1*a* and 5.3-1*b*. We will find the appropriate relationship required between these two circuits so that they are interchangeable, both providing the same response at terminals a–b.

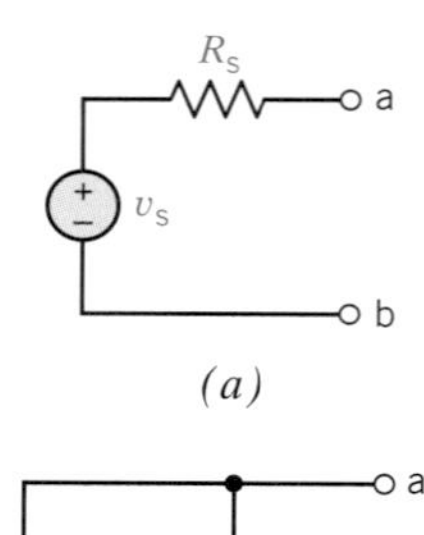

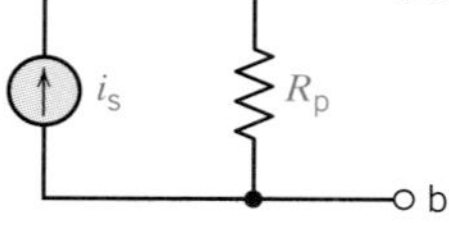

Figure 5.3-1 Two equivalent circuits.

A **source transformation** is a procedure for transforming one source into another while retaining the terminal characteristics of the original source.

A source transformation rests on the concept of equivalence. An *equivalent circuit* is one whose terminal characteristics remain identical to those of the original circuit. It is important to note that equivalence implies an identical effect at the terminals but not *within* the equivalent circuits themselves.

We want the circuit of Figure 5.3-1*a* to transform into that of Figure 5.3-1*b*. We then require that both circuits have the same characteristic for all values of an external resistor R connected between terminals a–b (Figures 5.3-2*a* and 5.3-2*b*). We will try the two extreme values $R = 0$ and $R = \infty$.

When the external resistance $R = 0$, we have a short circuit across terminals a–b. First, we require the short-circuit current to be the same for each circuit. The short-circuit current for Figure 5.3-2*a* is

$$i = \frac{v_s}{R_s} \tag{5.3-1}$$

The short-circuit current for Figure 5.3-2*b* is $i = i_s$. Therefore, we require that

$$i_s = \frac{v_s}{R_s} \tag{5.3-2}$$

For the open-circuit condition R is infinite, and from Figure 5.3-2*a* we have the voltage $v = v_s$. For the open-circuit voltage of Figure 5.3-2*b* we have

$$v = i_s R_p \tag{5.3-3}$$

Since v must be equal for both circuits to be equivalent, we require that

$$v_s = i_s R_p$$

Also, from Eq. 5.3-2 we require $i_s = v_s/R_s$. Therefore, we must have

$$v_s = \left(\frac{v_s}{R_s}\right) R_p$$

and, therefore, we require that

$$R_s = R_p \tag{5.3-4}$$

Equations 5.3-2 and 5.3-4 must be true simultaneously for both circuits for the two sources to be equivalent. Of course, we have proved that the two sources are equivalent at two values ($R = 0$ and $R = \infty$). We have not proved that the circuits are equal for all R, but we assert that the equality relationship holds for all R for these two circuits as we show below.

For the circuit of Figure 5.3-2*a* we use KVL to obtain

$$v_s = i\,R_s + v$$

Dividing by R_s gives

$$\frac{v_s}{R_s} = i + \frac{v}{R_s}$$

If we use KCL for the circuit of Figure 5.3-2*b*, we have

$$i_s = i + \frac{v}{R_p}$$

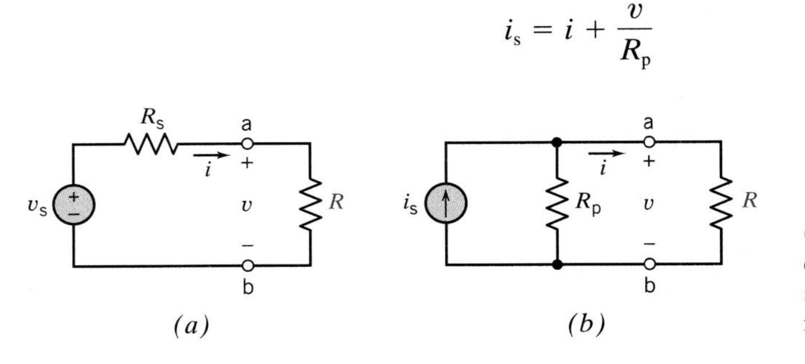

Figure 5.3-2
(*a*) Voltage source with an external resistor R. (*b*) Current source with an external resistance R.

Thus, the two circuits are equivalent when $i_s = v_s/R_s$ and $R_s = R_p$.

A voltage source v_s connected in series with a resistor R_s and a current source i_s connected in parallel with a resistor R_p are equivalent circuits provided that

$$R_p = R_s \text{ and } v_s = R_s i_s$$

Replacing a voltage source in series with a resistor by its equivalent circuit will not change the element currents or voltages in the rest of the circuit. Similarly, replacing a current source in parallel with a resistor by its equivalent circuit will not change the element currents or voltages in the rest of the circuit.

Source transformations are useful for circuit simplification and also may be useful in node or mesh analysis. The method of transforming one form of source into the other form is summarized in Figure 5.3-3.

Circuits are said to be *dual* when the characterizing equations of one network can be obtained from the other by simply interchanging v and i and interchanging G and R. There is a duality between resistance and conductance, current and voltage, and a series circuit and a parallel circuit. The two equivalent sources of Figure 5.3-3 are said to be dual circuits. Note that $R_s = 1/G_p$ is one element of duality. The equation for the voltage source with a short circuit is

$$i_s = G_p v_s$$

and the equation for the current source with an open circuit is

$$v_s = R_s i_s$$

Note that the voltage and the current are dual, the resistance R_s is the dual of the conductance G_p, and the short circuit is the dual of the open-circuit condition.

Therefore, *dual circuits* are defined by the same characterizing equations with v and i interchanged and G and R interchanged. When the set of transforms that converts one circuit into another also converts the second into the first, the circuits are said to be *duals*. Note that the transform of Figure 5.3-3*a* converts the first circuit into the second, while the transform of Figure 5.3-3*b* converts the second circuit into the first.

Dual circuits are two circuits such that the equations describing the first circuit, with v and i interchanged and R and G interchanged, describe the second circuit.

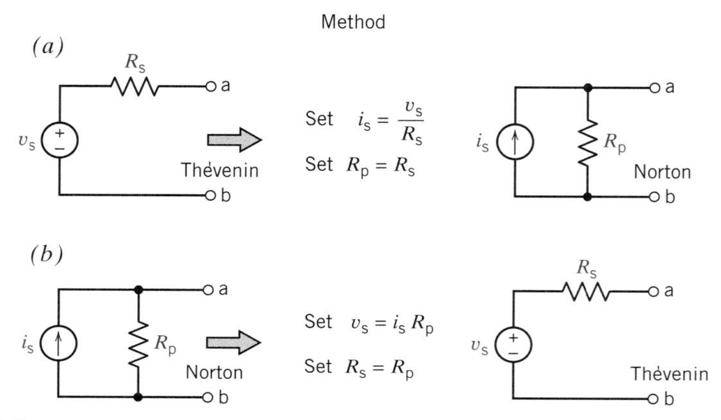

Figure 5.3-3 Method of source transformations.

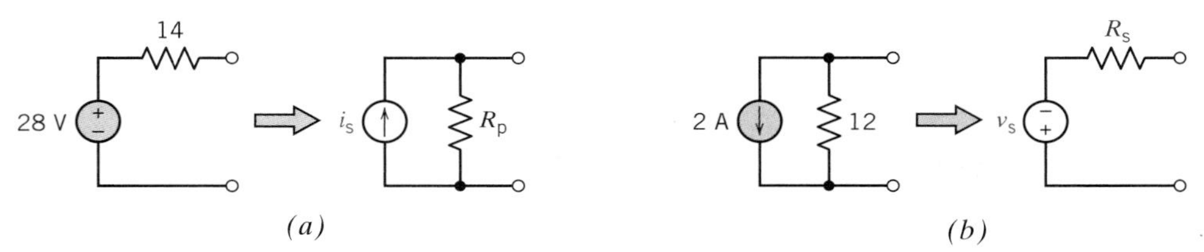

Figure 5.3-4 The circuits of Example 5.3-1. All resistances in ohms.

Example 5.3-1

Find the source transformation for the circuits shown in Figures 5.3-4*a* and 5.3-4*b*.

Solution

Using the method summarized in Figure 5.3-3, we note that the voltage source of Figure 5.3-4*a* can be transformed to a current source with $R_p = R_s = 14\ \Omega$. The current source is

$$i_s = \frac{v_s}{R_s} = \frac{28}{14} = 2\text{ A}$$

The resulting transformed source is shown on the right side of Figure 5.3-4*a*.

Starting with the current source of Figure 5.3-4*b*, we have $R_s = R_p = 12\ \Omega$. The voltage source is

$$v_s = i_s R_p = 2(12) = 24\text{ V}$$

The resulting transformed source is shown on the right side of Figure 5.3-4*b*. Note that the positive sign of the voltage source v_s appears on the lower terminal since the current source arrow points downward.

Example 5.3-2

A circuit is shown in Figure 5.3-5. Find the current *i* by reducing the circuit to the right of terminals a–b to its simplest form using source transformations.

Solution

The first step is to transform the 30 Ω series resistor and the 3-V source to a current source with a parallel resistance. First, we note that $R_p = R_s = 30\ \Omega$. The current source is

$$i_s = \frac{v_s}{R_s} = \frac{3}{30} = 0.1\text{ A}$$

as shown in Figure 5.3-6*a*. Combining the two parallel resistances in Figure 5.3-6*a*, we have $R_{p2} = 12\ \Omega$, as shown in Figure 5.3-6*b*.

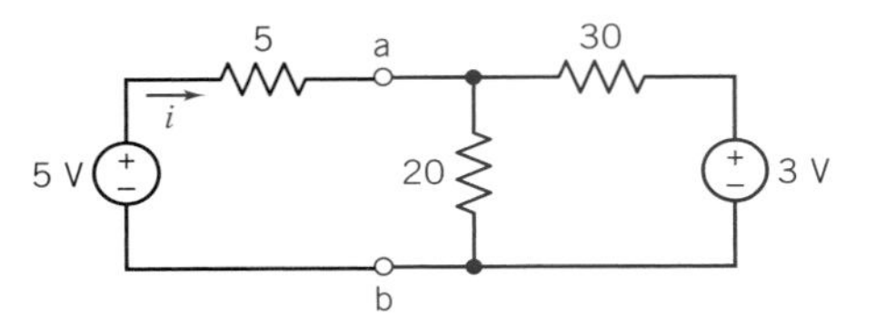

Figure 5.3-5
The circuit of Example 5.3-2. Resistances in ohms.

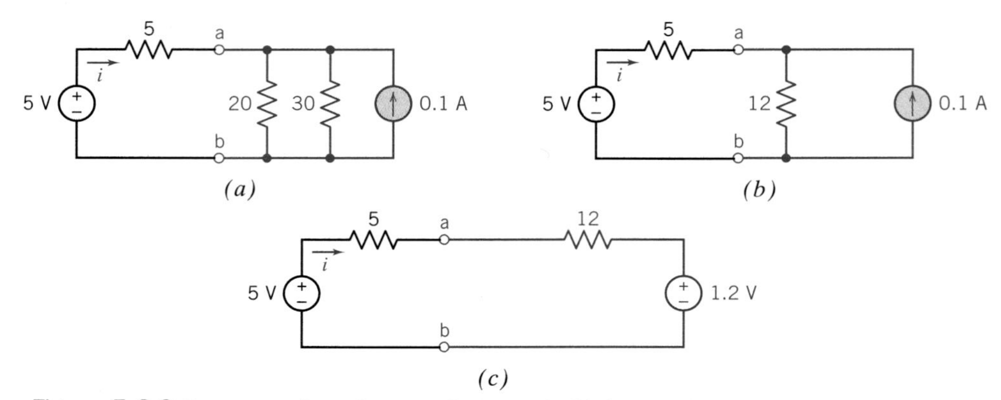

Figure 5.3-6 Source transformation steps for Example 5.3-2. All resistances in ohms.

The parallel resistance of 12 Ω and the current source of 0.1 A can be transformed to a voltage source in series with $R_{s2} = 12\ \Omega$, as shown in Figure 5.3-6*c*. The voltage source v_s is found as:

$$v_s = i_s R_{s2} = 0.1(12) = 1.2\ \text{V}$$

Source transformations do not disturb the currents and voltages in the rest of the circuit. Therefore, the current *i* in Figure 5.3-5 is equal to the current *i* in Figure 5.3-6*c*. The current *i* is found by using KVL around the loop of Figure 5.3-6*c*, yielding $i = 3.8/17 = 0.224$ A.

Exercise 5.3-1 Determine values of R and i_s so that the circuits shown in Figure E 5.3-1*a* and 5.3-1*b* are equivalent to each other due to a source transformation.

Answer: $R = 10\ \Omega$ and $i_s = 1.2$ A.

Exercise 5.3-2 Determine values of R and i_s so that the circuits shown in Figure E 5.3-2*a* and 5.3-2*b* are equivalent to each other due to a source transformation.

Hint: Notice that the polarity of the voltage source in Figure E 5.3-2*a* is not the same as in Figure E 5.3-1*a*.

Answer: $R = 10\ \Omega$ and $i_s = -1.2$ A.

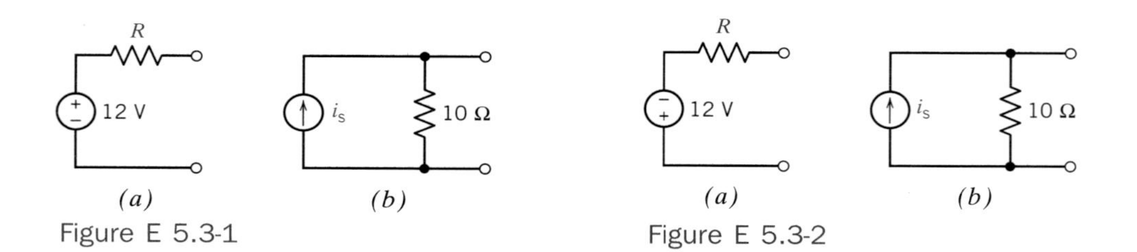

Figure E 5.3-1

Figure E 5.3-2

Exercise 5.3-3 Determine values of R and v_s so that the circuits shown in Figure E 5.3-3*a* and 5.3-3*b* are equivalent to each other due to a source transformation.

Answer: $R = 8\ \Omega$ and $v_s = 24$ V.

Exercise 5.3-4 Determine values of R and v_s so that the circuits shown in Figure E 5.3-4*a* and 5.3-4*b* are equivalent to each other due to a source transformation.

Hint: Notice that the polarity of the current source in Figure E 5.3-4*b* is not the same as in Figure E 5.3-3*b*.

Answer: $R = 8\ \Omega$ and $v_s = -24$ V.

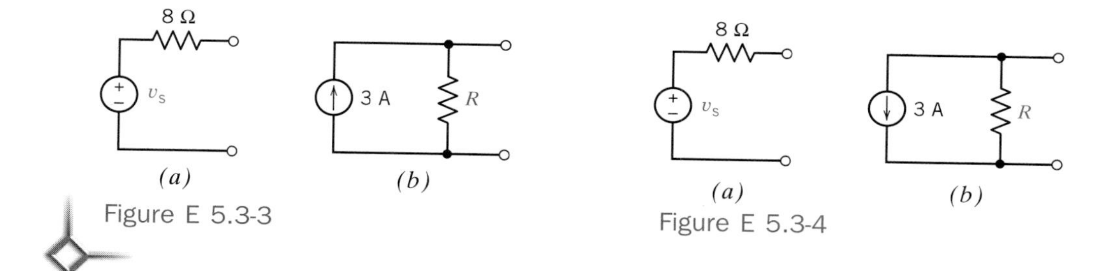

Figure E 5.3-3

Figure E 5.3-4

5.4 SUPERPOSITION

We discussed linear elements in Section 2.2 and noted that one of the properties of a linear element is that the principle of superposition holds. A linear element satisfies superposition when it satisfies the following response and excitation relationship:

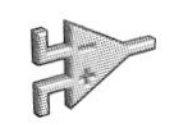

$$i_1 \longrightarrow v_1$$
$$i_2 \longrightarrow v_2 \tag{5.4-1}$$

then

$$i_1 + i_2 \longrightarrow v_1 + v_2$$

where the arrow implies the excitation causation and the resulting response. Thus, we can state that a device, if excited by current i_1, will exhibit response v_1. Similarly, an excitation i_2 will cause response v_2. Then if we use an excitation $i_1 + i_2$ we will find a response $v_1 + v_2$.

The principle of *superposition* states that for a linear circuit consisting of linear elements and independent sources, we can determine the total response by finding the response to each *independent* source with all other *independent* sources set to zero and then summing the individual responses. In this case, the response we seek may be a current or a voltage.

The **superposition principle** requires that the total effect of several causes acting simultaneously is equal to the sum of the effects of the individual causes acting one at a time.

The *superposition principle* may be restated as follows. In a linear circuit containing independent sources, the voltage across (or the current through) any element may be obtained by adding algebraically all the individual voltages (or currents) caused by each independent source acting alone, with all other independent voltage sources replaced by short circuits and all other independent current sources replaced by open circuits.

The principle of superposition requires that we deactivate (disable) all but one independent source and find the response due to that source. We then repeat the process by disabling all but a second source and determining the response. We find the response to each separate source, and then the total response is the sum of all the responses.

First, we note that when considering one independent source, we set the other independent sources to zero. Thus, an independent voltage source appears as a short circuit with zero voltage across it. Similarly, if an independent current source is set to zero, no current flows and it appears as an open circuit. Also, it is important to note that if a dependent source is present, it must remain active (unaltered) during the superposition process.

Example 5.4-1

Find the current measured by the ammeter in the circuit of Figure 5.4-1*a*.

Solution

An ideal ammeter is equivalent to a short circuit. In Figure 5.4-1*b* the ideal ammeter has been replaced by the equivalent short circuit and a label has been added to indicate the current measured by the ammeter, i_m.

The independent sources provide the inputs to a circuit. The current i_m is the response to two inputs, the voltage source voltage and the current source current. The principle of superposition tells us that we can determine the response to two inputs acting together by finding the response to each input acting separately and then adding the responses to the separate inputs.

First set the current source to zero; and then the current source appears as an open circuit as shown in Figure 5.4-2*a*. The current i_1 represents the portion of current i_m that is a response to the voltage source. In Figure 5.4-2*a* the 6 V of the voltage source appears across the series combination of the 3 Ω and 6 Ω resistors, therefore,

$$i_1 = \frac{6}{9}\text{ A}$$

For the second step, set the voltage source to zero, replacing it with a short circuit. Then we have the circuit of Figure 5.4-2*b*. The portion of the current i_m due to the current source is called i_2, and we obtain i_2 by the current divider principle as

$$i_2 = \frac{3}{3 + 6}2 = \frac{6}{9}\text{ A}$$

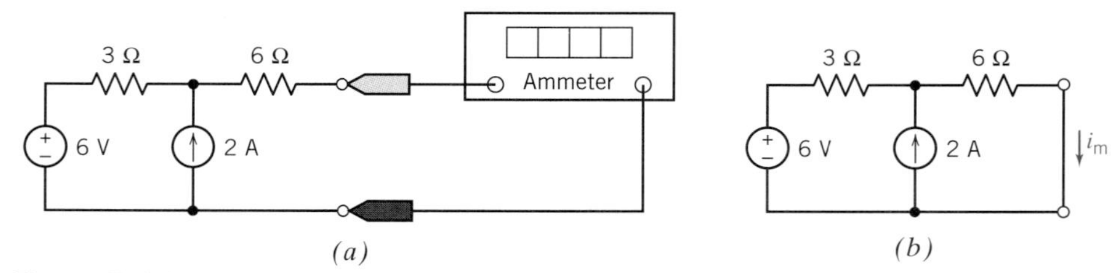

Figure 5.4-1 (*a*) A circuit containing two independent sources. (*b*) The circuit after the ideal ammeter has been replaced by the equivalent short circuit and a label has been added to indicate the current measured by the ammeter, i_m.

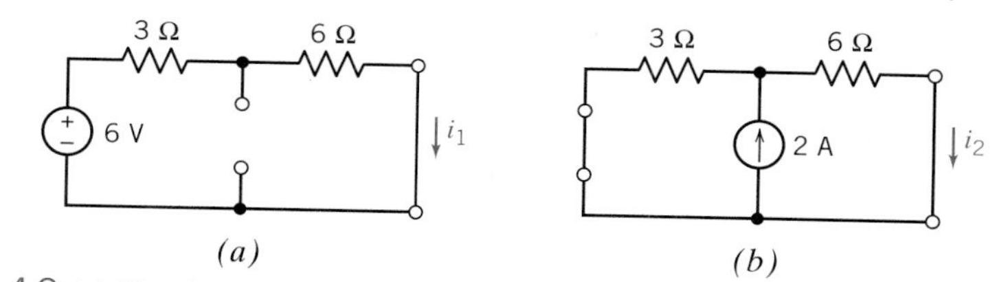

Figure 5.4-2 (*a*) Circuit of Example 5.4-1 with the current source set equal to zero. (*b*) Circuit of Example 5.4-1 with the voltage source set equal to zero.

The total current is then the sum of i_1 and i_2:

$$i_m = i_1 + i_2 = \frac{12}{9} \text{ A}$$

Exercise 5.4-1 Using the superposition principle, find the value of the voltage measured by the voltmeter in Figure E 5.4-1*a*.

Hint: Figure E 5.4-1*b* shows the circuit after the ideal voltmeter has been replaced by the equivalent open circuit and a label has been added to indicate the voltage measured by the voltmeter, v_m.

Answer: $v_m = \dfrac{20}{10 + 20 + 20}15 + 20\left(-\dfrac{10}{10 + (20 + 20)}2\right) = 6 + 20\left(-\dfrac{2}{5}\right) = -2 \text{ V}$

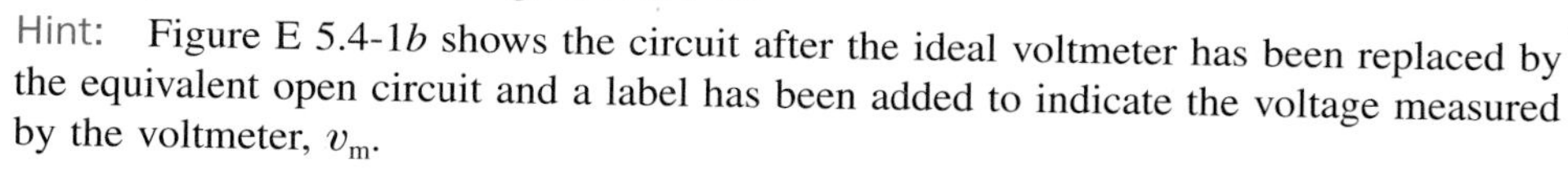

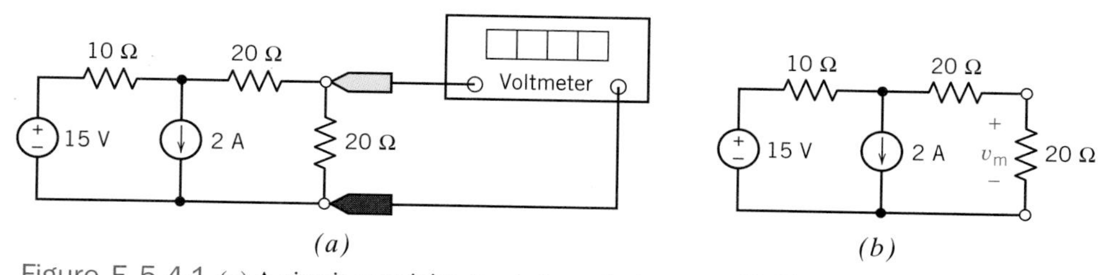

Figure E 5.4-1 (*a*) A circuit containing two independent sources. (*b*) The circuit after the ideal voltmeter has been replaced by the equivalent open circuit and a label has been added to indicate the voltage measured by the voltmeter, v_m.

Exercise 5.4-2 Using the superposition principle, find the value of the current measured by the ammeter in Figure E 5.4-2*a*.

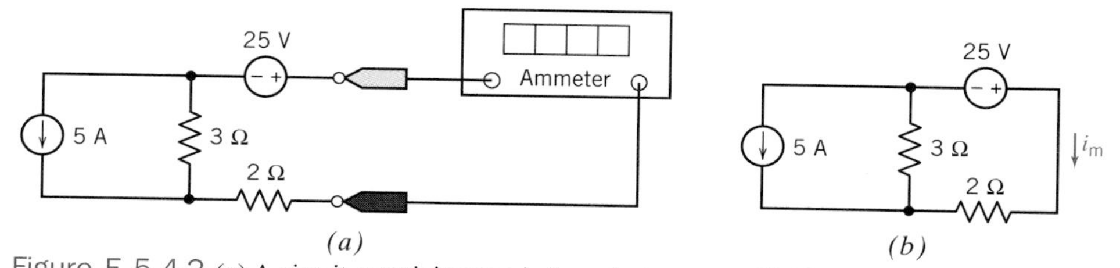

Figure E 5.4-2 (*a*) A circuit containing two independent sources. (*b*) The circuit after the ideal ammeter has been replaced by the equivalent short circuit and a label has been added to indicate the current measured by the ammeter, i_m.

Hint: Figure E 5.4-2b shows the circuit after the ideal ammeter has been replaced by the equivalent short circuit and a label has been added to indicate the current measured by the ammeter, i_{m}.

Answer: $i_m = \dfrac{25}{3+2} - \dfrac{3}{2+3}5 = 5 - 3 = 2\text{ A}$

Exercise 5.4-3 Using the superposition principle, find the value of the voltage measured by the voltmeter in Figure E 5.4-3a.

Hint: Figure E 5.4-3b shows the circuit after the ideal voltmeter has been replaced by the equivalent open circuit and a label has been added to indicate the voltage measured by the voltmeter, v_{m}.

Answer: $v_m = 3\left(\dfrac{3}{3+(3+3)}5\right) - \dfrac{3}{3+(3+3)}18 = 5 - 6 = -1\text{ V}$

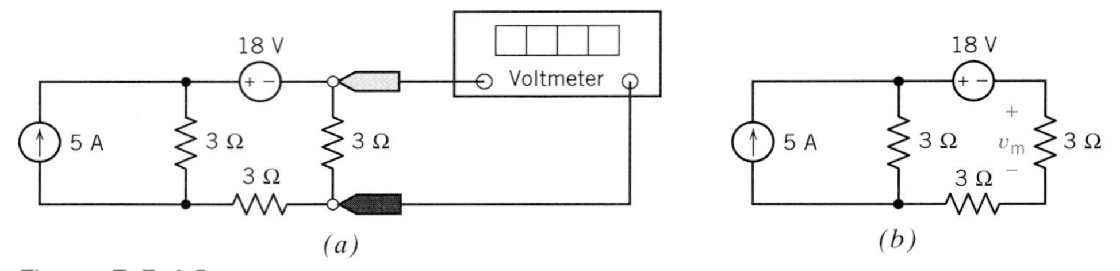

Figure E 5.4-3 (a) A circuit containing two independent sources. (b) The circuit after the ideal voltmeter has been replaced by the equivalent open circuit and a label has been added to indicate the voltage measured by the voltmeter, v_{m}.

In the following, let us use the superposition principle when the circuit contains both independent and dependent sources.

Example 5.4-2

Find the current i for the circuit of Figure 5.4-3. All resistances are in ohms.

Solution

We need to find the current i due to two independent sources. First, we will find the current resulting from the independent voltage source. The current source is replaced with an open circuit, and we have the circuit as shown in Figure 5.4-4a. The current i_1 represents the portion of current i resulting from the first source.

Kirchhoff's voltage law around the loop gives

$$-24 + (3+2)i_1 + 3i_1 = 0$$

or

$$i_1 = 3$$

When we set the independent voltage source to zero and determine the current i_2 due to the current source, we obtain the circuit of Figure 5.4-4b. Writing the KCL equation at node a, we obtain

$$-i_2 - 7 + \frac{v_{\text{a}} - 3\,i_2}{2} = 0 \qquad (5.4\text{-}2)$$

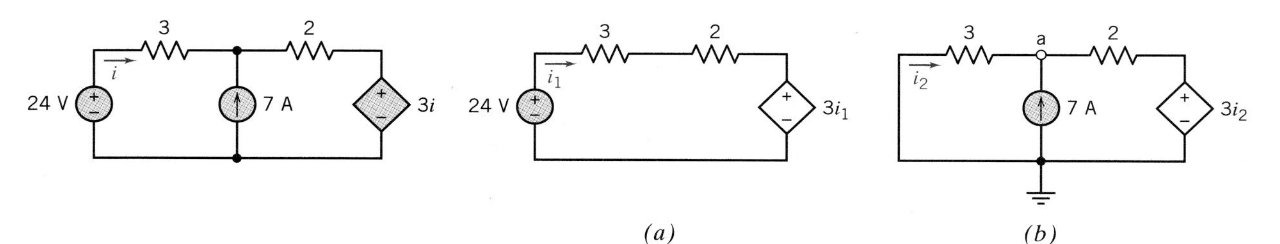

Figure 5.4-3 Circuit for Example 5.4-2.

Figure 5.4-4 Circuit with (*a*) the voltage source activated and the current source deactivated and (*b*) the current source activated and the voltage source deactivated. All resistances are in ohms.

Noting that $-i_2 = v_a/3$ in the left-hand branch, we can state that $v_a = -3i_2$. Substituting for v_a in Eq. 5.4-2, we obtain

$$-i_2 - 7 + \frac{-3\,i_2 - 3\,i_2}{2} = 0$$

Therefore

$$i_2 = -\frac{7}{4}$$

Thus, the total current is

$$i = i_1 + i_2 = 3 - \frac{7}{4} = \frac{5}{4}\,\text{A}$$

5.5 THÉVENIN'S THEOREM

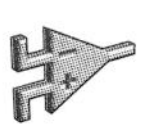

In the preceding section we saw that the analysis of a circuit may be significantly simplified by use of the superposition principle. A theorem was developed by M. L. Thévenin, a French engineer, who first published the principle in 1883. Thévenin, who is credited with the theorem, probably based his work on earlier work by Hermann von Helmholtz (see Figure 5.5-1).

The goal of Thévenin's theorem is to reduce some portion of a circuit to an equivalent source and a single element. Thévenin's theorem rests on the concept of equivalence. A circuit equivalent to another circuit exhibits identical characteristics at identical terminals.

This circuit simplification process can be illustrated by the circuit of Figure 5.5-2*a*. If we wish to determine the current or power delivered to R_L, the rest of the circuit can be

Figure 5.5-1 Hermann von Helmholtz (1821–1894), who is often credited with the basic work leading to Thévenin's theorem. Courtesy of the New York Public Library.

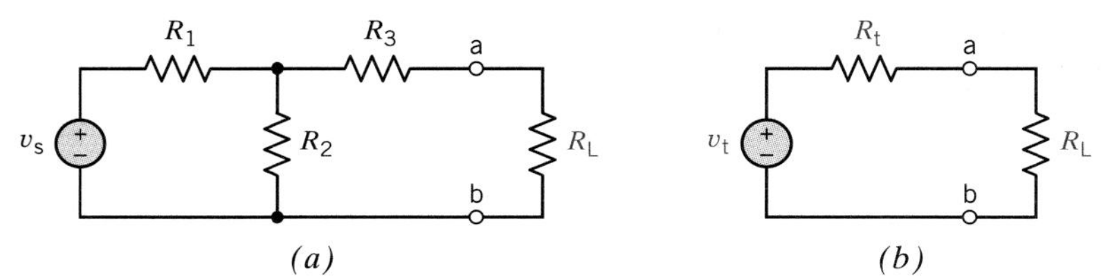

Figure 5.5-2 (*a*) Circuit and (*b*) a Thévenin circuit connected to a load resistor.

reduced to an equivalent circuit, shown in Figure 5.2-2*b*. The Thévenin equivalent circuit is shown to the left of terminals a–b and consists of a voltage v_t and a resistance R_t. Thévenin's principle is particularly useful when we wish to find the resulting current, voltage, or power delivered to a single element, especially when the element is variable. We reduce the rest of the circuit to R_t in series with a voltage source v_t, as shown in Figure 5.5-2*b*, and then later reconnect it to the element.

Thévenin's theorem may be stated as follows: Given any linear circuit, divide it into two circuits, A and B, each connected at the same pair of terminals. If either circuit contains a dependent source, it and its control variable must be in the same circuit. Then find the Thévenin equivalent for circuit A. Define v_{oc} as the open-circuit voltage of circuit A when circuit B is disconnected from the two terminals. Then the equivalent circuit of A is a voltage source v_{oc} in series with R_t, where R_t is the resistance looking into the terminals of circuit A with all its independent sources deactivated (set equal to zero).

Thévenin's theorem requires that, for any circuit of resistance elements and energy sources with an identified terminal pair, the circuit can be replaced by a series combination of an ideal voltage source v_t and a resistance R_t, where v_t is the open-circuit voltage at the two terminals and R_t is the ratio of the open-circuit voltage to the short-circuit current at the terminal pair.

Thévenin's theorem is summarized in Figure 5.5-3. If we examine the circuits of Figures 5.5-3*b* and 5.5-3*c*, they must exhibit the same circuit characteristics at the terminals. Since the circuit of Figure 5.5-3*b* must have the same open-circuit voltage at the two terminals as that of Figure 5.5-3*c*, we may state that $v_t = v_{oc}$, where v_{oc} is the open-circuit voltage of circuit A.

The resistance calculated looking into circuit A must be the same for the circuits of Figures 5.5-3*b* and 5.5-3*c*. If we choose to set all the independent sources of circuit A equal to zero, we may determine the equivalent resistance between the terminals, which is then required to be equal to R_t. Thévenin's theorem is often used to separate a circuit into a simple part and a second part where the components are nonlinear, unknown, or not yet specified. The simple part is selected to be as large as possible and is replaced by its Thévenin equivalent. Another use of Thévenin's theorem is to determine R_t of a circuit so that the load resistance can be selected for maximum power transfer.

In summary, Thévenin's equivalent circuit for circuit A is a voltage source $v_t = v_{oc}$ and a resistor R_t, where R_t is the resistance of circuit A seen at its terminals. (Many engineers use R_{th} instead of R_t as notation for the Thévenin resistance.) The Thévenin equivalent circuit is R_t in series with v_{oc}.

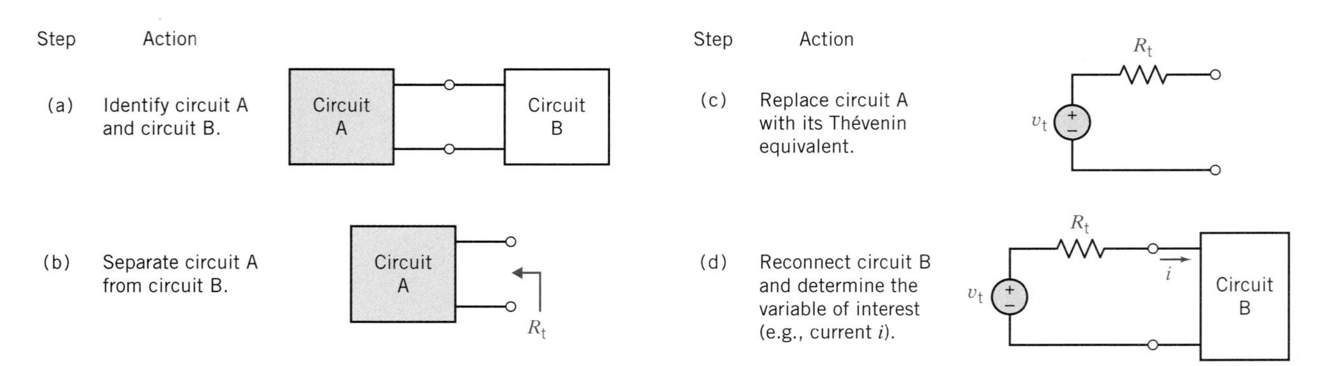

Figure 5.5-3 Summary of Thévenin circuit approach.

Example 5.5-1

Using Thévenin's theorem, find the current i through the resistor R in the circuit of Figure 5.5-4. All resistances are in ohms.

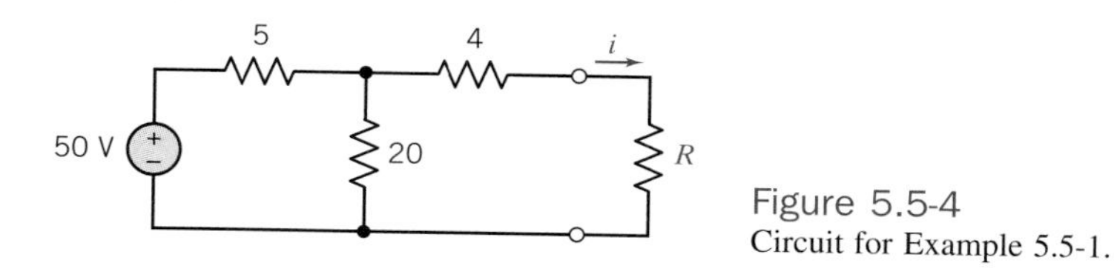

Figure 5.5-4
Circuit for Example 5.5-1.

Solution

Since we are interested in the current i, we identify the resistor R as circuit B. Then circuit A is as shown in Figure 5.5-5*a*. The Thévenin resistance R_t is found from Figure 5.5-5*b*, where we have deactivated the voltage source. We calculate the equivalent resistance looking into the terminals, obtaining $R_t = 8\ \Omega$.

The Thévenin voltage v_t is equal to v_{oc}, the open-circuit voltage. Using the voltage divider principle with the circuit of Figure 5.5-5*a*, we find $v_{oc} = 40$ V.

Reconnecting circuit B to the Thévenin equivalent circuit as shown in Figure 5.5-5*d*, we obtain

$$i = \frac{40}{R + 8} \text{ A}$$

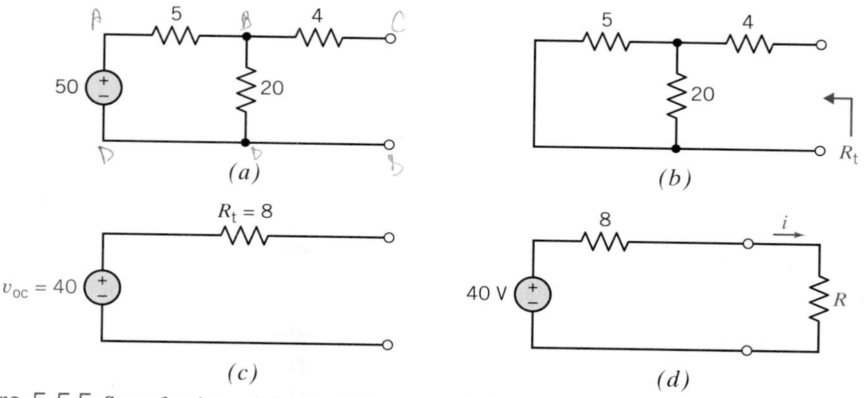

Figure 5.5-5 Steps for determining the Thévenin equivalent circuit for the circuit left of the terminals of Figure 5.5-4.

Example 5.5-2

Find the Thévenin equivalent circuit for the circuit shown in Figure 5.5-6.

Solution

One approach is to find the open-circuit voltage and the circuit's Thévenin equivalent resistance R_t. First, let us find the resistance R_t. Deactivating the sources results in a short

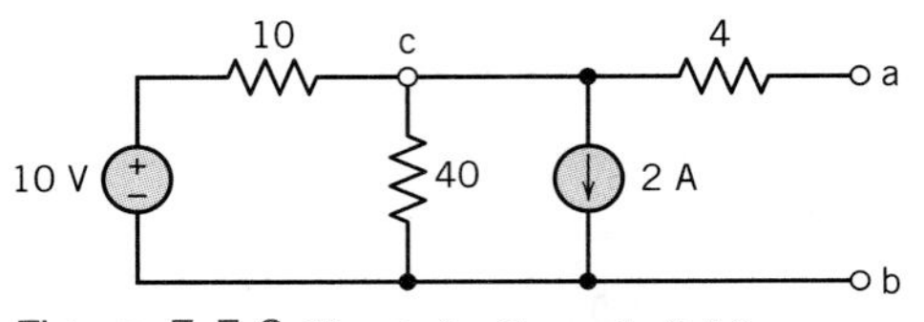

Figure 5.5-6 Circuit for Example 5.5-2. Resistances in ohms.

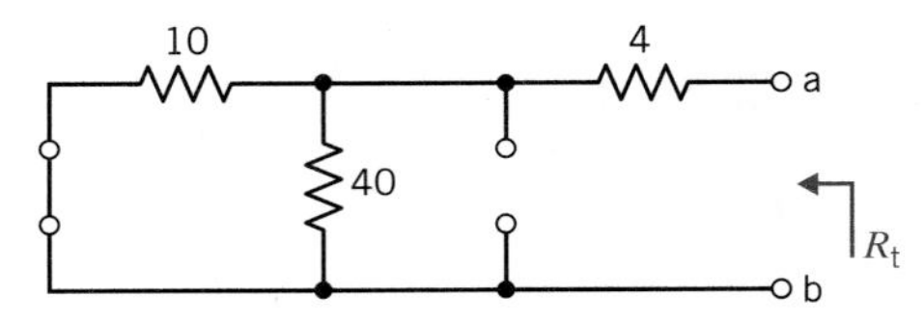

Figure 5.5-7 Circuit of Figure 5.5-6 with all the sources deactivated. Resistances in ohms.

circuit for the voltage source and an open circuit for the current source, as shown in Figure 5.5-7. Look into the circuit at terminals a–b to find R_t. The 10 Ω resistor in parallel with the 40 Ω resistor results in an equivalent resistance of 8 Ω. Adding 8 Ω to 4 Ω in series, we obtain

$$R_t = 12\ \Omega$$

Next, we wish to determine the open-circuit voltage at terminals a–b. Since no current flows through the 4 Ω resistor, the open-circuit voltage is identical to the voltage across the 40 Ω resistor, v_c. Using the bottom node as the reference node, we write KCL at node c of Figure 5.5-6 to obtain

$$\frac{v_c - 10}{10} + \frac{v_c}{40} + 2 = 0$$

Solving for v_c yields

$$v_c = -8\ \text{V}$$

Therefore, the Thévenin equivalent circuit is as shown in Figure 5.5-8.

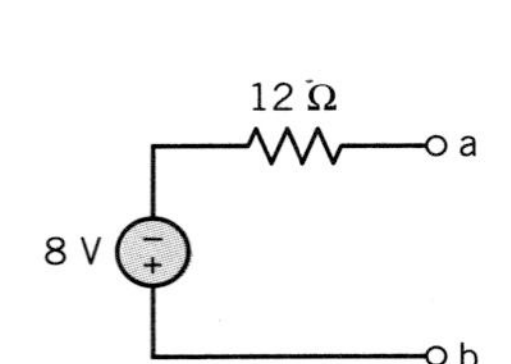

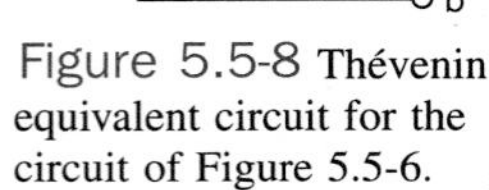
Figure 5.5-8 Thévenin equivalent circuit for the circuit of Figure 5.5-6.

Some circuits contain one or more dependent sources as well as independent sources. The presence of the dependent source prevents us from directly obtaining R_t from simple circuit reduction using the rules for parallel and series resistors.

A procedure for determining R_t is: (1) determine the open-circuit voltage v_{oc}, and (2) determine the short-circuit current i_{sc} when terminals a–b are connected by a short circuit, as shown in Figure 5.5-9; then

$$R_t = \frac{v_{oc}}{i_{sc}}$$

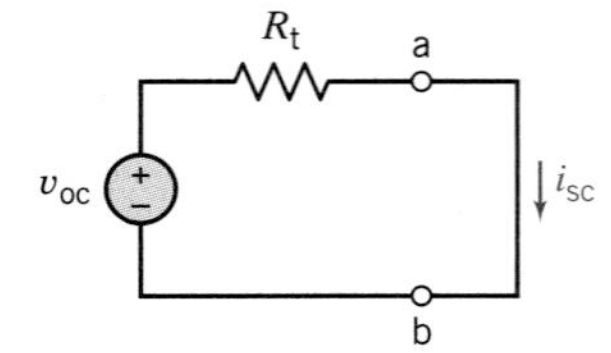

Figure 5.5-9 Thévenin circuit with a short circuit at terminals a–b.

This method is attractive since we already need the open-circuit voltage for the Thévenin equivalent circuit. We can show that $R_t = v_{oc}/i_{sc}$ by writing the KVL equation for the loop of Figure 5.5-9, obtaining

$$-v_{oc} + R_t\, i_{sc} = 0$$

Clearly, $R_t = v_{oc}/i_{sc}$.

Example 5.5-3

Find the Thévenin equivalent circuit for the circuit shown in Figure 5.5-10, which includes a dependent source. All resistances are in ohms.

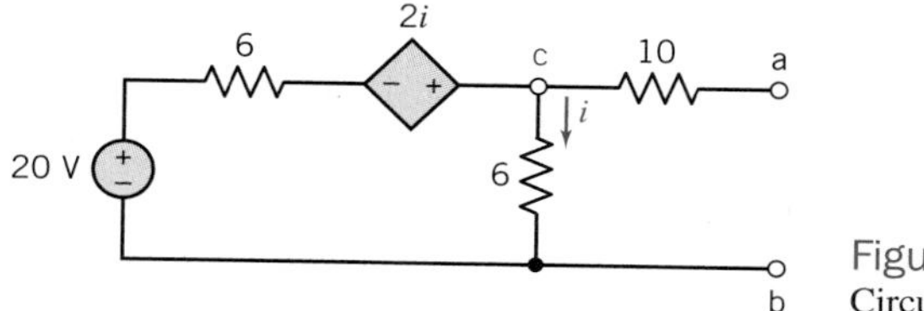

Figure 5.5-10
Circuit of Example 5.5-3.

Solution

First, we find the open-circuit voltage $v_{oc} = v_{ab}$. Writing KVL around the mesh of Figure 5.5-10 (using i as the mesh current), we obtain

$$-20 + 6i - 2i + 6i = 0$$

Therefore

$$i = 2 \text{ A}$$

Since there is no current flowing through the 10 Ω resistor, the open-circuit voltage is identical to the voltage across the resistor between terminals c and b. Therefore,

$$v_{oc} = 6i = 12 \text{ V}$$

The next step is to determine the short-circuit current for the circuit of Figure 5.5-11. Using the two mesh currents indicated, we have

$$-20 + 6i_1 - 2i + 6(i_1 - i_2) = 0$$

and

$$6(i_2 - i_1) + 10i_2 = 0$$

Substitute $i = i_1 - i_2$ and rearrange the two equations to obtain

$$10i_1 - 4i_2 = 20$$

and

$$-6i_1 + 16i_2 = 0$$

Therefore, we find that $i_2 = i_{sc} = 120/136$ A. The Thévenin resistance is

$$R_t = \frac{v_{oc}}{i_{sc}} = \frac{12}{120/136} = 13.6 \ \Omega$$

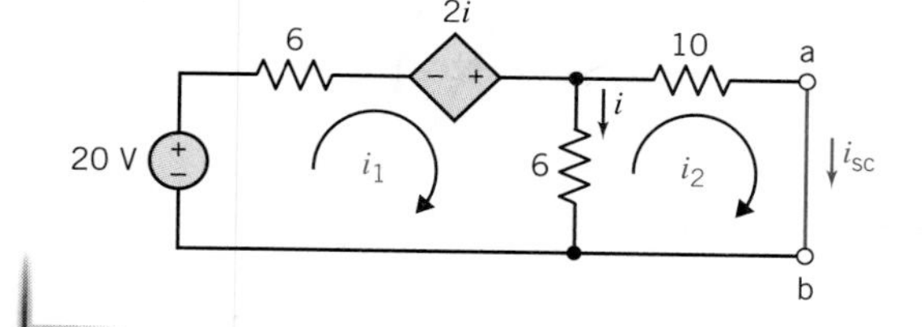

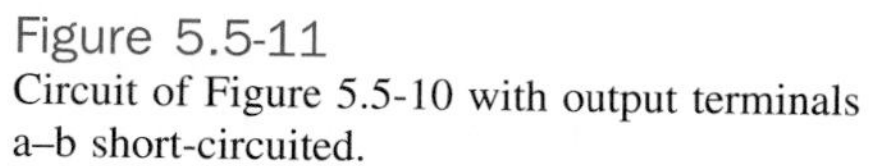
Figure 5.5-11
Circuit of Figure 5.5-10 with output terminals a–b short-circuited.

Example 5.5-4

Find the Thévenin equivalent circuit to the left of terminals a–b for the circuit of Figure 5.5-12. All resistances are in ohms.

Solution

First, we need to determine the open-circuit voltage $v_{ab} = v_{oc}$. Noting that the current source $v_{ab}/4$ provides a current source common to two meshes, we create a supermesh

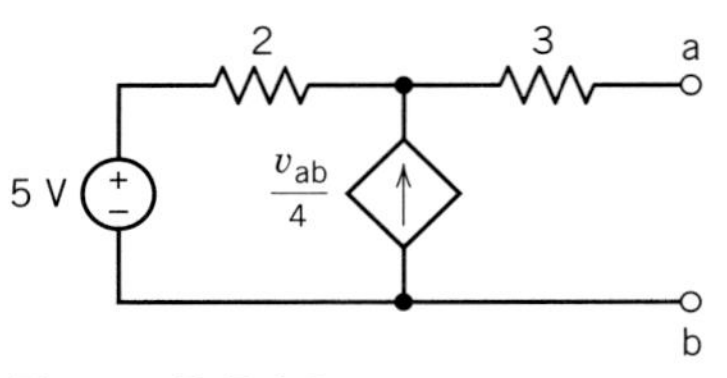

Figure 5.5-12 Circuit for Example 5.5-4.

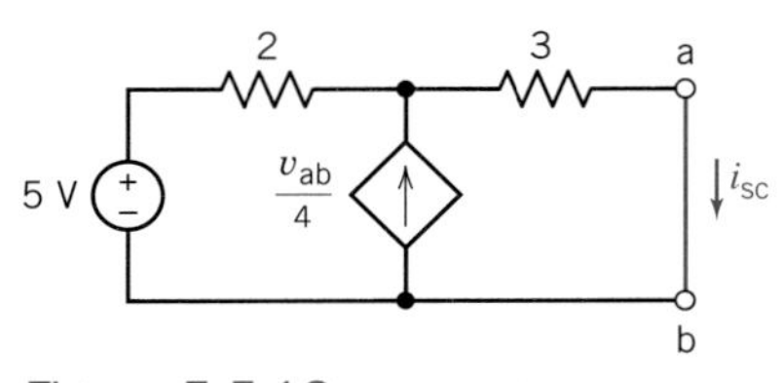

Figure 5.5-13 Circuit of Figure 5.5-12 with terminals a–b short-circuited. Resistances in ohms.

that includes v_{ab} and write one KVL equation around the periphery of the supermesh (see Table 4.7-1 to review the concept of supermesh). Also note that the current in the 3 Ω resistor is zero. The KVL equation around the periphery is

$$-5 + 2\left(\frac{-v_{ab}}{4}\right) + 3(0) + v_{ab} = 0$$

Therefore

$$v_{ab} = 10 \text{ V}$$

To find the short-circuit current, we establish a short circuit across a–b, as shown in Figure 5.5-13. Since $v_{ab} = 0$, the current source is set to zero (i.e., replaced by an open circuit). Then, using KVL around the periphery of the supermesh, we have

$$-5 + (2 + 3)\, i_{sc} = 0$$

or

$$i_{sc} = 1 \text{ A}$$

Therefore, the Thévenin resistance is

$$R_t = \frac{v_{oc}}{i_{sc}} = 10\ \Omega$$

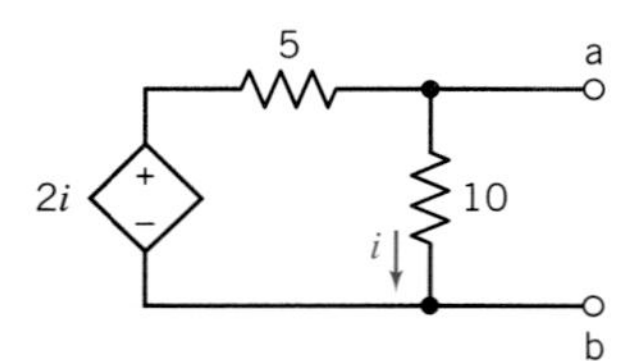

Figure 5.5-14 A circuit for which we seek its Thévenin equivalent. Resistances in ohms.

Another arrangement we may confront is a circuit containing no independent sources but one or more dependent sources. Consider the circuit without independent sources shown in Figure 5.5-14. We wish to determine the Thévenin equivalent circuit. Thus, we will determine v_{oc} and R_t at the terminals a–b.

Since the circuit has no independent sources, $i = 0$ when terminals a–b are open. Therefore, $v_{oc} = 0$. Similarly, we find that $i_{sc} = 0$.

It remains to determine R_t. Since $v_{oc} = 0$ and $i_{sc} = 0$, R_t cannot be determined from $R_t = v_{oc}/i_{sc}$. Therefore, we choose to connect a current source of 1 A at terminals a–b, as shown in Figure 5.5-15. Then, determining v_{ab}, the Thévenin resistance is

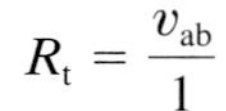

$$R_t = \frac{v_{ab}}{1}$$

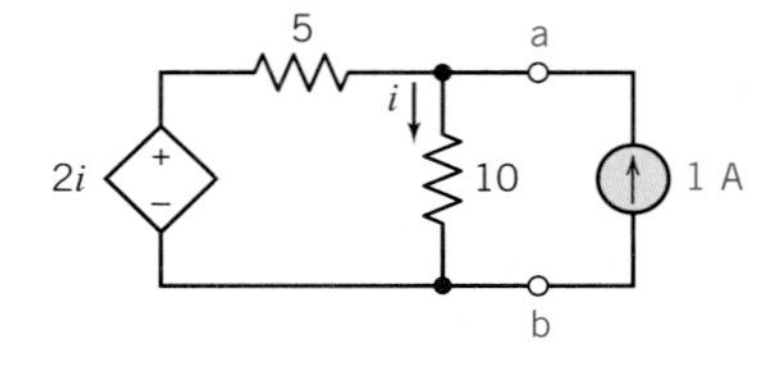

Figure 5.5-15 Circuit of Figure 5.5-14 with a 1-A source connected at terminals a–b. Resistances in ohms.

Writing KCL at a and setting b as the reference, we obtain

$$\frac{v_a - 2i}{5} + \frac{v_a}{10} - 1 = 0$$

For the 10 Ω resistor,

$$i = \frac{v_a}{10}$$

and therefore we have

$$\frac{v_a - 2(v_a/10)}{5} + \frac{v_a}{10} = 1$$

or

$$v_a = \frac{50}{13}\ \text{V}$$

The Thévenin resistance is $R_t = v_a/1$ or

$$R_t = \frac{50}{13}\ \Omega$$

The Thévenin equivalent is shown in Figure 5.5-16.

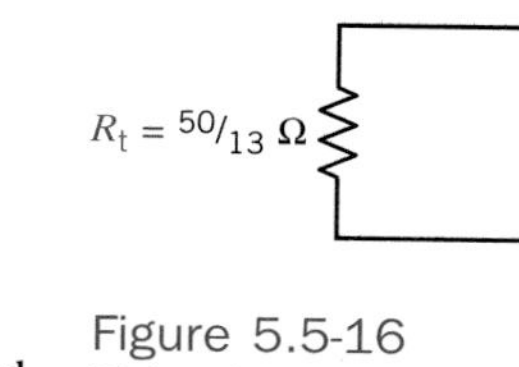

Figure 5.5-16 Thévenin equivalent circuit for the circuit of Figure 5.5-14.

The three common methods of finding a Thévenin equivalent circuit are summarized in Table 5.5-1. When separating a circuit containing dependent sources into two subcircuits to make a Thévenin equivalent, remember that the dependent source and its control voltage or current must be in the same subcircuit.

A laboratory procedure for determining the Thévenin equivalent of a black box circuit (see Figure 5.5-17*a*) is to measure i and v for two or more values of v_s and a fixed value of R. For the circuit of Figure 5.5-17*b* we replace the test circuit with its Thévenin equivalent obtaining

$$v = v_{oc} + i\,R_t \tag{5.5-1}$$

The procedure is to measure v and i for a fixed R and several values of v_s. For example, let $R = 10\ \Omega$ and consider the two measurement results

$$(1)\quad v_s = 49\ \text{V} : i = 0.5\ \text{A},\ v = 44\ \text{V}$$

and

$$(2)\quad v_s = 76\ \text{V} : i = 2\ \text{A},\quad v = 56\ \text{V}$$

Then we have two simultaneous equations (using Eq. 5.5-1):

$$44 = v_{oc} + 0.5\,R_t$$
$$56 = v_{oc} + 2\,R_t$$

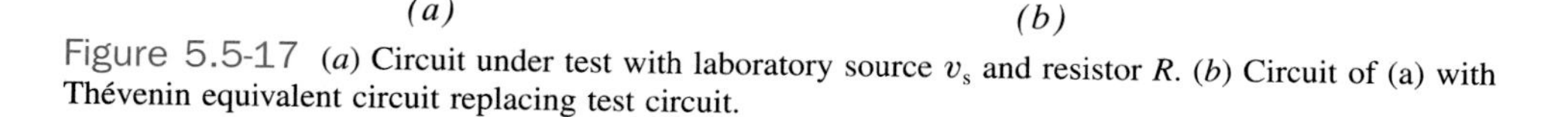

(*a*) (*b*)

Figure 5.5-17 (*a*) Circuit under test with laboratory source v_s and resistor R. (*b*) Circuit of (a) with Thévenin equivalent circuit replacing test circuit.

Table 5.5-1 **Methods of Finding a Thévenin Equivalent Circuit**

Number of Method	If the Circuit Contains:	Thevenin Equivalent Circuit
1	Resistors and independent sources	(a) Connect an open circuit between terminals a and b. Find $v_{oc} = v_{ab}$, the voltage across the open circuit. (b) Deactivate the independent sources. Find R_t by circuit resistance reduction.
2	Resistors and independent and dependent sources **or** Resistors and independent sources	(a) Connect an open circuit between terminals a and b. Find $v_{oc} = v_{ab}$, the voltage across the open circuit. (b) Connect a short circuit between terminals a and b. Find i_{sc}, the current directed from a to b in the short circuit. (c) Connect a 1-A current source from terminal b to terminal a. Determine v_{ab}. Then $R_t = v_{ab}/1$. or $R_t = v_{oc}/i_{sc}$.
3	Resistors and dependent sources (no independent sources)	(a) Note that $v_{oc} = 0$. (b) Connect a 1-A current source from terminal b to terminal a. Determine v_{ab}. Then $R_t = v_{ab}/1$.

Solving these simultaneous equations, we get $R_t = 8\ \Omega$ and $v_{oc} = 40$ V, thus obtaining the Thévenin equivalent of the black box circuit.

Exercise 5.5-1 Determine values of R_t and v_{oc} that cause the circuit shown in Figure 5.5-1*b* to be the Thévenin equivalent circuit of the circuit in Figure E 5.5-1*a*.

Answer: $R_t = 8\ \Omega$ and $v_{oc} = 2$ V.

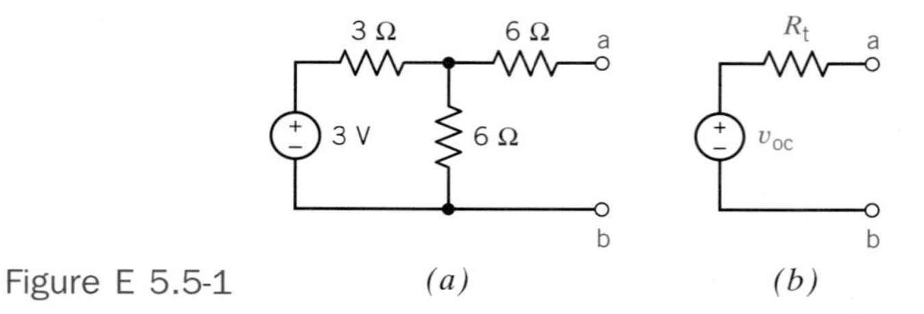

Figure E 5.5-1

Exercise 5.5-2 Determine values of R_t and v_{oc} that cause the circuit shown in Figure 5.5-2*b* to be the Thévenin equivalent circuit of the circuit in Figure E 5.5-2*a*.

Answer: $R_t = 3\ \Omega$ and $v_{oc} = -6$ V.

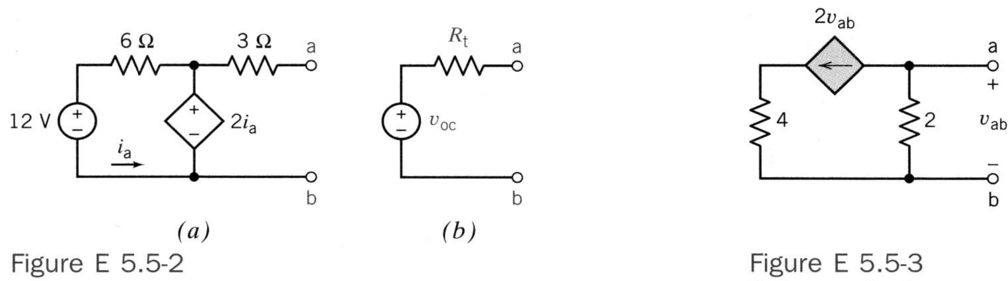

(*a*) (*b*)

Figure E 5.5-2

Figure E 5.5-3 Resistances in ohms.

Exercise 5.5-3 Find the Thévenin equivalent for the circuit shown in Figure E 5.5-3.

Answer: $v_{oc} = 0$, $R_t = 2/5\ \Omega$

5.6 NORTON'S EQUIVALENT CIRCUIT

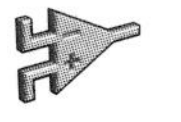

An American engineer, E. L. Norton at Bell Telephone Laboratories, proposed an equivalent circuit for circuit A of Figure 5.5-3 using a current source and an equivalent resistance. The Norton equivalent circuit is a dual of the Thévenin equivalent circuit. Norton published his method in 1926, 43 years after Thévenin. The Norton equivalent is the source transformation of the Thévenin equivalent.

Norton's theorem may be stated as follows: Given any linear circuit, divide it into two circuits, A and B. If either A or B contains a dependent source, its controlling variable must be in the same circuit. Consider circuit A and determine its short-circuit current i_{sc} at its terminals. Then the equivalent circuit of A is a current source i_{sc} in parallel with a resistance R_n, where R_n is the resistance looking into circuit A with all its independent sources deactivated.

Norton's theorem requires that, for any circuit of resistance elements and energy sources with an identified terminal pair, the circuit can be replaced by a parallel combination of an ideal current source i_{sc} and a conductance G_n, where i_{sc} is the short-circuit current at the two terminals and G_n is the ratio of the short-circuit current to the open-circuit voltage at the terminal pair.

Figure 5.6-1 Norton equivalent circuit for a linear circuit A.

We therefore have the Norton circuit for circuit A as shown in Figure 5.6-1. Since this is the dual of the Thévenin circuit, it is clear that $R_n = R_t$ and $v_{oc} = R_t\, i_{sc}$. The Norton equivalent is simply the source transformation of the Thévenin equivalent.

Example 5.6-1

Find the Norton equivalent circuit for the circuit of Figure 5.6-2.

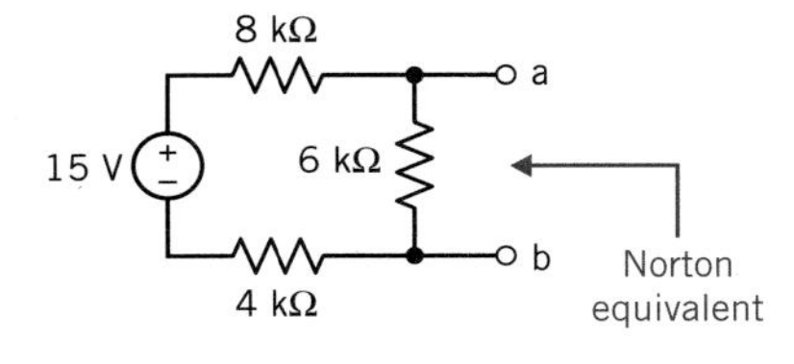

Figure 5.6-2 Circuit of Example 5.6-1.

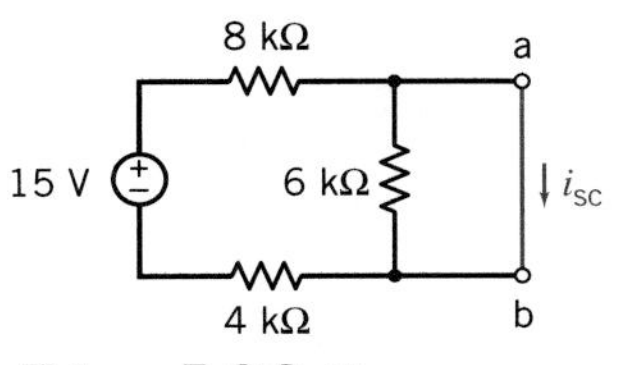

Figure 5.6-3 Short circuit connected to output terminals.

Solution

Since the circuit contains only an independent source, we can deactivate the source and find R_n by circuit reduction. Replacing the voltage source by a short circuit, we have a 6 kΩ resistor in parallel with (8 kΩ + 4 kΩ) = 12 kΩ. Therefore,

$$R_n = \frac{6 \times 12}{6 + 12} = 4\text{ k}\Omega$$

To determine i_{sc} we short circuit the output terminals with the voltage source activated as shown in Figure 5.6-3. Writing KCL at node a we have

$$-\frac{15\text{ V}}{12\text{ k}\Omega} + i_{sc} = 0$$

or

$$i_{sc} = 1.25\text{ mA}$$

Thus, the Norton equivalent (Figure 5.6-1) has $R_n = 4$ kΩ and $i_{sc} = 1.25$ mA.

Example 5.6-2

Find the Norton equivalent circuit for the circuit of Figure 5.6-4.

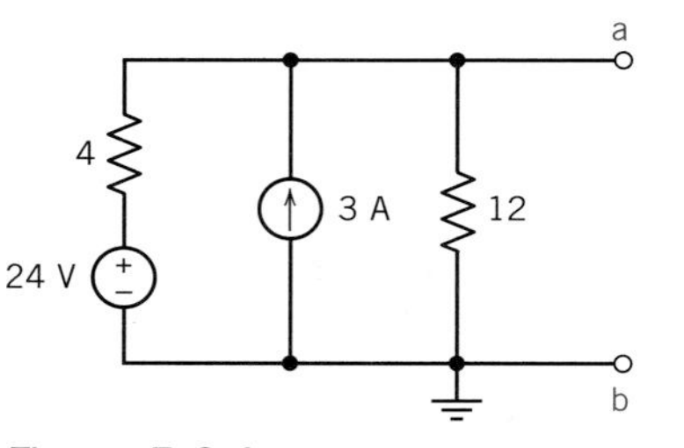

Figure 5.6-4 Circuit of Example 5.6-2. Resistances in ohms.

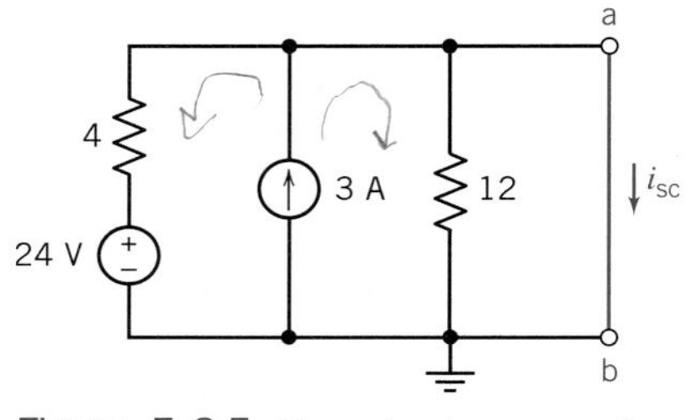

Figure 5.6-5 Short circuit connected to terminals a–b of the circuit of Figure 5.6-4. Resistances in ohms.

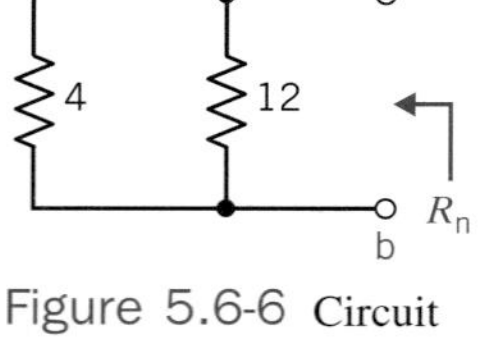

Figure 5.6-6 Circuit of Figure 5.6-4 with its sources deactivated. The voltage source becomes a short circuit, and the current source is replaced by an open circuit. Resistances in ohms.

Solution

First, determine the current i_{sc} for the short-circuit condition shown in Figure 5.6-5. Writing KCL at a, we obtain

$$-\frac{24}{4} - 3 + i_{sc} = 0$$

Note that no current flows in the 12 Ω resistor since it is in parallel with a short circuit. Also, because of the short circuit, the 24-V source causes 24 V to appear across the 4 Ω resistor. Therefore,

$$i_{sc} = \frac{24}{4} + 3 = 9\text{ A}$$

Now determine the equivalent resistance $R_n = R_t$ by deactivating the sources in the circuit as shown in Figure 5.6-6. Clearly, $R_n = 3$ Ω. Thus, we obtain the Norton equivalent circuit as shown in Figure 5.6-7.

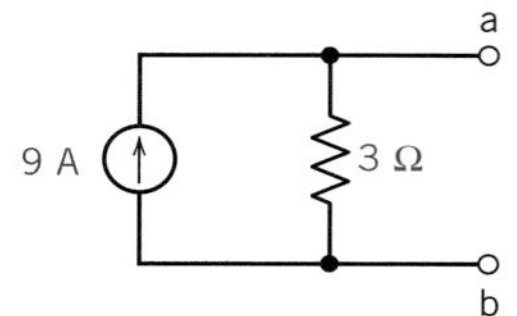

Figure 5.6-7 Norton equivalent of the circuit of Figure 5.6-4.

Table 5.6-1 **Methods of Finding a Norton Equivalent Circuit**

Number of Method	If the Circuit Contains:	Norton Equivalent Circuit
1	Resistors and independent sources	(a) Connect a short circuit between terminals a and b. Find i_{sc}, the current directed from a to b in the short circuit. (b) Deactivate the independent sources. Find $R_n = R_t$ by circuit resistance reduction.
2	Resistors and independent and dependent sources **or** Resistors and independent sources	(a) Connect an open circuit between terminals a and b. Find $v_{oc} = v_{ab}$, the voltage across the open circuit. (b) Connect a short circuit between terminals a and b. Find i_{sc}, the current directed from a to b in the short circuit. (c) Connect a 1-A current source from terminal b to terminal a. Determine v_{ab}. Then $R_n = R_t = v_{ab}/1$. or $R_n = R_t = v_{oc}/i_{sc}$.
3	Resistors and dependent sources (no independent sources)	(a) Note that $i_{sc} = 0$. (b) Connect a 1-A current source from terminal b to terminal a. Determine v_{ab}. Then $R_n = R_t = v_{ab}/1$.

We may encounter circuits with dependent sources as well as independent sources and wish to obtain the Norton equivalent. Since the Norton equivalent is the dual of the Thévenin equivalent, we may develop a table of methods parallel to that summarizing the Thévenin methods (Table 5.5-1). Table 5.6-1 summarizes the three methods of determining the Norton equivalent.

Example 5.6-3

Find the Norton equivalent to the left of terminals a–b for the circuit of Figure 5.6-8.

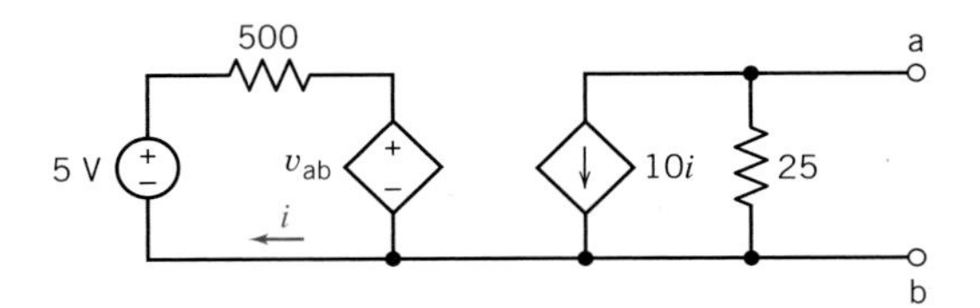

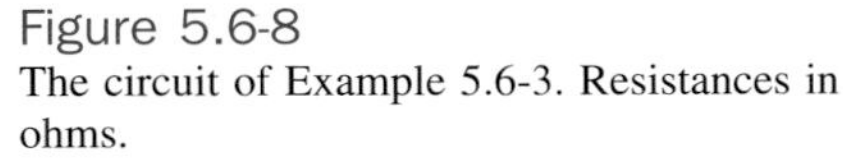
Figure 5.6-8
The circuit of Example 5.6-3. Resistances in ohms.

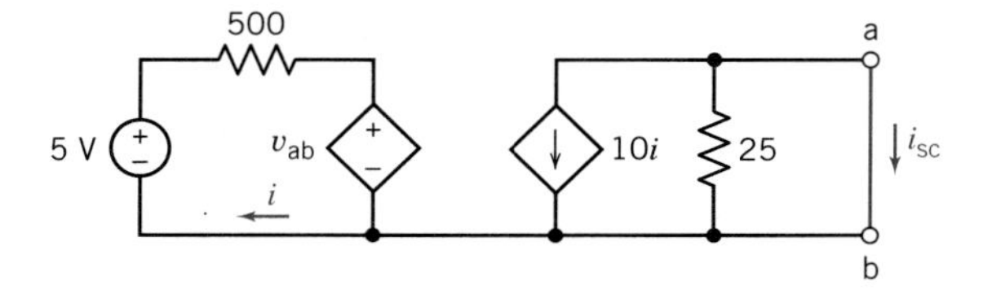

Figure 5.6-9 Circuit of Figure 5.6-8 with a short circuit at the terminals a–b. Resistances in ohms.

Solution

First, we need to determine the short-circuit current i_{sc} using Figure 5.6-9. Note that $v_{ab} = 0$ when the terminals are short-circuited. Then,

$$i = 5/500 = 10 \text{ mA}$$

Therefore, for the right-hand portion of the circuit

$$i_{sc} = -10i = -100 \text{ mA}$$

Now, to obtain R_t, we need $v_{oc} = v_{ab}$ from Figure 5.6-8, where i is the current in the first (left-hand) mesh. Writing the mesh current equation, we have

$$-5 + 500i + v_{ab} = 0$$

Also, for the right-hand mesh of Figure 5.6-8 we note that

$$v_{ab} = -25(10i) = -250i$$

Therefore,

$$i = \frac{-v_{ab}}{250}$$

Substituting i into the first mesh equation, we obtain

$$500\left(\frac{-v_{ab}}{250}\right) + v_{ab} = 5$$

Therefore,

$$v_{ab} = -5 \text{ V}$$

and

$$R_t = \frac{v_{ab}}{i_{sc}} = \frac{-5}{-0.1} = 50\ \Omega$$

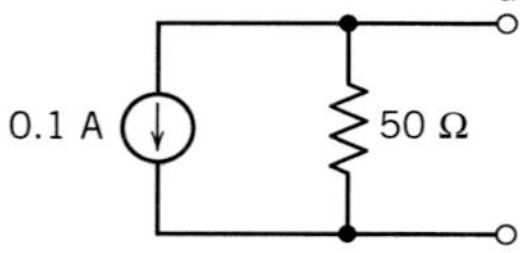

Figure 5.6-10 The Norton equivalent circuit for Example 5.6-3.

The Norton equivalent circuit is shown in Figure 5.6-10.

Exercise 5.6-1 Determine values of R_t and i_{sc} that cause the circuit shown in Figure E 5.6-1*b* to be the Norton equivalent circuit of the circuit in Figure E 5.6-1*a*.

Answer: $R_t = 8\ \Omega$ and $i_{sc} = 0.25$ A.

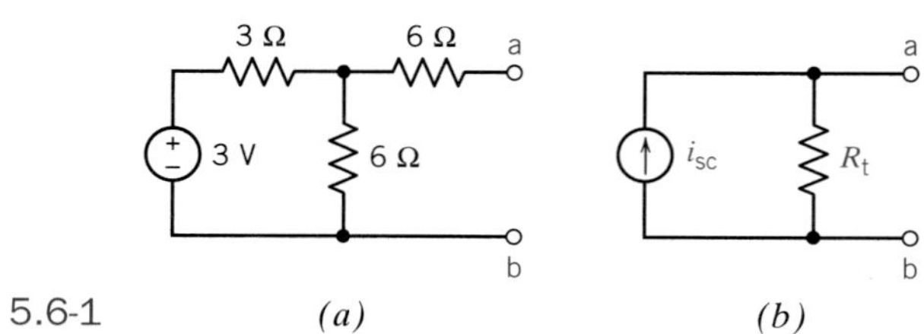

Figure E 5.6-1

Exercise 5.6-2 Determine values of R_t and i_{sc} that cause the circuit shown in Figure E 5.6-2*b* to be the Norton equivalent circuit of the circuit in Figure E 5.6-2*a*.

Answer: $R_t = 3\ \Omega$ and $i_{sc} = -2$ A.

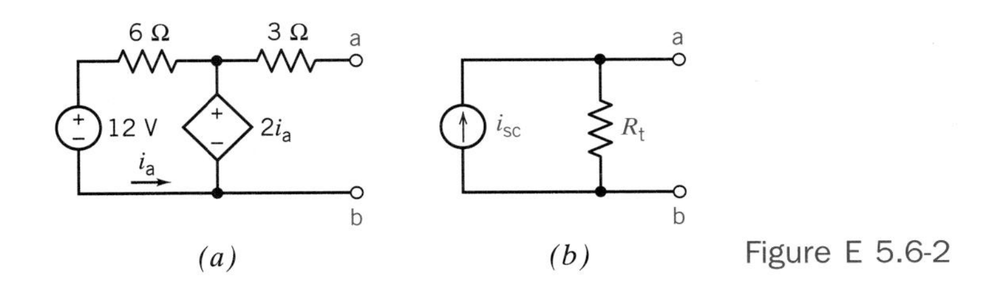

Figure E 5.6-2

Exercise 5.6-3 Use Norton's theorem to formulate a general expression for the current i in terms of the variable resistance R shown in Figure E 5.6-3.

Answer: $i = 20/(8 + R)$ A

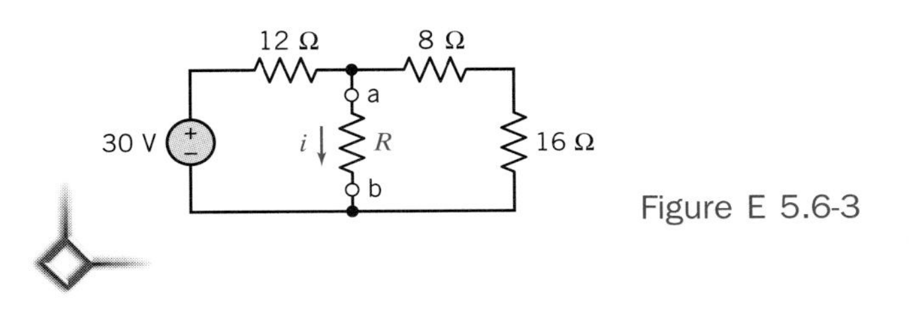

Figure E 5.6-3

5.7 MAXIMUM POWER TRANSFER

Many applications of circuits require that the maximum power available from a source be transferred to a load resistor R_L. Consider the circuit A shown in Figure 5.7-1, terminated with a load R_L. As demonstrated in Section 5.5, circuit A can be reduced to its Thévenin equivalent, as shown in Figure 5.7-2.

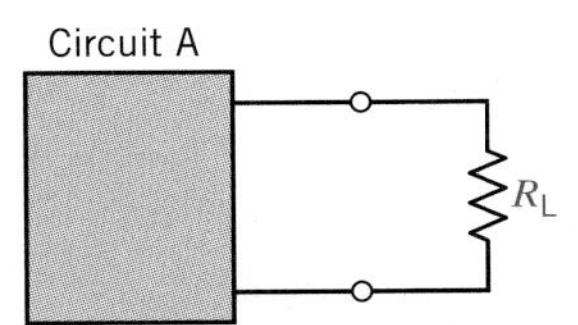

Figure 5.7-1 Circuit A contains resistors and independent and dependent sources. The load is the resistor R_L.

The general problem of power transfer can be discussed in terms of efficiency and effectiveness. Power utility systems are designed to transport the power to the load with the greatest efficiency by reducing the losses on the power lines. Thus, the effort is concentrated on reducing R_t, which would represent the resistance of the source plus the line resistance. Clearly, the idea of using superconducting lines that would exhibit no line resistance is exciting to power engineers.

In the case of signal transmission, as in the electronics and communications industries, the problem is to attain the maximum signal strength at the load. Consider the signal received at the antenna of an FM radio receiver from a distant station. It is the engineer's goal to design a receiver circuit so that the maximum power ultimately ends up at the output of the amplifier circuit connected to the antenna of your FM radio. Thus, we may represent the FM antenna and amplifier by the Thévenin equivalent shown in Figure 5.7-2.

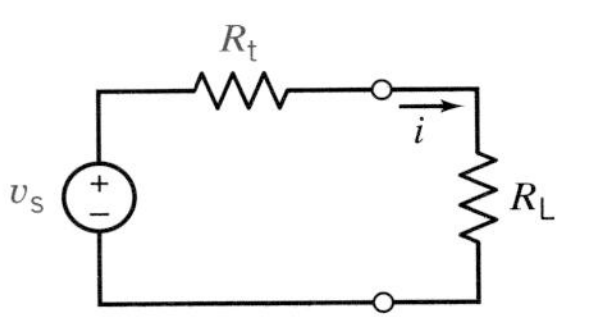

Figure 5.7-2 The Thévenin equivalent is substituted for circuit A. Here we use v_s for the Thévenin source voltage.

Let us consider the general circuit of Figure 5.7-2. We wish to find the value of the load R_L such that the maximum power is delivered to it. First, we need to find the power from

$$p = i^2 R_L$$

Since the current i is

$$i = \frac{v_s}{R_L + R_t}$$

we find that the power is

$$p = \left(\frac{v_s}{R_L + R_t}\right)^2 R_L \tag{5.7-1}$$

Assuming that v_s and R_t are fixed for a given source, the maximum power is a function of R_L. To find the value of R_L that maximizes the power, we use the differential calculus to find where the derivative dp/dR_L equals zero. Taking the derivative, we obtain

$$\frac{dp}{dR_L} = v_s^2 \frac{(R_t + R_L)^2 - 2(R_t + R_L)R_L}{(R_L + R_t)^4}$$

The derivative is zero when

$$(R_t + R_L)^2 - 2(R_t + R_L)R_L = 0 \tag{5.7-2}$$

or

$$(R_t + R_L)(R_t + R_L - 2R_L) = 0 \tag{5.7-3}$$

Solving Eq. 5.7-3, we obtain

$$R_L = R_t \tag{5.7-4}$$

To confirm that Eq. 5.7-4 is a maximum, it should be shown that $d^2p/dR_L^2 < 0$. Therefore, the maximum power is transferred to the load when R_L is equal to the Thévenin equivalent resistance R_t.

The maximum power, when $R_L = R_t$, is then obtained by substituting $R_L = R_t$ in Eq. 5.7-1 to yield

$$p_{max} = \frac{v_s^2 R_L}{(2 R_L)^2} = \frac{v_s^2}{4 R_L}$$

The power delivered to the load will differ from the maximum attainable as the load resistance R_L departs from $R_L = R_t$. The power attained as R_L varies from R_t is portrayed in Figure 5.7-3.

The **maximum power transfer** theorem states that the maximum power delivered by a source represented by its Thévenin equivalent circuit is attained when the load R_L is equal to the Thévenin resistance R_t.

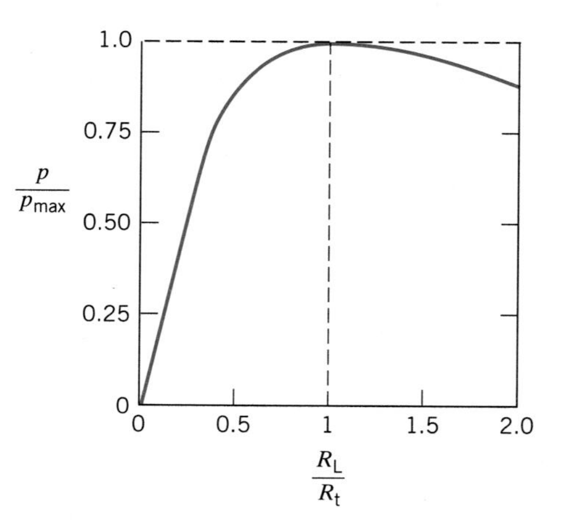

Figure 5.7-3
Power actually attained as R_L varies in relation to R_t.

The *efficiency of power transfer* is defined as the ratio of the power delivered to the load, p_{out}, to the power supplied by the source, p_{in}. Therefore, we have the efficiency η as

$$\eta = p_{out}/p_{in}$$

For maximum power transfer, when $R_s = R_L$, we have

$$p_{in} = v_s i = v_s\left(\frac{v_s}{R_s + R_L}\right) = \frac{v_s^2}{2R_L}$$

The output power delivered to the load was found above to be

$$p_{out} = p_{max} = \frac{v_s^2}{4R_L}$$

Therefore, $p_{out}/p_{in} = 1/2$, and only 50 percent efficiency can be achieved at maximum power transfer conditions. Circuits may have efficiencies less than 50 percent at the maximum power condition.

We may also use Norton's equivalent circuit to represent the circuit A. We then have a circuit with a load resistor R_L as shown in Figure 5.7-4. The current i may be obtained from the current divider principle to yield

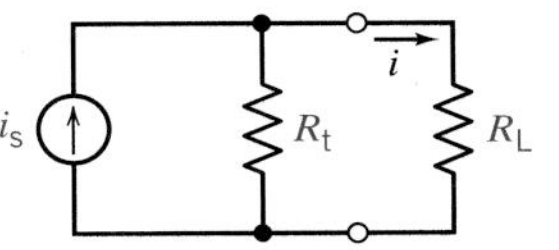

Figure 5.7-4 Norton's equivalent circuit representing the source circuit and a load resistor R_L. Here we use i_s as the Norton source current.

$$i = \frac{R_t}{R_t + R_L} i_s$$

Therefore, the power p is

$$p = i^2 R_L = \frac{i_s^2 R_t^2 R_L}{(R_t + R_L)^2} \qquad (5.7\text{-}5)$$

Using calculus, we can show that the maximum power occurs when

$$R_L = R_t \qquad (5.7\text{-}6)$$

Then the maximum power delivered to the load is

$$p_{max} = \frac{R_L i_s^2}{4} \qquad (5.7\text{-}7)$$

Example 5.7-1

Find the load R_L that will result in maximum power delivered to the load for the circuit of Figure 5.7-5. Also determine the p_{max}.

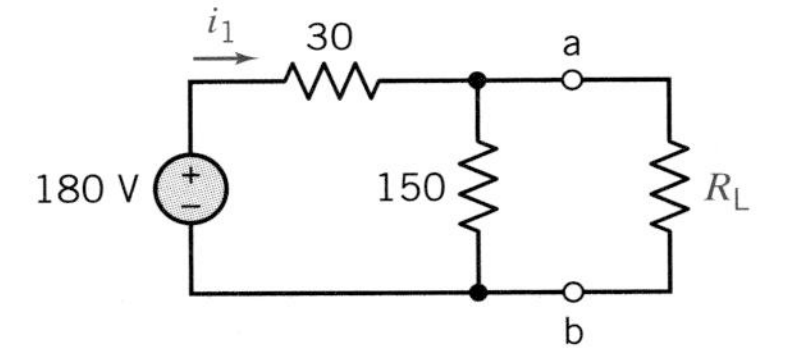

Figure 5.7-5 Circuit for Example 5.7-1. Resistances in ohms.

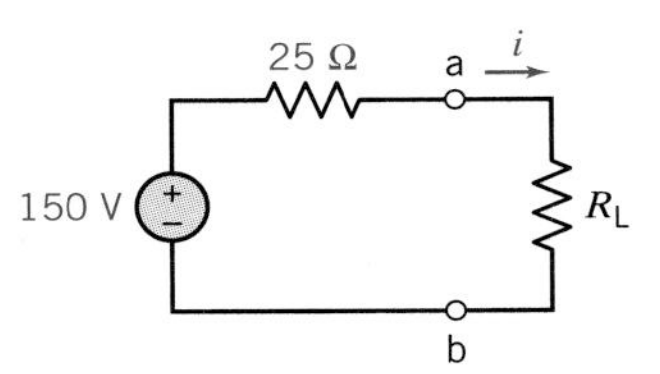

Figure 5.7-6 Thévenin equivalent circuit connected to R_L for Example 5.7-1.

Solution

First, we determine the Thévenin equivalent circuit for the circuit to the left of terminals a–b. Disconnect the load resistor. The Thévenin voltage source v_t is

$$v_t = \frac{150}{180} \times 180 = 150 \text{ V}$$

The Thévenin resistance R_t is

$$R_t = \frac{30 \times 150}{30 + 150} = 25 \ \Omega$$

The Thévenin circuit connected to the load resistor is shown in Figure 5.7-6. Maximum power transfer is obtained when $R_L = R_t = 25 \ \Omega$.

Then the maximum power is

$$p_{max} = \frac{v_s^2}{4 R_L} = \frac{(150)^2}{4 \times 25} = 225 \text{ W}$$

The Thévenin source v_t actually provides a total power of

$$p_s = v_t i = 150 \times 3 = 450 \text{ W}$$

Thus, we note that one-half the power is dissipated by the resistance R_t.

The actual source of the circuit of Figure 5.7-5 is 180 V. This source delivers a power $p = 180 i_1$ where i_1 is the current through the source when $R_L = 25 \ \Omega$. We readily calculate that $i_1 = 3.5$ A. Therefore, the actual source delivers 630 W to the total circuit, resulting in an efficiency of 35.7 percent.

If you prefer to use the Norton equivalent circuit, it is shown in Figure 5.7-4.

Example 5.7-2

Find the load R_L that will result in maximum power delivered to the load of the circuit of Figure 5.7-7a. Also determine p_{max} delivered.

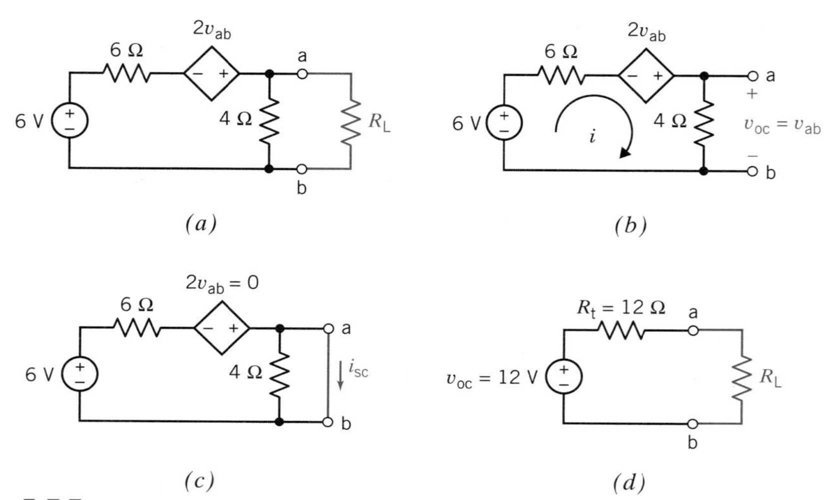

Figure 5.7-7 Determination of maximum power transfer to a load R_L.

Solution

We will use Method 2 of Table 5.5-1 to obtain the Thévenin equivalent circuit for the circuit of Figure 5.7-7*a*. First we find v_{oc} with the sources activated and the load resistor disconnected as shown in Figure 5.7-7*b*. The KVL gives

$$-6 + 10i - 2v_{ab} = 0$$

Also, we note that $v_{ab} = v_{oc} = 4i$. Therefore

$$10i - 8i = 6$$

or $i = 3$ A. Therefore, $v_{oc} = 4i = 12$ V.

To determine the short-circuit current, we add a short circuit as shown in Figure 5.7-7*c*. Writing KVL we have

$$-6 + 6i_{sc} = 0$$

Hence $i_{sc} = 1$ A.

Therefore, $R_t = v_{oc}/i_{sc} = 12\ \Omega$. The Thévenin equivalent circuit is shown in Figure 5.7-7*d* with the load resistor reconnected.

Maximum load power is achieved when $R_L = R_t = 12\ \Omega$. Then

$$p_{max} = \frac{v_{oc}^2}{4R_L} = \frac{12^2}{4(12)} = 3 \text{ W}$$

Exercise 5.7-1

Find the maximum power delivered to R_L for the circuit of Figure E 5.7-1 using a Thévenin equivalent circuit. All resistances in ohms.

Answer: 9 W

Exercise 5.7-2

Find the maximum power delivered to R_L for the circuit of Figure E 5.7-2 using a Norton equivalent circuit. All resistances in ohms.

Answer: 175 W

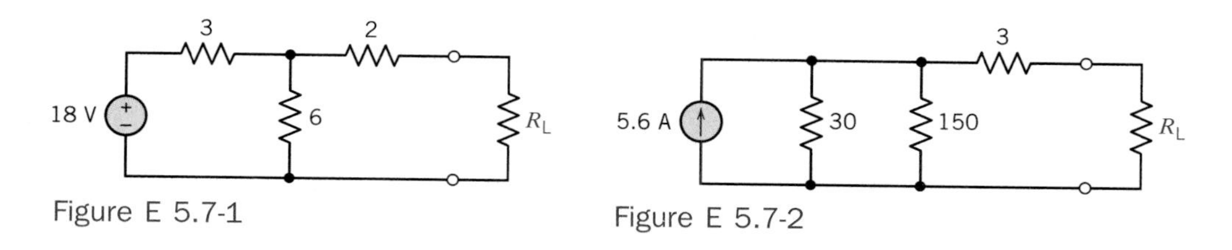

Figure E 5.7-1

Figure E 5.7-2

Exercise 5.7-3

For the circuit of Figure E 5.7-3, find the power delivered to the load when R_L is fixed and R_t may be varied between 1 Ω and 5 Ω. Select R_t so that maximum power is delivered to R_L.

Answer: 13.9 W

Exercise 5.7-4

A resistive circuit was connected to a variable resistor, and the power delivered to the resistor was measured as shown in Figure E 5.7-4. Determine the Thévenin equivalent circuit.

Partial Answer: $R_t = 20\ \Omega$

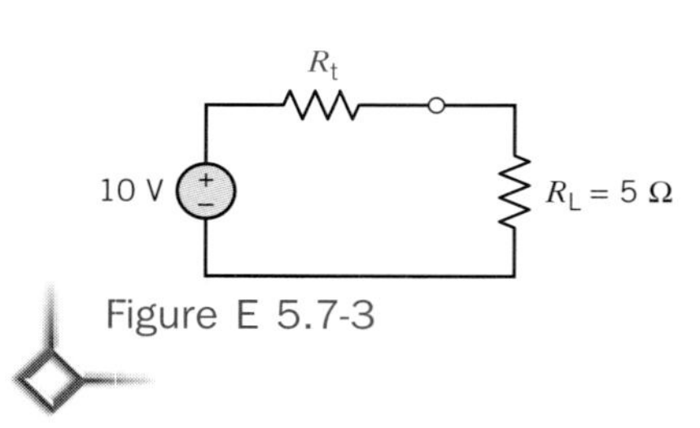

Figure E 5.7-3

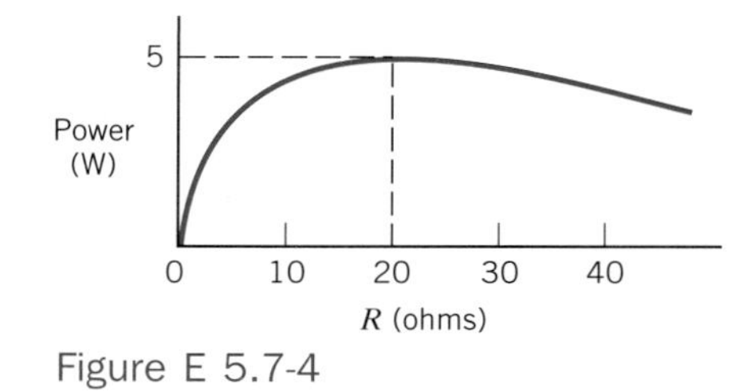

Figure E 5.7-4

5.8 FURTHER USE OF PSPICE FOR DC ANALYSIS[1]

In this section we will illustrate the utility of PSpice analysis when the circuit contains dependent sources and demonstrate the usefulness of the .DC command and the .PRINT command. We also will demonstrate the utility of the .TF statement.

First, let us demonstrate the utility of the .DC command, which permits the analysis of a circuit over a range of values of a source voltage or current. In this example we will change a voltage source over a specified range (often called a sweep analysis). Let us consider a circuit that contains a dependent source shown in Figure 5.8-1. The nodes are labeled with numerals within the circles, and the goal is to obtain the two mesh currents when V_1 equals two values: 10 V and 12 V. This type of analysis is useful; for example, the voltage source might change when it switches from standby to full on. See Appendix F for a review of the .DC statement. The PSpice program for the circuit is provided in Figure 5.8-2, where the VCVS is represented by E1. The controlling voltage is designated as the voltage v_{20} and the multiplier is 10.

The .DC statement sweeps the voltage V_1 between 10 V and 12 V in one increment of 2 V. The .PRINT statement (discussed in Appendix F) provides the output of the currents in R_1 and R_3 since these two currents are i_1 and i_2. The output is shown in Figure 5.8-3.

The .TF statement can be used to determine the dc transfer (gain ratio) function and the Thévenin equivalent of a circuit. The format is

```
.TF < Output Variable > < Input Source >
```

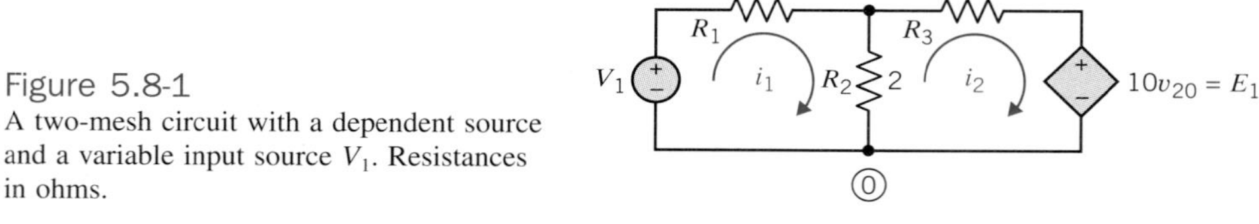

Figure 5.8-1
A two-mesh circuit with a dependent source and a variable input source V_1. Resistances in ohms.

[1]This section may be omitted or considered with a later chapter. See Appendix G for an introduction to PSpice.

```
**    TWO MESH WITH DEPENDENT SOURCE

*ELEM     NODE   NODE    VALUE
 R1        1      2        3
 R2        2      0        2
 R3        2      3        2
 V1        1      0       DC     10
*ELEM     NOD    NOD     CNOD   CNOD   MULT
 E1        3      0        2      0     10
 .DC       V1     10      12      2
 .PRINT   DC     I(R1)   I(R3)
 .END
```

Figure 5.8-2
The PSpice program for the circuit of Figure 5.8-1.

```
****     DC TRANSFER CURVES

*****************************************
  V1              I(R1)          I(R3)
   1.000E+01      3.636E+00       4.091E+00
   1.200E+01      4.364E+00       4.909E+00
```

Figure 5.8-3
Output for the PSpice program of Figure 5.8-2. The currents in R_1 and R_3 are provided for V_1 equal to 10 V and 12 V.

If v_o is the output variable and v_s is the input source, then the computer printout provides v_o/v_s, R_{in} seen by the input source, and the output resistance $R_o = R_t$ seen at v_o as shown in Figure 5.8-4. The .TF statement provides these three calculated values as the printout and a .PRINT statement is not necessary. The .TF calculation provides $v_o = v_t$ and R_t, the Thévenin equivalent circuit elements.

Let us use .TF analysis to determine the Thévenin equivalent circuit for the circuit of Example 5.5-3 as redrawn in Figure 5.8-5. The circuit is redrawn for PSpice analysis in Figure 5.8-6, where the CCVS is represented by H1 and the dummy source VM1 is used to provide i that controls H1. The 10 Ω resistor is omitted and will be added later to R_o of the circuit of Figure 5.8-6. The PSpice program for this circuit is shown in Figure 5.8-7. The output variable is $v_{30} = v_3$ referenced to the ground node.

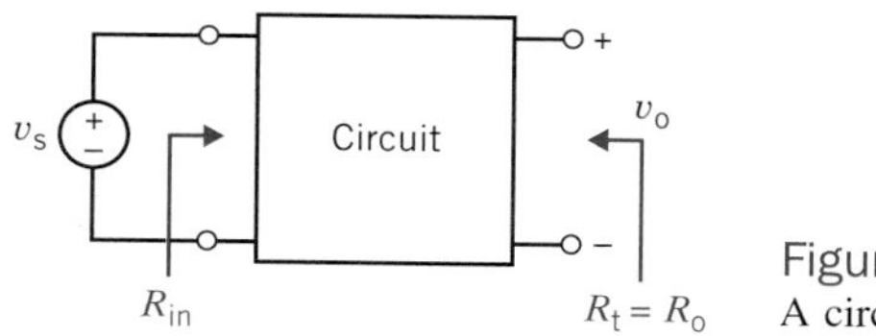

Figure 5.8-4
A circuit with an input source v_s.

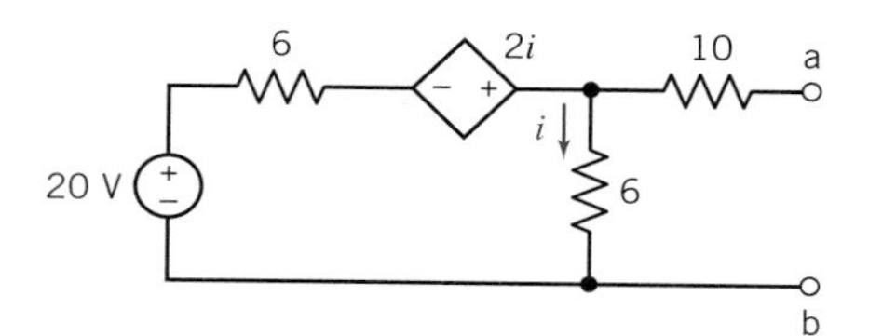

Figure 5.8-5 The circuit of Example 5.5-3. Resistances in ohms.

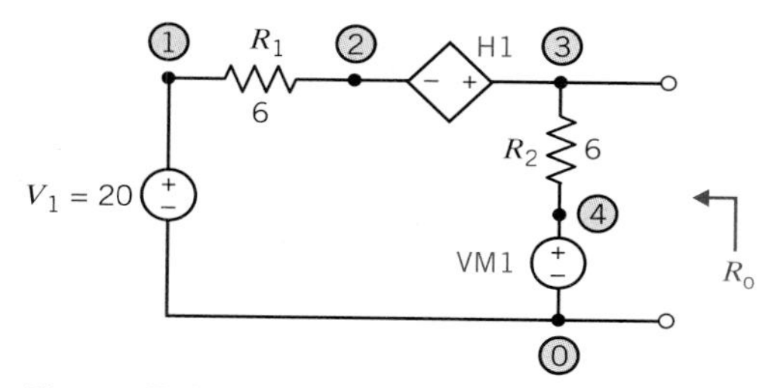

Figure 5.8-6 The circuit of Figure 5.8-5 redrawn for PSpice. Resistances in ohms.

```
**      THEVENIN  EQUIVALENT

V1       1       0       DC     20
R1       1       2       6
R2       3       4       6
VM1      4       0       DC      0
H1       3       2       VM1     2
.TF      V(3)    V1
.END
```

Figure 5.8-7 The PSpice program for Example 5.5-3. The dummy source VM1 generates the control current.

```
****       SMALL-SIGNAL  CHARACTERISTICS

        V(3)/V1  =  6.000E-01
        INPUT  RESISTANCE  AT  V1  =  1.000E+01
        OUTPUT  RESISTANCE  AT  V(3)  =  3.600E+00
```

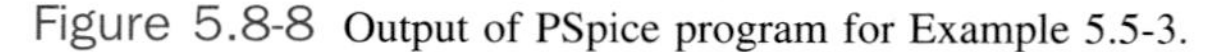

Figure 5.8-8 Output of PSpice program for Example 5.5-3.

The output of the calculation is shown in Figure 5.8-8, and $v_o = 0.6\, v_1 = 12$ V. The output resistance is $R_o = 3.6\ \Omega$ and thus $R_t = 3.6 + 10 = 13.6\ \Omega$. These calculations verify the results obtained in Example 5.3.

5.9 USING MATLAB TO DETERMINE THE THÉVENIN EQUIVALENT CIRCUIT

MATLAB can be used to reduce the work required to determine the Thévenin equivalent of a circuit such as the one shown in Figure 5.9-1*a*. First, connect a resistor, *R*, across the terminals of the network, as shown in Figure 5.9-1*b*. Next, write node or mesh equations to describe the circuit with the resistor connected across its terminals. In this case, the circuit in Figure 5.9-1*b* is represented by the mesh equations

$$\begin{aligned} 12 &= 28i_1 - 10i_2 - 8i_3 \\ 12 &= -10i_1 + 28i_2 - 8i_3 \\ 0 &= -8i_1 - 8i_2 + (16 + R)\, i_3 \end{aligned} \tag{5.9-1}$$

The current i in the resistor R is equal to the mesh current in the third mesh, that is,

$$i = i_3 \tag{5.9-2}$$

The mesh equations can be written using matrices such as

$$\begin{bmatrix} 28 & -10 & -8 \\ -10 & 28 & -8 \\ -8 & -8 & 16 + R \end{bmatrix} \begin{bmatrix} i_1 \\ i_2 \\ i_3 \end{bmatrix} = \begin{bmatrix} 12 \\ 12 \\ 0 \end{bmatrix} \tag{5.9-3}$$

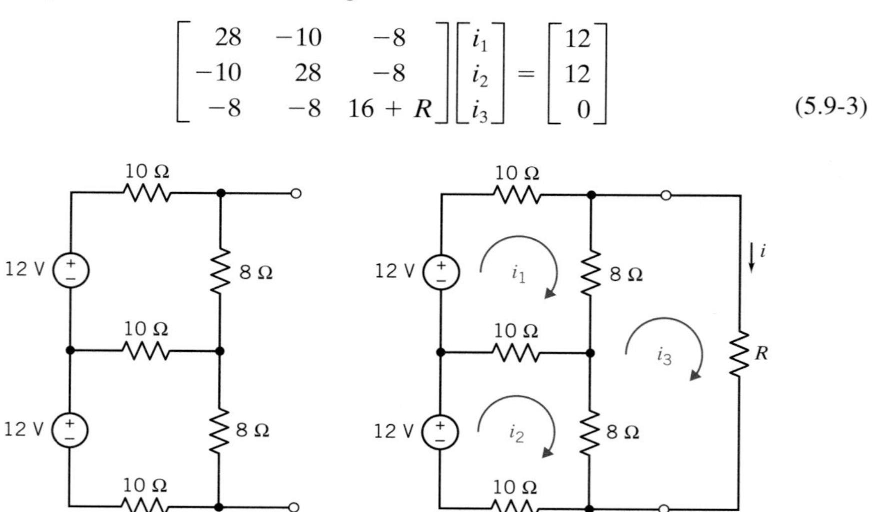

Figure 5.9-1 The circuit in (*b*) is obtained by connecting a resistor, *R*, across the terminals of the circuit in (*a*).

```
% ch5ex.m - MATLAB input file for Section 5-9

Z = [ 28   -10    -8;          %
     -10    28    -8;          % Mesh Equation
      -8    -8   16+R];        %
                               % Equation 5.9-3
V = [ 12;                      %
      12;                      %
       0 ];                    %


Im = Z\V;   % Calculate the mesh currents.

I = Im(3)   % Equation 5.9-2
```

Figure 5.9-2 MATLAB file used to solve the mesh equation representing the circuit shown in Figure 5.9-1*b*.

Notice that $i = i_3$ in Figure 5.9-1*b*.

Figure 5.9-2*a* shows a MATLAB file named ch5ex.m that solves Eq. 5.9-1. Figure 5.9-3 illustrates the use of this MATLAB file and shows that when $R = 6\ \Omega$, then $i = 0.7164$ A and that when $R = 12$ W, then $i = 0.5106$ A.

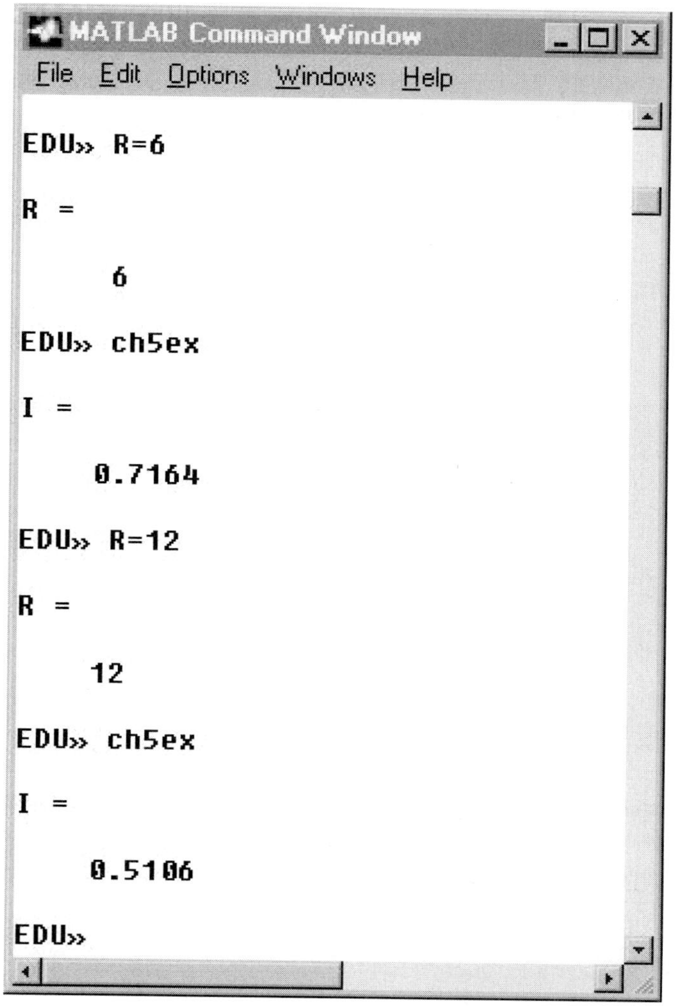

```
MATLAB Command Window
File Edit Options Windows Help

EDU>> R=6

R =

     6

EDU>> ch5ex

I =

    0.7164

EDU>> R=12

R =

    12

EDU>> ch5ex

I =

    0.5106

EDU>>
```

Figure 5.9-3 Computer screen showing the use of MATLAB to analyze the circuit shown in Figure 5.9-1.

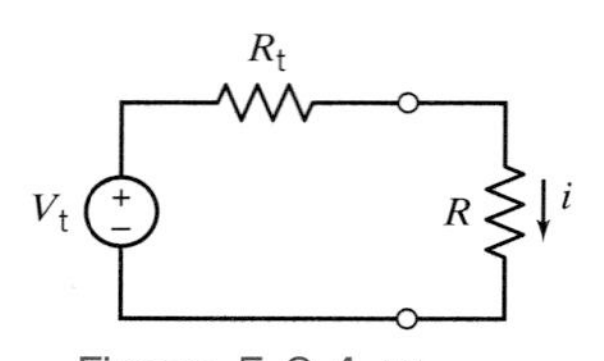

Figure 5.9-4 The circuit obtained by connecting a resistor, R, across the terminals of a Thévenin equivalent circuit.

Next, consider Figure 5.9-4 which shows a resistor R connected across the terminals of a Thévenin equivalent circuit. The circuit in Figure 5.9-4 is represented by the mesh equation

$$V_t = R_t i + Ri \tag{5.9-4}$$

As a matter of notation, let $i = i_a$ when $R = R_a$. Similarly, let $i = i_b$ when $R = R_b$. Equation 5.9-4 indicates that

$$\begin{aligned} V_t &= R_t i_a + R_a i_a \\ V_t &= R_t i_b + R_b i_b \end{aligned} \tag{5.9-5}$$

Equation 5.9-5 can be written using matrices as

$$\begin{bmatrix} R_a i_a \\ R_b i_b \end{bmatrix} = \begin{bmatrix} 1 & -i_a \\ 1 & -i_b \end{bmatrix} \begin{bmatrix} V_t \\ R_t \end{bmatrix} \tag{5.9-6}$$

Given i, R_a, i_b, and R_b, this matrix equation can be solved for V_t and R_t, the parameters of the Thévenin equivalent circuit. Figure 5.9-5 shows a MATLAB file that solves Eq. 5.9-6 using the values $i_b = 0.7164$ A, $R_b = 6\ \Omega$, $i_a = 0.5106$ A, $R_a = 12\ \Omega$. The resulting values of V_t and R_t are

$$V_t = 10.664 \text{ V} \quad \text{and} \quad R_t = 8.8863\ \Omega$$

```
% Find the Thevenin Equivalent of the circuit
% connected to the resister R.

Ra = 12;    ia = 0.5106;     % When R=Ra then i=ia

Rb = 6;     ib = 0.7164;     % When R=Rb then i=ib

A = [1 -ia;    %
      1 -ib];  %
               % Eqn 5.9-6
b = [Ra*ia;    %
      Rb*ib];  %

X = A\b;

Vt = X(1)   % Open-Circuit Voltage

Rt = X(2)   % Thevenin Resistance
```

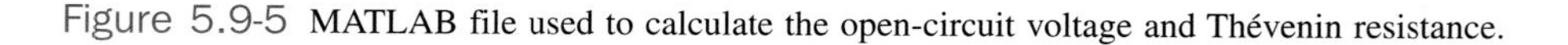

Figure 5.9-5 MATLAB file used to calculate the open-circuit voltage and Thévenin resistance.

5.10 VERIFICATION EXAMPLE

Suppose that the circuit shown in Figure 5.10-1a was built in the lab using $R = 2$ kΩ and that the voltage labeled v was measured to be $v = -1.87$ V. Next, the resistor labeled R was changed to $R = 5$ kΩ, and the voltage v was measured to be $v = -3.0$ V. Finally, the resistor was changed to $R = 10$ kΩ, and the voltage was measured to be $v = -3.75$ V. How can these measurements be checked to verify that they are consistent?

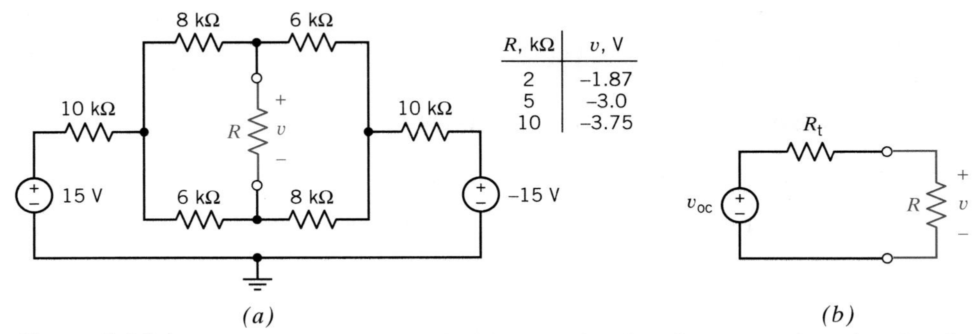

Figure 5.10-1 (*a*) A circuit with data obtained by measuring the voltage across the resistor *R* and (*b*) the circuit obtained by replacing the part of the circuit connected to *R* by its Thévenin equivalent circuit.

Solution

Let's replace the part of the circuit connected to the resistor *R* by its Thévenin equivalent circuit. Figure 5.10-1*b* shows the resulting circuit. Applying the voltage division principle to the circuit in Figure 5.10-1*b* gives

$$v = \frac{R}{R + R_t} v_{oc} \tag{5.10-1}$$

When $R = 2\ \text{k}\Omega$, then $v = -1.87$ V and Eq. 5.10-1 becomes

$$-1.87 = \frac{2000}{2000 + R_t} v_{oc} \tag{5.10-2}$$

Similarly, when $R = 5\ \text{k}\Omega$, then $v = -3.0$ V and Eq. 5.10-1 becomes

$$-3.0 = \frac{5000}{5000 + R_t} v_{oc} \tag{5.10-3}$$

Equations 5.10-2 and 5.10-3 constitute a set of two equations in two unknown, v_{oc} and R_t. Solving these equations gives $v_{oc} = -5$ V and $R_t = 3333\ \Omega$. Substituting these values into Eq. 5.10-1 gives

$$v = \frac{R}{R + 3333}(-5) \tag{5.10-4}$$

Equation 5.10-4 can be used to predict the voltage that would be measured if $R = 10\ \text{k}\Omega$. If the value of v obtained using Eq. 5.10-4 agrees with the measured value of v, then the measured data are consistent. Letting $R = 10\ \text{k}\Omega$ in Eq. 5.10-4 gives

$$v = \frac{10000}{10000 + 3333}(-5) = -3.75\ \text{V} \tag{5.10-5}$$

Since this value agrees with the measured value of v, the measured data are indeed consistent.

5.11 Design Challenge Solution

STRAIN GAUGE BRIDGE

Strain gauges are transducers that measure mechanical strain. Electrically, the strain gauges are resistors. The strain causes a change in resistance that is proportional to the strain.

Figure 5.11-1 shows four strain gauges connected in a configuration called a bridge. Strain gauge bridges are used to measure force or pressure (Doebelin 1966).

The bridge output is usually a small voltage. In Figure 5.11-1 an amplifer multiplies the bridge output, v_i, by a gain to obtain a larger voltage, v_o, which is displayed by the voltmeter.

Describe the Situation and the Assumptions

A strain gauge bridge is used to measure force. The strain gauges have been positioned so that the force will increase the resistance of two of the strain gauges while, at the same time, decreasing the resistance of the other two strain gauges.

The strain gauges used in the bridge have nominal resistances of $R = 120\ \Omega$. (The nominal resistance is the resistance when the strain is zero.) This resistance is expected to increase or decrease by no more than $2\ \Omega$ due to strain. This means that

$$-2\ \Omega \le \Delta R \le 2\ \Omega \tag{5.11-1}$$

The output voltage, v_o, is required to vary from -10 V to $+10$ V as ΔR varies from $-2\ \Omega$ to $2\ \Omega$.

State the Goal

Determine the amplifier gain, b, needed to cause v_o to be related to ΔR by

$$v_o = 5\frac{\text{volt}}{\text{ohm}} \cdot \Delta R \tag{5.11-2}$$

Generate a Plan

Use Thévenin's theorem to analyze the circuit shown in Figure 5.11-1 to determine the relationship between v_i and ΔR. Calculate the amplifier gain needed to satisfy Eq. 5.11-2.

Act on the Plan

We begin by finding the Thévenin equivalent of the strain gauge bridge. This requires two calculations: one to find the open-circuit voltage V_t and the other to find the Thévenin re-

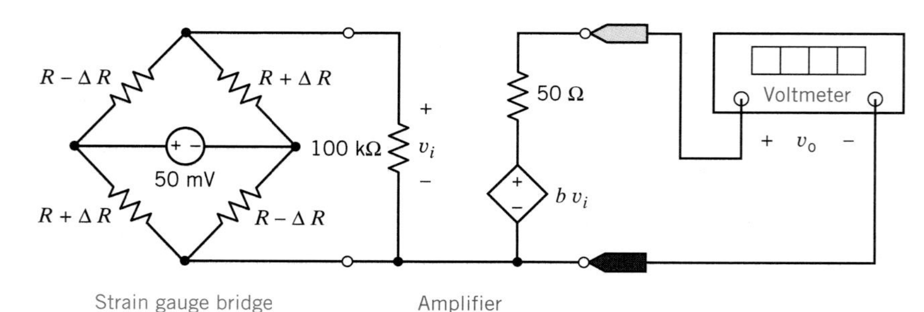

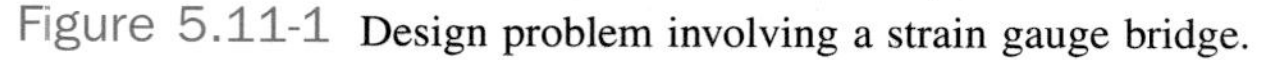

Figure 5.11-1 Design problem involving a strain gauge bridge.

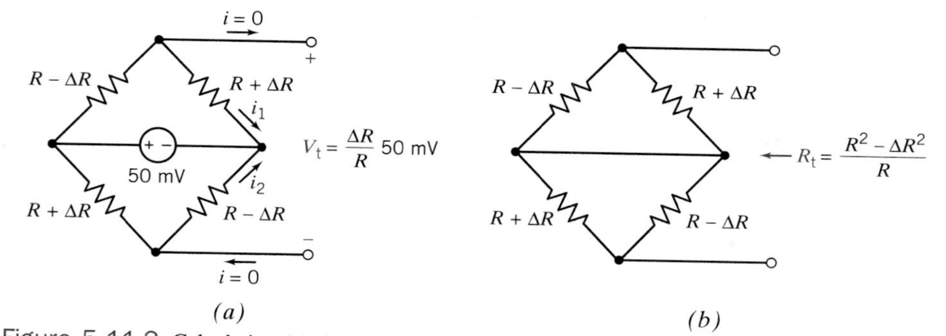

Figure 5.11-2 Calculating (*a*) the open-circuit voltage and (*b*) the Thévenin resistance of the strain gauge bridge.

sistance R_t. Figure 5.11-2*a* shows the circuit used to calculate V_t. Begin by finding the currents i_1 and i_2.

$$i_1 = \frac{50\text{ mV}}{(R - \Delta R) + (R + \Delta R)} = \frac{50\text{ mV}}{2R}$$

Similarly
$$i_2 = \frac{50\text{ mV}}{(R + \Delta R) + (R - \Delta R)} = \frac{50\text{ mV}}{2R}$$

Then
$$\begin{aligned} V_t &= (R + \Delta R)i_1 - (R - \Delta R)i_2 \\ &= (2\Delta R)\frac{50\text{ mV}}{2R} \\ &= \frac{\Delta R}{R}50\text{ mV} = \frac{50\text{ mV}}{120\ \Omega}\Delta R = (0.4167 \times 10^{-3})\Delta R \end{aligned} \tag{5.11-3}$$

Figure 5.11-2*b* shows the circuit used to calculate R_t. This figure shows that R_t is comprised of a series connection of two resistances, each of which is a parallel connection of two strain gauge resistances

$$\begin{aligned} R_t &= \frac{(R - \Delta R)(R + \Delta R)}{(R - \Delta R) + (R + \Delta R)} + \frac{(R + \Delta R)(R - \Delta R)}{(R + \Delta R) + (R - \Delta R)} \\ &= 2\frac{R^2 - \Delta R^2}{2R} \end{aligned}$$

Since R is much larger than ΔR, this equation can be simplified to

$$R_t = R$$

In Figure 5.11-3 the strain gauge bridge has been replaced by its Thévenin equivalent circuit. This simplification allows us to calculate v_i using voltage division

$$v_i = \frac{100\text{ k}\Omega}{100\text{ k}\Omega + R_t}V_t = 0.9988\ V_t = (0.4162 \times 10^{-3})\Delta R \tag{5.11-4}$$

Model the voltmeter as an ideal voltmeter. Then the voltmeter current is $i = 0$ as shown in Figure 5.11-3. Applying KVL to the right-hand mesh gives

$$v_o + 50(0) - bv_i = 0$$

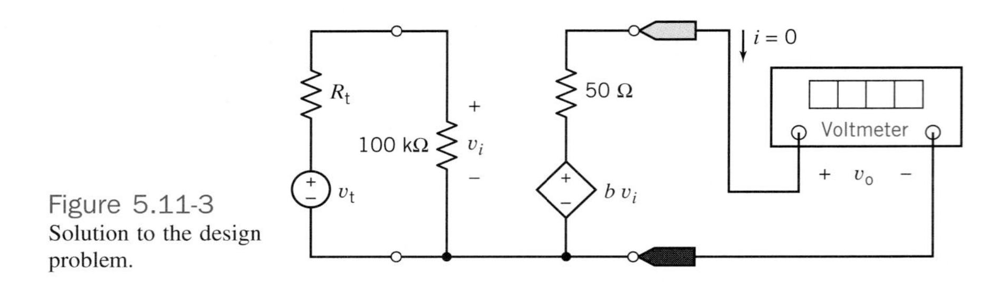

Figure 5.11-3
Solution to the design problem.

or
$$v_o = bv_i = b(0.4162 \times 10^{-3})\Delta R \tag{5.11-5}$$

Comparing Eq. 5.11-5 to Eq. 5.11-2 shows that the amplifier gain, b, must satisfy

$$b(0.4162 \times 10^{-3}) = 5$$

Hence, the amplifier gain is

$$b = 12{,}013$$

Verify the Proposed Solution

Substituting $b = 12{,}013$ into Eq. 5.11-5 gives

$$v_o = (12{,}013)(0.4162 \times 10^{-3})\Delta R = 4.9998\ \Delta R \tag{5.11-6}$$

which agrees with Eq. 5.11-2.

5.12 SUMMARY

- Electrical power is used for many applications such as transportation vehicles, heating and lighting, and communications.
- Source transformations, summarized in Table 5.12-1, are used to transform a circuit into an equivalent circuit. A voltage source v_{oc} in series with a resistor R_t can be transformed into a current source $i_{sc} = v_{oc}/R_t$ and a parallel resistor R_t. Conversely, a current source i_{sc} in parallel with a resistor R_t can be transformed into a voltage source $v_{oc} = R_t\, i_{sc}$ in series with a resistor R_t. The circuits in Table 5.12-1 are equivalent in the sense that the voltage and current of all circuit elements in the part of the circuit labeled "Circuit B" are unchanged by the source transformation.
- The superposition theorem permits us to determine the total response of a linear circuit to several independent sources by finding the response to each independent source separately and then adding the separate responses algebraically.
- Thévenin and Norton equivalent circuits, summarized in Table 5.12-2, are used to transform a circuit into a smaller, yet equivalent, circuit. First the circuit is separated into two parts, labeled as "Circuit A" and "Circuit B" in Table 5.12-2. "Circuit A" can be replaced by either its Thévenin equivalent circuit or its Norton equivalent circuit. The circuits in Table 5.15-2 are equivalent in the sense that the voltage and current of all circuit elements in the part of the circuit labeled "Circuit B" are unchanged by replacing "Circuit A" by either its Thévenin equivalent circuit or its Norton equivalent circuit.
- Procedures for calculating the parameters v_{oc}, i_{sc} and R_t of the Thévenin and Norton equivalent circuits are summarized in Tables 5.5-1 and 5.6-1.

Table 5.12-1 **Source Transformations**

Thévenin Circuit	Norton Circuit

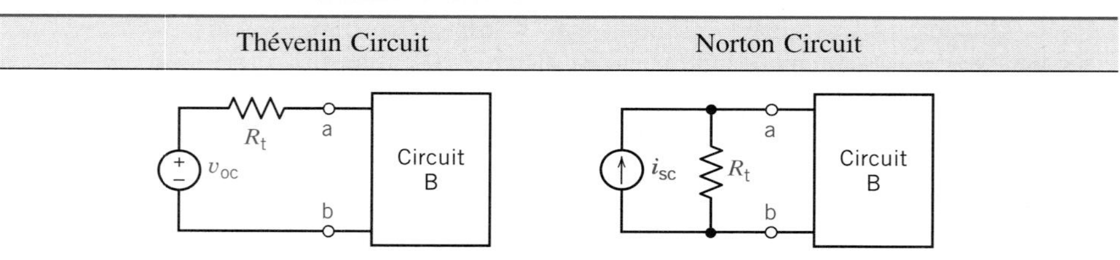

Table 5.12-2 **Thévenin and Norton Equivalent Circuits**

Original Circuit	Thévenin Equivalent Circuit	Norton Equivalent Circuit

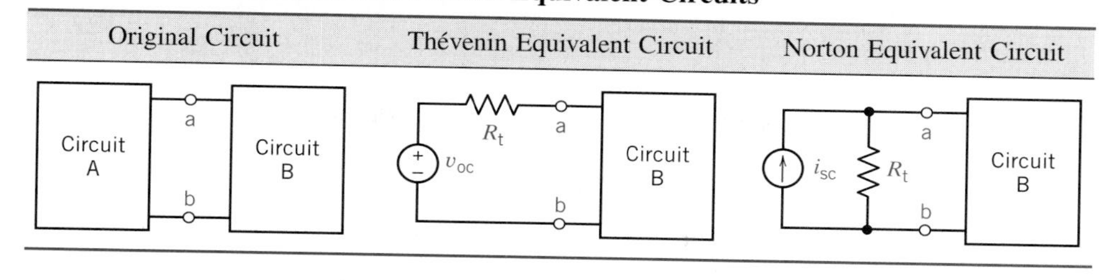

- The goal of many electronic and communications circuits is to deliver maximum power to a load resistor R_L. Maximum power is attained when R_L is set equal to the Thévenin resistance, R_t, of the circuit connected to R_L. This results in maximum power at the load when the series resistance R_t cannot be reduced.
- The circuit analysis program PSpice provides a command .TF <Output Variable> <Input Source> that can be used to determine the Thévenin or Norton equivalent of a circuit.
- The computer program MATLAB can be used to reduce the computational burden of calculating the parameters v_{oc}, i_{sc}, and R_t of the Thévenin and Norton equivalent circuits.

PROBLEMS

Section 5.3 Source Transformations

P 5.3-1 The circuit shown in Figure P 5.3-1*a* has been divided into two parts. The circuit shown in Figure P 5.3-1*b* was obtained by simplifying the part to the right of the terminals using source transformations. The part of the circuit to the left of the terminals was not changed.

(a) Determine the values of R_t and v_t in Figure P 5.3-1*b*.

(b) Determine the values of the current i and the voltage v in Figure P 5.3-1*b*. The circuit in Figure P 5.3-1*b* is equivalent to the circuit in Figure P 5.3-1*a*. Consequently, the current i and the voltage v in Figure P 5.3-1*a* have the same values as do the current i and the voltage v in Figure P 5.3-1*b*.

(c) Determine the value of the current i_a in Figure P 5.3-1*a*.

P 5.3-2 Consider the circuit of Figure P 5.3-2. Find i_a by simplifying the circuit (using source transformations) to a single-loop circuit so that you need to write only one KVL equation to find i_a.

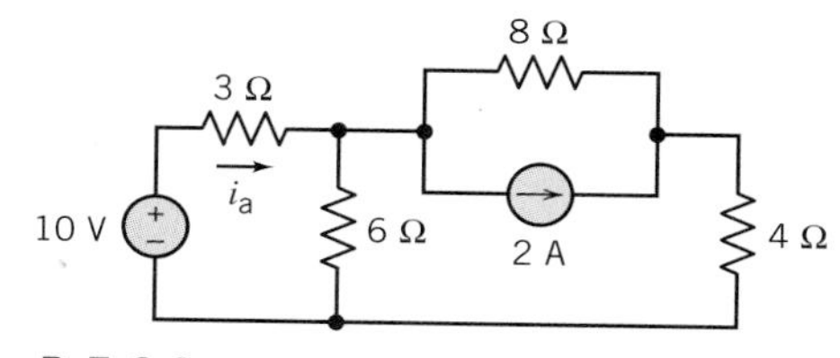

Figure P 5.3-2

P 5.3-3 Use source transformations to find the voltage v across the 1-mA current source for the circuit shown in Figure P 5.3-3.

Answer: $v = 3$ V

P 5.3-4 Find v_0 using source transformations if $i = 5/2$ A in the circuit shown in Figure P 5.3-4.

Hint: Reduce the circuit to a single mesh that contains the voltage source labeled v_o.

Answer: $v_o = 28$ V

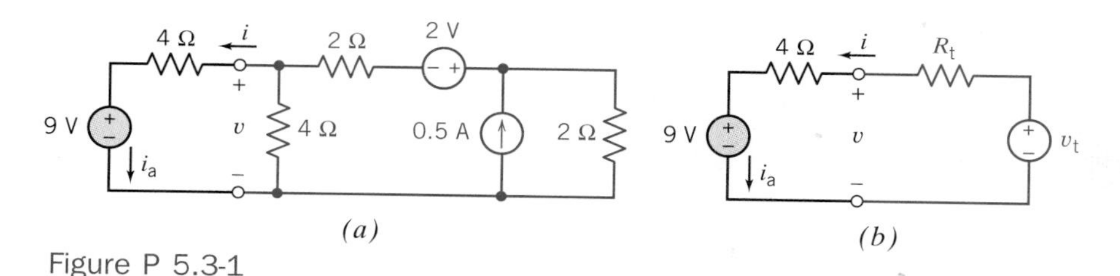

Figure P 5.3-1

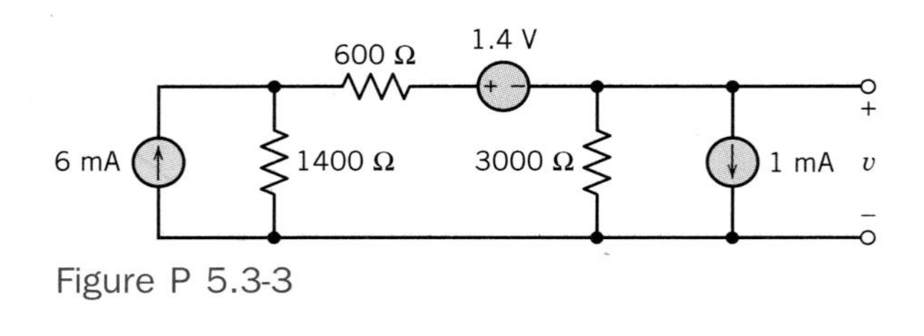

Figure P 5.3-3

P 5.3-5 Determine the value of the current i_a in the circuit shown in Figure P 5.3-5.

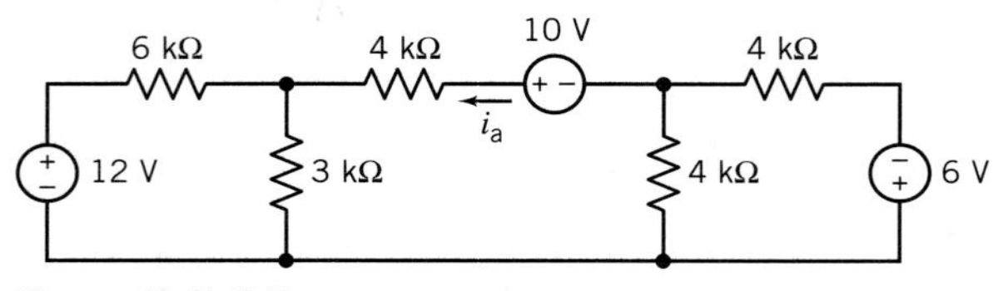

Figure P 5.3-5

P 5.3-6 For the circuit of Figure P 5.3-6, specify the resistance R_L that will cause current i to be 2 A.
Hint: Use source transformations to simplify the circuit connected to R_L.
Answer: $R_L = 2\ \Omega$

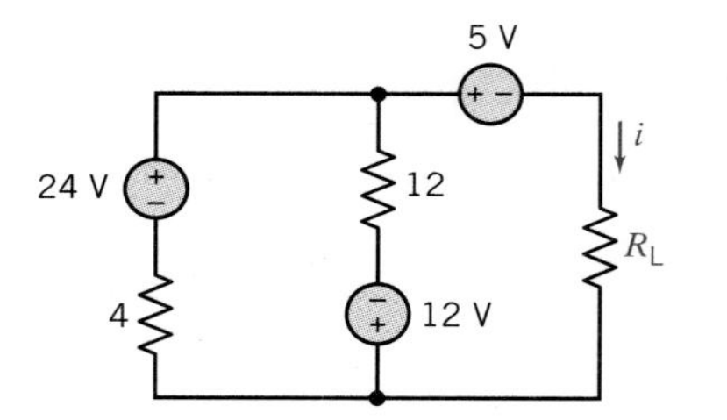

Figure P 5.3-6 Resistances in ohms.

Section 5.4 Superposition

P 5.4-1 Use superposition to find the current i_a in the circuit shown in Figure P 5.4-1.
Answer: $i_a = 1$ A

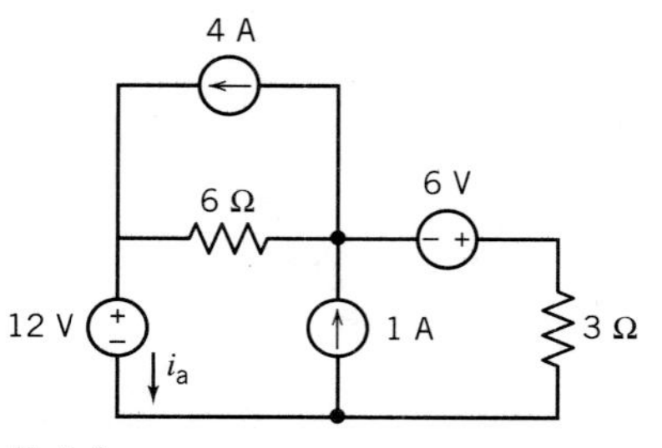

Figure P 5.4-1

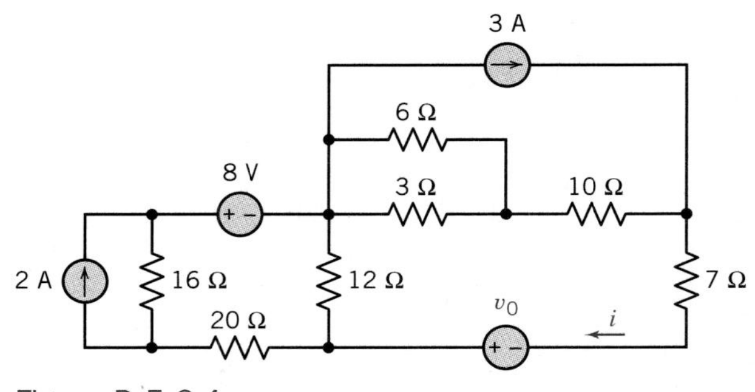

Figure P 5.3-4

P 5.4-2 Use superposition to find v for the circuit of Figure P 5.4-2.

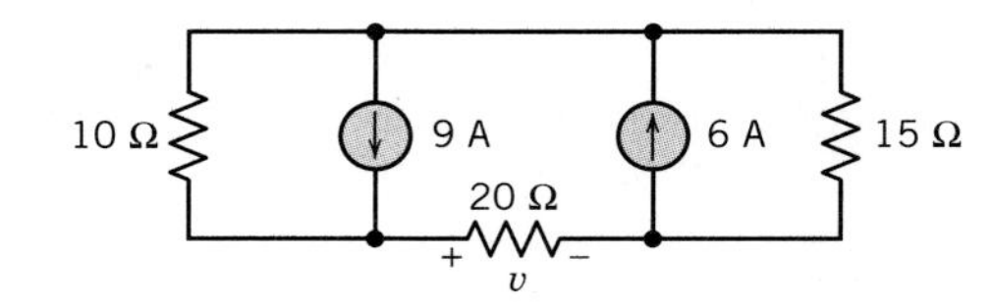

Figure P 5.4-2

P 5.4-3 Use superposition to find i for the circuit of Figure P 5.4-3.
Answer: $i = -2$ mA

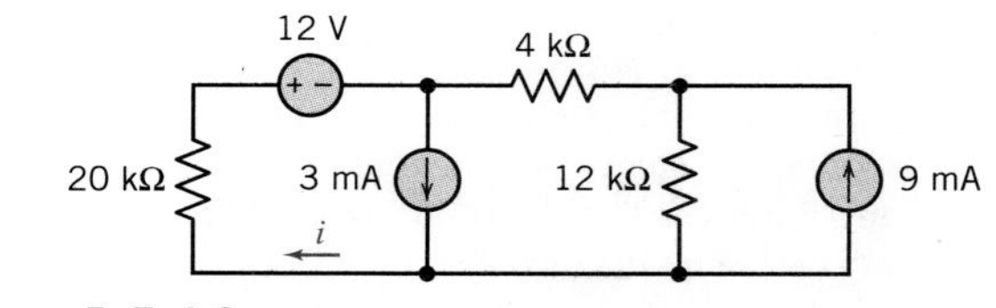

Figure P 5.4-3

P 5.4-4 Use superposition to find i for the circuit of Figure P 5.4-4.
Answer: $i = 3.5$ mA

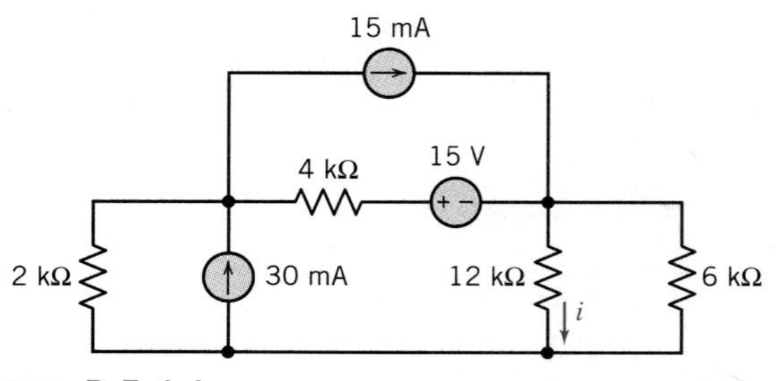

Figure P 5.4-4

P 5.4-5 Use superposition to find v_x for the circuit of Figure P 5.4-5.

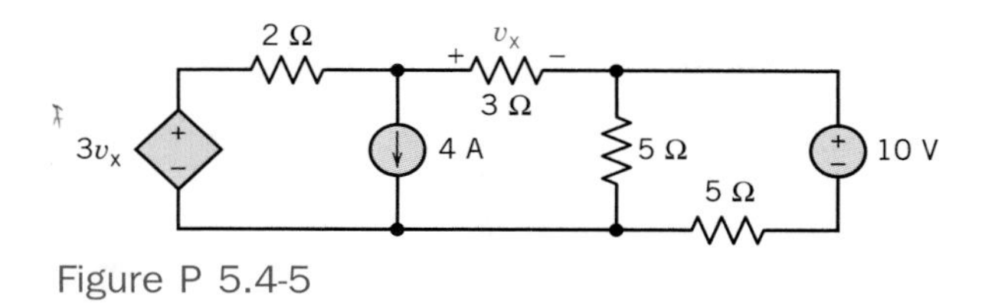

Figure P 5.4-5

P 5.4-6 Determine the current i of the circuit shown in Figure P 5.4-6.

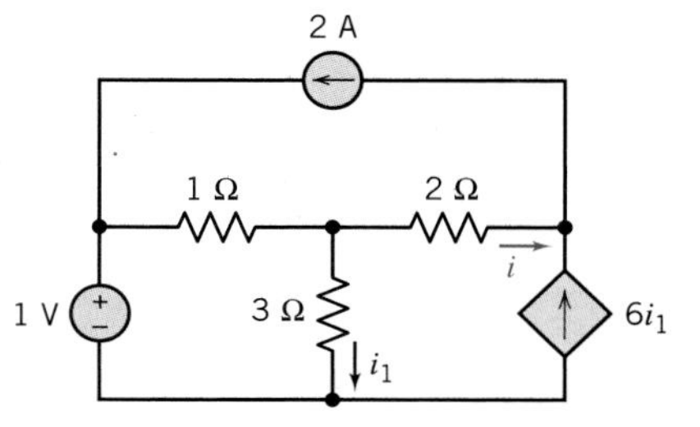

Figure P 5.4-6

P 5.4-7 For the circuit shown in Figure P 5.4-7, use superposition to find v in terms of the R's and source values.

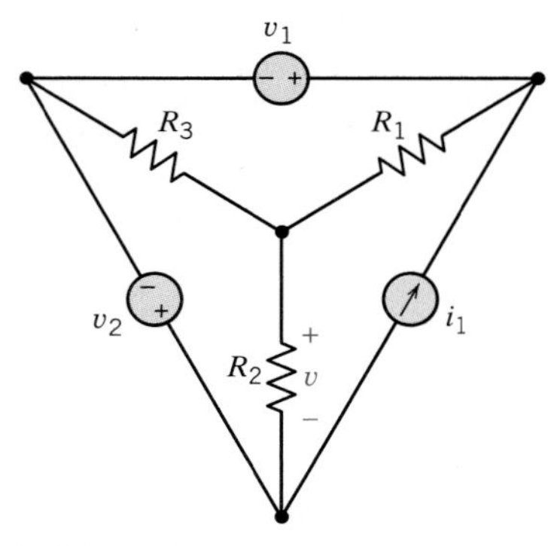

Figure P 5.4-7

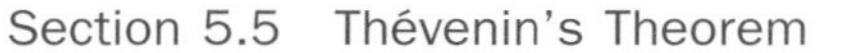

Section 5.5 Thévenin's Theorem

P 5.5-1 Determine values of R_t and v_{oc} that cause the circuit shown in Figure P 5.5-1b to be the Thévenin equivalent circuit of the circuit in Figure P 5.5-1a.

Hint: Use source transformations and equivalent resistances to reduce the circuit in Figure P 5.5-1a until it is the circuit in Figure P 5.5-1b.

Answer: $R_t = 5\ \Omega$ and $v_{oc} = 2$ V.

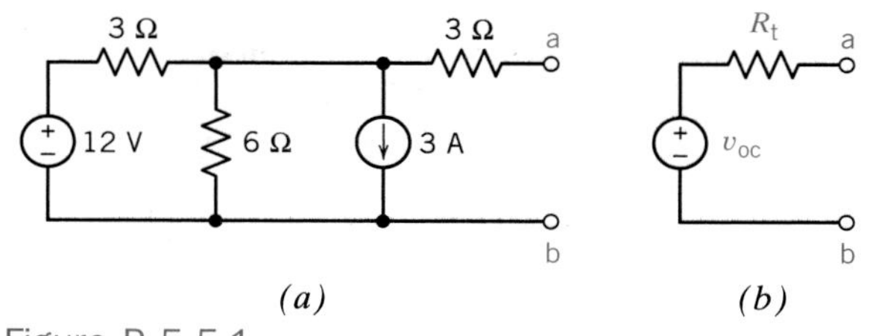

Figure P 5.5-1

P 5.5-2 Find the Thévenin equivalent circuit for the circuit of Figure P 5.5-2.

Answer: $R_t = 10\ \Omega$, $v_{oc} = -24$ V

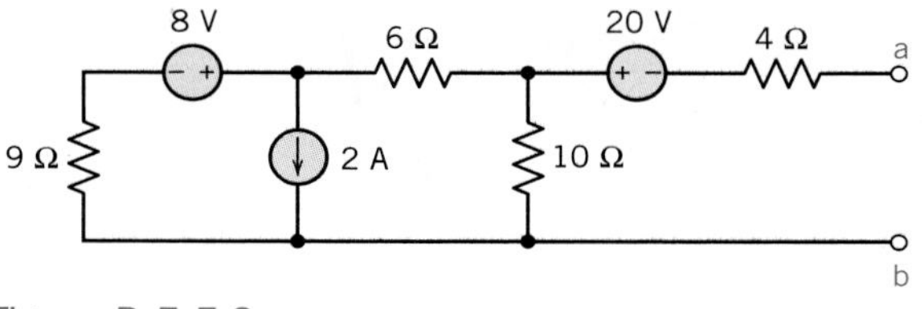

Figure P 5.5-2

P 5.5-3 Find the Thévenin equivalent for terminals a–b for the circuit of Figure P 5.5-3.

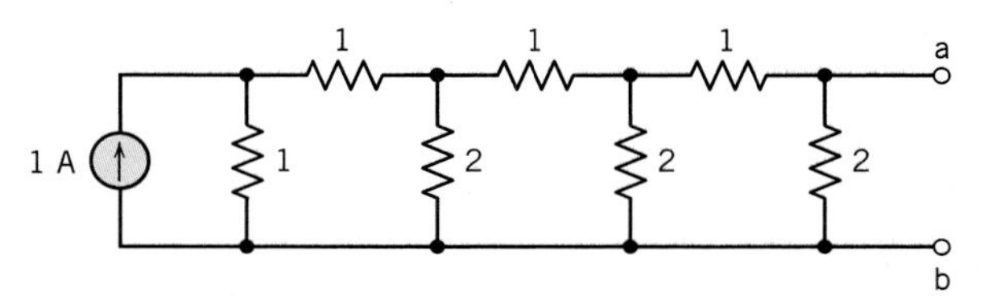

Figure P 5.5-3 All resistances in ohms.

P 5.5-4 Obtain the Thévenin equivalent for the circuit shown in Figure P 5.5-4. Resistances in ohms.

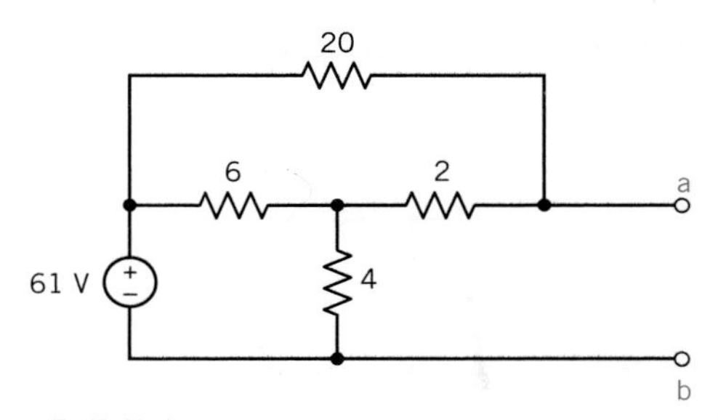

Figure P 5.5-4

P 5.5-5 Find the Thévenin equivalent circuit for the circuit of Figure P 5.5-5.

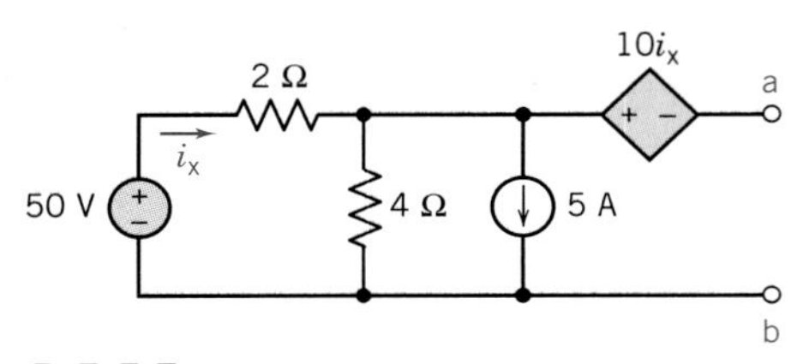

Figure P 5.5-5

P 5.5-6 Determine the Thévenin equivalent circuit for the circuit of Figure P 5.5-6 at the output terminals a–b.

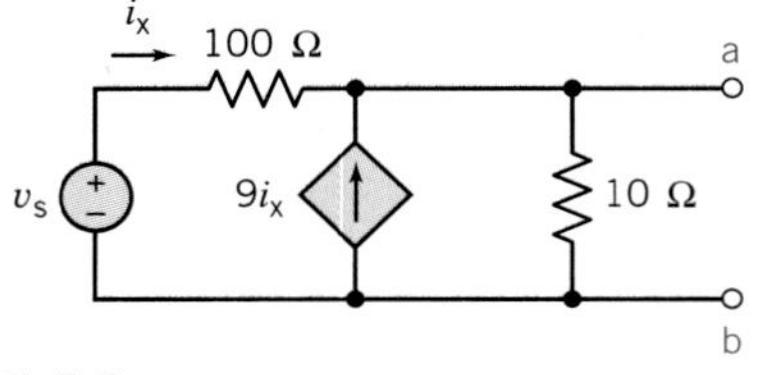

Figure P 5.5-6

P 5.5-7 The circuit shown in Figure P 5.5-7 has four unspecified circuit parameters: v_s, R_1, R_2, and d, where d is the gain of the CCCS.

(a) Show that the open circuit voltage, v_{oc}, the short circuit current, i_{sc}, and the Thévenin resistance, R_t, of this circuit are given by

$$v_{oc} = \frac{R_2(d+1)}{R_1 + (d+1)R_2} v_s,$$

$$i_{sc} = \frac{(d+1)}{R_1} v_s$$

and

$$R_t = \frac{R_1 R_2}{R_1 + (d+1)R_2}$$

(b) Let $R_1 = R_2 = 1\ \text{k}\Omega$. Determine the values of v_s and d required to cause $v_{oc} = 5$ V and $R_t = 625\ \Omega$.

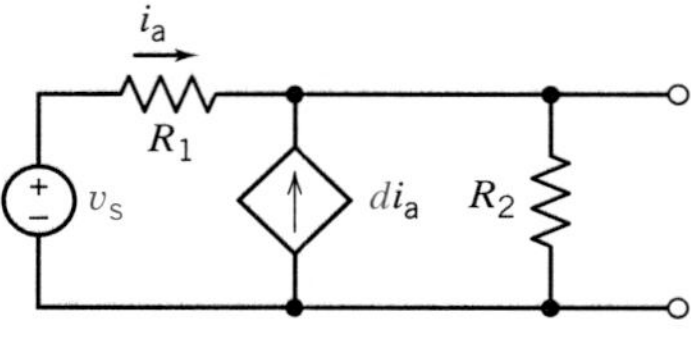

Figure P 5.5-7

P 5.5-8 Find R_t for the circuit of Figure P 5.5-8.
Answer: $R_t = 3\ \Omega$

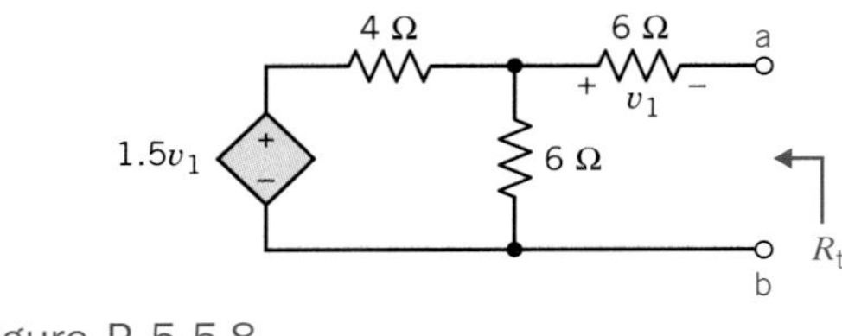

Figure P 5.5-8

P 5.5-9 A resistor, R, was connected to a circuit box as shown in Figure P 5.5-9. The voltage, v, was measured. The resistance was changed, and the voltage was measured again. The results are shown in the table. Determine the Thévenin equivalent of the circuit within the box and predict the voltage, v, when $R = 8\ \text{k}\Omega$.

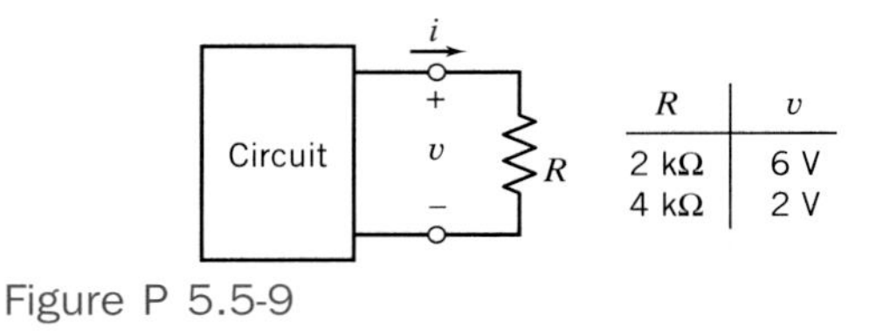

R	v
2 kΩ	6 V
4 kΩ	2 V

Figure P 5.5-9

P 5.5-10 A resistor, R, was connected to a circuit box as shown in Figure P 5.5-10. The current, i, was measured. The resistance was changed, and the current was measured again. The results are shown in the table.

(a) Specify the value of R required to cause $i = 2$ mA.

(b) Given that $R > 0$, determine the maximum possible value of the current i.

Hint: Use the data in the table to represent the circuit by a Thévenin equivalent.

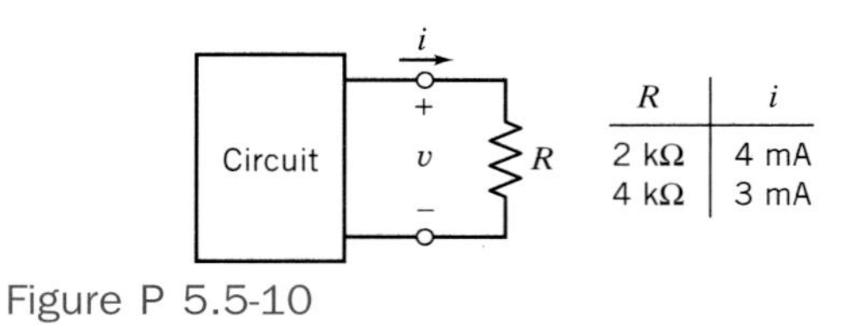

R	i
2 kΩ	4 mA
4 kΩ	3 mA

Figure P 5.5-10

P 5.5-11 Measurements made on terminals a–b of a linear circuit, Figure P 5.5-11a, which is known to be made up only of independent and dependent voltage sources and current sources and resistors, yield the current–voltage characteristics shown in Figure P 5.5-11b. Find the Thévenin equivalent circuit.

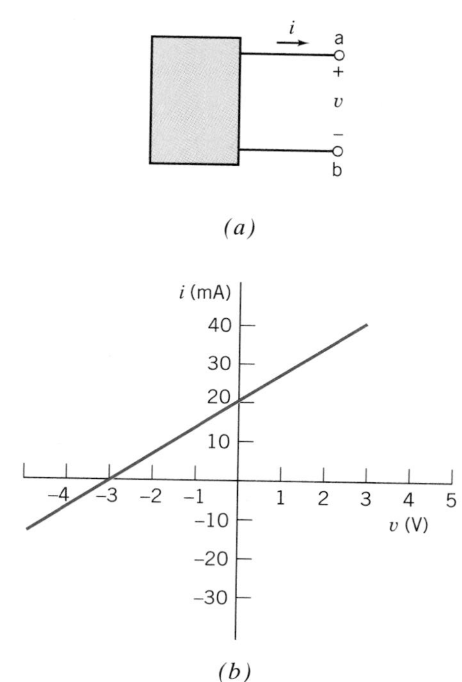

Figure P 5.5-11

P 5.5-12 For the circuit of Figure P 5.5-12, specify the resistance R that will cause current i_b to be 2 mA.
Hint: Find the Thévenin equivalent circuit of the circuit connected to R.

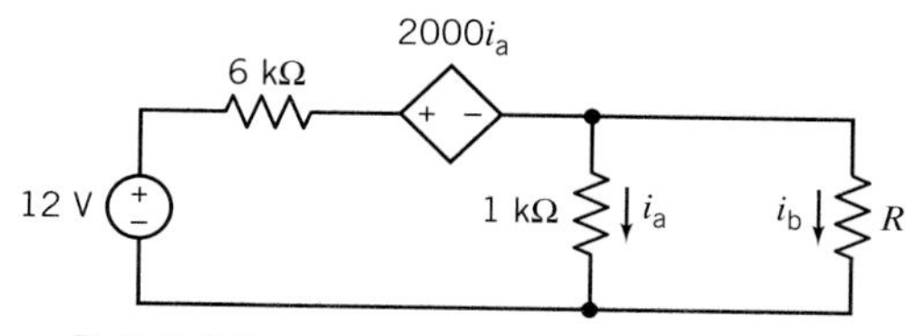

Figure P 5.5-12

P 5.5-13 For the circuit of Figure P 5.5-13, specify the value of the resistance R_L that will cause current i_L to be −2 A.
Answer: $R_L = 12\ \Omega$

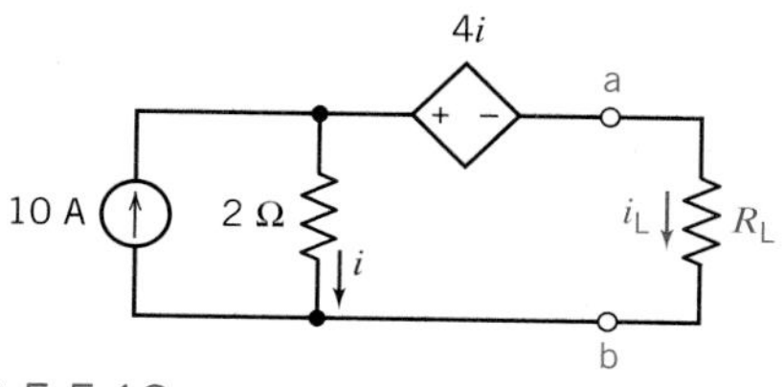

Figure P 5.5-13

P 5.5-14 Two black boxes are shown in Figure P 5.5-14. Box A contains the Thévenin equivalent of some linear circuit, and box B contains the Norton equivalent of the same circuit. With access to just the outsides of the boxes and their terminals, how can you determine which is which, using only one shorting wire?

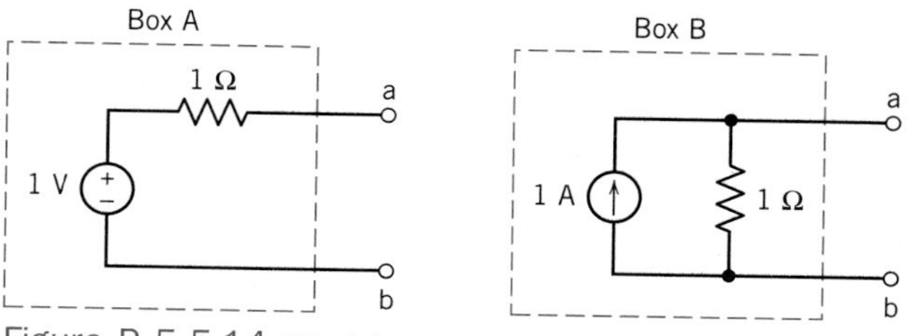

Figure P 5.5-14 Black boxes problem.

P 5.5-15 Using Thévenin's theorem, show that the circuit in Figure P 5.5-15*a* is equivalent to Figure P 5.5-15*b* where

$$R_b = \frac{R_1 R_2}{R_1 + R_2}$$

and

$$v = V_{cc} \frac{R_2}{R_1 + R_2}$$

The transistor Q is an electronic device.

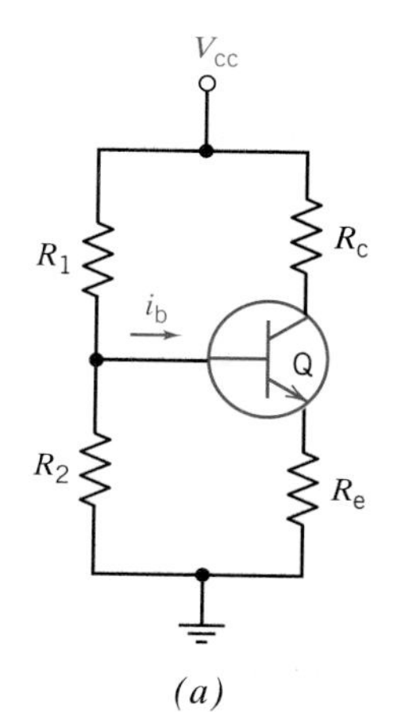

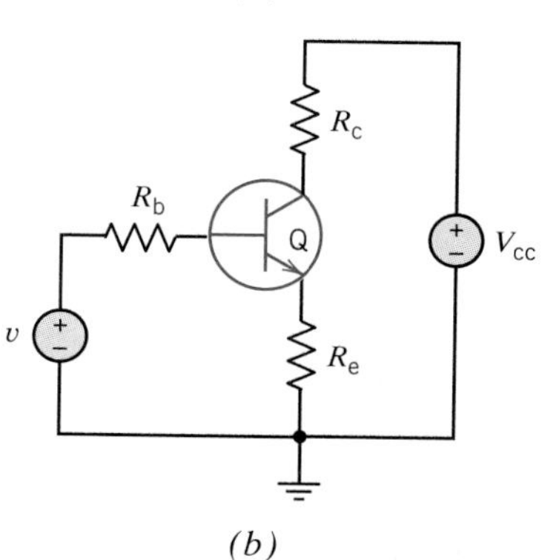

Figure P 5.5-15

P 5.5-16 For the circuit of Figure P 5.5-16, $R_{bc} = 10^6\ \Omega$, $R_{ce} = 100\ k\Omega$, $R_{bb} = 100\ \Omega$, $R_{be} = 2\ k\Omega$, and $g_m = 50$ mA/V.

(a) Find R_{in} (open C–E). (b) Find R_{out} (short B–E).
Answer: $R_{in} = 299\ \Omega$

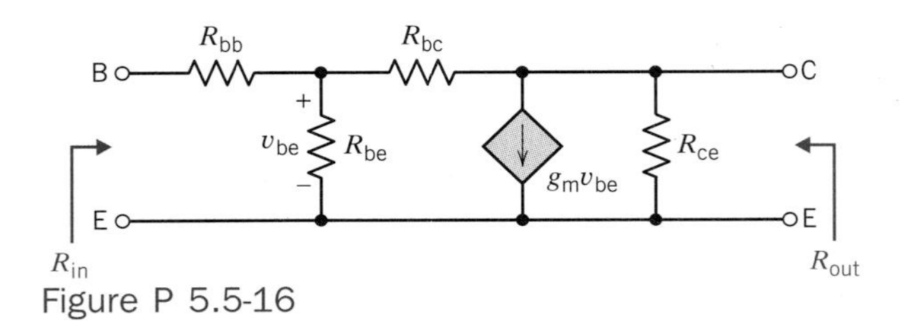

Figure P 5.5-16

P 5.5-17 The "tunnel diode" is a high-impurity-density *p-n* junction device. It can exhibit a negative resistance characteristic for certain current areas. Figure P 5.5-17 shows the *i-v* characteristic of a tunnel diode. In three ranges, it exhibits a linear *i-v* characteristic. Find the Thévenin equivalent circuit of the tunnel diode in these three areas.

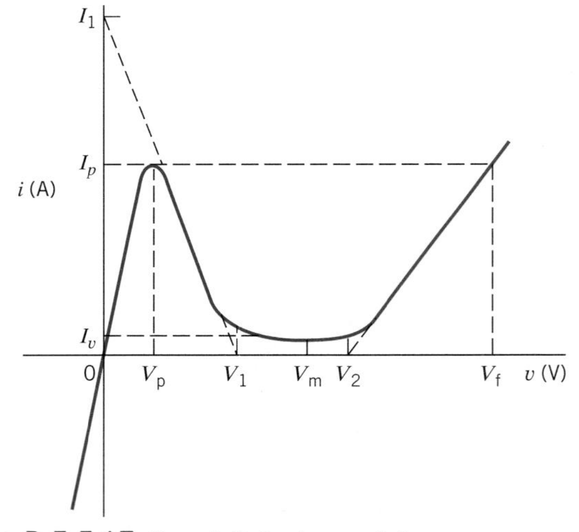

Figure P 5.5-17 Tunnel diode characteristic.

Section 5.6 Norton's Equivalent Theorem

P 5.6-1 Find the Norton equivalent circuit for the circuit shown in Figure P 5.6-1.
Answer: $R_t = 2\ \Omega$, $i_{sc} = -7.5$ A

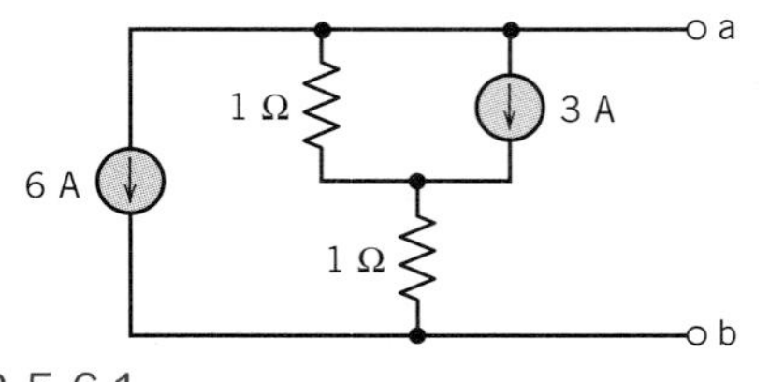

Figure P 5.6-1

P 5.6-2 Find the Norton equivalent circuit for the circuit shown in Figure P 5.6-2.

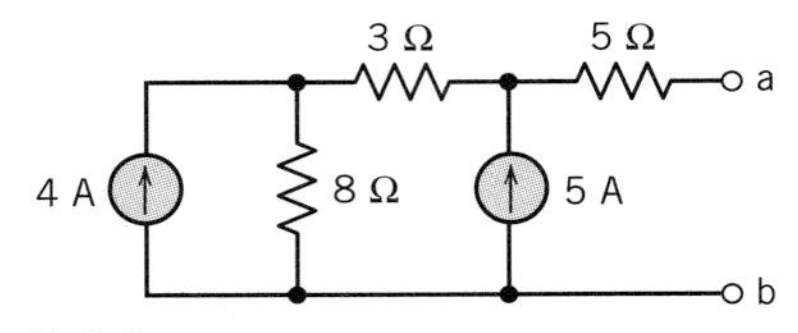

Figure P 5.6-2

P 5.6-3 Find R_t between terminals a–b and draw the Norton equivalent for the circuit shown in Figure P 5.6-3. The gain of the dependent source has units of V/mA.
Answer: $R_t = 3\ k\Omega$, $i_{sc} = 1$ mA

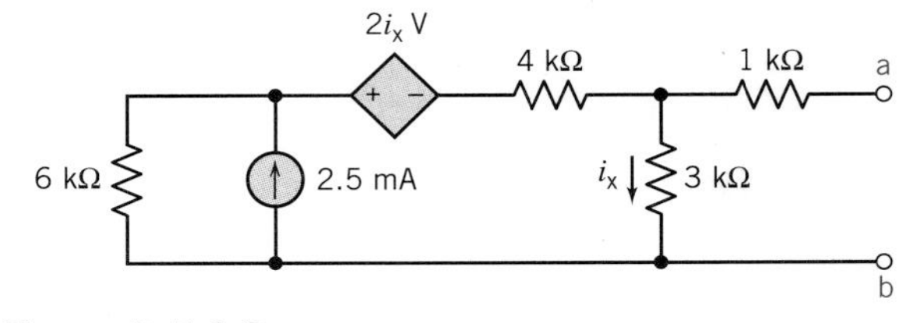

Figure P 5.6-3

P 5.6-4 Find the Norton equivalent circuit for the circuit shown in Figure P 5.6-4.

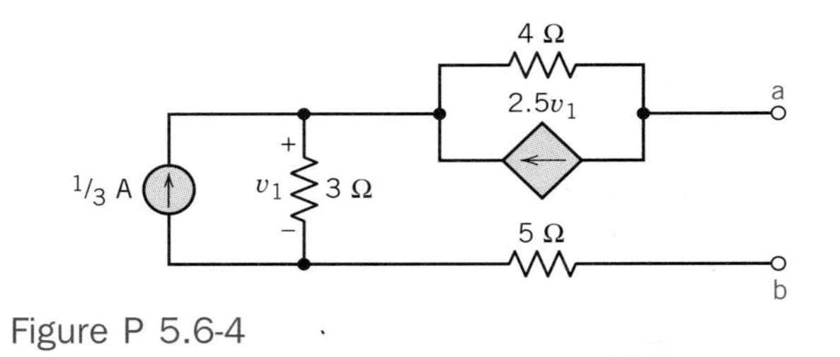

Figure P 5.6-4

Section 5.7 Maximum Power Transfer

P 5.7-1 The circuit model for a photovoltaic cell is given in Figure P 5.7-1 (Edelson 1992). The current i_s is proportional to the solar insolation (kW/m^2).
(a) Find the load resistance for maximum power transfer.
(b) Find the maximum power transferred when $i_s = 1$ A.

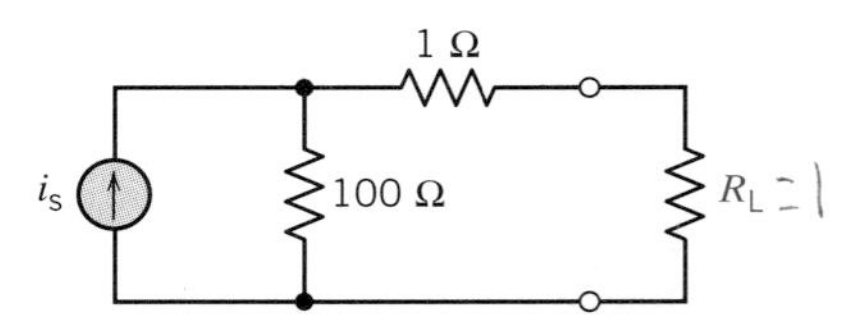

Figure P 5.7-1 Circuit model of photovoltaic cell.

P 5.7-2 For the circuit in Figure P 5.7-2, (a) find R such that maximum power is dissipated in R and (b) calculate the value of maximum power.
Answer: $R = 60\ \Omega$, $P_{max} = 54$ mW

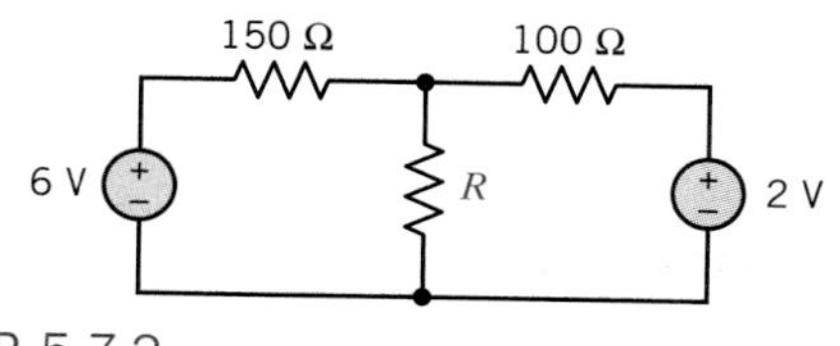

Figure P 5.7-2

P 5.7-3 For the circuit in Figure P 5.7-3, prove that for R_s variable and R_L fixed, the power dissipated in R_L is maximum when $R_s = 0$.

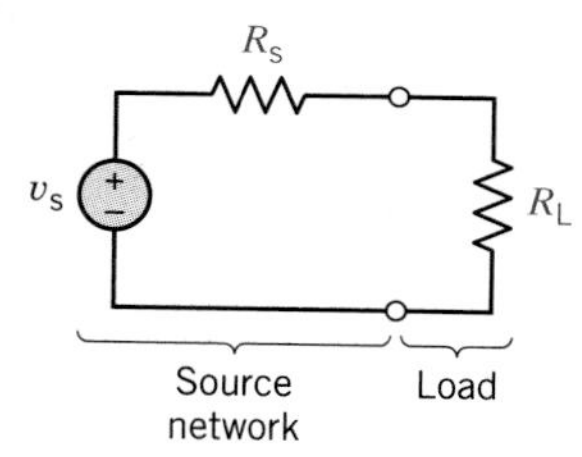

Figure P 5.7-3

P 5.7-4 Find the maximum power to the load R_L if the maximum power transfer condition is met for the circuit of Figure P 5.7-4.

Answer: max $p_L = 0.75$ W

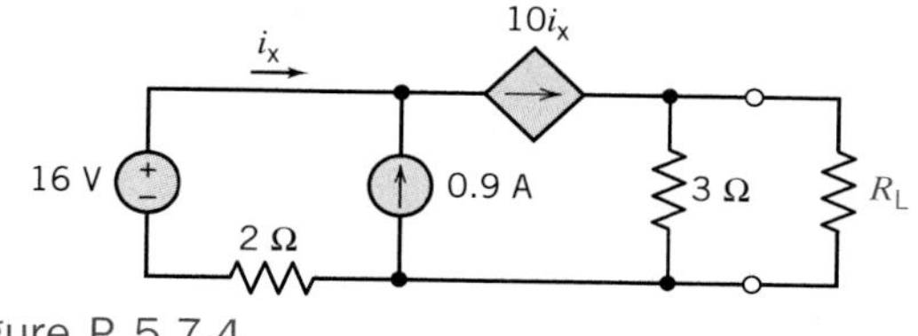

Figure P 5.7-4

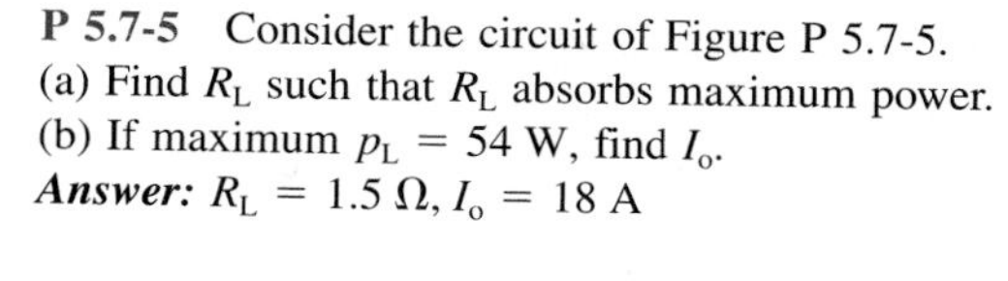

P 5.7-5 Consider the circuit of Figure P 5.7-5.
(a) Find R_L such that R_L absorbs maximum power.
(b) If maximum $p_L = 54$ W, find I_o.

Answer: $R_L = 1.5\ \Omega$, $I_o = 18$ A

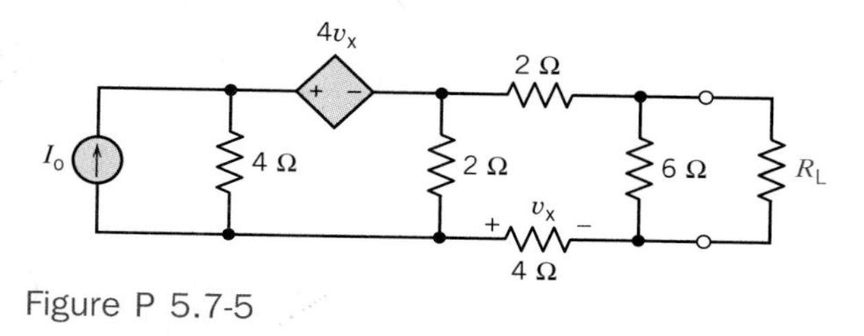

Figure P 5.7-5

P 5.7-6 Determine the maximum power that can be absorbed by a resistor, R, connected to terminals a–b of the circuit shown in Figure P 5.7-6. Specify the value of R.

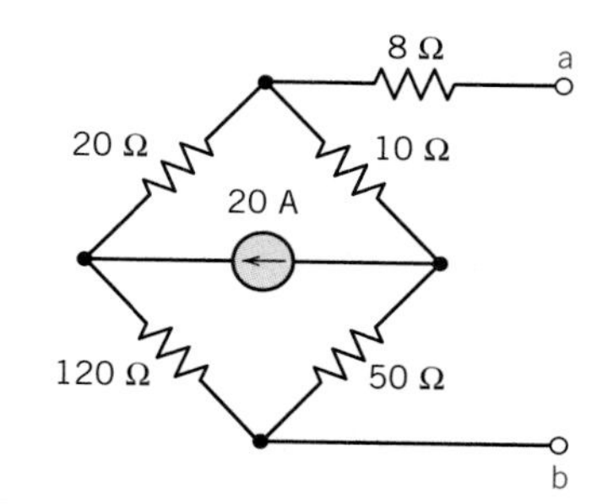

Figure P 5.7-6 Bridge circuit.

PSPICE PROBLEMS

SP 5.1 Determine the output voltage v for the circuit shown in Figure SP 5.1.

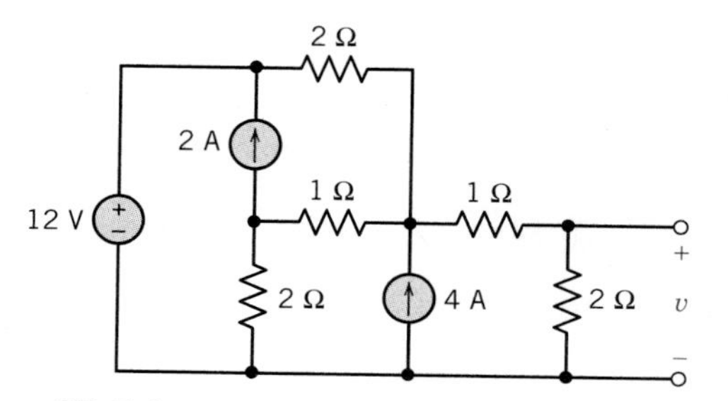

Figure SP 5.1

SP 5.2 Determine the current i for the circuit shown in Figure SP 5.2.

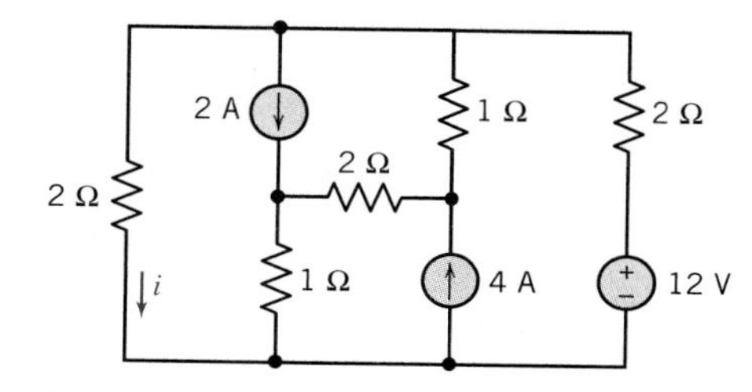

Figure SP 5.2

SP 5.3 Determine v for the circuit shown in Figure SP 5.3.

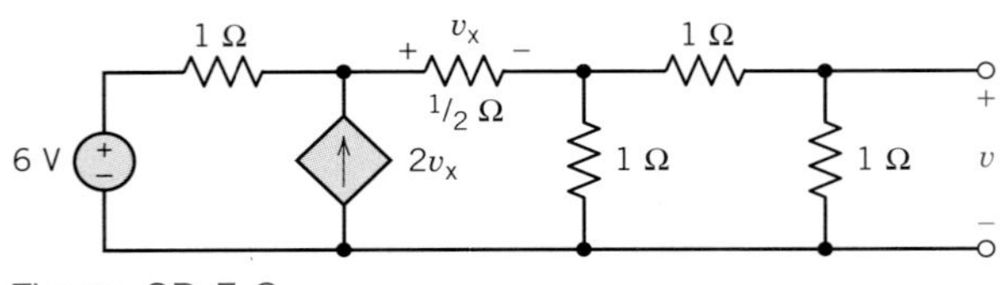

Figure SP 5.3

SP 5.4 A circuit with a constant-voltage source and a variable-current source is shown in Figure SP 5.4. Use PSpice to obtain a graphic plot of v_2 versus i_s when i_s is a constant I_o, which varies between 0 and 2 mA.

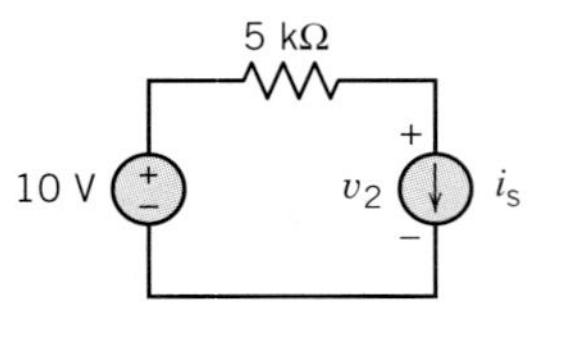

Figure SP 5.4

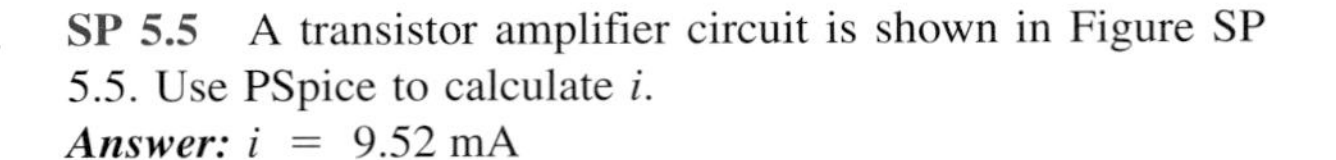

SP 5.5 A transistor amplifier circuit is shown in Figure SP 5.5. Use PSpice to calculate i.

Answer: $i = 9.52$ mA

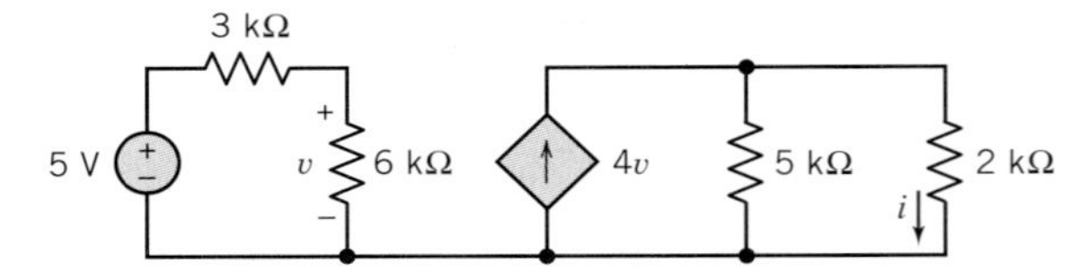

Figure SP 5.5 Transistor amplifier circuit.

SP 5.6 Use PSpice to determine the Thévenin equivalent of the circuit of Figure SP 5.6.

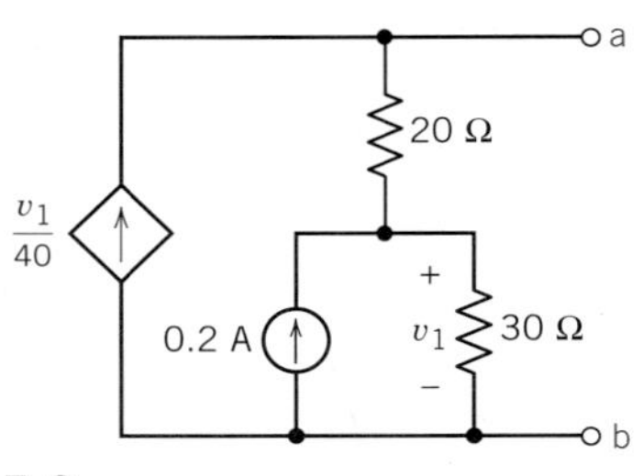

Figure SP 5.6

VERIFICATION PROBLEMS

VP 5-1 For the circuit of Figure VP 5.1 the current has been measured for three different values of R and is listed in the table. Are the data consistent?

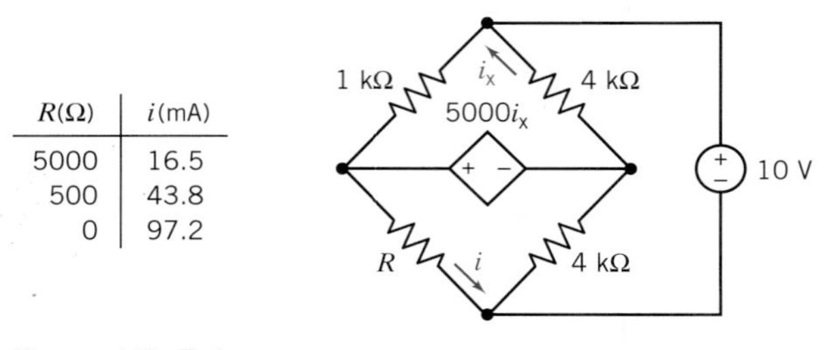

$R(\Omega)$	i(mA)
5000	16.5
500	43.8
0	97.2

Figure VP 5.1

VP 5-2 Your lab partner built the circuit shown in Figure VP 5.2 and measured the current i and voltage v corresponding to several values of the resistance R. The results are shown in the table in Figure VP 5.2. Your lab partner says that $R_L = 8000\ \Omega$ is required to cause $i = 1$ mA. Do you agree? Justify your answer.

VP 5-3 In preparation for lab, your lab partner determined the Thévenin equivalent of the circuit connected to R_L in Figure VP 5.3. She says that the Thévenin resistance is $R_t = \frac{6}{11}R$ and the open-circuit voltage is $v_{oc} = \frac{60}{11}$ V. In lab, you built the circuit using $R = 110\ \Omega$ and $R_L = 40\ \Omega$ and measured that $i = 54.5$ mA. Is this measurement consistent with the prelab calculations? Justify your answers.

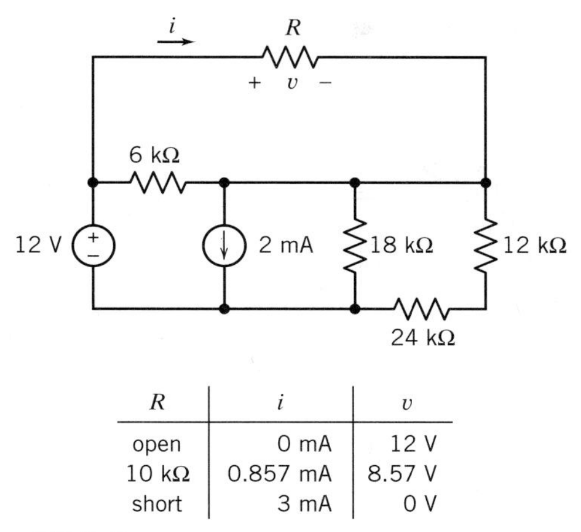

R	i	v
open	0 mA	12 V
10 kΩ	0.857 mA	8.57 V
short	3 mA	0 V

Figure VP 5.2

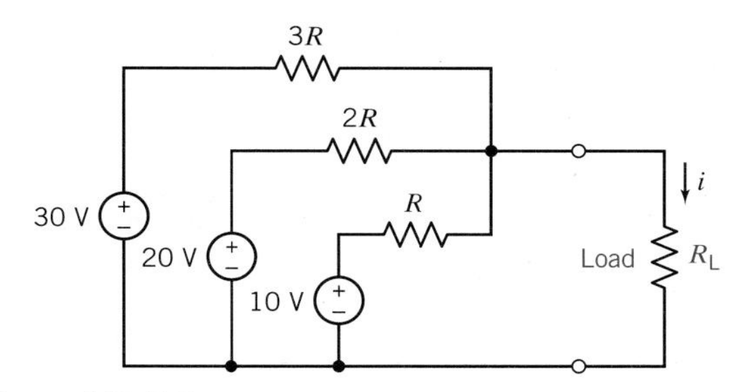

Figure VP 5.3

DESIGN PROBLEMS

DP 5-1 The circuit shown in Figure DP 5.1*a* has four unspecified circuit parameters: v_s, R_1, R_2, and R_3. To design this circuit, we must specify the values of these four parameters. The graph shown in Figure DP 5.1*b* describes a relationship between the current i and the voltage v.

Specify values of v_s, R_1, R_2, and R_3 that cause the current i and the voltage v in Figure DP 5.1*a* to satisfy the relationship described by the graph in Figure DP 5.1*b*.

First Hint: The equation representing the straight line in Figure DP 5.1*b* is $v = -R_t\, i + v_{oc}$. That is, the slope of the line is equal to -1 times the Thévenin resistance and the "v-intercept" is equal to the open circuit voltage.

Second Hint: There is more than one correct answer to this problem. Try setting $R_1 = R_2$.

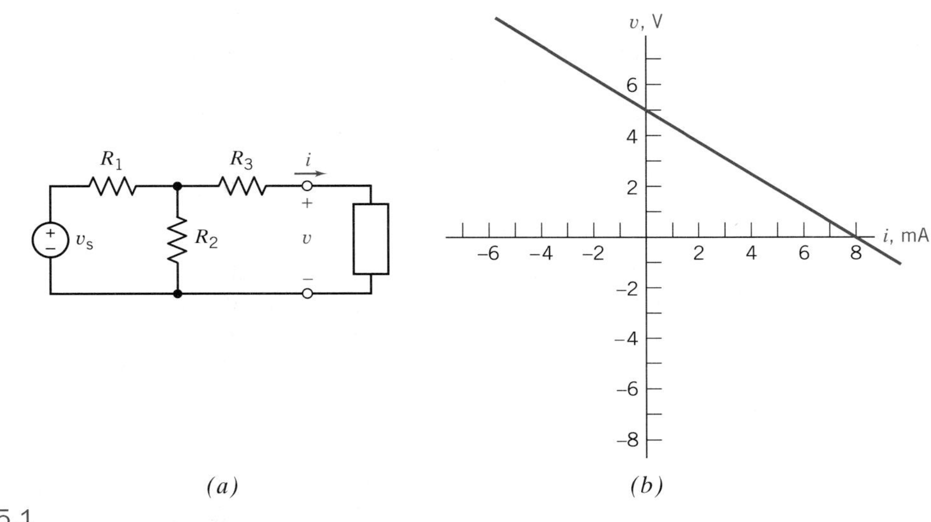

Figure DP 5.1

DP 5-2 The circuit shown in Figure DP 5.2*a* has four unspecified circuit parameters: i_s, R_1, R_2, and R_3. To design this circuit, we must specify the values of these four parameters. The graph shown in Figure DP 5.2*b* describes a relationship between the current i and the voltage v.

Specify values of i_s, R_1, R_2, and R_3 that cause the current i and the voltage v in Figure DP 5.2*a* to satisfy the relationship described by the graph in Figure DP 5.2*b*.

First Hint: Calculate the open circuit voltage, v_{oc}, and the Thévenin resistance, R_t, of the part of the circuit to the left of the terminals in Figure DP 5.2*a*.

Second Hint: The equation representing the straight line in Figure DP 5.2*b* is $v = -R_t i + v_{oc}$.

That is, the slope of the line is equal to -1 times the Thévenin resistance and the "v-intercept" is equal to the open circuit voltage.

Third Hint: There is more than one correct answer to this problem. Try setting both R_3 and $R_1 + R_2$ equal to twice the slope of the graph in Figure DP 5.2*b*.

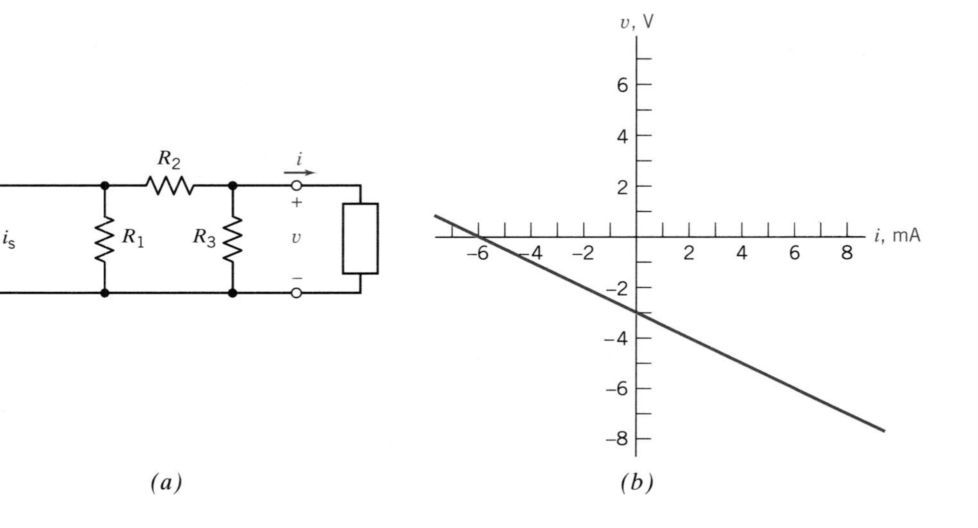

Figure DP 5.2

DP 5-3 The circuit shown in Figure DP 5.3*a* has four unspecified circuit parameters: v_s, R_1, R_2, and R_3. To design this circuit, we must specify the values of these four parameters. The graph shown in Figure DP 5.3*b* describes a relationship between the current i and the voltage v.

Is it possible to specify values of v_s, R_1, R_2 and R_3 that cause the current i and the voltage v in Figure DP 5.1*a* to satisfy the relationship described by the graph in Figure DP 5.3*b*? Justify your answer.

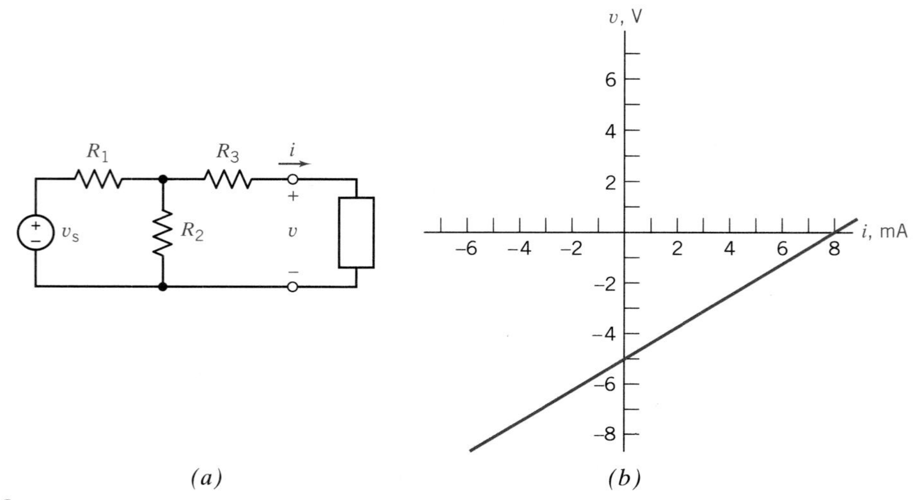

Figure DP 5.3

DP 5-4 The circuit shown in Figure DP 5.4*a* has four unspecified circuit parameters: v_s, R_1, R_2, and d, where d is the gain of the CCCS. To design this circuit, we must specify the values of these four parameters. The graph shown in Figure DP 5.4*b* describes a relationship between the current i and the voltage v.

Specify values of v_s, R_1, R_2, and d that cause the current i and the voltage v in Figure DP 5.4*a* to satisfy the relationship described by the graph in Figure DP 5.4*b*.

First Hint: The equation representing the straight line in Figure DP 5.4*b* is $v = -R_t i + v_{oc}$.
That is, the slope of the line is equal to -1 times the Thévenin resistance and the "v-intercept" is equal to the open circuit voltage.
Second Hint: There is more than one correct answer to this problem. Try setting $R_1 = R_2$.

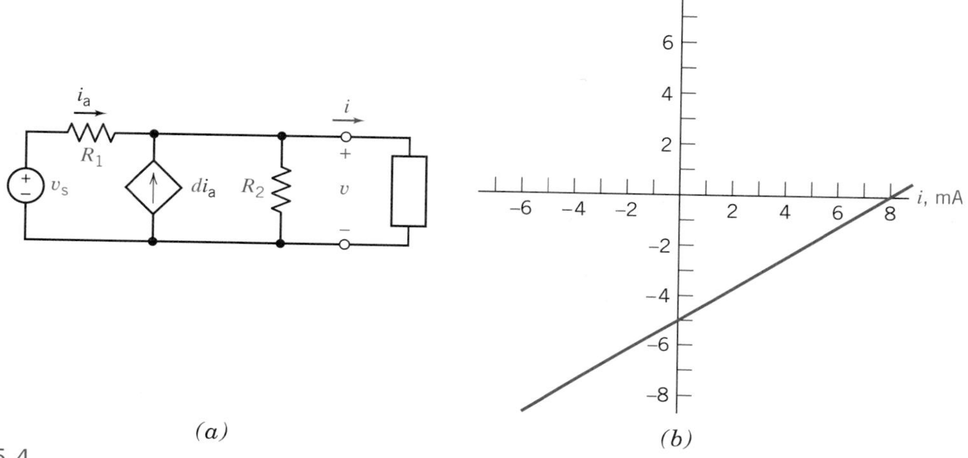

Figure DP 5.4

DP 5-5 A transmitter and antenna circuit is shown in Figure DP 5.5.

(a) Find the Thévenin equivalent circuit for the transmitter circuit (left of terminals a–b).

(b) Determine the value of β required to yield R_t of the transmitter so that it matches the load resistance, R_L.

(c) What value of β results in maximum power delivered to R_L?

(d) What practical engineering constraints could influence the value of β selected for the circuit?

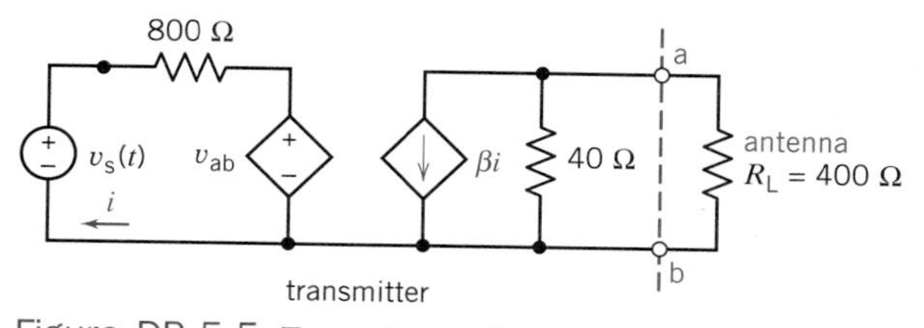

Figure DP 5.5 Transmitter and antenna circuit.

DP 5-6 There are many applications in dc, ac, and radio frequency (RF) circuits where power division is required. Consider the two-way resistive power divider circuit shown in Figure DP 5.6. The load resistances R_{L1} and R_{L2} represent two TV sets. If maximum power transfer to each TV set and equal power division are required at the interface a–b, what is the value of the resistor, R, that is necessary?

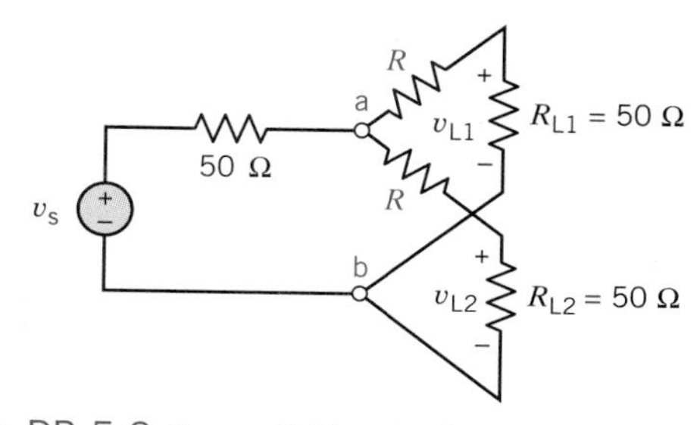

Figure DP 5.6 Power divider circuit.

CHAPTER 6

The Operational Amplifier

Preview

With the invention of the electron tube in 1907, the field of electronics was born. In this chapter we consider an active element called the operational amplifier, which has many useful applications in electronics.

When the operational amplifier is connected to resistors and capacitors, we obtain various configurations of amplifier circuits. Several useful forms of amplifier circuits are described and analyzed. We look at the time domain response of these circuits as well as their response when the input source is a steady-state sinusoid. The benefits of an amplifier circuit include isolation between input and output terminals and a power gain between the input and output terminals.

6.1 Design Challenge

TRANSDUCER INTERFACE CIRCUIT

A customer wants to automate a pressure measurement, which requires converting the output of the pressure transducer to a computer input. This conversion can be done using a standard integrated circuit called an *Analog-to-Digital Converter* (ADC) (Dorf 1998). The ADC requires an input voltage between 0 V and 10 V, while the pressure transducer output varies between −250 mV and 250 mV. Design a circuit to interface the pressure transducer with the ADC. That is, design a circuit that translates the range −250 mV to 250 mV to the range 0 V to 10 V.

Describe the Situation and the Assumptions

The situation is shown in Figure 6.1-1.

The specifications state that

$$-250 \text{ mV} \leq v_1 \leq 250 \text{ mV}$$
$$0 \text{ V} \leq v_2 \leq 10 \text{ V}$$

A simple relationship between v_2 and v_1 is needed so that information about the pressure is not obscured. Consider

$$v_2 = a \cdot v_1 + b$$

The coefficients, a and b, can be calculated by requiring that $v_2 = 0$ when $v_1 = -250$ mV and that $v_2 = 10$ V when $v_1 = 250$ mV, that is,

$$0 \text{ V} = a \cdot (-250 \text{ mV}) + b$$
$$10 \text{ V} = a \cdot 250 \text{ mV} + b$$

Solving these simultaneous equations gives $a = 20$ V and $b = 5$ V.

State the Goal

Design a circuit having input voltage v_1 and output voltage v_2. These voltages should be related by

$$v_2 = 20\, v_1 + 5 \quad \text{V}$$

In this chapter, we will see that such a circuit can be designed using operational amplifiers.

6.2 ELECTRONICS

As the use of radio began to expand in the 1920s, the industry associated with radio and electron tubes became important to electrical engineering. It was the invention of the triode vacuum tube by Lee De Forest in 1907 that enabled radio to grow. The use of the triode as an amplifier and detector demonstrated that radio circuits could readily be designed.

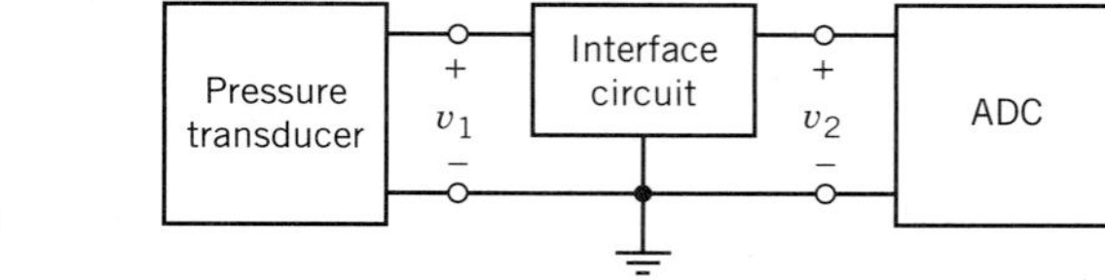

Figure 6.1-1
Interfacing a pressure transducer with an analog-to-digital converter (ADC).

By 1929 the vacuum tube became known as the electron tube. With the publication of *Electronics* magazine in 1930, a new word was born and an industry identified. *Electronics* is the engineering field and industry that uses electron devices in circuits and systems. During the first half of the century, electronics was dominated by the vacuum tube.

Engineers were interested in miniaturization of electronics and turned to experimentation with semiconductors, which led to the invention of the transistor.

The fundamental properties of semiconductors differentiate them from metals and insulators. A *semiconductor* is an electronic conductor with a resistivity in the range between the resistivity of metals and that of insulators. The first radio detectors used galena crystals, silicon, or silicon carbide crystals. With the development of the theory of quantum mechanics in the period 1926-1936, the understanding of semiconductors increased.

By the 1930s several early semiconductor devices had been built in the laboratory. The analogy between the semiconductor diode and the vacuum diode was obvious, and several people attempted to build a three-terminal semiconductor device.

In July 1945 a semiconductor research program was set up at Bell Laboratories. John Bardeen, William Shockley, and Walter Brattain, shown in Figure 6.2-1, were the inventors of the first transistor. One of the first transistors using a germanium crystal was assembled as shown in Figure 6.2-2. A *transistor* is an active semiconductor device with three or more terminals. Transistors became commercially available in 1954 as the manufacturing methods were perfected.

Texas Instruments (TI) was one of the early manufacturers of the transistor. One of the leading engineers at TI was Patrick E. Haggerty, who saw the opportunity for a miniature radio using transistors and small components. The goal was to build a miniature radio by late 1954 and sell it for $50. The first transistor radio built by TI is shown in Figure 6.2-3, and an inside view of the radio is shown in Figure 6.2-4. The radio was so popular that 100,000 were sold during 1955.

With the advent of the transistor, several electronic circuits became available. The motivation to miniaturize these electronic circuits was based on factors such as the complexity of circuits, the need for reliability, and the desire to reduce power consumption, weight,

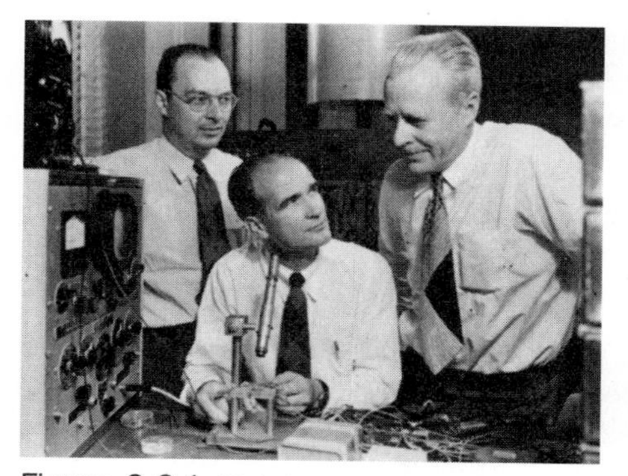

Figure 6.2-1 Nobel Prize winners John Bardeen, William Shockley, and Walter H. Brattain (left to right), shown at Bell Telephone Laboratories in 1948 with the apparatus used in the first investigations that led to the invention of the transistor. The trio received the 1956 Nobel Prize in physics for their invention of the transistor, which was announced by Bell Laboratories in 1948. Courtesy of Bell Telephone Laboratories.

Figure 6.2-2 The first transistors assembled by their inventors at Bell Laboratories (in 1947) were primitive by today's standards. Yet they revolutionized the electronics industry and changed our way of life. The first transistor, a "point-contact" type, amplified electrical signals by passing them through a solid semiconductor material, basically the same operation as performed by present "junction" transistors. The three terminal wires can be seen on the top of the transistor. The actual record of the first transistor operation was December 23, 1947. Courtesy of Bell Telephone Laboratories.

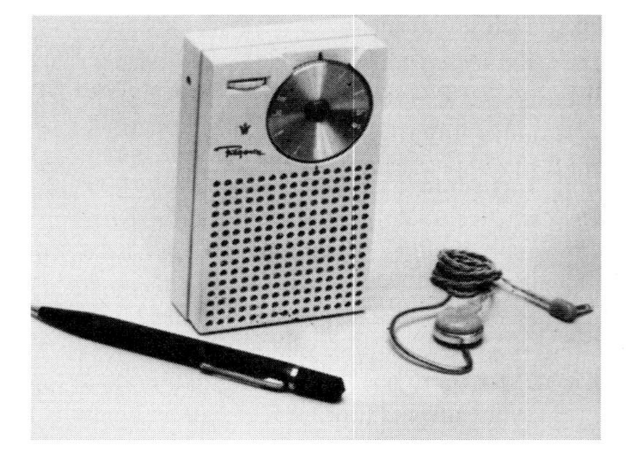

Figure 6.2-3 The first commercial transistor radio. The molded plastic case of the Regency TR-1 radio was designed to fit into the pocket of a man's dress shirt. It was 5 by 3 by $1\frac{1}{4}$ inches and sold for $49.95. This radio was introduced on October 18, 1954. Courtesy of Texas Instruments.

Figure 6.2-4 Inside view of the radio in Figure 6.2-3. The specially designed miniature components barely fit inside the 5 by 3 by $1\frac{1}{4}$ inch case. The four transistors, discrete resistors, capacitors, and other miniaturized components were mounted, along with the $2\frac{3}{4}$ inch speaker, in the front half of the plastic case. Courtesy of Texas Instruments.

and of course cost. By the late 1950s engineers were discussing the possibility that complete circuit functions could be formed within a single block of semiconductor. The first patent for an *integrated circuit* was filed on May 21, 1953, by Harwick Johnson of RCA.

By the 1960s integrated circuits were available for many electronic functions. An *integrated circuit* is defined as a combination of interconnected circuit elements inseparably associated on or within a continuous semiconductor (often called a chip). This chapter discusses the use and operation of one integrated electronic device, the operational amplifier.

The operational amplifier (op amp) was originally developed for analog computers. When op amps were used with associated resistors and capacitors, they could perform mathematical operations such as addition and integration. The earliest op amps were constructed of vacuum tubes by engineers in the early 1940s. They were used by George Philbrick as a building block for analog computers used to solve differential equations. One of the most widely used op amps, the μA741, became available as an integrated circuit in 1967. Op amps became popular because they are small, versatile, easy to use, and inexpensive.

6.3 THE OPERATIONAL AMPLIFIER

The *operational amplifier* is an active element with a high-gain ratio designed to be used with other circuit elements to perform a specified signal-processing operation. The μA741 operational amplifier is shown in Figure 6.3-1*a*. It has eight pin connections, whose functions are indicated in Figure 6.3-1*b*.

The operational amplifier shown in Figure 6.3-2 has five terminals. The names of these terminals are shown in both Figure 6.3-1*b* and Figure 6.3-2. Notice the plus and minus signs on the triangular part of the symbol of the operational amplifier. The plus sign identifies the noninverting input, and the minus sign identifies the inverting input.

The power supplies are used to bias the operational amplifier. In other words, the power supplies cause certain conditions that are required for the operational amplifier to function properly. Once the operational amplifiers have been biased, the power supplies have little effect on the way that the operational amplifiers work. It is inconvenient to include

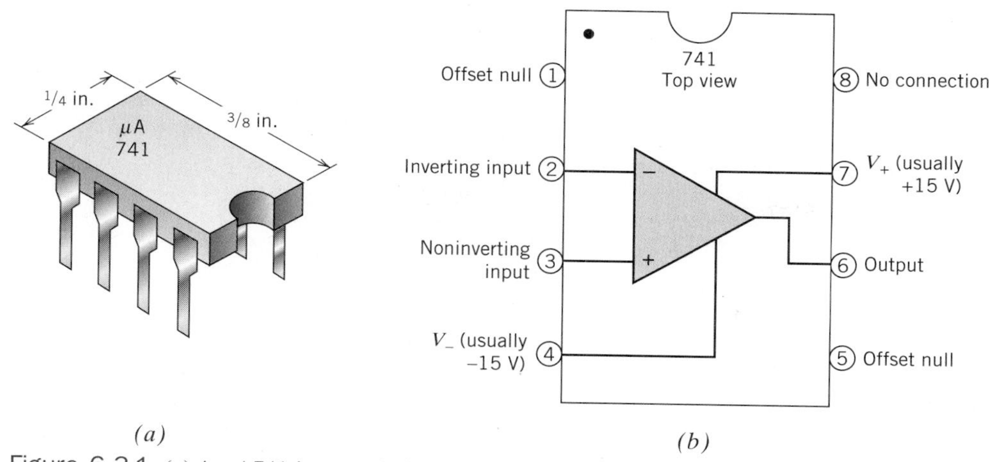

Figure 6.3-1 (*a*) A μA741 integrated circuit has eight connecting pins. (*b*) The correspondence between the circled pin numbers of the integrated circuit and the nodes of the operational amplifier.

the power supplies in drawings of operational amplifier circuits. These power supplies tend to clutter drawings of operational amplifier circuits, making them harder to read. Consequently, the power supplies are frequently omitted from drawings that accompany explanations of the function of operational amplifier circuits, such as the drawings found in textbooks. It is understood that power supplies are part of the circuit even though they are not shown. (Schematics, the drawings used to describe how to assemble a circuit, are a different matter.) The power supplies are shown in Figure 6.3-2, denoted as v_+ and v_-.

Since the power supplies are frequently omitted from the drawing of an operational amplifier circuit, it is easy to overlook the power supply currents. This mistake is avoided by careful application of Kirchhoff's Current Law (KCL). As a general rule, it is not helpful to apply KCL in a way that involves any power supply current. Two specific cases are of particular importance. First, the ground node in Figure 6.3-2 is a terminal of both power supplies. Both power supply currents would be involved if KCL was applied to the ground node. These currents must not be overlooked. It is best simply to refrain from applying KCL at the ground node of an operational amplifier circuit. Second, KCL requires that the sum of all currents into the operational amplifier be zero

$$i_1 + i_2 + i_o + i_+ + i_- = 0$$

Both power supply currents are involved in this equation. Once again, these currents must not be overlooked. It is best simply to refrain from applying KCL to sum the currents into an operational amplifier when the power supplies are omitted from the diagram.

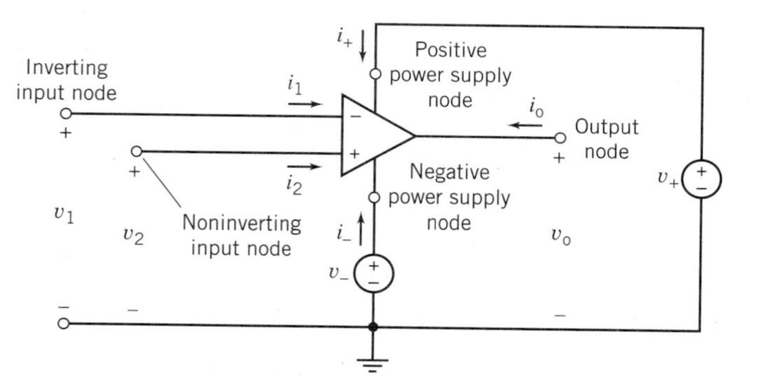

Figure 6.3-2
An op amp, including power supplies v_+ and v_-.

6.4 THE IDEAL OPERATIONAL AMPLIFIER

Operational amplifiers are complicated devices that exhibit both linear and nonlinear behavior. The operational amplifier output voltage and current, v_o and i_o, must satisfy three conditions in order for an operational amplifier to be linear, that is:

$$\begin{aligned} |v_o| &\le v_{sat} \\ |i_o| &\le i_{sat} \\ \left|\frac{dv_o(t)}{dt}\right| &\le SR \end{aligned} \tag{6.4-1}$$

The saturation voltage, v_{sat}, the saturation current, i_{sat}, and the slew rate limit, SR, are all parameters of an operational amplifier. For example, if a μA741 operational amplifier is biased using +15 V and −15 V power supplies, then

$$v_{sat} = 14\,\text{V}, \quad i_{sat} = 2\text{ mA}, \quad SR = 500{,}000\,\frac{\text{V}}{\text{s}} \tag{6.4-2}$$

These restrictions reflect the fact that operational amplifiers cannot produce arbitrarily large voltages or arbitrarily large currents or change output voltage arbitrarily quickly.

Figure 6.4-1 describes the *ideal operational amplifier.* The ideal operational amplifier is a simple model of an operational amplifier that is linear. The ideal operational amplifier is characterized by restrictions on its input currents and voltages. The currents into the input terminals of an ideal operational amplifier are zero. Consequently, in Figure 6.4-1

$$i_1 = 0 \quad \text{and} \quad i_2 = 0$$

The node voltages at the input nodes of an ideal operational amplifier are equal. Consequently, in Figure 6.4-1

$$v_2 = v_1$$

The ideal operational amplifier is a model of a linear operational amplifier, so the operational amplifier output current and voltage must satisfy the restrictions in Eq. 6.4-1. If they do not, then the ideal operational amplifier is not an appropriate model of the real operational amplifier. The output current and voltage depend on the circuit in which the operational amplifier is used. The ideal op amp conditions are summarized in Table 6.4-1.

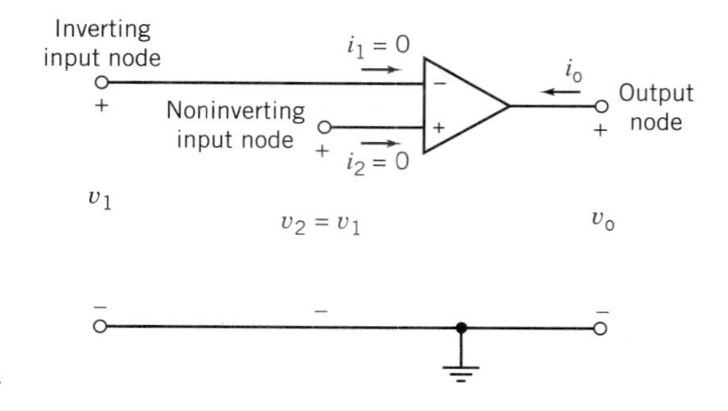

Figure 6.4-1
The ideal operational amplifier.

Table 6.4-1 **Operating Conditions for an Ideal Operational Amplifier**

Variable	Ideal Condition
Inverting node input current	$i_1 = 0$
Noninverting node input current	$i_2 = 0$
Voltage difference between inverting node voltage v_1 and noninverting node voltage v_2	$v_2 - v_1 = 0$

Example 6.4-1

Consider the circuit shown in Figure 6.4-2*a*. Suppose the operational amplifier is a μA741 operational amplifier. Model the operational amplifier as an ideal operational amplifier. Determine how the output voltage, v_o, is related to the input voltage, v_s.

Solution

Figure 6.4-2*b* shows the circuit when the operational amplifier of Figure 6.4-2*a* is modeled as an ideal operational amplifier.

1. The inverting input node and output node of the operational amplifier are connected by a short circuit, so the node voltages at these nodes are equal:

$$v_1 = v_o$$

2. The voltages at the inverting and noninverting nodes of an ideal op amp are equal

$$v_2 = v_1 = v_o$$

3. The currents into the inverting and noninverting nodes of an operational amplifier are zero, so

$$i_1 = 0 \quad \text{and} \quad i_2 = 0$$

4. The current in resistor R_s is $i_2 = 0$, so the voltage across R_s is 0 V. The voltage across R_s is $v_s - v_2 = v_s - v_o$, hence

$$v_s - v_o = 0$$

or

$$v_s = v_o$$

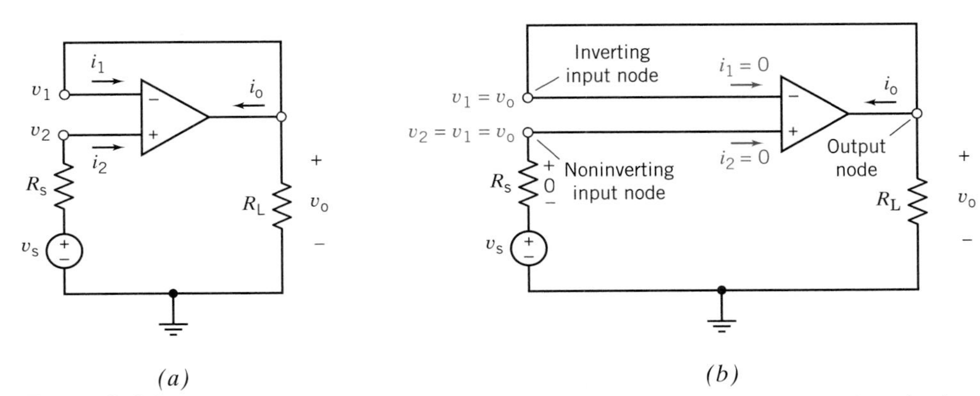

Figure 6.4-2 (*a*) The operational amplifier circuit for Example 6.4-1 and (*b*) an equivalent circuit showing the consequences of modeling the operational amplifier as an ideal operational amplifier. The voltages v_1, v_2, and v_o are node voltages and hence are the voltages at the node with respect to ground.

Does this solution satisfy the requirements of Eqs. 6.4-1 and 6.4-2? The output current of the operational amplifier must be calculated. Apply KCL at the output node of the operational amplifier to get

$$i_1 + i_o + \frac{v_o}{R_L} = 0$$

Since $i_1 = 0$,

$$i_o = -\frac{v_o}{R_L}$$

Now Eqs. 6.4-1 and 6.4-2 require

$$|v_s| \leq 14\,\text{V}$$
$$\left|\frac{v_s}{R_L}\right| \leq 2\text{ mA}$$
$$\left|\frac{d}{dt}v_s\right| \leq 500{,}000\,\frac{\text{V}}{\text{s}}$$

For example, when $v_s = 10$ V and $R_L = 20$ kΩ, then

$$|v_s| = 10\text{ V} < 14\text{ V}$$
$$\left|\frac{v_s}{R_L}\right| = \frac{10\,\text{V}}{20\text{ k}\Omega} = \frac{1}{2}\text{ mA} < 2\text{ mA}$$
$$\left|\frac{d}{dt}v_s\right| = 0 < 500{,}000\,\frac{\text{V}}{\text{s}}$$

This is consistent with the use of the ideal operational amplifier. On the other hand, when $v_s = 10$ V and $R_L = 2$ kΩ, then

$$\frac{v_s}{R_L} = 5\text{ mA} > 2\text{ mA}$$

so it is not appropriate to model the μA741 as an ideal operational amplifier when $v_s = 10$ V and $R_L = 2$ kΩ. When $v_s = 10$ V we require $R_L > 5$ kΩ in order to satisfy Eq. 6.4-1.

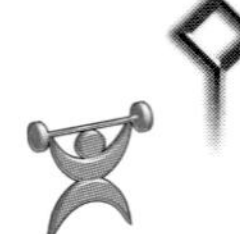

Exercise 6.4-1 Find the ratio v_o/v_s of the circuit shown in Figure E 6.4-1 where an ideal op amp is assumed.

Answer: $\frac{v_o}{v_s} = 1 + \frac{R_2}{R_1}$

Exercise 6.4-2 A noninverting op amp circuit is shown in Figure E 6.4-2.
(a) Find the voltage ratio v_o/v_s.
(b) Determine v_o/v_s when $R_2 \gg R_1$.

Answer: (a) $\frac{v_o}{v_s} = \frac{R_2}{R_1 + R_2}\left(1 + \frac{R_4}{R_3}\right)$

(b) $\frac{v_o}{v_s} = 1 + \frac{R_4}{R_3}$

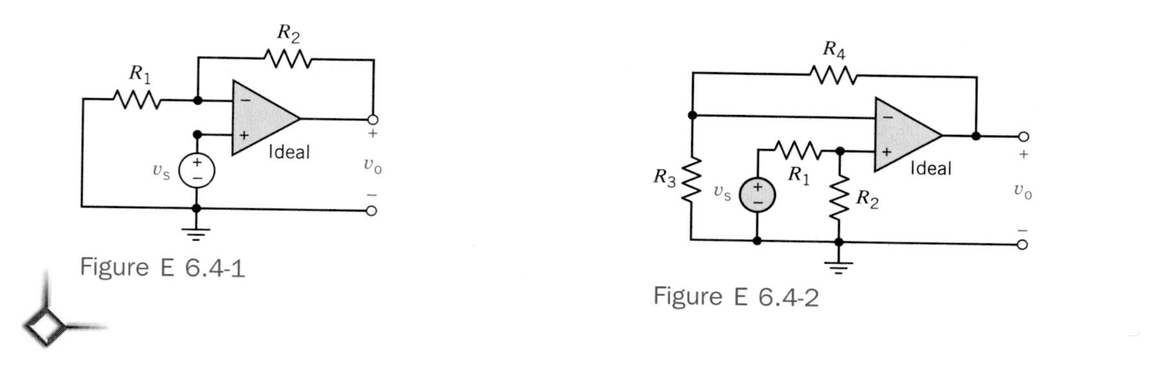

Figure E 6.4-1

Figure E 6.4-2

6.5 NODAL ANALYSIS OF CIRCUITS CONTAINING IDEAL OPERATIONAL AMPLIFIERS

It is convenient to use node equations to analyze circuits containing ideal operational amplifiers. There are three things to remember.

1. The node voltages at the input nodes of ideal operational amplifiers are equal. Thus, one of these two node voltages can be eliminated from the node equations. For example, in Figure 6.5-1, the voltages at the input nodes of the ideal operational amplifier are v_1 and v_2. Since

$$v_1 = v_2$$

v_2 can be eliminated from the node equations.

2. The currents in the input leads of an ideal operational amplifier are zero. These currents are involved in the KCL equations at the input nodes of the operational amplifier.
3. The output current of the operational amplifier is not zero. This current is involved in the KCL equations at the output node of the operational amplifier. Applying KCL at this node adds another unknown to the node equations. If the output current of the operational amplifier is not to be determined, then it is not necessary to apply KCL at the output node of the operational amplifier.

Example 6.5-1

The circuit shown in Figure 6.5-1 is called a difference amplifier. The operational amplifier has been modeled as an ideal operational amplifier. Use node equations to analyze this circuit and determine v_o in terms of the two source voltages, v_a and v_b.

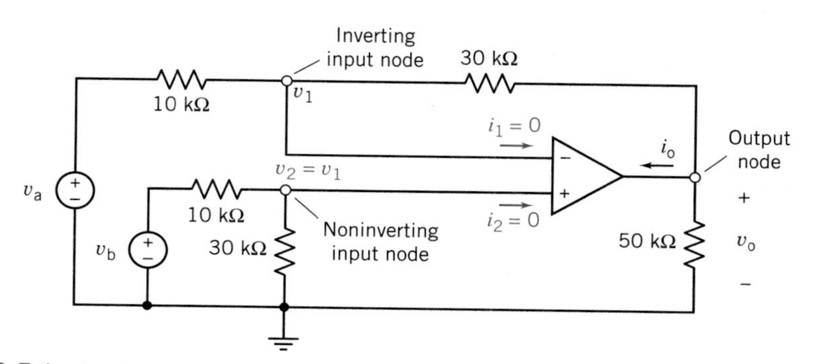

Figure 6.5-1 Circuit of Example 6.5-1.

Solution

The node equation at the noninverting node of the ideal operational amplifier is

$$\frac{v_2}{30\text{ k}\Omega} + \frac{v_2 - v_b}{10\text{ k}\Omega} + i_2 = 0$$

Since $v_2 = v_1$ and $i_2 = 0$, this equation becomes

$$\frac{v_1}{30\text{ k}\Omega} + \frac{v_1 - v_b}{10\text{ k}\Omega} = 0$$

Solving for v_1 we have

$$v_1 = 0.75 \cdot v_b$$

The node equation at the inverting node of the ideal operational amplifier is

$$\frac{v_1 - v_a}{10\text{ k}\Omega} + \frac{v_1 - v_o}{30\text{ k}\Omega} + i_1 = 0$$

Since $v_1 = 0.75 v_b$ and $i_1 = 0$, this equation becomes

$$\frac{0.75 \cdot v_b - v_a}{10\text{ k}\Omega} + \frac{0.75 \cdot v_b - v_o}{30\text{ k}\Omega} = 0$$

Solving for v_o we have

$$v_o = 3(v_b - v_a)$$

The difference amplifier takes its name from the fact that the output voltage, v_o, is a function of the difference, $v_b - v_a$, of the input voltages.

Example 6.5-2

Next, consider the circuit shown in Figure 6.5-2*a*. This circuit is called a bridge amplifier. The part of the circuit that is called a bridge is shown in Figure 6.5-2*b*. The operational amplifier and resistors R_5 and R_6 are used to amplify the output of the bridge. The operational amplifier in Figure 6.5-2*a* has been modeled as an ideal operational amplifier. As a consequence, $v_1 = 0$ and $i_1 = 0$ as shown. Determine the output voltage, v_o in terms of the source voltage, v_s.

Solution

Here is an opportunity to use Thévenin's theorem. Figure 6.5-2*c* shows the Thévenin equivalent of the bridge circuit. Figure 6.5-2*d* shows the bridge amplifier after the bridge has been replaced by its Thévenin equivalent. Figure 6.5-2*d* is simpler than Figure 6.5-2*a*. It is easier to write and solve the node equations representing Figure 6.5-2*d* than it is to write and solve the node equations representing Figure 6.5-2*a*. Thévenin's theorem assures us that the voltage v_o in Figure 6.5-2*d* is the same as the voltage v_o in Figure 6.5-2*a*.

Let us write node equations representing the circuit in Figure 6.5-2*d*. First notice that the node voltage v_a is given by (using KVL)

$$v_a = v_1 + v_{oc} + R_t i_1$$

Since $v_1 = 0$ and $i_1 = 0$,

$$v_a = v_{oc}$$

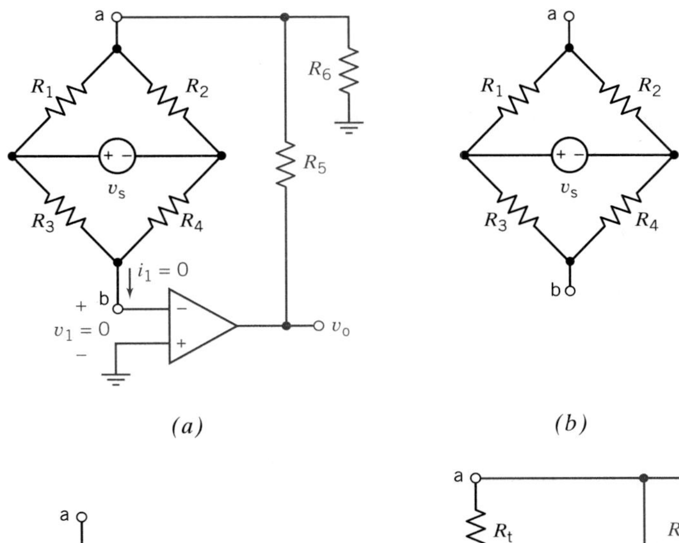

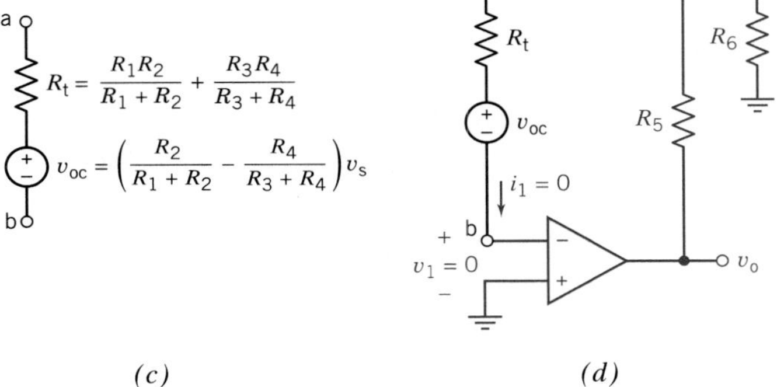

(c) (d)

Figure 6.5-2 (*a*) A bridge amplifier, including the bridge circuit. (*b*) The bridge circuit and, (*c*) its Thévenin equivalent circuit. (*d*) The bridge amplifier, including the Thévenin equivalent of the bridge.

Now writing the node equation at node a

$$i_1 + \frac{v_a - v_o}{R_5} + \frac{v_a}{R_6} = 0$$

Since $v_a = v_{oc}$ and $i_1 = 0$,

$$\frac{v_{oc} - v_o}{R_5} + \frac{v_{oc}}{R_6} = 0$$

Solving for v_o we have

$$v_o = \left(1 + \frac{R_5}{R_6}\right) v_{oc} = \left(1 + \frac{R_5}{R_6}\right)\left(\frac{R_2}{R_1 + R_2} - \frac{R_4}{R_3 + R_4}\right) v_s$$

Exercise 6.5-1 Find the relationship v_o/v_s for the circuit shown in Figure E 6.5-1.

Answer: $\frac{v_o}{v_s} = \frac{R_2}{R_1 + R_2}$

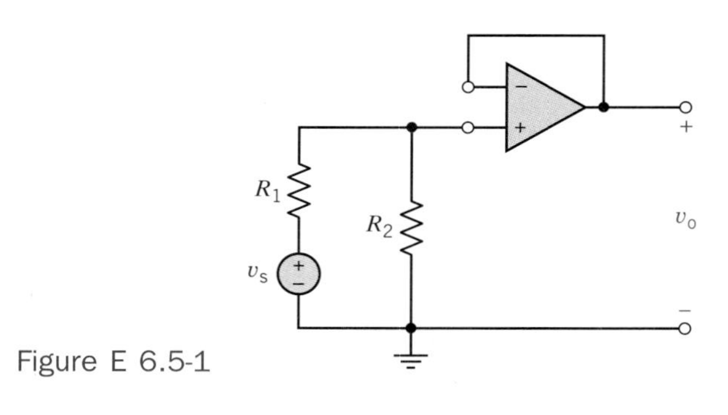

Figure E 6.5-1

6.6 DESIGN USING OPERATIONAL AMPLIFIERS

One of the early applications of operational amplifiers was to build circuits that performed mathematical operations. Indeed, the operational amplifier takes its name from this important application. Many of the operational amplifier circuits that perform mathematical operations are used so often that they have been given names. These names are part of an electrical engineer's vocabulary. Figure 6.6-1 shows several standard operational amplifier circuits. The next several examples show how to use Figure 6.6-1 to design simple operational amplifier circuits.

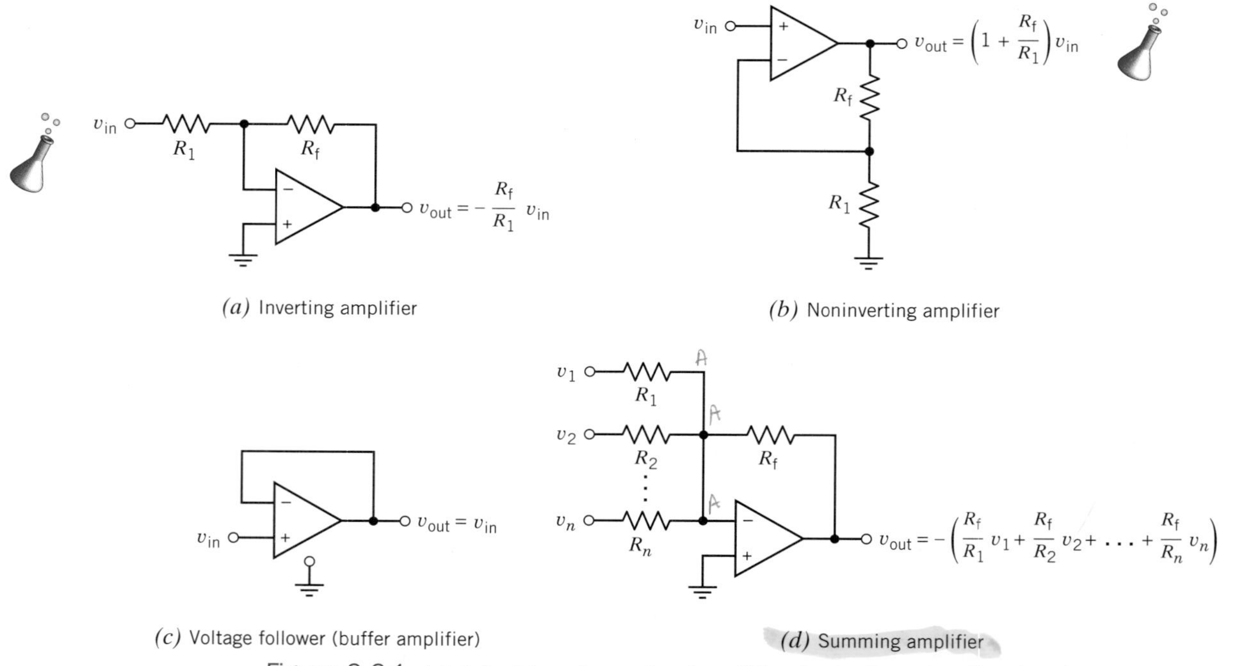

Figure 6.6-1 A brief catalog of operational amplifier circuits. Note that all node voltages are referenced to the ground node.

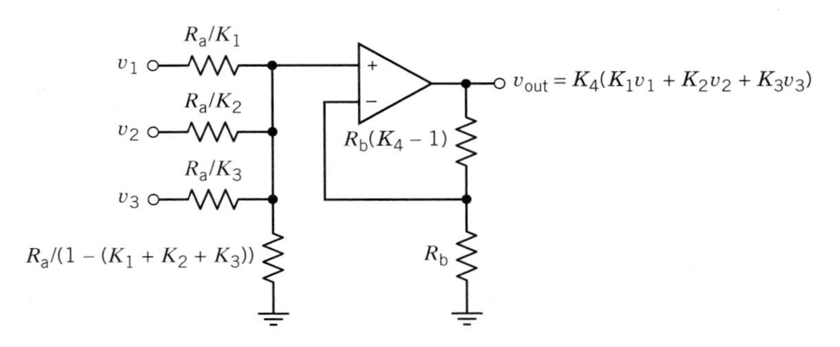

(e) Noninverting summing amplifier

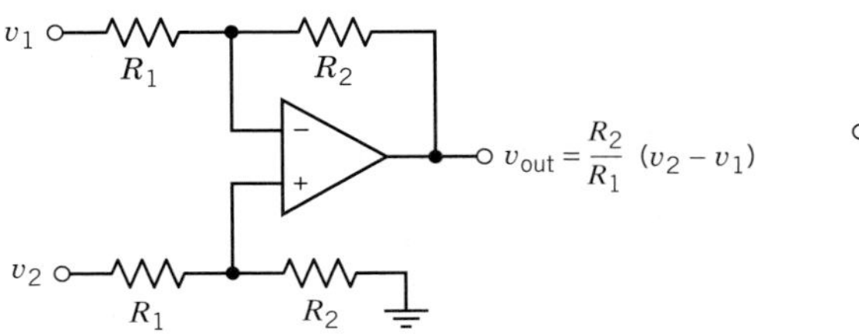

(f) Difference amplifier

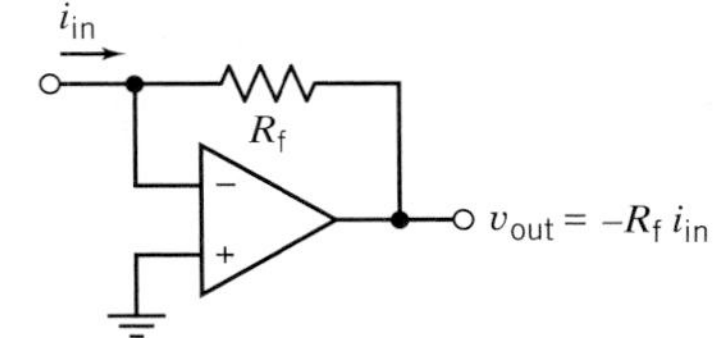

(g) Current-to-voltage converter

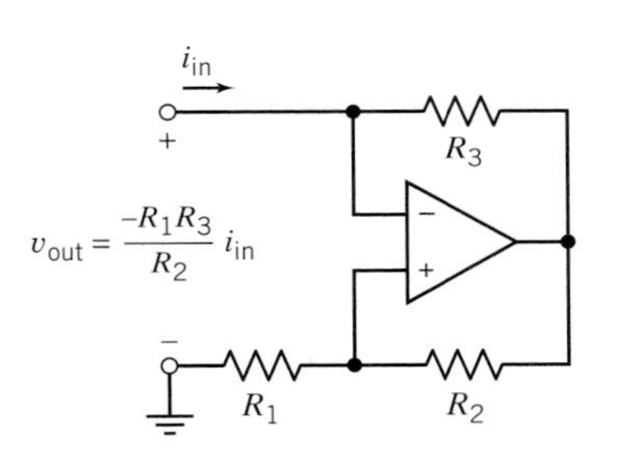

(h) Negative resistance convertor

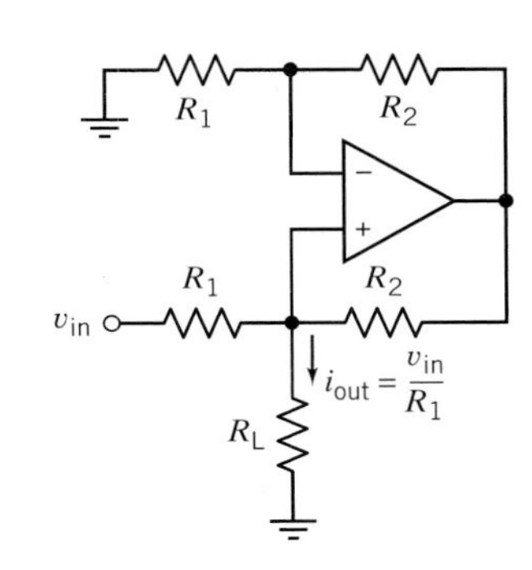

(i) Voltage-controlled current source (VCCS)

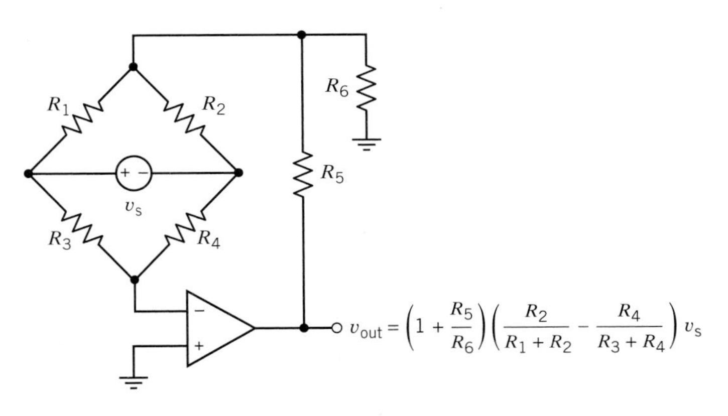

(j) Bridge amplifier

Figure 6.6-1 (cont.) A brief catalog of operational amplifier circuits.

Example 6.6-1

This example illustrates the use of a voltage follower to prevent loading. The voltage follower is shown in Figure 6.6-1*c*. Loading can occur when two circuits are connected. Consider Figure 6.6-2. In Figure 6.6-2*a* the output of Circuit #1 is the voltage v_a. In Figure 6.6-2*b*, Circuit #2 is connected to Circuit #1. The output of Circuit #1 is used as the input to Circuit #2. Unfortunately, connecting Circuit #2 to Circuit #1 can change the output of Circuit #1. This is called *loading*. Referring again to Figure 6.6-2, Circuit #2 is said to load Circuit #1 if $v_b \neq v_a$. The current i_b is called the load current. Circuit #1 is required to provide this current in Figure 6.6-2*b* but not in Figure 6.6-2*a*. This is the cause of the loading. The load current can be eliminated using a voltage follower as shown in Figure 6.6-2*c*. The voltage follower copies voltage v_a from the output of Circuit #1 to the input of Circuit #2 without disturbing Circuit #1.

Solution

As a specific example, consider Figure 6.6-3. The voltage divider shown in Figure 6.6-3*a* can be analyzed by writing a node equation at node a:

$$\frac{v_a - v_{in}}{20\ \text{k}\Omega} + \frac{v_a}{60\ \text{k}\Omega} = 0$$

Solving for v_a we have

$$v_a = \frac{3}{4} v_{in}$$

In Figure 6.6-3*b*, a resistor is connected across the output of the voltage divider. This circuit can be analyzed by writing a node equation at node a:

$$\frac{v_b - v_{in}}{20\ \text{k}\Omega} + \frac{v_b}{60\ \text{k}\Omega} + \frac{v_b}{30\ \text{k}\Omega} = 0$$

Solving for v_b, we have

$$v_b = \frac{1}{2} v_{in}$$

Since $v_b \neq v_a$, connecting the resistor directly to the voltage divider loads the voltage divider. This loading is caused by the current required by the 30 kΩ resistor. Without the voltage follower, the voltage divider must provide this current.

In Figure 6.6-3*c*, a voltage follower is used to connect the 30 kΩ resistor to the output of the voltage divider. Once again, the circuit can be analyzed by writing a node equation at node a:

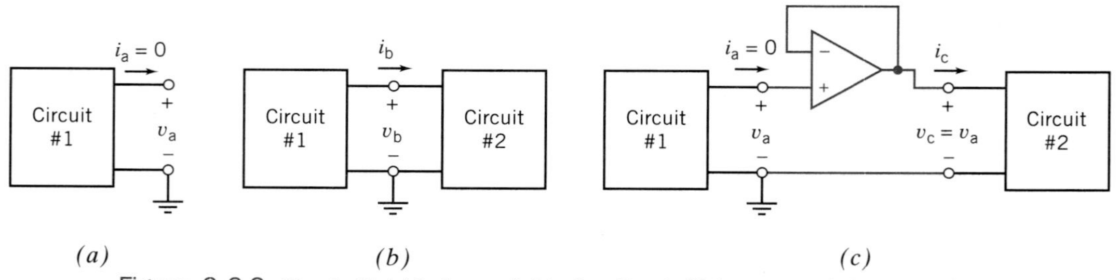

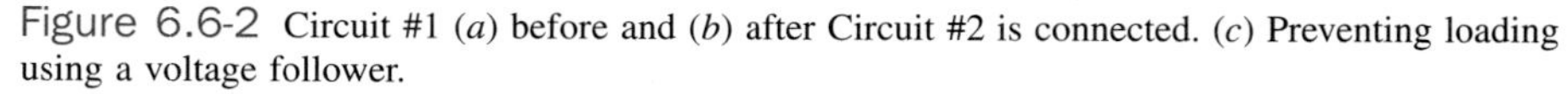

Figure 6.6-2 Circuit #1 (*a*) before and (*b*) after Circuit #2 is connected. (*c*) Preventing loading using a voltage follower.

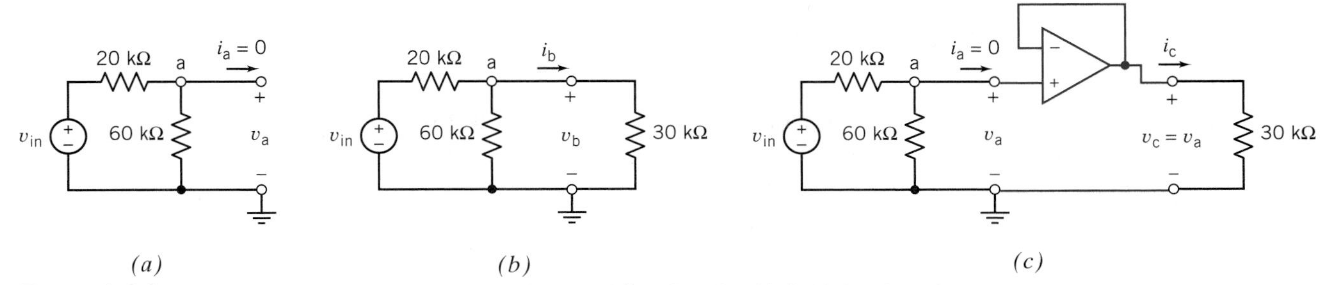

Figure 6.6-3 A voltage divider (*a*) before and (*b*) after a 30-kΩ resistor is added. (*c*) A voltage follower is added to prevent loading.

$$\frac{v_c - v_{in}}{20\ \text{k}\Omega} + \frac{v_c}{60\ \text{k}\Omega} = 0$$

Solving for v_c we have

$$v_c = \frac{3}{4} v_{in}$$

Since $v_c = v_a$, loading is avoided when the voltage follower is used to connect the resistor to the voltage divider. The voltage follower, not the voltage divider, provides the current required by the 30 kΩ resistor.

Example 6.6-2

A common application of operational amplifiers is to scale a voltage, that is, to multiply a voltage by a constant, K, so that

$$v_o = K\, v_{in}$$

This situation is illustrated in Figure 6.6-4*a*. The input voltage, v_{in}, is provided by an ideal voltage source. The output voltage, v_o, is the element voltage of a 100 kΩ resistor. This resistor, sometimes called a load resistor, represents the circuit that will use the voltage v_o as its input. Circuits that perform this operation are usually called amplifiers. The constant K is called the gain of the amplifier.

The required value of the constant K will determine which of the circuits is selected from Figure 6.6-1. There are four cases to consider: $K < 0$, $K > 1$, $K = 1$, and $0 < K < 1$.

Solution

Since resistor values are positive, the gain of the inverting amplifier, shown in Figure 6.6-1*a*, is negative. Accordingly, when $K < 0$ is required, an inverting amplifier is used. For example, suppose we require $K = -5$. From Figure 6.6-1*a*

$$-5 = -\frac{R_f}{R_1}$$

so

$$R_f = 5\, R_1$$

As a rule of thumb, it is a good idea to choose resistors in operational amplifier circuits that have values between 5 kΩ and 500 kΩ when possible. Choosing

$$R_1 = 10\ \text{k}\Omega$$

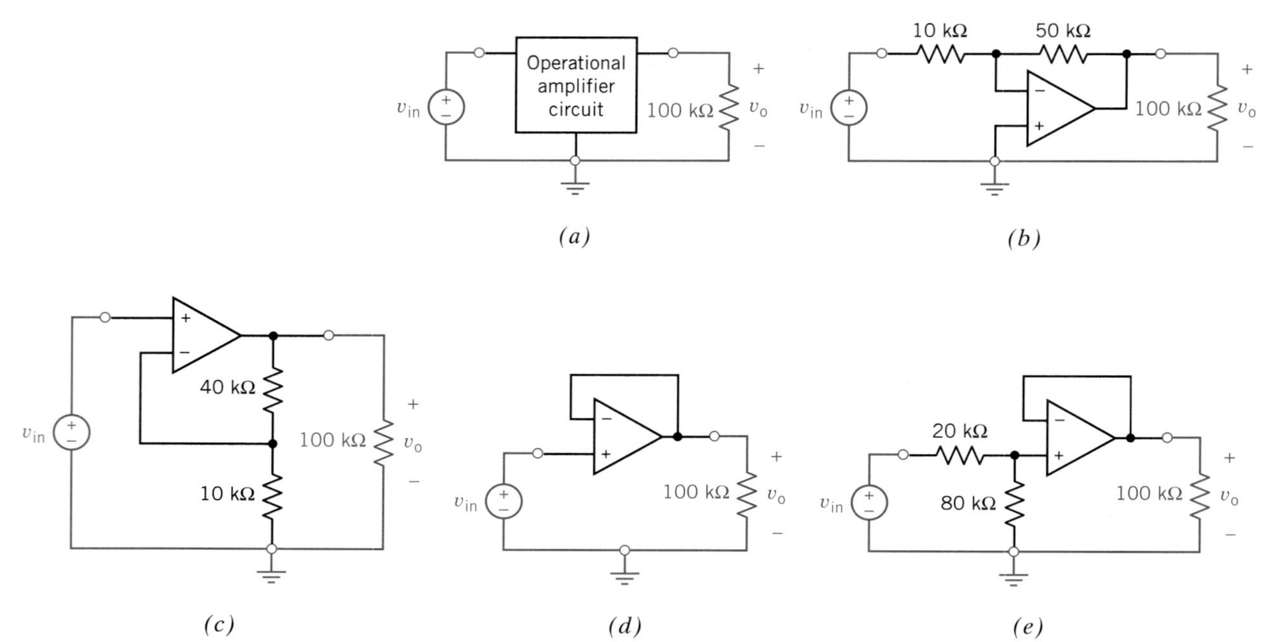

Figure 6.6-4 (*a*) An amplifier is required to make $v_o = K\,v_{in}$. The choice of amplifier circuit depends on the value of the gain *K*. Four cases are shown: (*b*) $K = -5$, (*c*) $K = 5$, (*d*) $K = 1$, and (*e*) $K = 0.8$.

gives

$$R_f = 50\ \text{k}\Omega$$

The resulting circuit is shown in Figure 6.6-4*b*.

Next suppose we require $K = 5$. The noninverting amplifier, shown in Figure 6.6-1*b*, is used to obtain gains greater than 1. From Figure 6.6-1*b*

$$5 = 1 + \frac{R_f}{R_1}$$

so

$$R_f = 4\,R_1$$

Choosing $R_1 = 10\ \text{k}\Omega$ gives $R_f = 40\ \text{k}\Omega$. The resulting circuit is shown in Figure 6.6-4*c*.

Consider using the noninverting amplifier of Figure 6.6-1*b* to obtain a gain $K = 1$. From Figure 6.6-1*b*

$$1 = 1 + \frac{R_f}{R_1}$$

so

$$\frac{R_f}{R_1} = 0$$

This can be accomplished by replacing R_f by a short circuit ($R_f = 0$) or by replacing R_1 by an open circuit ($R_1 = \infty$) or both. Doing both converts a noninverting amplifier into a voltage follower. The gain of the voltage follower is 1. In Figure 6.6-4*d* a voltage follower is used for the case $K = 1$.

There is no amplifier in Figure 6.6-1 that has a gain between 0 and 1. Such a circuit can be obtained using a voltage divider together with a voltage follower. Suppose we require $K = 0.8$. First, design a voltage divider to have an attenuation equal to K,

$$0.8 = \frac{R_2}{R_1 + R_2}$$

so

$$R_2 = 4 \cdot R_1$$

Choosing $R_1 = 20\ \text{k}\Omega$ gives $R_2 = 80\ \text{k}\Omega$. Adding a voltage follower gives the circuit shown in Figure 6.6-4*e*.

Example 6.6-3

Design a circuit having one output, v_o, and three inputs v_1, v_2, and v_3. The output must be related to the inputs by

$$v_o = 2v_1 + 3v_2 + 4v_3$$

In addition, the inputs are restricted to have values between -1 V and 1 V, that is,

$$|v_i| \le 1\ \text{V} \quad i = 1, 2, 3$$

Consider using an operational amplifier having $i_{sat} = 2$ mA and $v_{sat} = 15$ V, and design the circuit.

Solution

The required circuit must multiply each input by a separate positive number and add the results. The noninverting summer shown in Figure 6.6-1*e* can do these operations. This circuit is represented by six parameters: K_1, K_2, K_3, K_4, R_a, and R_b. Designing the non-inverting summer amounts to choosing values for these six parameters. Notice that $K_1 + K_2 + K_3 < 1$ is required to ensure that all of the resistors have positive values. Pick $K_4 = 10$ (a convenient value that is just a little larger than $2 + 3 + 4 = 9$). Then

$$v_o = 2v_1 + 3v_2 + 4v_3 = 10\,(0.2v_1 + 0.3v_2 + 0.4v_3)$$

That is $K_4 = 10$, $K_1 = 0.2$, $K_2 = 0.3$, and $K_3 = 0.4$. Figure 6.6-1*e* does not provide much guidance in picking values of R_a and R_b. Try $R_a = R_b = 100\ \Omega$. Then

$$(K_4 - 1)\,R_b = (10 - 1)100 = 900\ \Omega$$

Figure 6.6-5 shows the resulting circuit. It is necessary to check this circuit to ensure that it satisfies the specifications. Writing node equations

$$\frac{v_a - v_1}{500} + \frac{v_a - v_2}{333} + \frac{v_a - v_3}{250} + \frac{v_a}{1000} = 0$$

$$-\frac{v_o - v_a}{900} + \frac{v_a}{100} = 0$$

and solving these equations yield

$$v_o = 2v_1 + 3v_2 + 4v_3 \quad \text{and} \quad v_a = \frac{v_o}{10}$$

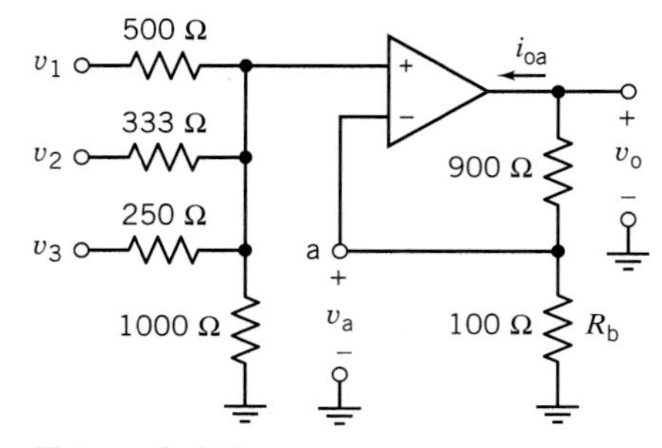

Figure 6.6-5 The proposed noninverting summing amplifier.

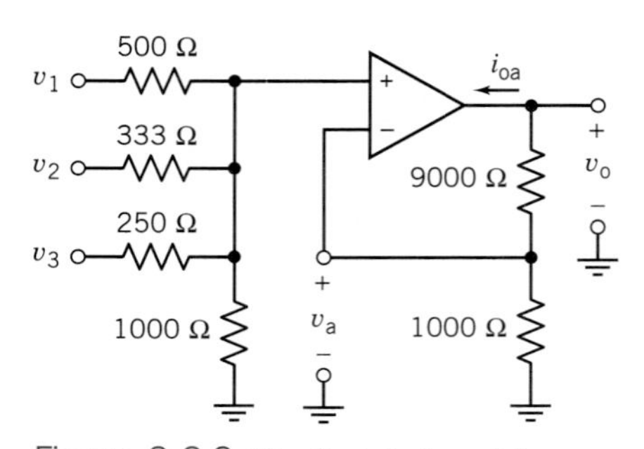

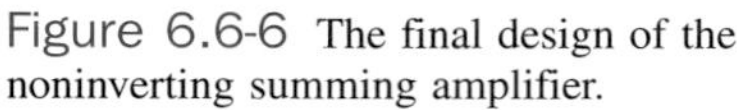

Figure 6.6-6 The final design of the noninverting summing amplifier.

The output current of the operational amplifier is given by

$$i_{oa} = \frac{v_a - v_o}{900} = -\frac{v_o}{1000} \tag{6.6-1}$$

How large can the output voltage be? We know that

$$|v_o| = |2v_1 + 3v_2 + 4v_3|$$

so

$$|v_o| \le 2|v_1| + 3|v_2| + 4|v_3| \le 9 \text{ V}$$

The operational amplifier output voltage will always be less than v_{sat}. That's good. Now what about the output current? Notice that $|v_o| \le 9$ V. From Eq. (6.6-1)

$$|i_{oa}| = \left|\frac{-v_o}{1000\ \Omega}\right| \le \left|\frac{-9\text{ V}}{1000\ \Omega}\right| = 9 \text{ mA}$$

The operational amplifier output current exceeds $i_{sat} = 2$mA. This is not allowed. Increasing R_b will reduce i_o. Try $R_b = 1000\ \Omega$. Then

$$(K_4 - 1)\,R_b = (10 - 1)\,1000 = 9000\ \Omega$$

This produces the circuit shown in Figure 6.6-6. Increasing R_a and R_b does not change the operational amplifier output voltage. As before

$$|v_o| \le 2|v_1| + 3|v_2| + 4|v_3| \le 9 \text{ V}$$

Increasing R_a and R_b does reduce the operational amplifier output current. Now

$$|i_{oa}| = \left|\frac{-v_o}{K_4 R_b}\right| \le \left|\frac{-9\text{ V}}{10000\ \Omega}\right| = 0.9 \text{ mA}$$

so $|i_{oa}| < 2$ mA and $|v_o| < 15$ V, as required.

Example 6.6-4

It is desired to design an inverting amplifier with an ideal operational amplifier and standard resistor values. The inverting amplifier, shown in Figure 6.6-1*a*, must have a gain of −23. The resistors R_1 and R_f must each be a single standard value resistor. The standard values are

$$R = r10^n$$

You Choose G

23

r_1 \ r_f	1.0	1.1	1.2	1.3	1.5	1.6	1.8	2.0	2.2	2.4	2.7	3.0	3.3	3.6	3.9	4.3	4.7	5.1	5.6	6.2	6.8	7.5	8.2	9.1
1	1.30	1.20	1.10	1.00	0.80	0.70	0.50	0.30	0.10	0.10	0.40	0.70	1.00	1.30	1.60	2.00	2.40	2.80	3.30	3.90	4.50	5.20	5.90	6.80
1.1	1.39	1.30	1.21	1.12	0.94	0.85	0.66	0.48	0.30	0.12	0.15	0.43	0.70	0.97	1.25	1.61	1.97	2.34	2.79	3.34	3.88	4.52	5.15	5.97
1.2	1.47	1.38	1.30	1.22	1.05	0.97	0.80	0.63	0.47	0.30	0.05	0.20	0.45	0.70	0.95	1.28	1.62	1.95	2.37	2.87	3.37	3.95	4.53	5.28
1.3	1.53	1.45	1.38	1.30	1.15	1.07	0.92	0.76	0.61	0.45	0.22	0.01	0.24	0.47	0.70	1.01	1.32	1.62	2.01	2.47	2.93	3.47	4.01	4.70
1.5	1.63	1.57	1.50	1.43	1.30	1.23	1.10	0.97	0.83	0.70	0.50	0.30	0.10	0.10	0.30	0.57	0.83	1.10	1.43	1.83	2.23	2.70	3.17	3.77
1.6	1.68	1.61	1.55	1.49	1.36	1.30	1.18	1.05	0.93	0.80	0.61	0.43	0.24	0.05	0.14	0.39	0.64	0.89	1.20	1.58	1.95	2.39	2.82	3.39
1.8	1.74	1.69	1.63	1.58	1.47	1.41	1.30	1.19	1.08	0.97	0.80	0.63	0.47	0.30	0.13	0.09	0.31	0.53	0.81	1.14	1.48	1.87	2.26	2.76
2	1.80	1.75	1.70	1.65	1.55	1.50	1.40	1.30	1.20	1.10	0.95	0.80	0.65	0.50	0.35	0.15	0.05	0.25	0.50	0.80	1.10	1.45	1.80	2.25
2.2	1.85	1.80	1.75	1.71	1.62	1.57	1.48	1.39	1.30	1.21	1.07	0.94	0.80	0.66	0.53	0.35	0.16	0.02	0.25	0.52	0.79	1.11	1.43	1.84
2.4	1.88	1.84	1.80	1.76	1.68	1.63	1.55	1.47	1.38	1.30	1.18	1.05	0.93	0.80	0.68	0.51	0.34	0.18	0.03	0.28	0.53	0.83	1.12	1.49
2.7	1.93	1.89	1.86	1.82	1.74	1.71	1.63	1.56	1.49	1.41	1.30	1.19	1.08	0.97	0.86	0.71	0.56	0.41	0.23	0.00	0.22	0.48	0.74	1.07
3	1.97	1.93	1.90	1.87	1.80	1.77	1.70	1.63	1.57	1.50	1.40	1.30	1.20	1.10	1.00	0.87	0.73	0.60	0.43	0.23	0.03	0.20	0.43	0.73
3.3	2.00	1.97	1.94	1.91	1.85	1.82	1.75	1.69	1.63	1.57	1.48	1.39	1.30	1.21	1.12	1.00	0.88	0.75	0.60	0.42	0.24	0.03	0.18	0.46
3.6	2.02	1.99	1.97	1.94	1.88	1.86	1.80	1.74	1.69	1.63	1.55	1.47	1.38	1.30	1.22	1.11	0.99	0.88	0.74	0.58	0.41	0.22	0.02	0.23
3.9	2.04	2.02	1.99	1.97	1.92	1.89	1.84	1.79	1.74	1.68	1.61	1.53	1.45	1.38	1.30	1.20	1.09	0.99	0.86	0.71	0.56	0.38	0.20	0.03
4.3	2.07	2.04	2.02	2.00	1.95	1.93	1.88	1.83	1.79	1.74	1.67	1.60	1.53	1.46	1.39	1.30	1.21	1.11	1.00	0.86	0.72	0.56	0.39	0.18
4.7	2.09	2.07	2.04	2.02	1.98	1.96	1.92	1.87	1.83	1.79	1.73	1.66	1.60	1.53	1.47	1.39	1.30	1.21	1.11	0.98	0.85	0.70	0.56	0.36
5.1	2.10	2.08	2.06	2.05	2.01	1.99	1.95	1.91	1.87	1.83	1.77	1.71	1.65	1.59	1.54	1.46	1.38	1.30	1.20	1.08	0.97	0.83	0.69	0.52
5.6	2.12	2.10	2.09	2.07	2.03	2.01	1.98	1.94	1.91	1.87	1.82	1.76	1.71	1.66	1.60	1.53	1.46	1.39	1.30	1.19	1.09	0.96	0.84	0.68
6.2	2.14	2.12	2.11	2.09	2.06	2.04	2.01	1.98	1.95	1.91	1.86	1.82	1.77	1.72	1.67	1.61	1.54	1.48	1.40	1.30	1.20	1.09	0.98	0.83
6.8	2.15	2.14	2.12	2.11	2.08	2.06	2.04	2.01	1.98	1.95	1.90	1.86	1.81	1.77	1.73	1.67	1.61	1.55	1.48	1.39	1.30	1.20	1.09	0.96
7.5	2.17	2.15	2.14	2.13	2.10	2.09	2.06	2.03	2.01	1.98	1.94	1.90	1.86	1.82	1.78	1.73	1.67	1.62	1.55	1.47	1.39	1.30	1.21	1.09
8.2	2.18	2.17	2.15	2.14	2.12	2.10	2.08	2.06	2.03	2.01	1.97	1.93	1.90	1.86	1.82	1.78	1.73	1.68	1.62	1.54	1.47	1.39	1.30	1.19
9.1	2.19	2.18	2.17	2.16	2.14	2.12	2.10	2.08	2.04	2.08	2.00	1.97	1.94	1.90	1.87	1.83	1.78	1.74	1.68	1.62	1.55	1.48	1.40	1.30

THIS DISPLAYS THE ERROR eij FOR THE COMBINATIONS OF r_1 AND r_f

Figure 6.6-7 The spreadsheet for all standard values of r_1 and r_f.

where

$r \in \{1.0, 1.1, 1.2, 1.3, 1.5, 1.6, 1.8, 2.0, 2.2, 2.4, 2.7, 3.0, 3.3, 3.6, 3.9, 4.3, 4.7, 5.1, 5.6, 6.2, 6.8, 7.5, 8.2, 9.1\}$

The operation of this inverting amplifier is characterized by

$$\frac{v_{out}}{v_s} = -\frac{R_f}{R_1} = -G$$

We need to select standard resistor values for R_1 and R_f so that $G = R_f/R_1$ is equal to 23. First, express the gain G as

$$G = g10^m$$

where $0 \leq g < 1$. Next, let

$$R_1 = r_1 10^n \quad \text{and} \quad R_f = r_f\, 10^{n+m}$$

Define the error as

$$e_{ij} = \left| g - \frac{r_f}{r_1} \right|$$

Use a spreadsheet computer program to set up a table that displays the error for each of the $24 \times 24 = 576$ possible combinations of r_1 and r_f.

The desired value of the gain is 23, and therefore $g = 2.3$ and $m = 1$. We then seek the minimum error where

$$e_{ij} = \left| 2.3 - \frac{r_f}{r_1} \right|$$

and use a spreadsheet to generate all combinations of r_1 and r_f as shown in Figure 6.6-7.

The spreadsheet identifies $r_f = 6.2$ and $r_1 = 2.7$ as providing minimum error, and then the actual gain, G_a, is

$$G_a = \frac{r_f \times 10^m}{r_1} = \frac{6.2 \times 10^1}{2.7} = 22.96$$

Exercise 6.6-1 Determine the ratio, v_o/v_{in}, of the circuit of Figure 6.6-1*b* when $R_f = 100\ \text{k}\Omega$ and $R_1 = 25\ \text{k}\Omega$.

Answer: $v_o/v_{in} = 5$

6.7 CHARACTERISTICS OF PRACTICAL OPERATIONAL AMPLIFIERS

The ideal operational amplifier is the simplest model of an operational amplifier. This simplicity is obtained by ignoring some imperfections of practical operational amplifiers. This section considers some of these imperfections and provides alternate operational amplifier models to account for these imperfections.

Consider the operational amplifier shown in Figure 6.7-1*a*. If this operational amplifier is ideal, then

$$i_1 = 0, i_2 = 0 \quad \text{and} \quad v_1 - v_2 = 0 \tag{6.7-1}$$

In contrast, the operational amplifier model shown in Figure 6.7-1*d* accounts for several nonideal parameters of practical operational amplifiers, namely:

- nonzero bias currents
- nonzero input offset voltage
- finite input resistance
- nonzero output resistance
- finite voltage gain

This model more accurately describes practical operational amplifiers than does the ideal operational amplifier. Unfortunately, the more accurate model of Figure 6.7-1*d* is much more complicated and much more difficult to use than the ideal operational amplifier. The models in Figures 6.7-1*b* and 6.7-1*c* provide a compromise. These models are more accurate than the ideal operational amplifier but easier to use than the model in Figure 6.7-1*d*. It will be convenient to have names for these models. The model in Figure 6.7-1*b* will be called "the offsets model" of the operational amplifier. Similarly, the model in Figure 6.7-1*c* will be called "the finite gain model" of the operational amplifier and the model in Figure 6.7-1*d* will be called "the offsets and finite gain model" of the operational amplifier.

The operational amplifier model shown in Figure 6.7-1*b* accounts for the nonzero bias current and nonzero input offset voltage of practical operational amplifiers but not the finite input resistance, the nonzero output resistance, or the finite voltage gain. This model

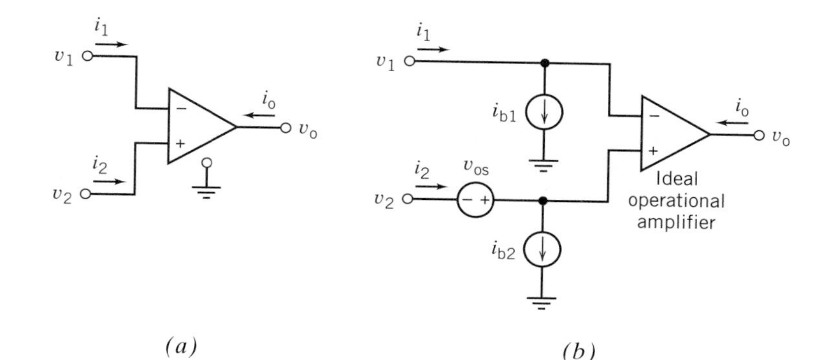

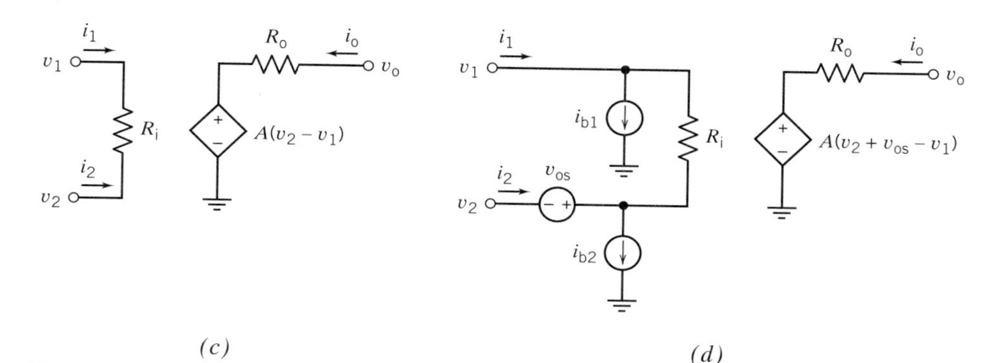

Figure 6.7-1 (*a*) An operational amplifier and (*b*) the offsets model of an operational amplifier. (*c*) The finite gain model of an operational amplifier. (*d*) The offsets and finite gain model of an operational amplifier.

consists of three independent sources and an ideal operational amplifier. In contrast to the ideal operational amplifier, the operational amplifier model that accounts for offsets is represented by the equations

$$i_1 = i_{b1},\ i_2 = i_{b2} \quad \text{and} \quad v_1 - v_2 = v_{os} \tag{6.7-2}$$

The voltage v_{os} is a small, constant voltage called the input offset voltage. The currents i_{b1} and i_{b2} are called the bias currents of the operational amplifier. They are small, constant currents. The difference between the bias currents is called the input offset current, i_{os}, of the amplifier:

$$i_{os} = i_{b1} - i_{b2}$$

Notice that when the bias currents and input offset voltage are all zero, Eq. 6.7-2 is the same as Eq. 6.7-1. In other words, the offsets model reverts to the ideal operational amplifier when the bias currents and input offset voltage are zero.

Frequently, the bias currents and input offset voltage can be ignored because they are very small. However, when the input signal to a circuit is itself small, the bias currents and input voltage can become important.

Manufacturers specify a maximum value for the bias currents, the input offset current, and the input offset voltage. For the μA741 the maximum bias current is specified to be 500 nA, the maximum input offset current is specified to be 200 nA, and the maximum input offset voltage is specified to be 5 mV. These specifications guarantee that

$$|i_{b1}| \leq 500 \text{ nA} \quad \text{and} \quad |i_{b2}| \leq 500 \text{ nA}$$
$$|i_{b1} - i_{b2}| \leq 200 \text{ nA}$$
$$|v_{os}| \leq 5 \text{ mV}$$

Table 6.7-1 shows the bias currents, offset current, and input offset voltage *typical* of several types of operational amplifier.

Table 6.7-1 **Selected Parameters of Typical Operational Amplifiers**

Parameter	Units	μA741	LF351	TL051C	OPA101 AM	OP-07E
Saturation voltage, v_{sat}	V	13	13.5	13.2	13	13
Saturation current, i_{sat}	mA	2	15	6	30	6
Slew rate, *SR*	V/μS	0.5	13	23.7	6.5	0.17
Bias current, i_b	nA	80	0.05	0.03	0.012	1.2
Offset current, i_{os}	nA	20	0.025	0.025	0.003	0.5
Input offset voltage, v_{os}	mV	1	5	0.59	0.1	0.03
Input resistance, R_i	MΩ	2	10^6	10^6	10^6	50
Output resistance, R_o	Ω	75	1000	250	500	60
Differential gain, *A*	V/mV	200	100	105	178	5000
Common mode rejection ratio, *CMRR*	V/mv	31.6	100	44	178	1413
Gain bandwidth product, *B*	MHz	1	4	3.1	20	0.6

Example 6.7-1

The inverting amplifier shown in Figure 6.7-2*a* contains a μA741 operational amplifier. This inverting amplifier designed in Example 6.6-2 has a gain of -5, that is,

$$v_o = -5 \cdot v_{in}$$

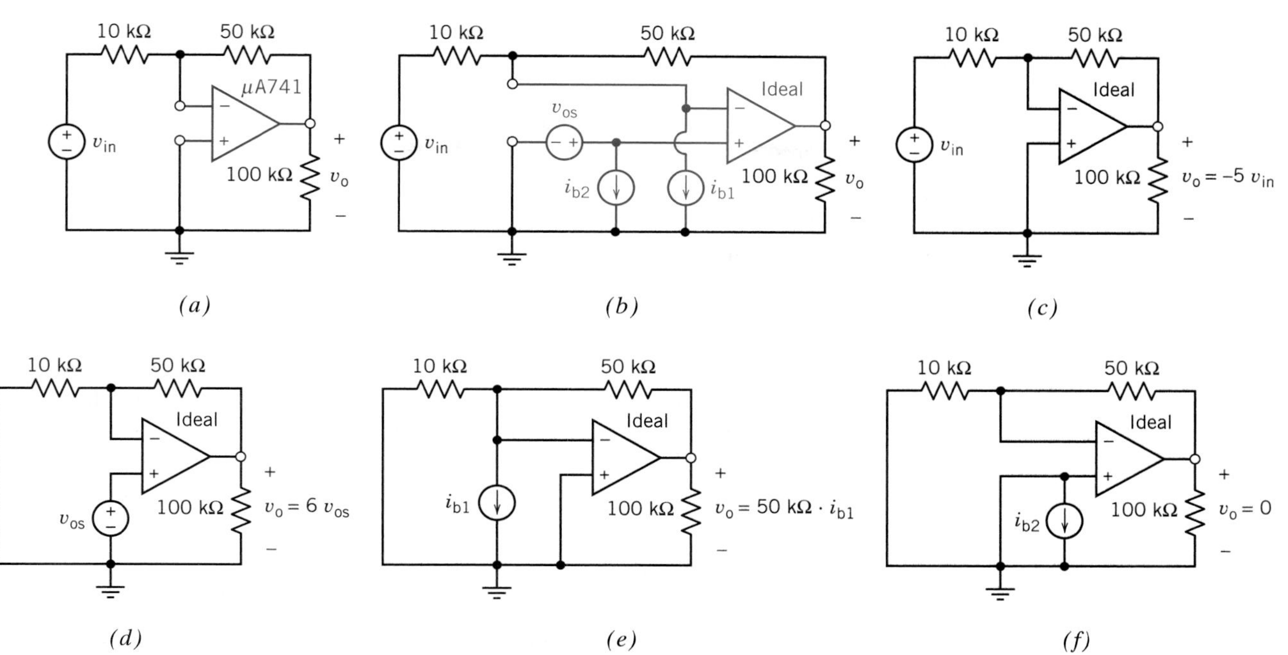

Figure 6.7-2 (*a*) An inverting amplifier and (*b*) an equivalent circuit that accounts for the input offset voltage and bias currents of the operational amplifier. (*c*)–(*f*) Analysis using superposition.

The design of the inverting amplifier is based on the ideal model of an operational amplifier and so did not account for the bias currents and input offset voltage of the μA741 operational amplifier. In this example, the offsets model of an operational amplifier will be used to analyze the circuit. This analysis will tell us what effect the bias currents and input offset voltage have on the performance of this circuit.

Solution

In Figure 6.7-2*b* the operational amplifier has been replaced by the offsets model of an operational amplifier. Notice that the operational amplifier in Figure 6.7-2*b* is the ideal operational amplifier that is part of the model of the operational amplifier used to account for the offsets. The circuit in Figure 6.7-2*b* contains four inputs that correspond to the four independent sources v_{in}, i_{b1}, i_{b2}, and v_{os}. (The input v_{in} is obtained by connecting a voltage source to the circuit. In contrast, the "inputs" i_{b1}, i_{b2}, and v_{os} are the results of imperfections of the operational amplifier. These inputs are part of the operational amplifier model and do not need to be added to the circuit.) Superposition can be used to good advantage in analyzing this circuit. Figures 6.7-2*c* through 6.7-2*f* illustrate this process. In each of these figures, all but one input has been set to zero, and the output due to that one input has been calculated.

Figure 6.7-2*c* shows the circuit used to calculate the response to v_{in} alone. The other inputs, i_{b1}, i_{b2}, and v_{os}, have all been set to zero. Recall that zero current sources act like open circuits and zero voltage sources act like short circuits. Figure 6.7-2*c* is obtained from Figure 6.7-2*b* by replacing the current sources i_{b1}, i_{b2} by open circuits and by replacing the voltage source v_{os} by a short circuit. The operational amplifier in Figure 6.7-2*c* is the ideal operational amplifier that is part of the offset model. Analysis of the inverting amplifier in Figure 6.7-2*c* gives

$$v_o = -5 \cdot v_{in}$$

Next consider Figure 6.7-2*d*. This circuit is used to calculate the response to v_{os} alone. The other inputs, v_{in}, i_{b1}, and i_{b2}, have all been set to zero. Figure 6.7-2*d* is obtained from Figure 6.7-2*b* by replacing the current sources i_{b1}, i_{b2} by open circuits and by replacing the voltage source v_{in} by a short circuit. Again, the operational amplifier is the ideal operational amplifier from the offsets model. The circuit in Figure 6.7-2*d* is one we have seen before; it is the noninverting amplifier (Figure 6.6-1*b*). Analysis of this noninverting amplifier gives

$$v_o = \left(1 + \frac{50\text{ k}\Omega}{10\text{ k}\Omega}\right) \cdot v_{os} = 6\, v_{os}$$

Next consider Figure 6.7-2*e*. This circuit is used to calculate the response to i_{b1} alone. The other inputs, v_{in}, v_{os}, and i_{b2}, have all been set to zero. Figure 6.7-2*e* is obtained from Figure 6.7-2*b* by replacing the current source i_{b2} by an open circuit and by replacing the voltage sources v_{in} and v_{os} by short circuits. Notice that the voltage across the 10 kΩ resistor is zero because this resistor is connected between the input nodes of the ideal operational amplifier. Ohm's law says that the current in the 10 kΩ resistor must be zero. The current in the 50 kΩ resistor is i_{b1}. Finally, paying attention to the reference directions,

$$v_o = 50\text{ k}\Omega \cdot i_{b1}$$

Figure 6.7-2*f* is used to calculate the response to i_{b2} alone. The other inputs, v_{in}, v_{os}, and i_{b1}, have all been set to zero. Figure 6.7-2*f* is obtained from Figure 6.7-2*b* by replacing the current source i_{b1} by an open circuit and by replacing the voltage sources v_{in} and v_{os} by short circuits. Replacing v_{os} by a short circuit inserts a short circuit across the current source i_{b2}. Again, the voltage across the 10 kΩ resistor is zero, so the current in the 10 kΩ resistor must be zero. Kirchhoff's Current Law shows that the current in the 50 kΩ resistor is also zero. Finally,

$$v_o = 0$$

The output caused by all four inputs working together is the sum of the outputs caused by each input working alone. Therefore

$$v_o = -5 \cdot v_{in} + 6 \cdot v_{os} + (50\text{ k}\Omega)\, i_{b1}$$

When the input of the inverting amplifier, v_{in}, is zero, the output v_o also should be zero. However, v_o is nonzero when we have a finite v_{os} or i_{b1}. Let

$$\text{output offset voltage} = 6 \cdot v_{os} + (50\text{ k}\Omega)\, i_{b1}$$

Then

$$v_o = -5 \cdot v_{in} + \text{output offset voltage}$$

Recall that when the operational amplifier is modeled as an ideal operational amplifier, analysis of this inverting amplifier gives

$$v_o = -5 \cdot v_{in}$$

Comparing these last two equations shows that bias currents and input offset voltage cause the output offset voltage. Modeling the operational amplifier as an ideal operational amplifier amounts to assuming that the output offset voltage is not important and thus ignoring it. Using the operational amplifier model that accounts for offsets is more accurate but also more complicated.

How large is the output offset voltage of this inverting amplifier? The input offset voltage of a μA741 operational amplifier will be at most 5 mV, and the bias current will be at most 500 nA, so

$$\text{output offset voltage} \leq 6 \cdot 5\text{ mV} + (50\text{ k}\Omega)\, 500\text{ nA} = 55\text{ mV}$$

We note that we can only ignore the effect of the offset voltage when $|5v_{in}| > 500$ mV or $|v_{in}| > 100$ mV. The output offset error can be reduced by using a better operational amplifier, that is, one that guarantees smaller bias currents and input offset voltage.

Now, let us turn our attention to different parameters of practical operational amplifiers. The operational amplifier model shown in Figure 6.7-1*c* accounts for the finite input resistance, the nonzero output resistance, and the finite voltage gain of practical operational amplifiers but not the nonzero bias current and nonzero input offset voltage. This model consists of two resistors and a VCVS.

The finite gain model reverts to an ideal operational amplifier when the gain, A, becomes infinite. To see that this is so notice that in Figure 6.7-1*c*

$$v_o = A(v_2 - v_1) + R_o i_o$$

so

$$v_2 - v_1 = \frac{v_o - R_o i_o}{A}$$

The models in Figure 6.7-1, as well as the model of the ideal operational amplifier, are valid only when v_o and i_o satisfy Eq. 6.4-1. Therefore,

$$|v_o| \le v_{sat} \quad \text{and} \quad |i_o| \le i_{sat}$$

Then

$$|v_2 - v_1| \le \frac{v_{sat} + R_o i_{sat}}{A}$$

Therefore,

$$\lim_{A \to \infty} (v_2 - v_1) = 0$$

Next, since

$$i_1 = -\frac{v_2 - v_1}{R_i} \quad \text{and} \quad i_2 = \frac{v_2 - v_1}{R_i}$$

we conclude that

$$\lim_{A \to \infty} i_1 = 0 \quad \text{and} \quad \lim_{A \to \infty} i_2 = 0$$

Thus, i_1, i_2, and $v_2 - v_1$ satisfy Eq. 6.7-1. In other words, the finite gain model of the operational amplifier reverts to the ideal operational amplifier as the gain becomes infinite. The gain for practical op amps ranges from 100,000 to 10^7.

Example 6.7-2

In Figure 6.7-3 a voltage follower is used as a buffer amplifier. Analysis based on the ideal operational amplifier shows that the gain of the buffer amplifier is

$$\frac{v_o}{v_s} = 1$$

What effects will the input resistance, output resistance, and finite voltage gain of a practical operational amplifier have on the performance of this circuit? To answer this ques-

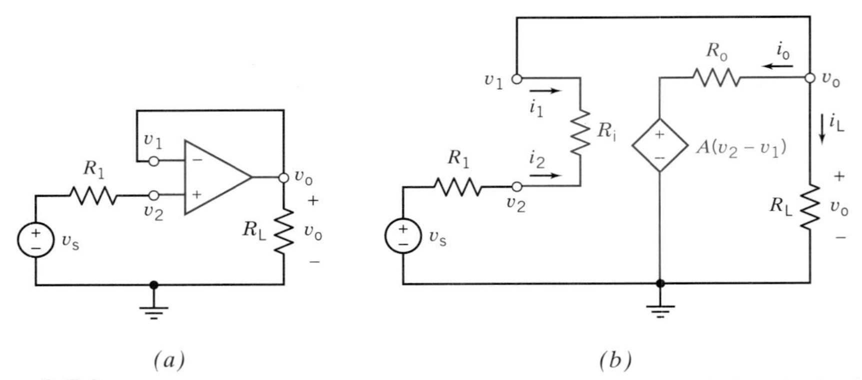

Figure 6.7-3 (*a*) A voltage follower used as a buffer amplifier and (*b*) an equivalent circuit with the operational amplifier model that accounts for finite voltage gain.

tion, replace the operational amplifier by the operational amplifier model that accounts for finite voltage gain. This gives the circuit shown in Figure 6.7-3*b*.

Solution

To be specific, suppose $R_1 = 1\ \text{k}\Omega$; $R_L = 10\ \text{k}\Omega$; and the parameters of the practical operational amplifier are $R_i = 100\ \text{k}\Omega$, $R_o = 100\ \Omega$, and $A = 10^5$ V/V.

Suppose that $v_o = 10$ V. We can find the current, i_L, in the output resistor as

$$i_L = \frac{v_o}{R_L} = \frac{10\ \text{V}}{10^4\ \Omega} = 10^{-3}\ \text{A}$$

Apply KCL at the top node of R_L to get

$$i_1 + i_o + i_L = 0$$

It will turn out that i_1 will be much smaller than both i_o and i_L. It is useful to make the approximation that $i_1 = 0$. (We will check this assumption later in this example.) Then

$$i_o = -i_L$$

Next, apply KVL to the mesh consisting of the VCVS, R_o, and R_L to get

$$-A(v_2 - v_1) - i_o R_o + i_L R_L = 0$$

Combining the last two equations and solving for $(v_2 - v_1)$ gives

$$v_2 - v_1 = \frac{i_L(R_o + R_L)}{\text{A}} = \frac{10^{-3}(100 + 10{,}000)}{10^5} = 1.01 \times 10^{-4}\ \text{V}$$

Now i_1 can be calculated using Ohm's law,

$$i_1 = \frac{v_1 - v_2}{R_i} = \frac{-1.01 \times 10^{-4}\ \text{V}}{100\ \text{k}\Omega} = -1.01 \times 10^{-9}\ \text{A}$$

This justifies our earlier assumption that i_1 is negligible compared with i_o and i_L.

Applying KVL to the outside loop gives

$$-v_s - i_1 R_1 - i_1 R_i + v_o = 0$$

Now let us do some algebra to determine v_s.

$$v_s = v_o - i_1(R_1 + R_i) = V_o + i_2\,(R_1 + R_i)$$
$$= v_o + \frac{v_2 - v_1}{R_i} \times (R_1 + R_i)$$
$$= v_o + \frac{i_L(R_o + R_L)}{A} \times \frac{R_1 + R_i}{R_i}$$
$$= v_o + \frac{v_o}{R_L} \times \frac{R_o + R_L}{A} \times \frac{R_1 + R_i}{R_i}$$

The gain of this circuit is

$$\frac{v_o}{v_s} = \frac{1}{1 + \frac{1}{A} \times \frac{R_o + R_L}{R_L} \times \frac{R_i + R_1}{R_i}}$$

This equation shows that the gain will be approximately 1 when A is very large, $R_o \ll R_L$, and $R_1 \ll R_i$. In this example, for the specified A, R_o, and R_i we have

$$\frac{v_o}{v_s} = \frac{1}{1 + \frac{1}{10^5} \times \frac{100 + 10{,}000}{10{,}000} \times \frac{10^5 + 1000}{10^5}} = \frac{1}{1.00001} = 0.99999$$

Thus, the input resistance, output resistance, and voltage gain of the practical operational amplifier have only a small, essentially negligible, combined effect on the performance of the buffer amplifier.

Table 6.7-1 lists two other parameters of practical operational amplifiers that have not yet been mentioned. They are the *Common Mode Rejection Ratio (CMRR)* and the gain bandwidth product. Consider first the common mode rejection ratio. In the finite gain model, the voltage of the dependent source is

$$A(v_2 - v_1)$$

In practice, we find that dependent source voltage is more accurately expressed as

$$A(v_2 - v_1) + A_{cm}\left(\frac{v_1 + v_2}{2}\right)$$

where $v_2 - v_1$ is called the differential input voltage

$\frac{v_1 + v_2}{2}$ is called the common mode input voltage

and A_{cm} is called the common mode gain

The gain A is sometimes called the differential gain to distinguish it from A_{cm}. The common mode rejection ratio is defined to be the ratio of A to A_{cm}

$$\text{CMRR} = \frac{A}{A_{cm}}$$

The dependent source voltage can be expressed using A and CMRR as

$$A(v_2 - v_1) + A_{cm}\frac{v_1 + v_2}{2} = A(v_2 - v_1) + \frac{A}{\text{CMRR}}\,\frac{v_1 + v_2}{2}$$

$$= A\left[\left(1 + \frac{1}{2\,\text{CMRR}}\right)v_2 - \left(1 - \frac{1}{2\,\text{CMRR}}\right)v_1\right]$$

CMRR can be added to the finite gain model by changing the voltage of the dependent source. The appropriate change is

$$\text{Replace } A(v_2 - v_1) \text{ by } A\left[\left(1 + \frac{1}{2\,\text{CMRR}}\right)v_2 - \left(1 - \frac{1}{2\,\text{CMRR}}\right)v_1\right]$$

This change will make the model more accurate but also more complicated. Table 6.7-1 shows that CMRR is typically very large. For example, a typical LF351 operational amplifier has $A = 100$ V/mV and CMRR $= 100$ V/mV. This means that

$$A\left[\left(1 + \frac{1}{2\,\text{CMRR}}\right)v_2 - \left(1 - \frac{1}{2\,\text{CMRR}}\right)v_1\right] = 100{,}000.5\,v_2 - 99{,}999.5\,v_1$$

compared to

$$A(v_2 - v_1) = 100{,}000\,v_2 - 100{,}000\,v_1$$

In most cases, negligible error is caused by ignoring the CMRR of the operational amplifier. The CMRR does not need to be considered unless accurate measurements of very small differential voltages must be made in the presence of very large common mode voltages.

Next we consider the gain bandwidth product of the operational amplifier. The finite gain model indicates that the gain, A, of the operational amplifier is a constant. Suppose

$$v_1 = 0 \quad \text{and} \quad v_2 = M \sin \omega t$$

so that

$$v_2 - v_1 = M \sin \omega t$$

The voltage of the dependent source in the finite gain model will be

$$A(v_2 - v_1) = A \cdot M \sin \omega t$$

The amplitude, $A \cdot M$, of this sinusoidal voltage does not depend on the frequency, ω. Practical operational amplifiers do not work this way. The gain of a practical amplifier is a function of frequency, say $A(\omega)$. For many practical amplifiers, $A(\omega)$ can be adequately represented as

$$A(\omega) = \frac{B}{j\omega}$$

It is not necessary to know now how this function behaves. Functions of this sort will be discussed in Chapter 13. For now it is enough to realize that the parameter B is used to describe the dependence of the operational amplifier gain on frequency. The parameter B is called the gain bandwidth product of the operational amplifier.

Exercise 6.7-1 The input offset voltage of a *typical* μA741 amplifier is 1 mV and the bias current is 80 nA. Suppose the operational amplifier in Figure 6.7-2*a* is a typical μA741. Show that the output offset voltage of the inverting amplifier will be at most 10 mV.

Exercise 6.7-2 Suppose the 10 kΩ resistor in Figure 6.7-2*a* is changed to 2 kΩ and the 50 kΩ resistor is changed to 10 kΩ. (These changes will not change the gain of the inverting amplifier. It will still be -5.) Show that the *maximum* output offset voltage is reduced to 35 mV. (Use $i_b = 500$ nA and $v_{os} = 5$ mV to calculate the *maximum* output offset voltage that could be caused by the μA741 amplifier.)

Exercise 6.7-3 Suppose the μA741 operational amplifier in Figure 6.7-2*a* is replaced with a *typical* OPA101AM operational amplifier. Show that the output offset voltage of the inverting amplifier will be at most 0.6 mV.

Exercise 6.7-4

(a) Determine the voltage ratio v_o/v_s for the op amp circuit shown in Figure E 6.7-4.

(b) Calculate v_o/v_s for a practical op amp with $A = 10^5$, $R_o = 100\ \Omega$, and $R_i = 500$ kΩ. The circuit resistors are $R_s = 10$ kΩ, $R_f = 50$ kΩ, and $R_a = 25$ kΩ.

Answer: (b) $v_o/v_s = -2$

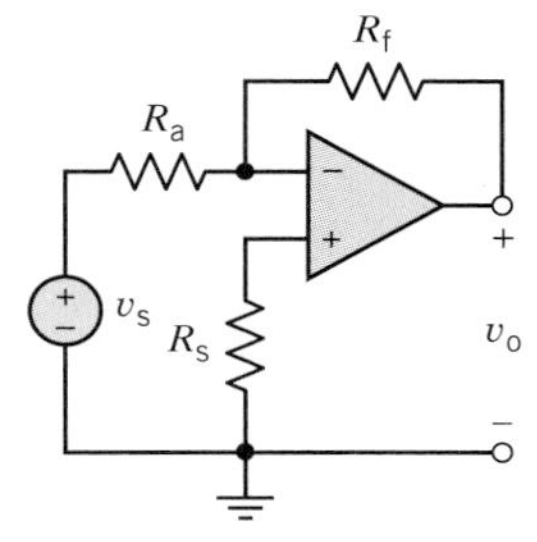

Figure E 6.7-4

6.8 THE OPERATIONAL AMPLIFIER AND PSPICE

Frequently, engineers design circuits based on a simple model of an operational amplifier, such as the ideal operational amplifier. These circuits are then checked by analyzing them using a more accurate, more complicated operational amplifier model. Fortunately, PSpice takes the drudgery out of this analysis.

PSpice does not recognize operational amplifiers as devices. Instead, the PSpice "subcircuit" feature includes operational amplifiers in a PSpice input file. The syntax for a subcircuit is

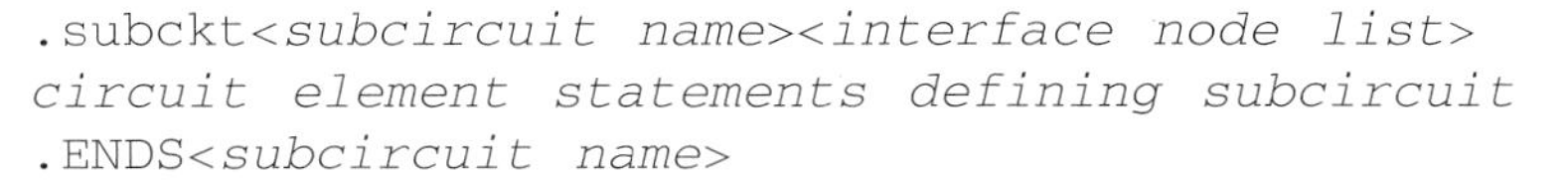

```
.subckt<subcircuit name><interface node list>
circuit element statements defining subcircuit
.ENDS<subcircuit name>
```

Table 6.8-1 provides subcircuits that implement the models of operational amplifiers that have been discussed in this chapter. The statement

```
Xname<interface node list><subcircuit name>
```

inserts a subcircuit into the circuit. (Xname is a PSpice device name that begins with the letter X. We will use XOAk to name the *k*th operational amplifier, for example, XOA1, XOA2, XOA3, . . .)

Example 6.8-1

The circuit shown in Figure 6.8-1*a* consists of two summing amplifiers. Assuming ideal operational amplifiers, we can use analysis of this circuit to determine v_o for the ideal and a practical op amp.

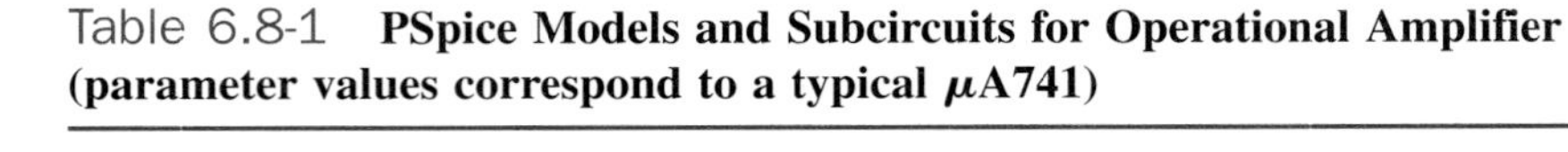

Table 6.8-1 **PSpice Models and Subcircuits for Operational Amplifier (parameter values correspond to a typical μA741)**

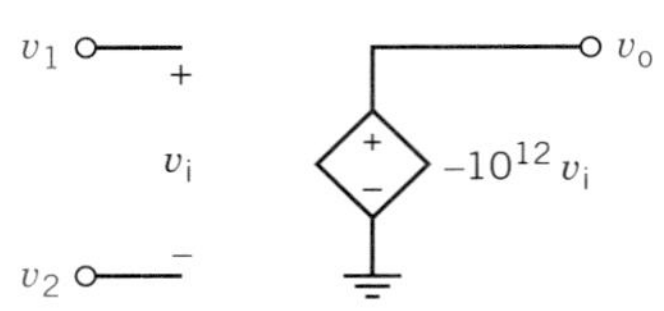

Ideal operational amplifier

```
.subckt ideal_op_amp 1 2 3
* op amp nodes listed in order: − + o
E  3 0  1 2  −1G
.ends ideal_op_amp
```

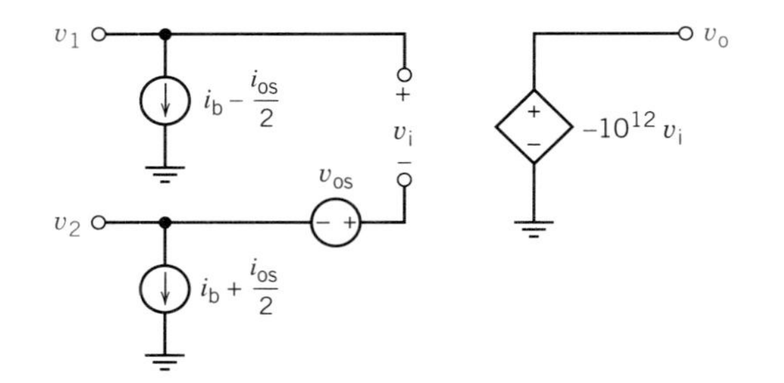

The offsets model of an operational amplifier

```
.subckt offsets_op_amp 1 2 4
* op amp nodes listed in order: − + o
Ib1   1  0  70nA
Ib2   2  0  90nA
Vos   3  2  ImV
Ri    1  3  1G
E     4  0  1  3  −1G
.ends offsets_op_amp
```

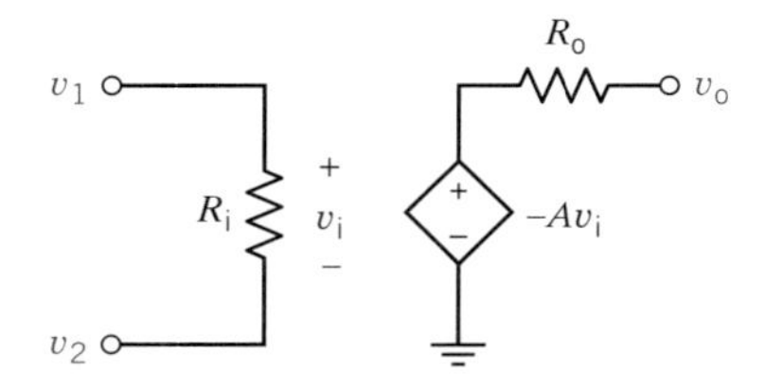

The finite gain model of an operational amplifier

```
.subckt finite_gain_op_amp 1 2 4
* op amp nodes listed in order: − + o
Ri    1  2  2MEG
E     3  0  1  2  −200000
Ro    4  3  75
.ends finite_gain_op_amp
```

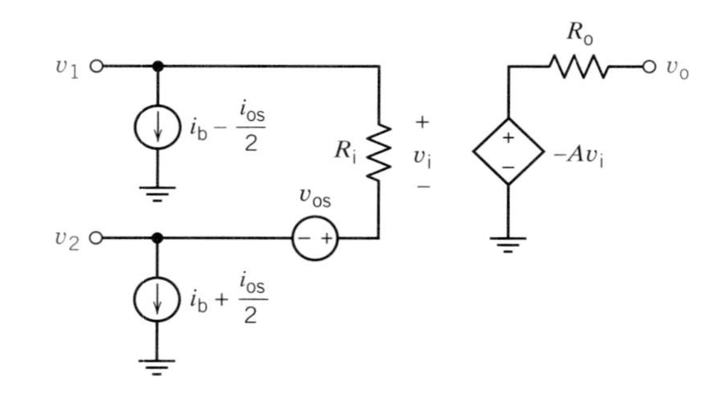

The offsets and finite gain model of an operational amplifier

```
.subckt op_amp 1  2  5
* op amp nodes listed in order: − + o
Ib1   1  0  70nA
Ib2   2  0  90nA
Vos   3  2  1mV
Ri    1  3  2MEG
E     4  0  1  3  −200000
Ro    4  5  75
.ends op_amp
```

Solution

We have v_4 as:

$$v_4 = -\left(\frac{100\ \text{k}\Omega}{50\ \text{k}\Omega} v_1 + \frac{100\ \text{k}\Omega}{25\ \text{k}\Omega} v_2\right) = -(2v_1 + 4v_2)$$

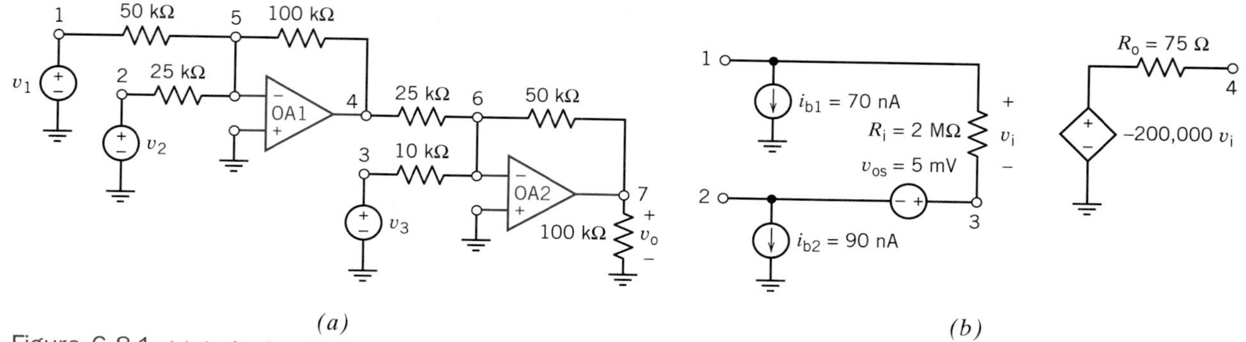

Figure 6.8-1 (*a*) A circuit using operational amplifiers and (*b*) a PSpice model of a typical μA741 operational amplifier.

where v_4 is the node voltage at node 4. Similarly

$$v_o = -\left(\frac{50\ \text{k}\Omega}{10\ \text{k}\Omega}v_3 + \frac{50\ \text{k}\Omega}{25\ \text{k}\Omega}v_4\right) = -(5v_3 + 2v_4)$$

Combining these equations yields

$$\begin{aligned} v_o &= -(5v_3 - 2(2v_1 + 4v_2)) \\ &= 4v_1 + 8v_2 - 5v_3 \end{aligned}$$

How will this circuit perform when the operational amplifiers are typical μA741s? This question can be answered by replacing each operational amplifier by the model that accounts for both offsets and finite gain and then analyzing the resulting circuit. Fortunately, PSpice can be used to do this analysis. Figure 6.8-2 shows a PSpice input file that corresponds to the circuit in Figure 6.8-1*a*. The lines

```
XOA1     5     0     4     uA741_op_amp
```

and

```
XOA2     6     0     7     uA741_op_amp
```

indicate that XOA1 and XOA2 are the device names of the operational amplifiers and that both operational amplifiers are to be modeled using the subcircuit named "uA741_op_amp." The operational amplifier nodes are listed in the same order as given in the .subckt statement. That means that node 5 is the inverting input node, node 0 is the noninverting input node, and node 4 is the output node of OA1. Similarly, node 6 is the inverting input node, node 0 is the noninverting input node, and node 7 is the output node of OA2.

PSpice will replace each operational amplifier with a copy of the subcircuit "uA741_op_amp." In doing so, the devices and nodes of the subcircuit will be renamed by appending the device names of the operational amplifiers. For example, the device

```
Ib1 1 0 70nA
```

in subcircuit uA741_op_amp becomes

```
XOA1.Ib1 XOA1.1 0 70nA
```

when the subcircuit uA741_op_amp is substituted for XOA1 and

```
XOA2.Ib1 XOA2.1 0 70nA
```

when the subcircuit uA741_op_amp is substituted for XOA2. Let us notice some things about node numbers. First, node 0 always represents the ground node. Node 0 is the

```
Two Summing Amplifiers

V1      1    0    200mV
V2      2    0    125mV
V3      3    0    250mV
R1      1    5    50k
R2      2    5    25k
R3      5    4    100k
XOA1    5    0    4 uA741_op_amp
R4      4    6    25k
R5      3    6    10k
R6      6    7    50k
R7      7    0    100k
XOA2    6    0    7 uA741_op_amp

.subckt uA741_op_amp 1  2  5
*                    -  +  o
Ib1     1    0    70nA
Ib2     2    0    90nA
Vos     3    2    1mV
Ri      1    3    2Meg
E       4    0    1  3  -200000
Ro      4    5    75
.ends uA741_op_amp

.end
```

Figure 6.8-2 PSpice input file for the summing amplifier.

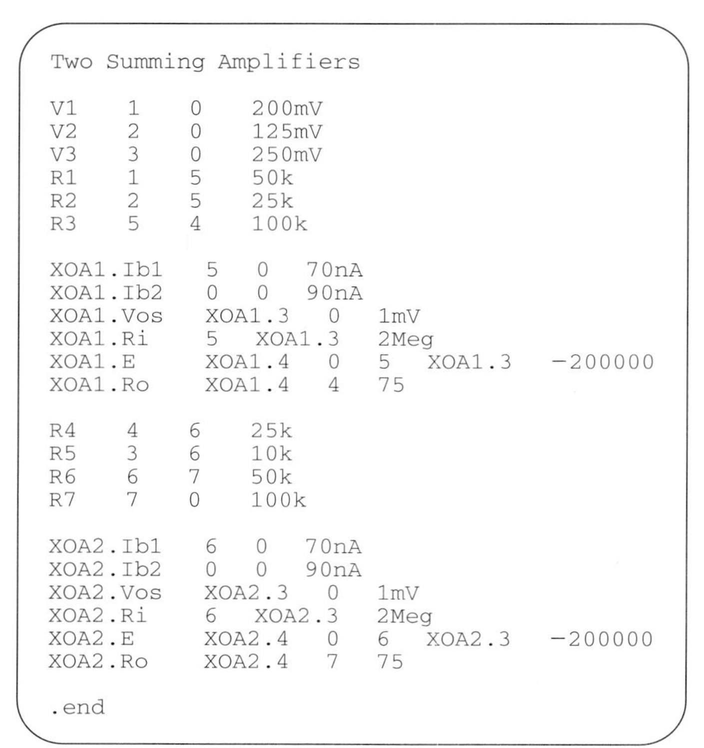

```
Two Summing Amplifiers

V1    1    0    200mV
V2    2    0    125mV
V3    3    0    250mV
R1    1    5    50k
R2    2    5    25k
R3    5    4    100k

XOA1.Ib1    5    0    70nA
XOA1.Ib2    0    0    90nA
XOA1.Vos    XOA1.3    0    1mV
XOA1.Ri     5    XOA1.3    2Meg
XOA1.E      XOA1.4    0    5    XOA1.3    -200000
XOA1.Ro     XOA1.4    4    75

R4    4    6    25k
R5    3    6    10k
R6    6    7    50k
R7    7    0    100k

XOA2.Ib1    6    0    70nA
XOA2.Ib2    0    0    90nA
XOA2.Vos    XOA2.3    0    1mV
XOA2.Ri     6    XOA2.3    2Meg
XOA2.E      XOA2.4    0    6    XOA2.3    -200000
XOA2.Ro     XOA2.4    7    75

.end
```

Figure 6.8-3 PSpice list showing substitution of subcircuits.

same node throughout the PSpice file, including inside subcircuits. For this reason, it was not necessary to include node 0 in the list of interface nodes in the .subckt statement.

Second, it is not necessary to keep the node numbers of the subcircuit separate from the node numbers of the main circuit. For example, consider node 4 of the subcircuit uA741_op_amp. This node becomes XOA1.4 when the subcircuit is substituted for XOA1 and XOA2.4 when the subcircuit is substituted for XOA2. Neither of these node numbers will be confused with node 4 of the main circuit.

Third, some nodes have two numbers—one as part of the main circuit and one as part of a subcircuit. In particular, the nodes in the statements

```
XOA1  5  0  4  uA741_op_amp
```

and

```
.subckt  uA741_op_amp  1  2  5
```

are listed in the same order. So nodes 5, 0, and 4 will correspond to nodes OA1.1, OA1.2, and OA1.5 when uA741_op_amp is substituted for XOA1. PSpice will use the node numbers from the main circuit rather than the node numbers from the subcircuit. This means that

```
XOA1.Ib1  XOA1.1  0  70nA
```

becomes

```
XOA1.Ib1  5  0  70nA
```

```
NODE        VOLTAGE     NODE        VOLTAGE

(    1)     .2000       (    2)      .1250
(    3)     .2500       (    4)     -.8860
(    5)     .0010       (    6)   997.3E-06
(    7)     .5334       (XOA1.3)     .0010
(XOA1.4)   -.8893       (XOA2.3)    -.0010
(XOA2.4)    .5346
```

Figure 6.8-4
PSpice output for the summing amplifier.

and

```
XOA2.Ib1 XOA2.1 0 70nA
```

becomes

```
XOA2.Ib1 6 0 70nA
```

Figure 6.8-3 shows the PSpice device list after the subcircuits have been substituted for the operational amplifiers. PSpice does not show this list to the user but does use the device names and node names from the list in output files and error messages.

Figure 6.8-4 shows PSpice output corresponding to the input file in Figure 6.8-2. Notice that $v_1 = 200$ mV, $v_2 = 125$ mV, and $v_3 = 250$ mV, so it is expected that $v_4 = -900$ mV and $v_o = 550$ mV if the operational amplifiers are ideal. PSpice shows that $v_4 = -886$ mV and $v_o = 534.6$ mV when the operational amplifiers are typical μA741s.

Exercise 6.8-1 Replace the subcircuit uA741_op_amp in Figure 6.8-2 with the subcircuit ideal_op_amp from Table 6.8-1. Make corresponding changes to the lines describing XOA1 and XOA2. Use PSpice to confirm that $v_4 = -900$ mV and $v_o = 550$ mV when the operational amplifiers are ideal.

Exercise 6.8-2 Modify the subcircuit uA741_op_amp to obtain a subcircuit TL051_op_amp that represents a typical TL051 operational amplifier. (See Table 6.7-1.) Use PSpice to find the node voltages v_o and v_4 of the summing amplifier in Figure 6.8-1*a* when $v_1 = 200$ mV, $v_2 = 125$ mV, $v_3 = 250$ mV, and the operational amplifiers are all a typical TL051.

Exercise 6.8-3 Redesign the summing amplifier of Figure 6.8-1*a* by dividing all the resistor values by 10. Notice that the performance of the summing amplifier depends on ratios of resistor values. Dividing all resistances by 10 does not change these ratios. The performance of the summing amplifier should be unchanged. Use PSpice to confirm that $v_4 = -900$ mV and $v_o = 55$ mV when the operational amplifiers are ideal and $v_1 = 200$ mV, $v_2 = 125$ mV, and $v_3 = 250$ mV. Use PSpice to find the node voltages v_o and v_4 when the operational amplifiers are all a typical uA741.

6.9 ANALYSIS OF OP AMP CIRCUITS USING MATLAB

Figure 6.9-1 shows an inverting amplifier. Model the operational amplifier as an ideal op amp. Then the output voltage of the inverting amplifier is related to the input voltage by

$$v_o(t) = -\frac{R_2}{R_1} v_s(t) \tag{6.9-1}$$

Suppose that $R_1 = 2\ \text{k}\Omega$, $R_2 = 50\ \text{k}\Omega$, and $v_s = -4\cos(2000\pi t)$ V. Using these values in Eq. 6.9-1 gives $v_o(t) = 100\cos(2000\pi t)$ V. This is not a practical answer. It's likely that the operational amplifier saturates, and, therefore, the ideal op amp is not an appropriate model of the operational amplifier. When voltage saturation is included in the model of the operational amplifier, the inverting amplifier is described by

$$v_o(t) = \begin{cases} v_{sat} & \text{when } -\dfrac{R_2}{R_1} v_s(t) > v_{sat} \\ -\dfrac{R_2}{R_1} \mathrm{v}_s(t) & \text{when } -v_{sat} < -\dfrac{R_2}{R_1} \mathrm{v}_s(t) < v_{sat} \\ -v_{sat} & \text{when } -\dfrac{R_2}{R_1} v_s(t) < -v_{sat} \end{cases} \tag{6.9-2}$$

where v_{sat} denotes the saturation voltage of the operational amplifier. Equation 6.9-2 is a more accurate, but more complicated, model of the inverting amplifier than Eq. 6.9-1. Of course, we prefer the simpler model, and we use the more complicated model only when we have reason to believe that answers based on the simpler model are not accurate.

Figures 6.9-2 and 6.9-3 illustrate the use of MATLAB to analyze the inverting amplifier when the operational amplifier model includes voltage saturation. Figure 6.9-2 shows the MATLAB input file, and Figure 6.9-3 shows the resulting plot of the input and output voltages of the inverting amplifier.

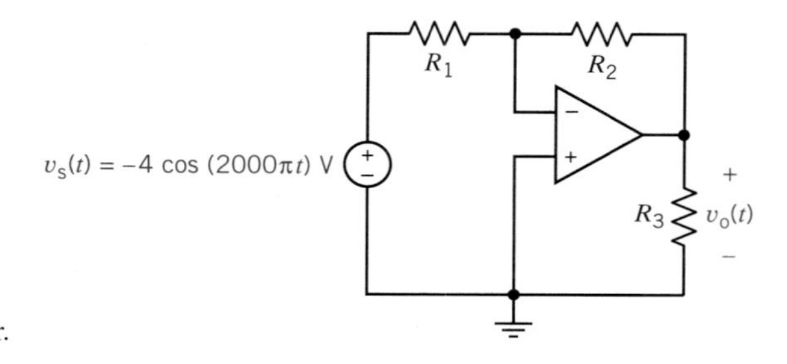

Figure 6.9-1
An inverting amplifier.

```
% Saturate.m simulates op amp voltage saturation
%---------------------------------------------------------------
% Enter values of the parameters that describe the circuit.
%---------------------------------------------------------------
                              % circuit parameters
R1=2e3;                       % resistance, ohms
R2=50e3;                      % resistance, ohms
R3=20e3;                      % resistance, ohms

                              % op amp parameter
vsat=15;                      % saturation voltage, V

                              % source parameters
M=4;                          % amplitude, V
f=1000;                       % frequency, Hz
w=2*pi*f;                     % frequency, rad/s
theta=(pi/180)*180;           % phase angle, rad

%---------------------------------------------------------------
% Divide the time interval (0, tf) into N increments
%---------------------------------------------------------------
tf=2/f;                       % final time
N=200;                        % number of incerments
t=0:tf/N:tf;                  % time, s

%---------------------------------------------------------------
% at each time t=k*(tf/N), calculate vo from vs
%---------------------------------------------------------------
vs = M*cos(w*t+theta);        % input voltage

for k=1:length(vs)

     if       (-(R2/R1)*vs(k) < -vsat)  vo(k) = -vsat; % ------
     elseif   (-(R2/R1)*vs(k) >  vsat)  vo(k) =  vsat; % eqn.
     else     vo(k) = -(R2/R1)*vs(k);                  % 6.9-2
     end                                               % ------

end

%---------------------------------------------------------------
% Plot Vo and vs versus t
%---------------------------------------------------------------
plot(t, vo, t, vs)            % plot the transfer characteristic
axis([0 tf -20 20])
xlabel('time, s')
ylabel('vo(t), V')
```

Figure 6.9-2 MATLAB input file corresponding to the circuit shown in Figure 6.9-1.

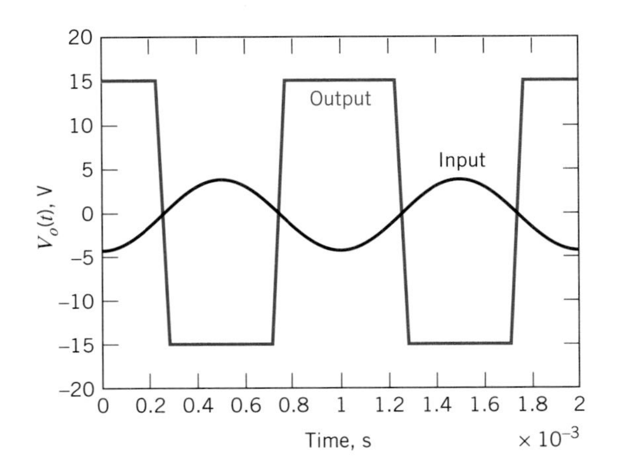

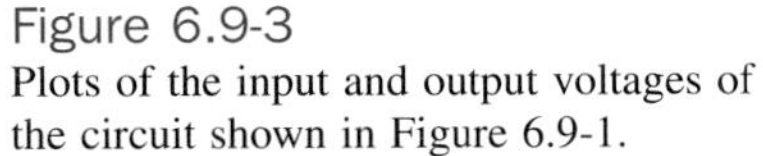
Figure 6.9-3 Plots of the input and output voltages of the circuit shown in Figure 6.9-1.

6.10 VERIFICATION EXAMPLE

The circuit in Figure 6.10-1a was analyzed by writing and solving the following set of simultaneous equations

$$\frac{v_6}{10} + i_5 = 0$$
$$10\, i_5 = v_4$$
$$\frac{v_4}{10} + i_3 = i_2$$
$$3 = 5i_2 + 10i_3$$
$$20i_3 = v_6$$

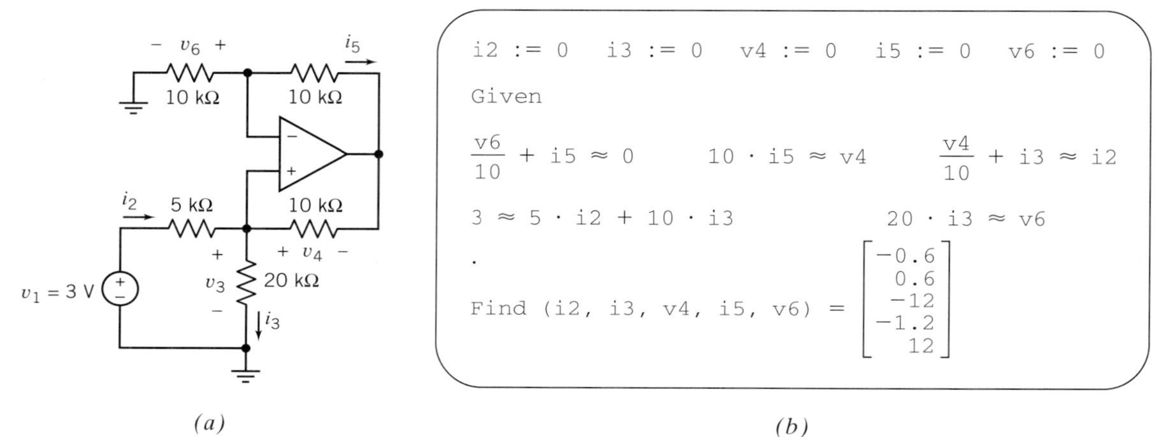

Figure 6.10-1 (a) An example circuit and (b) computer analysis using MathCad.

(These equations use units of volts, milliamps, and kohms.) A computer and the program Mathcad were used to solve these equations as shown in Figure 6.10-1b. The solution of these equations indicates that

$$i_2 = -0.6 \text{ mA}, \quad i_3 = 0.6 \text{ mA}, \quad v_4 = -12 \text{ V},$$
$$i_5 = 1.2 \text{ mA}, \quad \text{and} \quad v_6 = 12 \text{ V}$$

Are these voltages and currents correct?

Solution

Consider the voltage v_3. Using Ohm's law

$$v_3 = 20i_3 = 20\ (0.6) = 12 \text{ V}$$

Remember that resistances are in kohms and currents in milliamps. Applying KVL to the mesh consisting of the voltage source and the 5 kΩ and 20 kΩ resistors gives

$$v_3 = 3 - 5i_2 = 3 - 5(-0.6) = 6 \text{ V}$$

Clearly, v_3 cannot be both 12 and 6, so the values obtained for i_2, i_3, v_4, i_5, and v_6 cannot all be correct. Checking the simultaneous equations, we find that a resistor value has been entered incorrectly. The KVL equation corresponding to the mesh consisting of the voltage source and the 5 kΩ and 20 kΩ resistors should be

$$3 = 5i_2 + 20i_3$$

Note that $10i_3$ was incorrectly used in the fourth line of the Mathcad program of Figure 6.10-1. After making this correction, i_2, i_3, v_4, i_5, and v_6 are calculated to be

$$i_2 = -0.2 \text{ mA}, \quad i_3 = 0.2 \text{ mA}, \quad v_4 = -4 \text{ V},$$
$$i_5 = 0.4 \text{ mA}, \quad \text{and} \quad v_6 = 4 \text{ V}$$

Now

$$v_3 = 20i_3 = 20\,(0.2) = 4$$

and

$$v_3 = 3 - 5i_2 = 3 - 5\,(-0.2) = 4$$

This agreement suggests that the new values of i_2, i_3, v_4, i_5, and v_6 are correct. As an additional check, consider v_5. First, Ohm's law gives

$$v_5 = 10i_5 = 10\,(-0.4) = -4$$

Next, applying KVL to the loop consisting of the two 10 kΩ resistors and the input of the operational amplifier gives

$$v_5 = 0 + v_4 = 0 + (-4) = -4$$

This increases our confidence that the new values of i_2, i_3, v_4, i_5, and v_6 are correct.

6.11 Design Challenge Solution

TRANSDUCER INTERFACE CIRCUIT

A customer wants to automate a pressure measurement, which requires converting the output of the pressure transducer to a computer input. This conversion can be done using a standard integrated circuit called an analog-to-digital converter (ADC) (Dorf 1998). The ADC requires an input voltage between 0 V and 10 V, while the pressure transducer output varies between −250 mV and 250 mV. Design a circuit to interface the pressure transducer with the ADC. That is, design a circuit that translates the range −250 mV to 250 mV to the range 0 V to 10 V.

Describe the Situation and the Assumptions

The situation is shown in Figure 6.11-1.

The specifications state that

$$-250 \text{ mV} \leq v_1 \leq 250 \text{ mV}$$
$$0 \text{ V} \leq v_2 \leq 10 \text{ V}$$

A simple relationship between v_2 and v_1 is needed so that information about the pressure is not obscured. Consider

$$v_2 = a \cdot v_1 + b$$

The coefficients, a and b, can be calculated by requiring that $v_2 = 0$ when $v_1 = -250$ mV and that $v_2 = 10$ V when $v_1 = 250$ mV, that is,

$$0\text{ V} = a \cdot -250\text{ mV} + b$$
$$10\text{ V} = a \cdot 250\text{ mV} + b$$

Solving these simultaneous equations gives $a = 20$ V and $b = 5$ V.

State the Goal

Design a circuit having input voltage v_1 and output voltage v_2. These voltages should be related by

$$v_2 = 20v_1 + 5\text{ V} \tag{6.11-1}$$

Generate a Plan

Figure 6.11-2 shows a plan (or a structure) for designing the interface circuit. The operational amplifiers are biased using +15 V and −15 V power supplies. The constant 5 V input is generated from the 15 V power supply by multiplying by a gain of 1/3. The input voltage, v_1, is multiplied by a gain of 20. The summer (adder) adds the outputs of the two amplifiers in order to obtain v_2.

Each block in Figure 6.11-2 will be implemented using an operational amplifier circuit.

Act on the Plan

Figure 6.11-3 shows one proposed interface circuit. Some adjustments have been made to the plan. The summer is implemented using the inverting summing amplifier from Figure 6.6-1*d*. The inputs to this inverting summing amplifier must be $-20v_i$ and −5 V instead of $20v_i$ and 5 V. Consequently, an inverting amplifier is used to multiply v_1 by −20. A voltage follower is used to prevent the summing amplifier from loading the voltage divider. To make the signs work out correctly, the −15 V power supply provides the input to the voltage divider.

The circuit shown in Figure 6.11-3 is not the only circuit that solves this design challenge. There are several circuits that implement

$$v_2 = 20v_1 + 5\text{ V}$$

We will be satisfied with having found one circuit that does the job.

Verify the Proposed Solution

The circuit shown in Figure 6.11-3 was simulated using PSpice. The result of this simulation is the plot of the v_2 versus v_1 shown in Figure 6.11-4. Since this plot shows a straight line, v_2 is related to v_1 by the equation of a straight line

$$v_2 = mv_1 + b$$

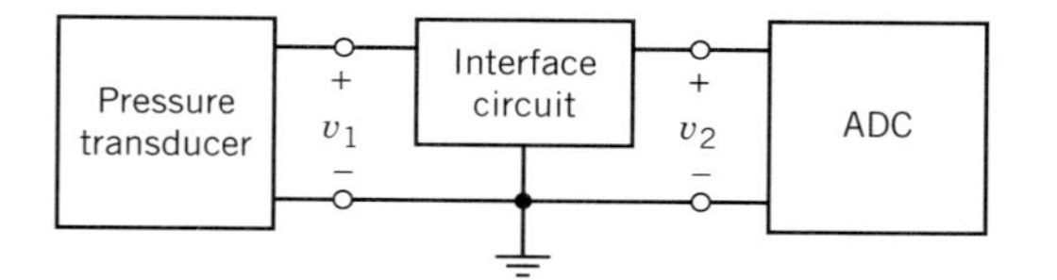

Figure 6.11-1
Interfacing a pressure transducer with an analog-to-digital converter (ADC).

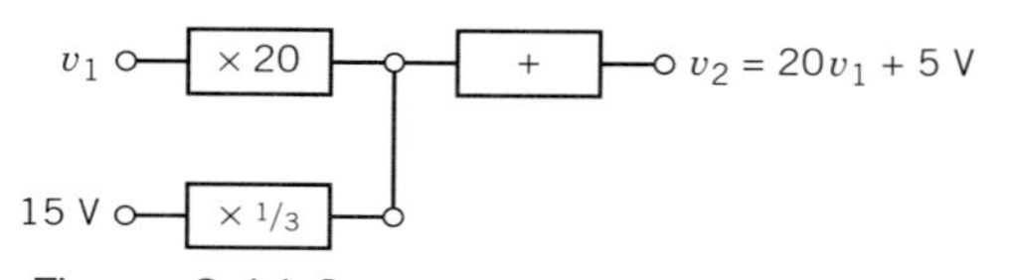

Figure 6.11-2
A structure (or plan) for the interface circuit.

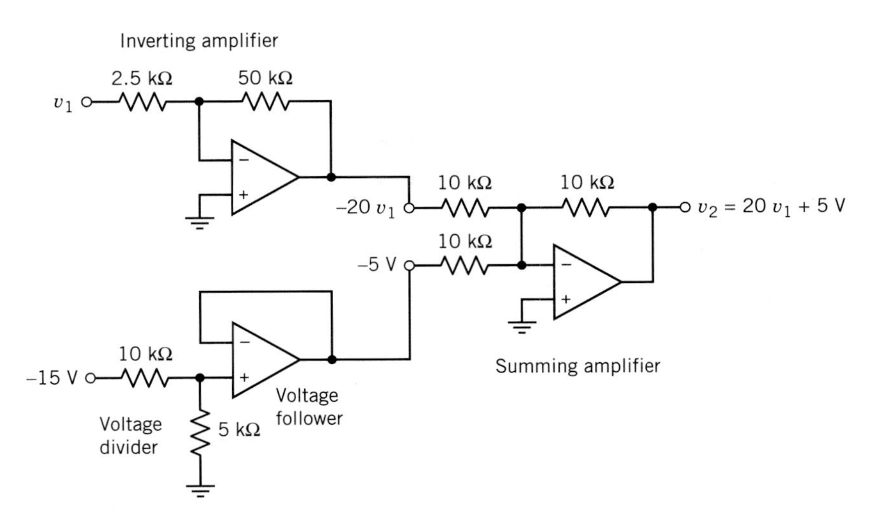

Figure 6.11-3 One implementation of the interface circuit.

where m is the slope of the line and b is the intercept of the line with the vertical axis. Two points on the line have been labeled to show that $v_2 = 10.002$ V when $v_1 = 0.250$ V and that $v_2 = 0.0047506$ V when $v_1 = -0.250$ V. The slope, m, and intercept, b, can be calculated from these points. The slope is given by

$$m = \frac{10.002 - (0.0047506)}{0.250 - (-0.250)} = 19.994$$

The intercept is given by

$$b = 10.002 - 19.994 \times .0250 = 5.003$$

Thus

$$v_2 = 19.994 v_1 + 5.003 \tag{6.11-2}$$

Comparing Eqs. 6.11-1 and 6.11-2 verifies that the proposed solution is indeed correct.

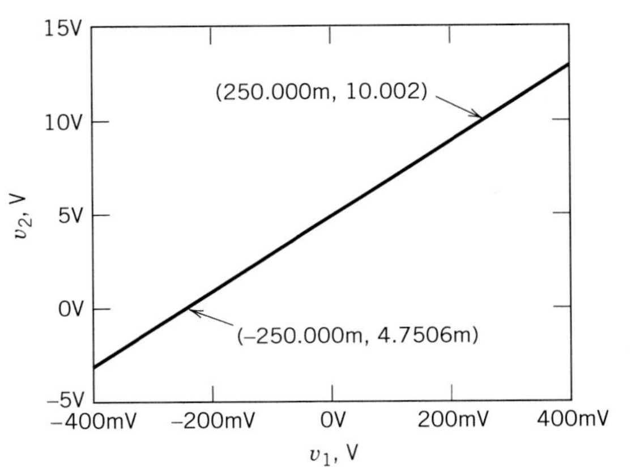

Figure 6.11-4 PSpice simulation of the circuit shown in Figure 6.11-3.

6.12 SUMMARY

- With the invention of the electron tube in 1907, the field of electronics was born. In 1947 the transistor was invented, which revolutionized the field of electronics. By the 1960s integrated electronic circuits such as the operational amplifier were available.
- Several models are available for operational amplifiers. Simple models are easy to use. Accurate models are more complicated. The simplest model of the operational amplifier is the ideal operational amplifier.
- The currents into the input terminals of an ideal operational amplifier are zero, and the voltages at the input nodes of an ideal operational amplifier are equal.
- It is convenient to use node equations to analyze circuits that contain ideal operational amplifiers.
- Operational amplifiers are used to build circuits that perform mathematical operations. Many of these circuits have been used so often that they have been given names. The inverting amplifier gives a response of the form $v_o = -Kv_i$ where K is a positive constant. The noninverting amplifier gives a response of the form $v_o = Kv_i$ where K is a positive constant. Another useful operational amplifier circuit is the noninverting amplifier with a gain of $K = 1$, often called a voltage follower or buffer. The output of the voltage follower faithfully follows the input voltage. The voltage follower reduces loading by isolating its output terminal from its input terminal.
- Figure 6.6-1 is a catalog of some frequently used operational amplifier circuits.
- Practical operational amplifiers have properties that are not included in the ideal operational amplifier. These include the input offset voltage, bias current, dc gain, input resistance, and output resistance. More complicated models are needed to account for these properties.
- PSpice can be used to reduce the drudgery of analyzing operational amplifier circuits with complicated models.

PROBLEMS

Section 6.4 The Ideal Operational Amplifier

P 6.4-1 Determine the value of voltage measured by the voltmeter in Figure P 6.4-1. Assume that the operational amplifier is ideal.
Answer: −4 V

P 6.4-2 Find v_o and i_o for the circuit of Figure P 6.4-2. Assume an ideal op amp.

P 6.4-3 Find v_o and i_o for the circuit of Figure P 6.4-3. Assume an ideal op amp.
Answer: $v_o = -30$ V, $i_o = 3.5$ mA

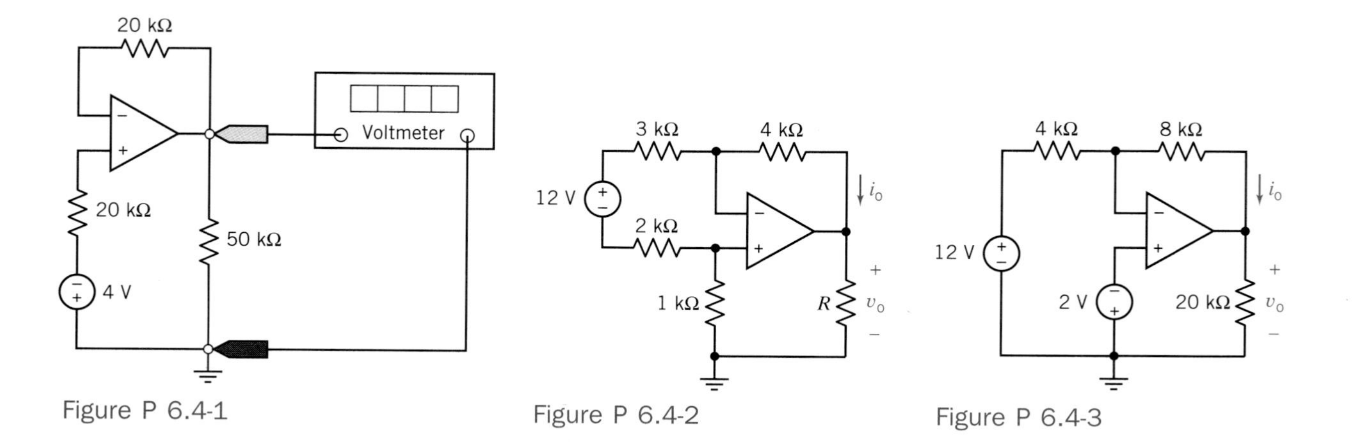

Figure P 6.4-1

Figure P 6.4-2

Figure P 6.4-3

P 6.4-4 Find v and i for the circuit of Figure P 6.4-4. Assume an ideal op amp.

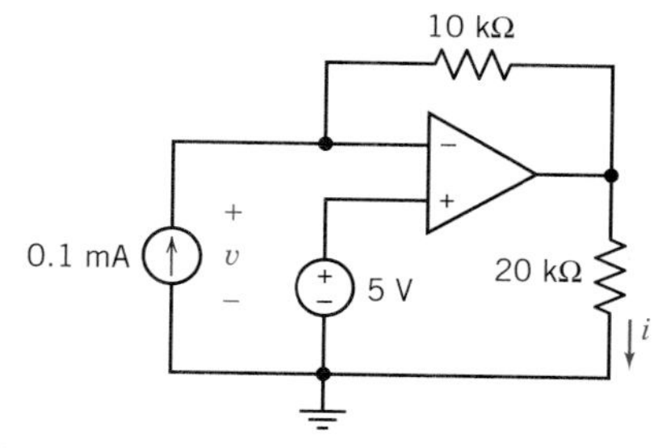

Figure P 6.4-4

P 6.4-5 Find v_o and i_o for the circuit of Figure P 6.4-5. Assume an ideal op amp.
Answer: $v_o = -15$ V, $i_o = 7.5$ mA

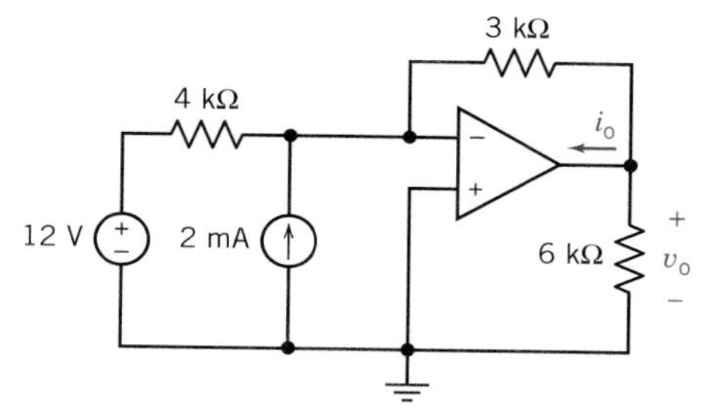

Figure P 6.4-5

P 6.4-6 Determine the value of voltage measured by the voltmeter in Figure P 6.4-6. Assume that the operational amplifier is ideal.
Answer: 7.5 V

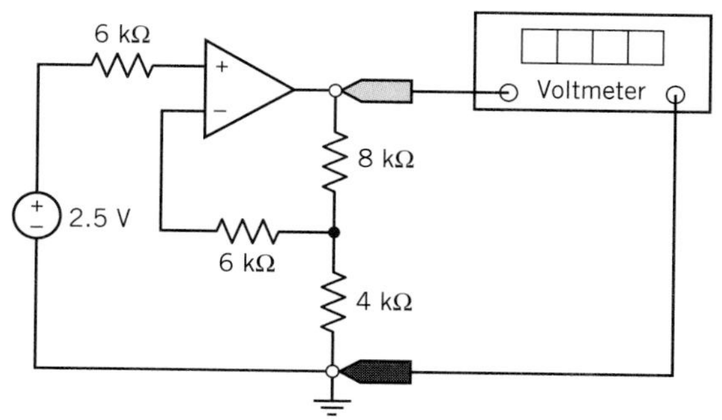

Figure P 6.4-6

P 6.4-7 Find v_o and i_o for the circuit of Figure P 6.4-7. Assume an ideal op amp.

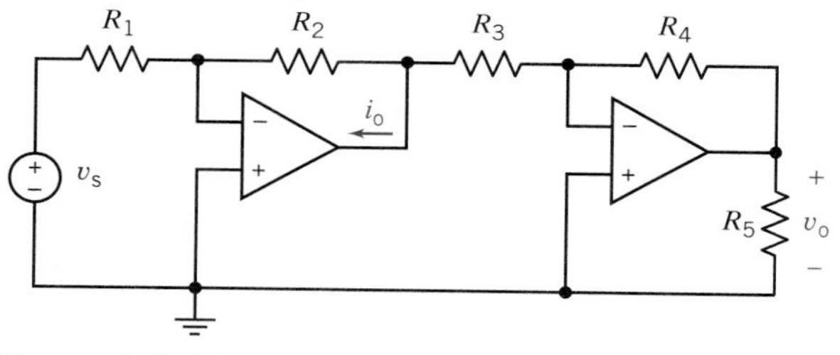

Figure P 6.4-7

P 6.4-8 An operational amplifier can be used to convert a current to a voltage, v_o, as shown in Figure P 6.4-8. Find the ratio v_o/i_s, using an ideal operational amplifier.

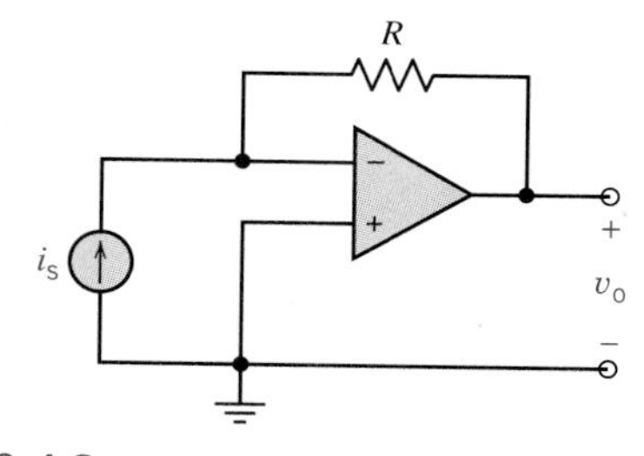

Figure P 6.4-8 A current-to-voltage converter.

P 6.4-9 Determine the current i_o for the circuit shown in Figure P 6.4-9. Assume that the operational amplifiers are ideal.
Answer: $i_o = 2.5$ mA

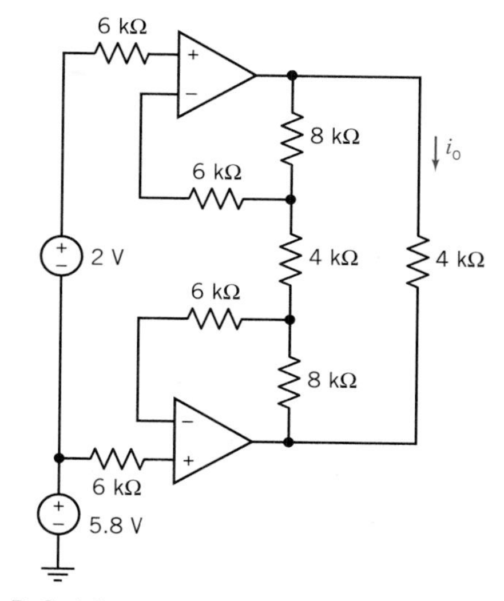

Figure P 6.4-9

P 6.4-10 Determine the voltage v_o for the circuit shown in Figure P 6.4-10. Assume that the operational amplifier is ideal.
Answer: $v_o = -8$ V

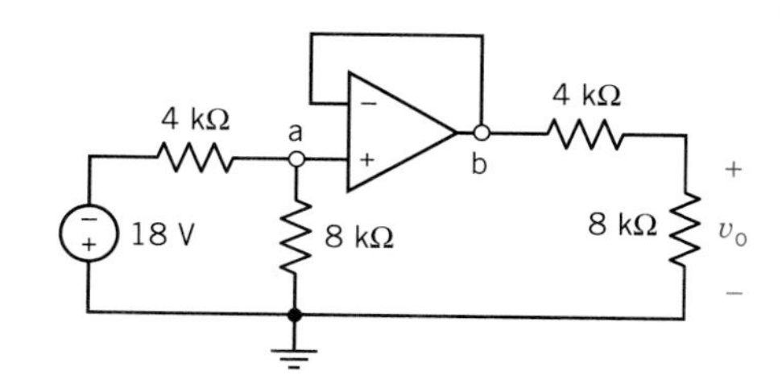

Figure P 6.4-10

Section 6.5 Nodal Analysis of Circuits Containing Ideal Operational Amplifiers

P 6.5-1 A circuit with two sources is shown in Figure P 6.5-1. Find v_o when the operational amplifier is assumed to be ideal.
Answer: $v_o = -2$ V

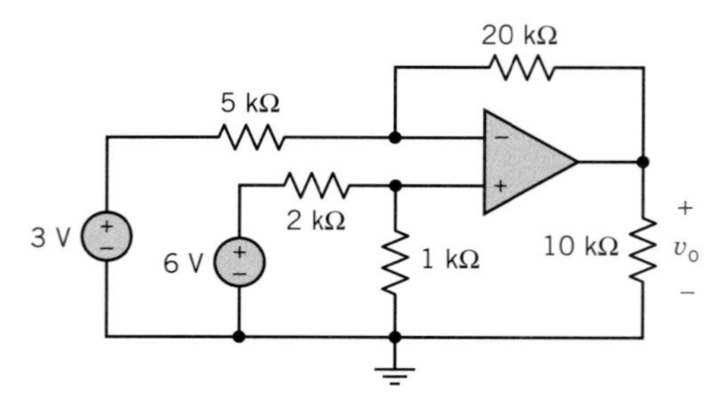

Figure P 6.5-1

P 6.5-2 Determine the node voltages for the circuit shown in Figure P 6.5-2. Assume that the operational amplifier is ideal.
Answer: $v_a = 2$ V, $v_b = -0.25$ V, $v_c = -5$ V, $v_d = -2.5$ V, and $v_e = -0.25$ V.

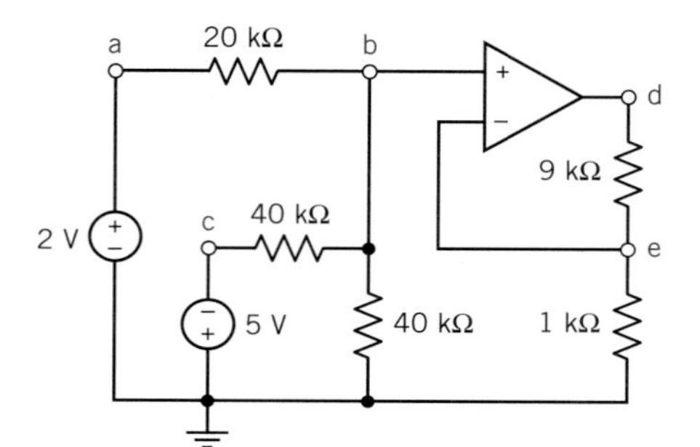

Figure P 6.5-2

P 6.5-3 Find v_o and i_o for the circuit of Figure P 6.5-3. Assume an ideal op amp.
Answer: $v_o = -4$V, $i_o = 1.33$ mA

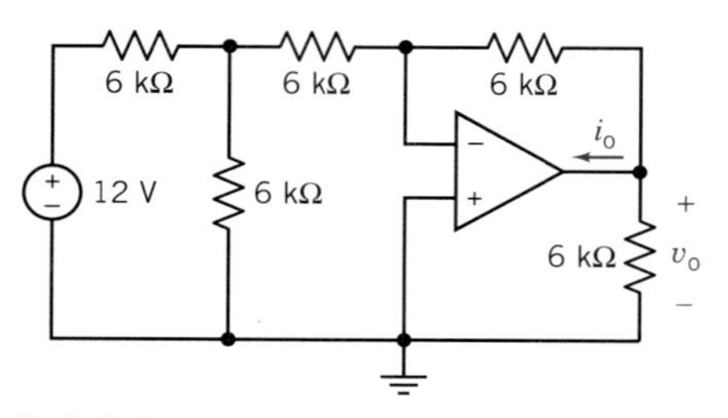

Figure P 6.5-3

P 6.5-4 If $R_1 = 4.8$ kΩ and $R_2 = R_4 = 30$ kΩ, find v_o/v_s for the circuit shown in Figure P 6.5-4 when $R_3 = 1$ kΩ.
Answer: $v_o/v_s = -200$

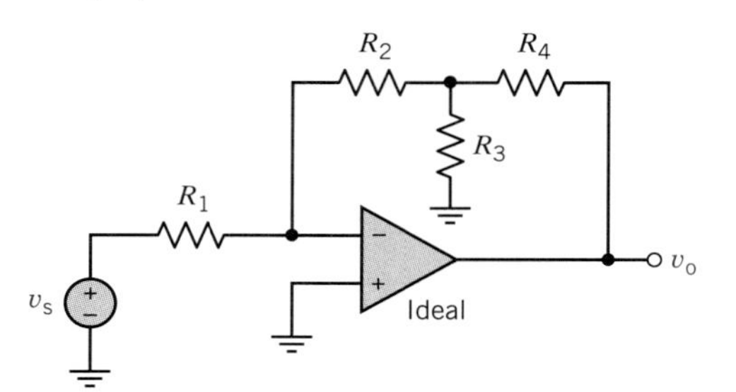

Figure P 6.5-4

P 6.5-5 Find v_o for the circuit shown in Figure P 6.5-5. Assume an ideal operational amplifier.

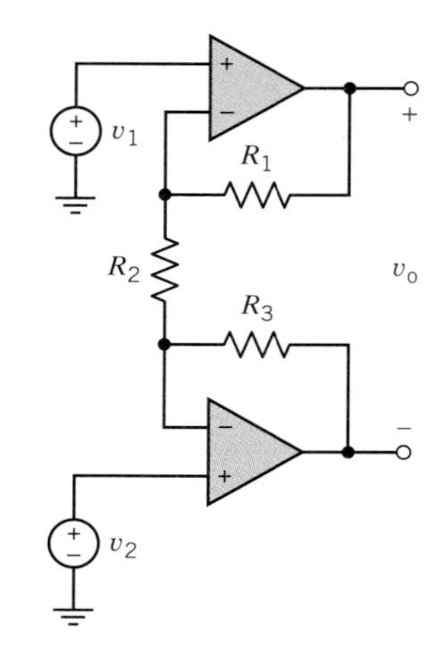

Figure P 6.5-5

P 6.5-6 Find v_o and i_o for the circuit of Figure P 6.5-6. Assume ideal op amps.

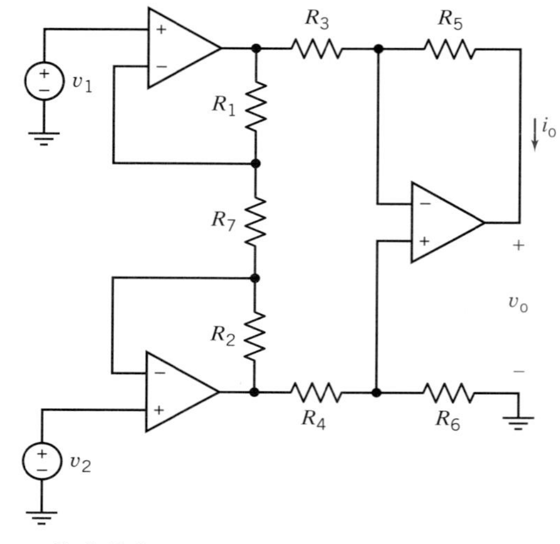

Figure P 6.5-6

P 6.5-7 Determine the node voltages for the circuit shown in Figure P 6.5-7. Assume that the operational amplifier is ideal.
Answer: $v_a = -0.75$ V, $v_b = 0$ V, and $v_c = -0.9375$ V.

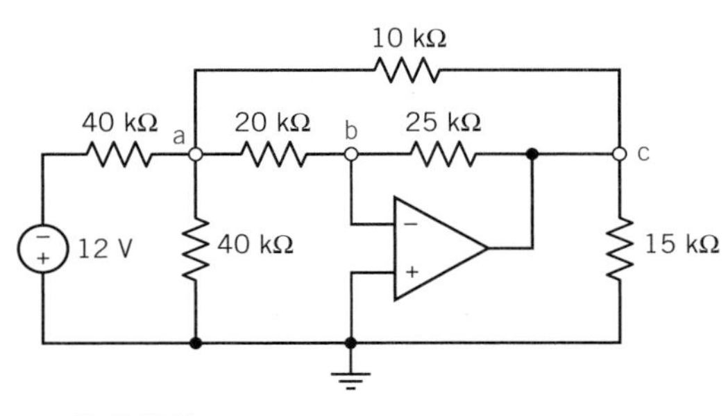

Figure P 6.5-7

P 6.5-8 Find v_o and i_o for the circuit shown in Figure P 6.5-8. Assume an ideal operational amplifier.

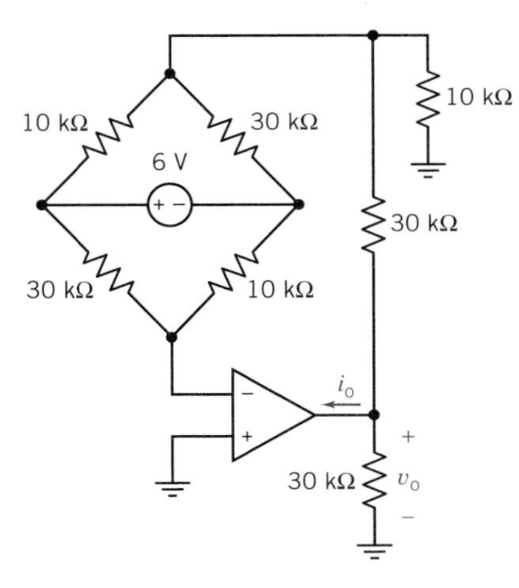

Figure P 6.5-8

P 6.5-9 Find v_o and i_o for the circuit shown in Figure P 6.5-9. Assume an ideal operational amplifier.

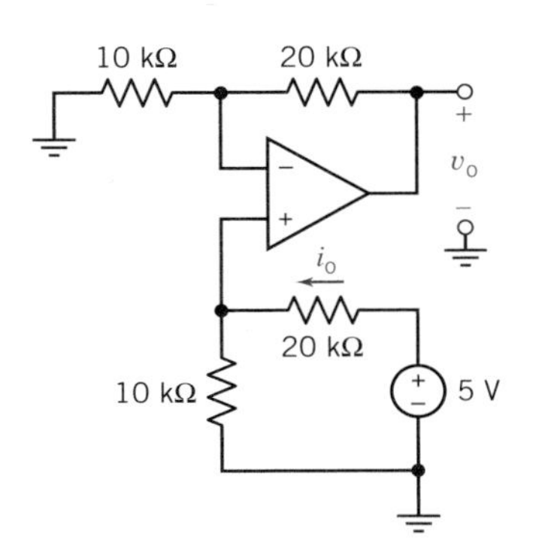

Figure P 6.5-9

P 6.5-10 Determine the node voltages for the circuit shown in Figure P 6.5-10. Assume that the operational amplifiers are ideal.
Answer: $v_a = -12$ V, $v_b = -4$ V, $v_c = -4$ V, $v_d = -4$ V, $v_e = -3.2$ V, $v_f = -4.8$ V, and $v_g = -3.2$ V.

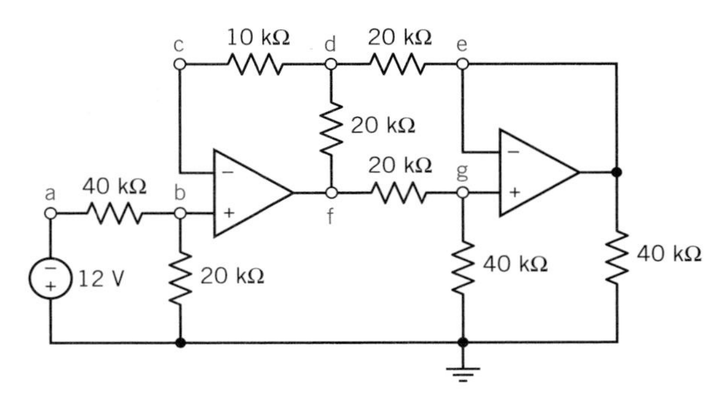

Figure P 6.5-10

P 6.5-11 The circuit shown in Figure P 6.5-11 includes a simple strain gauge. The resistor R changes its value by ΔR when it is twisted or bent. Derive a relation for the voltage gain v_o/v_s and show that it is proportional to the fractional change in R, namely $\Delta R/R_o$.

Answer: $v_o = -v_s \dfrac{R_o}{R_o + R_1} \dfrac{\Delta R}{R_o}$

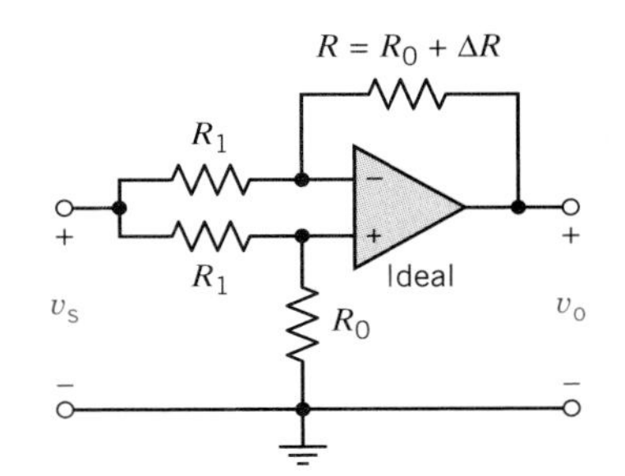

Figure P 6.5-11 A strain gauge circuit.

Section 6.6 Design Using Operational Amplifiers

P 6.6-1 Design the operational amplifier circuit in Figure P 6.6-1 so that

$$v_{out} = r \cdot i_{in}$$

where

$$r = 20 \frac{\text{V}}{\text{mA}}$$

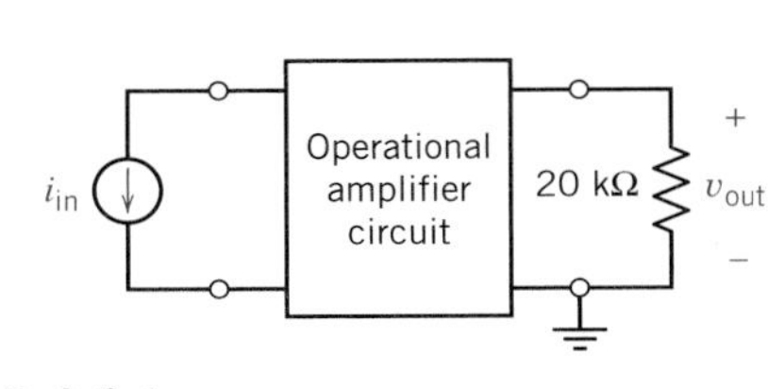

Figure P 6.6-1

P 6.6-2 Design the operational amplifier circuit in Figure P 6.6-2, so that

$$i_{out} = g \cdot v_{in}$$

where

$$g = 2\,\frac{\text{mA}}{\text{V}}$$

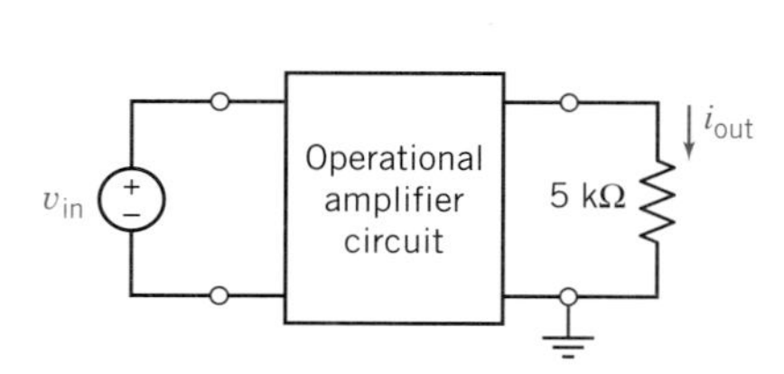

Figure P 6.6-2

P 6.6-3 Design the operational amplifier circuit in Figure P 6.6-3 so that

$$v_{out} = 5 \cdot v_1 + 2 \cdot v_2$$

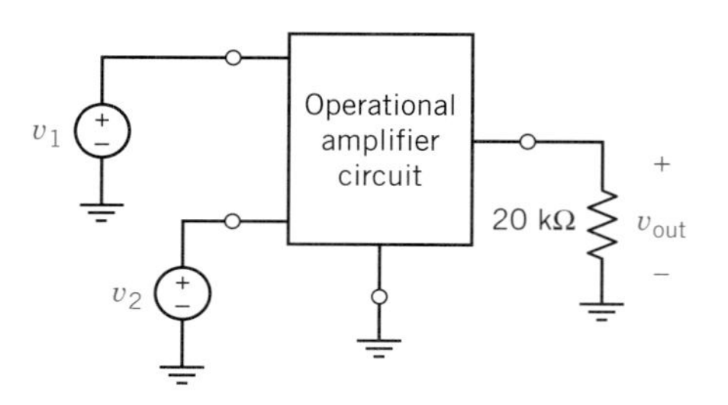

Figure P 6.6-3

P 6.6-4 Design the operational amplifier circuit in Figure P 6.6-3 so that

$$v_{out} = 5 \cdot (v_1 - v_2)$$

P 6.6-5 Design the operational amplifier circuit in Figure P 6.6-3 so that

$$v_{out} = 5 \cdot v_1 - 2 \cdot v_2$$

P 6.6-6 The voltage divider shown in Figure P 6.6-6 has a gain of

$$\frac{v_{out}}{v_{in}} = \frac{-10\ \text{k}\Omega}{5\ \text{k}\Omega + (-10\ \text{k}\Omega)} = 2$$

Design an operational amplifier circuit to implement the -10 kΩ resistor.

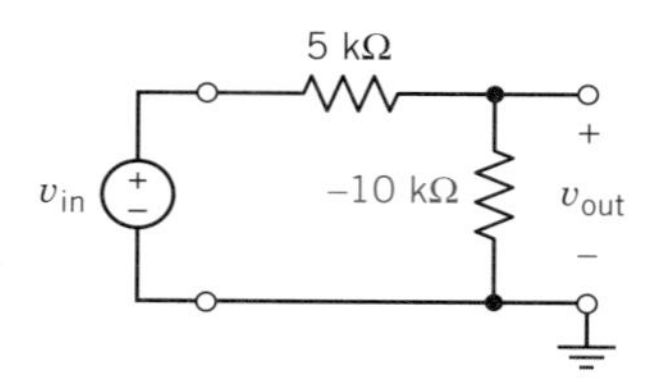

Figure P 6.6-6 A circuit with a negative resistor.

P 6.6-7 Design the operational amplifier circuit in Figure P 6.6-7 so that

$$i_{in} = 0 \quad \text{and} \quad v_{out} = 3 \cdot v_{in}$$

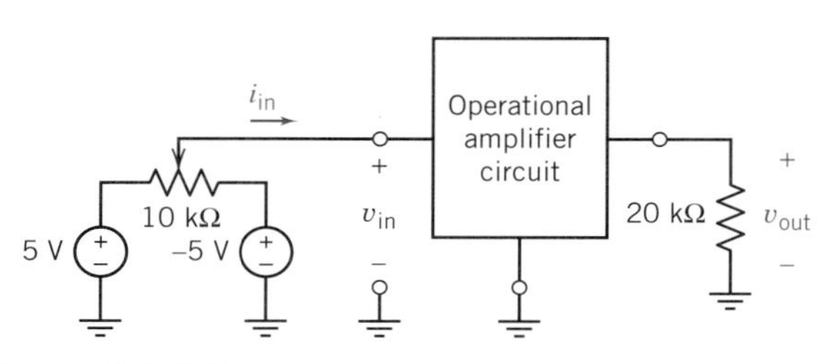

Figure P 6.6-7

P 6.6-8 A voltage-subtracting circuit is shown in Figure P 6.6-8. (a) Show that v_o can be expressed as

$$v_o = \frac{1 + R_2/R_1}{1 + R_3/R_4}\,v_2 - \frac{R_2}{R_1}\,v_1$$

(b) Design a circuit with an output $v_o = 4\,v_2 - 11\,v_1$.
Answer: $R_1 = 10$ kΩ, $R_2 = 110$ kΩ, $R_3 = 20$ kΩ, and $R_4 = 10$ kΩ (This answer is not unique.)

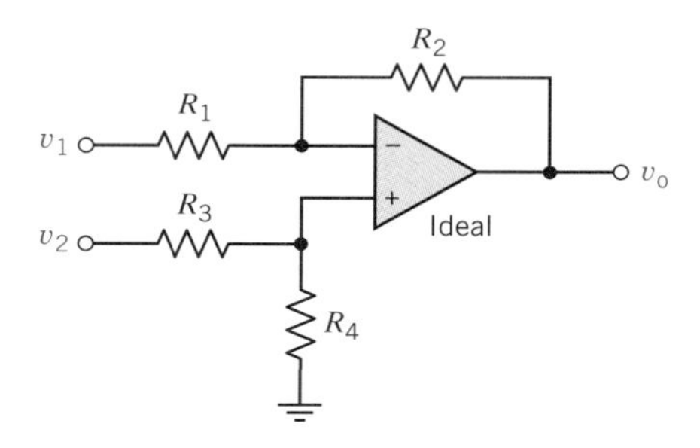

Figure P 6.6-8 A voltage subtracting circuit.

P 6.6-9 Design an operational amplifier circuit with output $v_o = 6\,v_1 + 2\,v_2$, where v_1 and v_2 are input voltages. Assume an ideal op amp.

P 6.6-10 Determine the voltage v_o for the circuit shown in Figure P 6.6-10. Assume that the operational amplifier is ideal.
Hint: Use superposition.
Answer: $v_o = (-3)(3) + (4)(-4) + (4)(8) = -7$ V.

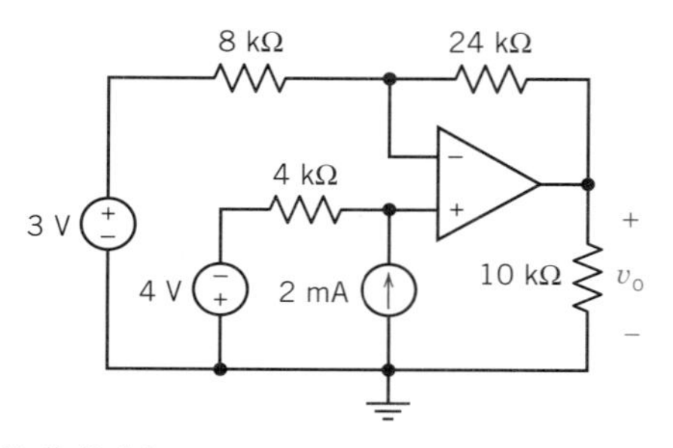

Figure P 6.6-10

P 6.6-11 For the op-amp circuit shown in Figure P 6.6-11, find and list all the possible voltage gains that can be achieved by connecting the resistor terminals to either the input or the output voltage terminals. Assume an ideal op amp.

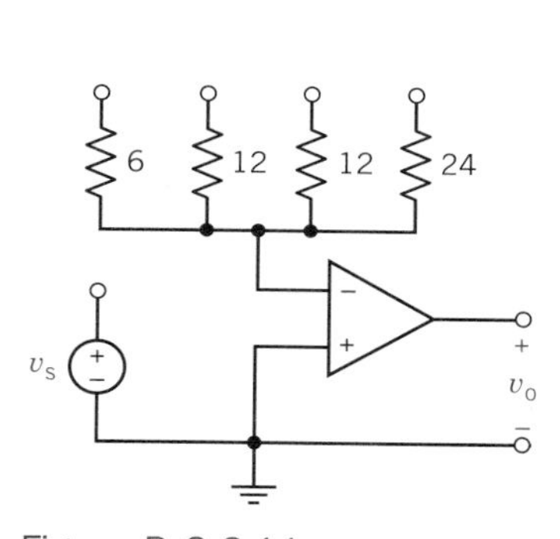

Figure P 6.6-11 Resistances in kΩ.

P 6.6-12 Assuming that all the op amps are ideal, determine v_o in terms of v_1 and v_2 for the circuit shown in Figure P 6.6-12.

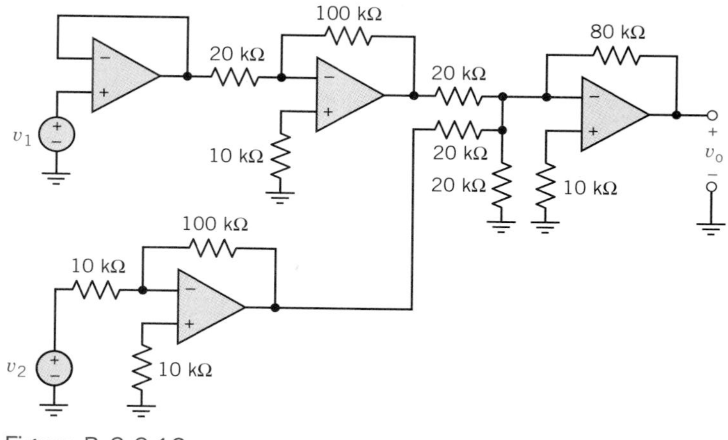

Figure P 6.6-12

P 6.6-13 Determine the voltage v_o as a function of v_a and v_b for the circuit shown in Figure P 6.6-13. Assume that the operational amplifiers are ideal.

Answer: $v_o = -3\,v_a + 6\,v_b$

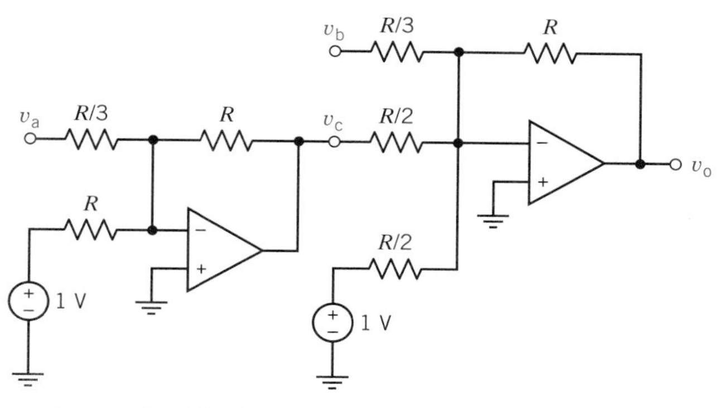

Figure P 6.6-13

P 6.6-14 The circuit shown in Figure P 6.6-14 is called the inverted R-2R ladder digital-to-analog converter (DAC). The input to this circuit is a binary code represented by $b_1\, b_2 \cdots b_n$, where b_i is either 1 or 0. Each switch shown in the figure is controlled by only one of the digits of the binary code. If $b_i = 1$, the switch will be at the left position, whereas if $b_i = 0$, the switch will be at the right position. Depending on the position of the switch, each current I_i is diverted either to true ground bus (adding to I^+) or the virtual ground bus (adding to I^-).

(a) Show that $I = V_R/R$ regardless of the digital input code.

(b) Show that the output voltage can be expressed as

$$V_o = -\frac{R_f}{R} V_R (b_1 2^{-1} + b_2 2^{-2} + \cdots + b_{n-1} 2^{-n+1} + b_n 2^{-n})$$

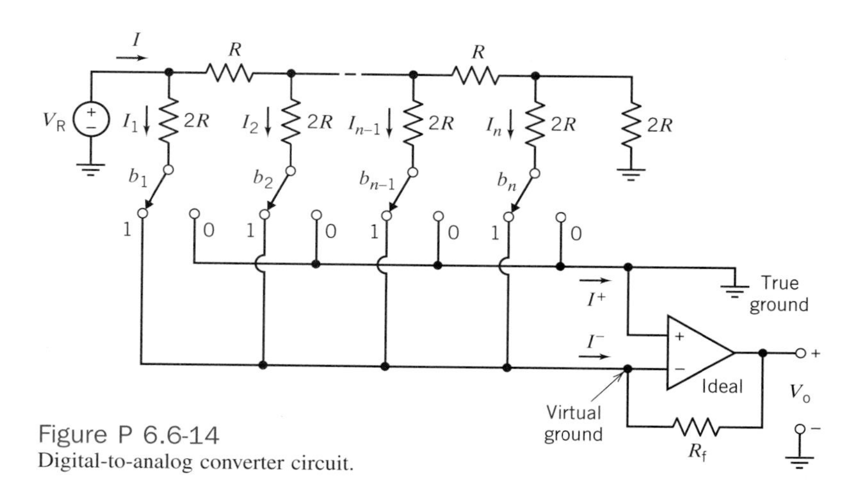

Figure P 6.6-14
Digital-to-analog converter circuit.

(c) Show that $I^+ + I^- = (1 - 2^{-n})\, V_R/R$ regardless of the binary code.
(d) Given $R_f = R = 10\ \text{k}\Omega$, $V_R = -16$ V and assuming a four-digit binary input code, find the output voltage V_o for each combination of the input code, ranging from 0000 to 1111. Explain the relationship between the output and the input code.

Section 6.7 Characteristics of the Practical Operational Amplifier

P 6.7-1 Consider the inverting amplifier shown in Figure P 6.7-1. The operational amplifier is a typical OP-07E (Table 6.7-1). Use the offsets model of the operational amplifier to calculate the output offset voltage. (Recall that the input, v_{in}, is set to zero when calculating the output offset voltage.)
Answer: 0.45 mV

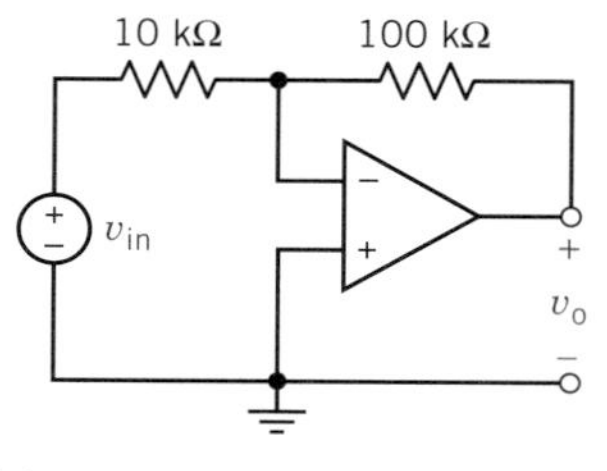

Figure P 6.7-1

P 6.7-2 Consider the noninverting amplifier shown in Figure P 6.7-2. The operational amplifier is a typical LF351 (Table 6.7-1). Use the offsets model of the operational amplifier to calculate the output offset voltage. (Recall that the input, v_{in}, is set to zero when calculating the output offset voltage.)

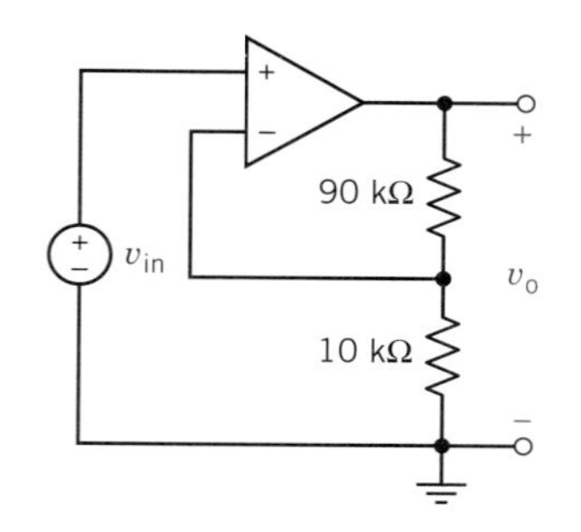

Figure P 6.7-2

P 6.7-3 Consider the inverting amplifier shown in Figure P 6.7-3. Use the finite gain model of the operational amplifier (Figure 6.7-1*c*) to calculate the gain of the inverting amplifier. Show that

$$\frac{v_o}{v_{in}} = \frac{R_{in}(R_o - AR_2)}{(R_1 + R_{in})(R_o + R_2) + R_1 R_{in}(1 + A)}$$

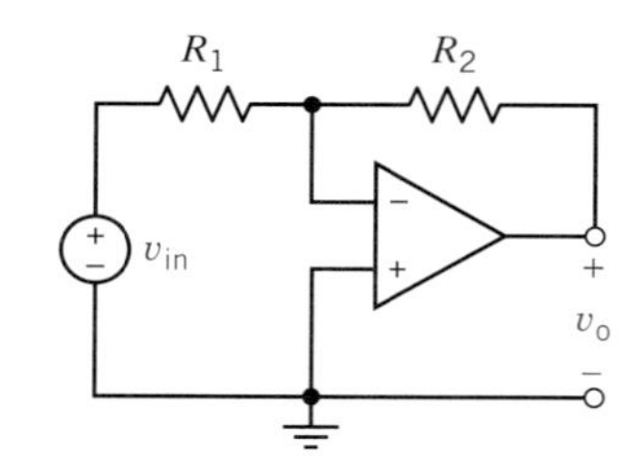

Figure P 6.7-3

P 6.7-4 Consider the inverting amplifier shown in Figure P 6.7-3. Suppose the operational amplifier is ideal, $R_1 = 5\ \text{k}\Omega$, and $R_2 = 50\ \text{k}\Omega$. The gain of the inverting amplifier will be

$$\frac{v_o}{v_{in}} = -10$$

Use the results of Problem P 6.7-3 to find the gain of the inverting amplifier in each of the following cases:
(a) The operational amplifier is ideal, but 2 percent resistors are used and $R_1 = 5.1\ \text{k}\Omega$ and $R_2 = 49\ \text{k}\Omega$.
(b) The operational amplifier is represented using the finite gain model with $A = 200{,}000$, $R_i = 2\ \text{M}\Omega$, and $R_o = 75\Omega$; $R_1 = 5\ \text{k}\Omega$ and $R_2 = 50\ \text{k}\Omega$.
(c) The operational amplifier is represented using the finite gain model with $A = 200{,}000$, $R_i = 2\ \text{M}\Omega$, and $R_o = 75\Omega$; $R_1 = 5.1\ \text{k}\Omega$ and $R_2 = 49\ \text{k}\Omega$.

P 6.7-5 The circuit in Figure P 6.7-5 is called a difference amplifier and is used for instrumentation circuits. The output of a measuring element is represented by the common mode signal v_{cm} and the differential signal $(v_n + v_p)$. Using an ideal operational amplifier, show that

$$v_o = -\frac{R_4}{R_1}(v_n + v_p)$$

when

$$\frac{R_4}{R_1} = \frac{R_3}{R_2}$$

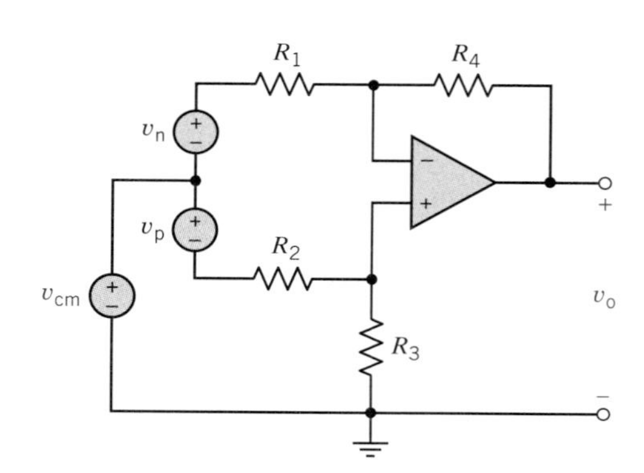

Figure P 6.7-5 A direct-coupled difference amplifier.

P 6.7-6 A circuit has three inputs v_1, v_2, and v_3 with $|v_i| \leq 1$ V for $i = 1, 2, 3$. The op amps have $i_{sat} = 2$ mA and $v_{sat} = 15$ V. Design an op amp circuit to produce
(a) $v_o = -(4v_1 + 2v_2)$
(b) $v_o = 4v_1 + 2v_2 - 3v_3$

P 6.7-7 An op amp circuit is shown in Figure P 6.7-7. It is desired to determine the output voltage for (a) an ideal op amp and (b) an op amp with $A = 10^4$, $R_i = 200$ kΩ, and $R_o = 5$ kΩ. Compare the results.

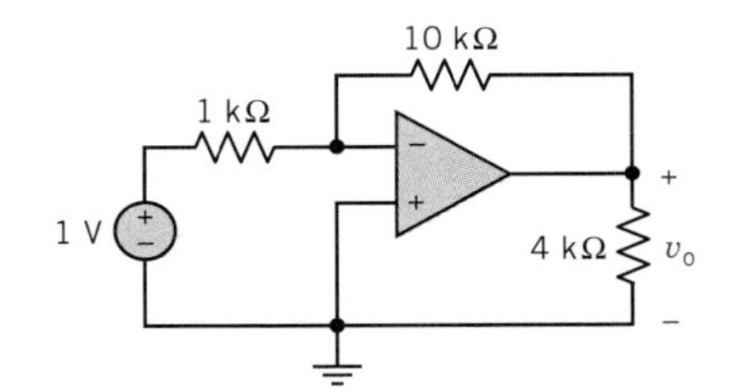

Figure P 6.7-7

PSPICE PROBLEMS

SP 6-1 Use PSpice to find v_o and i_o for the circuit of Figure SP 6.1 when $R_1 = 7$ kΩ, $R_2 = 98$ kΩ, $R_3 = 10$ kΩ, $R_4 = 20$ kΩ, $R_L = 2$ kΩ, and $v_s = 0.1$ V.

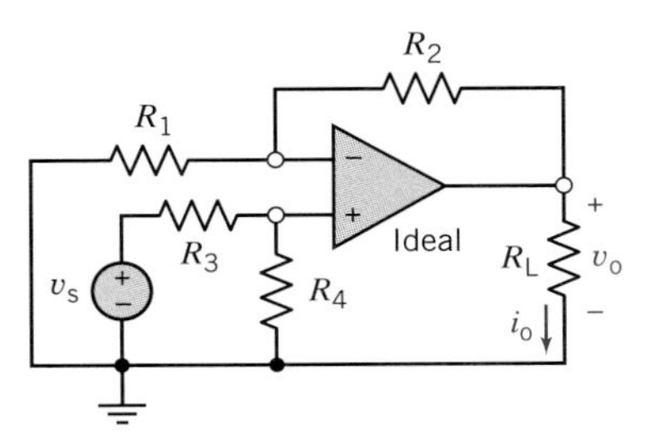

Figure SP 6.1

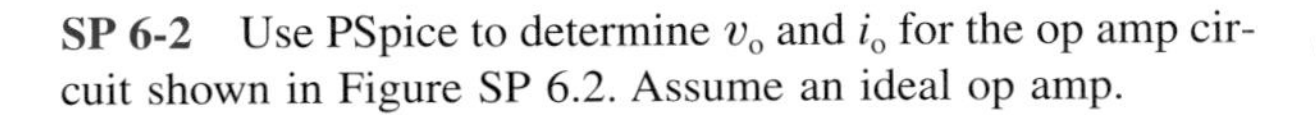

SP 6-2 Use PSpice to determine v_o and i_o for the op amp circuit shown in Figure SP 6.2. Assume an ideal op amp.

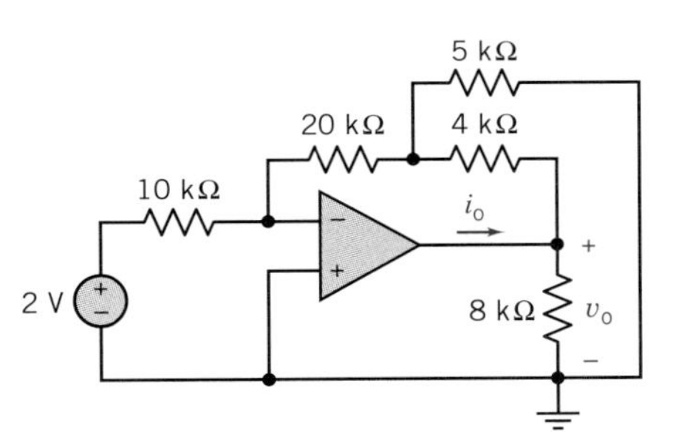

Figure SP 6.2

SP 6-3 A circuit with its nodes identified is shown in Figure SP 6.3. Determine v_{34}, v_{23}, v_{50}, and i_o.

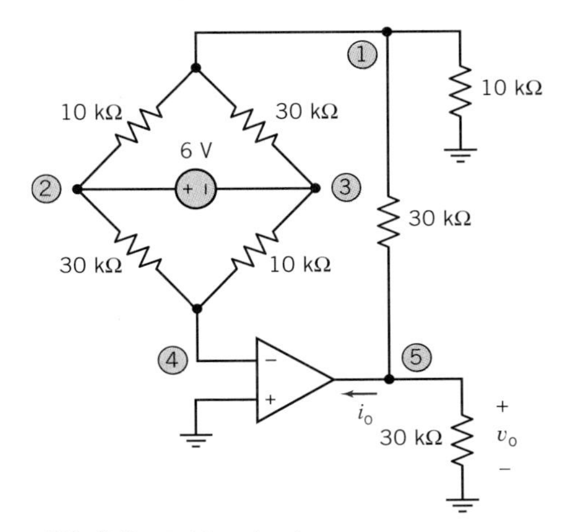

Figure SP 6.3 Bridge circuit.

SP 6-4 Use PSpice to analyze the VCCS shown in Figure SP 6.4. Consider two cases:
(a) The operational amplifier is ideal.
(b) The operational amplifier is a typical μA741 represented by the "offsets and finite gain" model.

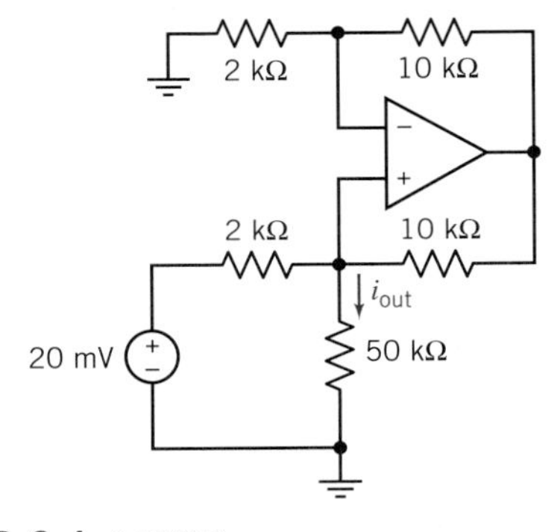

Figure SP 6.4 A VCCS.

SP 6-5 Use PSpice to analyze the noninverting summing amplifier shown in Figure SP 6.5. Consider two cases:
(a) The operational amplifier is ideal.
(b) The operational amplifier is a typical μA741 represented by the "offsets and finite gain" model.

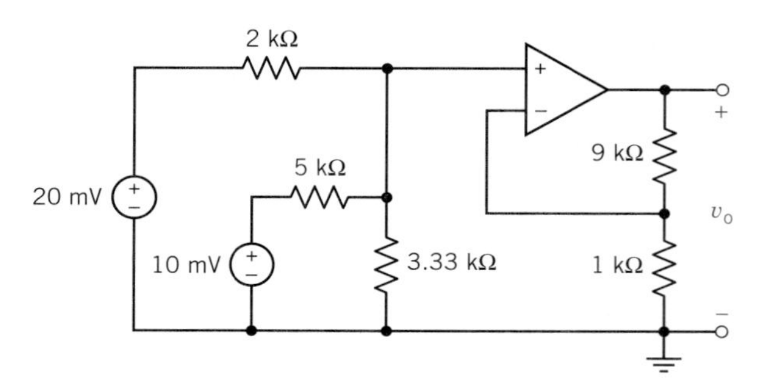

Figure SP 6.5

VERIFICATION PROBLEMS

VP 6-1 Analysis of the circuit in Figure VP 6.1 shows that $i_o = -1$ ma and $v_o = 7$ V. Is this analysis correct?
Hint: Is KCL satisfied at the output node of the op amp?

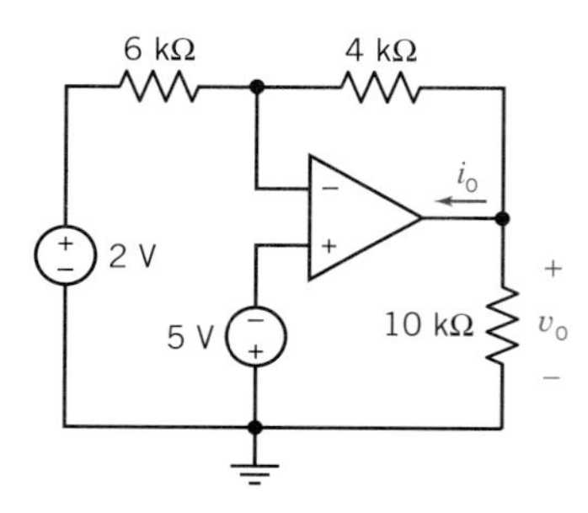

Figure VP 6.1

VP 6-2 Your lab partner measured the output voltage of the circuit shown in Figure VP 6.2 to be $v_o = 9.6$ V. Is this the correct output voltage for this circuit?
Hint: Ask your lab partner to check the polarity of the voltage that she measured.

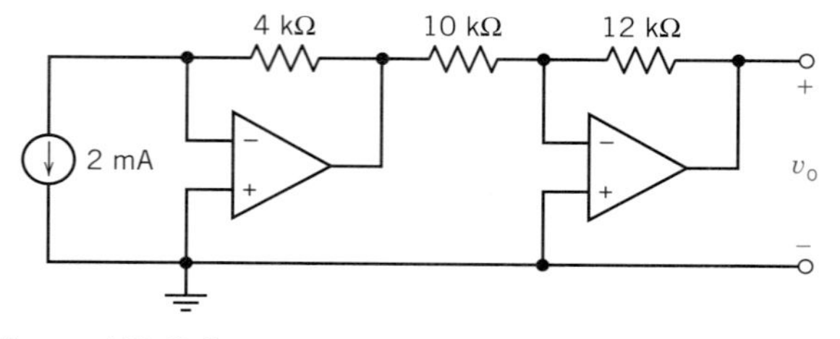

Figure VP 6.2

VP 6-3 Nodal analysis of the circuit shown in Figure VP 6.3 indicates that $v_o = -12$ V. Is this analysis correct?
Hint: Redraw the circuit to identify an inverting amplifier and a noninverting amplifier.

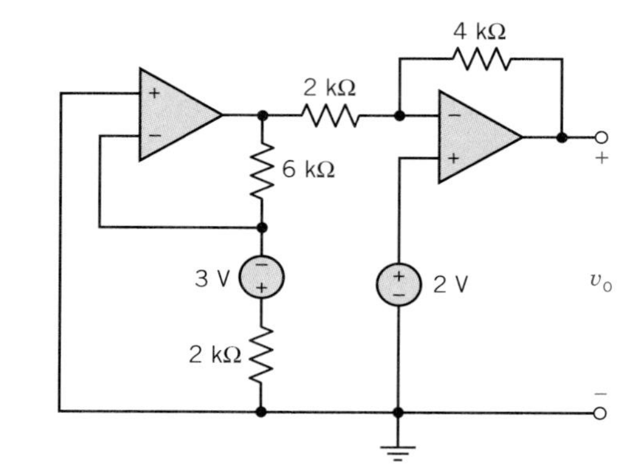

Figure VP 6.3

VP 6-4 Computer analysis of the circuit in Figure VP 6.4 indicates that the node voltages are $v_a = -5$ V, $v_b = 0$ V, $v_c = 2$ V, $v_d = 5$ V, $v_e = 2$ V, $v_f = 2$ V, and $v_g = 11$ V. Is this analysis correct? Justify your answer. Assume that the operational amplifier is ideal.
Hint: Verify that the resistor currents indicated by these node voltages satisfy KCL at nodes b, c, d, and f.

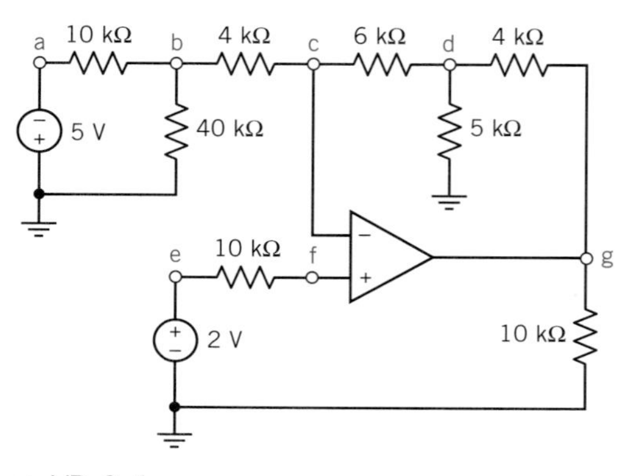

Figure VP 6.4

VP 6-5 Computer analysis of the noninverting summing amplifier shown in Figure VP 6.5 indicates that the node voltages are $v_a = 2$ V, $v_b = -0.25$ V, $v_c = -5$ V, $v_d = -2.5$ V, and $v_e = -0.25$ V.
(a) Is this analysis correct?
(b) Does this analysis verify that the circuit is a noninverting summing amplifier? Justify your answers. Assume that the operational amplifier is ideal.
1st Hint: Verify that the resistor currents indicated by these node voltages satisfy KCL at nodes b and e.
2nd Hint: Compare to Figure 6.6-1*e* to see that $R_a = 10$ kΩ and $R_b = 1$ kΩ. Determine K_1, K_2, and K_4 from the resistance values. Verify that $v_d = K_4(K_1 v_a + K_2 v_c)$.

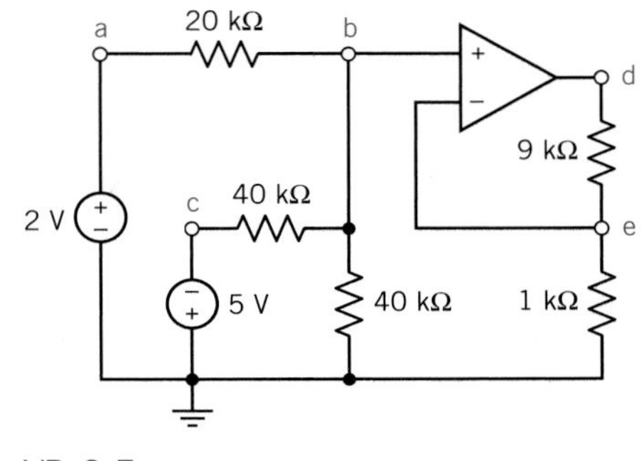

Figure VP 6.5

DESIGN PROBLEMS

DP 6-1 Design the operational amplifier circuit in Figure DP 6.1 so that

$$i_{\text{out}} = \frac{1}{4} \cdot i_{\text{in}}$$

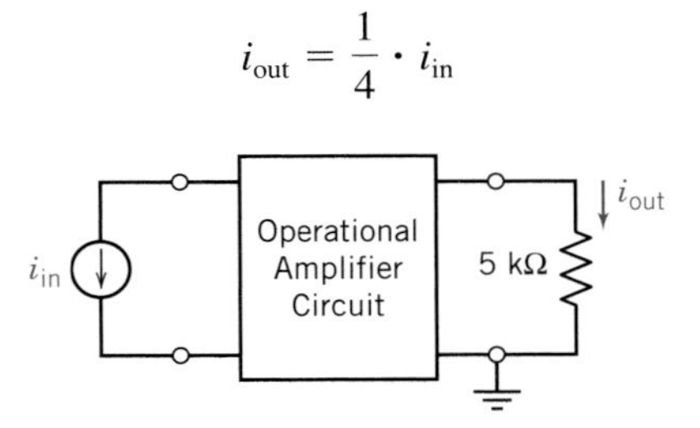

Figure DP 6.1

DP 6-2 Figure DP 6.2*a* shows a circuit that has one input, v_i, and one output, v_o. Figure DP 6.2*b* shows a graph that specifies a relationship between v_o and v_i. Design a circuit having input, v_i, and output, v_o, that have the relationship specified by the graph in Figure DP 6.2*b*.
Hint: A constant input is required. Assume that a 5 V source is available.

DP 6-3 Design a circuit having input, v_i, and output, v_o, that are related by the equations (a) $v_o = 12v_i + 6$, (b) $v_o = 12v_i - 6$, (c) $v_o = -12v_i + 6$, and (d) $v_o = -12v_i + 6$.
Hint: A constant input is required. Assume that a 5 V source is available.

DP 6-4 Design a circuit having three inputs, v_1, v_2, v_3, and two outputs, v_a, v_b, that are related by the equation

$$\begin{bmatrix} v_a \\ v_b \end{bmatrix} = \begin{bmatrix} 12 & 3 & -2 \\ 8 & -6 & 0 \end{bmatrix} \begin{bmatrix} v_1 \\ v_2 \\ v_3 \end{bmatrix} + \begin{bmatrix} 2 \\ -4 \end{bmatrix}$$

Hint: A constant input is required. Assume that a 5 V source is available.

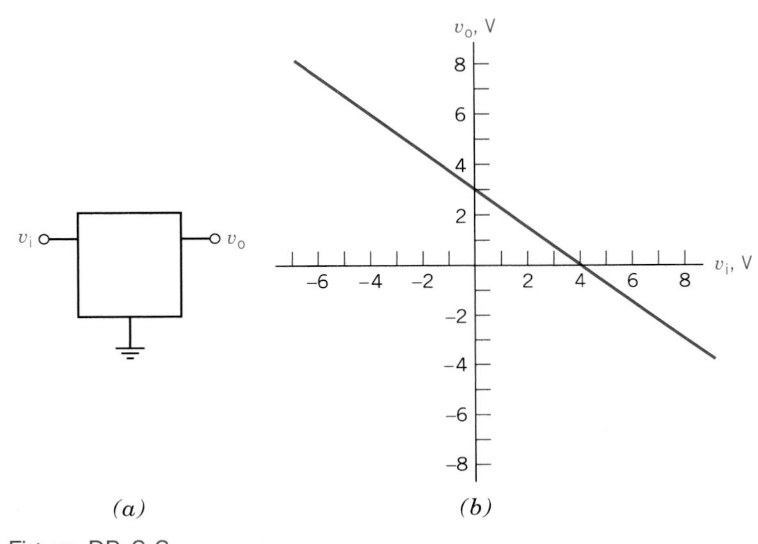

Figure DP 6.2

DP 6-5 A microphone has an unloaded voltage $v_s = 20$ mV, as shown in Figure DP 6.5*a*. An op amp is available as shown in Figure DP 6.5*b*. It is desired to provide an output voltage of 4 V. Design an inverting circuit and a noninverting circuit and contrast the input resistance at terminals x-y seen by the microphone. Which configuration would you recommend in order to achieve good performance in spite of changes in the microphone resistance R_s?

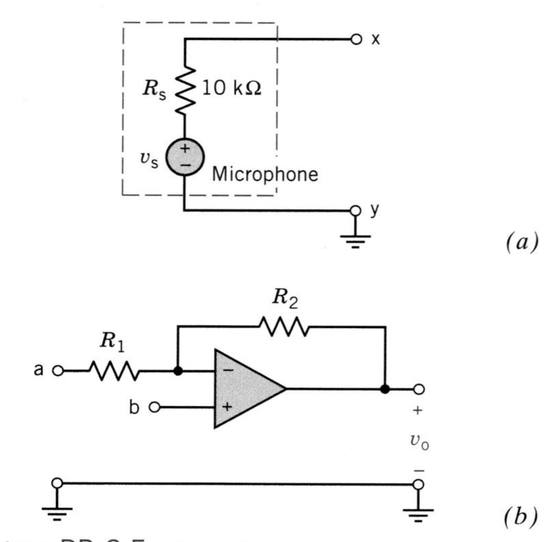

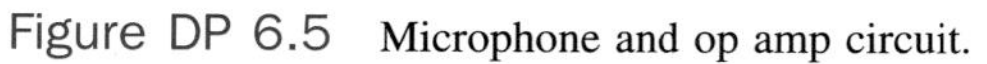

Figure DP 6.5 Microphone and op amp circuit.

CHAPTER 7

Energy Storage Elements

Preview

The energy storage in electric circuit elements is another aspect in the development of flexible and useful circuits. In this chapter we describe the characteristics of two energy storage elements: the inductor and the capacitor. These two terminal elements have been widely used by electrical engineers for over 100 years.

We may describe the storage of energy in electrical elements as analogous to the storage of information on a tablet or in a file drawer. As will be seen, the stored energy can later be retrieved and used for complex purposes. With the addition of inductors and capacitors to the familiar resistor, we will be ready to construct important and useful electric circuits. Since these circuits often contain one or more switches that open or close at a specified time, we also consider the effects of these switch changes on the circuit's behavior.

7.1 Design Challenge

INTEGRATOR AND SWITCH

This design challenge involves an integrator and a voltage-controlled switch.

An integrator is a circuit that performs the mathematical operation of integration. The output of an integrator, say $v_o(t)$, is related to the input of the integrator, say $v_s(t)$, by the equation

$$v_o(t_2) = K \cdot \int_{t_1}^{t_2} v_s(t)\, dt + v_o(t_1) \tag{7.1-1}$$

The constant K is called the gain of the integrator.

Integrators have many applications. One application of an integrator is to measure an interval of time. Suppose $v_s(t)$ is a constant voltage, V_s. Then

$$v_o(t_2) = K \cdot (t_2 - t_1) \cdot V_s + v_o(t_1) \tag{7.1-2}$$

This equation indicates that the output of the integrator at time t_2 is a measure of the time interval $t_2 - t_1$.

Switches can be controlled electronically. Figure 7.1-1 illustrates an electronically controlled SPST switch. The symbol shown in Figure 7.1-1*a* is sometimes used to emphasize that a switch is controlled electronically. The node voltage $v_c(t)$ is called the control voltage. Figure 7.1-1*b* shows a typical control voltage. This voltage-controlled switch is closed when $v_c(t) = v_H$ and open when $v_c(t) = v_L$. The switch shown in Figure 7.1-1 is open before time t_1. It closes at time t_1 and stays closed until time t_2. The switch opens at time t_2 and remains open.

Consider Figure 7.1-2. The voltage $v_c(t)$ controls the switch. The integrator converts the time interval $t_2 - t_1$ to a voltage that is displayed using the voltmeter. The time interval to be measured could be as small as 5 ms or as large as 200 ms. The challenge is to design the integrator. The available components include:

- standard 2% resistors (see Appendix E);
- 1 μF, 0.2 μF and 0.1 μF capacitors;
- operational amplifiers;
- +15-V and −15-V power supplies;
- 1 kΩ, 10 kΩ, and 100 kΩ potentiometers;
- voltage-controlled SPST switches.

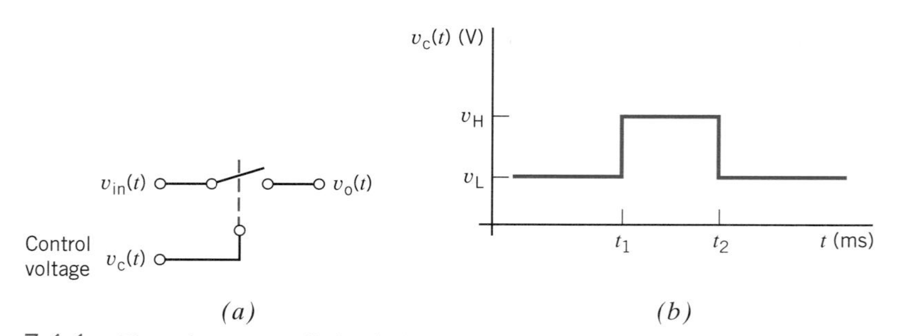

Figure 7.1-1 The voltage-controlled switch. (*a*) Switch symbol. (*b*) Typical control voltage.

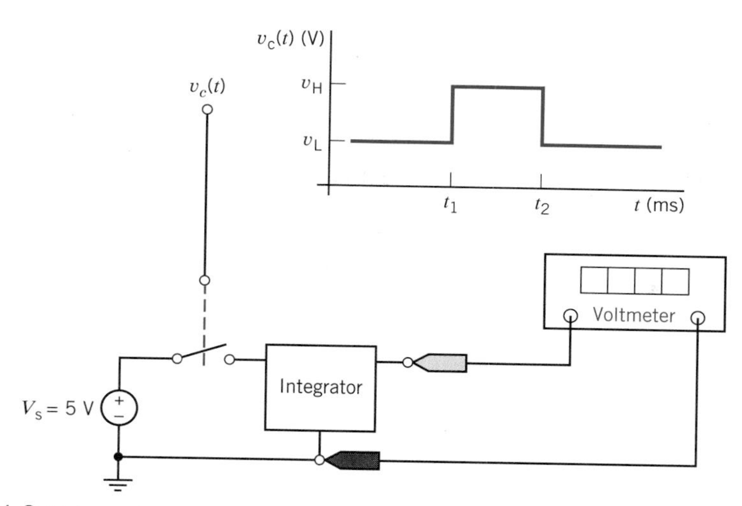

Figure 7.1-2 Using an integrator to measure an interval of time.

Describe the Situation and the Assumptions

It is convenient to set the integrator output to zero at time t_1. The relationship between the integrator output voltage and the time interval should be simple. Accordingly, let

$$v_o(t_2) = \frac{10\text{ V}}{200\text{ ms}} \cdot (t_2 - t_1) \tag{7.1-3}$$

Figure 7.1-2 indicates that $V_s = 5$ V. Comparing Eqs. 7.1-2 and 7.1-3 yields

$$K \cdot V_s = \frac{10\text{ V}}{200\text{ ms}} \quad \text{and therefore} \quad K = 10\frac{1}{s} \tag{7.1-4}$$

State the Goal

Design an integrator satisfying both

$$K = 10\frac{1}{s} \quad \text{and} \quad v_o(t_1) = 0 \tag{7.1-5}$$

An energy storage element—either a capacitor or an inductor—is needed to make an integrator. Energy storage elements are described in this chapter. We will return to this design challenge at the end of this chapter, after learning about capacitors and inductors.

7.2 ELECTRIC ENERGY STORAGE DEVICES

Storage of electric energy in devices has been pursued from the time of the Leyden jar. Part of the energy stored in these devices can later be released and supplied to a load.

In 1746 Pieter van Musschenbrock, professor of physics at Leyden, in Holland, stored a charge in a bottle containing water. This charge could then be released to provide a discharge or shock. The Leyden jar, the first manmade capacitor, had arrived, and it provided the first means of storing electric charge. It was shown that the charge stored was inversely proportional to the thickness of the glass and directly proportional to the surface area of the conductors. For a while the glass was thought to be essential, but that was proved wrong in 1762 when the first parallel-plate capacitor was made from two large boards covered with metal foil. It was the Leyden jar, used singly or grouped into banks of capacitors, that became the standard laboratory equipment.

With the development of the early capacitor, the concept of charge storage was explored further by Charles-Augustin de Coulomb and others as they developed the early theory of electricity. The study of electricity moved to quantitative description with the work of Coulomb, who succeeded in describing the ideas of electrostatics. Henry Cavendish proved the inverse-square law for electrostatics in 1772–1773 and performed detailed measurements on capacitance and conductivity. However, he did not publish his findings, and they came to full light only when Maxwell published them in 1879.

Many scientists were also interested in the theory of magnetic force. Hans Christian Oersted, a professor at the University of Copenhagen, discovered the magnetic field associated with an electric current. Oersted established the fundamental fact that a compass needle was affected only when a current was flowing nearby and not by the mere presence of a voltage or charge. He concluded that the magnetic field was circular and was dispersed in the space around the wire. Oersted, who first published his discovery in Latin in a pamphlet dated July 21, 1820, is shown in Figure 7.2-1.

The results of Oersted's discovery spread quickly, and at a public demonstration in September 1820 one member of the audience was André-Marie Ampère. Within weeks, Ampère published a paper on the mutual action of an electric current on a magnet and showed that two coils carrying currents behaved like magnets.

Michael Faraday and Sir Humphrey Davy repeated Oersted's experiments in 1821 at the British Royal Institution. Faraday's experiments with magnetism and electricity continued for decades. Faraday constructed an iron ring with two coils wound on opposite sides. On August 29, 1831, he connected one coil to a battery and the other coil to a galvanometer. He noted the transitory nature of the current induced in the second coil, which occurred only when the current in the first coil was started or stopped by connecting or disconnecting the battery. Faraday read an account of his work to the Royal Society on November 24, 1831, and published it in 1832. Faraday is shown in Figure 7.2-2. To honor Faraday, the unit of capacitance is named the farad.

Figure 7.2-1 Hans C. Oersted (1777–1851), the first person to observe the magnetic effects of an electric current. Courtesy of Burndy Library.

Figure 7.2-2 Michael Faraday's electrical discoveries were not his only legacy; his published account of them inspired much of the scientific work of the later nineteenth century. His *Experimental Researches in Electricity* remains one of the greatest accounts of scientific work ever written. Courtesy of Burndy Library.

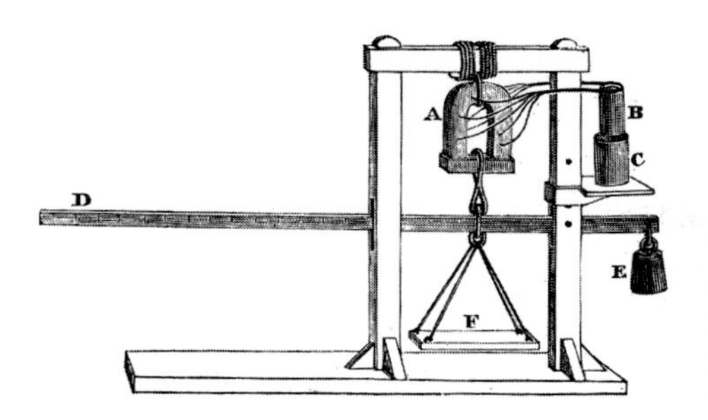

Figure 7.2-3
Joseph Henry's electromagnet. Direct current from the voltaic pile (B–C) was applied to a coil wound around an iron horseshoe core (A) to produce a powerful electromagnet. From Joseph Henry, *Galvanic Multiplier,* 1831. Courtesy of Burndy Library.

During the same period, Joseph Henry, an American, was exploring the concepts of electromagnetism. During the years 1827 to 1831, Henry studied the effects of magnetism with electromagnets such as the one shown in Figure 7.2-3. Although Faraday deserves the credit for the discovery of electromagnetic induction between two coils, it was Henry who first discovered self-induction with a single coil. Henry noted the principle of self-induction when a vivid spark was produced as a long coil of wire was disconnected from a battery. Henry was honored by the use of his name for the unit of self-induction. In 1847 he became the first director of the Smithsonian Institution, where he served for 32 years.

7.3 CAPACITORS

A capacitor is a two-terminal element that is a model of a device consisting of two conducting plates separated by a nonconducting material. Electric charge is stored on the plates, as shown in Figure 7.3-1, and the area between the plates is filled with a dielectric material. The capacitance value is proportional to the dielectric constant and surface area of the dielectric and is inversely proportional to its thickness. To obtain greater capacitance, a very thin structure of large area is required. For this configuration we may state the capacitance C as

$$C = \frac{\epsilon A}{d}$$

where ϵ is the dielectric constant, A the area of plates, and d the space between plates. Dielectric constants of some materials are given in Table 7.3-1. (The dielectric constant is

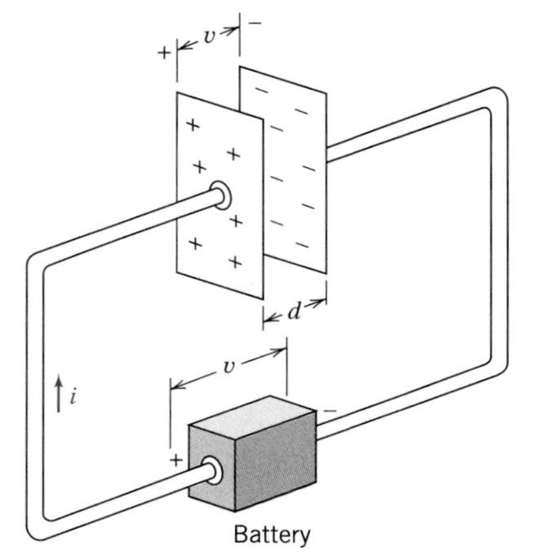

Figure 7.3-1
Capacitor connected to a battery.

Table 7.3-1 Relative Dielectric Constant, ϵ_r

Material	$\epsilon_r = \epsilon/\epsilon_0$
Glass	7
Nylon	2
Bakelite	5

ϵ_0 = permittivity of free space = 8.85×10^{-12} farad/meter. Dielectric constant is also called relative permittivity.

the property that determines the energy stored per unit volume for unit voltage difference across a capacitor.) Another common term for dielectric constant is permittivity (Dorf 1998).

We describe the charge $+q$ on one plate as identical to $-q$ on the other plate. The energy to move the charge q to the positive plate from the other plate is provided by the battery voltage v. The capacitor has been charged to the voltage v, which will be proportional to the charge q. Thus, we write

$$q = Cv \tag{7.3-1}$$

where C is the constant of proportionality, which is called the capacitance. The unit of capacitance is coulomb per volt and is called farad (F) in honor of Faraday. A capacitor is a linear element as long as it retains the relationship represented by Eq. 7.3-1.

Capacitance is a measure of the ability of a device to store energy in the form of separated charge or an electric field.

A capacitor can be used to model charge storage in any device. Most capacitors used in electronic equipment are linear. However, some are nonlinear. The charge associated with some reverse-biased semiconductor diodes, for example, varies as the 2/3 power of voltage because the distance across the junction, *d*, is a function of voltage. Unless otherwise stated, we will assume for our purposes that capacitors are linear.

When we first connect a battery to the capacitor shown in Figure 7.3-1, a current flows while the charges flow from one plate to the other. Since we use the conventional current i of positive charge flow, we can represent i as shown in Figure 7.3-1. Since the current is

$$i = \frac{dq}{dt}$$

we differentiate Eq. 7.3-1 to obtain

$$i = C\frac{dv}{dt} \tag{7.3-2}$$

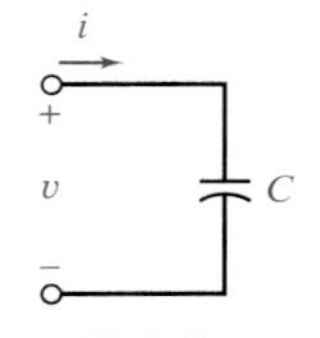

Figure 7.3-2 Circuit symbol of a capacitor.

Equation 7.3-2 is the current–voltage relationship for the model of a capacitor and can readily be shown to be a linear relation. Use Eq. 7.3-2 and the properties of superposition and homogeneity as described in Section 2.3 to show the capacitor relation is that of a linear element.

As the current flows toward the left-hand plate, it causes that plate (or its terminal) to have a positive voltage relative to the right-hand plate. The circuit symbol for a capacitor is shown in Figure 7.3-2. Again, the passive sign convention assumes that the current flows into the positive terminal of a capacitor as shown in Figure 7.3-2.

In summary, we may define a *capacitor* as a two-terminal element whose primary purpose is to introduce capacitance into an electric circuit. Capacitance is defined as the ratio of the charge stored to the voltage difference between the two conducting plates or wires, $C = q/v$.

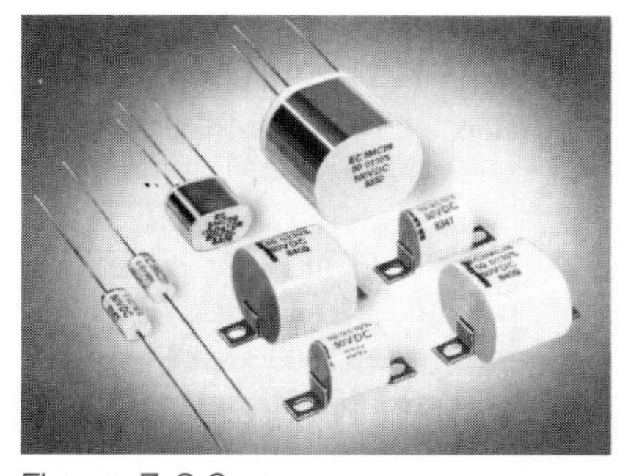

Figure 7.3-3 Miniature metal film capacitors ranging from 1 mF to 50 mF. Courtesy of Electronic Concepts Inc.

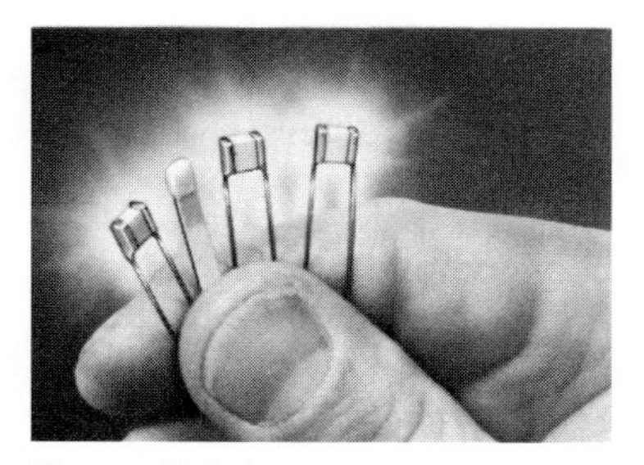

Figure 7.3-4 Miniature hermetically sealed polycarbonate capacitors ranging from 1 μF to 50 μF. Courtesy of Electronic Concepts Inc.

Capacitors use various dielectrics and are built in several forms (Trotter 1988). Some common capacitors use impregnated paper for a dielectric, whereas others use mica sheets, ceramics, and organic and metal films. Miniature metal film capacitors are shown in Figure 7.3-3. Miniature hermetically sealed polycarbonate capacitors are shown in Figure 7.3-4.

Capacitors come in a wide range of values. Two pieces of insulated wire about an inch long when twisted together will have a capacitance of about 1 pF. On the other hand, a power supply capacitor about an inch in diameter and a few inches long may have a capacitance of 0.01 F.

Reflecting on Eq. 7.3-2, we note that a circuit will have a current, i, that depends on the derivative of the voltage across the capacitor, v. If the voltage is constant, then $i = 0$. If the voltage in Figure 7.3-2 is

$$v = Kt$$

then, since K is a constant

$$i = C\frac{dv}{dt} = CK$$

Let us find the current through a capacitor when $v = 5 \sin t$ V in Figure 7.3-1:

$$i = C\frac{dv}{dt} = 5C \cos t \text{ A}$$

Example 7.3-1

Find the current for a capacitor $C = 1$ mF when the voltage across the capacitor is represented by the signal shown in Figure 7.3-5.

Solution

The voltage (with units of volts) is given by

$$\begin{aligned} v &= 0 && t \le 0 \\ &= 10t && 0 < t \le 1 \\ &= 20 - 10t && 1 < t \le 2 \\ &= 0 && t > 2 \end{aligned}$$

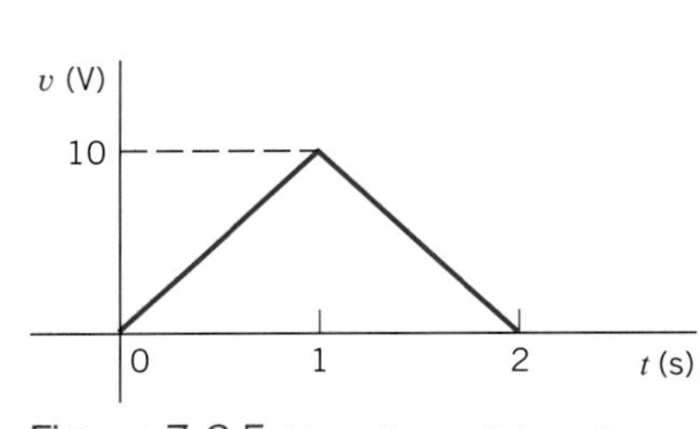

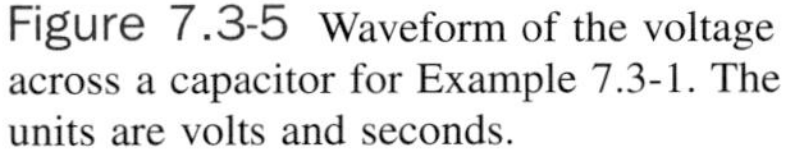
Figure 7.3-5 Waveform of the voltage across a capacitor for Example 7.3-1. The units are volts and seconds.

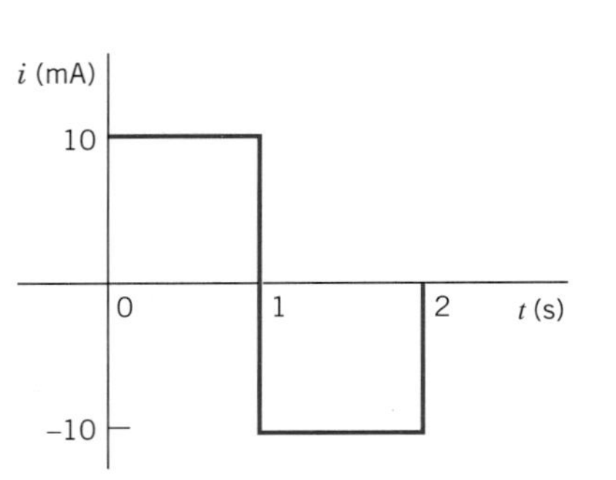

Figure 7.3-6 Current for Example 7.3-1.

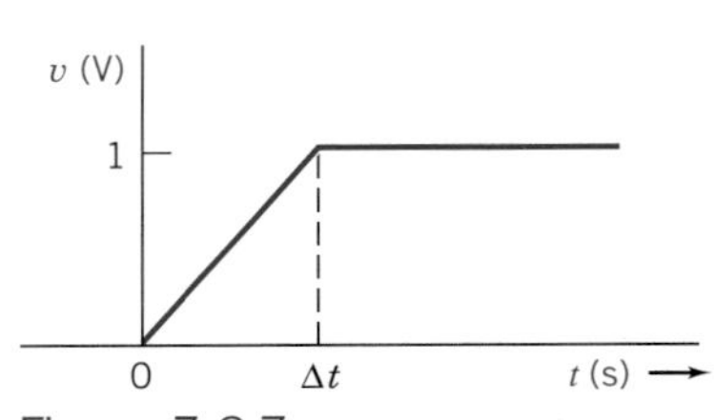

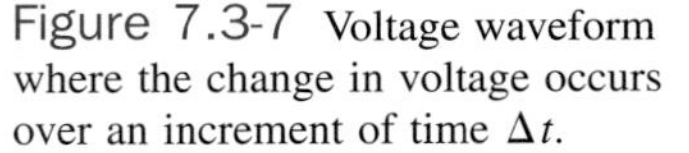
Figure 7.3-7 Voltage waveform where the change in voltage occurs over an increment of time Δt.

Then, since $i = C\,dv/dt$, where $C = 10^{-3}$ F, we obtain

$$
\begin{aligned}
i &= 0 & t &< 0 \\
&= 10^{-2} & 0 < t &< 1 \\
&= -10^{-2} & 1 < t &< 2 \\
&= 0 & t &> 2
\end{aligned}
$$

Therefore, the resulting current is a series of two pulses of magnitudes 10^{-2} A and -10^{-2} A, respectively, as shown in Figure 7.3-6.

Now consider the waveform shown in Figure 7.3-7, where the voltage changes from a constant voltage of zero to another constant voltage of 1 over an increment of time Δt. Since $i = C\,dv/dt$, we obtain

$$
\begin{aligned}
i &= 0 & t &< 0 \\
&= C(1/\Delta t) & 0 < t &< \Delta t \\
&= 0 & t &> \Delta t
\end{aligned}
$$

Thus, we obtain a pulse of height equal to $C/\Delta t$. As Δt decreases, the current will increase. Clearly, Δt cannot decline to zero or we would experience an infinite current. An infinite current is an impossibility, since it would require infinite power and an instantaneous movement of charge to occur at the capacitor terminals. According to the requirements of conservation of charge, the amount of the charge cannot change instantaneously. Thus, an instantaneous ($\Delta t = 0$) change of voltage across the capacitor is not possible.

The principle of conservation of charge states that the amount of electric charge cannot change instantaneously. Thus, $q(t)$ must be continuous over time. Recall that $q(t) = Cv(t)$. Thus, the voltage across the capacitor cannot change instantaneously; that is, we cannot have a discontinuity in $v(t)$.

The voltage across a **capacitor** cannot change instantaneously.

Now let us find the voltage $v(t)$ in terms of the current $i(t)$ by integrating both sides of Eq. 7.3-2. We obtain

$$v = \frac{1}{C}\int_{-\infty}^{t} i\,d\tau \tag{7.3-3}$$

This equation says that the capacitor voltage $v(t)$ can be found by integrating the capacitor current from time $\tau = -\infty$ until time $\tau = t$. To do so requires that we know the value of the capacitor current from time $\tau = -\infty$ until time $\tau = t$. Often, we don't know the value of the current all the way back to $\tau = -\infty$. Instead, we break the integral up into two parts

$$v = \frac{1}{C}\int_{t_0}^{t} i\,d\tau + \frac{1}{C}\int_{-\infty}^{t_0} i\,d\tau = \frac{1}{C}\int_{t_0}^{t} i\,d\tau + v(t_0) \qquad (7.3\text{-}4)$$

This equation says that the capacitor voltage $v(t)$ can be found by integrating the capacitor current from some convenient time $\tau = t_0$ until time $\tau = t$, provided that we also know the capacitor voltage at time t_0. Now we are only required to know the capacitor current from time $\tau = t_0$ until time $\tau = t$. The time t_0 is called the **initial time,** and the capacitor voltage $v(t_0)$ is called the **initial condition.** Frequently, it is convenient to select $t_0 = 0$ as the initial time.

Example 7.3-2

Find the voltage v for a capacitor $C = 1/2$ F when the current is as shown in Figure 7.3-8 and $v(t_0) = v(0) = 0$.

Solution

First, we write the equation for $i(t)$ as

$$\begin{aligned} i &= 0 & t &\le 0 \\ &= t & 0 &< t \le 1 \\ &= 1 & 1 &< t \le 2 \\ &= 0 & 2 &< t \end{aligned}$$

Then since

$$v = \frac{1}{C}\int_0^t i\,d\tau$$

and $C = 1/2$, we have

$$\begin{aligned} v &= 0 & t &\le 0 \\ &= 2\int_0^t \tau\,d\tau & 0 &< t \le 1 \\ &= 2\int_1^t (1)d\tau + v(1) & 1 &< t \le 2 \\ &= v(2) & 2 &< t \end{aligned}$$

with units of volts. Therefore, for $0 < t \le 1$, we have

$$v = t^2$$

For the period $1 < t \le 2$, we note that $v(1) = 1$ and therefore we have

$$v = 2(t - 1) + 1 = (2t - 1)\text{ V}$$

The resulting voltage waveform is shown in Figure 7.3-9. The voltage changes as t^2 during the first 1 s, changes linearly with t during the period from 1 to 2 s, and stays constant equal to 3 V after $t = 2$ s.

Actual capacitors have some resistance associated with them. Fortunately, it is easy to include approximate resistive effects in the circuit models. In capacitors the dielectric material between the plates is not a perfect insulator and has some small conductivity. This can be represented by a very high resistance in parallel with the capacitor. Ordinary ca-

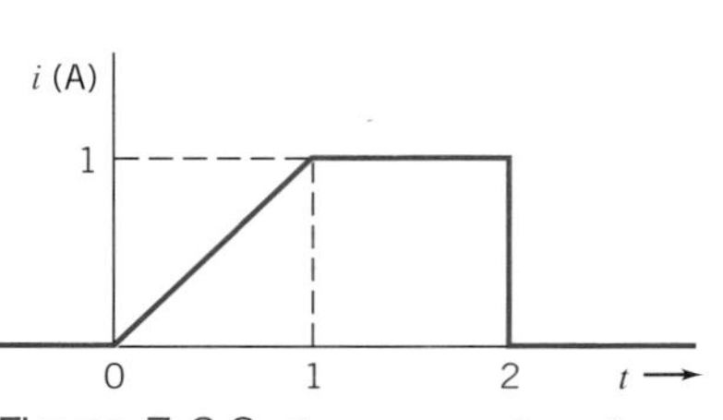

Figure 7.3-8 Current waveform for Example 7.3-2. The units are in amperes and seconds.

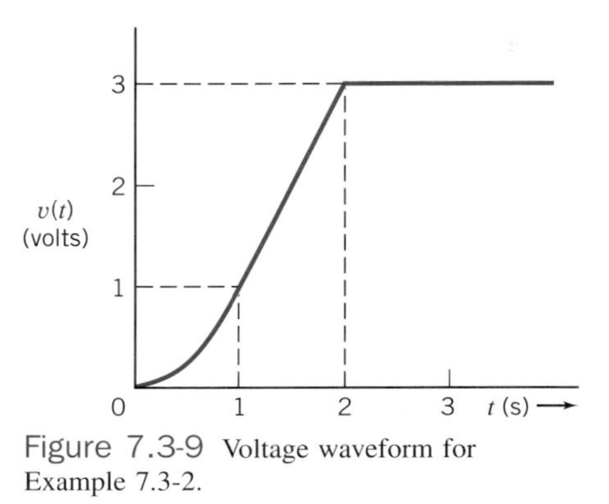

Figure 7.3-9 Voltage waveform for Example 7.3-2.

pacitors can hold a charge for hours, and the parallel resistance is then hundreds of megaohms. For this reason, the resistance associated with a capacitor is usually ignored. It is important to note that the voltage waveform cannot change instantaneously, but the current waveform may do so.

Exercise 7.3-1 Determine the current $i(t)$ for $t>0$ for the circuit of Figure E 7.3-1*b* when $v_s(t)$ is the voltage shown in Figure E 7.3-1*a*.

Hint: Determine $i_C(t)$ and $i_R(t)$ separately, then use KCL.

Answer: $$i(t) = \begin{cases} 2t-2 & 2<t<4 \\ 7-t & 4<t<8 \\ 0 & \text{otherwise} \end{cases}$$

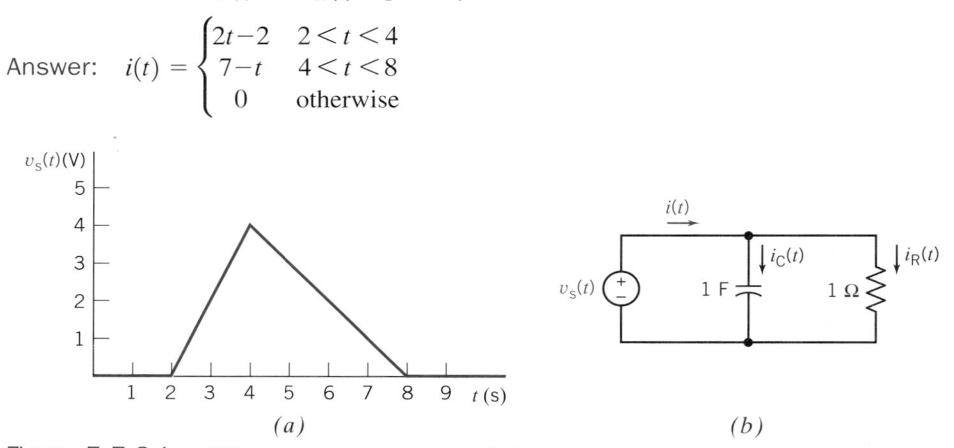

Figure E 7.3-1 (*a*) The voltage source voltage. (*b*) The circuit.

Exercise 7.3-2 Determine the voltage $v(t)$ for $t>0$ for the circuit of Figure E 7.3-2*b* when $i_s(t)$ is the current shown in Figure E 7.3-2*a*. The capacitor voltage at time $t=0$ is $v(0) = -12$ V.

Answer: $$v(t) = \begin{cases} 3\displaystyle\int_0^t 4d\tau - 12 = 12t - 12 & \text{for} \quad 0 < t < 4 \\ 3\displaystyle\int_4^t (-2)d\tau + 36 = 60 - 6t & \text{for} \quad 4 < t < 10 \\ 3\displaystyle\int_{10}^t 0d\tau + 0 = 0 & \text{for} \quad 10 < t \end{cases}$$

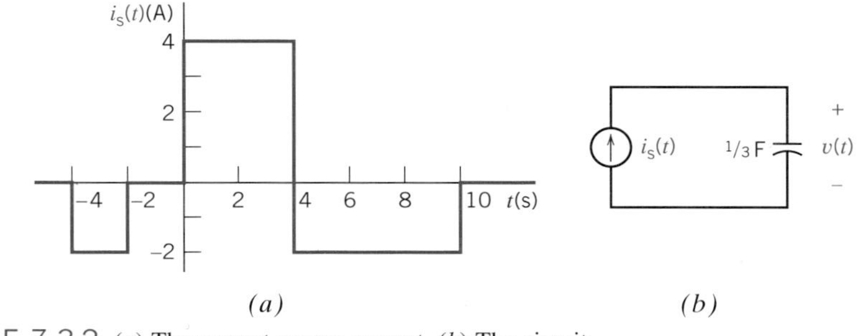

Figure E 7.3-2 (*a*) The current source current. (*b*) The circuit.

7.4 ENERGY STORAGE IN A CAPACITOR

Consider a capacitor that has been connected to a battery of voltage v. A current flows and a charge is stored on the plates of the capacitor, as shown in Figure 7.3-1. Eventually, the voltage across the capacitor is a constant, and the current through the capacitor is zero. The capacitor has stored energy by virtue of the separation of charges between the capacitor plates. These charges have an electrical force acting on them.

The forces acting on the charges stored in a capacitor are said to result from an electric field. An *electric field* is defined as the force acting on a unit positive charge in a specified region. Since the charges have a force acting on them along a direction x, we recognize that the energy required originally to separate the charges is now stored by the capacitor in the electric field.

The energy stored in a capacitor is

$$w_c(t) = \int_{-\infty}^{t} vi\, d\tau$$

Remember that v and i are both functions of time and could be written as $v(t)$ and $i(t)$. Since

$$i = C\frac{dv}{dt}$$

we have

$$\begin{aligned} w_c &= \int_{-\infty}^{t} vC\frac{dv}{d\tau}\, d\tau = C\int_{v(-\infty)}^{v(t)} v\, dv \\ &= \frac{1}{2}Cv^2\Big|_{v(-\infty)}^{v(t)} \end{aligned}$$

Since the capacitor was uncharged at $t = -\infty$, set $v(-\infty) = 0$. Therefore,

$$w_c(t) = \frac{1}{2}C\,v^2(t)\ \text{J} \tag{7.4-1}$$

Therefore, as a capacitor is being charged and $v(t)$ is changing, the energy stored, w_c, is changing. Note that $w_c(t) \geq 0$ for all $v(t)$, so the element is said to be passive (see Section 2–4).

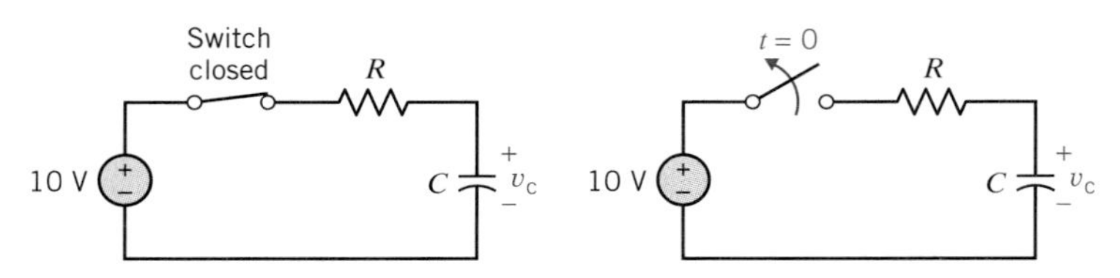

Figure 7.4-1 A circuit (*a*) where the capacitor is charged and $v_c = 10$ V and (*b*) the switch is opened at $t = 0$.

Since $q = Cv$, we may rewrite Eq. 7.4-1 as

$$w_c = \frac{1}{2C} q^2(t) \text{ J} \tag{7.4-2}$$

The capacitor is a storage element that stores but does not dissipate energy. For example, consider a 100-mF capacitor that has a voltage of 100 V across it. The energy stored is

$$w_c = \frac{1}{2} Cv^2 = \frac{1}{2}(0.1)(100)^2 = 500 \text{ J}$$

As long as the capacitor remains unconnected to any other element, the energy of 500 J remains stored. Now if we connect the capacitor to the terminals of a resistor, we expect a current to flow until all the energy is dissipated as heat by the resistor. After all the energy dissipates, the current is zero and the voltage across the capacitor is zero.

As noted in the previous section, the requirement of conservation of charge implies that the voltage on a capacitor is continuous. Thus, the *voltage and charge on a capacitor cannot change instantaneously.* This statement is summarized by the equation

$$v(0^+) = v(0^-)$$

where the time just prior to $t = 0$ is called $t = 0^-$ and the time immediately after $t = 0$ is called $t = 0^+$. The time between $t = 0^-$ and $t = 0^+$ is infinitely small. Nevertheless, the voltage will not change abruptly.

To illustrate the continuity of voltage for a capacitor, consider the circuit shown in Figure 7.4-1. For the circuit shown in Figure 7.4-1*a* the switch has been closed for a long time and the capacitor voltage has become $v_c = 10$ V. At time $t = 0$ we open the switch as shown in Figure 7.4-1*b*. Since the voltage on the capacitor is continuous,

$$v_c(0^+) = v_c(0^-) = 10 \text{ V}$$

Example 7.4-1

A 10-mF capacitor is charged to 100 V, as shown in the circuit of Figure 7.4-2. Find the energy stored by the capacitor and the voltage of the capacitor at $t = 0^+$ after the switch is opened.

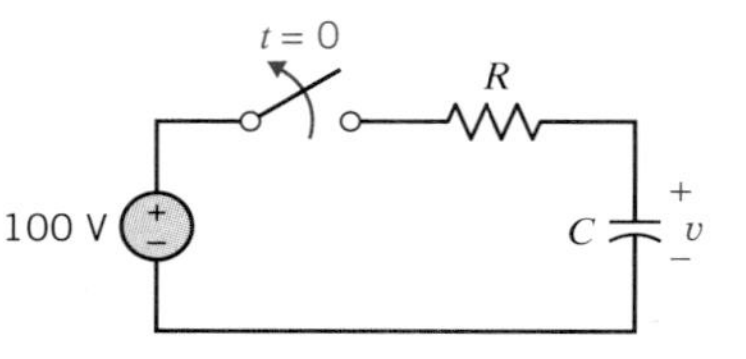

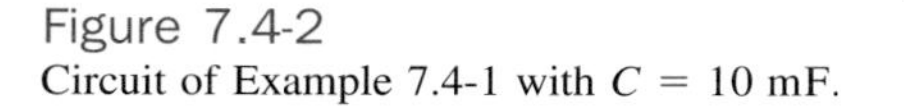
Figure 7.4-2
Circuit of Example 7.4-1 with $C = 10$ mF.

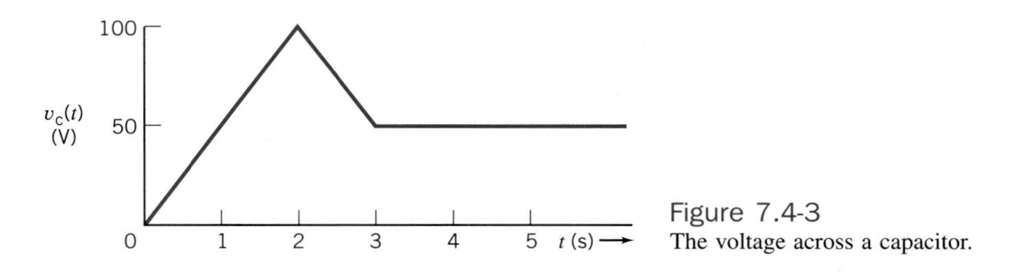

Figure 7.4-3
The voltage across a capacitor.

Solution

The voltage of the capacitor is $v = 100$ at $t = 0^-$. Since the voltage at $t = 0^+$ cannot change from the voltage at $t = 0^-$, we have

$$v(0^+) = v(0^-) = 100 \text{ V}$$

The energy stored by the capacitor at $t = 0^+$ is

$$w_c = \frac{1}{2} C v^2 = \frac{1}{2} (10^{-2})(100)^2 = 50 \text{ J}$$

Example 7.4-2

The voltage across a 5-mF capacitor varies as shown in Figure 7.4-3. Determine and plot the capacitor current, power, and energy.

Solution

The current is determined from $i_c = C\, dv/dt$ and is shown in Figure 7.4-4*a*. The power is $v(t)i(t)$—the product of the current curve (Figure 7.4-4*a*) and the voltage curve (Figure 7.4-3)—and is shown in Figure 7.4-4*b*. The capacitor receives energy during the first two seconds and then delivers energy for the period $2 < t < 3$.

The energy is $\omega = \int p\, dt$ and can be found as the area under the $p(t)$ curve. The curve for the energy is shown in Figure 7.4-4*c*. Note that the capacitor increasingly stores energy from $t = 0$ s to $t = 2$ s, reaching a maximum energy of 25 J. Then the capacitor delivers a total energy of 18.75 J to the external circuit from $t = 2$ s to $t = 3$ s. Finally, the capacitor holds a constant energy of 6.25 J after $t = 3$ s.

Exercise 7.4-1 A 200-μF capacitor has been charged to 100 V. Find the energy stored by the capacitor. Find the capacitor voltage at $t = 0^+$ if $v(0^-) = 100$ V.

Answer: $w = 1$ J; $v(0^+) = 100$ V

Exercise 7.4-2 A constant current $i = 2$ A flows into a capacitor of 100 μF after a switch is closed at $t = 0$. The voltage of the capacitor was equal to zero at $t = 0^-$. Find the energy stored at (a) $t = 1$ s and (b) $t = 100$ s.

Answer: $w(1) = 20$ kJ; $w(100) = 200$ MJ

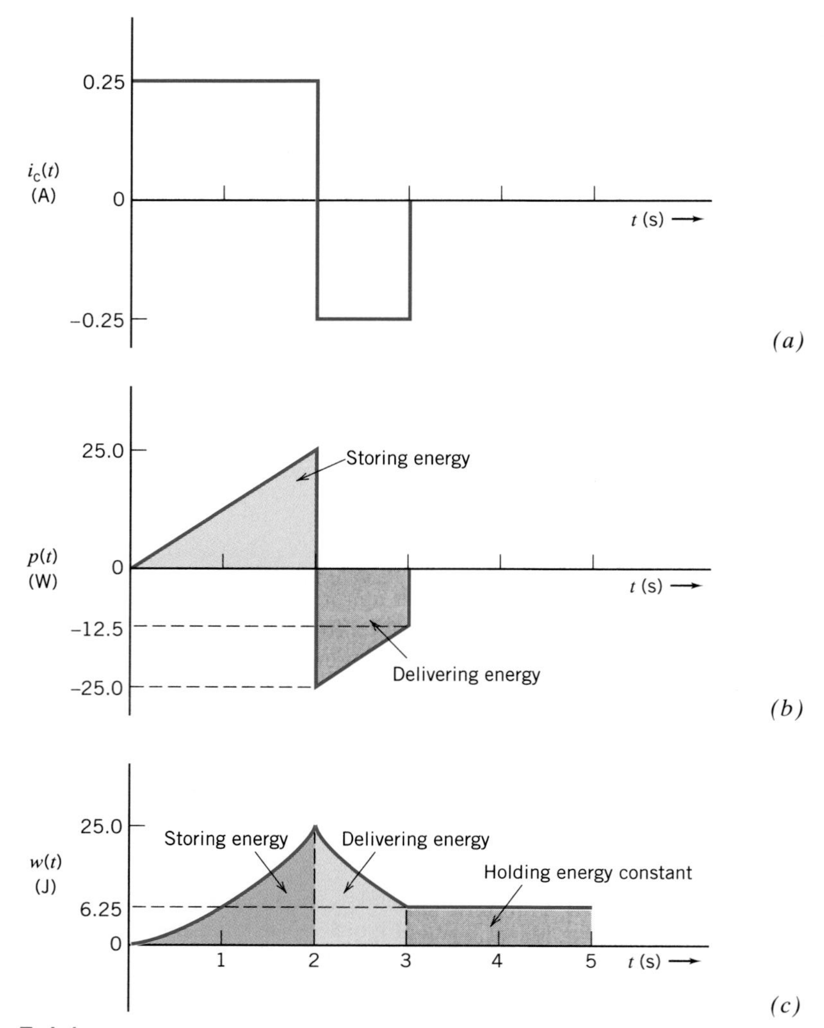

Figure 7.4-4 The current, power, and energy of the capacitor of Example 7.4-2.

Exercise 7.4-3 The initial capacitor voltage of the circuit shown in Figure E 7.4-3 is $v_c(0^-) = 3$ V. Determine (a) the voltage $v(t)$ and (b) the energy stored in the capacitor at $t = 0.2$ s and $t = 0.8$ s when

$$i(t) = \begin{cases} 3e^{5t}\ \text{A} & 0 < t < 1 \\ 0 & t \geq 1\ \text{s} \end{cases}$$

Answers: (a) $18e^{5t}$ V, $0 \leq t < 1$
(b) $w(0.2) = 6.65$ J, $w(0.8) = 2.68$ kJ

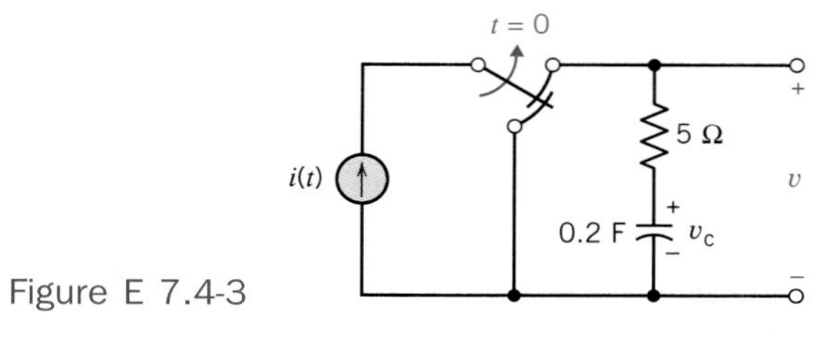

Figure E 7.4-3

7.5 SERIES AND PARALLEL CAPACITORS

First, let us consider the parallel connection of N capacitors as shown in Figure 7.5-1. We wish to determine the equivalent circuit for the N parallel capacitors as shown in Figure 7.5-2.

Using KCL, we have

$$i = i_1 + i_2 + i_3 + \cdots + i_N$$

Since

$$i_n = C_n \frac{dv}{dt}$$

and v appears across each capacitor, we obtain

$$\begin{aligned} i &= C_1 \frac{dv}{dt} + C_2 \frac{dv}{dt} + C_3 \frac{dv}{dt} + \cdots + C_N \frac{dv}{dt} \\ &= (C_1 + C_2 + C_3 + \cdots + C_N) \frac{dv}{dt} \\ &= \left(\sum_{n=1}^{N} C_n \right) \frac{dv}{dt} \end{aligned} \qquad (7.5\text{-}1)$$

For the equivalent circuit shown in Figure 7.5-2,

$$i = C_p \frac{dv}{dt} \qquad (7.5\text{-}2)$$

Comparing Eqs. 7.5-1 and 7.5-2, it is clear that

$$C_p = C_1 + C_2 + C_3 + \cdots + C_N = \sum_{n=1}^{N} C_n$$

Thus, the equivalent capacitance of a set of N parallel capacitors is simply the sum of the individual capacitances. It must be noted that all the parallel capacitors will have the same initial condition, $v(0)$. Note the similarity with the results we found in Section 3.5 for parallel conductances, where $G_P = \Sigma\, G_n$.

Now let us determine the equivalent capacitance C_S of a set of N series-connected capacitances, as shown in Figure 7.5-3. The equivalent circuit for the series of capacitors is shown in Figure 7.5-4.

Using KVL for the loop of Figure 7.5-3, we have

$$v = v_1 + v_2 + v_3 + \cdots + v_N \qquad (7.5\text{-}3)$$

Since, in general,

$$v_n = \frac{1}{C_n} \int_{t_0}^{t} i \, d\tau + v_n(t_0)$$

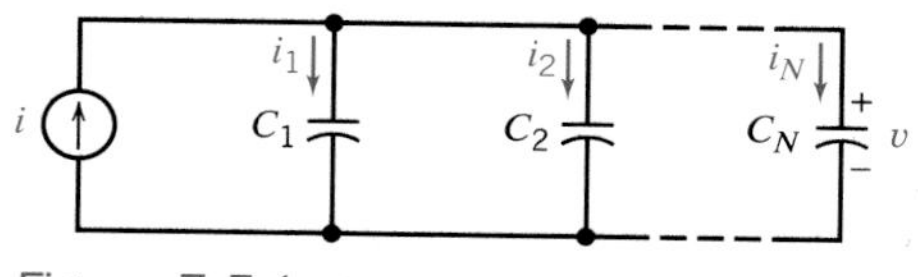

Figure 7.5-1 Parallel connection of N capacitors.

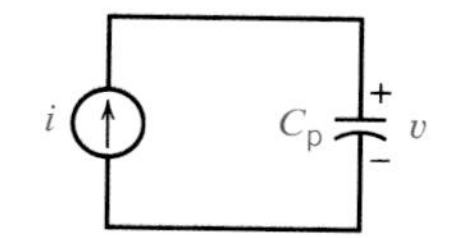

Figure 7.5-2 Equivalent circuit for N parallel capacitors.

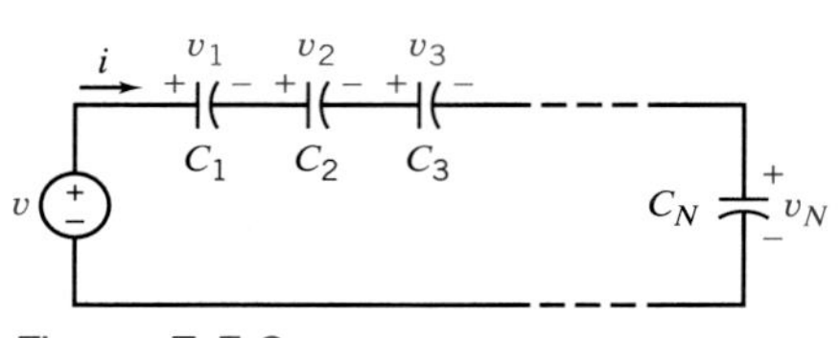

Figure 7.5-3 Series connection of N capacitors.

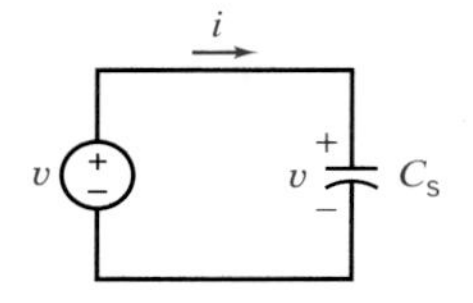

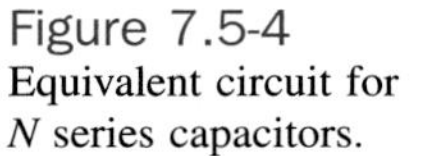
Figure 7.5-4 Equivalent circuit for N series capacitors.

where i is common to all capacitors, we obtain

$$v = \frac{1}{C_1}\int_{t_0}^{t} i\,d\tau + v_1(t_0) + \cdots + \frac{1}{C_N}\int_{t_0}^{t} i\,d\tau + v_N(t_0)$$

$$= \left(\frac{1}{C_1} + \frac{1}{C_2} + \cdots + \frac{1}{C_N}\right)\int_{t_0}^{t} i\,d\tau + \sum_{n=1}^{N} v_n(t_0)$$

$$= \sum_{n=1}^{N}\frac{1}{C_n}\int_{t_0}^{t} i\,d\tau + \sum_{n=1}^{N} v_n(t_0) \tag{7.5-4}$$

From Eq. 7.5-3 we note that at $t = t_0$

$$v(t_0) = v_1(t_0) + v_2(t_0) + \cdots + v_N(t_0) = \sum_{n=1}^{N} v_n(t_0) \tag{7.5-5}$$

Substituting Eq. 7.5-5 into Eq. 7.5-4, we obtain

$$v = \left(\sum_{n=1}^{N}\frac{1}{C_n}\right)\int_{t_0}^{t} i\,d\tau + v(t_0) \tag{7.5-6}$$

Using KVL for the loop of the equivalent circuit of Figure 7.5-4 yields

$$v = \frac{1}{C_s}\int_{t_0}^{t} i\,d\tau + v(t_0) \tag{7.5-7}$$

Comparing Eqs. 7.5-6 and 7.5-7, we find that

$$\frac{1}{C_s} = \sum_{n=1}^{N}\frac{1}{C_n} \tag{7.5-8}$$

For the case of two series capacitors, Eq. 7.5-8 becomes

$$\frac{1}{C_s} = \frac{1}{C_1} + \frac{1}{C_2}$$

or

$$C_s = \frac{C_1 C_2}{C_1 + C_2} \tag{7.5-9}$$

Example 7.5-1

Find the equivalent capacitance for the circuit of Figure 7.5-5 when $C_1 = C_2 = C_3 =$ 2 mF, $v_1(0) = 10$ V, and $v_2(0) = v_3(0) = 20$ V.

Solution

Since C_2 and C_3 are in parallel, we replace them with C_p, where

$$C_p = C_2 + C_3 = 4 \text{ mF}$$

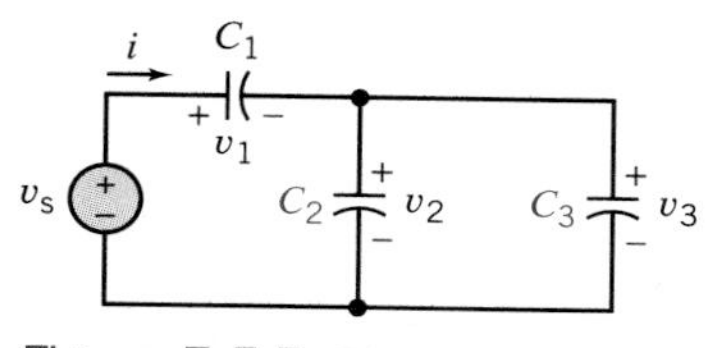

Figure 7.5-5 Circuit for Example 7.5-1.

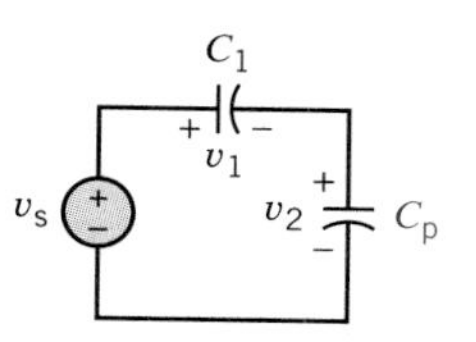

Figure 7.5-6 Circuit resulting from Figure 7.5-5 by replacing C_2 and C_3 with C_p.

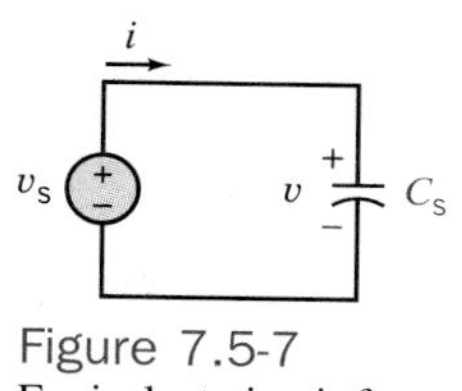

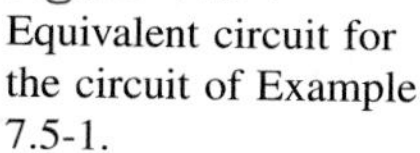

Figure 7.5-7 Equivalent circuit for the circuit of Example 7.5-1.

The voltage at $t = 0$ across the equivalent capacitance C_p is equal to the voltage across C_2 or C_3, which is $v_2(0) = v_3(0) = 20$ V. As a result of replacing C_2 and C_3 with C_p, we obtain the circuit shown in Figure 7.5-6.

We now want to replace the series of two capacitors C_1 and C_p with one equivalent capacitor. Using the relationship of Eq. 7.5-9, we obtain

$$C_s = \frac{C_1 C_p}{C_1 + C_p} = \frac{(2 \times 10^{-3})(4 \times 10^{-3})}{(2 \times 10^{-3}) + (4 \times 10^{-3})} = \frac{8}{6} \text{ mF}$$

The voltage at $t = 0$ across C_s is

$$v(0) = v_1(0) + v_p(0)$$

where $v_p(0) = 20$ V, the voltage across the capacitance C_p at $t = 0$. Therefore, we obtain

$$v(0) = 10 + 20 = 30 \text{ V}$$

Thus, we obtain the equivalent circuit shown in Figure 7.5-7.

Exercise 7.5-1 Find the equivalent capacitance for the circuit of Figure E 7.5-1.

Answer: $C_{eq} = 4$ mF

Exercise 7.5-2 Find the relationship for the division of current between two parallel capacitors as shown in Figure E 7.5-2.

Answer: $i_n = iC_n/(C_1 + C_2),\ n = 1, 2$

Exercise 7.5-3 Determine the equivalent capacitance C_{eq} for the circuit shown in Figure E 7.5-3.

Answer: 10/19 mF

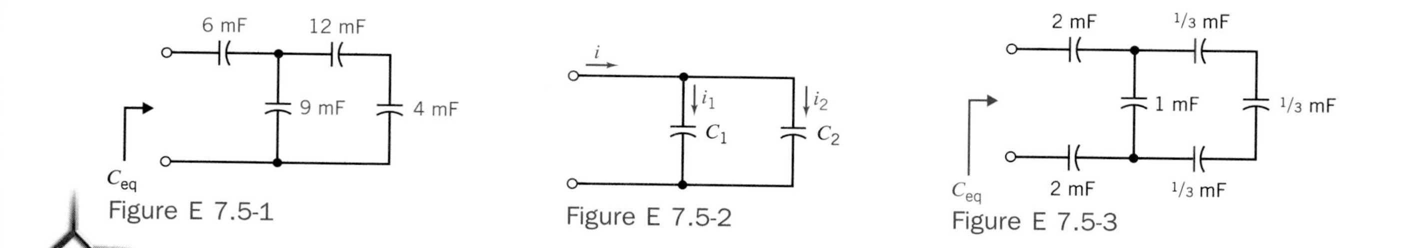

Figure E 7.5-1 Figure E 7.5-2 Figure E 7.5-3

7.6 INDUCTORS

A wire may be shaped as a multiturn coil, as shown in Figure 7.6-1. If we use a current source i_s, we find that the voltage across the coil is proportional to the rate of change of the current $i = i_s$, the current through the inductor. This proportional relationship may be expressed by the equation

$$v = L\frac{di}{dt} \tag{7.6-1}$$

where L is the constant of proportionality called *inductance* and is measured in henrys (H).

We define an *inductor* as a two-terminal element consisting of a winding of N turns for introducing inductance into an electric circuit. Inductance is defined as the property of an electric device by which a time-varying current through the device produces a voltage across it.

An ideal inductor is a coil wound with resistanceless wire. When current exists in the wire, energy is stored in the magnetic field around the coil. A constant current i in the coil results in zero voltage across the coil. A current that varies with time produces a self-induced voltage. Note that an abrupt (or instantaneous) change in current is impossible since an infinite voltage would be required.

Single-layer coils wound in a helix are often called solenoids. An example is shown in Figure 7.6-2. If the length of the coil is greater than one-half the diameter and the core is of nonferromagnetic material, the inductance of the coil is given by

$$L = \frac{\mu_0 N^2 A}{l + 0.45d}\ \text{H}$$

where N is the number of turns; A the cross-sectional area in square meters; l the length in meters; d the diameter in meters; and $\mu_0 = 4\pi \times 10^{-7}$ H/m, a constant known as the permeability of free space.

Iron cores have higher permeabilities than air, thereby concentrating the magnetic flux. This effect increases the inductance of the coil.

The force experienced by two neighboring current-carrying wires can be described in terms of the existence of a magnetic field, which can be described in terms of magnetic

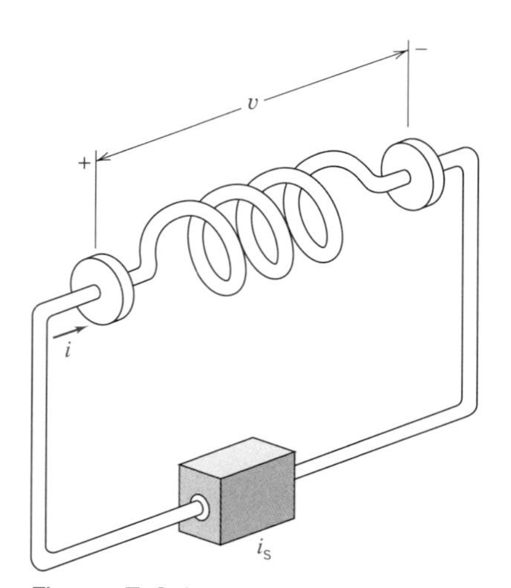

Figure 7.6-1 Coil of wire connected to a current source.

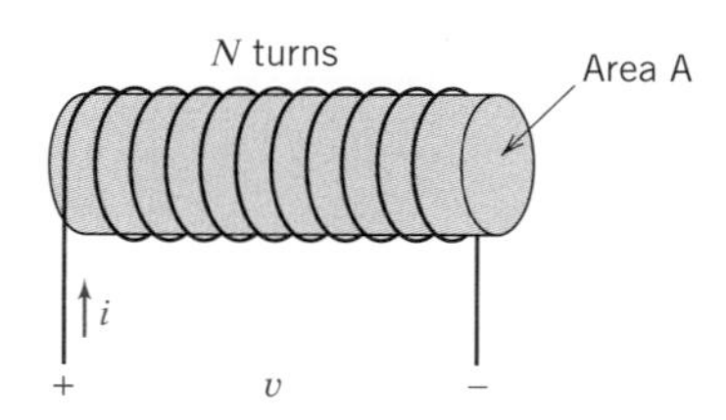

Figure 7.6-2 Coil wound as a tight helix on a core of area A.

flux that forms a loop around the coil, as shown in Figure 7.6-3. A *magnetic flux* $\phi(t)$ is associated with a current i in a coil. In this case we have an N-turn coil, and each flux line passes through all turns. Then the total flux is said to be $N\phi$.

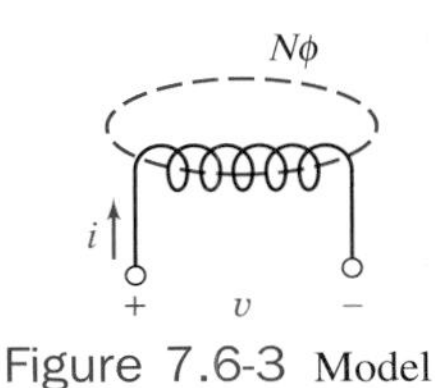

Figure 7.6-3 Model of the inductor.

Let us assume that the coil of wire has N turns and the core material has a relatively high permeability so that the magnetic flux ϕ is concentrated within the area A. According to Faraday, the changing flux creates an induced voltage in each turn equal to the derivative of the flux ϕ, so the total voltage v across N turns is

$$v = N\frac{d\phi}{dt} \tag{7.6-2}$$

However, since the total flux $N\phi$ is proportional to the current i in the coil, we have

$$N\phi = Li \tag{7.6-3}$$

where L, inductance, is the constant of proportionality. Substituting Eq. 7.6-3 into 7.6-2, we obtain

$$v = L\frac{di}{dt} \tag{7.6-4}$$

The circuit symbol for an inductor is shown in Figure 7.6-4. The passive sign convention for an inductor requires the current to flow into the positive terminal as shown in Figure 7.6-2. From the perspective of modeling electrical devices, the capacitor is a circuit element often used to model the effect of electric fields. Correspondingly, the inductor models the effects of magnetic fields. A magnetic field is a state, produced either by current flow in a wire or by a permanent magnet, that can induce voltage in a conductor when the magnetic field changes.

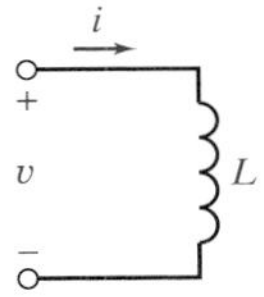

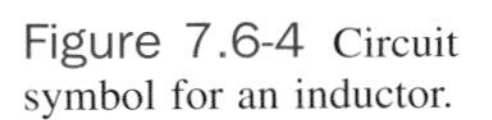

Figure 7.6-4 Circuit symbol for an inductor.

Inductance is a measure of the ability of a device to store energy in the form of a magnetic field.

Inductors include the actual resistance of the copper wire used in the coil. For this reason, inductors are far from ideal elements and are typically modeled by an ideal inductance in series with a small resistance.

An example of a coil with a large inductance is shown in Figure 7.6-5. Inductors are wound in various forms, as shown in Figure 7.6-6. Practical inductors have inductances ranging from 1 μH to 10 H.

Examining Eq. 7.6-4, we note that if the current i is constant, then the voltage across the inductor is zero. As the current changes more rapidly, the voltage will increase. Let us consider the voltage of an inductor when the current changes at $t = 0$ from zero to a

Figure 7.6-5 Coil with a large inductance. Courtesy of MuRata Company.

Figure 7.6-6 Elements with inductances arranged in various forms of coils. Courtesy of Dale Electronic Inc.

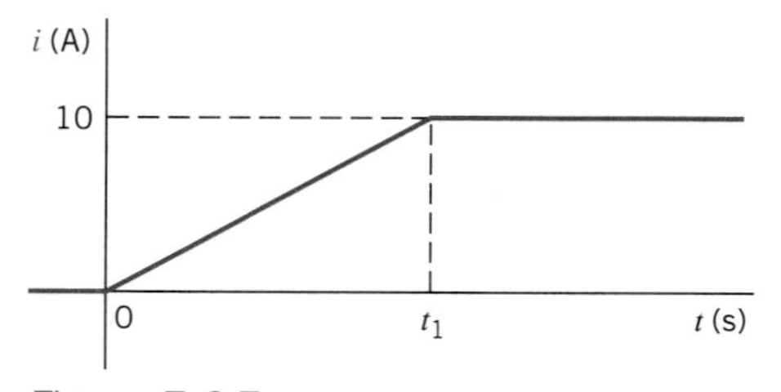

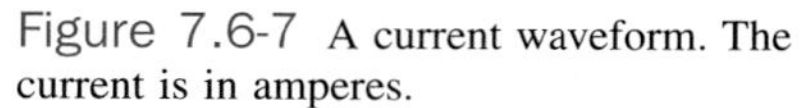
Figure 7.6-7 A current waveform. The current is in amperes.

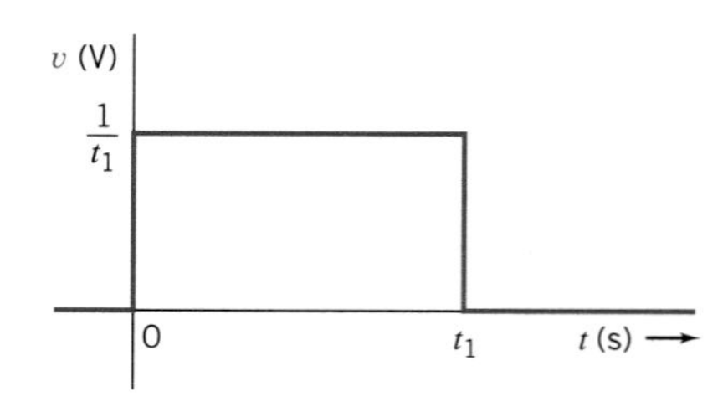

Figure 7.6-8 Voltage response for the current waveform of Figure 7.6-7 when $L = 0.1$ H.

constantly increasing current and eventually levels off as shown in Figure 7.6-7. Let us determine the voltage of the inductor. We may describe the current (in amperes) by

$$
\begin{aligned}
i &= 0 && t \le 0 \\
&= \frac{10t}{t_1} && 0 < t < t_1 \\
&= 10 && t \ge t_1
\end{aligned}
$$

Let us consider a 0.1-H inductor and find the voltage waveform. Since $v = L\,di/dt$, we have (in volts)

$$
\begin{aligned}
v &= 0 && t \le 0 \\
&= \frac{1}{t_1} && 0 < t < t_1 \\
&= 0 && t \ge t_1
\end{aligned}
$$

The resulting voltage pulse waveform is shown in Figure 7.6-8. Note that as t_1 decreases, the magnitude of the voltage increases. Clearly, we cannot let $t_1 = 0$, since the voltage required would then become infinite and we would require infinite power at the terminals of the inductor. Thus, instantaneous changes in the current through an inductor are not possible.

Faraday, Ampére, and Oersted developed the concept of magnetic flux $\phi(t)$ associated with the current in the inductor. For a linear inductor $\phi(t) = Mi(t)$, where M is a constant. Just as the principle of conservation of charge applies to a capacitor, the principle of conservation of flux applies to the inductor. Thus, the flux $\phi(t)$ cannot have discontinuities, and therefore $i(t)$ through the inductor cannot have discontinuities.

The current in an inductance cannot change instantaneously.

The limiting requirements on the current through the inductor and the voltage across a capacitor are summarized in Table 7.6-1.

Table 7.6-1 Limiting Requirements for Inductors and Capacitors

	Limiting Requirements
Capacitor	The voltage across a capacitor may not change instantaneously (change discontinuously).
Inductor	The current in an inductor may not change instantaneously (change discontinuously).

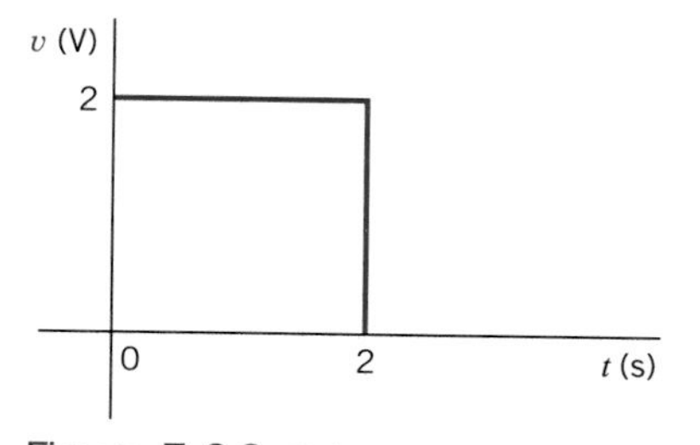

Figure 7.6-9 Voltage waveform for an inductor (in volts).

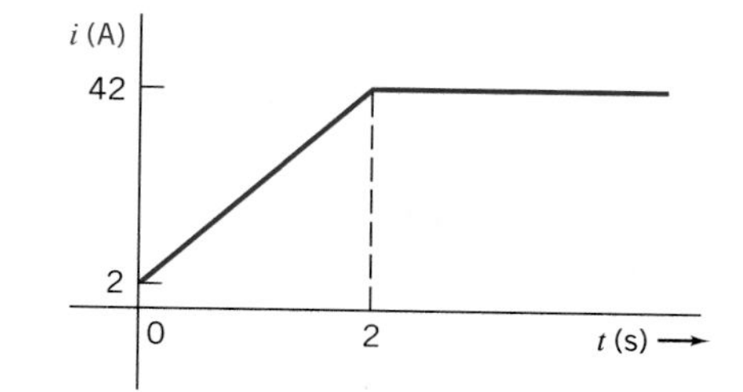

Figure 7.6-10 Current waveform for an inductor $L = 0.1$ H corresponding to the voltage waveform of Figure 7.6-9.

The current in an inductor in terms of the voltage across it may be determined by integrating the relationship

$$v = L\frac{di}{dt} \tag{7.6-5}$$

from t_0 to t. We obtain from Eq. 7.6-5 that

$$di = \frac{v}{L}\,dt$$

Integrating, we have

$$i = \frac{1}{L}\int_{t_0}^{t} v\,d\tau + i(t_0) \tag{7.6-6}$$

where $i(t_0)$ is the current that accumulates from $t = -\infty$ to t_0. Normally, we select $t_0 = 0$.

For example, let us consider the voltage waveform shown in Figure 7.6-9 for an inductor when $L = 0.1$ H and $i(t_0) = 2$ A. Since $v(t) = 2$ V between $t = 0$ and $t = 2$, we have

$$i = 10\int_{0}^{t} (2)\,d\tau + i(t_0) = 20t + 2\text{ A}$$

This current waveform is shown in Figure 7.6-10.

Example 7.6-1

Find the voltage across an inductor, $L = 0.1\text{H}$, when the current in the inductor is

$$i = 20te^{-2t}\text{A}$$

for $t > 0$ and $i(0) = 0$.

Solution

The voltage for $t > 0$ is

$$\begin{aligned} v &= L\frac{di}{dt} = (0.1)\frac{d}{dt}(20te^{-2t}) \\ &= 2(-2te^{-2t} + e^{-2t}) = 2e^{-2t}(1 - 2t)\text{ V} \end{aligned}$$

The voltage is equal to 2 V when $t = 0$, as shown in Figure 7.6-11*b*. The current waveform is shown in Figure 7.6-11*a*. Note that the current reaches a maximum value and the voltage is zero at $t = 0.5$ s.

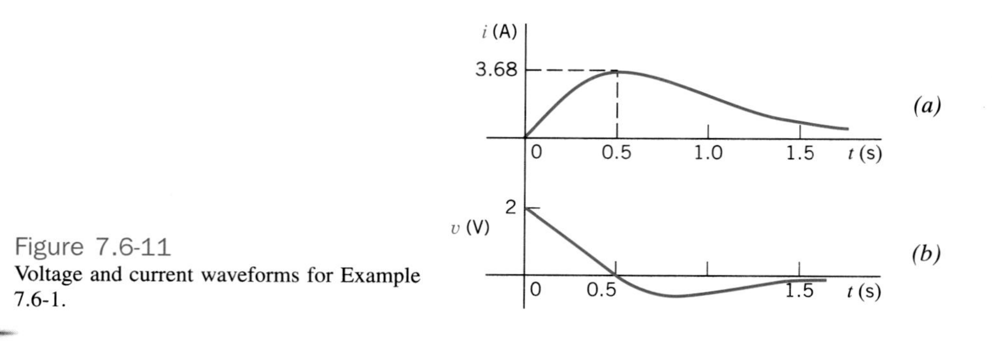

Figure 7.6-11
Voltage and current waveforms for Example 7.6-1.

Exercise 7.6-1 Determine the voltage $v(t)$ for $t > 0$ for the circuit of Figure E 7.6-1*b* when $i_s(t)$ is the current shown in Figure E 7.6-1*a*.

Hint: Determine $v_L(t)$ and $v_R(t)$ separately, then use KVL.

Answer: $v(t) = \begin{cases} 2t - 2 & 2 < t < 4 \\ 7 - t & 4 < t < 8 \\ 0 & \text{otherwise} \end{cases}$

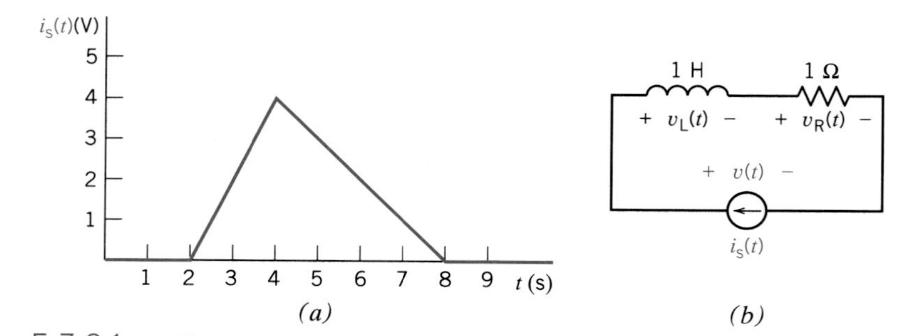

Figure E 7.6-1 (*a*) The current source current. (*b*) The circuit.

Exercise 7.6-2 Determine the current $i(t)$ for $t > 0$ for the circuit of Figure E 7.6-2*b* when $v_s(t)$ is the voltage shown in Figure E 7.6-2*a*. The inductor current at time $t = 0$ is $i(0) = -12$ A.

Answer: $$i(t) = \begin{cases} 3\int_0^t 4d\tau - 12 = 12t - 12 & \text{for} \quad 0 < t < 4 \\ 3\int_4^t -2d\tau + 36 = 60 - 6t & \text{for} \quad 4 < t < 10 \\ 3\int_{10}^t 0d\tau + 0 = 0 & \text{for} \quad 10 < t \end{cases}$$

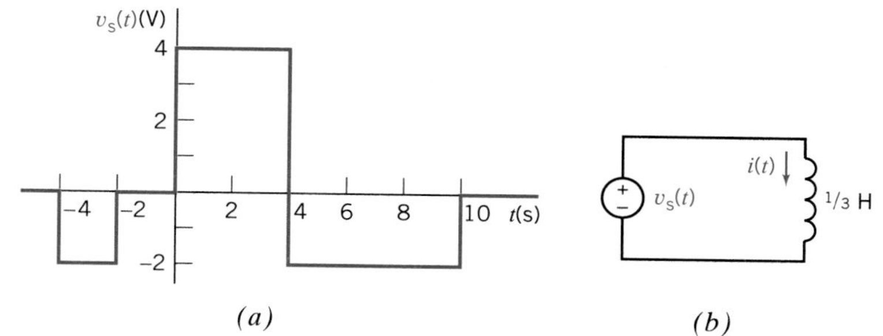

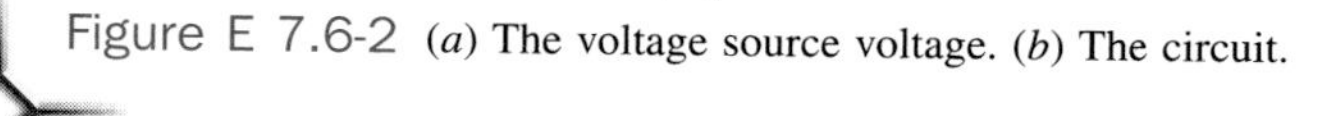
Figure E 7.6-2 (*a*) The voltage source voltage. (*b*) The circuit.

7.7 ENERGY STORAGE IN AN INDUCTOR

The power in an inductor is

$$p = vi = \left(L\frac{di}{dt}\right)i \tag{7.7-1}$$

The energy stored in the inductor is stored in the magnetic field and is

$$w = \int_{t_0}^{t} p\,d\tau = L\int_{i(t_0)}^{i(t)} i\,di$$

Integrating the current between $i(t_0)$ and $i(t)$, we obtain

$$w = \frac{L}{2}\left[i^2(t)\right]_{i(t_0)}^{i(t)} = \frac{L}{2}i^2(t) - \frac{L}{2}i^2(t_0) \tag{7.7-2}$$

Usually, we select $t_0 = -\infty$ for the inductor and then the current $i(-\infty) = 0$. Then we have

$$w = \frac{1}{2}Li^2 \tag{7.7-3}$$

Note that $w(t) \geq 0$ for all $i(t)$, so the inductor is a passive element. The inductor does not generate or dissipate energy but only stores energy. It is important to note that inductors and capacitors are fundamentally different from other devices considered in earlier chapters in that they have memory.

Example 7.7-1

Find the current in an inductor, $L = 0.1\,\text{H}$, when the voltage across the inductor is

$$v = 10te^{-5t}\,\text{V}$$

Assume that the current is zero for $t \leq 0$.

Solution

The voltage as a function of time is shown in Figure 7.7-1*a*. Note that the voltage reaches a maximum at $t = 0.2$ s.

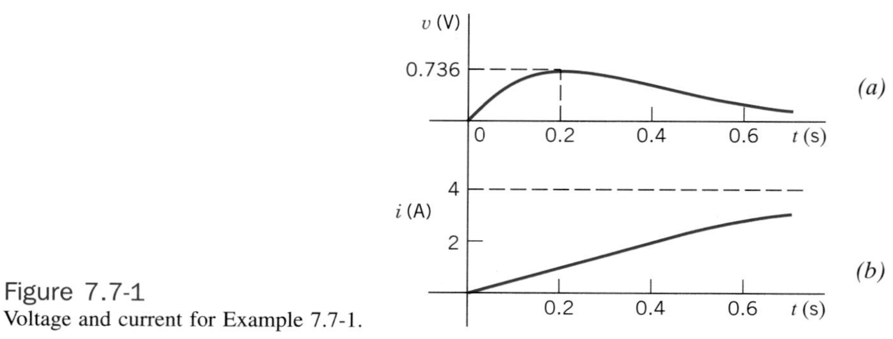

Figure 7.7-1
Voltage and current for Example 7.7-1.

The current is

$$i = \frac{1}{L}\int_0^t v\,d\tau + i\,(t_0)$$

Since the voltage is zero for $t < 0$, the current in the inductor at $t = 0$ is $i(0) = 0$. Then we have

$$i = 10\int_0^t 10\tau e^{-5\tau} d\tau = 100\left[\frac{-e^{-5\tau}}{25}(1 + 5\tau)\right]_0^t = 4(1 - e^{-5t}(1 + 5t))\text{ A}$$

The current as a function of time is shown in Figure 7.7-1*b*.

Example 7.7-2

Find the power and energy for an inductor of 0.1 H when the current and voltage are as shown in Figures 7.7-2*a* and 7.7-2*b*.

Solution

First, we write the expression for the current and the voltage. The current is

$$\begin{aligned} i &= 0 & t &< 0 \\ &= 20t & 0 &\le t \le 1 \\ &= 20 & 1 &< t \end{aligned}$$

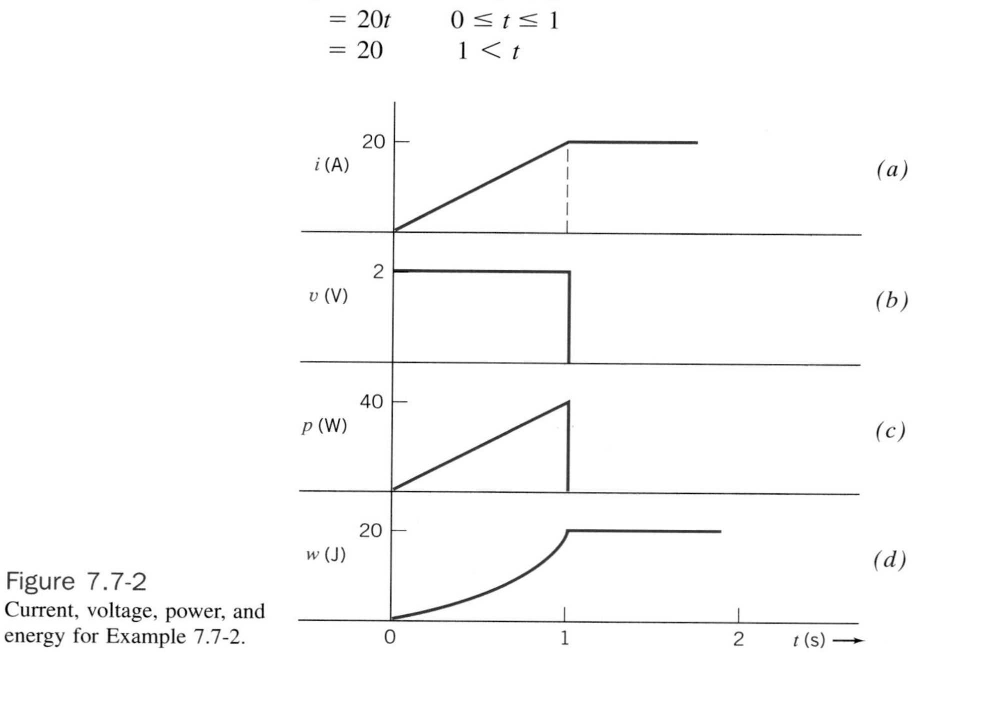

Figure 7.7-2
Current, voltage, power, and energy for Example 7.7-2.

The voltage is expressed as

$$\begin{aligned} v &= 0 \qquad t < 0 \\ &= 2 \qquad 0 < t < 1 \\ &= 0 \qquad 1 < t \end{aligned}$$

You may verify the voltage by using $v = L\,di/dt$. Then the power is

$$p = vi = 40t \text{ W}$$

for $0 \le t < 1$ and zero for all other time.

The energy, in joules, is then

$$\begin{aligned} w &= \frac{1}{2} Li^2 \\ &= 0.05(20t)^2 \quad 0 \le t \le 1 \\ &= 0.05(20)^2 \quad 1 < t \end{aligned}$$

and zero for all $t < 0$.

The power and energy are shown in Figures 7.7-2*c* and 7.7-2*d*.

Example 7.7-3

Find the power and the energy stored in a 0.1-H inductor when $i = 20te^{-2t}$ A and $v = 2e^{-2t}(1 - 2t)$ V for $t \ge 0$ and $i = 0$ for $t < 0$. (See Example 7.6-1.)

Solution

The power is

$$\begin{aligned} p = iv &= (20te^{-2t})[2e^{-2t}(1 - 2t)] \\ &= 40te^{-4t}(1 - 2t) \text{ W} \qquad t > 0 \end{aligned}$$

The energy is then

$$\begin{aligned} w &= \frac{1}{2} Li^2 = 0.05(20te^{-2t})^2 \\ &= 20t^2e^{-4t} \text{ J} \qquad t > 0 \end{aligned}$$

Note that w is positive for all values of $t > 0$. The energy stored in the inductor is shown in Figure 7.7-3.

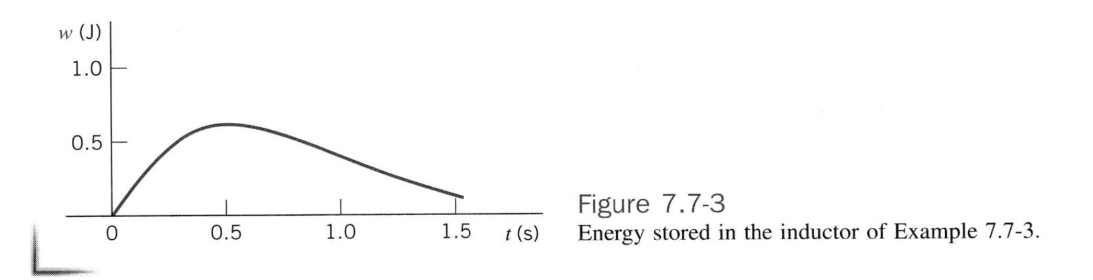

Figure 7.7-3
Energy stored in the inductor of Example 7.7-3.

Exercise 7.7-1 The current in an inductor, $L = 1/4$ H, is $i = 4te^{-t}$ A for $t \ge 0$ and $i = 0$ for $t < 0$. Find the voltage, power, and energy in this inductor.

Partial Answer: $w = 2t^2 e^{-2t}$ J

Exercise 7.7-2 The current through the inductor of a television tube deflection circuit is shown in Figure E 7.7-2 when $L = 1/2$ H. Find the voltage, power, and energy in the inductor.

Partial Answer: $p = 2t$ for $0 \le t < 1$
$= 2(t - 2)$ for $1 < t < 2$
$= 0$ for other t

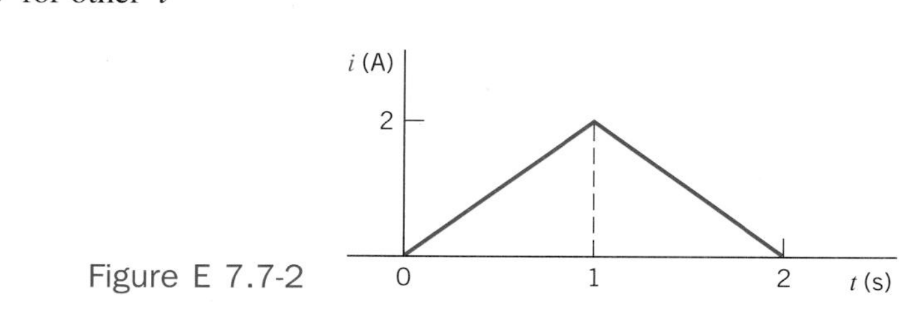

Figure E 7.7-2

7.8 SERIES AND PARALLEL INDUCTORS

A series and parallel connection of inductors can be reduced to an equivalent simple inductor. Consider a series connection of N inductors as shown in Figure 7.8-1. The voltage across the series connection is

$$\begin{aligned} v &= v_1 + v_2 + \cdots + v_N \\ &= L_1\frac{di}{dt} + L_2\frac{di}{dt} + \cdots + L_N\frac{di}{dt} \\ &= \left(\sum_{n=1}^{N} L_n\right)\frac{di}{dt} \end{aligned}$$

Since the equivalent series inductor L_s, as shown in Figure 7.8-2, is represented by

$$v = L_s\frac{di}{dt}$$

we require that

$$L_s = \sum_{n=1}^{N} L_n \tag{7.8-1}$$

Thus, an equivalent inductor for a series of inductors is the sum of the N inductors.

Now, consider the set of N inductors in parallel, as shown in Figure 7.8-3. The current i is equal to the sum of the currents in the N inductors:

$$i = \sum_{n=1}^{N} i_n$$

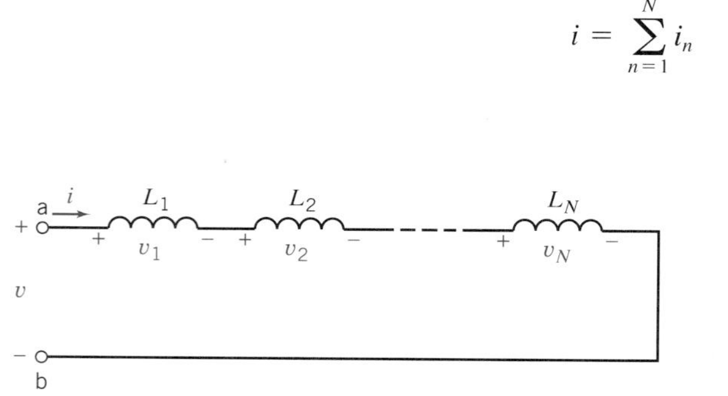

Figure 7.8-1
Series of N inductors.

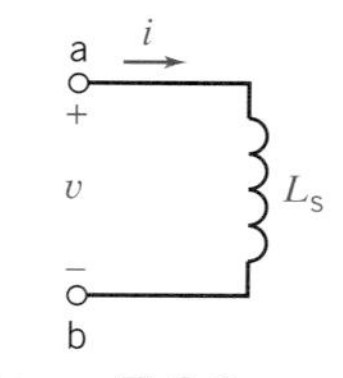

Figure 7.8-2
Equivalent inductor L_s for N series inductors.

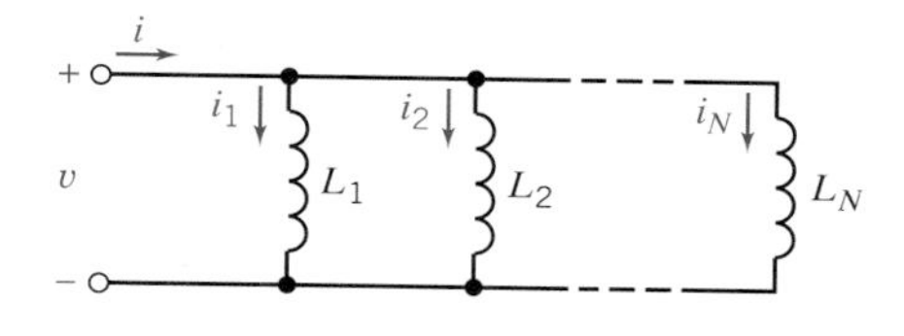

Figure 7.8-3
Connection of N parallel inductors.

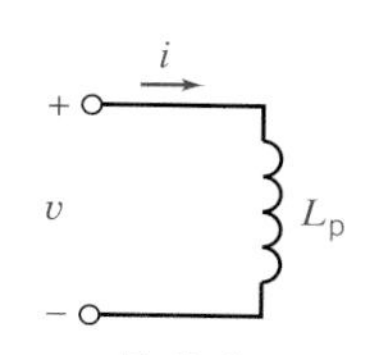

Figure 7.8-4
Equivalent inductor L_p for the connection of N parallel inductors.

However, since

$$i_n = \frac{1}{L_n}\int_{t_0}^{t} v\, d\tau + i_n(t_0)$$

we may obtain the expression

$$i = \sum_{n=1}^{N} \frac{1}{L_n}\int_{t_0}^{t} v\, d\tau + \sum_{n=1}^{N} i_n(t_0) \qquad (7.8\text{-}2)$$

The equivalent inductor L_p, as shown in Figure 7.8-4, is represented by the equation

$$i = \frac{1}{L_p}\int_{t_0}^{t} v\, d\tau + i(t_0) \qquad (7.8\text{-}3)$$

When Eqs. 7.8-2 and 7.8-3 are set equal to each other, we have

$$\frac{1}{L_p} = \sum_{n=1}^{N} \frac{1}{L_n} \qquad (7.8\text{-}4)$$

and

$$i(t_0) = \sum_{n=1}^{N} i_n(t_0) \qquad (7.8\text{-}5)$$

Example 7.8-1

Find the equivalent inductance for the circuit of Figure 7.8-5. All the inductor currents are zero at t_0.

Solution

First, we find the equivalent inductance for the 5-mH and 20-mH inductors in parallel. From Eq. 7.8-4 we obtain

$$\frac{1}{L_p} = \frac{1}{L_1} + \frac{1}{L_2}$$

or

$$L_p = \frac{L_1 L_2}{L_1 + L_2} = \frac{5 \times 20}{5 + 20} = 4 \text{ mH}$$

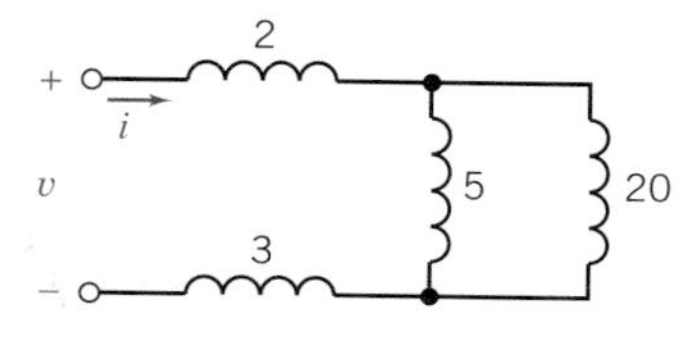

Figure 7.8-5
The circuit of Example 7.8-1. All inductances in millihenries.

This equivalent inductor is in series with the 2-mH and 3-mH inductors. Therefore, using Eq. 7.8-1, we obtain

$$L_{\mathrm{eq}} = \sum_{n=1}^{N} L_n = 2 + 3 + 4 = 9 \text{ mH}$$

Exercise 7.8-1 Find the equivalent inductance of the circuit of Figure E 7.8-1.

Answer: $L_{\mathrm{eq}} = 14$ mH

Exercise 7.8-2 Find the equivalent inductance of the circuit of Figure E 7.8-2.

Answer: $L_{\mathrm{eq}} = 4$ mH

Exercise 7.8-3 Determine the current ratio i_1/i for the circuit shown in Figure E 7.8-3. Assume that the initial currents are zero at t_0.

Answer: $\dfrac{i_1}{i} = \dfrac{L_2}{L_1 + L_2}$

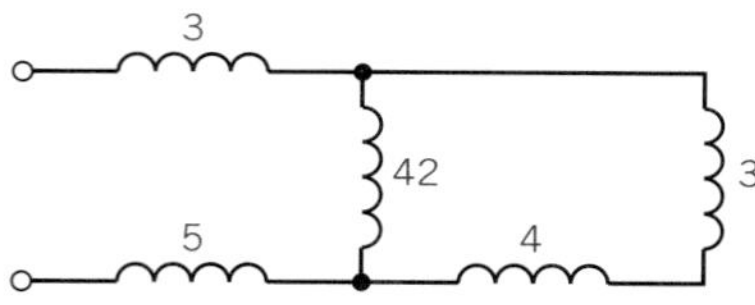

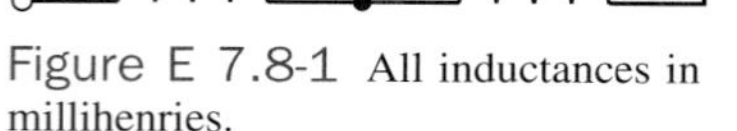

Figure E 7.8-1 All inductances in millihenries.

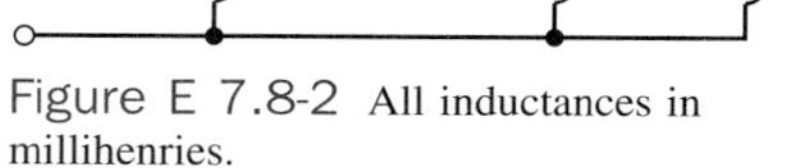

Figure E 7.8-2 All inductances in millihenries.

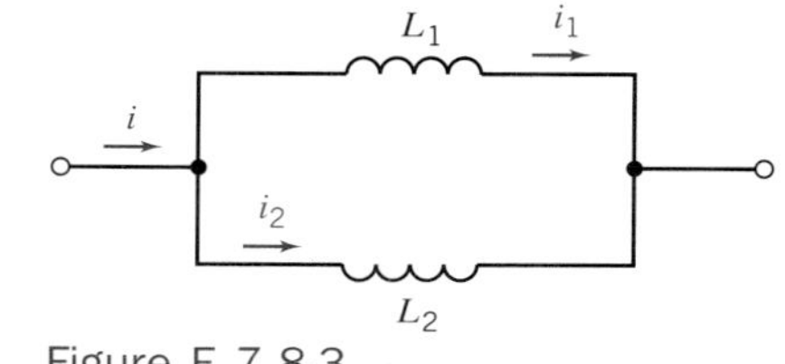

Figure E 7.8-3

7.9 INITIAL CONDITIONS OF SWITCHED CIRCUITS

In this section we concentrate on finding the change in selected variables in a circuit when a switch is thrown from open to closed or vice versa. The time of throwing the switch is considered to be $t = 0$, and we want to determine the value of the variable at $t = 0^-$ and $t = 0^+$, immediately before and after throwing the switch. Thus, a *switched circuit* is an electric circuit with one or more switches that open or close at time $t_0 = 0$.

We are particularly interested in the change in the current and voltage of energy storage elements after the switch is thrown since these variables along with the sources will dictate the behavior of the circuit for $t > 0$. In Table 7.9-1 we summarize the important characteristics of the behavior of an inductor and a capacitor. Note that we assume $t_0 = 0$. Recall that an instantaneous change in the inductor current and the capacitor voltage is not permitted. However, it is possible to change instantaneously an inductor's voltage and a capacitor's current.

We will assume that the switches in a circuit have been in position for a long time at $t = 0$, the switching time. Thus, we say the circuit conditions are in *steady state* at the

Table 7.9-1 Characteristics of Energy Storage Elements

Variable	Inductors	Capacitors
Passive sign convention	i, L, $+\ v\ -$	i, C, $+\ v\ -$
Voltage	$v = L\dfrac{di}{dt}$	$v = \dfrac{1}{C}\displaystyle\int_0^t i\,d\tau + v(0)$
Current	$i = \dfrac{1}{L}\displaystyle\int_0^t v\,d\tau + i(0)$	$i = C\dfrac{dv}{dt}$
Power	$p = Li\dfrac{di}{dt}$	$p = Cv\dfrac{dv}{dt}$
Energy	$w = \dfrac{1}{2}Li^2$	$w = \dfrac{1}{2}Cv^2$
An instantaneous change is not permitted for the element's:	Current	Voltage
Will permit an instantaneous change in the element's:	Voltage	Current
This element acts as a: (see note below)	Short circuit to a constant current into its terminals	Open circuit to a constant voltage across its terminals

Note: Assumes that the element is in a circuit with a steady-state condition.

time of switching. The steady-state response of a circuit is that which exists after a long time following any switching operation. Furthermore, when a circuit is excited by only dc sources, and the circuit attains steady state, all the branch currents and voltages are constant.

When a constant current flows into an inductor, the voltage, $v = L\,di/dt$, is zero across the element and the inductor appears as a short circuit.

An **inductance** behaves as a short circuit to a dc current.

Similarly, if a constant voltage is applied across a capacitor, it appears as an open circuit since $i = C\,dv/dt$ is equal to zero.

A **capacitance** behaves as an open circuit to a dc voltage.

First, let us consider a circuit with an inductor as shown in Figure 7.9-1. The symbol for the switch implies it is open at $t = 0^-$ and then closes at $t = 0$. Before the switch is thrown, we assume that the circuit has attained steady-state conditions. The source current at $t = 0^-$ divides between i_1 and i_L so that

$$i_1 + i_L = i_s$$

Note that i_L is a constant current, so the inductor voltage will be zero. Then using the current divider principle,

$$i_L = \frac{R_1}{R_1 + R_2}\, i_s$$

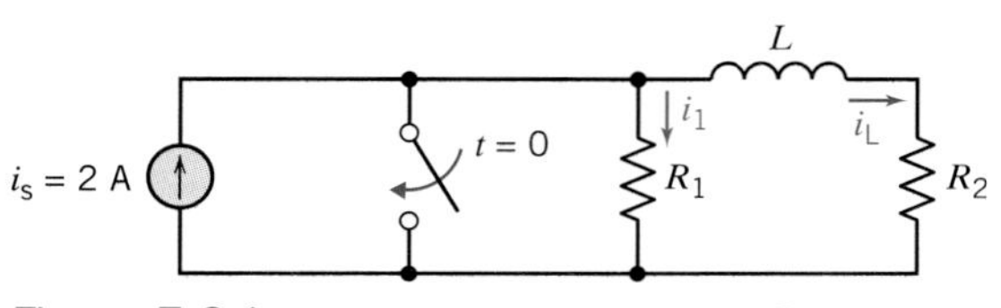

Figure 7.9-1 An *RL* circuit. $R_1 = R_2 = 1\ \Omega$. The switch is open for $t < 0$ and is closed at $t = 0$.

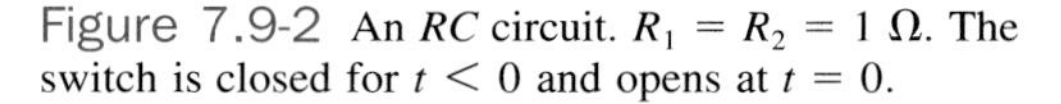

Figure 7.9-2 An *RC* circuit. $R_1 = R_2 = 1\ \Omega$. The switch is closed for $t < 0$ and opens at $t = 0$.

Since $i_s = 2\text{A}$, we have the current at $t = 0^-$. When $R_1 = R_2 = 1$, we obtain

$$i_L(0^-) = \left(\frac{1}{2}\right)2 = 1\ \text{A}$$

Since the current cannot change instantaneously for the inductor, we have

$$i_L(0^+) = i_L(0^-) = 1\ \text{A}$$

However, note that the current in the resistor can change instantaneously. Prior to $t = 0$ we have

$$i_1(0^-) = 1\ \text{A}$$

After the switch is thrown, we require that the voltage across R_1 be equal to zero because of the switched short circuit, and therefore

$$i_1(0^+) = 0$$

Thus, the current in the resistor changes abruptly from 1 to 0.

Now let us consider the circuit with a capacitor shown in Figure 7.9-2. Prior to $t = 0$, the switch has been closed for a long time. Since the source is a constant, the current in the capacitor is zero for $t < 0$ because the capacitor appears as an open circuit in a steady-state condition. Therefore, the voltage across the capacitor for $t < 0$ can be obtained from the voltage divider principle with R_1 and R_2 so that

$$v_c = \frac{R_2}{R_1 + R_2} v_s$$

Therefore, at $t = 0^-$, when $v_s = 10$ and $R_1 = R_2 = 1\ \Omega$, we obtain

$$v_c(0^-) = \left(\frac{1}{2}\right)10 = 5\ \text{V}$$

However, since the voltage across a capacitor cannot change instantaneously, we have

$$v_c(0^-) = v_c(0^+) = 5\ \text{V}$$

When the switch is opened at $t = 0$, the source is removed from the circuit, but $v_c(0^+)$ remains equal to 5 V.

If a circuit has both a capacitor and an inductor, we consider the current and voltage prior to $t = 0$ and find $v_c(0^-)$ and $i_L(0^-)$. As an example, consider the circuit shown in Figure 7.9-3. We assume that the switch has been closed a long time and steady-state conditions exist. In order to find $v_c(0^-)$ and $i_L(0^-)$ we replace the capacitor by an open circuit and the inductor by a short circuit, as shown in Figure 7.9-4. Immediately, we note that

$$i_L(0^-) = \frac{10}{5} = 2\ \text{A}$$

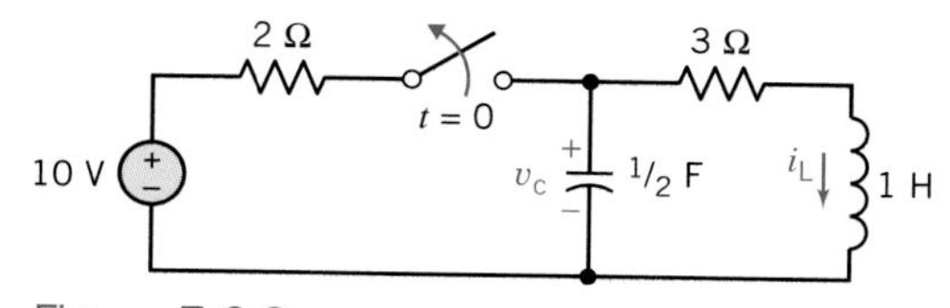

Figure 7.9-3 Circuit with an inductor and a capacitor. The switch is closed for a long time prior to opening at $t = 0$.

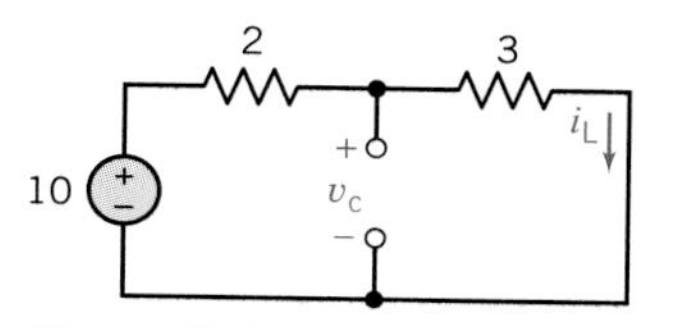

Figure 7.9-4 Circuit of Figure 7.9-3 for $t < 0$.

The capacitor is an open circuit as shown but does have a voltage v_c across its terminals. Using the voltage divider principle, we note that

$$v_c(0^-) = \left(\frac{3}{5}\right) 10 = 6 \text{ V}$$

Then we note that

$$v_c(0^+) = v_c(0^-) = 6 \text{ V}$$

and

$$i_L(0^+) = i_L(0^-) = 2 \text{ A}$$

Example 7.9-1

Find $i_L(0^+)$, $v_c(0^+)$, $dv_c(0^+)/dt$, and $di_L(0^+)/dt$ for the circuit of Figure 7.9-5. We will use $dv_c(0^+)/dt$ to denote $dv_c(t)/dt|_{t=0^+}$.

Assume that switch 1 has been open and switch 2 has been closed for a long time and steady-state conditions prevail at $t = 0^-$.

Solution

First, we redraw the circuit for $t = 0^-$ by replacing the inductor with a short circuit and the capacitor with an open circuit, as shown in Figure 7.9-6. Then we note that

$$i_L(0^-) = 0$$

and

$$v_c(0^-) = -2 \text{ V}$$

Therefore, we have

$$i_L(0^+) = i_L(0^-) = 0$$

and

$$v_c(0^+) = v_c(0^-) = -2 \text{ V}$$

In order to find $dv_c(0^+)/dt$ and $di_L(0^+)/dt$ we throw the switch at $t = 0$ and redraw the circuit of Figure 7.9-5, as shown in Figure 7.9-7. (We did not draw the current source since its switch is open.)

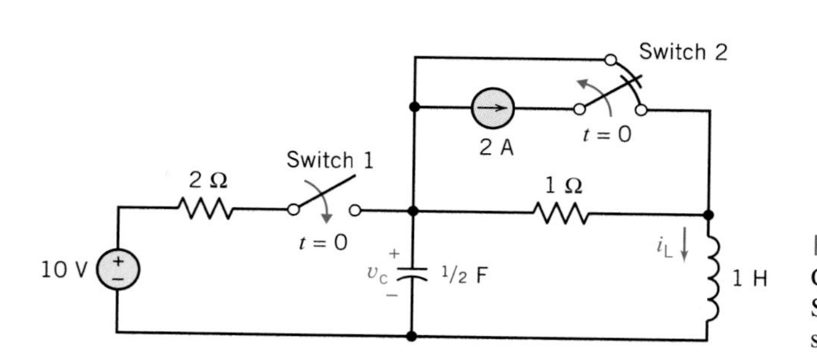

Figure 7.9-5 Circuit for Example 7.9-1. Switch 1 closes at $t = 0$ and switch 2 opens at $t = 0$.

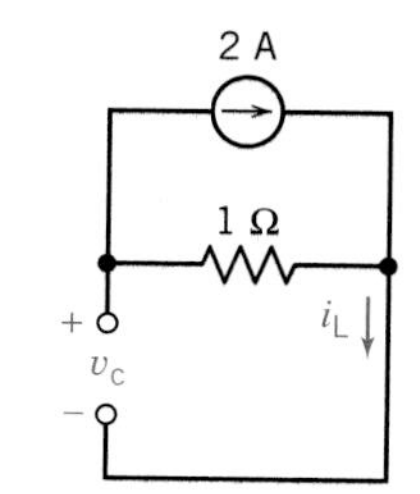

Figure 7.9-6 Circuit of Figure 7.9-5 at $t = 0^-$.

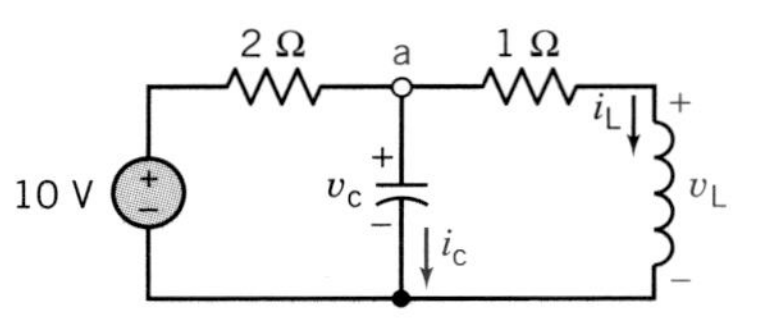

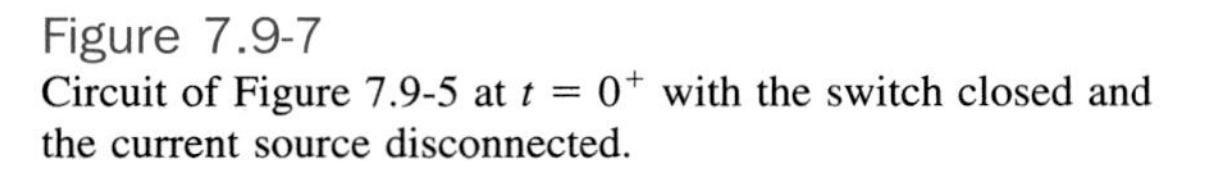

Figure 7.9-7
Circuit of Figure 7.9-5 at $t = 0^+$ with the switch closed and the current source disconnected.

Since we wish to find $dv_c(0^+)/dt$, we recall that

$$i_c = C\frac{dv_c}{dt}$$

so

$$\frac{dv_c(0^+)}{dt} = \frac{i_c(0^+)}{C}$$

Similarly, since for the inductor

$$v_L = L\frac{di_L}{dt}$$

we may obtain $di_L(0^+)/dt$ as

$$\frac{di_L(0^+)}{dt} = \frac{v_L(0^+)}{L}$$

Using KVL for the right-hand mesh of Figure 7.9-7, we obtain

$$v_L - v_c + 1i_L = 0$$

Therefore, at $t = 0^+$

$$v_L(0^+) = v_c(0^+) - i_L(0^+) = -2 - 0 = -2\text{ V}$$

Hence, we obtain

$$\frac{di_L(0^+)}{dt} = -2\text{ A/s}$$

Similarly, to find i_c we write KCL at node a to obtain

$$i_c + i_L + \frac{v_c - 10}{2} = 0$$

Consequently, at $t = 0^+$

$$i_c(0^+) = \frac{10 - v_c(0^+)}{2} - i_L(0^+) = 6 - 0 = 6\text{ A}$$

Accordingly,

$$\frac{dv_c(0^+)}{dt} = \frac{i_c(0^+)}{C} = \frac{6}{1/2} = 12\text{ V/s}$$

Thus, we found that at the switching time $t = 0$, the current in the inductor and the voltage of the capacitor remained constant. However, the inductor voltage did change instantaneously from $v_L(0^-) = 0$ to $v_L(0^+) = -2$ V, and we determined that $di_L(0^+)/dt = -2$A/s. Also, the capacitor current changed instantaneously from $i_c(0^-) = 0$ to $i_c(0^+) = 6$A and we found that $dv_c(0^+)/dt = 12$ V/s.

7.10 THE OPERATIONAL AMPLIFIER AND *RC* CIRCUITS

The circuit elements connected to the operational amplifier are not limited to resistors. It is useful to consider the connection of a combination of resistors and capacitors to an operational amplifier. First, let us consider the *RC* circuit and operational amplifier as shown in Figure 7.10-1.

Applying KVL to the mesh containing the independent voltage source gives

$$v_s = iR + v_1$$

Applying KCL at node a gives

$$i_c = i - i_1 = \frac{v_s - v_1}{R} - i_1$$

Applying KVL to the loop consisting of the operational amplifier input, capacitor, and operational amplifier output gives

$$v_o = v_1 - \frac{1}{C}\int_{-\infty}^{t} i_c \, dt$$

$$= v_1 - \frac{1}{C}\int_{-\infty}^{t} \left(\frac{v_s - v_1}{R} - i_1\right) dt$$

When the operational amplifier is modeled as an ideal operational amplifier, then $v_1 = 0$ and $i_1 = 0$, so

$$v_o = -\frac{1}{C}\int_{-\infty}^{t} \frac{v_s}{R} \, dt$$

$$= -\left(\frac{1}{RC}\int_{0}^{t} v_s \, dt - v_o(0)\right)$$

Therefore, the output voltage of this circuit is proportional to the integral of the input voltage. This circuit is commonly called an *integrator.* If we choose $RC = 1$, then

$$v_o = -\int_{-\infty}^{t} v_s \, dt = -\left(\int_{0}^{t} v_s \, dt + v_c(0)\right)$$

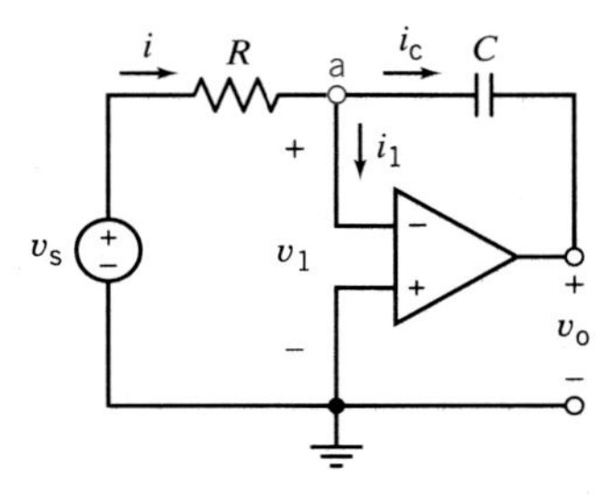

Figure 7.10-1 An integrator implemented using an operational amplifier.

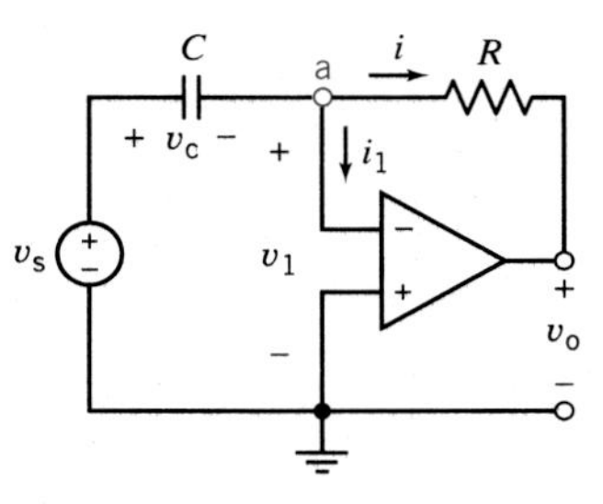

Figure 7.10-2 A differentiator implemented using an operational amplifier.

Next, consider the circuit shown in Figure 7.10-2. Applying KVL to the mesh containing the independent voltage source gives

$$v_s = v_c + v_1$$

Applying KCL at node a gives

$$i + i_1 = C\frac{dv_c}{dt} = C\frac{d(v_s - v_1)}{dt}$$

Applying KVL to the loop consisting of the operational amplifier input, resistor, and operational amplifier output gives

$$v_o = v_1 - Ri = v_1 - R\left(C\frac{d(v_s - v_1)}{dt} - i_1\right)$$

When the operational amplifier is modeled as an ideal operational amplifier, then $v_1 = 0$ and $i_1 = 0$, so

$$v_o = -RC\frac{dv_s}{dt}$$

Therefore, the output voltage of this circuit is proportional to the derivative of the input voltage. This circuit is commonly called a *differentiator.* Caution is advised in the use of this circuit, since it can accentuate the noise present within the circuit.

Exercise 7.10-1 For the integrator circuit of Figure 7.10-1, we have $R = 1\ \text{k}\Omega$, $C = 500\ \mu\text{F}$, and the initial capacitor voltage is $v_c(0) = 2$ V. When $v_s = 5$ V for $t > 0$ and $v_s = 0$ for $t < 0$, determine and sketch the output voltage $v_o(t)$. Assume the circuit is in steady state at $t = 0^-$.

Answer: $v_o = 2 - 10t$ V, $t \geq 0$

Exercise 7.10-2 For the differentiator circuit of Figure 7.10-2 we have $R = 20\ \text{k}\Omega$, $C = 50\ \mu\text{F}$, and the initial capacitor voltage is $v_c(0) = 0$. Assume the circuit is in steady state at $t = 0^-$. When $v_s(t) = 0.2 \sin 10t$ V, for $t \geq 0$ determine $v_o(t)$.

Answer: $v_o(t) = -2 \cos 10t$ V, $t \geq 0$

7.11 USING MATLAB TO PLOT CAPACITOR OR INDUCTOR VOLTAGE AND CURRENT

Suppose that the current in a 2-F capacitor is

$$i(t) = \begin{cases} 4 & t < 2 \\ t + 2 & 2 < t < 6 \\ 20 - 2t & 6 < t < 14 \\ -8 & t > 14 \end{cases} \tag{7.11-1}$$

where the units of current are A and the units of time are s. When the initial capacitor voltage is $v(0) = -5$ V, the capacitor voltage can be calculated using

$$v(t) = \frac{1}{2}\int_0^t i(\tau)\, d\tau - 5 \tag{7.11-2}$$

Equation 7.11-1 indicates that $i(t) = 4$ A while $t < 2$ s. Using this current in Eq. 7.11-2 gives

$$v(t) = \frac{1}{2}\int_0^t 4 d\tau - 5 = 2t - 5 \tag{7.11-3}$$

when $t < 2$ s. Next, Eq. 7.11-1 indicates that $i(t) = t + 2$ A while $2 < t < 6$ s. Using this current in Eq. 7.11-2 gives

$$v(t) = \frac{1}{2}\left(\int_2^t (t + 2)d\tau + \int_0^2 4\, d\tau\right) - 5 = \frac{1}{2}\int_2^t (t + 2)\, d\tau - 1 = \frac{t^2}{4} + t - 4 \tag{7.11-4}$$

when $2 < t < 6$ s. Continuing in this way, we calculate

$$v(t) = \frac{1}{2}\left(\int_6^t (20 - 2t)\, d\tau + \int_2^6 (t + 2)\, d\tau + \int_0^2 4\, d\tau\right) - 5 \tag{7.11-5}$$

$$= \frac{1}{2}\int_6^t (20 - 2t)\, d\tau + 11 = -\frac{t^2}{2} + 10t - 31$$

when $6 < t < 14$ s and

$$v(t) = \frac{1}{2}\left(\int_{14}^t - 8 d\tau + \int_6^{14} (20 - 2t)\, d\tau + \int_2^6 (t + 2)\, d\tau + \int_0^2 4 d\tau\right) - 5$$

$$= \frac{1}{2}\int_{14}^t - 8 d\tau + 11 = 67 - 4t \tag{7.11-6}$$

when $t > 14$ s.

Equations (7.11-3) through (7.11-6) can be summarized as

$$v(t) = \begin{cases} 2t - 5 & t < 2 \\ \dfrac{t^2}{4} + t - 4 & 2 < t < 6 \\ -\dfrac{t^2}{2} + 10\,t - 31 & 6 < t < 14 \\ 67 - 4\,t & t > 14 \end{cases} \tag{7.11-7}$$

Equations (7.11-1) and (7.11-7) provide an analytic representation of the capacitor current and voltage. MATLAB provides a convenient way to obtain graphical representation of these functions. Figures 7.11-1*a* and 7.11-1*b* show MATLAB input files that represent the capacitor current and voltage. Notice that the MATLAB input file representing the current, Figure 7.11-1*a*, is very similar to Eq. 7.11-1, while the MATLAB input file representing the voltage, Figure 7.11-1*b*, is very similar to Eq. 7.11-7. Figure 7.11-1*c* shows the MATLAB input file used to plot the capacitor current and voltage. Figure 7.11-2 shows the resulting plots of the capacitor current and voltage.

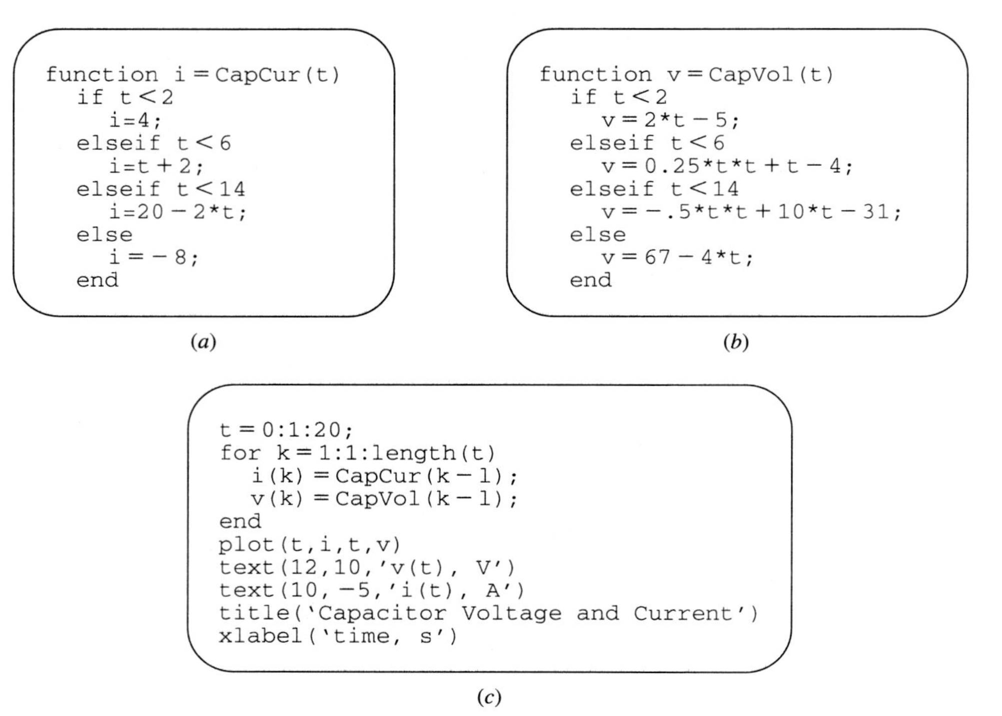

```
function i = CapCur(t)
  if t<2
    i=4;
  elseif t<6
    i=t + 2;
  elseif t<14
    i=20 - 2*t;
  else
    i = - 8;
  end
```

(*a*)

```
function v = CapVol(t)
  if t<2
    v = 2*t - 5;
  elseif t<6
    v = 0.25*t*t + t - 4;
  elseif t<14
    v = -.5*t*t + 10*t - 31;
  else
    v = 67 - 4*t;
  end
```

(*b*)

```
t = 0:1:20;
for k=1:1:length(t)
  i(k) = CapCur(k-1);
  v(k) = CapVol(k-1);
end
plot(t,i,t,v)
text(12,10,'v(t), V')
text(10, -5,'i(t), A')
title('Capacitor Voltage and Current')
xlabel('time, s')
```

(*c*)

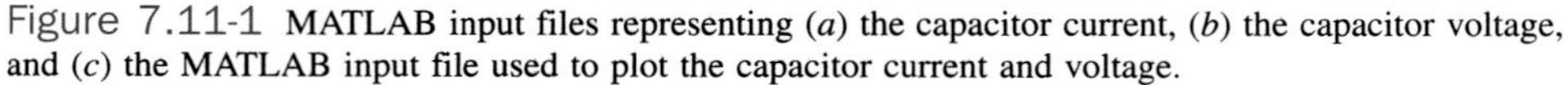

Figure 7.11-1 MATLAB input files representing (*a*) the capacitor current, (*b*) the capacitor voltage, and (*c*) the MATLAB input file used to plot the capacitor current and voltage.

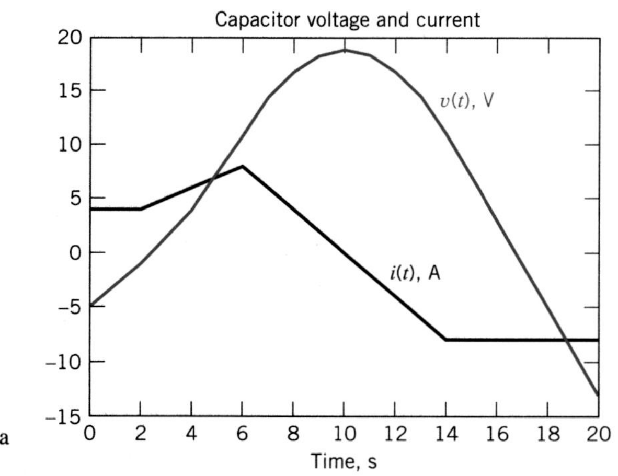

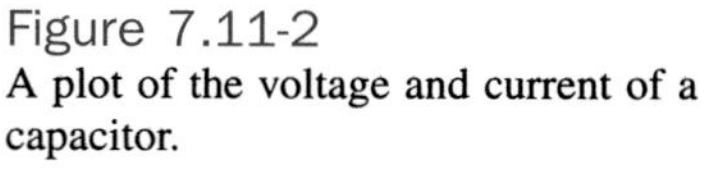

Figure 7.11-2
A plot of the voltage and current of a capacitor.

7.12 VERIFICATION EXAMPLE

A homework solution indicates that the current and voltage of a 2-F capacitor are

$$i(t) = \begin{cases} 4 & t < 2 \\ t + 2 & 2 < t < 6 \\ 20 - 2t & 6 < t < 14 \\ -8 & t > 14 \end{cases} \tag{7.12-1}$$

and

$$v(t) = \begin{cases} 2t - 5 & t < 2 \\ \dfrac{t^2}{4} + t - 4 & 2 < t < 6 \\ -\dfrac{t^2}{2} + 10\,t - 21 & 6 < t < 14 \\ 67 - 4\,t & t > 14 \end{cases} \tag{7.12-2}$$

where the units of current are A, the units of voltage are V, and the units of time are s. How can we check this homework solution to see if it is correct?

Solution

The capacitor voltage cannot change instantaneously. The capacitor voltage is given by

$$v(t) = 2t - 5 \tag{7.12-3}$$

when $t < 2$ s and by

$$v(t) = \frac{t^2}{4} + t - 4 \tag{7.12-4}$$

when $2 < t < 6$ s. Since the capacitor voltage cannot change instantaneously, Eqs. 7.12-3 and 7.12-4 must both give the same value for $v(2)$, the capacitor voltage at time $t = 2$ s. Solving Eq. 7.12-3 gives

$$v(2) = 2(2) - 5 = -1 \text{ V}$$

Also solving Eq. 7.12-4 gives

$$v(2) = \frac{2^2}{4} + 2 - 4 = -1 \text{ V}$$

These values agree so we haven't found an error. Next let's check $v(6)$, the capacitor voltage at time $t = 6$ s. The capacitor voltage is given by

$$v(t) = -\frac{t^2}{2} + 10t - 21 \tag{7.12-5}$$

when $6 < t < 14$ s. Equations 7.12-4 and 7.12-5 must both give the same value for $v(6)$. Solving Eq. 7.12-4 gives

$$v(6) = \frac{6^2}{4} + 6 - 4 = 11 \text{ V},$$

while solving Eq. 7.12-5 gives

$$v(6) = -\frac{6^2}{2} + 10(6) - 21 = 21 \text{ V}$$

These values don't agree. That means that $v(t)$ changes instantaneously at $t = 6$ s, and so $v(t)$ cannot be the voltage across the capacitor. The homework solution is not correct.

7.13 Design Challenge Solution

INTEGRATOR AND SWITCH

This design challenge involves an integrator and a voltage-controlled switch.

An integrator is a circuit that performs the mathematical operation of integration. The output of an integrator, say $v_o(t)$, is related to the input of the integrator, say $v_s(t)$, by the equation

$$v_o(t_2) = K \cdot \int_{t_1}^{t_2} v_s(t)\,dt + v_o(t_1) \tag{7.13-1}$$

The constant K is called the gain of the integrator.

Integrators have many applications. One application of an integrator is to measure an interval of time. Suppose $v_s(t)$ is a constant voltage, V_s. Then

$$v_o(t_2) = K \cdot (t_2 - t_1) \cdot V_s + v_o(t_1) \tag{7.13-2}$$

This equation indicates that the output of the integrator at time t_2 is a measure of the time interval $t_2 - t_1$.

Switches can be controlled electronically. Figure 7.13-1 illustrates an electronically controlled SPST switch. The symbol shown in Figure 7.13-1*a* is sometimes used to emphasize that a switch is controlled electronically. The node voltage $v_c(t)$ is called the control voltage. Figure 7.13-1*b* shows a typical control voltage. This voltage-controlled switch is closed when $v_c(t) = v_H$ and open when $v_c(t) = v_L$. The switch shown in Figure 7.13-1 is open before time t_1. It closes at time t_1 and stays closed until time t_2. The switch opens at time t_2 and remains open.

Consider Figure 7.13-2. The voltage $v_c(t)$ controls the switch. The integrator converts the time interval $t_2 - t_1$ to a voltage that is displayed using the voltmeter. The time interval to be measured could be as small as 5 ms or as large as 200 ms. The challenge is to design the integrator. The available components include:

- standard 2% resistors (see Appendix E)
- 1 μF, 0.2 μF, and 0.1 μF capacitors
- operational amplifiers
- +15-V and −15-V power supplies
- 1 kΩ, 10 kΩ, and 100 kΩ potentiometers
- voltage-controlled SPST switches

Define the Situation

It is convenient to set the integrator output to zero at time t_1. The relationship between the integrator output voltage and the time interval should be simple. Accordingly, let

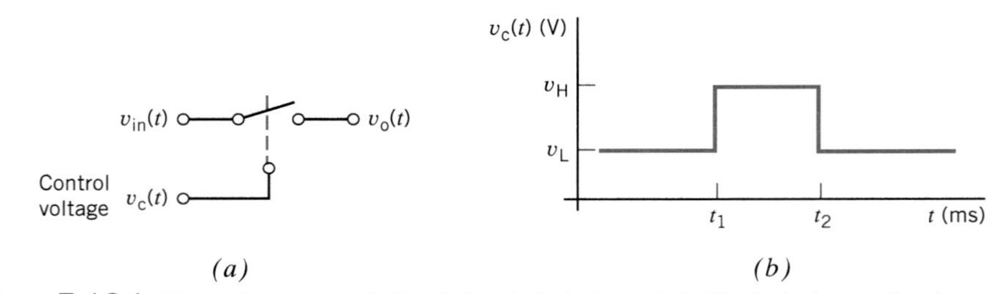

Figure 7.13-1 The voltage-controlled switch. (*a*) Switch symbol. (*b*) Typical control voltage.

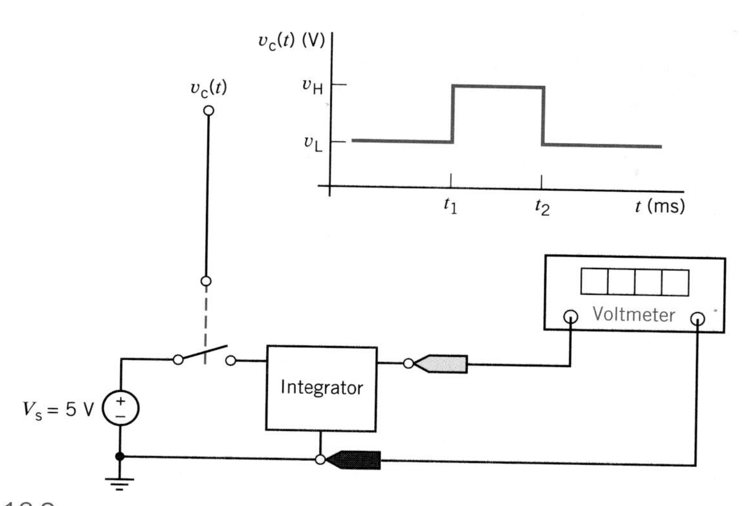

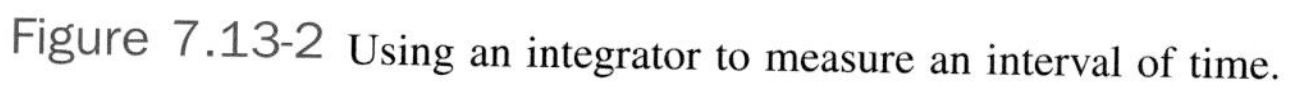
Figure 7.13-2 Using an integrator to measure an interval of time.

$$v_o(t_2) = \frac{10\text{ V}}{200\text{ ms}} \cdot (t_2 - t_1) \tag{7.13-3}$$

Figure 7.13-2 indicates that $V_s = 5$ V. Comparing Eqs. 7.13-2 and 7.13-3 yields

$$K \cdot V_s = \frac{10\text{ V}}{200\text{ ms}} \quad \text{and therefore} \quad K = 10\frac{1}{\text{s}} \tag{7.13-4}$$

State the Goal

Design an integrator satisfying both

$$K = 10\frac{1}{\text{s}} \quad \text{and} \quad v_o(t_1) = 0 \tag{7.13-5}$$

Generate a Plan

Let us use the integrator described in Section 7.10. Adding a switch as shown in Figure 7.13-3 satisfies the condition $v_o(t_1) = 0$. The analysis performed in Section 7.10 showed that

$$v_o(t_2) = -\frac{1}{RC} \cdot \int_{t_1}^{t_2} v_s(t)\, dt \tag{7.13-6}$$

so R and C must be selected to satisfy

$$\frac{1}{RC} = K = 10\frac{1}{\text{s}} \tag{7.13-7}$$

Take Action on the Plan

Any of the available capacitors would work. Select $C = 1\,\mu\text{F}$. Then

$$R = \frac{1}{10\frac{\text{V}}{\text{s}} \cdot 1\,\mu\text{F}} = 100\text{ k}\Omega \tag{7.13-8}$$

The final design is shown in Figure 7.13-4.

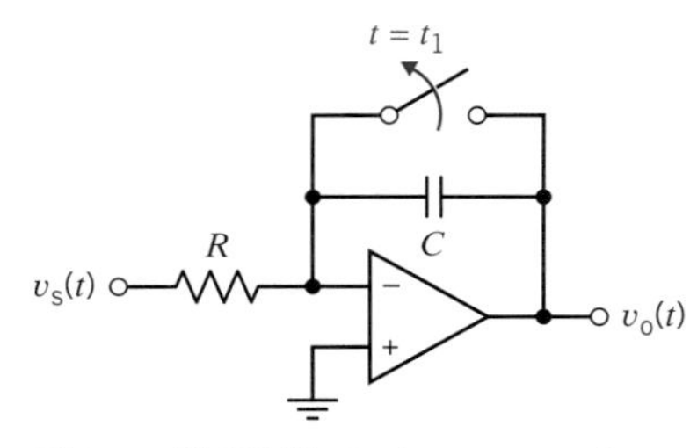

Figure 7.13-3 An integrator using an operational amplifier.

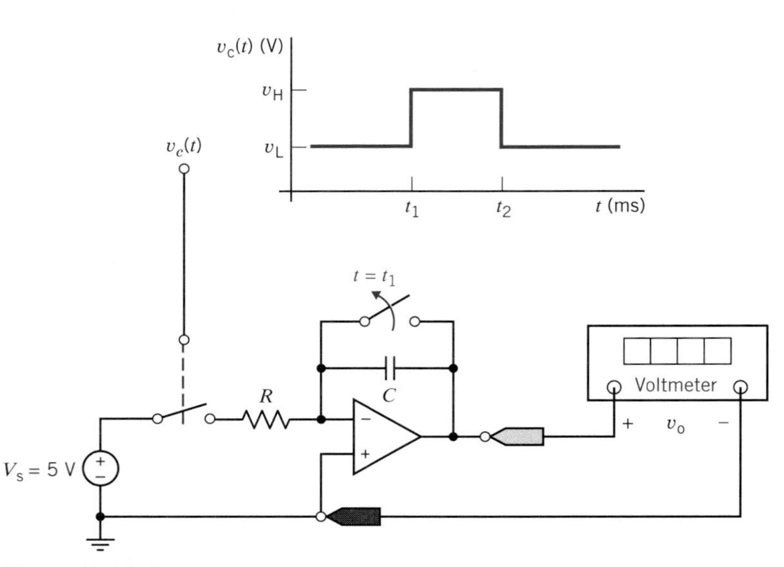

Figure 7.13-4 Using an operational amplifier integrator to measure an interval of time.

Verify the Proposed Solution

The output voltage of the integrator is given by

$$v_0(t) = -\frac{1}{RC}\int_{t_1}^{t} v_s(\tau)\,d\tau + v_0(0) = \frac{-1}{(100 \cdot 10^3)(10^{-6})}\int_{t_1}^{t} 5\,d\tau = -50(t - t_1)$$

where the units of voltage are V and the units of time are s. The interval of time can be calculated from the output voltage using

$$-(t - t_1) = \frac{v_0(t)}{50}$$

For example, an output voltage of -4 V indicates a time interval of $\frac{4}{50}$ s $= 80$ ms.

7.14 SUMMARY

- Table 7.14-1 summarizes the element equations for capacitors and inductors. (Notice that the voltage and current referred to in these equations adhere to the passive convention.) Unlike the circuit elements we encountered in previous chapters, the element equations for capacitors and inductors involve derivatives and integrals.
- Circuits that contain capacitors and/or inductors are able to store energy. The energy stored in the electric field of a capacitor is equal to $\frac{1}{2}Cv^2(t)$ where $v(t)$ is the voltage across the capacitor. The energy stored in the magnetic field of a inductor is equal to $\frac{1}{2}Li^2(t)$ where $i(t)$ is the current in the inductor.
- Circuits that contain capacitors and/or inductors have memory. The voltages and currents in that circuit at a particular time depend, not only on other voltages and currents at that same instant of time, but also on previous values of those currents and voltages. For example, the voltage across a capacitor at time t_1 depends on the voltage across that capacitor at an earlier time t_0 and also on the value of the capacitor current between t_0 and t_1.
- A set of series or parallel capacitors can be reduced to an equivalent capacitor. A set of series or parallel inductors can readily be reduced to an equivalent inductor. Table 7.14-2 summarizes the equations required to do so.
- In the absence of unbounded currents, the voltage across a capacitor cannot change instantaneously. Similarly, in the absence of unbounded voltages, the current in an inductor cannot change instantaneously. In contrast, the current in a capacitor and voltage across an inductor are both able to change instantaneously.

Table 7.14-1 Element Equations for Capacitors and Inductors

Capacitor	Inductor
$+\ v(t)\ -$, $i(t)$, C	$+\ v(t)\ -$, $i(t)$, L
$i(t) = C\dfrac{d}{dt}v(t)$	$i(t) = \dfrac{1}{L}\displaystyle\int_{t_0}^{t} v(\tau)\,d\tau + i(t_0)$
$v(t) = \dfrac{1}{C}\displaystyle\int_{t_0}^{t} i(\tau)\,d\tau + v(t_0)$	$v(t) = L\dfrac{d}{dt}i(t)$

Table 7.14-2 Parallel and Series Capacitors and Inductors

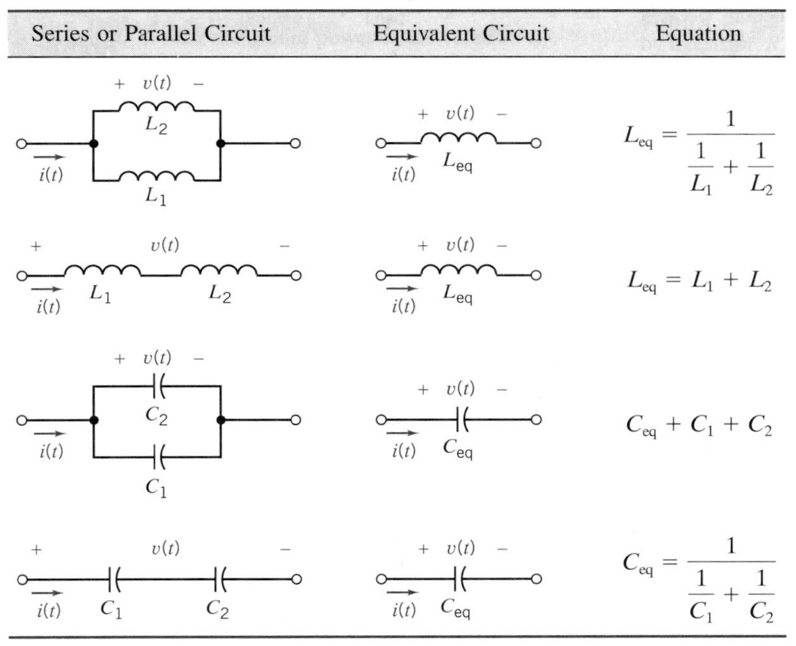

Series or Parallel Circuit	Equivalent Circuit	Equation
$+\ v(t)\ -$, L_2, $i(t)$, L_1	$+\ v(t)\ -$, $i(t)$, L_{eq}	$L_{eq} = \dfrac{1}{\dfrac{1}{L_1} + \dfrac{1}{L_2}}$
$+$, $v(t)$, $-$, $i(t)$, L_1, L_2	$+\ v(t)\ -$, $i(t)$, L_{eq}	$L_{eq} = L_1 + L_2$
$+\ v(t)\ -$, C_2, $i(t)$, C_1	$+\ v(t)\ -$, $i(t)$, C_{eq}	$C_{eq} + C_1 + C_2$
$+$, $v(t)$, $-$, $i(t)$, C_1, C_2	$+\ v(t)\ -$, $i(t)$, C_{eq}	$C_{eq} = \dfrac{1}{\dfrac{1}{C_1} + \dfrac{1}{C_2}}$

- We sometimes consider circuits that contain capacitors and inductors and have only constant inputs. (The voltages of the independent voltage sources and currents of the independent current sources are all constant.) When such a circuit is at steady state, all the currents and voltages in that circuit will be constant. In particular, the voltage across any capacitor will be constant. The current in that capacitor will be zero due to the derivative in the equation for the capacitor current. Similarly, the voltage across any inductor will be zero. Consequently, the capacitors will act like open circuits and the inductors will act like short circuits. Notice that this situation only occurs when all of the inputs to the circuit are constant.
- An op amp and a capacitor can be used to make circuits that perform the mathematical operations of integration and differentiation. Appropriately, these important circuits are called the integrator and the differentiator.
- The element voltages and currents in a circuit containing capacitors and inductors can be complicated functions of time. MATLAB is useful for plotting these functions.

PROBLEMS

Section 7.3 Capacitors and Their *v-i* Equation

P 7.3-1 A 15-μF capacitor has a voltage of 5 V across it at $t = 0$. If a constant current of 25 mA flows through the capacitor, how long will it take for the capacitor to charge up to 150 μC?
Answer: $t = 3$ ms

P 7.3-2 The voltage, $v(t)$, across a capacitor and current, $i(t)$, in that capacitor adhere to the passive convention. Determine the current, $i(t)$, when the capacitance is $C = 0.125$ F and the voltage is $v(t) = 12\cos(2t + 30°)$ V.

Hint:
$$\frac{d}{dt}A\cos(\omega t + \theta) = -A\sin(\omega t + \theta)\cdot\frac{d}{dt}(\omega t + \theta)$$
$$= -A\omega\sin(\omega t + \theta)$$
$$= A\omega\cos\left(\omega t + \left(\theta + \frac{\pi}{2}\right)\right)$$

Answer: $i(t) = 3\cos(2t + 120°)$ A

P 7.3-3 The voltage, $v(t)$, across a capacitor and current, $i(t)$, in that capacitor adhere to the passive convention. Determine the capacitance when the voltage is $v(t) = 12\cos(500t - 45°)$ V and the current is $i(t) = 3\cos(500t + 45°)$ mA.
Answer: $C = 0.5\,\mu$F

P 7.3-4 Determine $v(t)$ for the circuit shown in Figure P 7.3-4*a* when the $i_s(t)$ is as shown in Figure P 7.3-4*b* and $v_o(0^-) = -1$mV.

P 7.3-5 The voltage, $v(t)$, and current, $i(t)$, of a 1-F capacitor adhere to the passive convention. Also, $v(0) = 0$ V and $i(0) = 0$ A. (a) Determine $v(t)$ when $i(t) = x(t)$, where $x(t)$ is shown in Figure P 7.3-5 and $i(t)$ has units of A. (b) Determine $i(t)$ when $v(t) = x(t)$, where $x(t)$ is shown in Figure P 7.3-5 and $v(t)$ has units of V.

Hint: $x(t) = 4t - 4$ V when $1 < t < 2$, and $x(t) = -4t + 12$ V when $2 < t < 3$.

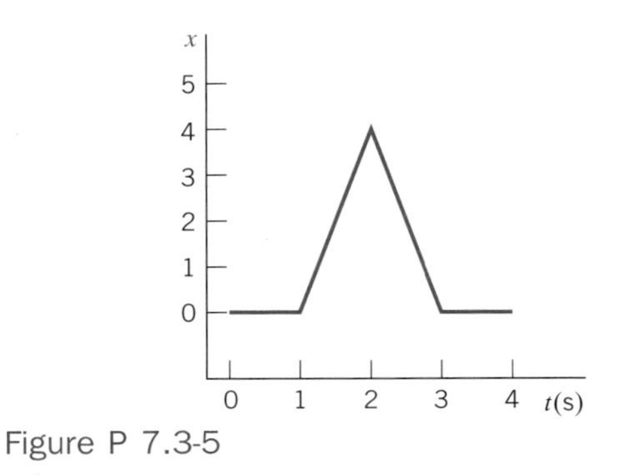

Figure P 7.3-5

P 7.3-6 The voltage, $v(t)$, and current, $i(t)$, of a 0.5-F capacitor adhere to the passive convention. Also, $v(0) = 0$ V and $i(0) = 0$ A. (a) Determine $v(t)$ when $i(t) = x(t)$, where $x(t)$ is shown in Figure P 7.3-6 and $i(t)$ has units of A. (b) Determine $i(t)$ when $v(t) = x(t)$, where $x(t)$ is shown in Figure P 7.3-6 and $v(t)$ has units of V.
Hint: $x(t) = 0.2t - 0.4$ V when $2 < t < 6$.

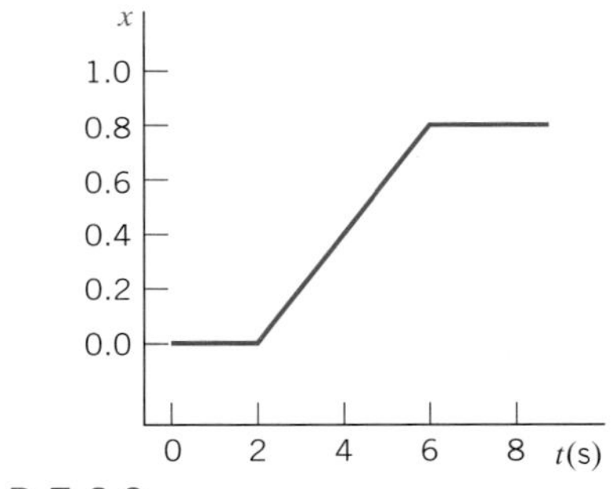

Figure P 7.3-6

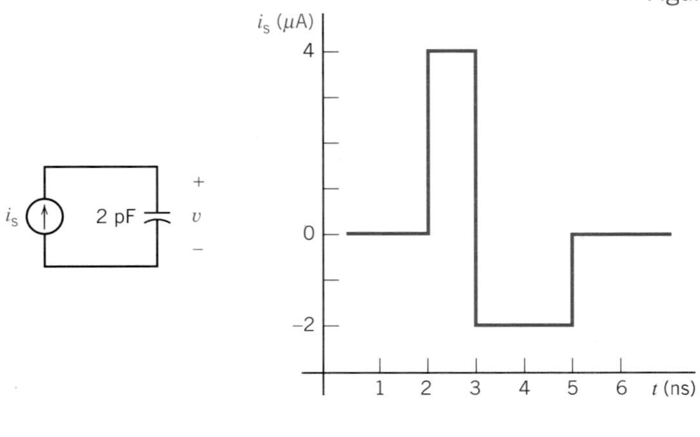

Figure P 7.3-4 (*a*) Circuit and (*b*) waveform of current source.

P 7.3-7 The voltage across a 40-μF capacitor is 25 V at $t_0 = 0$. If the current through the capacitor as a function of time is given by $i(t) = 6e^{-6t}$ mA for $t < 0$, find $v(t)$ for $t > 0$.
Answer: $v(t) = 50 - 25e^{-6t}$ V

P 7.3-8 Find i for the circuit of Figure P 7.3-8 if $v = 5(1 - 2e^{-2t})$V.

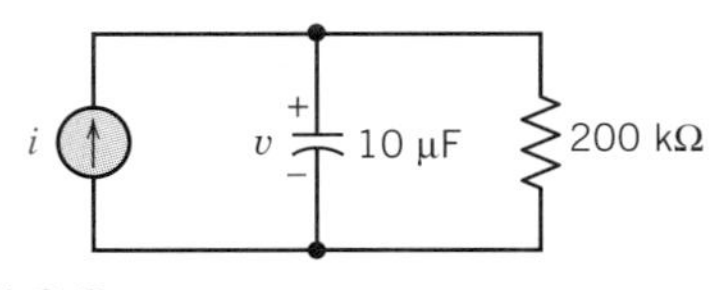

Figure P 7.3-8

Section 7.4 Energy Storage in a Capacitor

P 7.4-1 The current, i, through a capacitor is shown in Figure P 7.4-1. When $v(0) = 0$ and C = 0.5 F, determine and plot $v(t)$, $p(t)$, and $w(t)$ for 0 s $< t <$ 6 s.

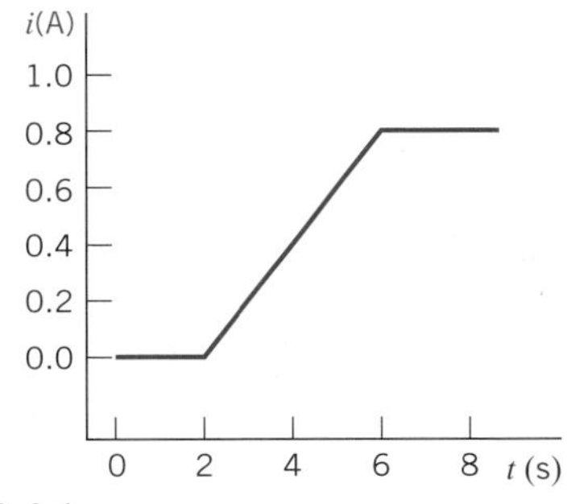

Figure P 7.4-1

P 7.4-2 In a pulse power circuit the voltage of a 10-μF capacitor is zero for $t < 0$ and

$$v = 5(1 - e^{-4000t})\text{ V} \qquad t \geq 0$$

Determine i_c and the energy stored in the capacitor at t = 0 ms and t = 10 ms.

P 7.4-3 If $v_c(t)$ is given by the waveform shown in Figure P 7.4-3, sketch the capacitor current for -1 s $\leq t \leq$ 2 s. Sketch the power and the energy for the capacitor over the same time interval when C = 1 mF.

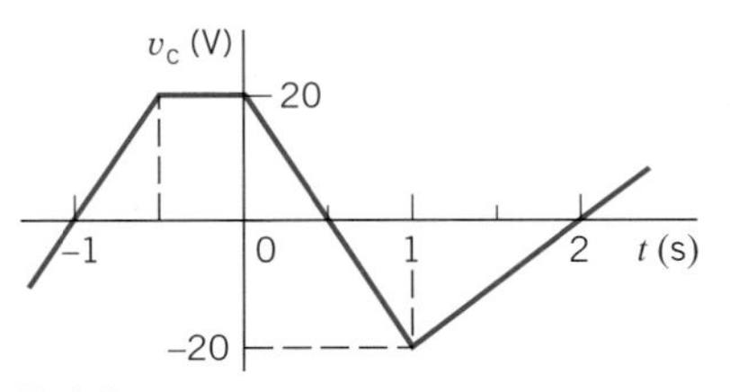

Figure P 7.4-3

P 7.4-4 The current through a 2-μF capacitor is 50 cos $(10t + \pi/6)$ μA for all time. The average voltage across the capacitor is zero. What is the maximum value of the energy stored in the capacitor? What is the first nonnegative value of t at which the maximum energy is stored?

P 7.4-5 The energy stored by a 1-mF capacitor used in a laser power supply is given as $w = 4e^{-10t}$ J for $t \geq 0$. Find the capacitor voltage and current at t = 0.1 s.

P 7.4-6 A capacitor is used in the electronic flash unit of a camera. A small battery with a constant voltage of 6 V is used to charge a capacitor with a constant current of 10 μA. How long does it take to charge the capacitor when C = 10 μF? What is the stored energy?

P 7.4-7 If a capacitor can store energy, as does a battery, could not a capacitor be used to power an electric train? Such a capacitor (electrolytic) would be about 1 cm^3 per 100 μF, for the voltage level required. Suppose we start with an initial voltage of 500 V and can extract all the energy from the capacitor to drive the car for one hour. The auto drive train requires 1.5 kW. How big would the capacitor have to be? Is this practical?

Section 7.5 Series and Parallel Capacitors

P 7.5-1 Find the current $i(t)$ for the circuit of Figure P 7.5-1.
Answer: $i(t) = -1.2 \sin 100t$ mA

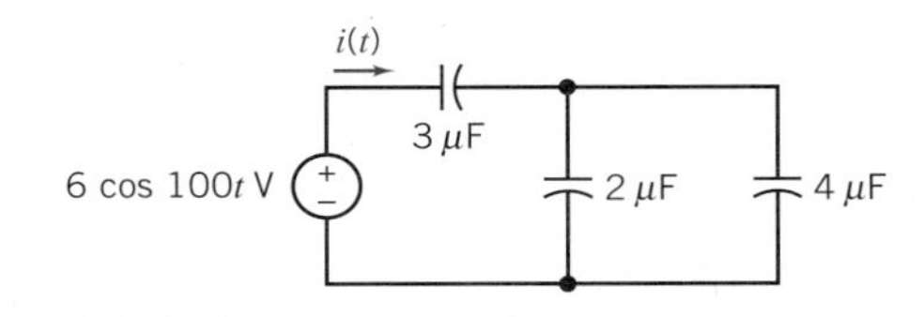

Figure P 7.5-1

P 7.5-2 Find the current $i(t)$ for the circuit of Figure P 7.5-2.
Answer: $i(t) = -1.5e^{-250t}$ mA

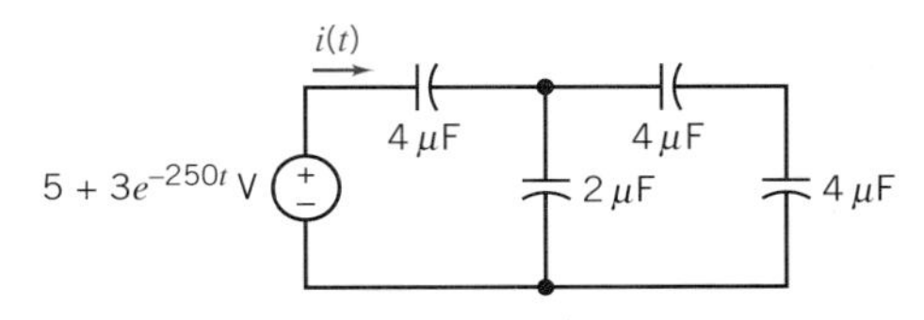

Figure P 7.5-2

P 7.5-3 The circuit of Figure P 7.5-3 contains five identical capacitors. Find the value of the capacitance C.
Answer: C = 10 μF

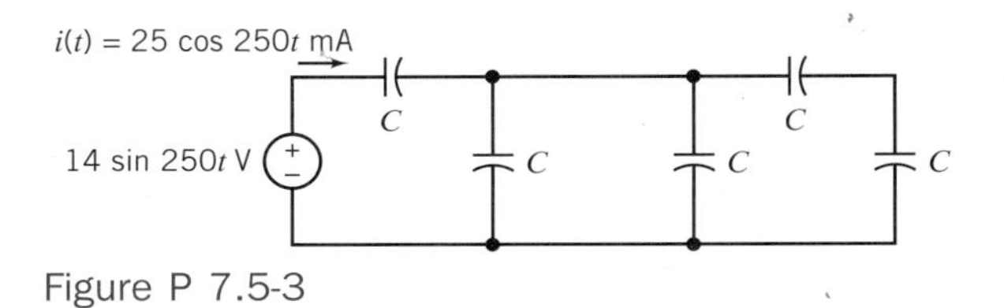

Figure P 7.5-3

Section 7.6 Inductors

P 7.6-1 Nikola Tesla (1857-1943) was an American electrical engineer who experimented with electric induction. Tesla built a large coil with a very large inductance, shown in Figure P 7.6-1. The coil was connected to a source current

$$i_s = 100 \sin 400t \text{ A}$$

so that the inductor current $i_L = i_s$. Find the voltage across the inductor and explain the discharge in the air shown in the figure. Assume that $L = 200$ H, and the average discharge distance is 2 m. Note that the dielectric strength of air is 3×10^6 V/m.

Figure P 7.6-1 Nikola Tesla sits impassively as alternating current induction coils discharge millions of volts with a roar audible 10 miles away (about 1910). Courtesy of Burndy Library.

P 7.6-2 The model of an electric motor consists of a series combination of a resistor and inductor. A current $i(t) = 4te^{-t}$ A flows through the series combination of a 10 Ω resistor and 0.1-henry inductor. Find the voltage across the combination.
Answer: $v(t) = 0.4e^{-t} + 39.6te^{-t}$ V

P 7.6-3 The voltage, $v(t)$, and current, $i(t)$ of a 1-H inductor adhere to the passive convention. Also, $v(0) = 0$ V and $i(0) = 0$ A. (a) Determine $v(t)$ when $i(t) = x(t)$, where $x(t)$ is shown in Figure P 7.6-3 and $i(t)$ has units of A. (b) Determine $i(t)$ when $v(t) = x(t)$, where $x(t)$ is shown in Figure P 7.6-3 and $v(t)$ has units of V.

Hint: $x(t) = 4t - 4$ V when $1 < t < 2$, and $x(t) = -4t + 12$ V when $2 < t < 3$.

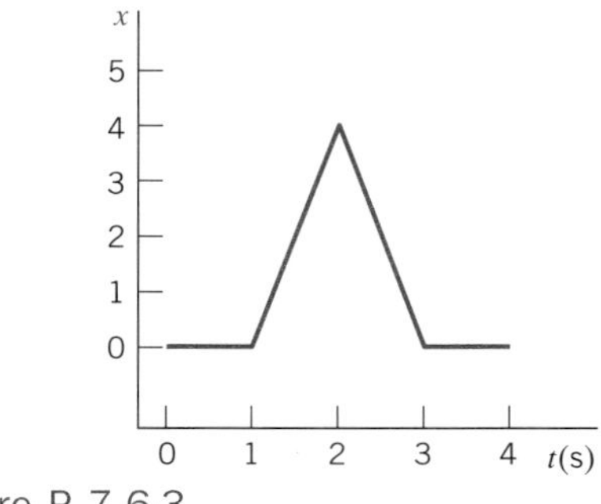

Figure P 7.6-3

P 7.6-4 The voltage, $v(t)$, across an inductor and current, $i(t)$, in that inductor adhere to the passive convention. Determine the voltage, $v(t)$, when the inductance is $L = 250$ mH and the current is $i(t) = 120 \sin(500t - 30°)$ mA.

Hint:
$$\frac{d}{dt} A \sin(\omega t + \theta) = A \cos(\omega t + \theta) \cdot \frac{d}{dt}(\omega t + \theta)$$
$$= A\omega \cos(\omega t + \theta)$$
$$= A\omega \sin\left(\omega t + \left(\theta + \frac{\pi}{2}\right)\right)$$

Answer: $v(t) = 15 \sin(500t + 60°)$ V

P 7.6-5 Determine $i_L(t)$ for $t > 0$ when $i_L(0) = -2$ μA for the circuit of Figure P 7.6-5*a* when $v_s(t)$ is as shown in Figure P 7.6-5*b*.

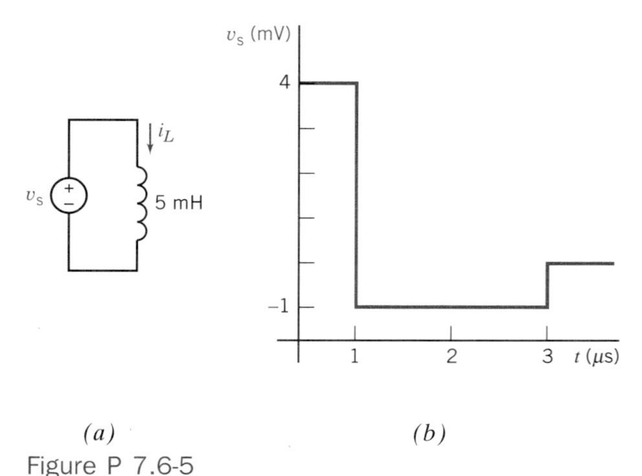

Figure P 7.6-5

P 7.6-6 Determine $v(t)$ for $t > 0$ for the circuit of Figure P 7.6-6*a* when $i_L(0) = 0$ and i_s is as shown in Figure P 7.6-6*b*.

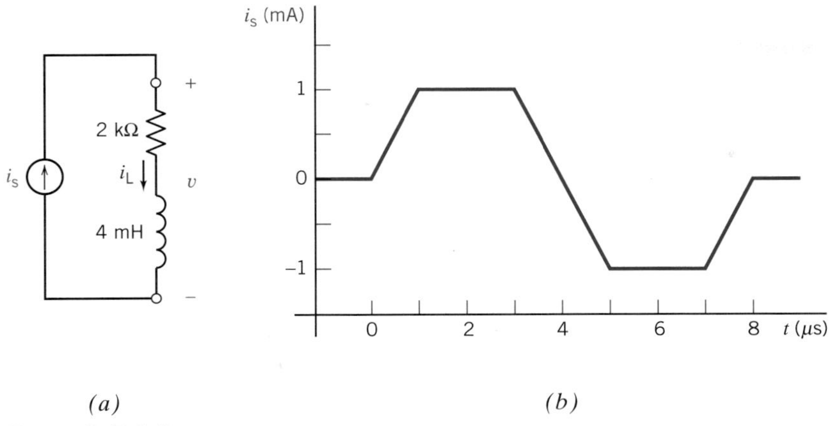

Figure P 7.6-6

P 7.6-7 Find R of the circuit shown in Figure P 7.6-7 if $v_1 = 1e^{-200t}$ V for $t \geq 0$.
Answer: $R = 80\ \Omega$

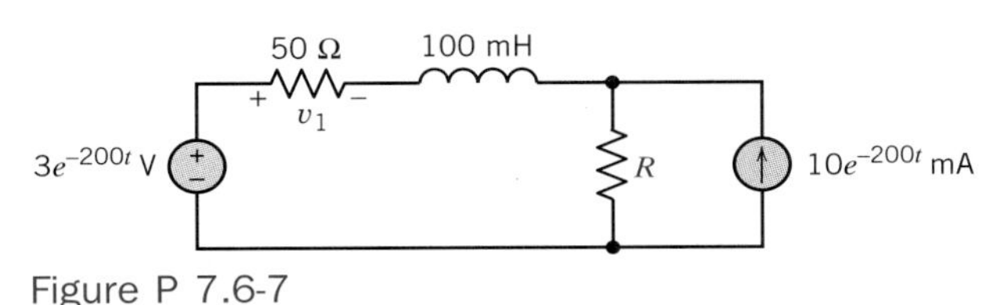

Figure P 7.6-7

P 7.6-8 The voltage, $v(t)$, and current, $i(t)$, of a 0.5-H inductor adhere to the passive convention. Also, $v(0) = 0$ V and $i(0) = 0$ A. (a) Determine $v(t)$ when $i(t) = x(t)$, where $x(t)$ is shown in Figure P 7.6-8 and $i(t)$ has units of A. (b) Determine $i(t)$ when $v(t) = x(t)$, where $x(t)$ is shown in Figure P 7.6-8 and $v(t)$ has units of V.
Hint: $x(t) = 0.2t - 0.4$ V when $2 < t < 6$.

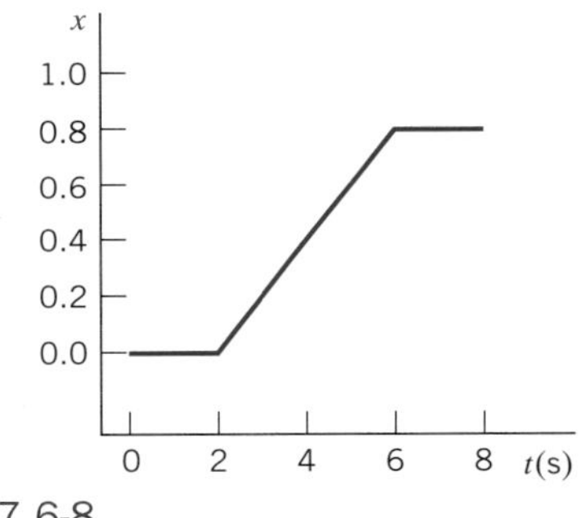

Figure P 7.6-8

Section 7.7 Energy Storage in an Inductor

P 7.7-1 The current, $i(t)$, in a 100-mH inductor connected in a telephone circuit changes according to

$$i(t) = \begin{cases} 0 & t \leq 0 \\ 4t & 0 < t < 1 \\ 4 & t \geq 1 \end{cases}$$

where the units of time are ms and the units of current are mA. Determine the power, $p(t)$, absorbed by the inductor and the energy, $w(t)$, stored in the inductor.

Answer: $p(t) = \begin{cases} 0 & t \leq 0 \\ 1.6t & 0 < t < 1 \\ 0 & t \geq 1 \end{cases}$

and $w(t) = \begin{cases} 0 & t \leq 0 \\ 0.8t^2 & 0 < t < 1 \\ 0.8 & t \geq 1 \end{cases}$

The units of $p(t)$ are W and the units of $w(t)$ are J.

P 7.7-2 The current, $i(t)$, in a 5 H inductor is

$$i(t) = \begin{cases} 0 & t \leq 0 \\ 4 \sin 2t & t \geq 0 \end{cases}$$

where the units of time are s and the units of current are A. Determine the power, $p(t)$, absorbed by the inductor and the energy, $w(t)$, stored in the inductor.
Hint: 2(cos A)(sin B) = sin(A+B) + sin(A-B)

P 7.7-3 The voltage, $v(t)$, across a 25-mH inductor used in a fusion power experiment is

$$v(t) = \begin{cases} 0 & t \leq 0 \\ 6 \cos 100t & t \geq 0 \end{cases}$$

where the units of time are s and the units of voltage are V. The current in this inductor is zero before the voltage changes

at $t = 0$. Determine the power, $p(t)$, absorbed by the inductor and the energy, $w(t)$, stored in the inductor.
Hint: $2(\cos A)(\sin B) = \sin(A+B) + \sin(A-B)$
Answer: $p(t) = 7.2 \sin 200t$ W and $w(t) = 3.6\,[1 - \cos 200t]$ mJ

Section 7.8 Series and Parallel Inductors

P 7.8-1 Find the current $i(t)$ for the circuit of Figure P 7.8-1.
Answer: $i(t) = 15 \sin 100t$ mA

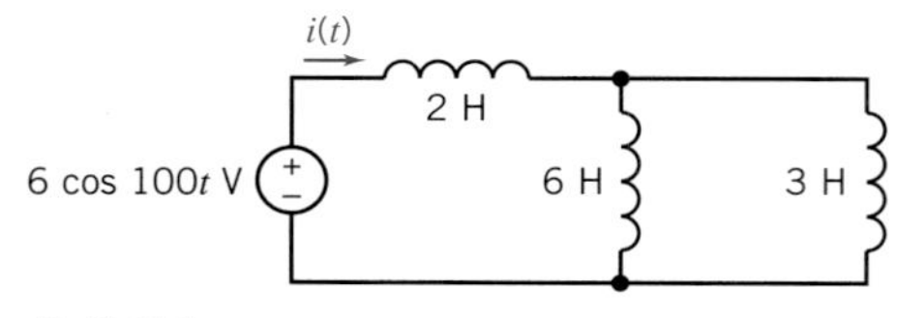

Figure P 7.8-1

P 7.8-2 Find the voltage $v(t)$ for the circuit of Figure P 7.8-2.
Answer: $v(t) = -6\, e^{-250t}$ mV

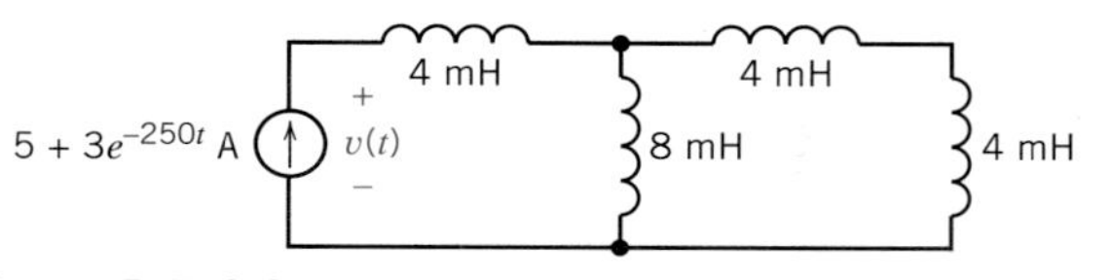

Figure P 7.8-2

P 7.8-3 The circuit of Figure P 7.8-3 contains four identical inductors. Find the value of the inductance L.
Answer: $L = 2.86$ H

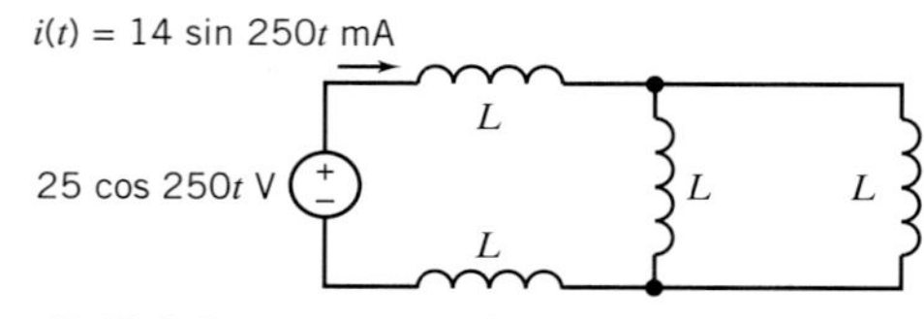

Figure P 7.8-3

Section 7.9 Initial Conditions of Switched Circuits

P 7.9-1 The switch in Figure P 7.9-1 has been open for a long time before closing at time $t = 0$. Find $v_C(0+)$ and $i_L(0+)$, the values of the capacitor voltage and inductor current immediately after the switch closes. Let $v_C(\infty)$ and $i_L(\infty)$ denote the values of the capacitor voltage and inductor current after the switch has been closed for a long time. Find $v_C(\infty)$ and $i_L(\infty)$.
Answer: $v_C(0+) = 12$ V, $i_L(0+) = 0$, $v_C(\infty) = 4$ V, and $i_L(\infty) = 1$ mA

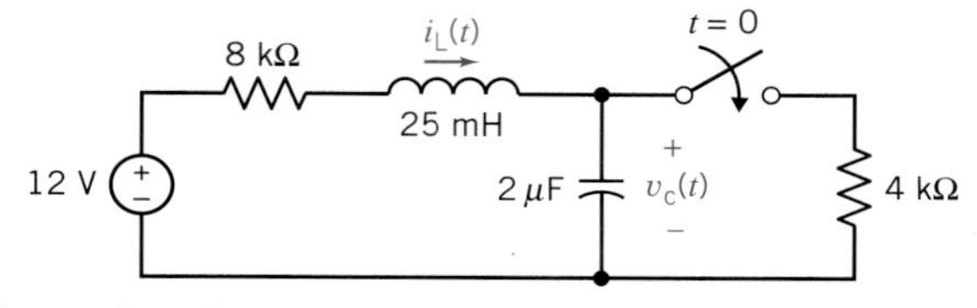

Figure P 7.9-1

P 7.9-2 The switch in Figure P 7.9-2 has been open for a long time before closing at time $t = 0$. Find $v_C(0+)$ and $i_L(0+)$, the values of the capacitor voltage and inductor current immediately after the switch closes. Let $v_C(\infty)$ and $i_L(\infty)$ denote the values of the capacitor voltage and inductor current after the switch has been closed for a long time. Find $v_C(\infty)$ and $i_L(\infty)$.
Answer: $v_C(0+) = 6$ V, $i_L(0+) = 1$ mA, $v_C(\infty) = 3$ V, and $i_L(\infty) = 1.5$ mA

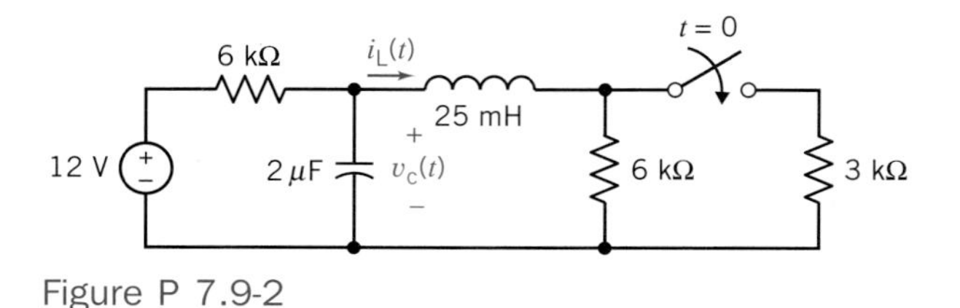

Figure P 7.9-2

P 7.9-3 The switch in Figure P 7.9-3 has been open for a long time before closing at time $t = 0$. Find $v_C(0+)$ and $i_L(0+)$, the values of the capacitor voltage and inductor current immediately after the switch closes. Let $v_C(\infty)$ and $i_L(\infty)$ denote the values of the capacitor voltage and inductor current after the switch has been closed for a long time. Find $v_C(\infty)$ and $i_L(\infty)$.
Answer: $v_C(0+) = 0$ V, $i_L(0+) = 0$, $v_C(\infty) = 8$ V, and $i_L(\infty) = 0.5$ mA

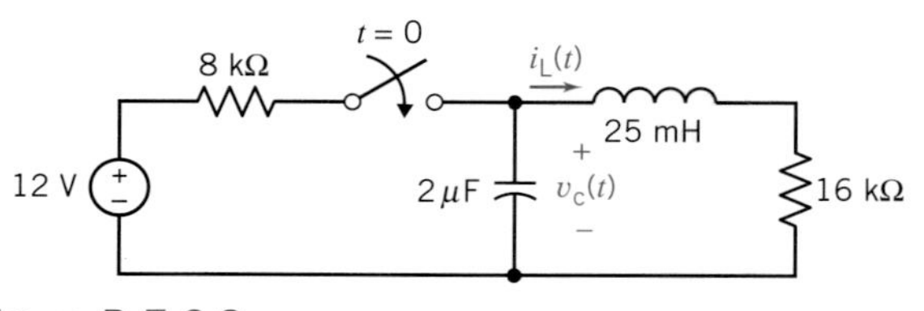

Figure P 7.9-3

P 7.9-4 Find $v_C(0^+)$ and $dv_c(0^+)/dt$ if $v(0^-) = 15$ V for the circuit of Figure P 7.9-4.

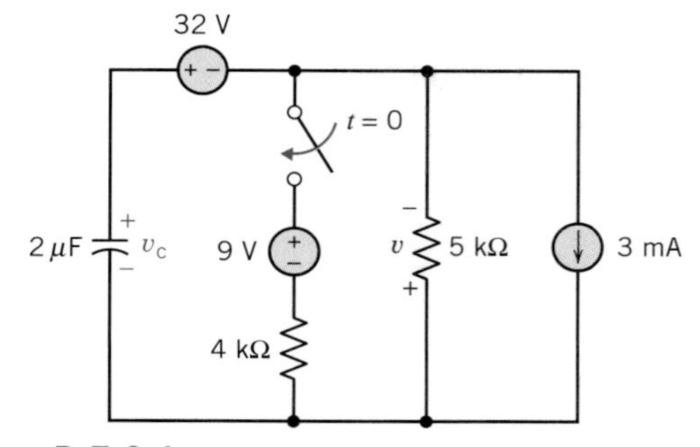

Figure P 7.9-4

P 7.9-5 For the circuit shown in Figure P 7.9-5, find $dv_c(0^+)/dt$, $di_L(0^+)/dt$, and $i(0^+)$ if $v(0^-) = 16$ V. Assume that the switch was closed for a long time prior to $t = 0$.

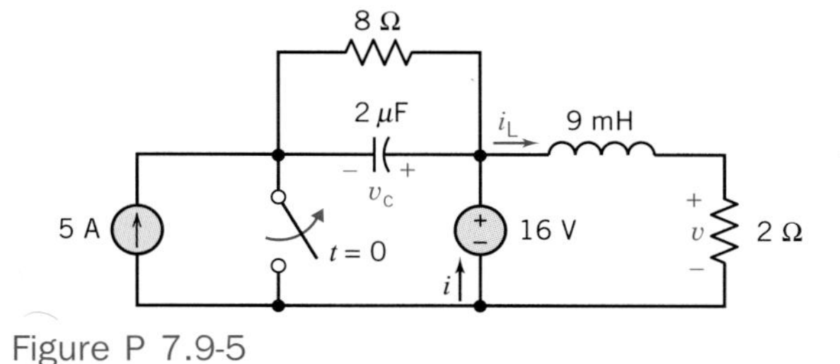

Figure P 7.9-5

P 7.9-6 For the circuit of Figure P 7.9-6, determine the current and voltage of each passive element at $t = 0^-$ and $t = 0^+$. The current source is $i_s = 0$ for $t < 0$ and $i_s = 4$ A for $t \geq 0$.

Figure P 7.9-6

Section 7.10 The Operational Amplifier and *RC* Circuits

P 7.10-1 Find the output voltage $v_o(t)$ for the circuit of Figure P 7.10-1 when

$$v_s(t) = \begin{cases} 0 & t < 0 \\ 12 \cos 100t \quad \text{V} & t \geq 0 \end{cases}$$

and the initial condition is $v_o(0) = 0$.
Answer: $v_o(t) = -3 \sin 100t$ V

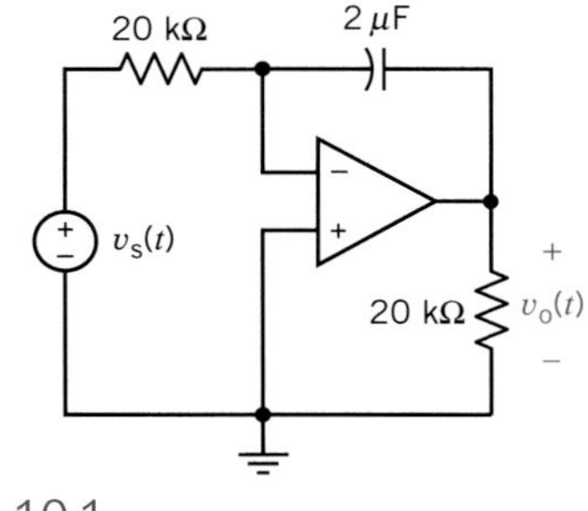

Figure P 7.10-1

P 7.10-2 Find the output voltage $v_o(t)$ for the circuit of Figure P 7.10-2 when

$$v_s(t) = \begin{cases} 0 & t < 0 \\ -4 \text{ V} & 0 \leq t < 3 \text{ ms} \\ 0 & t \geq 3 \text{ ms} \end{cases}$$

and the initial condition is $v_o(0) = 0$.

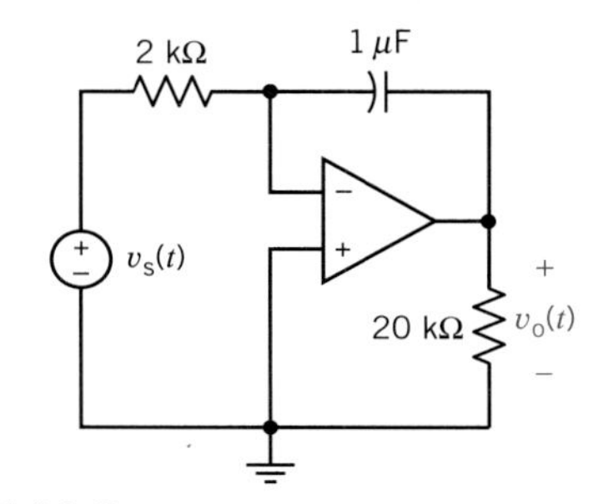

Figure P 7.10-2

P 7.10-3 Determine values of R and C for the circuit of Figure P 7.10-3 so that

$$v_o(t) = \begin{cases} 0 & t < 0 \\ 250t & 0 \leq t < 20 \text{ ms} \\ 5 & t \geq 20 \text{ ms} \end{cases}$$

when

$$v_s(t) = \begin{cases} 0 & t < 0 \\ -5 & 0 \leq t < 20 \text{ ms} \\ 0 & t \geq 20 \text{ ms} \end{cases}$$

Both $v_s(t)$ and $v_o(t)$ have units of V.
Answer: $R = 20$ kΩ and $C = 1$ μF

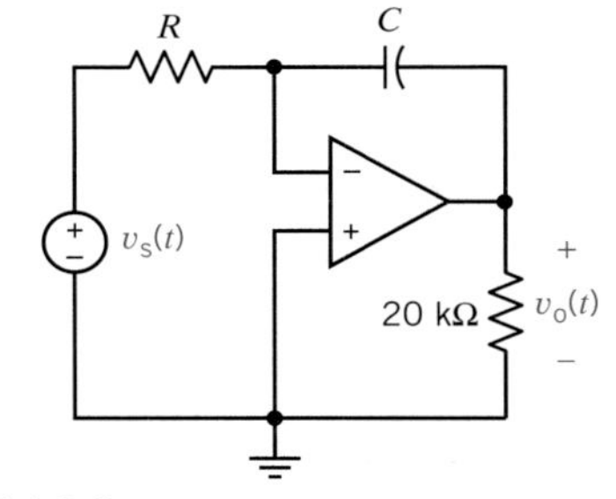

Figure P 7.10-3

VERIFICATION PROBLEMS

VP 7-1 A homework solution indicates that the current and voltage of a 100-H inductor are

$$i(t) = \begin{cases} .025 & t < 1 \\ -\dfrac{t}{25} + 0.065 & 1 < t < 3 \\ \dfrac{t}{50} - 0.115 & 3 < t < 9 \\ 0.065 & t > 9 \end{cases}$$

$$\text{and } v(t) = \begin{cases} 0 & t < 1 \\ -4 & 1 < t < 3 \\ 2 & 3 < t < 9 \\ 0 & t > 9 \end{cases}$$

where the units of current are A, the units of voltage are V, and the units of time are s. Verify that the inductor current does not change instantaneously.

VP 7-2 A homework solution indicates that the current and voltage of a 100-H inductor are

$$i(t) = \begin{cases} -\dfrac{t}{200} + 0.025 & t < 1 \\ -\dfrac{t}{100} + 0.03 & 1 < t < 4 \\ \dfrac{t}{100} - 0.03 & 4 < t < 9 \\ 0.015 & t > 9 \end{cases}$$

$$\text{and } v(t) = \begin{cases} -1 & t < 1 \\ -2 & 1 < t < 4 \\ 1 & 4 < t < 9 \\ 0 & t > 9 \end{cases}$$

where the units of current are A, the units of voltage are V, and the units of time are s. Is this homework solution correct? Justify your answer.

DESIGN PROBLEMS

DP 7-1 Select the resistance R for the circuit shown in Figure DP 7.1 so that $v(0) = 20$ V and $i(0) = 5$ A. Assume that the switch has been closed for a long time.

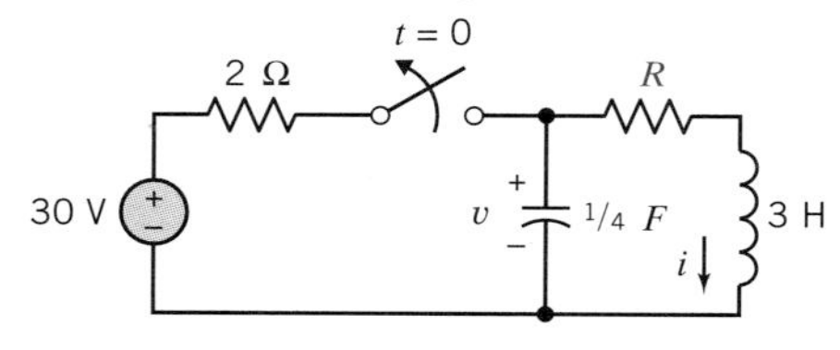

Figure DP 7.1

DP 7-2 A laser pulse power circuit is shown in Figure DP 7.2. It is required that $v(0) = 7.4$ V and $i(0) = 3.7$ A. Determine the required resistance R. Assume that the switch has been closed for a long time before it is opened at $t = 0$.

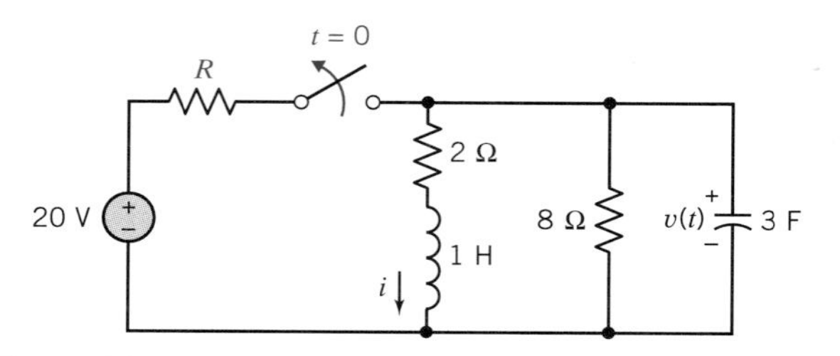

Figure DP 7.2 Laser pulse power circuit.

DP 7-3 A high-speed flash unit for sports photography requires a flash voltage $v(0^+) = 3$ V and

$$\left.\frac{dv(t)}{dt}\right|_{t=0^+} = 24 \text{ V/s}$$

The flash unit uses the circuit shown in Figure DP 7.3. Switch 1 has been closed a long time, and switch 2 has been open a long time at $t = 0$. Actually, the long time in this case is 3 s. Determine the required battery voltage, V_B, when $C = 1/8$ F.

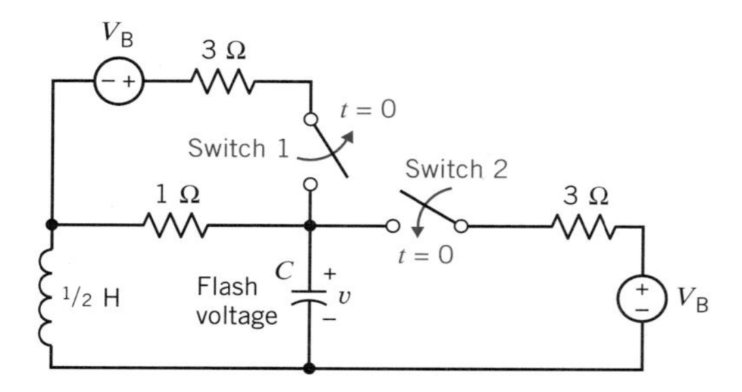

Figure DP 7.3 High-speed flash unit circuit.

DP 7-4 For the circuit shown in Figure DP 7.4, select a value of R so that the energy stored in the inductor is equal to the energy stored in the capacitor at steady state.

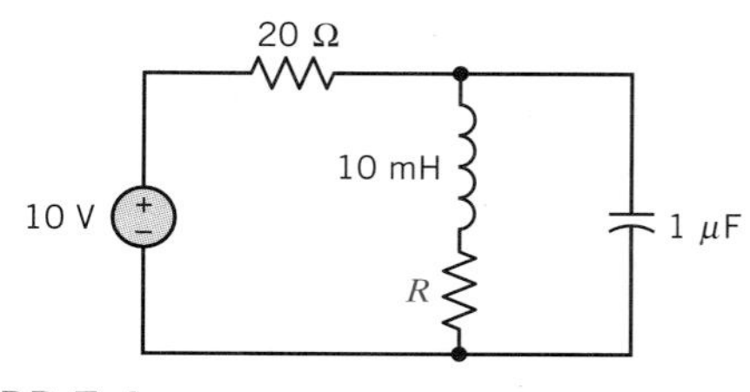

Figure DP 7.4

CHAPTER 8

The Complete Response of *RL* and *RC* Circuits

Preview

Circuits that contain capacitors and inductors can be represented by differential equations. The order of the differential equation is usually equal to the number of capacitors plus the number of inductors in the circuit. Circuits that contain only one inductor and no capacitors or only one capacitor and no inductors can be represented by a first-order differential equation. These circuits are called **first-order circuits.**

Thévenin and Norton equivalent circuits simplify the analysis of first-order circuits by showing that all first-order circuits are equivalent to one of two simple first-order circuits. Figure 8.0-1 shows how this is accomplished. In Figure 8.0-1*a*, a first-order circuit is partitioned into two parts. One part is the single capacitor or inductor that we expect to find in a first-order circuit. The other part is the rest of the circuit—everything except that capacitor or inductor. The next step, shown in Figure 8.0-1*b*, depends on whether the energy storage element is a capacitor or an inductor. If it is a capacitor, then the rest of the circuit is replaced by its Thévenin equivalent circuit. The result is

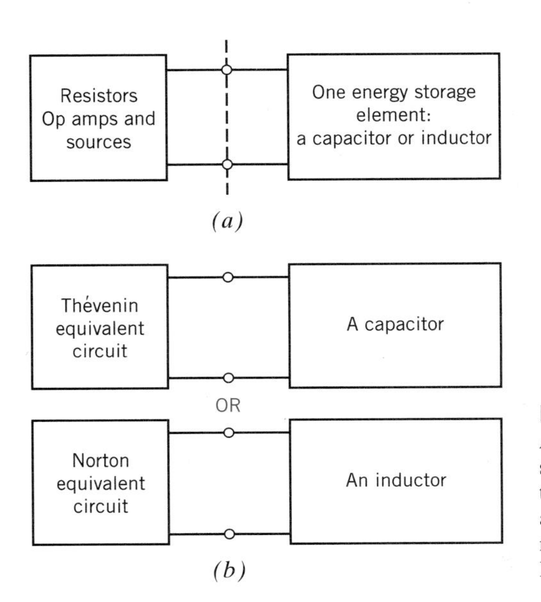

Figure 8.0-1
A plan for analyzing first-order circuits. (*a*) First, separate the energy storage element from the rest of the circuit. (*b*) Next, replace the circuit connected to a capacitor by its Thévenin equivalent circuit, or replace the circuit connected to an inductor by its Norton equivalent circuit.

a simple first-order circuit—a series circuit consisting of a voltage source, a resistor, and a capacitor. On the other hand, if the energy storage element is an inductor, then the rest of the circuit is replaced by its Norton equivalent circuit. The result is another simple first-order circuit—a parallel circuit consisting of a current source, a resistor, and an inductor. Indeed, all first-order circuits are equivalent to one of these two simple first-order circuits.

Consider the first-order circuit shown in Figure 8.0-2*a*. The input to this circuit is the voltage $v_s(t)$. The output, or response, of this circuit is the voltage across the capacitor. This circuit is at steady state before the switch is closed at time $t = 0$. Closing the switch disturbs this circuit. Eventually, the disturbance dies out and the circuit is again at steady state. The steady-state condition with the switch closed will probably be different from the steady-state condition with the switch open. Figure 8.0-2*b* shows a plot of the capacitor voltage versus time.

When the input to a circuit is sinusoidal, the steady-state response is also sinusoidal. Furthermore, the frequency of the response sinusoid must be the same as the frequency of the input sinusoid. The circuit shown in Figure 8.0-2*a* is at steady state before the switch is closed. The steady-state capacitor voltage will be

$$v(t) = B\cos(1000t + \phi), \quad t < 0 \tag{8.0-1}$$

The switch closes at time $t = 0$. The value of the capacitor voltage at the time the switch closes is

$$v(o) = B\cos(\phi), \quad t = 0 \tag{8.0-2}$$

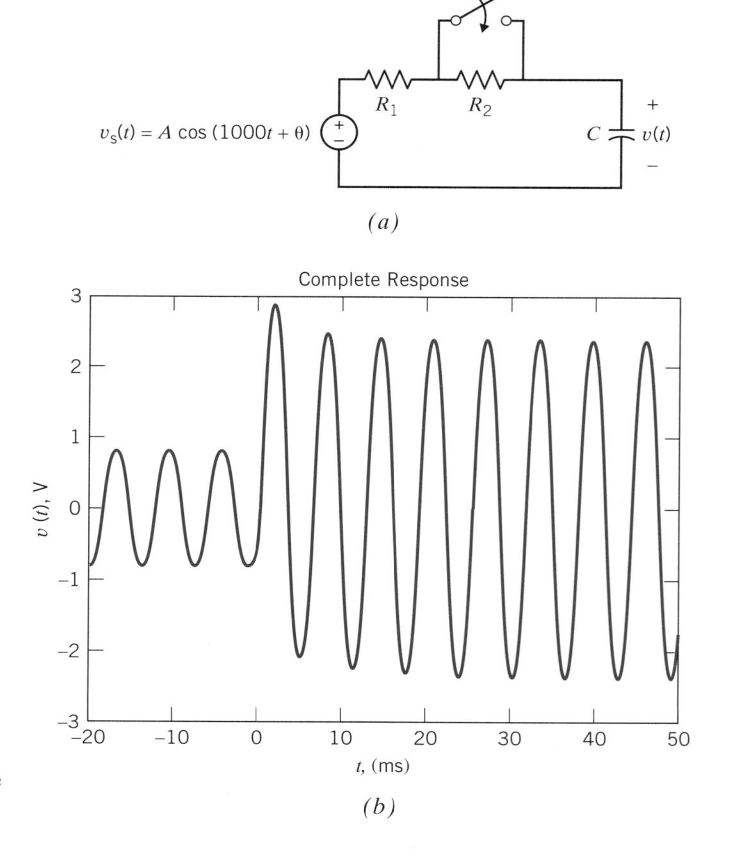

Figure 8.0-2
(*a*) A circuit and (*b*) its complete response.

After the switch closes, the response will consist of two parts: a transient part that eventually dies out and a steady-state part. The steady-state part of the response will be sinusoidal and will have the frequency of the input. For a first-order circuit, the transient part of the response is exponential. Indeed, we consider first-order circuits separately in order to take advantage of the simple form of the transient response of these circuits. After the switch is closed, the capacitor voltage is

$$v(t) = Ke^{-\frac{t}{\tau}} + M\cos(1000t + \delta) \tag{8.0-3}$$

Notice that $Ke^{-\frac{t}{\tau}}$ goes to zero as t becomes large. This is the transient part of the response. The transient part of the response dies out, leaving the steady-state response, $M\cos(1000t + \delta)$.

As a matter of vocabulary, the "transient part of the response" is frequently shortened to the **transient response** and the "steady-state part of the response" is shortened to the "steady-state response." The response, $v(t)$, given by Eq. 8.0-3, is called the **complete response** to contrast it with the transient and steady-state responses.

$$\text{Complete Response} = \text{Transient Response} + \text{Steady-State Response}$$

(The term *transient response* is used in two different ways by electrical engineers. Sometimes it refers to the "transient part of the complete response," and at other times it refers to a complete response, which includes a transient part. In particular, PSpice uses the term transient response to refer to the complete response. This can be confusing, so the term transient response must be used carefully.)

In general, the complete response of a first-order circuit can be represented as the sum of two parts, the **natural response** and the **forced response:**

$$\text{Complete Response} = \text{Natural Response} + \text{Forced Response}$$

The natural response is the general solution of the differential equation representing the first-order circuit, when the input is set to zero. The forced response is a particular solution of the differential equation representing the circuit.

The complete response of a first-order circuit will depend on an initial condition, usually a capacitor voltage or an inductor current at a particular time. Let t_o denote the time at which the initial condition is given. The natural response of a first-order circuit will be of the form

$$\text{Natural Response} = Ke^{-\frac{t-t_o}{\tau}}$$

When $t_o = 0$, as in Figure 8.0-2, then

$$\text{Natural Response} = Ke^{-\frac{t}{\tau}}$$

The constant K in the natural response depends on the initial condition, for example, the capacitor voltage at time t_0.

In this chapter, we will consider three cases. In these cases the input to the circuit after the disturbance will be (1) a constant, for example,

$$v_s(t) = V_o$$

or (2) an exponential, for example,

$$v_s(t) = V_o e^{-\frac{t}{\tau}}$$

or (3) a sinusoid, for example,

$$v_s(t) = V_o\cos(\omega t + \theta)$$

These three cases are special because the forced response will have the same form as the input. For example, in Figure 8.0-2, both the forced response and the input are sinusoidal, and the frequency of the forced response is the same as the frequency of the input. For other inputs, the forced response may not have the same form as the input. For example, when the input is a square wave, the forced response is not a square wave.

In the cases when the input is a constant or a sinusoid, the forced response is also called the steady-state response and the natural response is called the transient response.

Here is our plan for finding the complete response of first-order circuits:

Step 1: Find the forced response before the disturbance. Evaluate this response at time $t = t_0$ to obtain the initial condition of the energy storage element.

Step 2: Find the forced response after the disturbance.

Step 3: Add the natural response $= Ke^{-\frac{t}{\tau}}$ to the forced response to get the complete response. Use the initial condition to evaluate the constant K.

8.1 Design Challenge

A COMPUTER AND PRINTER

It is frequently necessary to connect two pieces of electronic equipment together so that the output from one device can be used as the input to another device. For example, this situation occurs when a printer is connected to a computer as shown in Figure 8.1-1*a*. This situation is represented more generally by the circuit shown in Figure 8.1-1*b*. Here a cable is used to connect two circuits. The driver (source) is used to send a signal through the cable to the receiver. Let us replace the driver, cable, and receiver with simple models. Model the driver as a voltage source, the cable as an *RC* circuit, and the receiver as an open circuit. The values of resistance and capacitance used to model the cable will depend on the length of the cable. For example, when RG58 coaxial cable is used,

$$R = r \cdot L \quad \text{where} \quad r = 0.54\,\frac{\Omega}{\text{m}}$$

and

$$C = c \cdot L \quad \text{where} \quad c = 88\,\frac{\text{pF}}{\text{m}}$$

and L is the length of the cable in meters. Figure 8.1-1*c* shows the equivalent circuit.

Suppose that the circuits connected by the cable are digital circuits. The driver will send 1's and 0's to the receiver. These 1's and 0's will be represented by voltages. The output of the driver will be one voltage, called V_{OH} (voltage out high), to represent logic 1 and another voltage, V_{OL} (voltage out low), to represent a logic 0. For example, one popular type of logic, called TTL (transistor-transistor logic) logic, uses $V_{OH} = 2.4$ V and $V_{OL} = 0.4$ V. The receiver uses two different voltages, V_{IH} and V_{IL}, to represent 1's and 0's. (This is done to provide noise immunity, but that is another story.) The receiver will interpret its input, v_b, to be a logic 1 whenever $v_b > V_{IH}$ and to be a logic 0 whenever $v_b < V_{IL}$. (Voltages between V_{IH} and V_{IL} will only occur during transitions between logic 1 and logic 0. These voltages will sometimes be interpreted as logic 1 and other times as logic 0.) TTL logic uses $V_{IH} = 2.0$ V and $V_{IL} = 0.8$ V.

Figure 8.1-2 shows what happens when the driver output changes from logic 0 to logic 1. Before time t_0,

$$v_a = V_{OL} \quad \text{and} \quad v_b < V_{IL} \quad \text{for} \quad t \le t_0$$

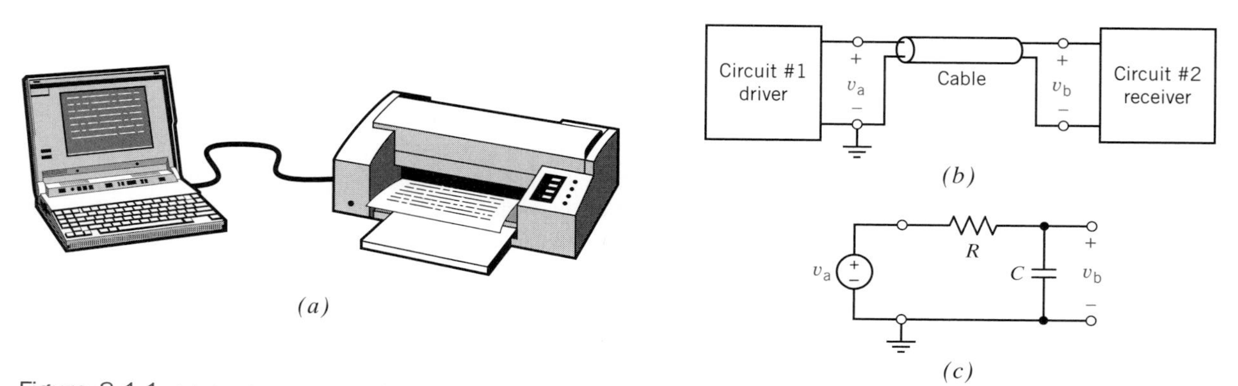

Figure 8.1-1 (*a*) A printer connected to a laptop computer. (*b*) Two circuits connected by a cable and (*c*) an equivalent circuit.

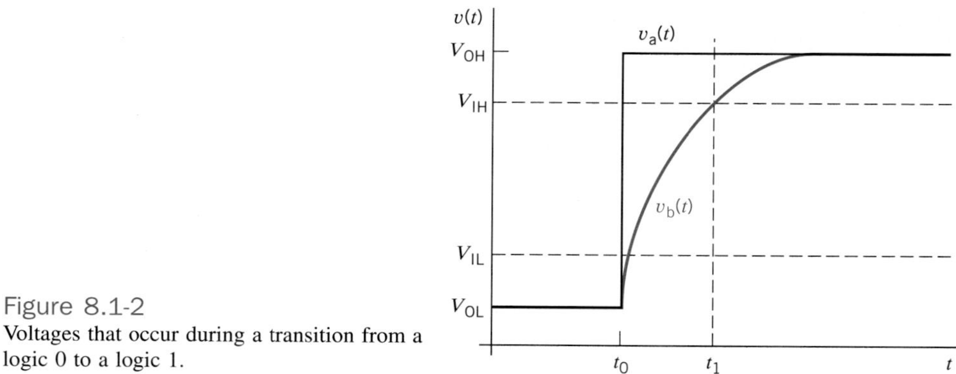

Figure 8.1-2
Voltages that occur during a transition from a logic 0 to a logic 1.

In words, a logic 0 is sent and received. The driver output switches to V_{OH} at time t_0. The receiver input, v_b, makes this transition more slowly. Not until time t_1 does the receiver input become large enough to be interpreted as a logic 1. That is,

$$v_b > V_{IH} \quad \text{for} \quad t \geq t_1$$

The time that it takes for the receiver to recognize the transition from logic 0 to logic 1

$$\Delta t = t_1 - t_0$$

is called the delay. This delay is important because it puts a limit on how fast 1's and 0's can be sent from the driver to the receiver. To ensure that the 1's and 0's are received reliably, each 1 and each 0 must last for at least Δt. The rate at which 1's and 0's are sent from the driver to the receiver is inversely proportional to the delay.

Suppose two TTL circuits are connected using RG58 coaxial cable. What restriction must be placed on the length of the cable to ensure that the delay, Δt, is less than 2 ns?

This chapter describes procedures for solving this type of problem. We will return to this problem at the end of the chapter.

8.2 SIGNALS AND COMMUNICATIONS

Attempts to communicate effectively have been pursued since society was first organized. A principal application of electric circuits is the communication of messages. Typically, the message is available in the form of *electric signals,* which may be a voltage or current varying in time in a manner that conveys information. The varying current in a telephone receiver is an example of an electric signal.

Perhaps the form of signal best known to schoolchildren is that used by Paul Revere to inform the patriot leaders of the advance of the British army. Revere, in 1775, used the visual signal method: "one lantern if by land, two if by sea."

A *signal* is defined as a real-valued function of time. By real valued, we mean that for any fixed value of time, the value of the signal at that time is a real number. Electric signals or waveforms are processed by electric circuits to yield the desired output form. Several signal waveforms are shown in Table 8.2-1. Frequently, electric signals are displayed using an oscilloscope as shown in Figure 8.2-1.

Information can be represented using electrical signals. Electrical communication devices process, transmit, and receive information by processing, transmitting, and receiving those electrical signals. Examples of electrical communication devices include underwater cable, telegraph, telephone, radio, television, and computers. A chronology of

Table 8.2-1 **Common Signal Waveforms**

Name	Equation	Waveform
1. Continuous or dc	$v(t) = V_0$	
2. Step	$v(t) = 0 \quad t < 0$ $= V_0 \quad t > 0$	
3. Decaying exponential	$v(t) = 0 \quad t < 0$ $= V_0 e^{-at} \quad t > 0$	
4. Ramp	$v(t) = 0 \quad t < 0$ $= Kt \quad t \geq 0$	
5. Pulse	$v(t) = V_0 \quad 0 < t < t_1$ $= 0$ elsewhere	
6. Sinusoid	$v(t) = V_0 \sin(\omega t + \theta)$	

Note: Voltage is used to designate the waveform equation; current could be used equally as well.

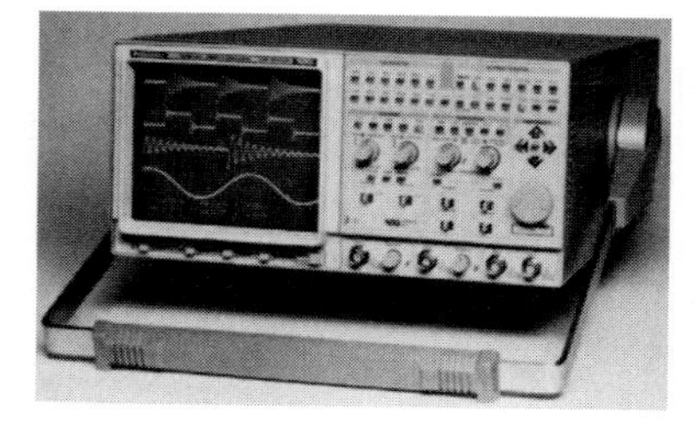

Figure 8.2-1
Common forms of electrical signals shown on the screen of an oscilloscope. Courtesy of Panasonic Industrial Co.

Table 8.2-2 **Chronology of Advances in Electrical Communication**

Year	Event
1838–1866	The birth of telegraphy. Morse perfects his telegraph. Transatlantic cables installed by Cyrus Field.
1864	*A Dynamical Theory of the Electromagnetic Field,* by James Clerk Maxwell, establishes the concept of electromagnetic radiation.
1876–1899	The birth of telephony. Acoustic transducer perfected by Alexander Graham Bell. First telephone exchange, in New Haven, Connecticut, with eight lines (1878).
1887–1907	Wireless telegraphy. Heinrich Hertz verifies Maxwell's theory. Demonstrations by Marconi and Popov. Marconi patents a complete wireless telegraph system (1897).
1904–1920	Radio. Lee De Forest invents the audion (triode based on Fleming's valve). First commercial broadcasting station, KDKA, opens in Pittsburgh in 1920.
1943–1946	John W. Mauchly and J. Presper Eckert build the ENIAC, the world's first true electronic digital computer, at the University of Pennsylvania.

important advances in electrical communications for the period 1838–1946 is given in Table 8.2-2.

As computers became more powerful and inexpensive, digital representation of electric signals became widespread. Digital representations of analog electric signals are obtained by sampling the analog signal and then encoding the samples as binary numbers. These binary numbers in turn are represented by electric signals. For example, 5 volts might represent the binary number 1, while 0 volts represents the binary number 0. Figure 8.2-2 illustrates this procedure. The analog signal, $v_a(t)$, is sampled to obtain the signal, $v_s(t)$, shown in Figure 8.2-2*a*. The sampled signal is then encoded as a sequence of binary numbers. Representing 1s by 5 volts and 0s by 0 volts produces the digital signal, $v_d(t)$, shown in Figure 8.2-2*b*.

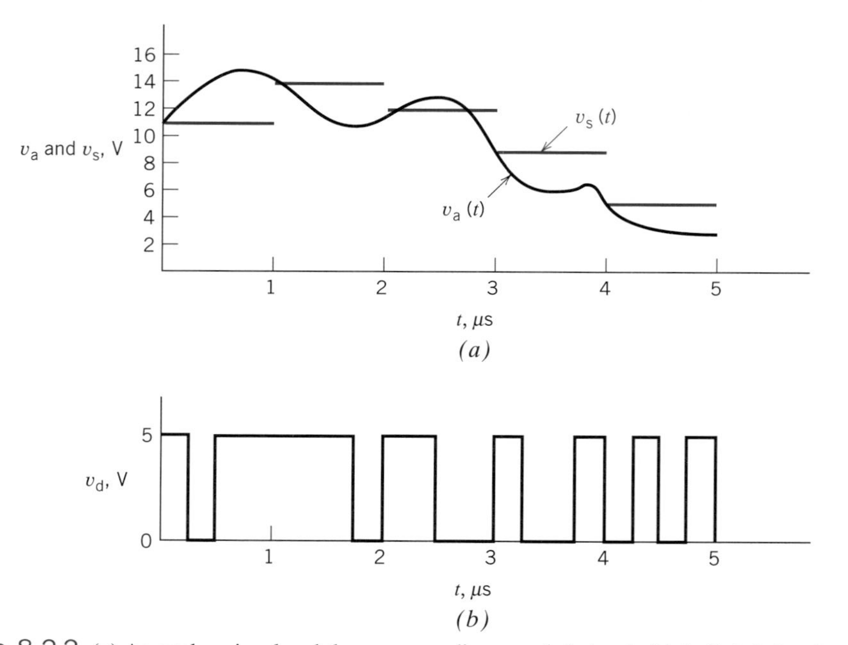

Figure 8.2-2 (*a*) An analog signal and the corresponding sampled signal. (*b*) A digital signal obtained by encoding the sampled signal.

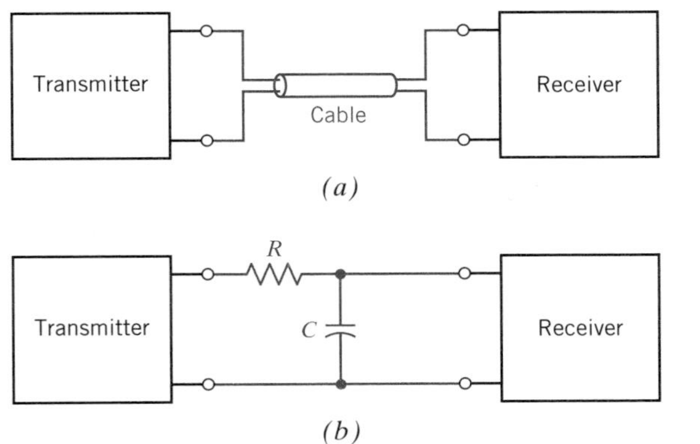

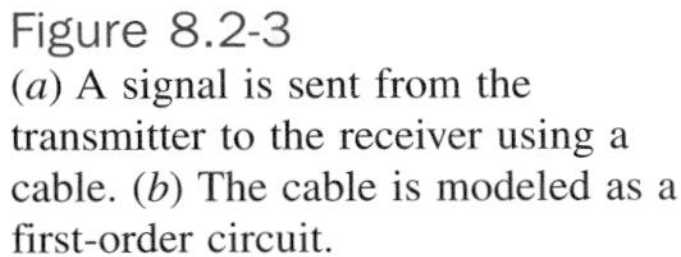

Figure 8.2-3
(*a*) A signal is sent from the transmitter to the receiver using a cable. (*b*) The cable is modeled as a first-order circuit.

The information represented by the digital signal can be transmitted by one circuit and received by another circuit. This situation occurs when a computer is connected to a printer using a cable, when two integrated circuits are connected by a path on a circuit board, or when two transistors inside an integrated circuit are connected by a conductor. In Figure 8.2-3*a*, two circuits are connected by a cable. A simple model of the cable is a first-order circuit. In Figure 8.2-3*b* the cable from Figure 8.2-3*a* has been replaced by this simple model. Analysis and design of first-order circuits like the one in Figure 8.2-3*b* is the subject of this chapter.

8.3 THE RESPONSE OF A FIRST-ORDER CIRCUIT TO A CONSTANT INPUT

In this section we find the complete response of a first-order circuit when the input to the circuit is constant after time t_o. Figure 8.3-1 illustrates this situation. In Figure 8.3-1*a* we find a first-order circuit that contains a single capacitor and no inductors. This circuit is at steady state before the switch closes, disturbing the steady state. The time at which steady state is disturbed is denoted as t_o. In Figure 8.3-1*a*, $t_o = 0$. Closing the switch removes the resistor R_1 from the circuit. (A closed switch is modeled by a short circuit. A short circuit in parallel with a resistor is equivalent to a short circuit.) After the switch closes, the circuit can be represented as shown in Figure 8.3-1*b*. In Figure 8.3-1*b* the part of the circuit that is connected to the capacitor has been replaced by its Thévenin equivalent circuit. Therefore

$$V_{oc} = \frac{R_3}{R_2 + R_3} V_s \quad \text{and} \quad R_{TH} = \frac{R_2 R_3}{R_2 + R_3}$$

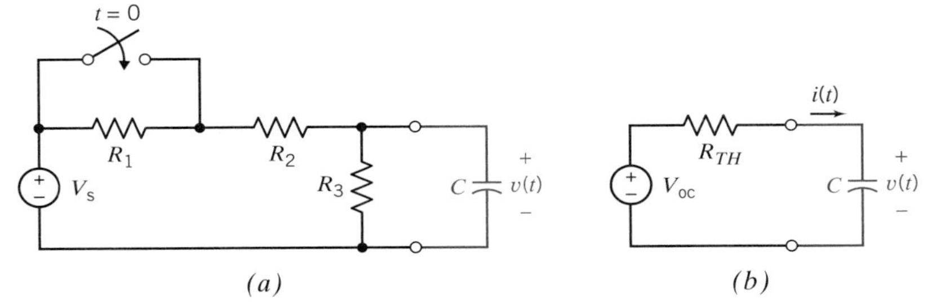

Figure 8.3-1 (*a*) A first-order circuit containing a capacitor. (*b*) After the switch closes, the circuit connected to the capacitor is replaced by its Thévenin equivalent circuit.

Let's represent the circuit in Figure 8.3-1*b* by a differential equation. The capacitor current is given by

$$i(t) = C\frac{d}{dt}v(t)$$

The same current, $i(t)$, passes through the resistor. Apply KVL to Figure 8.3-1*b* to get

$$V_{oc} = R_{TH}\,i(t) + v(t) = R_{TH}\left(C\frac{d}{dt}v(t)\right) + v(t)$$

Therefore,

$$\frac{d}{dt}v(t) + \frac{v(t)}{R_{TH}C} = \frac{V_{oc}}{R_{TH}C} \tag{8.3-1}$$

The highest order derivative in this equation is first order, so this is a first order differential equation.

Next, let's turn our attention to the circuit shown in Figure 8.3-2*a*. This circuit contains a single inductor and no capacitors. This circuit is at steady state before the switch closes at time $t_o = 0$, disturbing the steady state. After the switch closes, the circuit can be represented as shown in Figure 8.3-2*b*. In Figure 8.3-2*b* the part of the circuit that is connected to the inductor has been replaced by its Norton equivalent circuit. We calculate

$$I_{sc} = \frac{V_s}{R_2} \quad \text{and} \quad R_{TH} = \frac{R_2R_3}{R_2 + R_3}$$

Let's represent the circuit in Figure 8.3-2*b* by a differential equation. The inductor voltage is given by

$$v(t) = L\frac{d}{dt}i(t)$$

The voltage, $v(t)$, appears across the resistor. Apply KCL to the top node in Figure 8.3-2*b* to get

$$I_{sc} = \frac{v(t)}{R_{TH}} + i(t) = \frac{L\dfrac{d}{dt}i(t)}{R_{TH}} + i(t)$$

Therefore,

$$\frac{d}{dt}i(t) + \frac{R_{TH}}{L}i(t) = \frac{R_{TH}}{L}I_{sc} \tag{8.3-2}$$

As before, this is a first-order differential equation.

Equations 8.3-1 and 8.3-2 have the same form. That is

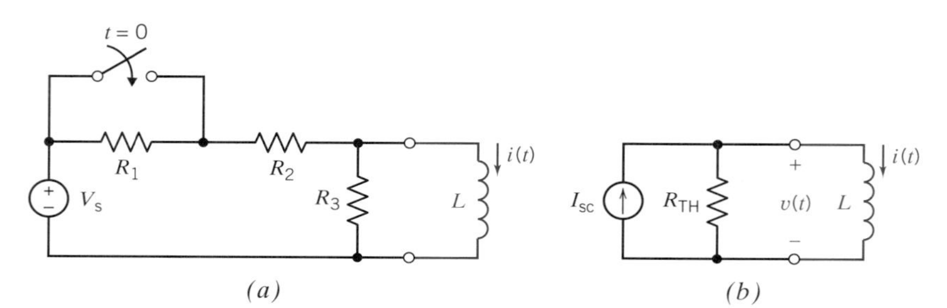

Figure 8.3-2 (*a*) A first-order circuit containing an inductor. (*b*) After the switch closes, the circuit connected to the inductor is replaced by its Norton equivalent circuit.

$$\frac{d}{dt}x(t) + \frac{x(t)}{\tau} = K \qquad (8.3\text{-}3)$$

The parameter τ is called the time constant. We will solve this differential equation by separating the variables and integrating. Then we will use the solution of Eq. 8.3-3 to obtain solutions of Eqs. 8.3-1 and 8.3-2.

We may rewrite Eq. 8.3-3 as

$$\frac{dx}{dt} = \frac{K\tau - x}{\tau}$$

or, separating the variables,

$$\frac{dx}{x - K\tau} = -\frac{dt}{\tau}$$

Forming the indefinite integral, we have

$$\int \frac{dx}{x - K\tau} = -\frac{1}{\tau}\int dt + D$$

where D is a constant of integration. Performing the integration, we have

$$\ln(x - K\tau) = -\frac{t}{\tau} + D$$

Solving for x gives

$$x(t) = K\tau + Ae^{-\frac{t}{\tau}}$$

where $A = e^D$, which is determined from the initial condition, $x(0)$. To find A, let $t = 0$. Then

$$x(0) = K\tau + Ae^{-\frac{0}{\tau}} = K\tau + A$$

or

$$A = x(0) - K\tau$$

Therefore, we obtain

$$x(t) = K\tau + [x(0) - K\tau]e^{-\frac{t}{\tau}} \qquad (8.3\text{-}4)$$

Since

$$x(\infty) = \lim_{t\to\infty} x(t) = K\tau$$

Eq. 8.3-4 can be written as

$$x(t) = x(\infty) + [x(0) - x(\infty)]e^{-\frac{t}{\tau}}$$

Taking the derivative of $x(t)$ with respect to t leads to a procedure for measuring or calculating the time constant.

$$\frac{d}{dt}x(t) = -\frac{1}{\tau}[x(0) - x(\infty)]e^{-\frac{t}{\tau}}$$

Now let $t = 0$ to get

$$\left.\frac{d}{dt}x(t)\right|_{t=0} = -\frac{1}{\tau}[x(0) - x(\infty)]$$

or

$$\tau = \frac{x(\infty) - x(0)}{\left.\frac{d}{dt}x(t)\right|_{t=0}} \qquad (8.3\text{-}5)$$

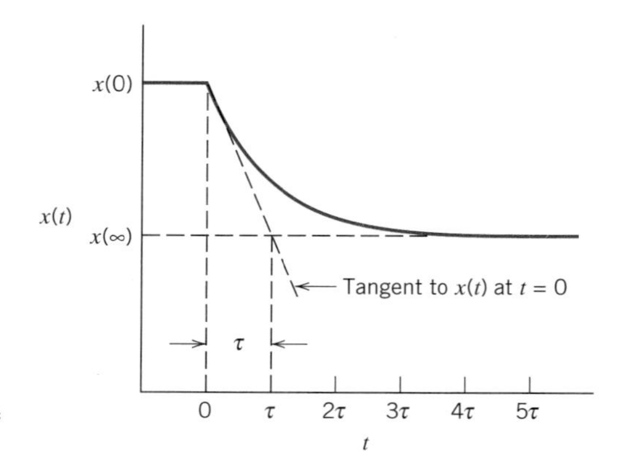

Figure 8.3-3
A graphical technique for measuring the time constant of a first-order circuit.

Figure 8.3-3 shows a plot of $x(t)$ versus t. We can determine the values of (1) the slope of the plot at time $t = 0$, (2) the initial value of $x(t)$, and (3) the final value of $x(t)$ from this plot. Equation 8.3-5 can be used to calculate the time constant from these values. Equivalently, Figure 8.3-3 shows how to measure the time constant from a plot of $x(t)$ versus t.

Next, we apply these results to the RC circuit in Figure 8.3-1. Comparing Eqs. 8.3-1 and 8.3-3, we see that

$$x(t) = v(t), \tau = R_{TH}C, \quad \text{and} \quad K = \frac{V_{oc}}{R_{TH}C}$$

Making these substitutions in Eq. 8.3-4 gives

$$v(t) = V_{oc} + (v(0) - V_{oc})e^{-\frac{t}{R_{TH}C}} \tag{8.3-6}$$

The second term on the right-hand side of Eq. 8.3-6 dies out as t increases. This is the transient or natural response. At $t = 0$, $e^{-0} = 1$. Letting $t = 0$ in Eq. 8.3-6 gives $v(0) = v(0)$, as required. When $t = 5\tau$, $e^{-5} = 0.0067 \approx 0$ so at time $t = 5$ the capacitor voltage will be

$$v(5\tau) = 0.9933\, V_{oc} + 0.0067\, v(0) \approx V_{oc}$$

This is the steady-state or forced response. The forced response is of the same form, a constant, as the input to the circuit. The sum of the natural and forced responses is the complete response.

$$\text{Complete Response} = v(t), \quad \text{Forced Response} = V_{oc},$$

and

$$\text{Natural Response} = (v(0) - V_{oc})e^{-\frac{t}{R_{TH}C}}$$

Next, compare Eqs. 8.3-2 and 8.3-3 to find the solution of the RL circuit in Figure 8.3-2. We see that

$$x(t) = i(t), \quad \tau = \frac{L}{R_{TH}}, \quad \text{and} \quad K = \frac{L}{R_{TH}} I_{sc}$$

Making these substitutions in Eq. 8.3-4 gives

$$i(t) = I_{sc} + (i(0) - I_{sc})e^{-\frac{R_{TH}}{L}t} \tag{8.3-7}$$

Again, the complete response is the sum of the forced (steady-state) response and the transient (natural) response.

$$\text{Complete Response} = i(t), \quad \text{Forced Response} = I_{sc},$$

and

$$\text{Natural Response} = (i(0) - I_{sc})e^{-\frac{R_{TH}}{L}t}$$

Example 8.3-1

Find the capacitor voltage after the switch opens in the circuit shown in Figure 8.3-4*a*. What is the value of the capacitor voltage 50 ms after the switch opens?

Solution

The 2-volt voltage source forces the capacitor voltage to be 2 volts until the switch opens. Since the capacitor voltage cannot change instantaneously, the capacitor voltage will be 2 volts immediately after the switch opens. Therefore, the initial condition is

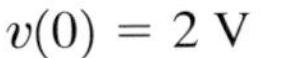

$$v(0) = 2 \text{ V}$$

Figure 8.3-4*b* shows the circuit after the switch opens. Comparing this circuit to the *RC* circuit in Figure 8.3-1*b*, we see that

$$R_{TH} = 10 \text{ k}\Omega \quad \text{and} \quad V_{oc} = 8 \text{ V}$$

The time constant for this first-order circuit containing a capacitor is

$$\tau = R_{TH}\,C = (10 \times 10^3)(2 \times 10^{-6}) = 20 \times 10^{-3} = 20 \text{ ms}$$

Substituting these values into Eq. 8.3-6 gives

$$v(t) = 8 - 6e^{-\frac{t}{20}} \text{ V} \tag{8.3-8}$$

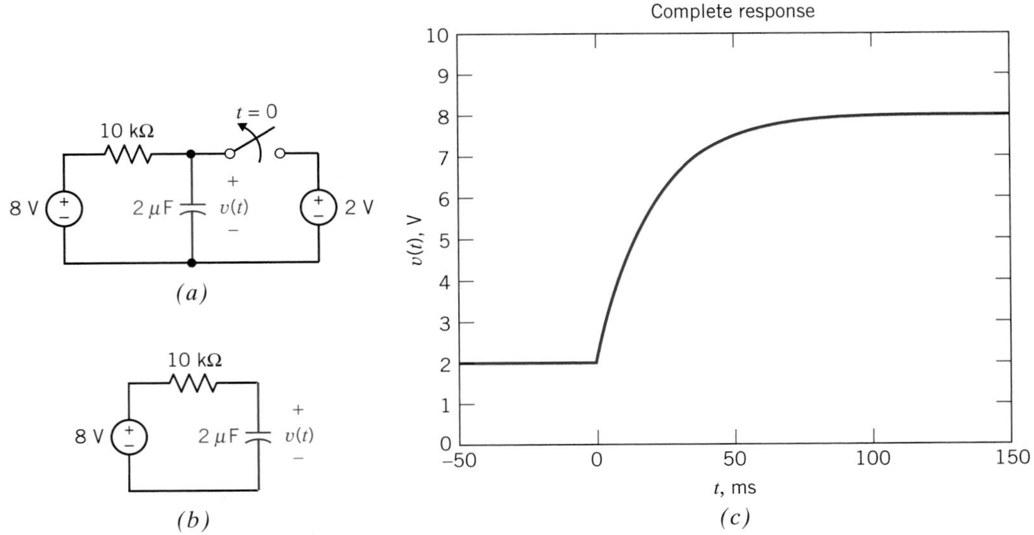

Figure 8.3-4 (*a*) A first-order circuit and (*b*) an equivalent circuit that is valid after the switch opens. (*c*) A plot of the complete response, $v(t)$, given in Eq. 8.3-8.

where t has units of ms. To find the voltage 50 ms after the switch opens, let $t = 50$. Then

$$v(50) = 8 - 6e^{-\frac{50}{20}} = 7.51 \text{ V}$$

Figure 8.3-4*c* shows a plot of the capacitor voltage as a function of time.

Example 8.3-2

Find the inductor current after the switch closes in the circuit shown in Figure 8.3-5*a*. How long will it take for the inductor current to reach 2 mA?

Solution

The inductor current will be 0 A until the switch closes. Since the inductor current cannot change instantaneously, it will be 0 A immediately after the switch closes. Therefore, the initial condition is

$$i(0) = 0$$

Figure 8.3-5*b* shows the circuit after the switch closes. Comparing this circuit to the *RL* circuit in Figure 8.3-2*b*, we see that

$$R_{TH} = 1000\ \Omega \quad \text{and} \quad I_{sc} = 4 \text{ mA}$$

The time constant for this first-order circuit containing an inductor is

$$\tau = \frac{L}{R_{TH}} = \frac{5 \times 10^{-3}}{1000} = 5 \times 10^{-6} = 5\,\mu\text{s}$$

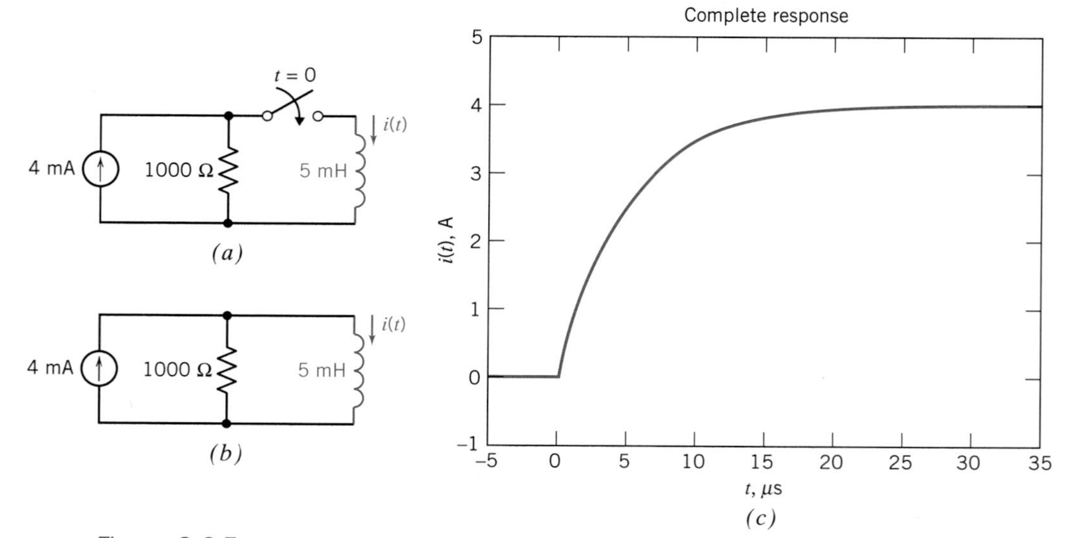

Figure 8.3-5 (*a*) A first-order circuit and (*b*) an equivalent circuit that is valid after the switch closes. (*c*) A plot of the complete response, $i(t)$, given by Eq. 8.3-9.

Substituting these values into Eq. 8.3-7 gives

$$i(t) = 4 - 4e^{-\frac{t}{5}}\,\text{mA} \tag{8.3-9}$$

where t has units of microseconds. To find the time when the current reaches 2 mA, substitute $i(t) = 2$ mA. Then

$$2 = 4 - 4e^{-\frac{t}{5}}\,\text{mA}$$

Solving for t gives

$$t = -5 \times \ln\left(\frac{2-4}{-4}\right) = 3.47\ \mu\text{s}$$

Figure 8.3-5c shows a plot of the inductor current as a function of time.

Example 8.3-3

The switch in Figure 8.3-6a has been open for a long time, and the circuit has reached steady state before the switch closes at time $t = 0$. Find the capacitor voltage after the switch closes.

Solution

The switch has been open for a long time before it closes at time $t = 0$. The circuit will have reached steady state before the switch closes. Since the input to this circuit is a constant, all the element currents and voltages will be constant when the circuit is at steady state. In particular, the capacitor voltage will be constant. The capacitor current will be

$$i(t) = C\frac{d}{dt}v(t) = C\frac{d}{dt}(\text{a constant}) = 0$$

The capacitor voltage is unknown, but the capacitor current is zero. In other words, the capacitor acts like an open circuit when the input is constant and the circuit is at steady state. (By a similar argument, inductors act like short circuits when the input is constant and the circuit is at steady state.)

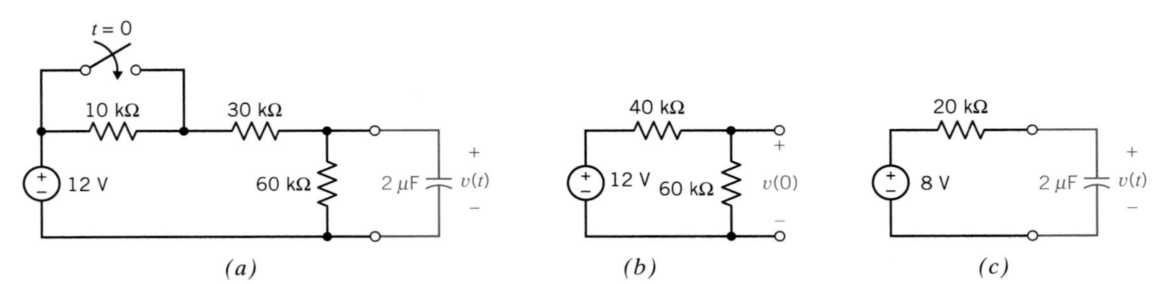

Figure 8.3-6 (a) A first-order circuit. The equivalent circuit for (b) $t < 0$ and (c) $t > 0$.

Figure 8.3-6*b* shows the appropriate equivalent circuit while the switch is open. An open switch acts like an open circuit, thus the 10 kΩ and 30 kΩ resistors are in series. They have been replaced by an equivalent 40 kΩ resistor. The input to the circuit is a constant (12 volts), and the circuit is at steady state, therefore the capacitor acts like an open circuit. The voltage across this open circuit is the capacitor voltage. Since we are interested in the initial condition, the capacitor voltage has been labeled as $v(0)$. Analyzing the circuit in Figure 8.3-6*b* using voltage division gives

$$v(0) = \frac{60 \times 10^3}{40 \times 10^3 + 60 \times 10^3} 12 = 7.2 \text{ V}$$

Figure 8.3-6*c* shows the appropriate equivalent circuit after the switch closes. Closing the switch shorts out the 10 kΩ resistor, removing it from the circuit. (A short circuit in parallel with any resistor is equivalent to a short circuit.) The part of the circuit that is connected to the capacitor has been replaced by its Thévenin equivalent circuit. After the switch is closed

$$V_{OC} = \frac{60 \times 10^3}{30 \times 10^3 + 60 \times 10^3} 12 = 8 \text{ V}$$

and
$$R_{TH} = \frac{30 \times 10^3 \times 60 \times 10^3}{30 \times 10^3 + 60 \times 10^3} = 20 \times 10^3 = 20 \text{ k}\Omega$$

and the time constant is

$$\tau = R_{TH} \times C = (20 \times 10^3) \times (2 \times 10^{-6}) = 40 \times 10^{-3} = 40 \text{ ms}$$

Substituting these values into Eq. 8.3-6 gives

$$v(t) = 8 - 0.8e^{-\frac{t}{40}} \text{ V}$$

where t has units of ms.

Example 8.3-4

The switch in Figure 8.3-7*a* has been open for a long time, and the circuit has reached steady state before the switch closes at time $t = 0$. Find the inductor current after the switch closes.

Solution

Figure 8.3-7*b* shows the appropriate equivalent circuit while the switch is open. The 100 Ω and 200 Ω resistors are in series and have been replaced by an equivalent 300 Ω resistor. The input to the circuit is a constant (12 volts), and the circuit is at steady state, therefore the inductor acts like a short circuit. The current in this short circuit is the inductor current. Since we are interested in the initial condition, the initial inductor current has been labeled as $i(0)$. This current can be calculated using Ohm's law

$$i(0) = \frac{12}{300} = 40 \text{ mA}$$

Figure 8.3-7*c* shows the appropriate equivalent circuit after the switch closes. Closing the switch shorts out the 100 Ω resistor, removing it from the circuit. The part of the cir-

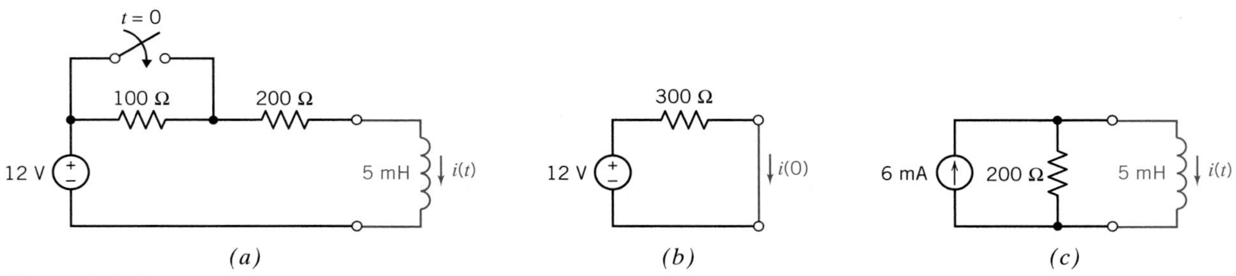

Figure 8.3-7 (*a*) A first-order circuit. The equivalent circuit for (*b*) $t < 0$ and (*c*) $t > 0$.

cuit that is connected to the inductor has been replaced by its Norton equivalent circuit. After the switch is closed

$$I_{sc} = \frac{12}{200} = 60\text{ mA} \quad \text{and} \quad R_{TH} = 200\ \Omega$$

and the time constant is

$$\tau = \frac{L}{R_{TH}} = \frac{5 \times 10^{-3}}{200} = 25 \times 10^{-6} = 25\ \mu\text{s}$$

Substituting these values into Eq. 8.3-7 gives

$$i(t) = 60 - 20e^{-\frac{t}{25}}\text{ mA}$$

where t has units of microseconds.

Example 8.3-5

The circuit in Figure 8.3-8*a* is at steady state before the switch opens. Find the current $i(t)$ after the switch opens.

Solution

The response or output of a circuit can be any element current or voltage. Frequently, the response is not the capacitor voltage or inductor current. In Figure 8.3-8*a*, the

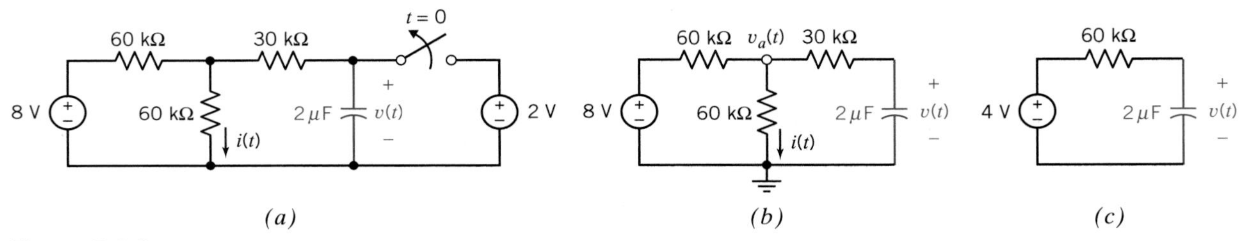

Figure 8.3-8 (*a*) A first-order circuit, (*b*) the circuit after the switch opens, and (*c*) the equivalent circuit after the switch opens.

response is the current $i(t)$ in a resistor rather than the capacitor voltage. In this case, two steps are required to solve the problem. First, find the capacitor voltage using the methods already described in this chapter. Once the capacitor voltage is known, write node or mesh equations to express the response in terms of the input and the capacitor voltage.

First we find the capacitor voltage. Before the switch opens, the capacitor voltage is equal to the voltage of the 2-volt source. The initial condition is

$$v(0) = 2 \text{ V}$$

Figure 8.3-8*b* shows the circuit as it will be after the switch is opened. The part of the circuit connected to the capacitor has been replaced by its Thévenin equivalent circuit in Figure 8.3-8*c*. The parameters of the Thévenin equivalent circuit are

$$V_{oc} = \frac{60 \times 10^3}{60 \times 10^3 + 60 \times 10^3} 8 = 4 \text{ V}$$

and

$$R_{TH} = 30 \times 10^3 + \frac{60 \times 10^3 \times 60 \times 10^3}{60 \times 10^3 + 60 \times 10^3} = 60 \times 10^3 = 60 \text{ k}\Omega$$

The time constant is

$$\tau = R_{TH} \times C = (60 \times 10^3) \times (2 \times 10^{-6}) = 120 \times 10^{-3} = 120 \text{ ms}$$

Substituting these values into Eq. 8.3-6 gives

$$v(t) = 4 - 2e^{-\frac{t}{120}} \text{ V}$$

where t has units of ms.

Now that the capacitor voltage is known, we return to the circuit in Figure 8.3-8*b*. Notice that the node voltage at the middle node at the top of the circuit has been labeled as $v_a(t)$. The node equation corresponding to this node is

$$\frac{v_a(t) - 8}{60 \times 10^3} + \frac{v_a(t)}{60 \times 10^3} + \frac{v_a(t) - v(t)}{30 \times 10^3} = 0$$

Substituting the expression for the capacitor voltage gives

$$\frac{v_a(t) - 8}{60 \times 10^3} + \frac{v_a(t)}{60 \times 10^3} + \frac{v_a(t) - (4 - 2e^{-\frac{t}{120}})}{30 \times 10^3} = 0$$

or

$$v_a(t) - 8 + v_a(t) + 2[v_a(t) - (4 - 2e^{-\frac{t}{120}})] = 0$$

Solving for $v_a(t)$, we get

$$v_a(t) = \frac{8 + 2(4 - 2e^{-\frac{t}{120}})}{4} = 4 - e^{-\frac{t}{120}} \text{ V}$$

Finally, we calculate $i(t)$ using Ohm's law

$$i(t) = \frac{v_a(t)}{60 \times 10^3} = \frac{4 - e^{-\frac{t}{120}}}{60 \times 10^3} = 66.7 - 16.7e^{-\frac{t}{120}} \ \mu\text{A}$$

Exercise 8.3-1 The circuit shown in Figure E 8.3-1 is at steady state before the switch closes at time $t = 0$ s. Determine the capacitor voltage, $v(t)$, for $t > 0$ s.

Answer: $v(t) = 2 + e^{-2.5t}$ V for $t > 0$

Exercise 8.3-2 The circuit shown in Figure E 8.3-2 is at steady state before the switch closes at time $t = 0$ s. Determine the inductor current, $i(t)$, for $t > 0$ s.

Answer: $i(t) = \frac{1}{4} + \frac{1}{12}e^{-1.33t}$ A for $t > 0$

Exercise 8.3-3 The circuit in Figure E 8.3-3 is at steady state before the switch closes. Find the inductor current after the switch closes.

Hint: $i(0) = 0.1\ A,\ \ I_{sc} = 0.3\ A,\ \ R_{TH} = 40\ \Omega$

Answer: $i(t) = 0.3 - 0.2e^{-2t}$ A $t \geq 0$

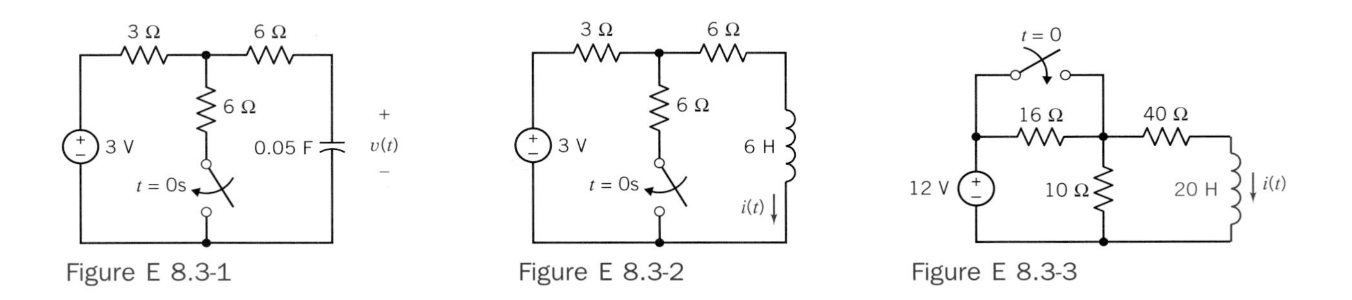

Figure E 8.3-1 Figure E 8.3-2 Figure E 8.3-3

Exercise 8.3-4 The circuit in Figure E 8.3-4 is at steady state before the switch closes. Find the capacitor voltage after the switch closes.

Hint: $v(0) = 12$ V, $V_{oc} = 12$ V

Answer: $v(t) = 12.0$ V

Exercise 8.3-5 The circuit shown in Figure E 8.3-5 is at steady state before the switch closes. The response of the circuit is the voltage $v(t)$. Find $v(t)$ for $t > 0$.

Hint: After the switch closes, the inductor current is $i(t) = 0.2\,(1 - e^{-1.8t})$ A

Answer: $v(t) = 8 + e^{-1.8t}$ V

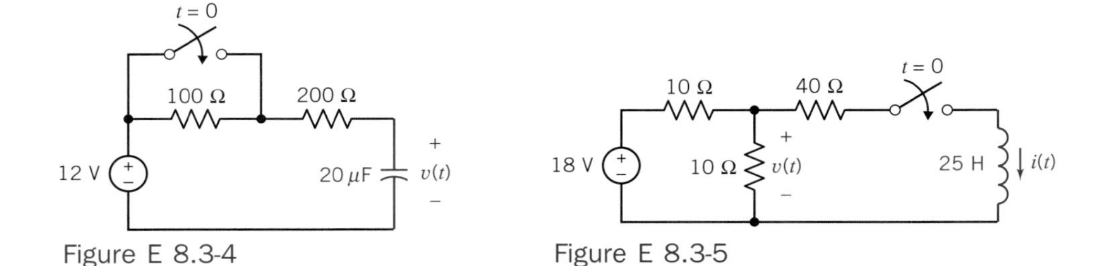

Figure E 8.3-4 Figure E 8.3-5

Exercise 8.3-6 The circuit shown in Figure E 8.3-6 is at steady state before the switch closes. The response of the circuit is the voltage $v(t)$. Find $v(t)$ for $t > 0$.

Answer: $v(t) = 37.5 - 97.5e^{-6400t}$ V

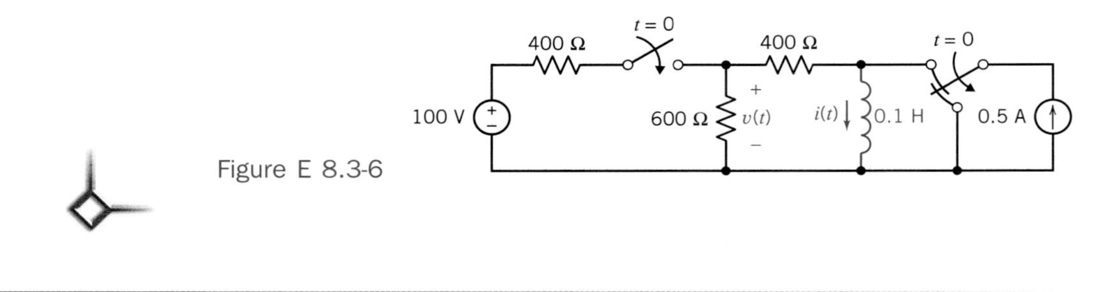

Figure E 8.3-6

8.4 SEQUENTIAL SWITCHING

Often circuits contain several switches that are not switched at the same time. For example, a circuit may have two switches where the first switch changes state at time $t = 0$ and the second switch closes at $t = 1$ ms.

Sequential switching occurs when a circuit contains two or more switches that change state at different instants.

Circuits with sequential switching can be solved using the methods described in the previous sections, based on the fact that inductor currents and capacitor voltages do not change instantaneously.

As an example of sequential switching, consider the circuit shown in Figure 8.4-1*a*. This circuit contains two switches—one that changes state at time $t = 0$, and a second that closes at $t = 1$ ms. Suppose this circuit has reached steady state before the switch changes state at time $t = 0$. Figure 8.4-1*b* shows the equivalent circuit that is appropriate for $t < 0$. Since the circuit is at steady state and the input is constant, the inductor acts like a short circuit and the current in this short is the inductor current. The short circuit forces the voltage across the resistor to be zero, so the current in the resistor is also zero. As a result, all of the source current flows in the short circuit and

$$i(t) = 10 \text{ A} \quad t < 0$$

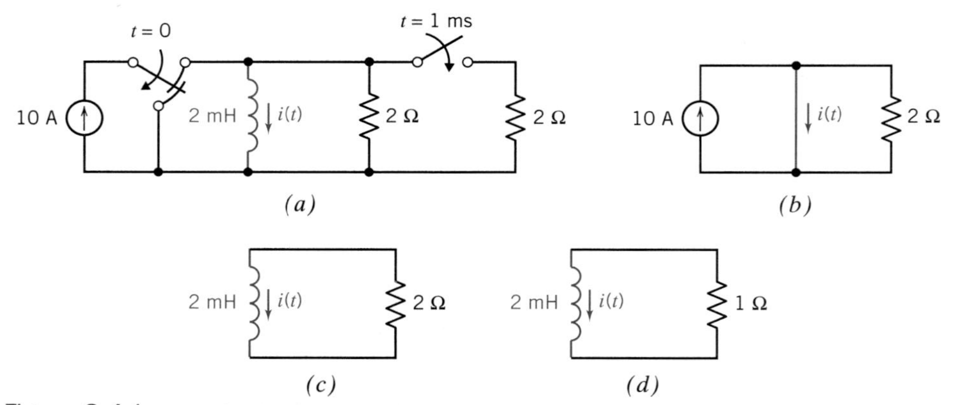

Figure 8.4-1 (*a*) A circuit with sequential switching. (*b*) The equivalent circuit before $t = 0$. (*c*) The equivalent circuit for $0 < t < 1$ ms. (*d*) The equivalent circuit after $t = 1$ ms.

The inductor current will be 10 A immediately before the switch changes state at time $t = 0$. We express this as

$$i(0^-) = 10 \text{ A}$$

Since the inductor current does not change instantaneously, the inductor current will also be 10 Amps immediately after the switch changes state. That is,

$$i(0^+) = 10 \text{ A}$$

This is the initial condition that is used to calculate the inductor current after $t = 0$. Figure 8.4-1*c* shows the equivalent circuit that is appropriate after one switch changes state at time $t = 0$ and before the other switch closes at time $t = 1\,\text{ms}$. We see that the Norton equivalent of the part of the circuit connected to the inductor has the parameters

$$I_{sc} = 0 \text{ A} \quad \text{and} \quad R_{TH} = 2\ \Omega$$

The time constant of this first-order circuit is

$$\tau = \frac{L}{R_{TH}} = \frac{2 \times 10^{-3}}{2} = 1 \times 10^{-3} = 1\,\text{ms}$$

The inductor current is

$$i(t) = i(0)e^{-\frac{t}{\tau}} = 10e^{-t} \text{ A}$$

for $0 < t < 1\,\text{ms}$. Notice that t has units of ms. Immediately before the other switch closes at time $t = 1$ ms, the inductor current will be

$$i(1^-) = 10e^{-1} = 3.68 \text{ A}$$

Since the inductor current does not change instantaneously, inductor current will also be 3.68 A immediately after the switch changes state. That is,

$$i(1^+) = 3.68 \text{ A}$$

This is the initial condition that is used to calculate the inductor current after the switch closes at time $t = 1\,\text{ms}$. Figure 8.4-1*d* shows the appropriate equivalent circuit. We see that the Norton equivalent of the part of the circuit connected to the inductor has the parameters

$$I_{sc} = 0 \text{ A} \quad \text{and} \quad R_{TH} = 1\ \Omega$$

The time constant of this first-order circuit is

$$\tau = \frac{L}{R_{TH}} = \frac{2 \times 10^{-3}}{1} = 2 \times 10^{-3} = 2\,\text{ms}$$

The inductor current is

$$i(t) = i(t_0)e^{-\frac{t-t_0}{\tau}} = 3.68e^{-\frac{t-1}{2}} \text{ A}$$

for $1\,\text{ms} < t$. Once again, t has units of ms. Also, t_0 denotes the time when the switch changes state, 1 ms in this example.

Figure 8.4-2 shows a plot of the inductor current. The time constant changes when the second switch closes. As a result, the slope of the plot changes at $t = 1\,\text{ms}$. Immediately before the switch closes, the slope is -3.68 A/ms. Immediately after the switch closes, the slope becomes $-3.68/2$ A/ms.

In some applications, switching occurs at prescribed voltage values rather than at prescribed times. Figure 8.4-3 shows a device, called a comparator, that can be used to

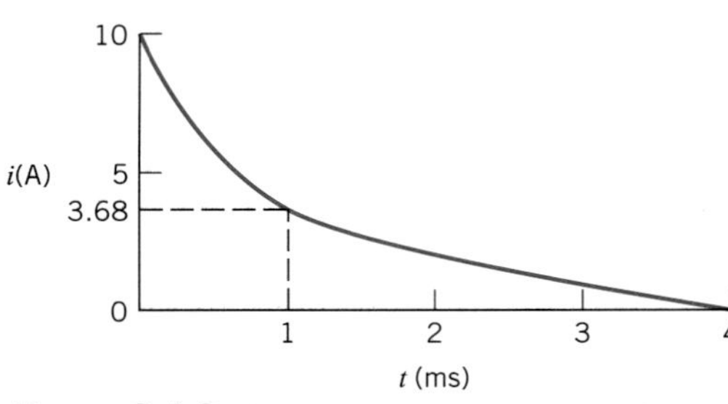

Figure 8.4-2 Current waveform for $t \geq 0$. The exponential has a different time constant for $0 \leq t < t_1$ and for $t \geq t_1$ where $t_1 = 1$ ms.

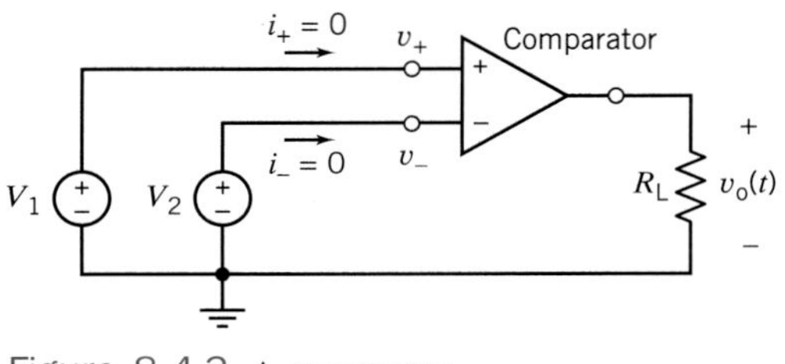

Figure 8.4-3 A comparator.

accomplish this kind of switching. (The symbol for the comparator is the same as the symbol of an op amp, so we add the label "comparator" to avoid confusion.) The input currents of the comparator are zero. The output voltage of the comparator will be one of two values, depending on the relative values of its input voltages. The relationship is

$$v_o(t) = \begin{cases} V_H & if \quad v_+ > v_- \\ V_L & if \quad v_+ < v_- \end{cases}$$

In this equation, v_o is the node voltage at the comparator output, v_+ and v_- are node voltages at the comparator input nodes, and V_H and V_L are the only two values that $v_o(t)$ can have. For example, if $V_H = 5$ volts and $V_L = 0$ volts, then $v_o(t)$ will be 5 volts whenever $v_+ > v_-$ and $v_o(t)$ will be 0 volts whenever $v_+ < v_-$.

In Figure 8.4-4, a comparator is used to compare the capacitor voltage to a threshold voltage, V_T. Suppose

$$V_A > V_T > v_c(0)$$

The input voltages of the comparator are

$$v_+ = v_c(t) \quad \text{and} \quad v_- = V_T$$

so the output voltage of the comparator is

$$v_o(t) = \begin{cases} V_H & if \quad v_c(t) > V_T \\ V_L & if \quad v_c(t) < V_T \end{cases}$$

The comparator output voltage will be $v_o(t) = V_L$, while $v_c(t) < V_T$ and $v_o(t) = V_H$ when $v_c(t) > V_T$. The comparator output voltage will switch from V_L to V_H at the instant when $v_c(t) = V_T$.

We know that the capacitor voltage of this first-order circuit will be

$$v_c(t) = V_A + (v_c(0) - V_A)e^{-\frac{t}{RC}}$$

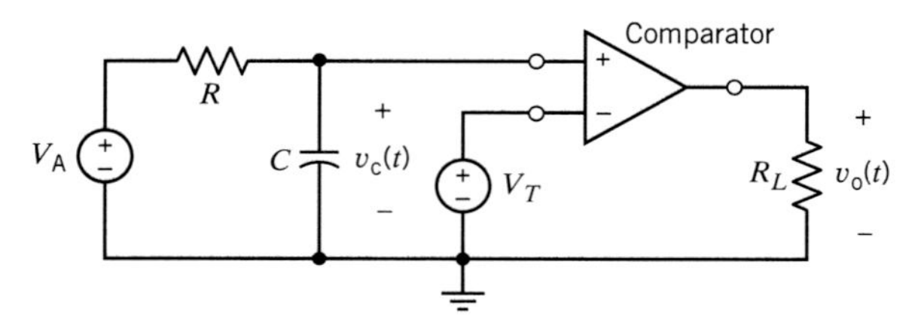

Figure 8.4-4 A comparator is used to compare the capacitor voltage, $v_c(t)$, to a threshold voltage, V_T.

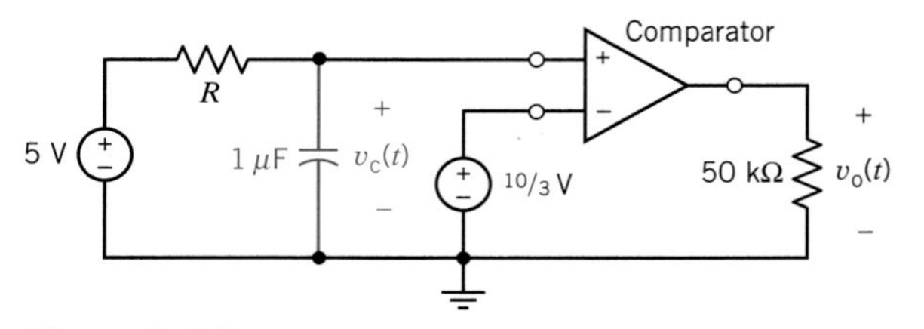

Figure 8.4-5 The initial capacitor voltage is $v_c(0) = 1.667$ volts, and the comparator output is to switch from $V_L = 0$ to $V_H = 5$ volts at time $t_1 = 1$ ms.

Let t_1 denote the time when the comparator output voltage switches from V_L to V_H. Then $v_c(t_1) = V_T$, so

$$V_T = V_A + [v_c(0) - V_A]e^{-\frac{t_1}{RC}}$$

Solving for t_1 gives

$$t_1 = RC \ln\left(\frac{v_c(0) - V_A}{V_T - V_A}\right) \qquad (8.4\text{-}1)$$

Example 8.4-1

Consider the circuit shown in Figure 8.4-5. The initial value of the capacitor voltage is $v_c(0) = 1.667$ volts. What value of resistance, R, is required if the comparator is to switch from $V_L = 0$ to $V_H = 5$ volts at time $t_1 = 1$ ms?

Solution

The circuit in Figure 8.4-5 is a specific example of the circuit in Figure 8.4-4 so we can use Eq. 8.4-1. Substituting the given values into Eq. 8.4-1, we get

$$1 \times 10^{-3} = R(1 \times 10^{-6}) \ln\left(\frac{\frac{5}{3} - 5}{\frac{10}{3} - 5}\right) = R(1 \times 10^{-6}) \ln(2)$$

Then, solving for R

$$R = \frac{1 \times 10^{-3}}{\ln(2) \times 10^{-6}} = 1.44 \text{ k}\Omega$$

Example 8.4-2

In Figure 8.4-6 a comparator is used to compare the resistor voltage, $v_R(t)$, to a threshold voltage, V_T. Suppose

$$V_A > V_T > R\, i_L(0)$$

Determine the time t_1 when the comparator output voltage switches from V_L to V_H.

Solution

The current into each comparator input is zero. This fact has two consequences. First, the resistor current is equal to the inductor current so

$$v_R(t) = R\, i_L(t)$$

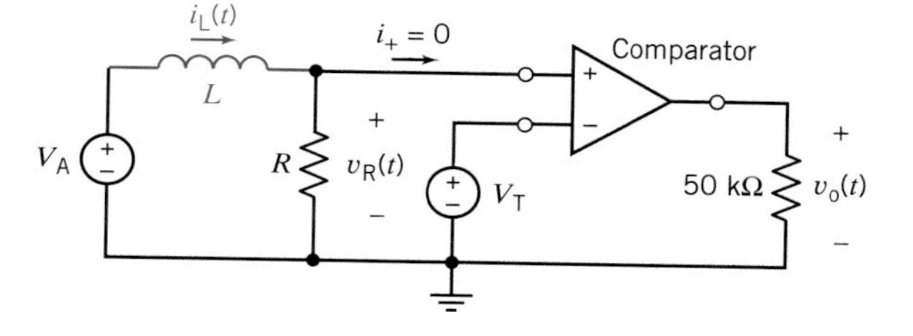

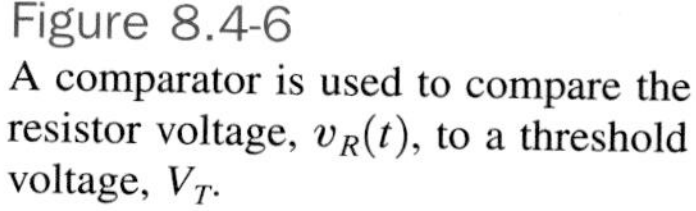

Figure 8.4-6
A comparator is used to compare the resistor voltage, $v_R(t)$, to a threshold voltage, V_T.

Second, the comparator does not disturb the first-order circuit consisting of the voltage source, resistor, and inductor. The inductor current is

$$i_L(t) = \frac{V_A}{R} + \left(i_L(0) - \frac{V_A}{R}\right)e^{-\frac{R}{L}t}$$

Next, t_1 is the time when $R\,i_L(t_1) = V_T$, so

$$V_T = V_A + (Ri_L(0) - V_A)e^{-\frac{R}{L}t_1}$$

Solving for t_1 gives

$$t_1 = \frac{L}{R}\ln\left(\frac{Ri_L(0) - V_A}{V_T - V_A}\right)$$

Exercise 8.4-1 Consider the circuit shown in Figure E 8.4-1. The initial value of the capacitor voltage is $v_c(0) = 1.5$ volts. What value of the threshold voltage, V_T, is required if the comparator is to switch from $V_L = 0$ to $V_H = 5$ volts at time $t_1 = 1$ ms?

Answer: $V_T = 2.88\ V$

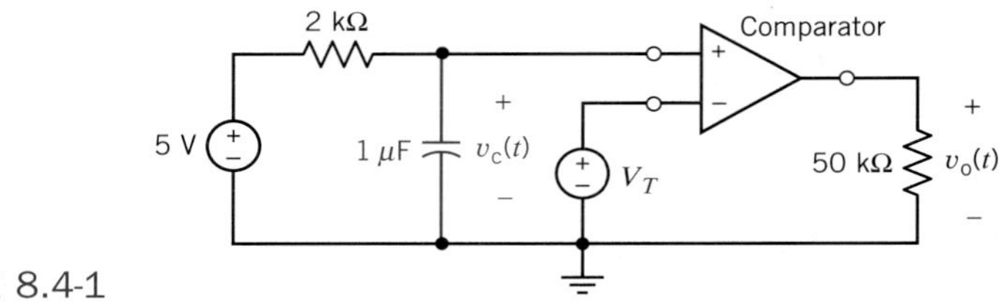

Figure E 8.4-1

Exercise 8.4-2 Consider the circuit shown in Figure E 8.4-2. The initial value of the inductor current is $i_L(0) = 1$ mA. What value of the inductance, L, is required if the comparator is to switch from $V_L = 0$ to $V_H = 5$ volts at time $t_1 = 10$ ms?

Hint: First show that $i_L(t) = 10 - 9e^{-\frac{500}{L}t}$ mA.

Answer: L = 8.5 H

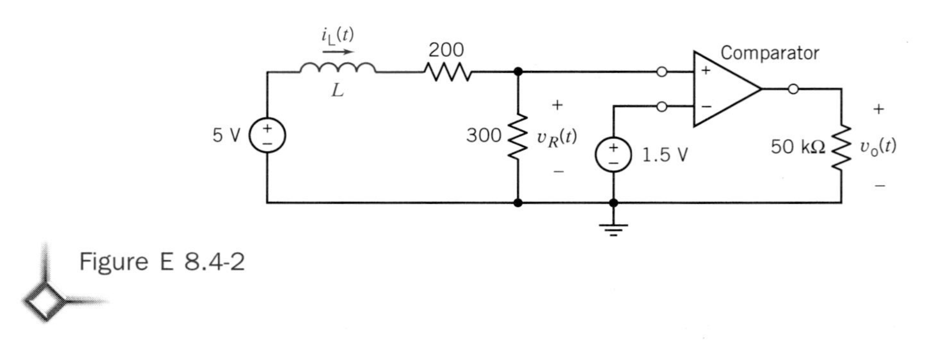

Figure E 8.4-2

8.5 STABILITY OF FIRST-ORDER CIRCUITS

We have shown that the natural response of a first-order circuit is

$$x_n(t) = Ke^{-\frac{t}{\tau}}$$

and that the complete response is the sum of the natural and forced responses

$$x(t) = x_n(t) + x_f(t)$$

When $\tau > 0$ the natural response vanishes as $t \rightarrow 0$, leaving the forced response. In this case, the circuit is said to be *stable*. When $\tau < 0$, the natural response grows without bound as $t \rightarrow 0$. The forced response becomes negligible compared to the natural response. The circuit is said to be *unstable*. When a circuit is stable, the forced response depends on the input to the circuit. That means that the forced response contains information about the input. When the circuit is unstable, the forced response is negligible and this information is lost. In practice, the natural response of an unstable circuit is not unbounded. This response will grow until something happens to change the circuit. Perhaps that change will be saturation of an op amp or of a dependent source. Perhaps that change will be the destruction of a circuit element. In most applications, the behavior of unstable circuits is undesirable and is to be avoided.

How can we design first-order circuits to be stable? Recalling that $\tau = R_{TH}C$ or $\tau = \frac{L}{R_{TH}}$, we see that $R_{TH} > 0$ is required to make a first-order circuit be stable. This condition will always be satisfied whenever the part of the circuit connected to the capacitor or inductor consists of only resistors and independent sources. Such circuits are guaranteed to be stable. In contrast, a first-order circuit that contains op amps or dependent sources may be unstable.

Example 8.5-1

The first-order circuit shown in Figure 8.5-1*a* is at steady state before the switch closes at $t = 0$. This circuit contains a dependent source and so may be unstable. Find the capacitor voltage, $v(t)$, for $t > 0$.

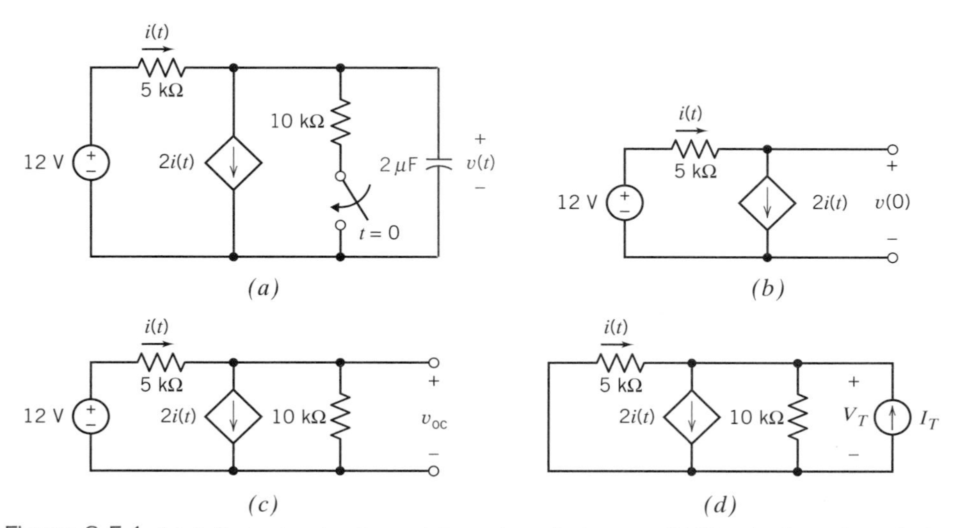

Figure 8.5-1 (*a*) A first-order circuit containing a dependent source. (*b*) The circuit used to calculate the initial condition. (*c*) The circuit used to calculate V_{oc}. (*d*) The circuit used to calculate R_{TH}.

Solution

The input to the circuit is a constant so the capacitor acts like an open circuit at steady state. We calculate the initial condition from the circuit in Figure 8.5-2*b*. Applying KCL to the top node of the dependent current source, we get

$$-i + 2i = 0$$

Therefore, $i = 0$. Consequently, there is no voltage drop across the resistor and

$$v(0) = 12\ V$$

Next, we determine the Thévenin equivalent circuit for the part of the circuit connected to the capacitor. This requires two calculations. First, calculate the open-circuit voltage using the circuit in Figure 8.5-1*c*. Writing a KVL equation for the loop consisting of the two resistors and the voltage source, we get

$$12 = (5 \times 10^3) \times i + (10 \times 10^3) \times (i - 2i)$$

Solving for the current, we find

$$i = -2.4\ \text{mA}$$

Applying Ohm's law to the 10 kΩ resistor, we get

$$V_{oc} = (10 \times 10^3) \times (i - 2i) = 24\ \text{V}$$

Now calculate the Thévenin resistance using the circuit shown in Figure 8.5-1*d*. Apply KVL to the loop consisting of the two resistors to get

$$0 = (5 \times 10^3) \times i + (10 \times 10^3) \times (I_T + i - 2i)$$

Solving for the current,

$$i = 2I_T$$

Applying Ohm's law to the 10 kΩ resistor, we get

$$V_T = 10 \times 10^3 \times (I_T + i - 2i) = -10 \times 10^3 \times I_T$$

The Thévenin resistance is given by

$$R_{TH} = \frac{V_T}{I_T} = -10\ \text{k}\Omega$$

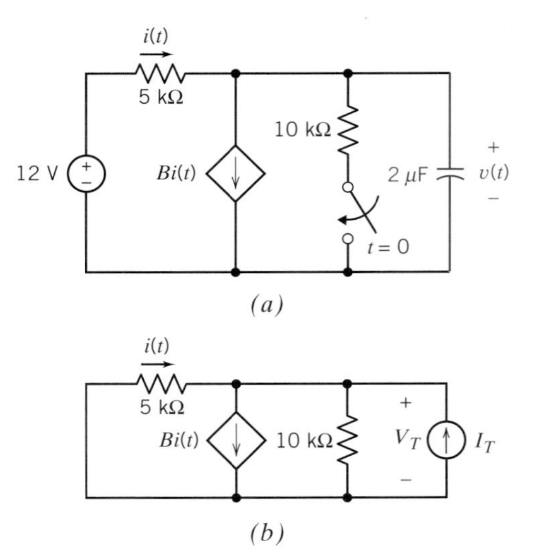

Figure 8.5-2
(*a*) A first-order circuit containing a dependent source. (*b*) The circuit used to calculate the Thévenin resistance of the part of the circuit connected to the capacitor.

The time constant is

$$\tau = R_{TH} C = -20 \text{ ms}$$

This circuit is unstable. The complete response is

$$v(t) = 24 - 12e^{\frac{t}{20}}$$

The capacitor voltage *decreases* from $v(0) = 12$ rather than *increasing* toward $v_f = 24$. It's not appropriate to refer to the forced response as a steady-state response when the circuit is unstable.

Example 8.5-2

The circuit considered in Example 8.5-1 has been redrawn in Figure 8.5-2*a*, with the gain of the dependent source represented by the variable *B*. What restrictions must be placed on the gain of the dependent source to ensure that it is stable? Design this circuit to have a time constant of +20 ms.

Solution

Figure 8.5-2*b* shows the circuit used to calculate R_{TH}. Applying KVL to the loop consisting of the two resistors

$$5 \times 10^3 \times i + V_T = 0$$

Solving for the current gives

$$i = -\frac{V_T}{5 \times 10^3}$$

Applying KCL to the top node of the dependent source, we get

$$-i + Bi + \frac{V_T}{10 \times 10^3} - I_T = 0$$

Combining these equations, we get

$$\left(\frac{1 - B}{5 \times 10^3} + \frac{1}{10 \times 10^3}\right) V_T - I_T = 0$$

The Thévenin resistance is given by

$$R_{TH} = \frac{V_T}{I_T} = -\frac{10 \times 10^3}{2B - 3}$$

The condition $B < 3/2$ is required to ensure that R_{TH} is positive and the circuit is stable.

To obtain a time constant of +20 ms requires

$$R_{TH} = \frac{\tau}{C} = \frac{20 \times 10^{-3}}{2 \times 10^{-6}} = 10 \times 10^3 = 10 \text{ k}\Omega$$

which in turn requires

$$10 \times 10^3 = -\frac{10 \times 10^3}{2B - 3}$$

Therefore $B = 1$. This suggests that we can fix the unstable circuit by decreasing the gain of the dependent source from 2 A/A to 1 A/A.

8.6 THE UNIT STEP SOURCE

In the preceding section, we considered circuits where the sources are switched in and out of the circuit at $t = t_0$. At the instant t_0 when these sources are applied to or disconnected from the circuit, many of the currents and voltages within the circuit change abruptly.

The application of a battery (constant source) by means of two switches, as shown in Figure 8.6-1, may be considered equivalent to a source that is zero up to t_0 and equal to the voltage V_0 thereafter.

We define the *unit step forcing function* as a function of time that is zero for $t < t_0$ and unity for $t > t_0$. At $t = t_0$ the magnitude changes from zero to one. We represent the unit step function by $u(t - t_0)$ where

$$u(t - t_0) = \begin{cases} 0 & t < t_0 \\ 1 & t > t_0 \end{cases} \tag{8.6-1}$$

The value of $u(t - t_0)$ is not defined at $t = t_0$, where it switches instantaneously from a magnitude of zero to one. The unit step function is shown in Figure 8.6-2. We will often consider $t_0 = 0$.

The unit step function is dimensionless. If we wish to represent the voltage of Figure 8.6-1, we use the representation

$$v(t) = V_0 u(t - t_0)$$

which implies the application of a voltage V_0 at $t = t_0$.

An exact one-switch equivalent of the source $v = V_0 u(t - t_0)$ requires that v be zero for $t < t_0$. As shown in Figure 8.6-3, v is defined as zero for $t < t_0$ and jumps to V_0 at $t = t_0$.

Henceforth, we will use the symbol as shown in Figure 8.6-4 to represent the step voltage source. The *step response* of a circuit is the response of the circuit to the sudden application of a constant source $v = V_0 u(t - t_0)$ when all the initial conditions of the circuit are equal to zero.

It is worth noting that $u(-t)$ simply implies that we have a value of 1 for $t < 0$, so that

$$u(-t) = \begin{cases} 1 & t < 0 \\ 0 & t > 0 \end{cases}$$

For the case where the change occurs at t_0, we have

$$u(t_0 - t) = \begin{cases} 1 & t < t_0 \\ 0 & t > t_0 \end{cases}$$

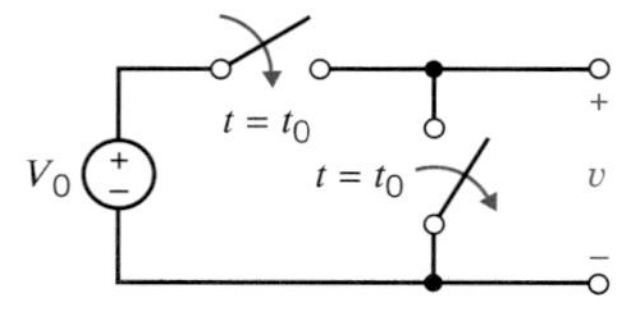

Figure 8.6-1 Application of a constant-voltage source at $t = t_0$ using two switches both acting at $t = t_0$.

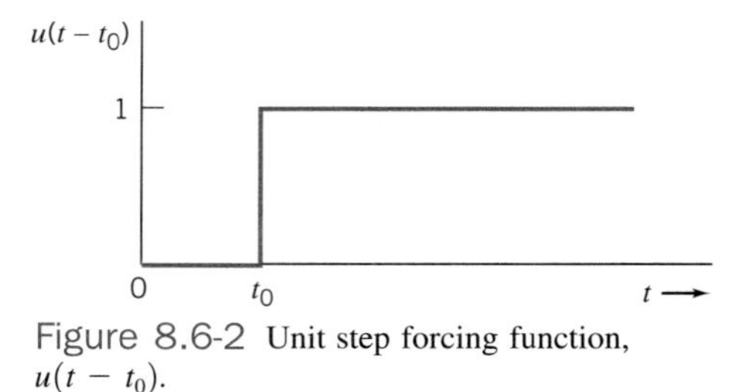

Figure 8.6-2 Unit step forcing function, $u(t - t_0)$.

Let us consider the *pulse* source

$$v(t) = \begin{cases} 0 & t < t_0 \\ V_0 & t_0 < t < t_1 \\ 0 & t_1 < t \end{cases}$$

which is shown in Figure 8.6-5*a*. As shown in Figure 8.6-5*b*, the pulse can be obtained from two-step voltage sources, the first of magnitude V_0 occurring at $t = t_0$ and the second equal to $-V_0$ occurring at $t = t_1$. Thus, the two-step sources of magnitude V_0 shown in Figure 8.6-6 will yield the desired pulse. We have $v(t) = V_0 u(t - t_0) - V_0 u(t - t_1)$ to provide the pulse. Notice how easy it is to use two-step function symbols to represent this pulse source. The pulse is said to have a duration of $(t_1 - t_0)$s.

We recognize that the unit step function is an ideal model. No real element can switch instantaneously at $t = t_0$. However, if it switches in a very short time (say, 1 ns), we can consider the switching as instantaneous for most of the medium-speed circuits. As long as the switching time is small compared to the time constant of the circuit, it can be ignored.

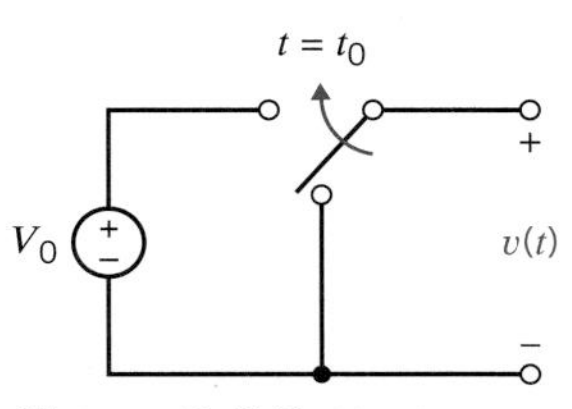

Figure 8.6-3 Single-switch equivalent circuit for the step voltage source.

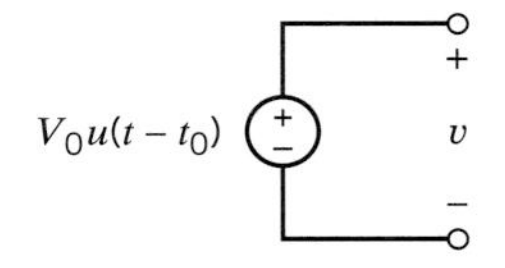

Figure 8.6-4 Symbol for the step voltage source of magnitude V_0 applied at $t = t_0$.

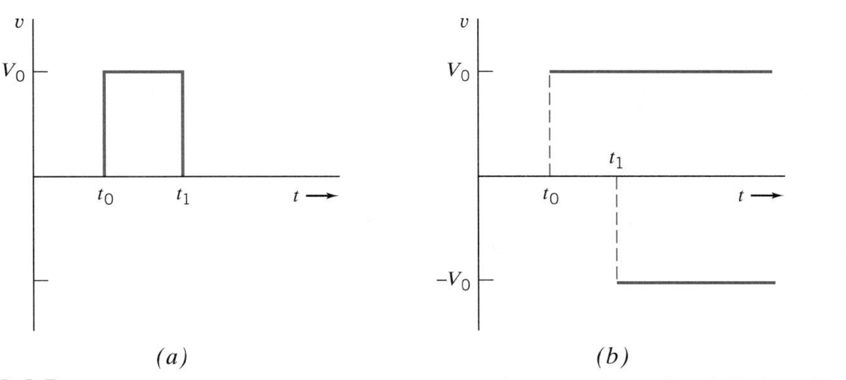

Figure 8.6-5 (*a*) Rectangular voltage pulse. (*b*) Two-step voltage waveforms that yield the voltage pulse.

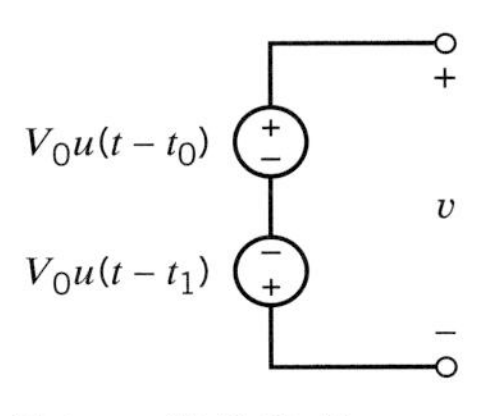

Figure 8.6-6 Two-step voltage sources that yield a rectangular voltage pulse, $v(t)$ with a magnitude of V_0 and a duration of $(t_1 - t_0)$ where $t_0 < t_1$.

Let us consider the application of a pulse to an *RL* circuit as shown in Figure 8.6-7. Here we let $t_0 = 0$ for convenience. The duration of the pulse is t_1 s.

A **pulse signal** has a constant nonzero value for a time duration of $\Delta t = t_1 - t_0$.

The pulse is applied to the *RL* circuit when $i(0) = 0$.

We wish to determine i for the pulse source. Since the circuit is linear, we may use the principle of superposition so that

$$i = i_1 + i_2$$

where i_1 is the response to $V_0 u(t)$ and i_2 is the response to $V_0 u(t - t_1)$.

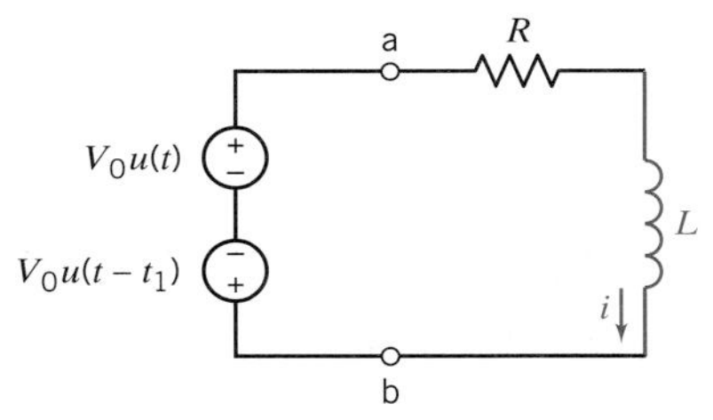

Figure 8.6-7
Pulse source connected to an *RL* circuit. The pulse is generated by the combination of two-step voltage sources each with units in volts.

We recall that the response of an *RL* circuit to a constant forcing function applied at $t = t_n$ is $\left(\text{see Eq. 8.3-7 with } i(0) = 0 \text{ and } I_{sc} = \dfrac{V_0}{R}\right)$

$$i = \frac{V_0}{R}\left(1 - e^{-\frac{t-t_n}{\tau}}\right) \text{A when } t > t_n$$

where $\tau = L/R$. Consequently, we may add the two solutions to the two-step sources, carefully noting $t_0 = 0$ and t_1 as the start of each response. Therefore,

$$i_1 = \frac{V_0}{R}\left(1 - e^{-\frac{t}{\tau}}\right) \text{A when } t \geq 0$$

and

$$i_2 = \frac{-V_0}{R}\left(1 - e^{-\frac{t-t_1}{\tau}}\right) \text{A when } t > t_1$$

Adding the responses, we have

$$i = \begin{cases} \dfrac{V_0}{R}\left(1 - e^{-\frac{t}{\tau}}\right) & 0 < t \leq t_1 \\ \dfrac{V_0}{R} e^{-\frac{t}{\tau}}\left(e^{\frac{t_1}{\tau}} - 1\right) & t > t_1 \end{cases}$$

Of course, the current i is zero for $t < 0$. The response i is shown in Figure 8.6-8. The magnitude of the response at $t = t_1$ is

$$i(t_1) = \frac{V_0}{R}(1 - e^{-t_1/\tau}) \text{ A}$$

If t_1 is greater than τ, the response will approach V_0/R before starting its decline as shown in Figure 8.6-8. The response at $t = 2t_1$ is

$$\begin{aligned} i(2t_1) &= \frac{V_0}{R} e^{-2\frac{t_1}{\tau}}\left(e^{\frac{t_1}{\tau}} - 1\right) \\ &= \frac{V_0}{R}\left(e^{-\frac{t_1}{\tau}} - e^{-2\frac{t_1}{\tau}}\right) \text{A} \end{aligned}$$

Exercise 8.6-1 Find v for the circuit shown in Figure E 8.6-1 when $v(0^-) = 0$, $I_0 = 1$ A, and $t_1 = 0.5$ s. Plot the response for $0 < t < 1$ s.

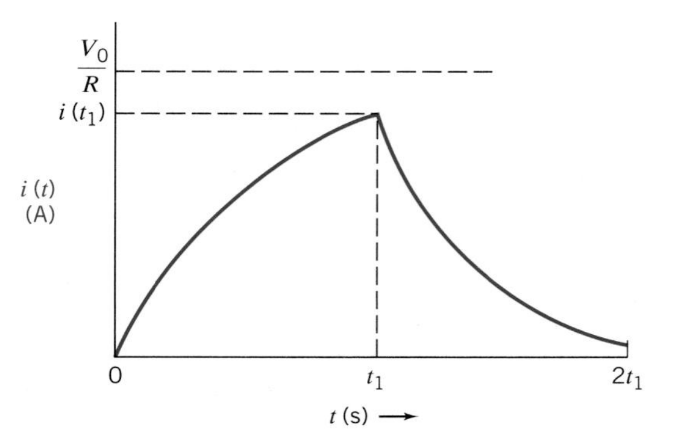

Figure 8.6-8
Response of the *RL* circuit shown in Figure 8.6-7.

Exercise 8.6-2 A pulse is transmitted along a telephone line represented by the circuit shown in Figure E 8.6-2. Find the received voltage $v(t)$. The length of the pulse is 100 ms and, therefore, $t_1 = 100$ ms. The line is represented by $C = 0.1\,\mu$F and $R = 200$ kΩ.

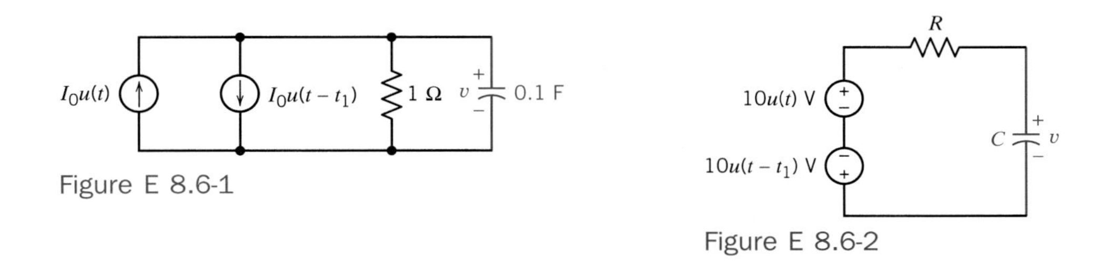

Figure E 8.6-1

Figure E 8.6-2

Exercise 8.6-3 The circuit shown in Figure E 8.6-3*a* has a current source as shown in Figure E 8.6-3*b*. Determine the current $i(t)$ in the inductor.

Answer: $i(t) = \begin{cases} 5(1 - e^{-10t}) \text{ A} & t \le 0.2 \text{ s} \\ 4.32e^{-10(t-0.2)} \text{ A} & t \ge 0.2 \text{ s} \end{cases}$

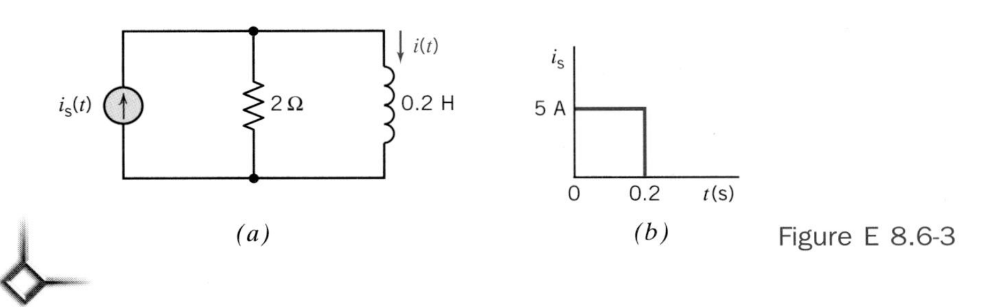

Figure E 8.6-3

8.7 THE RESPONSE OF A FIRST-ORDER CIRCUIT TO A NONCONSTANT SOURCE

In the previous sections we wisely used the fact that the forced response to a constant source will be a constant itself. It now remains to determine what the response will be when the forcing function is not a constant.

The differential equation described by an *RL* or *RC* circuit is represented by the general form

$$\frac{dx(t)}{dt} + ax(t) = y(t) \tag{8.7-1}$$

where $y(t)$ is a constant only when we have a constant current or constant voltage source and where $a = 1/\tau$ is the reciprocal of the time constant.

In this section we introduce the *integrating factor method,* which consists of multiplying Eq. 8.7-1 by a factor that makes the left-hand side a perfect derivative, and then integrating both sides.

Consider the derivative of a product of two terms such that

$$\frac{d}{dt}(xe^{at}) = \frac{dx}{dt}e^{at} + axe^{at} = \left(\frac{dx}{dt} + ax\right)e^{at} \tag{8.7-2}$$

The term within the parentheses on the right-hand side of Eq. 8.7-2 is exactly the form on the left-hand side of Eq. 8.7-1.

Therefore, if we multiply both sides of Eq. 8.7-1 by e^{at}, the left-hand side of the equation can be represented by the perfect derivative $d(xe^{at})/dt$. Carrying out these steps, we show that

$$\left(\frac{dx}{dt} + ax\right)e^{at} = ye^{at}$$

or

$$\frac{d}{dt}\left(xe^{at}\right) = ye^{at}$$

Integrating both sides of the second equation, we have

$$xe^{at} = \int ye^{at}\, dt + K$$

where K is a constant of integration. Therefore, solving for $x(t)$, we multiply by e^{-at} to obtain

$$x = e^{-at}\int ye^{at}\, dt + Ke^{-at} \tag{8.7-3}$$

For the case where the source is a constant so that $y(t) = M$, we have

$$\begin{aligned} x &= e^{-at}M\int e^{at}\, dt + Ke^{-at} \\ &= \frac{M}{a} + Ke^{-at} \\ &= x_{\mathrm{f}} + x_{\mathrm{n}} \end{aligned}$$

where the natural response is $x_{\mathrm{n}} = Ke^{-at}$ and the forced response is $x_{\mathrm{f}} = M/a$, a constant.

Now consider the case where $y(t)$, the forcing function, is not a constant. Considering Eq. 8.7-3, we see that the natural response remains, $x_{\mathrm{n}} = Ke^{-at}$. However, the forced response is

$$x_{\mathrm{f}} = e^{-at}\int y(t)e^{at}\, dt$$

Thus, the forced response will be dictated by the form of $y(t)$. Let us consider the case where $y(t)$ is an exponential function so that $y(t) = e^{bt}$. We assume that $(a + b)$ is not equal to zero. Then we have

$$\begin{aligned} x_{\mathrm{f}} &= e^{-at}\int e^{bt}e^{at}\, dt \\ &= e^{-at}\int e^{(a+b)t}\, dt \\ &= \frac{1}{a+b}e^{-at}e^{(a+b)t} \\ &= \frac{e^{bt}}{a+b} \end{aligned} \tag{8.7-4}$$

Therefore, the forced response of an *RL* or *RC* circuit to an exponential forcing function is of the same form as the forcing function itself. When $a + b$ is not equal to zero, we assume that the forced response will be of the same form as the forcing function itself, and we try to obtain the relationship that will be satisfied under those conditions.

Example 8.7-1

Find the current i for the circuit of Figure 8.7-1*a* for $t > 0$ when

$$v_s = 10e^{-2t}u(t) \text{ V}$$

Assume the circuit is in steady state at $t = 0^-$.

Solution

Since the forcing function is an exponential, we expect an exponential for the forced response i_f. Therefore, we expect i_f to be

$$i_f = Be^{-2t}$$

for $t > 0$. Writing KVL around the loop, we have

$$L\frac{di}{dt} + Ri = v_s$$

or

$$\frac{di}{dt} + 4i = 10e^{-2t}$$

for $t > 0$. Substituting $i_f = Be^{-2t}$, we have

$$-2Be^{-2t} + 4Be^{-2t} = 10e^{-2t}$$

or

$$(-2B + 4B)e^{-2t} = 10e^{-2t}$$

Hence, $B = 5$ and

$$i_f = 5e^{-2t}$$

The natural response can be obtained by considering the circuit shown in Figure 8.7-1*b*. This is the equivalent circuit that is appropriate after the switch has opened. The part of the circuit that is connected to the inductor has been replaced by its Norton equivalent circuit. The natural response is

$$i_n = Ae^{-\frac{R_{TH}}{L}t} = Ae^{-4t}$$

The complete response is

$$i = i_n + i_f = Ae^{-4t} + 5e^{-2t}$$

The constant A can be determined from the value of the inductor current at time $t = 0$. The initial inductor current, $i(0)$, can be obtained by considering the circuit shown in Figure 8.7-1*c*. This is the equivalent circuit that is appropriate before the switch opens. Because $v_s(t) = 0$ for $t < 0$ and a zero voltage source is a short circuit, the voltage source

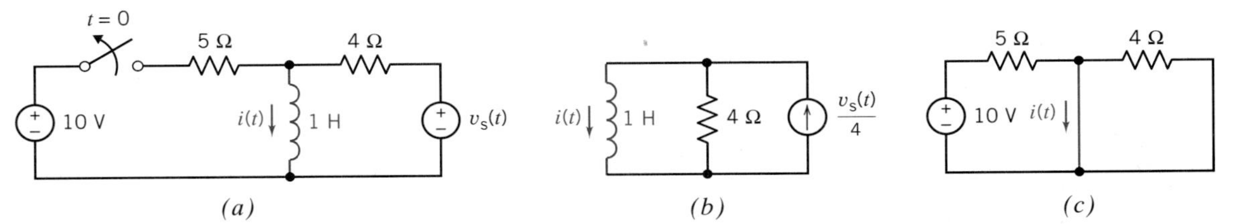

Figure 8.7-1 (*a*) A circuit with a nonconstant source, (*b*) the appropriate equivalent circuit after the switch opens; and (*c*) the appropriate equivalent circuit before the switch opens.

at the right side of the circuit has been replaced by a short circuit. Also, because the circuit is at steady state before the switch opens and the only input is the constant 10-volt source, the inductor acts like a short circuit. The current in the short circuit that replaces the inductor is the initial condition, $i(0)$. From Figure 8.7-1*c*

$$i(0) = \frac{10}{5} = 2 \text{ A}$$

Therefore, at $t = 0$

$$i(0) = Ae^{-4\times 0} + 5e^{-2\times 0} = A + 5$$

or

$$2 = A + 5$$

or $A = -3$. Therefore,

$$i = (-3e^{-4t} + 5e^{-2t}) \text{ A} \quad t > 0$$

The voltage source of Example 8.7-1 is a decaying exponential of the form

$$v_s = 10e^{-2t}u(t) \text{ V}$$

This source is said to be *aperiodic* (nonperiodic). A periodic source is one that repeats itself exactly after a fixed length of time. Thus, the signal $f(t)$ is *periodic* if there is a number T such that for all t

$$f(t + T) = f(t) \tag{8.7-5}$$

The smallest positive number T that satisfies Eq. 8.7-5 is called the *period.* The period defines the duration of one complete cycle of $f(t)$. Thus, any source for which there is no value of T satisfying Eq. 8.7-5 is said to be aperiodic. An example of a periodic source is 10 sin $2t$, which we consider in Example 8.7-2. The period of this sinusoidal source is π s.

Example 8.7-2

Find the response $v(t)$ for $t > 0$ for the circuit of Figure 8.7-2*a*. The initial voltage $v(0) = 0$, and the current source is $i_s = (10 \sin 2t)u(t)$ A.

Solution

Since the forcing function is a sinusoidal function, we expect that v_f is of the same form. Writing KCL at node a, we obtain

$$C\frac{dv}{dt} + \frac{v}{R} = i_s$$

or

$$0.5\frac{dv}{dt} + \frac{v}{4} = 10 \sin 2t \tag{8.7-6}$$

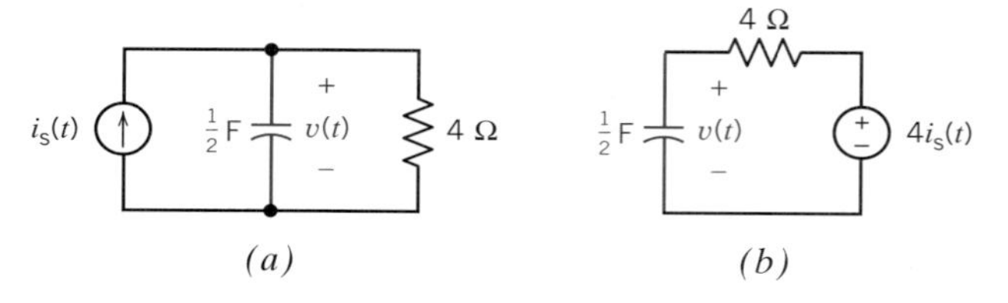

Figure 8.7-2
(*a*) A circuit with a nonconstant source. (*b*) The equivalent circuit for $t > 0$.

for $t > 0$. We assume that v_f will consist of the sinusoidal function sin $2t$ and its derivatives.

Examining Eq. 8.7-6, $v_f/4$ plus $0.5\, dv_f/dt$ must equal 10 sin $2t$. However, $d(\sin 2t)/dt = 2\cos 2t$. Therefore, the trial v_f needs to contain both sin $2t$ and cos $2t$ terms. Thus, we try the proposed solution

$$v_f = A\sin 2t + B\cos 2t$$

The derivative of v_f is then

$$\frac{dv_f}{dt} = 2A\cos 2t - 2B\sin 2t$$

Substituting v_f and dv_f/dt into Eq. 8.7-6, we obtain

$$(A\cos 2t - B\sin 2t) + \frac{1}{4}(A\sin 2t + B\cos 2t) = 10\sin 2t$$

Therefore, equating sin $2t$ terms and cos $2t$ terms, we obtain

$$\left(\frac{A}{4} - B\right) = 10 \quad \text{and} \quad \left(A + \frac{B}{4}\right) = 0$$

Solving for A and B, we obtain

$$A = \frac{40}{17} \quad \text{and} \quad B = \frac{-160}{17}$$

Consequently,

$$v_f = \frac{40}{17}\sin 2t - \frac{160}{17}\cos 2t$$

It is necessary that v_f be made up of sin $2t$ and cos $2t$ since the solution has to satisfy the differential equation. Of course, the derivative of sin ωt is $\omega \cos \omega t$.

The natural response can be obtained by considering the circuit shown in Figure 8.7-2b. This is the equivalent circuit that is appropriate for $t > 0$. The part of the circuit connected to the capacitor has been replaced by its Thévenin equivalent circuit. The natural response is

$$v_n = De^{-\frac{t}{R_{TH}C}} = De^{-\frac{t}{2}}$$

The complete response is then

$$v = v_n + v_f = De^{-t/2} + \frac{40}{17}\sin 2t - \frac{160}{17}\cos 2t$$

Since $v(0) = 0$, we obtain at $t = 0$

$$0 = D - \frac{160}{17}$$

or

$$D = \frac{160}{17}$$

Then the complete response is

$$v = \left(\frac{160}{17}e^{-t/2} + \frac{40}{17}\sin 2t - \frac{160}{17}\cos 2t\right)\text{V}$$

Table 8.7-1 **Forced Response to a Forcing Function**

Forcing Function, $y(t)$	Forced Response, $x_f(t)$
1. Constant $y(t) = M$	$x_f = N$, a constant
2. Exponential $y(t) = Me^{-bt}$	$x_f = Ne^{-bt}$
3. Sinusoid $y(t) = M \sin(\omega t + \theta)$	$x_f = A \sin \omega t + B \cos \omega t$

A special case for the forced response of a circuit may occur when the forcing function is a damped exponential where we have $y(t) = e^{-bt}$. Referring back to Eq. 8.7-4, we can show that

$$x_f = \frac{e^{-bt}}{a - b}$$

when $y(t) = e^{-bt}$. Note that here we have e^{-bt} while we used e^{bt} for Eq. 8.7-4. For the special case where $a = b$, we have $a - b = 0$, and this form of the response is indeterminate. For this special case, we must use $x_f = te^{-bt}$ as the forced response. The solution x_f, for the forced response, when $a = b$, will satisfy the original differential equation (8.7-1). Thus, when the natural response already contains a term of the same form as the forcing function, we need to multiply the assumed form of the forced response by t.

The forced response to selected forcing functions is summarized in Table 8.7-1. We note that if a circuit is linear, at steady state and excited by a single sinusoidal source having frequency ω, then all the element currents and voltages are sinusoids having frequency ω.

Exercise 8.7-1 The electrical power plant for the orbiting space station shown in Figure E 8.7-1*a* uses photovoltaic cells to store energy in batteries. The charging circuit is modeled by the circuit shown in Figure E 8.7-1*b*, where $v_s = 10 \sin 20t$ V. If $v(0^-) = 0$, find $v(t)$ for $t > 0$.

Answer: $v = 4e^{-10t} - 4 \cos 20t + 2 \sin 20t$ V

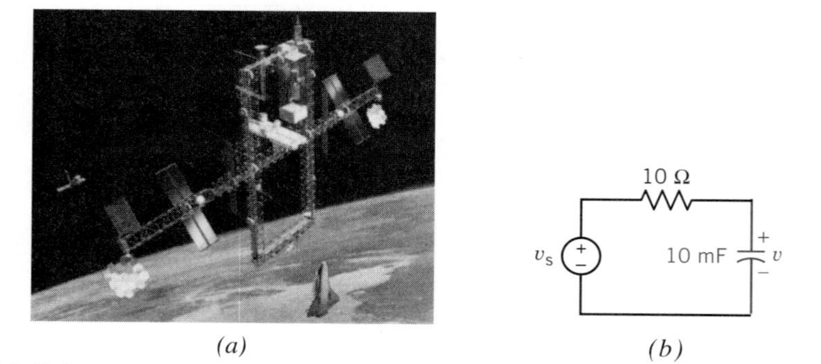

Figure E 8.7-1 (*a*) The NASA space station design shows the longer habitable modules that would house an orbiting scientific laboratory. (*b*) The circuit for energy storage for the laboratories. Photograph courtesy of the National Aeronautics and Space Administration.

Exercise 8.7-2 Find the current i in the circuit of Figure E 8.7-2 for $t > 0$ when $i_s = 10e^{-5t}u(t)$ A and $i(0^-) = 0$.

Answer: $i = 10.53(e^{-5t} - e^{-100t})$ A

Exercise 8.7-3 An experimenter is working in her laboratory with an electromagnet as shown in Figure E 8.7-3. She notices that whenever she turns off the electromagnet, a big spark appears at the switch contacts. Explain the occurrence of the spark. Suggest a way to suppress the spark by adding one element.

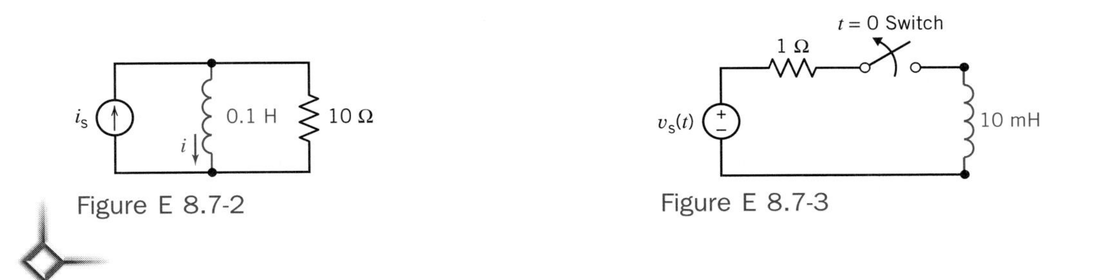

Figure E 8.7-2

Figure E 8.7-3

8.8 DIFFERENTIAL OPERATORS

In this section we introduce the differential operator s.

An **operator** is a symbol that represents a mathematical operation. We can define a differential *operator* s such that

$$sx = \frac{dx}{dt} \quad \text{and} \quad s^2x = \frac{d^2x}{dt^2}$$

Thus, the operator s denotes differentiation of the variable with respect to time. The utility of the operator s is that it can be treated as an algebraic quantity. This permits the replacement of differential equations with algebraic equations, which are easily handled.

Use of the s operator is particularly attractive when higher order differential equations are involved. Then we use the s operator so that

$$s^n x = \frac{d^n x}{dt^n} \quad \text{for} \quad n \geq 0$$

We assume that $n = 0$ represents no differentiation so that

$$s^0 = 1$$

which implies $s^0 x = x$.

Because integration is the inverse of differentiation, we define

$$\frac{1}{s}x = \int_{-\infty}^{t} x\, d\tau \tag{8.8-1}$$

The operator $1/s$ must be shown to satisfy the usual rules of algebraic manipulations. Of these rules, the commutative multiplication property presents the only difficulty. Thus, we require

$$s \cdot \frac{1}{s} = \frac{1}{s} \cdot s = 1 \tag{8.8-2}$$

Is this true for the operator s? First, we examine Eq. 8.8-1. Multiplying Eq. 8.8-1 by s yields

$$s \cdot \frac{1}{s} x = \frac{d}{dt} \int_{-\infty}^{t} x \, d\tau$$

or

$$x = x$$

as required. Now we try the reverse order by multiplying sx by the integration operator to obtain

$$\frac{1}{s} sx = \int_{-\infty}^{t} \frac{dx}{d\tau} d\tau = x(t) - x(-\infty)$$

Therefore

$$\frac{1}{s} sx = x$$

only when $x(-\infty) = 0$. From a physical point of view, we require that all capacitor voltages and inductor currents be zero at $t = -\infty$. Then the operator $1/s$ can be said to satisfy Eq. 8.8-2 and can be manipulated as an ordinary algebraic quantity.

Differential operators can be used to find the natural solution of a differential equation. For example, consider the first-order differential equation

$$\frac{d}{dt} x(t) + ax(t) = by(t) \tag{8.8-3}$$

The natural solution of this differential equation is

$$x_n(t) = K e^{st} \tag{8.8-4}$$

The homogeneous form of a differential equation is obtained by setting the forcing function equal to zero. The forcing function in Eq. 8.8-3 is $y(t)$. The homogeneous form of this equation is

$$\frac{d}{dt} x(t) + ax(t) = 0 \tag{8.8-5}$$

To see that $x_n(t)$ is a solution of the homogeneous form of the differential equation, we substitute Eq. 8.8-4 into Eq. 8.8-5.

$$\frac{d}{dt}(K e^{st}) + a(K e^{st}) = s K e^{st} + a K e^{st} = 0$$

To obtain the parameter s in Eq. 8.8-4, replace $\frac{d}{dt}$ in Eq. 8.8-5 by the differential operator s. This results in

$$sx + ax = (s + a)x = 0 \tag{8.8-6}$$

This equation has two solutions: $x = 0$ and $s = -a$. The solution $x = 0$ isn't useful, so we use the solution $s = -a$. Substituting this solution into Eq. 8.8-4 gives

$$x_n(t) = K e^{-at}$$

This is the same expression for the natural response that we obtained earlier in this chapter, by other methods. That's reassuring but not new. Differential operators will be quite useful when we analyze circuits that are represented by second- and higher order differential equations.

As a second application of differential operators, consider using the computer program MATLAB to find the complete response of a first-order circuit. Differential operators are used to describe differential equations to MATLAB. As an example, consider the circuit shown in Figure 8.8-1a. To represent this circuit by a differential equation, apply KVL to get

$$10 \times 10^3\left(1 \times 10^{-6} \frac{d}{dt} v(t)\right) + v(t) - 4\cos(100t) = 0$$

or

$$0.01 \frac{d}{dt} v(t) + v(t) = 4\cos(100t) \tag{8.8-7}$$

In the syntax used by MATLAB, the differential operator is represented by D instead of s. Replace $\frac{d}{dt}$ in Eq. 8.8-7 by the differential operator D to get

$$0.01Dv + v = 4\cos(100t)$$

Entering the MATLAB commands

```
v = dsolve('0.01*Dv + v = 4*cos(100*t)', 'v(0) = -8')
ezplot(v, [0, .2])
```

tells MATLAB to solve the differential equation using the initial condition $v(0) = -8$ volts and then plot the result. (The function named dsolve determines the symbolic solution of ordinary differential equations. This function is provided with the student edition of version 4 of MATLAB.) MATLAB responds by providing the complete solution of the differential equation

```
v = 2.*cos(100*t) + 2.*sin(100*t)-10.*exp(-100.*t)
```

and the plot of $v(t)$ versus t shown in Figure 8.8-1b.

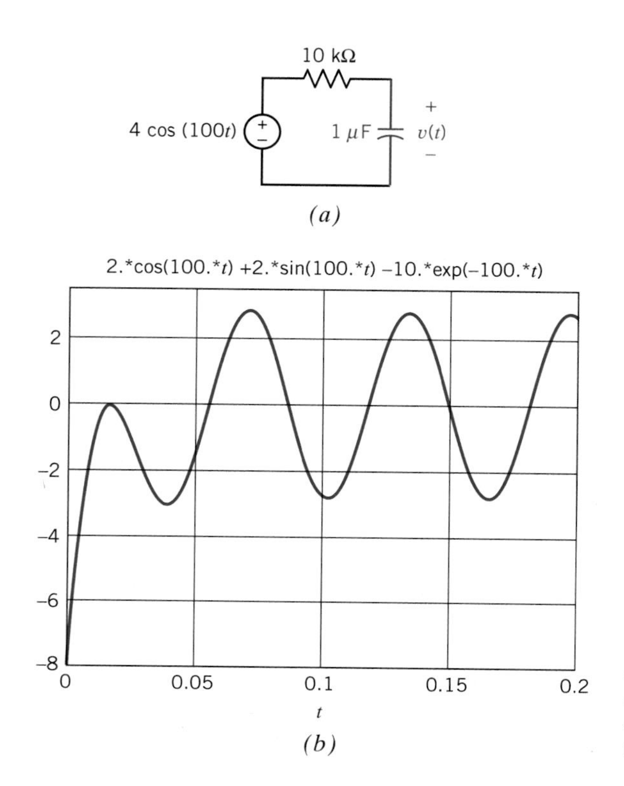

Figure 8.8-1
(a) A first-order circuit with a sinusoidal input and (b) a plot of its complete response produced using MATLAB.

8.9 THE COMPLETE RESPONSE USING PSPICE

In this section we consider the use of the transient analysis statement (.TRAN) in order to obtain the response of an *RC* or *RL* circuit to a forcing function with an initial condition. Let us consider the *RC* circuit shown in Figure 8.9-1*a*. The initial capacitor voltage is 5 V, and the magnitude of the step voltage input is 15 V. The circuit redrawn for PSpice is shown in Figure 8.9-1*b*, where $v(0) = -5$ V. We use the piecewise linear input source V1 to generate a step as described in Appendix F-6. The piecewise linear waveform starts with a value of zero at $t = 0$ and jumps to 15 V after 1 ps and remains at 15 V. The PSpice program is shown in Figure 8.9-1*c*. The output of the PSpice simulation is the plot of the complete response, $v(t)$, shown in Figure 8.9-2.

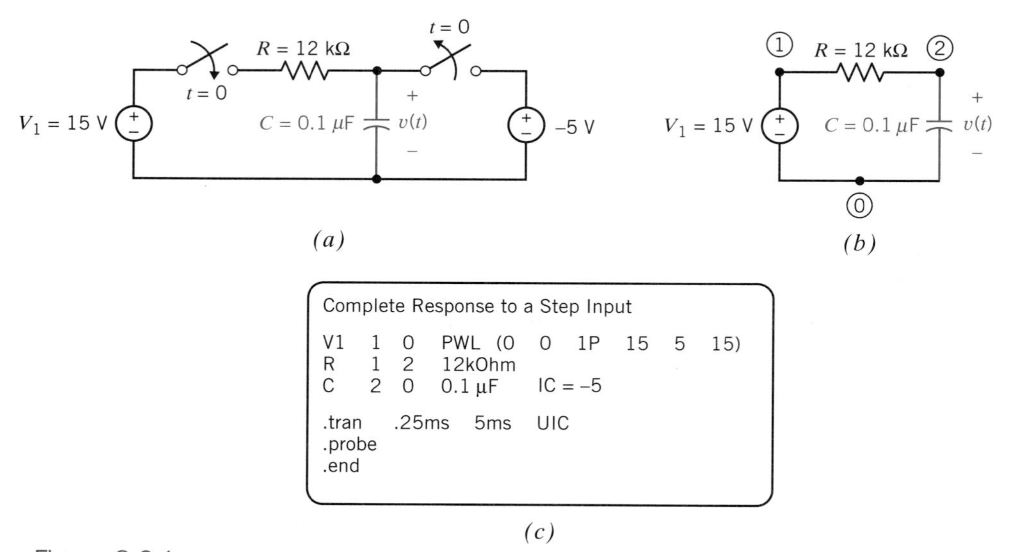


```
Complete Response to a Step Input
V1   1  0   PWL (0   0   1P   15   5   15)
R    1  2   12kOhm
C    2  0   0.1 µF      IC = -5
.tran   .25ms   5ms   UIC
.probe
.end
```

(*c*)

Figure 8.9-1 (*a*) An *RC* circuit. (*b*) The circuit redrawn for Pspice. (*c*) The Pspice input file.

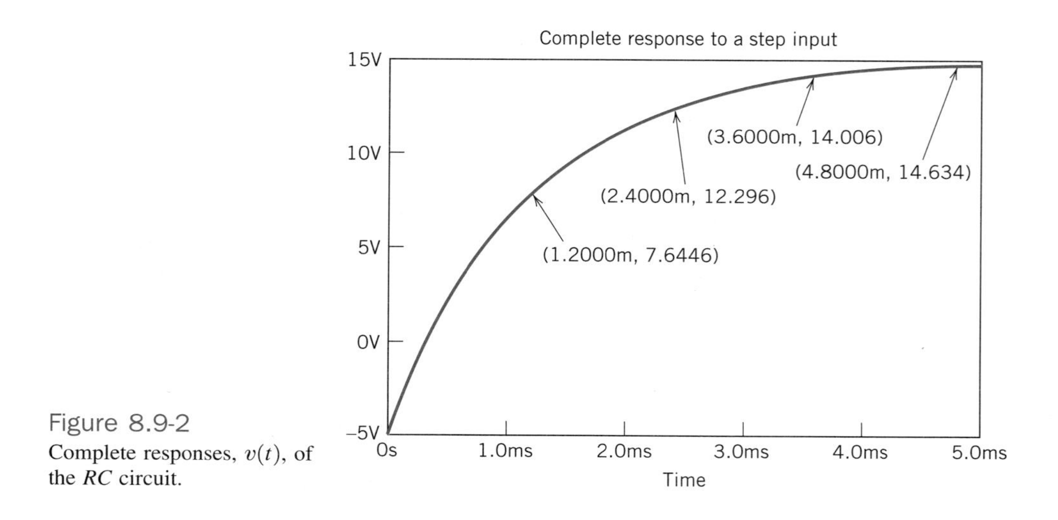

Figure 8.9-2
Complete responses, $v(t)$, of the *RC* circuit.

8.10 VERIFICATION EXAMPLES

Example 8.10-1

Consider the circuit and corresponding transient response shown in Figure 8.10-1. How can we tell if the transient response is correct? There are three things that need to be verified: the initial voltage, $v_0(t_0)$; the final voltage, $v_0(\infty)$; and the time constant, τ.

Solution

Consider first the initial voltage, $v_0(t_0)$. (In this example, $t_0 = 10\ \mu s$.) Before time $t_0 = 10\ \mu s$, the switch is closed and has been closed long enough for the circuit to reach steady state, that is, for any transients to have died out. To calculate $v_0(t_0)$ we simplify the circuit in two ways. First, replace the switch with a short circuit because the switch is closed. Second, replace the inductor with a short circuit because inductors act like short circuits when all the inputs are constants and the circuit is at steady state. The resulting circuit is shown in Figure 8.10-2a. After replacing the parallel 300 Ω and 600 Ω resistors with the equivalent 200 Ω resistor, the initial voltage is calculated using voltage division as

$$v_0(t_0) = \frac{200}{200 + 200} 8 = 4 \text{ V}$$

Next consider the final voltage, $v_0(\infty)$. In this case, the switch is open and the circuit has reached steady state. Again, the circuit is simplified in two ways. The switch is replaced with an open circuit because the switch is open. The inductor is replaced by a short circuit because inductors act like short circuits when all the inputs are constants and the

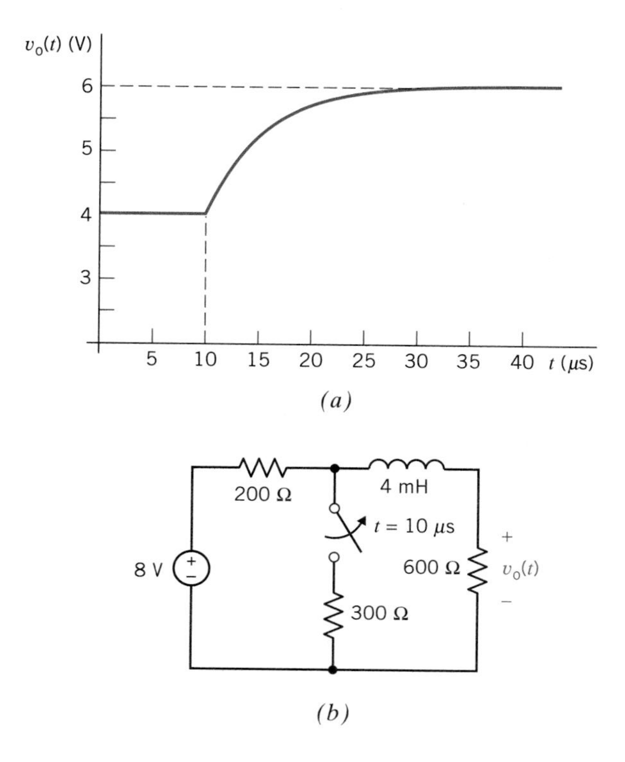

Figure 8.10-1
(a) A transient response and (b) the corresponding circuit.

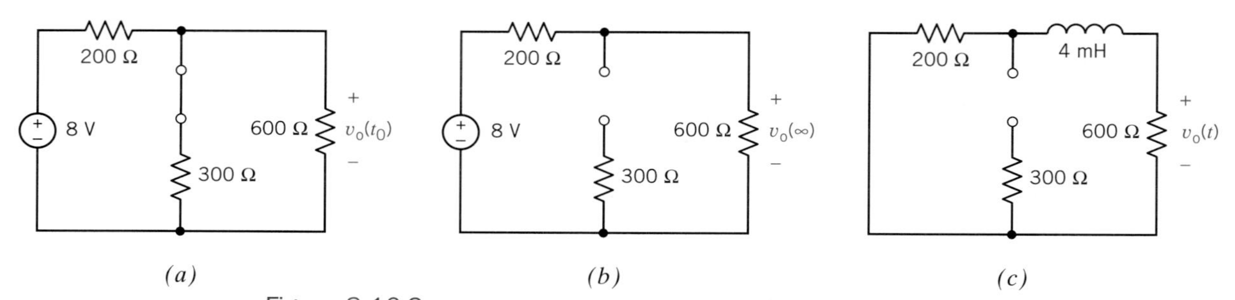

Figure 8.10-2 Circuits used to calculate the (a) initial voltage, (b) final voltage, and (c) time constant.

circuit is at steady state. The simplified circuit is shown in Figure 8.10-2*b*. The final voltage is calculated using voltage division as

$$v_0(\infty) = \frac{600}{200 + 600} 8 = 6 \text{ V}$$

The time constant is calculated from the circuit shown in Figure 8.10-2*c*. This circuit has been simplified by setting the input to zero (a zero voltage source acts like a short circuit) and replacing the switch by an open circuit. The time constant is

$$\tau = \frac{4 \times 10^{-3}}{200 + 600} = 5 \times 10^{-6} = 5 \ \mu\text{s}$$

Figure 8.10-3 shows how the initial voltage, final voltage, and time constant can be determined from the plot of the transient response. (Recall that a procedure for determining the time constant graphically was illustrated in Figure 8.3-3.) Since the values of $v_0(t_0)$, $v_0(\infty)$, and τ obtained from the transient response are the same as the values obtained by analyzing the circuit, we conclude that the transient response is indeed correct.

Example 8.10-2

Consider the circuit and corresponding transient response shown in Figure 8.10-4. How can we tell if the transient response is correct? Four things need to be verified: the steady-state capacitor voltage when the switch is open; the steady-state capacitor voltage when the switch is closed; the time constant when the switch is open; and the time constant when the switch is closed.

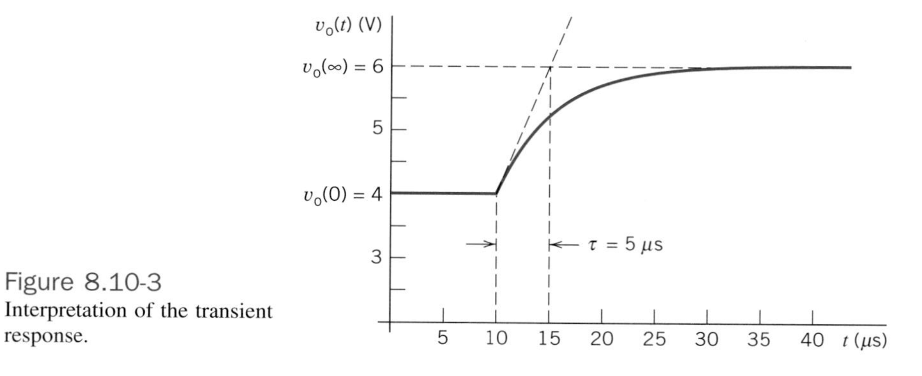

Figure 8.10-3 Interpretation of the transient response.

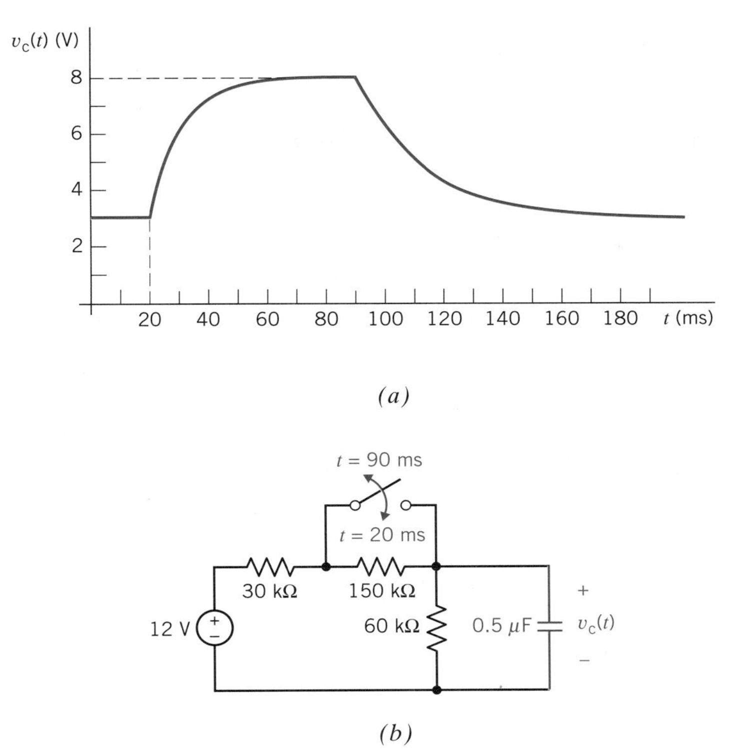

Figure 8.10-4 (*a*) A transient response and (*b*) the corresponding circuit.

Solution

Figure 8.10-5*a* shows the circuit used to calculate the steady-state capacitor voltage when the switch is open. The circuit has been simplified in two ways. First, the switch has been replaced with an open circuit. Second, the capacitor has been replaced with an open circuit because capacitors act like open circuits when all the inputs are constants and the circuit is at steady state. The steady-state capacitor voltage is calculated using voltage division as

$$v_c(\infty) = \frac{60}{60 + 30 + 150} 12 = 3 \text{ V}$$

Figure 8.10-5*b* shows the circuit used to calculate the steady-state capacitor voltage when the switch is closed. Again, this circuit has been simplified in two ways. First, the switch has been replaced with a short circuit. Second, the capacitor has been replaced with an open circuit. The steady-state capacitor voltage is calculated using voltage division as

$$v_c(\infty) = \frac{60}{60 + 30} 12 = 8 \text{ V}$$

Figure 8.10-5*c* shows the circuit used to calculate the time constant when the switch is open. This circuit has been simplified in two ways. First, the switch has been replaced with an open circuit. Second, the input has been set to zero (a zero voltage source acts like a short circuit). Notice that 180 kΩ in parallel with 60 kΩ is equivalent to 45 kΩ. The time constant is

$$\tau = (45 \times 10^3) \cdot (0.5 \times 10^{-6}) = 22.5 \times 10^{-3} = 22.5 \text{ ms}$$

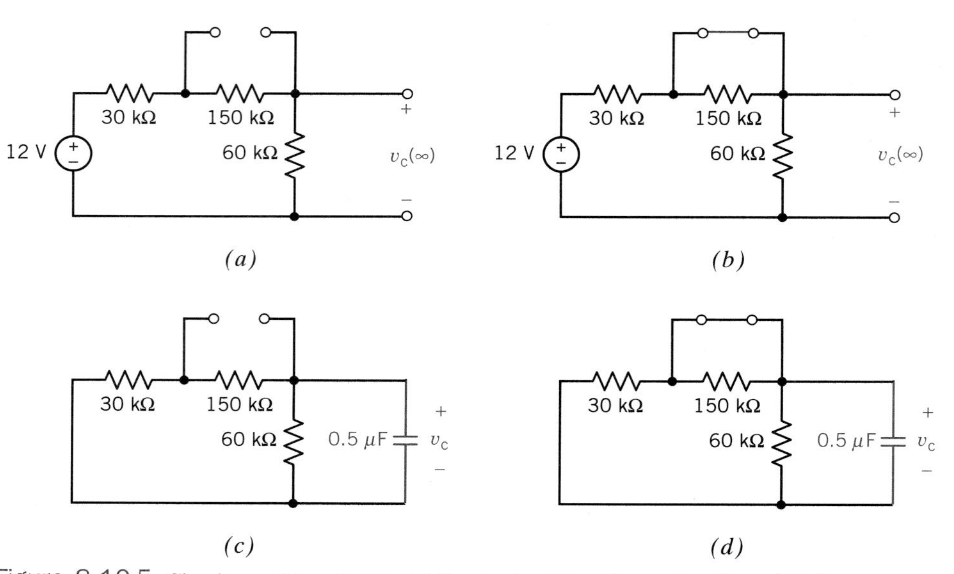

Figure 8.10-5 Circuits used to calculate (*a*) the steady-state voltage when the switch is open, (*b*) the steady-state voltage when the switch is closed, (*c*) the time constant when the switch is open, and (*d*) the time constant when the switch is closed.

Figure 8.10-5*d* shows the circuit used to calculate the time constant when the switch is closed. The switch has been replaced with a short circuit, and the input has been set to zero. Notice that 30 kΩ in parallel with 60 kΩ is equivalent to 20 kΩ. The time constant is

$$\tau = (20 \times 10^3) \cdot (0.5 \times 10^{-6}) = 10^{-2} = 10 \text{ ms}$$

Having done these calculations, we expect the capacitor voltage to be 3 V until the switch closes at $t = 20$ ms. The capacitor voltage will then increase exponentially to 8 V, with a time constant equal to 10 ms. The capacitor voltage will remain 8 V until the switch opens at $t = 90$ ms. The capacitor voltage will then decrease exponentially to 3 V, with a time constant equal to 22.5 ms. Figure 8.10-6 shows that the transient response satisfies this description. We conclude that the transient response is correct.

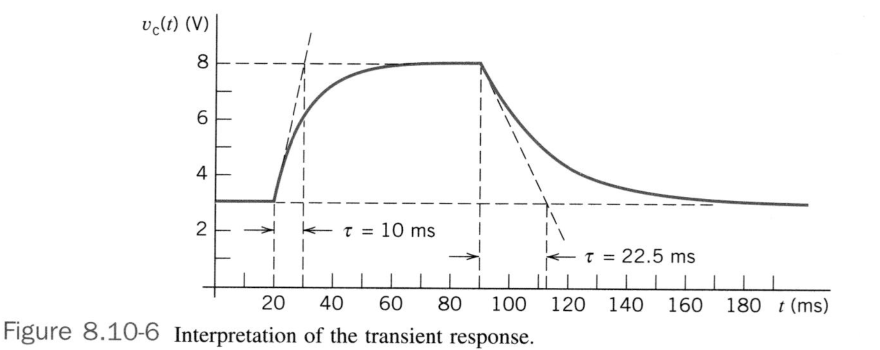

Figure 8.10-6 Interpretation of the transient response.

8.11 Design Challenge Solution

A COMPUTER AND PRINTER

It is frequently necessary to connect two pieces of electronic equipment together so that the output from one device can be used as the input to another device. For example, this situation occurs when a printer is connected to a computer as shown in Figure 8.11-1*a*. This situation is represented more generally by the circuit shown in Figure 8.11-1*b*. The driver is used to send a signal through the cable to the receiver. Let us replace the driver, cable, and receiver with simple models. Model the driver as a voltage source, the cable as an *RC* circuit, and the receiver as an open circuit. The values of resistance and capacitance used to model the cable will depend on the length of the cable. For example, when RG58 coaxial cable is used,

$$R = r \cdot L \quad \text{where} \quad r = 0.54\ \frac{\Omega}{\text{m}}$$

and

$$C = c \cdot L \quad \text{where} \quad c = 88\ \frac{\text{pF}}{\text{m}}$$

and L is the length of the cable in meters, Figure 8.11-1*c* shows the equivalent circuit.

Suppose that the circuits connected by the cable are digital circuits. The driver will send 1's and 0's to the receiver. These 1's and 0's will be represented by voltages. The output of the driver will be one voltage, V_{OH}, to represent logic 1 and another voltage, V_{OL}, to represent a logic 0. For example, one popular type of logic, called TTL logic, uses $V_{OH} = 2.4$ V and $V_{OL} = 0.4$ V. (TTL stands for Transistor-Transistor Logic.) The receiver uses two different voltages, V_{IH} and V_{IL}, to represent 1's and 0's. (This is done to provide noise immunity, but that is another story.) The receiver will interpret its input, v_b, to be a logic 1 whenever $v_b > V_{IH}$ and to be a logic 0 whenever $v_b < V_{IL}$. (Voltages between V_{IH} and V_{IL} will occur only during transitions between logic 1 and logic 0. These voltages will sometimes be interpreted as logic 1 and other times as logic 0.) TTL logic uses $V_{IH} = 2.0$ V and $V_{IL} = 0.8$ V.

Figure 8.11-2 shows what happens when the driver output changes from logic 0 to logic 1. Before time t_0

$$v_a = V_{OL} \quad \text{and} \quad v_b < V_{IL} \quad \text{for} \quad t \le t_0$$

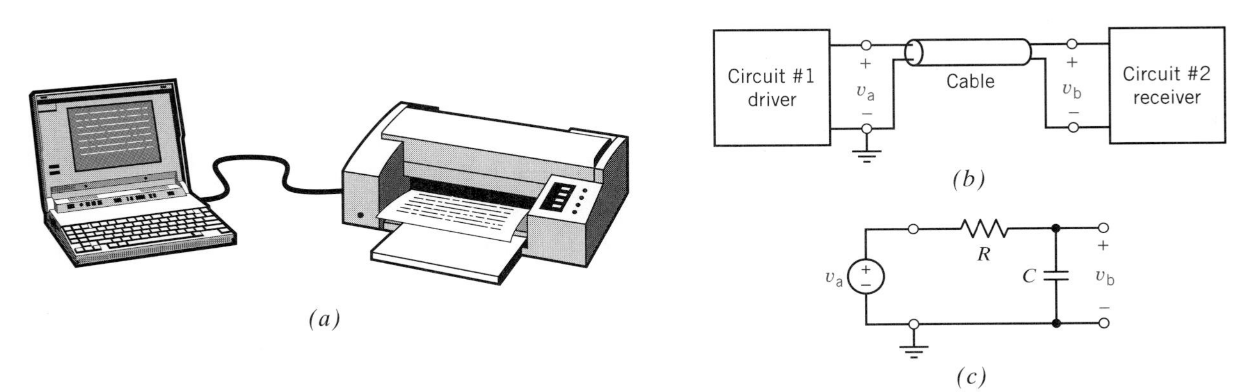

Figure 8.11-1 (*a*) A printer connected to a laptop computer. (*b*) Two circuits connected by a cable and (*c*) an equivalent circuit.

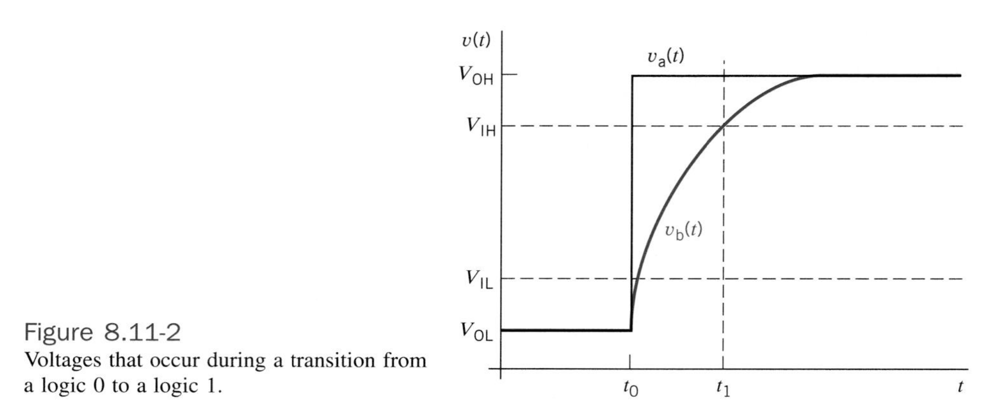

Figure 8.11-2
Voltages that occur during a transition from a logic 0 to a logic 1.

In words, a logic 0 is sent and received. The driver output switches to V_{OH} at time t_0. The receiver input, v_b, makes this transition more slowly. Not until time t_1 does the receiver input become large enough to be interpreted as a logic 1. That is,

$$v_b > V_{IH} \quad \text{for } t \geq t_1$$

The time that it takes for the receiver to recognize the transition from logic 0 to logic 1

$$\Delta t = t_1 - t_0$$

is called the delay. This delay is important because it puts a limit on how fast 1's and 0's can be sent from the driver to the receiver. To ensure that the 1's and 0's are received reliably, each 1 and each 0 must last at least Δt. The rate at which 1's and 0's are sent from the driver to the receiver is inversely proportional to the delay.

Suppose two TTL circuits are connected using RG58 coaxial cable. What restriction must be placed on the length of the cable to ensure that the delay, Δt, is less than 2 ns?

Describe the Situation and the Assumptions

The voltage $v_b(t)$ is the capacitor voltage of an *RC* circuit. The *RC* circuit is at steady state just before time t_0.

The input to the RC circuit is $v_a(t)$. Before time t_0, $v_a(t) = V_{OL} = 0.4$ V. At time t_0, $v_a(t)$ changes abruptly. After time t_0, $v_a(t) = V_{OH} = 2.4$ V.

Before time t_0, $v_b(t) = V_{OL} = 0.4$ V. After time t_0, $v_b(t)$ increases exponentially. Eventually $v_b(t) = V_{OH} = 2.4$ V.

The time constant of the *RC* circuit is

$$\tau = R \cdot C = rcL^2 = 47.52 \times 10^{-12} \cdot L^2$$

where L is the cable length in meters.

State the Goal

Calculate the maximum value of the cable length, L, for which $v_b > V_{IH} = 2.0$ V by time $t = t_0 + \Delta t$, where $\Delta t = 2$ ns.

Generate a Plan

Calculate the voltage $v_b(t)$ in Figure 8.11-1*b*. The voltage $v_b(t)$ will depend on the length of the cable, L, because the time constant of the *RC* circuit is a function of L. Set $v_b = V_{IH}$ at time $t = t_0 + \Delta t$. Solve the resulting equation for the length of the cable.

Act on the Plan

Using the notation introduced in this chapter,

$$v_b(0) = V_{OL} = 0.4 \text{ V}$$

$$v_b(\infty) = V_{OH} = 2.4 \text{ V}$$

and

$$\tau = 47.52 \times 10^{-12} \cdot L^2$$

Using Eq. 8.3-6, the voltage $v_b(t)$ is expressed as

$$v_b(t) = V_{OH} + (V_{OL} - V_{OH})e^{-(t-t_0)/\tau}$$

The capacitor voltage, v_b, will be equal to V_{IH} at time $t_1 = t_0 + \Delta t$, so

$$V_{IH} = V_{OH} + (V_{OL} - V_{OH})e^{-\Delta t/\tau}$$

Solving for the delay, Δt, gives

$$\Delta t = -\tau \ln\left[\frac{V_{IH} - V_{OH}}{V_{OL} - V_{OH}}\right]$$

$$= -47.52 \times 10^{-12} \cdot L^2 \cdot \ln\left[\frac{V_{IH} - V_{OH}}{V_{OL} - V_{OH}}\right]$$

In this case

$$L = \sqrt{\frac{-\Delta t}{47.52 \times 10^{-12} \cdot \ln\left[\frac{V_{IH} - V_{OH}}{V_{OL} - V_{OH}}\right]}}$$

and therefore

$$L = \sqrt{\frac{-2 \cdot 10^{-9}}{47.52 \times 10^{-12} \cdot \ln\left[\frac{2.0 - 2.4}{0.4 - 2.4}\right]}} = 5.11\text{m} = 16.8 \text{ ft}$$

Verify the Result

When $L = 5.11$ m, then

$$R = 0.54 \times 5.11 = 2.76 \ \Omega$$

and

$$C = (88 \times 10^{-12}) \times 5.11 = 450 \text{ pF}$$

so

$$\tau = 2.76 \times (450 \times 10^{-12}) = 1.24 \text{ ns}$$

Finally,

$$\Delta t = -1.24 \times 10^{-9} \times \ln\left[\frac{2.0 - 2.4}{0.4 - 2.4}\right] = 1.995 \text{ ns}$$

Since $\Delta t < 2$ ns, the specifications have been satisfied but with no margin for error.

8.12 SUMMARY

- Voltages and currents can be used to encode, store, and process information. When a voltage or current is used to represent information, that voltage or current is called a signal. Electric circuits that process that information are called signal processing circuits.
- Circuits that contain energy-storing elements, that is, capacitors and inductors, are represented by differential equations rather than algebraic equations. Analysis of these circuits requires the solution of differential equations.
- In this chapter we restricted our attention to first-order circuits. First-order circuits contain one energy storage element and are represented by first-order differential equations, which are reasonably easy to solve. We solved first-order differential equations using the method called separation of variables.

- The *complete response* of a circuit is the sum of the *natural response* and *the forced response*. The natural response is the general solution of the differential equation that represents the circuit when the input is set to zero. The forced response is the particular solution of the differential equation representing the circuit.
- The complete response can be separated into the *transient response* and the *steady state response*. The transient response vanishes with time leaving the steady-state response. When the input to the circuit is either a constant or a sinusoid, the steady-state response can be used as the forced response.
- The term *transient response* sometimes refers to the "transient part of the complete response" and other times to a complete response which includes a transient part. In particular, PSpice uses the term *transient response* to refer to the complete response. Since this can be confusing, the term must be used carefully.
- The *step response* of a circuit is the response when the input is equal to a unit step function and all the initial conditions of the circuit are equal to zero.
- We used Thévenin and Norton equivalent circuits to reduce the problem of analyzing any first-order circuit to the problem of analyzing one of two simple first-order circuits. One of the simple first-order circuits is a series circuit consisting of a voltage source, a resistor, and a capacitor. The other is a parallel circuit consisting of a current source, a resistor, and an inductor. Table 8.12-1 summarizes the equations used to determine the complete response of a first-order circuit.
- The parameter τ in the first-order differential equation $\frac{d}{dt}x(t) + \frac{x(t)}{\tau} = K$ is called the time constant. The time constant τ is the time for the response of a first-order circuit to complete 63 percent of the transition from initial value to final value.
- Stability is a property of well-behaved circuits. It is easy to tell if a first-order circuit is stable. A first-order circuit is stable if, and only if, its time constant is not negative, that is, $\tau \geq 0$.

Table 8.12-1 **Summary of First Order Circuits.**

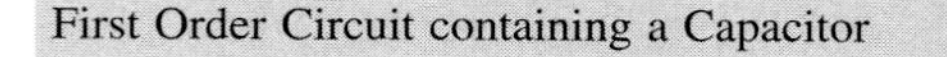

First Order Circuit containing a Capacitor	First Order Circuit containing an Inductor
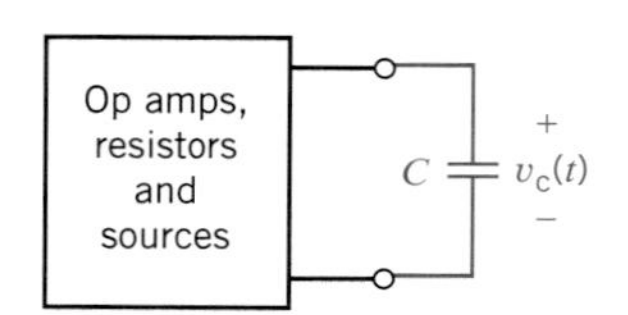	Op amps, resistors and sources L $i_L(t)$
Replace the circuit consisting of op amps, resistors and sources by its Thévenin equivalent circuit:	Replace the circuit consisting of op amps, resistors and sources by its Norton equivalent circuit:
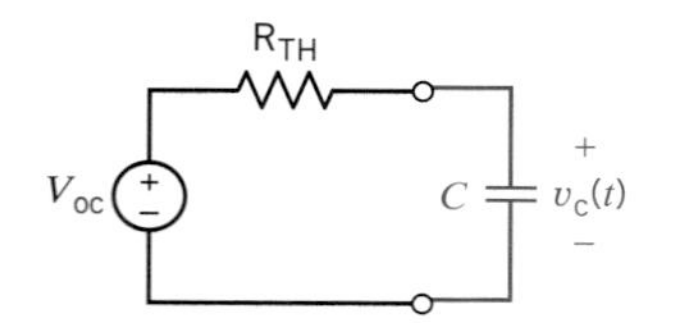	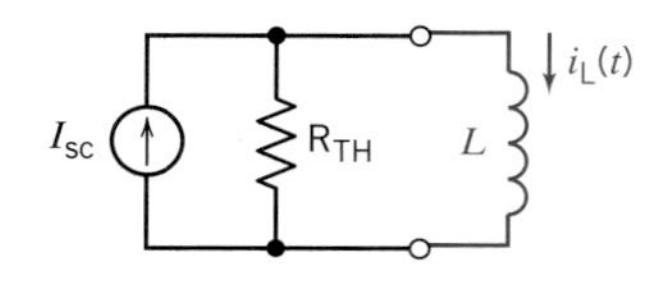
The capacitor voltage is:	The inductor current is:
$v(t) = V_{oc} + (v(0) - V_{oc})\, e^{-\frac{t}{R_{TH}C}}$	$i(t) = I_{sc} + (i(0) - I_{sc})\, e^{-\frac{R_{TH}}{L}t}$
where the time constant, τ, is	where the time constant, τ, is
$\tau = R_{TH}C$	$\tau = \frac{L}{R_{TH}}$
and the initial condition, $v(0)$, is the capacitor voltage at time $t = 0$.	and the initial condition, $i(0)$, is the inductor current at time $t = 0$.

PROBLEMS

Section 8.3 The Response of a First-Order Circuit to a Constant Input

P 8.3-1 A circuit containing a single inductor is at steady state until time $t = 0$. Before $t = 0$, the inductor current is $i_L(t) = 3$ mA. The circuit is disturbed at time $t = 0$. Figure P 8.3-1 shows the circuit after $t = 0$. Find the inductor current after time $t = 0$.

Answer: $i_L(t) = 5 - 2e^{-\frac{t}{3.75}}$ mA after $t = 0$, where t is in ms

P 8.3-2 A circuit containing a single capacitor is at steady state until time $t = 0$. Before $t = 0$, the capacitor voltage is $v_c(t) = 8$ V. The circuit is disturbed at time $t = 0$. Figure P 8.3-2 shows the circuit after $t = 0$. Find the capacitor voltage after time $t = 0$.

Answer: $v_c(t) = 4 + 4\,e^{-\frac{t}{0.67}}$ V after $t = 0$, where t is in ms.

P 8.3-3 The circuit shown in Figure P 8.3-3 is at steady state before the switch closes at time $t = 0$ s. Determine the capacitor voltage, $v(t)$, for $t > 0$ s.

Answer: $v(t) = -6 + 18e^{-6.67t}$ V for $t > 0$

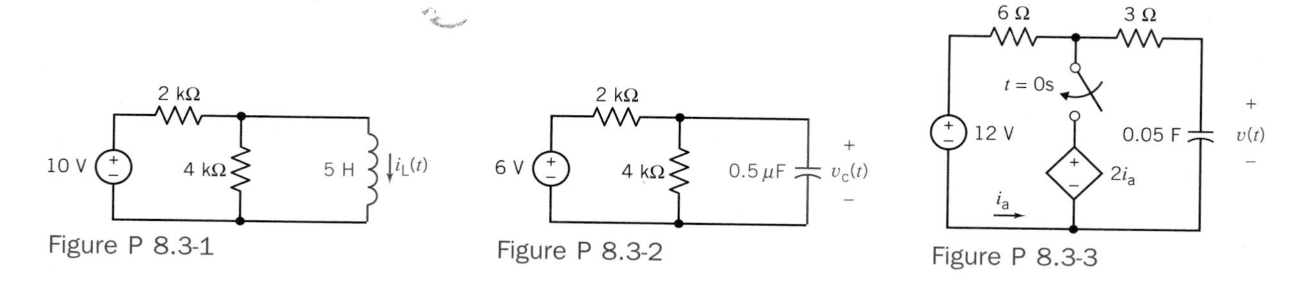

Figure P 8.3-1

Figure P 8.3-2

Figure P 8.3-3

P 8.3-4 The circuit shown in Figure P 8.3-4 is at steady state before the switch closes at time $t = 0$ s. Determine the inductor current, $i(t)$, for $t > 0$ s.

Answer: $i(t) = -2 + \frac{10}{3}e^{-0.5t}$ A for $t > 0$

P 8.3-5 The circuit shown in Figure P 8.3-5 is at steady state before the switch opens at time $t = 0$ s. Determine the voltage, $v_o(t)$, for $t > 0$ s.

Answer: $v_o(t) = 10 - 10e^{-12.5t}$ V for $t > 0$

P 8.3-6 The circuit shown in Figure P 8.3-6 is at steady state before the switch opens at time $t = 0$ s. Determine the voltage, $v_o(t)$, for $t > 0$ s.

Answer: $v_o(t) = 10e^{-4000t}$ V for $t > 0$

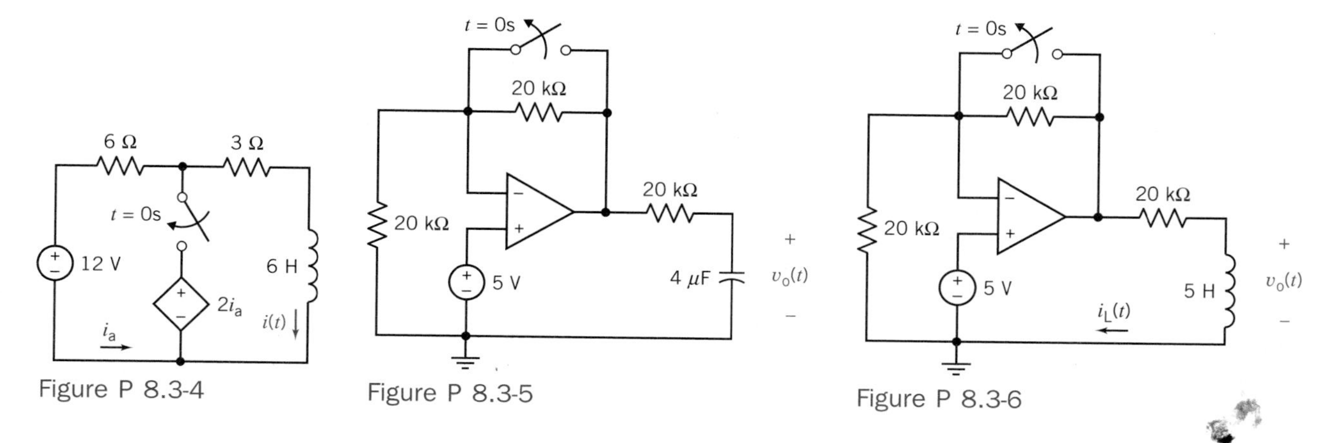

Figure P 8.3-4

Figure P 8.3-5

Figure P 8.3-6

P 8.3-7 The circuit shown in Figure P 8.3-7 is disturbed at time $t = 0$ when the switch opens. The circuit changes when the switch opens. Before time $t = 0$, the circuit is excited by a constant input. After time $t = 0$, the changed circuit is excited by the same constant input. This circuit has two steady-state responses, $v_C(t)$, one before $t = 0$ and one after $t = 0$. Find both of these steady-state responses.

Answer: $v_C(t) = 2$ V before $t = 0$ and $v_c(t) = 4$ V after $t = 0$

P 8.3-8 The circuit shown in Figure P 8.3-8 is disturbed at time $t = 0$ when the switch closes. The circuit changes when the switch closes. Before time $t = 0$, the circuit is excited by a constant input. After time $t = 0$, the changed circuit is excited by the same constant input. This circuit has two steady-state responses, $i_L(t)$, one before $t = 0$ and one after $t = 0$. Find both of these steady-state responses.

Answer: $i_L(t) = 2$ mA before $t = 0$ and $i_L(t) = 4$ mA after $t = 0$

P 8.3-9 The circuit shown in Figure P 8.3-9 is disturbed at time $t = 0$ when the switch closes. The circuit changes when the switch closes. Before time $t = 0$, the circuit is excited by a constant input. After time $t = 0$, the changed circuit is excited by the same constant input. This circuit has two steady-state responses, $i_L(t)$, one before $t = 0$ and one after $t = 0$. Find both of these steady-state responses.

Answer: $i_L(t) = 2$ mA before $t = 0$ and $i_L(t) = 2$ mA after $t = 0$

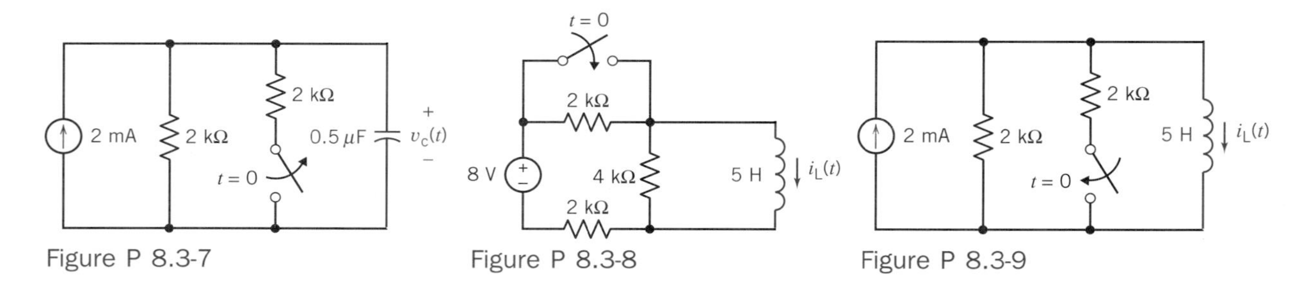

Figure P 8.3-7 Figure P 8.3-8 Figure P 8.3-9

P 8.3-10 The circuit shown in Figure P 8.3-10 is disturbed at time $t = 0$ when the switch opens. The circuit changes when the switch opens. Before time $t = 0$, the circuit is excited by a constant input. After time $t = 0$, the changed circuit is excited by the same constant input. This circuit has two steady-state responses, $v_C(t)$, one before $t = 0$ and one after $t = 0$. Find both of these steady-state responses.

Answer: $v_C(t) = 5.33$ V before $t = 0$ and $v_C(t) = 4$ V after $t = 0$

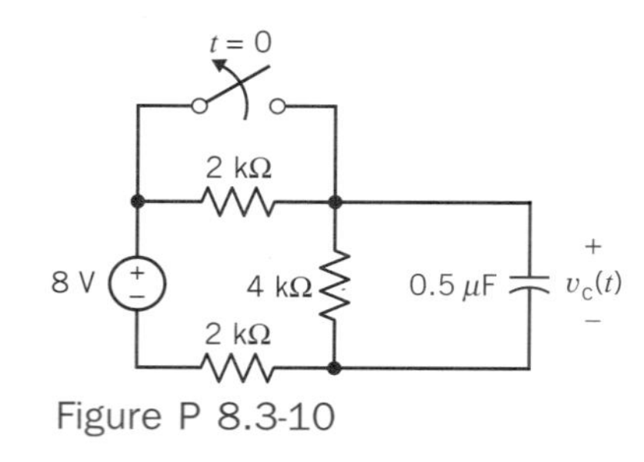

Figure P 8.3-10

(*a*)

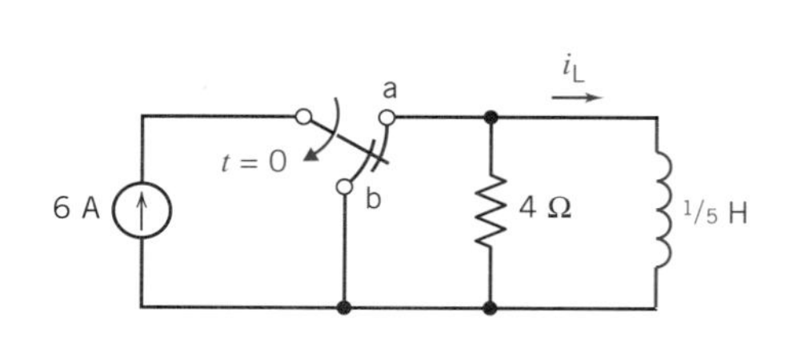

(*b*)

Figure P 8.3-11 (*a*) Astronaut Dale Gardner using the manned maneuvering unit to dock with the Westar VI satellite. Courtesy of NASA. (*b*) Model of the apogee capture device. Assume that the switch has been in position for a long time at $t = 0^-$.

P 8.3-11 Figure P 8.3-11*a* shows astronaut Dale Gardner using the manned maneuvering unit to dock with the spinning Westar VI satellite on November 14, 1984. Gardner used a large tool called the apogee capture device (ACD) to stabilize the satellite and capture it for recovery, as shown in Figure P 8.3-11*a*. The ACD can be modeled by the circuit of Figure P 8.3-11*b*. Find the inductor current i_L for $t > 0$.

Answer: $i_L(t) = 6e^{-20t}$ A

P 8.3-12 An electronic flash of a camera uses a small battery to charge a capacitor. Then, when the flash is activated, the capacitor is switched across the flashbulb. Assume that the battery is a 6-V battery that should not be operated with a current above 100 μA. The capacitor is to be selected. (a) Draw a circuit model that will represent the charging and discharging action. (b) It is desired to charge the capacitor within 5 s and to discharge it within 1/2 s. Select the appropriate values for the elements in the circuit. Assume that the value of the bulb resistance is 10 kΩ. Assume that the capacitor is charged or discharged in five time constants.

P 8.3-13 A security alarm for an office building door is modeled by the circuit of Figure P 8.3-13. The switch represents the door interlock, and v is the alarm indicator voltage. Find $v(t)$ for $t > 0$ for the circuit of Figure P 8.3-13. The switch has been closed for a long time at $t = 0^-$.

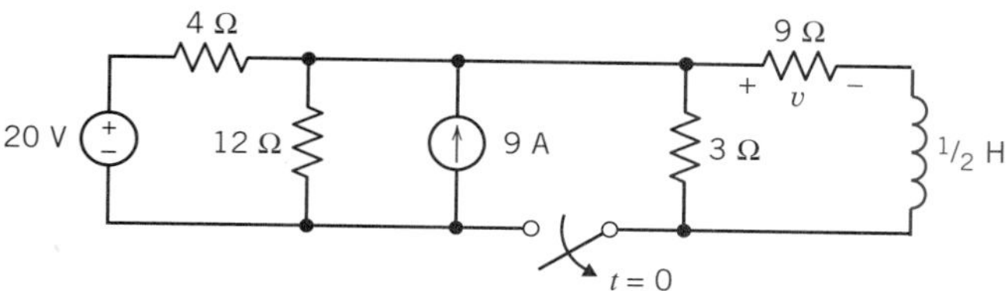

Figure P 8.3-13 A security alarm circuit.

P 8.3-14 The circuit shown in Figure P 8.3-14 is at steady state before the switch opens at time $t = 0$ s. Determine the voltage, $v_o(t)$, for $t > 0$ s.

Answer: $v_0(t) = -10 + 10e^{-12.5t}$ V for $t > 0$

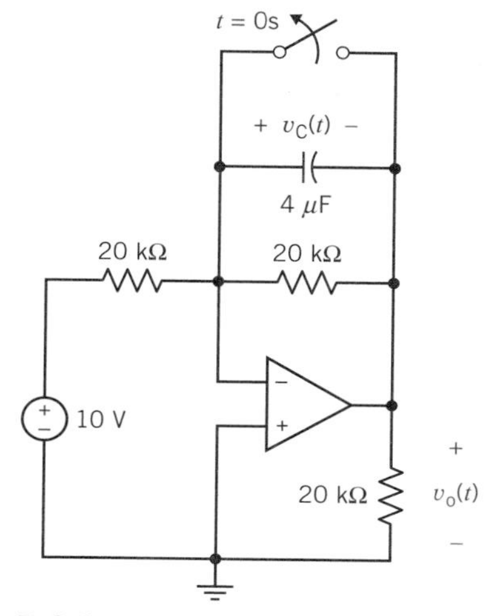

Figure P 8.3-14

Section 8.4 Sequential Switching

P 8.4-1 The circuit shown in Figure P 8.4-1 is at steady state before the switch closes at time $t = 0$ s. The switch remains closed for 1.5 s and then opens. Determine the capacitor voltage, $v(t)$, for $t > 0$ s.

Hint: Determine $v(t)$ when the switch is closed. Evaluate $v(t)$ at time $t = 1.5$ to get $v(1.5)$. Use $v(1.5)$ as the initial condition to determine $v(t)$ after the switch opens again.

Answer: $v(t) = \begin{cases} 5 + 5e^{-0.5t}\text{ V} & \text{for } 0 < t < 1.5\text{ s} \\ 10 - 2.64e^{-0.25(t-1.5)}\text{ V} & \text{for } 1.5\text{ s} < t \end{cases}$

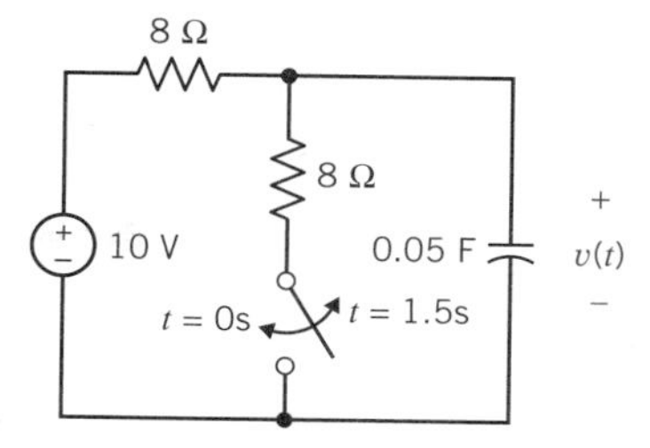

Figure P 8.4-1

P 8.4-2 The circuit shown in Figure P 8.4-2 is at steady state before the switch closes at time $t = 0$ s. The switch remains closed for 1.5 s and then opens. Determine the inductor current, $i(t)$, for $t > 0$ s.

Answer: $i(t) = \begin{cases} 2 + e^{-0.5t}\text{ A} & \text{for } 0 < t < 1.5\text{ s} \\ 3 - 0.53\,e^{-0.667(t-1.5)}\text{ A} & \text{for } 1.5\text{ s} < t \end{cases}$

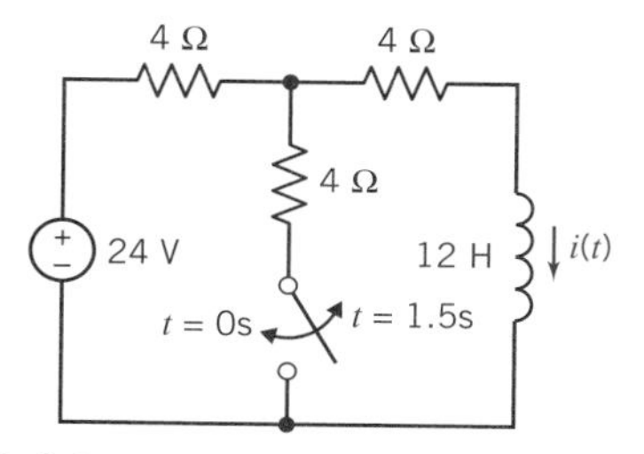

Figure P 8.4-2

P 8.4-3 Find $i(t)$ for $t > 0$ for the circuit shown in Figure P 8.4-3. The circuit is in steady state at $t = 0^-$.

Answer: $i(t) = 2/3e^{-6t}$ A for $0 \le t \le 51$ ms
$i(t) = 1.47e^{-14(t-0.051)}$ A for $t > 51$ ms

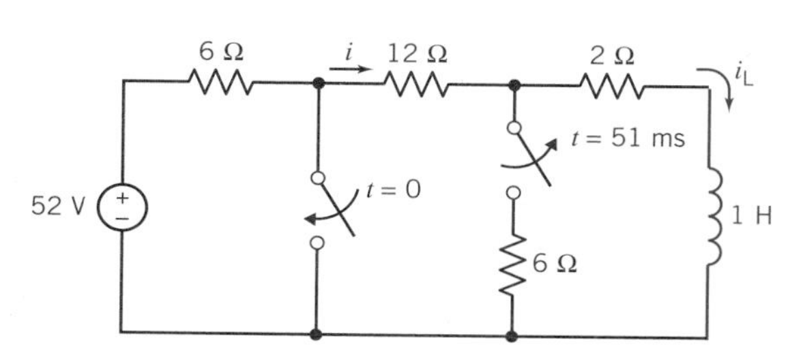

Figure P 8.4-3

P 8.4-4 Cardiac pacemakers are used by people to maintain regular heart rhythm when they have a damaged heart. The circuit of a pacemaker can be represented as shown in Figure P 8.4-4. The resistance of the wires, R, can be neglected since $R < 1$ mΩ. The heart's load resistance, R_L, is 1 kΩ. The first switch is activated at $t = t_0$ and the second switch is activated at $t_1 = t_0 + 10$ms. This cycle is repeated every second. Find $v(t)$ for $t_0 \le t \le 1$. Note that it is easiest to consider $t_0 = 0$ for this calculation. The cycle repeats by switch 1 returning to position a and switch 2 returning to its open position.

Hint: Use $q = Cv$ to determine $v(0^-)$ for the $100-\mu$F capacitor.

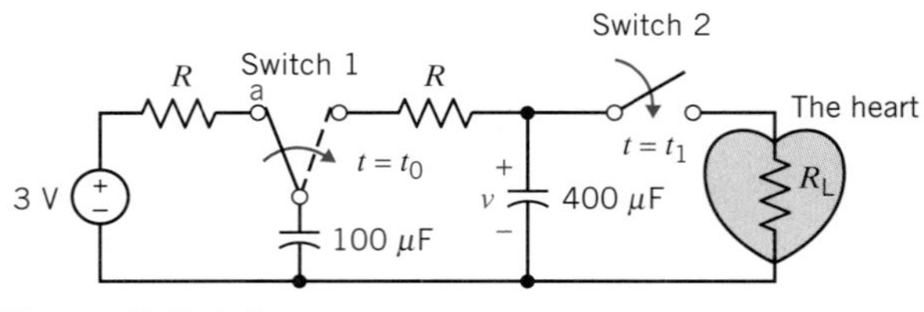

Figure P 8.4-4

P 8.4-5 Determine and sketch $i(t)$ for the circuit shown in Figure P 8.4-5. Calculate the time required for $i(t)$ to reach 99 percent of its final value.

In practice, a solenoid is not directly shorted to turn it off but is shorted through a device called a diode with a voltage drop of 0.7 V, as this is easier to implement. Why would you not simply open-circuit the coil to achieve zero current?

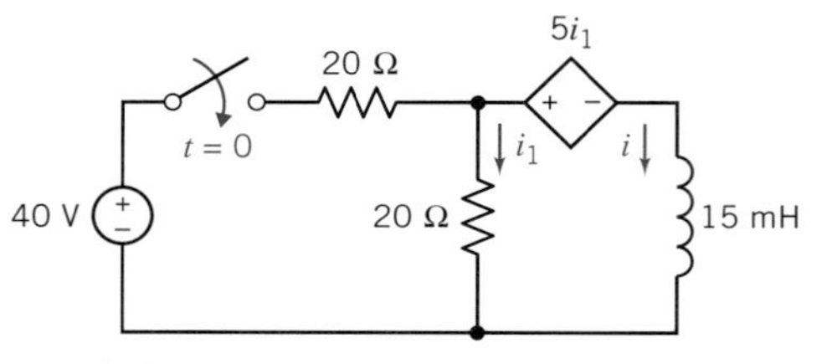

Figure P 8.4-5

P 8.4-6 An electronic flash on a camera uses the circuit shown in Figure P 8.4-6. Harold E. Edgerton invented the electronic flash in 1930. A capacitor builds a steady-state voltage and then discharges it as the shutter switch is pressed. The discharge produces a very brief light discharge. Determine the elapsed time t_1 to reduce the capacitor voltage to one-half of its initial voltage. Find the current at $t = t_1$.

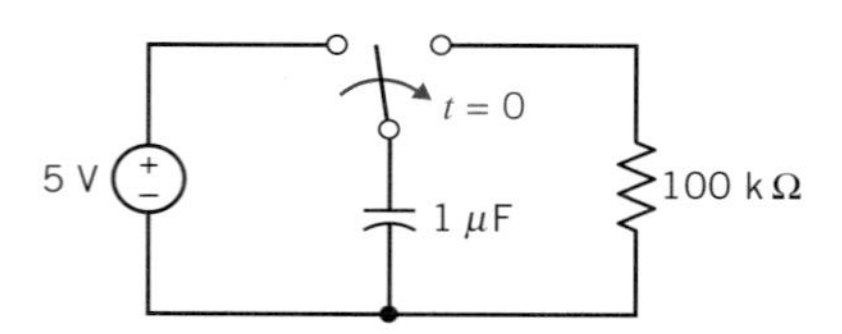

Figure P 8.4-6 Electronic flash circuit.

P 8.4-7 Sequential switching is used repetitively to generate communication signals. For the circuit shown in Figure P 8.4-7, switch a has been in position 1 and switch b has been open for a long time. At $t = 0$, switch a moves to position 2. Then, 100 ms after switch a moves, switch b closes. Find the capacitor voltage v for $t \geq 0$.

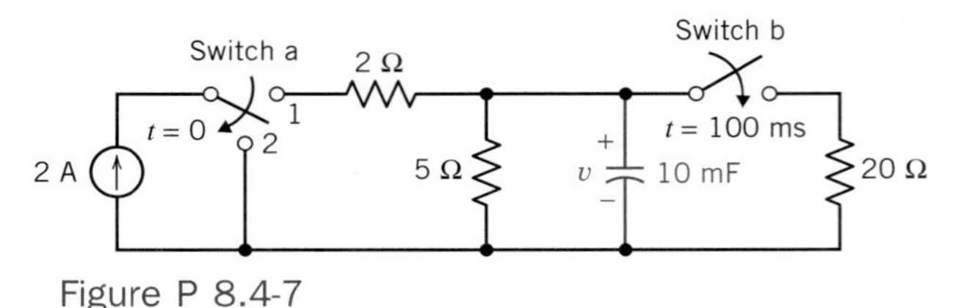

Figure P 8.4-7

P 8.4-8 For the circuit of Figure P 8.4-8, determine $v_c(t)$ and $v_x(t)$ for $t > 0$ when $C = 0.2\text{F}$. Assume that the circuit is in steady state when $t = 0^-$.

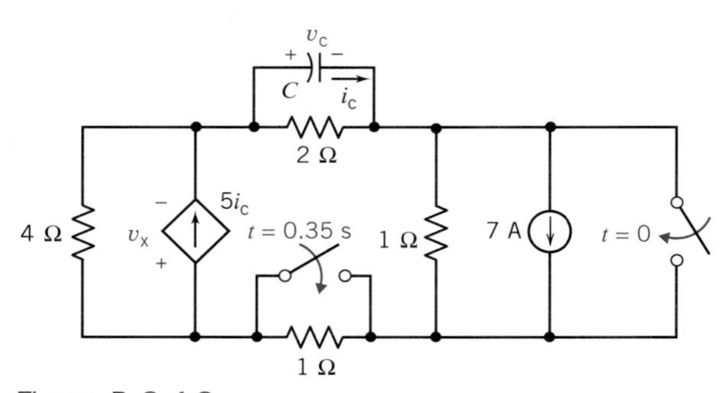

Figure P 8.4-8

Section 8.5 Stability of First-Order Circuits

P 8.5-1 The circuit in Figure P 8.5-1 contains a current-controlled voltage source. What restriction must be placed on the gain, R, of this dependent source in order to guarantee stability?

Answer: R $< 400\Omega$

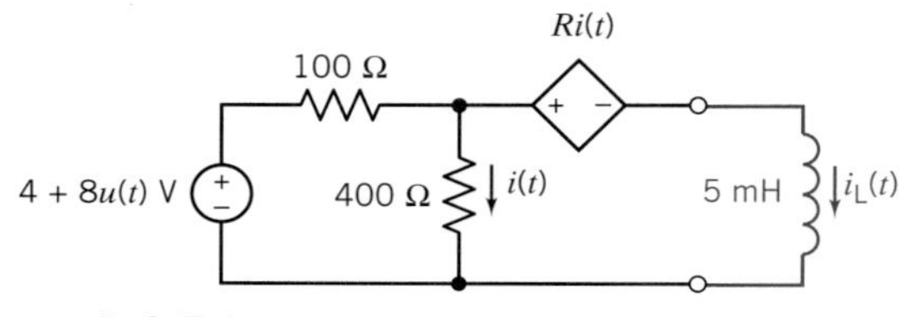

Figure P 8.5-1

P 8.5-2 The circuit in Figure P 8.5-2 contains a voltage-controlled voltage source. What restriction must be placed on the gain, A, of this dependent source in order to guarantee stability?

Answer: A < 5

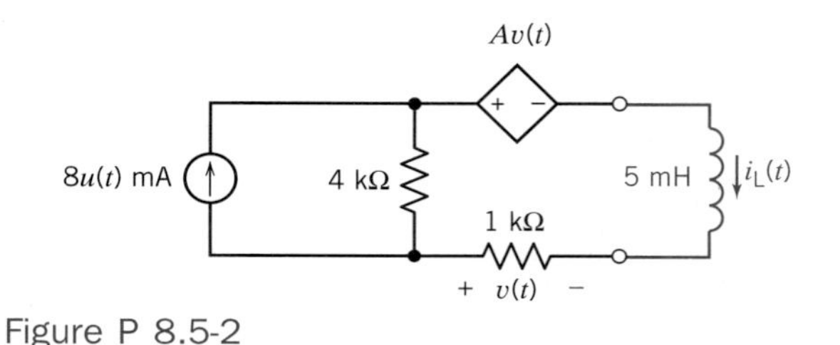

Figure P 8.5-2

P 8.5-3 The circuit in Figure P 8.5-3 contains a current-controlled current source. What restriction must be placed on the gain, B, of this dependent source in order to guarantee stability?

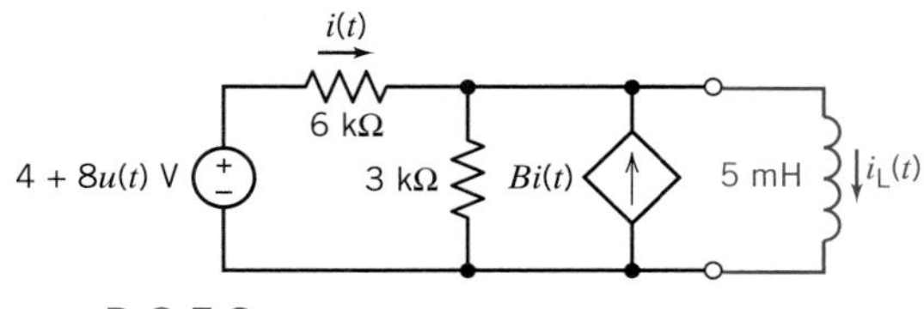

Figure P 8.5-3

P 8.5-4 The circuit in Figure P 8.5-4 contains a voltage-controlled voltage source. What restriction must be placed on the gain, A, of this dependent source in order to guarantee stability?

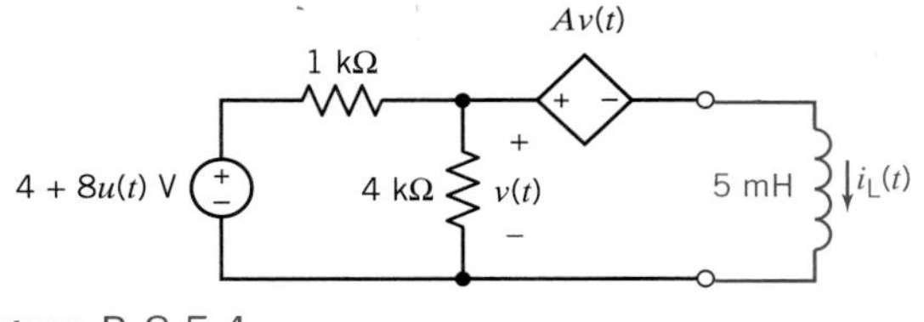

Figure P 8.5-4

Section 8.6 The Unit Step Response

P 8.6-1 The input to the circuit shown in Figure P 8.6-1 changes at time $t = 0$. Before time $t = 0$, the circuit is excited by a constant input. After time $t = 0$, the circuit is excited by a different constant input. The response of this circuit is the inductor current, $i_L(t)$. This circuit has two steady-state responses, one before $t = 0$ and one after $t = 0$. Find both of these steady-state responses.

Answer: $i_L(t) = -2$ mA before $t = 0$ and $i_L(t) = 3$ mA after $t = 0$

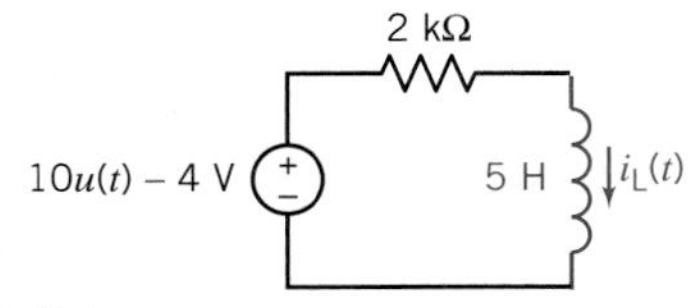

Figure P 8.6-1

P 8.6-2 The input to the circuit shown in Figure P 8.6-2 changes at time $t = 0$. Before time $t = 0$, the circuit is excited by a constant input. After time $t = 0$, the circuit is excited by a different constant input. The response of this circuit is the inductor current, $i_L(t)$. This circuit has two steady-state responses, one before $t = 0$ and one after $t = 0$. Find both of these steady-state responses.

Answer: $i_L(t) = 3$ mA before $t = 0$ and $i_L(t) = 2$ mA after $t = 0$

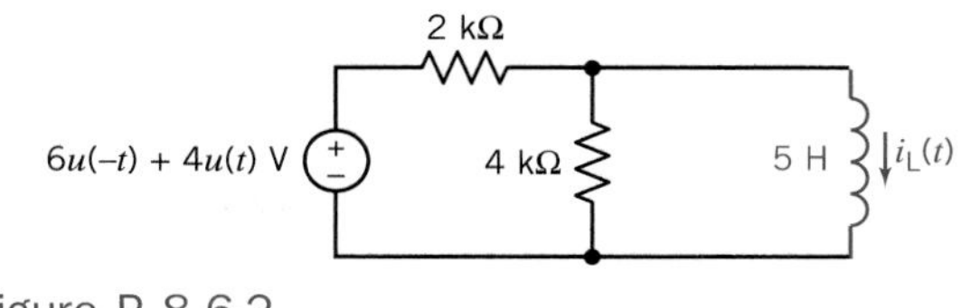

Figure P 8.6-2

P 8.6-3 The input to the circuit shown in Figure P 8.6-3 changes at time $t = 0$. Before time $t = 0$, the circuit is excited by a constant input. After time $t = 0$, the circuit is excited by a different constant input. The response of this circuit is the capacitor voltage, $v_C(t)$. This circuit has two steady-state responses, one before $t = 0$ and one after $t = 0$. Find both of these steady-state responses.

Answer: $v_C(t) = 8$ V before $t = 0$ and $v_C(t) = 12$ V after $t = 0$.

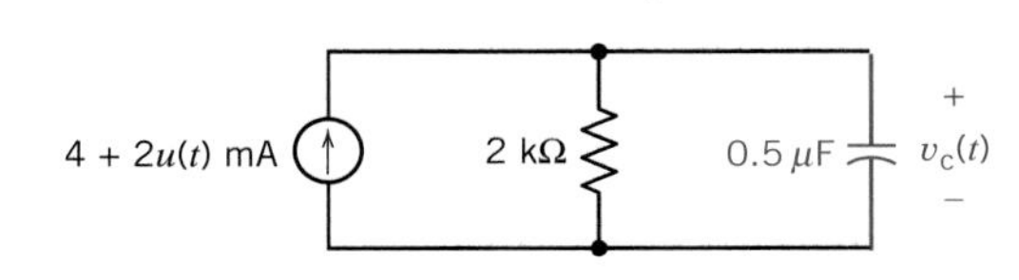

Figure P 8.6-3

P 8.6-4 The input to the circuit shown in Figure P 8.6-4 changes at time $t = 0$. Before time $t = 0$, the circuit is excited by a constant input. After time $t = 0$, the circuit is excited by a different constant input. The response of this circuit is the capacitor voltage $v_C(t)$. This circuit has two steady-state responses, one before $t = 0$ and one after $t = 0$. Find both of these steady-state responses.

Answer: $v_C(t) = 8$ V before $t = 0$ and $v_C(t) = 4$ V after $t = 0$

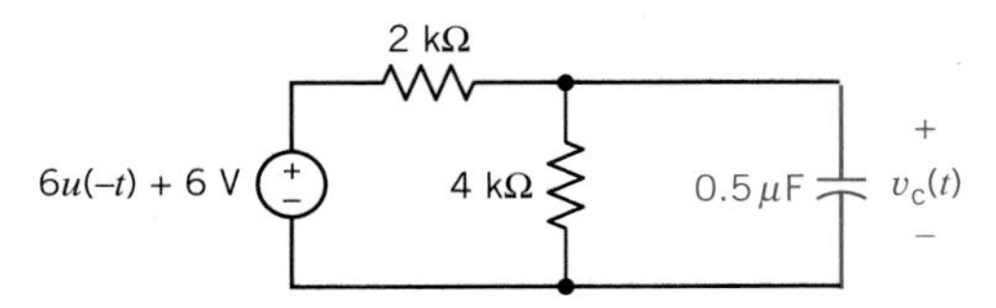

Figure P 8.6-4

P 8.6-5 Find $i(t)$ for $t > 0$ for the circuit of Figure P 8.6-5. Assume the circuit is in steady state at $t = 0^-$.

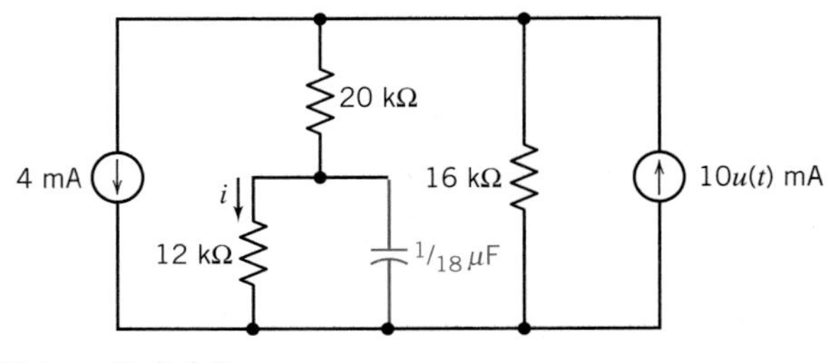

Figure P 8.6-5

P 8.6-6 Find the step response $v_c(t)$ of the circuit shown in Figure P 8.6-6 when $v_s = 20u(t)$ V. The initial voltage $v_c(0)$ is zero.

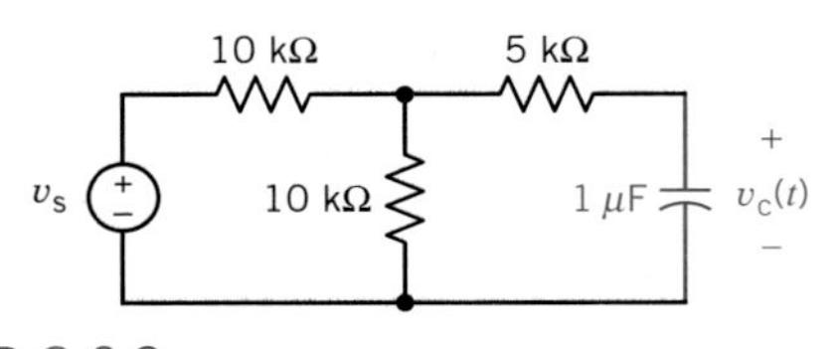

Figure P 8.6-6

P 8.6-7 Find $v(t)$ for $t > 0$ for the circuit shown in Figure P 8.6-7 when $v_s = 15e^{-t}[u(t) - u(t-1)]$ V. The circuit has reached steady state before $t = 0$.

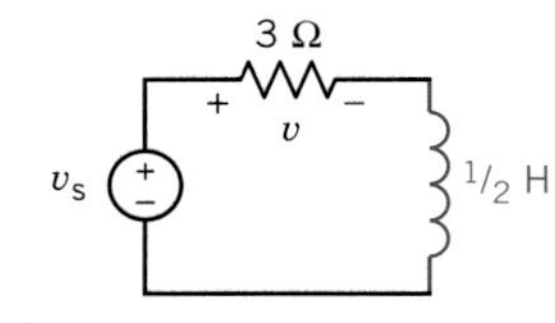

Figure P 8.6-7

P 8.6-8 Use step functions to represent the signal of Figure P 8.6-8.

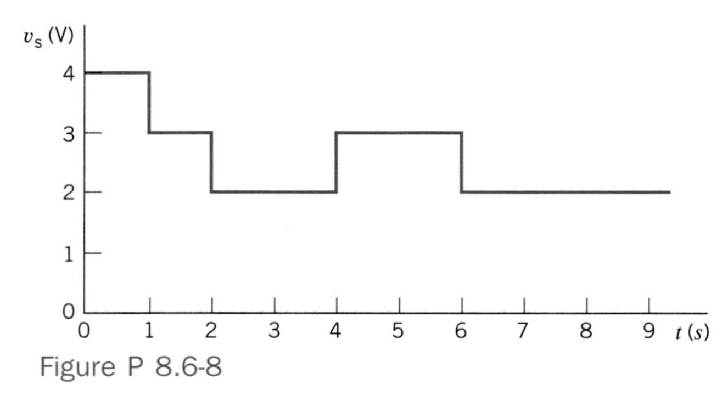

Figure P 8.6-8

P 8.6-9 The initial voltage of the capacitor of the circuit shown in Figure P 8.6-9 is zero. Determine the voltage $v(t)$ when the source is a pulse, described by

$$v_s = \begin{cases} 0 & t < 1 \text{ s} \\ 4 \text{ V} & 1 < t < 2 \text{ s} \\ 0 & t > 2 \text{ s} \end{cases}$$

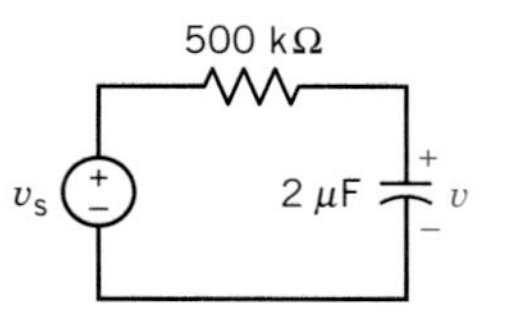

Figure P 8.6-9

P 8.6-10 Studies of an artificial insect are being used to understand the nervous system of animals (Beer 1991). A model neuron in the nervous system of the artificial insect is shown in Figure P 8.6-10. A series of pulses, called synapses, is required. The switch generates a pulse by opening at $t = 0$ and closing at $t = 0.5$ s. Assume that the circuit is in steady state and that $v(0^-) = 10$ V. Determine the voltage $v(t)$ for $0 < t < 2$ s.

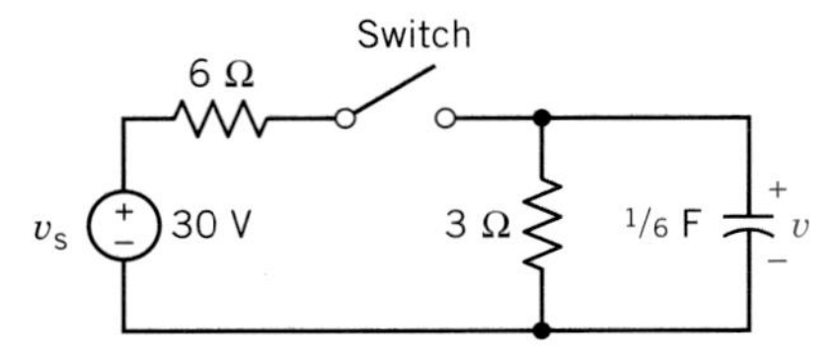

Figure P 8.6-10 Neuron circuit model.

P 8.6-11 An electronic circuit can be used to replace the springs and levers normally used to detonate a shell in a handgun (Jurgen 1989). The electric trigger would eliminate the clicking sensation, which may cause a person to misaim. The proposed trigger uses a magnet and a solenoid with a trigger switch. The circuit of Figure P 8.6-11 represents the trigger circuit with $i_s(t) = 40\,[u(t) - u(t - t_0)]$ A where $t_0 = 1$ ms. Determine and plot $v(t)$ for $0 < t < 0.3$ s.

Answer: $v = \begin{cases} 480(1-e^{-1000t}) & 0 < t < 1 \text{ ms} \\ 480(1-e^{-1})e^{-1000(t-t_0)} & t > 1 \text{ ms}, t_0 = 1 \text{ ms} \end{cases}$

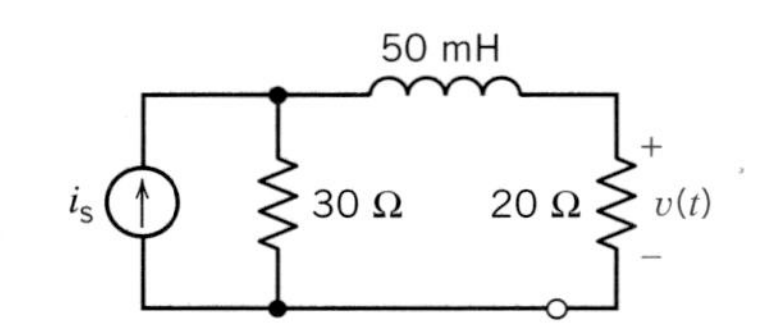

Figure P 8.6-11 Electric trigger circuit for handgun.

P 8.6-12 Determine $v_C(t)$ for $t > 0$ for the circuit of Figure P 8.6-12.

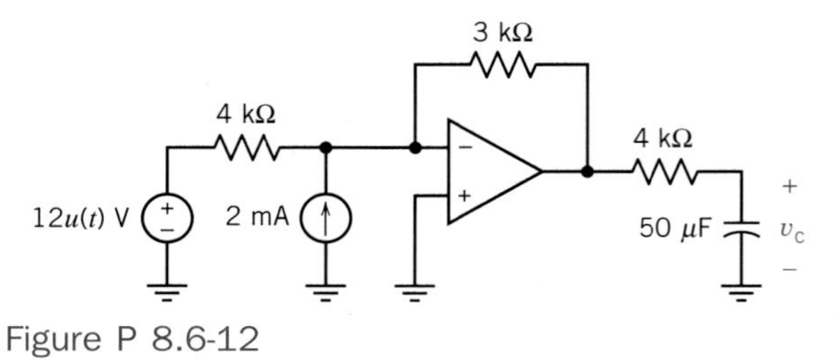

Figure P 8.6-12

Section 8.7 The Response of an *RL* or *RC* Circuit to a Nonconstant Source

P 8.7-1 Find $v_C(t)$ for $t > 0$ for the circuit shown in Figure P 8.7-1 when $v_1 = 8e^{-5t}u(t)$ V. Assume the circuit is in steady state at $t = 0^-$.

Answer: $v_C(t) = 4e^{-9t} + 18e^{-5t}$ V

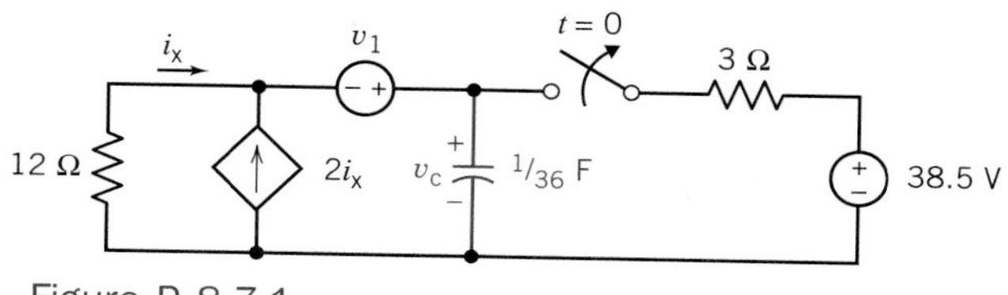

Figure P 8.7-1

P 8.7-2 Find $v(t)$ for $t > 0$ for the circuit shown in Figure P 8.7-2. Assume steady state at $t = 0^-$.
Answer: $v(t) = 20e^{-10t/3} - 12e^{-2t}$ V

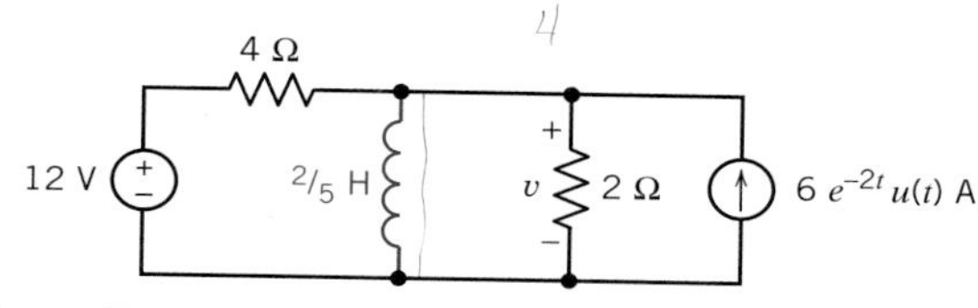

Figure P 8.7-2

P 8.7-3 Find $v(t)$ for $t > 0$ for the circuit shown in Figure P 8.7-3 when $v_1 = (25 \sin 4000t)u(t)$ V. Assume steady state at $t = 0^-$.

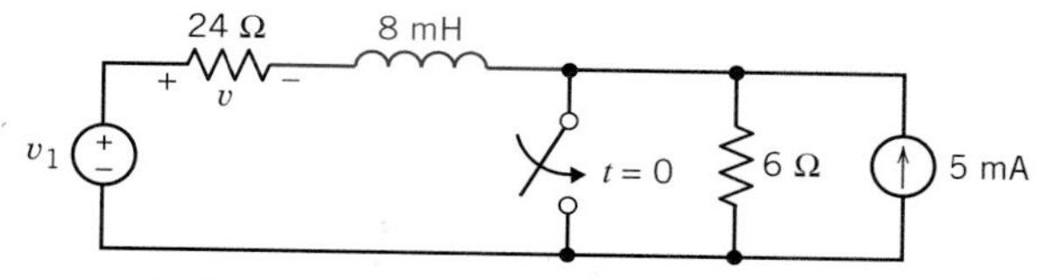

Figure P 8.7-3

P 8.7.4 Find $v_C(t)$ for $t > 0$ for the circuit shown in Figure P 8.7-4 when $i_s = [2 \cos 2t]u(t)$ mA.

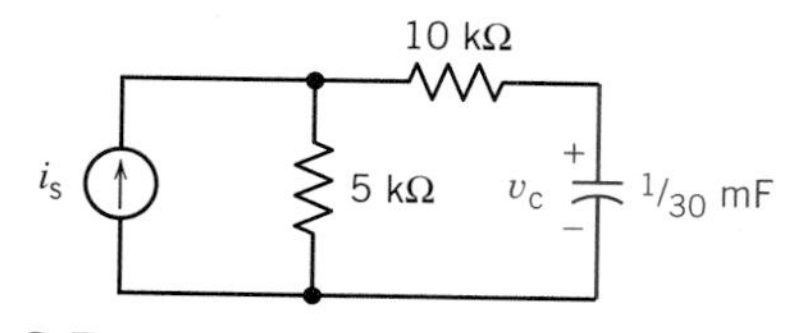

Figure P 8.7-4

P 8.7-5 Many have witnessed the use of an electrical megaphone for amplification of speech to a crowd. A model of a microphone and speaker is shown in Figure P 8.7-5*a*, and the circuit model is shown in Figure P 8.7-5*b*. Find $v(t)$ for $v_s = 10(\sin 100t)u(t)$, which could represent a person whistling or singing a pure tone.

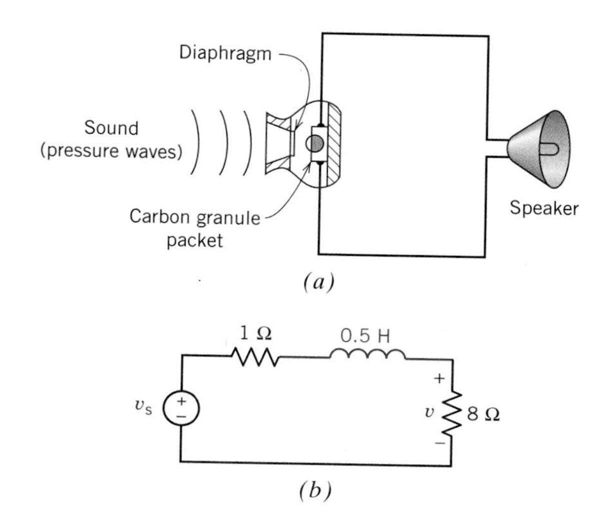

Figure P 8.7-5 Megaphone circuit.

P 8.7-6 A lossy integrator is shown in Figure P 8.7-6. It can be seen that the lossless capacitor of the ideal integrator circuit has been replaced with a model for the lossy capacitor, namely, a lossless capacitor in parallel with a 1 kΩ resistor. If $v_s = 15e^{-2t}u(t)$ V and $v_o(0) = 10$ V, find $v_o(t)$ for $t > 0$. Assume an ideal op amp.

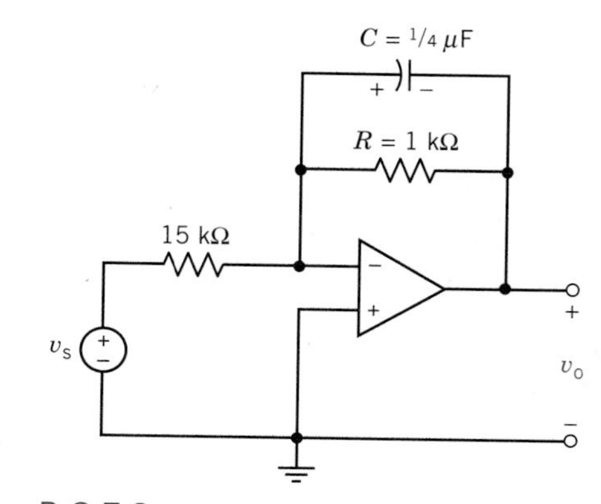

Figure P 8.7-6 Integrator circuit.

P 8.7-7 Most television sets use magnetic deflection in the cathode ray tube. To move the electron beam across the screen, it is necessary to have a ramp of current as shown in Figure P 8.7-7*a*, to flow through the deflection coil. The deflection coil circuit is shown in Figure P 8.7-7*b*. Find the waveform v_1 that will generate the current ramp, i_L.

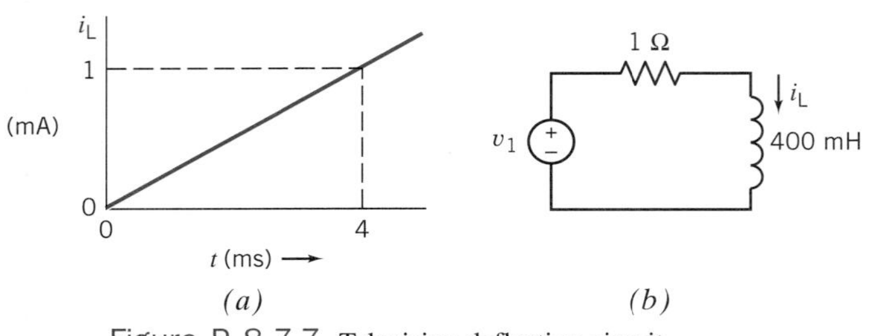

Figure P 8.7-7 Television deflection circuit.

P 8.7-8 Determine $v(t)$ for the circuit shown in Figure P 8.7-8.

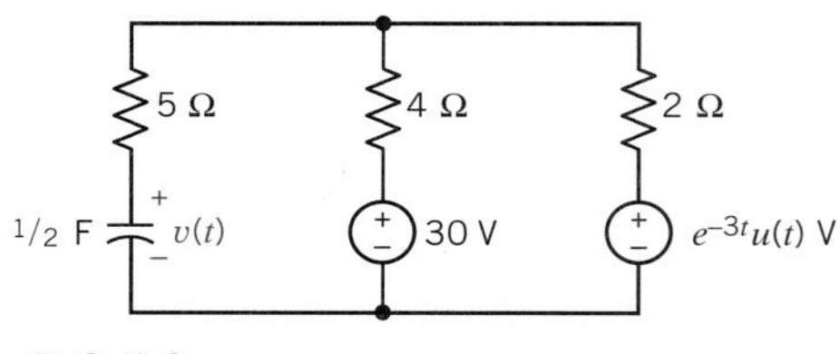

Figure P 8.7-8

P 8.7-9 (a) In the circuit of Figure P 8.7-9, given $v_{c1}(0) = 10$ V, $v_{c2}(0) = 20$ V, find $v_o(t)$ in terms of $v_1(t)$ and $v_2(t)$ for $t > 0$. (b) If $v_1(t) = 10e^{-2000t}$ V and $v_2(t) = 20e^{-1000t}$ V, find $v_0(t)$ for $t > 0$.

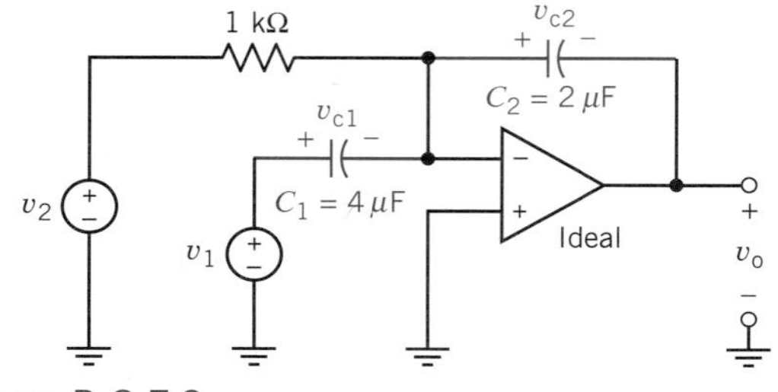

Figure P 8.7-9

P 8.7-10 For the circuit shown in Figure P 8.7-10, find $v_C(t)$, $t \geq 0$, when $v_C(0^-) = 3$ V.
Answer: $v_C(t) = 4 - e^{-250t}$ V, $t \geq 0$

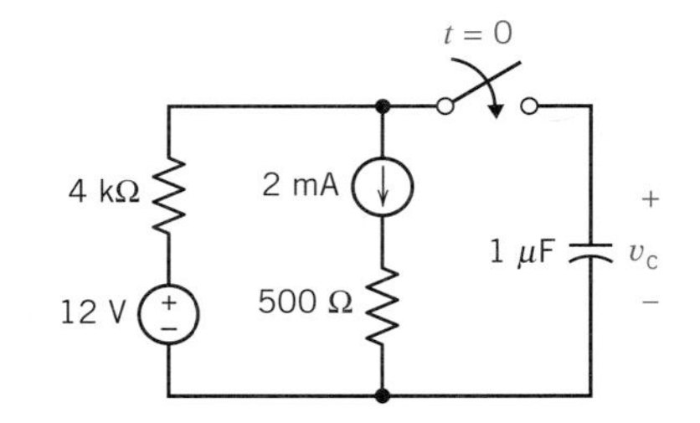

Figure P 8.7-10

P 8.7-11 Determine $v(t)$ for the circuit shown in Figure P 8.7-11*a* when v_s varies as shown in Figure P 8.7-11*b*. The initial capacitor voltage is $v_C(0) = 0$.

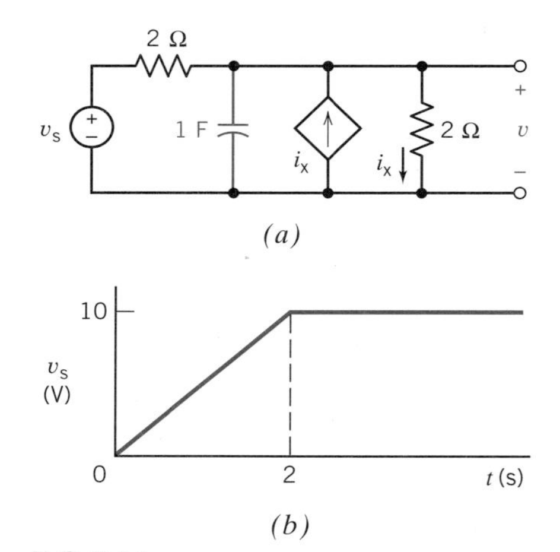

Figure P 8.7-11

P 8.7-12 The electron beam, which is used to "draw" signals on an oscilloscope, is moved across the face of a cathode-ray tube (CRT) by a force exerted on electrons in the beam. The basic system is shown in Figure P 8.7-12*a*. The force is created from a time-varying, ramp-type voltage applied across the vertical or the horizontal plates. As an example, consider the simple circuit of Figure P 8.7-12*b* for horizontal deflection where the capacitance between the plates is *C*.

(a) Derive an expression for the voltage across the capacitance. If $v(t) = kt$ and $R_s = 625\ \text{k}\Omega$, $k = 1000$, and $C = 2000$ pF, compute v_c as a function of time. Sketch $v(t)$ and $v_c(t)$ on the same graph for time less than 10 ms. Does the voltage across the plates track the input voltage?

(b) Describe the deflection of the electron beam if the force is $F = qE$, where q is the charge and E is the electric field; E is the ratio of the voltage across the plates to the spacing of the plates ($E = v/S$). Assume a zero initial condition for the capacitor.

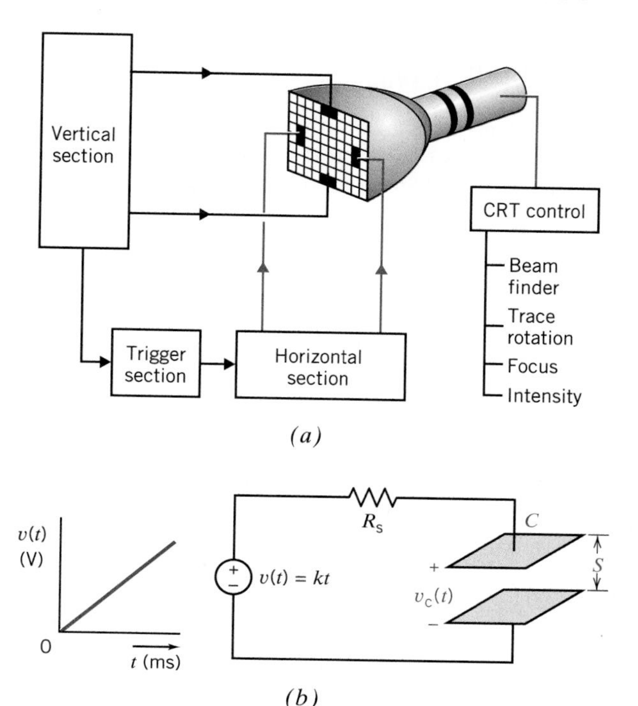

Figure P 8.7-12 Cathode-ray tube beam circuit.

PSPICE PROBLEMS

SP 8-1 Determine and plot $i(t)$ for the circuit of Figure SP 8.1.

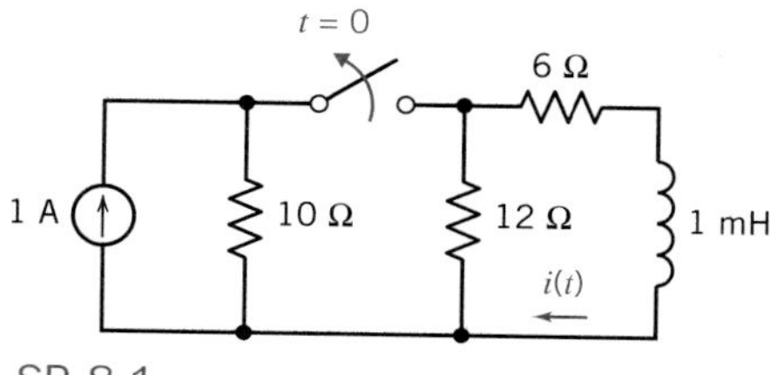

Figure SP 8.1

SP 8-2 An RC circuit is shown in Figure SP 8.2 with $R = 1\ \text{k}\Omega$ and $C = 20\ \mu\text{F}$. Determine the time, t_1, when $v(t_1) = 4.00$ V, given $v(0^-) = 12$ V.

Figure SP 8.2

SP 8-3 For the circuit shown in Figure SP 8.3 find $v_c(t)$, $t \geq 0$, when $v_c(0^-) = 5$ V. Plot $v_c(t)$ for five time constants.

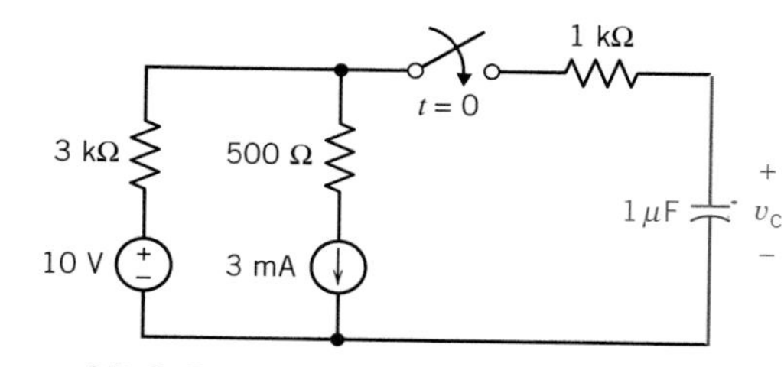

Figure SP 8.3

SP 8-4 An RC circuit as shown in Figure SP 8.4 has $v(0) = 0$. It is desired to plot the response of the circuit for four time constants: 2, 4, 8, and 16 ms. Select the appropriate value of R and use a PSpice program to plot the step response for the four time constants on one graphical plot.

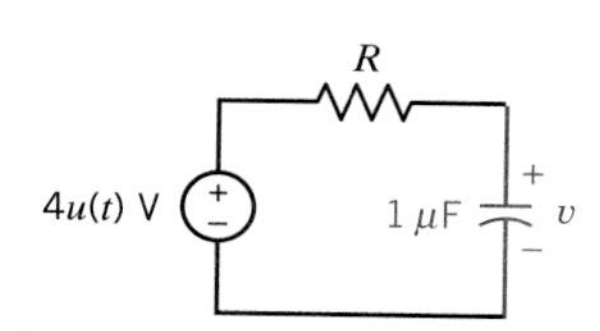

Figure SP 8.4

VERIFICATION PROBLEMS

VP 8-1 Figure VP 8.1 shows the transient response of a first-order circuit. This transient response was obtained using the computer program PSpice. A point on this transient response has been labeled. The label indicates a time and the capacitor voltage at that time. Placing the circuit diagram on the plot suggests that the plot corresponds to the circuit. Verify that the plot does indeed represent the voltage of the capacitor in this circuit.

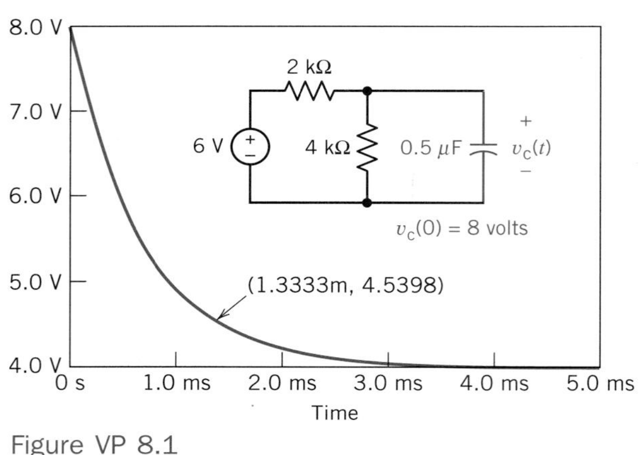

Figure VP 8.1

VP 8-2 Figure VP 8.2 shows the transient response of a first-order circuit. This transient response was obtained using the computer program PSpice. A point on this transient response has been labeled. The label indicates a time and the inductor current at that time. Placing the circuit diagram on the plot suggests that the plot corresponds to the circuit. Verify that the plot does indeed represent the current of the inductor in this circuit.

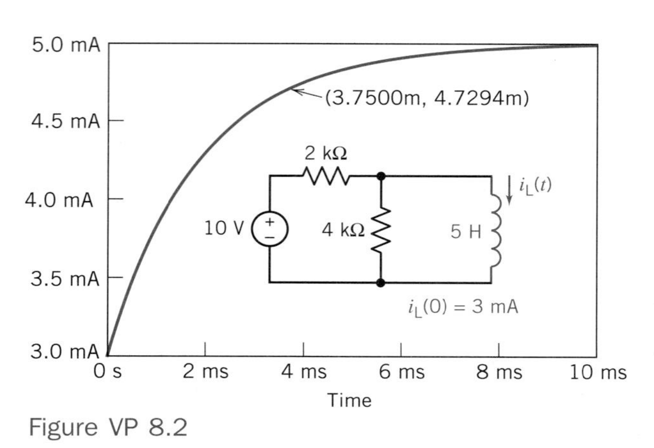

Figure VP 8.2

VP 8-3 Figure VP 8.3 shows the transient response of a first-order circuit. This transient response was obtained using the computer program PSpice. A point on this transient response has been labeled. The label indicates a time and the inductor current at that time. Placing the circuit diagram on the plot suggests that the plot corresponds to the circuit. Specify that value of the inductance, *L*, required to cause the current of the inductor in this circuit to be accurately represented by this plot.

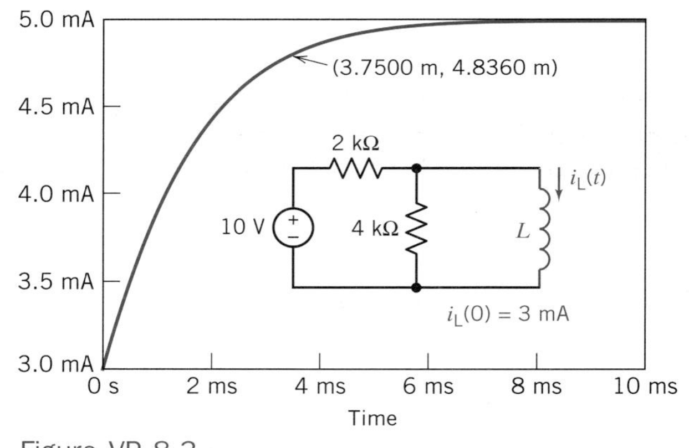

Figure VP 8.3

VP 8-4 Figure VP 8.4 shows the transient response of a first-order circuit. This transient response was obtained using the computer program PSpice. A point on this transient response has been labeled. The label indicates a time and the capacitor voltage at that time. Assume that this circuit has reached steady state before time $t = 0$. Placing the circuit diagram on the plot suggests that the plot corresponds to the circuit. Specify values of A, B, R1, R2, and C that cause the voltage across the capacitor in this circuit to be accurately represented by this plot.

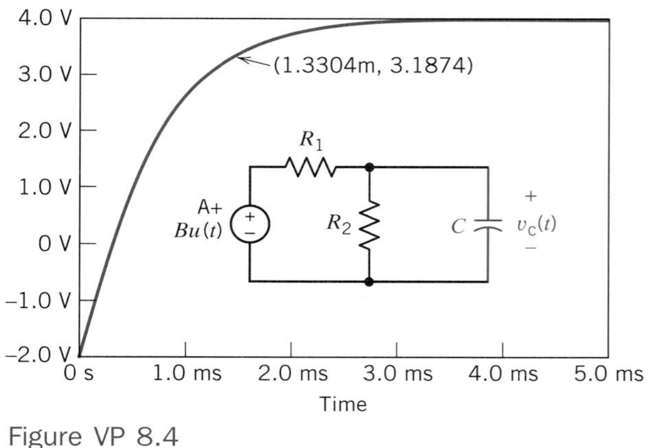

Figure VP 8.4

DESIGN PROBLEMS 

DP 8-1 Design the circuit in Figure DP 8.1 so that $v(t)$ makes the transition from $v(t) = 6\text{V}$ to $v(t) = 10\text{V}$ in 10 ms after the switch is closed. Assume that the circuit is at steady state before the switch is closed. Also, assume that the transition will be complete after 5 time constants.

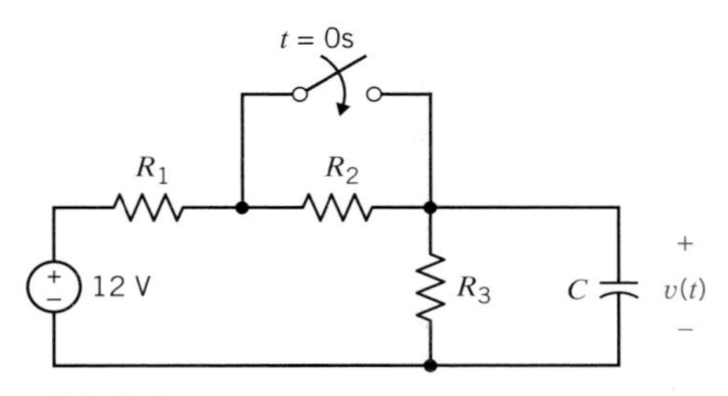

Figure DP 8.1

DP 8-2 Design the circuit in Figure DP 8.2 so that $i(t)$ makes the transition from $i(t) = 1\,\text{mA}$ to $i(t) = 4\,\text{mA}$ in 10 ms after the switch is closed. Assume that the circuit is at steady state before the switch is closed. Also, assume that the transition will be complete after 5 time constants.

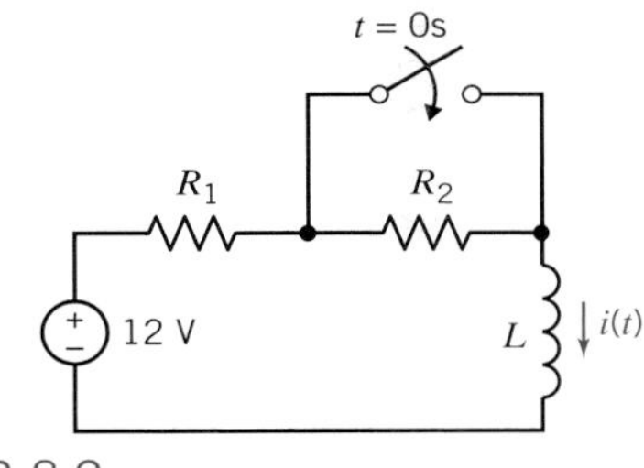

Figure DP 8.2

DP 8-3 The switch in Figure DP 8.3 closes at time 0, $2\Delta t$, $4\Delta t, \ldots 2k\Delta t$ and opens at times $\Delta t, 3\Delta t, 5\Delta t \ldots (2k+1)\Delta t$. When the switch closes $v(t)$ makes the transition from $v(t) = 0\,\text{V}$ to $v(t) = 5\,\text{V}$. Conversely, when the switch opens $v(t)$ makes the transition from $v(t) = 5\,\text{V}$ to $v(t) = 0\,\text{V}$. Suppose we require that $\Delta t = 5\tau$ so that one transition is complete before the next one begins. (a) Determine the value of C required so that $\Delta t = 1\ \mu\text{s}$. (b) How large must Δt be when $C = 2\ \mu\text{F}$?

Answer: (a) C = 4pF (b) $\Delta t = 0.5\text{s}$.

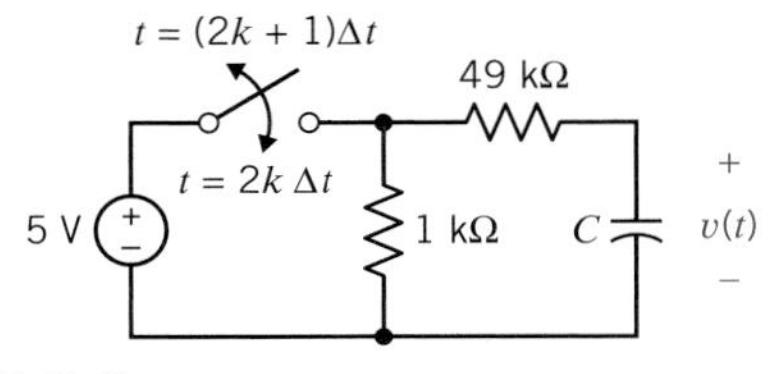

Figure DP 8.3

DP 8-4 The switch in Figure DP 8.3 closes at time 0, $2\Delta t$, $4\Delta t, \ldots 2k\Delta t$ and opens at times $\Delta t, 3\Delta t, 5\Delta t \ldots (2k+1)\Delta t$. When the switch closes $v(t)$ makes the transition from $v(t) = 0\,\text{V}$ to $v(t) = 5\,\text{V}$. Conversely, when the switch opens $v(t)$ makes the transition from $v(t) = 5\,\text{V}$ to $v(t) = 0\,\text{V}$. Suppose we require that one transition is 95% complete before the next one begins. (a) Determine the value of C required so that $\Delta t = 1\ \mu\text{s}$. (b) How large must Δt be when $C = 2\ \mu\text{F}$?

Hint: Show that $\Delta t = -\tau \ln(1\text{-}k)$ is required for the transition to be $100k\%$ complete.

Answer: (a) C = 6.67 pF (b) $\Delta t = 0.3$ s

DP 8-5 A laser trigger circuit is shown in Figure DP 8.5. In order to trigger the laser, we require $60\,\text{mA} < |i| < 180\ \text{mA}$ for $0 < t < 200\ \mu\text{s}$. Determine a suitable value for R_1 and R_2.

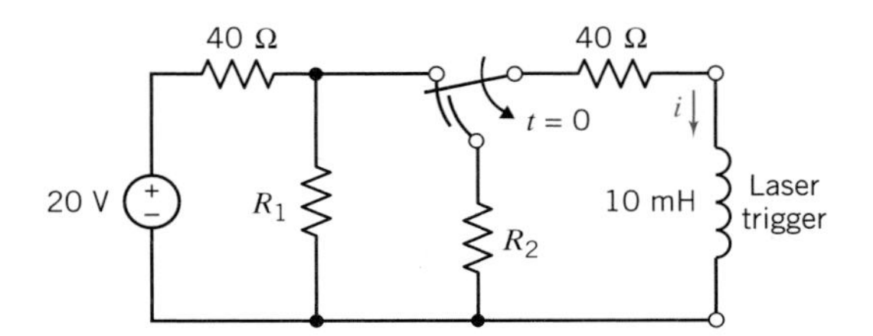

Figure DP 8.5 Laser trigger circuit.

DP 8-6 Fuses are used to open a circuit when excessive current flows (Wright 1990). One fuse is designed to open when the power absorbed by R exceeds 10 W for 0.5 s. The source represents the turn-on condition for the load where $v_s = A[u(t) - u(t - 0.75)]$ V. Assume that $i_L(0^-) = 0$. The goal is to achieve the maximum current while not opening the fuse. Determine an appropriate value of A and sketch the current waveform. The circuit is shown in Figure DP 8.6.

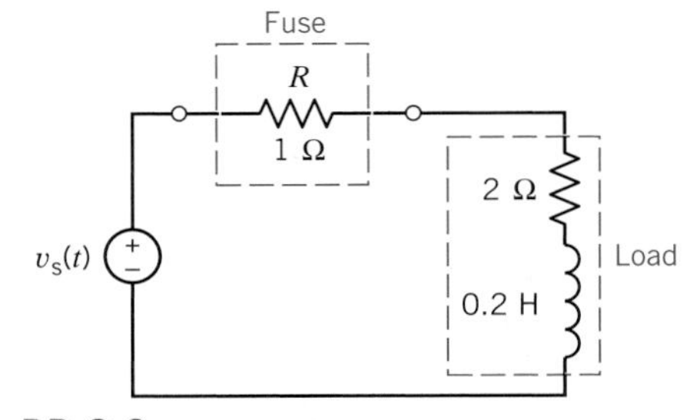

Figure DP 8.6 Fuse circuit.

CHAPTER 9

The Complete Response of Circuits with Two Energy Storage Elements

Preview

In the preceding chapter we determined the natural response and the forced response of circuits with one energy storage element. In this chapter we proceed to determine the complete response $x(t)$ of a circuit with two energy storage elements. A circuit with two energy storage elements is described by a second-order differential equation in terms of $x(t)$.

We describe three methods of obtaining the second-order differential equation: (1) the direct method, (2) the operator method, and (3) the state variable method. Then, using the differential equation, we proceed to find the natural response x_n and the forced response x_f.

Although we focus on circuits with two energy storage elements, the methods described can be used for circuits with three or more energy storage elements.

9.1 Design Challenge

AUTO AIRBAG IGNITER

Problem

Airbags are now widely used for driver protection in automobiles. A pendulum is used to switch a charged capacitor to the inflation ignition device, as shown in Figure 9.1-1. The automobile airbag is inflated by an explosive device that is ignited by the energy absorbed by the resistive device represented by R. In order to inflate, it is required that the energy dissipated in R is at least 1 J. It is required that the ignition device trigger within 0.1 s. Select the L and C that meet the specifications.

This problem involves a circuit that contains *two* energy storage elements. This type of circuit is the subject of this chapter. We will return to the problem at the end of the chapter.

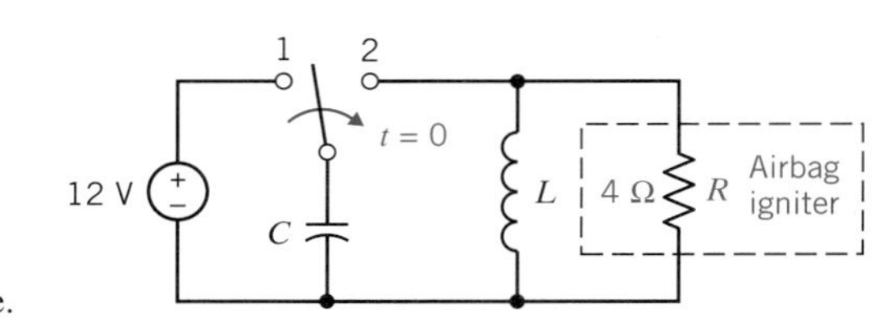

Figure 9.1-1
An automobile airbag ignition device.

9.2 COMMUNICATIONS AND POWER SYSTEMS

The objective of an electrical circuit is to transmit an electrical signal or electrical power. An *electrical system* is an interconnection of electrical elements and circuits to achieve a desired objective. Consider the electrical system shown in Figure 9.2-1. This system has a source and transmits the source power or signal to a receiver and ultimately to the end user.

In a communication system the source is an input signal such as a voltage signal. The transmitter converts the signal to a suitable form for the transmission medium. Then the output of the transmitter proceeds through the medium until it reaches the receiver, where it is converted into a form that is useful for the end user. The Marconi receiver is shown in Figure 9.2-2.

An example of a well-known communication system is the early railway electric telegraph for communication from one railway station to another. The message is converted to Morse code, and the transmitter key is opened and closed to generate a series of short and long pulses. These electrical pulses travel along the telegraph wire until they reach a telegraph receiver at another railway station, where they are received and decoded for use by the reader.

Another example of a communication system is the radio, which converts spoken words into electrical waveforms. The transmitter then couples these waveforms to the transmission medium, in this case the earth's atmosphere. At the receiver, an antenna couples the medium to the receiver and eventually converts these electrical waveforms back to audio signals for the listener.

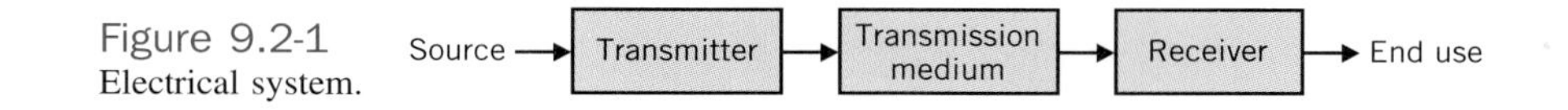

Figure 9.2-1
Electrical system.

Figure 9.2-2
Marconi with the receiving apparatus used at Signal Hill, 1901. Courtesy of the IEEE Center for the History of Electrical Engineering.

The goal of an efficient communication system is to achieve the best transmission and reception of the original signal so that the user may accurately discern the message. Thus, the goal is to provide the user with a faithful, undistorted output waveform that is true to the input signal. The chronology of electrical communication developments is provided in Table 9.2-1.

Now let us consider the goal of an electric power system. A power plant, such as a hydroelectric power station, generates large amounts of electric power. It is common to generate as much as 30 MW to 70 MW continuously at a hydroelectric plant. Since these plants are usually distant from customers in the cities, the electrical power must be transmitted over long pairs of wires to the receiver (see Figure 9.2-1). The transmitter, in this case, must efficiently couple the power to the transmission medium (wires). Similarly, the receiver must efficiently couple the power to local distribution circuits for ultimate use by the end user.

Thus, the goal of the communication system is faithful undistorted transmission, while the goal of the electric power system is to transmit the energy efficiently, with minimum loss of energy in the transmission medium (the wires).

Both communication and power systems contain capacitance and inductance throughout their circuits. Thus, with both energy storage elements present, it is necessary to consider circuits that have both elements and determine the forced and natural responses of the circuits. It is the purpose of this charter to consider the behavior of circuits that contain both capacitors and inductors.

Table 9.2-1 **Twentieth-Century Chronology of Electrical Communication**

1920	Birth of commercial radio. KDKA, Pittsburgh, the first radio broadcasting station.
1923–1938	Birth of television. Philo T. Farnsworth and Vladimir Zworykin propose television systems. Experimental broadcasts begin.
1939	First commercial television broadcast in New York City.
1938–1945	World War II. Radar and microwave systems developed.
1955	John R. Pierce proposes satellite communication systems.
1962	Telstar I launched as first communication satellite.
1983	First optical fiber telephone communication system installed in Chicago.
1984	Cellular mobile telephone systems introduced.
1995	Internet established.

9.3 DIFFERENTIAL EQUATION FOR CIRCUITS WITH TWO ENERGY STORAGE ELEMENTS

In Chapter 8 we considered circuits that contained only one energy storage element, and these could be described by a first-order differential equation. In this section we consider the description of circuits with two irreducible energy storage elements that are described by a second-order differential equation. Later, we will consider circuits with three or more irreducible energy storage elements that are described by a third-order (or higher) differential equation. We use the term *irreducible* to indicate that all parallel or series connections or other reducible combinations of like storage elements have been reduced to their irreducible form. Thus, for example, all parallel capacitors have been reduced to one capacitor, C_p.

In the following paragraphs we use two methods to obtain the second-order differential equation for circuits with two energy storage elements. Then in the next section we obtain the solution to these second-order differential equations.

First, let us consider the circuit shown in Figure 9.3-1, which consists of a parallel combination of a resistor, an inductor, and a capacitor. Writing the nodal equation at the top node, we have

$$\frac{v}{R} + i + C\frac{dv}{dt} = i_s \tag{9.3-1}$$

Then we write the equation for the inductor as

$$v = L\frac{di}{dt} \tag{9.3-2}$$

Substitute Eq. 9.3-2 into Eq. 9.3-1, obtaining

$$\frac{L}{R}\frac{di}{dt} + i + CL\frac{d^2i}{dt^2} = i_s \tag{9.3-3}$$

which is the second-order differential equation we seek. Solve this equation for $i(t)$. If $v(t)$ is required, use Eq. 9.3-2 to obtain it.

This method of obtaining the second-order differential equation may be called the *direct method* and is summarized in Table 9.3-1.

In Table 9.3-1 the circuit variables are called x_1 and x_2. In any example, x_1 and x_2 will be specific element currents or voltages. When we analyzed the circuit of Figure 9.3-1, we used $x_1 = v$ and $x_2 = i$. In contrast, to analyze the circuit of Figure 9.3-2 we will use $x_1 = i$ and $x_2 = v$, where i is the inductor current and v is the capacitor voltage.

Now let us consider the *RLC* series circuit shown in Figure 9.3-2 and use the direct method to obtain the second-order differential equation. We chose $x_1 = i$ and $x_2 = v$. First we seek an equation for $\frac{dx_1}{dt} = \frac{di}{dt}$. Writing KVL around the loop, we have

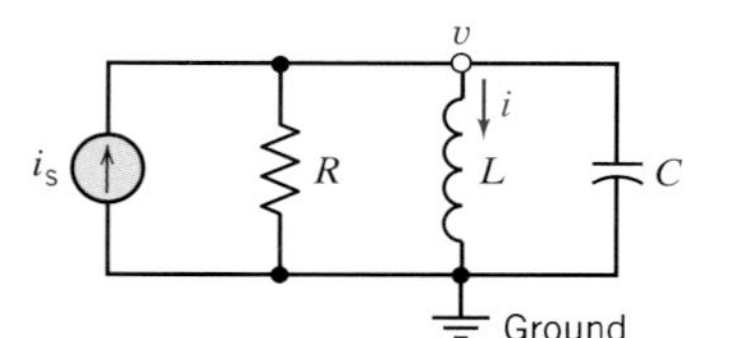

Figure 9.3-1 An *RLC* circuit with a current source.

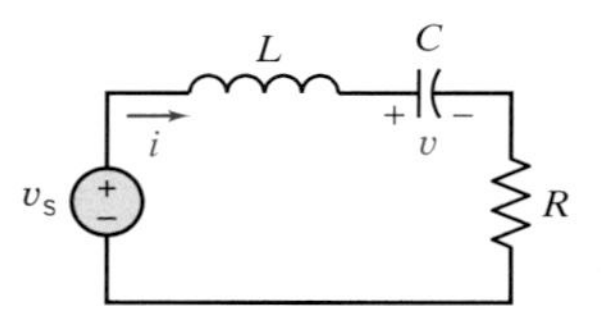

Figure 9.3-2 An *RLC* series circuit.

Table 9.3-1 **The Direct Method for Obtaining the Second-Order Differential Equation of a Circuit**

Step 1	Identify the first and second variables x_1 and x_2. These variables are capacitor voltages and/or inductor currents.
Step 2	Write one first-order differential equation, obtaining $\frac{dx_1}{dt} = f(x_1, x_2)$. (Eq. 1)
Step 3	Obtain an additional first-order differential equation in terms of the second variable so that $\frac{dx_2}{dt} = Kx_1$ or $x_1 = \frac{1}{K}\frac{dx_2}{dt}$. (Eq. 2)
Step 4	Substitute the equation of Step 3 into the equation of Step 2, thus obtaining a second-order differential equation in terms of x_2.

$$L\frac{di}{dt} + v + Ri = v_s \tag{9.3-4}$$

where v is the capacitor voltage. This equation may be written as

$$\frac{di}{dt} + \frac{v}{L} + \frac{R}{L}i = \frac{v_s}{L} \tag{9.3-5}$$

Recall $v = x_2$, and obtain an equation in terms of $\frac{dx_2}{dt}$. Since

$$C\frac{dv}{dt} = i \tag{9.3-6}$$

or

$$C\frac{dx_2}{dt} = x_1 \tag{9.3-7}$$

substitute Eq. 9.3-6 into Eq. 9.3-5 to obtain the desired second-order differential equation:

$$C\frac{d^2v}{dt^2} + \frac{v}{L} + \frac{RC}{L}\frac{dv}{dt} = \frac{v_s}{L} \tag{9.3-8}$$

Equation 9.3-8 may be rewritten as

$$\frac{d^2v}{dt^2} + \frac{R}{L}\frac{dv}{dt} + \frac{1}{LC}v = \frac{v_s}{LC} \tag{9.3-9}$$

Another method of obtaining the second-order differential equation describing a circuit is called the *operator method.* First, we obtain differential equations describing node voltages or mesh currents and use operators to obtain the differential equation for the circuit.

As a more complicated example of a circuit with two energy storage elements, consider the circuit shown in Figure 9.3-3. This circuit has two inductors and can be described by the mesh currents as shown in Figure 9.3-3. The mesh equations are

$$L_1\frac{di_1}{dt} + R(i_1 - i_2) = v_s \tag{9.3-10}$$

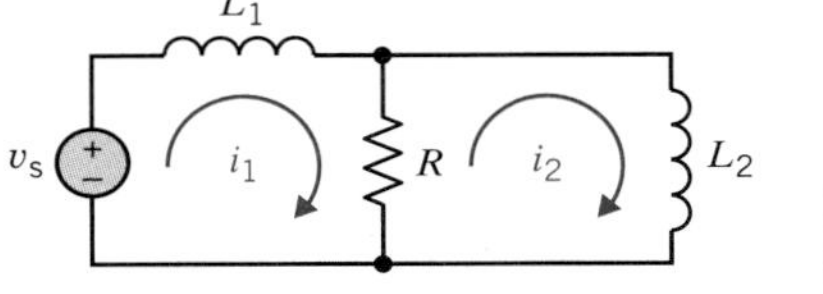

Figure 9.3-3
Circuit with two inductors.

and
$$R(i_2 - i_1) + L_2 \frac{di_2}{dt} = 0 \tag{9.3-11}$$

Now, let us use $R = 1\ \Omega, L_1 = 1$ H, and $L_2 = 2$ H. Then we have

$$\frac{di_1}{dt} + i_1 - i_2 = v_s$$

and
$$i_2 - i_1 + 2\frac{di_2}{dt} = 0 \tag{9.3-12}$$

In terms of i_1 and i_2, we may rearrange these equations as

$$\frac{di_1}{dt} + i_1 - i_2 = v_s \tag{9.3-13}$$

and
$$-i_1 + i_2 + 2\frac{di_2}{dt} = 0 \tag{9.3-14}$$

It remains to obtain one second-order differential equation. This is done in the second step of the operator method. The differential operator s, where $s = d/dt$, is used to transform differential equations into algebraic equations. Upon replacing d/dt by s, Eqs. 9.3-13 and 9.3-14 become

$$si_1 + i_1 - i_2 = v_s$$

and
$$-i_1 + i_2 + 2si_2 = 0$$

These two equations may be rewritten as

$$(s+1)i_1 - i_2 = v_s$$

and
$$-i_1 + (2s+1)i_2 = 0$$

We may use Cramer's rule (see Appendix A) to solve for i_2, obtaining

$$i_2 = \frac{1v_s}{(s+1)(2s+1) - 1}$$

Therefore,
$$(2s^2 + 3s)i_2 = v_s$$

and the differential equation is

$$2\frac{d^2i_2}{dt^2} + 3\frac{di_2}{dt} = v_s \tag{9.3-15}$$

The operator method for obtaining the second-order differential equation is summarized in Table 9.3-2.

Example 9.3-1

Find the differential equation for the current i_2 for the circuit of Figure 9.3-4.

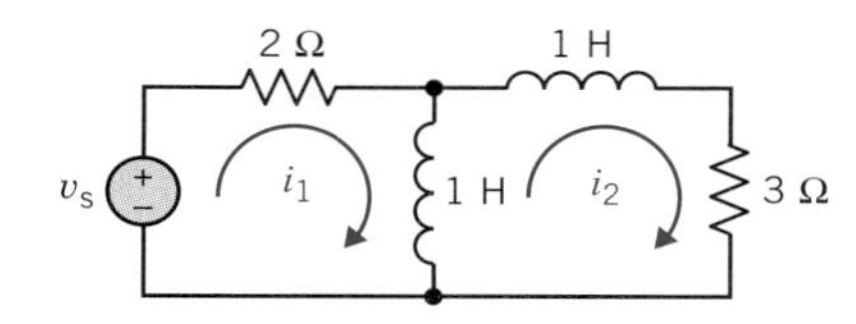

Figure 9.3-4
Circuit for Example 9.3-1.

Table 9.3-2 **Operator Method for Obtaining the Second-Order Differential Equation of a Circuit**

Step 1	Identify the variable x_1 for which the solution is desired.
Step 2	Write one differential equation in terms of the desired variable x_1 and a second variable x_2.
Step 3	Obtain an additional equation in terms of the second variable and the first variable.
Step 4	Use the operator $s = d/dt$ and $1/s = \int dt$ to obtain two algebraic equations in terms of s and the two variables x_1 and x_2.
Step 5	Using Cramer's rule, solve for the desired variable so that $x_1 = f(s, \text{sources}) = P(s)/Q(s)$, where $P(s)$ and $Q(s)$ are polynomials in s.
Step 6	Rearrange the equation of Step 5 so that $Q(s)x_1 = P(s)$.
Step 7	Convert the operators back to derivatives for the equation of Step 6 to obtain the second-order differential equation.

Solution

Write the two mesh equations using KVL to obtain

$$2i_1 + \frac{di_1}{dt} - \frac{di_2}{dt} = v_s$$

$$-\frac{di_1}{dt} + 3\,i_2 + 2\frac{di_2}{dt} = 0$$

Using the operator $s \equiv d/dt$, we have

$$(2 + s)i_1 - si_2 = v_s$$

and

$$-si_1 + (3 + 2s)i_2 = 0$$

Using Cramer's rule to solve for i_2, we obtain

$$i_2 = \frac{sv_s}{(2 + s)(3 + 2s) - s^2} = \frac{sv_s}{s^2 + 7s + 6} \tag{9.3-16}$$

Rearranging Eq. 9.3-16, we obtain

$$(s^2 + 7s + 6)i_2 = sv_s \tag{9.3-17}$$

Therefore, the differential equation for i_2 is

$$\frac{d^2i_2}{dt^2} + 7\frac{di_2}{dt} + 6\,i_2 = \frac{dv_s}{dt} \tag{9.3-18}$$

Example 9.3-2

Find the differential equation for the voltage v for the circuit of Figure 9.3-5.

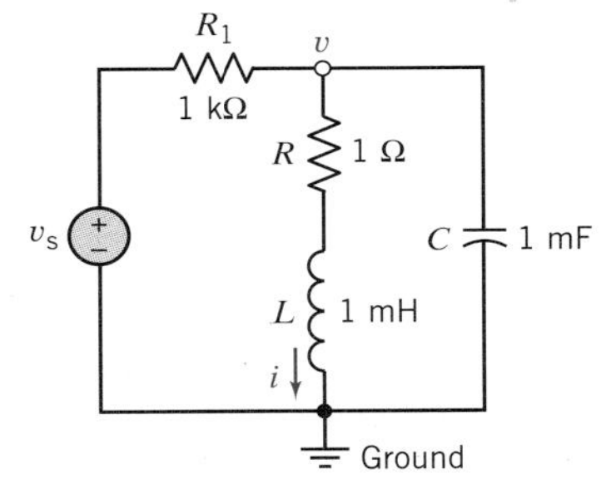

Figure 9.3-5
The *RLC* circuit for Example 9.3-2.

Solution

The KCL node equation at the upper node is

$$\frac{v - v_s}{R_1} + i + C\frac{dv}{dt} = 0 \tag{9.3-19}$$

Since we wish to determine the equation in terms of v, we need a second equation in terms of the current i. Write the equation for the current through the branch containing the inductor as

$$Ri + L\frac{di}{dt} = v \tag{9.3–20}$$

Using the operator $s = d/dt$, we have the two equations

$$\frac{v}{R_1} + Csv + i = \frac{v_s}{R_1}$$

and

$$-v + Ri + Lsi = 0$$

Substituting the parameter values and rearranging, we have

$$(10^{-3} + 10^{-3}s)v + i = 10^{-3}v_s$$

and

$$-v + (10^{-3}s + 1)i = 0$$

Using Cramer's rule, solve for v to obtain

$$v = \frac{(s + 1000)v_s}{(s + 1)(s + 1000) + 10^6} = \frac{(s + 1000)v_s}{s^2 + 1001s + 1001 \times 10^3}$$

Therefore, we have

$$(s^2 + 1001s + 1001 \times 10^3)v = (s + 1000)v_s$$

or the differential equation we seek is

$$\frac{d^2v}{dt^2} + 1001\frac{dv}{dt} + 1001 \times 10^3\, v = \frac{dv_s}{dt} + 1000\, v_s$$

Exercise 9.3-1 Find the second-order differential equation for the circuit shown in Figure E 9.3-1 in terms of i using the direct method.

Answer: $\dfrac{d^2i}{dt^2} + \dfrac{1}{2}\dfrac{di}{dt} + i = \dfrac{1}{2}\dfrac{di_s}{dt}$

Exercise 9.3-2 Find the second-order differential equation for the circuit shown in Figure E 9.3-2 in terms of v using the operator method.

Answer: $\dfrac{d^2v}{dt^2} + 2\dfrac{dv}{dt} + 2\,v = 2\dfrac{di_s}{dt}$

Exercise 9.3-3 Find the second differential equation for i_2 for the circuit of Figure E 9.3-3 using the operator method. Recall that the operator for the integral is $1/s$.

Answer: $3\dfrac{d^2i_2}{dt^2} + 4\dfrac{di_2}{dt} + 2\,i_2 = \dfrac{d^2v_s}{dt^2}$

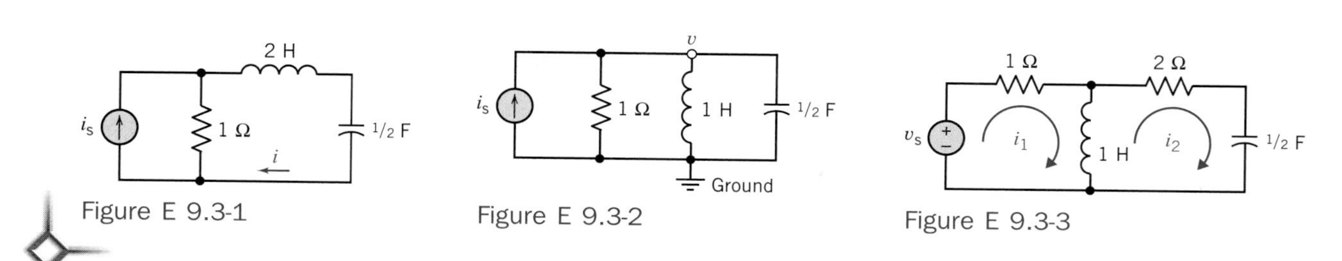

Figure E 9.3-1

Figure E 9.3-2

Figure E 9.3-3

9.4 SOLUTION OF THE SECOND-ORDER DIFFERENTIAL EQUATION—THE NATURAL RESPONSE

In the preceding section we found that a circuit with two irreducible energy storage elements can be represented by a second-order differential equation of the form

$$a_2\frac{d^2x}{dt^2} + a_1\frac{dx}{dt} + a_0 x = f(t)$$

where the constants a_2, a_1, a_0 are known and the forcing function $f(t)$ is specified.

The complete response $x(t)$ is given by

$$x = x_n + x_f \tag{9.4-1}$$

where x_n is the natural response and x_f is the forced response. The natural response satisfies the unforced differential equation when $f(t) = 0$. The forced response x_f satisfies the differential equation with the forcing function present.

The natural response of a circuit, x_n, will satisfy the equation

$$a_2\frac{d^2x_n}{dt^2} + a_1\frac{dx_n}{dt} + a_0 x_n = 0 \tag{9.4-2}$$

Since x_n and its derivatives must satisfy the equation, we postulate the exponential solution

$$x_n = Ae^{st} \tag{9.4-3}$$

where A and s are to be determined. The exponential is the only function that is proportional to all of its derivatives and integrals and therefore is the natural choice for the solution of a differential equation with constant coefficients. Substituting Eq. 9.4-3 in Eq. 9.4-2 and differentiating where required, we have

$$a_2As^2e^{st} + a_1Ase^{st} + a_0Ae^{st} = 0 \tag{9.4-4}$$

Since $x_n = Ae^{st}$, we may rewrite Eq. 9.4-4 as

$$a_2s^2x_n + a_1sx_n + a_0x_n = 0$$

or

$$(a_2s^2 + a_1s + a_0)x_n = 0$$

Since we do not accept the trivial solution, $x_n = 0$, it is required that

$$(a_2s^2 + a_1s + a_0) = 0 \tag{9.4-5}$$

This equation, in terms of s, is called a *characteristic equation*. It is readily obtained by replacing the derivative by s and the second derivative by s^2. Clearly, we have returned to the familiar operator

$$s^n \equiv \frac{d^n}{dt^n}$$

The **characteristic equation** is derived from the governing differential equation for a circuit by setting all independent sources to zero value and assuming an exponential solution.

Figure 9.4-1 Oliver Heaviside (1850–1925). Photograph courtesy of the Institution of Electrical Engineers.

Oliver Heaviside (1850–1925), shown in Figure 9.4-1, advanced the theory of operators for the solution of differential equations.

The solution of the quadratic equation (9.4-5) has two roots s_1 and s_2, where

$$s_1 = \frac{-a_1 + \sqrt{a_1^2 - 4a_2 a_0}}{2a_2} \tag{9.4-6}$$

and

$$s_2 = \frac{-a_1 - \sqrt{a_1^2 - 4a_2 a_0}}{2a_2} \tag{9.4-7}$$

When there are two distinct roots, there are two solutions such that

$$x_n = A_1 e^{s_1 t} + A_2 e^{s_2 t} \tag{9.4-8}$$

While there are indeed two solutions to the second-order differential equation, their sum is also a solution, *since the equation is linear*. Furthermore, the *general solution* must consist of as many terms, each with an arbitrary coefficient, as the order of the equation to satisfy the fundamental theorem of differential equations. We will delay considering the special case when $s_1 = s_2$.

The **roots** of the characteristic equation contain all the information necessary for determining the character of the natural response.

Example 9.4-1

Find the natural response of the circuit current i_2 shown in Figure 9.4-2. Use operators to formulate the differential equation and obtain the response in terms of two arbitrary constants.

Solution

Writing the two mesh equations, we have

$$12\, i_1 + 2\frac{di_1}{dt} - 4\, i_2 = v_s$$

and

$$-4\, i_1 + 4\, i_2 + 1\frac{di_2}{dt} = 0$$

Using the operator $s = d/dt$, we obtain

$$(12 + 2s)\, i_1 - 4\, i_2 = v_s \tag{9.4-9}$$

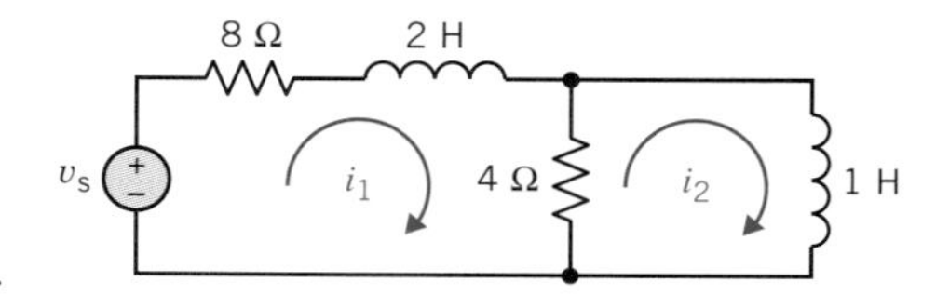

Figure 9.4-2
Circuit of Example 9.4-1.

$$-4\,i_1 + (4 + s)\,i_2 = 0 \tag{9.4-10}$$

Using Cramer's rule and solving for i_2, we have

$$i_2 = \frac{4\,v_s}{(12 + 2s)(4 + s) - 16} = \frac{4\,v_s}{2s^2 + 20s + 32} = \frac{2\,v_s}{s^2 + 10s + 16}$$

Therefore,
$$(s^2 + 10s + 16)i_2 = 2v_s$$

Note that $(s^2 + 10s + 16) = 0$ is the characteristic equation and may be determined directly by evaluating the determinant for Eqs. 9.4-9 and 9.4-10. Thus, the roots of the characteristic equation are $s_1 = -2$ and $s_2 = -8$. Therefore, the natural response is

$$x_n = A_1 e^{-2t} + A_2 e^{-8t}$$

where $x = i_2$. The roots s_1 and s_2 are the *characteristic roots* and are often called the *natural frequencies*. The reciprocals of the magnitude of the real characteristic roots are the *time constants*. The time constants of this circuit are 1/2 s and 1/8 s.

Exercise 9.4-1 Find the characteristic equation and the natural frequencies for the circuit shown in Figure E 9.4-1.

Answer: $s^2 + 7s + 10 = 0$

$s_1 = -2$

$s_2 = -5$

Exercise 9.4-2 German automaker Volkswagen, in its bid to make more efficient cars, has come up with an auto whose engine saves energy by shutting itself off at stoplights. The "stop–start" system springs from a campaign to develop cars in all its world markets that use less fuel and pollute less than vehicles now on the road. The stop–start transmission control has a mechanism that senses when the car does not need fuel: coasting downhill and idling at an intersection. The engine shuts off, but a small starter flywheel keeps turning so that power can be quickly restored when the driver touches the accelerator.

A model of the stop–start circuit is shown in Figure E 9.4-2. Determine the characteristic equation and the natural frequencies for the circuit.

Answer: $s^2 + 20s + 400 = 0$

$s = -10 \pm j17.3$

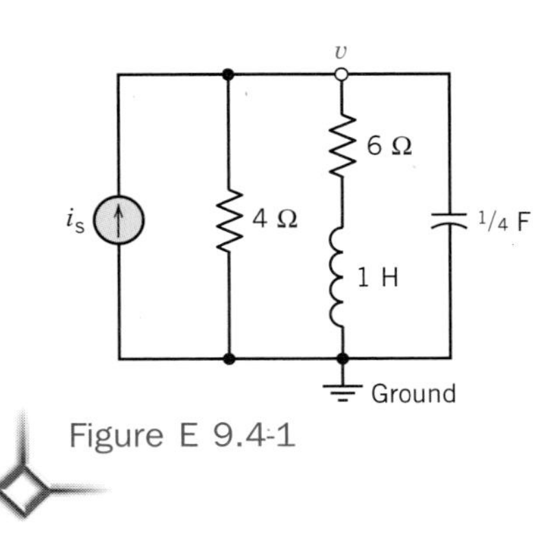

Figure E 9.4-1

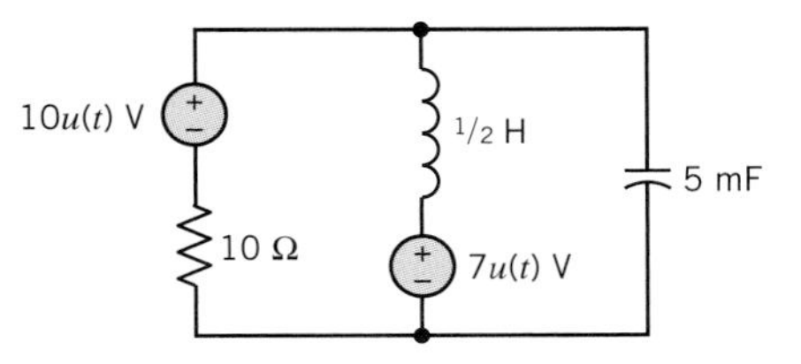

Figure E 9.4-2 Stop–start circuit.

9.5 NATURAL RESPONSE OF THE UNFORCED PARALLEL *RLC* CIRCUIT

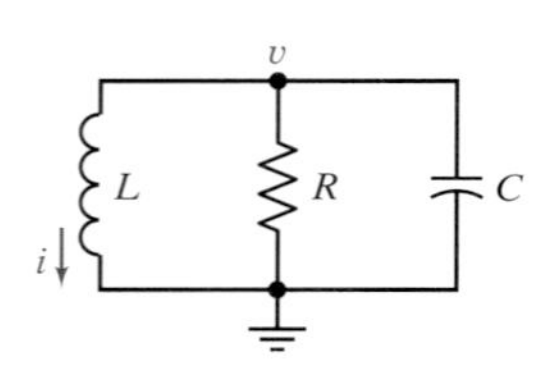

Figure 9.5-1 Parallel *RLC* circuit.

In this section we consider the (unforced) natural response of the parallel *RLC* circuit shown in Figure 9.5-1. We choose to examine the parallel *RLC* circuit to illustrate the three forms of the natural response. An analogous discussion of the series *RLC* circuit could be presented, but it is omitted since the purpose is not to obtain the solution to specific circuits but rather to illustrate the general method.

Write the KCL at the node to obtain

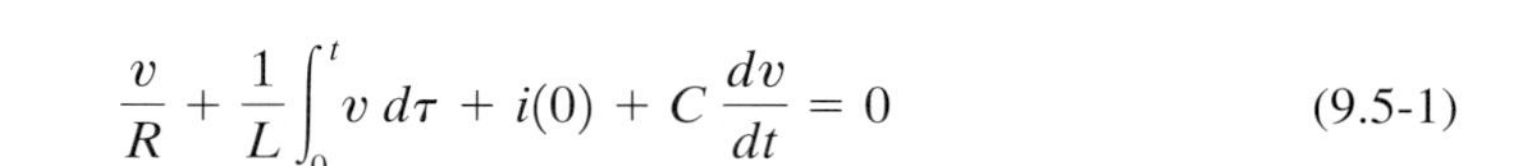

$$\frac{v}{R} + \frac{1}{L}\int_0^t v\,d\tau + i(0) + C\frac{dv}{dt} = 0 \qquad (9.5\text{-}1)$$

Taking the derivative of Eq. 9.5-1, we have

$$C\frac{d^2v}{dt^2} + \frac{1}{R}\frac{dv}{dt} + \frac{1}{L}v = 0 \qquad (9.5\text{-}2)$$

A **second-order circuit** has a homogeneous differential equation containing a second-order derivative term due to the presence of two independent energy storage elements.

Using the operator s, we obtain the characteristic equation

$$s^2 + \frac{1}{RC}s + \frac{1}{LC} = 0 \qquad (9.5\text{-}3)$$

The two roots of the characteristic equation are

$$s_1 = -\frac{1}{2RC} + \left[\left(\frac{1}{2RC}\right)^2 - \frac{1}{LC}\right]^{1/2} \qquad (9.5\text{-}4)$$

$$s_2 = -\frac{1}{2RC} - \left[\left(\frac{1}{2RC}\right)^2 - \frac{1}{LC}\right]^{1/2} \qquad (9.5\text{-}5)$$

When s_1 is not equal to s_2, the solution to the second-order differential Eq. 9.5-2 is

$$v_n = A_1e^{s_1t} + A_2e^{s_2t} \qquad (9.5\text{-}6)$$

The roots of the characteristic equation may be rewritten as

$$s_1 = -\alpha + \sqrt{\alpha^2 - \omega_0^2} \qquad (9.5\text{-}7)$$

$$s_2 = -\alpha - \sqrt{\alpha^2 - \omega_0^2} \qquad (9.5\text{-}8)$$

where $\alpha = 1/2RC$ and $\omega_0^2 = 1/LC$. Normally, ω_0 is called the *resonant frequency*. The concept of resonant frequency is expanded in subsequent chapters.

The roots of the characteristic equation assume three possible conditions:

1. Two real and distinct roots when $\alpha^2 > \omega_0^2$.
2. Two real equal roots when $\alpha^2 = \omega_0^2$.
3. Two complex roots when $\alpha^2 < \omega_0^2$.

When the two roots are real and distinct, the circuit is said to be *overdamped*. When the roots are both real and equal, the circuit is *critically damped*. When the two roots are complex conjugates, the circuit is said to be *underdamped*.

Let us determine the natural response for the overdamped *RLC* circuit of Figure 9.5-1 when the initial conditions are $v(0)$ and $i(0)$ for the capacitor and the inductor, respectively. Notice that because the circuit in Figure 9.5-1 has no input, $v_n(0)$ and $v(0)$ are both names for the same voltage. Then, at $t = 0$ for Eq. 9.5-6, we have

$$v_n(0) = A_1 + A_2 \tag{9.5-9}$$

Since A_1 and A_2 are both unknown, we need one more equation at $t = 0$. Rewriting Eq. 9.5-1 at $t = 0$, we have[1]

$$\frac{v(0)}{R} + i(0) + C\frac{dv(0)}{dt} = 0$$

Since $i(0)$ and $v(0)$ are known, we have

$$\frac{dv(0)}{dt} = -\frac{v(0)}{RC} - \frac{i(0)}{C} \tag{9.5-10}$$

Thus, we now know the initial value of the derivative of v. Taking the derivative of Eq. 9.5-6 and setting $t = 0$, we obtain

$$\frac{dv_n(0)}{dt} = s_1 A_1 + s_2 A_2 \tag{9.5-11}$$

Using Eqs. 9.5-10 and 9.5-11, we obtain a second equation in terms of the two constants as

$$s_1 A_1 + s_2 A_2 = -\frac{v(0)}{RC} - \frac{i(0)}{C} \tag{9.5-12}$$

Using Eqs. 9.5-9 and 9.5-12, we may obtain A_1 and A_2.

Example 9.5-1

Find the natural response of $v(t)$ for $t > 0$ for the parallel *RLC* circuit shown in Figure 9.5-1 when $R = 2/3\ \Omega$, $L = 1$ H, $C = 1/2$ F, $v(0) = 10$ V, and $i(0) = 2$ A.

Solution

The characteristic equation is

$$s^2 + \frac{1}{RC}s + \frac{1}{LC} = 0$$

or

$$s^2 + 3s + 2 = 0$$

Therefore, the roots of the characteristic equation are

$$s_1 = -1$$

and

$$s_2 = -2$$

Then the natural response is

$$v_n = A_1 e^{-t} + A_2 e^{-2t} \tag{9.5-13}$$

The initial capacitor voltage is $v(0) = 10$, so we have

$$v_n(0) = A_1 + A_2$$

or

$$10 = A_1 + A_2 \tag{9.5-14}$$

[1] *Note:* $\dfrac{dv(0)}{dt}$ means $\left.\dfrac{dv(t)}{dt}\right|_{t=0}$

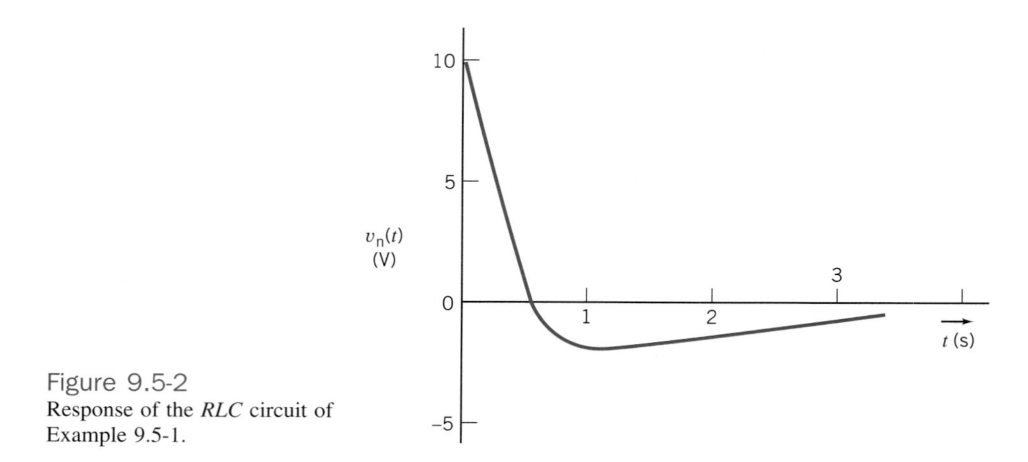

Figure 9.5-2
Response of the *RLC* circuit of Example 9.5-1.

We use Eq. 9.5-12 to obtain the second equation for the unknown constants. Then

$$s_1 A_1 + s_2 A_2 = -\frac{v(0)}{RC} - \frac{i(0)}{C}$$

or

$$-A_1 - 2A_2 = -\frac{10}{1/3} - \frac{2}{1/2}$$

Therefore, we have

$$-A_1 - 2A_2 = -34 \tag{9.5-15}$$

Solving Eqs. 9.5-14 and 9.5-15 simultaneously, we obtain $A_2 = 24$ and $A_1 = -14$. Therefore, the natural response is

$$v_n = (-14e^{-t} + 24e^{-2t}) \text{ V}$$

The natural response of the circuit is shown in Figure 9.5-2.

Exercise 9.5-1 Find the natural response of the *RLC* circuit of Figure 9.5-1 when $R = 6\ \Omega$, $L = 7$ H, and $C = 1/42$ F. The initial conditions are $v(0) = 0$ and $i(0) = 10$ A.

Answer: $v_n(t) = -84(e^{-t} - e^{-6t})$ V

Exercise 9.5-2 The circuit shown in Figure E 9.5-2 is used to detect smokers in airplanes who surreptitiously light up in nonsmoking areas before they can take a single puff. The sensor activates the switch, and the change in the voltage $v(t)$ activates a light at the flight attendant's station. Determine the natural response $v(t)$.

Answer: $v(t) = -1.16e^{-2.7t} + 1.16e^{-37.3t}$ V

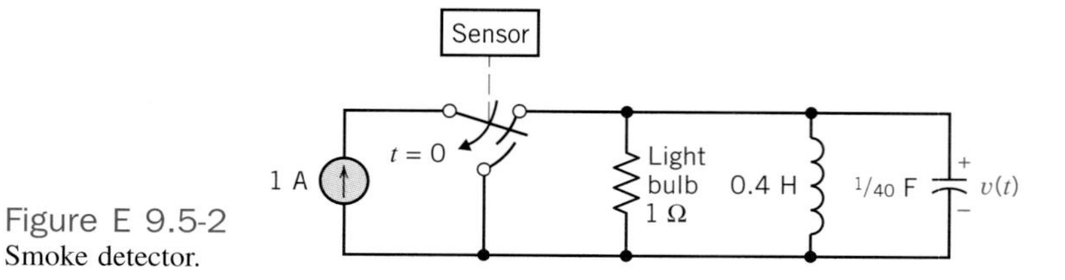

Figure E 9.5-2
Smoke detector.

9.6 NATURAL RESPONSE OF THE CRITICALLY DAMPED UNFORCED PARALLEL *RLC* CIRCUIT

Again we consider the parallel *RLC* circuit, and here we will determine the special case when the characteristic equation has two equal real roots. Two real equal roots occur when $\alpha^2 = \omega_0^2$ where

$$\alpha = \frac{1}{2RC} \quad \text{and} \quad \omega_0^2 = \frac{1}{LC}$$

Let us assume that $s_1 = s_2$ and proceed to find $v_n(t)$. We write the natural response as the sum of two exponentials as

$$v_n = A_1 e^{s_1 t} + A_2 e^{s_1 t} = A_3 e^{s_1 t} \tag{9.6-1}$$

where $A_3 = A_1 + A_2$. Since the two roots are equal, we have only one undetermined constant, but we still have two initial conditions to satisfy. Clearly, Eq. 9.6-1 is not the total solution for the natural response of a critically damped circuit. We need the solution that will contain two arbitrary constants, so with some foreknowledge we try the solution

$$x_n = g(t)e^{s_1 t}$$

where $g(t)$ is a polynomial in t. Let us try

$$g(t) = A_2 + A_1 t$$

Substituting $x_n = g(t)e^{s_1 t}$ into the original differential equation and applying the initial conditions, we will obtain the solution for A_1 and A_2.

Therefore, for two simultaneous equal roots of the characteristic equation, we try the solution

$$v_n = e^{s_1 t}(A_1 t + A_2) \tag{9.6-2}$$

Let us consider a parallel *RLC* circuit where $L = 1$ H, $R = 1\ \Omega$, $C = 1/4$ F, $v(0) = 5$ V, and $i(0) = -6$ A. The characteristic equation for the circuit is

$$s^2 + \frac{1}{RC}s + \frac{1}{LC} = 0$$

or

$$s^2 + 4s + 4 = 0$$

The two roots are then $s_1 = s_2 = -2$. Using Eq. 9.6-2 for the natural response, we have

$$v_n = e^{-2t}(A_1 t + A_2) \tag{9.6-3}$$

Since $v_n(0) = 5$, we have at $t = 0$

$$5 = A_2$$

Now, to obtain A_1, we proceed to find the derivative of v_n and evaluate it at $t = 0$. The derivative of v_n is found by differentiating Eq. 9.6-3 to obtain

$$\frac{dv}{dt} = -2A_1 te^{-2t} + A_1 e^{-2t} - 2A_2 e^{-2t} \tag{9.6-4}$$

Evaluating Eq. 9.6-4 at $t = 0$, we have

$$\frac{dv(0)}{dt} = A_1 - 2A_2$$

Again, we may use Eq. 9.5-10 so that

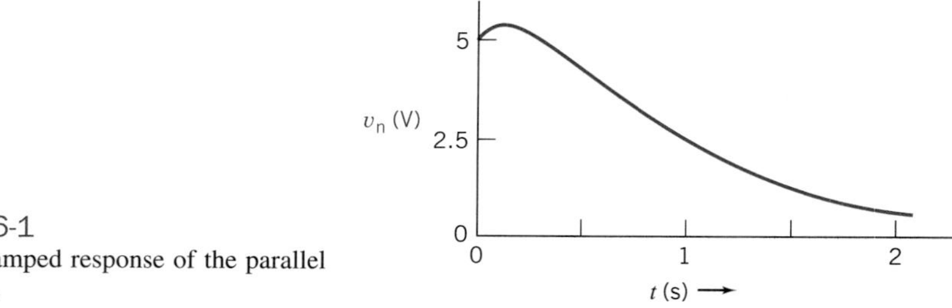

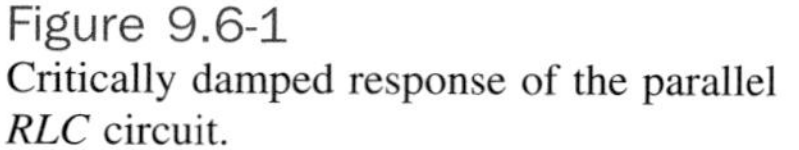

Figure 9.6-1
Critically damped response of the parallel *RLC* circuit.

$$\frac{dv(0)}{dt} = -\frac{v(0)}{RC} - \frac{i(0)}{C}$$

or

$$\frac{dv(0)}{dt} = \frac{-5}{1/4} - \frac{-6}{1/4} = 4$$

Therefore, $A_1 = 14$ and the natural response is

$$v_n = e^{-2t}(14t + 5) \text{ V}$$

The critically damped natural response of this *RLC* circuit is shown in Figure 9.6-1.

Exercise 9.6-1 A parallel *RLC* circuit has $R = 10\ \Omega$, $C = 1$ mF, $L = 0.4$ H, $v(0) = 8$ V, and $i(0) = 0$. Find the natural response $v_n(t)$ for $t > 0$.

Answer: $v_n(t) = e^{-50t}(8 - 400t)$ V

9.7 NATURAL RESPONSE OF AN UNDERDAMPED UNFORCED PARALLEL *RLC* CIRCUIT

The characteristic equation of the parallel *RLC* circuit will have two complex conjugate roots when $\alpha^2 < \omega_0^2$. This condition is met when

$$LC < (2RC)^2$$

or when

$$L < 4R^2C$$

Recall that

$$v_n = A_1e^{s_1 t} + A_2e^{s_2 t} \tag{9.7-1}$$

where

$$s_{1,2} = -\alpha \pm \sqrt{\alpha^2 - \omega_0^2}$$

When

$$\omega_0^2 > \alpha^2$$

we have

$$s_{1,2} = -\alpha \pm j\sqrt{\omega_0^2 - \alpha_0^2}$$

where

$$j = \sqrt{-1}$$

See Appendix B for a review of complex numbers.

The complex roots lead to an oscillatory-type response. We define the square root $\sqrt{\omega_0^2 - \alpha^2}$ as ω_d, which we will call the *damped resonant frequency.* The factor α, called the *damping coefficient,* determines how quickly the oscillations subside. Then the roots are

$$s_{1,2} = -\alpha \pm j\,\omega_d$$

Therefore, the natural response is

$$v_n = A_1 e^{-\alpha t} e^{j\omega_d t} + A_2 e^{-\alpha t} e^{-j\omega_d t}$$

or

$$v_n = e^{-\alpha t}(A_1 e^{j\omega_d t} + A_2 e^{-j\omega_d t}) \tag{9.7-2}$$

Let us use the Euler identity[2]

$$e^{\pm j\omega t} = \cos \omega t \pm j \sin \omega t \tag{9.7-3}$$

Let $\omega = \omega_d$ in Eq. 9.7-3 and substitute into Eq. 9.7-2 to obtain

$$\begin{aligned} v_n &= e^{-\alpha t}(A_1 \cos \omega_d t + jA_1 \sin \omega_d t + A_2 \cos \omega_d t - jA_2 \sin \omega_d t) \\ &= e^{-\alpha t}[(A_1 + A_2)\cos \omega_d t + j(A_1 - A_2)\sin \omega_d t] \end{aligned} \tag{9.7-4}$$

Since the unknown constants A_1 and A_2 remain arbitrary, we replace $(A_1 + A_2)$ and $j(A_1 - A_2)$ with new arbitrary (yet unknown) constants B_1 and B_2. A_1 and A_2 must be complex conjugates so that B_1 and B_2 are real numbers. Therefore, Eq. 9.7-4 becomes

$$v_n = e^{-\alpha t}(B_1 \cos \omega_d t + B_2 \sin \omega_d t) \tag{9.7-5}$$

where B_1 and B_2 will be determined by the initial conditions, $v(0)$ and $i(0)$.

The natural underdamped response is oscillatory with a decaying magnitude. The rapidity of decay depends on α, and the frequency of oscillation depends on ω_d.

Let us find the general form of the solution for B_1 and B_2 in terms of the initial conditions when the circuit is unforced. Then at $t = 0$ we have

$$v_n(0) = B_1$$

In order to find B_2, we evaluate the first derivative of v_n at $t = 0$. Therefore, the derivative is

$$\frac{dv_n}{dt} = e^{-\alpha t}[(\omega_d B_2 - \alpha B_1)\cos \omega_d t - (\omega_d B_1 + \alpha B_2)\sin \omega_d t]$$

and at $t = 0$ we obtain

$$\frac{dv_n(0)}{dt} = \omega_d B_2 - \alpha B_1 \tag{9.7-6}$$

Recall that we found earlier that Eq. 9.5-10 provides $dv(0)/dt$ for the parallel *RLC* circuit as

$$\frac{dv_n(0)}{dt} = -\frac{v(0)}{RC} - \frac{i(0)}{C} \tag{9.7-7}$$

Therefore, we use Eqs. 9.7-6 and 9.7-7 to obtain

$$\omega_d B_2 = \alpha B_1 - \frac{v(0)}{RC} - \frac{i(0)}{C} \tag{9.7-8}$$

[2]See Appendix D for a discussion of Euler's identity.

Example 9.7-1

Consider the parallel *RLC* circuit when $R = 25/3\ \Omega$, $L = 0.1$ H, $C = 1$ mF, $v(0) = 10$ V, and $i(0) = -0.6$ A. Find the natural response $v_n(t)$ for $t > 0$.

Solution

First, we determine α^2 and ω_0^2 to determine the form of the response. Consequently, we obtain

$$\alpha = \frac{1}{2RC} = 60$$

and

$$\omega_0^2 = \frac{1}{LC} = 10^4$$

Therefore, $\omega_0^2 > \alpha^2$ and the natural response is underdamped. We proceed to determine the damped resonant frequency ω_d as

$$\omega_d = (\omega_0^2 - \alpha^2)^{1/2} = (10^4 - 3.6 \times 10^3)^{1/2} = 80 \text{ rad/s}$$

Hence, the characteristic roots are

$$s_1 = -\alpha + j\omega_d = -60 + j80$$

and

$$s_2 = -\alpha - j\omega_d = -60 - j80$$

Consequently, the natural response is obtained from Eq. 9.7-5 as

$$v_n(t) = B_1 e^{-60t} \cos 80t + B_2 e^{-60t} \sin 80t$$

Since $v(0) = 10$, we have

$$B_1 = v(0) = 10$$

We can use Eq. 9.7-8 to obtain B_2 as

$$B_2 = \frac{\alpha}{\omega_d} B_1 - \frac{v(0)}{\omega_d RC} - \frac{i(0)}{\omega_d C}$$

$$= \frac{60 \times 10}{80} - \frac{10}{80 \times 25/3000} - \frac{-0.6}{80 \times 10^{-3}} = 7.5 - 15.0 + 7.5 = 0$$

Therefore, the natural response is

$$v_n(t) = 10e^{-60t} \cos 80\,t \text{ V}$$

A sketch of this response is shown in Figure 9.7-1. While the response is oscillatory in form because of the cosine function, it is damped by the exponential function, e^{-60t}.

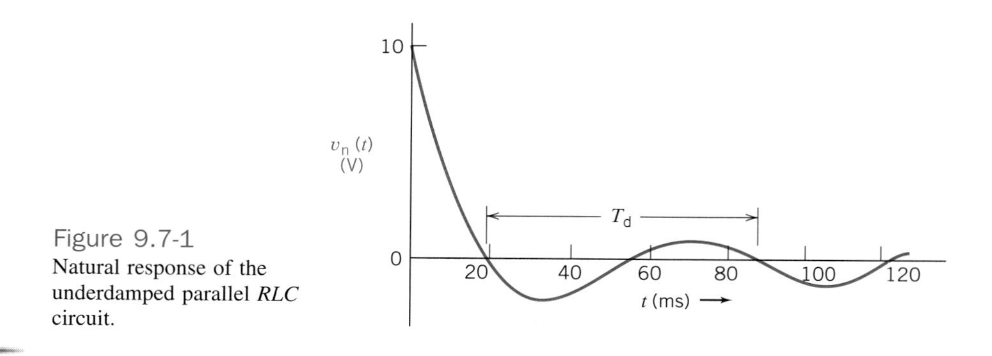

Figure 9.7-1
Natural response of the underdamped parallel *RLC* circuit.

The *period of the damped oscillation* is the time interval, denoted as T_d, expressed as

$$T_d = \frac{2\pi}{\omega_d} \text{ s} \tag{9.7-9}$$

The natural response of an underdamped circuit is not a pure oscillatory response, but it does exhibit the form of an oscillatory response. Thus, we may approximate T_d by the period between the first and third zero-crossings, as shown in Figure 9.7-1. Therefore, the frequency in hertz is

$$f_d = \frac{1}{T_d} \text{ Hz}$$

The period of the oscillation of the circuit of Example 9.7-1 is

$$T_d = \frac{2\pi}{80} = 79 \text{ ms}$$

Exercise 9.7-1 A parallel *RLC* circuit has $R = 62.5\ \Omega$, $L = 10$ mH, $C = 1\ \mu$F, $v(0) = 10$ V, and $i(0) = 80$ mA. Find the natural response $v_n(t)$.

Answer: $v_n(t) = e^{-8000t}[10 \cos 6000t - 26.7 \sin 6000t]$ V

9.8 FORCED RESPONSE OF AN *RLC* CIRCUIT

The forced response of an *RLC* circuit described by a second-order differential equation must satisfy the differential equation and contain no arbitrary constants. As we noted earlier, the response to a forcing function will often be of the same form as the forcing function. Again, we consider the differential equation for the second-order circuit as

$$\frac{d^2x}{dt^2} + a_1\frac{dx}{dt} + a_0x = f(t) \tag{9.8-1}$$

The forced response x_f must satisfy Eq. 9.8-1. Therefore, substituting x_f, we have

$$\frac{d^2x_f}{dt^2} + a_1\frac{dx_f}{dt} + a_0x_f = f(t) \tag{9.8-2}$$

We need to determine x_f so that x_f and its first and second derivatives all satisfy Eq. 9.8-2.

If the forcing function is a constant, we expect the forced response also to be a constant since the derivatives of a constant are zero. If the forcing function is of the form $f(t) = Be^{-at}$, then the derivatives of $f(t)$ are all exponentials of the form Qe^{-at}, and we expect

$$x_f = De^{-at}$$

If the forcing function is a sinusoidal function, we can expect the forced response to be a sinusoidal function. If $f(t) = A \sin \omega_0 t$, we will try

$$x_f = M \sin \omega_0 t + N \cos \omega_0 t = Q \sin (\omega_0 t + \theta)$$

Table 9.8-1 summarizes selected forcing functions and their associated assumed solutions.

Table 9.8-1

Forcing Function	Assumed Solution
K	A
Kt	$At + B$
Kt^2	$At^2 + Bt + C$
$K \sin \omega t$	$A \sin \omega t + B \cos \omega t$
Ke^{-at}	Ae^{-at}

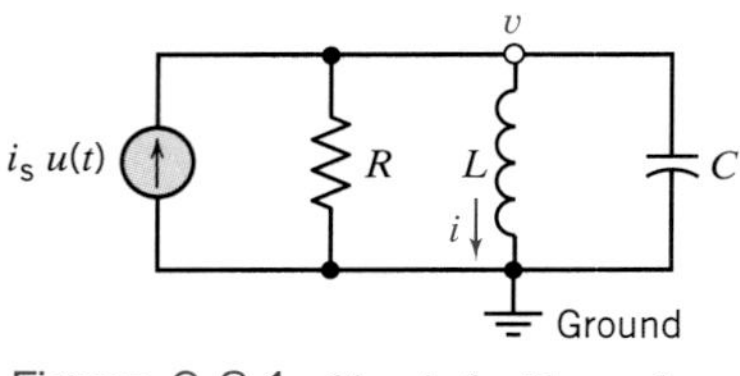

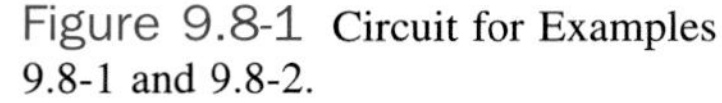

Figure 9.8-1 Circuit for Examples 9.8-1 and 9.8-2.

Example 9.8-1

Find the forced response for the inductor current i_f for the parallel RLC circuit shown in Figure 9.8-1 when $i_s = 8e^{-2t}$ A. Let $R = 6\ \Omega$, $L = 7$ H, and $C = 1/42$ F.

Solution

The source current is applied at $t = 0$ as indicated by the unit step function $u(t)$. The KCL equation at the upper node is

$$i + \frac{v}{R} + C\frac{dv}{dt} = i_s \tag{9.8-3}$$

We wish to obtain the second-order differential equation in terms of i using Eq. 9.8-3. We note that

$$v = L\frac{di}{dt} \tag{9.8-4}$$

and

$$\frac{dv}{dt} = L\frac{d^2i}{dt^2} \tag{9.8-5}$$

Substituting Eqs. 9.8-4 and 9.8-5 into Eq. 9.8-3, we have

$$i + \frac{L}{R}\frac{di}{dt} + CL\frac{di^2}{dt^2} = i_s$$

Then we divide by LC and rearrange to obtain the familiar second-order differential equation

$$\frac{d^2i}{dt^2} + \frac{1}{RC}\frac{di}{dt} + \frac{1}{LC}i = \frac{i_s}{LC} \tag{9.8-6}$$

Substituting the component values and the source i_s, we obtain

$$\frac{d^2i}{dt^2} + 7\frac{di}{dt} + 6\,i = 48e^{-2t} \tag{9.8-7}$$

We wish to obtain the forced response, so we assume that the response will be

$$i_f = Be^{-2t} \tag{9.8-8}$$

where B is to be determined. Substituting the assumed solution, Eq. 9.8-8, into the differential equation, we have

$$4Be^{-2t} + 7(-2Be^{-2t}) + 6\,Be^{-2t} = 48e^{-2t}$$

or

$$(4 - 14 + 6)Be^{-2t} = 48e^{-2t}$$

Therefore, $B = -12$ and

$$i_f = -12e^{-2t} \text{ A}$$

Example 9.8-2

Find the forced response i_f of the circuit of Example 9.8-1 when $i_s = I_0$, where I_0 is a constant.

Solution

Since the source is a constant applied at $t = 0$, we expect the forced response to be a constant also. As a first method, we will use the differential equation to find the forced response. Second, we will demonstrate the alternative method that uses the steady-state behavior of the circuit to find i_f.

The differential equation with the constant source is obtained from Eq. 9.8-6 as

$$\frac{d^2i}{dt^2} + 7\frac{di}{dt} + 6i = 6\,I_0$$

Again, we assume that the forced response is $i_f = D$, a constant. Since the first and second derivatives of the assumed forced response are zero, we have

$$6\,D = 6\,I_0$$

or

$$D = I_0$$

Therefore,

$$i_f = I_0$$

Another approach is to determine the steady-state response i_f of the circuit of Figure 9.8-1 by drawing the steady-state circuit model. The inductor is represented by a short circuit and the capacitor is represented by an open circuit, as shown in Figure 9.8-2. Clearly, since the steady-state model of the inductor is a short circuit, all the source current flows through the inductor in steady state and

$$i_f = I_0$$

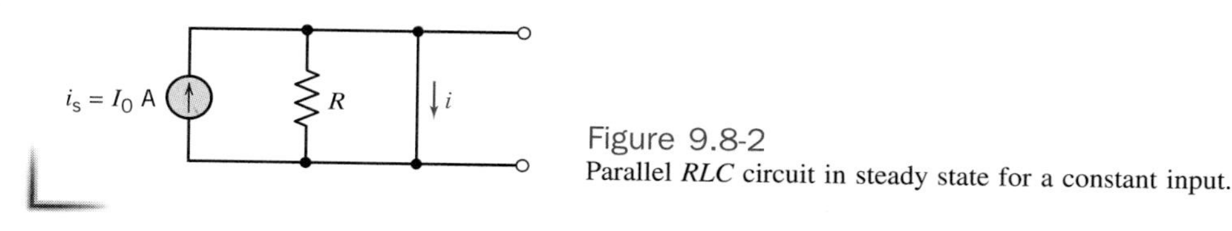

Figure 9.8-2
Parallel *RLC* circuit in steady state for a constant input.

The two previous examples showed that it is relatively easy to obtain the response of the circuit to a forcing function. However, we are sometimes confronted with a special case where the form of the forcing function is the same as the form of one of the components of the natural response.

Again, consider the circuit of Examples 9.8-1 and 9.8-2 (Figure 9.8-1) when the differential equation is

$$\frac{d^2i}{dt^2} + 7\frac{di}{dt} + 6\,i = 6\,i_s \tag{9.8-9}$$

The characteristic equation of the circuit is

$$s^2 + 7s + 6 = 0$$

or
$$(s + 1)(s + 6) = 0$$

Thus, the natural response is

$$i_n = A_1 e^{-t} + A_2 e^{-6t} \tag{9.8-10}$$

Consider the special case where $i_s = 3e^{-6t}$. Then we at first expect the forced response to be

$$i_f = Be^{-6t} \tag{9.8-11}$$

However, the forced response and one component of the natural response would then both have the form De^{-6t}. Will this work? Let's try substituting Eq. 9.8-11 into the differential equation (9.8-9). We then obtain

$$36Be^{-6t} - 42Be^{-6t} + 6Be^{-6t} \neq 18e^{-6t}$$

or
$$0 \neq 18e^{-6t}$$

which is an impossible solution. Therefore, we need another form of the forced response when one of the natural response terms has the same form as the forcing function.

Let us try the forced response

$$i_f = B\,t\,e^{-6t} \tag{9.8-12}$$

Then, substituting Eq. 9.8-12 into Eq. 9.8-9, we have

$$B(-6g - 6g + 36tg) + 7B(g - 6tg) + 6Btg = 18g \tag{9.8-13}$$

where $g = g(t) = e^{-6t}$. Simplifying Eq. 9.8-13, we have

$$B = -\frac{18}{5}$$

Therefore,
$$i_f = -\frac{18}{5}\,t\,e^{-6t}$$

In general, if the forcing function is of the same form as one of the components of the natural response, x_{n1}, we will use

$$x_f = t^p x_{n1}$$

where the integer p is selected so that the x_f is not duplicated in the natural response. Use the lowest power, p, of t that is not duplicated in the natural response.

Exercise 9.8-1 A circuit is described for $t > 0$ by the equation

$$\frac{d^2v}{dt^2} + 5\frac{dv}{dt} + 6v = v_s$$

Find the forced response v_f for $t > 0$ when (a) $v_s = 8$ V, (b) $v_s = 3e^{-4t}$ V, and (c) $v_s = 2e^{-2t}$ V.

Answer: (a) $v_f = 8/6$ V
(b) $v_f = \frac{3}{2}e^{-4t}$ V
(c) $v_f = 2te^{-2t}$ V

Exercise 9.8-2 A circuit is described for $t > 0$ by the equation

$$\frac{d^2i}{dt^2} + 9\frac{di}{dt} + 20\,i = 6\,i_s$$

where $i_s = 6 + 2t$ A. Find the forced response i_f.

Answer: $i_f = 1.53 + 0.6t$ A

9.9 COMPLETE RESPONSE OF AN *RLC* CIRCUIT

We have succeeded in finding the natural response and the forced response of a circuit described by a second-order differential equation. We wish to proceed to determine the complete response for the circuit.

We know that the *complete response* is the sum of the natural response and the forced response; thus

$$x = x_n + x_f$$

We may then obtain the complete response along with its unspecified constants by evaluating $x(t)$ at $t = 0$ and dx/dt at $t = 0$ to determine these constants.

Let us consider the series *RLC* circuit of Figure 9.3-2 with a differential equation (9.3-8) as

$$LC\frac{d^2v}{dt^2} + RC\frac{dv}{dt} + v = v_s$$

When $L = 1$ H, $C = \frac{1}{6}$ F, and $R = 5\ \Omega$, we obtain

$$\frac{d^2v}{dt^2} + 5\frac{dv}{dt} + 6\,v = 6\,v_s \tag{9.9-1}$$

We let $v_s = \dfrac{2e^{-t}}{3}$ V, $v(0) = 10$ V, and $dv(0)/dt = -2$ V/s.

We will first determine the form of the natural response and then determine the forced response. Adding these responses, we have the complete response with two unspecified constants. We will then use the initial conditions to specify these constants to obtain the complete response.

To obtain the natural response, we write the characteristic equation using operators as

$$s^2 + 5s + 6 = 0$$

or

$$(s + 2)(s + 3) = 0$$

Therefore, the natural response is

$$v_n = A_1e^{-2t} + A_2\,e^{-3t}$$

The forced response is obtained by examining the forcing function and noting that its exponential response has a different time constant than the natural response, so we may write

$$v_f = Be^{-t} \tag{9.9-2}$$

We can determine B by substituting Eq. 9.9-2 into Eq. 9.9-1. Then we have

$$Be^{-t} + 5(-Be^{-t}) + 6(Be^{-t}) = 4\,e^{-t}$$

or

$$B = 2$$

The complete response is then

$$v = v_n + v_f = A_1 e^{-2t} + A_2 e^{-3t} + 2e^{-t}$$

In order to find A_1 and A_2, we use the initial conditions. At $t = 0$ we have $v(0) = 10$, so we obtain

$$10 = A_1 + A_2 + 2 \qquad (9.9\text{-}3)$$

From the fact that $dv/dt = -2$ at $t = 0$, we have

$$-2A_1 - 3A_2 - 2 = -2 \qquad (9.9\text{-}4)$$

Solving Eqs. 9.9-3 and 9.9-4 by Cramer's rule, we have $A_1 = 24$ and $A_2 = -16$. Therefore,

$$v = 24\,e^{-2t} - 16e^{-3t} + 2\,e^{-t}\ \text{V}$$

Example 9.9-1

Find the complete response $v(t)$ for $t > 0$ for the circuit of Figure 9.9-1. Assume the circuit is in steady state at $t = 0^-$.

Solution

First, we determine the initial conditions of the circuit. At $t = 0^-$ we have the circuit model shown in Figure 9.9-2, where we replace the capacitor with an open circuit and the inductor with a short circuit. Then the voltage is

$$v(0^-) = 6\ \text{V}$$

and the inductor current is

$$i(0^-) = 1\ \text{A}$$

After the switch is thrown, we can write the KVL for the right-hand mesh of Figure 9.9-1 to obtain

$$-v + \frac{di}{dt} + 6\,i = 0 \qquad (9.9\text{-}5)$$

The KCL equation at node a will provide a second equation in terms of v and i as

$$\frac{v - v_s}{4} + i + \frac{1}{4}\frac{dv}{dt} = 0 \qquad (9.9\text{-}6)$$

Equations 9.9-5 and 9.9-6 may be rearranged as

$$\left(\frac{di}{dt} + 6\,i\right) - v = 0 \qquad (9.9\text{-}7)$$

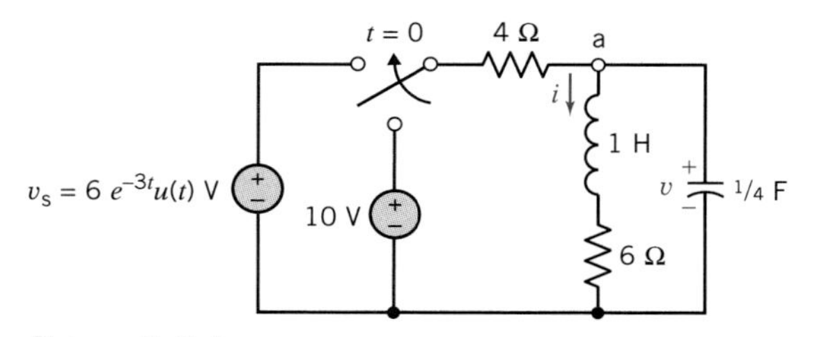

Figure 9.9-1 Circuit of Example 9.9-1.

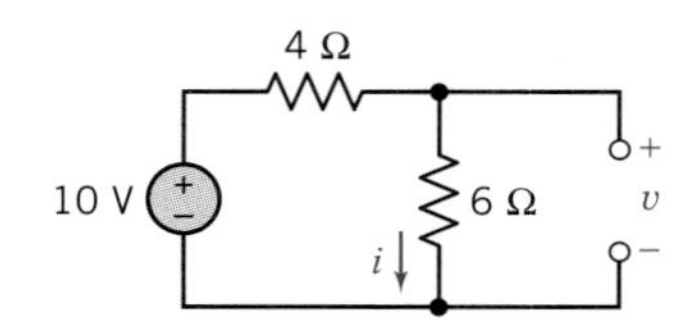

Figure 9.9-2 Circuit of Example 9.9-1 at $t = 0^-$.

$$i + \left(\frac{v}{4} + \frac{1}{4}\frac{dv}{dt}\right) = \frac{v_s}{4} \tag{9.9-8}$$

We will use operators so that $s = d/dt$, $s^2 = d^2/dt^2$, and $1/s = \int dt$. Then we obtain

$$(s + 6)i - v = 0 \tag{9.9-9}$$

$$i + \frac{1}{4}(s + 1)\,v = v_s/4 \tag{9.9-10}$$

The characteristic equation is obtained from Cramer's rule as the determinant Δ where

$$\Delta = \frac{1}{4}(s + 6)(s + 1) + 1$$

Set the determinant to zero to obtain

$$(s + 6)(s + 1) + 4 = 0$$

or

$$s^2 + 7s + 10 = 0$$

Therefore, the roots of the characteristic equation are

$$s_1 = -2 \quad \text{and} \quad s_2 = -5$$

To find the second-order differential equation describing the circuit, we use Cramer's rule for Eqs. 9.9-9 and 9.9-10 to solve for v in order to obtain

$$v = \frac{(s + 6)(v_s/4)}{\Delta} = \frac{(s + 6)v_s}{s^2 + 7s + 10}$$

Of course, this equation can be rewritten as

$$(s^2 + 7s + 10)v = (s + 6)v_s$$

and hence the second-order differential equation is

$$\frac{d^2v}{dt^2} + 7\frac{dv}{dt} + 10v = \frac{dv_s}{dt} + 6v_s \tag{9.9-11}$$

The natural response v_n is

$$v_n = A_1 e^{-2t} + A_2 e^{-5t}$$

The forced response is assumed to be of the form

$$v_f = Be^{-3t} \tag{9.9-12}$$

Substituting v_f into the differential equation, we have

$$9Be^{-3t} - 21Be^{-3t} + 10Be^{-3t} = -18e^{-3t} + 36e^{-3t}$$

Therefore,

$$B = -9$$

and

$$v_f = -9e^{-3t}$$

The complete response is then

$$v = v_n + v_f = A_1 e^{-2t} + A_2\, e^{-5t} - 9e^{-3t} \tag{9.9-13}$$

Since $v(0) = 6$, we have

$$v(0) = 6 = A_1 + A_2 - 9$$

or

$$A_1 + A_2 = 15 \tag{9.9-14}$$

We also know that $i(0) = 1$ A. We can use Eq. 9.9-8 to determine $dv(0)/dt$ and then evaluate the derivative of Eq. 9.9-13 at $t = 0$. Equation 9.9-8 states that

$$\frac{dv}{dt} = -4\,i - v + v_s$$

At $t = 0$ we have

$$\frac{dv(0)}{dt} = -4\,i(0) - v(0) + v_s(0) = -4 - 6 + 6 = -4$$

Let us take the derivative of Eq. 9.9-13 to obtain

$$\frac{dv}{dt} = -2A_1e^{-2t} - 5A_2e^{-5t} + 27e^{-3t}$$

At $t = 0$ we obtain

$$\frac{dv(0)}{dt} = -2A_1 - 5A_2 + 27$$

Since $dv(0)/dt = -4$, we have

$$2\,A_1 + 5A_2 = 31 \tag{9.9-15}$$

Solving Eqs. 9.9-15 and 9.9-14 simultaneously, we obtain

$$A_1 = \frac{44}{3} \quad \text{and} \quad A_2 = \frac{1}{3}$$

Therefore,

$$v = \frac{44}{3}e^{-2t} + \frac{1}{3}e^{-5t} - 9e^{-3t}\ \text{V}$$

Note that we used the capacitor voltage and the inductor current as the unknowns. This is very convenient, since you will normally have the initial conditions of these variables. These variables, v_c and i_L, are known as the *state variables*. We will consider this approach more fully in the next section.

Exercise 9.9-1 Find the differential equation for $v_c(t)$ in the circuit of Figure E 9.9-1 using the direct method. Find $v_c(t)$ for time $t > 0$ for each of the following sets of component values: (a) $C = 1$ F, $L = 0.25$ H, $R_1 = R_2 = 1.309\ \Omega$, (b) $C = 1$ F, $L = 1$ H, $R_1 = 3\ \Omega$, $R_2 = 1\ \Omega$, (c) $C = 0.125$ F, $L = 0.5$ H, $R_1 = 1\ \Omega$, $R_2 = 4\ \Omega$.

Answers: (a) $v_c(t) = \frac{1}{2} - e^{-2t} + \frac{1}{2}e^{-4t}$ V (b) $v_c(t) = \frac{1}{4} - \left(\frac{1}{4} + \frac{1}{2}t\right)e^{-2t}$ V
(c) $v_c(t) = 0.8 - e^{-2t}(0.8\cos 4t + 0.4\sin 4t)$ V

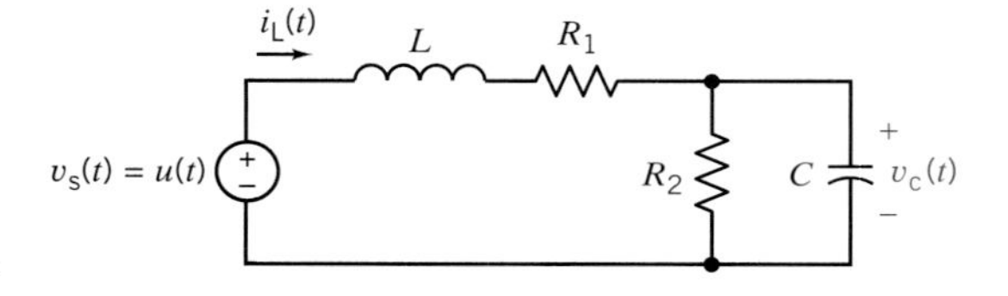

Figure E 9.9-1

Exercise 9.9-2 Find the differential equation for $v_o(t)$ in the circuit of Figure E 9.9-2 using the direct method. Find $v_o(t)$ for time $t > 0$ for each of the following sets of component values: (a) $C = 1$ F, $L = 0.25$ H, $R_1 = R_2 = 1.309\ \Omega$, (b) $C = 1$ F, $L = 1$ H, $R_1 = 1\ \Omega$, $R_2 = 3\ \Omega$, (c) $C = 0.125$ F, $L = 0.5$ H, $R_1 = 4\ \Omega$, $R_2 = 1\ \Omega$.

Answers: (a) $v_o(t) = \frac{1}{2} - e^{-2t} + \frac{1}{2}e^{-4t}$ V (b) $v_o(t) = \frac{3}{4} - \left(\frac{3}{4} + \frac{3}{2}t\right)e^{-2t}$ V
(c) $v_c(t) = 0.2 - e^{-2t}(0.2 \cos 4t + 0.1 \sin 4t)$ V

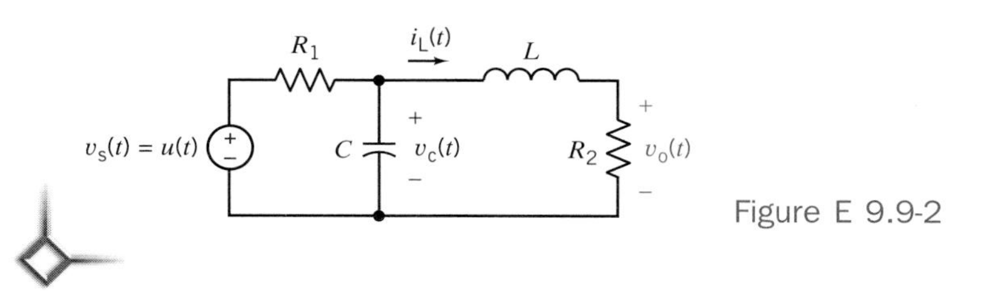

Figure E 9.9-2

9.10 STATE VARIABLE APPROACH TO CIRCUIT ANALYSIS

The *state variables* of a circuit are a set of variables associated with the energy of the energy storage elements of the circuit. Thus, they describe the complete response of a circuit to a forcing function and the circuit's initial conditions. Here the word "state" means condition, as in "state of the union." We will choose as the state variables those variables that describe the energy storage of the circuit. Thus, we will use the independent capacitor voltages and the independent inductor currents.

Consider the circuit shown in Figure 9.10-1. The two energy storage elements are C_1 and C_2, and the two capacitors cannot be reduced to one. We expect the circuit to be described by a second-order differential equation. However, let us first obtain the two first-order differential equations that describe the response for $v_1(t)$ and $v_2(t)$, which are the state variables of the circuit. If we know the value of the state variables at one time and the value of the input variables thereafter, we can find the value of any state variable for any subsequent time.

Writing the KCL at nodes 1 and 2, we have

$$\textit{node 1:} \qquad C_1 \frac{dv_1}{dt} = \frac{v_a - v_1}{R_1} + \frac{v_2 - v_1}{R_2} \tag{9.10-1}$$

$$\textit{node 2:} \qquad C_2 \frac{dv_2}{dt} = \frac{v_b - v_2}{R_3} + \frac{v_1 - v_2}{R_2} \tag{9.10-2}$$

Equations 9.10-1 and 9.10-2 can be rewritten as

$$\frac{dv_1}{dt} + \frac{v_1}{C_1 R_1} + \frac{v_1}{C_1 R_2} - \frac{v_2}{C_1 R_2} = \frac{v_a}{C_1 R_1} \tag{9.10-3}$$

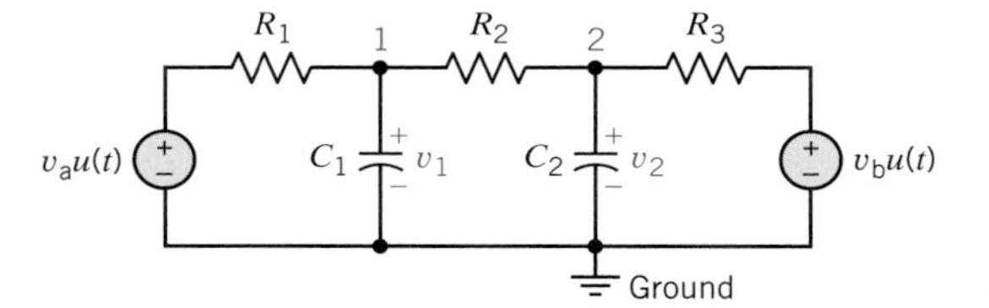

Figure 9.10-1
Circuit with two energy storage elements.

$$\frac{dv_2}{dt} + \frac{v_2}{C_2R_3} + \frac{v_2}{C_2R_2} - \frac{v_1}{C_2R_2} = \frac{v_b}{C_2R_3} \quad (9.10\text{-}4)$$

Assume that $C_1R_1 = 1$, $C_1R_2 = 1$, $C_2R_3 = 1$, and $C_2R_2 = 1/2$. Then we have

$$\frac{dv_1}{dt} + 2v_1 - v_2 = v_a \quad (9.10\text{-}5)$$

and

$$-2\,v_1 + \frac{dv_2}{dt} + 3v_2 = v_b \quad (9.10\text{-}6)$$

Using operators, we have

$$(s + 2)v_1 - v_2 = v_a$$

$$-2\,v_1 + (s + 3)v_2 = v_b$$

If we wish to solve for v_1, we use Cramer's rule to obtain

$$v_1 = \frac{(s + 3)v_a + v_b}{(s + 2)(s + 3) - 2} \quad (9.10\text{-}7)$$

The characteristic equation is obtained from the denominator and has the form

$$s^2 + 5s + 4 = 0$$

The characteristic roots are $s = -4$ and $s = -1$. The second-order differential equation can be obtained by rewriting Eq. 9.10-7 as

$$(s^2 + 5s + 4)v_1 = (s + 3)v_a + v_b$$

Then the differential equation for v_1 is

$$\frac{d^2v_1}{dt^2} + 5\frac{dv_1}{dt} + 4v_1 = \frac{dv_a}{dt} + 3v_a + v_b \quad (9.10\text{-}8)$$

We now proceed to obtain the natural response

$$v_{1n} = A_1e^{-t} + A_2e^{-4t}$$

and the forced response, which depends on the form of the forcing function. For example, if $v_a = 10$ V and $v_b = 6$ V, v_{1f} will be a constant (see Table 9.8-1). We obtain v_{1f} by substituting v_a and v_b into Eq. 9.10-8, obtaining

$$4v_{1f} = 3v_a + v_b$$

or

$$4v_{1f} = 30 + 6 = 36$$

Therefore,

$$v_{1f} = 9$$

Then

$$v_1 = v_{1n} + v_{1f} = A_1e^{-t} + A_2e^{-4t} + 9 \quad (9.10\text{-}9)$$

We will usually know the initial conditions of the energy storage elements. For example, if we know that $v_1(0) = 5$ V and $v_2(0) = 10$ V, we first use $v_1(0) = 5$ along with Eq. 9.10-9 to obtain

$$v_1(0) = A_1 + A_2 + 9$$

and, therefore,

$$A_1 + A_2 = -4 \quad (9.10\text{-}10)$$

Now we need the value of dv_1/dt at $t = 0$. Referring back to Eq. 9.10-5, we have

Table 9.10-1 **State Variable Method of Circuit Analysis**

1. Identify the state variables as the independent capacitor voltages and inductor currents.
2. Determine the initial conditions at $t = 0$ for the capacitor voltages and the inductor currents.
3. Obtain a first-order differential equation for each state variable using KCL or KVL.
4. Use the operator s to substitute for d/dt.
5. Obtain the characteristic equation of the circuit by noting that it can be obtained by setting the determinant of Cramer's rule equal to zero.
6. Determine the roots of the characteristic equation, which then determines the form of the natural response.
7. Obtain the second-order (or higher order) differential equation for the selected variable x by Cramer's rule.
8. Determine the forced response x_f by assuming an appropriate form of x_f and determining the constant by substituting the assumed solution in the second-order differential equation.
9. Obtain the complete solution $x = x_n + x_f$.
10. Use the initial conditions on the state variables along with the set of first-order differential equations (step 3) to obtain $dx(0)/dt$.
11. Using $x(0)$ and $dx(0)/dt$ for each state variable, find the arbitrary constants $A_1, A_2, \ldots A_n$ to obtain the complete solution $x(t)$.

$$\frac{dv_1}{dt} = v_a + v_2 - 2\, v_1$$

Therefore, at $t = 0$ we have

$$\frac{dv_1(0)}{dt} = v_a(0) + v_2(0) - 2v_1(0) = 10 + 10 - 2(5) = 10$$

The derivative of the complete solution, Eq. 9.10-9, at $t = 0$ is

$$\frac{dv_1(0)}{dt} = -A_1 - 4\, A_2$$

Therefore,

$$A_1 + 4\,A_2 = -10 \tag{9.10-11}$$

Solving Eqs. 9.10-10 and 9.10-11, we have

$$A_1 = -2 \quad \text{and} \quad A_2 = -2$$

Therefore,

$$v_1(t) = -2e^{-t} - 2e^{-4t} + 9\text{ V}$$

As you encounter circuits with two or more energy storage elements, you should consider using the state variable method of describing a set of first-order differential equations.

The **state variable method** uses a first-order differential equation for each state variable to determine the complete response of a circuit.

A summary of the state variable method is given in Table 9.10-1. We will use this method in Example 9.10-1.

Example 9.10-1

Find $i(t)$ for $t > 0$ for the circuit shown in Figure 9.10-2 when $R = 3\ \Omega$, $L = 1$ H, $C = 1/2$ F, and $i_s = 2e^{-3t}$ A. Assume steady state at $t = 0^-$.

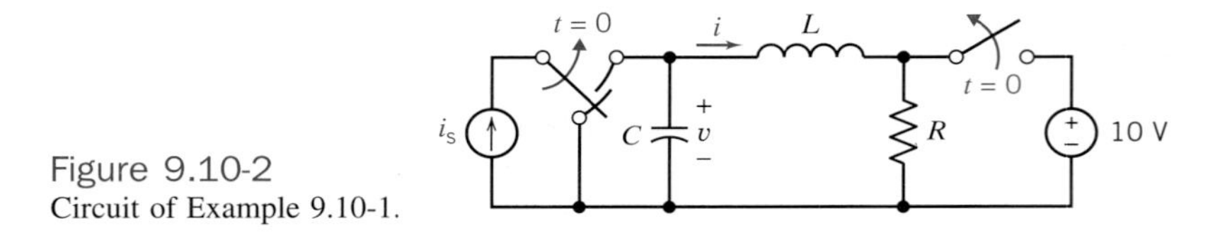

Figure 9.10-2
Circuit of Example 9.10-1.

Solution

First, we identify the state variables as i and v. The initial conditions at $t = 0$ are obtained by considering the circuit with the 10-V source connected for a long time at $t = 0^-$. Then $v(0) = 10$ V and $i(0) = 0$ A. At $t = 0$, the voltage source is disconnected and the current source is connected.

The first differential equation is obtained by using KVL around the *RLC* mesh to obtain

$$L\frac{di}{dt} + Ri = v$$

The second differential equation is obtained by using KCL at the node at the top of the capacitor to get

$$C\frac{dv}{dt} + i = i_s$$

We may rewrite these two first-order differential equations as

$$\frac{di}{dt} + \frac{R}{L}i - \frac{v}{L} = 0$$

and

$$\frac{dv}{dt} + \frac{i}{C} = \frac{i_s}{C}$$

Substituting the component values, we have

$$\frac{di}{dt} + 3i - v = 0 \tag{9.10-12}$$

and

$$\frac{dv}{dt} + 2i = 2i_s \tag{9.10-13}$$

Using the operator $s = d/dt$, we have

$$(s + 3)\, i - v = 0 \tag{9.10-14}$$

$$2\, i + sv = 2i_s \tag{9.10-15}$$

Therefore, the characteristic equation obtained from the determinant is

$$(s + 3)s + 2 = 0$$

or

$$s^2 + 3s + 2 = 0$$

Thus, the roots of the characteristic equation are

$$s_1 = -2 \quad \text{and} \quad s_2 = -1$$

Since we wish to solve for $i(t)$ for $t > 0$, we use Cramer's rule to solve Eqs. 9.10-14 and 9.10-15 for i, obtaining

$$i = \frac{2i_s}{s^2 + 3s + 2}$$

Therefore, the differential equation is

$$\frac{d^2i}{dt^2} + 3\frac{di}{dt} + 2i = 2i_s \qquad (9.10\text{-}16)$$

The natural response is

$$i_n = A_1 e^{-t} + A_2 e^{-2t}$$

We assume the forced response is of the form

$$i_f = Be^{-3t}$$

Substituting i_f into Eq. 9.10-16, we have

$$(9Be^{-3t}) + 3(-3Be^{-3t}) + 2\,Be^{-3t} = 2(2e^{-3t})$$

or

$$9B - 9B + 2B = 4$$

Therefore, $B = 2$ and

$$i_f = 2e^{-3t}$$

The complete response is

$$i = A_1 e^{-t} + A_2 e^{-2t} + 2e^{-3t}$$

Since $i(0) = 0$,

$$0 = A_1 + A_2 + 2 \qquad (9.10\text{-}17)$$

We need to obtain $di(0)/dt$ from Eq 9.10-12, which we repeat here as

$$\frac{di}{dt} + 3i - v = 0$$

Therefore, at $t = 0$ we have

$$\frac{di(0)}{dt} = -3i(0) + v(0) = 10$$

The derivative of the complete response at $t = 0$ is

$$\frac{di(0)}{dt} = -A_1 - 2A_2 - 6$$

Since $di(0)/dt = 10$, we have

$$-A_1 - 2A_2 = 16$$

and repeating Eq. 9.10-17, we have

$$A_1 + A_2 = -2$$

Adding these two equations, we determine that $A_1 = 12$ and $A_2 = -14$. Then we have the complete solution for i as

$$i = 12\,e^{-t} - 14e^{-2t} + 2\,e^{-3t}\text{ A}$$

We recognize that the state variable method is particularly powerful for finding the response of energy storage elements in a circuit. This is also true if we encounter higher

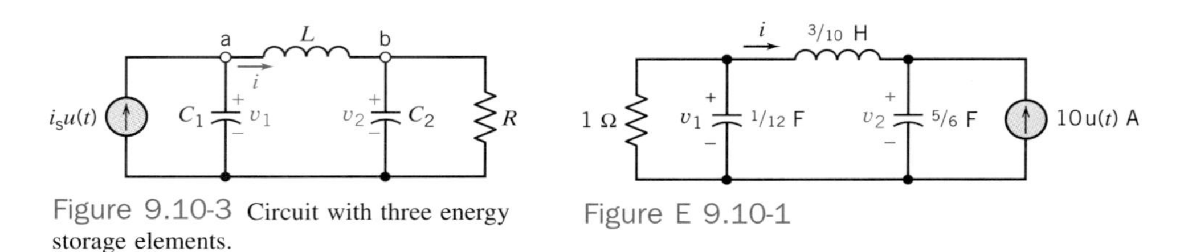

Figure 9.10-3 Circuit with three energy storage elements.

Figure E 9.10-1

order circuits with three or more energy storage elements. For example, consider the circuit shown in Figure 9.10-3. The state variables are v_1, v_2, and i. Two first-order differential equations are obtained by writing the KCL equations at node a and node b. Then a third first-order differential equation is obtained by writing the KVL around the middle mesh containing i. The solution for one or more of these variables can then be obtained by proceeding with the state variable method summarized in Table 9.10-1.

Exercise 9.10-1 Find $v_2(t)$ for $t > 0$ for the circuit of Figure E 9.10-1. Assume there is no initial stored energy.

Answer: $v_2(t) = -15e^{-2t} + 6e^{-4t} - e^{-6t} + 10$ V

9.11 ROOTS IN THE COMPLEX PLANE

We have observed that the character of the natural response of a second-order system is determined by the roots of the characteristic equation. Let us consider the roots of a parallel *RLC* circuit. The characteristic equation 9.5-3 is

$$s^2 + \frac{s}{RC} + \frac{1}{LC} = 0$$

and the roots are (9.5-7 and 9.5-8)

$$s = -\alpha \pm \sqrt{\alpha^2 - \omega_0^2}$$

where $\alpha = 1/2RC$ and $\omega_0^2 = 1/LC$. When $\omega_0 > \alpha$, the roots are complex and

$$s = -\alpha \pm j\sqrt{\omega_0^2 - \alpha^2} = -\alpha \pm j\omega_d \tag{9.11-1}$$

In general, roots are located in the complex plane, the location being defined by coordinates measured along the real or σ axis and the imaginary or $j\omega$-axis. This is referred to as the s-plane or, since s has the units of frequency, as the *complex frequency plane.* Where the roots are real, negative, and distinct, the response is the sum of two decaying exponentials and is said to be overdamped. Where the roots are complex conjugates, the natural response is an exponentially decaying sinusoid and is said to be underdamped or oscillatory.

Now, let us show the location of the roots of the characteristic equation for the four conditions: (a) undamped, $\alpha = 0$; (b) underdamped, $\alpha < \omega_o$; (c) critically damped, $\alpha = \omega_0$;

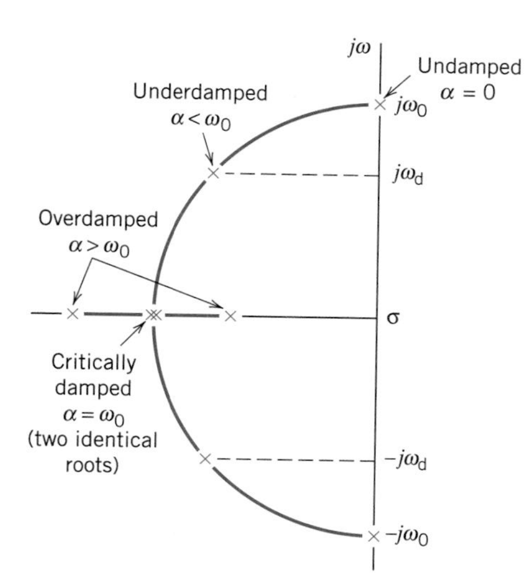

Figure 9.11-1
The complete s-plane showing the location of the two roots, s_1 and s_2, of the characteristic equation in the left-hand portion of the s-plane. The roots are designated by the $\times$ symbol.

and (d) overdamped, $\alpha > \omega_0$. These four conditions lead to root locations on the s-plane as shown in Figure 9.11-1. When $\alpha = 0$, the two complex roots are $\pm j\omega_0$. When $\alpha < \omega_0$, the roots are $s = -\alpha \pm j\omega_d$. When $\alpha = \omega_0$, there are two roots at $s = -\alpha$. Finally, when $\alpha > \omega_0$, there are two real roots, $s = -\alpha \pm \sqrt{\alpha^2 - \omega_0^2}$.

A summary of the root locations, the type of response, and the form of the response $v(t)$ for $v(0) = 1$ V and $i(0) = 0$ is presented in Table 9.11-1.

Exercise 9.11-1 A parallel *RLC* circuit has $L = 0.1$ H and $C = 100$ mF. Determine the roots of the characteristic equation and plot them on the s-plane when (a) $R = 0.4\ \Omega$ and (b) $R = 1.0\ \Omega$.

Answer: (a) $s = -5, -20$ (Figure E 9.11-1)

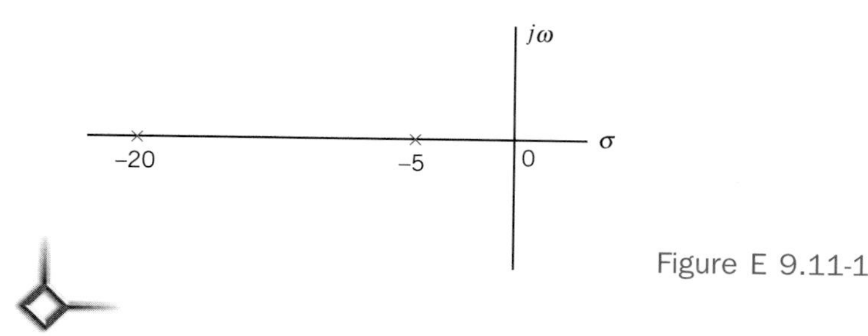

Figure E 9.11-1

9.12 PSPICE ANALYSIS OF THE *RLC* CIRCUIT

In this section we use PSpice analysis methods to obtain the transient response of *RLC* circuits. First, let us consider the *RLC* series circuit shown in Figure 9.3-2 and redrawn in PSpice format in Figure 9.12-1 with $v_s(t) = 0$. Let us consider that $i(0) = 0$ A and $v(0) = 10$ V are established at $t = 0$. Using Eq. 9.3-8, we have

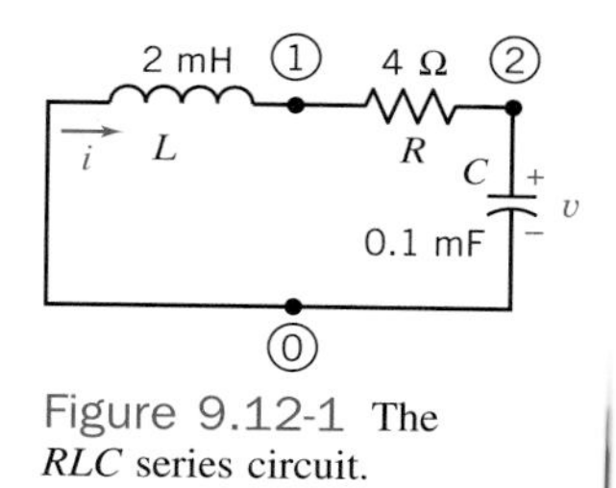

Figure 9.12-1 The *RLC* series circuit.

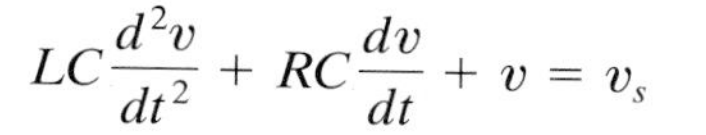

$$LC\frac{d^2v}{dt^2} + RC\frac{dv}{dt} + v = v_s$$

Table 9.11-1 **The Natural Response of a Parallel RLC Circuit**

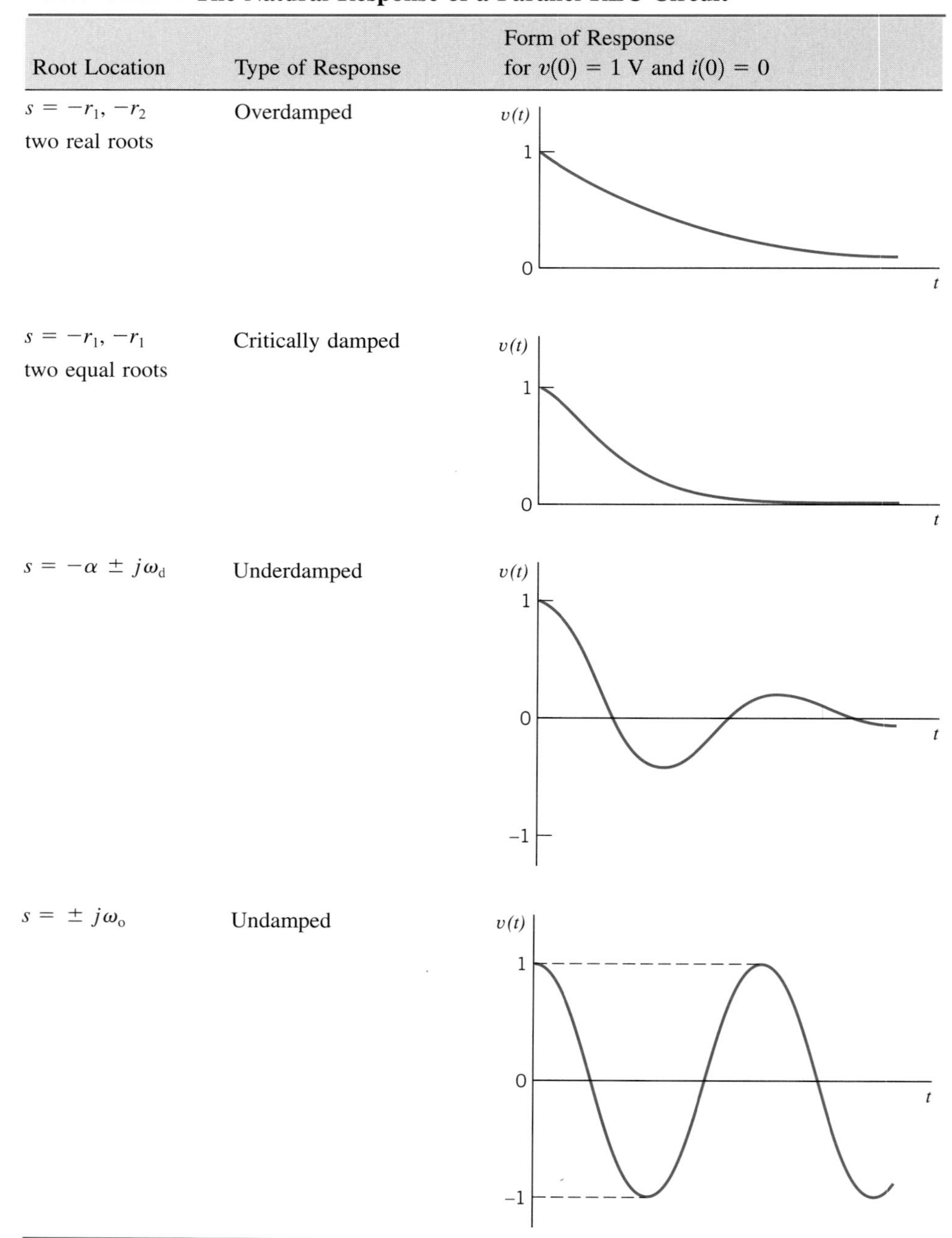

Root Location	Type of Response	Form of Response for $v(0) = 1$ V and $i(0) = 0$
$s = -r_1, -r_2$ two real roots	Overdamped	
$s = -r_1, -r_1$ two equal roots	Critically damped	
$s = -\alpha \pm j\omega_d$	Underdamped	
$s = \pm j\omega_o$	Undamped	

Note: $v(t)$ is the capacitor voltage and $i(t)$ is the inductor current

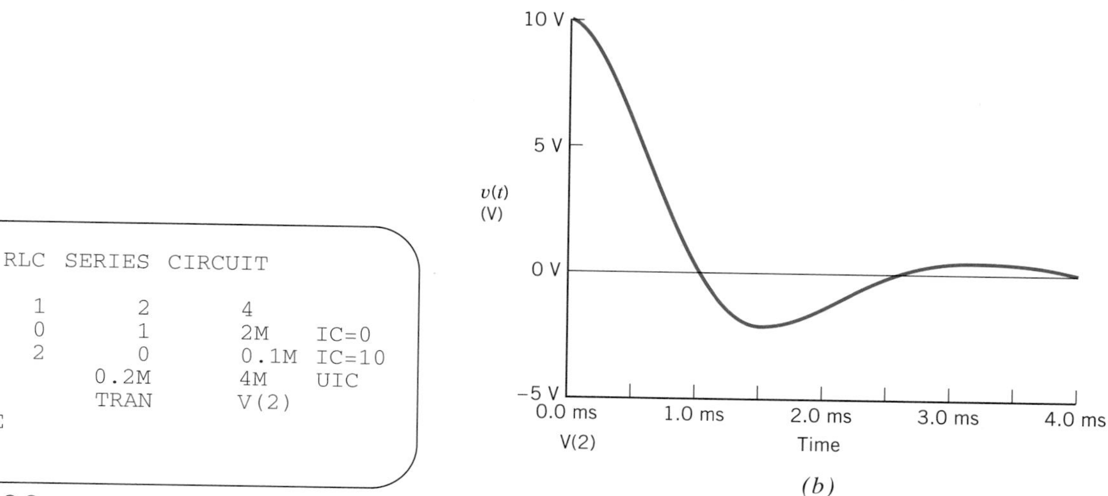

```
**        RLC  SERIES  CIRCUIT

R          1       2        4
L          0       1        2M     IC=0
C          2       0        0.1M   IC=10
.TRAN            0.2M       4M     UIC
.PLOT            TRAN       V(2)
.PROBE
.END
```


Figure 9.12-2 The PSpice program for the *RLC* series circuit.

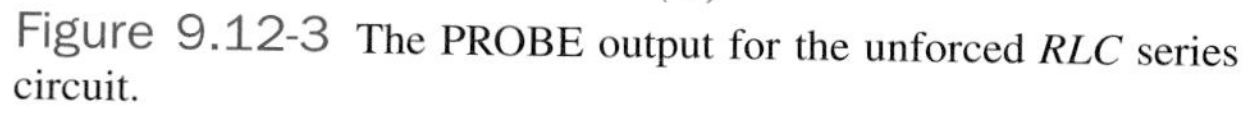

Figure 9.12-3 The PROBE output for the unforced *RLC* series circuit.

Let $s = d/dt$ and substitute the values of the parameters to obtain the unforced equation

$$(2 \times 10^{-7}s^2 + 4 \times 10^{-4}s + 1)v = 0$$

or

$$s^2 + 2000s + 5 \times 10^6 = 0$$

The roots of the characteristic equation are complex conjugates, and the circuit response is underdamped and oscillatory. We wish to determine the circuit response.

The PSpice program is shown in Figure 9.12-2, incorporating the initial conditions. The transient response, $v(t)$, is shown in Figure 9.12-3. Note that the voltage $v(t)$ attains a negative value and eventually decays to zero after 4 ms.

As an example of the solution of an *RLC* circuit using PSpice with a forcing function and initial conditions, reconsider Example 9.10-1. Reviewing Figure 9.10-2, we find that $v(0) = 10$ V and $i(0) = 0$. Figure 9.10-2 is redrawn for $t \geq 0$ in a PSpice format as shown in Figure 9.12-4. The input signal is $i_s = 2e^{-3t}$ A. The EXP input is used in the PSpice program shown in Figure 9.12-5. We plot the capacitor voltage as shown in Figure 9.12-6.

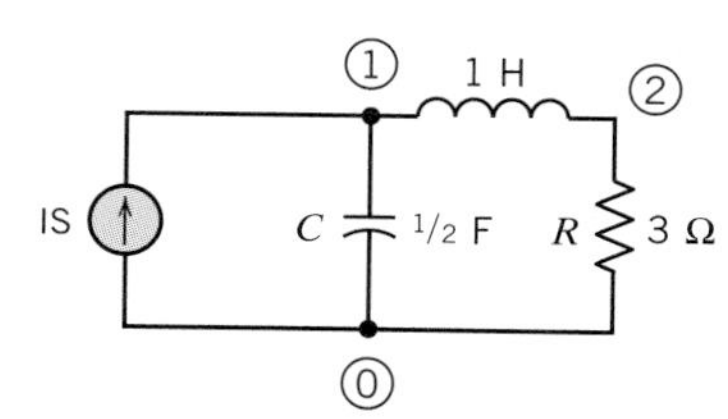

Figure 9.12-4 The *RLC* circuit of Example 9.10-1 redrawn for PSpice for $t \geq 0$ with a current source input.

```
**        FORCED RLC CIRCUIT
R   2    0 3
C   1    0 0.5     IC = 10
L   1    2 1       IC = 0
IS  0    1 EXP   (0  2 0  1E-6  0  .333)
.TRAN 0.2  5 UIC
.PLOT TRAN V(1)
.PROBE
.END
```

Figure 9.12-5 The PSpice program for the *RLC* circuit with an exponential input current waveform.

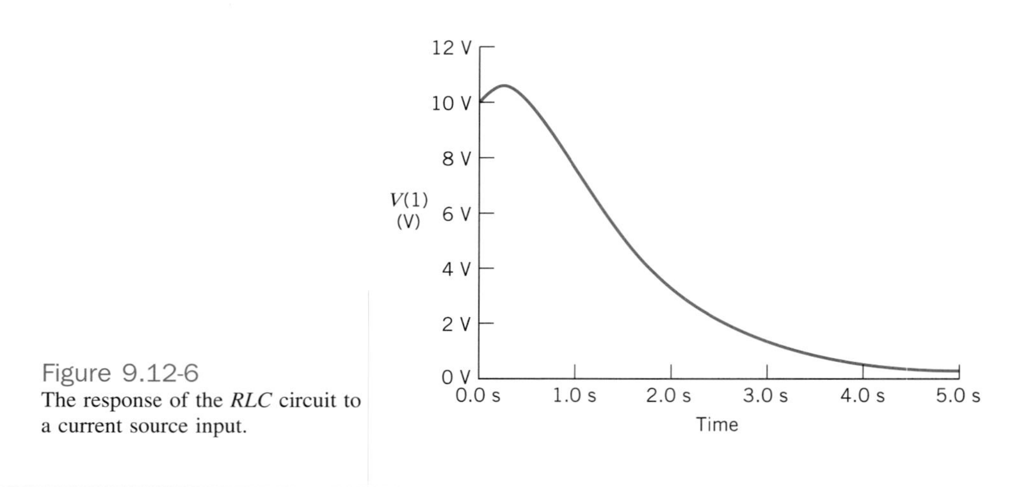

Figure 9.12-6
The response of the *RLC* circuit to a current source input.

9.13 VERIFICATION EXAMPLE

Figure 9.13-1*b* shows an *RLC* circuit. The voltage, $v_s(t)$, of the voltage source is the square wave shown in Figure 9.13-1*a*. Figure 9.13-2 shows a plot of the inductor current, $i(t)$, which was obtained by simulating this circuit using PSpice. Verify that the plot of $i(t)$ is correct.

Solution

Several features of the plot can be checked. The plot indicates that steady-state values of the inductor current are $i(\infty) = 0$ and $i(\infty) = 200$ mA and that the circuit is underdamped. In addition, some points on the response have been labeled to give the corresponding values of time and current. These values can be used to check the value of the damped resonant frequency, ω_d.

If the voltage of the voltage source were a constant, $v_s(t) = V_s$, then the steady-state inductor current would be

$$i(t) = \frac{V_s}{100}$$

Thus, we expect the steady-state inductor current to be $i(\infty) = 0$ when $V_s = 0$ V and to be $i(\infty) = 200$ mA when $V_s = 20$ V. The plot in Figure 9.13-2 shows that the steady-state values of the inductor current are indeed $i(\infty) = 0$ and $i(\infty) = 200$ mA.

The plot in Figure 9.13-2 shows an underdamped response. The *RLC* circuit will be underdamped if

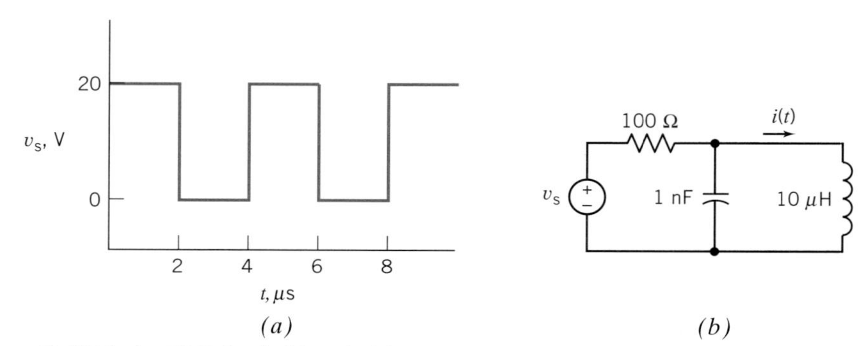

Figure 9.13-1 An *RLC* circuit (*b*) excited by a square wave (*a*).

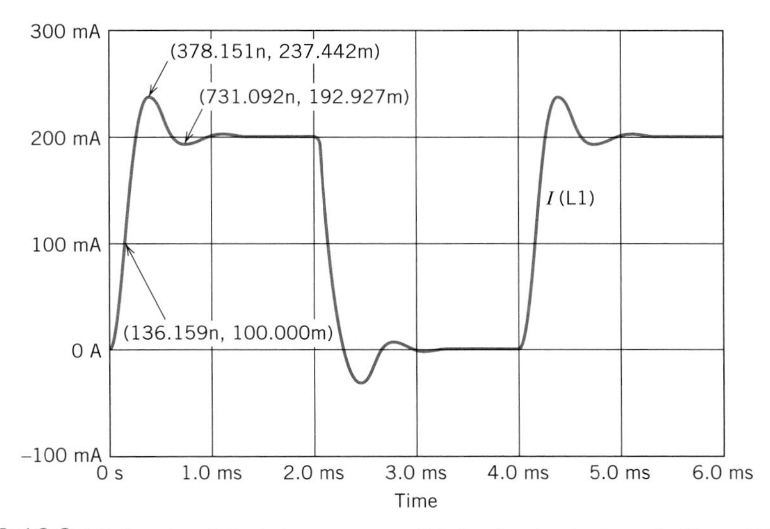

Figure 9.13-2 PSpice plot of the inductor current, $i(t)$, for the circuit shown in Figure 9.13-1.

$$10^{-5} = L < 4R^2C = 4 \times 100^2 \times 10^{-9}$$

Since this inequality is satisfied, the circuit is indeed underdamped, as indicated by the plot.

The damped resonant frequency, ω_d, is given by

$$\omega_d = \sqrt{\frac{1}{LC} - \left(\frac{1}{2RC}\right)^2} = \sqrt{\frac{1}{10^{-5} \times 10^{-9}} - \left(\frac{1}{2 \times 100 \times 10^{-9}}\right)^2} = 8.66 \times 10^6$$

The plot indicates that the plot has a maxima at 378 ns and a minima at 731 ns. Therefore, the period of the damped oscillation is

$$T_d = 2(731 \times 10^{-9} - 378 \times 10^{-9}) = 363 \times 10^{-9}$$

The damped resonant frequency, ω_d, is related to T_d by Eq. 9.7-9. Therefore

$$\omega_d = \frac{2\pi}{T_d} = \frac{2\pi}{363 \times 10^{-9}} = 8.65 \times 10^6$$

The value of ω_d obtained from the plot agrees with the value obtained from the circuit.

We conclude that the plot is correct.

9.14 Design Challenge Solution

AUTO AIRBAG IGNITER

Problem

Airbags are now widely used for driver protection in automobiles. A pendulum is used to switch a charged capacitor to the inflation ignition device, as shown in Figure 9.14-1. The automobile airbag is inflated by an explosive device that is ignited by the energy absorbed by the resistive device represented by R. In order to inflate, it is required that the energy dissipated in R is at least 1 J. It is required that the ignition device trigger within 0.1 s. Select the L and C that meet the specifications.

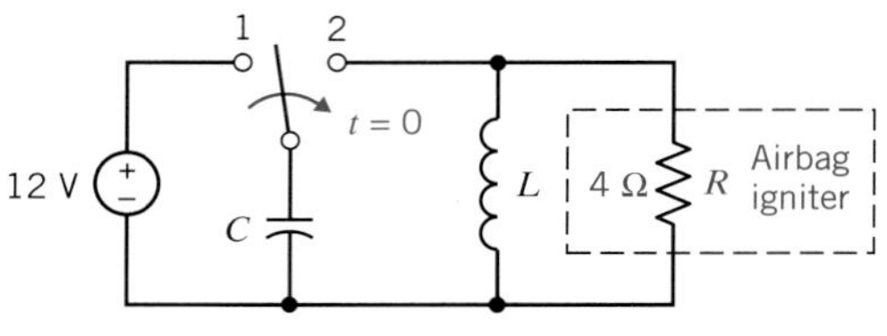

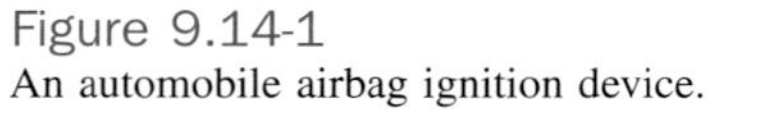
Figure 9.14-1
An automobile airbag ignition device.

Describe the Situation

1. The switch is changed from position 1 to position 2 at $t = 0$.
2. The switch was connected to position 1 for a long time.
3. A parallel *RLC* circuit occurs for $t \geq 0$.

State the Goal

Select L and C so that the energy stored in the capacitor is quickly delivered to the resistive device R.

Generate a Plan

1. Select L and C so that an underdamped response is obtained with a period of less than or equal to 0.4 s ($T \leq 0.4$ s).
2. Solve for $v(t)$ and $i(t)$ for the resistor R.

Act on the Plan

We assume that the initial capacitor voltage is $v(0) = 12$ V and $i_L(0) = 0$ since the switch is in position 1 for a long time prior to $t = 0$. The response of the parallel *RLC* circuit for an underdamped response is of the form (Eq. 9.7-5)

$$v(t) = e^{-\alpha t}(B_1 \cos \omega_d t + B_2 \sin \omega_d t) \tag{9.14-1}$$

as discussed in Section 9.7. This natural response is obtained when $\alpha^2 < \omega_0^2$ or $L < 4R^2C$. We choose an underdamped response for our design but recognize that an overdamped or critically damped response may satisfy the circuit's design objectives. Furthermore, we recognize that the parameter values selected below represent only one acceptable solution.

Since we want a rapid response, we will select $\alpha = 2$ (a time constant of 1/2 s) where $\alpha = 1/2RC$. Therefore, we have

$$C = \frac{1}{2R\alpha} = \frac{1}{16}\text{ F}$$

Recall that $\omega_0^2 = 1/LC$ and it is required that $\alpha^2 < \omega_0^2$. Since we want a rapid response, we select the natural frequency ω_0 so that (recall $T \approx 0.4$ s)

$$\omega_0 = \frac{2\pi}{T} = \frac{2\pi}{0.4} = 5\pi$$

Therefore, we obtain

$$L = \frac{1}{\omega_0^2 C} = \frac{1}{25\pi^2(1/16)} = 0.065\text{ H}$$

Thus, we will use $C = 1/16$ F and $L = 65$ mH. We then find that $\omega_d = 15.58$ rad/s and, using Eq. 9.7-5, we have

$$v(t) = e^{-2t}(B_1 \cos \omega_d t + B_2 \sin \omega_d t) \tag{9.14-2}$$

Then $B_1 = v(0) = 12$ and

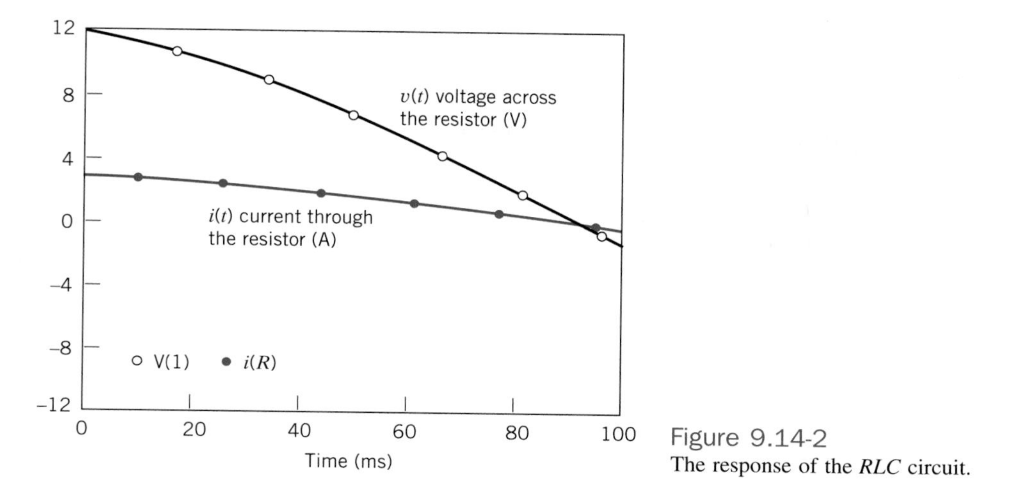

Figure 9.14-2
The response of the *RLC* circuit.

$$\omega_d B_2 = \alpha B_1 - \frac{B_1}{RC} = (2 - 4)12 = -24$$

Therefore, $B_2 = -24/15.58 = -1.54$. Since $B_2 \ll B_1$, we can approximate Eq. 9.14-2 as

$$v(t) \cong 12e^{-2t} \cos \omega_d t \text{ V}$$

The power is then

$$p = \frac{v^2}{R} = 36e^{-4t} \cos^2 \omega_d t \text{ W}$$

Verify the Proposed Solution

The actual voltage and current for the resistor R are shown in Figure 9.14-2 for the first 100 ms. If we sketch the product of v and i for the first 100 ms, we obtain a linear approximation declining from 36 W at $t = 0$ to 0 W at $t = 95$ ms. The energy absorbed by the resistor over the first 100 ms is then

$$w \cong \tfrac{1}{2}(36)(0.1 \text{ s}) = 1.8 \text{ J}$$

Therefore, the airbag will trigger in less than 0.1 s and our objective is achieved.

9.15 SUMMARY

- Second-order circuits are circuits that are represented by a second order differential equation, for example

$$\frac{d^2}{dt^2}x(t) + 2\alpha\frac{d}{dt}x(t) + \omega_o^2 x(t) = f(t)$$

where $x(t)$ is the output current or voltage of the circuit and $f(t)$ is the input to the circuit. The output of the circuit, also called the response of the circuit, can be the current or voltage of any device in the circuit. The output is frequently chosen to be the current of an inductor or the voltage of a capacitor. The input to the circuit is provided by the voltages of independent voltage sources and/or currents of independent current sources. The coefficients of this differential equation have names: α is called the damping coefficient and ω_o is called the resonant frequency.

- Obtaining the differential equation to represent an arbitrary circuit can be challenging. This chapter presents three methods for obtaining that differential equation: the direct method (Section 9.3), the operator method (Section 9.3), and the state variable method (Section 9.10).
- The characteristic equation of a second-order circuit is

$$s^2 + 2\alpha s + \omega_o^2 = 0$$

This second order equation has two solutions, s_1 and s_2. These solutions are called the natural frequencies of the second-order circuit.

- Second-order circuits are characterized as overdamped, critically damped, or underdamped. A second-order circuit is overdamped when s_1 and s_2 are real and unequal, or, equivalently, $\alpha > \omega_0$. A second-order circuit is critically damped when s_1 and s_2 are real and equal, or, equivalently, $\alpha = \omega_0$. A second-order circuit is underdamped when s_1 and s_2 are complex conjugates, or, equivalently, $\alpha < \omega_0$.

• Table 9.15-1 describes the natural frequencies of overdamped, underdamped, and critically damped parallel and series *RLC* circuits.

• The complete response of a second-order circuit is the sum of the natural response and the forced response

$$x = x_n + x_f$$

The form of the natural response depends on the natural frequencies of the circuit as summarized in Table 9.15-2. The form of the forced response depends on the input to the circuit as summarized in Table 9.15-3.

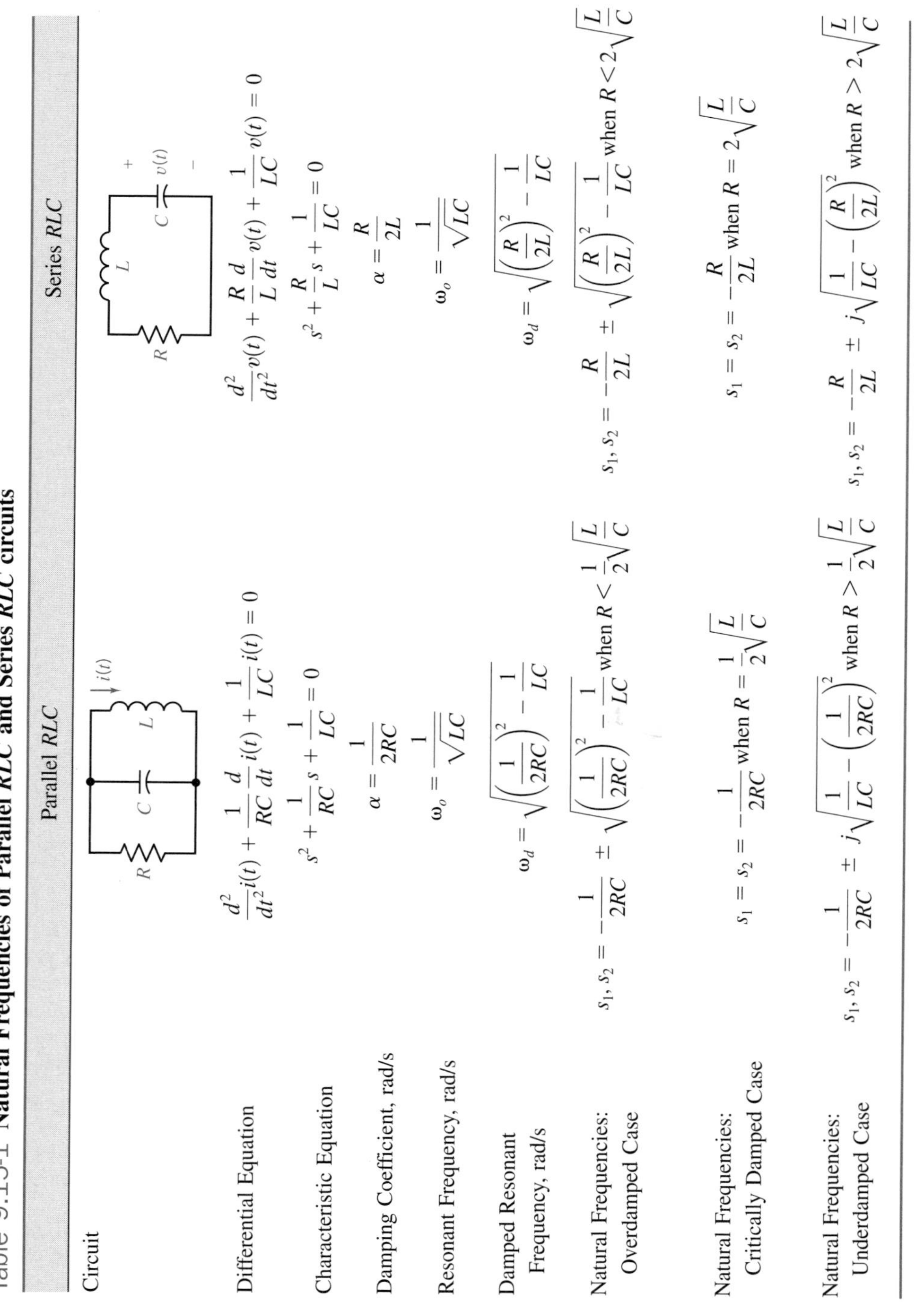

Table 9.15-1 **Natural Frequencies of Parallel *RLC* and Series *RLC* circuits**

	Parallel *RLC*	Series *RLC*
Circuit	(circuit: R, C, L in parallel; $i(t)$)	(circuit: R, L, C in series; $+\ v(t)\ -$)
Differential Equation	$\frac{d^2}{dt^2}i(t) + \frac{1}{RC}\frac{d}{dt}i(t) + \frac{1}{LC}i(t) = 0$	$\frac{d^2}{dt^2}v(t) + \frac{R}{L}\frac{d}{dt}v(t) + \frac{1}{LC}v(t) = 0$
Characteristic Equation	$s^2 + \frac{1}{RC}s + \frac{1}{LC} = 0$	$s^2 + \frac{R}{L}s + \frac{1}{LC} = 0$
Damping Coefficient, rad/s	$\alpha = \frac{1}{2RC}$	$\alpha = \frac{R}{2L}$
Resonant Frequency, rad/s	$\omega_o = \frac{1}{\sqrt{LC}}$	$\omega_o = \frac{1}{\sqrt{LC}}$
Damped Resonant Frequency, rad/s	$\omega_d = \sqrt{\left(\frac{1}{2RC}\right)^2 - \frac{1}{LC}}$	$\omega_d = \sqrt{\left(\frac{R}{2L}\right)^2 - \frac{1}{LC}}$
Natural Frequencies: Overdamped Case	$s_1, s_2 = -\frac{1}{2RC} \pm \sqrt{\left(\frac{1}{2RC}\right)^2 - \frac{1}{LC}}$ when $R < \frac{1}{2}\sqrt{\frac{L}{C}}$	$s_1, s_2 = -\frac{R}{2L} \pm \sqrt{\left(\frac{R}{2L}\right)^2 - \frac{1}{LC}}$ when $R < 2\sqrt{\frac{L}{C}}$
Natural Frequencies: Critically Damped Case	$s_1 = s_2 = -\frac{1}{2RC}$ when $R = \frac{1}{2}\sqrt{\frac{L}{C}}$	$s_1 = s_2 = -\frac{R}{2L}$ when $R = 2\sqrt{\frac{L}{C}}$
Natural Frequencies: Underdamped Case	$s_1, s_2 = -\frac{1}{2RC} \pm j\sqrt{\frac{1}{LC} - \left(\frac{1}{2RC}\right)^2}$ when $R > \frac{1}{2}\sqrt{\frac{L}{C}}$	$s_1, s_2 = -\frac{R}{2L} \pm j\sqrt{\frac{1}{LC} - \left(\frac{R}{2L}\right)^2}$ when $R > 2\sqrt{\frac{L}{C}}$

Table 9.15-2 **Natural Response of Second-Order Circuits**

Case	Natural Frequencies	Natural Response, x_n
Overdamped	$s_1, s_2 = -\alpha \pm \sqrt{\alpha^2 - \omega_o^2}$	$A_1 e^{-s_1 t} + A_2 e^{-s_2 t}$
Critically Damped	$s_1, s_2 = -\alpha$	$(A_1 + A_2 t)e^{-\alpha t}$
Underdamped	$s_1, s_2 = -\alpha \pm j\sqrt{\omega_o^2 - \alpha^2} = -\alpha \pm j\omega_d$	$(A_1 \cos \omega_d t + A_2 \sin \omega_d t)e^{-\alpha t}$

Table 9.15-3 **Forced Response of Second-Order Circuits**

	Input, $f(t)$	Forced Response, x_f
Constant	K	A
Ramp	$K\,t$	$A + B\,t$
Sinusoid	$K \cos \omega t$, $K \sin \omega t$, or $K \cos (\omega + \theta)$	$A \cos \omega t + B \sin \omega t$
Exponential	$K\,e^{-bt}$	Ae^{-bt}

PROBLEMS

Section 9.3 Differential Equation for Circuits with Two Energy Storage Elements

P 9.3-1 Find the differential equation for the circuit shown in Figure P 9.3-1 using the direct method.

P 9.3-2 Find the differential equation for the circuit shown in Figure P 9.3-2 using the operator method.

Answer: $$\frac{d^2}{dt^2} i_L(t) + 11000 \frac{d}{dt} i_L(t) + 1.1 \times 10^8 i_L(t) = 10^8 i_s(t)$$

P 9.3-3 Find the differential equation for $i_L(t)$ for $t > 0$ for the circuit of Figure P 9.3-3.

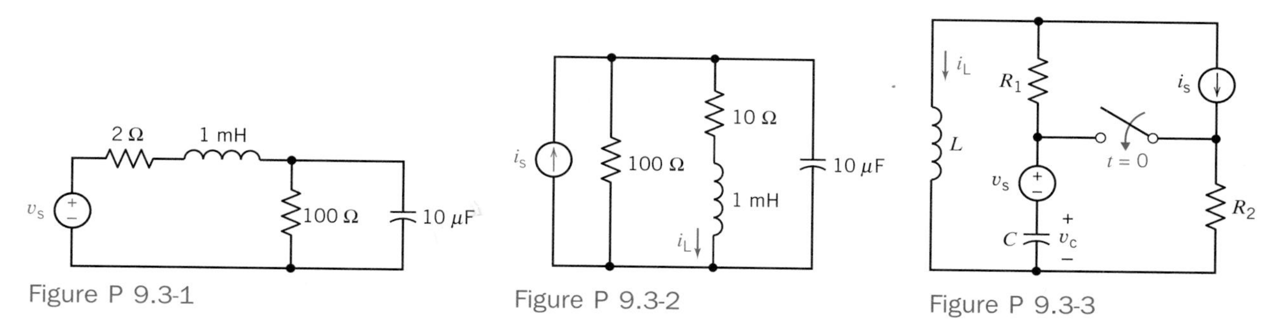

Figure P 9.3-1

Figure P 9.3-2

Figure P 9.3-3

Section 9.4 Solution of the Second-Order Differential Equation—The Natural Response

P 9.4-1 Find the characteristic equation and its roots for the circuit of Figure P 9.3-2.

P 9.4-2 Find the characteristic equation and its roots for the circuit of Figure P 9.4-2.

Answer: $s^2 + 400s + 3 \times 10^4 = 0$

roots: $s = -300, -100$

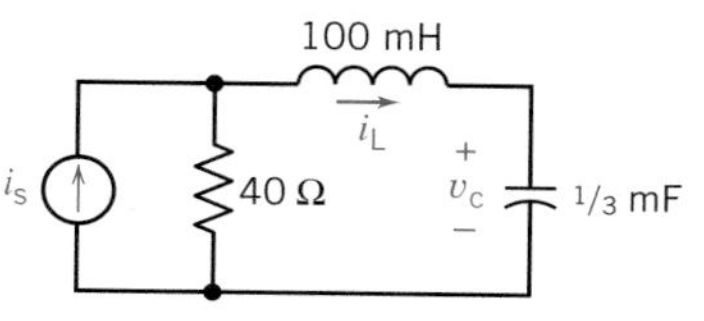

Figure P 9.4-2

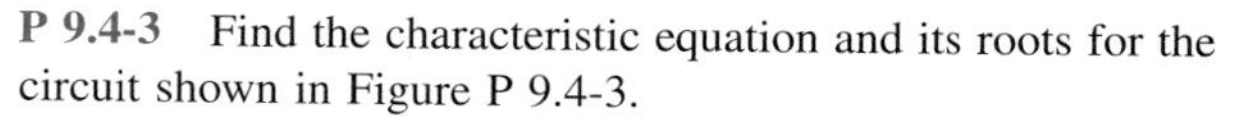

P 9.4-3 Find the characteristic equation and its roots for the circuit shown in Figure P 9.4-3.

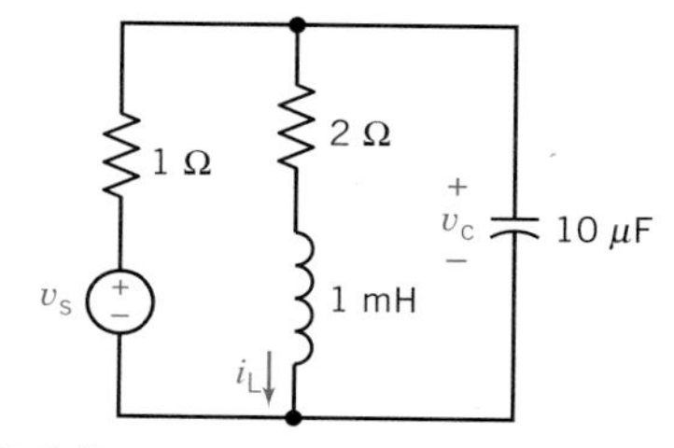

Figure P 9.4-3

Section 9.5 Natural Response of the Unforced Parallel *RLC* Circuit

P 9.5-1 Determine $v(t)$ for the circuit of Figure P 9.5-1 when $L = 1$ H and $v_s = 0$ for $t \geq 0$. The initial conditions are $v(0) = 6$ V and $dv/dt(0) = -3000$ V/s.
Answer: $v(t) = -2e^{-100t} + 8e^{-400t}$ V

P 9.5-2 An *RLC* circuit is shown in Figure P 9.5-2 where $v(0) = 2$ V. The switch has been open for a long time before closing at $t = 0$. Determine and plot $v(t)$.

P 9.5-3 Determine $i_1(t)$ and $i_2(t)$ for the circuit of Figure P 9.5-3 when $i_1(0) = i_2(0) = 11$ A.

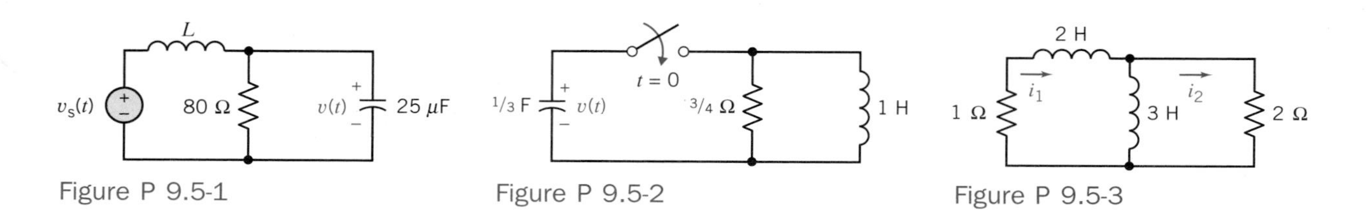

Figure P 9.5-1

Figure P 9.5-2

Figure P 9.5-3

Section 9.6 Natural Response of the Critically Damped Unforced Parallel *RLC* Circuit

P 9.6-1 Find $v_c(t)$ for $t > 0$ for the circuit shown in Figure P 9.6-1.
Answer: $v_c(t) = (3 + 6000t)e^{-2000t}$ V

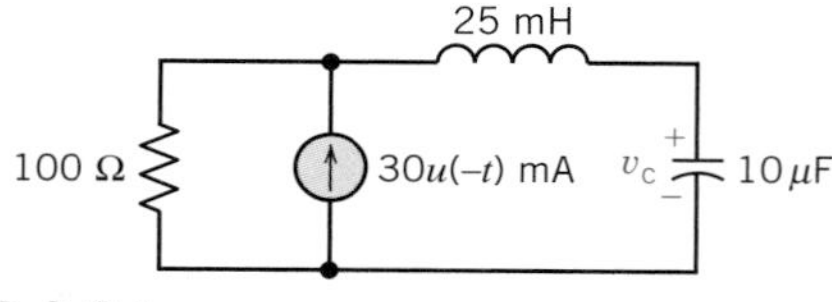

Figure P 9.6-1

P 9.6-2 Find $v_c(t)$ for $t > 0$ for the circuit of Figure P 9.6-2. Assume steady-state conditions exist at $t = 0^-$.
Answer: $v_c(t) = -8te^{-2t}$ V

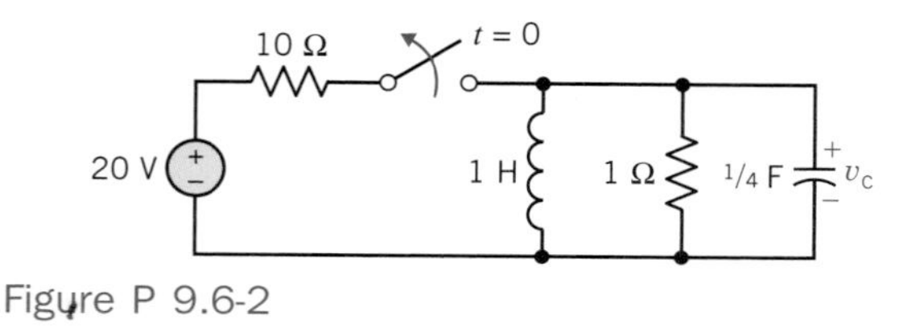

Figure P 9.6-2

P 9.6-3 Police often use stun guns to incapacitate potentially dangerous felons. The hand-held device provides a series of high-voltage, low-current pulses. The power of the pulses is far below lethal levels, but it is enough to cause muscles to contract and put the person out of action. The device provides a pulse of up to 50,000 V, and a current of 1 mA flows through an arc. A model of the circuit for one period is shown in Figure P 9.6-3. Find $v(t)$ for $0 < t < 1$ ms. The resistor R represents the spark gap. Select C so that the response is critically damped.

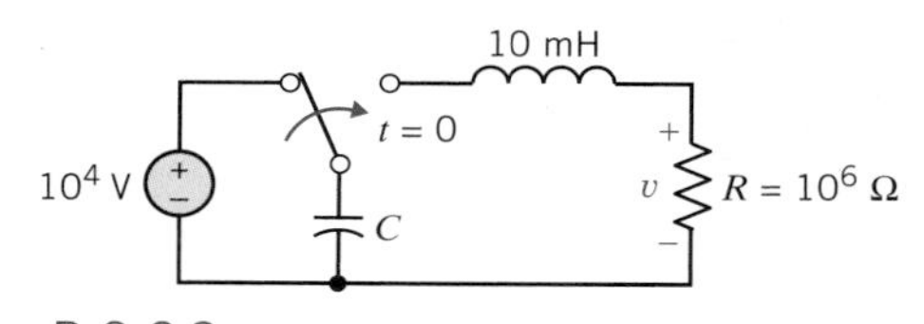

Figure P 9.6-3

P 9.6-4 Reconsider Problem 9.5-1 when $L = 640$ mH and the other parameters and conditions remain the same.
Answer: $v(t) = (6 - 1500t)e^{-250t}$ V

P 9.6-5 An automobile ignition uses an electromagnetic trigger. The *RLC* trigger circuit shown in Figure P 9.6-5 has a step input of 6 V, and $v(0) = 2$ V and $i(0) = 0$. The resistance R must be selected from $2\ \Omega < R < 7\ \Omega$ so that the current $i(t)$ exceeds 0.6 A for greater than 0.5 s in order to activate the trigger. A critically damped response $i(t)$ is required to avoid oscillations in the trigger current. Select R and determine and plot $i(t)$.

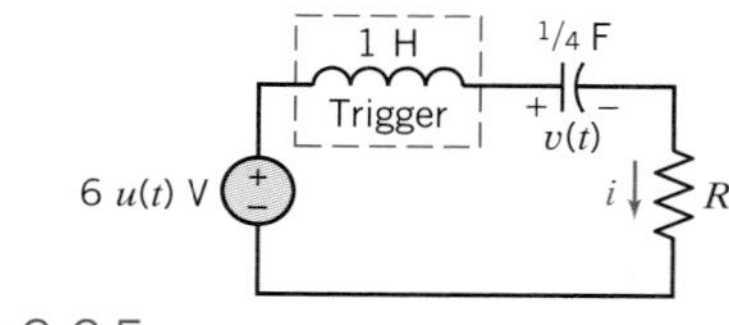

Figure P 9.6-5

Section 9.7 Natural Response of an Underdamped Unforced Parallel *RLC* Circuit

P 9.7-1 A communication system from a space station uses short pulses to control a robot operating in space. The transmitter circuit is modeled in Figure P 9.7-1. Find the output voltage $v_c(t)$ for $t > 0$. Assume steady-state conditions at $t = 0^-$.
Answer: $v_c(t) = e^{-400t}[3 \cos 300t + 4 \sin 300t]$ V

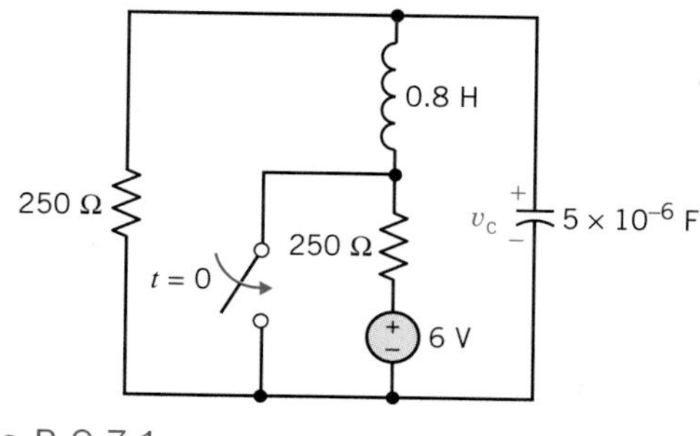

Figure P 9.7-1

P 9.7-2 The switch of the circuit shown in Figure P 9.7-2 is opened at $t = 0$. Determine and plot $v(t)$ when $C = 1/4$ F. Assume steady state at $t = 0^-$.
Answer: $v(t) = -4e^{-2t} \sin 2t$ V

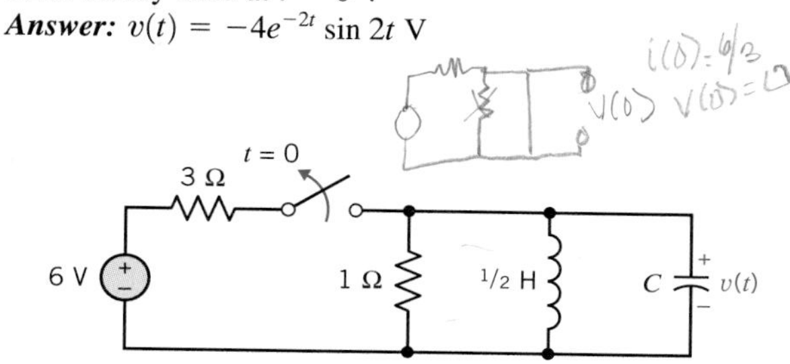

Figure P 9.7-2

P 9.7-3 A 240-W power supply circuit is shown in Figure P 9.7-3*a*. This circuit employs a large inductor and a large capacitor. The model of the circuit is shown in Figure P 9.7-3*b*. Find $i_L(t)$ for $t > 0$ for the circuit of Figure P 9.7-3*b*. Assume steady-state conditions exist at $t = 0^-$.
Answer: $i_L(t) = e^{-2t}(-4 \cos t + 2 \sin t)$ A

(*a*)

(*b*)

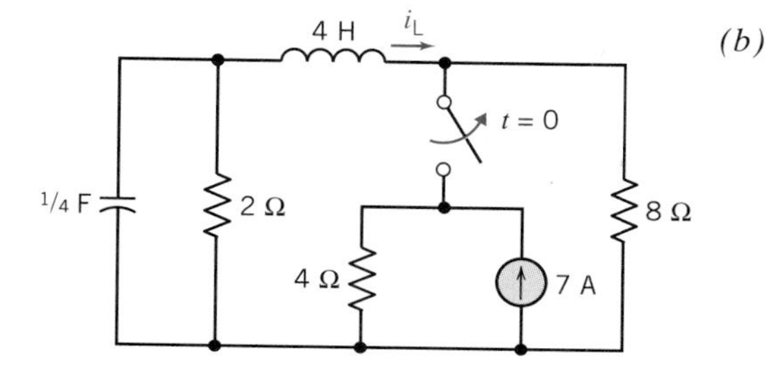

Figure P 9.7-3 (*a*) A 240-W power supply. Courtesy of Kepco, Inc. (*b*) Model of the power supply circuit.

P 9.7-4 The natural response of a parallel *RLC* circuit is measured and plotted as shown in Figure P 9.7-4. Using this chart, determine an expression for $v(t)$.
Hint: Notice that $v(t) = 260$ mV at $t = 5$ ms and that $v(t) = -200$ mV at $t = 7.5$ ms. Also, notice that the time between the first and third zero crossings is 5 ms.
Answer: $v(t) = 544e^{-276t} \sin 1257t$ V

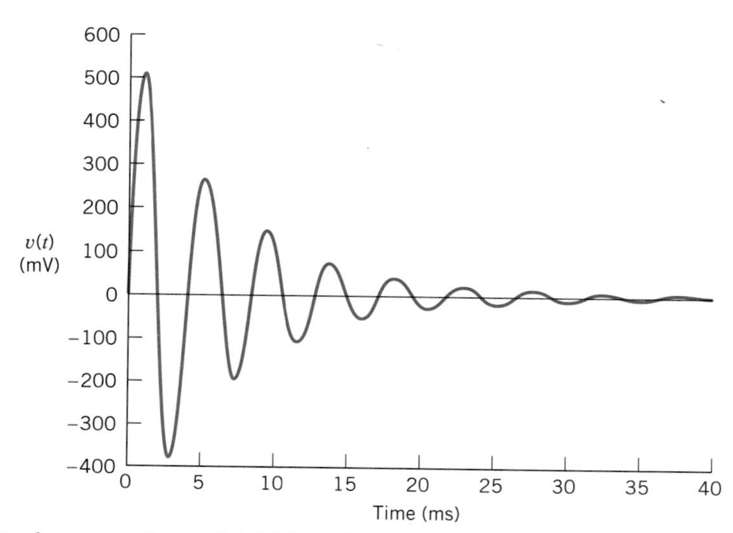

Figure P 9.7-4 The natural response of a parallel *RLC* circuit.

P 9.7-5 The photovoltaic cells of the proposed space station shown in Figure P 9.7-5*a* provide the voltage $v(t)$ of the circuit shown in Figure P 9.7-5*b*. The space station passes behind the shadow of Earth (at $t = 0$) with $v(0) = 2$ V and $i(0) = 1/10$ A. Determine and sketch $v(t)$ for $t > 0$.

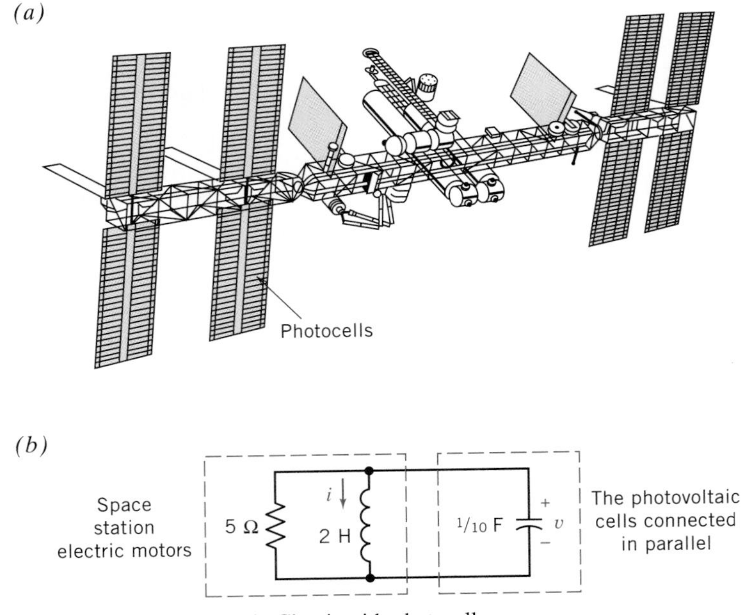

Figure P 9.7-5 *(a)* Photocells on space station. *(b)* Circuit with photocells.

Section 9.8 Forced Response of an *RLC* Circuit

P 9.8-1 Determine the forced response for the inductor current i_f when (a) $i_s = 1$ A, (b) $i_s = 0.5t$ A, and (c) $i_s = 2e^{-250t}$ A for the circuit of Figure P 9.8-1.

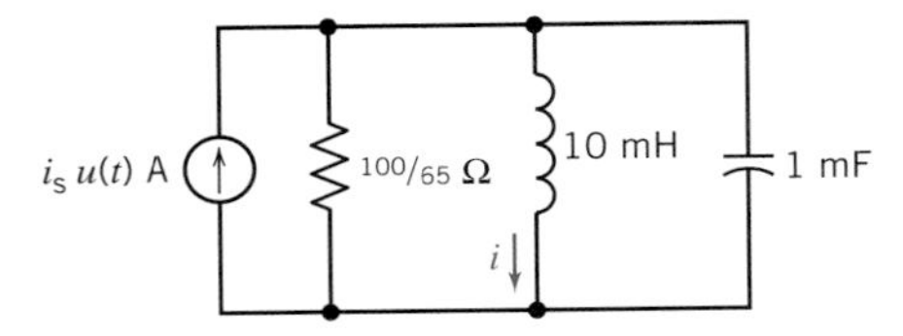

Figure P 9.8-1

P 9.8-2 Determine the forced response for the capacitor voltage, v_f, for the circuit of Figure P 9.8-2 when (a) $v_s = 2$ V, (b) $v_s = 0.2t$ V, and (c) $v_s = 1e^{-30t}$ V.

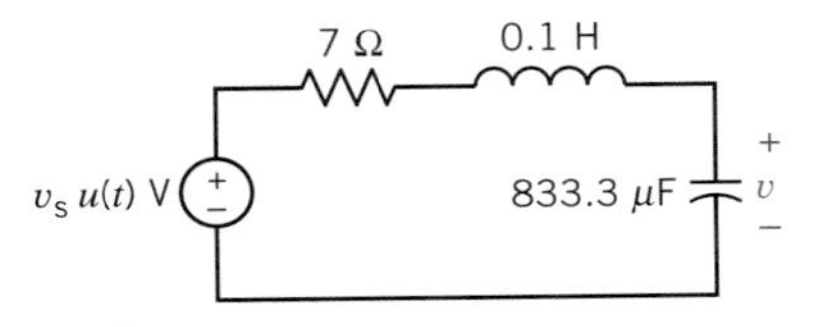

Figure P 9.8-2

Section 9.9 Complete Response of an *RLC* Circuit

P 9.9-1 Find $i(t)$ for $t > 0$ using the direct method for the circuit of Figure P 9.9-1.

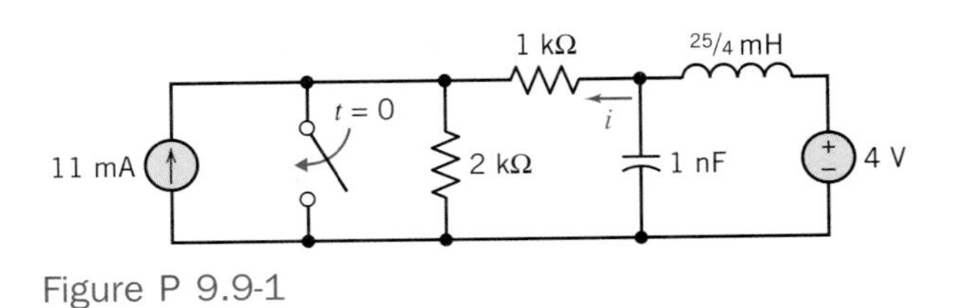

Figure P 9.9-1

P 9.9-2 Find $i_L(t)$ for $t > 0$ for the circuit shown in Figure P 9.9-2. Use the direct method.

Hint: Show that $1 = \frac{d^2}{dt^2}i_L(t) + 5\frac{d}{dt}i_L(t) + 5i_L(t)$ for $t > 0$.

Answer: $i_L(t) = 0.246e^{-3.62t} - 0.646e^{-1.38t} + 0.2$ A for $t > 0$.

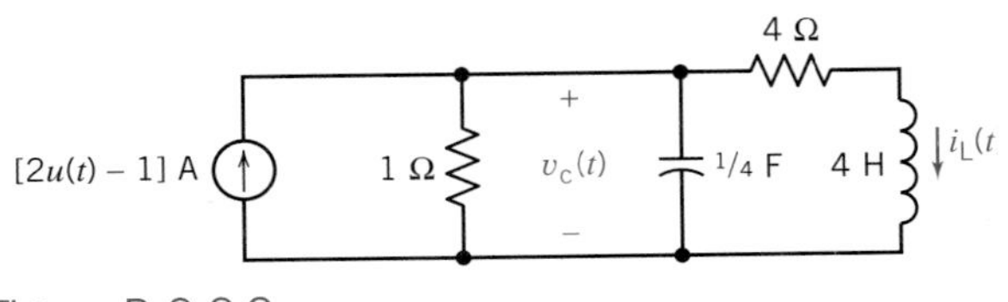

Figure P 9.9-2

P 9.9-3 Find $v_1(t)$ for $t > 0$ for the circuit shown in Figure P 9.9-3. Assume steady-state conditions at $t = 0^-$.
Answer: $v_1(t) = -12e^{-4\times10^3 t} + 2e^{-2.4\times10^9 t} + 20$ V

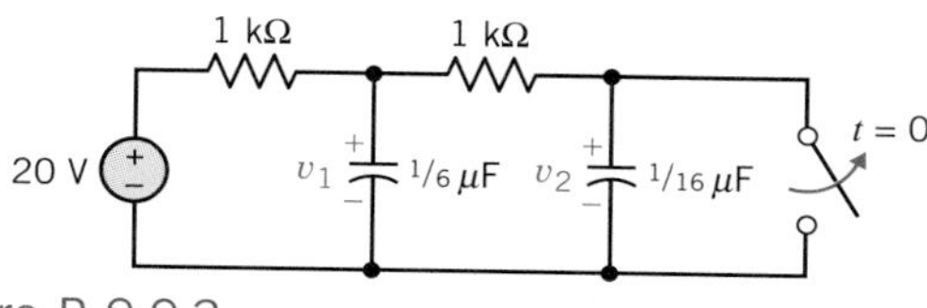

Figure P 9.9-3

P 9.9-4 Find $v(t)$ for $t > 0$ for the circuit shown in Figure P 9.9-4 when $v(0) = 1$ V and $i_L(0) = 0$.
Answer: $v = 25e^{-3t} - \frac{1}{17}[429e^{-4t} - 21\cos t + 33\sin t]$ V

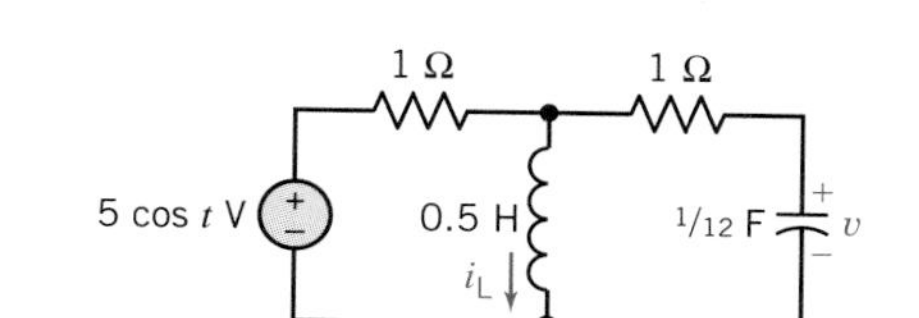

Figure P 9.9-4

P 9.9-5 Find $v(t)$ for $t > 0$ for the circuit of Figure P 9.9-5 using the method of operators.
Answer: $v(t) = [-16e^{-t} + 16e^{-3t} + 8]u(t) + [16e^{-(t-2)} - 16e^{-3(t-2)} - 8]u(t-2)$ V

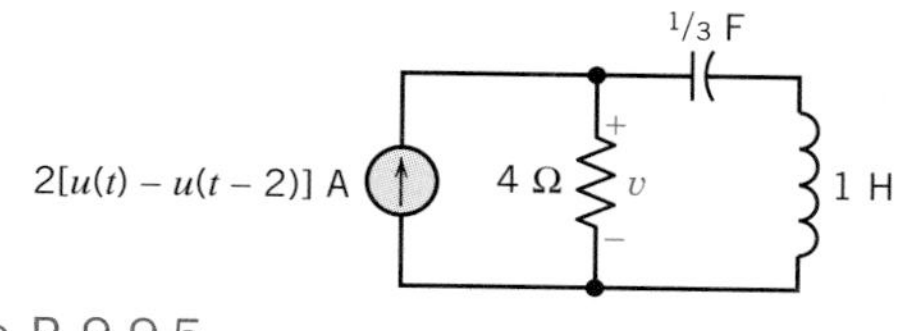

Figure P 9.9-5

P 9.9-6 An experimental space station power supply system is modeled by the circuit shown in Figure P 9.9-6. Find $v_c(t)$ for $t > 0$. Assume steady-state conditions at $t = 0^-$.

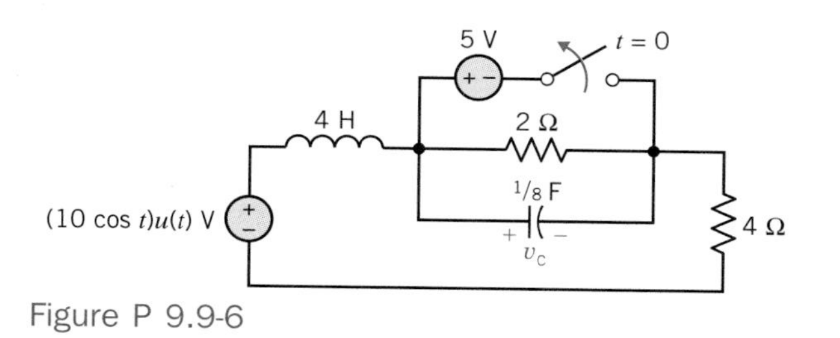

Figure P 9.9-6

P 9.9-7 Find $v_c(t)$ for $t > 0$ in the circuit of Figure P 9.9-7 when (a) $C = 1/18$ F, (b) $C = 1/10$ F, and (c) $C = 1/20$ F.
Answer: (a) $v_c(t) = 8e^{-3t} + 24te^{-3t} - 8$ V
(b) $v_c(t) = 10e^{-t} - 2e^{-5t} - 8$ V
(c) $v_c(t) = e^{-3t}(8\cos t + 24\sin t) - 8$ V

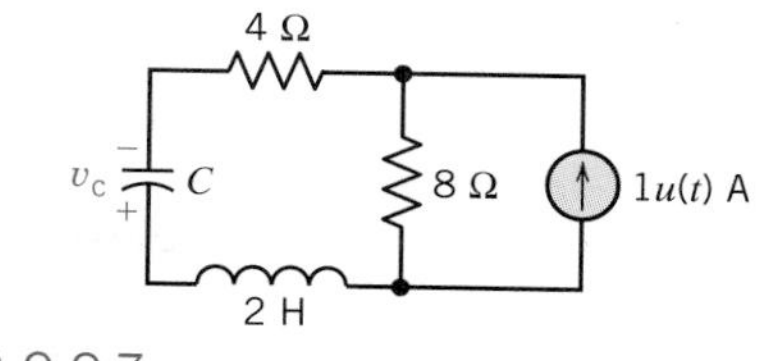

Figure P 9.9-7

P 9.9-8 Find $i_L(t)$ for $t > 0$ for the circuit shown in Figure P 9.9-8. Use the direct method.
Hint: Show that $2 = \frac{d^2}{dt^2}v_C(t) + 6\frac{d}{dt}v_C(t) + 2v_C(t)$ for $t > 0$.

Answer: $v_C(t) = 0.123e^{-5.65t} + 0.877e^{-0.35t} + 1$ V for $t > 0$.

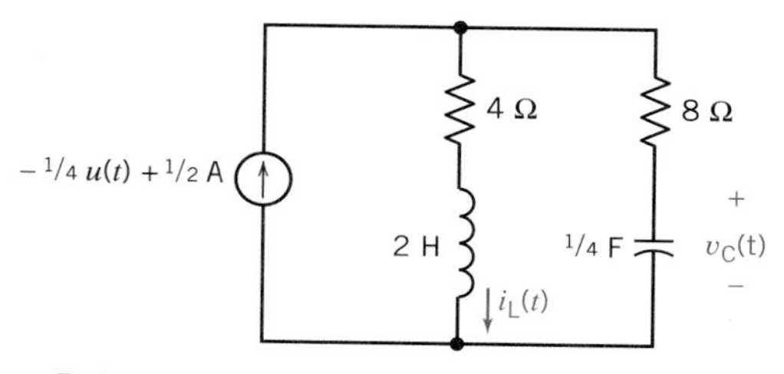

Figure P 9.9-8

P 9.9-9 In Figure P 9.9-9, determine the inductor current $i(t)$ when $i_s = 5u(t)$ A. Assume that $i(0) = 0$, $v_c(0) = 0$.
Answer: $i(t) = 5 + e^{-2t}[-5\cos 5t - 2\sin 5t]$ A

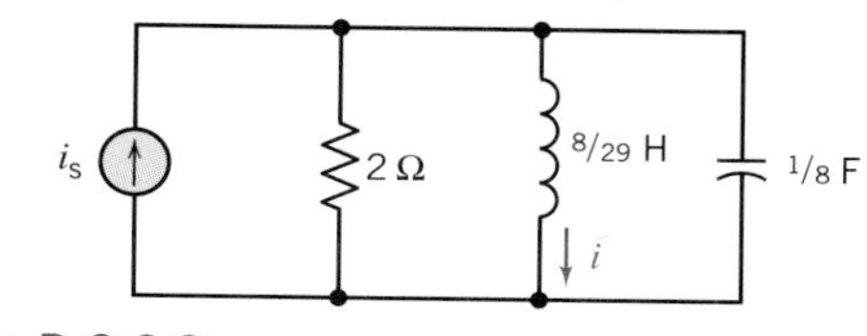

Figure P 9.9-9

P 9.9-10 Railroads widely use automatic identification of railcars. When a train passes a tracking station, a wheel detector activates a radio-frequency module. The module's antenna, as shown in Figure P 9.9-10*a*, transmits and receives a signal that bounces off a transponder on the locomotive. A trackside processor turns the received signal into useful information consisting of the train's location, speed, and direction of travel. The railroad uses this information to schedule locomotives, trains, crews, and equipment more efficiently.

One proposed transponder circuit is shown in Figure P 9.9-10*b* with a large transponder coil of $L = 5$ H. Determine $i(t)$ and $v(t)$. The received signal is $i_s = 9 + 3e^{-2t}u(t)$ A.

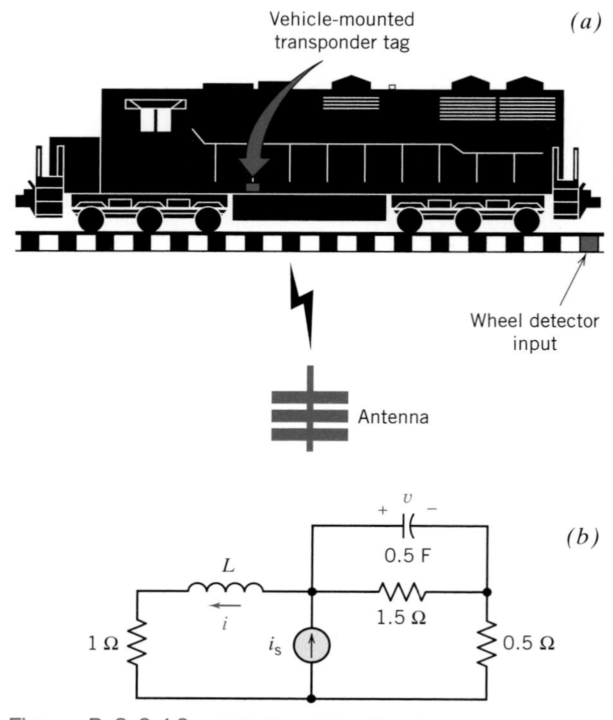

Figure P 9.9-10 (*a*) Railroad identification system. (*b*) Transponder circuit.

Section 9.10 State Variable Approach to Circuit Analysis

P 9.10-1 Find $v(t)$ for $t > 0$ using the state variable method of Section 9.10 when $C = 1/5$ F in the circuit of Figure P 9.10-1. Sketch the response for $v(t)$ for $0 < t < 10$ s.
Answer: $v(t) = -25e^{-t} + e^{-5t} + 24$ V

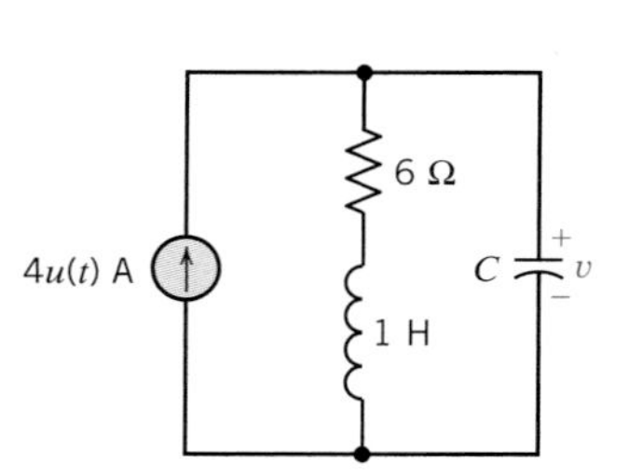

Figure P 9.10-1

P 9.10-2 Repeat Problem 9.10-1 when $C = 1/10$ F. Sketch the response for $v(t)$ for $0 < t < 3$ s.
Answer: $v(t) = e^{-3t}(-24 \cos t - 32 \sin t) + 24$ V

P 9.10-3 Determine the current $i(t)$ and the voltage $v(t)$ for the circuit of Figure P 9.10-3.
Answer: $i(t) = (3.08e^{-2.57t} - 0.08e^{-97.4t} - 6)$ A

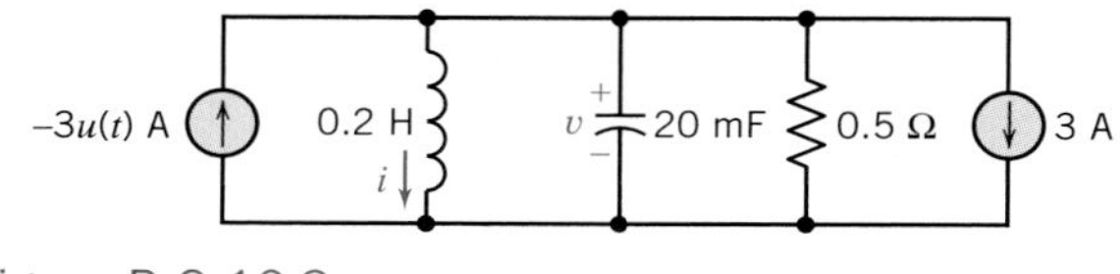

Figure P 9.10-3

P 9.10-4 Clean-air laws are pushing the auto industry toward the development of electric cars. One proposed vehicle using an ac motor is shown in Figure P 9.10-4*a*. The motor-controller circuit is shown in Figure P 9.10-4*b* with $L = 100$ mH and $C = 10$ mF. Using the state equation approach, determine $i(t)$ and $v(t)$ where $i(t)$ is the motor-control current. The initial conditions are $v(0) = 10$ V and $i(0) = 0$.

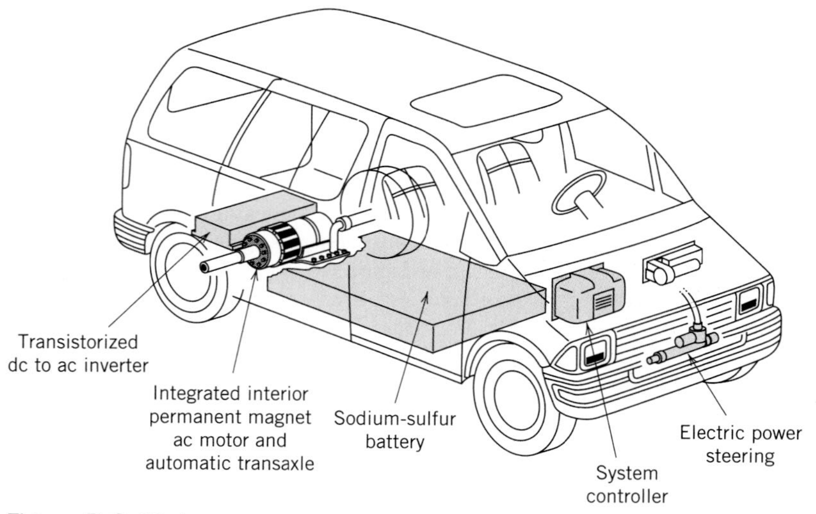

Figure P 9.10-4*a* Electric vehicle.

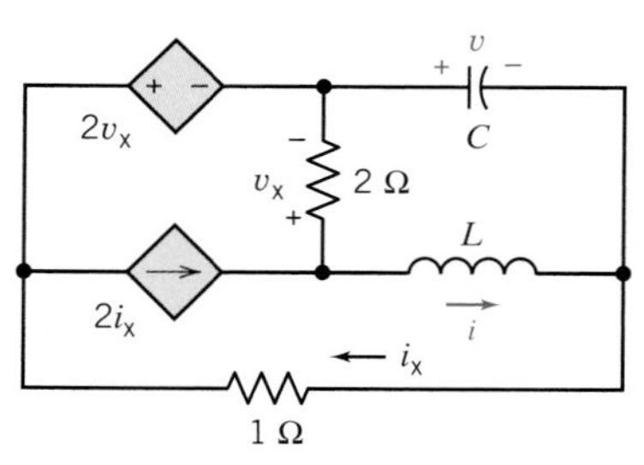

Figure P 9.10-4*b* Motor controller circuit.

P 9.10-5 Studies of an artificial insect are being used to understand the nervous system of animals (Beer 1991). A model neuron in the nervous system of the artificial insect is shown in Figure P 9.10-5. The input signal v_s, is used to generate a series of pulses, called synapses. The switch generates a pulse by opening at $t = 0$ and closing at $t = 0.5$ s. Assume that the circuit is in steady state and that $v(0^-) = 10$ V. Determine the voltage $v(t)$ for $0 < t < 2$ s.

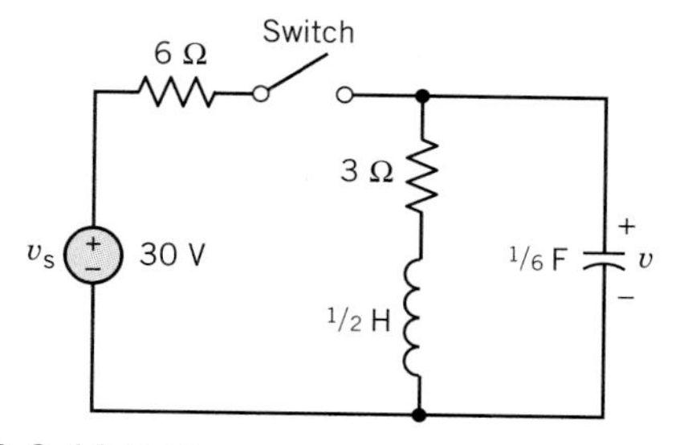

Figure P 9.10-5 Newton circuit model.

Section 9.11 Roots in the Complex Plane

P 9.11-1 For the circuit of Figure P 9.11-1, determine the roots of the characteristic equation and plot the roots on the s-plane.

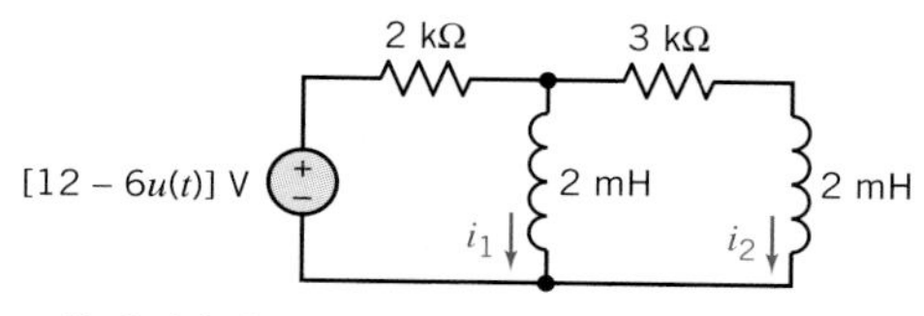

Figure P 9.11-1

P 9.11-2 For the circuit of Figure P 9.7-1 determine the roots of the characteristic equation and plot the roots on the s-plane.

P 9.11-3 For the circuit of Figure P 9.11-3 determine the roots of the characteristic equation and plot the roots on the s-plane.

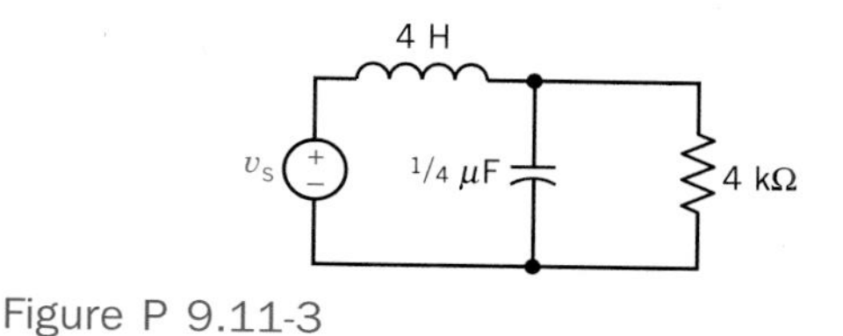

Figure P 9.11-3

P 9.11-4 An *RLC* circuit is shown in Figure P 9.11-4.
(a) Obtain the two-node voltage equations using operators.
(b) Obtain the characteristic equation for the circuit.
(c) Show the location of the roots of the characteristic equation in the s-plane.
(d) Determine $v(t)$ for $t > 0$.

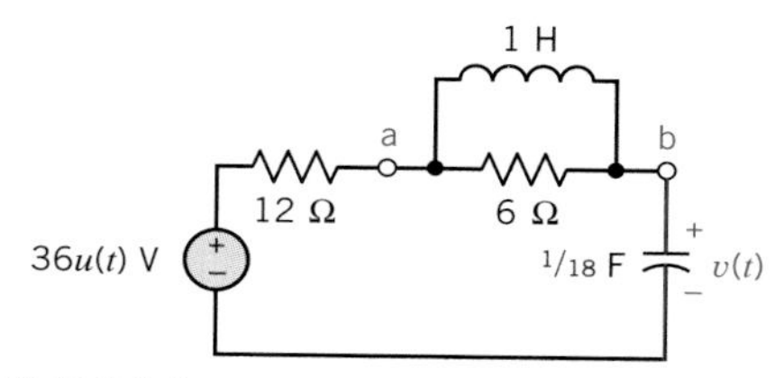

Figure P 9.11-4

PSPICE PROBLEMS

SP 9-1 Determine and plot $v(t)$ for Example 9.5-1.

SP 9-2 A simple amplifier is represented by the circuit model shown in Figure SP 9.2. Determine $v(t)$ when $v_s = 5u(t)$.

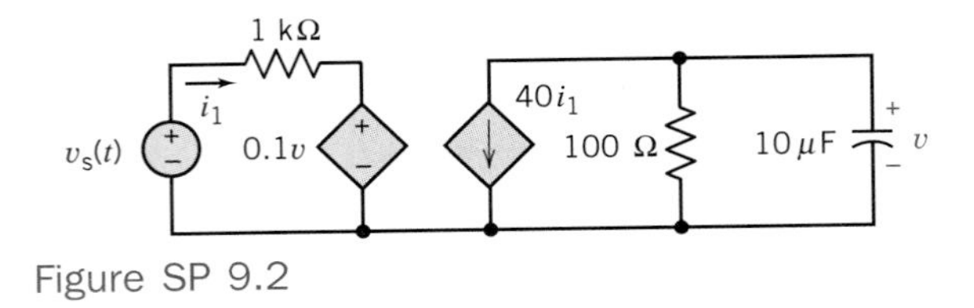

Figure SP 9.2

SP 9-3 Determine and plot $v(t)$ for Problem P 9.7-3.

SP 9-4 Determine and plot the natural response $v(t)$ of the circuit of Figure SP 9.4 when $i_s = 3e^{-6t}$ A.

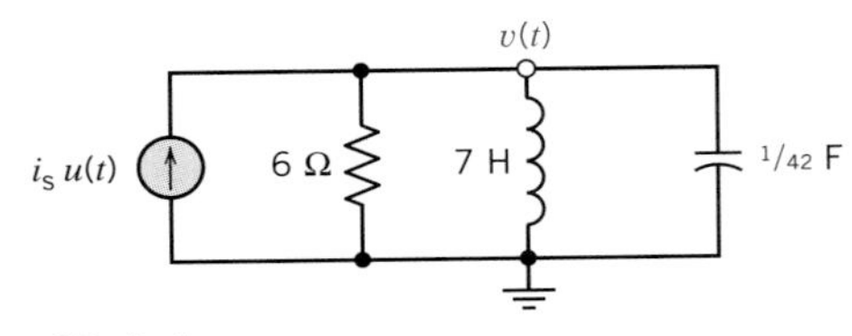

Figure SP 9.4

SP 9-5 Determine and plot the capacitor voltage $v(t)$ for $0 < t < 300$ μs for the circuit shown in Figure SP 9.5*a*. The sources are pulses as shown in Figure SP 9.5*b* and SP 9.5*c*.

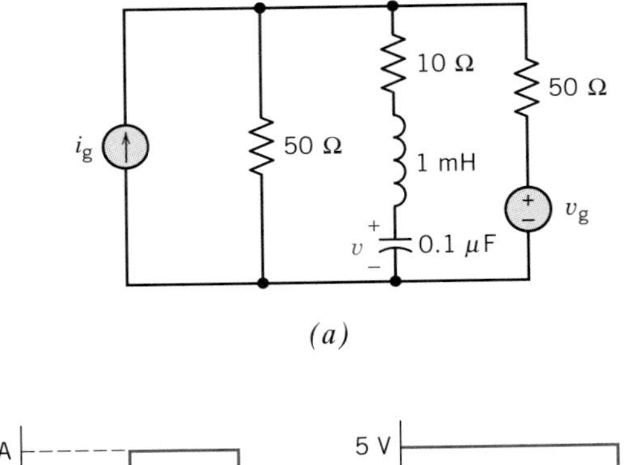

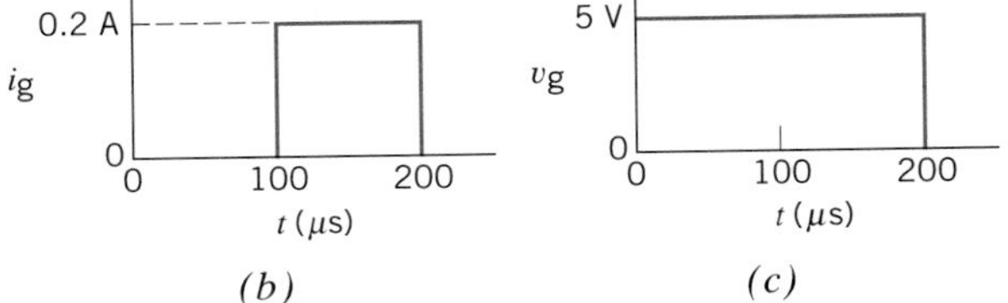

Figure SP 9.5 (*a*) Circuit, (*b*) current pulse, and (*c*) voltage pulse.

SP 9-6 Determine and plot $v(t)$ for the circuit of Figure SP 9.6 when $v_s(t) = 5u(t)$ V. Plot $v(t)$ for $0 < t < 0.25$ s.

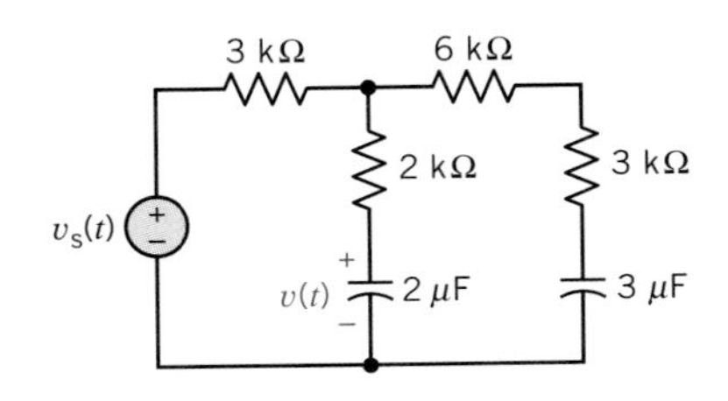

Figure SP 9.6

VERIFICATION PROBLEMS

VP 9-1 Figure VP 9.1*b* shows an *RLC* circuit. The voltage, $v_s(t)$, of the voltage source is the square wave shown in Figure VP 9.1*a*. Figure VP 9.1*c* shows a plot of the inductor current, $i(t)$, which was obtained by simulating this circuit using PSpice. Verify that the plot of $i(t)$ is correct.

Answer: The plot is correct.

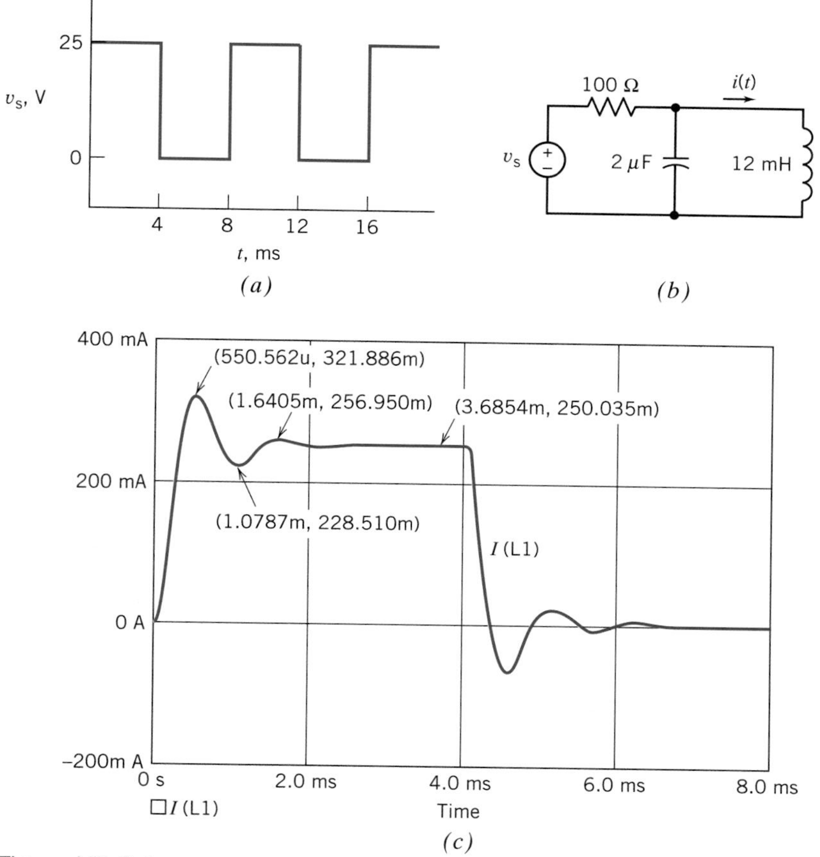

Figure VP 9.1

VP 9-2 Figure VP 9.2*b* shows an *RLC* circuit. The voltage, $v_s(t)$, of the voltage source is the square wave shown in Figure VP 9.2*a*. Figure VP 9.2*c* shows a plot of the inductor current, $i(t)$, which was obtained by simulating this circuit using PSpice. Verify that the plot of $i(t)$ is correct.
Answer: The plot is not correct.

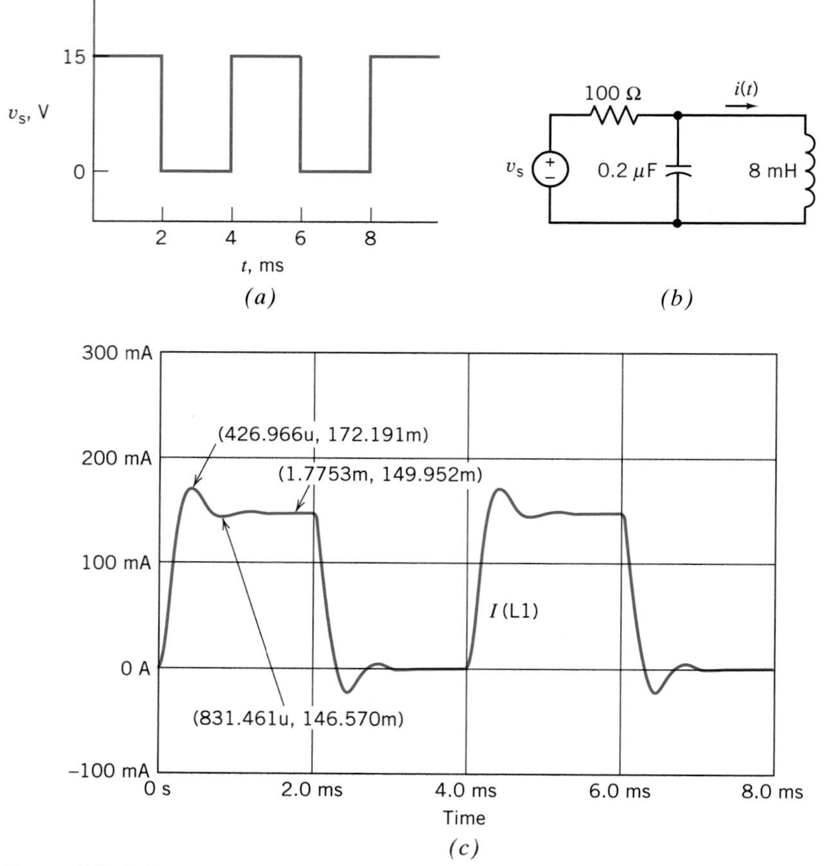

Figure VP 9.2

DESIGN PROBLEMS

DP 9-1 Design the circuit shown in Figure DP 9.1 so that

$$v_c(t) = \frac{1}{2} + A_1e^{-2t} + A_2e^{-4t}\text{ V} \quad \text{for } t > 0$$

Determine the values of the unspecified constants, A_1 and A_2. ***Hint:*** The circuit is overdamped and the natural frequencies are 2 and 4 rad/sec.

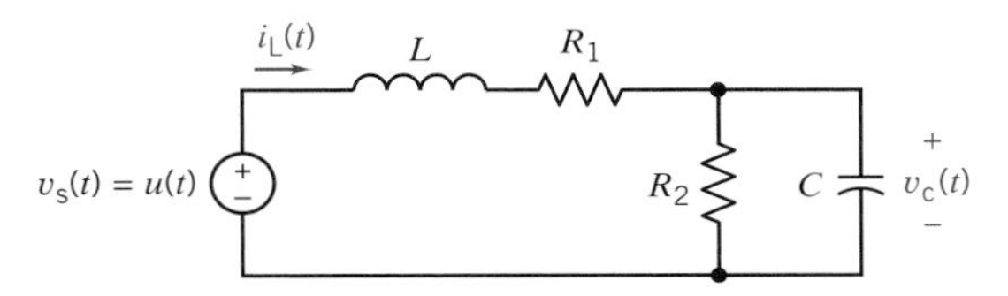

Figure DP 9.1

DP 9-2 Design the circuit shown in Figure DP 9.1 so that

$$v_c(t) = \frac{1}{4} + (A_1 + A_2t)e^{-4t}\text{ V} \quad \text{for } t > 0$$

Determine the values of the unspecified constants, A_1 and A_2. ***Hint:*** The circuit is critically damped and the natural frequencies are both 2 rad/sec.

DP 9-3 Design the circuit shown in Figure DP 9.1 so that

$$v_c(t) = 0.8 + e^{-2t}(A_1 \cos 4t + A_2 \sin 4t)\text{ V} \quad \text{for } t > 0$$

Determine the values of the unspecified constants, A_1 and A_2. ***Hint:*** The circuit is underdamped and the damped resonant frequency is 4 rad/sec and the damping coefficient is 2.

DP 9-4 Show that the circuit shown in Figure DP 9.1 cannot be designed so that

$$v_c(t) = 0.5 + e^{-2t}(A_1 \cos 4t + A_2 \sin 4t)\text{ V} \quad \text{for } t > 0$$

Hint: Show that such a design would require $\frac{1}{RC} + 10RC = 4$ where $R = R_1 = R_2$. Next, show that $\frac{1}{RC} + 10RC = 4$ would require the value of RC to be complex.

DP 9-5 Design the circuit shown in Figure DP 9.5 so that

$$v_o(t) = \frac{1}{2} + A_1e^{-2t} + A_2e^{-4t}\text{ V} \quad \text{for } t > 0$$

Determine the values of the unspecified constants, A_1 and A_2. ***Hint:*** The circuit is overdamped and the natural frequencies are 2 and 4 rad/sec.

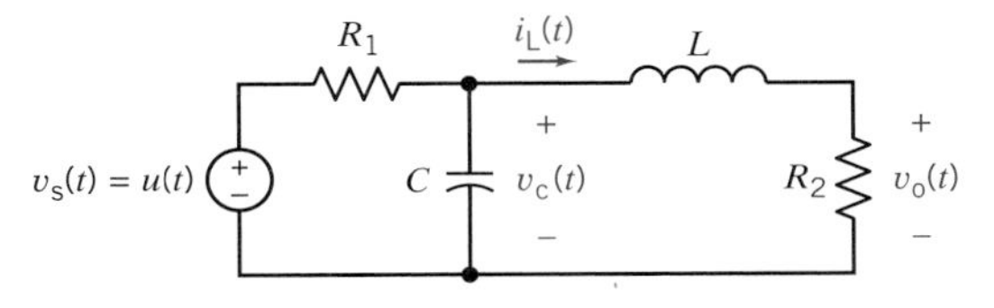

Figure DP 9.5

DP 9-6 Design the circuit shown in Figure DP 9.5 so that

$$v_o(t) = \frac{3}{4} + (A_1 + A_2t)e^{-4t}\text{ V} \quad \text{for } t > 0$$

Determine the values of the unspecified constants, A_1 and A_2. ***Hint:*** The circuit is critically damped and the natural frequencies are both 2 rad/sec.

DP 9-7 Design the circuit shown in Figure DP 9.5 so that

$$v_c(t) = 0.2 + e^{-2t}(A_1 \cos 4t + A_2 \sin 4t)\text{ V} \quad \text{for } t > 0$$

Determine the values of the unspecified constants, A_1 and A_2. ***Hint:*** The circuit is underdamped and the damped resonant frequency is 4 rad/sec and the damping coefficient is 2.

DP 9-8 Show that the circuit shown in Figure DP 9.5 cannot be designed so that

$$v_c(t) = 0.5 + e^{-2t}(A_1 \cos 4t + A_2 \sin 4t)\text{ V} \quad \text{for } t > 0$$

Hint: Show that such a design would require $\frac{1}{RC} + 10RC = 4$ where $R = R_1 = R_2$. Next, show that $\frac{1}{RC} + 10RC = 4$ would require the value of RC to be complex.

DP 9-9 A fluorescent light uses cathodes (coiled tungsten filaments coated with an electron-emitting substance) at each end that send current through mercury vapors sealed in the tube. Ultraviolet radiation is produced as electrons from the cathodes knock mercury electrons out of their natural orbits. Some of the displaced electrons settle back into orbit, throwing off the excess energy absorbed in the collision. Almost all of this energy is in the form of ultraviolet radiation. The ultraviolet rays, which are invisible, strike a phosphor coating on the inside of the tube. The rays energize the electrons in the phosphor atoms, and the atoms emit white light. The conversion of one kind of light into another is known as fluorescence.

One form of a fluorescent lamp is represented by the *RLC* circuit shown in Figure DP 9.9. Select L so that the current $i(t)$ reaches a maximum at approximately $t = 0.5$ s. Determine the maximum value of $i(t)$. Assume that the switch was in position 1 for a long time before switching to position 2 at $t = 0$.

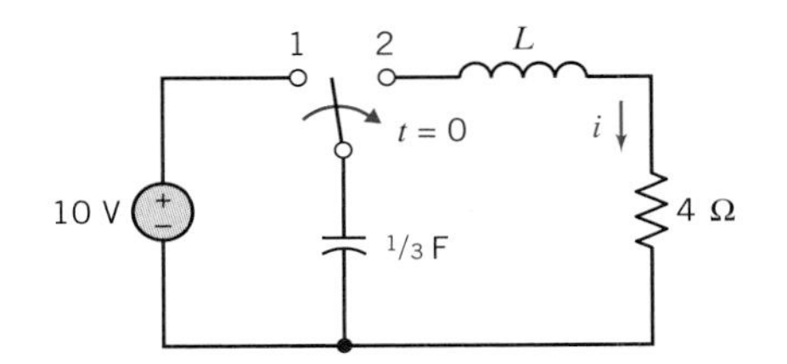

Figure DP 9.9 Fluorescent lamp circuit.

CHAPTER 10

Sinusoidal Steady-State Analysis

Preview

The wide use of alternating current voltage sources requires us to study the analysis of circuits in the sinusoidal steady state. In this chapter, we are interested primarily in the steady-state response of circuits to these sinusoidal sources. The need for the analysis of sinusoidal steady-state behavior led engineers to develop the concept of the phasor and impedance, which linearly relates the phasor current and phasor voltage of a circuit element. With this concept we can proceed to use all the circuit analysis methods we developed carefully in Chapter 5 for resistive circuits.

10.1 Design Challenge

OP AMP CIRCUIT

Figure 10.1-1*a* shows two sinusoidal voltages, one labeled as input and the other labeled as output. We want to design a circuit that will transform the input sinusoid into the output sinusoid. Figure 10.1-1*b* shows a candidate circuit. We must first determine if this circuit can do the job. Then, if it can, we will design the circuit, that is, specify the required values of R_1, R_2, and C.

In this chapter we will learn how to find the steady-state response of a linear circuit when the input is a sinusoid. We will see that phasors can be used to describe the way a circuit transforms a sinusoidal input into a sinusoidal output. With these tools we will be able to solve this problem, to which we will return at the end of this chapter.

10.2 ALTERNATING CURRENT BECOMES STANDARD

In the late 1800s the primary use for electricity was illumination, with Edison's direct current system operating in New York and New Jersey. As electricity became an important energy source, inventors such as Michael Faraday and Joseph Henry demonstrated small electric motors. However, it was clear that reliance on batteries had hindered the development of widespread use of electric power. Thus began the battle of alternating current (ac) versus direct current (dc).

Alternating current was generated in a power plant by a rotating machine called an alternator. Direct current was generated by a machine called a dynamo. Thus, it became a significant task for electrical engineers in the late 1800s to decide which system would be superior for future use.

Direct current seemed to offer some advantages. Batteries could be used as emergency backup when dynamos failed or could supply power during periods of low demand. Also, dynamos could be run in parallel to add supply as power demand increased. In 1890 the parallel use of alternators was thought to be difficult because of synchronization problems.

The Frankfurt Exhibition of 1891 was largely responsible for thrusting the case for ac transmission into the consciousness of European and American electrical engineers. Frank-

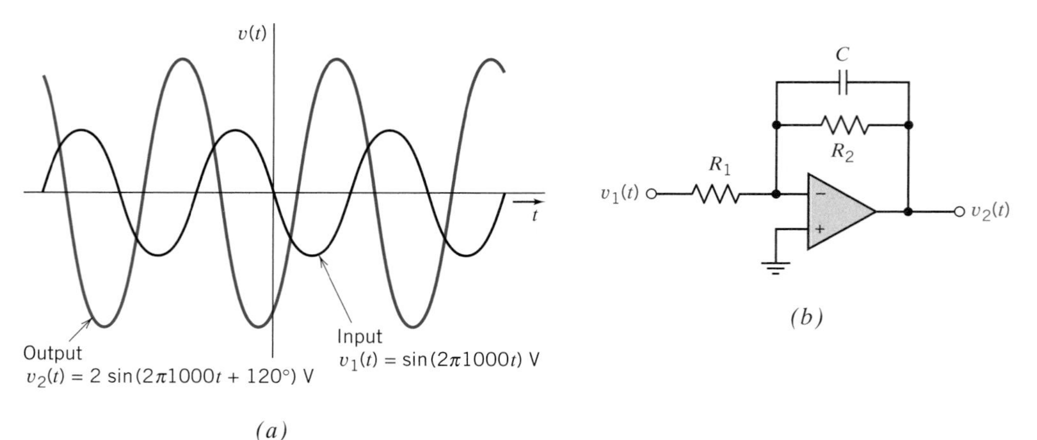

Figure 10.1-1 (*a*) Input and output voltages. (*b*) Proposed circuit.

furt had been ready to establish a municipal dc system, whereas Cologne had just become the first German city to adopt an ac system.

The battle of the systems was in full swing, and ac was demonstrated to be worthy of consideration at the Frankfurt Exhibition. When the exhibition opened, lights shone and motors ran on power generated by alternators at a dam on the Neckar River, 176 km away. The power was transmitted at 15 kV, and then a transformer was used to reduce the voltage to the 110 V used at the exhibition. A transformer is a device with two or more coils (windings) used to transfer ac power between circuits, usually with changed values of voltage and current. Transformers are considered in Chapter 11.

In March 1892 the Cataract Construction Company issued a request for proposals to build a large power plant at Niagara Falls, New York, and transmit the power to Buffalo, New York, some 20 miles distant. However, by 1893 the Cataract directors had decided to proceed with their own plan. But they had to choose: Was it to be an ac or a dc system?

The main advantage of ac over dc is efficiency of transmission. Alternating voltages could be transformed to high voltage, thus reducing the loss on the transmission line. If a line has a wire resistance R and the power transmitted is VI, then power lost in the lines is I^2R. Thus, if the transmitted voltage could be set at a high level and the current I kept low, the line losses would be minimized.

Many engineers defended the dc system. Thomas Edison, for example, was firmly against the adoption of ac systems. He emphasized the greater reliability of dc and the potential hazard of high ac voltages. However, after 1891 the safety issue died with the safe operation of a 19-km ac transmission line from Willamette Falls to Portland, Oregon, which operated at 4 kV. The Frankfurt line mentioned earlier operated at 30 kV. Thus, the Cataract Company began the construction of an ac power system at Niagara, and the ac versus dc issue was settled.

In 1893, when General Electric Company acquired a small firm, it acquired the services of a young electrical engineer, Charles P. Steinmetz, shown in Figure 10.2-1. Steinmetz had completed his doctoral dissertation in mathematics at the University of Breslau but had left for Switzerland and eventually the United States.

Steinmetz's claim to fame was his symbolic method of steady-state ac calculation—the $j\omega$ notation—and the theory of electrical transients. McGraw-Hill published the first of his many textbooks, *Theory and Calculation of AC Phenomena,* in 1897. Four years later, he was elected president of the American Institute of Electrical Engineers in recognition of his leadership.

One of Steinmetz's greatest insights occurred in 1893, when he discussed a paper by A. E. Kennelly on impedance. Steinmetz, noting that $a + jb = r(\cos\phi + j\sin\phi)$, showed the equivalence of the complex notation and the polar form in ac circuit calculations. Thus, the use of complex numbers to solve ac circuit problems was born.

The subject of this chapter is the analysis of the steady-state response using complex numbers as originally introduced by Steinmetz in 1893.

Figure 10.2-1
Charles P. Steinmetz (1865–1923), the developer of the mathematical analytical tools for studying ac circuits. Courtesy of General Electric Co.

10.3 SINUSOIDAL SOURCES

In electrical engineering, sinusoidal forcing functions are particularly important since power sources and communication signals are usually transmitted as sinusoids or modified sinusoids. The forcing function causes the forced response, and the natural response is caused by the internal dynamics of the circuit. The natural response will normally decay after some period of time, but the forced response continues indefinitely. Therefore, in this chapter we are interested primarily in the steady-state response of a circuit to the sinusoidal forcing function.

We consider the forcing function

$$v_s = V_m \sin \omega t \tag{10.3-1}$$

or, in the case of a current source,

$$i_s = I_m \sin \omega t \tag{10.3-2}$$

The amplitude of the sinusoid is V_m, and the radian frequency is ω (rad/s). The sinusoid is a *periodic function* defined by the property

$$x(t + T) = x(t)$$

for all t and where T is the period of oscillation.

The reciprocal of T defines the *frequency* or number of cycles per second, denoted by f, where

$$f = \frac{1}{T}$$

The frequency f is in cycles per second, more commonly referred to as hertz (Hz) in honor of the scientist Heinrich Hertz, shown in Figure 10.3-1. Therefore, the angular (radian) frequency to the sinusoidal function is

Figure 10.3-1
Heinrich R. Hertz (1857–1894). Courtesy of the Institution of Electrical Engineers.

$$\omega = 2\pi f = \frac{2\pi}{T} \text{ (rad/s)}$$

For the voltage source of Eq. 10.3-1, the maximum value is V_m. If the sinusoidal voltage has an associated *phase angle* ϕ, in radians, the voltage source is

$$v_s = V_m \sin(\omega t + \phi) \tag{10.3-3}$$

The sinusoidal voltage of Eq. 10.3-3 is represented by Figure 10.3-2. Normally, we consider ωt and ϕ in radians, but many use degrees as the convention for the total angle. Either approach is acceptable as long as it is clear which notation you are using.

Since, conventionally, the angle ϕ may be expressed in degrees, you will encounter the notation

$$v_s = V_m \sin(4t + 30^\circ)$$

or alternatively

$$v_s = V_m \sin\left(4t + \frac{\pi}{6}\right)$$

This angular inconsistency will not deter us as long as we recognize that in the actual calculation of $\sin \theta$, θ must be in degrees or radians as our calculator or math tables require.

In addition, it is worth noting that

$$V_m \sin(\omega t + 30^\circ) = V_m \cos(\omega t - 60^\circ)$$

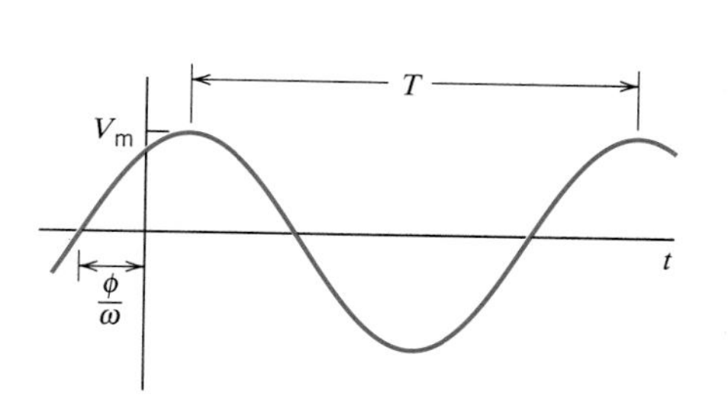

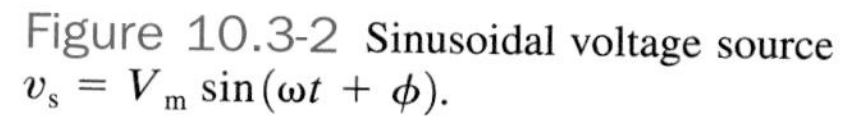
Figure 10.3-2 Sinusoidal voltage source $v_s = V_m \sin(\omega t + \phi)$.

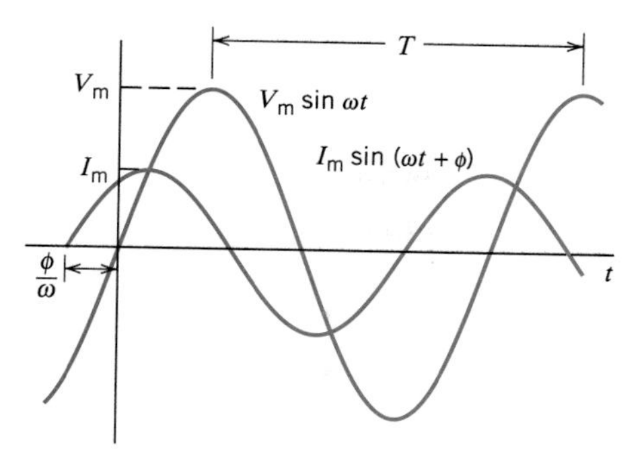

Figure 10.3-3 Voltage and current of a circuit element.

This relationship can be deduced using the trigonometric formulas summarized in Appendix B.

If a circuit has a voltage across an element as

$$v = V_m \sin \omega t$$

and a current flows through the element

$$i = I_m \sin(\omega t + \phi)$$

we have the v and the i shown in Figure 10.3-3. We say that the current *leads* the voltage by ϕ radians. Examining Figure 10.3-3, we note that the current reaches its peak value before the voltage and thus is said to lead the voltage. Thus, a point on $i(t)$ is reached in time ahead of the corresponding point of $v(t)$. An alternative form is to state that the voltage lags the current by ϕ radians.

Consider a sine waveform with

$$v = 2 \sin(3t + 20°)$$

and the associated current waveform

$$i = 4 \sin(3t - 10°)$$

Clearly, the voltage v leads the current i by 30°, or $\pi/6$ radians.

Example 10.3-1

The voltage across an element is $v = 3 \cos 3t$ V, and the associated current through the element is $i = -2 \sin(3t + 10°)$ A. Determine the phase relationship between the voltage and current.

Solution

First, we need to convert the current to a cosine form with a positive magnitude so that it can be contrasted with the voltage. In order to determine a phase relationship, it is necessary to express both waveforms in a consistent form.

Since $-\sin \omega t = \sin(\omega t + \pi)$, we have

$$i = 2 \sin(3t + 180° + 10°)$$

Also, we note that

$$\sin \theta = \cos(\theta - 90°)$$

Therefore, $i = 2\cos(3t + 180° + 10° - 90°) = 2\cos(3t + 100°)$

Recall that $v = 3\cos 3t$. Therefore, the current leads the voltage by 100°.

In the preceding chapter we learned that the response of a circuit to a forcing function

$$v_s = V_0 \cos \omega t$$

will result in a forced response of the form

$$v_f = A\cos \omega t + B\sin \omega t \tag{10.3-4}$$

Equation 10.3-4 may also be written as

$$v_f = \sqrt{A^2 + B^2}\left(\frac{A}{\sqrt{A^2 + B^2}}\cos \omega t + \frac{B}{\sqrt{A^2 + B^2}}\sin \omega t\right)$$

Consider the triangle shown in Figure 10.3-4 and note that

$$\sin \theta = \frac{B}{\sqrt{A^2 + B^2}}$$

and

$$\cos \theta = \frac{A}{\sqrt{A^2 + B^2}}$$

Then we have for v_f

$$v_f = \sqrt{A^2 + B^2}\,(\cos \theta \cos \omega t + \sin \theta \sin \omega t) \tag{10.3-5}$$

However, we can also write Eq. 10.3-5 as

$$v_f = \sqrt{A^2 + B^2}\cos(\omega t - \theta) \tag{10.3-6}$$

where $\theta = \tan^{-1} B/A$ since $A > 0$.

Therefore, if a circuit has a forcing function $V_0 \cos \omega t$, the resulting forced response is

$$v_f = A\cos \omega t + B\sin \omega t = \sqrt{A^2 + B^2}\cos(\omega t - \theta)$$

where $\theta = \tan^{-1} B/A$ when $A > 0$ and $\theta = 180° + \tan^{-1} B/A$ when $A < 0$. We will use this result in the next section.

Example 10.3-2

A current has the form $i = -6\cos 2t + 8\sin 2t$. Find the current restated in the form of Eq. 10.3-6.

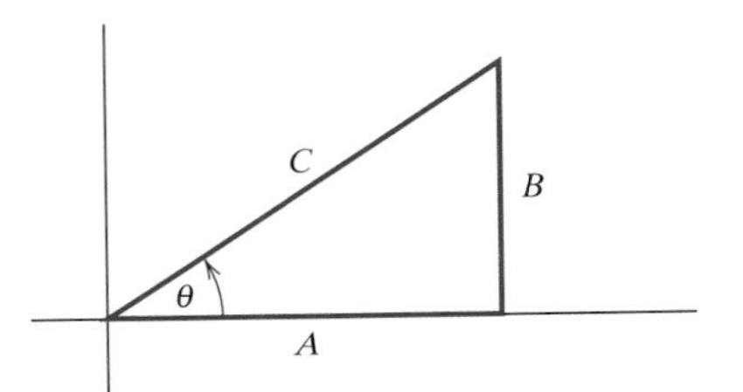

Figure 10.3-4 Triangle for A and B of Eq. 10.3-4, where $C = \sqrt{A^2 + B^2}$.

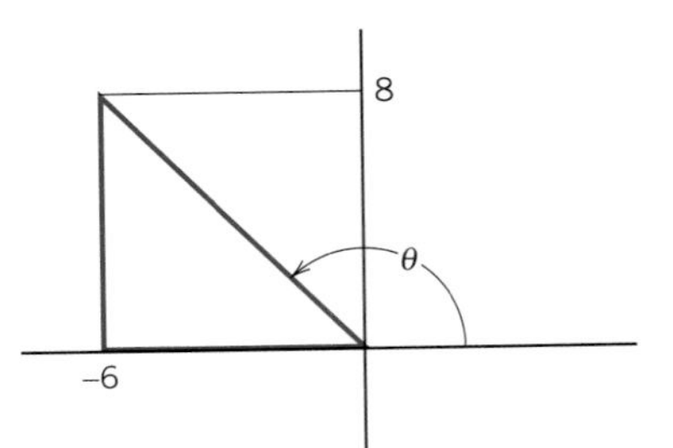

Figure 10.3-5 The A-B triangle for Example 10.3-2.

Solution

The triangle for A and B is shown in Figure 10.3-5. Since the coefficient A is equal to -6 and B is $+8$, we have the angle θ shown. Therefore,

$$\theta = 180° + \tan^{-1}\left(\frac{8}{-6}\right) = 180° - 53.1° = 126.9°$$

Hence

$$i = 10\cos(2t - 126.9°)$$

Exercise 10.3-1 A voltage is $v = 6\cos(4t + 30°)$. (a) Find the period of oscillation. (b) State the phase relation to the associated current $i = 8\cos(4t - 70°)$.

Answer: (a) $T = 2\pi/4$.
(b) The voltage leads the current by 100°.

Exercise 10.3-2 A voltage is $v = 3\cos 4t + 4\sin 4t$. Find the voltage in the form of Eq. 10.3-6.

Answer: $v = 5\cos(4t - 53°)$

Exercise 10.3-3 A current is $i = 12\sin 5t - 5\cos 5t$. Find the current in the form of Eq. 10.3-6.

Answer: $i = 13\cos(5t - 112.6°)$

10.4 STEADY-STATE RESPONSE OF AN *RL* CIRCUIT FOR A SINUSOIDAL FORCING FUNCTION

As an example of the task of determining the steady-state response of a circuit, let us consider the *RL* circuit shown in Figure 10.4-1. We will use the forcing function

$$v_s = V_m \cos \omega t$$

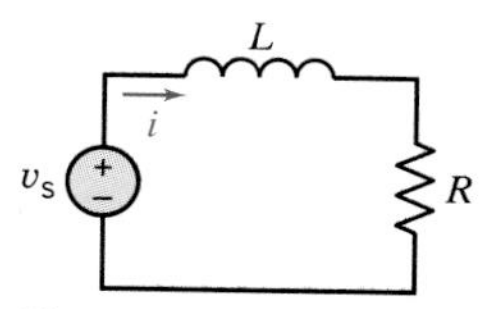

Figure 10.4-1 An *RL* circuit.

We wish to determine only the forced response i_f since we will take the long-run view and look for the steady-state response. Regardless of the initial value of the current and provided that the root of the characteristic equation is in the left-hand s-plane, the response will become purely sinusoidal as $t \to \infty$. This response is called the sinusoidal steady-state response.

The governing differential equation of the *RL* circuit is given by

$$L\frac{di}{dt} + Ri = V_m \cos \omega t \tag{10.4-1}$$

Following the method of the previous chapter, we assume that

$$i_f = A\cos \omega t + B\sin \omega t \tag{10.4-2}$$

At this point, since we are only solving for the forced response, we drop the subscript f notation. Substituting the assumed solution of Eq. 10.4-2 into the differential equation and completing the derivative, we have

$$L(-\omega A \sin \omega t + \omega B \cos \omega t) + R(A\cos \omega t + B \sin \omega t) = V_m \cos \omega t$$

Equating the coefficients of cos ωt, we obtain

$$\omega LB + RA = V_m$$

Next, equating the coefficients of sin ωt, we obtain

$$-\omega LA + RB = 0$$

Solving for A and B, we have

$$A = \frac{RV_m}{R^2 + \omega^2 L^2}$$

and

$$B = \frac{\omega L V_m}{R^2 + \omega^2 L^2}$$

The response to the sinusoidal input is then

$$i = A \cos \omega t + B \sin \omega t$$

or

$$i = \frac{V_m}{Z} \cos(\omega t - \beta)$$

where

$$Z = \sqrt{R^2 + \omega^2 L^2}$$

and

$$\beta = \tan^{-1} \frac{\omega L}{R}$$

Thus, the forced (steady-state) response is of the form

$$i = I_m \cos(\omega t + \phi)$$

where

$$I_m = \frac{V_m}{Z}$$

and

$$\phi = -\beta$$

In this case, we have found only the steady-state response of a circuit with one energy storage element. Clearly, this approach can be quite complicated if the circuit has several storage elements.

Steinmetz saw a way beyond this differential equation approach. The method is discussed in the next section. It was an insight that, in many ways, enabled electrical engineering to leap forward to new achievements.

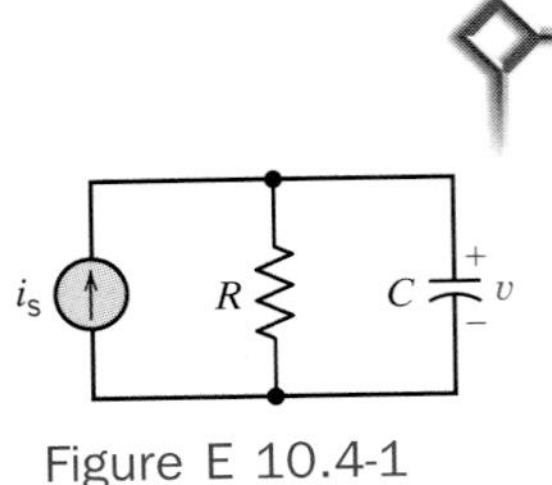

Figure E 10.4-1

Exercise 10.4-1 Find the forced response v for the RC circuit shown in Figure E 10.4-1 when $i_s = I_m \cos \omega t$.

Answer: $v = (R\, I_m / P) \cos(\omega t - \theta)$, $P = \sqrt{1 + \omega^2 R^2 C^2}$, $\theta = \tan^{-1}(\omega R C)$

Exercise 10.4-2 Find the forced response $i(t)$ for the RL circuit of Figure 10.4-1 when $R = 2\ \Omega$, $L = 1$ H, and $v_s = 10 \cos 3t$ V.

Answer: $i = 2.77 \cos(3t - 56.3°)$ A

10.5 COMPLEX EXPONENTIAL FORCING FUNCTION

Upon reviewing the preceding section, we see that the input to the circuit was of the form

$$v_s = V_\text{m} \cos \omega t$$

and the response was

$$i = \frac{V_\text{m}}{Z} \cos(\omega t - \beta)$$

Thus, the response to a sinusoidal input is also sinusoidal and has the same frequency as the input but has a different magnitude and phase angle than the original voltage source.

It is useful to consider the exponential signal

$$v_\text{e} = V_\text{m} e^{j\omega t} \tag{10.5-1}$$

Using Euler's equation we can relate the exponential signal to a sinusoidal signal

$$v_\text{s} = V_\text{m} \cos \omega t = \text{Re}\{V_\text{m} e^{j\omega t}\} = \text{Re}\{v_\text{e}\}$$

The notation $\text{Re}\{a + jb\}$ is read as the real part of the complex number $(a + jb)$. For example,

$$\text{Re}\{a + jb\} = a$$

Let us try the exponential source v_e of Eq. 10.5-1 with the differential equation of the RL circuit shown in Figure 10.5-1

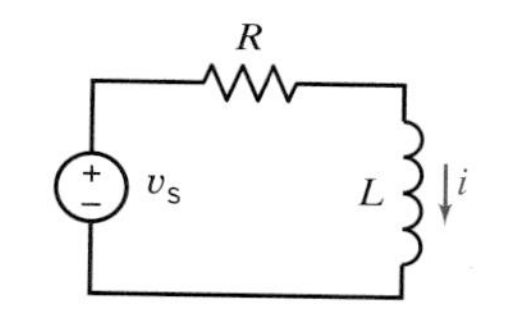

Figure 10.5-1

$$L \frac{di_e}{dt} + Ri_e = v_e \tag{10.5-2}$$

where i_e is the response to the exponential input. Since the source is an exponential, we try the solution

$$i_e = Ae^{j\omega t} \tag{10.5-3}$$

and substitute into Eq. 10.5-2 to obtain

$$(j\omega L + R)Ae^{j\omega t} = V_\text{m} e^{j\omega t}$$

Hence,[1]

$$A = \frac{V_\text{m}}{R + j\omega L} = \frac{V_\text{m}}{Z} e^{-j\beta}$$

where

$$\beta = \tan^{-1} \frac{\omega L}{R}$$

and

$$Z = \sqrt{R^2 + \omega^2 L^2}$$

Therefore, substituting for A, we have

$$i_e = \frac{V_\text{m}}{Z} e^{-j\beta} e^{j\omega t} \tag{10.5-4}$$

Again noting that the original forcing function was

$$v_s = \text{Re}\{V_m e^{j\omega t}\} = V_m \cos \omega t$$

[1]Note: See Appendix B for a review of complex numbers.

we expect that

$$i = \text{Re}\{i_e\} = \text{Re}\left\{\frac{V_\text{m}}{Z}e^{-j\beta}e^{j\omega t}\right\}$$

Accordingly, $i = \dfrac{V_\text{m}}{Z}\text{Re}\{e^{-j\beta}e^{j\omega t}\} = \dfrac{V_\text{m}}{Z}\text{Re}\{e^{j(\omega t-\beta)}\} = \dfrac{V_\text{m}}{Z}\cos(\omega t - \beta)$

In general we are finding the sinusoidal response

$$i = I_\text{m}\cos(\omega t - \beta) = \text{Re}\left\{\frac{V_\text{m}}{Z}e^{j(\omega t-\beta)}\right\}$$

to the sinusoidal excitation

$$v_s = V_\text{m}\cos\omega t = \text{Re}\{V_\text{m}e^{j\omega t}\}$$

We have learned that this response is readily obtained by using the complex exponential excitation, $\text{Re}\{V_\text{m}e^{j\omega t}\}$.

As an example, let us find the steady-state response for the *RLC* circuit shown in Figure 10.5-2. This circuit is represented by the differential equation

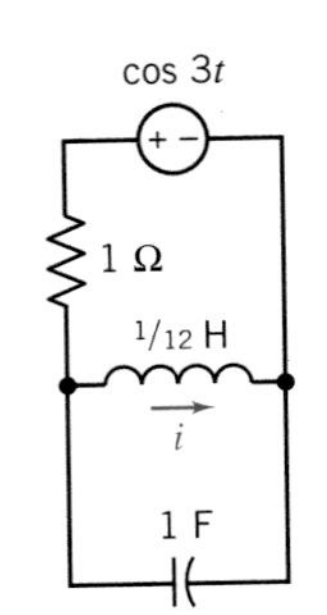

Figure 10.5-2

$$\frac{d^2i}{dt^2} + \frac{di}{dt} + 12i = 12\cos 3t \tag{10.5-5}$$

First, replace the real excitation by the complex exponential excitation

$$v_e = 12e^{j3t}$$

Then we have Eq. 10.5-5 restated as

$$\frac{d^2i_e}{dt^2} + \frac{di_e}{dt} + 12i_e = 12e^{j3t} \tag{10.5-6}$$

We expect the response to the exponential input to be of the form

$$i_e = Ae^{j3t} \tag{10.5-7}$$

The first and second derivatives of the i_e of Eq. 10.5-7 are

$$\frac{di_e}{dt} = j3Ae^{j3t}$$

and

$$\frac{d^2i_e}{dt^2} = -9Ae^{j3t}$$

Substituting into Eq. 10.5-6, we have

$$(-9 + j3 + 12)Ae^{j3t} = 12e^{j3t} \tag{10.5-8}$$

Solving for A, we obtain

$$A = \frac{12}{3 + j3} = \frac{12(3 - j3)}{(3 + j3)(3 - j3)} = \frac{12(3 - j3)}{18} = 2\sqrt{2}\angle -45°$$

Therefore, $\quad i_e = Ae^{j3t} = 2\sqrt{2}\,e^{-j(\pi/4)}e^{j3t} = 2\sqrt{2}\,e^{j(3t-\pi/4)}$

Recall that Euler's identity is $e^{j\phi} = \cos\phi + j\sin\phi$. Thus, the desired answer for the steady-state current is[2]

$$i(t) = \text{Re}\{i_e\} = \text{Re}\{2\sqrt{2}e^{j(3t-\pi/4)}\} = 2\sqrt{2}\cos(3t - 45°)$$

[2]See Appendix D for a discussion of Euler's equation.

Table 10.5-1 **Use of the Complex Exponential Excitation to Determine a Circuit's Steady-State Response to a Sinusoidal Source**

1. Write the excitation (forcing function) as a cosine waveform with a phase angle so that $y_s = Y_m \cos(\omega t + \phi)$, where y_s is a current source i_s or a voltage source v_s in the circuit.
2. Recall Euler's identity, which is

$$e^{j\alpha} = \cos \alpha + j \sin \alpha$$

where $\alpha = \omega t + \phi$ in this case.

3. Introduce the complex excitation so that for a voltage source, for example, we have

$$v_s = Re\{V_m e^{j(\omega t + \phi)}\}$$

where $V_m e^{j(\omega t + \phi)}$ is a complex exponential excitation.

4. Use the complex excitation and the differential equation along with the assumed response $x_e = Ae^{j(\omega t + \phi)}$, where A is to be determined. Note that A will normally be a complex quantity.
5. Determine the constant $A = Be^{-j\beta}$, so that

$$x_e = Ae^{j(\omega t + \phi)} = Be^{j(\omega t + \phi - \beta)}$$

6. Recognize that the desired response is

$$x(t) = \text{Re}\{x_e\} = B \cos(\omega t + \phi - \beta)$$

Note that we have changed from $\pi/4$ radians to 45°, which are interchangeable and equivalent. Both degree and radian notation are acceptable and interchangeable.

Compare Eqs. 10.5-6 and 10.5-8. Eq. 10.5-6 is a differential equation, which is what we expect for the equation representing a circuit that contains capacitors or inductors. In contrast, Eq. 10.5-8 is an algebraic equation, involving addition and multiplication but not integration or differentiation. The coefficients of Eq. 10.5-8 are complex numbers while the coefficients of equation 10.5-6 are real numbers. Algebraic equations are easier to solve than differential equations, so we prefer to solve Eq. 10.5-8, even though it contains complex coefficients.

We have developed a straightforward method for determining the steady-state response of a circuit to a sinusoidal excitation. The process is as follows: (1) instead of applying the actual forcing function, we apply a complex exponential forcing function, and (2) we then obtain the complex response whose real part is the desired response. The process is summarized in Table 10.5-1. Let us use this process in another example.

Example 10.5-1

Find the response i of the RL circuit of Figure 10.5-1 when $R = 2\ \Omega$, $L = 1$ H, and $v_s = 10 \sin 3t$ V.

Solution

First, we will rewrite the voltage source so that it is expressed as a cosine waveform as follows:

$$v_s = 10 \sin 3t = 10 \cos(3t - 90°)$$

Using the complex excitation, we have

$$v_e = 10e^{j(3t - 90°)}$$

Introduce the complex excitation into the circuit's differential equation, which is

$$L\frac{di_e}{dt} + R\,i_e = v_e$$

obtaining

$$\frac{di_e}{dt} + 2\,i_e = 10e^{j(3t-90^\circ)}$$

Assume that the response is

$$i_e = Ae^{j(3t-90^\circ)} \tag{10.5-9}$$

where A is a complex quantity to be determined. Substituting the assumed solution, Eq. 10.5-9, into the differential equation and taking the derivative, we have

$$j3Ae^{j(3t-90^\circ)} + 2Ae^{j(3t-90^\circ)} = 10e^{j(3t-90^\circ)}$$

Therefore,

$$j3A + 2A = 10$$

or

$$A = \frac{10}{j3+2} = \frac{10}{\sqrt{9+4}}\,e^{-j\beta}$$

where

$$\beta = \tan^{-1}\frac{3}{2} = 56.3^\circ$$

Then the solution is

$$i_e = Ae^{j(3t-90^\circ)} = \frac{10}{\sqrt{13}}e^{-j56.3^\circ}\,e^{j(3t-90^\circ)} = \frac{10}{\sqrt{13}}e^{j(3t-146.3^\circ)}$$

Consequently, the actual response is

$$i(t) = \mathrm{Re}\{i_e\} = \frac{10}{\sqrt{13}}\cos(3t - 146.3^\circ)\text{ A}$$

Steinmetz observed the process we just utilized and decided to formulate a method for solving the sinusoidal steady-state response of circuits using complex number notation. The development of this approach is the subject of the next section.

Exercise 10.5-1
Find a and b when

$$\frac{10}{a+jb} = 2.36e^{j45}$$

Answer: $a = 3, b = -3$

Exercise 10.5-2
Find A and θ when

$$[A\angle\theta](-3+j8) = j32$$

Answer: $A = 3.75, \theta = -20.56^\circ$

10.6 THE PHASOR CONCEPT

A sinusoidal current or voltage at a given frequency is characterized by its amplitude and phase angle. For example, the current response in the *RL* circuit considered in Example 10.5-1 was

$$i(t) = \mathrm{Re}\{I_m e^{j(\omega t + \phi - \beta)}\}$$
$$= I_m \cos(\omega t + \phi - \beta)$$

The magnitude I_m and the phase angle $(\phi - \beta)$, along with knowledge of ω, completely specify the response. Thus, we may write $i(t)$ as

$$i(t) = \mathrm{Re}\{I_m e^{j(\phi - \beta)} e^{j\omega t}\}$$

However, we note that the complex factor $e^{j\omega t}$ remained unchanged throughout all our previous calculations. Thus, the information we seek is represented by

$$\mathbf{I} = I_m e^{j(\phi - \beta)} = I_m \underline{/\phi - \beta} \qquad (10.6\text{-}1)$$

where **I** is called a *phasor*. A phasor is a complex number that represents the magnitude and phase of a sinusoid. The term phasor is used instead of vector because the angle is time based rather than space based. A phasor may be written in exponential form, polar form, or rectangular form.

The **phasor concept** may be used when the circuit is linear, the steady-state response is sought, and all independent sources are sinusoidal and have the same frequency.

A real sinusoidal current, where $\theta = \phi - \beta$, is written as

$$i(t) = I_m \cos(\omega t + \theta)$$

It can be represented by

$$i(t) = \mathrm{Re}\{I_m e^{j(\omega t + \theta)}\}$$

We then decide to drop the notation Re and the redundant $e^{j\omega t}$ to obtain the phasor representation

$$\mathbf{I} = I_m e^{j\theta} = I_m \underline{/\theta}$$

This abbreviated representation is the *phasor notation*. Phasor quantities are complex and thus are printed in boldface in this book. You may choose to use the underline notation as follows:

$$\underline{I} = I_m \underline{/\theta}$$

Although we have dropped or suppressed the complex frequency $e^{j\omega t}$, we continue to note that we are in the complex frequency form and are performing calculations in the *frequency domain*. We have transformed the problem from the time domain to the frequency domain by the use of phasor notation. A transform is a means of encoding to simplify a calculation process. One example of a mathematical transform is the logarithmic transform.

A **transform** is a change in the mathematical description of a physical variable to facilitate computation.

The actual steps involved in transforming a function in the time domain to the frequency domain are summarized in Table 10.6-1. Since it is easy to move through these steps, we usually jump directly from step 1 to step 4.

Table 10.6-1 **Transformation from the Time Domain to the Frequency Domain**

1. Write the function in the time domain, $y(t)$, as a cosine waveform with a phase angle ϕ as

$$y(t) = Y_{\mathrm{m}} \cos(\omega t + \phi)$$

2. Express the cosine waveform as the real part of a complex quantity by using Euler's identity so that

$$y(t) = \mathrm{Re}\{Y_{\mathrm{m}} e^{j(\omega t + \phi)}\}$$

3. Drop the real part notation.
4. Suppress the $e^{j\omega t}$, while noting the value of ω for later use, obtaining the phasor

$$\mathbf{Y} = Y_{\mathrm{m}} e^{j\phi} = Y_{\mathrm{m}} \angle \phi$$

For example, let us determine the phasor notation for

$$i = 5 \sin(100t + 120°)$$

We have chosen to use cosine functions as the standard for phasor notation. Thus, we express the current as a cosine waveform:

$$i = 5 \cos(100t + 30°)$$

At this point, it is easy to see that the information we require is the amplitude and the phase. Thus, the phasor is

$$\mathbf{I} = 5 \angle 30°$$

Of course, the reverse process from phasor notation to time notation is exactly the reverse of the steps required to go from the time to the phasor notation. Thus, if we have a voltage in phasor notation:

$$\mathbf{V} = 24 \angle 125°$$

the time domain notation is

$$v(t) = 24 \cos(\omega t + 125°)$$

where the frequency ω was noted in the original statement of the circuit formulation. This transformation from the frequency domain to the time domain is summarized in Table 10.6-2.

Table 10.6-2 **Transformation from the Frequency Domain to the Time Domain**

1. Write the phasor in exponential form as

$$\mathbf{Y} = Y_{\mathrm{m}} e^{j\beta}$$

2. Reinsert the factor $e^{j\omega t}$ so that you have

$$Y_{\mathrm{m}} e^{j\beta} e^{j\omega t}$$

3. Reinsert the real part operator Re as

$$\mathrm{Re}\{Y_{\mathrm{m}} e^{j\beta} e^{j\omega t}\}$$

4. Use Euler's identity to obtain the time function

$$y(t) = \mathrm{Re}\{Y_{\mathrm{m}} e^{j(\omega t + \beta)}\} = Y_{\mathrm{m}} \cos(\omega t + \beta)$$

A **phasor** is a transformed version of a sinusoidal voltage or current waveform and consists of the magnitude and phase angle information of the sinusoid.

The phasor method uses the transformation from the time domain to the frequency domain to more easily obtain the sinusoidal steady-state solution of the differential equation. Consider the *RL* circuit of Figure 10.6-1. We wish to find the solution for the steady-state current i when the voltage source is $v_s = V_m \cos \omega t$ V and $\omega = 100$ rad/s.

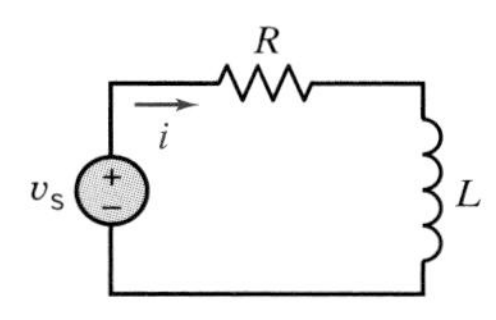

Figure 10.6-1 *RL* circuit.

Also, for this circuit let $R = 200\ \Omega$ and $L = 2$ H. Then we may write the differential equation as

$$L\frac{di}{dt} + Ri = v_s \tag{10.6-2}$$

Since

$$v_s = V_m \cos(\omega t + \phi) = \text{Re}\{V_m e^{j(\omega t + \phi)}\} \tag{10.6-3}$$

we will use the assumed solution

$$i = I_m \cos(\omega t + \beta) = \text{Re}\{I_m e^{j(\omega t + \beta)}\} \tag{10.6-4}$$

Therefore, we may substitute Eqs. 10.6-3 and 10.6-4 into Eq. 10.6-2 and suppress the Re notation to obtain

$$(j\omega L I_m + R I_m) e^{j(\omega t + \beta)} = V_m e^{j(\omega t + \phi)}$$

Suppress the $e^{j\omega t}$ to obtain

$$(j\omega L + R) I_m e^{j\beta} = V_m e^{j\phi}$$

Now we recognize the phasors

$$\mathbf{I} = I_m e^{j\beta}$$

and

$$\mathbf{V}_S = V_m e^{j\phi}$$

Therefore, in phasor notation we have

$$(j\omega L + R)\mathbf{I} = \mathbf{V}_S$$

Solving for **I**, we have

$$\mathbf{I} = \frac{\mathbf{V}_S}{j\omega L + R} = \frac{\mathbf{V}_S}{j200 + 200}$$

for $\omega = 100$, $L = 2$, and $R = 200$. Therefore, since

$$\mathbf{V}_S = V_m \angle 0^\circ$$

we have

$$\mathbf{I} = \frac{V_m}{283\angle 45^\circ} = \frac{V_m}{283}\angle -45^\circ$$

Using the method of Table 10.6-2, we may transform this result back to the time domain to obtain the steady-state time solution as

$$i(t) = \frac{V_m}{283}\cos(100t - 45^\circ)\ \text{A}$$

It is clear that we can use phasors directly to obtain a linear algebraic equation expressed in terms of the phasors and complex numbers and then solve for the phasor variable of interest. After obtaining the phasor we desire, we simply transform it back to the time domain to obtain the steady-state solution.

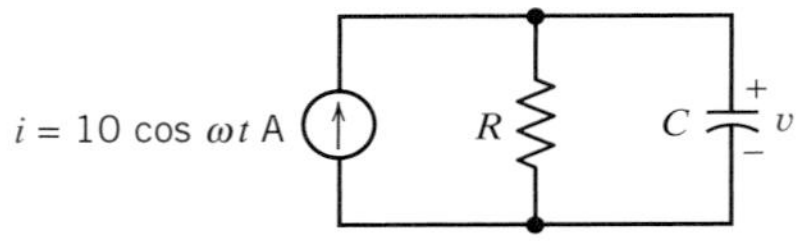

Figure 10.6-2
An *RC* circuit with a sinusoidal current source.

Example 10.6-1

Find the steady-state voltage v for the *RC* circuit shown in Figure 10.6-2 when $i = 10 \cos \omega t$ A, $R = 1\ \Omega$, $C = 10$ mF, and $\omega = 100$ rad/s.

Solution

First, we find the phasor representation of the source current as

$$\mathbf{I} = I_m \angle 0° = 10\angle 0° \tag{10.6-5}$$

We seek to find the voltage v by first obtaining the phasor **V.**

Write the node voltage differential equation for the circuit to obtain

$$\frac{v}{R} + C\frac{dv}{dt} = i \tag{10.6-6}$$

Since
$$i = 10\,\text{Re}\{e^{j\omega t}\}$$

and
$$v = V_m \text{Re}\{e^{j(\omega t + \phi)}\}$$

we substitute into Eq. 10.6-6 and suppress the Re notation to obtain

$$\frac{V_m}{R} e^{j(\omega t + \phi)} + j\omega C\, V_m e^{j(\omega t + \phi)} = 10e^{j\omega t}$$

We now suppress the $e^{j\omega t}$ and obtain

$$\left(\frac{1}{R} + j\omega C\right) V_m e^{j\phi} = 10e^{j0°}$$

Recalling the phasor representation of Eq. 10.6-5, we have

$$\left(\frac{1}{R} + j\omega C\right)\mathbf{V} = \mathbf{I}$$

Since $R = 1$, $C = 10^{-2}$, and $\omega = 100$, we have

$$(1 + j1)\mathbf{V} = \mathbf{I}$$

or
$$\mathbf{V} = \frac{\mathbf{I}}{1 + j1}$$

Therefore,
$$\mathbf{V} = \frac{10}{\sqrt{2}\angle 45°} = \frac{10}{\sqrt{2}}\angle -45°$$

Transforming from the phasor notation back to the steady-state time solution, we have

$$v = \frac{10}{\sqrt{2}} \cos(100t - 45°)\ \text{V}$$

Exercise 10.6-1 Express the current i as a phasor.
(a) $i = 4\cos(\omega t - 80^\circ)$ (b) $i = 10\cos(\omega t + 20^\circ)$
(c) $i = 8\sin(\omega t - 20^\circ)$

Answers: (a) $4\angle -80^\circ$
(b) $10\angle +20^\circ$
(c) $8\angle -110^\circ$

Exercise 10.6-2 Find the steady-state voltage v represented by the phasor
(a) $\mathbf{V} = 10\angle -140^\circ$ (b) $\mathbf{V} = 80 + j75$

Answer: (a) $v = 10\cos(\omega t - 140^\circ)$
(b) $109.7\cos(\omega t + 43.2^\circ)$

Exercise 10.6-3 Find the response v for the circuit shown in Figure E 10.6-3 when $i_s = 10\cos 100t$ A.

Answer: $v = 7.071\cos(100t - 45^\circ)$

Exercise 10.6-4 Find the current $i(t)$ for the RLC circuit of Figure E 10.6-4 when $v_s = 4\cos 100t$ V.

Answer: $i(t) = 2\sqrt{2}\cos(100t + 45^\circ)$ A

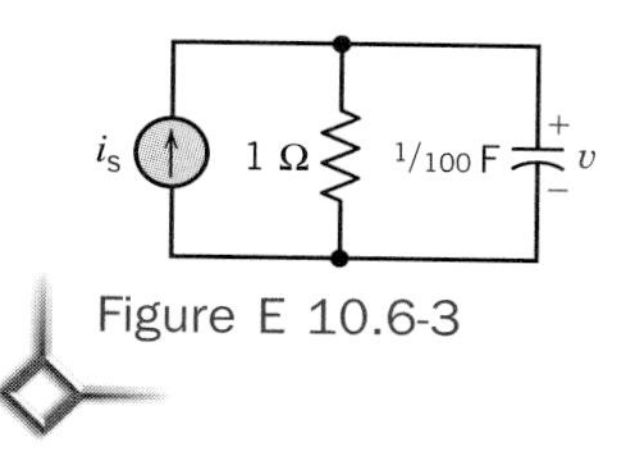

Figure E 10.6-3

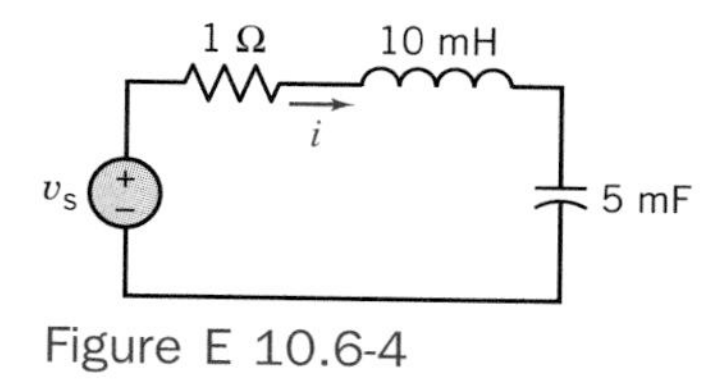

Figure E 10.6-4

10.7 PHASOR RELATIONSHIPS FOR *R*, *L*, AND *C* ELEMENTS

In the preceding section we found that the phasor representation is actually a transformation from the time domain into the frequency domain. With this transform, we have converted the solution of a differential equation into the solution of an algebraic equation.

In this section we determine the relationship between the phasor voltage and the phasor current of the elements: R, L, and C. We use the transformation from the time to the frequency domain and then solve for the phasor relationship for a specified element. We are using the method of Table 10.6-2 as recorded in the last section.

Let us begin with the resistor, as shown in Figure 10.7-1*a*. The voltage–current relationship in the time domain is

$$v = Ri \tag{10.7-1}$$

Now consider the steady-state voltage

$$v = V_m \cos(\omega t + \phi)$$

Then

$$v = \text{Re}\{V_m e^{j(\omega t + \phi)}\} \tag{10.7-2}$$

(*a*)

(*b*)

Figure 10.7-1
(*a*) The v-i time domain relationship for R.
(*b*) The frequency domain relationship for R.

Assume that the current source is of the form

$$i = \text{Re}\{I_{\text{m}} e^{j(\omega t + \beta)}\} \tag{10.7-3}$$

Then substitute Eqs. 10.7-2 and 10.7-3 into Eq. 10.7-1 and suppress the Re notation to obtain

$$V_{\text{m}} e^{j(\omega t + \phi)} = R\, I_{\text{m}} e^{j(\omega t + \beta)}$$

Suppress $e^{j\omega t}$ to obtain

$$V_{\text{m}} e^{j\phi} = R\, I_{\text{m}} e^{j\beta}$$

Therefore, we note that $\beta = \phi$ and

$$\mathbf{V} = R\mathbf{I} \tag{10.7-4}$$

Since $\beta = \phi$, the current and voltage waveforms are in phase. This phasor relationship is shown in Figure 10.7-1*b*.

For example, if the voltage across a resistor is $v = 10 \cos 10t$, we know that the current will be

$$i = \frac{10}{R} \cos 10t$$

in the time domain.

In the frequency domain, we first note that the voltage is

$$\mathbf{V} = 10\angle 0^\circ$$

Then, using the phasor relationship of the resistor, Eq. 10.7-4, we have

$$\mathbf{I} = \frac{\mathbf{V}}{R} = \frac{10\angle 0^\circ}{R}$$

Then, obtaining the time domain relationship for i, we have

$$i = \frac{10}{R} \cos 10t$$

Now consider the inductor as shown in Figure 10.7-2*a*. The time domain voltage–current relationship is

$$v = L\frac{di}{dt} \tag{10.7-5}$$

Again, we use the complex voltage as

$$v = \text{Re}\{V_{\text{m}} e^{j(\omega t + \phi)}\} \tag{10.7-6}$$

and assume that the current is

$$i = \text{Re}\{I_{\text{m}} e^{j(\omega t + \beta)}\} \tag{10.7-7}$$

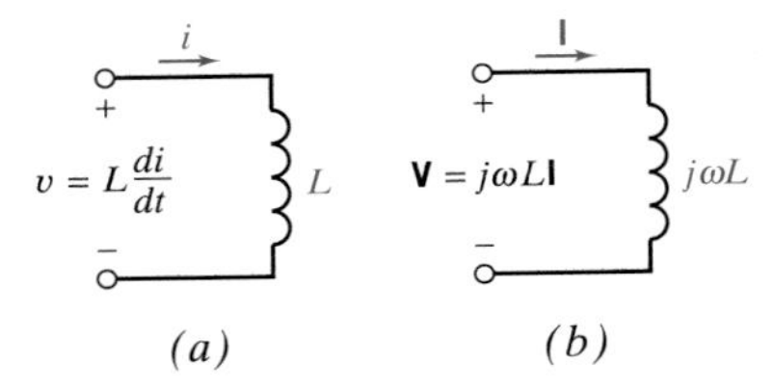

Figure 10.7-2
(*a*) The time domain v-i relationship for an inductor.
(*b*) The frequency domain relationship for an inductor.

Substituting Eqs. 10.7-6 and 10.7-7 into Eq. 10.7-5 and suppressing the Re notation, we have

$$V_m e^{j\phi} e^{j\omega t} = L \frac{d}{dt}\{I_m e^{j\omega t} e^{j\beta}\}$$

Taking the derivative, we have

$$V_m e^{j\phi} e^{j\omega t} = j\omega L\, I_m e^{j\omega t} e^{j\beta}$$

Now suppressing the $e^{j\omega t}$, we have

$$V_m e^{j\phi} = j\omega L\, I_m e^{j\beta} \tag{10.7-8}$$

or

$$\mathbf{V} = j\omega L\mathbf{I} \tag{10.7-9}$$

This phasor relationship is shown in Figure 10.7-2*b*. Since $j = e^{j90°}$, Eq. 10.7-8 can also be written as

$$V_m e^{j\phi} = \omega L\, I_m e^{j90°} e^{j\beta}$$

Therefore,

$$\phi = \beta + 90°$$

Thus, the voltage leads the current by exactly 90°.

As an illustration, consider an inductor of 2 H with $\omega = 100$ rad/s and with voltage $v = 10\cos(\omega t + 50°)$ V. Then the phasor voltage is

$$\mathbf{V} = 10\underline{/50°}$$

and the phasor current is

$$\mathbf{I} = \frac{\mathbf{V}}{j\omega L}$$

Since $\omega L = 200$, we have

$$\mathbf{I} = \frac{V}{j200} = \frac{10\underline{/50°}}{200\underline{/90°}} = 0.05\underline{/-40°} \text{ A}$$

Then the current expressed in the time domain is

$$i = 0.05\cos(100t - 40°) \text{ A}$$

Therefore, the current lags the voltage by 90°.

Finally, let us consider the case of the capacitor, as shown in Figure 10.7-3*a*. The current–voltage relationship is

$$i = C\frac{dv}{dt} \tag{10.7-10}$$

We assume that the voltage is

$$\begin{aligned} v &= V_m \cos(\omega t + \phi) \\ &= \mathrm{Re}\{V_m e^{j(\omega t + \phi)}\} \end{aligned} \tag{10.7-11}$$

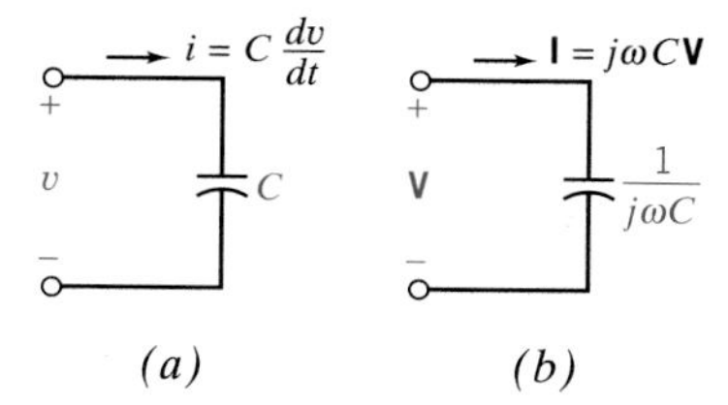

Figure 10.7-3
(*a*) The time domain *v*-*i* relationship for a capacitor. (*b*) The frequency domain relationship for a capacitor.

and the current is of the form

$$i = \mathrm{Re}\{I_m e^{j(\omega t + \beta)}\} \tag{10.7-12}$$

Suppress the Re notation in Eqs. 10.7-11 and 10.7-12 and substitute them into Eq. 10.7-10 to obtain

$$I_m e^{j(\omega t + \beta)} = C\frac{d}{dt}(V_m e^{j(\omega t + \phi)})$$

Taking the derivative, we have

$$I_m e^{j\omega t} e^{j\beta} = j\omega\, C\, V_m e^{j\omega t} e^{j\phi}$$

Suppressing the $e^{j\omega t}$, we obtain

$$I_m e^{j\beta} = j\omega\, C\, V_m e^{j\phi}$$

or

$$\mathbf{I} = j\omega\, C\,\mathbf{V} \tag{10.7-13}$$

This phasor relationship is shown in Figure 10.7-3*b*. Since $j = e^{j90^\circ}$, the current leads the voltage by 90°. As an example, consider a voltage $v = 100 \cos \omega t$ V and let us find the current when $\omega = 1000$ rad/s and $C = 1$ mF. Since

$$\mathbf{V} = 100\underline{/\ 0^\circ}$$

we have

$$\begin{aligned}\mathbf{I} &= j\omega\, C\,\mathbf{V}\\ &= (\omega\, C\, e^{j90^\circ})100\, e^{j0} = (1e^{j90^\circ})100 = 100\underline{/\ 90^\circ}\end{aligned}$$

Therefore, transforming this phasor into the time domain, we have

$$i = 100\cos(\omega t + 90^\circ)\ \mathrm{A}$$

We can rewrite Eq. 10.7-13 as

$$\mathbf{V} = \frac{1}{j\omega\, C}\mathbf{I} \tag{10.7-14}$$

Using this form, we summarize the phasor equations for sources and the resistor, inductor, and capacitor in Table 10.7-1, where the phasor voltage is expressed in its relationship to the phasor current.

Exercise 10.7-1 A current in an element is $i = 5\cos 100t$ A. Find the steady-state voltage $v(t)$ across the element for (a) a resistor of 10 Ω, (b) an inductor $L = 10$ mH, and (c) a capacitor $C = 1$ mF.

Answer: (a) $50 \cos 100t$ V
(b) $5\cos(100t + 90^\circ)$ V
(c) $50\cos(100t - 90^\circ)$ V

Exercise 10.7-2 A capacitor $C = 10\ \mu$F has a steady-state voltage across it of $v = 100\cos(500t + 30^\circ)$ V. Find the steady-state current in the capacitor.

Answer: $i = 0.5\cos(500t + 120^\circ)$A

Table 10.7-1 **Time Domain and Frequency Domain Relationships**

Element	Time Domain	Frequency Domain
Current source	$i(t) = A\cos(\omega t + \theta)$	$\mathbf{I}(\omega) = Ae^{j\theta}$
Voltage source	$v(t) = B\cos(\omega t + \phi)$	$\mathbf{V}(\omega) = Be^{j\phi}$
Resistor	$i(t)$, R, $v(t)$; $v(t) = R\,i(t)$	$\mathbf{I}(\omega)$, R, $\mathbf{V}(\omega)$; $\mathbf{V}(\omega) = R\,\mathbf{I}(\omega)$
Capacitor	$i(t)$, C, $v(t)$; $v(t) = \frac{1}{C}\int_{-\infty}^{t} i(\tau)\,d\tau$	$\mathbf{I}(\omega)$, $\frac{1}{j\omega C}$, $\mathbf{V}(\omega)$; $\mathbf{V}(\omega) = \frac{1}{j\omega C}\,\mathbf{I}(\omega)$
Inductor	$i(t)$, L, $v(t)$; $v(t) = L\frac{d}{dt}i(t)$	$\mathbf{I}(\omega)$, $j\omega L$, $\mathbf{V}(\omega)$; $\mathbf{V}(\omega) = j\omega L\,\mathbf{I}(\omega)$
CCVS	$i_c(t)$, $i(t)$; $v(t) = K\,i_c(t)$	$\mathbf{I}_c(\omega)$, $\mathbf{I}(\omega)$; $\mathbf{V}(\omega) = K\,\mathbf{I}_c(\omega)$
Ideal Op-amp	$i_1 = 0$, $i_2 = 0$, 0, $i(t)$, $v(t)$	$\mathbf{I}_1 = 0$, $\mathbf{I}_2 = 0$, 0, $\mathbf{I}(\omega)$, $\mathbf{V}(\omega)$

Exercise 10.7-3 The voltage $v(t)$ and current $i(t)$ for an element are shown in Figure E 10.7-3. Determine if the element is an inductor or a capacitor.

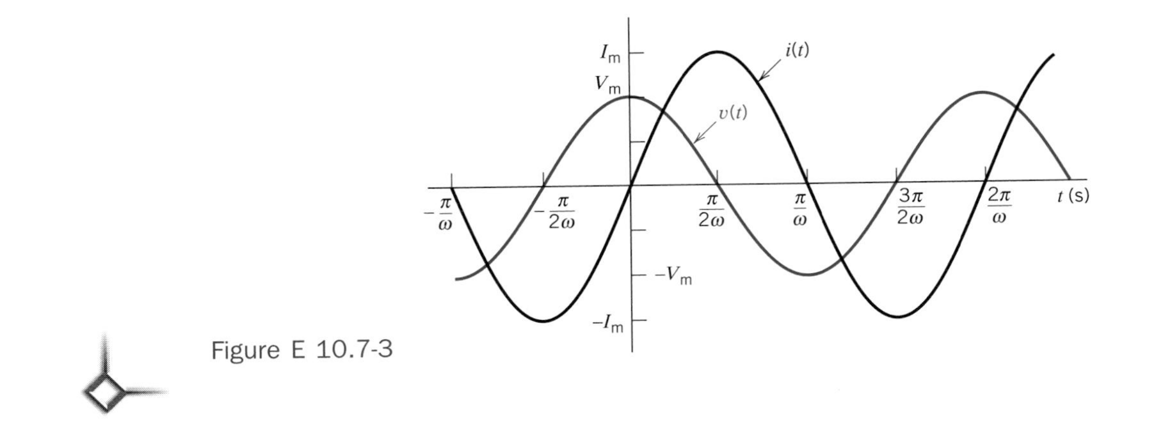

Figure E 10.7-3

10.8 IMPEDANCE AND ADMITTANCE

The relationships in the frequency domain for the phasor current and phasor voltage of a capacitor, inductor, and resistor are summarized in Table 10.7-1. These relationships appear similar to Ohm's law for resistors.

We will define the *impedance* of an element as the ratio of the phasor voltage to the phasor current, which we denote by **Z**. Therefore,

$$\mathbf{Z} = \frac{\mathbf{V}}{\mathbf{I}} \tag{10.8-1}$$

This is called Ohm's law in phasor notation.

Since $\mathbf{V} = V_m \angle \phi$ and $\mathbf{I} = I_m \angle \beta$, we have

$$\mathbf{Z} = \frac{V_m \angle \phi}{I_m \angle \beta} = \frac{V_m}{I_m} \angle \phi - \beta \tag{10.8-2}$$

Thus, the impedance is said to have a magnitude $|\mathbf{Z}|$ and a phase angle $\theta = \phi - \beta$. Therefore,

$$|\mathbf{Z}| = \frac{V_m}{I_m} \tag{10.8-3}$$

and

$$\theta = \phi - \beta \tag{10.8-4}$$

Impedance has a role similar to the role of resistance in resistive circuits. Also, since it is a ratio of volts to amperes, impedance has units of ohms. Impedance is the ratio of two phasors; however, it is not a phasor itself. Impedance is a complex number that relates one phasor **V** to the other phasor **I** as

$$\mathbf{V} = \mathbf{Z}\mathbf{I} \tag{10.8-5}$$

The phasors **V** and **I** may be transformed to the time domain to yield the steady-state voltage or current, respectively. Impedance has no meaning in the time domain, however.

With the concept of impedance we can solve for the behavior of sinusoidally excited circuits, using complex algebra, in the same way we solved resistive circuits.

Since the impedance is a complex number, it may be written in several forms, as follows:

$$\begin{aligned} \mathbf{Z} &= |\mathbf{Z}| \angle \theta && \longrightarrow \text{polar form} \\ &= Z\, e^{j\theta} && \longrightarrow \text{exponential form} \\ &= R + jX && \longrightarrow \text{rectangular form} \end{aligned} \quad (10.8\text{-}6)$$

where R is the real part and X is the imaginary part of the complex number $\mathbf{Z}$. We introduce the notation, in Eq. 10.8-6, $|\mathbf{Z}| = Z$. Thus, the magnitude of the impedance can be written as Z (not boldface). The $R = \text{Re}\, \mathbf{Z}$ is called the resistive part of the impedance, and $X = \text{Im}\, \mathbf{Z}$ is called the reactive part of the impedance. Both R and X are measured in ohms.

We also note that the magnitude of the impedance is

$$Z = \sqrt{R^2 + X^2} \quad (10.8\text{-}7)$$

and the phase angle is

$$\theta = \tan^{-1} \frac{X}{R} \quad (10.8\text{-}8)$$

These relationships are summarized graphically in the complex plane, in Figure 10.8-1. As an example, let us consider

$$\mathbf{Z} = 2 + j2$$

Then,

$$Z = \sqrt{8}$$

and

$$\theta = 45^\circ$$

The three elements R, L, and C are uniquely represented by an impedance that follows from their **V–I** relationship. For a resistor we have

$$\mathbf{V} = R\,\mathbf{I}$$

and, therefore,

$$\mathbf{Z} = R \quad (10.8\text{-}9)$$

For the inductor we have

$$\mathbf{V} = j\omega L\,\mathbf{I}$$

and, therefore,

$$\mathbf{Z} = j\omega L \quad (10.8\text{-}10)$$

Finally, for the capacitor we have

$$\mathbf{V} = \frac{\mathbf{I}}{j\omega C}$$

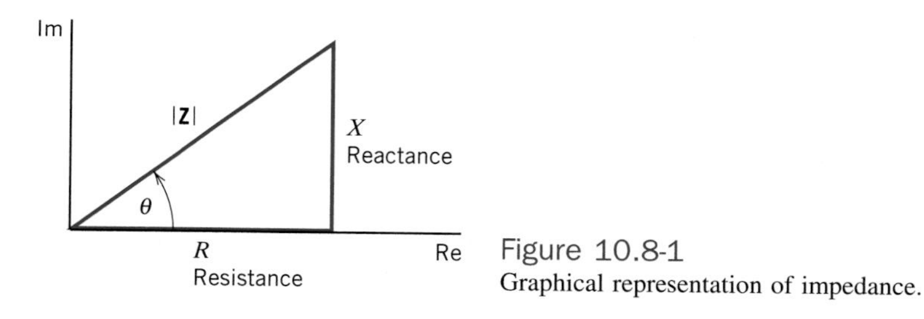

Figure 10.8-1
Graphical representation of impedance.

so that

$$\mathbf{Z} = \frac{1}{j\omega C} \tag{10.8-11}$$

The impedances for R, L, and C are used in Table 10.7-1 to represent resistors, inductors, and capacitors in the frequency domain. The unit for an impedance is ohms.

The reciprocal of impedance is called the *admittance* and is denoted by $\mathbf{Y}$:

$$\mathbf{Y} = \frac{1}{\mathbf{Z}} \tag{10.8-12}$$

Admittance is analogous to conductance for resistive circuits. The units of admittance are siemens, abbreviated as S. Recalling from Eq. 10.8-6 that $\mathbf{Z} = Z\underline{/\theta}$, we have

$$\mathbf{Y} = \frac{1}{|\mathbf{Z}|\underline{/\theta}} = |\mathbf{Y}|\underline{/-\theta} \tag{10.8-13}$$

Therefore, $|\mathbf{Y}| = 1/|\mathbf{Z}|$ and the angle of $\mathbf{Y}$ is $-\theta$. We may also write the magnitude relation as $Y = 1/Z$.

Using the form

$$\mathbf{Z} = R + jX$$

we obtain

$$\mathbf{Y} = \frac{1}{R + jX} = \frac{R - jX}{R^2 + X^2} = G + jB \tag{10.8-14}$$

Note that G is not simply the reciprocal of R, nor is B the reciprocal of X. The real part of admittance, G, is called the *conductance,* and the imaginary part, B, is called the *susceptance*. The units of G and B are siemens.

The impedance of an element is $\mathbf{Z} = R + jX$. The element is inductive if the reactive part X is positive, capacitive if X is negative. Since $\mathbf{Y}$ is the reciprocal of $\mathbf{Z}$ and $\mathbf{Y} = G + jB$, one can also say that if B is positive the element is capacitive and that a negative B indicates an inductive element.

Let us consider the capacitor $C = 1$ mF and find its impedance and admittance. The impedance of a capacitor is

$$\mathbf{Z} = \frac{1}{j\omega C}$$

Therefore, in addition to the value of $C = 1$ mF, we need the frequency ω. If we consider the case $\omega = 100$ rad/s, we obtain

$$\mathbf{Z} = \frac{1}{j0.1} = \frac{10}{j} = -10j = 10\underline{/-90^\circ}\ \Omega$$

To find the admittance, we note that

$$\mathbf{Y} = \frac{1}{\mathbf{Z}} = j\omega C = j0.1 = 0.1\underline{/90^\circ}\ \text{S}$$

Exercise 10.8-1 Figure E 10.8-1*a* shows a circuit represented in the time domain. Figure 10.8-1*b* shows the same circuit represented in the frequency domain, using pha-

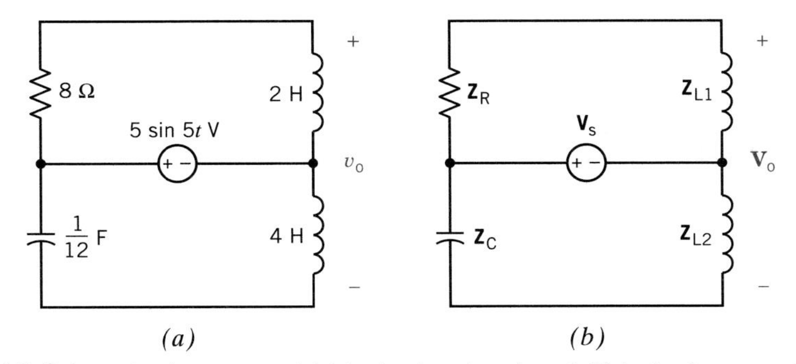

Figure E 10.8-1 A circuit represented (*a*) in the time domain and (*b*) in the frequency domain.

sors and impedances. $\mathbf{Z}_R$, $\mathbf{Z}_C$, $\mathbf{Z}_{L1}$, and $\mathbf{Z}_{L2}$ are the impedances corresponding to the resistor, capacitor, and two inductors in Figure 10.8-1*a*. $\mathbf{V}_s$ is the phasor corresponding to the voltage of the voltage source. Determine $\mathbf{Z}_R$, $\mathbf{Z}_C$, $\mathbf{Z}_{L1}$, $\mathbf{Z}_{L2}$, and $\mathbf{V}_s$.

Hint: $5 \sin 5t = 5 \cos(5t - 90°)$.

Answer:

$$\mathbf{Z}_R = 8\ \Omega, \mathbf{Z}_C = \frac{1}{j5\frac{1}{12}} = \frac{2.4}{j} = \frac{j2.4}{j \times j} = -j2.4\ \Omega, \mathbf{Z}_{L1} = j5(2) = j10\ \Omega,$$

$$\mathbf{Z}_{L2} = j5(4) = j20\ \Omega \text{ and } \mathbf{V}_s = 5\underline{/-90°}\text{ V}.$$

Exercise 10.8-2 Figure E 10.8-2*a* shows a circuit represented in the time domain. Figure 10.8-2*b* shows the same circuit represented in the frequency domain, using phasors and impedances. $\mathbf{Z}_R$, $\mathbf{Z}_C$, $\mathbf{Z}_{L1}$, and $\mathbf{Z}_{L2}$ are the impedances corresponding to the resistor, capacitor, and two inductors in Figure 10.8-2*a*. $\mathbf{I}_s$ is the phasor corresponding to the current of the current source. Determine $\mathbf{Z}_R$, $\mathbf{Z}_C$, $\mathbf{Z}_{L1}$, $\mathbf{Z}_{L2}$, and $\mathbf{I}_s$.

Answer: $$\mathbf{Z}_R = 8\ \Omega, \mathbf{Z}_C = \frac{1}{j3\frac{1}{12}} = \frac{4}{j} = \frac{j4}{j \times j} = -j4\ \Omega, \mathbf{Z}_{L1} = j3(2) = j6\ \Omega,$$

$$\mathbf{Z}_{L2} = j3(4) = j12\ \Omega \text{ and } \mathbf{I}_s = 4\underline{/15°}\text{ V}.$$

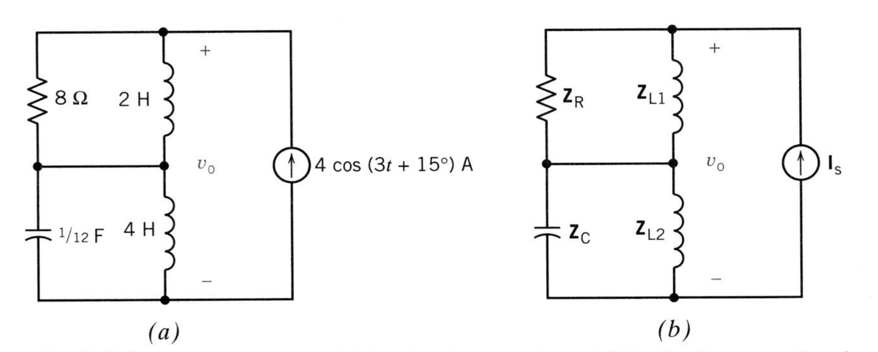

Figure E 10.8-2 A circuit represented (*a*) in the time domain and (*b*) in the frequency domain.

10.9 KIRCHHOFF'S LAWS USING PHASORS

Kirchhoff's current law and voltage law were considered earlier in the time domain. Consider the KVL around a closed path, which requires that

$$v_1 + v_2 + v_3 + \cdots + v_n = 0 \quad (10.9\text{-}1)$$

For sinusoidal steady-state voltages, we may write the equation in terms of cosine waveforms as

$$V_{m_1}\cos(\omega t + \theta_1) + V_{m_2}\cos(\omega t + \theta_2) + \cdots + V_{m_n}\cos(\omega t + \theta_n) = 0 \quad (10.9\text{-}2)$$

All the information concerning each voltage v_n is incorporated in the magnitude and phase, V_{m_n} and θ_n (assuming we keep note of ω, which is the same for each term). Equation 10.9-2 can be rewritten, using Euler's identity, as

$$\text{Re}\{V_{m_1}e^{j\theta_1}e^{j\omega t}\} + \cdots + \text{Re}\{V_{m_n}e^{j\theta_n}e^{j\omega t}\} = 0$$

or

$$\text{Re}\{V_{m_1}e^{j\theta_1}e^{j\omega t} + \cdots + V_{m_n}e^{j\theta_n}e^{j\omega t}\} = 0$$

We can factor out the $e^{j\omega t}$ to obtain

$$\text{Re}\{(V_{m_1}e^{j\theta_1} + \cdots + V_{m_n}e^{j\theta_n})e^{j\omega t}\} = 0$$

Writing $V_{m_p}e^{j\theta_p}$ as $\mathbf{V}_p$, we have

$$\text{Re}(\mathbf{V}_1 + \mathbf{V}_2 + \cdots + \mathbf{V}_n)e^{j\omega t} = 0$$

Since $e^{j\omega t}$ cannot equal zero, we require that

$$\mathbf{V}_1 + \mathbf{V}_2 + \cdots + \mathbf{V}_n = 0 \quad (10.9\text{-}3)$$

Therefore, we have the important result that the sum of the *phasor voltages* in a closed path is zero. Thus, Kirchhoff's voltage law holds in the frequency domain with phasor voltages. Using a similar process, one can show that Kirchhoff's current law holds in the frequency domain for phasors, so that at a node we have

$$\mathbf{I}_1 + \mathbf{I}_2 + \cdots + \mathbf{I}_n = 0 \quad (10.9\text{-}4)$$

Since both the KVL and the KCL hold in the frequency domain, it is easy to conclude that all the techniques of analysis we developed for resistive circuits hold for phasor currents and voltages. For example, we can use the principle of superposition, source transformations, Thévenin and Norton equivalent circuits, and node voltage and mesh current analysis. All these methods apply as long as the circuit is linear.

First, let us consider impedances connected in series, as shown in Figure 10.9-1. The phasor current $\mathbf{I}$ flows through each impedance. Applying KVL, we can write

$$\mathbf{V}_1 + \mathbf{V}_2 + \cdots + \mathbf{V}_n = \mathbf{V}$$

Since $\mathbf{V}_j = \mathbf{Z}_j\mathbf{I}_j$, we have

$$(\mathbf{Z}_1 + \mathbf{Z}_2 + \cdots + \mathbf{Z}_n)\mathbf{I} = \mathbf{V}$$

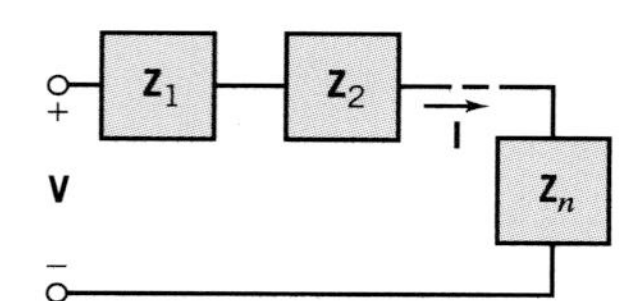

Figure 10.9-1 Impedances in series.

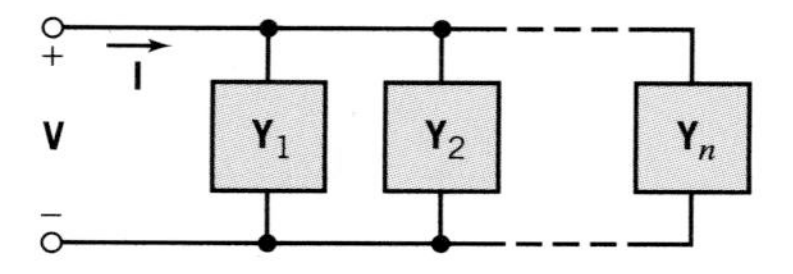

Figure 10.9-2 Admittances in parallel.

Therefore, the equivalent impedance seen at the input terminals is

$$\mathbf{Z}_{eq} = \mathbf{Z}_1 + \mathbf{Z}_2 + \cdots + \mathbf{Z}_n \tag{10.9-5}$$

Thus, the equivalent impedance for a series of impedances is the sum of the individual impedances.

Consider the set of parallel admittances shown in Figure 10.9-2. It can easily be shown that the equivalent admittance $\mathbf{Y}_{eq}$ is

$$\mathbf{Y}_{eq} = \mathbf{Y}_1 + \mathbf{Y}_2 + \cdots + \mathbf{Y}_n \tag{10.9-6}$$

In the case of two parallel admittances, we have

$$\mathbf{Y}_{eq} = \mathbf{Y}_1 + \mathbf{Y}_2$$

and the corresponding equivalent impedance is

$$\mathbf{Z}_{eq} = \frac{1}{\mathbf{Y}_{eq}} = \frac{1}{\mathbf{Y}_1 + \mathbf{Y}_2} = \frac{\mathbf{Z}_1\mathbf{Z}_2}{\mathbf{Z}_1 + \mathbf{Z}_2} \tag{10.9-7}$$

Similarly, the current divider and voltage divider rules hold for phasor currents and voltages.

Example 10.9-1

Consider the RLC circuit shown in Figure 10.9-3 where $R = 9\ \Omega$, $L = 10$ mH, and $C = 1$ mF. Find the steady-state current i using phasors.

Solution

First, we redraw the circuit in phasor form as shown in Figure 10.9-3b. Then, we write KVL around the mesh to obtain

$$R\mathbf{I} + \mathbf{Z}_2\mathbf{I} + \mathbf{Z}_3\mathbf{I} = \mathbf{V}_s \tag{10.9-8}$$

where

$$\mathbf{Z}_2 = j\omega L$$

and

$$\mathbf{Z}_3 = \frac{1}{j\omega C}$$

Since $\omega = 100$, $L = 10$ mH, and $C = 1$ mF, we have

$$\mathbf{Z}_2 = j1$$

and

$$\mathbf{Z}_3 = -j10$$

Therefore, Eq. 10.9-8 becomes

$$(9 + j1 - j10)\mathbf{I} = \mathbf{V}_s \tag{10.9-9}$$

or

$$\mathbf{I} = \frac{\mathbf{V}_s}{9 - j9}$$

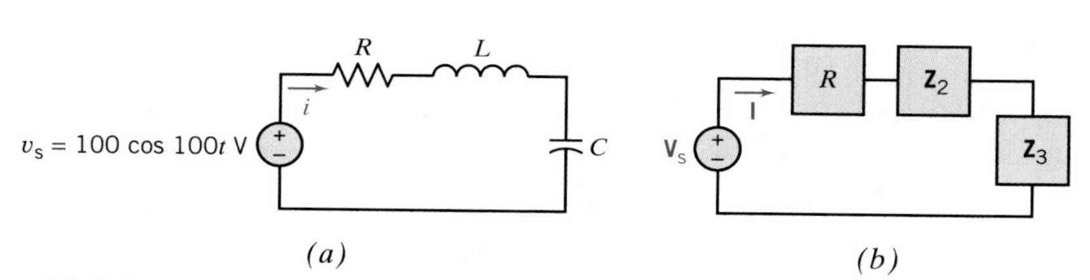

Figure 10.9-3 (a) An RLC series circuit. (b) Circuit in phasor form.

Since $\mathbf{V}_s = 100\underline{/0^\circ}$, we obtain the phasor current as

$$\mathbf{I} = \frac{100\underline{/0^\circ}}{9\sqrt{2}\underline{/-45^\circ}} = 7.86\underline{/45^\circ}$$

Therefore, the steady-state current in the time domain is

$$i = 7.86\cos(100t + 45^\circ)\text{ A}$$

Example 10.9-2

Find the steady-state voltage v for the circuit of Figure 10.9-4a.

Solution

First represent the circuit in the frequency domain, using impedances and phasors. The impedance of the inductor is

$$j\omega L = j1000(10 \times 10^{-3}) = j10\ \Omega$$

The impedance of the capacitor is

$$\frac{1}{j\omega C} = \frac{1}{j1000(100 \times 10^{-6})} = \frac{10}{j} = -j10\ \Omega$$

The phasor representation of the input current is

$$10\underline{/0^\circ} = 10\text{ A}$$

Figure 10.9-4b shows the frequency domain representation of the circuit. The phasor voltage $\mathbf{V}$ can be obtained by applying Kirchhoff's current law at the top node of the circuit in Figure 10.9-4b to get

$$\frac{\mathbf{V}}{10} + \frac{\mathbf{V}}{10 + j10} + \frac{\mathbf{V}}{-j10} = 10 \tag{10.9-10}$$

or

$$\frac{\mathbf{V}}{10} + \frac{\mathbf{V}}{10 + j10}\left(\frac{10 - j10}{10 - j10}\right) + \frac{\mathbf{V}}{-j10} = 0.1\mathbf{V} + (0.05 - j0.05)\mathbf{V} + j0.1\mathbf{V} = 10$$

Solving for $\mathbf{V}$, we have

$$\mathbf{V} = \frac{10}{0.158\underline{/18.4^\circ}} = 63.3\underline{/-18.4^\circ}$$

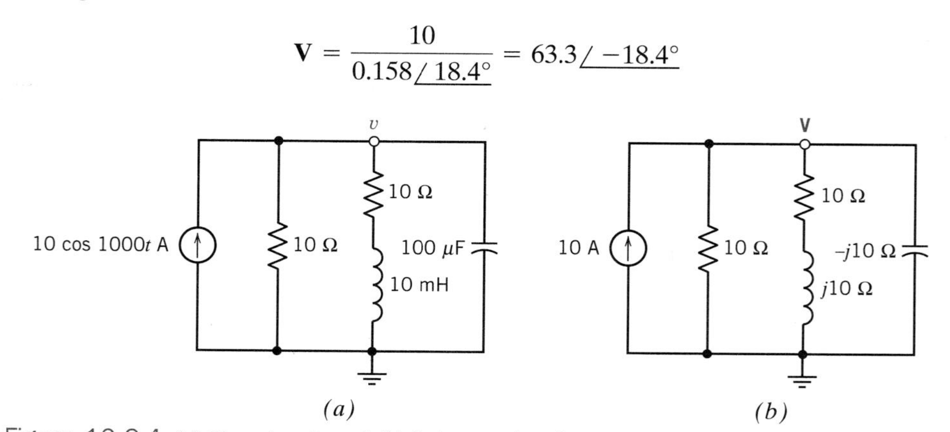

Figure 10.9-4 (a) Time domain and (b) frequency domain representation of the circuit for Example 10.9-2.

Therefore, we have the steady-state voltage as

$$v = 63.3 \cos(1000t - 18.4°)\ \text{V}$$

Exercise 10.9-1 Determine the voltage $v(t)$ for the circuit of Figure E 10.9-1.

Hint: Analyze the circuit in the frequency domain, using impedances and phasors. Use voltage division, twice. Add the results.

Answer: $v(t) = 3.58 \cos(5t + 47.2°)$ V.

Exercise 10.9-2 Determine the voltage $v(t)$ for the circuit of Figure E 10.9-2.

Hint: Analyze the circuit in the frequency domain, using impedances and phasors. Replace parallel impedances with an equivalent impedance, twice. Apply KVL.

Answer: $v(t) = 14.4 \cos(3t - 22°)$ V.

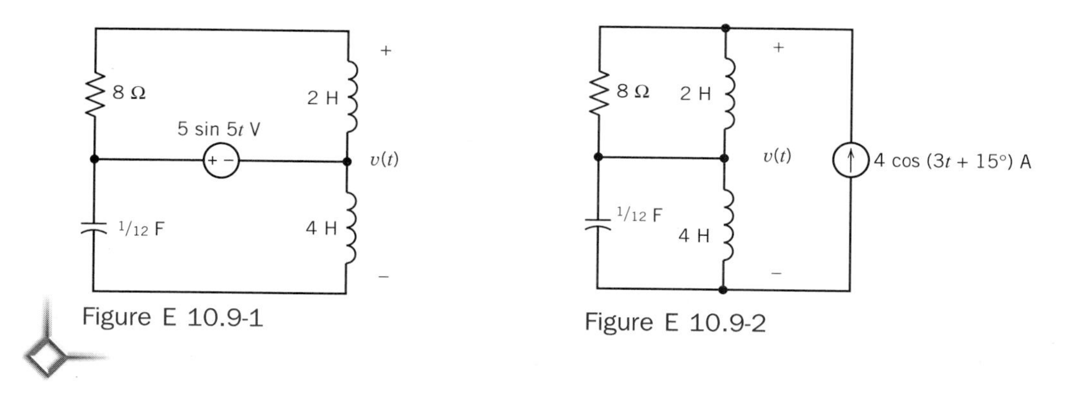

Figure E 10.9-1

Figure E 10.9-2

10.10 NODE VOLTAGE AND MESH CURRENT ANALYSIS USING PHASORS

Circuit analysis in the frequency domain follows the same procedure as we utilized for resistive circuits; however, we use impedances and phasors instead of resistances and time functions. Since Ohm's law can be used in the frequency domain, we use the relationship $\mathbf{V} = \mathbf{ZI}$ for the passive elements and proceed to use the node voltage and mesh current techniques.

As an example of the node voltage method using phasors, consider the circuits of Figure 10.10-1 when $i_s = I_m \cos \omega t$. For a specified ω and for specified L and C, we can obtain the impedance for the L and C elements. When $\omega = 1000$ rad/s and $C = 100\ \mu$F, we obtain

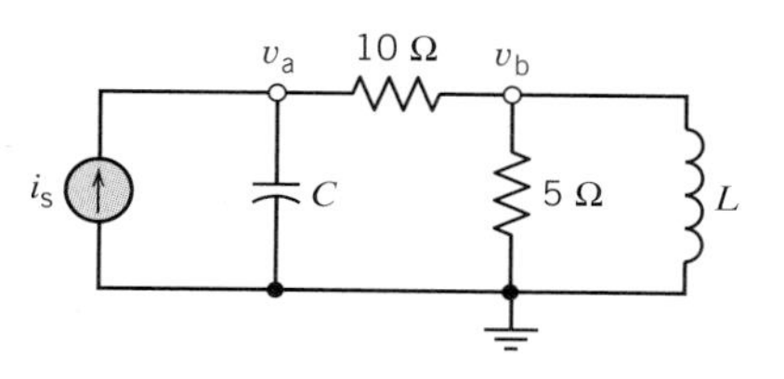

Figure 10.10-1
Circuit for which we wish to determine v_a and v_b.

$$\mathbf{Z}_1 = \frac{1}{j\omega C} = -j10$$

When $L = 5$ mH for the inductor, we have the impedance

$$\mathbf{Z}_L = j\omega L = j5$$

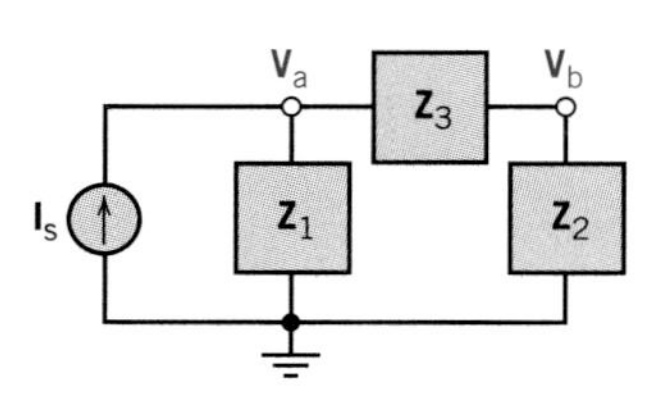

Figure 10.10-2
Circuit equivalent to that of Figure 10.10-1 in phasor form.

Then, we may redraw the circuit shown in Figure 10.10-1 using the phasor format shown in Figure 10.10-2. Clearly, $\mathbf{Z}_3 = 10\ \Omega$ and $\mathbf{Z}_2$ is obtained from the parallel combination of the 5 Ω resistor and the inductor's impedance $\mathbf{Z}_L$. Rather than obtaining $\mathbf{Z}_2$, let us determine $\mathbf{Y}_2$, which is readily found by adding the two parallel admittances as follows:

$$\mathbf{Y}_2 = \frac{1}{5} + \frac{1}{\mathbf{Z}_L} = \frac{1}{5} + \frac{1}{j5} = \frac{1}{5}(1 - j)$$

Using KCL at node a, we have

$$\frac{\mathbf{V}_a}{\mathbf{Z}_1} + \frac{\mathbf{V}_a - \mathbf{V}_b}{\mathbf{Z}_3} = \mathbf{I}_s \tag{10.10-1}$$

At node b, we have

$$\frac{\mathbf{V}_b}{\mathbf{Z}_2} + \frac{\mathbf{V}_b - \mathbf{V}_a}{\mathbf{Z}_3} = 0 \tag{10.10-2}$$

Rearranging Eqs. 10.10-1 and 10.10-2, we obtain

$$(\mathbf{Y}_1 + \mathbf{Y}_3)\mathbf{V}_a + (-\mathbf{Y}_3)\mathbf{V}_b = \mathbf{I}_s \tag{10.10-3}$$

$$(-\mathbf{Y}_3)\mathbf{V}_a + (\mathbf{Y}_2 + \mathbf{Y}_3)\mathbf{V}_b = 0 \tag{10.10-4}$$

where we use the admittance $\mathbf{Y}_n = 1/\mathbf{Z}_n$ and $\mathbf{I}_s = I_m \angle 0°$

We find that Eqs. 10.10-3 and 10.10-4 are similar to the node voltage equations we found in Chapter 4 for resistive circuits. In this case, however, we obtain the node voltage equations in terms of phasor currents, phasor voltages, and complex impedances and admittances.

In general, we may state that for circuits containing only admittances and independent sources, KCL at node k requires that the coefficient of $\mathbf{V}_k$ be the sum of the admittances at node k and the coefficients of the other terms are the negative of the admittance between those nodes and the kth node.

Let us proceed to solve for $\mathbf{V}_a$ for the circuit shown in Figures 10.10-1 and 10.10-2 when $I_m = 10$ A. Substituting the admittances into Eqs. 10.10-3 and 10.10-4, we have

$$\left(\frac{1}{-j10} + \frac{1}{10}\right)\mathbf{V}_a + \frac{-1}{10}\mathbf{V}_b = 10 \tag{10.10-5}$$

$$\frac{-1}{10}\mathbf{V}_a + \left[\frac{1}{5}(1 - j) + \frac{1}{10}\right]\mathbf{V}_b = 0 \tag{10.10-6}$$

We then use Cramer's rule to solve for $\mathbf{V}_a$, obtaining

$$\mathbf{V}_a = \frac{100(3 - 2j)}{4 + j}$$

$$= \frac{100(3 - 2j)(4 - j)}{17} = \frac{100}{17}(10 - 11j) = 87.5\angle -47.7°$$

Therefore, we have the steady-state voltage v_a:

$$v_a = 87.5 \cos(1000t - 47.7°) \text{ V}$$

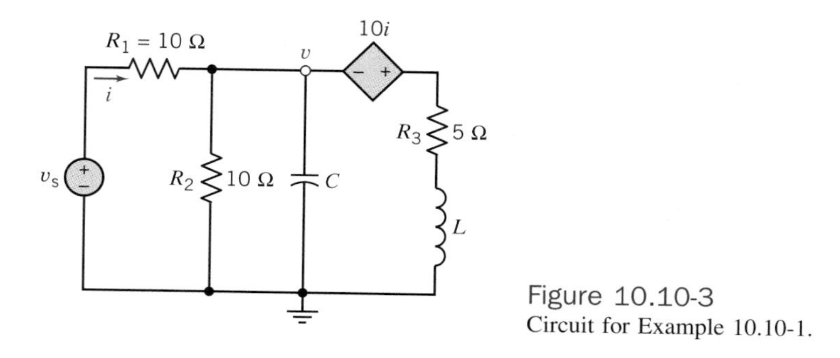

Figure 10.10-3
Circuit for Example 10.10-1.

The general nodal analysis methods of Chapter 4 may be utilized here, where we are careful to note that we use complex impedances and admittances and phasor voltages and currents. Once we have determined the desired phasor currents or voltages, we transform them back to the time domain to obtain the steady-state sinusoidal current or voltage desired. We use the concept of a supernode, if necessary, and include the effect of a dependent source, if required.

Example 10.10-1

A circuit is shown in Figure 10.10-3 with $\omega = 10$ rad/s, $L = 0.5$ H, and $C = 10$ mF. Find the node voltage v in its sinusoidal steady-state form when $v_s = 10 \cos \omega t$ V.

Solution

The circuit has a dependent source between two nodes, so we identify a supernode as shown in Figure 10.10-4, where we also show the impedance for each element in phasor form. For example, the impedance of the inductor is $\mathbf{Z}_L = j\omega L = j5$. Similarly, the impedance for the capacitor is

$$\mathbf{Z}_c = \frac{1}{j\omega C} = \frac{10}{j} = -j10$$

First, we note that $\mathbf{Y}_1 = 1/R_1 = 1/10$. We now wish to bring together the two parallel admittances for R_2 and C to yield one admittance $\mathbf{Y}_2$ as shown in Figure 10.10-5. We then obtain

$$\mathbf{Y}_2 = \frac{1}{R_2} + \frac{1}{\mathbf{Z}_c} = \frac{1}{10} + \frac{j}{10} = \frac{1}{10}(1 + j)$$

We may obtain $\mathbf{Y}_3$ for the series resistance and inductance as

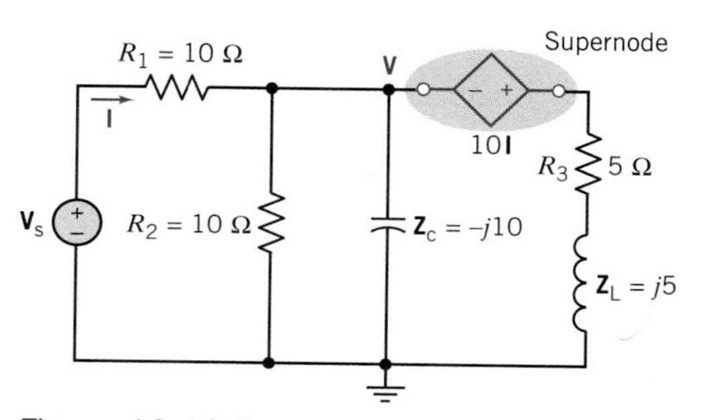

Figure 10.10-4 Circuit for Example 10.10-1 in phasor notation.

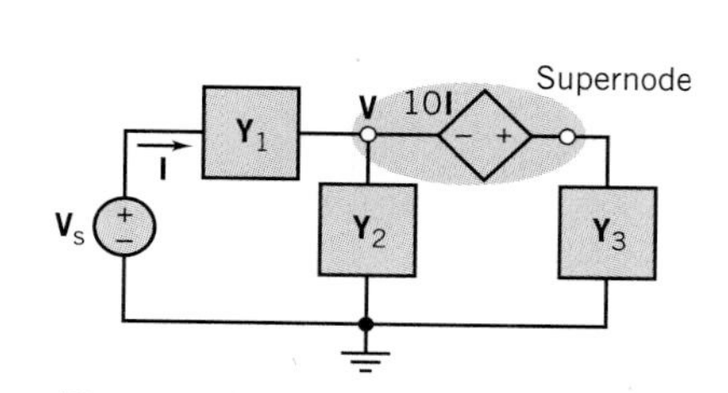

Figure 10.10-5 Circuit for Example 10.10-1 with three admittances and the supernode identified.

$$\mathbf{Y}_3 = \frac{1}{\mathbf{Z}_3}$$

where $\mathbf{Z}_3 = R_3 + \mathbf{Z}_L = 5 + j5$. Therefore, we have

$$\mathbf{Y}_3 = \frac{1}{5 + j5} = \frac{1}{50}(5 - j5)$$

Writing the KCL at the supernode of Figure 10.10-5, we have

$$\mathbf{Y}_1(\mathbf{V} - \mathbf{V}_s) + \mathbf{Y}_2\mathbf{V} + \mathbf{Y}_3(\mathbf{V} + 10\mathbf{I}) = 0 \tag{10.10-7}$$

Furthermore, we note that

$$\mathbf{I} = \mathbf{Y}_1(\mathbf{V}_s - \mathbf{V}) \tag{10.10-8}$$

Substituting Eq. 10.10-8 into Eq. 10.10-7, we obtain

$$\mathbf{Y}_1(\mathbf{V} - \mathbf{V}_s) + \mathbf{Y}_2\mathbf{V} + \mathbf{Y}_3[\mathbf{V} + 10\mathbf{Y}_1(\mathbf{V}_s - \mathbf{V})] = 0$$

Rearranging, we have

$$(\mathbf{Y}_1 + \mathbf{Y}_2 + \mathbf{Y}_3 - 10\mathbf{Y}_1\mathbf{Y}_3)\,\mathbf{V} = (\mathbf{Y}_1 - 10\mathbf{Y}_1\mathbf{Y}_3)\mathbf{V}_s$$

Therefore
$$\mathbf{V} = \frac{(\mathbf{Y}_1 - 10\mathbf{Y}_1\mathbf{Y}_3)\mathbf{V}_s}{\mathbf{Y}_1 + \mathbf{Y}_2 + \mathbf{Y}_3 - 10\mathbf{Y}_1\mathbf{Y}_3}$$

Since $\mathbf{V}_s = 10\angle 0^\circ$, we have

$$\mathbf{V} = \frac{(\frac{1}{10} - \frac{1}{50}(5 - j5))10}{\frac{1}{10} + \frac{1}{10}(1 + j)} = \frac{1 - (1 - j)}{\frac{1}{10}(2 + j)} = \frac{10j}{2 + j}$$

Therefore, we obtain

$$v = \frac{10}{\sqrt{5}}\cos(10t + 63.4^\circ)\text{ V}$$

The processes of node voltage and mesh current analysis using phasors for determining the steady-state sinusoidal response of a circuit are recorded in Tables 10.10-1 and 10.10-2, respectively.

Mesh current analysis, using the method of Table 10.10-2, is relatively straightforward. Once you have the impedance of each element, you may readily write the KVL equations for each mesh.

Example 10.10-2

Find the steady-state sinusoidal current i_1 for the circuit of Figure 10.10-6 when $v_s = 10\sqrt{2}\cos(\omega t + 45^\circ)$ V and $\omega = 100$ rad/s. Also, $L = 30$ mH and $C = 5$ mF.

Solution

First, we transform the source voltage to phasor form to obtain

$$\mathbf{V}_s = 10\sqrt{2}\angle 45^\circ = 10 + 10j$$

We then select the two mesh currents as $\mathbf{I}_1$ and $\mathbf{I}_2$, as shown in Figure 10.10-7. Since the frequency of the source is $\omega = 100$, we find that the inductance has an impedance of

$$\mathbf{Z}_L = j\omega L = j3$$

Table 10.10-1 **Node Voltage Analysis Using the Phasor Concept to Find the Sinusoidal Steady-State Node Voltages**

1. Convert the independent sources to phasor form.
2. Select the nodes and the reference node and label the node voltages in the time domain, v_n, and the corresponding phasor voltage, $\mathbf{V}_n$.
3. If the circuit contains only independent current sources, proceed to step 5; otherwise proceed to step 4.
4. If the circuit contains a voltage source, select one of the following three cases and the associated method:

Case	Method
a. The voltage source connects node q and the reference node.	Set $\mathbf{V}_q = \mathbf{V}_s$ and proceed.
b. The voltage source lies between two nodes.	Create a supernode including both nodes.
c. The voltage source in series with an impedance lies between node d and the ground with its positive terminal at node d.	Replace the voltage source and series impedance with a parallel combination of an admittance $\mathbf{Y}_1 = 1/\mathbf{Z}_1$ and a current source $\mathbf{I}_1 = \mathbf{V}_s\mathbf{Y}_1$ entering node d.

5. Using the known frequency of the sources, ω, find the impedance of each element in the circuit.
6. For each branch at a given node, find the equivalent admittance of that branch, $\mathbf{Y}_n$.
7. Write KCL at each node.
8. Solve for the desired node voltage $\mathbf{V}_a$ using Cramer's rule.
9. Convert the phasor voltage $\mathbf{V}_a$ back to the time domain form.

Table 10.10-2 **Mesh Current Analysis Using the Phasor Concept to Find the Sinusoidal Steady-State Mesh Currents**

1. Convert the independent sources to phasor form.
2. Select the mesh currents and label the currents in the time domain, i_n, and the corresponding phasor currents, $\mathbf{I}_n$.
3. If the circuit contains only independent voltage sources, proceed to step 5; otherwise proceed to step 4.
4. If the circuit contains a current source, select one of the following two cases and the associated method:

Case	Method
a. The current source appears as an element of only one mesh, n.	Equate the mesh current $\mathbf{I}_n$ to the current of the source, accounting for the direction of the current source.
b. The current source is common to two meshes.	Create a supermesh as the periphery of the two meshes. In step 6 write one KVL equation around the periphery of the supermesh. Also record the constraining equation incurred by the current source.

5. Using the known frequency of the sources, ω, find the impedance of each element in the circuit.
6. Write KVL for each mesh.
7. Solve for the desired mesh current $\mathbf{I}_n$ using Cramer's rule.
8. Convert the phasor current $\mathbf{I}_n$ back to the time domain form.

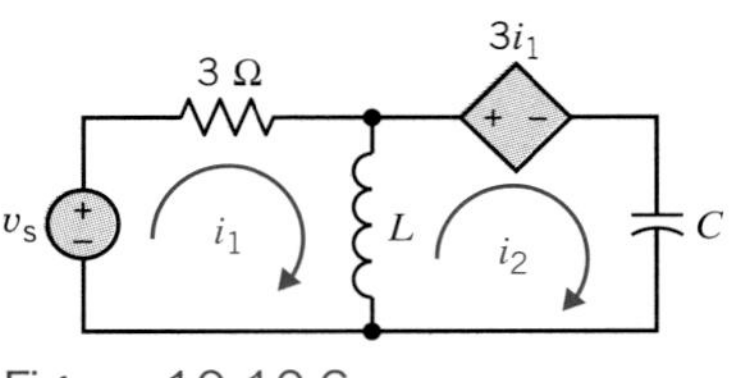

Figure 10.10-6 Circuit of Example 10.10-2.

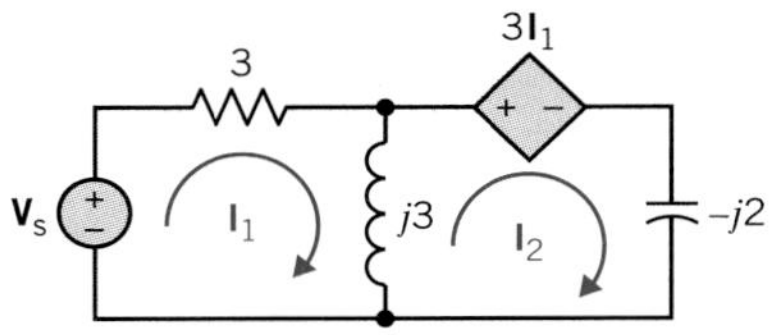

Figure 10.10-7 Circuit of Example 10.10-2 with phasors and impedances.

The capacitor has an impedance of

$$\mathbf{Z}_c = \frac{1}{j\omega C} = \frac{1}{j(\frac{1}{2})} = -j2\Omega$$

We can then summarize the circuit's phasor currents and the impedance of each element by redrawing the circuit in terms of phasors, as shown in Figure 10.10-7. Now we can write the KVL equations for each mesh, obtaining

$$\text{mesh 1:} \quad (3 + j3)\mathbf{I}_1 - j3\mathbf{I}_2 = \mathbf{V}_s$$

$$\text{mesh 2:} \quad (3 - j3)\mathbf{I}_1 + (j3 - j2)\mathbf{I}_2 = 0$$

Solving for $\mathbf{I}_1$ using Cramer's rule, we have

$$\mathbf{I}_1 = \frac{(10 + j10)j}{\Delta}$$

where the determinant is

$$\Delta = (3 + j3)(j) + j3(3 - j3) = 6 + 12j$$

Therefore, we have

$$\mathbf{I}_1 = \frac{10j - 10}{6 + 12j}$$

Continuing, we obtain

$$\mathbf{I}_1 = \frac{10(j - 1)}{6(1 + 2j)} = \frac{10(\sqrt{2}\angle 135^\circ)}{6(\sqrt{5}\angle 63.4^\circ)} = 1.05\angle 71.6^\circ$$

Thus, the steady-state time response is

$$i_1 = 1.05\cos(100t + 71.6^\circ)\ \text{A}$$

Example 10.10-3

Find the steady-state current i_1 when the voltage source is $v_s = 10\sqrt{2}\cos(\omega t + 45^\circ)$ V and the current source is $i_s = 3\cos\omega t$ A for the circuit of Figure 10.10-8. The circuit of the figure provides the impedance in ohms for each element at the specified ω.

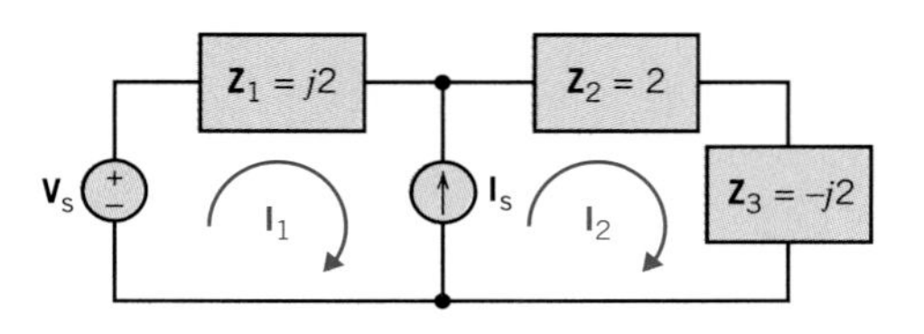

Figure 10.10-8 Circuit of Example 10.10-3.

Solution

First, we transform the independent sources into phasor form. The voltage source is

$$\mathbf{V}_s = 10\sqrt{2}\underline{/45^\circ} = 10(1 + j)\,V$$

and the current source is

$$\mathbf{I}_s = 3\underline{/0^\circ}\,A$$

We note that the current source connects the two meshes and provides a constraining equation

$$\mathbf{I}_2 - \mathbf{I}_1 = \mathbf{I}_s \qquad (10.10\text{-}9)$$

Creating a supermesh around the periphery of the two meshes, we write one KVL equation, obtaining

$$\mathbf{I}_1\mathbf{Z}_1 + \mathbf{I}_2(\mathbf{Z}_2 + \mathbf{Z}_3) = \mathbf{V}_s \qquad (10.10\text{-}10)$$

Since we wish to solve for $\mathbf{I}_1$, we will use $\mathbf{I}_2$ from Eq. 10.10-9 and substitute it into Eq. 10.10-10, obtaining

$$\mathbf{I}_1\mathbf{Z}_1 + (\mathbf{I}_s + \mathbf{I}_1)(\mathbf{Z}_2 + \mathbf{Z}_3) = \mathbf{V}_s$$

Rearranging, we have

$$(\mathbf{Z}_1 + \mathbf{Z}_2 + \mathbf{Z}_3)\mathbf{I}_1 = \mathbf{V}_s - (\mathbf{Z}_2 + \mathbf{Z}_3)\mathbf{I}_s$$

Therefore, we have

$$\mathbf{I}_1 = \frac{\mathbf{V}_s - (\mathbf{Z}_2 + \mathbf{Z}_3)\mathbf{I}_s}{\mathbf{Z}_1 + \mathbf{Z}_2 + \mathbf{Z}_3}$$

Substituting the impedances and the sources, we have

$$\mathbf{I}_1 = \frac{(10 + j10) - (2 - j2)3}{2} = 2 + j8 = 8.25\underline{/76^\circ}\,A$$

Thus, we obtain

$$i_1 = 8.25\cos(\omega t + 76^\circ)\text{ A}$$

Exercise 10.10-1 A circuit has the form shown in Figure E 10.10-1 when $i_{s1} = 1\cos 100t$ A and $i_{s2} = 0.5\cos(100t - 90^\circ)$ A. Find the voltage v_a in the time domain.

Answer: $v_a = \sqrt{5}\cos(100t - 63.5^\circ)$ V

Exercise 10.10-2 Use mesh current analysis for the circuit of Figure E 10.10-2 to find the steady-state voltage across the inductor, v_L, when $v_{s1} = 20\cos\omega t$ V, $v_{s2} = 30\cos(\omega t - 90^\circ)$V, and $\omega = 1000$ rad/s.

Answer: $v_L = 24\sqrt{2}\cos(\omega t + 82^\circ)$ V

Exercise 10.10-3 Determine the node phasor voltages at terminals a and b for the circuit of Figure E 10.10-3 when $\mathbf{V}_s = j50$ V and $\mathbf{V}_1 = j30$ V.

Answer: $\mathbf{V}_a = 14.33\underline{/-71.75^\circ}$ V, $\mathbf{V}_b = 36.67\underline{/83^\circ}$ V

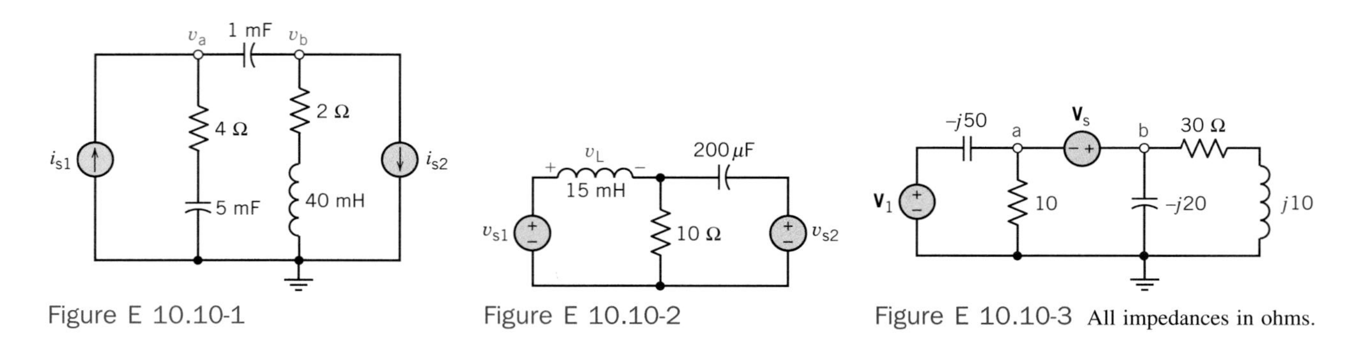

Figure E 10.10-1

Figure E 10.10-2

Figure E 10.10-3 All impedances in ohms.

10.11 SUPERPOSITION, THÉVENIN AND NORTON EQUIVALENTS, AND SOURCE TRANSFORMATIONS

Circuits in the frequency domain with phasor currents and voltages and impedances are analogous to the resistive circuits we considered earlier. Since they are linear, we expect that the principle of superposition and the source transformation method will hold. Furthermore, we can define Thévenin and Norton equivalent circuits in terms of impedance or admittance.

First, let us consider the *superposition principle,* which may be restated as follows: For a linear circuit containing two or more independent sources, any circuit voltage or current may be calculated as the algebraic sum of all the individual currents or voltages caused by each independent source acting alone.

If a linear circuit is excited by several sinusoidal sources all having the same frequency, ω, then superposition *may* be used. If a linear circuit is excited by several sources all having different frequencies, then superposition *must* be used.

The superposition principle is particularly useful if a circuit has two or more sources acting at different frequencies. Clearly, the circuit will have one set of impedance values at one frequency and a different set of impedance values at another frequency. We can determine the phasor response at each frequency. Then we find the time response corresponding to each phasor response and add them. Note that superposition, in the case of sources operating at two or more frequencies, applies to time responses only. We cannot superpose the phasor responses.

Example 10.11-1

Using the superposition principle, find the steady-state current i for the circuit shown in Figure 10.11-1 when $v_s = 10 \cos 10t$ V, $i_s = 3$ A, $L = 1.5$ H, and $C = 10$ mF.

Solution

The principle of superposition says that the response to the voltage source and current source acting together is equal to the sum of the response to the voltage source acting

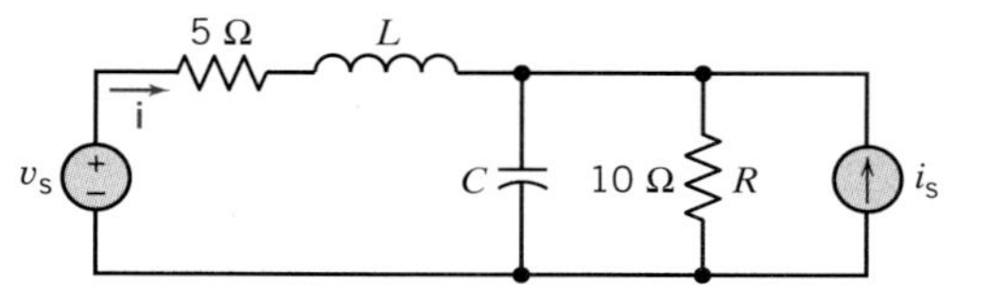

Figure 10.11-1
Circuit of Example 10.11-1.

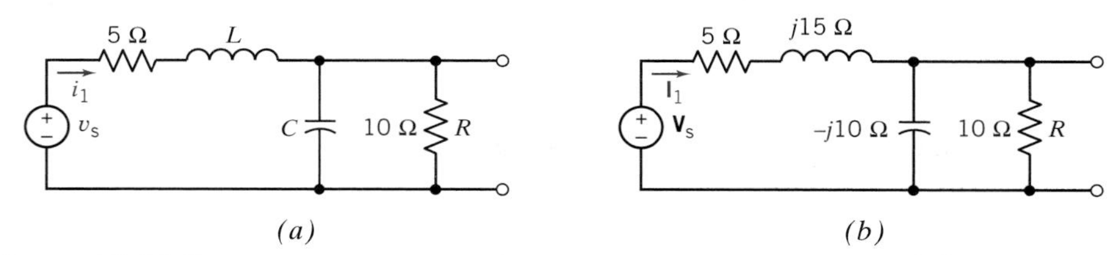

Figure 10.11-2 (*a*) Circuit for Example 10.11-1 for the voltage source acting alone. (*b*) Representation in the frequency domain.

alone plus the response to the current source acting alone. Let i_1 denote the response to the voltage source acting alone. Figure 10.11-2*a* shows the circuit that is used to calculate i_1. In Figure 10.11-2*b*, this circuit has been represented in the frequency domain using impedances and phasors. Similarly, let i_2 denote the response to the voltage source acting alone. Figure 10.11-3*a* shows the circuit that is used to calculate i_2. In Figure 10.11-3*b*, this circuit has been represented in the frequency domain.

The first step is to convert the independent sources into phasor form noting that the sources operate at different frequencies. For the voltage source operating at $\omega = 10$, we have

$$\mathbf{V}_s = 10\angle 0^\circ \text{ V}$$

We note that the current source is a direct current, so we can state that $\omega = 0$ for the current source. The phasor form of the current source is then

$$\mathbf{I}_s = 3\angle 0^\circ \text{ A}$$

The second step is to convert the circuit to phasor form with the impedance of each element shown as in Figure 10.11-2*b*.

Now let us determine the phasor current $\mathbf{I}_1$, which is the component of current $\mathbf{I}$ due to the voltage source. We remove the current source, replacing it with an open circuit across the 10 Ω resistor. Then we may find the current $\mathbf{I}_1$ due to the first source as

$$\mathbf{I}_1 = \frac{\mathbf{V}_s}{5 + j\omega L + \mathbf{Z}_p} \tag{10.11-1}$$

where $\mathbf{Z}_p$ is the impedance of the capacitor and the 10 Ω resistance in parallel. Recall that $\omega = 10$ and $C = 10$ mF. Therefore, since $\mathbf{Z}_c = -j10$, we have

$$\mathbf{Z}_p = \frac{\mathbf{Z}_c R}{R + \mathbf{Z}_c} = \frac{(-j10)10}{10 - j10} = 5(1 - j)\ \Omega$$

Substituting $\mathbf{Z}_p$ and $\omega L = 15$ into Eq. 10.11-1, we have

$$\mathbf{I}_1 = \frac{10\angle 0^\circ}{5 + j15 + (5 - j5)} = \frac{10}{10 + j10} = \frac{10}{\sqrt{200}}\angle -45^\circ \text{ A}$$

Therefore, the time domain current resulting from the voltage source is

$$i_1 = 0.71\cos(10t - 45^\circ) \text{ A}$$

Now let us determine the phasor current $\mathbf{I}_2$ due to the current source. Setting the voltage source to zero results in a short circuit. Since $\omega = 0$ for the dc source, the capacitor impedance becomes an open circuit because $\mathbf{Z}_C = 1/j\omega C = \infty$. The inductor's impedance becomes a short circuit because $\mathbf{Z}_L = j\omega L = 0$. Hence, we obtain the circuit shown in Figure 10.11-3*b*. We see that we have returned to a familiar resistive circuit for a dc source. Then the response due to the current source is

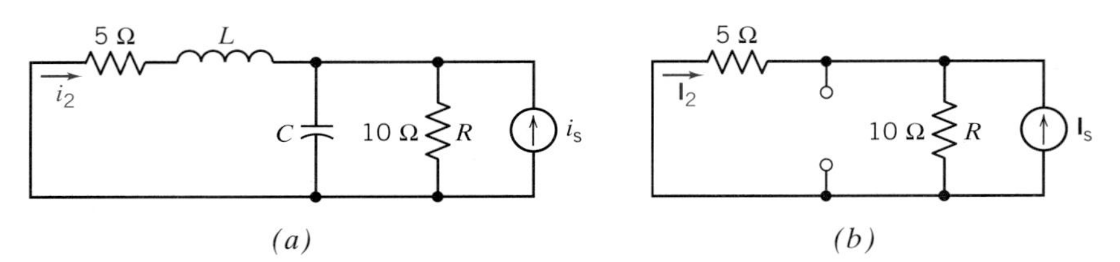

Figure 10.11-3 (*a*) Circuit for Example 10.11-1 for the current source acting alone. (*b*) Representation in the frequency domain.

$$\mathbf{I}_2 = -\frac{10}{15}3 = -2 \text{ A}$$

Therefore, using the principle of superposition, the total steady-state current is $i = i_1 + i_2$ or

$$i = 0.71 \cos(10t - 45°) - 2 \text{ A}$$

Now let us consider the *source transformations* for frequency domain (phasor) circuits. The techniques considered for resistive circuits discussed in Chapter 5 can readily be extended. The source transformation is concerned with transforming a voltage source and its associated series impedance to a current source and its associated parallel impedance, or vice versa, as shown in Figure 10.11-4. The method of transforming from one source to another source is summarized in Figure 10.11-5.

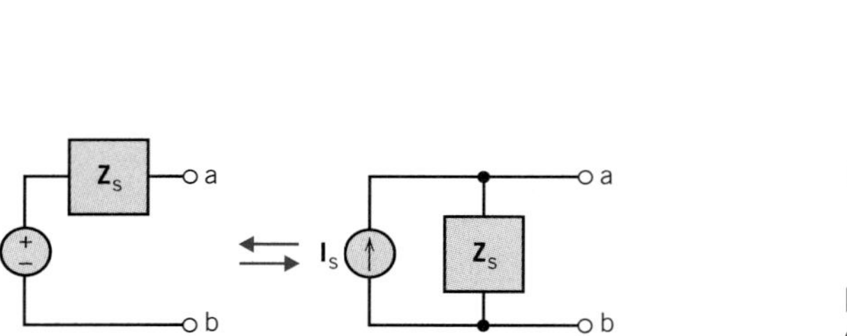

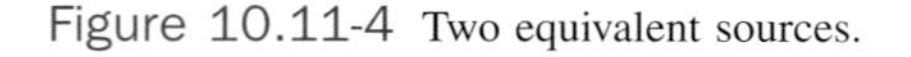

Figure 10.11-4 Two equivalent sources.

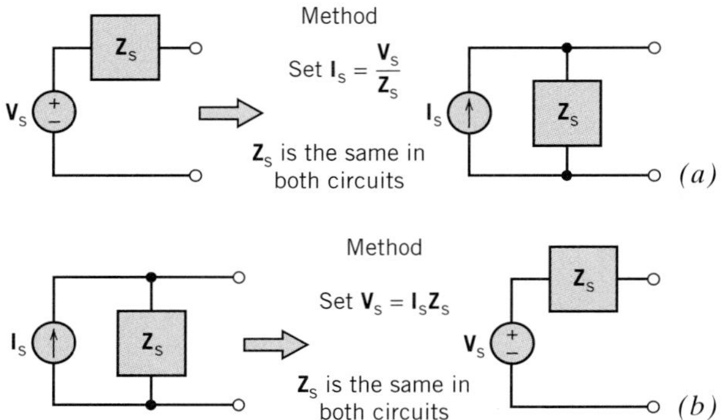

Figure 10.11-5 Method of source transformations. (*a*) Converting a voltage source to a current source. (*b*) Converting a current source to a voltage source.

Example 10.11-2

A circuit has a voltage source v_s in series with two elements, as shown in Figure 10.11-6. Determine the phasor equivalent current source form when $v_s = 10 \cos(\omega t + 45°)$ V and $\omega = 100$ rad/s.

Solution

First, we determine the equivalent current source as

$$\mathbf{I}_s = \frac{\mathbf{V}_s}{\mathbf{Z}_s}$$

Since $\mathbf{Z}_s = 10 + j10$ and $\mathbf{V}_s = 10\underline{/45°}$, we obtain

$$\mathbf{I}_s = \frac{10\underline{/45°}}{\sqrt{200}\underline{/45°}} = \frac{10}{\sqrt{200}}\underline{/0°} \text{A}$$

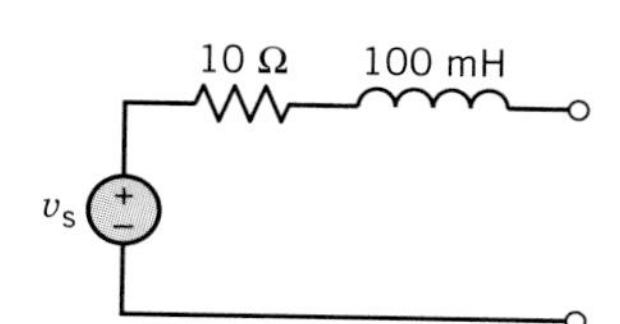

Figure 10.11-6 Circuit of Example 10.11-2.

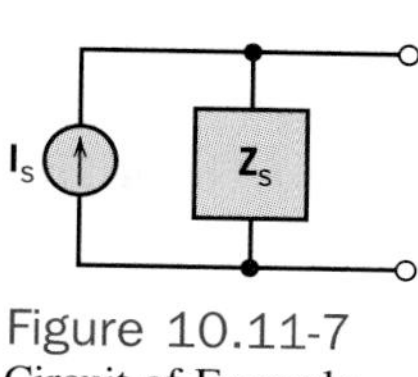

Figure 10.11-7 Circuit of Example 10.11-2 transformed to a current source where $\mathbf{Z}_s = 10 + j10\ \Omega$ and $\mathbf{I}_s = 1/\sqrt{2}$ A.

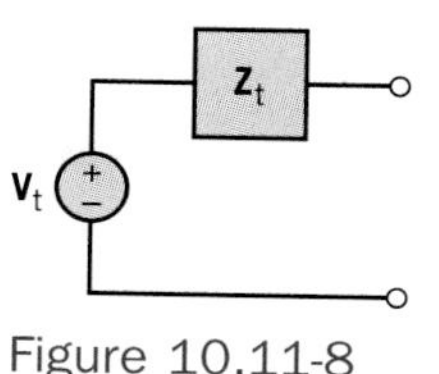

Figure 10.11-8 The Thévenin equivalent circuit.

The equivalent current source circuit is shown in Figure 10.11-7.

Thévenin's and Norton's theorems apply to phasor current or voltages and impedances in the same way that they do for resistive circuits. The Thévenin theorem is used to obtain an equivalent circuit as discussed in Chapter 5. The Thévenin equivalent circuit is shown in Figure 10.11-8.

A procedure for determining the Thévenin equivalent circuit is as follows:

1. Identify a separate circuit portion of a total circuit.
2. Determine the Thévenin voltage $\mathbf{V}_t = \mathbf{V}_{oc}$, the open-circuit voltage at the terminals.
3. (a) Find $\mathbf{Z}_t$ by deactivating all the independent sources and reducing the circuit to an equivalent impedance; (b) if the circuit has one or more dependent sources, then either short circuit the terminals and determine $\mathbf{I}_{sc}$ from which $\mathbf{Z}_t = \mathbf{V}_{oc}/\mathbf{I}_{sc}$; or (c) deactivate the independent sources, attach a current source at the terminals, and determine both $\mathbf{V}$ and $\mathbf{I}$ at the terminals from which $\mathbf{Z}_t = \mathbf{V}/\mathbf{I}$.

Example 10.11-3

Find the Thévenin equivalent circuit for the circuit shown in Figure 10.11-9 when $\mathbf{Z}_1 = 1 + j$ and $\mathbf{Z}_2 = -j1$.

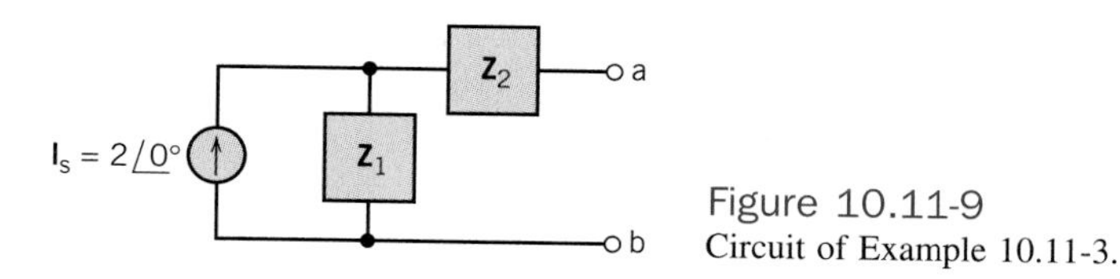

Figure 10.11-9 Circuit of Example 10.11-3.

Solution

The open-circuit voltage is

$$\mathbf{V}_{oc} = \mathbf{I}_s \mathbf{Z}_1 = (2\angle 0°)(1 + j) = 2\sqrt{2}\angle 45°\ \text{V}$$

The impedance $\mathbf{Z}_t$ is found by deactivating the current source by replacing it with an open circuit. Then we have $\mathbf{Z}_1$ in series with $\mathbf{Z}_2$, so that

$$\mathbf{Z}_t = \mathbf{Z}_1 + \mathbf{Z}_2 = (1 + j) - j = 1\ \Omega$$

Example 10.11-4

Find the Thévenin equivalent circuit of the circuit shown in Figure 10.11-10 in phasor form.

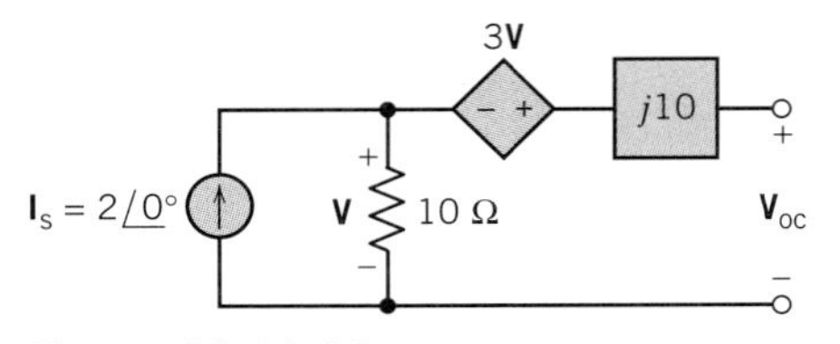

Figure 10.11-10 Circuit of Example 10.11-4.

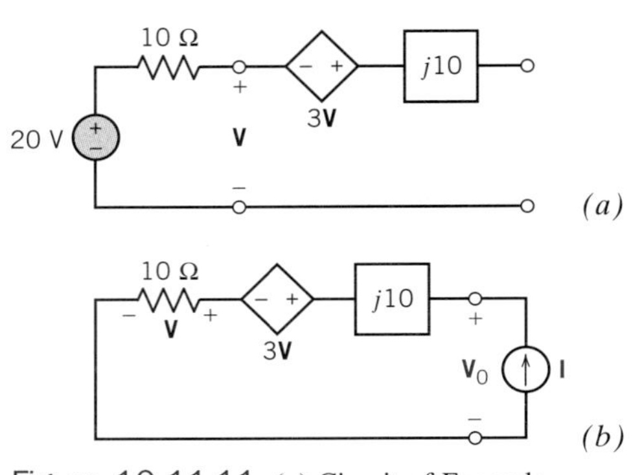

Figure 10.11-11 (*a*) Circuit of Example 10.11-4 with an open circuit at the output and the current source transformed to a voltage source. (*b*) Circuit with a test current source connected at the output terminal.

Solution

The Thévenin voltage $\mathbf{V}_t = \mathbf{V}_{oc}$, so we first determine $\mathbf{V}_{oc}$. Note that with the open circuit,

$$\mathbf{V} = 10\ \mathbf{I}_s = 20 \text{ V}$$

Then, for the mesh on the right, using KVL, we have

$$\mathbf{V}_{oc} = 3\mathbf{V} + \mathbf{V} = 4\mathbf{V} = 80\angle 0° \text{ V}$$

Examining the circuit of Figure 10.11-10, we transform the current source and 10 Ω resistance to the voltage source and 10 Ω series resistance as shown in Figure 10.11-11*a*. When the voltage source is deactivated and a current source is connected at the terminals as shown in Figure 10.11-11*b*, KVL gives

$$\mathbf{V}_o = j10\ \mathbf{I} + 4\ \mathbf{V} = (j10 + 40)\ \mathbf{I}$$

Therefore,

$$\mathbf{Z}_t = 40 + j10\ \Omega$$

Now let us consider the procedure for finding the Norton equivalent circuit. The steps are similar to those used for the Thévenin equivalent since $\mathbf{Z}_t$ in series with the Thévenin voltage is equal to the Norton impedance in parallel with the Norton current source. The Norton equivalent circuit is shown in Figure 10.11-12.

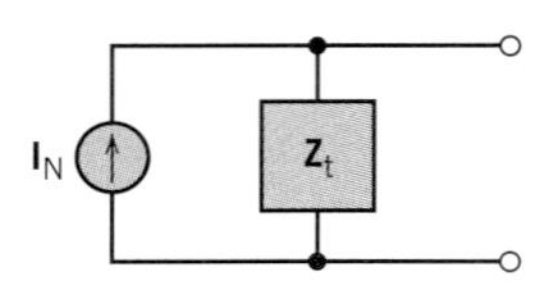

Figure 10.11-12 The Norton equivalent circuit expressed in terms of a phasor current and an impedance.

In order to determine the Norton circuit, we follow the procedure as follows:

1. Identify a separate circuit portion of a total circuit.
2. The Norton current $\mathbf{I}_N$ is the current through a short circuit at the terminals, so $\mathbf{I}_N = \mathbf{I}_{sc}$.
3. Find $\mathbf{Z}_t$ by (a) deactivating all the independent sources and reducing the circuit to an equivalent impedance, *or* (b) if the circuit has one or more dependent sources, find the open-circuit voltage at the terminals, $\mathbf{V}_{oc}$, so that

$$\mathbf{Z}_t = \frac{\mathbf{V}_{oc}}{\mathbf{I}_{sc}}$$

Example 10.11-5

Find the Norton equivalent of the circuit shown in Figure 10.11-13 in phasor and impedance forms. Assume that $\mathbf{V}_s = 100\angle 0°$ V.

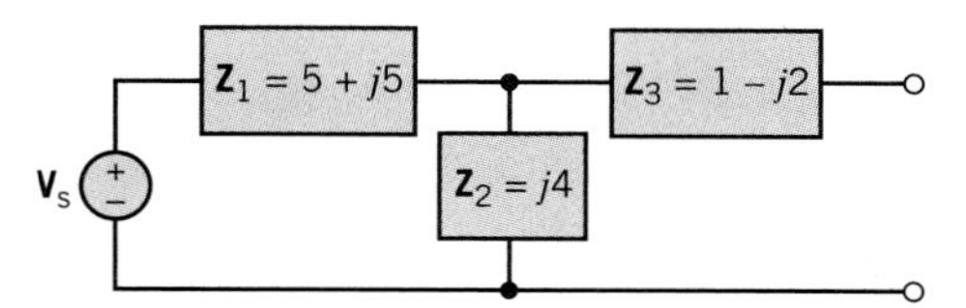

Figure 10.11-13 Circuit of Example 10.11-5.

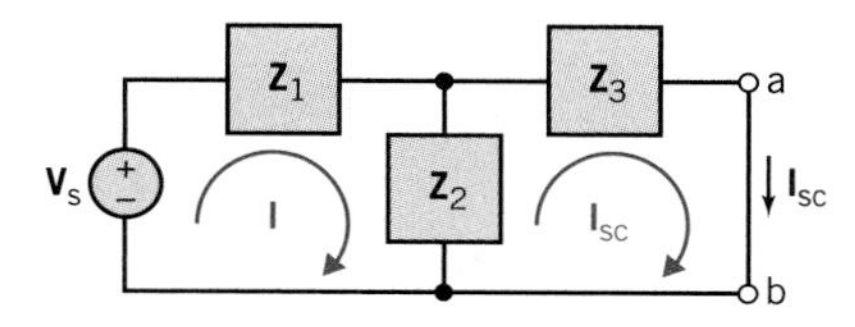

Figure 10.11-14 Circuit of Example 10.11-5 with a short circuit at terminals a–b.

Solution

First, let us find the equivalent impedance by deactivating the voltage source by replacing it with a short circuit. Since $\mathbf{Z}_1$ appears in parallel with $\mathbf{Z}_2$, we have

$$\mathbf{Z}_t = \mathbf{Z}_3 + \frac{\mathbf{Z}_1\mathbf{Z}_2}{\mathbf{Z}_1 + \mathbf{Z}_2} = (1 - j2) + \frac{(5 + j5)(j4)}{(5 + j5) + (j4)}$$

$$= (1 - j2) + \frac{20}{53}(2 + j7) = \frac{93}{53} + j\frac{34}{53} = \frac{1}{53}(93 + j34)$$

We now proceed to determine the Norton equivalent current source by determining the current flowing through a short circuit connected at terminals a–b, as shown in Figure 10.11-14.

We will use mesh currents to find $\mathbf{I}_{sc}$ as shown in Figure 10.11-14. The two mesh KVL equations are

$$\text{mesh 1: } (\mathbf{Z}_1 + \mathbf{Z}_2)\mathbf{I} + (-\mathbf{Z}_2)\mathbf{I}_{sc} = \mathbf{V}_s$$
$$\text{mesh 2: } (-\mathbf{Z}_2)\mathbf{I} + (\mathbf{Z}_2 + \mathbf{Z}_3)\mathbf{I}_{sc} = 0$$

Using Cramer's rule, we find $\mathbf{I}_N = \mathbf{I}_{sc}$ as follows:

$$\mathbf{I}_{sc} = \frac{\mathbf{Z}_2\mathbf{V}_s}{(\mathbf{Z}_1 + \mathbf{Z}_2)(\mathbf{Z}_2 + \mathbf{Z}_3) - \mathbf{Z}_2^2} = \frac{(j4)100}{(5 + j9)(1 + j2) - (-16)}$$

$$= \frac{j400}{3 + j19} = \frac{400}{370}(19 + 3j)\,\text{A}$$

Exercise 10.11-1 Determine value of $\mathbf{V}_t$ and $\mathbf{Z}_t$ such that the circuit shown in Figure E 10.11-1b is the Thévenin equivalent circuit of the circuit shown in Figure E 10.11-1a.

Answer: $\mathbf{V}_t = 3.58 \angle 47^\circ$ and $\mathbf{Z}_t = 4.9 + j1.2\ \Omega$

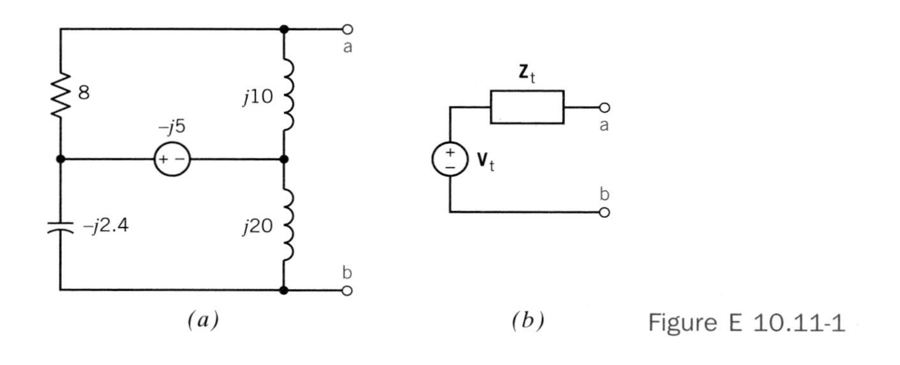

Figure E 10.11-1

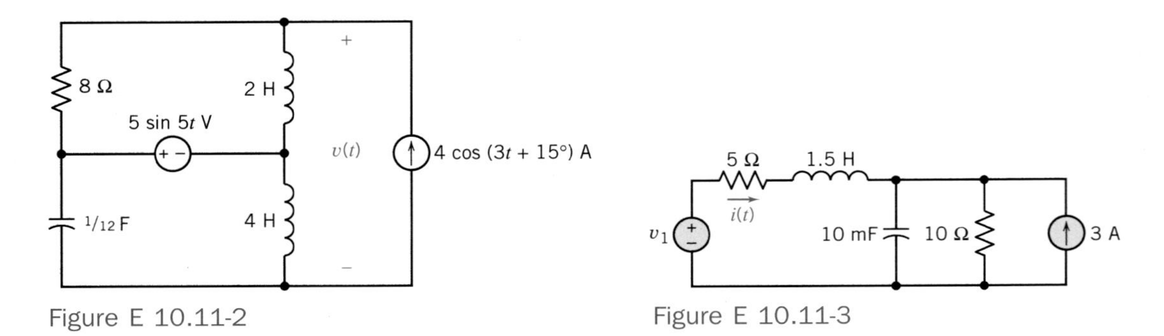

Figure E 10.11-2

Figure E 10.11-3

Exercise 10.11-2 Determine the voltage $v(t)$ for the circuit of Figure E 10.11-2.

Hint: Use superposition.

Answer: $v(t) = 3.58\cos(5t + 47.2°) + 14.4\cos(3t - 22°)$ V.

Exercise 10.11-3 Using the principle of superposition, determine $i(t)$ of the circuit shown in Figure E 10.11-3 when $v_1 = 10\cos 10t$ V.

Answer: $i = -2 + 0.71\cos(10t - 45°)$ A

10.12 PHASOR DIAGRAMS

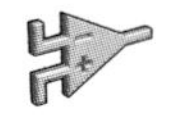

Phasors representing the voltage or current of a circuit are time quantities transformed or converted into the frequency domain. Phasors are complex numbers and can be portrayed in a complex plane. The relationship of phasors on a complex plane is called a *phasor diagram*.

Let us consider an *RLC* series circuit as shown in Figure 10.12-1. The impedance of each element is also identified in the diagram. Since the current flows through all elements and is common to all, we take **I** as the reference phasor.

$$\mathbf{I} = I\angle 0°$$

Then the voltage phasors are

$$\mathbf{V}_R = R\mathbf{I} = RI\angle 0° \qquad (10.12\text{-}1)$$

$$\mathbf{V}_L = j\omega L\,\mathbf{I} = \omega LI\angle 90° \qquad (10.12\text{-}2)$$

$$\mathbf{V}_c = \frac{-j\mathbf{I}}{\omega C} = \frac{I}{\omega C}\angle -90° \qquad (10.12\text{-}3)$$

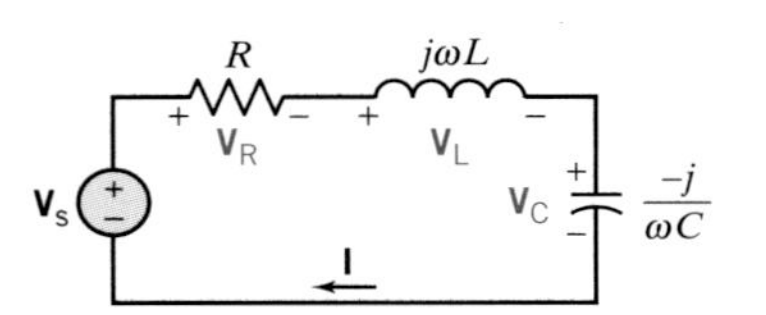

Figure 10.12-1 An *RLC* circuit.

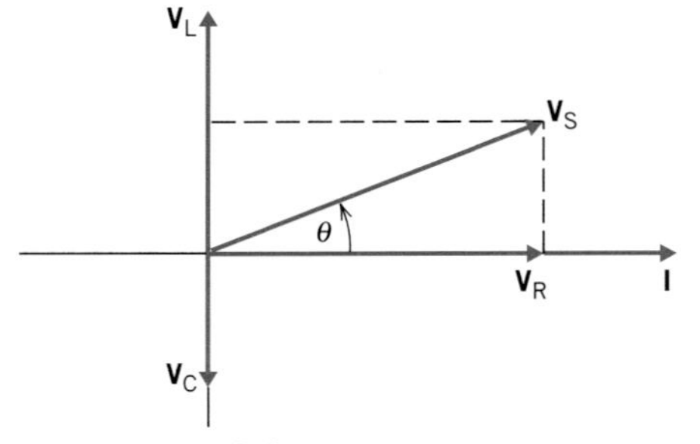

Figure 10.12-2 Phasor diagram for the *RLC* circuit of Figure 10.12-1.

These phasors are shown in the phasor diagram of Figure 10.12-2. Note that KVL for this circuit requires that

$$\mathbf{V}_s = \mathbf{V}_R + \mathbf{V}_L + \mathbf{V}_c$$

A **phasor diagram** is a graphical representation of phasors and their relationship on the complex plane.

The current $\mathbf{I}$ and the voltage across the resistor are in phase. The inductor voltage leads the current by 90°, and the capacitor voltage lags the current by 90°. For a given L and C there will be a frequency ω that results in

$$|\mathbf{V}_L| = |\mathbf{V}_c|$$

Referring to Eqs. 10.12-2 and 10.12-3, this equality of voltage magnitudes occurs when

$$\omega L = \frac{1}{\omega C}$$

or

$$\omega^2 = \frac{1}{LC}$$

When $\omega^2 = 1/LC$, the magnitudes of the inductor voltage and capacitor voltage are equal. Since they are out of phase by 180°, they cancel, and the resulting condition is

$$\mathbf{V}_s = \mathbf{V}_R$$

and then both $\mathbf{V}_s$ and $\mathbf{V}_R$ are in phase with $\mathbf{I}$. This condition is called *resonance*.

Exercise 10.12-1 Consider the *RLC* series circuit of Figure 10.12-1 when $L = 1$ mH and $C = 1$ mF. Find the frequency ω when the current, source voltage, and $\mathbf{V}_R$ are all in phase.

Answer: $\omega = 1000$ rad/s.

Exercise 10.12-2 Draw the phasor diagram for the circuit of Figure E 10.12-2 when $\mathbf{V} = V\underline{/0°}$. Show each current on the diagram.

Exercise 10.12-3 The circuit shown in Figure E 10.12-3 contains a sinusoidal current source of $25\,\underline{/0°}$ A. An ammeter reads the magnitude of the current. Ammeter A_1 reads 15 A, and ammeter A_2 reads 6 A. Find the reading of ammeter A_3.

Hint: The ammeter measures the magnitude of the current through the ammeter.

Answer: 26 A.

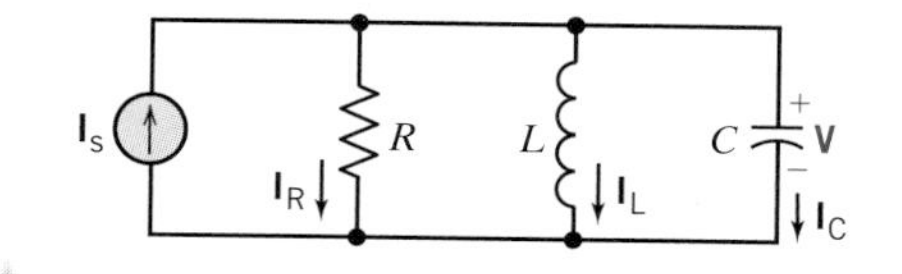

Figure E 10.12-2

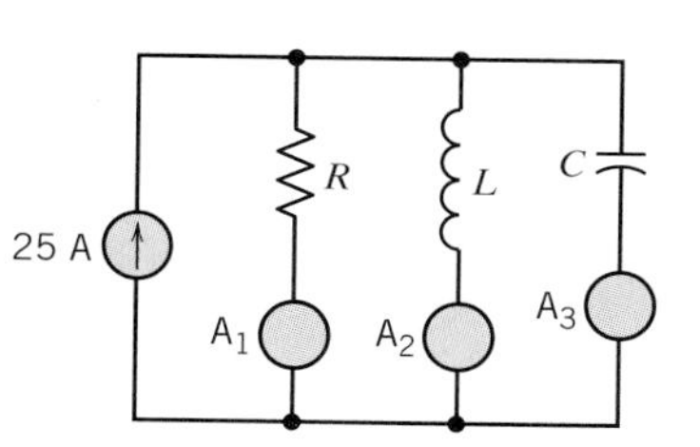

Figure E 10.12-3

10.13 SINUSOIDAL STEADY-STATE ANALYSIS USING PSPICE

We can readily determine the sinusoidal steady-state response of a circuit using the .AC analysis command of PSpice (see Appendix F-7). This is often called ac analysis and the calculation is obtained for a specific frequency. The form of the .AC statement is

```
.AC <sweep type> <n> <start freq> <end freq>
```

For phasor analysis, start freq = end freq and we choose LIN (linear) for the sweep type. Then n is the number of frequency points that will be output, which is one for the single frequency chosen. Thus, if we wish to determine the phasor calculation at a frequency of 100 Hz, we use

```
.AC LIN 1 100 100
```

This statement requires the frequency stated in hertz. The output of the calculation is obtained by using the statement

```
.PRINT AC <ac output variables>
```

Let us use PSpice to determine the solution of Example 10.9-2. The circuit of Figure 10.9-4 is redrawn in Figure 10.13-1 where $i_s = 10 \cos 1000t$. The PSpice program is shown in Figure 10.13-2. The current source is designated as IS and has a magnitude of 10 and a phase of 0°. PSpice uses $M \sin(\omega t + \phi)$ as the sinusoidal input signal for transient analysis. However, for .AC analysis, we can designate the input source as a cosine or sine function and PSpice determines the angle of the output variable, given the assumed form of the input source. Note that the .AC statement calculates the response at $f = 1000/2\pi = 159.15$ Hz as required by PSpice.

We then call for a printout of the magnitude and phase of the voltage at node 1, V(1). The simulation results are provided, and we have VM(1) = 63.28 V and VP(1) = −18.4°. Therefore, the capacitor voltage is

$$v = 63.3 \cos(1000t - 18.4°)$$

as we determined in Example 10.9-2.

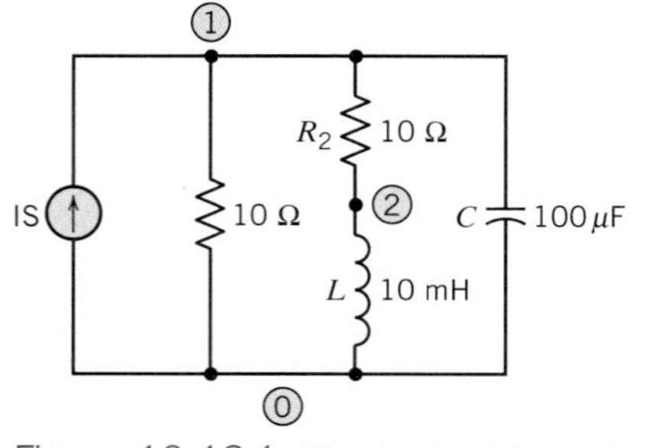

Figure 10.13-1 The circuit of Example 10.9-2 formatted for PSpice.

```
           EXAMPLE 10.9-2 BY SPICE

IS         0      1          AC         10   0
R1         1      0          10
R2         1      2          10
L1         2      0          10M
C1         1      0    .END  0.1 M

.AC        LIN    1          159.15     159.15
.PRINT     AC                VM(1)      VP(1)
```

Figure 10.13-2 The PSpice program for Example 10.9-2.

10.14 PHASOR CIRCUITS AND THE OPERATIONAL AMPLIFIER

The discussion in the prior sections considered the behavior of operational amplifiers and their associated circuits in the time domain. In this section we consider the behavior of operational amplifiers and associated *RLC* circuits in the frequency domain using phasors.

Figure 10.14-1 shows two frequently used operational amplifier circuits, the inverting amplifier, and the noninverting amplifier. These circuits are represented using impedances

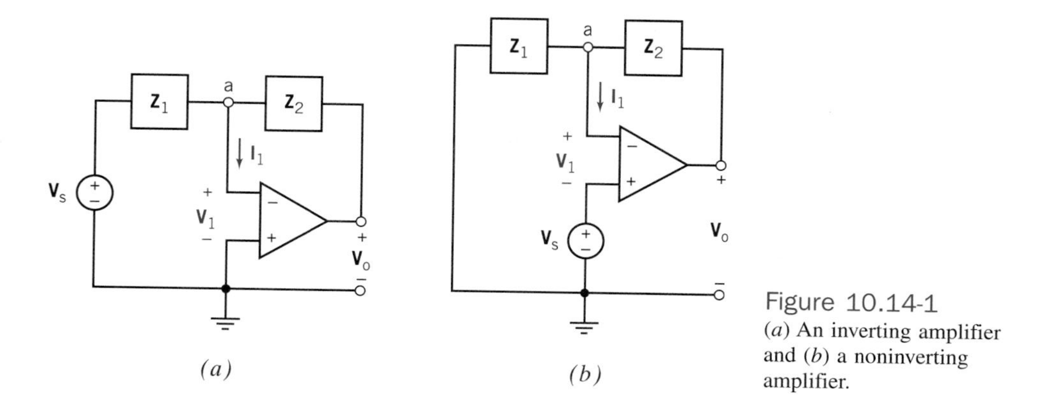

Figure 10.14-1
(*a*) An inverting amplifier and (*b*) a noninverting amplifier.

and phasors. This representation is appropriate when the input is sinusoidal and the circuit is at steady state. $\mathbf{V}_s$ is the phasor corresponding to a sinusoidal input voltage, and $\mathbf{V}_o$ is the phasor representing the resulting sinusoidal output voltage. Both circuits involve two impedances, $\mathbf{Z}_1$ and $\mathbf{Z}_2$.

Now let us proceed to determine the ratio of output-to-input voltage, $\mathbf{V}_o/\mathbf{V}_s$, for the inverting amplifier shown in Figure 10.14-1*a*. This circuit can be analyzed by writing the node equation at node a as

$$\frac{\mathbf{V}_s - \mathbf{V}_1}{\mathbf{Z}_1} + \frac{\mathbf{V}_o - \mathbf{V}_1}{\mathbf{Z}_2} - \mathbf{I}_1 = 0 \tag{10.14-1}$$

When the operational amplifier is ideal, $\mathbf{V}_1$ and $\mathbf{I}_1$ are both 0. Then

$$\frac{\mathbf{V}_s}{\mathbf{Z}_1} + \frac{\mathbf{V}_o}{\mathbf{Z}_2} = 0 \tag{10.14-2}$$

Finally,

$$\frac{\mathbf{V}_o}{\mathbf{V}_s} = -\frac{\mathbf{Z}_2}{\mathbf{Z}_1} \tag{10.14-3}$$

Next, we will determine the ratio of output-to-input voltage, $\mathbf{V}_o/\mathbf{V}_s$, for the noninverting amplifier shown in Figure 10.14-1*b*. This circuit can be analyzed by writing the node equation at node a as

$$\frac{(\mathbf{V}_s + \mathbf{V}_1)}{\mathbf{Z}_1} - \frac{\mathbf{V}_o - (\mathbf{V}_s + \mathbf{V}_1)}{\mathbf{Z}_2} + \mathbf{I}_1 = 0 \tag{10.14-4}$$

When the operational amplifier is ideal, $\mathbf{V}_1$ and $\mathbf{I}_1$ are both 0. Then

$$\frac{\mathbf{V}_s}{\mathbf{Z}_1} - \frac{\mathbf{V}_o - \mathbf{V}_s}{\mathbf{Z}_2} = 0$$

Finally,

$$\frac{\mathbf{V}_o}{\mathbf{V}_s} = \frac{\mathbf{Z}_1 + \mathbf{Z}_2}{\mathbf{Z}_1} \tag{10.14-5}$$

Typically, impedances $\mathbf{Z}_1$ and $\mathbf{Z}_2$ are obtained using only resistors and capacitors. Of course, in theory we could use inductors, but their cost and size relative to capacitors result in little use of inductors with operational amplifiers.

An example of the inverting amplifier is shown in Figure 10.14-2.

The impedance $\mathbf{Z}_n$, where n is equal to 1 or 2, is a parallel R_nC_n impedance so that

$$\mathbf{Z}_n = \frac{R_n(-jX_n)}{R_n - jX_n} = \frac{R_n X_n (X_n - jR_n)}{R_n^2 + X_n^2} \tag{10.14-6}$$

where $X_n = 1/\omega C_n$. Using Eqs. 10.14-3 and 10.14-6, one may obtain the ratio $\mathbf{V}_o/\mathbf{V}_s$.

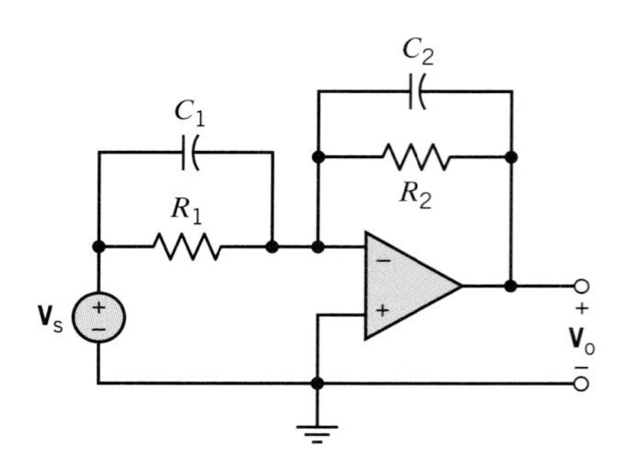

Figure 10.14-2
Operational amplifier with two *RC* circuits connected.

Example 10.14-1

Find the ratio $\mathbf{V}_o/\mathbf{V}_s$ for the circuit of Figure 10.14-2 when $R_1 = 1\ \text{k}\Omega$, $R_2 = 10\ \text{k}\Omega$, $C_1 = 0$ and $C_2 = 0.1\ \mu\text{F}$ for $\omega = 1000$ rad/s.

Solution

Since $C_1 = 0$, $\mathbf{Z}_1 = R_1 = 1\ \text{k}\Omega$. The impedance $\mathbf{Z}_2$ is

$$\mathbf{Z}_2 = \frac{R_2 X_2 (X_2 - jR_2)}{R_2^2 + X_2^2}$$

The reactance X_2 is

$$X_2 = \frac{1}{\omega C_2} = 10^4$$

Therefore, since $R_2 = 10^4$, we have

$$\mathbf{Z}_2 = \frac{(10^4)(10^4)(10^4 - j10^4)}{(10^4)^2 + (10^4)^2} = \frac{10^4(1-j)}{2}$$

Therefore,

$$\frac{\mathbf{V}_o}{\mathbf{V}_s} = -\frac{\mathbf{Z}_2}{\mathbf{Z}_1} = \frac{-10^4(1-j)}{2(10^3)} = -5(1-j)$$

Exercise 10.14-1 Find the ratio $\mathbf{V}_o/\mathbf{V}_s$ for the circuit shown in Figure 10.14-2 when $R_1 = R_2 = 1\ \text{k}\Omega$, $C_2 = 0$, $C_1 = 1\ \mu\text{F}$, and $\omega = 1000$ rad/s.

Answer: $\mathbf{V}_o/\mathbf{V}_s = -1 - j$

10.15 USING MATLAB FOR ANALYSIS OF STEADY-STATE CIRCUITS WITH SINUSOIDAL INPUTS

Analysis of steady-state linear circuits with sinusoidal inputs using phasors and impedances requires complex arithmetic. MATLAB can be used to reduce the effort required to do this complex arithmetic. Consider the circuit shown in Figure 10.15-1*a*. The input to this circuit, $v_s(t)$, is a sinusoidal voltage. At steady state, the output, $v_o(t)$, will also be a sinusoidal voltage as shown in Figure 10.15-1*a*. This circuit can be represented in the frequency domain

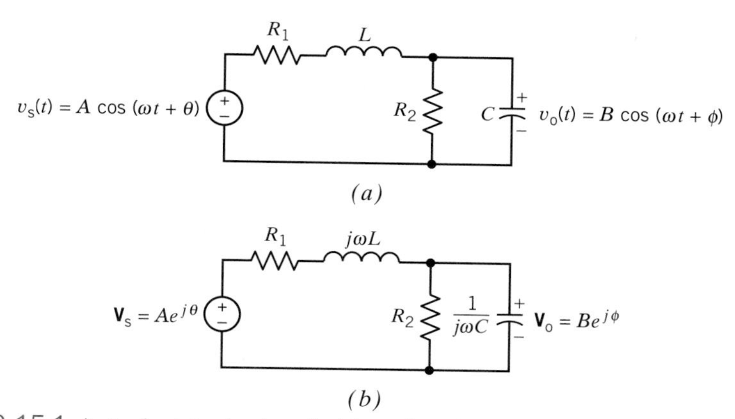

Figure 10.15-1 A steady-state circuit excited by a sinusoidal input voltage. This circuit is represented both (*a*) in the time domain and (*b*) in the frequency domain.

using phasors and impedances as shown in Figure 10.15-1*b*. Analysis of this circuit proceeds as follows. Let $\mathbf{Z}_1$ denote the impedance of the series combination of R_1 and $j\omega L$. That is,

$$\mathbf{Z}_1 = R_1 + j\omega L \tag{10.15-1}$$

Next, let Y_2 denote the admittance of the parallel combination of R_2 and $1/j\omega\, C$. That is,

$$\mathbf{Y}_2 = \frac{1}{R_2} + j\omega\, C \tag{10.15-2}$$

Let Z_2 denote the corresponding impedance, that is,

$$\mathbf{Z}_2 = \frac{1}{\mathbf{Y}_2} \tag{10.15-3}$$

Finally, V_o is calculated from V_s using voltage division. That is,

$$\mathbf{V}_o = \frac{\mathbf{Z}_2}{\mathbf{Z}_1 + \mathbf{Z}_2}\mathbf{V}_s \tag{10.15-4}$$

Figure 10.15-2 shows a MATLAB input file that uses Equations 10.15-1 through 10.15-3 to find the steady-state response of the circuit shown in Figure 10.15-1. Equation 10.15-4 is used to calculate V_o. Next $B = |V_o|$ and $\phi = \angle V_o$ are calculated and used to determine the magnitude and phase angle of the sinusoidal output voltage. Notice that MATLAB, not the user, does the complex arithmetic needed to solve these equations. Finally, MATLAB produces the plot shown in Figure 10.15-3, which displays the sinusoidal input and output voltages in the time domain.

10.16 VERIFICATION EXAMPLES

Example 10.16-1

It is known that

$$\frac{10}{R - j4} = A\angle 53^\circ$$

A computer program states that $A = 2$. Verify this result. (Notice that values are given to only two significant figures.)

```
%-------------------------------------------------
%       Describe the input voltage source.
%-------------------------------------------------
w = 2;
A = 12;
theta = (pi/180)*60;
Vs = A*exp(j*theta)
%-------------------------------------------------
%Describe the resistors, inductor and capacitor.
%-------------------------------------------------
R1 = 6;
L = 4;
R2 = 12;
C = 1/24;
%-------------------------------------------------
% Calculate the equivalent impedances of the
%  series resistor and inductor and of the
%       parallel resistor and capacitor
%-------------------------------------------------
Z1 = R1 + j*w*L               % Eqn 10.15-1
Y2 = 1/R2 + j*w*C;            % Eqn 10.15-2
Z2 = 1 / Y2                   % Eqn 10.15-3
%-------------------------------------------------
%  Calculate the phasor corresponding to the
%                 output voltage.
%-------------------------------------------------
Vo = Vs * Z2/(Z1 + Z2) % Eqn 10.15-4
B = abs(Vo);
phi = angle(Vo);
%-------------------------------------------------
%
%-------------------------------------------------
T = 2*pi/w;
tf = 2*T; N = 100; dt = tf/N;
t = 0 : dt : tf;
%-------------------------------------------------
%      Plot the input and output voltages.
%-------------------------------------------------
for k = 1 : 101
    vs(k) = A * cos(w * t(k) + theta);
    vo(k) = B * cos(w * t(k) + phi);
end
plot (t, vs, t, vo)
```

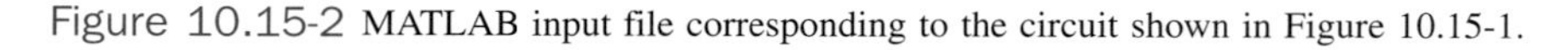

Figure 10.15-2 MATLAB input file corresponding to the circuit shown in Figure 10.15-1.

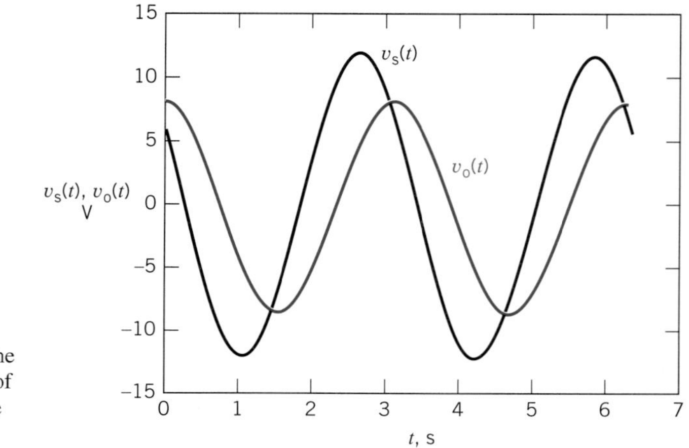

Figure 10.15-3 MATLAB plots showing the input and output voltages of the circuit shown in Figure 10.15-1.

Solution

The equation for the angle is

$$-\tan^{-1}\left(\frac{-4}{R}\right) = 53°$$

Then, we have

$$R = \frac{-4}{\tan(-53°)} = 3.014$$

Solving for A in terms of R, we obtain

$$A = \frac{10}{(R^2 + 16)^{1/2}} = 1.997$$

Therefore $A = 2$ is correct to two significant figures.

Example 10.16-2

Consider the circuit shown in Figure 10.16-1. Suppose we know that the capacitor voltages are

$$1.96\cos(100t - 101.3°)\text{ V} \quad \text{and} \quad 4.39\cos(100t - 37.88°)\text{ V}$$

but we do not know which voltage is $v_1(t)$ and which is $v_2(t)$. How can we match the voltages with the capacitors?

Solution

Let us guess that

$$v_1(t) = 1.96\cos(100t - 101.3°)$$

and

$$v_2(t) = 4.39\cos(100t - 37.88°)$$

and then check to see if this choice satisfies the node equations representing the circuit. These node equations are

$$\frac{10 - \mathbf{V}_1}{R_1} = j\omega\, C_1\mathbf{V}_1 + \frac{\mathbf{V}_1 - \mathbf{V}_2}{R_2}$$

and

$$j\omega\, C_2\mathbf{V}_2 = \frac{\mathbf{V}_1 - \mathbf{V}_2}{R_2}$$

where $\mathbf{V}_1$ and $\mathbf{V}_2$ are the phasors corresponding to $v_1(t)$ and $v_2(t)$. That is,

$$\mathbf{V}_1 = 1.96e^{-j101.3°} \quad \text{and} \quad \mathbf{V}_2 = 4.39e^{-j37.88°}$$

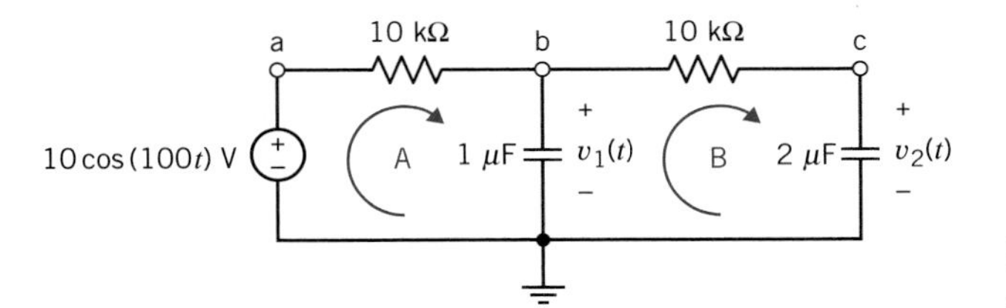

Figure 10.16-1
An example circuit.

Substituting the phasors $\mathbf{V}_1$ and $\mathbf{V}_2$ into the left-hand side of the first node equation gives

$$\frac{10 - 1.96e^{-j101.3}}{10 \times 10^3} = 0.001 + j1.92 \times 10^{-4}$$

Substituting the phasors $\mathbf{V}_1$ and $\mathbf{V}_2$ into the right-hand side of the first node equation gives

$$j \cdot 100 \times 10^{-6} \cdot 1.96e^{-j101.3} + \frac{1.96e^{-j101.3} - 4.39e^{-j37.88}}{10 \times 10^3}$$

$$= -19.3 \times 10^{-4} + j3.89 \times 10^{-5}$$

Since the right-hand side is not equal to the left-hand side, $\mathbf{V}_1$ and $\mathbf{V}_2$ do not satisfy the node equation. That means that the selected order of $v_1(t)$ and $v_2(t)$ are not correct. Instead, use the reverse order so that

$$v_1(t) = 4.39 \cos(100t - 37.88°)$$

and

$$v_2(t) = 1.96 \cos(100t - 101.3°)$$

Now the phasors $\mathbf{V}_1$ and $\mathbf{V}_2$ will be

$$\mathbf{V}_1 = 4.39e^{-j37.88°} \quad \text{and} \quad \mathbf{V}_2 = 1.96e^{-j101.3°}$$

Substituting the new values of the phasors $\mathbf{V}_1$ and $\mathbf{V}_2$ into the left-hand side of the first node equation gives

$$\frac{10 - 4.39e^{-j37.88}}{10 \times 10^3} = 6.353 \times 10^{-4} + j2.696 \times 10^{-4}$$

Substituting the new values of the phasors $\mathbf{V}_1$ and $\mathbf{V}_2$ into the right-hand side of the first node equation gives

$$j \cdot 100 \cdot 10^{-6} \cdot 4.39e^{-j37.88} + \frac{4.39e^{-j37.88} - 1.96e^{-j101.3}}{10 \times 10^3}$$

$$= +\ 6.545 \times 10^{-4} + j2.69 \times 10^{-4}$$

Since the right-hand side is very close to equal to the left-hand side, $\mathbf{V}_1$ and $\mathbf{V}_2$ satisfy the first node equation. That means that $v_1(t)$ and $v_2(t)$ are probably correct. To be certain, we will also check the second node equation. Substituting the phasors $\mathbf{V}_1$ and $\mathbf{V}_2$ into the left-hand side of the second node equation gives

$$j \cdot 100 \cdot 2 \times 10^{-6} \cdot 1.96e^{-j101.3} = +3.84 \times 10^{-4} - j7.681 \times 10^{-5}$$

Substituting the phasors $\mathbf{V}_1$ and $\mathbf{V}_2$ into the right-hand side of the second node equation gives

$$\frac{4.39e^{-j37.88} - 1.96e^{-j101.3}}{10 \times 10^3} = 3.85 \times 10^{-4} - j7.735 \times 10^{-5}$$

Since the right-hand side is equal to the left-hand side, $\mathbf{V}_1$ and $\mathbf{V}_2$ satisfy the second node equation. Now we are certain that

$$v_1(t) = 4.39 \cos(100t - 37.88°) \text{ V}$$

and

$$v_2(t) = 1.96 \cos(100t - 101.3°) \text{ V}$$

10.17 Design Challenge Solution

OP AMP CIRCUIT

Figure 10.17-1*a* shows two sinusoidal voltages, one labeled as input and the other labeled as output. We want to design a circuit that will transform the input sinusoid into the output sinusoid. Figure 10.17-1*b* shows a candidate circuit. We must first determine if this circuit can do the job. Then, if it can, we will design the circuit, that is, specify the required values of R_1, R_2, and C.

Define the Situation and the Assumptions

The input and output sinusoids have different amplitudes and phase angles but the same frequency

$$f = 1000 \text{ Hz}$$

or, equivalently,

$$\omega = 2\pi 1000 \text{ rad/s}$$

We now know that this must be the case. When the input to a linear circuit is a sinusoid, the steady-state output will also be a sinusoid having the same frequency.

In this case, the input sinusoid is

$$v_1(t) = \sin(2\pi 1000 t) = \cos(2\pi 1000 t - 90°) \text{ V}$$

and the corresponding phasor is

$$\mathbf{V}_1 = 1 e^{-j90°} = 1\underline{/-90°} \text{ V}$$

The output sinusoid is

$$v_2(t) = 2\sin(2\pi 1000 t + 120°) = 2\cos(2\pi 1000 t + 30°) \text{ V}$$

and the corresponding phasor is

$$\mathbf{V}_2 = 2 e^{j30°} \text{ V}$$

The ratio of these phasors is

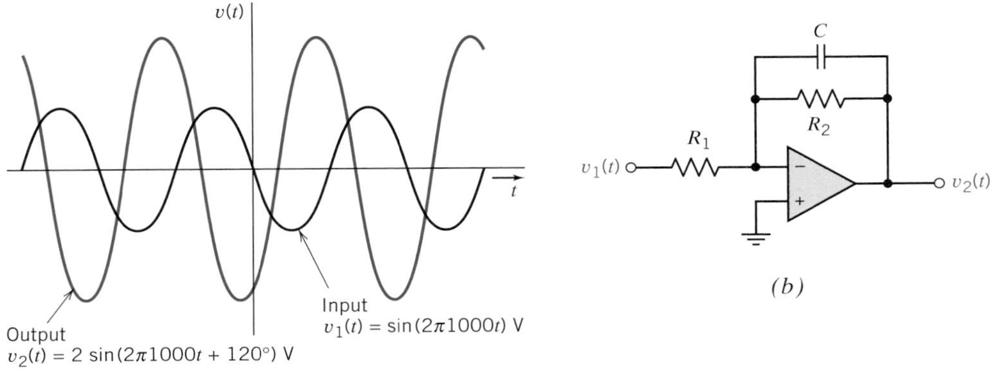

Figure 10.17-1 (*a*) Input and output voltages. (*b*) Proposed circuit.

$$\frac{\mathbf{V}_2}{\mathbf{V}_1} = \frac{2e^{j30°}}{1e^{-j90°}} = 2e^{j120°}$$

The magnitude of this ratio is called the gain, G, of the circuit used to transform the input sinusoid into the output sinusoid:

$$G = \left|\frac{\mathbf{V}_2}{\mathbf{V}_1}\right| = 2$$

The angle of this ratio is called the phase shift, θ, of the required circuit.

$$\theta = \angle\frac{\mathbf{V}_2}{\mathbf{V}_1} = 120°$$

Therefore, we need a circuit that has a gain of 2 and a phase shift of 120°.

State the Goal

Determine if it is possible to design the circuit shown in Figure 10.17-1*b* to have a gain of 2 and a phase shift of 120°. If it is possible, specify the appropriate values of R_1, R_2, and C.

Generate a Plan

Analyze the circuit shown in Figure 10.17-1*b* to determine the ratio of the output phasor to the input phasor, $\mathbf{V}_2/\mathbf{V}_1$. Determine if this circuit can have a gain of 2 and a phase shift of 120°. If so, determine the required values of R_1, R_2, and C.

Act on the Plan

The circuit in Figure 10.17-1*b* is a special case of the circuit shown in Figure 10.14-1. The impedance $\mathbf{Z}_1$ in Figure 10.14-1 corresponds to the resistor R_1 in Figure 10.17-1*b* and impedance $\mathbf{Z}_2$ corresponds to the parallel combination of resistor R_2 and capacitor C. That is,

$$\mathbf{Z}_1 = R_1$$

and

$$\mathbf{Z}_2 = \frac{R_2(1/j\omega C)}{R_2 + 1/j\omega C} = \frac{R_2}{1 + j\omega C R_2}$$

Then, using Eq. 10.14-3,

$$\frac{\mathbf{V}_2}{\mathbf{V}_1} = -\frac{\mathbf{Z}_2}{\mathbf{Z}_1} = -\frac{R_2/(1 + j\omega C R_2)}{R_1} = -\frac{R_2/R_1}{1 + j\omega C R_2}$$

The phase shift of the circuit in Figure 10.17-1*b* is given by

$$\theta = \angle\frac{\mathbf{V}_2}{\mathbf{V}_1} = \angle\left(-\frac{R_2/R_1}{1 + j\omega C R_2}\right) = 180° - \tan^{-1}\omega C R_2 \qquad (10.17\text{-}1)$$

What values of phase shift are possible? Notice that ω, C, and R_2 are all positive, which means that

$$0° \leq \tan^{-1}\omega C R_2 \leq 90°$$

Therefore, the circuit shown in Figure 10.17-1*b* can be used to obtain phase shifts between 90° and 180°. Hence, we can use this circuit to produce a phase shift of 120°.

The gain of the circuit in Figure 10.17-1*b* is given by

$$G = \left|\frac{\mathbf{V}_2}{\mathbf{V}_1}\right| = \left|-\frac{R_2/R_1}{1 + j\omega C R_2}\right|$$

$$= \frac{R_2/R_1}{\sqrt{1 + \omega^2 C^2 R_2^2}} = \frac{R_2/R_1}{\sqrt{1 + \tan^2(180° - \theta)}} \qquad (10.7\text{-}2)$$

Next, first solve Eq. 10.17-1 for R_2 and then Eq. 10.17-2 for R_1 to get

$$R_2 = \frac{\tan(180° - \theta)}{\omega C}$$

and

$$R_1 = \frac{R_2/G}{\sqrt{1 + \tan^2(180° - \theta)}}$$

These equations can be used to design the circuit. First, pick a convenient, readily available, and inexpensive value of the capacitor, say

$$C = 0.02\ \mu\text{F}$$

Next, calculate values of R_1 and R_2 from the values of ω, C, G, and θ. For $\omega = 6283$ rad/s, $C = 0.02\ \mu$F, $G = 2$, and $\theta = 120°$ we calculate

$$R_1 = 3446\ \Omega \quad \text{and} \quad R_2 = 13.78\ k\Omega$$

and the design is complete.

Verify the Proposed Solution

When $C = 0.02\ \mu$F, $R_1 = 3446\ \Omega$, and $R_2 = 13.78$ kΩ, the network function of the circuit is

$$\frac{\mathbf{V}_2}{\mathbf{V}_1} = -\frac{R_2/R_1}{1 + j\omega C R_2} = -\frac{4}{1 + j\omega(0.2756 \times 10^{-3})}$$

In this case, $\omega = 2\pi 1000$ and $\mathbf{V}_1 = 1\angle -90°$ so

$$\frac{\mathbf{V}_2}{\mathbf{V}_1} = -\frac{4}{1 + j(2\pi \times 10^3)(0.2756 \times 10^{-3})} = 2\angle 120°$$

as required by the specifications.

10.18 SUMMARY

- With the pervasive use of ac electric power in the home and industry, it is important for engineers to analyze circuits with sinusoidal independent sources.
- The steady-state response of a linear circuit to a sinusoidal input is itself a sinusoid having the same frequency as the input signal.
- Circuits that contain inductors and capacitors are represented by differential equations. When the input to the circuit is sinusoidal, the phasors and impedances can be used to represent the circuit in the frequency domain. In the frequency domain, the circuit is represented by algebraic equations. The original circuit, represented by a differential equation, is called the time domain representation of the circuit.
- The steady-state response of a linear circuit with a sinusoidal input is obtained as follows:
  1. Transform the circuit into the frequency domain using phasors and impedances.
  2. Represent the frequency domain circuit by algebraic equations, e.g. mesh or node equations.
  3. Solve the algebraic equations to obtain the response of the circuit.
  4. Transform the response into the time domain using phasors.
- Table 10.7-1 summarizes the relationships used to transform a circuit from the time domain to the frequency domain or visa versa.
- When a circuit contains several sinusoidal sources, we distinguish two cases.
  * When all of the sinusoidal sources have the same frequency, the response will be a sinusoid with that frequency and the problem can be solved in the same way that it would be if there was only one source.
  * When the sinusoidal sources have different frequencies, superposition is used to break the time domain circuit up into several circuits each with sinusoidal inputs all at the same frequency. Each of the separate circuits is analyzed separately and the responses are summed, *in the time domain*.
- MATLAB greatly reduces the computational burden associated with solving mesh or node equations having complex coefficients.

PROBLEMS

Section 10.3 Sinusoidal Sources

P 10.3-1 Express the following summations of sinusoids in the general form $A \sin(\omega t + \theta)$ by using trigonometric identities.

Answer: (a) $i(t) = 2\cos(6t + 120°) + 4\sin(6t - 60°)$
(b) $v(t) = 5\sqrt{2}\cos 8t + 10\sin(8t + 45°)$

P 10.3-2 A sinusoidal voltage has a maximum value of 100 V, and the value is 10 V at $t = 0$. The period is $T = 1$ ms. Determine $v(t)$.

P 10.3-3 A sinusoidal current is given as $i = 300\cos(1200\pi t + 55°)$ mA. Determine the frequency f and the value of the current at $t = 2$ ms.

P 10.3-4 Plot a graph of the voltage signal

$$v(t) = 15\cos(628t + 45°)\text{ mV}$$

P 10.3-5 Represent each of the signals shown in Figure P 10.3-5 by a function of the form $A\cos(\omega t + \theta)$.

Answer: (a) $v_s(t) = 10\cos(1900t + 30°)$ V,
(b) $v(t) = 10\cos(1260t + 30°)$ V

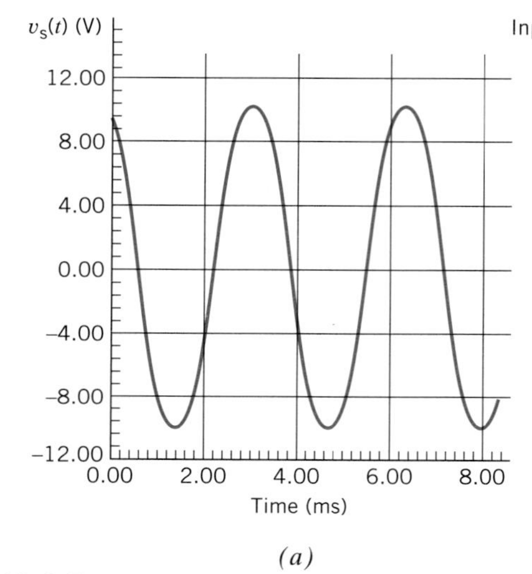

(a)

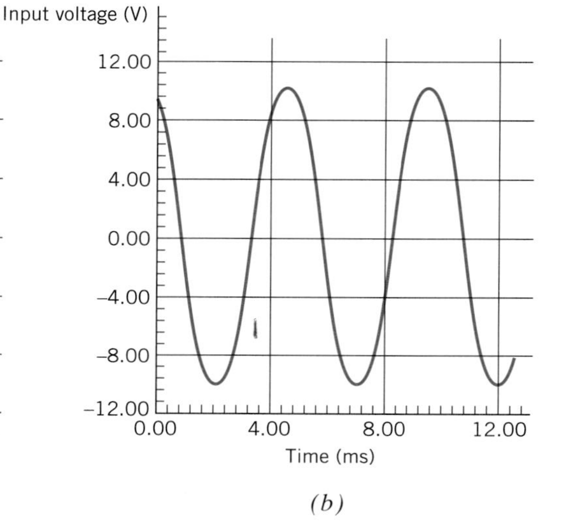

(b)

Figure P 10.3-5

Section 10.4 Steady-State Response of an *RL* Circuit for a Sinusoidal Forcing Function

P 10.4-1 Find the forced response i for the circuit of Figure P 10.4-1 when $v_s(t) = 10\cos(300t)$ V.

Answer: $i(t) = 1.24\cos(300t - 68°)$

P 10.4-2 Find the forced response v for the circuit of Figure P 10.4-2 when $i_s(t) = 0.5\cos\omega t$ A and $\omega = 1000$ rad/s.

P 10.4-3 Find the forced response $i(t)$ for the circuit of Figure P 10.4-3.

Answer: $i(t) = 2\cos(4t + 45°)$ mA

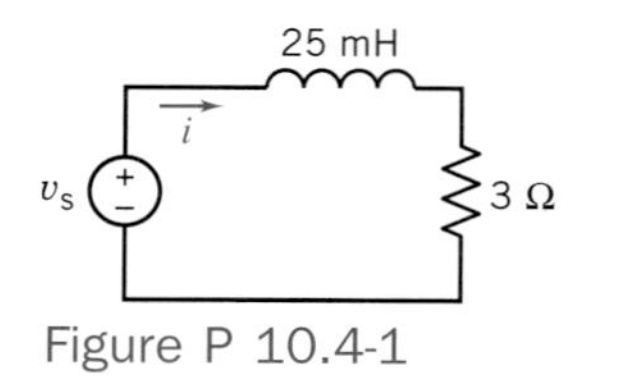

Figure P 10.4-1

Figure P 10.4-2

Figure P 10.4-3

Section 10.5 Complex Exponential Forcing Function (a) Complex Numbers

P 10.5-1 Determine the polar form of the quantity

$$\frac{(5\angle 36.9°)(10\angle -53.1°)}{(4 + j3) + (6 - j8)}$$

Answer: $2\sqrt{5}\angle 10.36°$

P 10.5-2 Determine the polar and rectangular form of the expression

$$5\angle +81.87°\left(4 - j3 + \frac{3\sqrt{2}\angle -45°}{7 - j1}\right)$$

Answer: $28\angle +45° = 14\sqrt{2} + j14\sqrt{2}$

P 10.5-3 Given $\mathbf{A} = 3 + j7$, $\mathbf{B} = 6\angle 15°$, and $\mathbf{C} = 5e^{j2.3}$, find $(\mathbf{A}^*\mathbf{C}^*)/\mathbf{B}$.
Answer: $0.65 - j6.32$

P 10.5-4 Determine a and b when (angles in degrees)

$$(6\angle 120°)(-4 + j3 + 2e^{j15}) = a + jb$$

P 10.5-5 Find a, b, A, and θ as required (angles given in degrees).
(a) $A\,e^{j120} + jb = -4 + j3$
(b) $6e^{j120}(-4 + jb + 8e^{j\theta}) = 18$
(c) $(a + j4)j2 = 2 + Ae^{j60}$

Section 10.5 Complex Exponential Forcing Function (b) Response of a Circuit

P 10.5-6 For the circuit of Figure P 10.5-6 find $i(t)$ when $v_s(t) = 0.1\cos(\omega t + 90°)$ V and $\omega = 10^7$ rad/s.
Answer: $i(t) = 1\cos(\omega t + 90°)$ mA

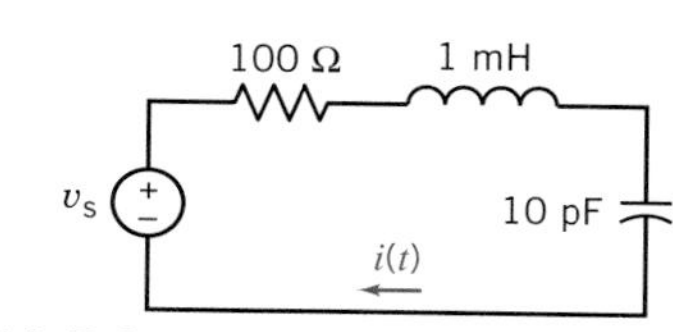

Figure P 10.5-6

P 10.5-7 For the circuit of Figure P 10.5-7 find $i(t)$ when $v_s(t) = 20\cos(\omega t + 45°)$ V when $\omega = 25$ Mrad/s.

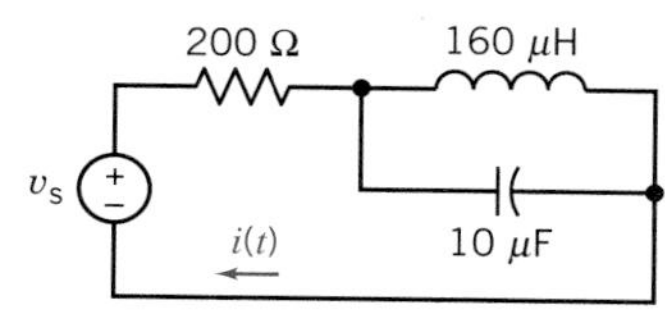

Figure P 10.5-7

Section 10.6 The Phasor Concept

P 10.6-1 For the circuit of Figure P 10.6-1 find $v(t)$ when $i_s(t) = 5\cos(\omega t - 120°)$ mA when $\omega = 10^5$ rad/s.
Answer: $v(t) = 1.5\cos(\omega t + 60°)$ V

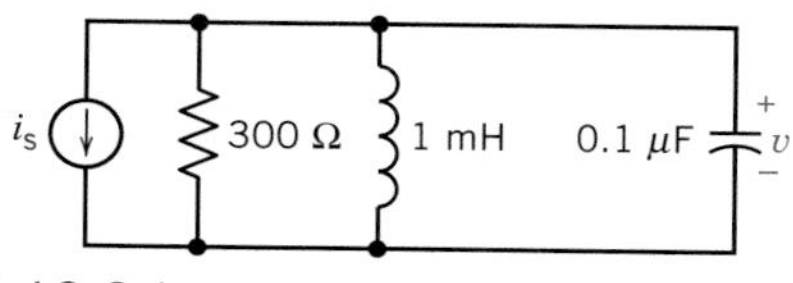

Figure P 10.6-1

P 10.6-2 For the circuit of Figure P 10.6-2 find $i(t)$ when $i_s = 25\cos(\omega t - 120°)$ mA and $\omega = 10^3$ rad/s.

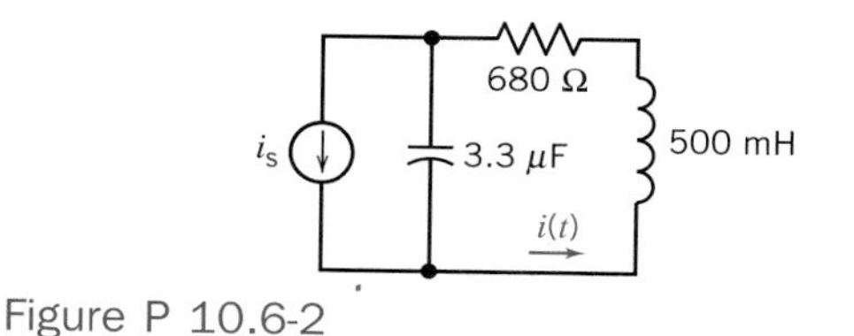

Figure P 10.6-2

P 10.6-3 For the circuit of Figure P 10.6-3, find $v(t)$ when $v_s = 2\sin 500t$ V.
Answer: $v(t) = 1.25\cos(500t - 141°)$ V

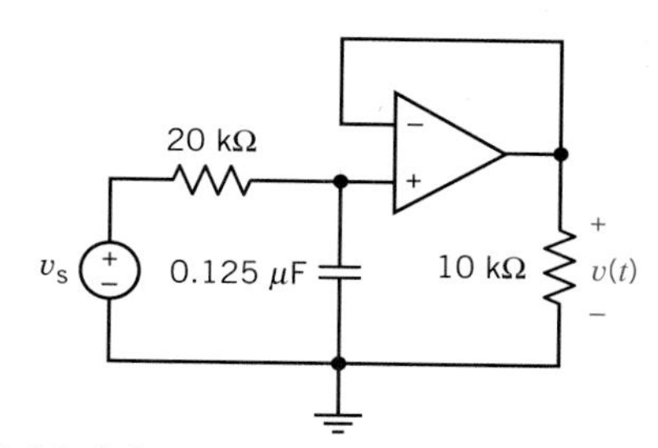

Figure P 10.6-3

Section 10.7 Phasor Relationships for *R*, *L*, and *C* Elements

P 10.7-1 Given the phasor $\mathbf{I} = 6 + j8$, rotate it in the phasor plane by the following amounts and express the final result in rectangular coordinates: (a) $-45°$; (b) $90°$.
Answer: (a) $\mathbf{I} = 7\sqrt{2} + j\sqrt{2}$
(b) $\mathbf{I} = -8 + j6$

P 10.7-2 Add the following voltage waveforms using phasors, obtaining the resultant voltage in the form $v_1 + v_2 = A\cos(\omega t + \phi)$.
(a) $v_1 = 3\cos(2t + 60°)$; $v_2 = 8\cos(2t - 22.5°)$
(b) $v_1 = 2\sqrt{2}\sin 4t$; $v_2 = 10\cos(4t + 30°)$

P 10.7-3 Find the two phasors $\mathbf{A}$ and $\mathbf{B}$ so that $|A| = 5\sqrt{2}$, $|B| = 4$ and so that $2\mathbf{A} + 5\mathbf{B} = j10(1 + \sqrt{3})$ and $\mathbf{B}$ leads $\mathbf{A}$ by 75°.

P 10.7-4 For the following voltage and current expressions, indicate whether the element is capacitive, inductive, or resistive and find the element value.
(a) $v = 15\cos(400t + 30°)$; $i = 3\sin(400t + 30°)$
(b) $v = 8\sin(900t + 50°)$; $i = 2\sin(900t + 140°)$
(c) $v = 20\cos(250t + 60°)$; $i = 5\sin(250t + 150°)$
Answer: (a) $L = 12.5$ mH
(b) $C = 277.77$ μF
(c) $R = 4\ \Omega$

P 10.7-5 Two voltages appear in series so $v = v_1 + v_2$. Find v when $v_1 = 150\cos(377t - \pi/6)$ V and $\mathbf{V}_2 = 200\angle +60°$ V.
Answer: $v(t) = 250\cos(377t + 23.1°)$ V

Section 10.8 Impedance and Admittance

P 10.8-1 Find $\mathbf{Z}$ and $\mathbf{Y}$ for the circuit of Figure P 10.8-1 operating at 10 kHz.

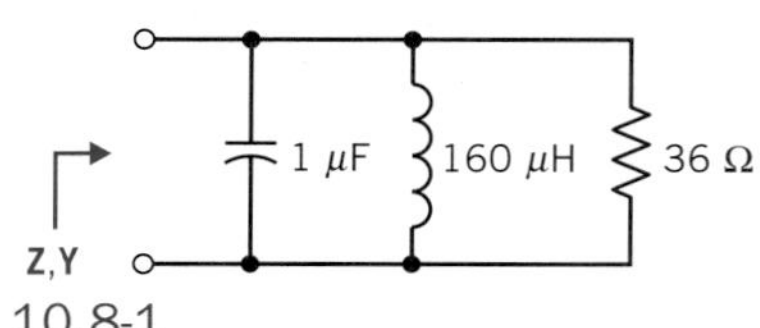

Figure P 10.8-1

P 10.8-2 Find R and L of the circuit of Figure P 10.8-2 when $v(t) = 10\cos(\omega t + 40°)$ V, $i(t) = 2\cos(\omega t + 15°)$ mA, and $\omega = 2 \times 10^6$ rad/s.
Answer: 4.532 kΩ, L = 1.057 mH

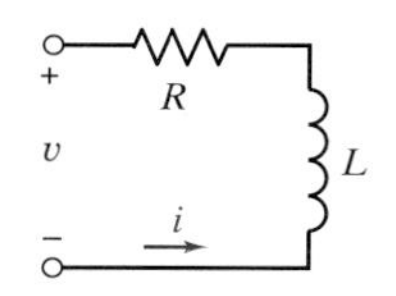

Figure P 10.8-2

P 10.8-3 Consider the circuit of Figure P 10.8-3 when $R = 6\ \Omega$, $L = 27$ mH, and $C = 22\ \mu$F. Determine the frequency f when the impedance **Z** is purely resistive, and find the input resistance at that frequency.

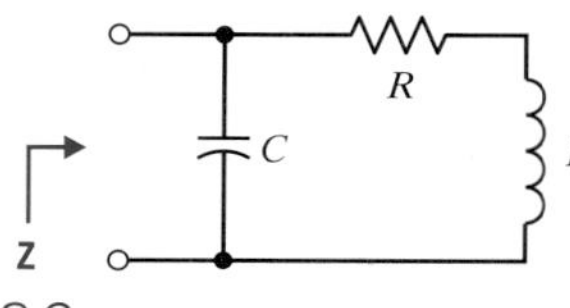

Figure P 10.8-3

P 10.8-4 Consider the circuit of Figure P 10.8-4 when $R = 10$ kΩ and $f = 1$ kHz. Find L and C so that $\mathbf{Z} = 100 + j0\ \Omega$.
Answer: L = 0.1587 H, C = 0.158 μF.

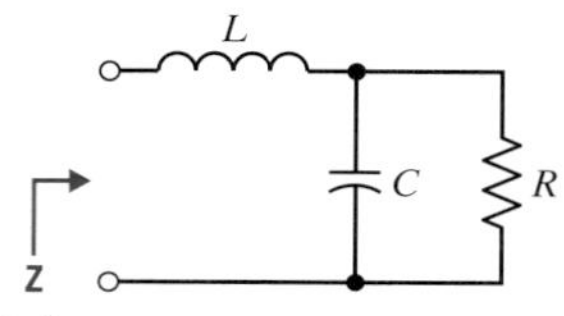

Figure P 10.8-4

P 10.8-5 For the circuit of Figure P 10.8-5 find the value of C required so that $\mathbf{Z} = 590.7\ \Omega$ when $f = 1$ MHz.
Answer: C = 0.27 nF

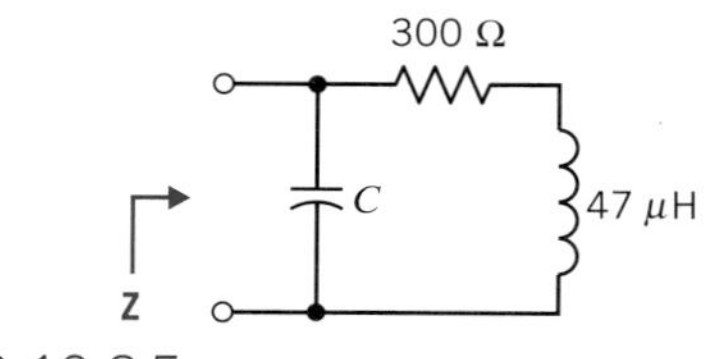

Figure P 10.8-5

Section 10.9 Kirchhoff's Laws Using Phasors

P 10.9-1 For the circuit shown in Figure P 10.9-1, find (a) the impedances $\mathbf{Z}_1$ and $\mathbf{Z}_2$ in polar form, (b) the total combined impedance in polar form, and (c) the steady-state current $i(t)$.
Answer: (a) $\mathbf{Z}_1 = 5\angle 53.1°$; $\mathbf{Z}_2 = 8\sqrt{2}\angle -45°$
(b) $\mathbf{Z}_1 + \mathbf{Z}_2 = 11.7\angle -20°$
(c) $i(t) = (8.55)\cos(1250t + 20°)$ A

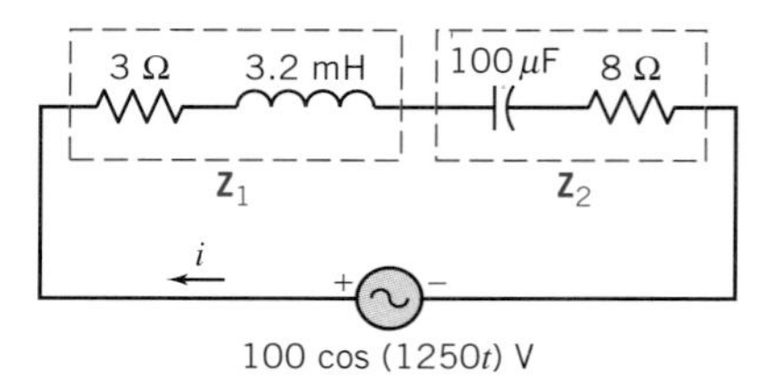

Figure P 10.9-1

P 10.9-2 Spinal cord injuries result in paralysis of the lower body and can cause loss of bladder control. Numerous electrical devices have been proposed to replace the normal nerve pathway stimulus for bladder control. Figure P 10.9-2 shows the model of a bladder control system where $v_s = 20\cos\omega t$ V and $\omega = 100$ rad/s. Find the steady-state voltage across the 10 Ω load resistor.
Answer: $v(t) = 10\sqrt{2}\cos(100t + 45°)$ V

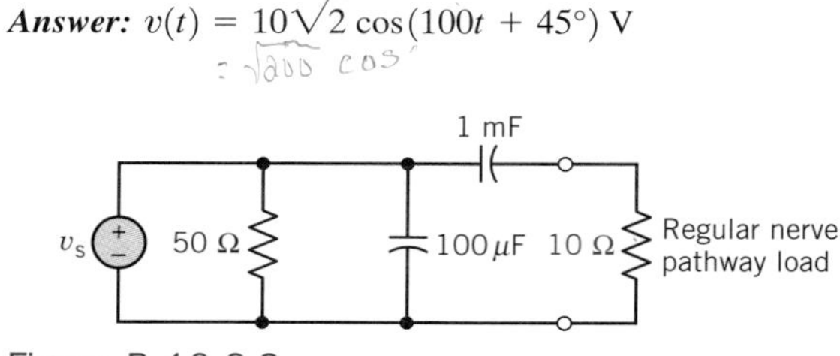

Figure P 10.9-2

P 10.9-3 There are 500 to 1000 deaths each year in the United States from electric shock. If a person makes a good contact with his hands, the circuit can be represented by Figure P 10.9-3, where $v_s = 160\cos\omega t$ V and $\omega = 2\pi f$. Find the steady-state current i flowing through the body when (a) $f = 60$ Hz and (b) $f = 400$ Hz.
Answer: (a) $i(t) = 0.53\cos(120\pi t + 5.9°)$,
(b) $i(t) = 0.625\cos(800\pi t + 59.9°)$ A

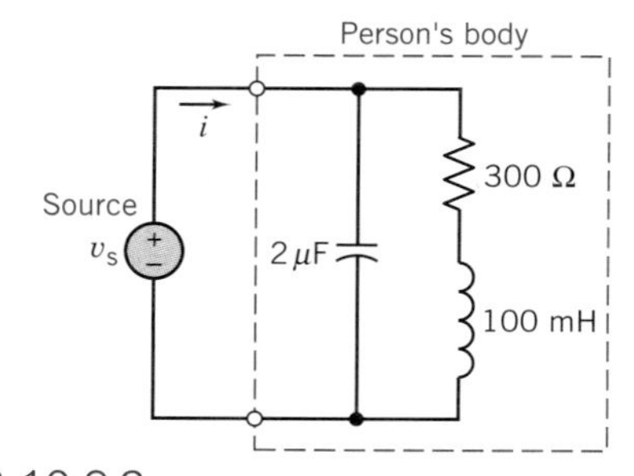

Figure P 10.9-3

P 10.9-4 Determine $i(t)$ of the *RLC* circuit shown in Figure P 10.9-4 when $v_s = 2\cos(4t + 30°)$ V.
Answer: $i(t) = 0.185\cos(4t - 26.3°)$ A

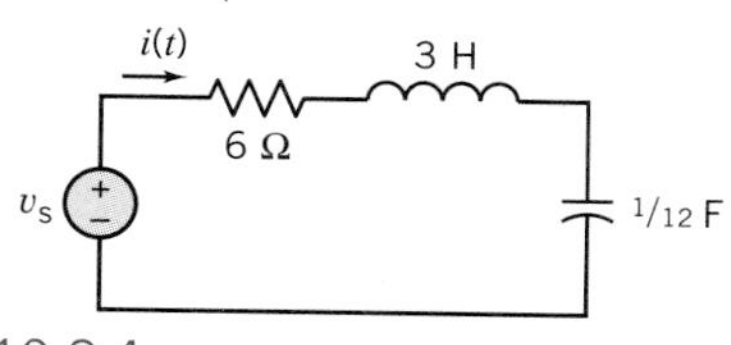

Figure P 10.9-4

P 10.9-5 The big toy from the hit movie *Big* is a child's musical fantasy come true—a sidewalk-sized piano. Like a hopscotch grid, this Christmas's hot toy invites anyone who passes to jump on, move about, and make music. The developer of the "toy" piano used a tone synthesizer and stereo speakers as shown in Figure P 10.9-5 (Gardner, 1988). Determine the current $i(t)$ for a tone at 796 Hz when $C = 10\ \mu$F.

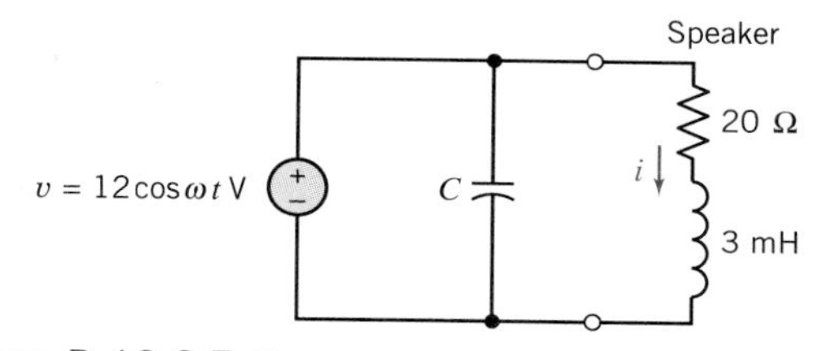

Figure P 10.9-5 Tone synthesizer.

P 10.9-6 Determine *B* and *L* for the circuit of Figure P 10.9-6 when $i(t) = B\cos(3t - 51.87°)$ A.
Answer: $B = 1.6$, $L = 2$ H.

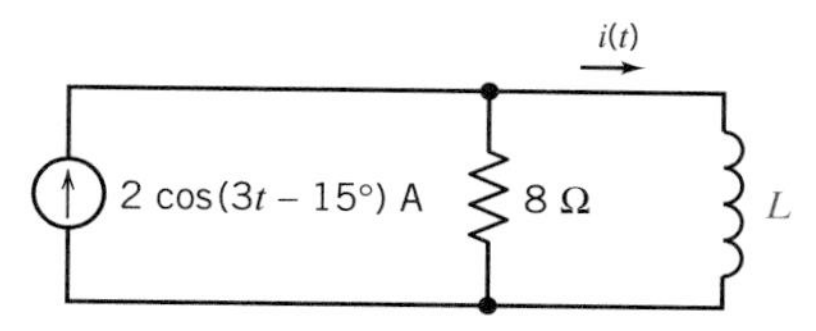

Figure P 10.9-6

P 10.9-7 Determine $i(t)$, $v(t)$, and L for the circuit shown in Figure P 10.9-7.
Answer: $i(t) = 1.34\cos(2t - 87°)$ A, $v(t) = 7.29\cos(2t - 24°)$ V, L = 4 H.

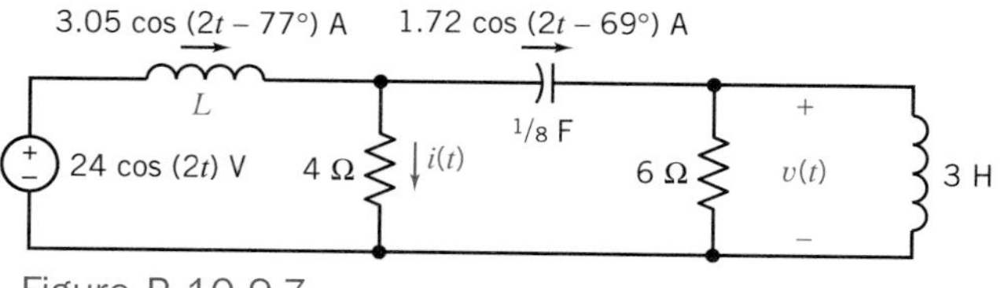

Figure P 10.9-7

Section 10-10 Node Voltage and Mesh Current Analysis Using Phasors

(a) Node Voltage Analysis

P 10.10-1 Find the phasor voltage $\mathbf{V}_c$ for the circuit shown in Figure P 10.10-1.

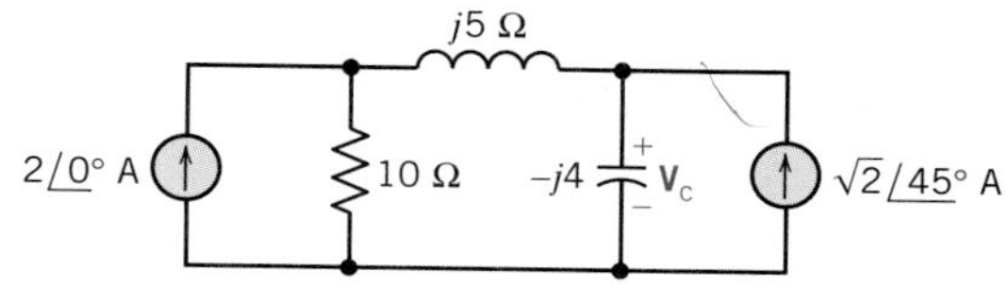

Figure P 10.10-1

P 10.10-2 For the circuit shown in Figure P 10.10-2, determine the phasor currents $\mathbf{I}_s$, $\mathbf{I}_c$, $\mathbf{I}_L$, and $\mathbf{I}_R$ if $\omega = 1000$ rad/s.
Answer: $\mathbf{I}_s = 0.347\angle -25.5°$ A
$\mathbf{I}_c = 0.461\angle 112.9°$ A
$\mathbf{I}_L = 0.720\angle -67.1°$ A
$\mathbf{I}_R = 0.230\angle 22.9°$ A

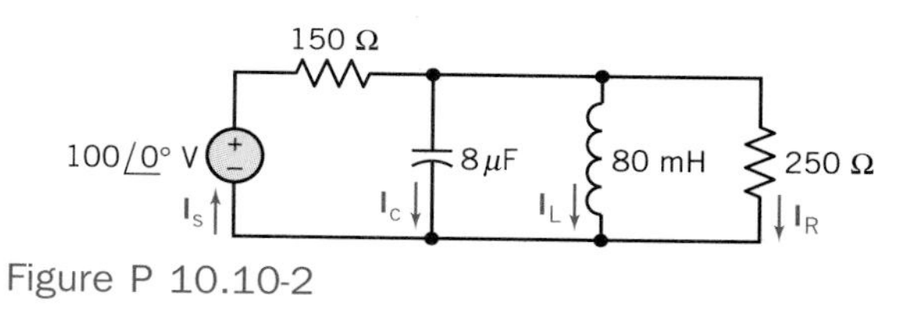

Figure P 10.10-2

P 10.10-3 Consider the circuit of Figure P 10.10-3. Using nodal analysis, find $\mathbf{V}_s$ when $\mathbf{I}$ in the 4 Ω resistor is $3\angle 45°$ A.

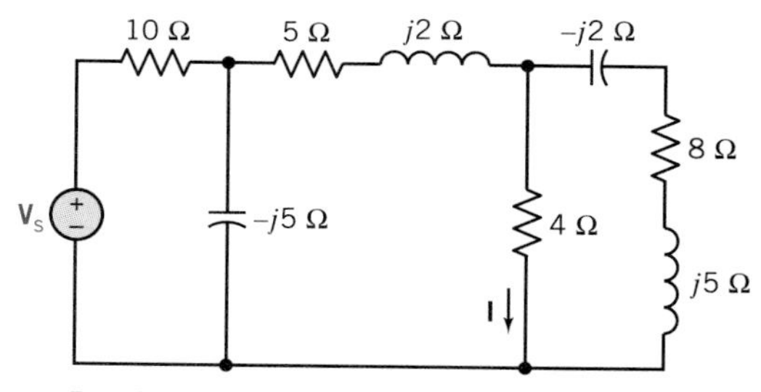

Figure P 10.10-3

P 10.10-4 Find the two node voltages $v_a(t)$ and $v_b(t)$ for the circuit of Figure P 10.10-4 when $v_s(t) = 1.2\cos 4000t$.
Answer: $v_a(t) = 1.97\cos(4000t - 171°)$ V
$v_b(t) = 2.21\cos(4000t - 144°)$ V

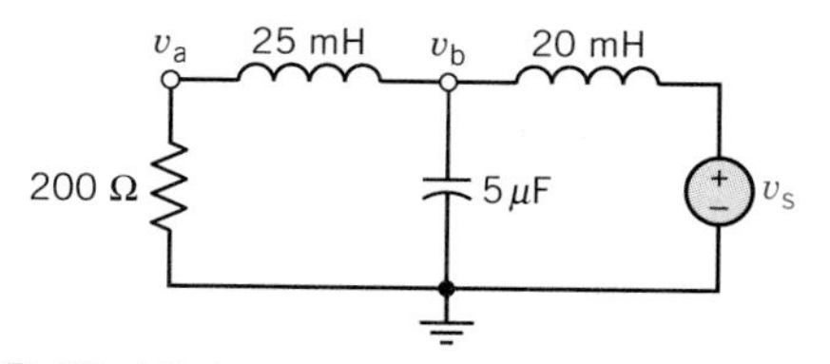

Figure P 10.10-4

P 10.10-5 The circuit shown in Figure P 10.10-5 has two sources: $v_s = 20\cos(\omega_0 t + 90°)$ V and $i_s = 6\cos\omega_0 t$ A where $\omega_0 = 10^5$ rad/s. Determine $v_0(t)$.
Answer: $v_0(t) = 16.31\cos(10^5 t + 71.5°)$ V

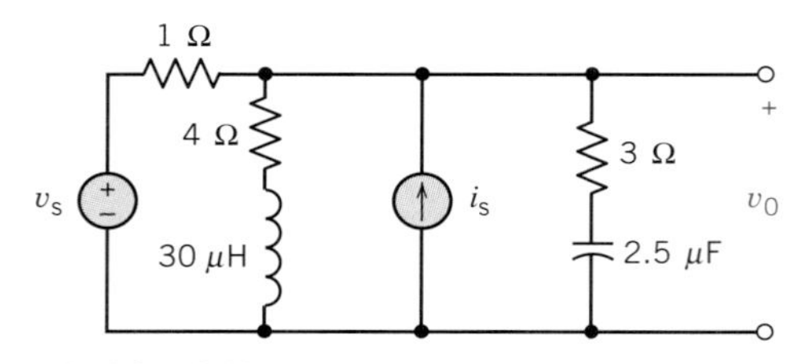

Figure P 10.10-5

P 10.10-6 Determine the voltage v_a for the circuit in Figure P 10.10-6 when $i_s = 20\cos(\omega t + 53.13°)$ A and $\omega = 10^4$ rad/s.
Answer: $v_a(t) = 339.4\cos(10^4 t + 45°)$ V

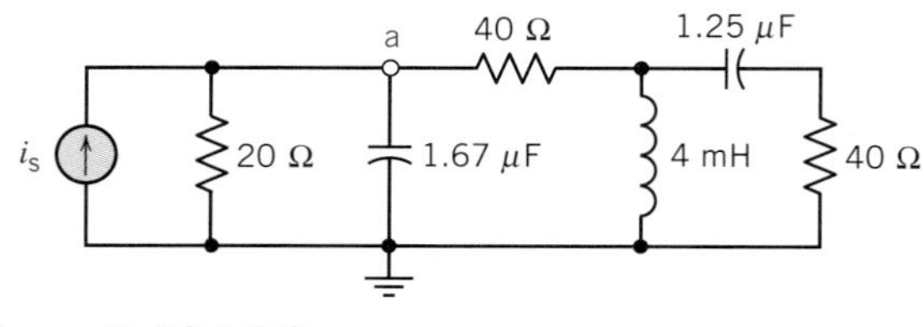

Figure P 10.10-6

P 10.10-7 A commercial airliner has sensing devices to indicate to the cockpit crew that each door and baggage hatch is closed. A device called a search coil magnetometer, also known as a proximity sensor, provides a signal indicative of the proximity of metal or other conducting material to an inductive sense coil. The inductance of the sense coil changes as the metal gets closer to the sense coil. The sense coil inductance is compared to a reference coil inductance with a circuit called a balanced inductance bridge (see Figure P 10.10-7). In the inductance bridge, a signal indicative of proximity is observed between terminals A and B by subtracting the voltage at B, v_B, from the voltage at A, v_A (Lenz, 1990).

The bridge circuit is excited by a sinusoidal voltage source $v_s = \sin(800\,\pi t)$ V. The two resistors, $R = 100\ \Omega$, are of equal resistance. When the door is open (no metal is present), the sense coil inductance, L_s, is equal to the reference coil inductance, $L_R = 40$ mH. In this case, what is the magnitude of the signal $\mathbf{V}_A - \mathbf{V}_B$?

When the airliner door is completely closed, $L_s = 60$ mH. With the door closed, what is the phasor representation of the signal $\mathbf{V}_A - \mathbf{V}_B$?

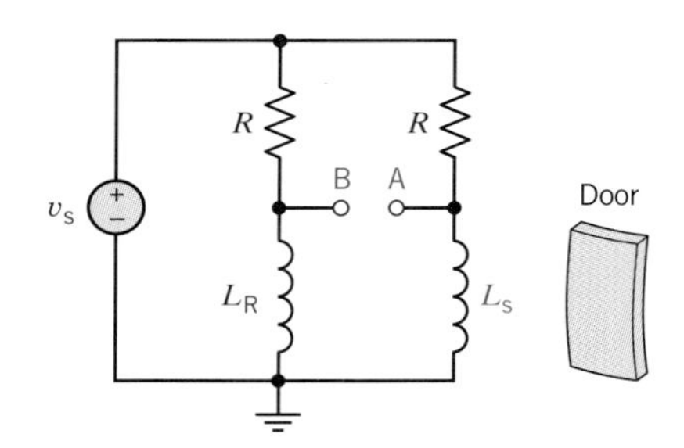

Figure P 10.10-7 Airline door sensing circuit.

P 10.10-8 Using a tiny diamond-studded burr operating at 190,000 rpm, cardiologists can remove life-threatening plaque deposits is coronary arteries. The procedure is fast, uncomplicated, and relatively painless (McCarty, 1991). The Rotablator, an angioplasty system, consists of an advancer/catheter, a guide wire, a console, and a power source. The advancer/catheter contains a tiny turbine that drives the flexible shaft that rotates the catheter burr. The model of the operational and control circuit is shown in Figure P 10.10-8. Determine $v(t)$, the voltage that drives the tip, when $v_s = \sqrt{2}\cos(40t + 135°)$ V.
Answer: $v(t) = 1.414\cos(40t - 135°)$ V

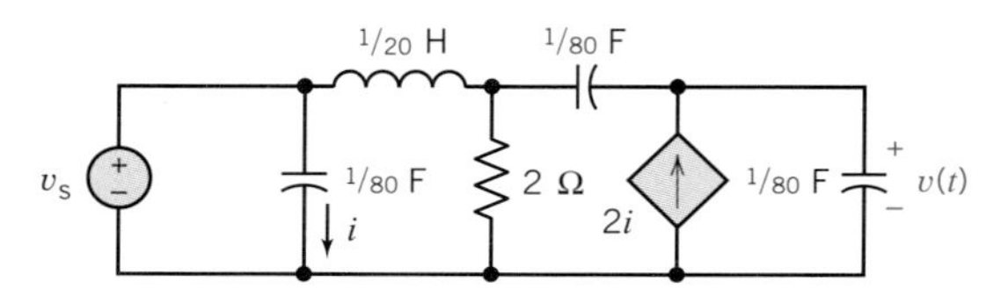

Figure P 10.10-8 Control circuit for Rotablator.

P 10.10-9 For the circuit of Figure P 10.10-9, it is known that

$$v_2(t) = 0.7571\cos(2t + 66.7°)\ \text{V}$$
$$v_3(t) = 0.6064\cos(2t - 69.8°)\ \text{V}$$

Determine $i_1(t)$.

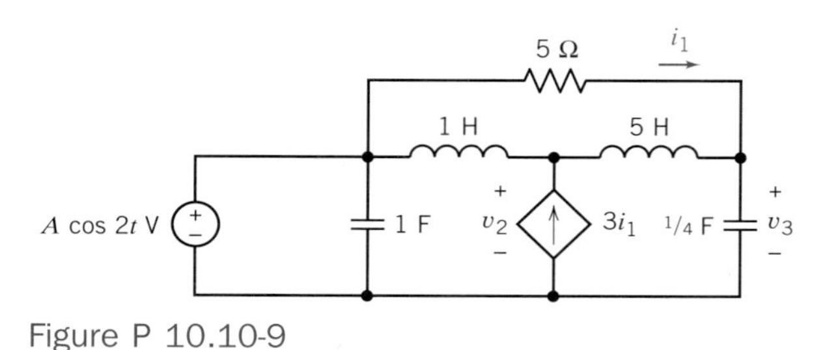

Figure P 10.10-9

Section 10.10 Node Voltage and Mesh Current Analysis Using Phasors
(b) Mesh Current Analysis

P 10.10-10 Determine $\mathbf{I}_1$, $\mathbf{I}_2$, $\mathbf{V}_L$, and $\mathbf{V}_c$ for the circuit of Figure P 10.10-10 using KVL and mesh analysis.

Answer: $\mathbf{I}_1 = 2.5\angle 29.0^\circ$ A
$\mathbf{I}_2 = 1.8\angle 105^\circ$ A
$\mathbf{V}_L = 16.3\angle 78.7^\circ$ V
$\mathbf{V}_c = 7.2\angle 15^\circ$ V

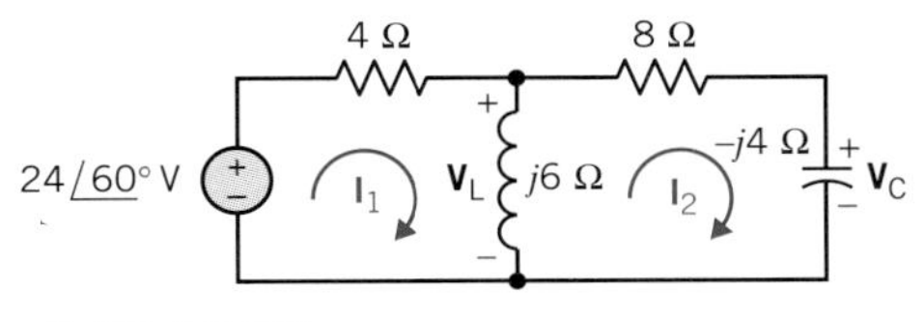

Figure P 10.10-10

P 10.10-11 The model of a high-frequency transistor amplifier is shown in Figure P 10.10-11 with a source voltage and a load resistor. The source voltage is $v_s = 10 \cos \omega t$ where $\omega = 10^8$ rad/s. Find the voltage across the load resistor.

Answer: $v_L(t) = 9.59 \times 10^4 \cos(10^8 t + 175.6^\circ)$ V

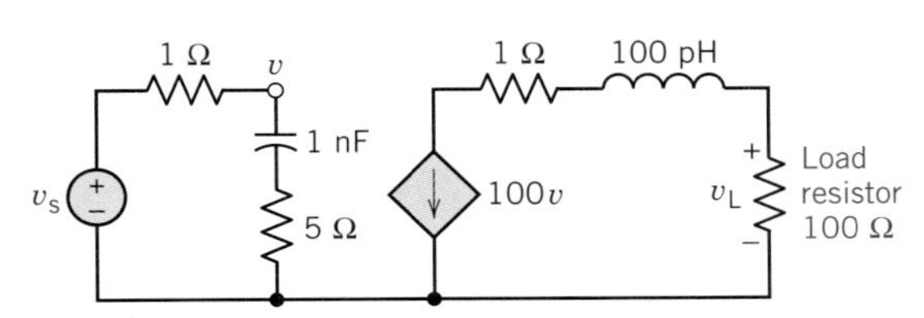

Figure P 10.10-11 Model of a high-frequency transistor amplifier.

P 10.10-12 Determine the current $i(t)$ for the circuit of Figure P 10.10-12 using mesh currents when $\omega = 1000$ rad/s.

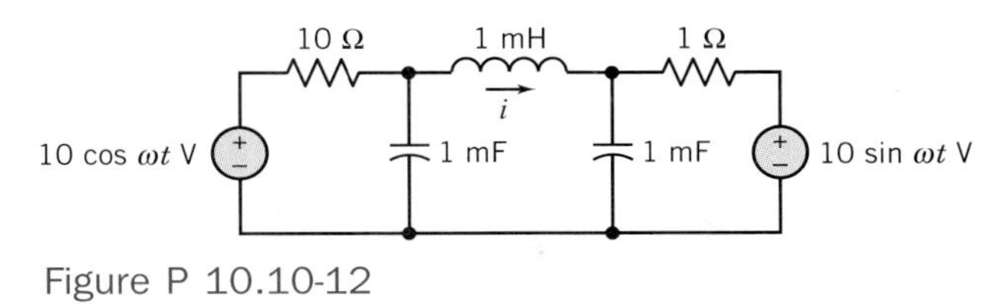

Figure P 10.10-12

P 10.10-13 The idea of using an induction coil in a lamp isn't new, but applying it in a commercially available product is. An induction coil in a bulb induces a high-frequency energy flow in mercury vapor to produce light. The lamp uses about the same amount of energy as a fluorescent bulb but lasts six times longer, with 60 times the life of a conventional incandescent bulb. The circuit model of the bulb and its associated circuit are shown in Figure P 10.10-13. Determine the voltage $v(t)$ across the $2\ \Omega$ resistor when $C = 40\ \mu$F, $L = 40\ \mu$H, $v_s = 10 \cos(\omega_0 t + 30^\circ)$, and $\omega_0 = 10^5$ rad/s.

Answer: $v(t) = 6.45 \cos(10^5 t + 44^\circ)$ V

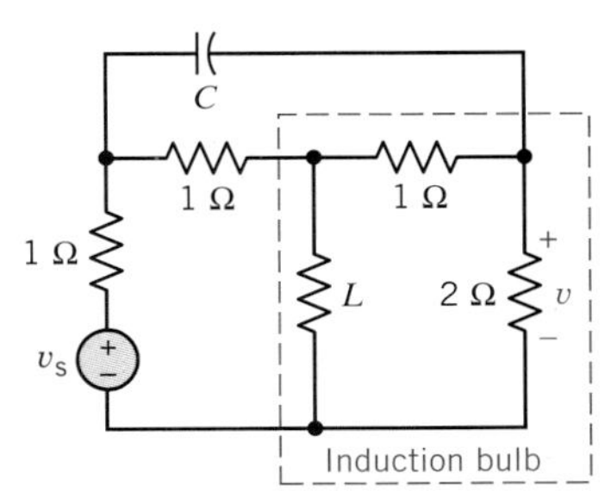

Figure P 10.10-13 Induction bulb circuit.

P 10.10-14 The development of coastal hotels in various parts of the world is a rapidly growing enterprise. The need for environmentally acceptable shark protection is manifest where these developments take place alongside shark-infested waters (Smith, 1991). One concept is to use an electrified line submerged in the water in order to deter the sharks, as shown in Figure P 10.10-14*a*. The circuit model of the electric fence is shown in Figure P 10.10-14*b* where the shark is represented by an equivalent resistance of $100\ \Omega$. Determine the current flowing through the shark's body, $i(t)$, when $v_s = 375 \cos 400t$ V.

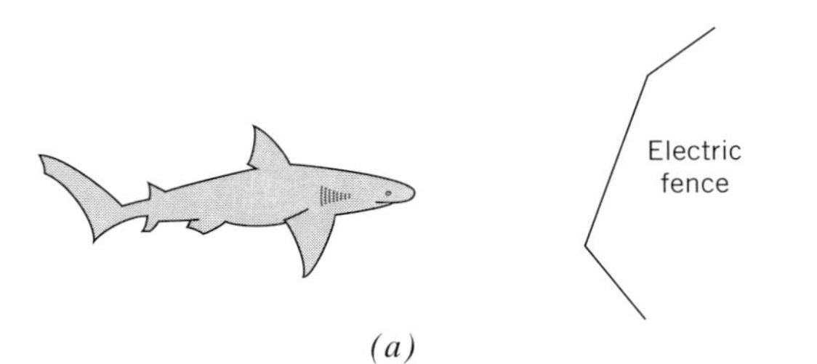

(*a*)

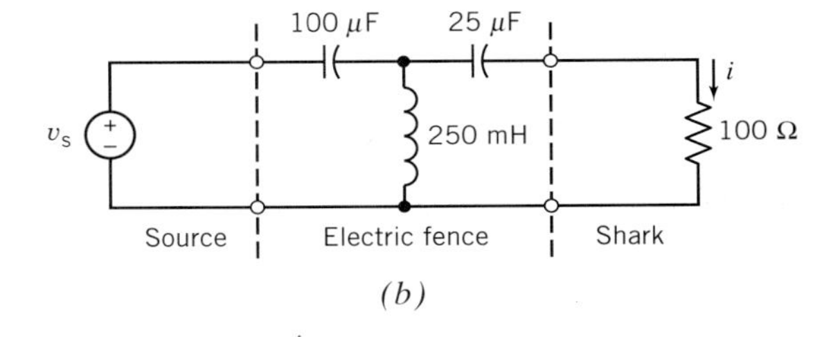

(*b*)

Figure P 10.10-14 Electric fence for repelling sharks.

Section 10.11 Superposition, Thévenin and Norton Equivalents, and Source Transformations
(a) Superposition Principle

P 10.11-1 For the circuit of Figure P 10.11-1 find $i(t)$ when $v_1 = 12 \cos(4000t + 45°)$ V and $v_2 = 5 \cos 3000t$ V.

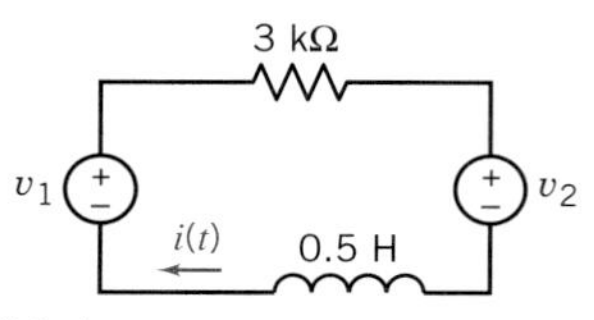

Figure P 10.11-1

P 10.11-2 Determine $i(t)$ of the circuit of Figure P 10.11-2. ***Hint:*** Replace the voltage source by a series combination of a dc voltage and a sinusoidal voltage source.
Answer: $i(t) = 0.166 \cos(4t - 135°) + 0.5$ mA

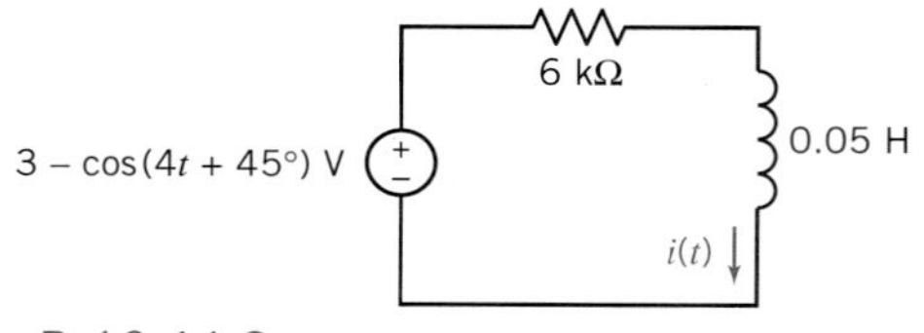

Figure P 10.11-2

P 10.11-3 Determine $i(t)$ for the circuit of Figure P 10.11-3.
Answer: $i(t) = 2 \cos(4t + 45°) - 0.833 \cos(3t - 90°)$ mA

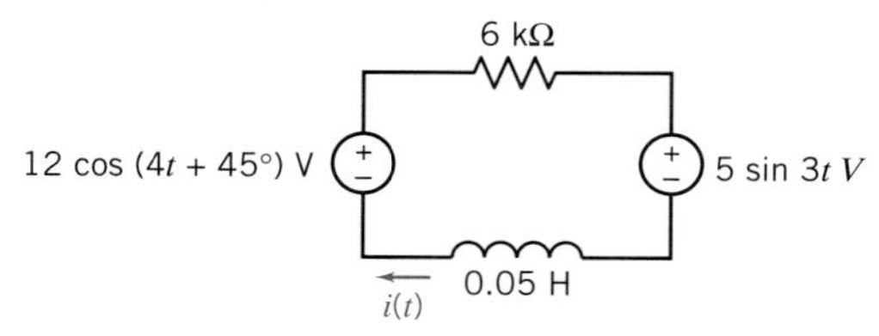

Figure P 10.11-3

Section 10.11 Superposition, Thévenin and Norton Equivalents, and Source Transformations
(b) Thévenin and Norton Equivalent Circuits

P 10.11-4 Determine the Thévenin equivalent circuit for the circuit shown in Figure P 10.11-4 when $v_s = 5 \cos (4000t - 30°)$.
Answer: $\mathbf{V}_t = 5.7\angle{-21.9°}$ V
$\mathbf{Z}_t = 23\angle{-81.9°}\ \Omega$

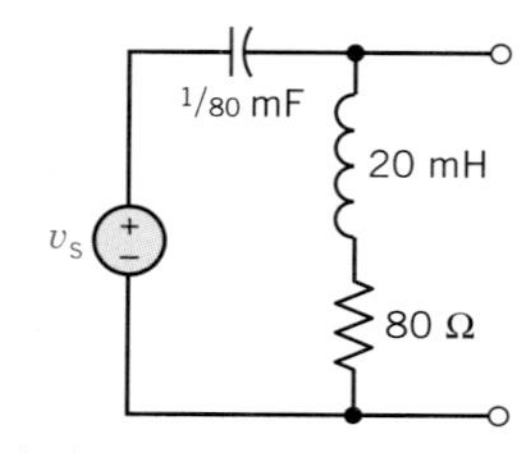

Figure P 10.11-4

P 10.11-5 Find $v_x(t)$ by first replacing the circuit to the left of terminals a–b in Figure P 10.11-5 with its Thévenin equivalent circuit when $C = 10$ mF.
Answer: $v_x(t) = 33.13 \cos(20t - 83.66°)$ V

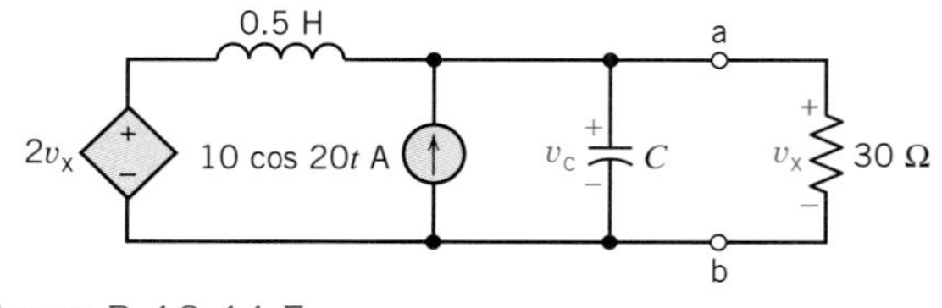

Figure P 10.11-5

P 10.11-6 Find the Thévenin equivalent circuit for the circuit shown in Figure P 10.11-6 using the mesh current method.
Answer: $\mathbf{V}_t = 3.71\angle{-16°}$ V
$\mathbf{Z}_t = 247\angle{-16°}\ \Omega$

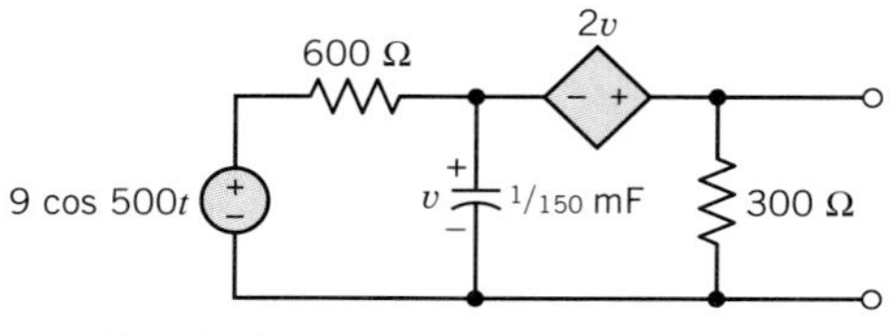

Figure P 10.11-6

P 10.11-7 For the circuit shown in Figure P 10.11-7, find the capacitance, C, if $v(t) = 100\sqrt{2} \cos(100t - 45°)$ V.
Answer: $C = 1/800$ F

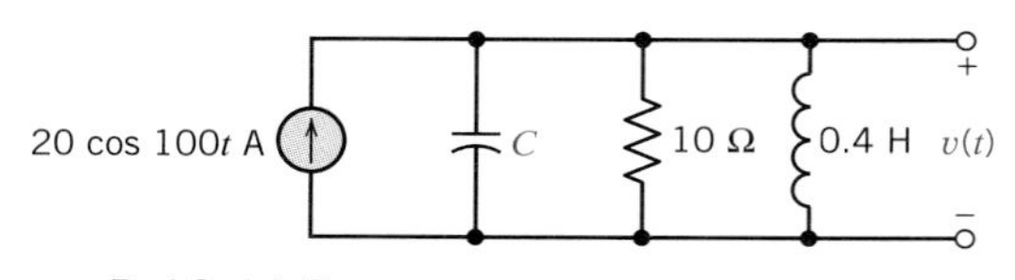

Figure P 10.11-7

P 10.11-8 A pocket-sized mini-disk CD player system has an amplifier circuit shown in Figure P 10.11-8 with a signal $v_s = 10 \cos(\omega t + 53.1°)$ at $\omega = 10{,}000$ rad/s. Determine the Thévenin equivalent at the output terminals a–b.

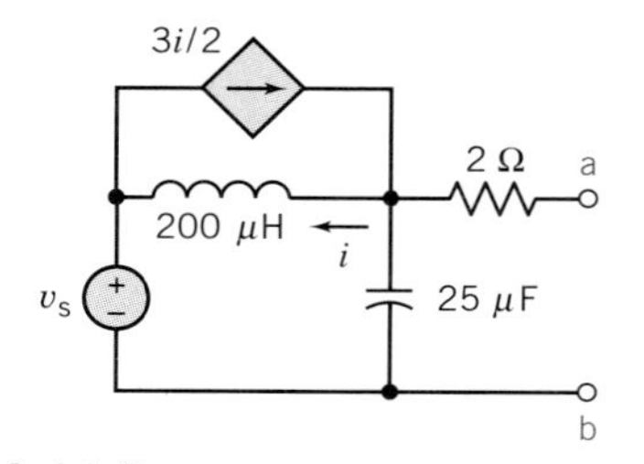

Figure P 10.11-8

P 10.11-9 In the analysis of guided waves on transmission lines, circuit theory is often used. One of the simplest unit-length circuit models of a transmission line is shown in Figure P 10.11-9. The transmission line is represented by L and C, and the load by R. A line has $L = 97.5$ nH and $C = 39$ pF and operates at 100 MHz. Determine the input impedance when $R = 25\ \Omega$ and when $R = 50\ \Omega$.

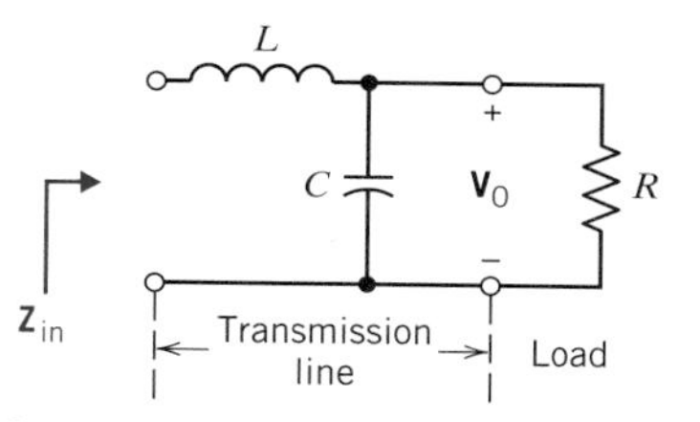

Figure P 10.11-9 Circuit model of transmission line.

P 10.11-10 An AM radio receiver uses the parallel RLC circuit shown in Figure P 10.11-10. Determine the frequency, f_0, at which the admittance **Y** is a pure conductance. What is the "number" of this station on the AM radio dial?

Answer: $f_0 = 800$ Hz, which corresponds to 80 on the AM radio dial.

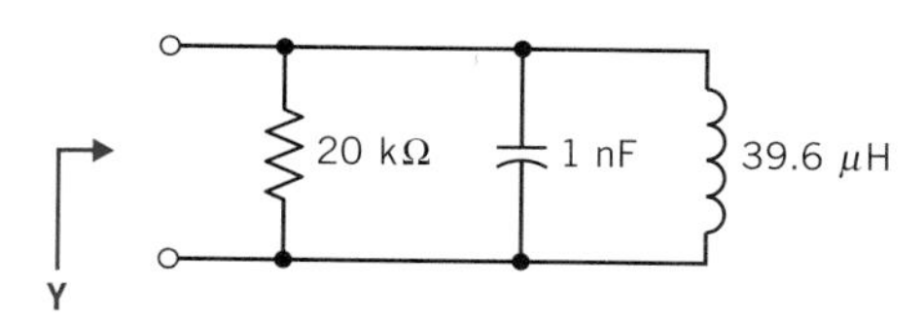

Figure P 10.11-10

P 10.11-11 A linear circuit is placed within a black box with only the terminals a–b available as shown in Figure P 10.11-11. There are three elements available in the laboratory: (1) a 50 Ω resistor, (2) a 2.5-μF capacitor, and (3) a 50-mH inductor. These three elements are placed across terminals a–b as the load $\mathbf{Z}_L$, and the magnitude of **V** is measured as (1) 25 V, (2) 100 V, and (3) 50 V, respectively. It is known that the sources within the box are sinusoidal with $\omega = 2 \times 10^3$ rad/s. Determine the Thévenin equivalent for the circuit in the box as shown in Figure P 10.11-11.

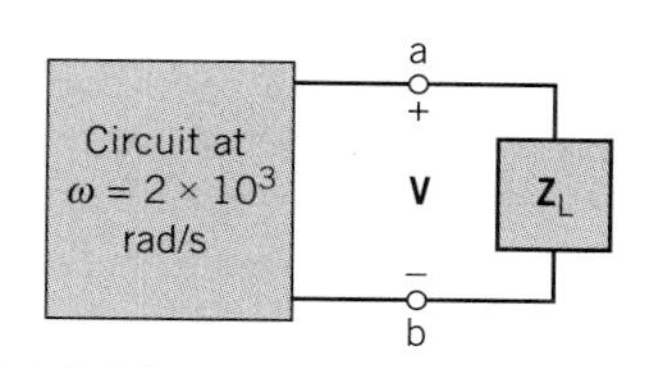

Figure P 10.11-11 A circuit within a black box is connected to a selected impedance $\mathbf{Z}_n$.

Section 10.11 Superposition, Thévenin and Norton Equivalents, and Source Transformations (c) Source Transformations

P 10.11-12 Consider the circuit of Figure P 10.11-12, where we wish to determine the current **I**. Use a series of source transformations to find a current source in parallel with an equivalent impedance, and then find the current in the 2 Ω resistor by current division.

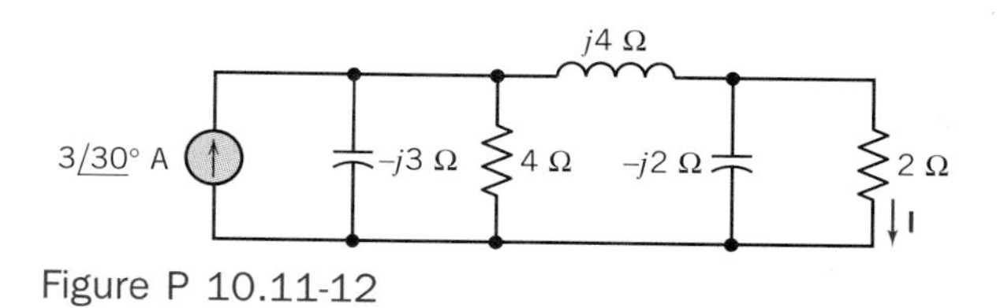

Figure P 10.11-12

P 10.11-13 For the circuit of Figure P 10.11-13 determine the current **I** using a series of source transformations. The source has $\omega = 25 \times 10^3$ rad/s.

Answer: $i(t) = 4\cos(25000t - 44°)$ mA

Figure P 10.11-13

Section 10.12 Phasor Diagrams

P 10.12-1 Using a phasor diagram determine **V** when $\mathbf{V} = \mathbf{V}_1 - \mathbf{V}_2 + \mathbf{V}_3^*$ and $\mathbf{V}_1 = 3 + j3$, $\mathbf{V}_2 = 4 + j2$, and $\mathbf{V}_3 = -3 - j2$. (Units are volts.)

Answer: $\mathbf{V} = 5\underline{/143.1°}$ V

P 10.12-2 Consider the series RLC circuit of Figure P 10.12-2 when $R = 10\ \Omega$, $L = 1$ mH, $C = 100\ \mu$F, and $\omega = 10^3$ rad/s. Find **I** and plot the phasor diagram.

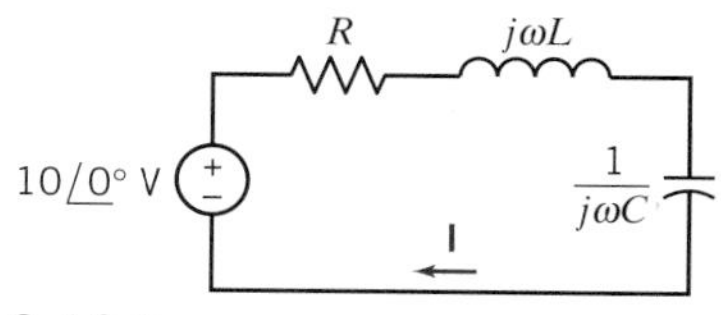

Figure P 10.12-2

P 10.12-3 Consider the signal

$$i(t) = 72\sqrt{3}\cos 8t + 36\sqrt{3}\sin(8t + 140°) + 144\cos(8t + 210°) + 25\cos(8t + \phi)$$

Using the phasor plane, for what value of ϕ does the $|I|$ attain its maximum?

Section 10.14 Phasor Circuits and the Operational Amplifier

P 10.14-1 Find the steady-state response $v_0(t)$ if $v_s(t) = \sqrt{2}\cos 1000t$ for the circuit of Figure P 10.14-1. Assume an ideal op amp.

Answer: $v_0(t) = 10\cos(1000t - 225°)$

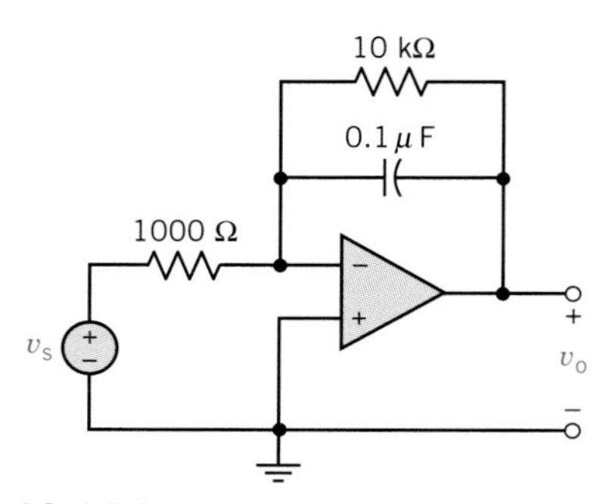

Figure P 10.14-1

P 10.14-2 Determine $\mathbf{V}_0/\mathbf{V}_s$ for the op amp circuit shown in Figure P 10.14-2. Assume an ideal op amp.

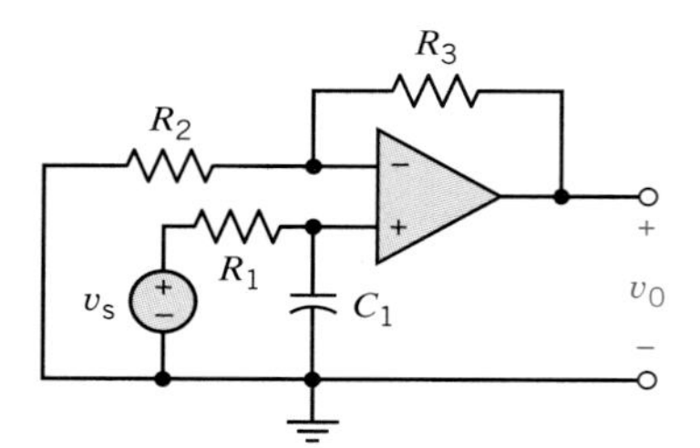

Figure P 10.14-2 Amplifier circuit for disk player.

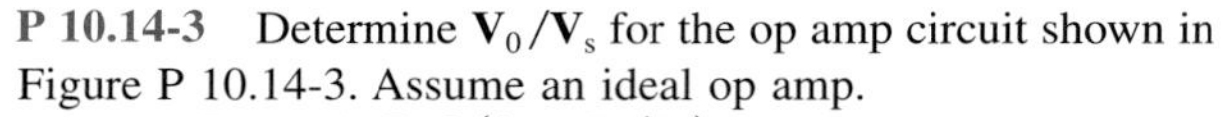

P 10.14-3 Determine $\mathbf{V}_0/\mathbf{V}_s$ for the op amp circuit shown in Figure P 10.14-3. Assume an ideal op amp.

Answer: $\dfrac{\mathbf{V}_0}{\mathbf{V}_s} = \dfrac{j\omega R_1 C_1(1 + R_3/R_2)}{1 + j\omega R_1 C_1}$

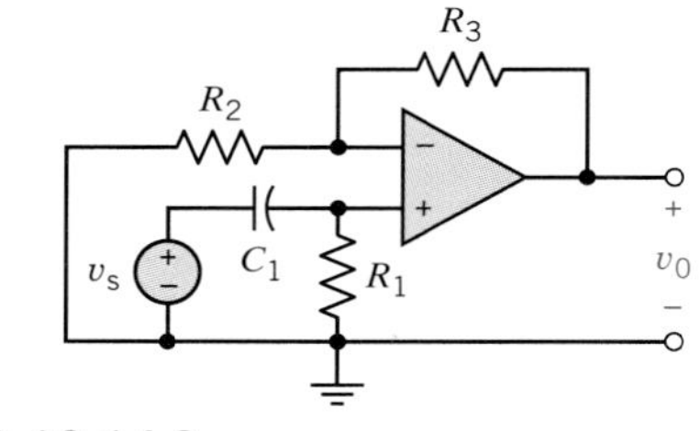

Figure P 10.14-3

P 10.14-4 For the circuit of Figure P 10.14-4 determine $v_0(t)$ when $v_s = 5\cos\omega t$ mV and $f = 10$ kHz.

Answer: $v_0 = 0.5\cos(\omega t - 89.5°)$ mV

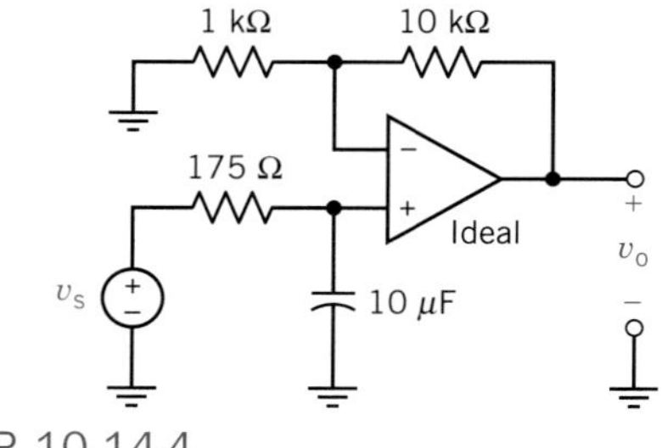

Figure P 10.14-4

PSPICE PROBLEMS

SP 10-1 Determine the current $i(t)$ for the circuit of Figure SP 10.1 when $v_s = 10\cos(6t + 45°)$ V and $i_s = 2\cos(6t + 60°)$ A.

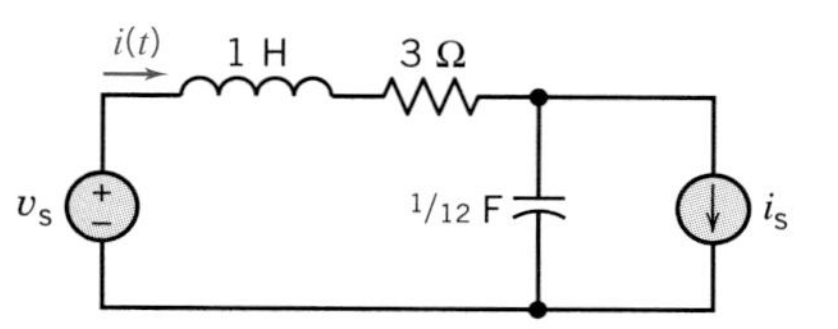

Figure SP 10.1

SP 10-2 Determine $v_0(t)$ for the circuit of Figure SP 10.2 when $v_s = 5\cos(3t - 30°)$ V.

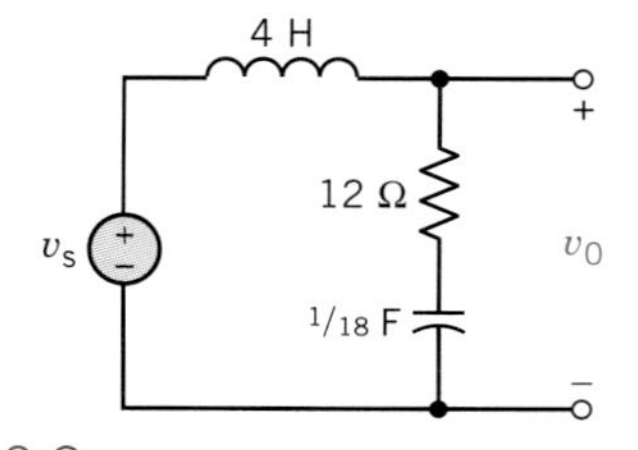

Figure SP 10.2

SP 10-3 Determine the current $i(t)$ for the circuit shown in Figure SP 10.3 when $v_s = 200 \cos \omega t$ V, $i_s = 8 \cos (\omega t + 90°)$ A, and $\omega = 1000$ rad/s.

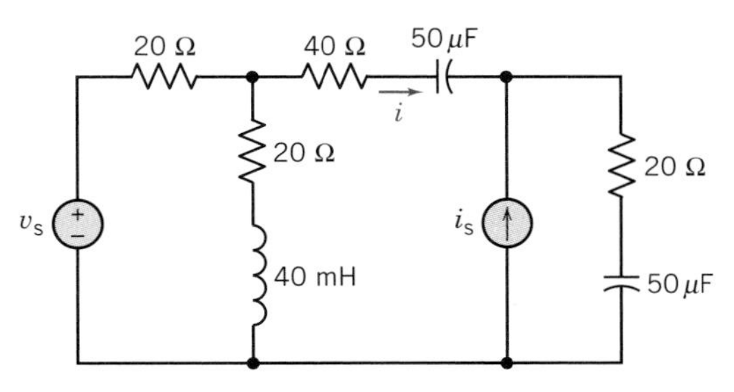

Figure SP 10.3

SP 10-4 Determine the output voltage v, for the circuit shown in Figure SP 10.4 when $v_s = 4 \cos \omega t$ V and $\omega = 2\pi \times 1000$ rad/s.

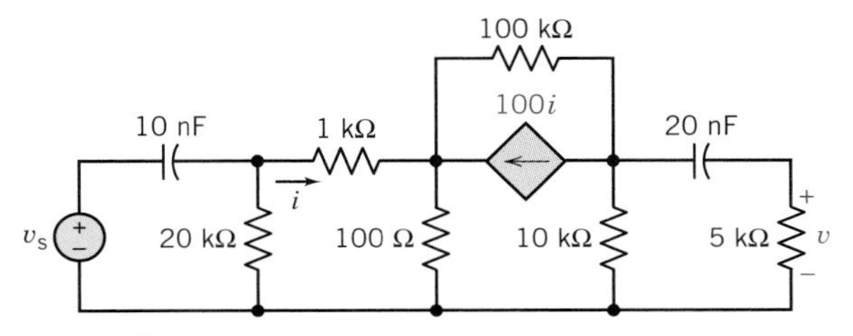

Figure SP 10.4

SP 10-5 Determine $i(t)$ for the circuit of Figure SP 10.5 when $i_s = 2 \cos (3t + 10°)$ A and $v_s = 3 \cos (2t + 30°)$ V.

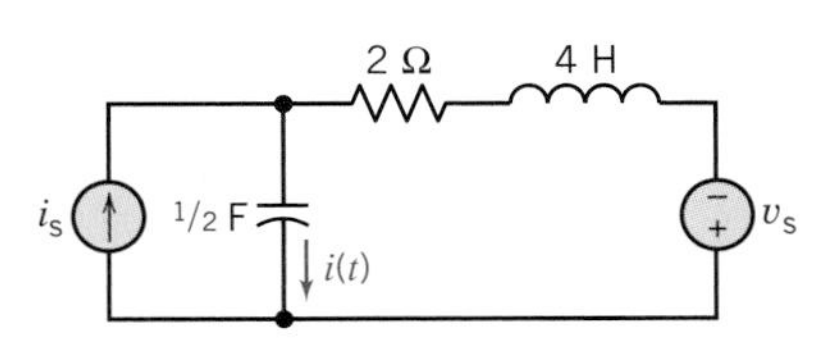

Figure SP 10.5

SP 10-6 Determine $v_a(t)$ and $i(t)$ for the circuit shown in Figure SP 10.6 when $v_s = 5 \cos 2t$ V and $i_s = 5 \cos 2t$ A.

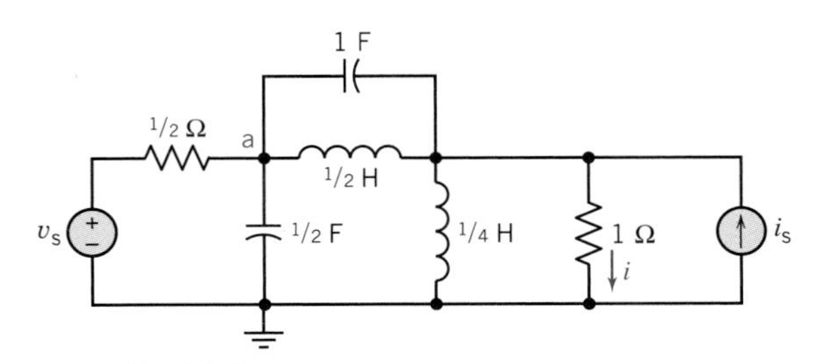

Figure SP 10.6

SP 10-7 Determine $i(t)$ for the circuit shown in Figure SP 10.7 when $v_s = 4 \cos 5000t$ V.

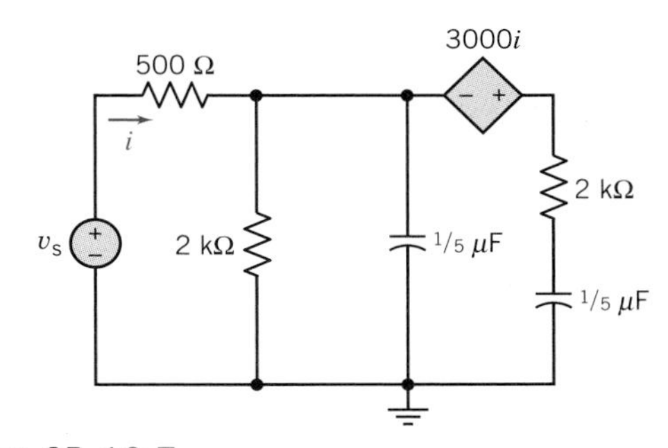

Figure SP 10.7

SP 10-8 Determine the impedance **Z** for the circuit as shown in Figure SP 10.8 at a frequency of 60 Hz.

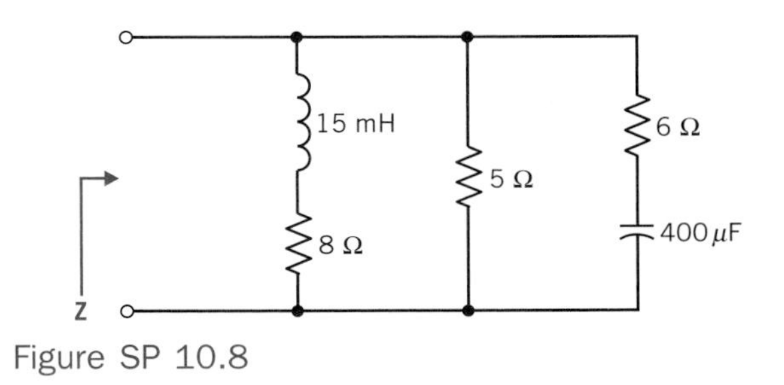

Figure SP 10.8

SP 10-9 Determine the current $i(t)$ for the circuit shown in Figure SP 10.9 when $v_s(t) = 120 \sin(\omega t + 30°)$ V and $f = 10$ kHz.

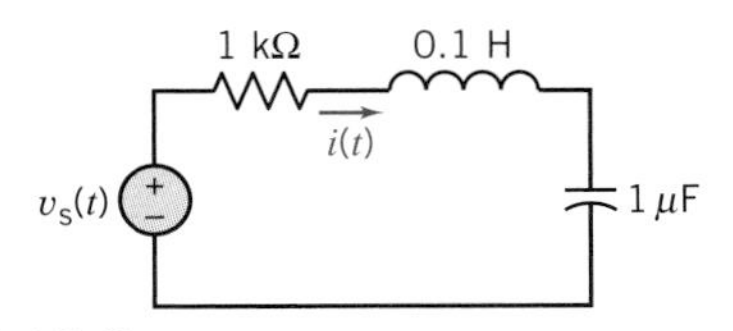

Figure SP 10.9

VERIFICATION PROBLEMS

VP 10-1 Computer analysis of the circuit in Figure VP 10.1 indicates that the values of the node voltages are $\mathbf{V}_1 = 20\underline{/-90^\circ}$ and $\mathbf{V}_2 = 44.7\underline{/-63.4^\circ}$. Are these values correct?

Hint: Calculate the current in each circuit element using the values of $\mathbf{V}_1$ and $\mathbf{V}_2$. Check to see if KCL is satisfied at each node of the circuit.

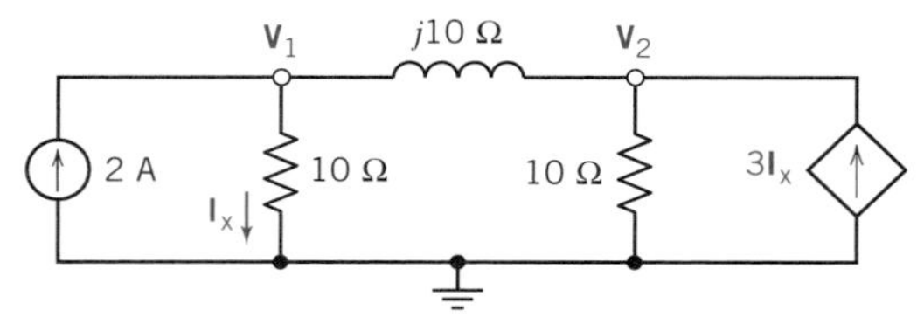

Figure VP 10.1

VP 10-2 Computer analysis of the circuit in Figure VP 10.2 indicates that the mesh currents are $i_1(t) = 0.39 \cos (5t + 39^\circ)$ A and $i_2(t) = 0.28 \cos(5t + 180^\circ)$ A. Is this analysis correct?

Hint: Represent the circuit in the frequency domain using impedances and phasors. Calculate the voltage across each circuit element using the values of $\mathbf{I}_1$ and $\mathbf{I}_2$. Check to see if KVL is satisfied for each mesh of the circuit.

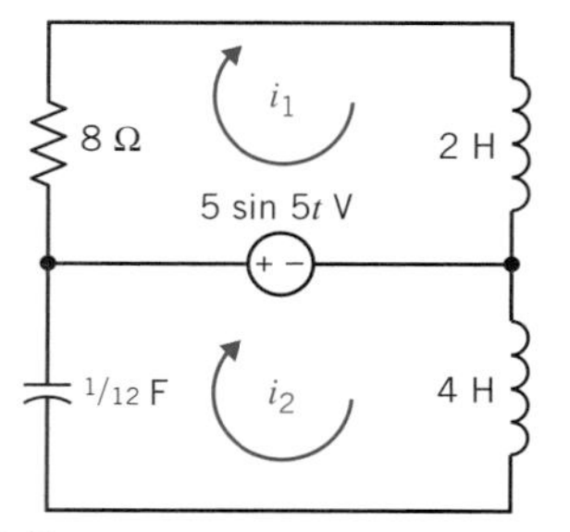

Figure VP 10.2

VP 10-3 Computer analysis of the circuit in Figure VP 10.3 indicates that the values of the node voltages are $v_1(t) = 19.2 \cos(3t + 68^\circ)$ V and $v_2(t) = 2.4 \cos(3t - 75^\circ)$ V. Is this analysis correct?

Hint: Represent the circuit in the frequency domain using impedances and phasors. Calculate the current in each circuit element using the values of $\mathbf{V}_1$ and $\mathbf{V}_2$. Check to see if KCL is satisfied at each node of the circuit.

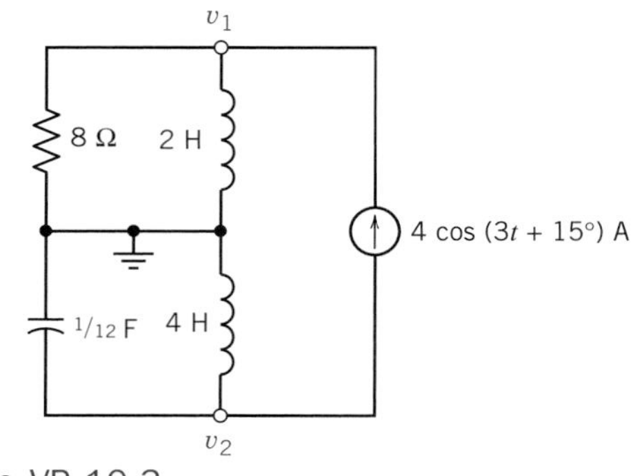

Figure VP 10.3

VP 10-4 A computer program reports that the currents of the circuit of Figure VP 10.4 are $\mathbf{I} = 0.2\underline{/53.1^\circ}$ A, $\mathbf{I}_1 = 632\underline{/-18.4^\circ}$ mA, and $\mathbf{I}_2 = 190\underline{/71.6^\circ}$ mA. Verify this result.

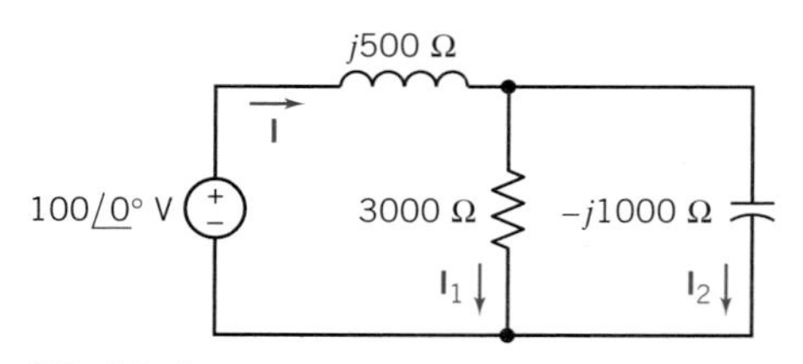

Figure VP 10.4

DESIGN PROBLEMS

DP 10-1 Design the circuit shown in Figure DP 10.1 to produce the specified output voltage $v_o(t)$ when provided with the given input voltage $v_i(t)$.

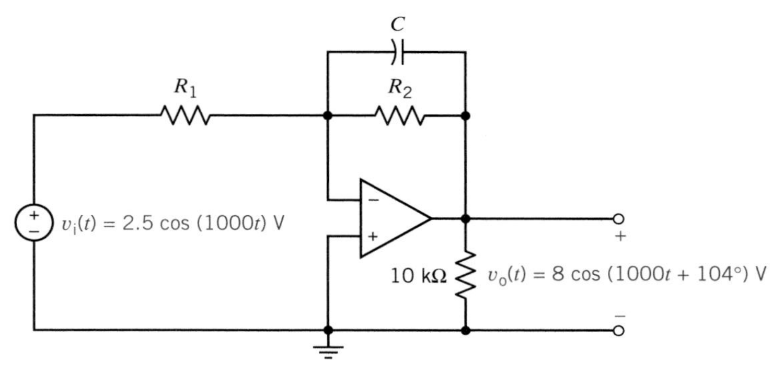

Figure DP 10.1

DP 10-2 Design the circuit shown in Figure DP 10.2 to produce the specified output voltage $v_o(t)$ when provided with the given input voltage $v_i(t)$.

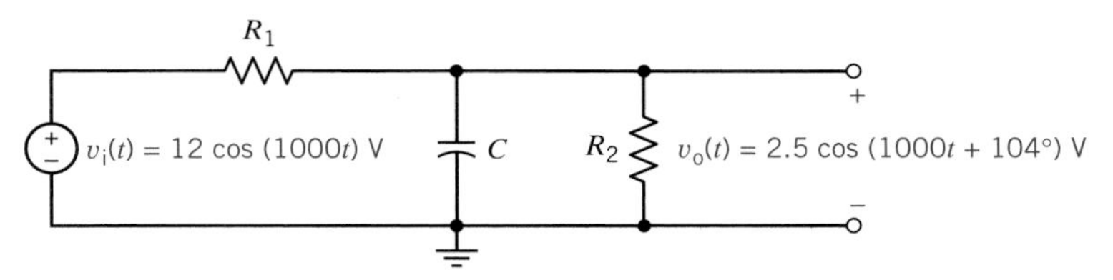

Figure DP 10.2

DP 10-3 Design the circuit shown in Figure DP 10.3 to produce the specified output voltage $v_o(t)$ when provided with the given input voltage $v_i(t)$.

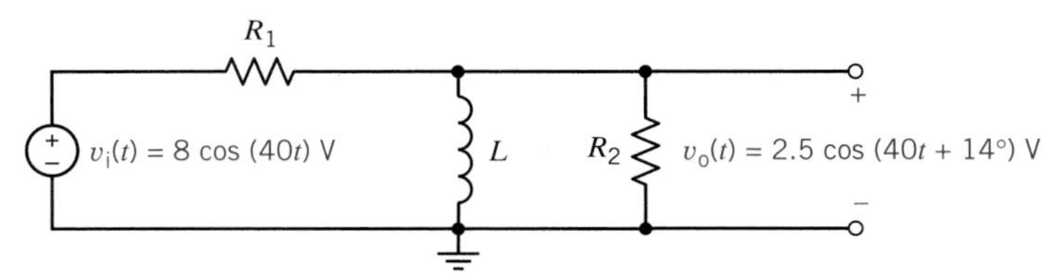

Figure DP 10.3

DP 10-4 Show that it is not possible to design the circuit shown in Figure DP 10.4 to produce the specified output voltage $v_o(t)$ when provided with the given input voltage $v_i(t)$.

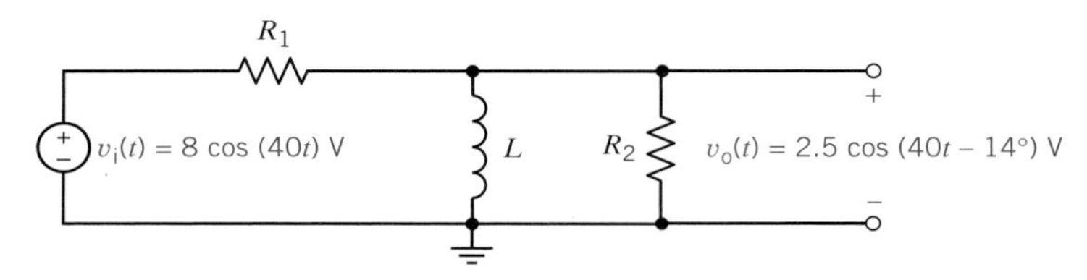

Figure DP 10.4

DP 10-5 A circuit with an unspecified R, L, and C is shown in Figure DP 10.5. The input source is $i_s = 10 \cos 1000t$ A, and the goal is to select the R, L, and C so that the node voltage is $v = 80 \cos(1000t - \theta)$ V where $-40° < \theta < 40°$.

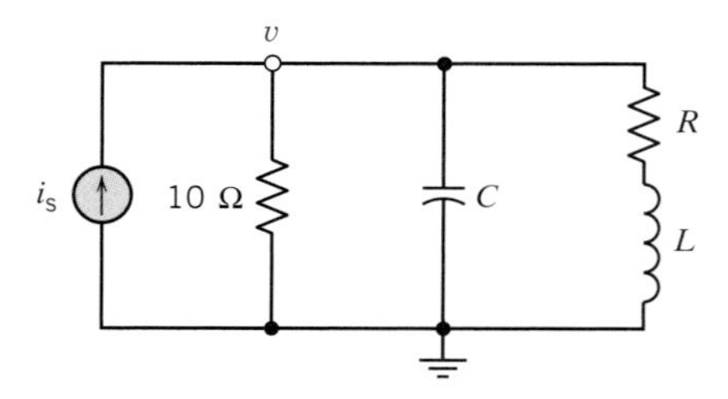

Figure DP 10.5

CHAPTER 11

AC Steady-State Power

Preview

We have seen how sinusoidal steady-state currents and voltages, described as ac, have become the predominant form of power delivery to the home and industry. In this chapter we develop appropriate relationships for describing the power delivered by ac current and voltage to a load impedance. We also reexamine the concepts of superposition and the maximum power theorem. Finally, we consider the concept of mutual inductance associated with an electric transformer, and we describe how a transformer enables us to increase or decrease a voltage or current flowing through a load.

11.1 Design Challenge

MAXIMUM POWER TRANSFER

The matching network in Figure 11.1-1 is used to interface the source with the load, which means that the matching network is used to connect the source to the load in a desirable way. In this case, the purpose of the matching network is to transfer as much power as possible to the load. This problem occurs frequently enough that it has been given a name, the maximum power transfer problem.

An important example of the application of maximum power transfer is the connection of a cellular phone or wireless radio transmitter to the cell's antenna. For example, the input impedance of a practical cellular telephone antenna is $\mathbf{Z} = (10 + j6.28)\ \Omega$ (Dorf 1998).

In this chapter we continue our study of circuits that are excited by sinusoidal sources. In particular, we will learn that phasors can be used to calculate the average power delivered to an element. We will return to the problem of designing the matching network at the end of the chapter. Then we will be ready to design this network to transfer maximum power to the load.

11.2 ELECTRIC POWER

Human civilization's progress has been enhanced by society's ability to control and distribute energy. Electricity serves as a carrier of energy to the user. Energy present in a fossil fuel or a nuclear fuel is converted to electric power in order to transport and readily distribute it to customers. By means of transmission lines, electric power is transmitted and distributed to essentially all the residences, industries, and commercial buildings in the United States and Canada.

Electric power may be readily transported with low attendant losses, and improved methods for safe handling of electric power have been developed over the past 80 years. Furthermore, methods of converting fossil fuels to electric power are well developed, economical, and safe. Means of converting solar and nuclear energy to electric power are currently in various stages of development or of proven safety. Geothermal energy, tidal energy, and wind energy may also be converted to electric power. The kinetic energy of falling water may readily be used to generate hydroelectric power.

In 1900 the United States consumed about 3.7 billion kWh of electric energy, for a per capita consumption of 49 kWh per person per year. By 1990 total consumption had grown to 2550 billion kWh, with a per capita consumption of 9481 kWh per year. The growth in the consumption of electric energy is shown in Table 11.2-1. The annual growth rate averaged 7.5 percent per year over the period 1900–1970. This growth rate decreased over

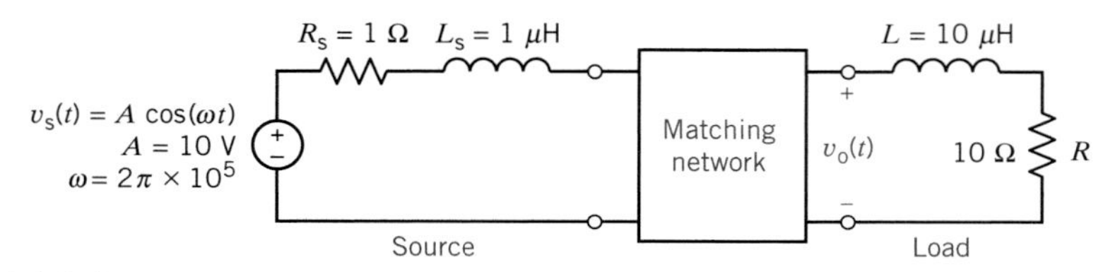

Figure 11.1-1 Design the matching network to transfer maximum power to the load where the load is the model of an antenna of a wireless communication system.

Table 11-2.1 **Electric Energy Use in the United States**

Year	Population (millions)	Generating Capacity (million kW)	Consumption (billion kWh)	Consumption per Person per Year (kWh)
1900	76	1	4	49
1920	106	20	58	540
1940	132	50	142	1074
1960	181	168	753	4169
1980	230	570	2126	9025
1990	242	699	2550	9481

the next 25 years as population leveled off, conservation practices were developed, and more efficient use of electricity was achieved.

As we noted earlier, the necessity of transmitting electrical power over long distances fostered the development of ac high-voltage power lines from power plant to end user. George Westinghouse and his company developed the transformer, which permitted the efficient transformation of ac voltages to high magnitudes. Westinghouse, who is shown in Figure 11.2-1, designed a transformer that could be produced in quantity by ordinary manufacturing processes. The first large-scale installation was made in Buffalo in 1886.

Frequencies were generally rather high; 133 Hz was common in the early days of ac power transmission. Nikola Tesla, the inventor of the ac induction motor, who is shown in Figure 11.2-2, found that 133 Hz was too high for his motor. By 1910, he had convinced electrical engineers that 60 Hz should be adopted in the United States.

Transmission over long distances required higher ac voltages. A 155-mile line at 110,000 V was put into operation in California in 1908. As needs grew, transmission voltages increased to 230,000 V, and the Hoover Dam project of the 1930s used 287 kV on a 300-mile transmission line. By the 1970s, voltages over 600 kV were used. A modern transmission line is shown in Figure 11.2-3.

Figure 11.2-1 George Westinghouse, 1846–1914, was one of the persons who championed the development of ac power and fostered the development and use of the practical ac transformer. Courtesy of Westinghouse Corporation.

Figure 11.2-2 Nikola Tesla, 1856–1943, joined George Westinghouse in the promotion of ac power in the late 1800s. Tesla was responsible for many inventions, including the ac induction motor, and was a contributor to the selection of 60 Hz as the standard ac frequency in the United States. Courtesy of Burndy Library.

Figure 11.2-3 AC power high-voltage transmission lines. Courtesy of Pacific Gas and Electric Company.

Figure 11.2-4 A large hydroelectric power plant. Courtesy of Hydro Quebec.

Figure 11.2-5 A large wind-power turbine and generator. Courtesy of *EPRI Journal.*

Electric energy generation uses original sources such as hydropower, coal, and nuclear energy. An example of a large hydroelectric power project is shown in Figure 11.2-4. A typical hydroelectric power plant can generate 1000 MW. On the other hand, many regions are turning to small generators such as the wind-power device shown in Figure 11.2-5. A typical wind-power machine may be capable of generating 75 kW.

During the period 1920–1960, there was extensive development of transmission tower equipment, insulators, and power lines. An example of an early high-voltage power laboratory is shown in Figure 11.2-6.

Figure 11.2-6 High-voltage laboratory of the Ohio Insulator Co. (Ohio Brass Co.): outdoor impulse gaps (left and right); test section of 220-V transmission towers (left). From A. O. Austin, "A Laboratory for Making Lightning" (paper presented at the International High Tension Congress, Paris, June 6–15, 1929). Courtesy of the Ohio Brass Co.

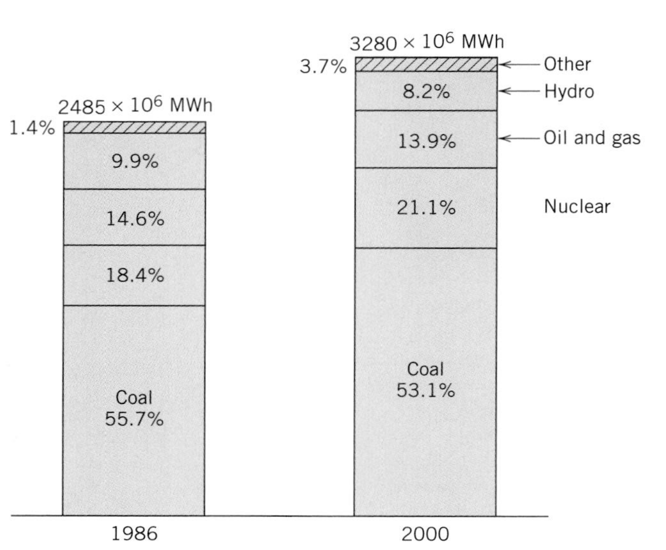

Figure 11.2-7 Sources of electric energy in 1986 and projected for 2000. Source: *World Resources,* World Resource Institute, Oxford University Press, NY, 2000.

A unique element of the American power system is its interconnectedness. Although the power system of the United States consists of many independent companies, it is interconnected by large transmission facilities. An electric utility is often able to save money by buying electricity from another utility and by transmitting the energy over the transmission lines of a third utility.

Coal fueled the largest percentage of the 2485×10^{12} watt-hours produced in the United States in 1986 (see Figure 11.2-7). Nuclear sources are projected to grow to 21 percent of the total by 2000.

11.3 INSTANTANEOUS POWER AND AVERAGE POWER

We are interested in determining the power generated and absorbed in a circuit or in an element of a circuit. Electrical engineers talk about several types of power, for example, instantaneous power, average power, and complex power. We will start with an examination of the instantaneous power, which is the product of the time domain voltage and current associated with one or more circuit elements. The instantaneous power is likely to be a complicated function of time. This prompts us to look for a simpler measure of the power generated and absorbed in a circuit element, such as the average power.

Consider the circuit element shown in Figure 11.3-1. Notice that the element voltage $v(t)$ and the element current $i(t)$ adhere to the passive convention. The **instantaneous power** delivered to this circuit element is the product of the voltage $v(t)$ and the current $i(t)$, so that

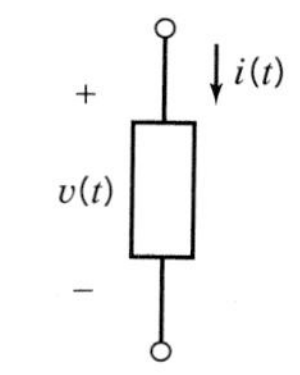

Figure 11.3-1 A circuit element.

$$p(t) = v(t)\, i(t) \tag{11.3-1}$$

The unit of power is watts (W). We can always calculate the instantaneous power since no restrictions have been placed on either $v(t)$ or $i(t)$. The instantaneous power can be a quite complicated function of t when $v(t)$ or $i(t)$ are themselves complicated functions of t.

Suppose that the voltage $v(t)$ is a periodic function having period T. That is,

$$v(t) = v(t + T)$$

since the voltage repeats every T seconds. Then, for a linear circuit, the current will also be a periodic function having the same period, so

$$i(t) = i(t + T)$$

Therefore, the instantaneous power is

$$\begin{aligned} p(t) &= v(t)\, i(t) \\ &= v(t + T)\, i(t + T) \end{aligned}$$

The *average value* of a periodic function is the integral of the time function over a complete period, divided by the period. We use a capital P to denote average power and a lowercase p to denote instantaneous power. Therefore, the average power P is given by

$$P = \frac{1}{T}\int_{t_0}^{t_0+T} p(t)\, dt \tag{11.3-2}$$

where t_0 is an arbitrary starting point in time.

Next, suppose that the voltage $v(t)$ is sinusoidal, that is,

$$v(t) = V_{\text{m}}(\cos \omega t + \theta_{\text{V}})$$

Then, for a linear circuit at steady state, the current will also be sinusoidal and will have the same frequency, so

$$i(t) = I_m(\cos \omega t + \theta_I)$$

The period and frequency of $v(t)$ and $i(t)$ are related by

$$\omega = \frac{2\pi}{T}$$

The instantaneous power delivered to the element is

$$p(t) = V_m I_m \cos(\omega t + \theta_V) \cos(\omega t + \theta_I)$$

Using the trigonometric identity (see Appendix C) for the product of two cosine functions,

$$p(t) = \frac{V_m I_m}{2}[\cos(\theta_V - \theta_I) + \cos(2\omega t + \theta_V + \theta_I)]$$

We see that the instantaneous power has two terms. The first term within the brackets is independent of time, and the second term varies sinusoidally over time at twice the radian frequency of $v(t)$.

The average power delivered to the element is

$$P = \frac{1}{T}\int_0^T \frac{V_m I_m}{2}[\cos(\theta_V - \theta_I) + \cos(2\omega t + \theta_V + \theta_I)]\, dt$$

where we have chosen $t_0 = 0$. Then we have

$$P = \frac{1}{T}\int_0^T \frac{V_m I_m}{2} \cos(\theta_V - \theta_I)\, dt + \frac{1}{T}\int_0^T \frac{V_m I_m}{2} \cos(2\omega t + \theta_V + \theta_I)\, dt$$

$$= \frac{V_m I_m \cos(\theta_V - \theta_I)}{2T}\int_0^T dt + \frac{V_m I_m}{2T}\int_0^T \cos(2\omega t + \theta_V + \theta_I)\, dt$$

The second integral is zero, since the average value of the cosine function over a complete period is zero. Then we have

$$P = \frac{V_m I_m}{2} \cos(\theta_V - \theta_I) \tag{11.3-3}$$

Example 11.3-1

Find the average power delivered to a resistor R when the current through the resistor is $i(t)$, as shown in Figure 11.3-2.

Solution

The current waveform repeats every T seconds and attains a maximum value of I_m. Using the period from $t = 0$ to $t = T$, we have

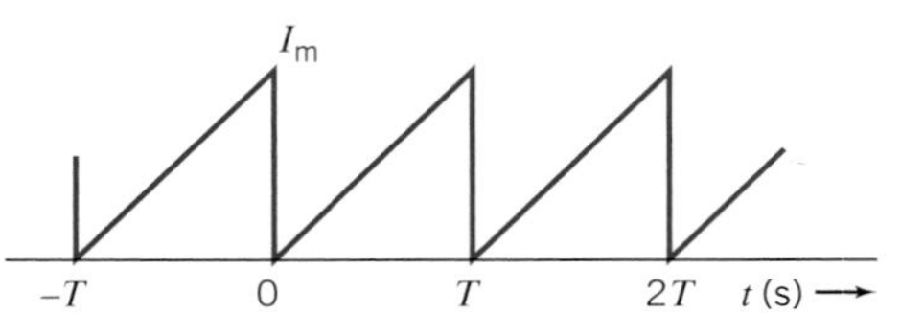

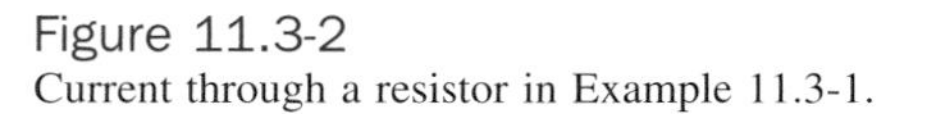
Figure 11.3-2
Current through a resistor in Example 11.3-1.

$$i = \frac{I_m}{T} t \quad 0 < t \leq T$$

Then the instantaneous power is

$$p = i^2 R = \frac{I_m^2 t^2}{T^2} R \quad 0 < t \leq T$$

It is sufficient to find the average power over $0 < t < T$ since the power is periodic with period T. Then the average power is

$$P = \frac{1}{T} \int_0^T \frac{I_m^2 R}{T^2} t^2 \, dt$$

Integrating, we have

$$P = \frac{I_m^2 R}{T^3} \int_0^T t^2 \, dt = \frac{I_m^2 R}{T^3} \frac{T^3}{3} = \frac{I_m^2 R}{3} \text{ W}$$

Example 11.3-2

The circuit shown in Figure 11.3-3 is at steady state. The mesh current is

$$i(t) = 721 \cos(100t - 41°) \text{ mA}$$

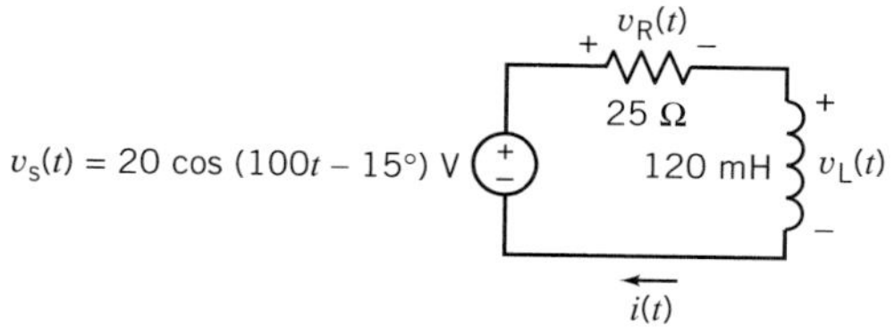

Figure 11.3-3
An *RL* circuit with a sinusoidal voltage source.

The element voltages are

$$v_s(t) = 20 \cos(100t - 15°) \text{ V}$$
$$v_R(t) = 18 \cos(100t - 41°) \text{ V}$$
$$v_L(t) = 8.66 \cos(100t + 49°) \text{ V}$$

Find the average power *delivered to* each device in this circuit.

Solution

Notice that $v_s(t)$ and $i(t)$ don't adhere to the passive convention. Thus, $v_s(t)\, i(t)$ is the power *delivered by* the voltage source. Therefore, the average power calculated using Eq. 11.3-3 is the average power *delivered by* the voltage source. The average power *delivered by* the voltage source is

$$P_s = \frac{(20)(0.721)}{2} \cos(-15° - (-41°)) = 6.5 \text{ W}$$

The average power *delivered to* the voltage source is -6.5 W.

Since $v_R(t)$ and $i(t)$ do adhere to the passive convention, the average power calculated using Eq. 11.3-3 is the average power *delivered to* the resistor. The power *delivered to* the resistor is

$$P_R = \frac{(18)(0.721)}{2} \cos(-41° - (-41°)) = 6.5 \text{ W}$$

The power *delivered to* the inductor is

$$P_L = \frac{(8.66)(0.721)}{2}\cos(49° - (-41°)) = 0 \text{ W}$$

Why is the average power delivered to the inductor equal to zero? The angle of the element voltage is 90° larger than the angle of the element current. Since $\cos(90°) = 0$, the average power delivered to the inductor is zero. The angle of the inductor voltage will always be 90° larger than the angle of the inductor current. Therefore, the average power delivered to any inductor is zero.

Exercise 11.3-1 An *RLC* circuit is shown in Figure E 11.3-1 with a voltage source $v_s = 7 \cos 10t$ V.
(a) Determine the instantaneous power delivered to the circuit by the voltage source.
(b) Find the instantaneous power delivered to the inductor.

Answer: (a) $p = 7.54 + 15.2 \cos(20t - 60.3°)$ W
(b) $p = 28.3 \cos(20t - 30.6°)$ W

Exercise 11.3-2 Determine the instantaneous power delivered to an element and sketch $p(t)$ when the element is (a) a resistance R and (b) an inductor L. The voltage across the element is $v(t) = V_m \cos(\omega t + \theta)$ V.

Answer: (a) $P_R = \dfrac{V_m^2}{2R}[1 + \cos(2\omega t + 2\theta)]$ W

(b) $P_L = \dfrac{V_m^2}{2\omega L}\cos(2\omega t + 2\theta - 90°)$ W

Exercise 11.3-3 (a) Find the average power delivered by the source to the circuit shown in Figure E 11.3-3.
(b) Find the power absorbed by resistor R_1.

Answer: (a) 30 W (b) 20 W

Figure E 11.3-1

Figure E 11.3-3

Exercise 11.3-4 Find the average power delivered to each impedance in the network shown in Figure E 11.3-4.

Answer: $P_1 = 234.4$ W, $P_2 = 78.1$ W

Exercise 11.3-5 (a) Determine the instantaneous power of an element, and sketch it, when the voltage across it and the current through it are

$$v(t) = \begin{cases} 2 \text{ V} & 0 < t \le 1.5 \\ 1 \text{ V} & 1.5 < t < 2 \end{cases}$$

$$i(t) = \begin{cases} 2t \text{ A} & 0 < t \le 1 \\ 2 \text{ A} & 1 < t < 2 \end{cases}$$

The period of the voltage and current is 2 s.
(b) Calculate the average power of the element.

Answer: (b) $P = 2.5$ W

Exercise 11.3-6 For the circuit of Figure E 11.3-6, determine the average power of each element and verify that the average power delivered by the source is equal to the sum of the average power absorbed by the other circuit elements. The source is $v_s = 5 \cos 2t$ V.

Answer: $P_S = P_{R_1} + P_{R_2} = 4.6$ W

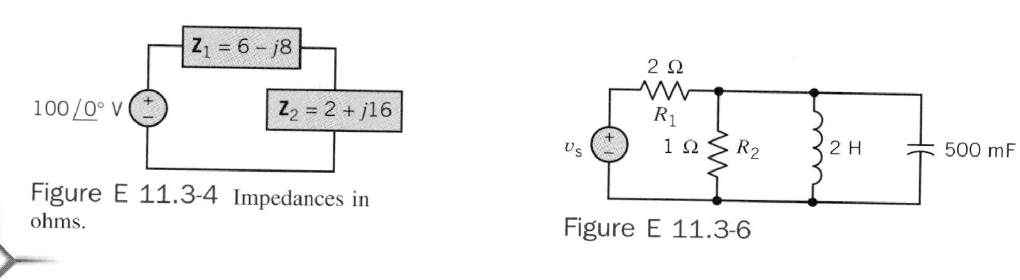

Figure E 11.3-4 Impedances in ohms.

Figure E 11.3-6

11.4 EFFECTIVE VALUE OF A PERIODIC WAVEFORM

The voltage available from a wall plug in a residence is said to be 110 V. Of course, this is not the average value of the sinusoidal voltage since we know that the average would be zero. It is also not the instantaneous value or the maximum value, V_m, of the voltage $v = V_m \cos \omega t$.

The effective value of a voltage is a measure of its effectiveness in delivering power to a load resistor. The concept of an *effective value* comes from a desire to have a sinusoidal voltage (or current) deliver to a load resistor the same average power as an equivalent dc voltage (or current). The goal is to find a dc V_{eff} (or I_{eff}) that will deliver the same average power to the resistor as would be delivered by a periodically varying source, as shown in Figure 11.4-1. The energy delivered in a period T is $W = PT$, where P is the average power.

The average power delivered to the resistor R by a periodic current is

$$P = \frac{1}{T}\int_0^T i^2 R\, dt \tag{11.4-1}$$

We select the period, T, of the periodic current as the integration interval.

The power delivered by a direct current is

$$P = I_{eff}^2 R \tag{11.4-2}$$

where I_{eff} is the dc current that will deliver the same power as the periodically varying current. That is, I_{eff} is defined as the steady (constant) current that is as *effective* in delivering power as the periodically varying current.

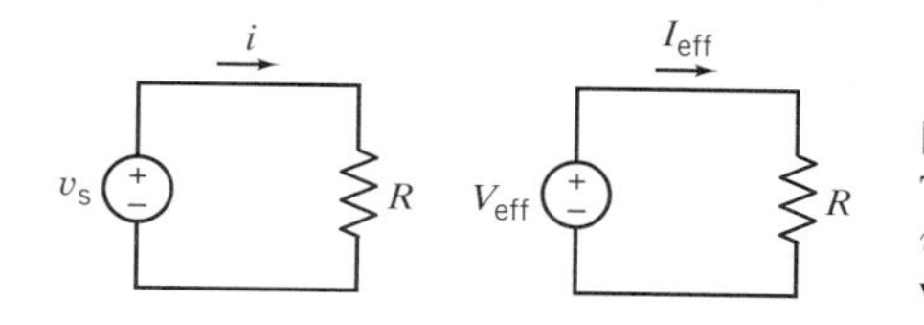

Figure 11.4-1
The goal is to find a dc voltage, V_{eff}, for a specified $v_s(t)$ that will deliver the same average power to R as would be delivered by the ac source.

We equate Eqs. 11.4-1 and 11.4-2, obtaining

$$I_{\text{eff}}^2 R = \frac{R}{T}\int_0^T i^2\, dt$$

Solving for I_{eff}, we have

$$I_{\text{eff}} = \left(\frac{1}{T}\int_0^T i^2\, dt\right)^{1/2} \tag{11.4-3}$$

We see that I_{eff} is the square root of the mean of the squared value. Thus, the effective current I_{eff} is commonly called the root-mean-square current I_{rms}.

The **effective value** of a current is the steady current (dc) that transfers the same average power as the given varying current.

Of course, the effective value of the voltage in a circuit is similarly found from the equation

$$V_{\text{eff}}^2 = V_{\text{rms}}^2 = \frac{1}{T}\int_0^T v^2\, dt$$

Thus

$$V_{\text{rms}} = \left(\frac{1}{T}\int_0^T v^2\, dt\right)^{1/2}$$

Now let us find the I_{rms} of a sinusoidally varying current $i = I_{\text{m}} \cos \omega t$. Using Eq. 11.4-3 and a trigonometric formula from Appendix C, we have

$$\begin{aligned} I_{\text{rms}} &= \left(\frac{1}{T}\int_0^T I_{\text{m}}^2 \cos^2\omega t\, dt\right)^{1/2} \\ &= \left(\frac{I_{\text{m}}^2}{T}\int_0^T \frac{1}{2}\,(1 + \cos 2\omega t)\, dt\right)^{1/2} \\ &= \frac{I_{\text{m}}}{\sqrt{2}} \end{aligned} \tag{11.4-4}$$

since the integral of cos $2\omega t$ is zero over the period T. Remember that Eq. 11.4-4 is true only for sinusoidal currents.

In practice, we must be careful to determine whether a sinusoidal voltage is expressed in terms of its effective value or its maximum value I_{m}. In the case of power transmission and use in the home, the voltage is said to be 110 V or 220 V, and it is understood that the voltage is the rms or effective value.

In electronics or communications circuits, the voltage could be described as 10 V, and the person is typically indicating the maximum or peak amplitude, V_{m}. Henceforth, we will use V_{m} as the peak value and V_{rms} as the rms value. Sometimes it is necessary to distinguish V_{rms} from V_{m} by the context in which the voltage is given.

Example 11.4-1

Find the effective value of the current for the sawtooth waveform shown in Figure 11.4-2.

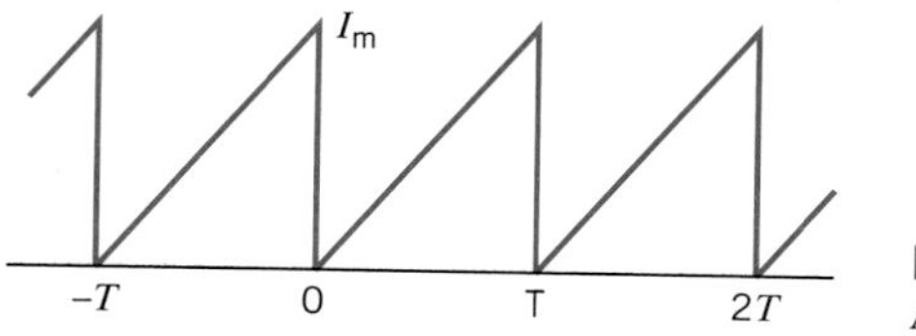

Figure 11.4-2
A sawtooth current waveform.

Solution
First, we will express the current waveform over the period $0 < t \le T$. The current is then

$$i = \frac{I_m}{T} t \qquad 0 < t \le T$$

The effective value is found from

$$I_{eff}^2 = \frac{1}{T}\int_0^T i^2\,dt = \frac{1}{T}\int_0^T \frac{I_m^2}{T^2} t^2\,dt = \frac{I_m^2}{T^3}\left[\frac{t^3}{3}\right]_0^T = \frac{I_m^2}{3}$$

Therefore, solving for I_{eff}, we have

$$I_{eff} = \frac{I_m}{\sqrt{3}}$$

Exercise 11.4-1 Find the effective value of the current waveform shown in Figure E 11.4-1.

Answer: $I_{eff} = 8.66$

Exercise 11.4-2 Find the effective value of the following currents: (a) $\cos 3t + \cos 3t$; (b) $\sin 3t + \cos(3t + 60°)$; (c) $2 \cos 3t + 3 \cos 5t$

Answer: (a) $\sqrt{2}$ (b) 0.366 (c) 2.55

Exercise 11.4-3 Calculate the effective value of the voltage across the resistance R of the circuit shown in Figure E 11.4-3 when $\omega = 100$ rad/s.

Hint: Use superposition.

Answer: $V_{eff} = 4.82$ V

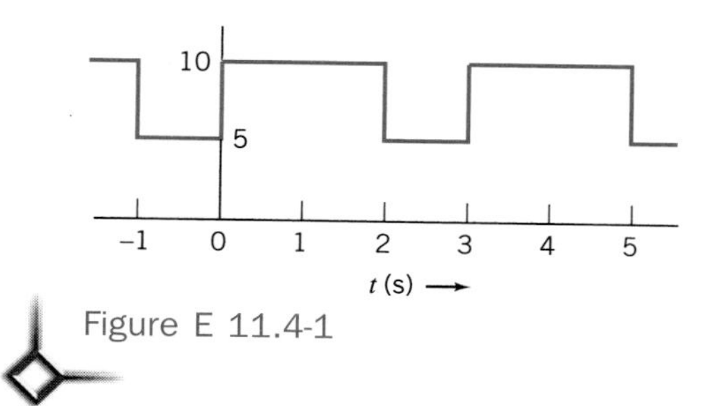

Figure E 11.4-1

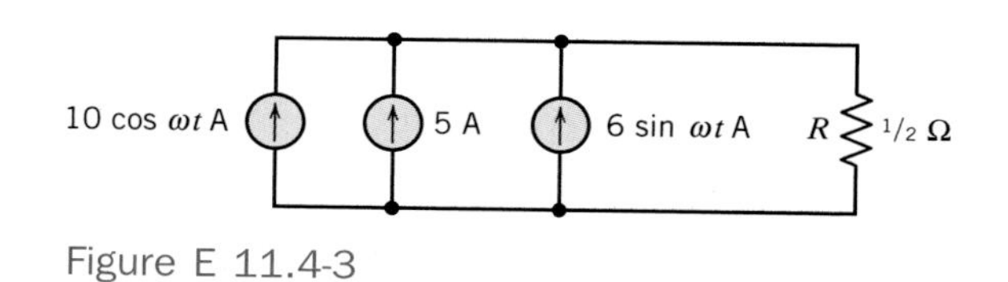

Figure E 11.4-3

11.5 COMPLEX POWER

Suppose that a linear circuit with a sinusoidal input is at steady state. All the element voltages and currents will be sinusoidal with the same frequency as the input. Such a circuit can be analyzed in the frequency domain, using phasors. Indeed, we can calculate the power generated or absorbed in a circuit or in any element of a circuit, in the frequency domain, using phasors.

Figure 11.5-1 represents the voltage and current of an element in both the time domain and the frequency domain. Notice that the element current and voltage adhere to the passive convention. In a previous section, the instantaneous power and the average power were calculated from the time domain representations of the element current and voltage, $v(t)$ or $i(t)$. In contrast, we now turn our attention to the frequency domain representations of the element current and voltage

$$\mathbf{I}(\omega) = I_m \angle \theta_I \quad \text{and} \quad \mathbf{V}(\omega) = V_m \angle \theta_V \tag{11.5-1}$$

The **complex power** delivered to the element is defined to be

$$\mathbf{S} = \frac{\mathbf{VI}^*}{2} = \frac{(V_m\angle\theta_V)(I_m\angle -\theta_I)}{2} = \frac{V_m I_m}{2}\angle(\theta_V - \theta_I) \tag{11.5-2}$$

where $\mathbf{I}^*$ denotes the complex conjugate of $\mathbf{I}$ (see Appendix B). The magnitude of $\mathbf{S}$

$$|\mathbf{S}| = \frac{V_m I_m}{2} \tag{11.5-3}$$

is called the **apparent power.**

Converting the complex power, $\mathbf{S}$, from polar to rectangular form gives

$$\mathbf{S} = \frac{V_m I_m}{2}\cos(\theta_V - \theta_I) + j\frac{V_m I_m}{2}\sin(\theta_V - \theta_I) \tag{11.5-4}$$

The real part of $\mathbf{S}$ is equal to the average power that we calculated previously in the time domain! (See Eq. 11.3-3.) Recall that the average power was denoted as P. We can represent the complex power as

$$\mathbf{S} = P + jQ \tag{11.5-5}$$

where

$$P = \frac{V_m I_m}{2}\cos(\theta_V - \theta_I) \tag{11.5-6}$$

is the **average power** and

$$Q = \frac{V_m I_m}{2}\sin(\theta_V - \theta_I) \tag{11.5-7}$$

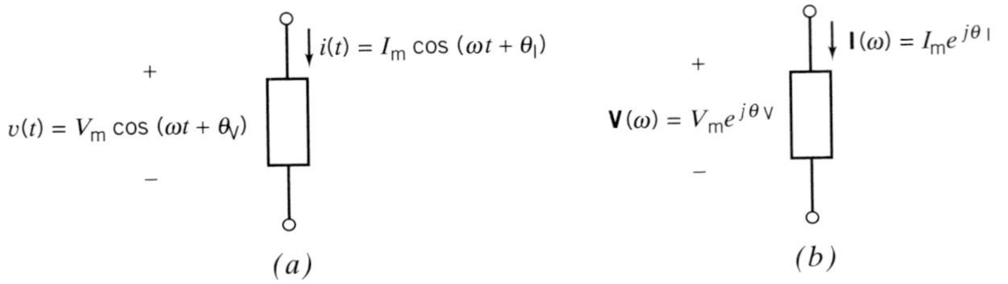

$i(t) = I_m \cos(\omega t + \theta_I)$

$v(t) = V_m \cos(\omega t + \theta_V)$

(a)

$\mathbf{I}(\omega) = I_m e^{j\theta_I}$

$\mathbf{V}(\omega) = V_m e^{j\theta_V}$

(b)

Figure 11.5-1 A linear circuit is excited by a sinusoidal input. The circuit has reached steady state. The element voltage and current can be represented in (a) the time domain or (b) the frequency domain.

is the **reactive power.** The complex power, average power, and reactive power are all the product of a voltage and a current. Nonetheless, it is conventional to use different units for these three types of power. We have already seen that the units of the average power are watts. The units of complex power are Volt-Amps (VA), and the units of reactive power are Volt-Amps Reactive (VAR). The formulas used to calculate power in the frequency domain are summarized in Table 11.5-1.

Let's return to Figure 11.5-1*b*. The impedance of the element can be expressed as

$$\mathbf{Z}(\omega) = \frac{\mathbf{V}(\omega)}{\mathbf{I}(\omega)} = \frac{V_m \angle\theta_V}{I_m \angle\theta_I} = \frac{V_m}{I_m}\angle(\theta_V - \theta_I) \tag{11.5-8}$$

Converting the impedance, $\mathbf{Z}$, from polar to rectangular form gives

$$\mathbf{Z}(\omega) = \frac{V_m}{I_m}\cos(\theta_V - \theta_I) + j\frac{V_m}{I_m}\sin(\theta_V - \theta_I) \tag{11.5-9}$$

We can represent the impedance as

$$\mathbf{Z}(\omega) = R + jX$$

where $R = \frac{V_m}{I_m}\cos(\theta_V - \theta_I)$ is the resistance and $X = \frac{V_m}{I_m}\sin(\theta_V - \theta_I)$ is the reactance.

The similarity between Eqs. 11.5-4 and 11.5-9 suggests that the complex power can be expressed in terms of the impedance

$$\begin{aligned}\mathbf{S} &= \frac{V_m I_m}{2}\cos(\theta_V - \theta_I) + j\frac{V_m I_m}{2}\sin(\theta_V - \theta_I) \\ &= \left(\frac{I_m^2}{2}\right)\frac{V_m}{I_m}\cos(\theta_V - \theta_I) + j\left(\frac{I_m^2}{2}\right)\frac{V_m}{I_m}\sin(\theta_V - \theta_I) \\ &= \left(\frac{I_m^2}{2}\right)\mathrm{Re}(\mathbf{Z}) + j\left(\frac{I_m^2}{2}\right)\mathrm{Im}(\mathbf{Z})\end{aligned} \tag{11.5-10}$$

In particular, the average power delivered to the element is given by

$$P = \left(\frac{I_m^2}{2}\right)\mathrm{Re}(\mathbf{Z}) \tag{11.5-11}$$

Table 11.5-1 **Frequency Domain Power Relationships**

Quantity	Relationship Using Peak Values	Relationship Using rms Values	Units
Element voltage, $v(t)$	$v(t) = V_m\cos(\omega t + \theta_V)$	$v(t) = V_{rms}\sqrt{2}\cos(\omega t + \theta_V)$	V
Element current, $i(t)$	$i(t) = I_m\cos(\omega t + \theta_I)$	$i(t) = I_{rms}\sqrt{2}\cos(\omega t + \theta_I)$	A
Complex power, $\mathbf{S}$	$\mathbf{S} = \frac{V_m I_m}{2}\cos(\theta_V - \theta_I)$	$\mathbf{S} = V_{rms} I_{rms}\cos(\theta_V - \theta_I)$	VA
	$+ j\frac{V_m I_m}{2}\sin(\theta_V - \theta_I)$	$+ jV_{rms} I_{rms}\sin(\theta_V - \theta_I)$	
Apparent power, $\lvert\mathbf{S}\rvert$	$\lvert\mathbf{S}\rvert = \frac{V_m I_m}{2}$	$\lvert\mathbf{S}\rvert = V_{rms} I_{rms}$	VA
Average power, P	$P = \frac{V_m I_m}{2}\cos(\theta_V - \theta_I)$	$P = V_{rms} I_{rms}\cos(\theta_V - \theta_I)$	W
Reactive power, Q	$Q = \frac{V_m I_m}{2}\sin(\theta_V - \theta_I)$	$Q = V_{rms} I_{rms}\sin(\theta_V - \theta_I)$	VAR

When the element is a resistor, then $\text{Re}(\mathbf{Z}) = R$

$$P_R = \left(\frac{I_m^2}{2}\right) R$$

When the element is a capacitor or an inductor, then $\text{Re}(\mathbf{Z}) = 0$; thus, the average power delivered to a capacitor or an inductor is zero.

Figure 11.5-2 summarizes Eqs. 11.5-4 and 11.5-9 using (a) the impedance triangle and (b) the power triangle.

Complex power is conserved. The sum of the complex power absorbed by all the elements of a circuit is zero. This fact can be expressed by the equation

$$\sum_{\substack{\text{all} \\ \text{elements}}} \frac{\mathbf{V}_k \mathbf{I}_k^*}{2} = 0 \qquad (11.5\text{-}12)$$

where $\mathbf{V}_k$ and $\mathbf{I}_k$ are the phasors corresponding to the element voltage and current of the kth element of the circuit. The phasors $\mathbf{V}_k$ and $\mathbf{I}_k$ must adhere to the passive convention so that $\mathbf{V}_k\mathbf{I}_k^*/2$ is the complex power *absorbed* by the kth branch. The summation in Eq. 11.5-12 adds up the complex powers in all elements of the circuit. When an element of the circuit is a source that is supplying power to the circuit $\mathbf{V}_k\mathbf{I}_k^*/2$ will be negative, indicating that positive complex power is being supplied rather than absorbed. Sometimes conservation of complex power is expressed as

$$\sum_{\text{sources}} \frac{\mathbf{V}_k \mathbf{I}_k^*}{2} = \sum_{\substack{\text{other} \\ \text{elements}}} \frac{\mathbf{V}_k \mathbf{I}_k^*}{2} \qquad (11.5\text{-}13)$$

where phasors $\mathbf{V}_k$ and $\mathbf{I}_k$ adhere to the passive convention for the "other elements" but do not adhere to the passive convention for the sources. When $\mathbf{V}_k$ and $\mathbf{I}_k$ do not adhere to the passive convention then $\mathbf{V}_k\mathbf{I}_k^*/2$ is the complex power *supplied* by the kth branch. We read Eq. 11.5-13 to say that the complex power supplied by the sources is equal to the complex power absorbed by the other elements of the circuit.

Equation 11.5-12 implies that both

$$\text{Re} \sum_{\text{sources}} \left(\frac{\mathbf{V}_k \mathbf{I}_k^*}{2}\right) = \sum_{\substack{\text{other} \\ \text{elements}}} \text{Re}\left(\frac{\mathbf{V}_k \mathbf{I}_k^*}{2}\right) = 0$$

and

$$\text{Im} \sum_{\text{sources}} \left(\frac{\mathbf{V}_k \mathbf{I}_k^*}{2}\right) = \sum_{\substack{\text{other} \\ \text{elements}}} \text{Im}\left(\frac{\mathbf{V}_k \mathbf{I}_k^*}{2}\right) = 0$$

Therefore,

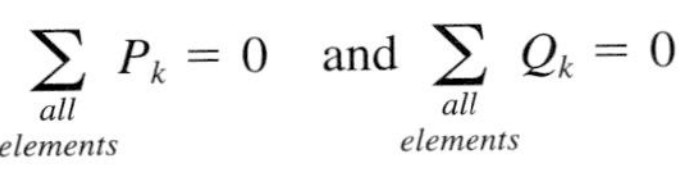

$$\sum_{\substack{\text{all} \\ \text{elements}}} P_k = 0 \quad \text{and} \quad \sum_{\substack{\text{all} \\ \text{elements}}} Q_k = 0$$

In other words, average power and reactive power are both conserved.

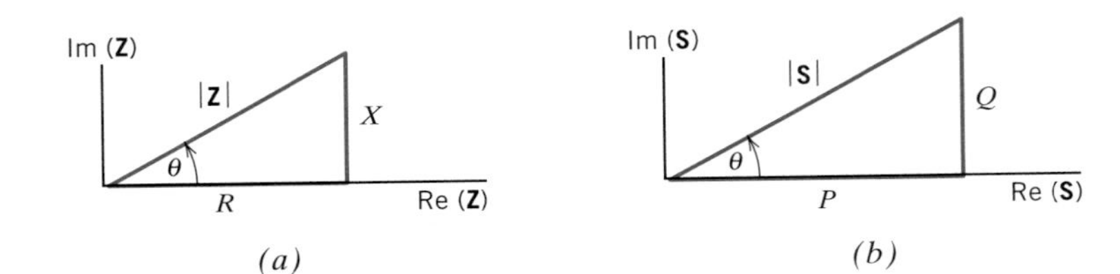

Figure 11.5-2 (a) The impedance triangle where $\mathbf{Z} = R + jX = Z\angle\theta$. (b) The complex power triangle where $\mathbf{S} = P + jQ$.

Example 11.5-1

Verify that complex power is conserved in the circuit of Figure 11.5-3 when $v_s = 100 \cos 1000t$ V.

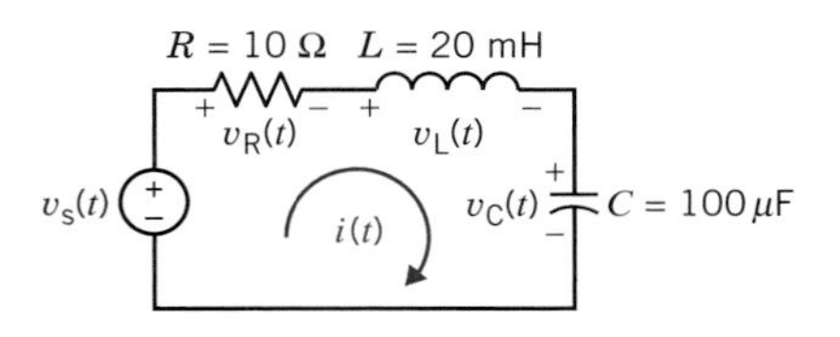

Figure 11.5-3
Circuit for Examples 11.5-1 and 11.5-2.

Solution

The phasor corresponding to the source voltage is

$$\mathbf{V}_s(\omega) = 100\angle 0 \text{ V}$$

Writing and solving a mesh equation, we find that the phasor corresponding to the mesh current is

$$\mathbf{I}(\omega) = \frac{\mathbf{V}_s(\omega)}{R + j\omega L - j\dfrac{1}{\omega C}} = \frac{100}{10 + j(1000)(0.02) - j\dfrac{1}{(1000)10^{-4}}} = 7.07\angle -45° \text{ A}$$

Ohm's law provides the phasors corresponding to the element voltages:

$$\mathbf{V}_R(\omega) = R\,\mathbf{I}(\omega) = 10(7.07\angle -45°) = 70.7\angle -45° \text{ V}$$
$$\mathbf{V}_L(\omega) = j\omega L\,\mathbf{I}(\omega) = j(1000)(0.02)(7.07\angle -45°)$$
$$= (20\angle 90°)(7.07\angle -45°) = 141.4\angle 45° \text{ V}$$
$$\mathbf{V}_C(\omega) = -j\frac{1}{\omega C}\mathbf{I}(\omega) = -j\frac{1}{(1000)(10^{-4})}(7.07\angle -45°)$$
$$= (10\angle -90°)(7.07\angle -45°) = 70.7\angle -135° \text{ V}$$

Consider the voltage source. The phasors $\mathbf{V}_S$ and $\mathbf{I}$ do not adhere to the passive convention. The complex power

$$\mathbf{S}_V = \frac{\mathbf{V}_S\mathbf{I}^*}{2} = \frac{100(7.07\angle -45°)^*}{2} = \frac{100(7.07\angle 45°)}{2} = \frac{100(7.07)}{2}\angle 45° = 353.5\angle 45° \text{ VA}$$

is the complex power supplied by the voltage source.

The phasors $\mathbf{I}$ and $\mathbf{V}_R$ do adhere to the passive convention. The complex power

$$\mathbf{S}_R = \frac{\mathbf{V}_R\mathbf{I}^*}{2} = \frac{(70.7\angle -45°)(7.07\angle -45°)^*}{2}$$
$$= \frac{(70.7\angle -45°)(7.07\angle 45°)}{2} = \frac{(70.7)(7.07)}{2}\angle(-45° + 45°) = 250\angle 0 \text{ VA}$$

is the complex power absorbed by the resistor. Similarly,

$$\mathbf{S}_L = \frac{\mathbf{V}_L\mathbf{I}^*}{2} = \frac{(141.4\angle 45°)(7.07\angle 45°)}{2} = \frac{(141.4)(7.07)}{2}\angle(45° + 45°) = 500\angle 90° \text{ VA}$$

is the complex power delivered to the inductor and

$$\mathbf{S}_C = \frac{\mathbf{V}_C\mathbf{I}^*}{2} = \frac{(70.7\angle -135°)(7.07\angle 45°)}{2} = \frac{(70.7)(7.07)}{2}\angle(-135° + 45°)$$
$$= 250\angle -90° \text{ VA}$$

is the complex power delivered to the capacitor.

To verify that complex power has been conserved, we calculate the complex power absorbed by the "other elements" and compare it to the complex power supplied by the source

$$\begin{aligned}\mathbf{S}_R + \mathbf{S}_L + \mathbf{S}_C &= 250\angle 0 + 500\angle 90 + 250\angle -90\\ &= (250 + j0) + (0 + j500) + (0 - j250)\\ &= 250 + j250 = 353.5\angle 45^\circ = \mathbf{S}_V\end{aligned}$$

As expected, the complex power supplied by the sources is equal to the complex power absorbed by the other elements of the circuit.

Example 11.5-2

Verify that average power is conserved in the circuit of Figure 11.5-3 when $v_s = 100 \cos 1000t$ V.

Solution

The phasor corresponding to the source voltage is

$$\mathrm{V}_s(\omega) = 100\angle 0 \text{ V}$$

Writing and solving a mesh equation, we find that the phasor corresponding to the mesh current is

$$\mathbf{I}(\omega) = \frac{\mathbf{V}_s(\omega)}{R + j\omega L - j\dfrac{1}{\omega C}} = \frac{100}{10 + j(1000)(0.02) - j\dfrac{1}{(1000)10^{-4}}} = 7.07\angle -45^\circ \text{ A}$$

The average power absorbed by the resistor, the capacitor, and the inductor can be calculated using

$$P = \left(\frac{I_{\mathrm{m}}^2}{2}\right)\mathrm{Re}(\mathbf{Z})$$

Since $\mathrm{Re}(\mathbf{Z}) = 0$ for the capacitor and for the inductor, the average power absorbed by each of these devices is zero. $\mathrm{Re}(\mathbf{Z}) = \mathrm{R}$ for the resistor so

$$P_R = \left(\frac{I_m^2}{2}\right)R = \frac{(7.07^2)}{2}10 = 250 \text{ W}$$

The average power supplied by the source is

$$P_V = \mathrm{Re}(\mathbf{S}_V) = \mathrm{Re}\left(\frac{\mathbf{V}_s\mathbf{I}^*}{2}\right) = \mathrm{Re}\left(\frac{100(7.07)}{2}\angle 45^\circ\right) = \mathrm{Re}(353.5\angle 45^\circ) = 250 \text{ W}$$

To verify that average power has been conserved, we calculate the average power absorbed by the "other elements" and compare it to the average power supplied by the source

$$P_R + P_L + P_C = 250 + 0 + 0 = 250 = P_V$$

As expected, the average power supplied by the sources is equal to the average power absorbed by the other elements of the circuit.

Exercise 11.5-1 Determine the average power delivered to each element of the circuit shown in Figure E 11.5-1. Verify that average power is conserved.

Answer: $4.39 + 0 = 4.39$ W

Exercise 11.5-2 Determine the complex power delivered to each element of the circuit shown in Figure E 11.5-2. Verify that complex power is conserved.

Answer: $6.606 + j5.248 - j3.303 = 6.606 + j1.982$ VA

Exercise 11.5-3 Design the circuit shown in Figure E 11.5-3, that is, specify values for R and L, so that the complex power delivered to the RL circuit is $8 + j6$ VA.

Answer: $R = 5.76\ \Omega$, $L = 2.16$ H.

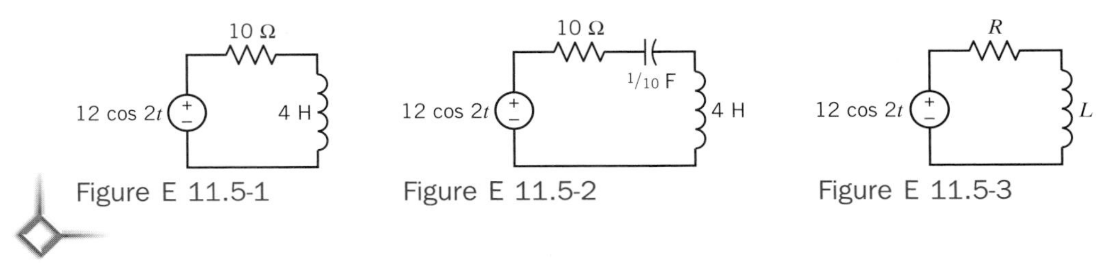

Figure E 11.5-1 Figure E 11.5-2 Figure E 11.5-3

11.6 POWER FACTOR

In this section, as in the previous section, we consider a linear circuit with a sinusoidal input that is at steady state. All the element voltages and currents will be sinusoidal and will have the same frequency as the input. Such a circuit can be analyzed in the frequency domain, using phasors. In particular, we can calculate the power generated or absorbed in a circuit or in any element of a circuit, in the frequency domain, using phasors.

Recall that in Section 11.5 we showed that the average power absorbed by the element shown in Figure 11.5-1 is

$$P = \frac{V_m I_m}{2} \cos(\theta_V - \theta_I)$$

and that the apparent power is

$$|\mathbf{S}| = \frac{V_m I_m}{2}$$

The ratio of the average power to the apparent power is called the **power factor** (pf). The power factor is calculated as

$$pf = \cos(\theta_V - \theta_I)$$

The angle $(\theta_V - \theta_I)$ is often referred to as the *power factor angle*. The average power absorbed by the element shown in Figure 11.5-1 can be expressed as

$$P = \frac{V_m I_m}{2} pf \tag{11.6-1}$$

The cosine is an even function, that is, $\cos(\theta) = \cos(-\theta)$. So

$$pf = \cos(\theta_V - \theta_I) = \cos(\theta_I - \theta_V)$$

This causes a small difficulty. We can't calculate $\theta_V - \theta_I$ from *pf* without some additional information. For example, suppose $pf = 0.8$. We calculate

$$36.87° = \cos^{-1}(0.8)$$

but that's not enough to determine $\theta_V - \theta_I$ uniquely. Since the cosine is even, both $\cos(36.87°) = 0.8$ and $\cos(-36.87°) = 0.8$, so either $\theta_V - \theta_I = 36.87°$ or $\theta_V - \theta_I = -36.87°$. This difficulty is resolved by labeling the power factor as *leading* or *lagging*. When $\theta_V - \theta_I > 0$, the power factor is said to be lagging, and when $\theta_V - \theta_I < 0$, the power factor is said to be leading. If the power factor is specified to be 0.8 leading, then $\theta_V - \theta_I = -36.87°$. On the other hand, if the power factor is specified to be 0.8 lagging, then $\theta_V - \theta_I = 36.87°$.

The significance of the power factor is illustrated by the circuit shown in Figure 11.6-1. This circuit models the transmission of electric power from a power utility company to a customer. The customer's load is connected to the power company's power plant by a transmission line. Typically, the customer requires power at a specified voltage. The power company must supply both the power used by the customer and the power absorbed by the transmission line. The power absorbed by the transmission line is lost; it doesn't do anybody any good, and we want to minimize it.

The circuit in Figure 11.6-2 models the transmission of electric power from a power utility company to a customer in the frequency domain, using impedances and phasors. Our objective is to find a way to reduce the power absorbed by the transmission line. In this situation, it is likely that we cannot change the transmission line, so we can't change R_l or $j\omega L_l$. Similarly, since the customer requires a specified average power at a specified voltage, we can't change V_m or P. In the following analysis we leave R_l, L_l, V_m, and P as variables for the sake of generality. We won't need to repeat the analysis later if we encounter a similar situation with a different customer and a different transmission line. We will see that it is possible to adjust the power factor by adding a compensating impedance to the customer's load. We will leave the power factor, *pf*, as a variable in our analysis because we plan to vary the power factor to reduce the power absorbed by the load.

The impedance of the line is

$$\mathbf{Z}_{\text{LINE}}(\omega) = \frac{R_l}{2} + j\omega\frac{L_l}{2} + \frac{R_l}{2} + j\omega\frac{L_l}{2} = R_l + j\omega L_l$$

The average power absorbed by the line is

$$P_{\text{LINE}} = \frac{I_m^2}{2}\text{Re}(\mathbf{Z}_{\text{LINE}}) = \frac{I_m^2}{2}R_l$$

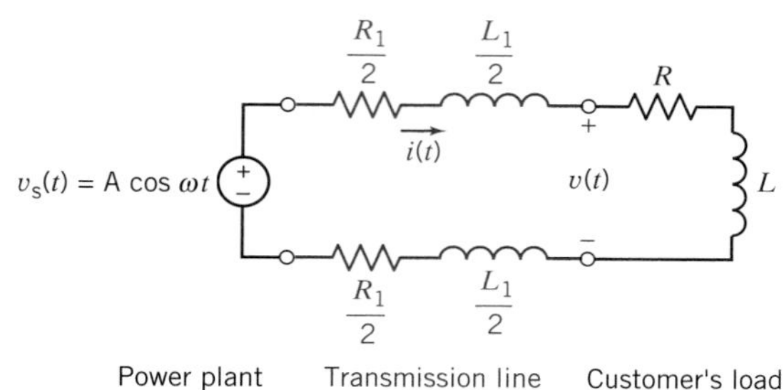

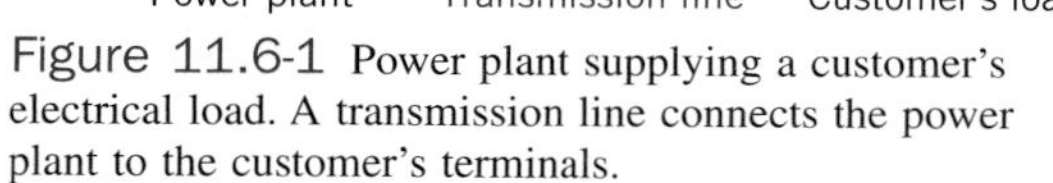

Figure 11.6-1 Power plant supplying a customer's electrical load. A transmission line connects the power plant to the customer's terminals.

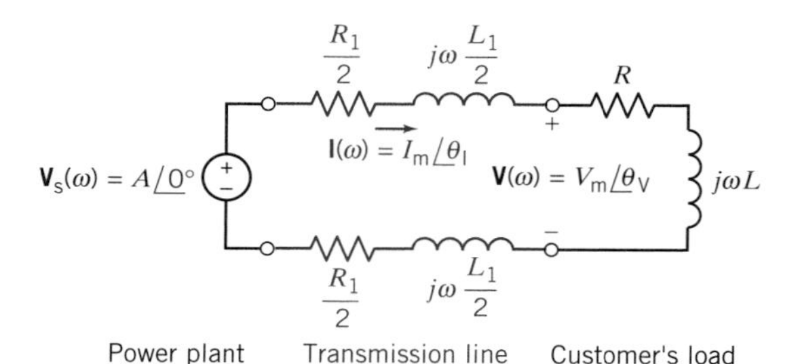

Figure 11.6-2 Frequency domain representation of the power plant supplying a customer's electrical load.

Since the customer requires power at a specified voltage, we will treat the voltage across the load, V_m, and the average power delivered to the load, P, as known quantities. Recall from Eq. 11.6-1 that

$$P = \frac{V_m I_m}{2} pf$$

Solving for I_m gives

$$I_m = \frac{2P}{V_m pf}$$

so

$$P_{LINE} = 2\left(\frac{P}{V_m pf}\right)^2 R_1$$

Increasing pf will reduce the power absorbed in the transmission line. The power factor is the cosine of an angle, so its maximum value is 1. Notice that $pf = 1$ occurs when $\theta_V = \theta_I$, that is, when the load appears to be resistive.

In Figure 11.6-3 a compensating impedance has been attached across the terminals of the customer's load. We plan to use this impedance to adjust the power factor of the customer's load. Since it is to the advantage of both the power company and the user to keep the power factor of a load as close to unity as feasible, we say that we are *correcting* the power factor of the load. We will denote the corrected power factor as pfc and the corresponding phase angle as θ_c. That is,

$$pfc = \cos \theta_C$$

We can represent the impedance of the load as

$$\mathbf{Z} = R + jX$$

Similarly, we can represent the impedance of the compensating impedance as

$$\mathbf{Z}_C = R_C + jX_C$$

Since $\mathbf{Z}$ is connected to draw a current $\mathbf{I}$, the power delivered to $\mathbf{Z}$ will remain P. The benefit of the parallel impedance is that the parallel combination appears as the load to the source and $\mathbf{I}_L$ is the current that flows through the transmission line. We want $\mathbf{Z}_C$ to absorb no average power. Therefore, we choose a reactive element so that

$$\mathbf{Z}_C = jX_C$$

The impedance of the parallel combination, $\mathbf{Z}_p$, is

$$\mathbf{Z}_P = \frac{\mathbf{Z}\,\mathbf{Z}_C}{\mathbf{Z} + \mathbf{Z}_C}$$

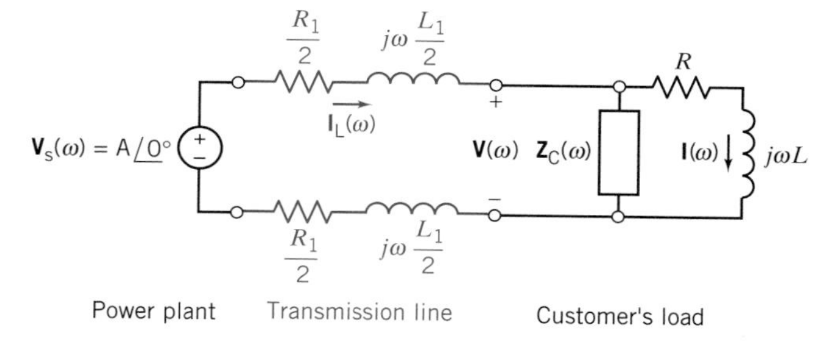

Figure 11.6-3 Power plant supplying a customer's electrical load. A compensating impedance has been added to the customer's load to correct the power factor.

The parallel impedance may be written as

$$\mathbf{Z}_P = R_P + jX_P = Z_P \angle \theta_P$$

and the power factor of the new combination is

$$pfc = \cos\theta_P = \cos\left(\tan^{-1}\frac{X_P}{R_P}\right) \tag{11.6-2}$$

where pfc is the corrected power factor and the corrected phase $\theta_c = \theta_p$. Some algebra is needed to calculate R_P and X_P.

$$\begin{aligned}\mathbf{Z}_P &= \frac{(R + jX)\, jX_C}{R + jX + jX_C} \\ &= \frac{RX_C^2 + j[R^2X_C + (X_C + X)\,X\,X_C]}{R^2 + (X + X_C)^2} \\ &= \frac{RX_C^2}{R^2 + (X + X_C)^2} + j\,\frac{R^2X_C + (X_C + X)\,X\,X_C}{R^2 + (X + X_C)^2}\end{aligned}$$

Therefore, the ratio of X_P to R_P is

$$\frac{X_P}{R_P} = \frac{R^2 + (X_C + X)\,X}{RX_C} \tag{11.6-3}$$

Equation 11.6-2 may be written as

$$\frac{X_P}{R_P} = \tan(\cos^{-1} pfc) \tag{11.6-4}$$

Combining Eqs. 11.6-3 and 11.6-4 and solving for X_C, we have

$$X_C = \frac{R^2 + X^2}{R\tan(\cos^{-1} pfc) - X} \tag{11.6-5}$$

We note that X_C may be positive or negative depending on the required pfc and the original R and X of the load. The factor $\tan(\cos^{-1}(pfc))$ will be positive if pfc is specified as lagging and negative if it is specified as leading.

Typically, we will find that the customer's load is inductive, and we will need a capacitive impedance $\mathbf{Z}_C$. Recall that for a capacitor, we have

$$\mathbf{Z}_C = \frac{-j}{\omega C} = jX_C \tag{11.6-6}$$

Note that we determine that X_C is typically negative. Combining Eqs. 11.6-5 and 11.6-6 gives

$$\frac{-1}{\omega C} = \frac{R^2 + X^2}{R\tan(\cos^{-1} pfc) - X}$$

Solving for ωC gives

$$\omega C = \frac{X - R\tan(\cos^{-1} pfc)}{R^2 + X^2} = \frac{R}{R^2 + X^2}\left(\frac{X}{R} - \tan(\cos^{-1} pfc)\right)$$

Let $\theta = \tan^{-1}\left(\dfrac{X}{R}\right)$. Then

$$\omega C = \frac{R}{R^2 + X^2}(\tan\theta - \tan\theta_C) \qquad (11.6\text{-}7)$$

where $\theta = \cos^{-1}(pf)$ and $\theta_c = \cos^{-1}(pfc)$.

Example 11.6-1

A customer's plant has two parallel loads connected to the power utility's distribution lines. The first load consists of 50 kW of heating and is resistive. The second load is a set of motors that operate at 0.86 lagging power factor. The motors' load is 100 kVA. Power is supplied to the plant at 10,000 volts rms. Determine the total current flowing from the utility's lines into the plant and the plant's overall power factor.

Solution

Figure 11.6-4*a* summarizes what is known about this power system.

First consider the heating load. Since this load is resistive, the reactive power is zero. Therefore,

$$\mathbf{S}_1 = P_1 = 50\ \text{kW}$$

Next consider the motors. The power factor is lagging, so $\theta_2 > 0°$.

$$\theta_2 = \cos^{-1}(pf_2) = \cos^{-1}(0.86) = 30.7°$$

The complex power absorbed by the motors is

$$\mathbf{S}_2 = |\mathbf{S}_2| \angle\theta_2 = 100\angle 30.7°\ \text{kVA}$$

The average power and reactive power absorbed by the motors is obtained by converting the complex power to rectangular form

$$\mathbf{S}_2 = |\mathbf{S}_2|\cos\theta_2 + \mathrm{j}|\mathbf{S}_2|\sin\theta_2 = 100\cos 30.7° + \mathrm{j}100\sin 30.7° = 86 + \mathrm{j}51\ \text{kVA}$$

Therefore

$$P_2 = 86\ \text{kW} \quad \text{and} \quad Q_2 = 51\ \text{kVAR}$$

The total complex power $\mathbf{S}$ delivered to the total load is the sum of the complex power delivered to each load,

$$\mathbf{S} = \mathbf{S}_1 + \mathbf{S}_2 = 50 + (86 + j51) = 136 + j51\ \text{kVA}$$

The average power and reactive power of the customer's load is

$$P = 136\ \text{kW} \quad \text{and} \quad Q = 51\ \text{kVAR}$$

To calculate the power factor of the customer's load, first convert $\mathbf{S}$ to polar form

$$\mathbf{S} = 145.2\angle 20.6°\ \text{kVA}$$

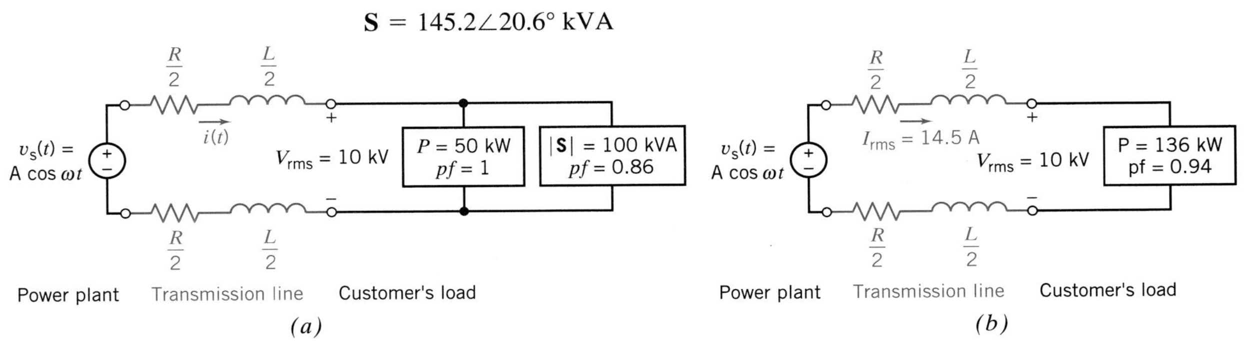

Figure 11.6-4 Power system for Example 11.6-1.

Then
$$pf = \cos(20.6^\circ) = 0.94 \text{ lagging}$$

The total current flowing from the utility's lines into the plant can be calculated from the apparent power absorbed by the customer's load and the voltage across the terminals of the customer's load. Recall that

$$|\mathbf{S}| = \frac{V_m I_m}{2} = V_{rms} I_{rms}$$

Solving for the current gives

$$I_{rms} = \frac{|\mathbf{S}|}{V_{rms}} = \frac{145200}{10^4} = 14.52 \text{ A rms}$$

Figure 11.6-4*b* summarizes the results of this example.

Example 11.6-2

A load as shown in Figure 11.6-5 has an impedance of $\mathbf{Z} = 100 + j100\ \Omega$. Find the parallel capacitance required to correct the power factor to (a) 0.95 lagging and (b) 1.0. Assume that the source is operating at $\omega = 377$ rad/s.

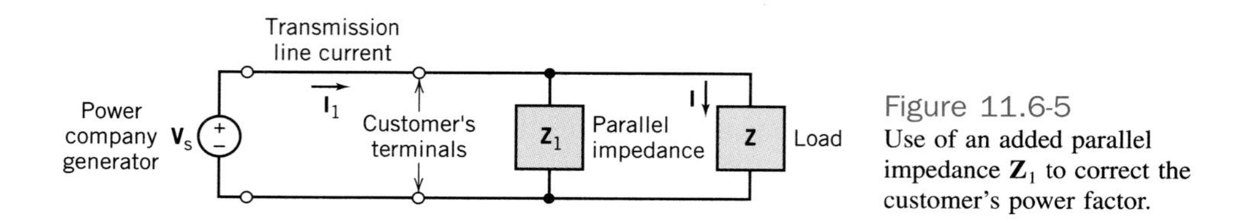

Figure 11.6-5 Use of an added parallel impedance $\mathbf{Z}_1$ to correct the customer's power factor.

Solution

The phase angle of the impedance is $\theta = 45^\circ$, so the original load has a lagging power factor with

$$\cos\theta = \cos 45^\circ = 0.707$$

First, we wish to correct the *pf* so that $pfc = 0.95$ lagging. Then, we use Eq. 11.6-5 as follows:

$$X_C = \frac{100^2 + 100^2}{100\tan(\cos^{-1} 0.95) - 100} = -297.9$$

The capacitor required is determined from

$$-\frac{1}{\omega C} = X_C$$

Therefore, since $\omega = 377$ rad/s

$$C = -\frac{1}{\omega X_C} = \frac{-1}{377(-297.9)} = 8.9\ \mu\text{F}$$

If we wish to correct the load to $pfc = 1$, we have

$$X_C = \frac{2 \times 10^4}{100\tan(\cos^{-1} 1) - 100} = -200$$

The capacitor required to correct the power factor to 1.0 is determined from

$$C = \frac{-1}{\omega X_C} = \frac{-1}{377(-200)} = 13.3\ \mu\text{F}$$

Since the uncorrected power factor is lagging, we can alternatively use Eq. 11.6-7 to determine C. For example, it follows that $pfc = 1$. Then $\theta_c = 0°$. Therefore,

$$\omega C = \frac{100}{2 \times 10^4}(\tan\theta - \tan\theta_c) = (5 \times 10^{-3})(\tan(45°) - \tan(0°)) = 5 \times 10^{-3}$$

and

$$C = \frac{5 \times 10^{-3}}{377} = 13.3\ \mu\text{F}$$

As expected, this is the same value of capacitance as was calculated using Eq.11.6-5.

Exercise 11.6-1 A circuit has a large motor connected to the ac power lines ($\omega = (2\pi)60 = 377$ rad/s). The model of the motor is a resistor of 100 Ω in series with an inductor of 5 H. Find the power factor of the motor.

Answer: $pf = 0.053$ lagging

Exercise 11.6-2 A circuit has a load impedance $\mathbf{Z} = 50 + j80\ \Omega$ as shown in Figure 11.6-5. Determine the power factor of the uncorrected circuit. Determine the impedance $\mathbf{Z}_C$ required to obtain a corrected power factor of 1.0.

Answer: $\mathbf{Z}_C = -j111.25\ \Omega$

Exercise 11.6-3 Determine the power factor for the total plant of Example 11.6-1 when the resistive heating load is decreased to 30 kW. The motor load and the supply voltage remain as described in Example 11.6-1.

Answer: $pf = 0.915$

Exercise 11.6-4 A 4-kW, 110-Vrms load, as shown in Figure 11.6-5, has a power factor of 0.82 lagging. Find the value of the parallel capacitor that will correct the power factor to 0.95 lagging when $\omega = 377$ rad/s.

Answer: $C = 0.324$ mF

11.7 THE POWER SUPERPOSITION PRINCIPLE

In this section, let us consider the case where the circuit contains two or more sources. For example, consider the circuit shown in Figure 11.7-1*a* with two sinusoidal voltage sources. The principle of superposition states that the response to both sources acting together is equal to the sum of the responses to each voltage source acting alone. The application of the principle of superposition is illustrated in Figure 11.7-1*b* where i_1 is the response to source 1 acting alone and the response i_2 is the response to source 2 acting alone. The total response is

$$i = i_1 + i_2 \tag{11.7-1}$$

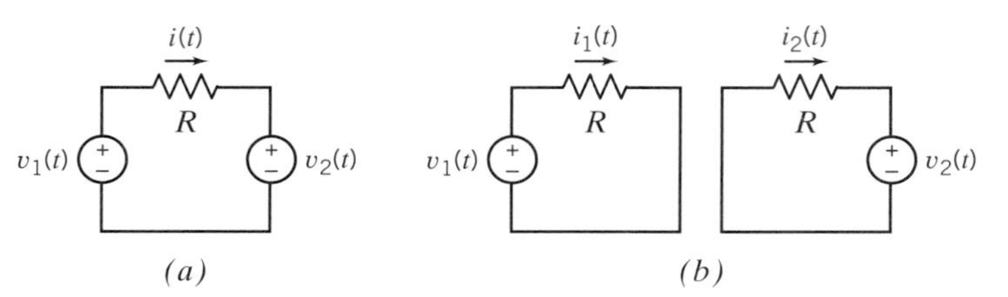

Figure 11.7-1 (*a*) A circuit with two sources. (*b*) Using superposition to calculate the resistor current as $i(t) = i_1(t) + i_2(t)$.

The instantaneous power is

$$p = i^2R = R(i_1 + i_2)^2 = R(i_1^2 + i_2^2 + 2i_1i_2)$$

where R is the resistance of the circuit. Then the average power is

$$\begin{aligned} P &= \frac{1}{T}\int_0^T p\,dt = \frac{R}{T}\int_0^T (i_1^2 + i_2^2 + 2i_1i_2)dt \\ &= \frac{R}{T}\int_0^T i_1^2\,dt + \frac{R}{T}\int_0^T i_2^2\,dt + \frac{2R}{T}\int_0^T i_1i_2\,dt = P_1 + P_2 + \frac{2R}{T}\int_0^T i_1i_2\,dt \end{aligned} \quad (11.7\text{-}2)$$

where P_1 is the average power due to v_1 and P_2 is the average power due to v_2. We will see that when v_1 and v_2 are sinusoids having different frequencies, then

$$\frac{2R}{T}\int_0^T i_1i_2\,dt = 0 \quad (11.7\text{-}3)$$

When Eq. 11.7-3 is satisfied, then Eq. 11.7-2 reduces to

$$P = P_1 + P_2 \quad (11.7\text{-}4)$$

This equation states that the average power delivered to the resistor by both sources acting together is equal to the sum of the average power delivered to the resistor by each voltage source acting alone. This is the principle of power superposition. Notice that the principle of power superposition is valid only when Eq. 11.7-3 is satisfied.

Now let us determine under what conditions Eq. 11.7-3 is satisfied. Let the radian frequency for the first source be $m\omega$, and let the radian frequency for the second source be $n\omega$. The currents can be represented by the general form

$$i_1 = I_1 \cos(m\omega t + \phi)$$

and

$$i_2 = I_2 \cos(n\omega t + \theta)$$

Let P_{12} denote the average power due to the product of the two currents. That is

$$P_{12} = \frac{2R}{T}\int_0^T i_1i_2\,dt = \frac{2R}{T}\int_0^T I_1I_2\cos(m\omega t + \phi)\cos(n\omega t + \theta)\,dt \quad (11.7\text{-}5)$$

When m and n are integers, we perform this integration to get

$$\begin{aligned} P_{12} &= \frac{2R}{T}\int_0^T I_1I_2\cos(m\omega t + \phi)\cos(n\omega t + \theta)\,dt \\ &= \frac{RI_1I_2}{T}\int_0^T (\cos((m - n)\omega t + (\phi - \theta)) + \cos((m + n)\omega t + (\phi + \theta)))\,dt \end{aligned}$$

$$= \begin{cases} 0, & m \neq n \\ RI_1I_2 \cos(\phi - \theta), & m = n \end{cases}$$

We note that the integral is equal to zero when $m \neq n$ and is equal to a nonzero quantity when $m = n$, where m and n are integers.

Therefore, we may state that the total average power delivered to a load is equal to the sum of the average power delivered by each source when the radian frequency of each source is an integer multiple of the other sources. However, when two or more sources have the same radian frequency, the total average power is not the sum of the average power due to each source.

Now let us further consider Eq. 11.7-5 when m and n are not integers. For convenience, let $m = 1, n = 1.5$, and $\phi = \theta = 0°$. Furthermore, since we wish to determine the average power when one of the cosine functions has a period that is not an integer multiple of period T, we must revert to the definition of average power over all time as

$$\begin{aligned} P_{12} &= \lim_{t\to\infty} \frac{1}{\mathrm{T}} \int_{-t/2}^{t/2} p\, dt \\ &= \lim_{t\to\infty} \frac{1}{\mathrm{T}} \int_{-t/2}^{t/2} 2\, R\, I_1 I_2 \cos \omega t \cos(1.5\omega t)\, dt \\ &= \lim_{t\to\infty} \frac{1}{\mathrm{T}} \int_{-t/2}^{t/2} R\, I_1 I_2 \left(\cos 0.5\omega t + \cos 2.5\omega t\right) dt \\ &= 0 \end{aligned}$$

since the integral of each cosine wave within the final form of the integral is zero.

Therefore, in summary, the *superposition of average power* states that the average power delivered to a circuit by several sinusoidal sources, acting together, is equal to the sum of the average power delivered to the circuit by each source acting alone, if, and only if, no two of the sources have the same frequency. Similar arguments show that superposition can be used to calculate the reactive power or the complex power delivered to a circuit by several sinusoidal sources, provided again that no two sources have the same frequency.

If two or more sources are operating at the same frequency, the principle of *power* superposition is not valid, but the principle of superposition remains valid. In this case, we use the principle of superposition to find each phasor current and then add the currents to obtain the total phasor current

$$\mathbf{I} = \mathbf{I}_1 + \mathbf{I}_2 + \cdots + \mathbf{I}_N$$

for N sources. Then we have the average power

$$P = \frac{I_\mathrm{m}^2 R}{2} \tag{11.7-6}$$

where $|\mathbf{I}| = I_\mathrm{m}$.

Example 11.7-1

The circuit in Figure 11.7-2 contains two sinusoidal sources. To illustrate power superposition, consider two cases:

(1) $v_A(t) = 12 \cos 3t$ V and $i_B(t) = 2 \cos 4t$ A

(2) $v_A(t) = 12 \cos 4t$ V and $i_B(t) = 2 \cos 4t$ A

Find the average power absorbed by the resistor.

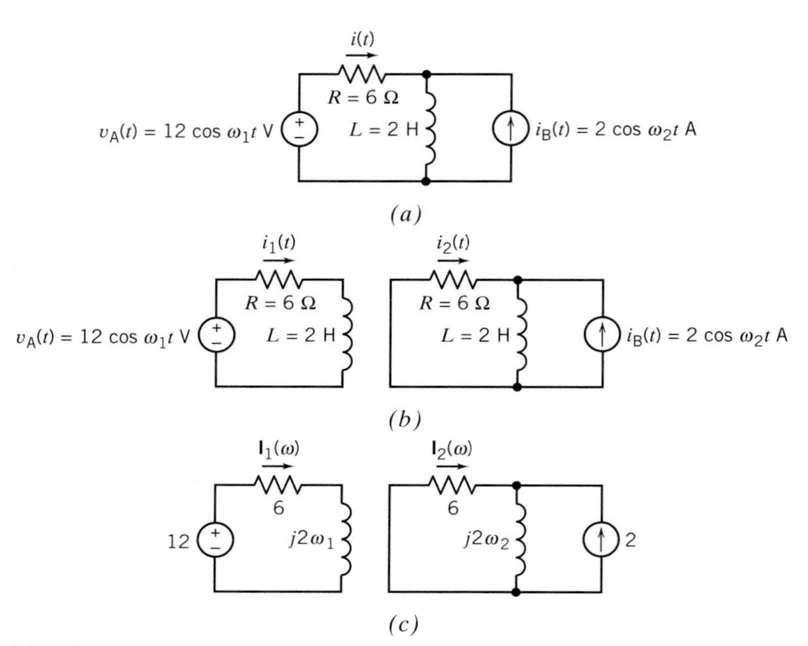

Figure 11.7-2 (*a*) A circuit with two sinusoidal sources. (*b*) Using superposition to find the response to each source separately. (*c*) Representing the circuit from (*b*) in the frequency domain.

Solution

The resistor current caused by both sources acting together is equal to the sum of the resistor current caused by the voltage source acting alone and the resistor current caused by the current source acting alone. The application of the principle of superposition is illustrated in Figure 11.7-2*b* where i_1 is the response to voltage source acting alone and the response i_2 is the response to current source acting alone. The total response is $i = i_1 + i_2$. In Figure 11.7-2*c*, the circuits from Figure 11.7-2*b* are represented in the frequency domain using impedances and phasors.

Now consider the two cases.

Case 1: Analysis of the circuits in Figure 11.7-2*c* gives

$$\mathbf{I}_1(\omega) = 1.414\angle -45° \quad \text{and} \quad \mathbf{I}_2(\omega) = 1.6\angle -143°$$

These phasors correspond to different frequencies and cannot be added. The corresponding time domain currents are

$$i_1(t) = 1.414\cos(3t - 45°)\text{ A} \quad \text{and} \quad i_2(t) = 1.6\cos(4t - 143°)\text{ A}$$

Using superposition, we find that the total current in the resistor is

$$i(t) = 1.414\cos(3t - 45°) + 1.6\cos(4t - 143°)\text{ A}$$

The average power could be calculated as

$$P = \frac{R}{T}\int_0^T i^2\,dt = \frac{R}{T}\int_0^T (1.414\cos(3t - 45°) + 1.6\cos(4t - 143°))^2\,dt$$

Since the two sinusoidal sources have different frequencies, the average power can be calculated more easily using power superposition.

$$P = P_1 + P_2 = \frac{1.414^2}{2}6 + \frac{1.6^2}{2}6 = 13.7\text{ W}$$

Notice that both superposition and power superposition were used in this case. First, superposition was used to calculate $\mathbf{I}_1(\omega)$ and $\mathbf{I}_2(\omega)$. Next, P_1 was calculated using $\mathbf{I}_1(\omega)$, and P_2 was calculated using $\mathbf{I}_2(\omega)$. Finally, power superposition was used to calculate P from P_1 and P_2.

Case 2: Analysis of the circuits in Figure 11.7-2*c* gives

$$\mathbf{I}_1(\omega) = 1.2\angle-53.1^\circ \quad \text{and} \quad \mathbf{I}_2(\omega) = 1.6\angle-143^\circ$$

Both of these phasors correspond to the same frequency, $\omega = 4$ rad/s. Therefore, these phasors can be added to obtain the phasor corresponding to $i(t)$.

$$\mathbf{I}(\omega) = \mathbf{I}_1(\omega) + \mathbf{I}_2(\omega) = (1.2\angle-53.1^\circ) + (1.6\angle-143^\circ) = 2.0\angle-106.3^\circ$$

The sinusoidal current corresponding to this phasor is

$$i(t) = 2.0\cos(4t - 106.3^\circ)\text{ A}$$

The average power absorbed by the resistor is

$$P = \frac{2.0^2}{2}6 = 12\text{ W}$$

Alternately, the time domain currents corresponding to $\mathbf{I}_1(\omega)$ and $\mathbf{I}_2(\omega)$ are

$$i_1(t) = 1.2\cos(4t - 53.1^\circ)\text{ A} \quad \text{and} \quad i_2(t) = 1.6\cos(4t - 143^\circ)\text{ A}$$

Using superposition, we find that the total current in the resistor is

$$i(t) = 1.2\cos(4t - 53.1^\circ) + 1.6\cos(4t - 143^\circ) = 2.0\cos(4t - 106.3^\circ)\text{ A}$$

So $P = 12$ W, as before.

Power superposition cannot be used in this case because the two sinusoidal sources have the same frequency.

Exercise 11.7-1 Determine the average power absorbed by the resistor in Figure 11.7-2*a* for these two cases:

(a) $v_A(t) = 12\cos 3t$ V and $i_B(t) = 2\cos 3t$ A

(b) $v_A(t) = 12\cos 4t$ V and $i_B(t) = 2\cos 3t$ A

Answer: (a) 12 W (b) 10.3 W

11.8 THE MAXIMUM POWER TRANSFER THEOREM

In Chapter 5, we proved that for a resistive network, maximum power is transferred from a source to a load when the load resistance is set equal to the Thévenin resistance of the Thévenin equivalent source. Now let us consider a circuit represented by a Thévenin equivalent circuit for a sinusoidal steady source, as shown in Figure 11.8-1, when the load is $\mathbf{Z}_L$.

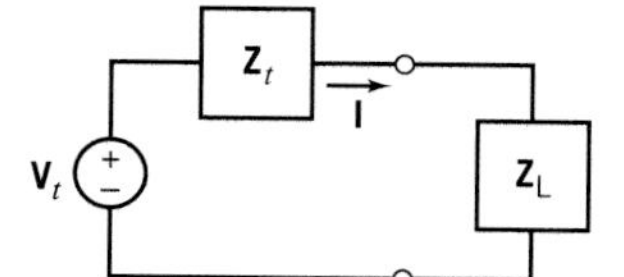

Figure 11.8-1 The Thévenin equivalent circuit with a load impedance.

We then have

$$\mathbf{Z}_t = R_t + jX_t$$

and

$$\mathbf{Z}_L = R_L + jX_L$$

The average power delivered to the load is

$$P = \frac{I_m^2}{2} R_L$$

The phasor current **I** is given by

$$\mathbf{I} = \frac{\mathbf{V}_t}{\mathbf{Z}_t + \mathbf{Z}_L} = \frac{\mathbf{V}_t}{(R_t + jX_t) + (R_L + jX_L)}$$

where we may select the values of R_L and X_L. The average power delivered to the load is

$$P = \frac{I_m^2 R_L}{2} = \frac{|\mathbf{V}_t|^2 R_L/2}{(R_t + R_L)^2 + (X_t + X_L)^2}$$

and we wish to maximize P. The term $(X_t + X_L)^2$ can be eliminated by setting $X_L = -X_t$. We have

$$P = \frac{|\mathbf{V}_t|^2 R_L}{2(R_t + R_L)^2}$$

The maximum is determined by taking the derivative dP/dR_L and setting it equal to zero. Then we find that $dP/dR_L = 0$ when $R_L = R_t$.

Consequently, we have

$$\mathbf{Z}_L = R_t - jX_t$$

Thus, the *maximum power transfer* from a circuit with a Thévenin equivalent circuit with an impedance $\mathbf{Z}_t$ is obtained when $\mathbf{Z}_L$ is set equal to $\mathbf{Z}_t^*$, the complex conjugate of $\mathbf{Z}_t$.

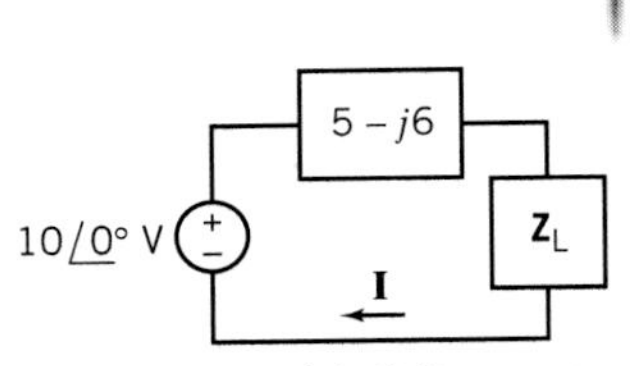

Figure 11.8-2 Circuit for Example 11.8-1. Impedances in ohms.

Example 11.8-1

Find the load impedance that transfers maximum power to the load and determine the maximum power quantity obtained for the circuit shown in Figure 11.8-2.

Solution

We select the load to be the complex conjugate of $\mathbf{Z}_t$ so that

$$\mathbf{Z}_L = \mathbf{Z}_t^* = 5 + j6$$

Then the maximum power transferred can be obtained by noting that

$$\mathbf{I} = \frac{10\angle 0°}{5 + 5} = 1\angle 0°$$

Therefore, the average power transferred to the load is

$$P = \frac{I_m^2}{2} R_L = \frac{(1)^2}{2} 5 = 2.5 \text{ W}$$

Exercise 11.8-1 For the circuit of Figure 11.8-1, find $\mathbf{Z}_L$ to obtain the maximum power transferred when the Thévenin equivalent circuit has $\mathbf{V}_t = 100\angle 0°$ V and $\mathbf{Z}_t = 10 + j14\ \Omega$. Also determine the maximum power transferred to the load.

Answer: $\mathbf{Z}_L = 10 - j14\ \Omega, \quad P = 125$ W

Exercise 11.8-2 A television receiver uses a cable to connect the antenna to the TV, as shown in Figure E 11.8-2, with $v_s = 4 \cos \omega t$ mV. The TV station is received at 52 MHz. Determine the average power delivered to each TV set if (a) the load impedance is $\mathbf{Z} = 300\ \Omega$; (b) two identical TV sets are connected in parallel with $\mathbf{Z} = 300\ \Omega$ for each set; (c) two identical sets are connected in parallel and $\mathbf{Z}$ is to be selected so that maximum power is delivered at each set.

Answer: (a) 9.6 nW (b) 4.9 nW (c) 5 nW

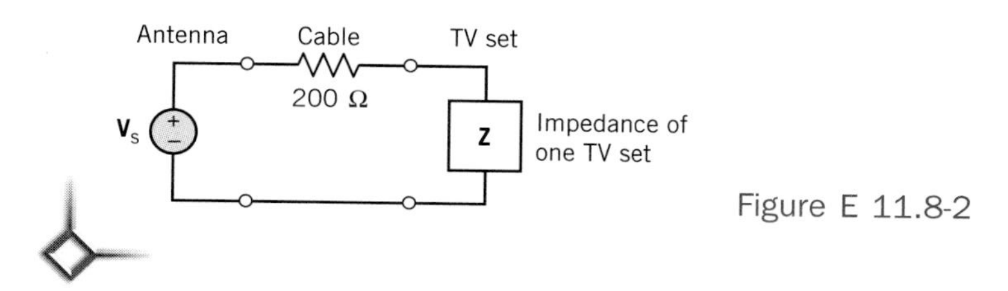

Figure E 11.8-2

11.9 COUPLED INDUCTORS

The concept of self-inductance was introduced in Chapter 7. We commonly use the term *inductance* for self-inductance, and we are familiar with circuits that have inductors. In this section we consider coupled inductors, which are useful in circuits with sinusoidal steady-state (ac) voltages and currents and are also widely used in electronic circuits.

Coupled inductors or *coupled coils* are a magnetic device that consists of two or more multiturn coils wound on a common core. Figure 11.9-1*a* shows two coils of wire wrapped around a magnetic core. These coils are said to be magnetically coupled. A voltage applied to one coil, as shown in Figure 11.9-1*a*, causes a voltage across the second coil. Here's why. The input voltage, $v_1(t)$, causes a current $i_1(t)$ in coil 1. The current and voltage are related by

$$v_1 = L_1 \frac{di_1}{dt} \tag{11.9-1}$$

where L_1 is the self-inductance of coil 1. The current $i_1(t)$ causes a flux in the magnetic core. This flux is related to the current by

$$\phi = c_1 N_1 i_1 \tag{11.9-2}$$

where c_1 is a constant that depends on the magnetic properties and geometry of the core and N_1 is the number of turns in coil 1. The number of turns in a coil indicates the number of times the wire is wrapped around the core. The flux, ϕ, is contained within the magnetic core. The core has a cross-sectional area A. The voltage across the coil 1 is related to the flux by

$$v_1 = N_1 \frac{d\phi}{dt} = N_1 \frac{d}{dt}(c_1 N_1 i_1) = c_1 N_1^2 \frac{di_1}{dt} \tag{11.9-3}$$

Comparing Eqs. 11.9-1 and 11.9-3 shows that

$$L_1 = c_1 N_1^2 \tag{11.9-4}$$

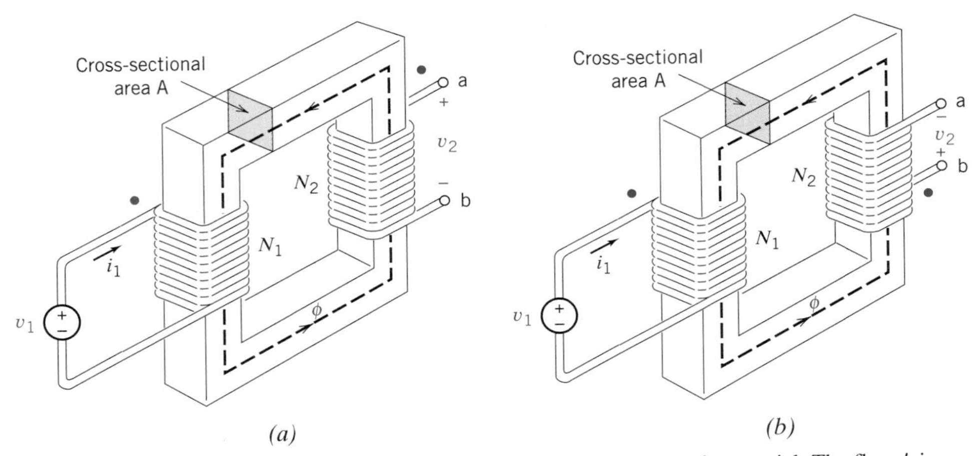

Figure 11.9-1 Two magnetically coupled coils mounted on a magnetic material. The flux ϕ is contained within the magnetic core.

A voltage, v_2, at the terminals of the second coil is induced by ϕ that flows through the second coil. This voltage is related to the flux by

$$v_2 = N_2 \frac{d\phi}{dt} = c_M N_1 N_2 \frac{di_1}{dt} = M \frac{di_1}{dt} \quad (11.9\text{-}5)$$

where c_M is a constant that depends on the magnetic properties and geometry of the core, N_2 is the number of turns in the second coil, and $M = c_M N_1 N_2$ is a positive number called the **mutual inductance.** The unit of mutual inductance is the henry, H.

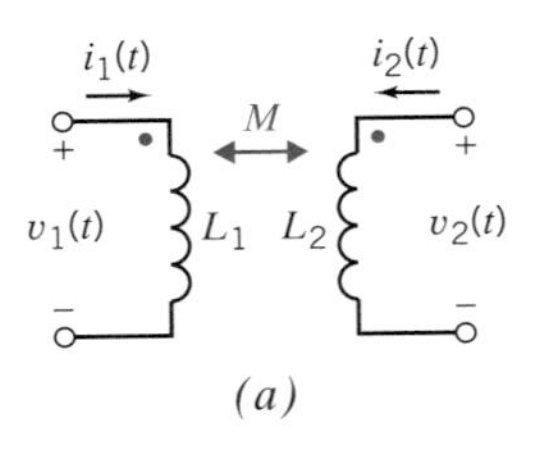

The polarity of the voltage v_2, compared to the polarity of v_1, depends on the way in which the coils are wrapped on the core. There are two distinct cases, and they are shown in Figure 11.9-1a and b. The difference between these two figures is the direction in which coil 2 is wrapped around the core. A **dot convention** is used to indicate the way the coils have been wrapped on the coil. Notice that one end of each coil is marked with a dot. The dots at the end of the coils indicate that the dotted ends have a positive voltage at the same time.

The circuit symbol that is used to represent coupled inductors is shown in Figure 11.9-2 with the dots shown and the mutual inductance identified as M. Two cases are shown in Figure 11.9-2. In Figure 11.9-2a both coil currents enter the dotted ends of the coils. In Figure 11.9-2b, one current, i_1, enters the dotted end of a coil, but the other current, i_2, enters the undotted end on the coil. In both cases, the reference directions of the voltage and current of each coil adhere to the passive convention.

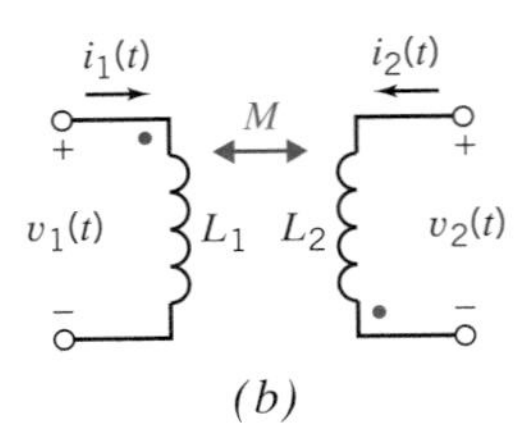

Figure 11.9-2 Circuit symbol for coupled inductors. In (a) both coil currents enter the dotted ends of the coils. In (b) one coil current enters the dotted end of the coil, but the other coil current enters the undotted end.

Suppose both coil currents enter the dotted ends of the coils, as in Figure 11.9-1a, or both coil currents enter the undotted ends of the coils. The voltage across the first coil, v_1, is related to the coil currents by

$$v_1 = L_1 \frac{di_1}{dt} + M \frac{di_2}{dt} \quad (11.9\text{-}6)$$

Similarly, the voltage across the second coil is related to the coil currents by

$$v_2 = L_2 \frac{di_2}{dt} + M \frac{di_1}{dt} \quad (11.9\text{-}7)$$

In contrast, suppose one coil current enters the dotted end of a coil while the other coil current enters the undotted end of a coil, as in Figure 11.9-2*b*. The voltage across the first coil, v_1, is related to the coil currents by

$$v_1 = L_1\frac{di_1}{dt} - M\frac{di_2}{dt} \tag{11.9-8}$$

Similarly, the voltage across the second coil is related to the coil currents by

$$v_2 = L_2\frac{di_2}{dt} - M\frac{di_1}{dt} \tag{11.9-9}$$

Thus, the mutual inductance can be seen to induce a voltage in a coil due to the current in the other coil.

Coupled inductors can be modeled using inductors (without coupling) and dependent sources. Figure 11.9-3 shows an equivalent circuit for coupled inductors.

The use of coupled inductors is usually limited to non-dc applications since coils behave as short circuits for a steady current.

Suppose that coupled inductors are part of a linear circuit with a sinusoidal input and that the circuit is at steady state. Such a circuit can be analyzed in the frequency domain, using phasors. The coupled inductors shown in Figure 11.9-2*a* are represented by the phasor equations

$$\mathbf{V}_1 = j\omega\, L_1\mathbf{I}_1 + j\omega\, M\,\mathbf{I}_2 \tag{11.9-10}$$

and

$$\mathbf{V}_2 = j\omega\, L_2\mathbf{I}_2 + j\omega\, M\,\mathbf{I}_1 \tag{11.9-11}$$

In contrast, the coupled inductors shown in Figure 11.9-2*b* are represented by the phasor equations

$$\mathbf{V}_1 = j\omega\, L_1\mathbf{I}_1 - j\omega\, M\,\mathbf{I}_2 \tag{11.9-12}$$

and

$$\mathbf{V}_2 = j\omega\, L_2\mathbf{I}_2 - j\omega\, M\,\mathbf{I}_1 \tag{11.9-13}$$

The inductances, L_1 and L_2, and mutual inductance, M, each depend on the magnetic properties and geometry of the core and the number of turns in the coils. Referring to Eqs. 11.9-4 and 11.9-5, we can write

$$L_1L_2 = (c_1N_1^2)(c_2N_2^2) = c_1c_2(N_1N_2)^2 = \left(\frac{c_M N_1 N_2}{k}\right)^2 = \frac{M^2}{k^2} \tag{11.9-14}$$

where the constant $k = c_M/\sqrt{c_1c_2}$ is called the **coupling coefficient.** Since the coupling coefficient depends on c_1, c_2, and c_M, it depends on the magnetic properties and geometry of the core. Solving Eq. 11.9-14 for the coupling coefficient gives

$$k = \frac{M}{\sqrt{L_1L_2}} \tag{11.9-15}$$

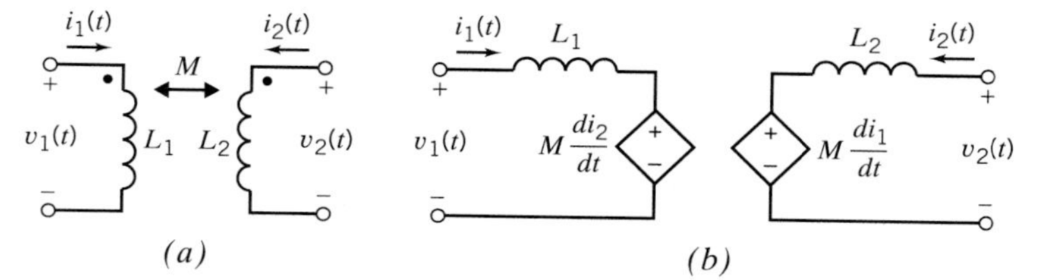

Figure 11.9-3 (*a*) Coupled inductors and (*b*) an equivalent circuit.

The instantaneous power absorbed by coupled inductors is

$$\begin{aligned} p(t) &= v_1(t)i_1(t) + v_2(t)i_2(t) \\ &= \left(L_1 \frac{d}{dt} i_1(t) \pm M \frac{d}{dt} i_2(t)\right) i_1(t) + \left(L_2 \frac{d}{dt} i_2(t) \pm M \frac{d}{dt} i_1(t)\right) i_2(t) \\ &= L_1 i_1(t) \frac{d}{dt} i_1(t) \pm M \frac{d}{dt}(i_1(t) i_2(t)) + L_2 i_2(t) \frac{d}{dt} i_2(t) \end{aligned} \tag{11.9-16}$$

where $-M$ is used if one current enters the undotted end of a coil while the other current enters the dotted end; otherwise $+M$ is used. The energy stored in the coupled inductors is calculated by integrating the power absorbed by the coupled inductors. The energy stored in coupled inductors is

$$w(t) = \int_{-\infty}^{t} p(\tau)\, d\tau = \tfrac{1}{2} L_1 i_1^2 + \tfrac{1}{2} L_2 i_2^2 \pm M\, i_1\, i_2 \tag{11.9-17}$$

where again $-M$ is used if one current enters the undotted end of a coil, while the other current enters the dotted end; otherwise $+M$ is used. We can use this equation to find how large a value M can attain in terms of L_1 and L_2. Since the transformer is a passive element, the energy stored must be greater than or equal to zero. The limiting quantity for M is obtained when $w = 0$ in Eq. 11.9-17. Then we have

$$\tfrac{1}{2} L_1 i_1^2 + \tfrac{1}{2} L_2 i_2^2 - M\, i_1\, i_2 = 0 \tag{11.9-18}$$

as the limiting condition for the case where one current enters the dotted terminal and the other current leaves the dotted terminal. Now add and subtract the term $i_1 i_2 = \sqrt{L_1 L_2}$ in the equation to generate a term that is a perfect square as follows:

$$\left(\sqrt{\frac{L_1}{2}}\, i_1 - \sqrt{\frac{L_2}{2}}\, i_2\right)^2 + i_1 i_2 \left(\sqrt{L_1 L_2} - M\right) = 0$$

The perfect square term can be positive or zero. Therefore, in order to have $w \geq 0$, we require that

$$\sqrt{L_1 L_2} \geq M \tag{11.9-19}$$

Thus, the maximum value of M is $\sqrt{L_1 L_2}$.

Therefore, the coupling coefficient of passive coupled inductors can be no larger than 1. In addition, the coupling coefficient cannot be negative since L_1, L_2, and M are all non-negative. When $k = 0$, no coupling exists. Therefore, the coupling coefficient must satisfy

$$0 \leq k \leq 1 \tag{11.9-20}$$

Most power system transformers have a k that approaches one, while k is low for radio circuits.

Figure 11.9-4a shows coupled inductors used as a transformer to connect a source to a load. The coil connected to the source is called the **primary coil** and the coil connected to the load is called the **secondary coil.** Circuit #2 is connected to Circuit #1 through the magnetic coupling of the transformer, but there is no electrical connection between these two circuits. For example, there is no path for current to flow from Circuit #1 to Circuit #2. In addition, no circuit element is incident to one node in Circuit #1 and incident to another node in Circuit #2.

Figure 11.9-4b shows a specific example of the situation shown in Figure 11.9-4a. The source is a single sinusoidal voltage source, and the load is a single impedance. The circuit has been represented in the frequency domain, using phasors and impedances. The

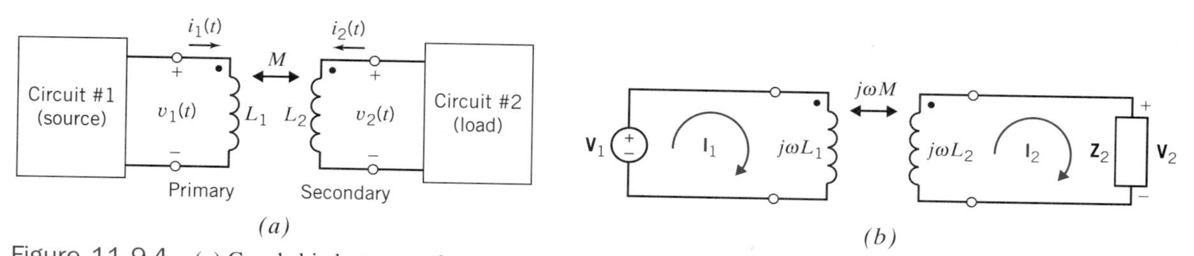

Figure 11.9-4 (*a*) Coupled inductors used as a transformer to magnetically couple two circuits and (*b*) a transformer used to magnetically couple a voltage source to an impedance.

circuit in Figure 11.9-4*b* can be analyzed by writing mesh equations. The two mesh equations are

$$j\omega L_1\mathbf{I}_1 - j\omega M\,\mathbf{I}_2 = \mathbf{V}_1$$
$$-j\omega M\,\mathbf{I}_1 + (j\omega L_2 + \mathbf{Z}_2)\,\mathbf{I}_2 = 0$$

Solving for $\mathbf{I}_2$ in terms of $\mathbf{V}_1$, we have

$$\mathbf{I}_2 = \left[\frac{j\omega M}{((j\omega)^2(L_1L_2 - M^2) + (j\omega L_1\mathbf{Z}_2))}\right]\mathbf{V}_1 \tag{11.9-21}$$

When the coupling coefficient of the coupled inductors is unity, then $M = \sqrt{L_1L_2}$ and Eq. 11.9-21 reduces to

$$\mathbf{I}_2 = \left[\frac{j\omega M}{j\omega L_1\mathbf{Z}_2}\right]\mathbf{V}_1 = \left[\frac{j\omega\sqrt{L_1L_2}}{j\omega L_1\mathbf{Z}_2}\right]\mathbf{V}_1 = \frac{\sqrt{L_2}}{\mathbf{Z}_2\sqrt{L_1}}\mathbf{V}_1 \tag{11.9-22}$$

The voltage across the impedance is given by

$$\mathbf{V}_2 = \mathbf{Z}_2\mathbf{I}_2 = \sqrt{\frac{L_2}{L_1}}\mathbf{V}_1 \tag{11.9-23}$$

The ratio of the inductances is related to the magnetic properties and geometry of the core and the number of turns in the coils. Referring to Eq. 11.9-4, we can write

$$\frac{L_2}{L_1} = \frac{c_2N_2^2}{c_1N_1^2}$$

When both coils are wound symmetrically on the same core, then $c_1 = c_2$. In this case,

$$\frac{L_2}{L_1} = \frac{N_2^2}{N_1^2} = n^2 \tag{11.9-24}$$

where n is called the **turns ratio** of the transformer. Combining Eqs. 11.9-23 and 11.9-24 gives

$$\mathbf{V}_2 = n\mathbf{V}_1 \tag{11.9-25}$$

where $\mathbf{V}_1$ is the voltage across the primary coil, $\mathbf{V}_2$ is the voltage across the secondary coil, and n is the turns ratio. A similar calculation shows that

$$\mathbf{I}_1 = -n\mathbf{I}_2 \tag{11.9-26}$$

where $\mathbf{I}_1$ is the current in the primary coil and $\mathbf{I}_2$ is the current in the secondary coil. Equations 11.9-25 and 11.9-26 represent the ideal transformer. The ideal transformer is the subject of the next section of this chapter. We conclude that a transformer is an ideal transformer when the coupling coefficient of the transformer coils is unity and both coils are wound symmetrically on the same core so that $c_1 = c_2$.

Example 11.9-1

Find the voltage $v_2(t)$ in the circuit as shown in Figure 11.9-5*a*. Is the transformer an ideal transformer?

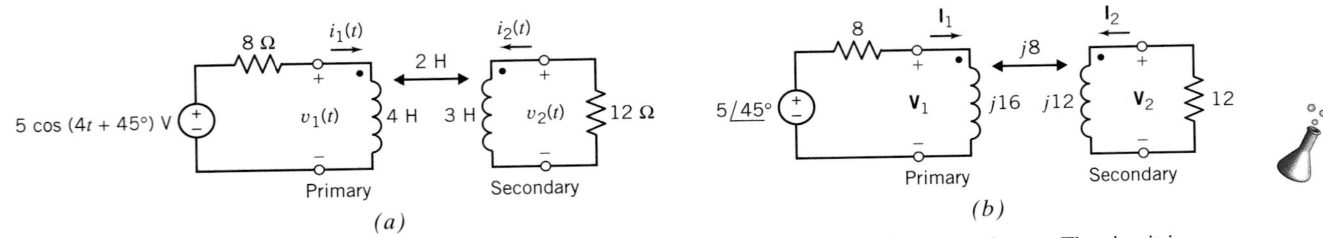

Figure 11.9-5 A circuit in which coupled inductors are used as a transformer. The circuit is represented (*a*) in the time domain and (*b*) in the frequency domain, using phasors and impedances.

Solution

First, represent the circuit in the frequency domain, using phasors and impedances, as shown in Figure 11.9-5*b*. Notice that the coil currents, $\mathbf{I}_1$ and $\mathbf{I}_2$, both enter the dotted end of the coils. Express the coil voltages as functions of the coil currents using the equations that describe the coupled inductors, Eqs. 11.9-14 and 11.9-15.

$$\mathbf{V}_1 = j16\,\mathbf{I}_1 + j8\,\mathbf{I}_2$$
$$\mathbf{V}_2 = j8\,\mathbf{I}_1 + j12\,\mathbf{I}_2$$

Next, write two mesh equations

$$5\angle 45^\circ = 8\,\mathbf{I}_1 + \mathbf{V}_1$$

and

$$\mathbf{V}_2 = -12\,\mathbf{I}_2$$

Substituting the equations for the coil voltages into the mesh equations gives

$$5\angle 45^\circ = 8\,\mathbf{I}_1 + (j16\,\mathbf{I}_1 + j8\,\mathbf{I}_2) = (8 + j16)\,\mathbf{I}_1 + j8\,\mathbf{I}_2$$

and

$$j8\,\mathbf{I}_1 + j12\,\mathbf{I}_2 = -12\,\mathbf{I}_2$$

Solving for $\mathbf{I}_2$ gives

$$\mathbf{I}_2 = 0.138\angle -141^\circ \text{ A}$$

Next, $\mathbf{V}_2$ is given by

$$\mathbf{V}_2 = -12\,\mathbf{I}_2 = 1.656\angle 39^\circ \text{ V}$$

Returning to the time domain,

$$v_2(t) = 1.656\cos(4t + 39^\circ) \text{ V}$$

The transformer is not ideal since $L_1 L_2 = 4 \cdot 3 > 2^2 = M^2$.

Exercise 11.9-1 Determine the voltage v_o for the circuit of Figure E 11.9-1.

Hint: Write a single mesh equation. The currents in the two coils are equal to each other and equal to the mesh current.

Answer: $v_o = 14\cos 4t$ V

Exercise 11.9-2 Determine the voltage v_o for the circuit of Figure E 11.9-2.

Hint: This exercise is the same as Exercise 11.9-1, except for the position of the dot on the vertical coil.

Answer: $v_o = 18 \cos 4t$ V

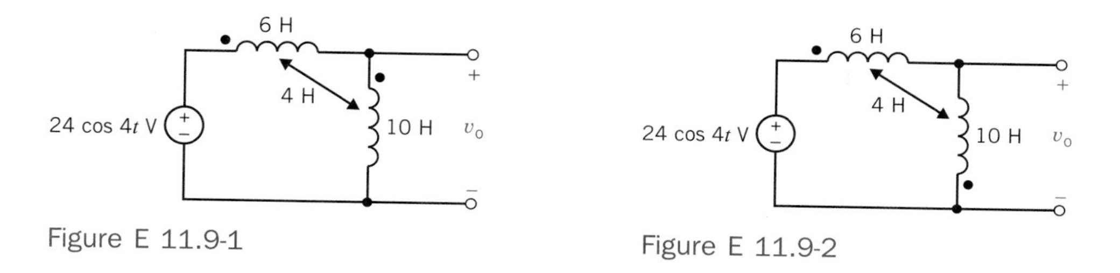

Figure E 11.9-1

Figure E 11.9-2

Exercise 11.9-3 Determine the current i_o for the circuit of Figure E 11.9-3.

Hint: The voltage across the vertical coil is zero because of the short circuit. The voltage across the horizontal coil induces a current in the vertical coil. Consequently, the current in the vertical coil is not zero.

Answer: $i_o = 1.909 \cos(4t - 90°)$ A

Exercise 11.9-4 Determine the current i_o for the circuit of Figure E 11.9-4.

Hint: This exercise is the same as Exercise 11.9-3, except for the position of the dot on the vertical coil.

Answer: $i_o = 0.818 \cos(4t - 90°)$ A

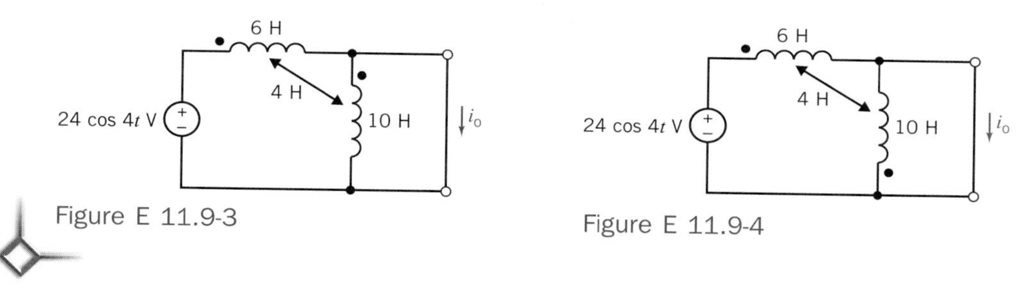

Figure E 11.9-3

Figure E 11.9-4

11.10 THE IDEAL TRANSFORMER

One major use of transformers is in ac power distribution. Transformers possess the ability to "step up" or "step down" ac voltages or currents. Transformers are used by power utilities to raise (step up) the voltage from 10 kV at a generating plant to 200 kV or higher for transmission over long distances. Then, at a receiving plant, transformers are used to reduce (step down) the voltage to 220 or 110 V for use by the customer (Coltman 1988).

In addition to power systems, transformers are commonly used in electronic and communication circuits. They provide the ability to raise or reduce voltages and to isolate one circuit from another.

One of the coils, typically drawn on the left of the diagram of a transformer, is designated as the *primary coil,* and the other is called the *secondary coil* or winding. The primary coil is connected to the energy source, and the secondary coil is connected to the load.

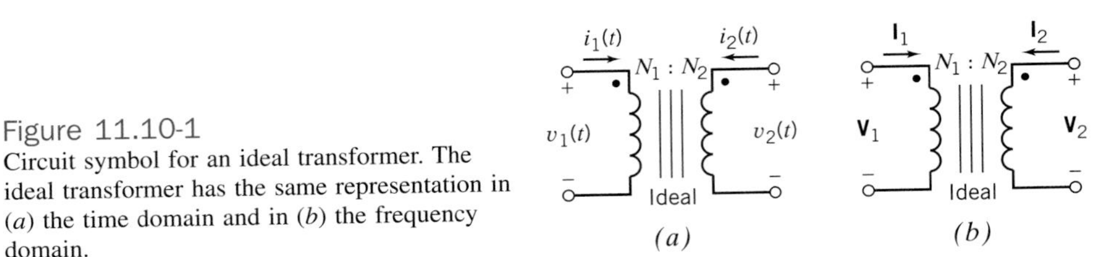

Figure 11.10-1
Circuit symbol for an ideal transformer. The ideal transformer has the same representation in (*a*) the time domain and in (*b*) the frequency domain.

An **ideal transformer** is a model of a transformer with a coupling coefficient equal to unity. These characteristics are usually present for specially designed iron-core transformers. They approximate the ideal transformer for a range of frequencies.

The symbol for the ideal transformer is shown in Figure 11.10-1. The operation of the ideal transformer is the same in the time domain as in the frequency domain. The time domain representation of the transformer is shown in Figure 11.10-1*a*. In the time domain, the two defining equations for an ideal transformer are

$$v_2(t) = nv_1(t) \tag{11.10-1}$$

and

$$i_1(t) = -ni_2(t) \tag{11.10-2}$$

where $n = N_2/N_1$ is called the *turns ratio* of the transformer.

The use of transformers is usually limited to non-dc applications since the primary and secondary windings behave as short circuits for a steady current.

The frequency domain representation of the transformer is shown in Figure 11.10-1*b*. In the frequency domain, the two defining equations for an ideal transformer are

$$\mathbf{V}_2 = n\mathbf{V}_1 \tag{11.10-3}$$

and

$$\mathbf{I}_1 = -n\mathbf{I}_2 \tag{11.10-4}$$

The vertical bars in Figure 11.10-1 indicate the iron core, and we write ideal with the transformer to ensure recognition of the ideal case. An ideal transformer can be modeled using dependent sources as shown in Figure 11.10-2.

Notice that the voltage and current of both coils of the transformer in Figure 11.10-1 adhere to the passive convention. The instantaneous power absorbed by the ideal transformer is

$$p(t) = v_1(t)i_1(t) + v_2(t)i_2(t) = v_1(t)(-ni_2(t)) + (nv_1(t))i_2(t) = 0 \tag{11.10-5}$$

The ideal transformer is said to be *loss*less since instantaneous power absorbed by it is zero. A similar argument shows that the ideal transformer absorbs zero complex power, zero average power, and zero reactive power.

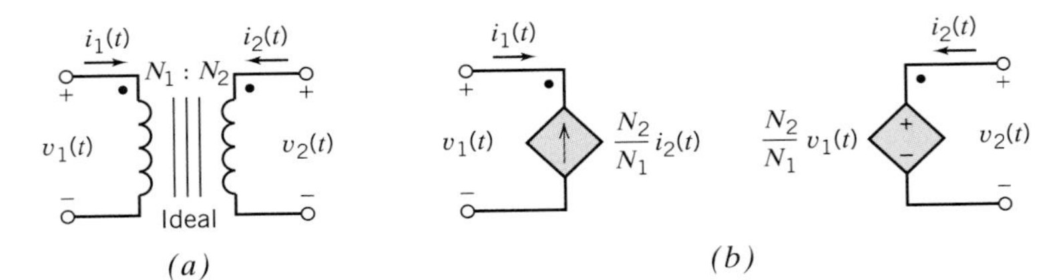

Figure 11.10-2 (*a*) Ideal transformer and (*b*) an equivalent circuit.

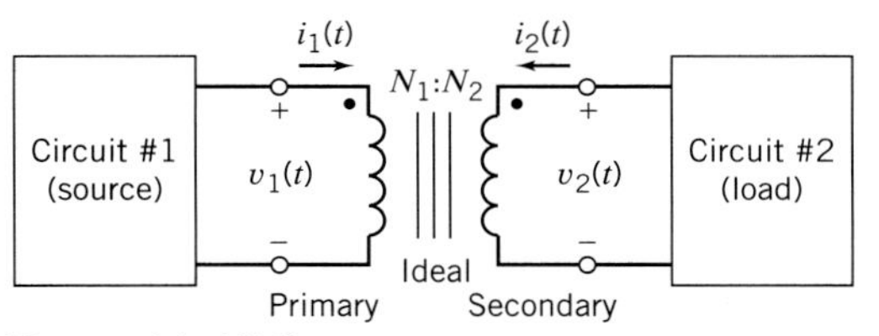

Figure 11.10-3 An ideal transformer used to magnetically couple two circuits.

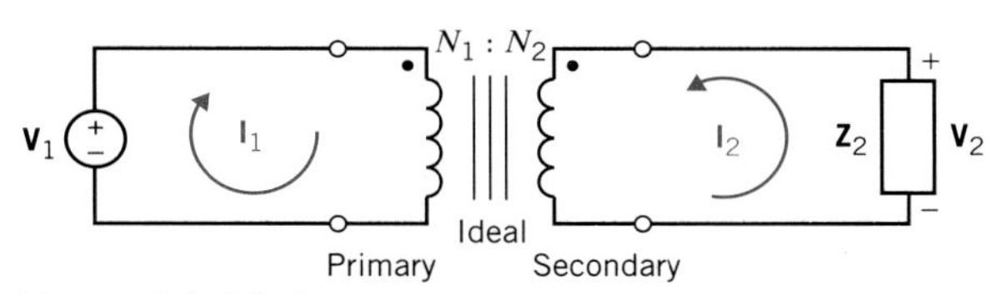

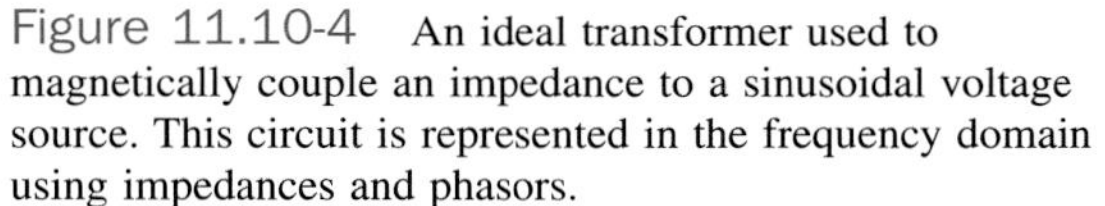
Figure 11.10-4 An ideal transformer used to magnetically couple an impedance to a sinusoidal voltage source. This circuit is represented in the frequency domain using impedances and phasors.

Figure 11.10-3 shows an ideal transformer that is used to connect a source to a load. The coil connected to the source is called the primary coil, and the coil connected to the load is called the secondary coil. Circuit #2 is connected to Circuit #1 through the magnetic coupling of the transformer, but there is no electrical connection between these two circuits. Because the ideal transformer is lossless, all of the power delivered to the ideal transformer by Circuit #1 is in turn delivered to Circuit #2 by the ideal transformer.

Let us consider the circuit of Figure 11.10-4, which has a load impedance $\mathbf{Z}_2$ magnetically coupled to a voltage source using an ideal transformer.

The input impedance of the circuit connected to the voltage source is

$$\mathbf{Z}_1 = \frac{\mathbf{V}_1}{\mathbf{I}_1}$$

$\mathbf{Z}_1$ is called the impedance seen at the primary of the transformer or the impedance seen by the voltage source.

The transformer is represented by the equations

$$\mathbf{V}_1 = \mathbf{V}_2/n$$

and

$$\mathbf{I}_1 = -n\,\mathbf{I}_2$$

where $n = N_2/N_1$ is the turns ratio of the transformer.

The current and voltage of the impedance, $\mathbf{I}_2$ and $\mathbf{V}_2$, do not adhere to the passive convention so

$$\mathbf{V}_2 = -\mathbf{Z}_2\,\mathbf{I}_2$$

Therefore, for $\mathbf{Z}_1$ we have

$$\mathbf{Z}_1 = \frac{\mathbf{V}_2/n}{-n\,\mathbf{I}_2} = \frac{1}{n^2}\left(-\frac{\mathbf{V}_2}{\mathbf{I}_2}\right) = \frac{1}{n^2}\,\mathbf{Z}_2$$

The source experiences the impedance $\mathbf{Z}_1$, which is equal to $\mathbf{Z}_2$ scaled by the factor $1/n^2$. We sometimes say that $\mathbf{Z}_1$ is the impedance $\mathbf{Z}_2$ reflected to the primary of the transformer.

Suppose we are going to connect a load impedance to a source. If we connect the load impedance directly to the source, then the source sees the load impedance $\mathbf{Z}_2$. In contrast, if we connect the load impedance to the source using an ideal transformer, the source sees the impedance $\mathbf{Z}_1$. In this context, we say that the transformer has changed the impedance seen by the source from $\mathbf{Z}_2$ to $\mathbf{Z}_1$.

Example 11.10-1

Often we can use an ideal transformer to represent a transformer that connects the output of a stereo amplifier, $\mathbf{V}_1$, to a stereo speaker, as shown in Figure 11.10-5. Find the re-

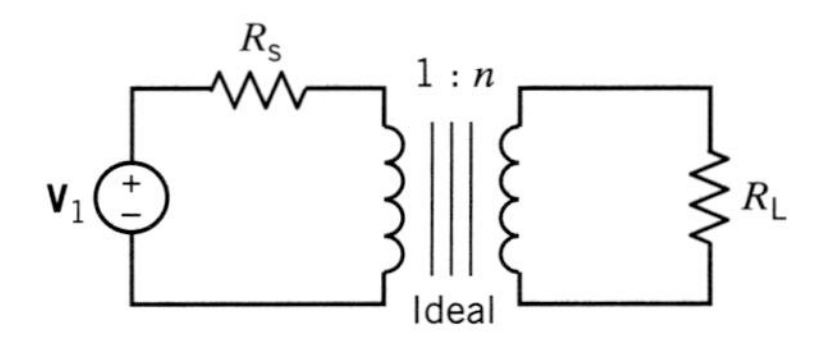

Figure 11.10-5
Output of an amplifier connected to a stereo speaker with resistance R_L.

quired turns ratio n when $R_L = 8\ \Omega$ and $R_s = 48\ \Omega$ if we wish to achieve maximum power transfer to the load.

Solution

The impedance seen at the primary due to R_L is

$$\mathbf{Z}_1 = \frac{R_L}{n^2} = \frac{8}{n^2}$$

To achieve maximum power transfer, we require that

$$\mathbf{Z}_1 = R_s$$

Since $R_s = 48\ \Omega$, we require that $\mathbf{Z}_1 = 48\ \Omega$, so

$$n^2 = \frac{8}{48} = \frac{1}{6}$$

and, therefore,

$$\left(\frac{N_2}{N_1}\right)^2 = \frac{1}{6}$$

or

$$N_1 = \sqrt{6}N_2$$

Exercise 11.10-1 Find $\mathbf{V}_1$ and $\mathbf{I}_1$ for the circuit of Figure E 11.10-1 when $n = 5$.

Exercise 11.10-2 Determine v_2 and i_2 for the circuit shown in Figure E 11.10-2 when $n = 2$. Note that i_2 does not enter the dotted terminal.

Answer: $v_2 = 0.68 \cos(10t + 47.7°)$ V
$i_2 = 0.34 \cos(10t + 42°)$ A

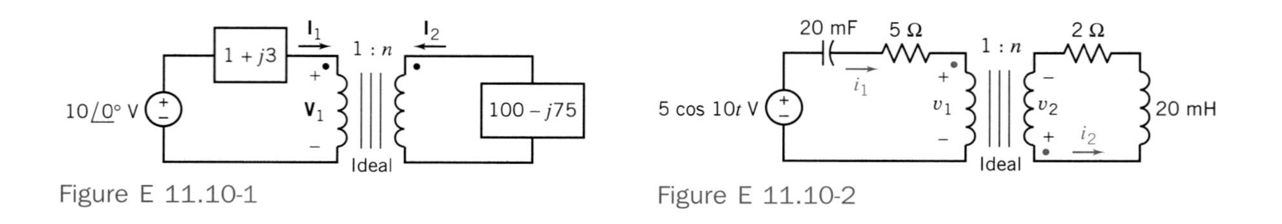

Figure E 11.10-1

Figure E 11.10-2

Exercise 11.10-3 Determine the impedance $\mathbf{Z}_{ab}$ for the circuit of Figure E 11.10-3. All the transformers are ideal.

Answer: $\mathbf{Z}_{ab} = 4.063\mathbf{Z}$

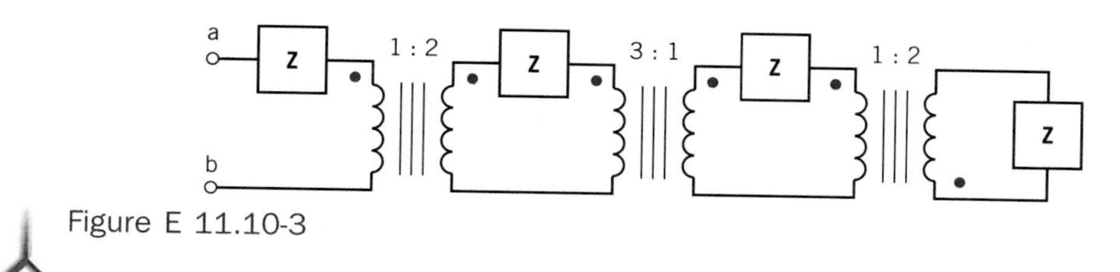

Figure E 11.10-3

11.11 TRANSFORMER CIRCUIT ANALYSIS USING PSPICE

Transformer circuits may be modeled using PSpice by using the coupled inductor statement as

K *<name>* L *<name>* L *<name>* *<coupling k>*

Therefore, the mutual inductor statement for the transformer shown in Figure 11.11-1 is

K1 L1 L2 0.5

where the coefficient of coupling is 0.5. This statement is used to couple the two individual inductors that model the transformer coils. Recall that the current must flow into the dotted terminals of the transformer as shown in Figure 11.11-1. Therefore, the dotted node of the inductor is listed first.

Consider the circuit shown in PSpice format in Figure 11.11-2. The source is $v_s = 110 \cos 10t$ V, and the coupling coefficient is 0.5. The dummy resistor R_d is included to provide an *electrical* connection between the coils of the transformer. (PSpice requires a dc path to ground from every node in the circuit.) The dummy resistance R_d will be set equal to 100 MΩ to approximate an open circuit. Then we obtain the PSpice program as shown in Figure 11.11-3.

Let us consider the transformer circuit shown in Figure 11.11-4, where $\omega = 1000$ rad/s, $R = 1\ \Omega$, and $R_L = 100\ \Omega$. For this transformer

$$k = \frac{M}{\sqrt{L_1 L_2}} = 1$$

Therefore, the coupling coefficient is equal to 1, and the transformer is an ideal transformer. The PSpice program to calculate $\mathbf{V}_2$ is shown in Figure 11.11-5. Note that we use a coupling coefficient of 0.999999 since PSpice requires that $0 \le k < 1$.

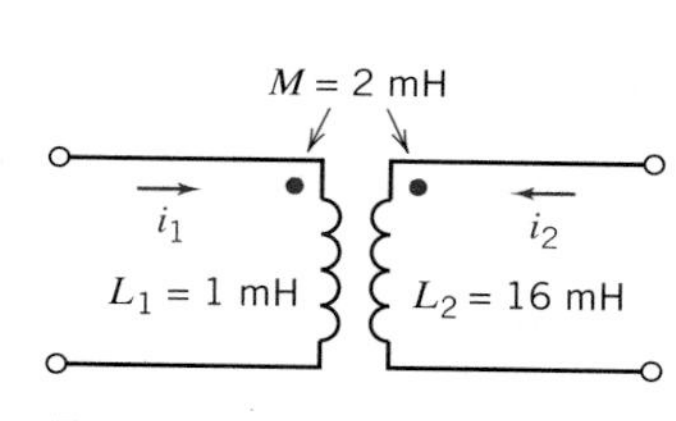

Figure 11.11-1

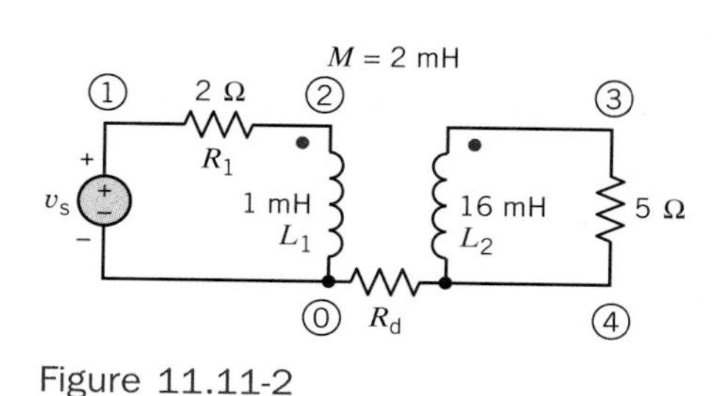

Figure 11.11-2

```
TRANSFORMER CIRCUIT
VS    1   0   AC   110   0
R1    1   2   2
L1    2   0   1M
L2    3   4   16M
R2    3   4   5
K1    L1   L2   0.5
RD    4   0   100 MEG
.AC   LIN   1   1.591   1.591
.PRINT   AC   VM (3,4)   VP (3,4)
.END
```

Figure 11.11-3 The PSpice program for the circuit of Figure 11.11-2.

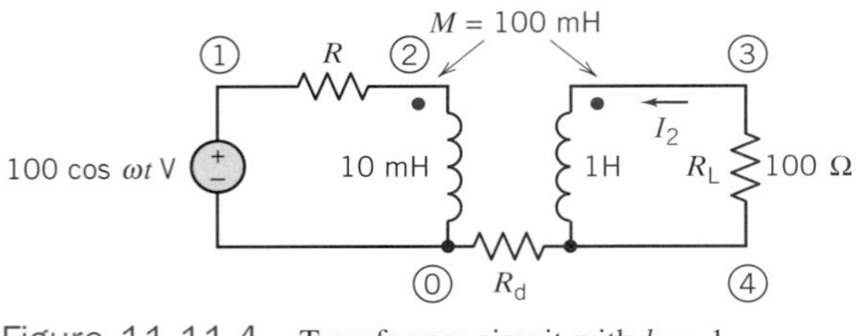

Figure 11.11-4 Transformer circuit with $k = 1$.

```
TRANSFORMER
VI   1   0   AC   100   0
R1   1   2   1
L1   2   0   10M
L2   3   4   1
RL   4   3   100
RD   4   0   1G
K1   L1   L2   .999999
.AC   LIN   1   1592   1592
.PRINT   AC   VM (3,4)   VP (3,4)
.END
```

Figure 11.11-5 PSpice program with $R = 1\ \Omega$ and $\omega = 10{,}000$ rad/s.

The ideal transformer is obtained when $k = 1$ and the reactances of the inductances are much greater than the other circuit impedances. The circuit of Figure 11.11-4 has $k = 1$ and will closely satisfy the impedance requirements when ω is increased to 10,000 rad/s and $R = 1\ \Omega$. Then, $\omega L_1 = 100$ and $\omega L_2 = 10{,}000$. Thus, we have $\omega L_1 \gg R$ and $\omega L_2 \gg R_L$ and the transformer appears to be ideal. If we rewrite the PSpice program for these conditions, we have the program shown in Figure 11.11-5. The output of this program indicates that the voltage across L_1 is

$$\mathbf{V}_1 = 50.01\angle 0.9^\circ$$

and that the voltage across L_2 is

$$\mathbf{V}_2 = 499.9\angle -0.7^\circ$$

For a perfectly ideal transformer, the output voltage across R_L is given by

$$\begin{aligned}\mathbf{V}_2 &= n\mathbf{V}_1\\ &= 10\mathbf{V}_1\end{aligned}$$

since $n^2 = L_2/L_1 = 100$ and $n = 10$. Thus, we see that the actual transformer acts as a nearly ideal transformer when $\omega = 10{,}000$ and the impedance of each inductance is large relative to the other circuit impedances.

The ideal transformer is a model of a transformer with a coupling coefficient equal to one and with primary and secondary reactances very large compared to the impedances

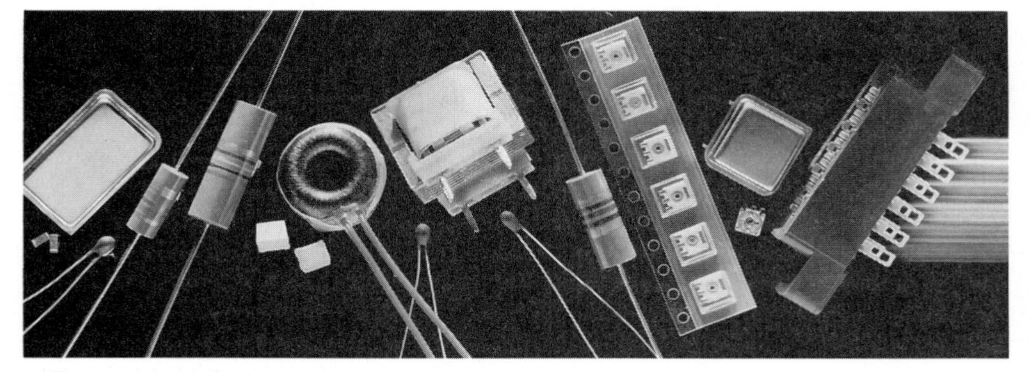

Figure 11.11-6 Components shown include oscillators, inductors, transformers, thermistors, chip potentiometers, and connectors. (Courtesy of Dale-Vishay Inc.)

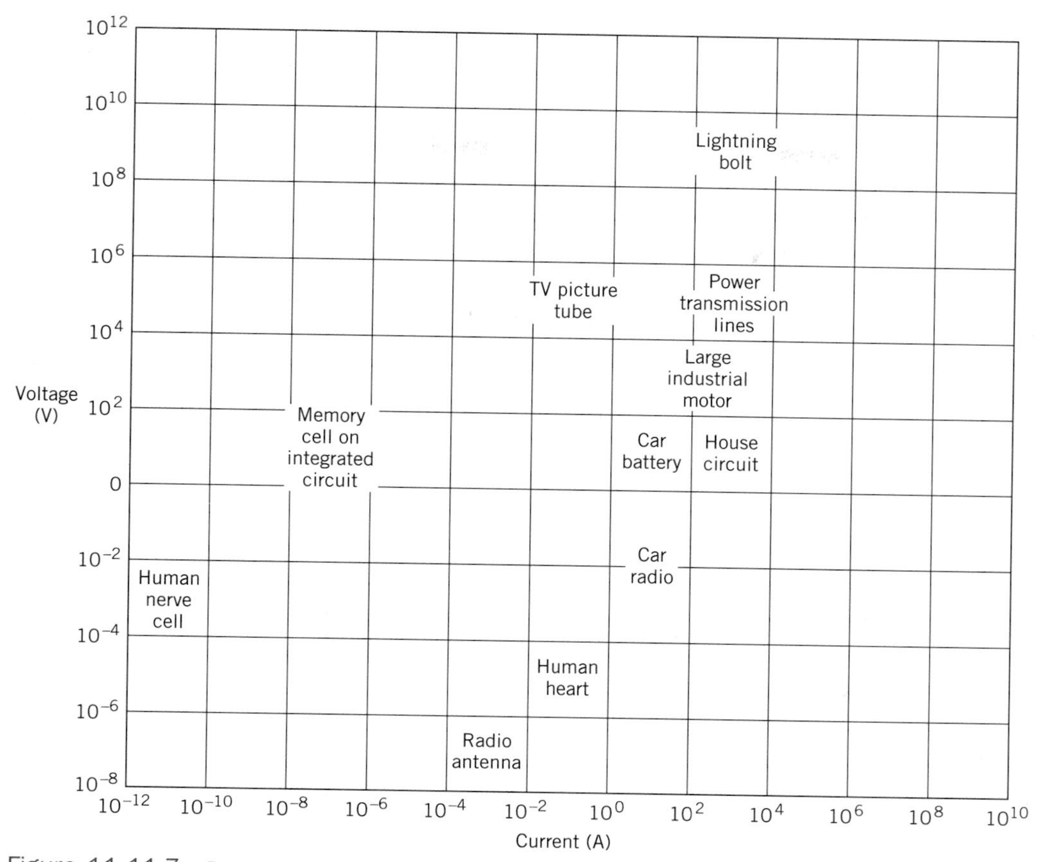

Figure 11.11-7 Power levels for selected electrical devices or phenomena.

connected to the terminals of the transformer. This ideal transformer model fits many practical applications in ac circuits. Various components including transformers are shown in Figure 11.11-6. The power levels for selected electrical devices or phenomena are shown in Figure 11.11-7.

11.12 VERIFICATION EXAMPLE

The circuit shown in Figure 11.12-1*a* has been analyzed using a computer, and the results are tabulated in Figure 11.12-1*b*. The labels Xp and Xs refer to the primary and secondary of the transformer. The passive convention is used for all elements, including the voltage sources, which means that

$$\frac{(30)(1.76)}{2}\cos(133^\circ - 0) = -18.00$$

is the average power *absorbed* by the voltage source. The average power *supplied* by the voltage source is $+18.00$ W.

How can we verify that the computer analysis of this circuit is indeed correct?

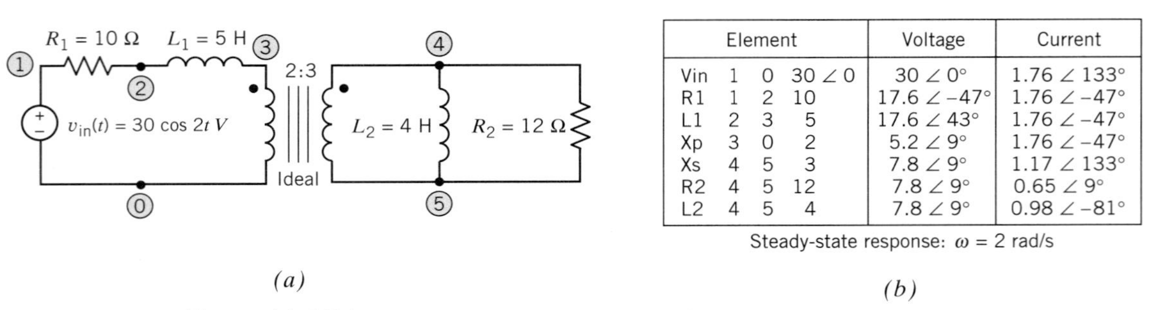

Element				Voltage	Current
Vin	1	0	30 ∠ 0	30 ∠ 0°	1.76 ∠ 133°
R1	1	2	10	17.6 ∠ −47°	1.76 ∠ −47°
L1	2	3	5	17.6 ∠ 43°	1.76 ∠ −47°
Xp	3	0	2	5.2 ∠ 9°	1.76 ∠ −47°
Xs	4	5	3	7.8 ∠ 9°	1.17 ∠ 133°
R2	4	5	12	7.8 ∠ 9°	0.65 ∠ 9°
L2	4	5	4	7.8 ∠ 9°	0.98 ∠ −81°

Steady-state response: $\omega = 2$ rad/s

(*a*) (*b*)

Figure 11.12-1 (*a*) A circuit and (*b*) the results from computer analysis for the circuit.

Solution

There are several things that can be easily checked.

1. The element current and voltage of each inductor should be 90° out of phase with each other so that the average power delivered to each inductor is zero. The element current and voltage of both L_1 and L_2 satisfy this condition.
2. An ideal transformer absorbs zero average power. The sum of the average power absorbed by the transformer primary and the secondary is

$$\frac{(5.2)(1.76)}{2}\cos(9° - (-47°)) + \frac{(7.8)(1.17)}{2}\cos(133° - 9°) = 2.56 + (-2.55) \approx 0 \text{ W}$$

so this condition is satisfied.

3. All of the power delivered to the primary of the transformer is in turn delivered to the load. In this example, the load consists of the inductor L_2 and the resistor R_2. Since the average power delivered to the inductor is zero, all the power delivered to the transformer primary should be delivered by the secondary to the resistor R_2. The power delivered to the transformer primary is

$$\frac{(5.2)(1.76)}{2}\cos(9° - (-47°)) = 2.56 \text{ W}$$

The power delivered to R_2 is

$$\frac{(7.8)(0.65)}{2}\cos(0) = 2.53 \text{ W}$$

There seems to be some roundoff error in the voltages and currents provided by the computer. Nonetheless, it seems reasonable to conclude that all the power delivered to the transformer primary is delivered by the secondary to the resistor R_2.

4. The average power supplied by the voltage source should be equal to the average power absorbed by the resistors. We have already calculated that the average power delivered by the voltage source is 18 W. The average power absorbed by the resistors is

$$\frac{(17.6)(1.76)}{2}\cos(0) + \frac{(7.8)(0.65)}{2}\cos(0) = 15.49 + (2.53) = 18.02 \text{ W}$$

so this condition is satisfied.

Since these four conditions are satisfied, we are confident that the computer analysis of the circuit is correct.

11.13 Design Challenge Solution

MAXIMUM POWER TRANSFER

The matching network in Figure 11.13-1 is used to interface the source with the load, which means that the matching network is used to connect the source to the load in a desirable way. In this case, the purpose of the matching network is to transfer as much power as possible to the load. This problem occurs frequently enough that it has been given a name, the maximum power transfer problem.

An important example of the application of maximum power transfer is the connection of a cellular phone or wireless radio transmitter to the cell's antenna. For example, the input impedance of a practical cellular telephone antenna is $\mathbf{Z} = (10 + j6.28)\ \Omega$ (Dorf 1998).

Describe the Situation and the Assumptions

The input voltage is a sinusoidal function of time. The circuit is at steady state. The matching network is to be designed to deliver as much power as possible to the load.

State the Goal

To achieve maximum power transfer, the matching network should "match" the load and source impedances. The source impedance is

$$\mathbf{Z}_S = R_S + j\omega L_S = 1 + j(2 \cdot \pi \cdot 10^5)(10^{-6}) = 1 + j0.628\ \Omega$$

For maximum power transfer, the impedance $\mathbf{Z}_{\text{in}}$, shown in Figure 11.13-2, must be the complex conjugate of $\mathbf{Z}_s$. That is,

$$\mathbf{Z}_{\text{in}} = \mathbf{Z}_S^* = 1 - j0.628$$

Generate a Plan

Let us use a transformer for the matching network as shown in Figure 11.13-3. The impedance $\mathbf{Z}_{\text{in}}$ will be a function of n, the turns ratio of the transformer. We will set $\mathbf{Z}_{\text{in}}$ equal to the complex conjugate of $\mathbf{Z}_s$ and solve the resulting equation to determine the turns ratio, n.

Act on the Plan

$$\mathbf{Z}_{\text{in}} = \frac{1}{n^2}(R + j\omega L) = \frac{1}{n^2}(10 + j6.28)$$

We require that

$$\frac{1}{n^2}(10 + j6.28) = 1 - j0.628$$

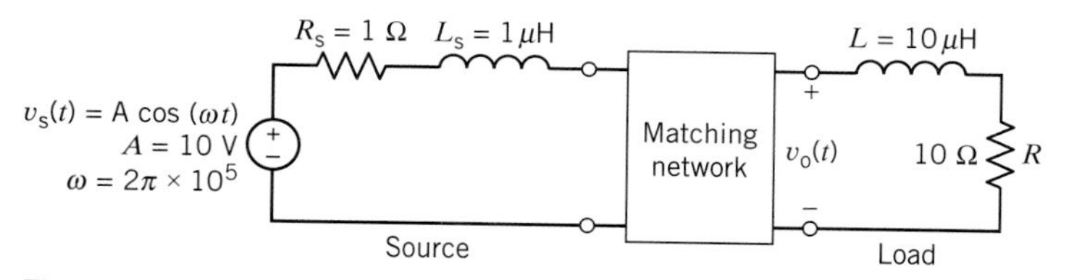

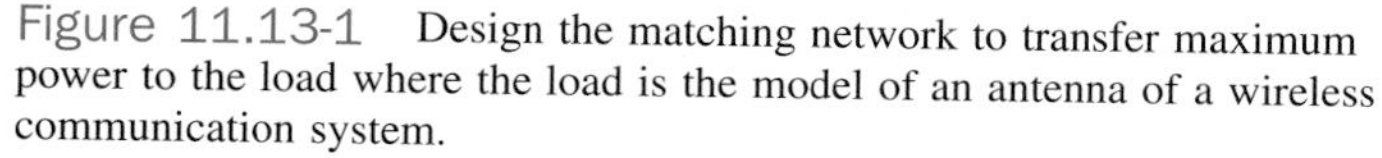
Figure 11.13-1 Design the matching network to transfer maximum power to the load where the load is the model of an antenna of a wireless communication system.

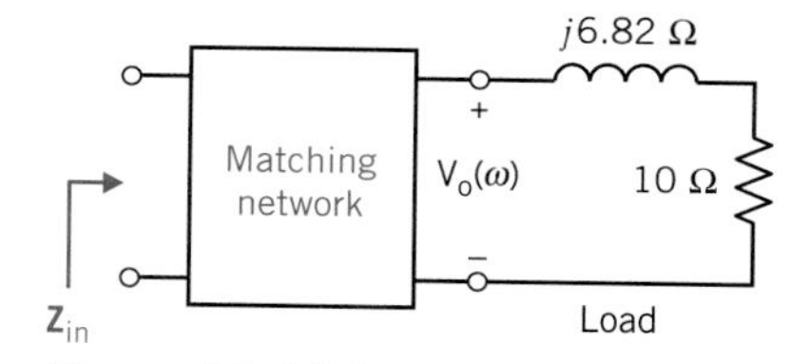

Figure 11.13-2 Z_{in} is the impedance seen looking into the matching network.

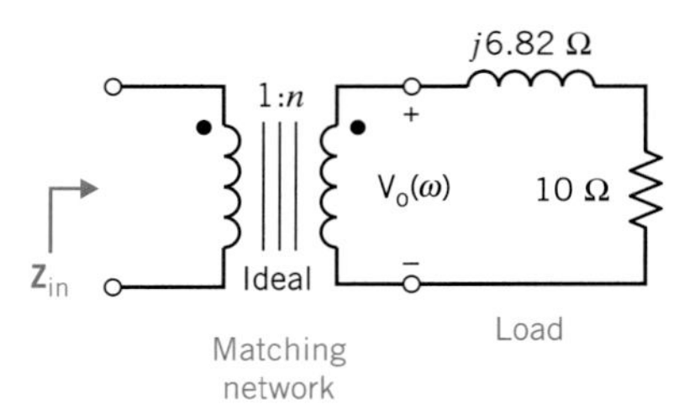

Figure 11.13-3 Using an ideal transformer as the matching network.

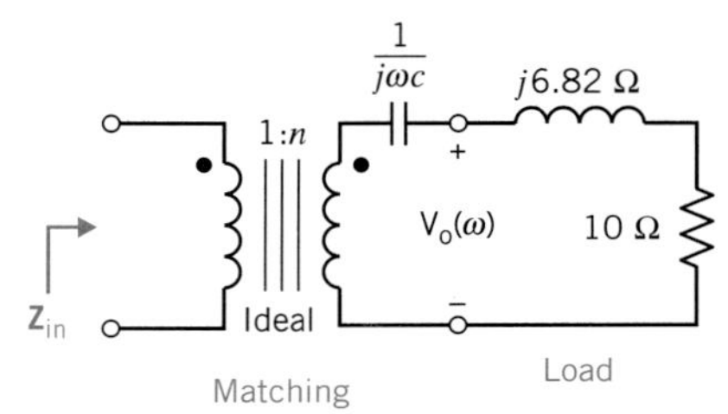

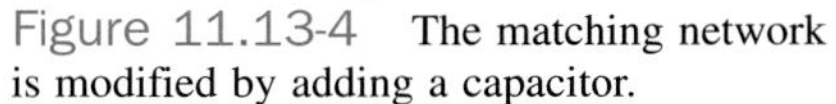

Figure 11.13-4 The matching network is modified by adding a capacitor.

This requires both

$$\frac{1}{n^2}10 = 1 \tag{11.13-1}$$

and

$$\frac{1}{n^2}\,6.28 = -0.628 \tag{11.13-2}$$

Selecting

$$n = 3.16$$

(e.g., $N_2 = 158$ and $N_1 = 50$) satisfies Eq. 11.13-1 but not 11.13-2. Indeed, no positive value of n will satisfy Eq. 11.13-2.

We need to modify the matching network to make the imaginary part of $\mathbf{Z}_{\text{in}}$ negative. This can be accomplished by adding a capacitor as shown in Figure 11.13-4. Then

$$\mathbf{Z}_{\text{in}} = \frac{1}{n^2}\left(R + j\omega L - j\frac{1}{\omega C}\right) = \frac{1}{n^2}\left(10 + j6.28 - j\frac{1}{2\pi \cdot 10^5 \cdot C}\right)$$

We require that

$$\frac{1}{n^2}\left(10 + j6.28 - j\frac{1}{2\pi \cdot 10^5 \cdot C}\right) = 1 - j0.628$$

This requires both

$$\frac{1}{n^2}10 = 1 \tag{11.13-3}$$

and

$$\frac{1}{n^2}\left(6.28 - \frac{1}{2\pi \cdot 10^5 \cdot C}\right) = -0.628 \tag{11.13-4}$$

First, solving Eq. 11.13-3 gives

$$n = 3.16$$

Next, solving Eq. 11.13-4 gives

$$C = 0.1267\ \mu\text{F}$$

and the design is complete.

Verify the Proposed Solution

When $n = 3.16$ and $C = 0.1267\ \mu$F, the input impedance of the matching network is

$$
\begin{aligned}
\mathbf{Z}_{\text{in}} &= \frac{1}{n^2}\left(R + j\omega L + \frac{1}{j\omega C}\right) \\
&= \frac{1}{3.16^2}\left(10 + j(2\pi \times 10^5)(10^{-5}) + \frac{1}{j(2\pi \times 10^5)(0.1267 \times 10^{-6})}\right) \\
&= 1 - j0.629
\end{aligned}
$$

as required.

11.14 SUMMARY

- With the adoption of ac power as the generally used conventional power for industry and the home, engineers became involved in analyzing ac power relationships.
- The instantaneous power delivered to this circuit element is the product of the element voltage and current. Let $v(t)$ and $i(t)$ be the element voltage and current, chosen to adhere to the passive convention. Then $p(t) = v(t)\,i(t)$ is the instantaneous power delivered to this circuit element. Instantaneous power is calculated in the time domain.
- The instantaneous power can be a quite complicated function of t. When the element voltage and current are periodic functions having the same period, T, it is convenient to calculate the average power $P = \frac{1}{T}\int_{t_0}^{t_0+T} i(t)\,v(t)\,dt$.
- The effective value of a current is the constant (dc) current that delivers the same average power to a 1 Ω resistor as the given varying current. The effective value of a voltage is the constant (dc) voltage that delivers the same average power as the given varying voltage.
- Consider a linear circuit with a sinusoidal input that has reached steady state. All the element voltages and currents will be sinusoidal with the same frequency as the input. Such a circuit can be analyzed in the frequency domain, using phasors and impedances. Indeed, we can calculate the power generated or absorbed in a circuit or in any element of a circuit, in the frequency domain, using phasors. Table 11.5-1 summarizes the equations used to calculate average power, complex power, or reactive power in the frequency domain.
- Since it is important to keep the current I as small as possible in the transmission lines, engineers strive to achieve a power factor close to one. The power factor is equal to $\cos\theta$, where θ is the phase angle difference between the sinusoidal steady-state load voltage and current. A purely reactive impedance in parallel with the load is used to correct the power factor.
- Finally, we considered the coupled coils and transformers. Coupled inductors and transformers exhibit mutual inductance, which relates the voltage in one coil to the change in current in another coil. The equations that describe coupled coils and transformers are collected in Tables 11.14-1 and 11.14-2.

Table 11.14-1 **Coupled Inductors**

Device Symbol (including reference directions of element voltages and currents)	Device Equations in the Time Domain	Device Equations in the Frequency Domain
$i_1(t)$, $i_2(t)$, M, L_1, L_2, $v_1(t)$, $v_2(t)$ (dots at top of both coils)	$v_1 = L_1\frac{di_1}{dt} + M\frac{di_2}{dt}$ $v_2 = L_2\frac{di_2}{dt} + M\frac{di_1}{dt}$	$\mathbf{V}_1 = j\omega L_1\mathbf{I}_1 + j\omega M\mathbf{I}_2$ $\mathbf{V}_2 = j\omega L_2\mathbf{I}_2 + j\omega M\mathbf{I}_1$
$i_1(t)$, $i_2(t)$, M, L_1, L_2, $v_1(t)$, $v_2(t)$ (dot at top of coil 1, bottom of coil 2)	$v_1 = L_1\frac{di_1}{dt} - M\frac{di_2}{dt}$ $v_2 = L_2\frac{di_2}{dt} - M\frac{di_1}{dt}$	$\mathbf{V}_1 = j\omega L_1\mathbf{I}_1 - j\omega M\mathbf{I}_2$ $\mathbf{V}_2 = j\omega L_2\mathbf{I}_2 - j\omega M\mathbf{I}_1$

Table 11.14-2 **Ideal Transformers**

Device Symbol (including reference directions of element voltages and currents)	Device Equations in the Frequency Domain
$i_1(t)$, $i_2(t)$, $N_1 : N_2$, $v_1(t)$, $v_2(t)$, Ideal (dots at top of both coils)	$\mathbf{V}_1 = \frac{N_1}{N_2}\mathbf{V}_2$ $\mathbf{I}_1 = -\frac{N_2}{N_1}\mathbf{I}_2$
$i_1(t)$, $i_2(t)$, $N_1 : N_2$, $v_1(t)$, $v_2(t)$, Ideal (dot at top of coil 1, bottom of coil 2)	$\mathbf{V}_1 = -\frac{N_1}{N_2}\mathbf{V}_2$ $\mathbf{I}_1 = \frac{N_2}{N_1}\mathbf{I}_2$

PROBLEMS

Section 11.3 Instantaneous Power and Average Power

P 11.3-1 An *RLC* circuit is shown in Figure P 11.3-1. Find the instantaneous power delivered to the inductor when $i_s = 1 \cos \omega t$ A and $\omega = 6283$ rad/s.

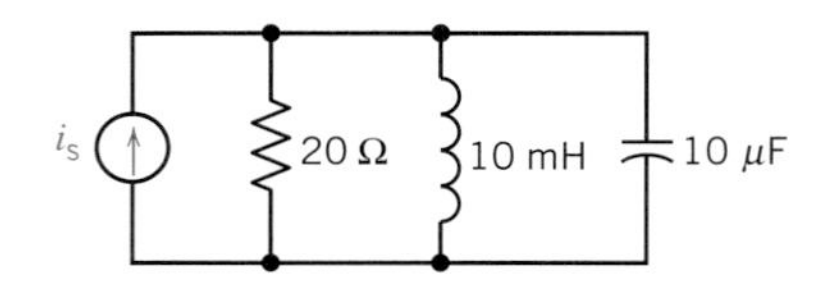

Figure P 11.3-1

P 11.3-2 Find the average power absorbed by the 0.6 kΩ resistor and the average power supplied by the current source for the circuit of Figure P 11.3-2.

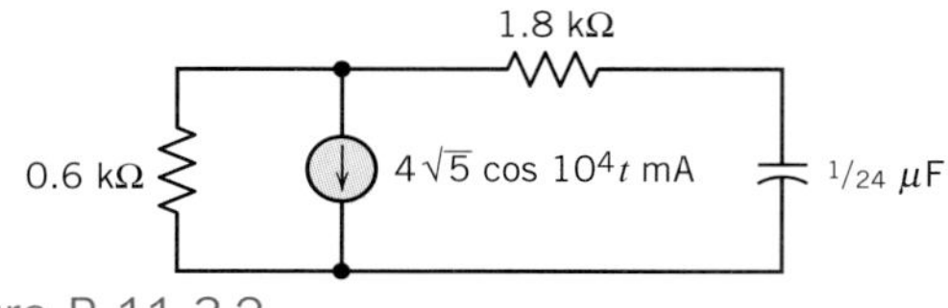

Figure P 11.3-2

P 11.3-3 For the circuit of Figure P 11.3-3, find (a) average power supplied by the independent source and (b) average power absorbed by the dependent source.

Hint: The gain of the dependent source is 5/2 V/mA = 2500 V/A.

Answer: (a) 2 mW (b) 0 W

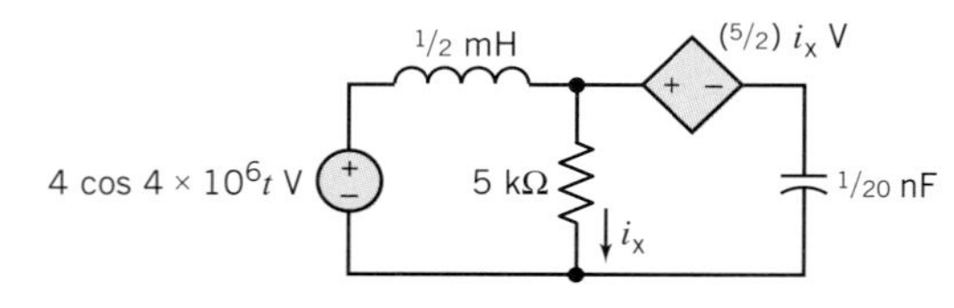

Figure P 11.3-3 Units of i_x are mA.

P 11.3-4 Use nodal analysis to find the average power absorbed by the 20 Ω resistor in the circuit of Figure P 11.3-4.

Answer: $P = 200$ W

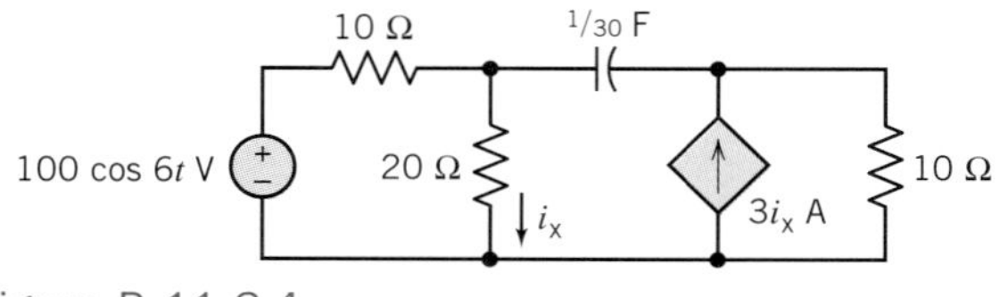

Figure P 11.3-4

P 11.3-5 Nuclear power stations have become very complex to operate, as illustrated by the training simulator for the operating room of the Pilgrim power plant shown in Figure P 11.3-5*a*. One control circuit has the model shown in Figure P 11.3-5*b*. Find the average power absorbed by each element.

Answer:
$P_{\text{source current}} = -12.8$ W
$P_{8\Omega} = 6.4$ W
$P_L = 0$ W
$P_{\text{voltage source}} = 6.4$ W

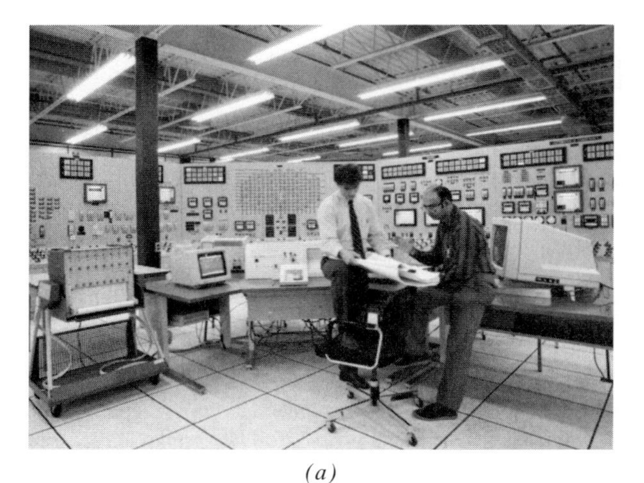

(*a*)

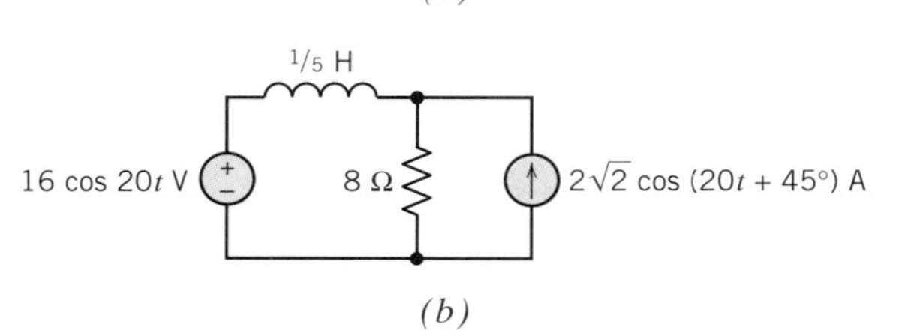

(*b*)

Figure P 11.3-5 (*a*) The simulation training room for the Pilgrim Power Station. The power station is located at Plymouth, Massachusetts, and generates 700 MW. It commenced operation in 1972. Courtesy of Boston Edison. (*b*) One control circuit of the reactor.

P 11.3-6 Find the average power absorbed by each element for the circuit of Figure P 11.3-6.

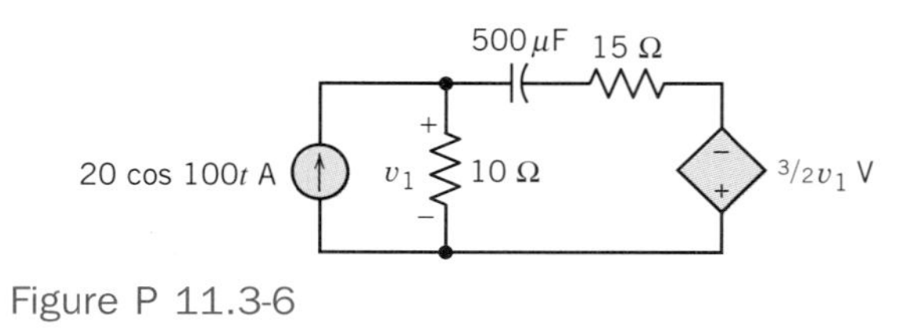

Figure P 11.3-6

P 11.3-7 Find the average power absorbed by the 6-kΩ resistor and supplied by the leftmost source in the circuit of Figure P 11.3-7.

Answers: 160 μW absorbed and 277 μW delivered.

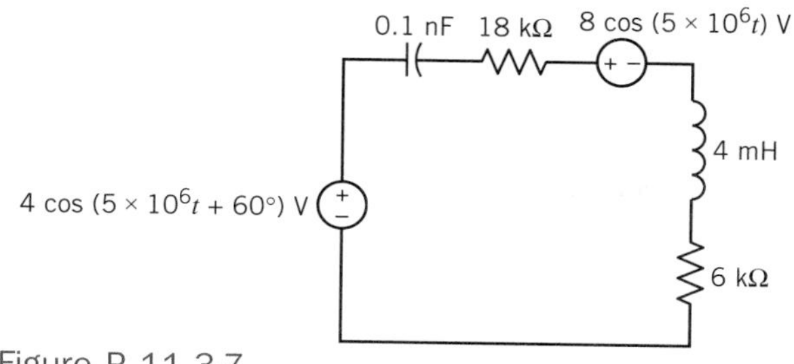

Figure P 11.3-7

P 11.3-8 A student experimenter in the laboratory encounters all types of electrical equipment. Some pieces of test equipment are battery-operated or operate at low voltage so that any hazard is minimal. Other types of equipment are isolated from electrical ground so that there is no problem if a grounded object makes contact with the circuit. Some types of test equipment, however, are supplied by voltages that can be hazardous or have dangerous voltage outputs. The standard power supply used in the United States for power and lighting in laboratories is the 120, grounded, 60-Hz sinusoidal supply. This supply provides power for much of the laboratory equipment, so an understanding of its operation is essential in its safe use (Bernstein 1991).

Consider the case where the experimenter has one hand on a piece of electrical equipment and the other hand on a ground connection, as shown in the circuit diagram of Figure P 11.3-8*a*. The hand-to-hand resistance is 200 Ω. Shocks with an energy of 30 J are hazardous to humans. Consider the model shown in Figure P 11.3-8*b*, which represents the human with *R*. Determine the energy delivered to the human in 1 s.

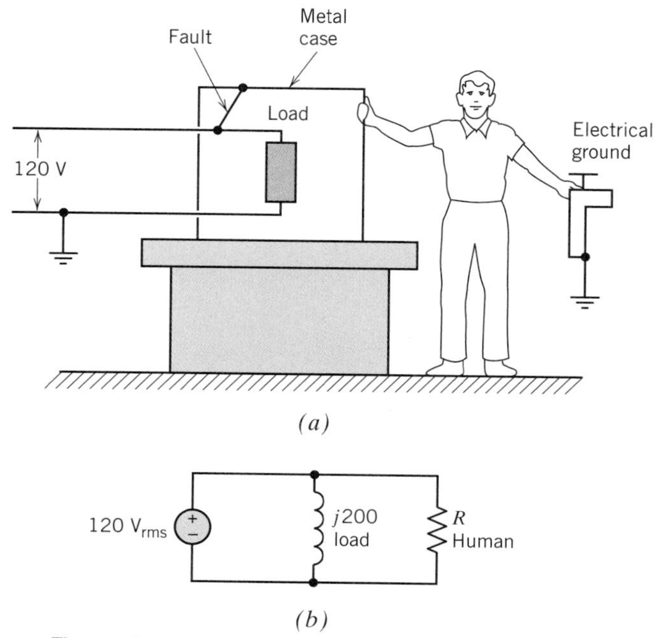

Figure P 11.3-8 Student experimenter touching an electrical device.

Section 11.4 Effective Value of a Periodic Waveform

P 11.4-1 Find the rms value of the current i for (a) $i = 2 - 4\cos 2t$ A, (b) $i = 3\sin \pi t + \sqrt{2}\cos \pi t$ A, and (c) $i = 2\cos 2t + 4\sqrt{2}\cos(2t + 45°) + 12\sin 2t$ A.

Answer: (a) $2\sqrt{3}$ (b) 2.35 A (c) $5\sqrt{2}$ A

P 11.4-2 Find the rms value for each of the waveforms shown in Figure P 11.4-2.

Answer: (a) $\sqrt{7/5}$ (b) and (c) 1/2

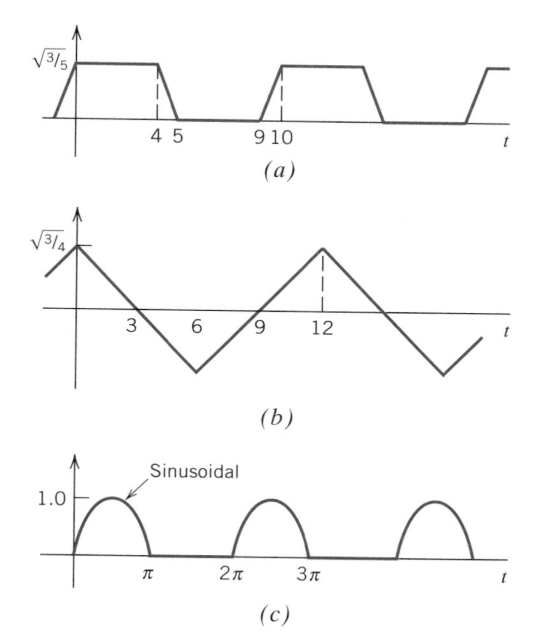

Figure P 11.4-2

P 11.4-3 Find the average and the rms value of the voltage waveform shown in Figure P 11.4-3.

Answer: $V_{aver} = 1.75$ V

$V_{rms} = 2.18$ V

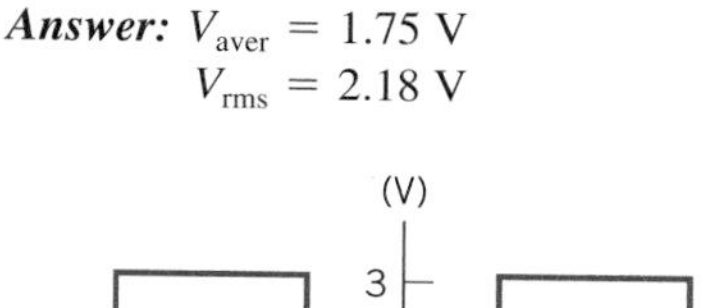
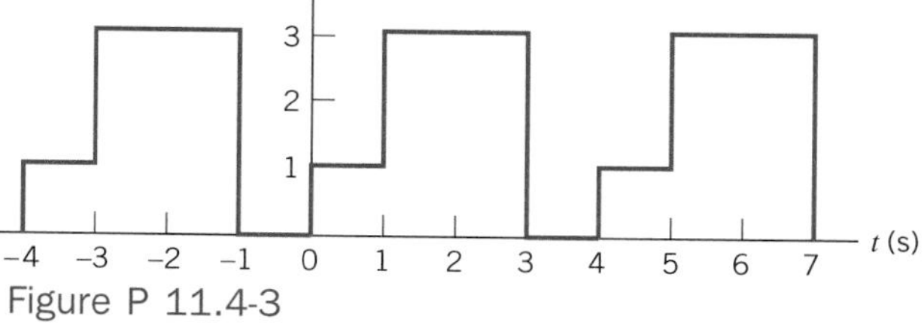

Figure P 11.4-3

P 11.4-4 Find the rms value for each of the waveforms of Figure P 11.4-4.

Answer: $V_{rms} = 1.225$ V

$I_{rms} = 5$ mA

P 11.4-5 Find the rms value of the voltage $v(t)$ shown in Figure P 11.4-5.

Answer: $V_{rms} = 4.24$ V

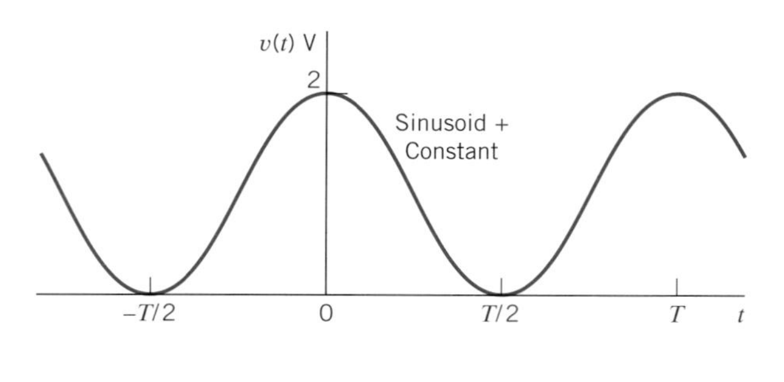

(a)

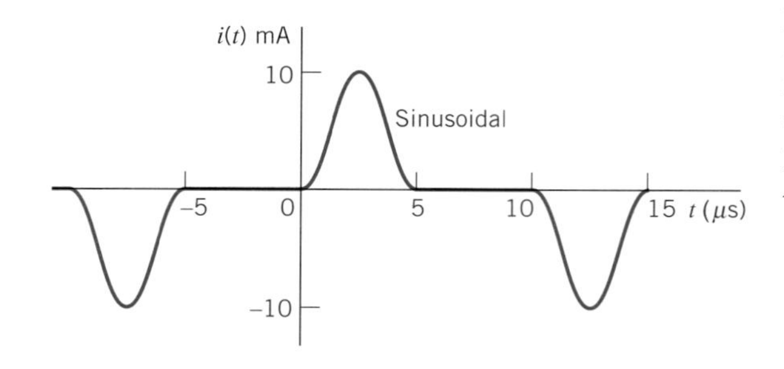

(b)

Figure P 11.4-4

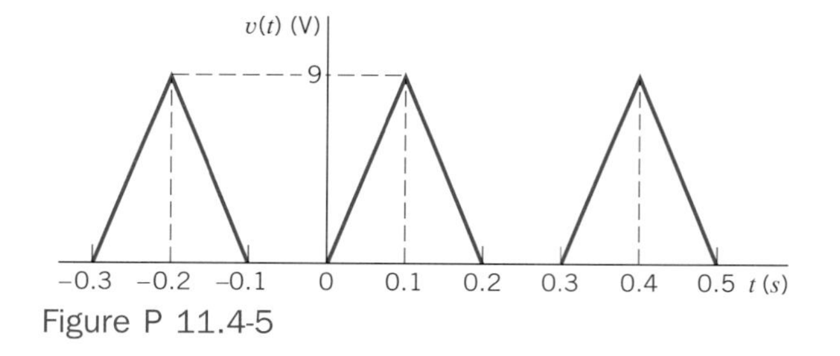

Figure P 11.4-5

Section 11.5 Complex Power

P 11.5-1 A coil is represented by a model consisting of an inductance L in series with a resistance R. A voltmeter reading rms voltage reads 26 V when a 2-A (rms) current is supplied to the coil. A wattmeter indicates that the average power delivered to the coil is 20 W. Determine L and R when $\omega = 377$.

Answer: $R = 5\ \Omega$, $L = 31.8$ mH

P 11.5-2 Assume the coil of Problem P 11.5-1 is connected to a 26-V (rms) source at the frequency $\omega = 377$. Find the complex power delivered to each element of this circuit. Verify the principle of conservation of complex power.

P 11.5-3 For the circuit of Figure P 11.5-3, find the value of the inductor L if the complex power supplied by the voltage source is $50/3\angle 53.13°$ VA.

Answer: $L = 2$ H

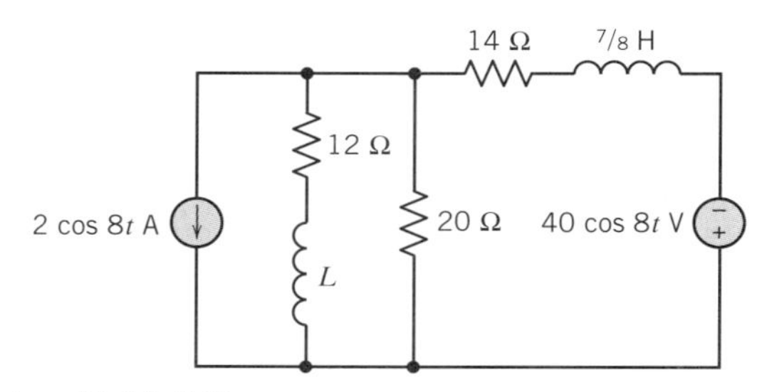

Figure P 11.5-3

P 11.5-4 Many engineers are working to develop photovoltaic power plants that provide ac power. An example of an experimental photovoltaic system is shown in Figure P 11.5-4*a*. A model of one portion of the energy conversion circuit is shown in Figure P 11.5-4*b*. Find the average, reactive, and complex power delivered by the dependent source.

Answer: $\mathbf{S} = +j8/9$ VA

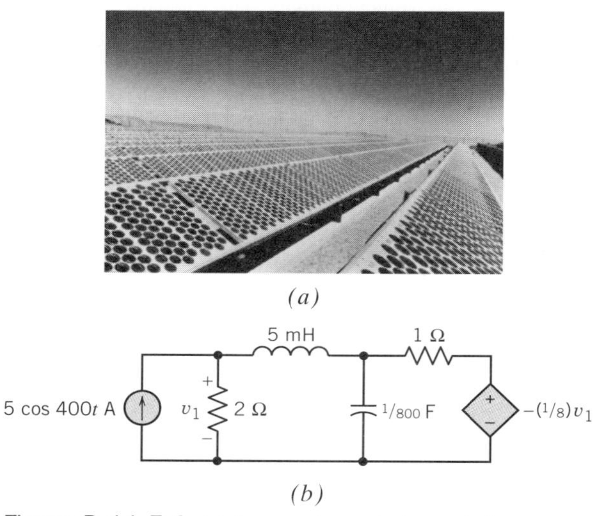

Figure P 11.5-4 (*a*) An experimental photovoltaic power plant. (*b*) Model of part of the energy conversion circuit. Courtesy of *EPRI Journal.*

P 11.5-5 For the circuit shown in Figure P 11.5-5, determine $\mathbf{I}$ and the complex power $\mathbf{S}$ delivered by the source when $\mathbf{V} = 50\angle 120°$ V rms.

Answer: $\mathbf{S} = 100 + j75$ VA

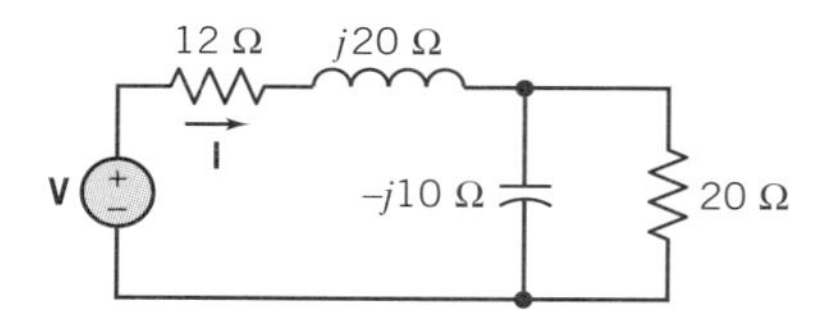

Figure P 11.5-5

P 11.5-6 For the circuit of Figure P 11.5-6, determine the complex power of the R, L, and C elements and show that the complex power delivered by the sources is equal to the complex power absorbed by the R, L, and C elements.

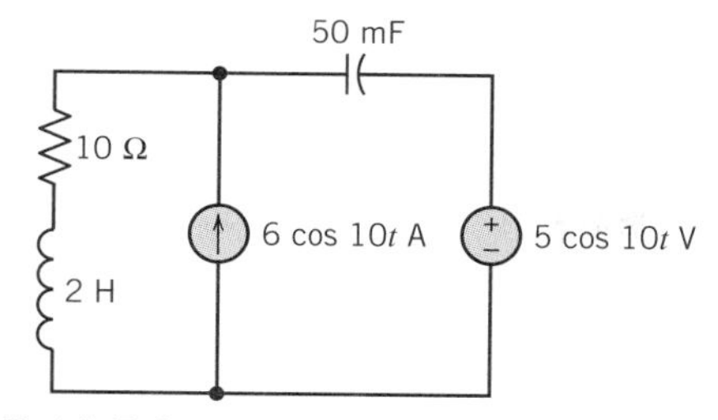

Figure P 11.5-6

P 11.5-7 A circuit is shown in Figure P 11.5-7 with an unknown impedance **Z.** However, it is known that $v(t) = 100\cos(100t + 20°)$ V and $i(t) = 25\cos(100t - 10°)$ A. (a) Find **Z.** (b) Find the power absorbed by the impedance. (c) Determine the type of element and its magnitude that should be placed across the impedance **Z** (connected to terminals a–b) so that the voltage $v(t)$ and the current entering the parallel elements are in phase.

Answers: *(a)* $4\angle 30°$ *(b)* 1082.5 W *(c)* 1.25 mF

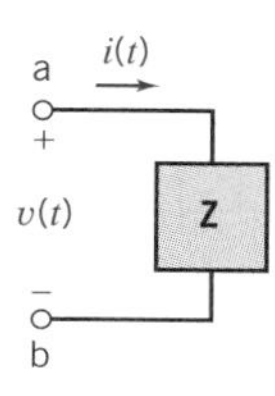

Figure P 11.5-7

P 11.5-8 Find the complex power delivered by the voltage source and the power factor seen by the voltage source for the circuit of Figure P 11.5-8.

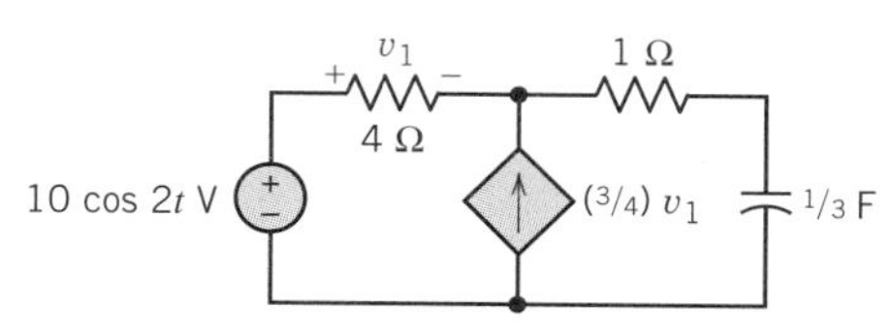

Figure P 11.5-8

P 11.5-9 The information below is related to tests performed on the MT46 Machine Tool Relay manufactured by Furnas Controls Co. The data are measured when the relay is energized from a 120-V rms, 60-Hz supply.

Inrush current	1.135 A rms
Seal current	0.2185 A rms
Inrush watts	46.0 W
Seal watts	5.0 W
Inrush VA	136.2 VA
Seal VA	26.22 VA

The inrush values describe the conditions at immediate switch-on of the relay, that is, as the relay armature is about to start moving. The seal values describe the conditions when the armature has moved to its final on position. Determine the series equivalent *RL* circuit representation of the relay (a) at initial switch-on and (b) with the armature sealed.

P 11.5-10 To encourage conservation of electricity, many electric utility companies are offering financial incentives to their customers to purchase more efficient appliances, lamps, and insulation. One utility, Pacific Gas and Electric Company (PG&E), headquartered in San Francisco, offers a rebate on customer purchases of low-power compact fluorescent lamps (Zorpette 1991).

One model of compact fluorescent lamp provides the same illumination as a 75-W incandescent lamp but consumes only 18 W. The equivalent resistance of the incandescent lamp is 192 Ω, and that of the compact fluorescent lamp is 800 Ω. Assume that the source resistance of the line voltage (generation source resistance plus transmission resistance) is 3 Ω. Derive expressions for power consumption in the lamp, for total power consumption (power consumed by the lamp and by the source resistance), and for power transfer efficiency (lamp power consumption divided by total power consumption). Evaluate these three expressions for the two lamps. Assume line voltage is 120 V rms. Which lamp causes the greatest power transfer to the load? Which lamp causes the greatest percentage of power delivered to the load? Which lamp causes the greatest total consumption of power? Which lamp is the most efficient? Explain what you mean by efficient.

Section 11.6 Power Factor

P 11.6-1 An industrial firm has two electrical loads connected in parallel across the power source. Power is supplied to the firm at 4000 V (rms). One load is 30 kW of heating use, and the other load is a set of motors that together operate as a load at 0.6 lagging power factor and at 150 kVA. Determine the total current and the plant power factor.

Answer: $I = 42.5$ A, pf $= 1/\sqrt{2}$

P 11.6-2 Two electrical loads are connected in parallel to a 440-V rms, 60-Hz supply. The first load is 12 kVA at 0.7 lagging power factor; the second load is 10 kVA at 0.8 lagging power factor. Find the average power, the apparent power, and the power factor of the two combined loads.

Answer: Total power factor = 0.75 lagging

P 11.6-3 The source of Figure P 11.6-3 delivers 50 VA with a power factor of 0.8 lagging. Find the unknown impedance **Z.**

Answer: $\mathbf{Z} = 6.39\angle 26.6°\ \Omega$

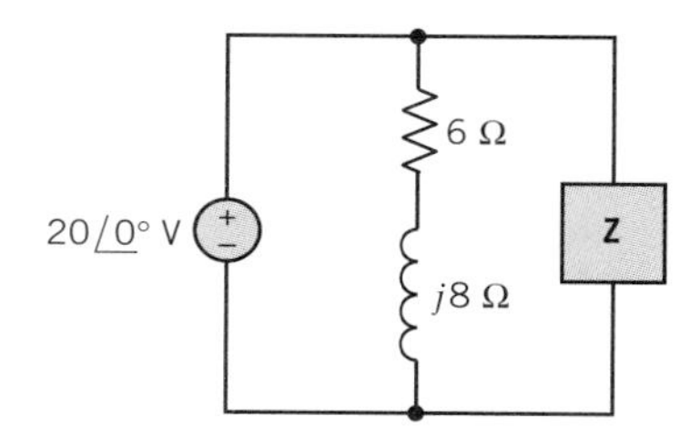

Figure P 11.6-3

P 11.6-4 Electric streetcars have been used in industrialized nations since the 1890s. A typical streetcar system will use the power distributed by overhead wires at 50 kV rms. It is desirable to keep the power factor equal to one. When 50 streetcars are running on a line, the average load is $\mathbf{Z}_L = 8 + j6\ \Omega$ and $\omega = 377$ rad/s. Find (a) the complex power delivered to the load, (b) the power factor of the load, and (c) the value of a capacitor connected in parallel that will correct the total power factor to 0.95 leading.

P 11.6-5 Manned space stations will require several continuously available ac power sources. Also, it is desired to keep the power factor close to 1. Consider the model of one communication circuit shown in Figure P 11.6-5. If an average power of 500 W is dissipated in the 20 Ω resistor, find (a) V_{rms}, (b) $I_{s\ rms}$, (c) the power factor seen by the source, and (d) $|\mathbf{V}_s|$.

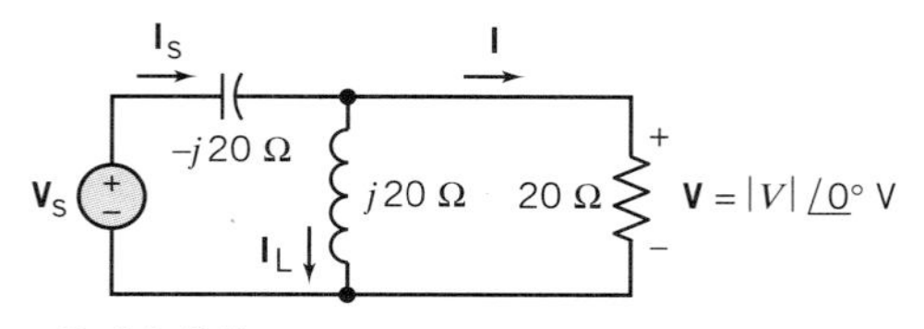

Figure P 11.6-5

P 11.6-6 Two impedances are supplied by $\mathbf{V} = 100\angle 160°$ V rms, as shown in Figure P 11.6-6, where $\mathbf{I} = 2\angle 190°$ A rms. The first load draws $P_1 = 23.2$ W and $Q_1 = 50$ VAR. Calculate $\mathbf{I}_1$, $\mathbf{I}_2$, the power factor of each impedance, and the total power factor of the circuit.

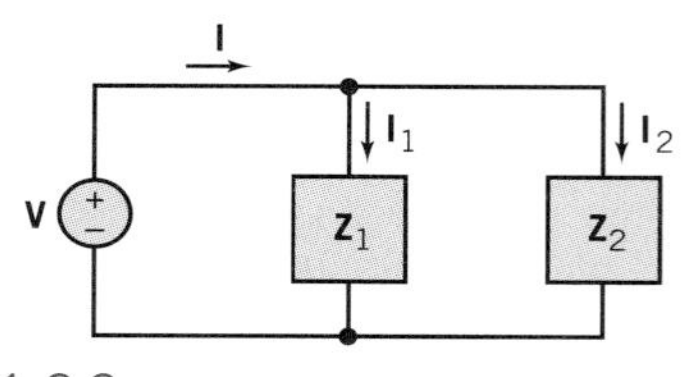

Figure P 11.6-6

P 11.6-7 A motor draws 1-A peak current at a 0.6 lagging power factor from a 120-V-rms 60-Hz source. The motor is modeled by a series resistance R and an inductance L. Determine (a) the complex power of the motor and (b) the values of R and L.

Answer: (a) $\mathbf{S} = 50.9 + j67.8$ VA (b) $R = 101.8\ \Omega$, $L = 0.36$ H

P 11.6-8 A genertor supplies two parallel loads at 220 V with a total power factor of 0.75 lagging. One load is 4800 VA at a power factor of 0.85 lagging. The second load absorbs 4 kW. What are the apparent power and power factor of the second load?

P 11.6-9 A residential electric supply three-wire circuit from a transformer is shown in Figure P 11.6-9*a*. The circuit model is shown in Figure P 11.6-9*b*. From its nameplate, the refrigerator motor is known to have a rated current of 8.5 A rms. It is reasonable to assume an inductive impedance angle of 45° for a small motor at rated load. Lamp and range loads are 100 W and 12 kW, respectively.

(a) Calculate the currents in line 1, line 2, and the neutral wire.

(b) Calculate: (i) P_{refrig}, Q_{refrig}, (ii) P_{lamp}, Q_{lamp}, and (iii) P_{total}, Q_{total}, $\mathbf{S}_{total}$, and overall power factor.

(c) The neutral connection resistance increases, because of corrosion and looseness, to 20 Ω. (This must be included as part of the neutral wire.) Use mesh analysis and calculate the voltage across the lamp.

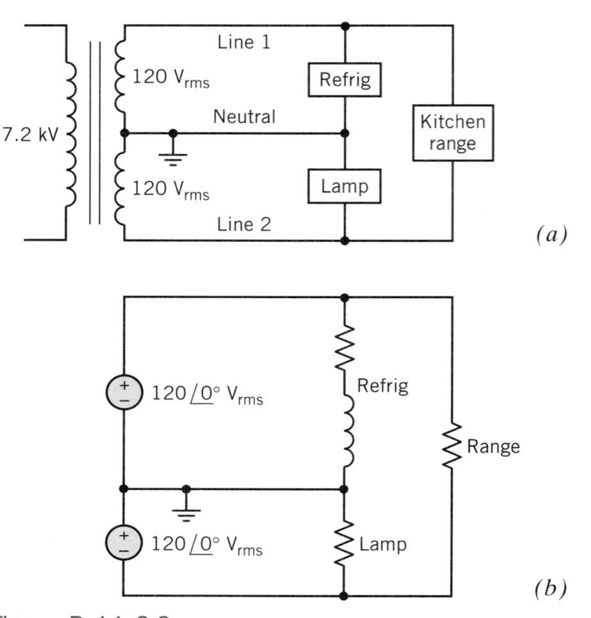

Figure P 11.6-9 Residential circuit with selected loads.

P 11.6-10 A motor connected to a 220-V supply line from the power company has a current of 7.6 A. Both the current and the voltage are rms values. The average power delivered to the motor is 1317 W. (a) Find the apparent power, the reactive power, and the power factor when $\omega = 377$ rad/s. (b) Find the capacitance of a parallel capacitor that will result in a unity power factor of the combination. (c) Find the current in the utility lines after the capacitor is installed.

Answer: (a) pf = 0.788 (b) $C = 56.5\ \mu$F (c) $I = 6.0$ A rms

P 11.6-11 Two loads are connected in parallel across a 1000-V (rms), 60-Hz source. One load absorbs 500 kW at 0.6 power factor lagging, and the second load absorbs 400 kW and 600 kVAR. Determine the value of the capacitor that should be added in parallel with the two loads to improve the overall power factor to 0.9 lagging.

Answer: $C = 2.2\ \mu$F

P 11.6-12 A voltage source with a complex internal impedance is connected to a load, as shown in Figure P 11.6-12. The load absorbs 1 kW of average power at 100 V rms with a power factor of 0.80 lagging. The source frequency is 200 rad/s. (a) Determine the source voltage $\mathbf{V}_1$. (b) Find the type of value of the element to be placed in parallel with the load so that maximum power is transferred to the load.

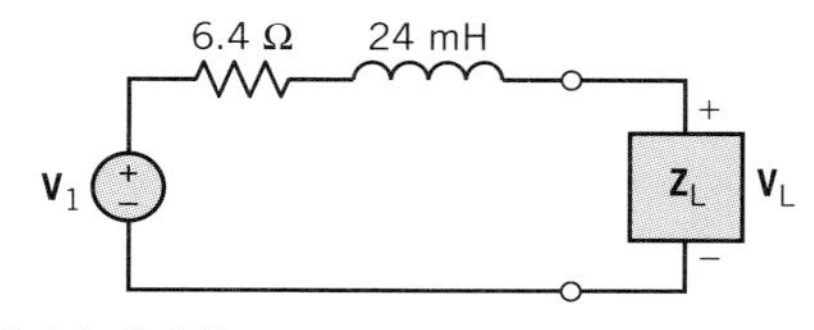

Figure P 11.6-12

P 11.6-13 A 100-kW induction motor, shown in Figure P 11.6-13, is receiving 100 kW at 0.8 power factor lagging. Determine the additional apparent power in kVA that is made available by improving the power factor to (a) 0.95 lagging and (b) 1.0. (c) Find the required reactive power in kVAR provided by a set of parallel capacitors for parts a and b. (d) Determine the ratio of kVA released to the kVAR of capacitors required for parts a and b alone. Set up a table recording the results of this problem for the two values of power factor attained.

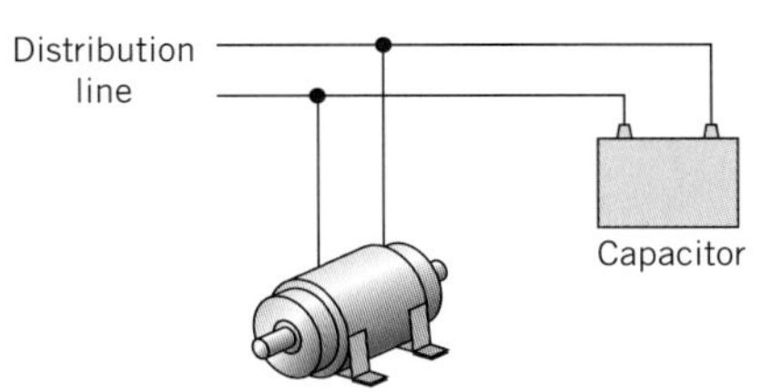

Figure P 11.6-13 Induction motor with parallel capacitor.

P 11.6-14 A resistor of 150 Ω and a parallel capacitor draw 0.2 A from a 24-V source operating at 400 Hz. The power factor is measured as 0.8 leading. Determine the capacitance of the parallel capacitor C.

P 11.6-15 Two loads are connected in parallel and supplied from a 7.2-kV rms 60 Hz source. The first load is 50 kVA at 0.9 lagging power factor, and the second load is 45 kW at 0.91 lagging power factor. Determine the kVAR rating and capacitance required to correct the overall power factor to 0.97 lagging.
Answer: $C = 1.01\ \mu F$

P 11.6-16 An industrial plant operates from a 500-V rms, 60-Hz source and has two loads connected in parallel to the source. The first load draws 48 kW at 0.6 lagging power factor. The second load draws 24 kW at 0.96 leading power factor. Select a parallel impedance to correct the overall power factor to unity.

Section 11.7 The Superposition Principle

P 11.7-1 Find the average power absorbed by the 2 Ω resistor in the circuit of Figure P 11.7-1.
Answer: $P = 413$ W

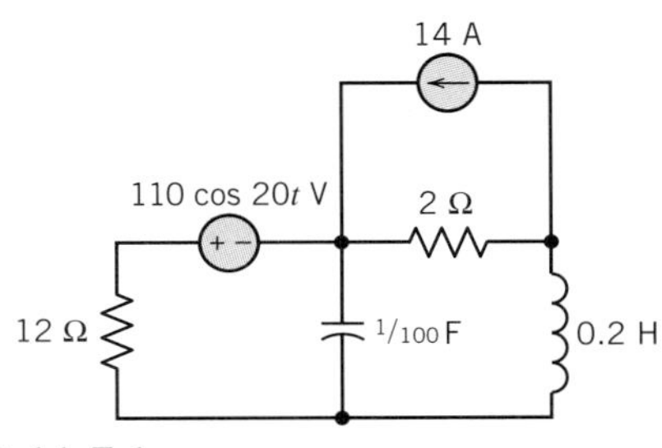

Figure P 11.7-1

P 11.7-2 Find the average power absorbed by the 8 Ω resistor in the circuit of Figure P 11.7-2.
Answer: $P = 22$ W

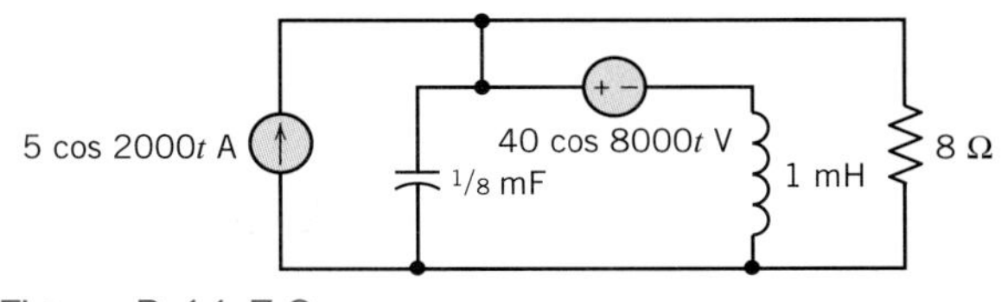

Figure P 11.7-2

P 11.7-3 For the circuit shown in Figure P 11.7-3, determine the average power absorbed by each resistor, R_1 and R_2. The voltage source is $v_s = 10 + 10\cos(5t + 40°)$ V, and the current source is $i_s = 4\cos(5t + 30°)$ A.

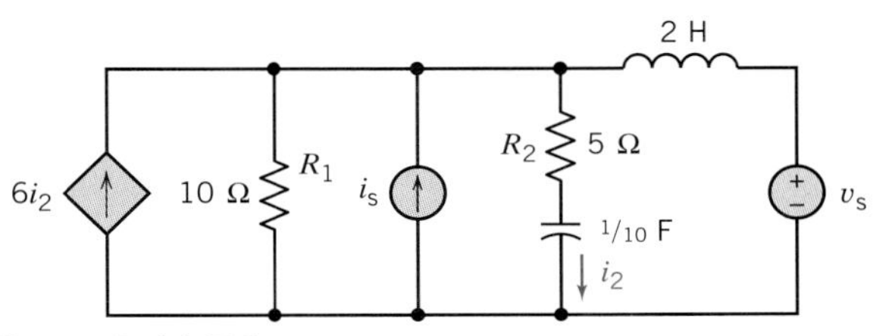

Figure P 11.7-3

P 11.7-4 For the circuit shown in Figure P 11.7-4, determine the effective value of the resistor voltage v_R and the capacitor voltage v_c.

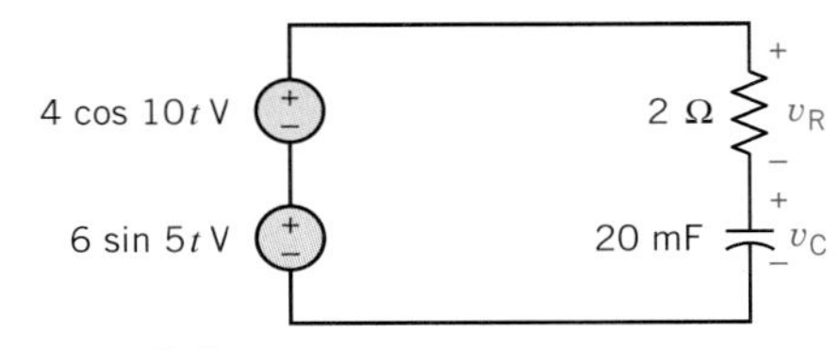

Figure P 11.7-4

Section 11.8 The Maximum Power Theorem

P 11.8-1 Determine values of R and L for the circuit shown in Figure P 11.8-1 that cause maximum power transfer to the load.
Answer: $R = 800\ \Omega$, $L = 1.6$ H

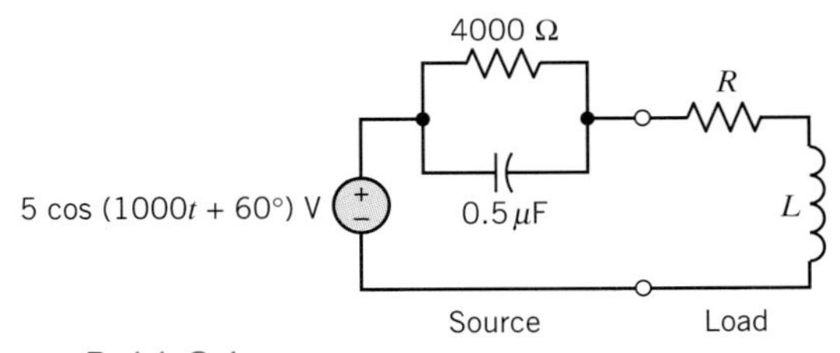

Figure P 11.8-1

P 11.8-2 Is it possible to choose R and L for the circuit shown in Figure P 11.8-2 so that the average power delivered to the load is 12 mW?
Answer: Yes

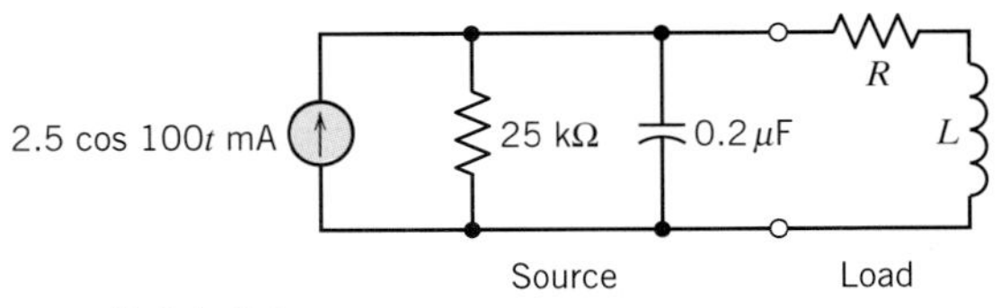

Figure P 11.8-2

P 11.8-3 The capacitor has been added to the load in the circuit shown in Figure P 11.8-3 in order to maximize the power absorbed by the 4000 Ω resistor. What value of capacitance should be used to accomplish that objective?
Answer: 0.1 μF

Figure P 11.8-3

P 11.8-4 What is the value of the average power delivered to the 2000 Ω resistor in the circuit shown in Figure 11.8-4? Can the average power delivered to the 2000 Ω resistor be increased by adjusting the value of the capacitance?
Answer: 8 mW. No.

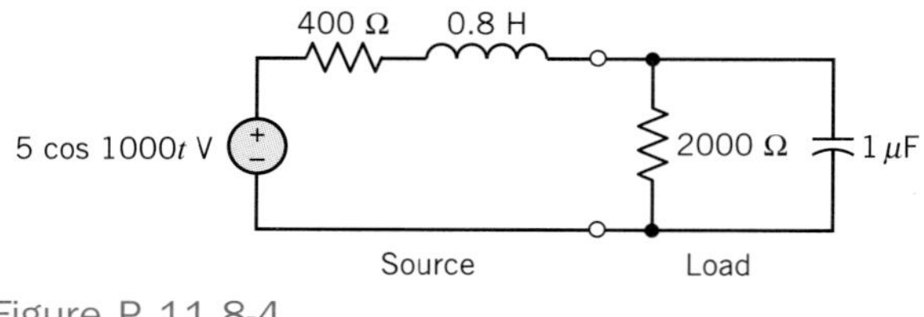

Figure P 11.8-4

P 11.8-5 What is the value of the resistance R in Figure P 11.8-5 that maximizes the average power delivered to the load?

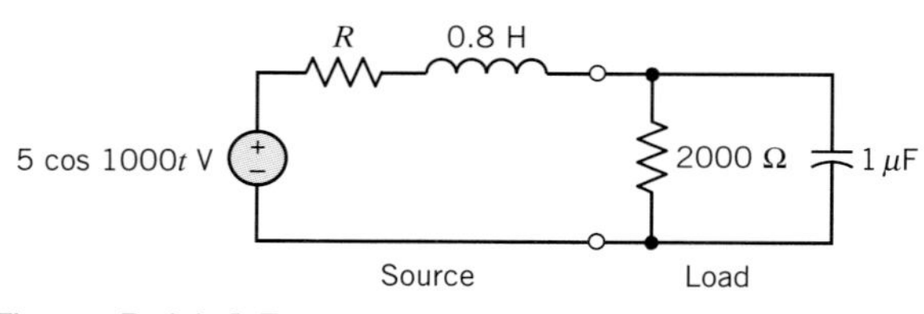

Figure P 11.8-5

P 11.8-6 For the circuit of Figure P 11.8-6, find the values of R and L such that R absorbs maximum power. What is this value of maximum power?
Answer: $R = 2$ kΩ, $L = 1$ mH

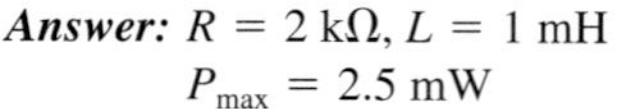
$P_{\max} = 2.5$ mW

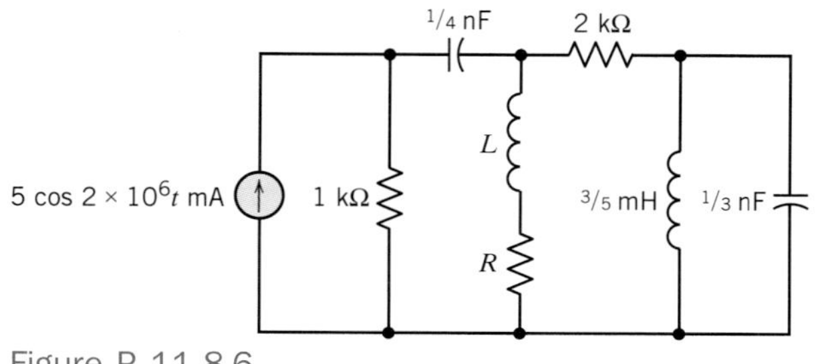

Figure P 11.8-6

P 11.8-7 Consider the circuit of Figure P 11.8-7 and find the value of L and R such that the maximum power transfer condition $\mathbf{Z}_L = \mathbf{Z}_t^*$ is met. What power is drawn by the load in this case?

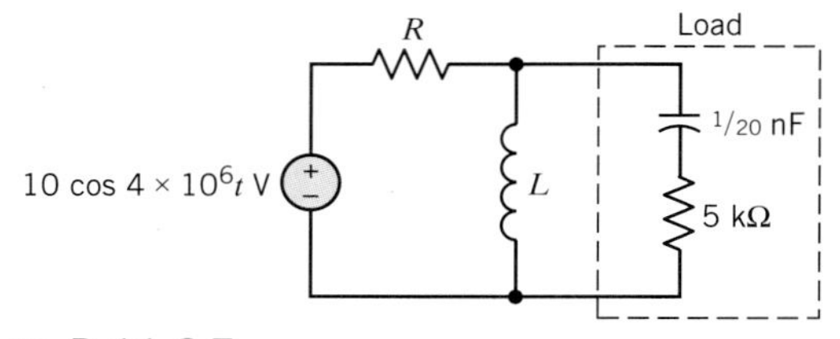

Figure P 11.8-7

P 11.8-8 The circuit shown in Figure P 11.8-8 operates at $\omega = 1000$ rad/s. Determine the appropriate values of R and C for maximum power transfer to the load. Determine the maximum power delivered to the load when $v_s = 100 \cos \omega t$ V.

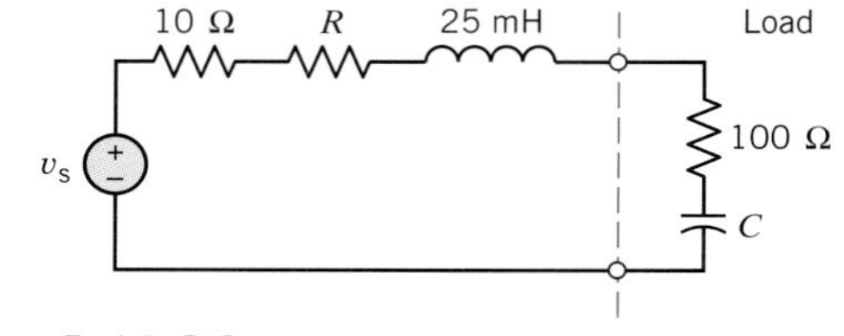

Figure P 11.8-8

P 11.8-9 Obtain the Thévenin equivalent circuit at terminals a–b for the transistor circuit shown in Figure P 11.8-9. Calculate the power delivered to the load connected at terminals a–b when the load is selected for the maximum power condition.

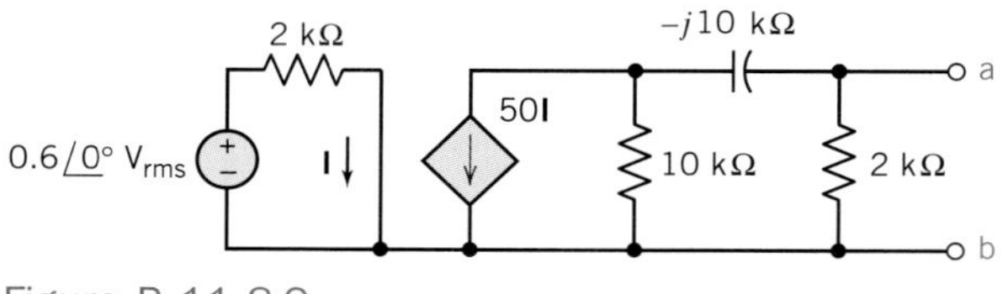

Figure P 11.8-9

P 11.8-10 (a) Determine the load impedance $\mathbf{Z}_{ab}$ that will absorb maximum power if it is connected to terminals a–b of the circuit shown in Figure P 11.8-10.
(b) Determine the maximum power absorbed by this load.
(c) Determine a model of the load and indicate the element values.

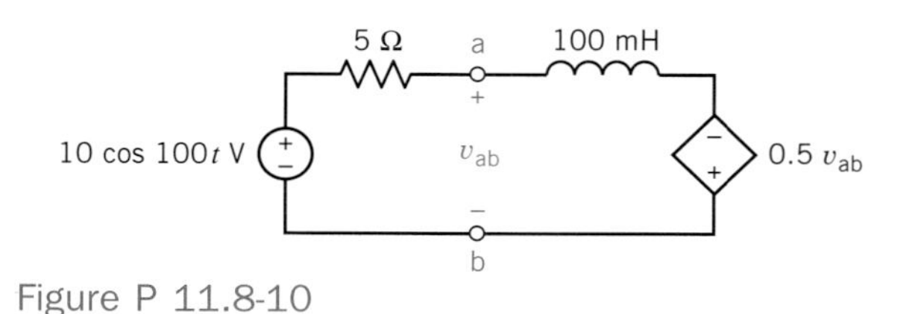

Figure P 11.8-10

Section 11.9 Coupled Inductors

P 11.9-1 Two magnetically coupled coils are connected as shown in Figure P 11.9-1. Using a source v_s connected to terminals a–b, show that an equivalent inductance at terminals a–b is $L_{ab} = L_1 + L_2 - 2M$.

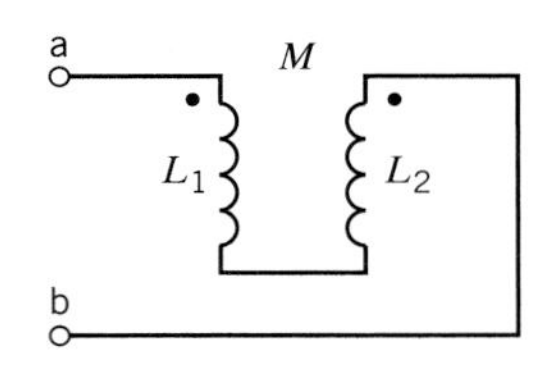

Figure P 11.9-1

P 11.9-2 Two magnetically coupled coils are shown connected in Figure P 11.9-2. Find the equivalent inductance L_{ab}.

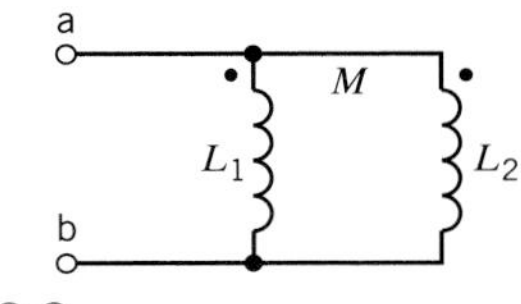

Figure P 11.9-2

P 11.9-3 The source voltage of the circuit shown in Figure P 11.9-3 is $v_s = 141.4 \cos 100t$ V. Determine $i_1(t)$ and $i_2(t)$.

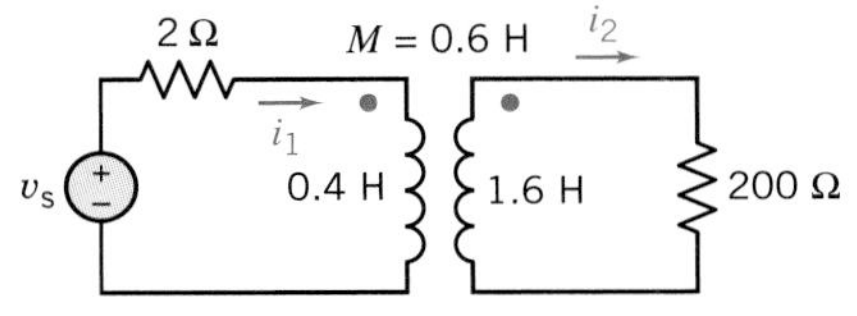

Figure P 11.9-3

P 11.9-4 A circuit with a mutual inductance is shown in Figure P 11.9-4. Find the voltage $\mathbf{V}_2$ when $\omega = 5000$.

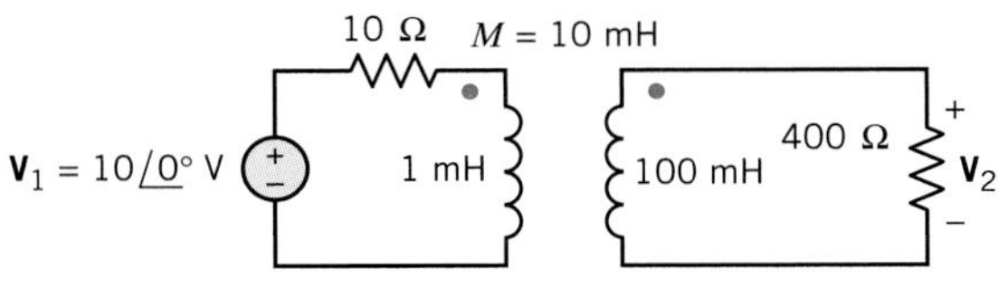

Figure P 11.9-4

P 11.9-5 Determine $v(t)$ for the circuit of Figure P 11.9-5 when $v_s = 10 \cos 30t$ V.
Answer: $v(t) = 23 \cos(30t + 9°)$ V

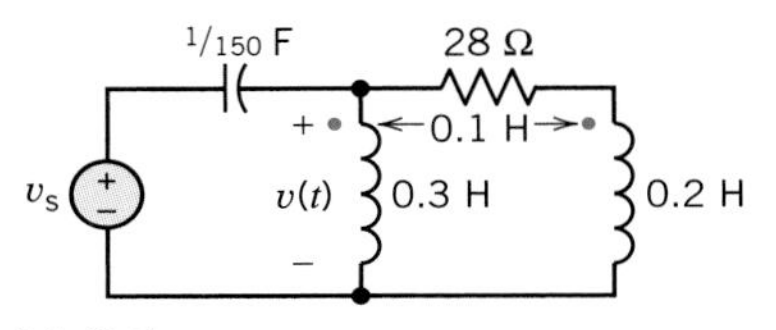

Figure P 11.9-5

P 11.9-6 Find the total energy stored in the circuit shown in Figure P 11.9-6 at $t = 0$ if the secondary winding is (a) open-circuited, (b) short-circuited, (c) connected to the terminals of a 7 Ω resistor.
Answer: (a) 15 J (b) 0 J (c) 5 J

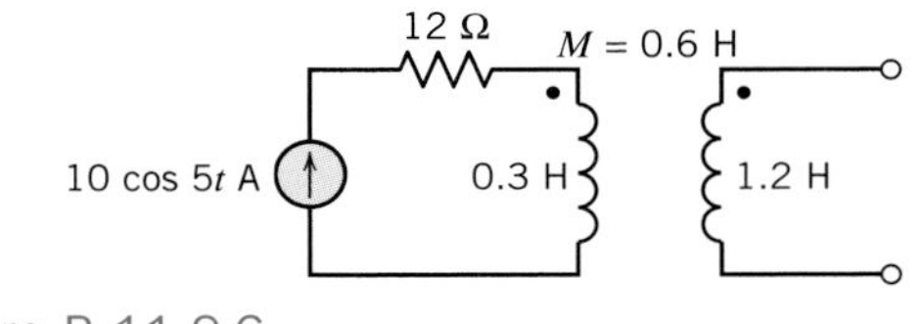

Figure P 11.9-6

P 11.9-7 Find the input impedance, $\mathbf{Z}$, of the circuit of Figure P 11.9-7 when $\omega = 1000$ rad/s.
Answer: $\mathbf{Z} = 8.4 \angle 14°\ \Omega$

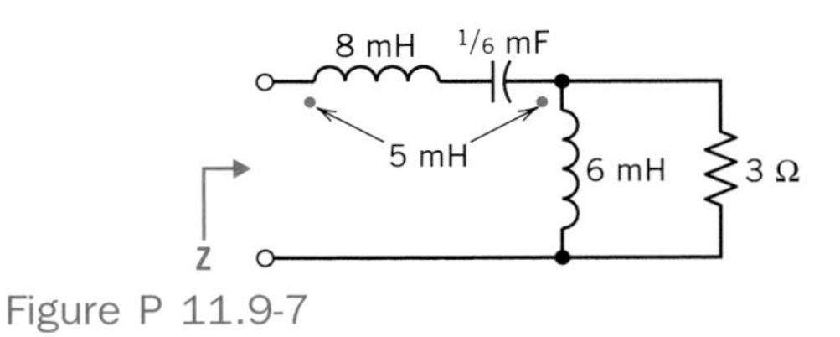

Figure P 11.9-7

P 11.9-8 A circuit with three mutual inductances is shown in Figure P 11.9-8. When $v_s = 10 \sin 2t$ V, $M_1 = 2$ H, and $M_2 = M_3 = 1$ H, determine the capacitor voltage $v(t)$.

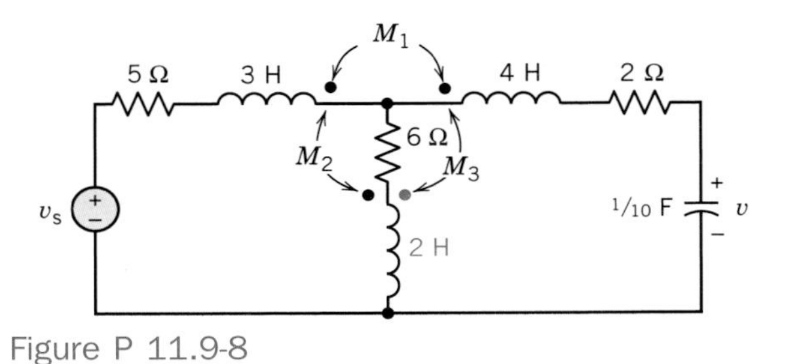

Figure P 11.9-8

Section 11.10 The Ideal Transformer

P 11.10-1 Find $\mathbf{V}_1$, $\mathbf{V}_2$, $\mathbf{I}_1$, and $\mathbf{I}_2$ for the circuit of Figure P 11.10-1, when $n = 5$.

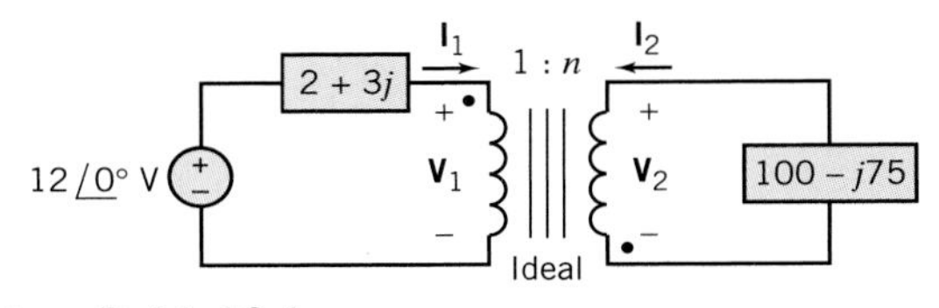

Figure P 11.10-1

P 11.10-2 A circuit with a transformer is shown in Figure P 11.10-2. (a) Determine the turns ratio. (b) Determine the value of R_{ab}. (c) Determine the current supplied by the voltage source.

Answer: (a) $n = 5$

(b) $R_{ab} = 400\ \Omega$

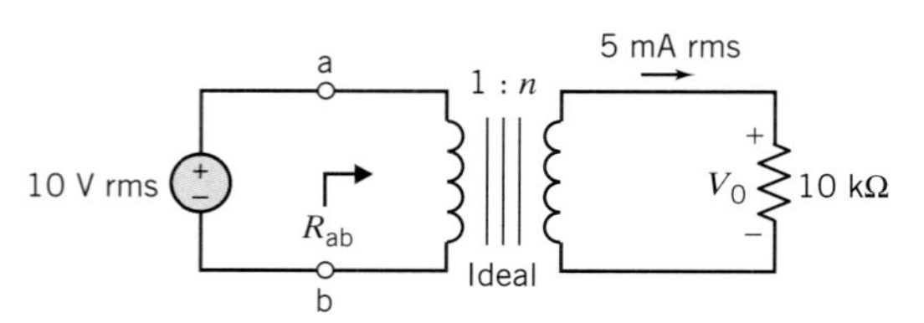

Figure P 11.10-2

P 11.10-3 Find the voltage $\mathbf{V}_c$ in the circuit shown in Figure P 11.10-3. Assume an ideal transformer. The turns ratio is $n = 1/3$.

Answer: $\mathbf{V}_c = 21.0\ \angle -105.3°$

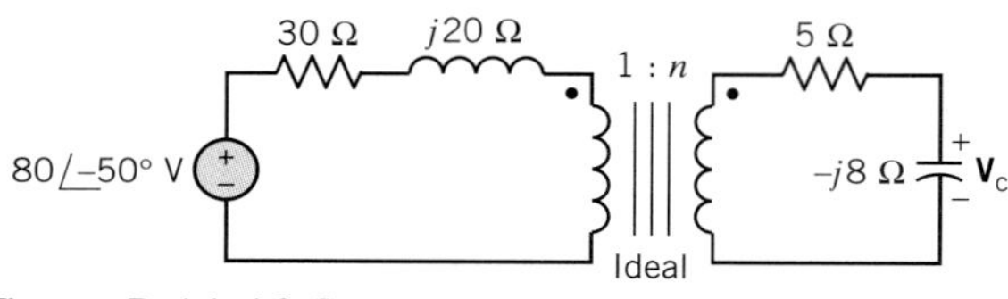

Figure P 11.10-3

P 11.10-4 An ideal transformer is connected in the circuit shown in Figure P 11.10-4, where $v_s = 50 \cos 1000t$ V and $n = N_2/N_1 = 5$. Calculate $\mathbf{V}_1$ and $\mathbf{V}_2$.

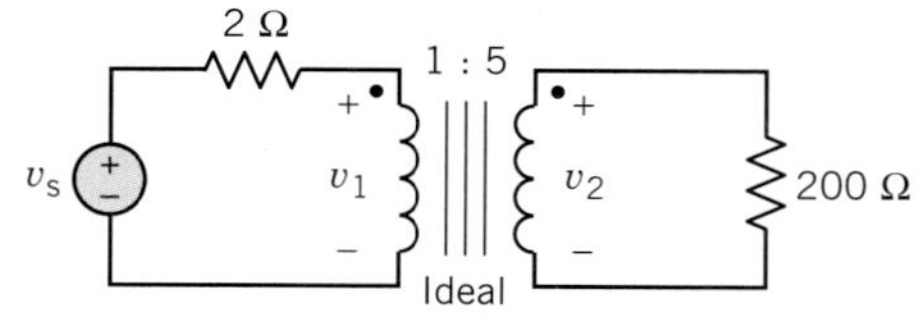

Figure P 11.10-4

P 11.10-5 The circuit of Figure P 11.10-5 is operating at 10^5 rad/s. Determine the inductance L and the turns ratio n to achieve maximum power transfer to the load.

Answer: $n = 2$

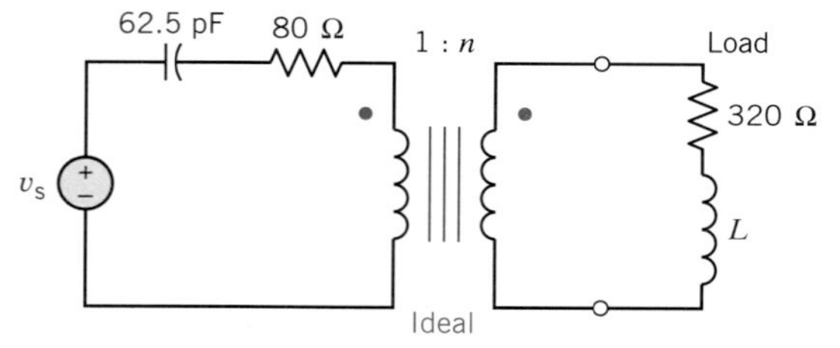

Figure P 11.10-5

P 11.10-6 Find the Thévenin equivalent at terminals a–b for the circuit of Figure P 11.10-6 when $v = 16 \cos 3t$ V.

Answer: $V_{oc} = 12$, $Z_{th} = 3.75\ \Omega$

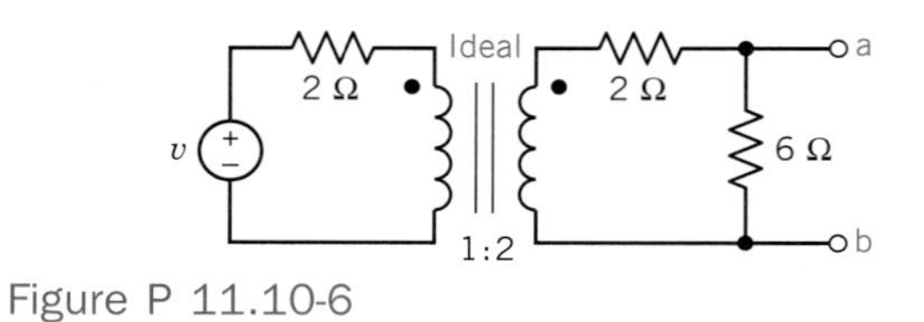

Figure P 11.10-6

P 11.10-7 Find the input impedance $\mathbf{Z}$ for the circuit of Figure P 11.10-7.

Answer: $\mathbf{Z} = 6\ \Omega$

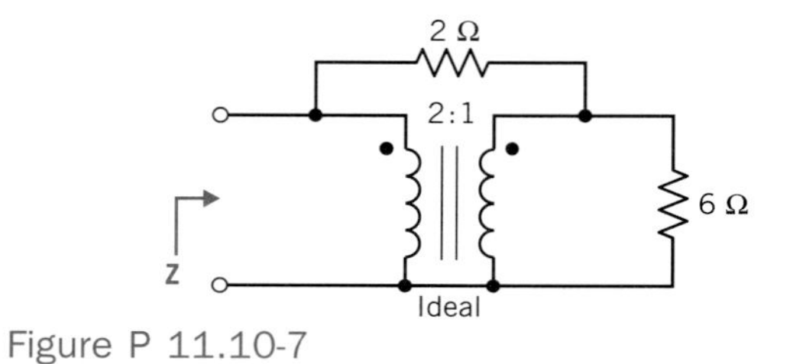

Figure P 11.10-7

P 11.10-8 Determine the node voltages for the circuit shown in Figure P 11.10-8 when $n = 4$.

Answer: $V_a = 12.5$ V; $V_b = 50$ V

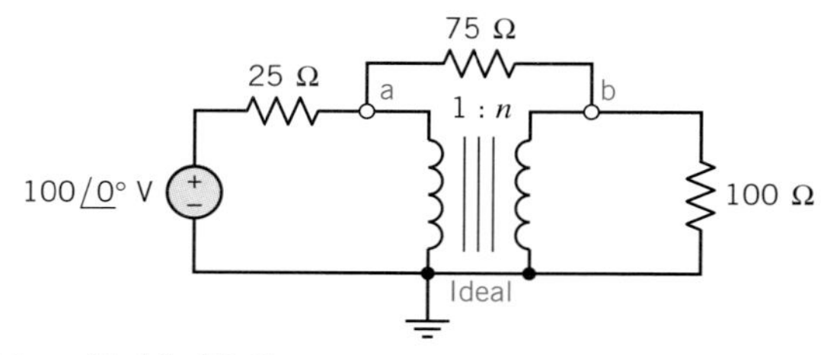

Figure P 11.10-8

P 11.10-9 A circuit with two ideal transformers is shown in Figure P 11.10-9, with $L = 2$ mH and $C = 100\ \mu$F operating at $\omega = 1$ krad/s. Select the turns ratio of each transformer in order to achieve maximum power to R_L.

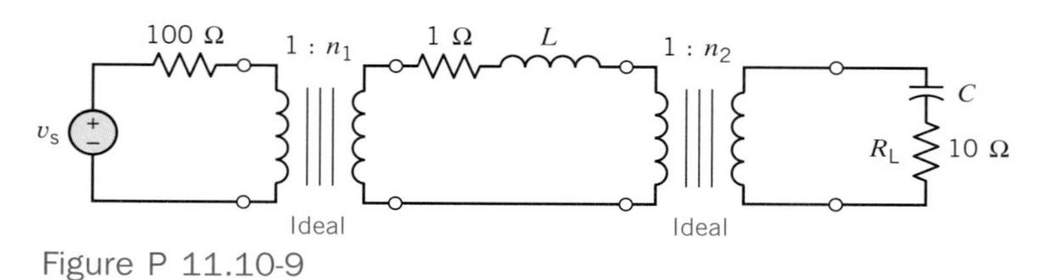

Figure P 11.10-9

P 11.10-10 In less-developed regions in mountainous areas, small hydroelectric generators are used to serve several residences (Mackay 1990). Assume each house uses an electric range and an electric refrigerator, as shown in Figure P 11.10-10. The generator is represented as $\mathbf{V}_s$ operating at 60 Hz and $\mathbf{V}_2 = 230\angle 0°$ V. Calculate the power consumed by six homes connected to the hydroelectric generator when $n = 5$.

P 11.10-11 An amplifier circuit with a transformer coupled output is shown in Figure P 11.10-11. It is desired to transfer maximum power to the output resistor R, which is equal to 100 Ω. Select the transformer turns ratio n and the compensator impedance $\mathbf{Z}$. Determine the value of the maximum power output when $v_s = 20 \cos 100t$ V.

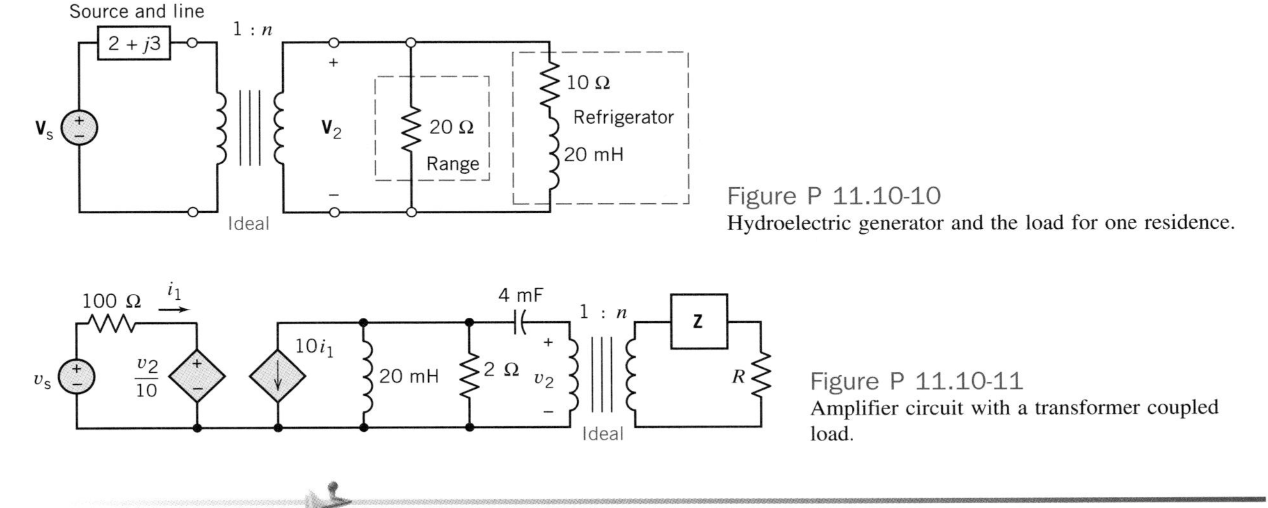

Figure P 11.10-10
Hydroelectric generator and the load for one residence.

Figure P 11.10-11
Amplifier circuit with a transformer coupled load.

PSPICE PROBLEMS

SP 11-1 For the circuit shown in Figure SP 11.1, determine the voltage, $v(t)$, when $v_s = 200 \sin(\omega t + 45°)$ V, $\omega = 10^4$ rad/s, and $R = 100$ Ω.

Answer: $70 \sin(\omega t + 7.2°)$ V

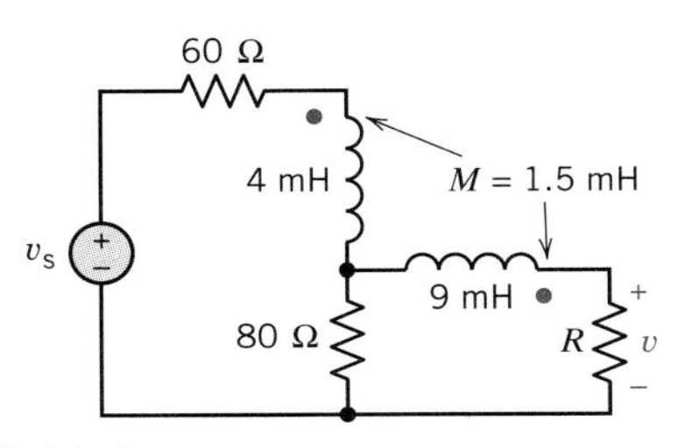

Figure SP 11.1

SP 11-2 For the circuit shown in Figure SP 11.2, determine $i(t)$ using a PSpice program. The two sources operate at $\omega = 500$ rad/s, $v_1 = 100 \cos \omega t$ V, and $v_2 = 100 \cos(\omega t - 90°)$ V.

Answer: $4 \cos(500t + 28.6°)$ A

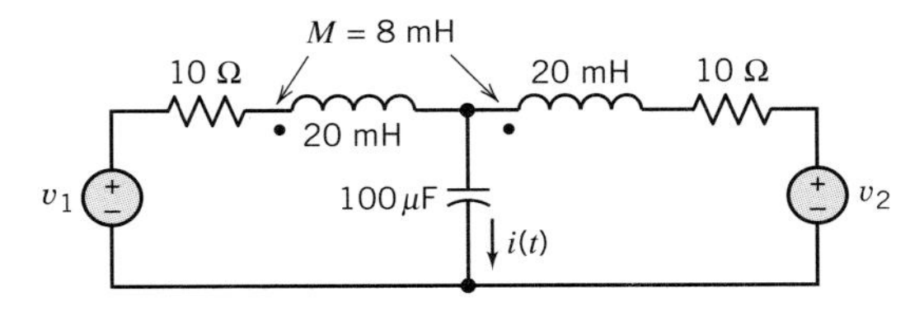

Figure SP 11.2

SP 11-3 The circuit with an iron core and with $k = 1$ is shown in Figure SP 11.3.

(a) Determine the ratio $\mathbf{V}_2/\mathbf{V}_1$ for $\omega = 1000$ rad/s analytically, by assuming that the transformer is ideal.

(b) Determine the ratio using PSpice.

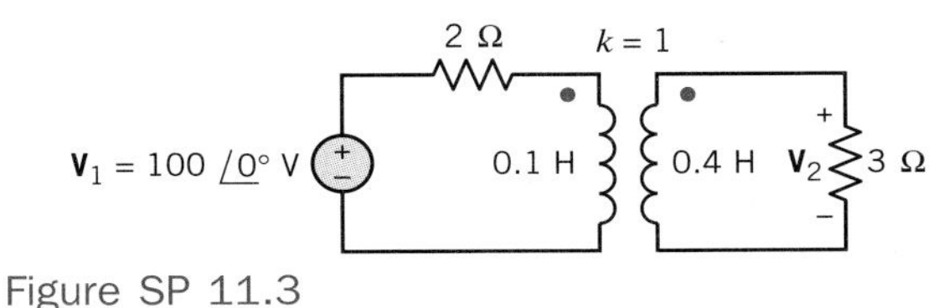

Figure SP 11.3

SP 11-4 Determine the currents i_1 and i_2 for the circuit shown in Figure SP 11.4 when $v_s(t) = 12 \sin 10t$ V and $i_s(t) = 6 \cos(10t + 45°)$ A.

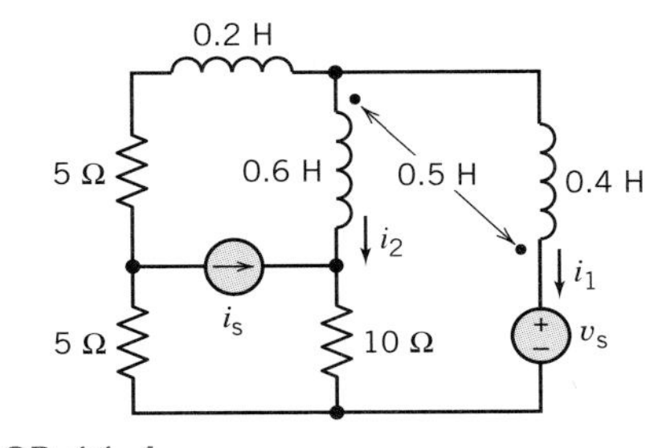

Figure SP 11.4

SP 11-5 A magnetic amplifier can be used to deliver large output power to a load R_L. Determine $i_o(t)$ for the circuit of Figure SP 11.5 when $v_s = 2 \cos 500t$ V and $R_L = 3\ \Omega$. Determine the output power to the load.

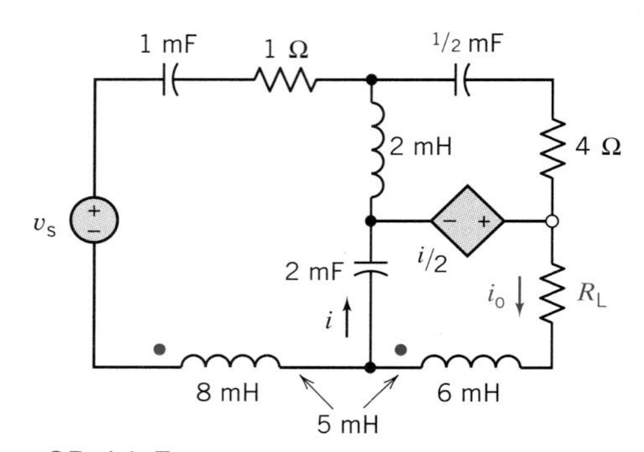

Figure SP 11.5 Magnetic amplifier circuit.

VERIFICATION PROBLEMS

VP 11-1 A laboratory report states that the average power is 214.6 W and the reactive power is 1,048 VAR delivered to the load shown in the circuit of Figure VP 11.1. Verify these values when $v_s = 115\sqrt{2} \cos(120\pi t)$ V.

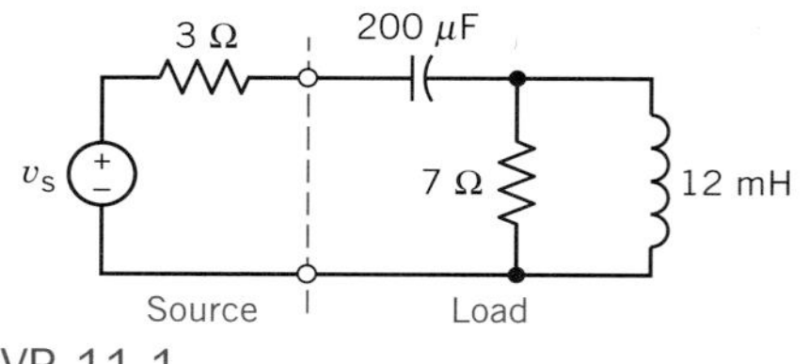

Figure VP 11.1

VP 11-2 A source and two loads are shown in Figure VP 11.2. One calculation reports that the power delivered to the total load is 5087 W. Check this report and calculate the overall power factor of the total load.

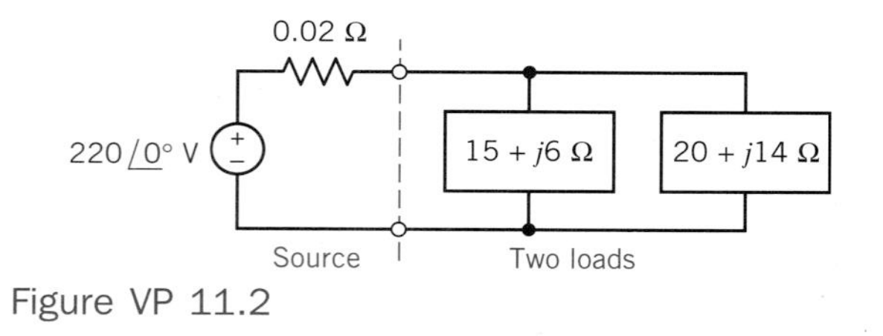

Figure VP 11.2

VP 11-3 A computer calculation states that the rms value of $v(t)$ of Figure VP 11.3 is 8.366 V. Verify this result.

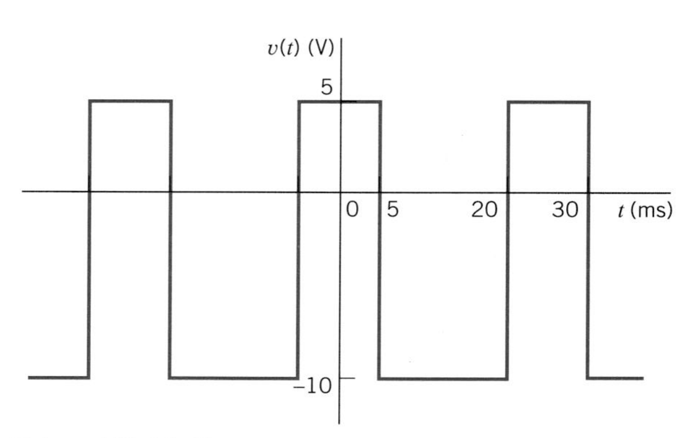

Figure VP 11.3

VP 11-4 A student report states that $\mathbf{Z}_{in} = j30\ \Omega$ for the circuit of Figure VP 11.4. Verify this result.

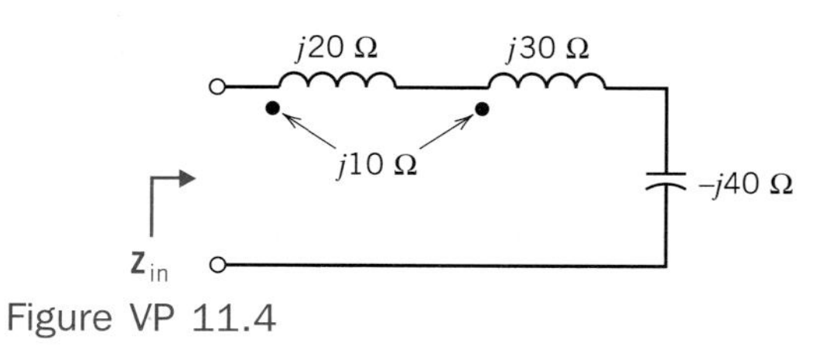

Figure VP 11.4

DESIGN PROBLEMS

DP 11-1 It is desired that the input impedance of the series *RLC* circuit shown in Figure DP 11.1 is $\mathbf{Z}_{in} = Z\angle\theta$ where $Z = 7.21\Omega$ and $\theta = 33.7°$. Select the inductance L when the frequency of the input source is 4 rad/s.

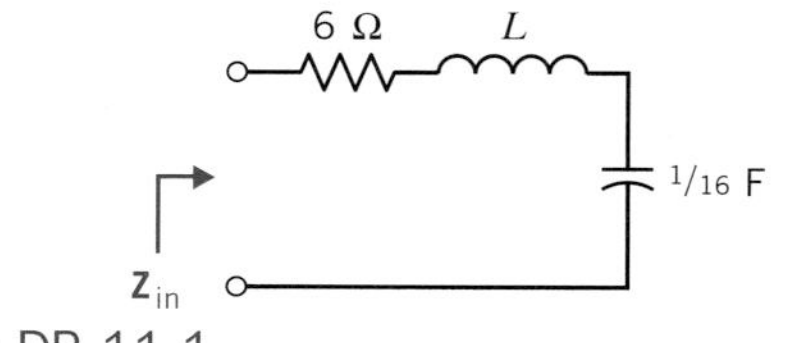

Figure DP 11.1

DP 11-2 A capacitor is in series with a 750 Ω resistor, as shown in Figure DP 11.2. The input is $v_s = 240 \cos \omega t$ V where $\omega = 2\pi(400)$. It is desired to limit the magnitude of the current to 0.2 A. Determine the required capacitance C.

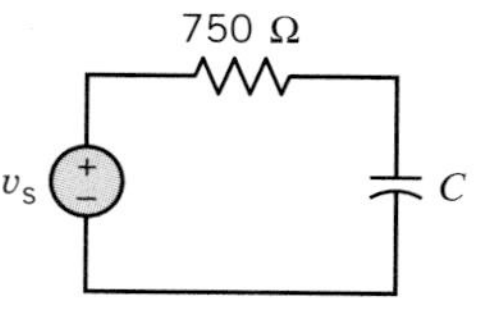

Figure DP 11.2

DP 11-3 The input admittance $\mathbf{Y}_{in} = Y\angle\theta$ of the circuit shown in Figure DP 11.3 is required to be $2.8 < Y < 2.9$ when the source frequency is 50 krad/s. Determine the required inductance L and the resulting angle θ.

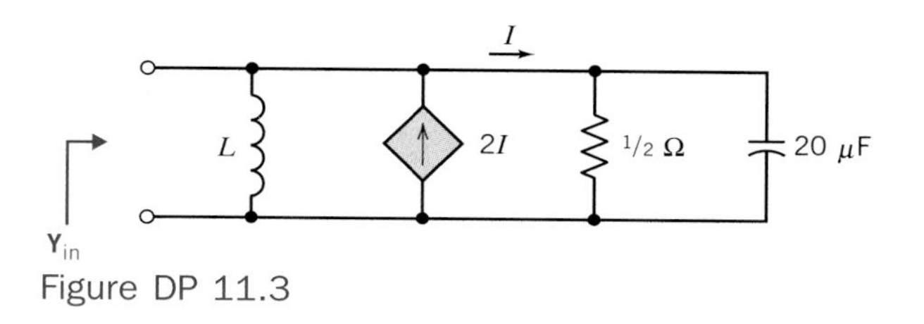

Figure DP 11.3

DP 11-4 Select the tuning capacitor C so that the input impedance of the circuit shown in Figure DP 11.4 is $\mathbf{Z}_{in} = Z\angle\theta$ where $4.2\Omega < Z < 4.6\Omega$. The frequency of operation is $\omega = 100$ rad/s.

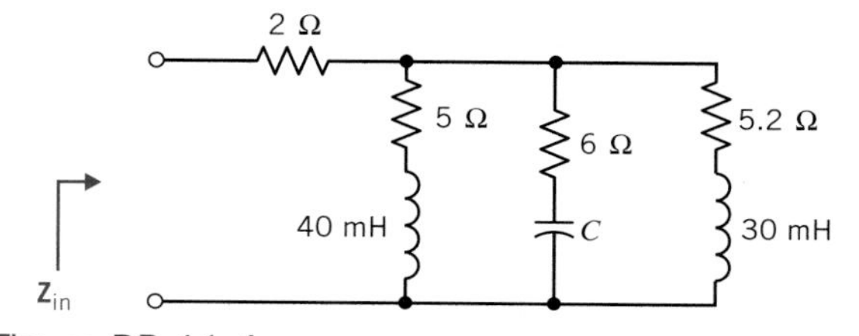

Figure DP 11.4

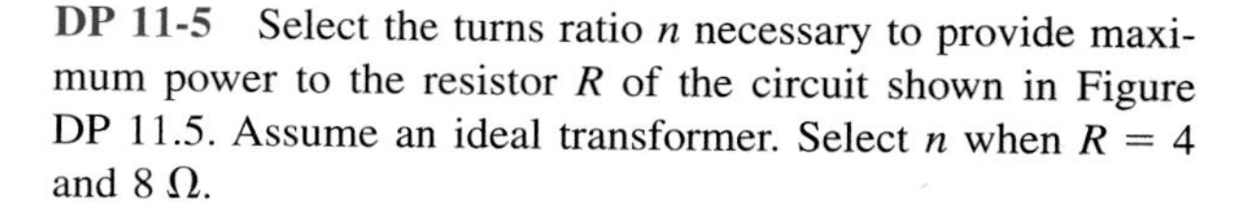

DP 11-5 Select the turns ratio n necessary to provide maximum power to the resistor R of the circuit shown in Figure DP 11.5. Assume an ideal transformer. Select n when $R = 4$ and 8 Ω.

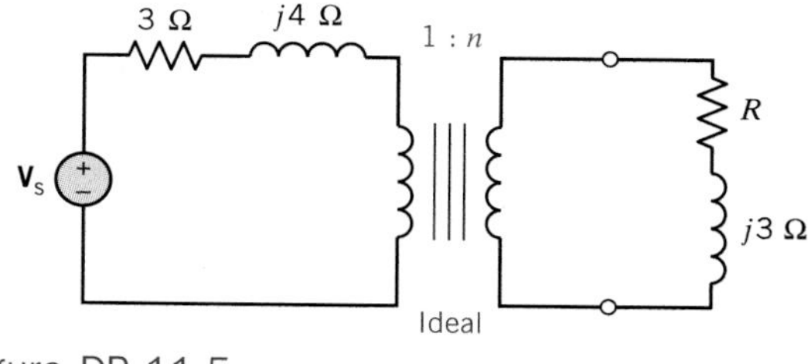

Figure DP 11.5

DP 11-6 A radio receiver operates at $\omega = 10^6$ rad/s with a load of a capacitor and resistor as shown in Figure DP 11.6. Select the cable R and L in order to deliver maximum power to the load. Determine the power delivered to the load when $v_s = V_1 \cos(\omega t + 30°)$ V and $V_1 = 0.045$ V.

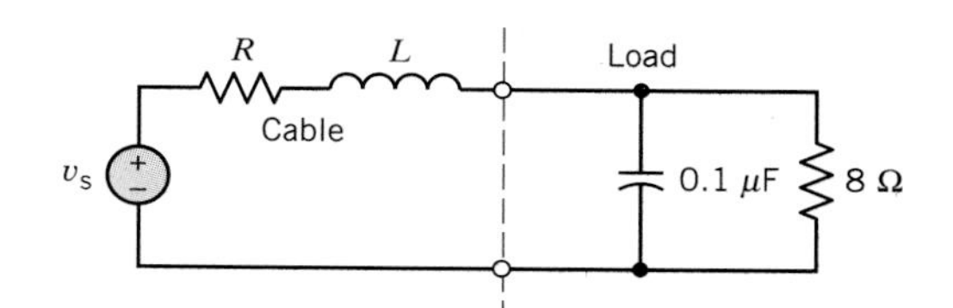

Figure DP 11.6 Radio receiver circuit.

DP 11-7 An amplifier in a short-wave radio operates at 100 kHz. The load $\mathbf{Z}_2$ is connected to a source through an ideal transformer, as shown in Figure DP 11.7. The load is a series connection 10 Ω resistance and 10-μH inductance. The $\mathbf{Z}_s$ consists of a 1 Ω resistance and a 1-μH inductance.

(a) Select an integer n in order to maximize the energy delivered to the load. Calculate $\mathbf{I}_2$ and the energy to the load.

(b) Add a capacitance C in series with Z_2 in order to improve the energy delivered to the load.

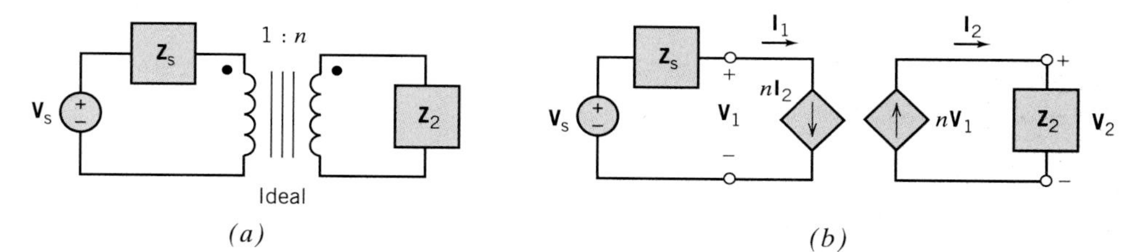

Figure DP 11.7 (*a*) Circuit with an ideal transformer. (*b*) Model of the circuit of Figure 12-25*a*. The direction of the dependent current source is downward due to $\mathbf{I}_2$ flowing out of the dotted terminal.

DP 11-8 A new electronic lamp (E-lamp) has been developed that uses a radio-frequency sinusoidal oscillator and a coil to transmit energy to a surrounding cloud of mercury gas as shown in Figure DP 11.8*a*. The mercury gas emits ultraviolet light which is transmitted to the phosphor coating, which, in turn, emits visible light. A circuit model of the E-lamp is shown in Figure DP 11.8*b*. The capacitance C and the resistance R are dependent on the lamp spacing design and the type of phosphor. Select R and C so that maximum power is delivered to R, which relates to the phosphor coating (Adler 1992). The circuit operates at $\omega_0 = 10^7$ rad/s.

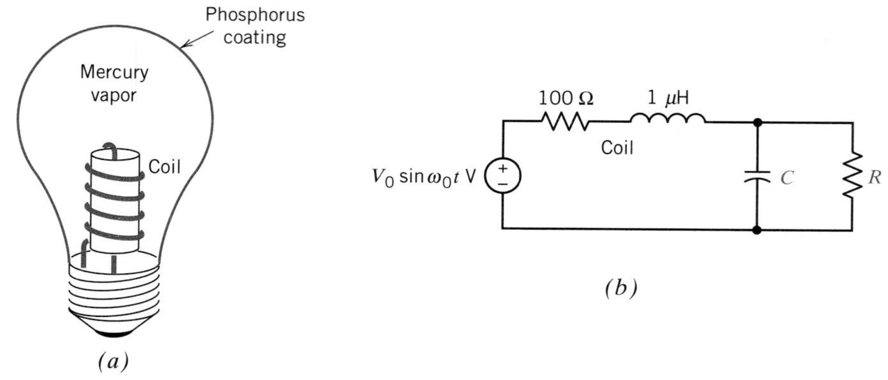

Figure DP 11.8 Electronic lamp.

CHAPTER 12

Three-Phase Circuits

Preview

In this chapter, we will begin to analyze *three-phase circuits*. These circuits will consist of three parts: a three-phase source, a three-phase load, and a transmission line. The three-phase source will consist of either three Y-connected sinusoidal voltage sources or three Δ-connected sinusoidal voltage sources. Similarly, the circuit elements that comprise the load will be connected to form either a Y or a Δ. The transmission line will be used to connect the source to the load and will consist of either three or four wires. These circuits are described using names that identify the way in which the source and the load are connected. For example, the circuit shown in Figure 12.4-1 has a Y-connected three-phase source and a Y-connected load. The circuit in Figure 12.4-1 is called a Y-to-Y circuit. The circuit in Figure 12.6-1 has a Y-connected three-phase source and a Δ-connected load. The circuit in Figure 12.6-1 is called a Y-to-Δ circuit.

Notice that the Y-to-Y circuit in Figure 12.4-1 has been represented in the frequency domain, using impedances and phasors. This is appropriate because the three voltage sources that comprise a three-phase source are sinusoidal sources having *the same frequency.* Analysis of three-phase circuits using phasors and impedances will determine the *steady-state response* of the three-phase circuit.

Before beginning our analysis of three-phase circuits, it is helpful to recall why it is advantageous to use phasors to find the steady-state response of linear circuits to sinusoidal inputs. Circuits that contain capacitors or inductors are represented by differential equations in the time domain. We can solve these *differential equations*, but it is a lot of work. Impedances and phasors are used to represent the circuit in the frequency domain. Linear circuits are represented by *algebraic equations* in the frequency domain. These algebraic equations involve complex numbers, but they are still easier to solve than the differential equations. Solving these algebraic equations provides the phasor corresponding to the output voltage or current. We know that the steady-state output voltage or current will be sinusoidal and will have the same frequency as the input sinusoid. The magnitude and phase angle of the phasor corresponding to the output voltage or current provide the magnitude and phase angle of the output sinusoid.

We will be particularly interested in the power the three-phase source delivers to the three-phase load. Table 12.0-1 summarizes the formulas that can be used to calculate the power delivered to an element when the element voltage and current adhere to the passive convention.

Table 12.0-1 **Frequency Domain Power Relationships**

Quantity	Relationship Using Peak Values	Relationship Using rms Values	Units
Element voltage, $v(t)$	$v(t) = V_m \cos(\omega t + \theta_V)$	$v(t) = V_{rms}\sqrt{2}\cos(\omega t + \theta_V)$	V
Element current, $i(t)$	$i(t) = I_m \cos(\omega t + \theta_I)$	$i(t) = I_{rms}\sqrt{2}\cos(\omega t + \theta_I)$	A
Complex power, $\mathbf{S}$	$\mathbf{S} = \frac{V_m I_m}{2}\cos(\theta_V - \theta_I) + j\frac{V_m I_m}{2}\sin(\theta_V - \theta_I)$	$\mathbf{S} = V_{rms} I_{rms}\cos(\theta_V - \theta_I) + jV_{rms} I_{rms}\sin(\theta_V - \theta_I)$	VA
Apparent power, $\lvert\mathbf{S}\rvert$	$\lvert\mathbf{S}\rvert = \frac{V_m I_m}{2}$	$\lvert\mathbf{S}\rvert = V_{rms} I_{rms}$	VA
Average power, P	$P = \frac{V_m I_m}{2}\cos(\theta_V - \theta_I)$	$P = V_{rms} I_{rms}\cos(\theta_V - \theta_I)$	W
Reactive power, Q	$Q = \frac{V_m I_m}{2}\sin(\theta_V - \theta_I)$	$Q = V_{rms} I_{rms}\sin(\theta_V - \theta_I)$	VAR

Table 12.0-1 also provides the equations for the sinusoidal element current and voltage. In the table, I_m and V_m are the magnitudes of the sinusoidal current and voltage, while I_{rms} and V_{rms} are the corresponding effective values of the current and voltage. Notice that the formulas for power in terms of I_{rms} and V_{rms} are simpler than the corresponding formulas in terms of I_m and V_m. In contrast, the equations giving the sinusoidal voltage and current are simpler when I_m and V_m are used. When engineers are interested primarily in power, they are likely to use I_{rms} and V_{rms}. On the other hand, when engineers are interested primarily in the sinusoidal currents and voltages, they are likely to use I_m and V_m. In this chapter, we are interested mainly in power and will use effective values.

12.1 Design Challenge

POWER FACTOR CORRECTION

Figure 12.1-1 shows a three-phase circuit. The capacitors are added to improve the power factor of the load. We need to determine the value of the capacitance C required to obtain a power factor of 0.9 lagging.

In this chapter we will describe three-phase circuits. In particular, we will show that balanced three-phase circuits can be analyzed by using "per-phase" equivalent circuits. This will enable us to calculate the capacitance required to correct the power factor of the load. We will return to this design problem at the end of the chapter.

12.2 TESLA AND POLYPHASE CIRCUITS

Nikola Tesla (1856–1943), greatly influenced the adoption of ac power. He first worked for Edison, who was quoted as saying, "My personal desire would be to prohibit entirely the use of alternating currents. They are as unnecessary as they are dangerous." Tesla left Edison and within five years had obtained ten patents for ac motors, employing his induction motor principle and his efficient polyphase power system. Manufacturer George Westinghouse bought these patents, and the two 3725-kW alternators that first supplied power at Niagara Falls in 1895 were based on Tesla's patents.

After much litigation, Tesla was awarded patent rights for his invention of the polyphase-current system to run an induction motor. Polyphase (as distinguished from single-phase) alternating current comes from coils in the generator, wound to produce two or more separate circuits delivering current to the terminals in such a way that they are in sequence as the machine rotates. During the following few decades, there were innumerable improvements in motors and a tremendous diversification of designs to run almost anything from a toy train to a ship.

However, because the early Westinghouse 133 1/3-Hz alternators, as alternating-current generators are often called, produced power at a frequency too high for efficient motor operation, 25- and 60-Hz alternators gradually became standard in the United States. At 60 Hz the human eye does not detect flicker in an incandescent lamp, while at 25 Hz a flicker is quite noticeable, especially in small lamps. So, for a time, 60 Hz was used when lighting loads predominated, and 25 Hz was used when power for motors was more important. In countries other than the United States, the standard lighting frequency became 50 Hz and the power frequency became 16 2/3 Hz. The large Westinghouse alternators

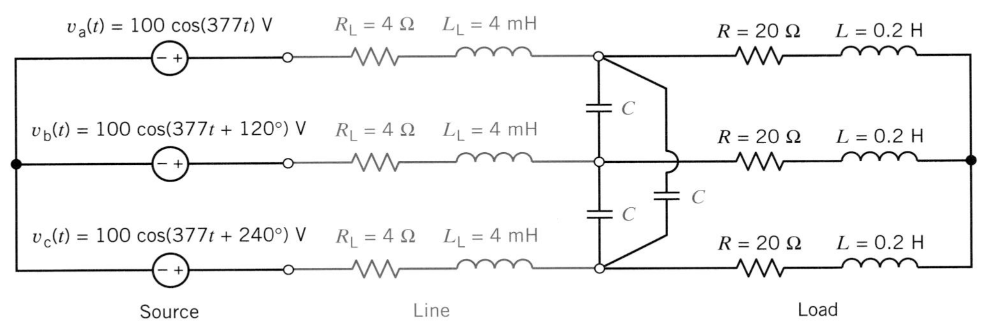

Figure 12.1-1 A balanced three-phase circuit.

at the Columbia Exposition in Chicago in 1893 were 60-Hz machines. The voltages of alternating and direct current were also becoming standardized in the 1890s, and by the end of the century, 110 V was the common American voltage for each phase of the polyphase circuits.

12.3 THREE-PHASE VOLTAGES

The generation and transmission of electrical power are more efficient in polyphase systems employing combinations of two, three, or more sinusoidal voltages. In addition, polyphase circuits and machines possess some unique advantages. For example, the power transmitted in a three-phase circuit is constant or independent of time rather than pulsating, as it is in a single-phase circuit. In addition, three-phase motors start and run much better than do single-phase motors. The most common form of polyphase system employs three balanced voltages, equal in magnitude and differing in phase by $360°/3 = 120°$.

An elementary ac generator consists of a rotating magnet and a stationary winding. The turns of the winding are spread along the periphery of the machine. The voltage generated in each turn of the winding is slightly out of phase with the voltage generated in its neighbor because it is cut by maximum magnetic flux density an instant earlier or later. The voltage produced in the first winding is $v_{aa'}$.

If the first winding were continued around the machine, the voltage generated in the last turn would be 180° out of phase with that in the first, and they would cancel, producing no useful effect. For this reason, one winding is commonly spread over no more than one-third of the periphery; the other two-thirds of the periphery can hold two more windings used to generate two other similar voltages. A simplified version of three windings around the periphery of a cylindrical drum is shown in Figure 12.3-1*a*. The three sinusoids (sinusoids are obtained with a proper winding distribution and magnet shape) generated by the three similar windings are shown in Figure 12.3-1*b*. Defining $v_{aa'}$ as the potential of terminal a with respect to terminal a', we describe the voltages as

$$\begin{aligned} v_{aa'} &= \sqrt{2}\,V \cos \omega t \\ v_{bb'} &= \sqrt{2}\,V \cos (\omega t - 120°) \\ v_{cc'} &= \sqrt{2}\,V \cos (\omega t - 240°) \end{aligned} \tag{12.3-1}$$

where V is the effective value.

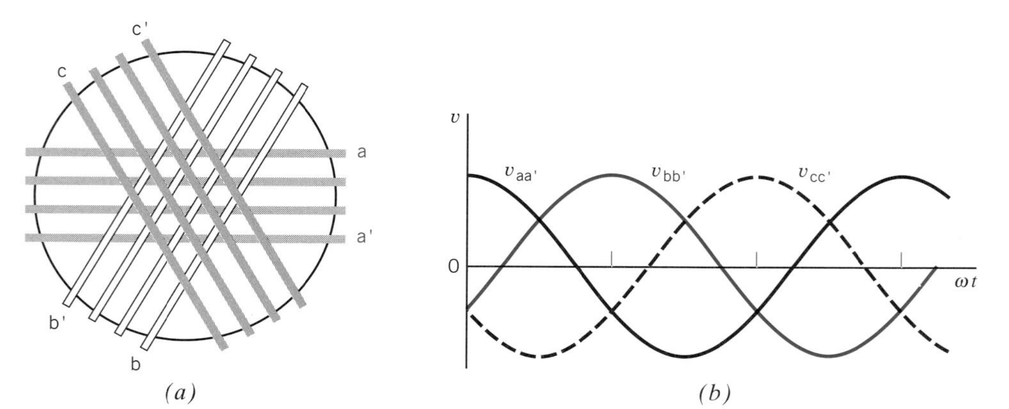

Figure 12.3-1 (*a*) The three windings on a cylindrical drum used to obtain three-phase voltages (end view). (*b*) Balanced three-phase voltages.

A **three-phase circuit** generates, distributes, and uses energy in the form of three voltages equal in magnitude and symmetric in phase.

The three similar portions of a three-phase system are called *phases*. Because the voltage in phase aa' reaches its maximum first, followed by that in phase bb' and then by that in phase cc', we say the phase rotation is *abc*. This is an arbitrary convention; for any given generator the phase rotation may be reversed by reversing the direction of rotation. The six-terminal ac generator is shown in Figure 12.3-2.

Using phasor notation, we may write Eq. 12.3-1 as

$$\begin{aligned} \mathbf{V}_{aa'} &= V \angle 0^\circ \\ \mathbf{V}_{bb'} &= V \angle -120^\circ \\ \mathbf{V}_{cc'} &= V \angle -240^\circ = V \angle 120^\circ \end{aligned} \tag{12.3-2}$$

The three voltages are said to be *balanced voltages* because they have identical amplitude, V, and frequency, ω, and are out of phase with each other by exactly 120°. The phasor diagram of the balanced three-phase voltages is shown in Figure 12.3-3. Examining Figure 12.3-3, we find

$$\mathbf{V}_{aa'} + \mathbf{V}_{bb'} + \mathbf{V}_{cc'} = 0 \tag{12.3-3}$$

For notational ease we henceforth use $\mathbf{V}_{aa'} = \mathbf{V}_a$, $\mathbf{V}_{bb'} = \mathbf{V}_b$, and $\mathbf{V}_{cc'} = \mathbf{V}_c$ as the three voltages.

The **positive phase sequence** is *abc*, as shown in Figure 12.3-3. The sequence *acb* is called the negative phase sequence, as shown in Figure 12.3-4.

Often the phase voltage in the Y connection is written as

$$\mathbf{V}_a = V_p \angle 0^\circ$$

where V_p is the magnitude of the phase voltage.

Referring to the generator of Figure 12.3-2, there are six terminals and three voltages, v_a, v_b, and v_c. We use phasor notation and assume that each phase winding provides a source voltage in series with a negligible impedance. Under these assumptions, there are two ways of interconnecting the three sources, as shown in Figure 12.3-5. The common terminal of the Y connection is called the *neutral terminal* and is labeled *n*. The neutral terminal may or may not be available for connection. Balanced loads result in no current in a neutral wire, and thus it is often not needed.

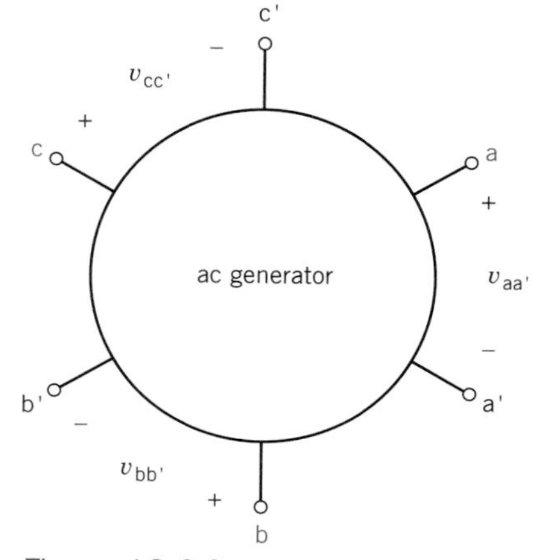

Figure 12.3-2 Generator with six terminals.

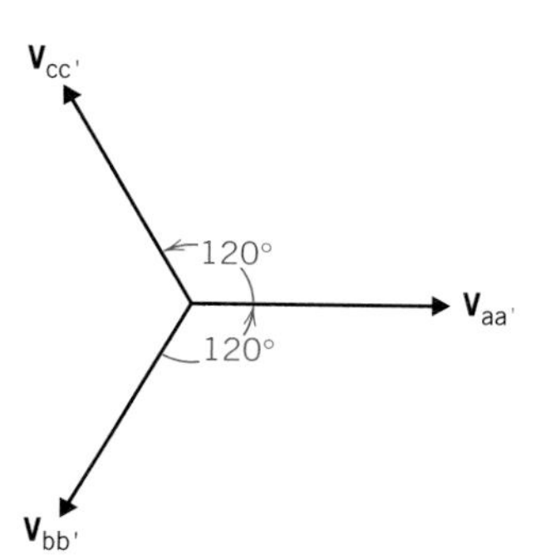

Figure 12.3-3 Phasor representation of the positive phase sequence of the balanced three-phase voltages.

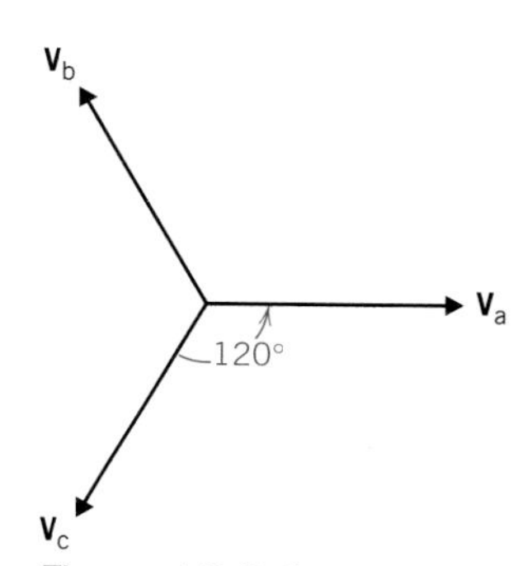

Figure 12.3-4 The negative phase sequence acb in the Y connection.

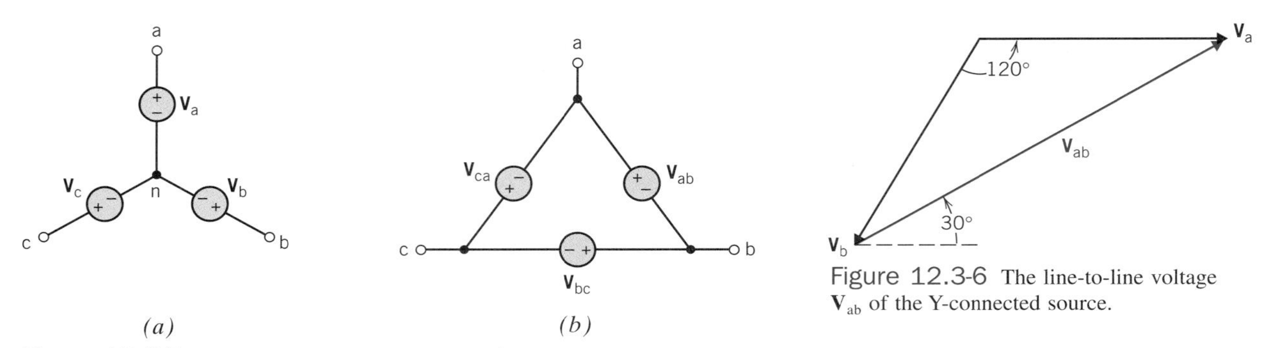

Figure 12.3-6 The line-to-line voltage $\mathbf{V}_{ab}$ of the Y-connected source.

(*a*) (*b*)

Figure 12.3-5 (*a*) Y-connected sources and (*b*) Δ-connected sources.

The connection shown in Figure 12.3-5*a* is called the Y connection, and the Δ connection is shown in Figure 12.3-5*b*. The Y connection selects terminals a', b', and c' and connects them together as neutral. Then the line-to-line voltage, $\mathbf{V}_{ab}$, of the Y-connected sources is

$$\mathbf{V}_{ab} = \mathbf{V}_a - \mathbf{V}_b \tag{12.3-4}$$

as is evident by examining Figure 12.3-5*a*. Since $\mathbf{V}_a = V_p \angle 0°$ *and* $\mathbf{V}_b = V_p \angle -120°$, we have

$$\begin{aligned} \mathbf{V}_{ab} &= V_p - V_p(-0.5 - j0.866) \\ &= V_p(1.5 + j0.866) \\ &= \sqrt{3}\, V_p \angle 30° \end{aligned} \tag{12.3-5}$$

This relationship is also demonstrated by the phasor diagram of Figure 12.3-6. Similarly,

$$\mathbf{V}_{bc} = \sqrt{3}\, V_p \angle -90° \tag{12.3-6}$$

and

$$\mathbf{V}_{ca} = \sqrt{3}\, V_p \angle -210° \tag{12.3-7}$$

Therefore, in a Y connection, the line-to-line voltage is $\sqrt{3}$ times the phase voltage and is displaced 30° in phase. The line current is equal to the phase current. We normally use V_p to remind us that it is the phase voltage.

Exercise 12.3-1 The Y-connected three-phase voltage source has $\mathbf{V}_c = 120 \angle -240°\ V$ rms. Find the line-to-line voltage $\mathbf{V}_{bc}$.

Answer: $207.8 \angle -90°$ V rms

12.4 THE Y-TO-Y CIRCUIT

Consider the Y-to-Y circuit shown in Figure 12.4-1. This three-phase circuit consists of three parts: a three-phase source, a three-phase load, and a transmission line. The three-phase source consists of three Y-connected sinusoidal voltage sources. The impedances that comprise the load are connected to form a Y. The transmission line used to connect the source to the load consists of four wires, including a wire connecting the neutral node of the source to the neutral node of the load. Figure 12.4-2 shows another Y-to-Y circuit.

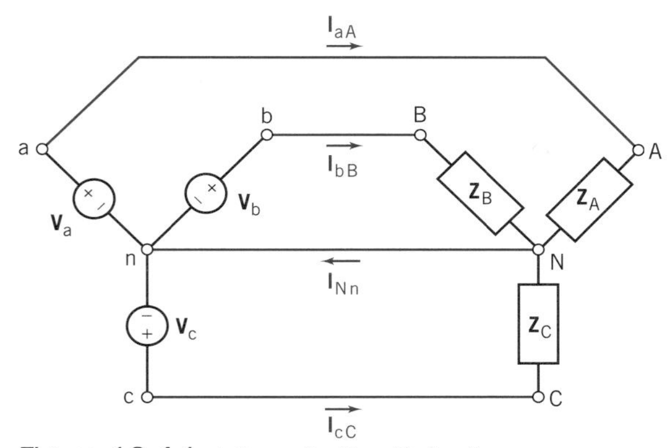

Figure 12.4-1 A four-wire Y-to-Y circuit.

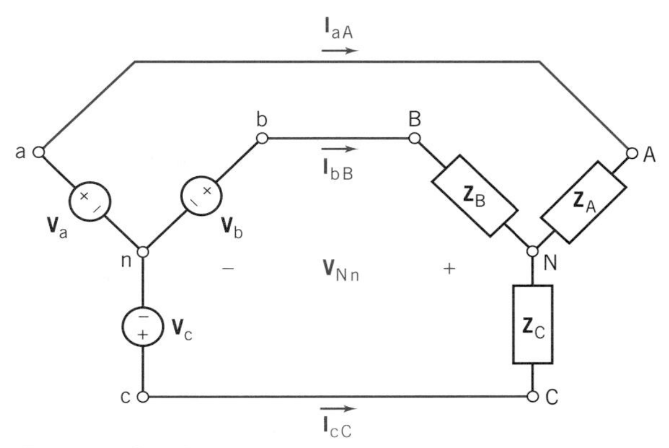

Figure 12.4-2 A three-wire Y-to-Y circuit.

In Figure 12.4-2, the three-phase source is connected to the load using three wires, without a wire connecting the neutral node of the source to the neutral node of the load. To distinguish between these circuits, the circuit in Figure 12.4-1 is called a four-wire Y-to-Y circuit, while the circuit in Figure 12.4-2 is called a three-wire Y-to-Y circuit.

Analysis of the four-wire Y-to-Y circuit in Figure 12.4-1 is relatively easy. Each impedance of the three-phase load is connected directly across a voltage source of the three-phase source. Therefore, the voltage across the impedance is known, and the line currents are easily calculated as

$$\mathbf{I}_{aA} = \frac{\mathbf{V}_a}{\mathbf{Z}_A}, \quad \mathbf{I}_{bB} = \frac{\mathbf{V}_b}{\mathbf{Z}_B}, \quad \text{and} \quad \mathbf{I}_{cC} = \frac{\mathbf{V}_c}{\mathbf{Z}_C} \tag{12.4-1}$$

The current in the wire connecting the neutral node of the source to the neutral node of the load is

$$\mathbf{I}_{Nn} = \mathbf{I}_{aA} + \mathbf{I}_{bB} + \mathbf{I}_{cC} = \frac{\mathbf{V}_a}{\mathbf{Z}_A} + \frac{\mathbf{V}_b}{\mathbf{Z}_B} + \frac{\mathbf{V}_c}{\mathbf{Z}_C} \tag{12.4-2}$$

The average power delivered by the three-phase source to the three-phase load is calculated by adding up the average power delivered to each impedance of the load.

$$P = P_A + P_B + P_C \tag{12.4-3}$$

where, for example, P_A is the average power absorbed by $\mathbf{Z}_A$. P_A is easily calculated once $\mathbf{I}_{aA}$ is known.

For convenience, let the phase voltages of the Y-connected source be

$$\mathbf{V}_a = V\angle 0° \text{ V rms}, \quad \mathbf{V}_b = V\angle -120° \text{ V rms}, \quad \text{and} \quad \mathbf{V}_c = V\angle 120° \text{ V rms}$$

When $\mathbf{Z}_A = \mathbf{Z}_B = \mathbf{Z}_C = \mathbf{Z} = Z\angle\theta$, the load is said to be a *balanced load*. In general, analysis of balanced three-phase circuits is easier than analysis of unbalanced three-phase circuits. The line currents in the balanced, four-wire Y-to-Y circuit are given by

$$\mathbf{I}_{aA} = \frac{\mathbf{V}_a}{\mathbf{Z}} = \frac{V\angle 0°}{Z\angle\theta}, \quad \mathbf{I}_{bB} = \frac{\mathbf{V}_b}{\mathbf{Z}} = \frac{V\angle -120°}{Z\angle\theta}, \quad \text{and} \quad \mathbf{I}_{cC} = \frac{\mathbf{V}_c}{\mathbf{Z}} = \frac{V\angle 120°}{Z\angle\theta}$$

Then

$$\mathbf{I}_{aA} = \frac{V}{Z}\angle -\theta, \quad \mathbf{I}_{bB} = \frac{V}{Z}\angle(-\theta - 120°), \quad \text{and} \quad \mathbf{I}_{cC} = \frac{V}{Z}\angle(-\theta + 120°) \tag{12.4-4}$$

The line currents have equal magnitudes and differ in phase by 120°. $\mathbf{I}_{bB}$ and $\mathbf{I}_{cC}$ can be calculated from $\mathbf{I}_{aA}$ by subtracting and adding 120° to the phase angle of $\mathbf{I}_{aA}$.

The current in the wire connecting the neutral node of the source to the neutral node of the load is

$$\mathbf{I}_{\mathrm{Nn}} = \mathbf{I}_{\mathrm{aA}} + \mathbf{I}_{\mathrm{bB}} + \mathbf{I}_{\mathrm{cC}} = \frac{V}{Z\angle\theta}(1\angle 0° + 1\angle -120° + 1\angle 120°)$$
$$= 0 \qquad (12.4\text{-}5)$$

There is no current in the wire connecting the neutral node of the source to the neutral node of the load.

Because effective values of the sinusoidal voltages and currents have been used instead of peak values, the appropriate formulas for power are those given in the "effective values" column of Table 12.0-1. The average power delivered to the load is

$$P = P_{\mathrm{A}} + P_{\mathrm{B}} + P_{\mathrm{C}} = V\frac{V}{Z}\cos(-\theta) + V\frac{V}{Z}\cos(-\theta) + V\frac{V}{Z}\cos(-\theta)$$
$$= 3\frac{V^2}{Z}\cos(\theta) \qquad (12.4\text{-}6)$$

where, for example, P_{A} is the average power absorbed by $\mathbf{Z}_{\mathrm{A}}$. Equal power is absorbed by each impedance of the three-phase load, $\mathbf{Z}_{\mathrm{A}}$, $\mathbf{Z}_{\mathrm{B}}$, and $\mathbf{Z}_{\mathrm{C}}$. It is not necessary to calculate P_{A}, P_{B}, and P_{C} separately. The average power delivered to the load can be determined by calculating P_{A} and multiplying by three.

Next, consider the three-wire Y-to-Y circuit shown in Figure 12.4-2. The phase voltages of the Y-connected source are $\mathbf{V}_{\mathrm{a}} = V\angle 0°$ V rms, $\mathbf{V}_{\mathrm{b}} = V\angle -120°$ V rms, and $\mathbf{V}_{\mathrm{c}} = V\angle 120°$ V rms. The first step in the analysis of this circuit is to calculate $\mathbf{V}_{\mathrm{Nn}}$, the voltage at the neutral node of the three-phase load with respect to the voltage at the neutral node of the three-phase source. (This step wasn't needed when the four-wire Y-to-Y circuit was analyzed because the fourth wire forced $\mathbf{V}_{\mathrm{Nn}} = 0$.) It is convenient to select node n, the neutral node of the three-phase source, to be the reference node. Then $\mathbf{V}_{\mathrm{a}}$, $\mathbf{V}_{\mathrm{b}}$, $\mathbf{V}_{\mathrm{c}}$, and $\mathbf{V}_{\mathrm{Nn}}$ are the node voltages of the circuit. Write a node equation at node N to get

$$0 = \frac{\mathbf{V}_{\mathrm{a}} - \mathbf{V}_{\mathrm{Nn}}}{\mathbf{Z}_{\mathrm{A}}} + \frac{\mathbf{V}_{\mathrm{b}} - \mathbf{V}_{\mathrm{Nn}}}{\mathbf{Z}_{\mathrm{B}}} + \frac{\mathbf{V}_{\mathrm{c}} - \mathbf{V}_{\mathrm{Nn}}}{\mathbf{Z}_{\mathrm{C}}}$$
$$= \frac{(V\angle 0°) - \mathbf{V}_{\mathrm{Nn}}}{\mathbf{Z}_{\mathrm{A}}} + \frac{(V\angle -120°) - \mathbf{V}_{\mathrm{Nn}}}{\mathbf{Z}_{\mathrm{B}}} + \frac{(V\angle 120°) - \mathbf{V}_{\mathrm{Nn}}}{\mathbf{Z}_{\mathrm{C}}} \qquad (12.4\text{-}7)$$

Solving for $\mathbf{V}_{\mathrm{Nn}}$ gives

$$\mathbf{V}_{\mathrm{Nn}} = \frac{(V\angle -120°)\mathbf{Z}_{\mathrm{A}}\mathbf{Z}_{\mathrm{C}} + (V\angle 120°)\mathbf{Z}_{\mathrm{A}}\mathbf{Z}_{\mathrm{B}} + (V\angle 0°)\mathbf{Z}_{\mathrm{B}}\mathbf{Z}_{\mathrm{C}}}{\mathbf{Z}_{\mathrm{A}}\mathbf{Z}_{\mathrm{C}} + \mathbf{Z}_{\mathrm{A}}\mathbf{Z}_{\mathrm{B}} + \mathbf{Z}_{\mathrm{B}}\mathbf{Z}_{\mathrm{C}}} \qquad (12.4\text{-}8)$$

Once $\mathbf{V}_{\mathrm{Nn}}$ has been determined, the line currents can be calculated using

$$\mathbf{I}_{\mathrm{aA}} = \frac{\mathbf{V}_{\mathrm{a}} - \mathbf{V}_{\mathrm{Nn}}}{\mathbf{Z}_{\mathrm{A}}}, \quad \mathbf{I}_{\mathrm{bB}} = \frac{\mathbf{V}_{\mathrm{b}} - \mathbf{V}_{\mathrm{Nn}}}{\mathbf{Z}_{\mathrm{B}}}, \quad \mathbf{I}_{\mathrm{cC}} = \frac{\mathbf{V}_{\mathrm{c}} - \mathbf{V}_{\mathrm{Nn}}}{\mathbf{Z}_{\mathrm{C}}} \qquad (12.4\text{-}9)$$

Analysis of the three-wire Y-to-Y circuit is much simpler when the circuit is balanced, that is, when $\mathbf{Z}_{\mathrm{A}} = \mathbf{Z}_{\mathrm{B}} = \mathbf{Z}_{\mathrm{C}} = \mathbf{Z} = Z\angle\theta$. When the circuit is balanced, Eq. 12.4-8 becomes

$$\mathbf{V}_{\mathrm{Nn}} = \frac{(V\angle -120°)\mathbf{ZZ} + (V\angle 120°)\mathbf{ZZ} + (V\angle 0°)\mathbf{ZZ}}{\mathbf{ZZ} + \mathbf{ZZ} + \mathbf{ZZ}}$$
$$= [(V\angle -120°) + (V\angle 120°) + (V\angle 0°)]/3$$
$$= 0 \qquad (12.4\text{-}10)$$

When a three-wire Y-to-Y is balanced, it is not necessary to write and solve a node equation to find $\mathbf{V}_{Nn}$ because $\mathbf{V}_{Nn}$ is known to be zero. Recall that $\mathbf{V}_{Nn} = 0$ in the four-wire Y-to-Y circuit. The balanced three-wire Y-to-Y circuit acts like the balanced four-wire Y-to-Y circuit. In particular, the line currents are given by Eq. 12.4-4, and the average power delivered to the load is given by Eq. 12.4-6.

Ideally, the transmission line connecting the load to the source can be modeled using short circuits. That's what was done in both Figures 12.4-1 and 12.4-2. Sometimes it is appropriate to model the lines connecting load to the source as impedances. For example, this is done when comparing the power that is delivered to the load to the power that is absorbed by the transmission line. Figure 12.4-3 shows a three-wire Y-to-Y circuit in which the transmission line is modeled by the line impedances $\mathbf{Z}_{aA}$, $\mathbf{Z}_{bB}$, and $\mathbf{Z}_{cC}$. The line impedances do not significantly complicate the analysis of the circuit since each line impedance is connected in series with a load impedance. After replacing series impedances by equivalent impedances, the analysis proceeds as before. If the circuit is not balanced, a node equation is written and solved to determine $\mathbf{V}_{Nn}$. Once $\mathbf{V}_{Nn}$ has been determined, the line currents can be calculated. Both the power delivered to the load and the power absorbed by the line can be calculated from the line currents and the load and line impedances.

Analysis of balanced Y-to-Y circuits is simpler than analysis of unbalanced Y-to-Y circuits in several ways:

1. $\mathbf{V}_{Nn} = 0$. It is not necessary to write and solve a node equation to determine $\mathbf{V}_{Nn}$.
2. The line currents have equal magnitudes and differ in phase by 120°. $\mathbf{I}_{bB}$ and $\mathbf{I}_{cC}$ can be calculated from $\mathbf{I}_{aA}$ by subtracting and adding 120° to the phase angle of $\mathbf{I}_{aA}$.
3. Equal power is absorbed by each impedance of the three-phase load, $\mathbf{Z}_A$, $\mathbf{Z}_B$, and $\mathbf{Z}_C$. It is not necessary to calculate P_A, P_B, and P_C separately. The average power delivered to the load can be determined by calculating P_A and multiplying by three.

The key to analysis of the balanced Y-to-Y circuit is calculation of the line current, $\mathbf{I}_{aA}$. The *per-phase equivalent circuit* provides the information needed to the line current, $\mathbf{I}_{aA}$. This equivalent circuit consists of the voltage source and impedances in the one phase of the three phases of the three-phase circuit. Figure 12.4-4 shows the per-phase equivalent circuit corresponding to the three-phase circuit shown in Figure 12.4-3. The neutral nodes, n and N, are connected by a short circuit in the per-phase equivalent circuit to indicate that $\mathbf{V}_{Nn} = 0$ in a balanced Y-to-Y circuit. The per-phase equivalent circuit can be used to analyze either three-wire or four-wire balanced Y-to-Y circuits, but it can only be used for *balanced* circuits.

The behavior of a balanced Y-to-Y circuit is summarized in Table 12.4-1.

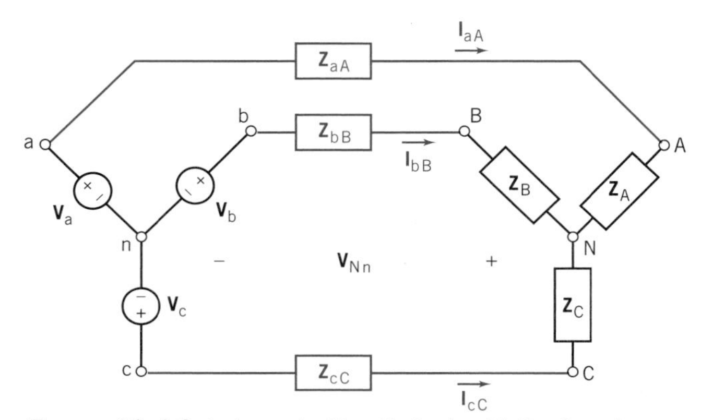

Figure 12.4-3 A three-wire Y-to-Y circuit with line impedances.

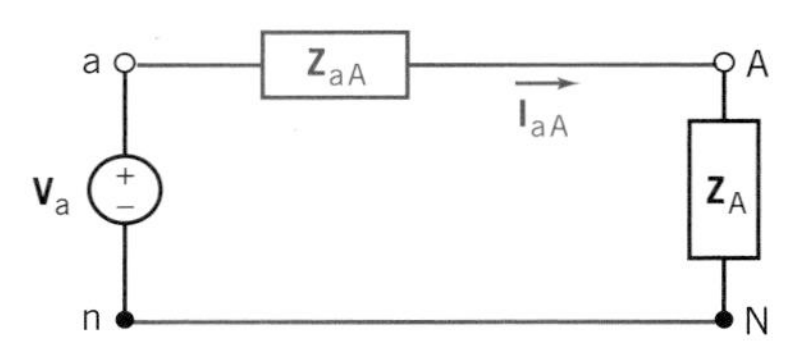

Figure 12.4-4 Per-phase equivalent circuit for the three-wire Y-to-Y circuit with line impedances.

Table 12.4-1 **The Balanced Y-to-Y Circuit**

Phase voltages	$\mathbf{V}_a = V_p\angle 0°$
	$\mathbf{V}_b = V_p\angle -120°$
	$\mathbf{V}_c = V_p\angle -240°$
Line-to-line voltages	$\mathbf{V}_{ab} = \mathbf{V}_L$
	$= \sqrt{3}\, V_p \angle 30°$
	$\mathbf{V}_{bc} = \sqrt{3}\, V_p \angle -90°$
	$\mathbf{V}_{ca} = \sqrt{3}\, V_p \angle -210°$
	$\mathbf{V}_L = \sqrt{3}\, V_p$
Currents	$\mathbf{I}_L = \mathbf{I}_p$ (line current = phase current)
	$\mathbf{I}_A = \dfrac{\mathbf{V}_a}{\mathbf{Z}_p} = \dfrac{\mathbf{V}_a}{\mathbf{Z}_Y} = I_p\angle -\theta$ with $\mathbf{Z}_p = Z\angle\theta$
	$\mathbf{I}_B = \mathbf{I}_A\angle -120°$
	$\mathbf{I}_C = \mathbf{I}_A\angle -240°$

p = phase, L = line.

Example 12.4-1

Determine the complex power delivered to the three-phase load of a four-wire Y-to-Y circuit such as the one shown in Figure 12.4-1. The phase voltages of the Y-connected source are $\mathbf{V}_a = 110\angle 0°$ V rms, $\mathbf{V}_b = 110\angle -120°$ V rms, and $\mathbf{V}_c = 110\angle 120°$ V rms. The load impedances are $\mathbf{Z}_A = 50 + j80\ \Omega$, $\mathbf{Z}_B = j50\ \Omega$, and $\mathbf{Z}_C = 100 + j25\ \Omega$.

Solution

The line currents of an *unbalanced* four-wire Y-to-Y circuit are calculated using Eq. 12.4-1. In this example

$$\mathbf{I}_{aA} = \frac{\mathbf{V}_a}{\mathbf{Z}_A} = \frac{110\angle 0°}{50 + j80},\quad \mathbf{I}_{bB} = \frac{\mathbf{V}_b}{\mathbf{Z}_B} = \frac{110\angle -120°}{j50},\quad \text{and}\quad \mathbf{I}_{cC} = \frac{\mathbf{V}_c}{\mathbf{Z}_C} = \frac{110\angle 120°}{100 + j25}$$

so

$$\mathbf{I}_{aA} = 1.16\angle -58°\ \text{A rms},\quad \mathbf{I}_{bB} = 2.2\angle 150°\ \text{A rms},\quad \text{and}\quad \mathbf{I}_{cC} = 1.07\angle 106°\ \text{A rms}$$

The complex power delivered to $\mathbf{Z}_A$ is

$$\mathbf{S}_A = \mathbf{I}_{aA}{}^*\mathbf{V}_a = (1.16\angle -58°)^*(110\angle 0°) = (1.16\angle 58°)(110\angle 0°) = 68 + j109\ \text{VA}$$

Similarly, we calculate the complex power delivered to $\mathbf{Z}_B$ and $\mathbf{Z}_C$ as

$$\mathbf{S}_B = (2.2\angle 150°)^*(110\angle -120°) = j242\ \text{VA}$$

and

$$\mathbf{S}_C = (107\angle 106°)^*(110\angle 120°) = 114 + j28\ \text{VA}$$

The total complex power delivered to the three-phase load is

$$\mathrm{S}_A + \mathrm{S}_B + \mathrm{S}_C = 182 + j379\ \text{VA}$$

Example 12.4-2

Determine the complex power delivered to the three-phase load of a four-wire Y-to-Y circuit such as the one shown in Figure 12.4-1. The phase voltages of the Y-connected source are $\mathbf{V}_a = 110\angle 0°$ V rms, $\mathbf{V}_b = 110\angle -120°$ V rms, and $\mathbf{V}_c = 110\angle 120°$ V rms. The load impedances are $\mathbf{Z}_A = \mathbf{Z}_B = \mathbf{Z}_C = 50 + j80\ \Omega$.

Solution

This example is similar to the previous example. The important difference is that this three-phase circuit is balanced. We need to calculate only one line current, $\mathbf{I}_{aA}$, and the complex power, $\mathbf{S}_A$, delivered to only one of the load impedances, $\mathbf{Z}_A$. The power delivered to the three-phase load is $3\mathbf{S}_A$. We begin by calculating $\mathbf{I}_{aA}$ as

$$\mathbf{I}_{aA} = \frac{\mathbf{V}_a}{\mathbf{Z}_A} = \frac{110\angle 0°}{50 + j80} = 1.16\angle -58° \text{ A rms}$$

The complex power delivered to $\mathbf{Z}_A$ is

$$\mathbf{S}_A = \mathbf{I}_{aA}{}^*\mathbf{V}_a = (1.16\angle -58°)^*(110\angle 0°) = (1.16\angle 58°)(110\angle 0°) = 68 + j109 \text{ VA}$$

The total power delivered to the three-phase load is

$$3\mathbf{S}_A = 204 + j326 \text{ VA}$$

(The currents $\mathbf{I}_{bB}$ and $\mathbf{I}_{cC}$ can also be calculated using Eq. 12.4-1. Verify that $\mathrm{I}_{bB} = 1.16\angle -177°$ A rms and $\mathbf{I}_{cC} = 1.16\angle 62°$ A rms. Notice that $\mathbf{I}_{bB}$ and $\mathbf{I}_{cC}$ can be calculated from $\mathbf{I}_{aA}$ by subtracting and adding 120° to the phase angle of $\mathbf{I}_{aA}$. Also, check that the complex power delivered to $\mathbf{Z}_B$ and to $\mathbf{Z}_C$ is equal to the complex power delivered to $\mathbf{Z}_A$. That is, $\mathbf{S}_B = 68 + j109$ VA and $\mathbf{S}_C = 68 + j109$ VA.)

Example 12.4-3

Determine the complex power delivered to the three-phase load of a three-wire Y-to-Y circuit such as the one shown in Figure 12.4-2. The phase voltages of the Y-connected source are $\mathbf{V}_a = 110\angle 0°$ V rms, $\mathbf{V}_b = 110\angle -120°$ V rms, and $\mathbf{V}_c = 110\angle 120°$ V rms. The load impedances are $\mathbf{Z}_A = 50 + j80\ \Omega$, $\mathbf{Z}_B = j50\ \Omega$, and $\mathbf{Z}_C = 100 + j25\ \Omega$.

Solution

This example seems similar to Example 12.4-1 but considers a three-wire Y-to-Y circuit instead of the four-wire circuit considered in Example 12.4-1. Since the circuit is unbalanced, $\mathbf{V}_{Nn}$ is not known. We begin by writing and solving a node equation to determine $\mathbf{V}_{Nn}$. The solution of that node equation is given in Eq. 12.4-8 to be

$$\mathbf{V}_{Nn} = \frac{(110\angle -120°)(50 + j80)(100 + j25) + (110\angle 120°)(50 + j80)(j50) + (110\angle 0°)(j50)(100 + j25)}{(50 + j80)(100 + j25) + (50 + j80)(j50) + (j50)(100 + j25)}$$
$$= 56\angle -151° \text{ V rms}$$

Now that $\mathbf{V}_{Nn}$ is known, the line currents are calculated as

$$\mathbf{I}_{aA} = \frac{\mathbf{V}_a - \mathbf{V}_{Nn}}{\mathbf{Z}_A} = \frac{110\angle 0° - 56\angle -151°}{50 + j80} = 1.71\angle -48° \text{ A rms}$$

$$\mathbf{I}_{bB} = \frac{\mathbf{V}_b - \mathbf{V}_{Nn}}{\mathbf{Z}_B} = 2.45\angle 3° \text{ A rms} \quad \text{and} \quad \mathbf{I}_{cC} = \frac{\mathbf{V}_c - \mathbf{V}_{Nn}}{\mathbf{Z}_C} = 1.19\angle 79° \text{ A rms}$$

The complex power delivered to $\mathbf{Z}_A$ is

$$\mathbf{S}_A = \mathbf{I}_{aA}^*\,\mathbf{V}_a = \mathbf{I}_{aA}{}^*(\mathbf{I}_{aA}\mathbf{Z}_A) = (1.71\angle -48°)^*(1.71\angle -48°)(50 + j80) = 146 + j234 \text{ VA}$$

Similarly, we calculate the complex power delivered to $\mathbf{Z}_B$ and $\mathbf{Z}_C$ as

$$\mathbf{S}_B = \mathbf{I}_{bB}^* (\mathbf{I}_{bB}\mathbf{Z}_B) = j94 \text{ VA} \quad \text{and} \quad \mathbf{S}_C = \mathbf{I}_{cC}^*(\mathbf{I}_{cC}\,\mathbf{Z}_C) = 141 + j35 \text{ VA}$$

The total complex power delivered to the three-phase load is

$$\mathbf{S}_A + \mathbf{S}_B + \mathbf{S}_C = 287 + j364 \text{ VA}$$

Example 12.4-4

Determine complex power delivered to the three-phase load of a three-wire Y-to-Y circuit such as the one shown in Figure 12.4-2. The phase voltages of the Y-connected source are $\mathbf{V}_a = 110\angle 0°$ V rms, $\mathbf{V}_b = 110\angle -120°$ V rms, and $\mathbf{V}_c = 110\angle 120°$ V rms. The load impedances are $\mathbf{Z}_A = \mathbf{Z}_B = \mathbf{Z}_C = 50 + j80\ \Omega$.

Solution

This example is similar to Example 12.4-3. The important difference is that this three-phase circuit is balanced, so $\mathbf{V}_{Nn} = 0$. It is not necessary to write and solve a node equation to determine $\mathbf{V}_{Nn}$.

Balanced three-wire Y-to-Y circuits and balanced four-wire Y-to-Y circuits are analyzed in the same way. We need calculate only one line current, $\mathbf{I}_{aA}$, and the complex power, $\mathbf{S}_a$, delivered to only one of the load impedances, $\mathbf{Z}_A$. The power delivered to the three-phase load is $3\mathbf{S}_a$.

The line current is calculated as

$$\mathbf{I}_{aA} = \frac{\mathbf{V}_a}{\mathbf{Z}_A} = \frac{110\angle 0°}{50 + j80} = 1.16\angle -58° \text{ A rms}$$

The total power delivered to the three-phase load is

$$3\mathbf{S}_A = 3\mathbf{I}_{aA}^*\,\mathbf{V}_a = 204 + j326 \text{ VA}$$

Example 12.4-5

Figure 12.4-5*a* shows a balanced three-wire Y-to-Y circuit. Determine average power delivered by the three-phase source, delivered to the three-phase load, and absorbed by the three-phase line.

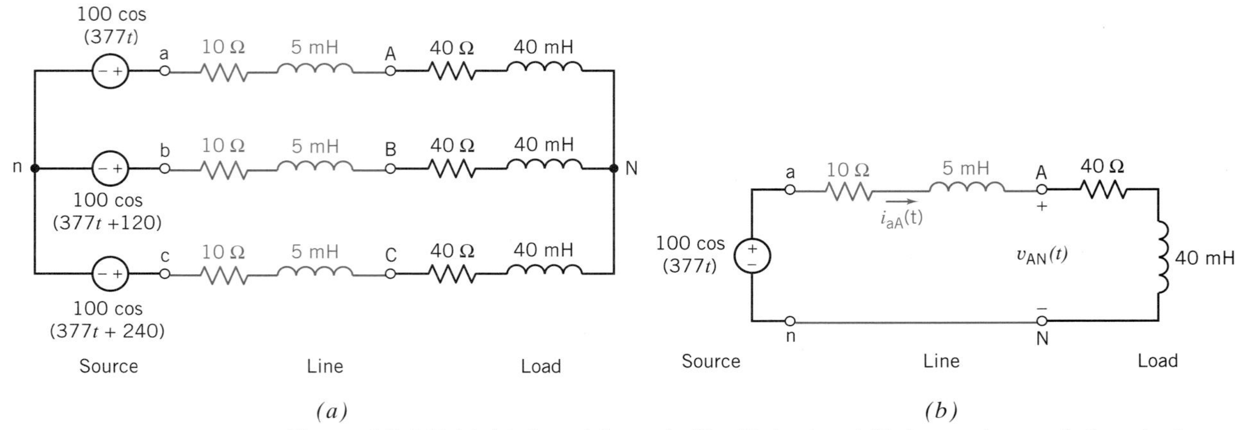

Figure 12.4-5 (*a*) A balanced three-wire Y-to-Y circuit and (*b*) the per-phase equivalent circuit.

Solution

The three-wire Y-to-Y circuit in Figure 12.4-5*a* looks different than the three-wire Y-to-Y circuit in Figure 12.4-2. One difference is cosmetic. The circuits are drawn differently, with all circuit elements drawn vertically or horizontally in Figure 12.4-5*a*. A more important difference is that the circuit in Figure 12.4-2 is represented in the frequency domain, using phasors and impedances, while the circuit in Figure 12.4-5*a* is represented in the time domain. Because the circuit is represented in the time domain, the magnitude, rather than the effective value, of source voltage is given.

Since this three-phase circuit is balanced, it can be analyzed using a per-phase equivalent circuit. Figure 12.4-5*b* shows the appropriate per-phase equivalent circuit.

The line current is calculated as

$$\mathbf{I}_{\mathrm{aA}}(\omega) = \frac{100}{50 + j(377)(0.045)} = 1.894\angle -18.7^\circ \text{ A}$$

The phase voltage at the load is

$$\mathbf{V}_{\mathrm{AN}}(\omega) = (40 + j(377)(0.04))\mathbf{I}_{\mathrm{aA}}(\omega) = 81\angle 2^\circ \text{ V}$$

Because peak values of the sinusoidal voltages and currents have been used instead of effective values, the appropriate formulas for power are those given in the "peak values" column of Table 12.0-1. The power delivered by the source is calculated as

$$\mathbf{I}_{\mathrm{aA}}(\omega) = 1.894\angle -18.7^\circ \text{ A} \quad \text{and} \quad \mathbf{V}_{\mathrm{an}}(\omega) = 100\angle 0^\circ \text{ V}$$

so

$$P_{\mathrm{a}} = \frac{(100)(1.894)}{2}\cos(18.7^\circ) = 89.7 \text{ W}$$

The power delivered to the load is calculated as

$$\mathbf{I}_{\mathrm{aA}}(\omega) = 1.894\angle -18.7^\circ \text{ A} \quad \text{and} \quad R_{\mathrm{A}} = 40\ \Omega \quad \text{so} \quad P_{\mathrm{A}} = \frac{1.894^2}{2}40 = 71.7 \text{ W}$$

The power lost in the line is calculated as

$$\mathbf{I}_{\mathrm{aA}}(\omega) = 1.894\angle -18.7^\circ \text{ A} \quad \text{and} \quad R_{\mathrm{aA}} = 10\ \Omega \quad \text{so} \quad P_{\mathrm{aA}} = \frac{1.894^2}{2}10 = 17.9 \text{ W}$$

A total of 80 percent of the power supplied by the source is delivered to the load. The other 20 percent is lost in the line.

Example 12.4-6

As noted in Example 12.4-5, 80 percent of the power supplied by the source is delivered to the load, and the other 20 percent is lost in the line. The loss in the line can be reduced by reducing the current in the line. Reducing the current in the load would reduce the power delivered to the load. Transformers provide a way of reducing the line current without reducing the load current.

In this example two three-phase transformers are added to the three-phase circuit considered in Example 12.4-5. A transformer at the source "steps up" the voltage and "steps down" the current. Conversely, a transformer at the load "steps down" the voltage and "steps up" the current. Since the turns ratio of these transformers are reciprocals of each other, the voltage and current at the load are unchanged. The current in the line will be reduced to reduce the power lost in line. The line voltage will increase. The higher line voltage will require increased insulation and increased attention to safety.

Figure 12.4-6*a* shows the per-phase equivalent circuit of the balanced three-wire Y-to-Y circuit that includes the two transformers. Determine the average power delivered by the three-phase source, delivered to the three-phase load, and absorbed by the three-phase line.

Solution

To analyze the per-phase equivalent circuit in Figure 12.4-6*a*, notice that

1. The secondary voltage of the left-hand transformer is 10 times the primary voltage, that is, 1000 cos (377*t*).
2. The impedance connected to the secondary of the right-hand transformer can be reflected to the primary of this transformer by multiplying by 100. The result is a 4000-ohm resistor in series with a 4-H inductor.

These observations lead to the one-mesh circuit shown in Figure 12.4-6*b*. The mesh current in this circuit is the line current of the three-phase circuit. This line current is calculated as

$$\mathbf{I}_{\mathrm{aA}}(\omega) = \frac{1000}{4010 + j(377)(4.005)} = 0.2334\angle -20.6^\circ \text{ A}$$

The current into the dotted end of the secondary of the left-hand transformer in Figure 12.4-6*a* is $-\mathrm{I}_{\mathrm{aA}}(\omega)$, so the current into the dotted end of the primary of this transformer is

$$\mathbf{I}_{\mathrm{a}}(\omega) = -10\,(-\mathbf{I}_{\mathrm{aA}}(\omega)) = 2.334\angle -20.6^\circ \text{ A}$$

The current into the dotted end of the primary of the right-hand transformer is $\mathrm{I}_{aA}(\omega)$, so the current into the dotted end of the secondary is

$$\mathbf{I}_{\mathrm{A}}\,(\omega) = -(-10\,\mathbf{I}_{\mathrm{aA}}(\omega)) = 2.334\angle 159^\circ \text{ A}$$

The phase voltage at the load is

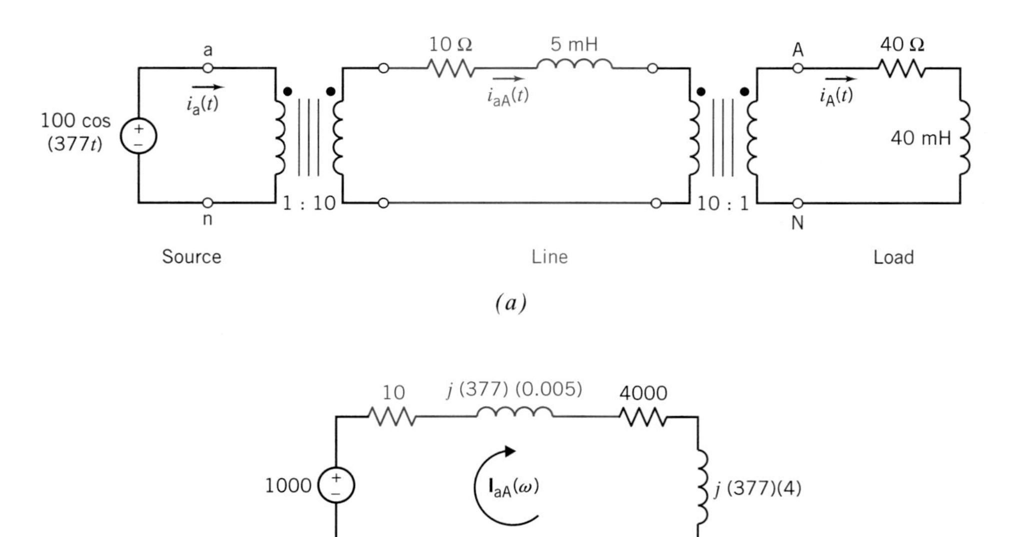

Figure 12.4-6 (*a*) A per-phase equivalent circuit for a balanced Y-to-Y circuit with step-up and step-down transformers and (*b*) the corresponding frequency domain circuit used to calculate the line current.

$$\mathbf{V}_{AN}(\omega) = (40 + j(377)(0.04))\mathbf{I}_A(\omega) = 99.77\angle 0° \text{ V}$$

The power delivered by the source is calculated as

$$\mathbf{I}_a(\omega) = 2.334\angle -20.6° \, A \quad \text{and}$$

$$\mathbf{V}_{an}(\omega) = 100\angle 0° \text{ V} \quad \text{so} \quad P_a = \frac{(100)(2.334)}{2}\cos(20.6°) = 109.2 \text{ W}$$

The power delivered to the load is calculated as

$$\mathbf{I}_A(\omega) = 2.334\angle -20.6° \text{ A} \quad \text{and} \quad R_A = 40\ \Omega \quad \text{so} \quad P_A = \frac{2.334^2}{2}40 = 108.95 \text{ W}$$

The power lost in the line is calculated as

$$\mathbf{I}_{aA}(\omega) = 0.2334\angle -20.6° \text{ A} \quad \text{and} \quad R_{aA} = 10\ \Omega \quad \text{so} \quad P_A = \frac{0.2334^2}{2}10 = 0.27 \text{ W}$$

Now 98 percent of the power supplied by the source is delivered to the load. Only 2 percent is lost in the line.

Exercise 12.4-1 Determine complex power delivered to the three-phase load of a four-wire Y-to-Y circuit such as the one shown in Figure 12.4-1. The phase voltages of the Y-connected source are $\mathbf{V}_a = 120\angle 0°$ V rms, $\mathbf{V}_b = 120\angle -120°$ V rms, and $\mathbf{V}_c = 120\angle 120°$ V rms. The load impedances are $\mathbf{Z}_A = 80 + j50\ \Omega$, $\mathbf{Z}_B = 80 + j80\ \Omega$, and $\mathbf{Z}_C = 100 - j25\ \Omega$.

Answers: $\mathbf{S}_A = 129 + j81$ VA, $\mathbf{S}_B = 90 + j90$ VA, $\mathbf{S}_C = 136 - j34$ VA, and $\mathbf{S} = 355 + j137$ VA

Exercise 12.4-2 Determine complex power delivered to the three-phase load of a four-wire Y-to-Y circuit such as the one shown in Figure 12.4-1. The phase voltages of the Y-connected source are $\mathbf{V}_a = 120\angle 0°$ V rms, $V_b = 120\angle -120°$ V rms, and $V_c = 120\angle 120°$ V rms. The load impedances are $\mathbf{Z}_A = \mathbf{Z}_B = \mathbf{Z}_C = 40 + j30\ \Omega$.

Answers: $\mathbf{S}_A = \mathbf{S}_B = \mathbf{S}_C = 230 + j173$ VA and $\mathbf{S} = 691 + j518$ VA.

Exercise 12.4-3 Determine complex power delivered to the three-phase load of a three-wire Y-to-Y circuit such as the one shown in Figure 12.4-2. The phase voltages of the Y-connected source are $\mathbf{V}_a = 120\angle 0°$ V rms, $\mathbf{V}_b = 120\angle -120°$ V rms, and $\mathbf{V}_c = 120\angle 120°$ V rms. The load impedances are $\mathbf{Z}_A = 80 + j50\ \Omega$, $\mathbf{Z}_B = 80 + j80\ \Omega$, and $\mathbf{Z}_C = 100 - j25\ \Omega$.

Intermediate Answer: $\mathbf{V}_{nN} = 28.89\angle -150.5$ V rms

Answer: $\mathbf{S} = 392 + j142$ VA

Exercise 12.4-4 Determine complex power delivered to the three-phase load of a three-wire Y-to-Y circuit such as the one shown in Figure 12.4-2. The phase voltages of the Y-connected source are $\mathbf{V}_a = 120\angle 0°$ V rms, $\mathbf{V}_b = 120\angle -120°$ V rms, and $\mathbf{V}_c = 120\angle 120°$ V rms. The load impedances are $\mathbf{Z}_A = \mathbf{Z}_B = \mathbf{Z}_C = 40 + j30\ \Omega$.

Answers: $\mathbf{S}_A = \mathbf{S}_B = \mathbf{S}_C = 230 + j173$ VA and $\mathbf{S} = 691 + j518$ VA

12.5 THE Δ-CONNECTED SOURCE AND LOAD

The Δ-connected source is shown in Figure 12.3-5*b*. This generator connection, however, is seldom used in practice because any slight imbalance in magnitude or phase of the three-phase voltages will not result in a zero sum. The result will be a large circulating current in the generator coils that will heat the generator and depreciate the efficiency of the generator. For example, consider the condition

$$\begin{aligned} \mathbf{V}_{ab} &= 120\angle 0^\circ \\ \mathbf{V}_{bc} &= 120.1\angle -121^\circ \\ \mathbf{V}_{ca} &= 120.2\angle 121^\circ \end{aligned} \tag{12.5-1}$$

If the total resistance around the loop is 1 Ω, we can calculate the circulating current as

$$\begin{aligned} \mathbf{I} &= (\mathbf{V}_{ab} + \mathbf{V}_{bc} + \mathbf{V}_{ca})/1 \\ &= 120 + 120.1(-0.515 - j0.857) + 120.2(-0.515 + j0.857) \\ &\cong 120 - 1.03(120.15) \\ &\cong -3.75 \text{ A} \end{aligned} \tag{12.5-2}$$

which would be unacceptable.

Therefore, we will consider only a Y-connected source as practical at the source side and consider both the Δ-connected load and the Y-connected load at the load side.

The Δ-to-Y and Y-to-Δ transformations convert Δ-connected loads to equivalent Y-connected loads and vice versa. These transformations are summarized in Table 12.5-1.

Table 12.5-1 **Y-to-Δ and Δ-to-Y Conversions**

Description	Circuit	Conversion Formulas (unbalanced)	Conversion Formulas (balanced)
Y-connected load	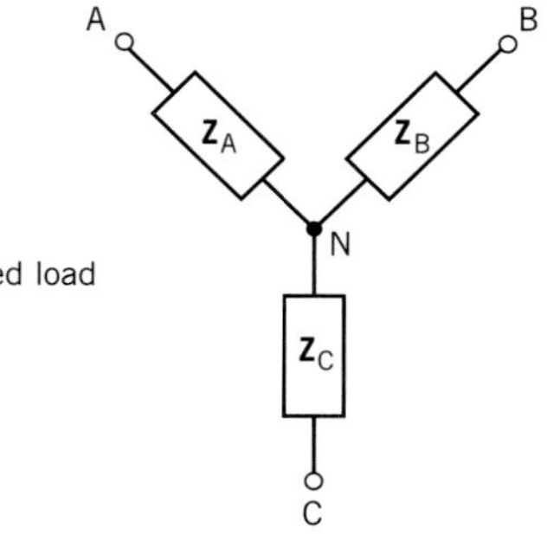	$\mathbf{Z}_A = \dfrac{\mathbf{Z}_1\mathbf{Z}_3}{\mathbf{Z}_1+\mathbf{Z}_2+\mathbf{Z}_3}$ $\mathbf{Z}_B = \dfrac{\mathbf{Z}_2\mathbf{Z}_3}{\mathbf{Z}_1+\mathbf{Z}_2+\mathbf{Z}_3}$ $\mathbf{Z}_C = \dfrac{\mathbf{Z}_1\mathbf{Z}_2}{\mathbf{Z}_1+\mathbf{Z}_2+\mathbf{Z}_3}$	When $\mathbf{Z}_1 = \mathbf{Z}_2 = \mathbf{Z}_3 = \mathbf{Z}_\Delta$ then $\mathbf{Z}_A = \mathbf{Z}_B = \mathbf{Z}_C = \dfrac{\mathbf{Z}_\Delta}{3}$
Δ-connected load	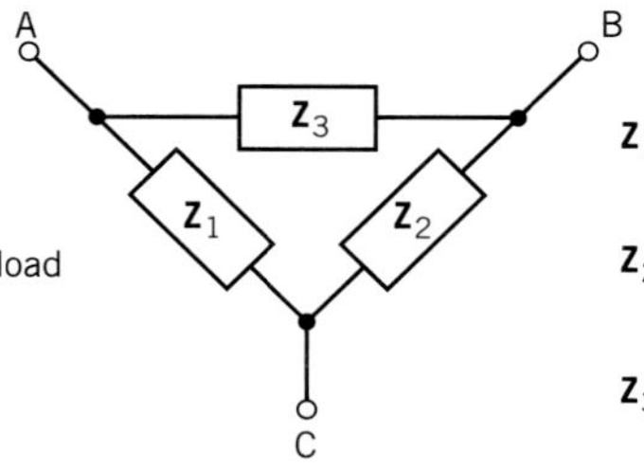	$\mathbf{Z}_1 = \dfrac{\mathbf{Z}_A\mathbf{Z}_B+\mathbf{Z}_B\mathbf{Z}_C+\mathbf{Z}_A\mathbf{Z}_C}{\mathbf{Z}_B}$ $\mathbf{Z}_2 = \dfrac{\mathbf{Z}_A\mathbf{Z}_B+\mathbf{Z}_B\mathbf{Z}_C+\mathbf{Z}_A\mathbf{Z}_C}{\mathbf{Z}_A}$ $\mathbf{Z}_3 = \dfrac{\mathbf{Z}_A\mathbf{Z}_B+\mathbf{Z}_B\mathbf{Z}_C+\mathbf{Z}_A\mathbf{Z}_C}{\mathbf{Z}_C}$	When $\mathbf{Z}_A = \mathbf{Z}_B = \mathbf{Z}_C = \mathbf{Z}_Y$ then $\mathbf{Z}_1 = \mathbf{Z}_2 = \mathbf{Z}_3 = 3\mathbf{Z}_Y$

Given the impedances $\mathbf{Z}_1, \mathbf{Z}_2, \mathbf{Z}_3$ of a Δ-connected load, Table 12.5-1 provides the formulas that are required to determine the impedances, $\mathbf{Z}_A, \mathbf{Z}_B, \mathbf{Z}_C$, of the equivalent Y-connected load. These three-phase loads are said to be equivalent because replacing the Δ-connected load by the Y-connected load will not change any of the voltages or currents of the three-phase source or three-phase line.

The Δ-to-Y and Y-to-Δ transformations are significantly simpler when the loads are balanced. Suppose the Δ-connected load is balanced, that is, $\mathbf{Z}_1 = \mathbf{Z}_2 = \mathbf{Z}_3 = \mathbf{Z}_\Delta$. The equivalent Y-connected load will also be balanced, so $\mathbf{Z}_A = \mathbf{Z}_B = \mathbf{Z}_C = \mathbf{Z}_Y$. Then, we have

$$\mathbf{Z}_Y = \frac{\mathbf{Z}_\Delta}{3} \tag{12.5-3}$$

Therefore, if we have a Y-connected source and a balanced Δ-connected load with $\mathbf{Z}_\Delta$, we convert the Δ load to a Y load with $\mathbf{Z}_Y = \mathbf{Z}_\Delta/3$. Then the line current is

$$\mathbf{I}_A = \frac{\mathbf{V}_a}{\mathbf{Z}_Y} = \frac{3\mathbf{V}_a}{\mathbf{Z}_\Delta} \tag{12.5-4}$$

Thus, we will consider only the Y-to-Y configuration. If the Y-to-Δ configuration is encountered, the Δ-connected load is converted to a Y-connected load equivalent, and the resulting currents and voltages are calculated.

Example 12.5-1

Figure 12.5-1*a* shows a three-phase load that consists of a parallel connection of a Y-connected and Δ-connected loads. Convert this load to an equivalent Y-connected load.

Solution

First convert the Y-connected load to a Δ-connected load as shown in Figure 12.5-1*b*. Notice, for example, that both of the Δ-connected loads in Figure 12.5-1*b* have an impedance connected between terminals A and B. These impedances are in parallel and can be replaced by a single equivalent impedance. Replace the parallel Δ-connected loads by a single equivalent Δ-connected load as shown in Figure 12.5-1*c*. Finally, convert the Δ-connected load to a Y-connected load as shown in Figure 12.5-1*d*.

12.6 THE Y-TO-Δ CIRCUIT

Now, let us consider the Y-to-Δ circuit as shown in Figure 12.6-1. Applying KCL at the nodes of the Δ-connected load shows that the relation between the line currents and phase currents is

$$\mathbf{I}_{aA} = \mathbf{I}_{AB} - \mathbf{I}_{CA}$$

$$\mathbf{I}_{bB} = \mathbf{I}_{BC} - \mathbf{I}_{AB}$$

and

$$\mathbf{I}_{cC} = \mathbf{I}_{CA} - \mathbf{I}_{BC} \tag{12.6-1}$$

The goal is to calculate the line and phase currents for the load.

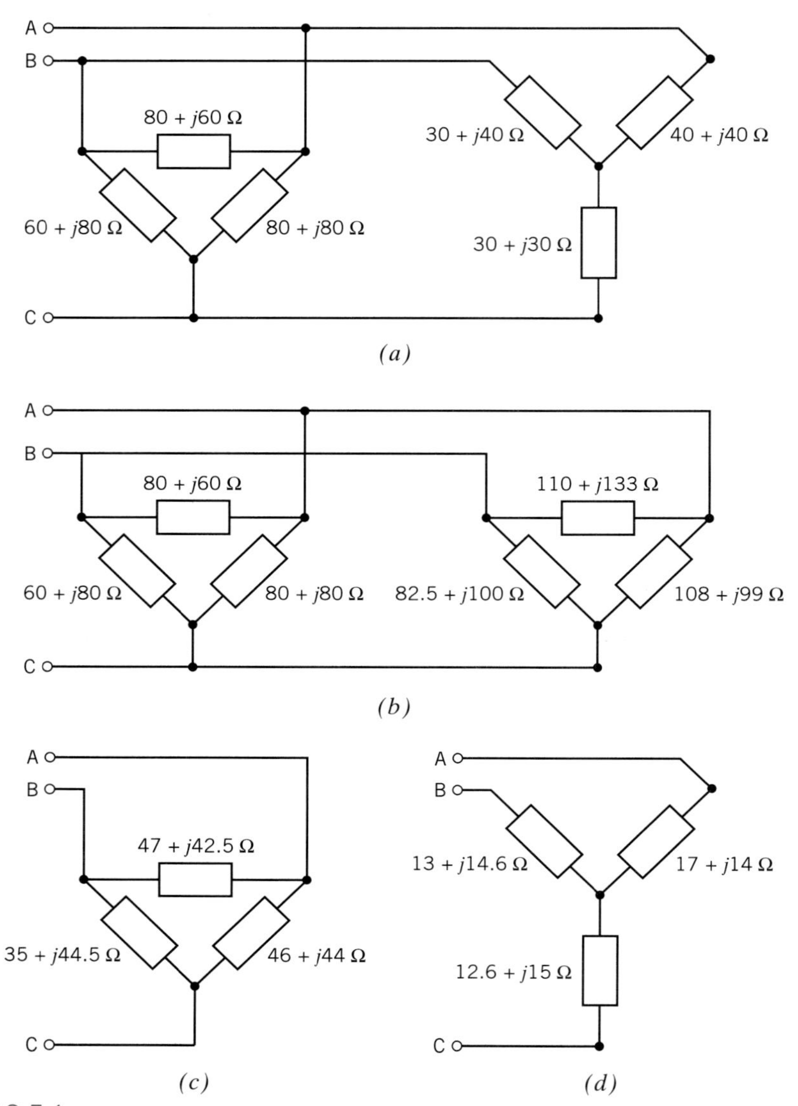

Figure 12.5-1 Example of Y-Δ conversions. (*a*) Parallel Y-connected and Δ-connected loads. (*b*) The Y-connected load is converted to a Δ-connected load. (*c*) The parallel Δ-connected loads are replaced by a single equivalent Δ-connected load. (*d*) The Δ-connected load is converted to a Y-connected load.

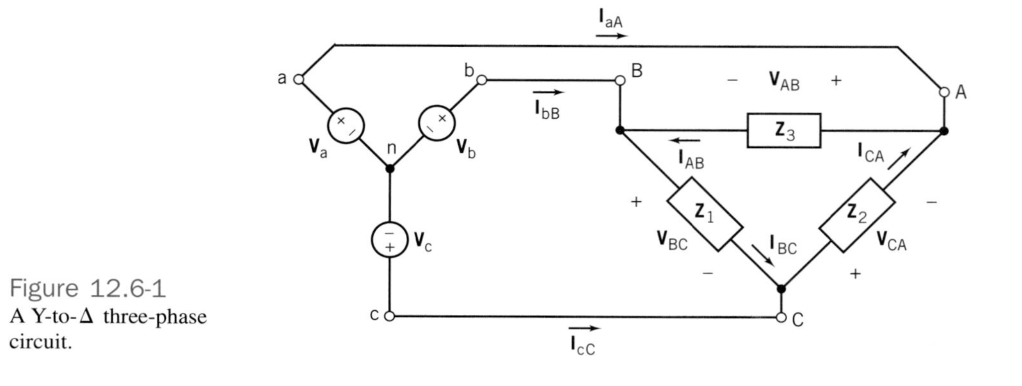

Figure 12.6-1
A Y-to-Δ three-phase circuit.

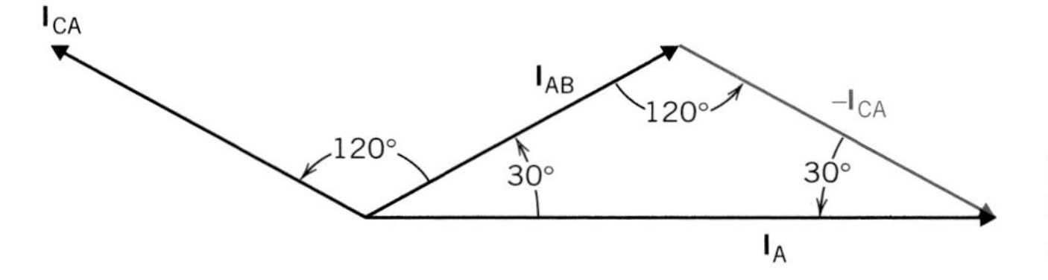

Figure 12.6-2
Phasor diagram for currents of a Δ load.

The phase currents in the Δ-connected load can be calculated from the line-to-line voltages. These line-to-line voltages appear directly across the impedances of the Δ-connected load. For example, $\mathbf{V}_{AB}$ appears across $\mathbf{Z}_3$, so

$$\mathbf{I}_{BA} = \frac{\mathbf{V}_{BA}}{\mathbf{Z}_3} \tag{12.6-2}$$

Similarly,

$$\mathbf{I}_{AC} = \frac{\mathbf{V}_{AC}}{\mathbf{Z}_2} \quad \text{and} \quad \mathbf{I}_{CB} = \frac{\mathbf{V}_{CB}}{\mathbf{Z}_1} \tag{12.6-3}$$

When the load is balanced, the phase currents in the load have the same magnitude and have phase angles that differ by 120°. For example, if the three-phase source has the *abc* sequence and $\mathbf{I}_{AB} = I\angle\phi$, then $\mathbf{I}_{CA} = I\angle(\phi + 120°)$. The line current $\mathbf{I}_{aA}$ is calculated as

$$\begin{aligned}
\mathbf{I}_{aA} &= \mathbf{I}_{AB} - \mathbf{I}_{CA} \\
&= I\cos\phi + jI\sin\phi - I\cos(\phi + 120°) - jI\sin(\phi + 120°) \\
&= -2I\sin(\phi + 60°)\sin(-60°) + j2I\cos(\phi + 60°)\sin(-60°) \\
&= \sqrt{3}\,I\,[\sin(\phi + 60°) - j\cos(\phi + 60°)] \\
&= \sqrt{3}\,I\,[\cos(\phi - 30°) + j\sin(\phi - 30°)] \\
&= \sqrt{3}\,I\angle(\phi - 30°)\text{ A}
\end{aligned} \tag{12.6-4}$$

Therefore,

$$|\,\mathbf{I}_{aA}\,| = \sqrt{3}\,|\,\mathbf{I}\,| \tag{12.6-5}$$

or

$$I_L = \sqrt{3}\,I_p$$

and the line current magnitude is $\sqrt{3}$ times the phase current magnitude. This result can also be obtained from the phasor diagram shown in Figure 12.6-2. In a Δ connection, the line current is $\sqrt{3}$ times the phase current and is displaced −30° in phase. The line-to-line voltage is equal to the phase voltage.

Example 12.6-1

Consider the three-phase circuit shown in Figure 12.6-1. The voltages of the Y-connected source are

$$\mathbf{V}_a = \frac{220}{\sqrt{3}}\angle -30°\text{ V rms}, \quad \mathbf{V}_b = \frac{220}{\sqrt{3}}\angle -150°\text{ V rms}, \quad \text{and} \quad \mathbf{V}_c = \frac{220}{\sqrt{3}}\angle 90°\text{ V rms}$$

The Δ-connected load is balanced. The impedance of each phase is $\mathbf{Z}_\Delta = 10\angle 50°\ \Omega$. Determine the phase and line currents.

Solution

The line-to-line voltages are calculated from the phase voltages of the source as

$$\mathbf{V}_{AB} = \mathbf{V}_a - \mathbf{V}_b = \frac{220}{\sqrt{3}}\angle -30^\circ - \frac{220}{\sqrt{3}}\angle -150^\circ = 220\angle 0^\circ \text{ V rms}$$

$$\mathbf{V}_{BC} = \mathbf{V}_b - \mathbf{V}_c = \frac{220}{\sqrt{3}}\angle -150^\circ - \frac{220}{\sqrt{3}}\angle 90^\circ = 220\angle -120^\circ \text{ V rms}$$

$$\mathbf{V}_{CA} = \mathbf{V}_c - \mathbf{V}_a = \frac{220}{\sqrt{3}}\angle 90^\circ - \frac{220}{\sqrt{3}}\angle -30^\circ = 220\angle -240^\circ \text{ V rms}$$

The phase voltages of a Δ-connected load are equal to the line-to-line voltages. The phase currents are

$$\mathbf{I}_{AB} = \frac{\mathbf{V}_{AB}}{\mathbf{Z}} = \frac{220\angle 0^\circ}{10\angle -50^\circ} = 22\angle 50^\circ \text{ A rms}$$

$$\mathbf{I}_{BC} = \frac{\mathbf{V}_{BC}}{\mathbf{Z}} = \frac{220\angle -120^\circ}{10\angle -50^\circ} = 22\angle -70^\circ \text{ A rms}$$

$$\mathbf{I}_{CA} = \frac{\mathbf{V}_{CA}}{\mathbf{Z}} = \frac{220\angle -240^\circ}{10\angle -50^\circ} = 22\angle -190^\circ \text{ A rms}$$

The line currents are

$$\mathbf{I}_{aA} = \mathbf{I}_{AB} - \mathbf{I}_{CA} = 22\angle 50^\circ - 22\angle -190^\circ$$
$$= 22\sqrt{3}\angle 20^\circ \text{ A rms}$$

Then $\mathbf{I}_{bB} = 22\sqrt{3}\angle -100^\circ$ A rms and $\mathbf{I}_{cC} = 22\sqrt{3}\angle -220^\circ$ A rms

The current and voltage relationships for a Δ load are summarized in Table 12.6-1.

Table 12.6-1 **The Current and Voltage for a Δ Load**

Phase voltages	$\mathbf{V}_{AB} = \mathbf{V}_{AB}\underline{/0^\circ} = V_p\underline{/0^\circ}$
Line-to-line voltages	$\mathbf{V}_{AB} = \mathbf{V}_L$ (line voltage = phase voltage)
Phase currents	$\mathbf{I}_{AB} = \frac{\mathbf{V}_{AB}}{\mathbf{Z}_p} = \frac{\mathbf{V}}{\mathbf{Z}_\Delta} = I_p\underline{/-\theta}$
	with $\mathbf{Z}_p = Z\underline{/\theta}$
	$\mathbf{I}_{BC} = \mathbf{I}_{AB}\underline{/-120^\circ}$
	$\mathbf{I}_{CA} = \mathbf{I}_{AB}\underline{/-240^\circ}$
Line currents	$\mathbf{I}_A = \sqrt{3}\, I_p\underline{/-\theta - 30^\circ}$
	$\mathbf{I}_B = \sqrt{3}\, I_p\underline{/-\theta - 150^\circ}$
	$\mathbf{I}_C = \sqrt{3}\, I_p\underline{/-\theta + 90^\circ}$
	$I_L = \sqrt{3}\, I_p$

L = line, p = phase.

Exercise 12.6-1 Consider the three-phase circuit shown in Figure 12.6-1. The voltages of the Y-connected source are

$$\mathbf{V}_a = \frac{360}{\sqrt{3}}\angle -30^\circ \text{ V rms}, \quad \mathbf{V}_b = \frac{360}{\sqrt{3}}\angle -150^\circ \text{ V rms}, \quad \text{and} \quad \mathbf{V}_c = \frac{360}{\sqrt{3}}\angle 90^\circ \text{ V rms}$$

The Δ-connected load is balanced. The impedance of each phase is $\mathbf{Z}_\Delta = 180\angle 45°\ \Omega$. Determine the phase and line currents when the line-to-line voltage is 360 V rms.

Partial Answers: $\mathbf{I}_{AB} = 2\angle 45°$ A rms, $\mathbf{I}_{aA} = 3.46\angle 15°$ A rms

12.7 BALANCED THREE-PHASE CIRCUITS

We have only two possible practical configurations for three-phase circuits, Y-to-Y and Y-to-Δ, and we can convert the latter to a Y-to-Y form. Thus, a practical three-phase circuit can always be converted to the Y-to-Y circuit.

Balanced circuits are easier to analyze than unbalanced circuits. Earlier, we saw that balanced three-phase Y-to-Y circuits can be analyzed using a per-phase equivalent circuit.

The circuit shown in Figure 12.7-1*a* is a balanced Y-to-Δ circuit. Figure 12.7-1*b* shows the equivalent Y-to-Y circuit where

$$\mathbf{Z}_Y = \frac{\mathbf{Z}_\Delta}{3}$$

This Y-to-Y circuit can be analyzed using the per-phase equivalent circuit shown in Figure 12.7-1*c*.

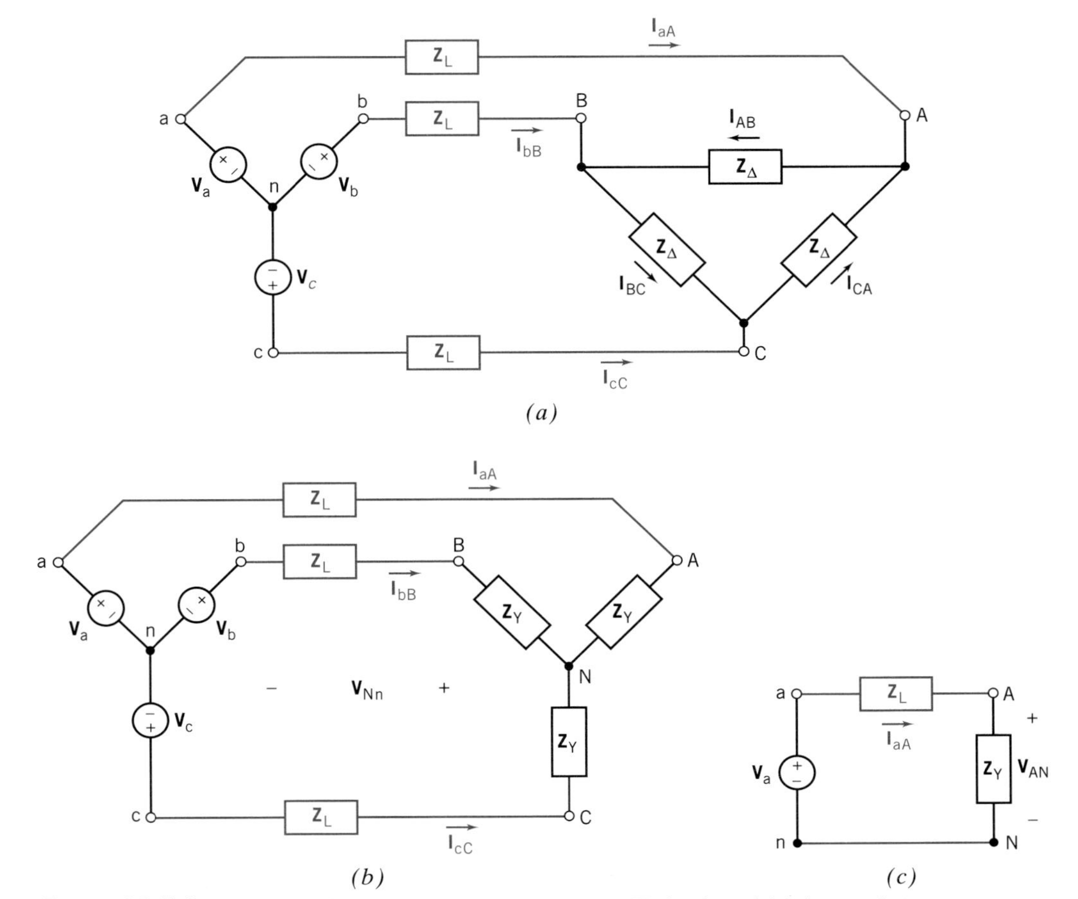

Figure 12.7-1 (*a*) A Y-to-Δ circuit, (*b*) the equivalent Y-to-Y circuit, and (*c*) the per-phase equivalent circuit.

Example 12.7-1

Figure 12.7-1*a* shows a balanced Y-to-Δ three-phase circuit. The phase voltages of the Y-connected source are $\mathbf{V}_a = 110\angle 0°$ V rms, $\mathbf{V}_b = 110\angle -120°$ V rms, and $\mathbf{V}_c = 110\angle 120°$ V rms. The line impedances are each $\mathbf{Z}_L = 10 + j5\ \Omega$. The impedances of the Δ-connected load are each $\mathbf{Z}_\Delta = 75 + j225\ \Omega$. Determine the phase currents in the Δ-connected load.

Solution

Convert the Δ-connected load to a Y-connected load using the Δ-to-Y transformation summarized in Table 12.5-1. The impedances of the balanced equivalent Y-connected load are

$$\mathbf{Z}_Y = \frac{75 + j225}{3} = 25 + j75\ \Omega$$

The per-phase equivalent circuit for the Y-to-Y circuit is shown in Figure 12.7-1*c*. The line current is given by

$$\mathbf{I}_{aA} = \frac{\mathbf{V}_a}{\mathbf{Z}_L + \mathbf{Z}_Y} = \frac{110\angle 0°}{(10 + j5) + (25 + j75)} = 1.26\angle -66° \text{ A rms} \quad (12.7\text{-}1)$$

The line current, $\mathbf{I}_{aA}$, calculated using the per-phase equivalent circuit, is also the line current, $\mathbf{I}_{aA}$, in the Y-to-Y circuit, as well as the line current, $\mathbf{I}_{aA}$, in the Y-to-Δ circuit. The other line currents in the balanced Y-to-Y circuit have the same magnitude but differ in phase angle by 120°. These line currents are

$$\mathbf{I}_{bB} = 1.26\angle -186° \text{ A rms} \quad \text{and} \quad \mathbf{I}_{cC} = 1.26\angle 54° \text{ A rms}$$

(To check the value of $\mathbf{I}_{bB}$, apply KVL to the loop in the Y-to-Y circuit that starts at node *n*, passes through nodes *b, B, N,* and returns to node *n*. The resulting KVL equation is

$$\mathbf{V}_b = \mathbf{Z}_L \mathbf{I}_{bB} + \mathbf{Z}_Y \mathbf{I}_{bB} + \mathbf{V}_{Nn}$$

Because the circuit is balanced, $\mathbf{V}_{Nn} = 0$. Solving for $\mathbf{I}_{bB}$ gives

$$\mathbf{I}_{bB} = \frac{\mathbf{V}_b}{\mathbf{Z}_L + \mathbf{Z}_Y} = \frac{110\angle -120°}{(10 + j5) + (25 + j75)} = 1.26\angle -186° \text{ A rms} \quad (12.7\text{-}2)$$

Comparing Eqs. 12.7-1 and 12.7-2 shows that the line currents in the balanced Y-to-Y circuit have the same magnitude but differ in phase angle by 120°.)

The line currents of the Y-to-Δ circuit in Figure 12.7-1*a* are equal to the line currents of the Y-to-Y circuit in Figure 12.7-1*b* because the Y-to-Δ and Y-to-Y circuits are equivalent.

The voltage $\mathbf{V}_{AN}$ in the per-phase equivalent circuit is

$$\mathbf{V}_{AN} = \mathbf{I}_{aA}\mathbf{Z}_Y = (1.26\angle -66°)(25 + j75) = 99.6\angle 5° \text{ V rms}$$

The voltage $\mathbf{V}_{AN}$ calculated using the per-phase equivalent circuit is also the phase voltage, $\mathbf{V}_{AN}$, of the Y-to-Y circuit. The other phase voltages of the balanced Y-to-Y circuit have the same magnitude but differ in phase angle by 120°. These phase voltages are

$$\mathbf{V}_{BN} = 99.6\angle -115° \text{ V rms} \quad \text{and} \quad \mathbf{V}_{CN} = 99.6\angle 125° \text{ V rms}$$

The line-to-line voltages of the Y-to-Y circuit are calculated as

$$\mathbf{V}_{AB} = \mathbf{V}_{AN} - \mathbf{V}_{BN} = 99.5\angle 5° - 99.5\angle -115° = 172\angle 35° \text{ V rms}$$
$$\mathbf{V}_{BC} = \mathbf{V}_{BN} - \mathbf{V}_{CN} = 99.5\angle -115° - 99.5\angle 125° = 172\angle -85° \text{ V rms}$$
$$\mathbf{V}_{CA} = \mathbf{V}_{CN} - \mathbf{V}_{AN} = 99.5\angle 125° - 99.5\angle 5° = 172\angle 155° \text{ V rms}$$

The phase voltages of a Δ-connected load are equal to the line-to-line voltages. The phase currents are

$$\mathbf{I}_{AB} = \frac{\mathbf{V}_{AB}}{\mathbf{Z}_\Delta} = \frac{172 \angle 35^\circ}{75 + j225} = 0.727 \angle -36^\circ \text{ A rms}$$

$$\mathbf{I}_{BC} = \frac{\mathbf{V}_{BC}}{\mathbf{Z}_\Delta} = \frac{172 \angle -85^\circ}{75 + j225} = 0.727 \angle -156^\circ \text{ A rms}$$

$$\mathbf{I}_{CA} = \frac{\mathbf{V}_{CA}}{\mathbf{Z}_\Delta} = \frac{172 \angle 155^\circ}{75 + j225} = 0.727 \angle 84^\circ \text{ A rms}$$

Exercise 12.7-1 Figure 12.7-1*a* shows a balanced Y-to-Δ three-phase circuit. The phase voltages of the Y-connected source are $\mathbf{V}_a = 110 \angle 0^\circ$ V rms, $\mathbf{V}_b = 110 \angle -120^\circ$ V rms, and $\mathbf{V}_c = 100 \angle 120^\circ$ V rms. The line impedances are each $\mathbf{Z}_L = 10 + j25\ \Omega$. The impedances of the Δ-connected load are each $\mathbf{Z}_\Delta = 150 + j270\ \Omega$. Determine the phase currents in the Δ-connected load.

Answers: $\mathbf{I}_{AB} = 0.49 \angle -32.5^\circ$ A rms, $\mathbf{I}_{BC} = 0.49 \angle -152.5^\circ$ A rms, $\mathbf{I}_{CA} = 0.49 \angle 87.5^\circ$ A rms

12.8 INSTANTANEOUS AND AVERAGE POWER IN A BALANCED THREE-PHASE LOAD

One advantage of three-phase power is the smooth flow of energy to the load. Consider a balanced load with resistance R. Then the *instantaneous power* is

$$p(t) = \frac{v_{ab}^2}{R} + \frac{v_{bc}^2}{R} + \frac{v_{ca}^2}{R} \tag{12.8-1}$$

where $v_{ab} = V \cos \omega t$ and the other two-phase voltages have a phase of ±120°, respectively. Furthermore,

$$\cos^2 \alpha t = (1 + \cos 2\alpha)/2$$

Therefore,

$$\begin{aligned} p(t) &= \frac{V^2}{2R}[1 + \cos 2\omega t + 1 + \cos 2(\omega t - 120^\circ) + 1 + \cos 2(\omega t - 240^\circ)] \\ &= \frac{3V^2}{2R} + \frac{V^2}{2R}[\cos 2\omega t + \cos(2\omega t - 240^\circ) + \cos(2\omega t - 480^\circ)] \\ &= \frac{3V^2}{2R} \end{aligned} \tag{12.8-2}$$

since the bracketed term is equal to zero for all time. Hence, the *instantaneous power* delivered to a balanced three-phase load is a constant.

The total *average power* delivered to a balanced three-phase load can be calculated using the per-phase equivalent circuit. Consider, again, Figure 12.7-1. Figure 12.7-1*a* shows a balanced Y-to-Δ circuit. Figure 12.7-1*b* shows the equivalent Y-to-Y circuit, obtained using the Δ-to-Y transformation summarized in Table 12.5-1. Figure 12.7-1*c* shows the per-phase equivalent circuit corresponding to the Y-to-Y circuit. The voltage $\mathbf{V}_{AN} = V_P \angle\theta_{AV}$ and the current $\mathbf{I}_{aA} = I_L \angle\theta_{IA}$ are obtained using per-phase equivalent circuit. The voltage $\mathbf{V}_{AN}$ and the current $\mathbf{I}_{aA}$ are the phase voltage and line current of the Y-connected load in Figure 12.7-1*b*. The total average power delivered to the balanced Y-connected load is given by

$$\begin{aligned} P_Y &= 3\,P_A \\ &= 3\,V_P I_L \cos(\theta_{AV} - \theta_{AI}) \\ &= 3\,V_P I_L \cos(\theta) \end{aligned} \tag{12.8-3}$$

where θ is the angle between the phase voltage and the line current, $\cos\theta$ is the power factor, and V_P and I_P are effective values of the phase voltage and line current.

It is easier to measure the line-to-line voltage and the line current of a circuit. Also recall that the line current equals the phase current and that the phase voltage is $V_P = V_L/\sqrt{3}$ for the Y-load configuration. Therefore,

$$\begin{aligned} P &= 3\frac{V_L}{\sqrt{3}} I_L \cos\theta \\ &= \sqrt{3}\, V_L\, I_L \cos\theta \end{aligned} \tag{12.8-4}$$

The total average power delivered to the Δ-connected load in Figure 12.7-1*c* is

$$\begin{aligned} P = 3P_{AB} = 3V_{AB}\, I_{AB} \cos\theta &= 3(\sqrt{3}\, V_p)\frac{I_L}{\sqrt{3}} \cos\theta \\ &= 3V_p I_L \cos\theta \end{aligned} \tag{12.8-5}$$

In summary, the total average power delivered to the Δ-connected load in Figure 12.7-1*a* is equal to the total average power delivered to the balanced Y-connected load in Figure 12.7-1*b*. That's appropriate because the two circuits are equivalent. Notice that the information required to calculate the power delivered to a balanced load, Y or Δ, is obtained from the per-phase equivalent circuit.

Example 12.8-1

Figure 12.7-1*a* shows a balanced Y-to-Δ three-phase circuit. The phase voltages of the Y-connected source are $\mathbf{V}_a = 110 \angle 0°$ V rms, $\mathbf{V}_b = 110 \angle -120°$ V rms, and $\mathbf{V}_c = 110 \angle 120°$ V rms. The line impedances are each $\mathbf{Z}_L = 10 + j5\ \Omega$. The impedances of the Δ-connected load are each $\mathbf{Z}_\Delta = 75 + j225\ \Omega$. Determine the average power delivered to the load.

Solution

This circuit was analyzed in Example 12.7-1. That analysis showed that

$$\mathbf{I}_{aA} = 1.26 \angle -66°\ \text{A rms}$$

and

$$\mathbf{V}_{AN} = 99.6 \angle 5°\ \text{V rms}$$

The total average power delivered to the load is given by Eq. 12.8-3 as

$$P = 3(99.6)\,(1.26) \cos(5° - (-66°)) = 122.6\ \text{W}$$

Exercise 12.8-1 Figure 12.7-1*a* shows a balanced Y-to-Δ three-phase circuit. The phase voltages of the Y-connected source are $\mathbf{V}_a = 110 \angle 0°$ V rms, $\mathbf{V}_b = 110 \angle -120°$ V rms, and $\mathbf{V}_c = 110 \angle 120°$ V rms. The line impedances are each $\mathbf{Z}_L = 10 + j25\ \Omega$. The impedances of the Δ-connected load are each $\mathbf{Z}_\Delta = 150 + j270\ \Omega$. Determine the average power delivered to the Δ-connected load.

Intermediate result: $\mathbf{I}_{aA} = 0.848 \angle -62.5°$ A rms and $\mathbf{V}_{AN} = 87.3 \angle -1.5°$ V rms

Answer: $P = 107.9$ W

12.9 TWO-WATTMETER POWER MEASUREMENT

For many load configurations, for example, a three-phase motor, the phase current or voltage is inaccessible. We may wish to measure power with a wattmeter connected to each phase. However, since the phases are not available, we measure the line currents and the line-to-line voltages. A wattmeter provides a reading of $V_L I_L \cos\theta$ where V_L and I_L are the rms magnitudes and θ is the angle between the line voltage, **V**, and the current, **I**. We choose to measure V_L and I_L, the line voltage and current, respectively. We will show that two wattmeters are sufficient to read the power delivered to the three-phase load, as shown in Figure 12.9-1. We use *cc* to denote current coil and *vc* to denote voltage coil.

Wattmeter 1 reads

$$P_1 = V_{AB} I_A \cos\theta_1 \tag{12.9-1}$$

and wattmeter 2 reads

$$P_2 = V_{CB} I_C \cos\theta_2 \tag{12.9-2}$$

For the *abc* phase sequence for a balanced load

$$\theta_1 = \theta_a + 30°$$

and

$$\theta_2 = \theta_a - 30° \tag{12.9-3}$$

where θ_a is the angle between the phase current and the phase voltage for phase *a* of the three-phase source.

Therefore

$$P = P_1 + P_2 = 2V_L I_L \cos\theta \cos 30°$$
$$= \sqrt{3}\, V_L I_L \cos\theta \tag{12.9-4}$$

which is the total average power of the three-phase circuit. The preceding derivation of Eq. 12.9-4 is for a balanced circuit; the result is good for any three-phase, three-wire load, even unbalanced or nonsinusoidal voltages.

The power factor angle, θ, of a balanced three-phase system may be determined from the reading of the two wattmeters shown in Figure 12.9-2.

The total power is obtained from Eqs. 12.9-1 through 12.9-3 as

$$P = P_1 + P_2 = V_L I_L[\cos(\theta + 30°) + \cos(\theta - 30°)]$$
$$= V_L I_L\, 2\cos\theta \cos 30° \tag{12.9-5}$$

Similarly,

$$P_1 - P_2 = V_L I_L(-2\sin\theta \sin 30°) \tag{12.9-6}$$

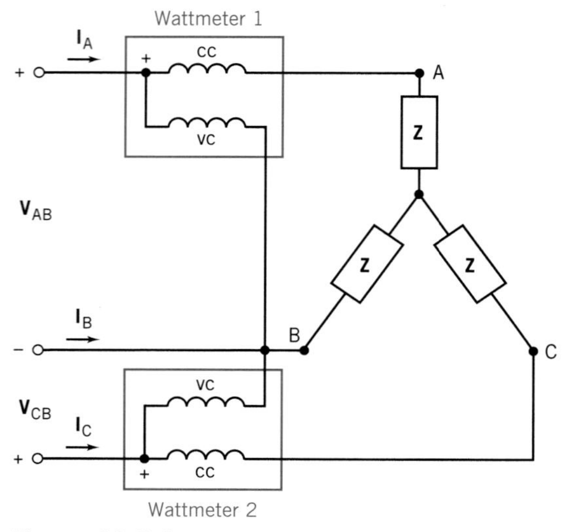

Figure 12.9-1 Two-wattmeter connection for a three-phase Y-connected load.

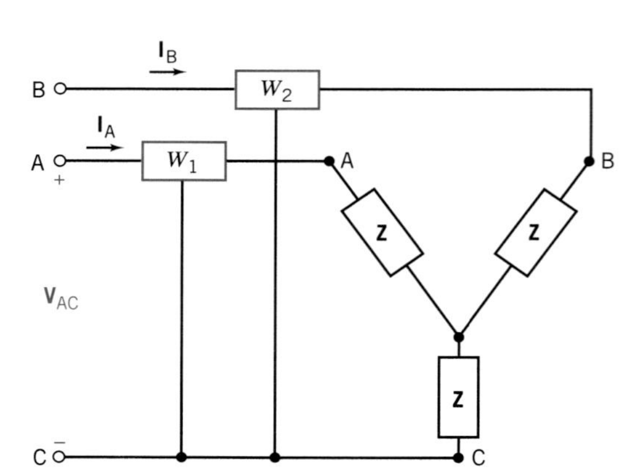

Figure 12.9-2 The two-wattmeter connection for Example 12.9-1.

Dividing Eq. 12.9-5 by Eq. 12.9-6, we obtain

$$\frac{P_1 + P_2}{P_1 - P_2} = \frac{2\cos\theta\cos 30^\circ}{-2\sin\theta\sin 30^\circ} = \frac{-\sqrt{3}}{\tan\theta}$$

Therefore,

$$\tan\theta = \sqrt{3}\,\frac{P_2 - P_1}{P_2 + P_1} \tag{12.9-7}$$

where θ = power factor angle.

Example 12.9-1

The two-wattmeter method is used, as shown in Figure 12.9-1, to measure the total power delivered to the Y-connected load when $Z = 10\angle 45^\circ\ \Omega$ and the supply line-to-line voltage is 220 V rms. Determine the reading of each wattmeter and the total power.

Solution

The phase voltage is (see Table 12.4-1)

$$\mathbf{V}_A = \frac{220}{\sqrt{3}}\angle -30^\circ \text{ V rms}$$

Then we obtain the line current as

$$\mathbf{I}_A = \frac{\mathbf{V}_A}{\mathbf{Z}} = \frac{220\angle -30^\circ}{10\sqrt{3}\angle 45^\circ} = 12.7\angle -75^\circ \text{ A rms}$$

Then the second line current is

$$\mathbf{I}_B = 12.7\angle -195^\circ \text{ A rms}$$

The voltage $\mathbf{V}_{AB} = 220\angle 0^\circ$ V rms, $\mathbf{V}_{CA} = 220\angle +120^\circ$ V rms, and $\mathbf{V}_{BC} = 220\angle -120^\circ$ V rms. The first wattmeter reads

$$P_1 = I_A V_{AC}\cos\theta_1 = 12.7(220)\cos 15^\circ = 2698 \text{ W}$$

Since $\mathbf{V}_{CA} = 220 \angle +120°$, then $\mathbf{V}_{AC} = 220 \angle -60°$. Therefore, the angle θ_1 lies between $\mathbf{V}_{AC}$ and $\mathbf{I}_A$ and is equal to 15°. The reading of the second wattmeter is

$$P_2 = I_B V_{BC} \cos \theta_2 = 12.7(220) \cos 75° = 723 \text{ W}$$

where θ_2 is the angle between $\mathbf{I}_B$ and $\mathbf{V}_{BC}$. Therefore, the total power is

$$P = P_1 + P_2 = 3421 \text{ W}$$

We note that all of the preceding calculations assume that the wattmeter itself absorbs negligible power.

Example 12.9-2

The two wattmeters in Figure 12.9-1 read $P_1 = 60$ kW and $P_2 = 180$ W, respectively. Find the power factor of the circuit.

Solution

From Eq. 12.9-7 we have

$$\tan \theta = \sqrt{3}\frac{P_2 - P_1}{P_2 + P_1} = \sqrt{3}\frac{120}{240} = \frac{\sqrt{3}}{2} = 0.866$$

Therefore, we have $\theta = 40.9°$ and the power factor is

$$pf = \cos \theta = 0.756$$

The positive angle, θ, indicates that the power factor is lagging. If θ is negative, then the power factor is leading.

Exercise 12.9-1 The line current to a balanced three-phase load is 24 A rms. The line-to-line voltage is 450 V rms, and the power factor of the load is 0.47 lagging. If two wattmeters are connected as shown in Figure 12.9-1, determine the reading of each meter and the total power to the load.

Answers: $P_1 = -371$ W, $P_2 = 9162$ W, $P = 8791$ W

Exercise 12.9-2 The two wattmeters are connected as shown in Figure 12.9-1 with $P_1 = 60$ kW and $P_2 = 40$ kW, respectively. Determine (a) the total power and (b) the power factor.

Answers: (a) 100 kW, (b) 0.945 leading

12.10 THE BALANCED THREE-PHASE CIRCUIT AND PSPICE

The analysis of three-phase balanced circuits may be facilitated by using the per-phase equivalent circuit. Once we have arrived at a per-phase circuit, we can use PSpice to solve for the phase and line currents and voltages.

Let us consider a Y-to-Y connected system with a transmission line resistance as shown in Figure 12.10-1. The per-phase equivalent circuit shown in Figure 12.10-2 includes the

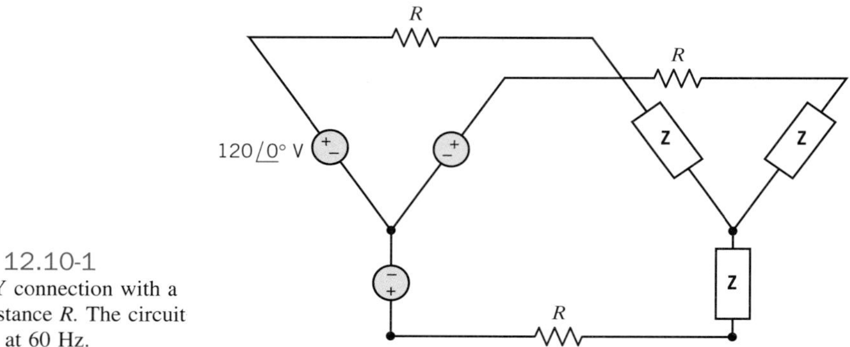

Figure 12.10-1
A Y-to-Y connection with a line resistance R. The circuit operates at 60 Hz.

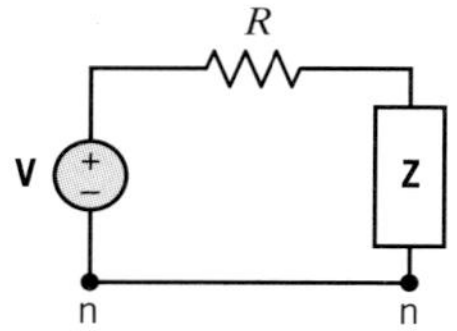

Figure 12.10-2
The per-phase equivalent circuit.

neutral wire but assumes that its resistance is zero since the neutral wire is only added in the equivalent circuit. Let us consider the case where $R = 1\ \Omega$ and $\mathbf{Z} = 9 + j\omega L$ where $f = 60$ Hz and $L = 26.5$ mH. The total impedance is $10 + j10$, and the line current is

$$\mathbf{I}_{\mathrm{A}} = \frac{120\ \angle 0^\circ}{10\sqrt{2}\ \angle 45^\circ} = 8.5\ \angle -45^\circ\ \mathrm{A}$$

To determine the line current using PSpice, we set up the PSpice program as shown in Figure 12.10-3. The output of the program yields the line current as IM(VA) = 8.5 and IP(VA) = 135°.

```
VA       1    0    AC   120    0
RLOSS    1    2    1
RLOAD    2    3    9
LLOAD    3    0    26.5M
.AC      LIN  1    60   60
.PRINT   AC        IM(VA)     IP(VA)
.END
```

Figure 12.10-3
The PSpice program for the per-phase circuit.

12.11 VERIFICATION EXAMPLES

Example 12.11-1

Figure 12.11-1*a* shows a balanced three-phase circuit. Computer analysis of this circuit produced the element voltages and currents tabulated in Figure 12.11-1*b*. Is this computer analysis correct?

Solution

Since the three-phase circuit is balanced, it can be analyzed by using a per-phase equivalent circuit. The appropriate per-phase equivalent circuit for this example is shown in Figure 12.11-2. This per-phase equivalent circuit can be analyzed by writing a single mesh equation:

$$10 = (9 + j12)\,\mathbf{I}_{\mathrm{L}}(\omega)$$

or

$$\mathbf{I}_{\mathrm{L}}(\omega) = 0.67e^{-j53^\circ}\ \mathrm{A}$$

where $\mathbf{I}_{\mathrm{L}}(\omega)$ is the phasor corresponding to the inductor current. The voltage across the inductor is given by

$$\mathbf{V}_{\mathrm{L}}(\omega) = j12\,\mathbf{I}_{\mathrm{L}}(\omega) = 8\,e^{j37^\circ}\ \mathrm{V}$$

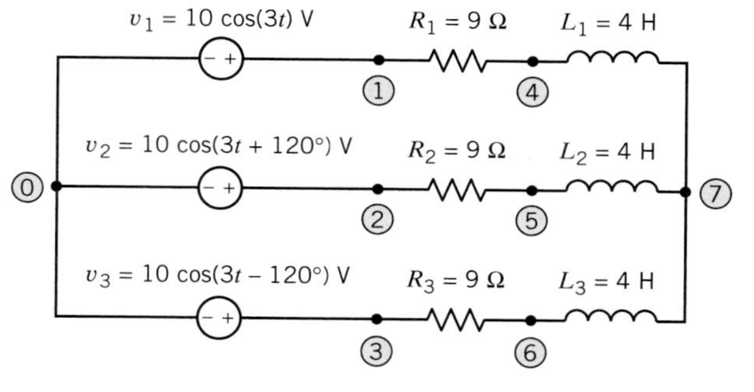

(a)

Element	Voltage	Current
V1 1 0 10 /0	10 /0	0.67 /127
V2 2 0 10 /120	10 /120	0.67 /-113
V3 3 0 10 /-120	10 /-120	0.67 /7
R1 1 4 9	6 /-53	0.67 /-53
R2 2 5 9	6 /67	0.67 /67
R3 3 6 9	6 /-173	0.67 /-173
L1 4 7 4	8 /37	0.67 /-53
L2 5 7 4	8 /157	0.67 /67
L3 6 7 4	8 /-83	0.67 /-173

(b)

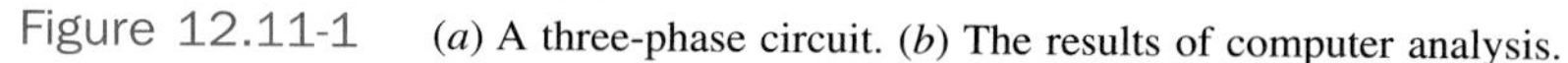

Figure 12.11-1 (a) A three-phase circuit. (b) The results of computer analysis.

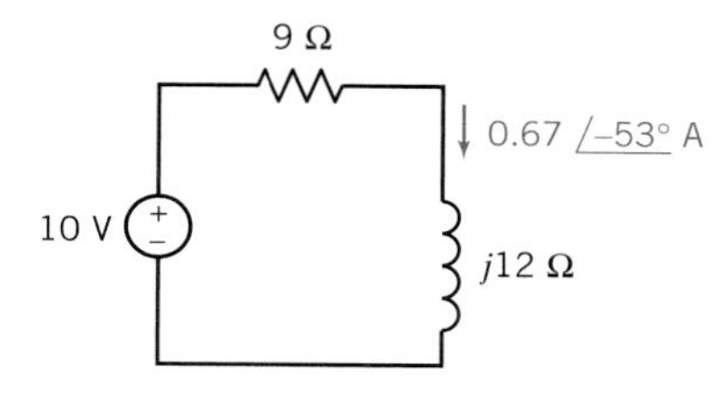

Figure 12.11-2
The per-phase equivalent circuit.

The voltage across the resistor is given by

$$\mathbf{V}_{\mathrm{R}}(\omega) = 9\,\mathbf{I}_{\mathrm{L}}(\omega) = 6\,e^{-j53^\circ}\ \mathrm{V}$$

These currents and voltages are the same as the values given in the computer analysis for the element currents and voltages of R_1 and L_1. We conclude that the computer analysis of the three-phase circuit is correct.

Example 12.11-2

Computer analysis of the circuit in Figure 12.11-3 shows that $\mathbf{V}_{\mathrm{Nn}}(\omega) = 12.67\angle 174.6^\circ$ V. This computer analysis did not use RMS values, so 12.67 is the magnitude of the sinusoidal voltage $v_{\mathrm{Nn}}(t)$ rather than the effective value. Verify that this voltage is correct.

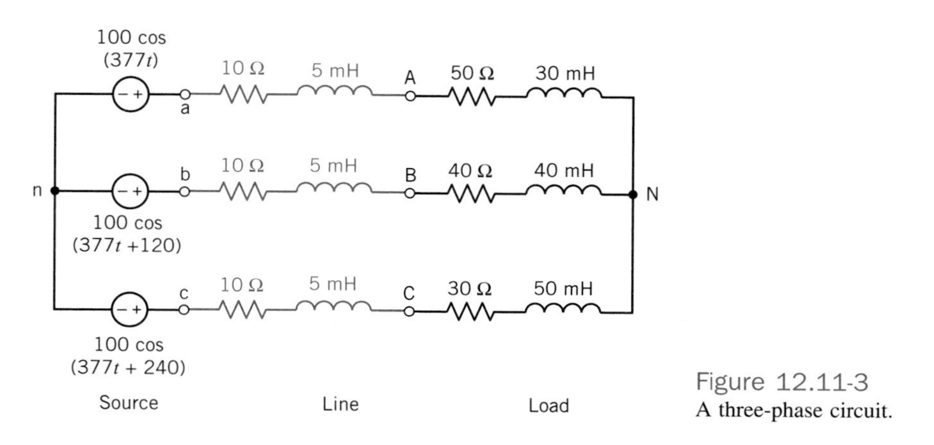

Figure 12.11-3
A three-phase circuit.

Solution

This result could be checked by writing and solving a node equation to calculate $\mathbf{V}_{\mathrm{Nn}}(\omega)$, but it is easier to check this result by verifying that KCL is satisfied at node N.

First, calculate the three line currents as

$$\mathbf{I}_{\mathrm{A}}(\omega) = \frac{100 - \mathbf{V}_{\mathrm{Nn}}(\omega)}{60 + j(377)(0.035)} = 1.833 \angle -13^\circ \text{ A}$$

$$\mathbf{I}_{\mathrm{B}}(\omega) = \frac{100\angle 120^\circ - \mathbf{V}_{\mathrm{Nn}}(\omega)}{50 + j(377)(0.045)} = 1.766 \angle 94.9^\circ \text{ A}$$

$$\mathbf{I}_{\mathrm{C}}(\omega) = \frac{100\angle 240^\circ - \mathbf{V}_{\mathrm{Nn}}(\omega)}{40 + j(377)(0.055)} = 2.118 \angle -140.5^\circ \text{ A}$$

Next apply KCL at node N to get

$$1.833 \angle -13^\circ + 1.766 \angle 95.9^\circ + 2.118 \angle -140.5^\circ = 0 \text{ A}$$

Since KCL is satisfied at node N, the given node voltage is correct.

We can also check that average power is conserved. Recall that peak values, rather than effective values, are being used in this example. First, determine the power delivered by the (three-phase) source:

$\mathbf{I}_{\mathrm{A}}(\omega) = 1.833 \angle -13^\circ$ and $\mathbf{V}_{\mathrm{an}}(\omega) = 100 \angle 0^\circ$ so

$$P_{\mathrm{a}} = \frac{(100)(1.833)}{2}\cos(0^\circ - (-13^\circ)) = 89.3 \text{ W}$$

$\mathbf{I}_{\mathrm{B}}(\omega) = 1.766 \angle 94.9^\circ$ and $\mathbf{V}_{\mathrm{bn}}(\omega) = 100 \angle 120^\circ$ so

$$P_{\mathrm{b}} = \frac{(100)(1.766)}{2}\cos(120^\circ - (94.9^\circ)) = 80 \text{ W}$$

$\mathbf{I}_{\mathrm{C}}(\omega) = 2.118 \angle -140.5^\circ$ and $\mathbf{V}_{\mathrm{cn}}(\omega) = 100 \angle 240^\circ$ so

$$P_{\mathrm{c}} = \frac{(100)(2.118)}{2}\cos(0^\circ + 140.5^\circ) = 99.2 \text{ W}$$

The power delivered by the source is 89.3 + 80 + 99.2 = 268.5 W.

Next, determine the power delivered to the (three-phase) load as

$\mathbf{I}_{\mathrm{A}}(\omega) = 1.833 \angle -13^\circ$ A and $R_{\mathrm{A}} = 50\ \Omega$ so $P_{\mathrm{A}} = \dfrac{1.833^2}{2} 50 = 84.0$ W

$\mathbf{I}_{\mathrm{B}}(\omega) = 1.766 \angle 94.9^\circ$ A and $R_{\mathrm{B}} = 40\ \Omega$ so $P_{\mathrm{B}} = \dfrac{1.766^2}{2} 40 = 62.4$ W

$\mathbf{I}_{\mathrm{C}}(\omega) = 2.118 \angle -140.5^\circ$ A and $R_{\mathrm{C}} = 30\ \Omega$ so $P_{\mathrm{C}} = \dfrac{2.118^2}{2} 30 = 67.3$ W

The power delivered to the load is 84 + 62.4 + 67.3 W.

Determine the power lost in the (three-phase) line as

$\mathbf{I}_{\mathrm{A}}(\omega) = 1.833 \angle -13^\circ$ A and $R_{\mathrm{aA}} = 10\ \Omega$ so $P_{\mathrm{aA}} = \dfrac{1.833^2}{2} 10 = 16.8$ W

$\mathbf{I}_{\mathrm{B}}(\omega) = 1.766 \angle 94.9^\circ$ A and $R_{\mathrm{bB}} = 10\ \Omega$ so $P_{\mathrm{bB}} = \dfrac{1.766^2}{2} 10 = 15.6$ W

$\mathbf{I}_{\mathrm{C}}(\omega) = 2.118 \angle -140.5^\circ$ A and $R_{\mathrm{cC}} = 10\ \Omega$ so $P_{\mathrm{cC}} = \dfrac{2.118^2}{2} 10 = 22.4$ W

The power lost in the line is 16.8 + 15.6 + 22.4 = 54.8 W.

The power delivered by the source is equal to the sum of the power lost in the line plus the power delivered to the load. Again, we conclude that the given node voltage is correct.

12.12 Design Challenge Solution

POWER FACTOR CORRECTION

Figure 12.12-1 shows a three-phase circuit. The capacitors are added to improve the power factor of the load. We need to determine the value of the capacitance C required to obtain a power factor of 0.9 lagging.

Describe the Situation and the Assumptions

1. The circuit is excited by sinusoidal sources all having the same frequency, 60 Hz or 377 rad/s. The circuit is at steady state. The circuit is a linear circuit. Phasors can be used to analyze this circuit.
2. The circuit is a balanced three-phase circuit. A per-phase equivalent circuit can be used to analyze this circuit.
3. The load consists of two parts. The part comprising resistors and inductors is connected as a Y. The part comprising capacitors is connected as a Δ. A Δ-to-Y transformation can be used to simplify the load.

The per-phase equivalent circuit is shown in Figure 12.12-2.

State the Goal

Determine the value of C required to correct the power factor to 0.9 lagging.

Generate a Plan

Power factor correction was considered in Chapter 11. A formula was provided for calculating the reactance, X_1, needed to correct the power factor of a load

$$X_1 = \frac{R^2 + X^2}{R\tan(\cos^{-1}\text{pfc}) - X}$$

where R and X are the real and imaginary parts of the load impedance before the power factor is corrected and pfc is the corrected power factor. After this equation is used to calculate X_1, the capacitance C can be calculated from X_1. Notice that X_1 will be the reactance of the equivalent Y-connected capacitors. We will need to calculate the Δ-connected capacitor equivalent of the Y-connected capacitor.

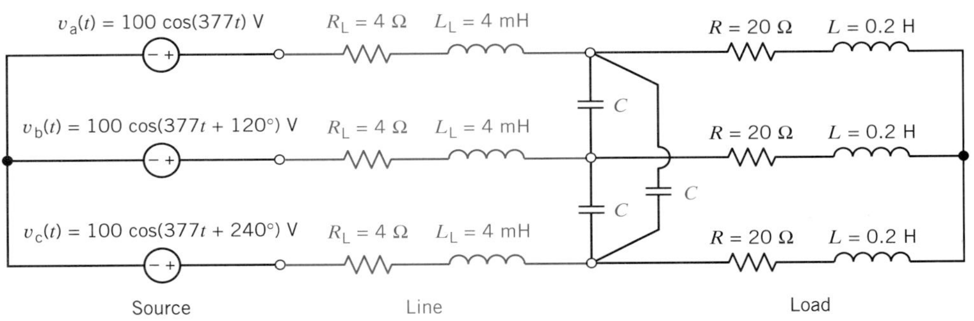

Figure 12.12-1 A balanced three-phase circuit.

Act on the Plan

We note that $\mathbf{Z} = R + jX = 20 + j75.4\ \Omega$. Therefore, the reactance, X_1, needed to correct the power factor is

$$X_1 = \frac{20^2 + 75.4^2}{20\tan(\cos^{-1}0.9) - 75.4} = -92.6$$

The Y-connected capacitor equivalent to the Δ-connected capacitor can be calculated from $\mathbf{Z}_Y = \mathbf{Z}_\Delta/3$. Therefore, the capacitance of the equivalent Y-connected capacitor is $3C$.

Finally, since $X_1 = 1/(3C\omega)$, we have

$$C = \frac{1}{\omega \cdot 3 \cdot X_1} = -\frac{1}{377 \cdot 3(-92.6)} = 9.458\ \mu\text{F}$$

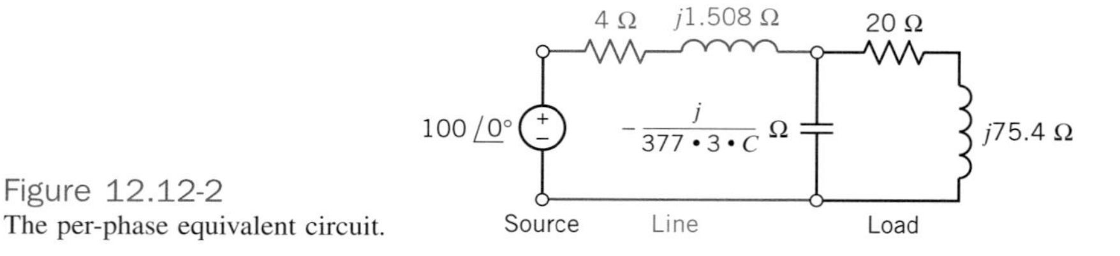

Figure 12.12-2
The per-phase equivalent circuit.

Verify the Proposed Solution

When $C = 9.458\ \mu\text{F}$, the impedance of one phase of the equivalent Y-connected load will be

$$\mathbf{Z}_Y = \frac{\dfrac{1}{j377 \times 3 \times C}(20 + j75.4)}{\dfrac{1}{j377 \times 3 \times C} + (20 + j75.4)} = 240.4 + j123.9$$

The value of the power factor is

$$pf = \cos\left(\tan^{-1}\left(\frac{123.9}{240.4}\right)\right) = 0.90$$

so the specifications have been satisfied.

12.13 SUMMARY

- The generation and transmission of electrical power are more efficient in three-phase systems employing three voltages of the same magnitude and frequency and differing in phase by 120° from each other.
- The three-phase source consists of either three Y-connected sinusoidal voltage sources or three Δ-connected sinusoidal voltage sources. Similarly, the circuit elements that comprise the load are connected to form either a Y or a Δ. The transmission line is used to connect the source to the load and consists of either three or four wires.
- Analysis of three-phase circuits using phasors and impedances will determine the *steady-state response* of the three-phase circuit. We are particularly interested in the power the three-phase source delivers to the three-phase load. Table 12.0-1 summarizes the formulas that are used to calculate the power delivered to an element when the element voltage and current adhere to the passive convention.
- The current in the neutral wire of a balanced Y-to-Y connection is zero; thus the wire may be removed if desired. The key to the analysis of the Y-to-Y circuit is the calculation of the line currents. When the circuit is not balanced, the first step in the analysis of this circuit is to calculate $\mathbf{V}_{Nn}$, the voltage at the neutral node of the three-phase load with respect to the voltage at the neutral node of the three-phase source. When the circuit is balanced, this step isn't needed because $\mathbf{V}_{Nn} = 0$. Once $\mathbf{V}_{Nn}$ is known, the line currents can be calculated. The

line current for a balanced Y-to-Y connection is $\mathbf{V}_a/\mathbf{Z}$ for phase a, and the other two currents are displaced by $\pm120°$ from $\mathbf{I}_A$.

- For a Δ load, we converted the Δ load to a Y-connected load by using the relation Δ-to-Y transformation. Then we proceeded with the Y-to-Y analysis.
- The line current for a balanced Δ load is $\sqrt{3}$ times the phase current and is displaced $-30°$ in phase. The line-to-line voltage of a Δ load is equal to the phase voltage.
- The power delivered to a balanced Y-connected load is $P_Y = \sqrt{3}\, V_{AB} I_A \cos\theta$ where V_{AB} is the line-to-line voltage, I_A is the line current, and θ is the angle between the phase voltage and the phase current ($\mathbf{Z}_Y = Z\angle\theta$).
- The two-wattmeter method of measuring three-phase power delivered to a load was described. Also, we considered the usefulness of the two-wattmeter method for determining the power factor angle of a three-phase system.

PROBLEMS

Section 12.3 Three-Phase Voltages

P 12.3-1 A balanced three-phase Y-connected load has one-phase voltage

$$\mathbf{V}_c = 277\angle45° \text{ V rms}$$

The phase sequence is *abc*. Find the line-to-line voltages $\mathbf{V}_{AB}$, $\mathbf{V}_{BC}$, and $\mathbf{V}_{CA}$. Draw a phasor diagram showing the phase and line voltages.

P 12.3-2 A three-phase system has a line-to-line voltage

$$\mathbf{V}_{BA} = 12{,}470\angle-35° \text{ V rms}$$

with a Y load. Find the phase voltages when the phase sequence is *abc*.

P 12.3-3 A three-phase system has a line-to-line voltage

$$\mathbf{V}_{ab} = 1500\angle30° \text{ V rms}$$

with a Y load. Determine the phase voltage.

Section 12.4 The Y-to-Y Circuit

P 12.4-1 Consider a three-wire Y-to-Y circuit. The voltages of the Y-connected source are $\mathbf{V}_a = \dfrac{208}{\sqrt{3}}\angle-30°$ V rms, $\mathbf{V}_b = \dfrac{208}{\sqrt{3}}\angle-150°$ V rms, and $\mathbf{V}_c = \dfrac{208}{\sqrt{3}}\angle90°$ V rms. The Y-connected load is balanced. The impedance of each phase is $\mathbf{Z} = 12\angle30°\ \Omega$.

(a) Find the phase voltages.

(b) Find the line currents and phase currents.

(c) Show the line currents and phase currents on a phasor diagram.

(d) Determine the power dissipated in the load.

P 12.4-2 A balanced three-phase Y-connected supply delivers power through a three-wire plus neutral-wire circuit in a large office building to a three-phase Y-connected load. The circuit operates at 60 Hz. The phase voltages of the Y-connected source are $\mathbf{V}_a = 120\angle0°$ V rms, $\mathbf{V}_b = 120\angle-120°$ V rms, and $\mathbf{V}_c = 120\angle120°$ V rms. Each transmission wire, including the neutral wire, has a 2 Ω resistance, and the balanced Y load has a 10 Ω resistance in series with 100 mH. Find the line voltage and the phase current at the load.

P 12.4-3 A Y-connected source and load are shown in Figure P 12.4-3. (a) Determine the rms value of the current $i_a(t)$. (b) Determine the average power delivered to the load.

P 12.4-4 An unbalanced Y-Y circuit is shown in Figure P 12.4-4. Find the average power delivered to the load.
Hint: $\mathbf{V}_{Nn}(\omega) = 27.4\angle-63.6$ V
Answer: 436.4 W

P 12.4-5 A balanced Y-Y circuit is shown in Figure P 12.4-5. Find the average power delivered to the load.

P 12.4-6 An unbalanced Y-Y circuit is shown in Figure P 12.4-6. Find the average power delivered to the load.
Hint: $\mathbf{V}_{Nn}(\omega) = 1.755\angle-29.5$ V
Answer: 436.4 W

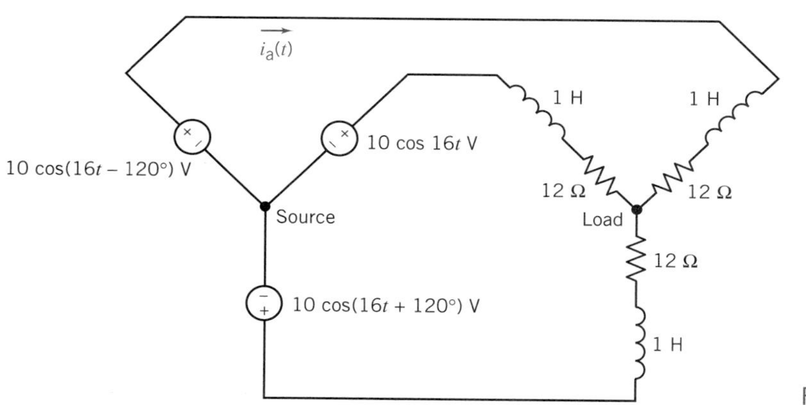

Figure P 12.4-3

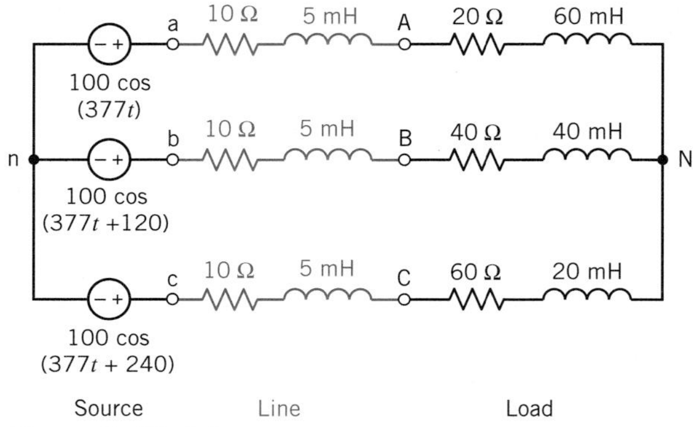

Figure P 12.4-4

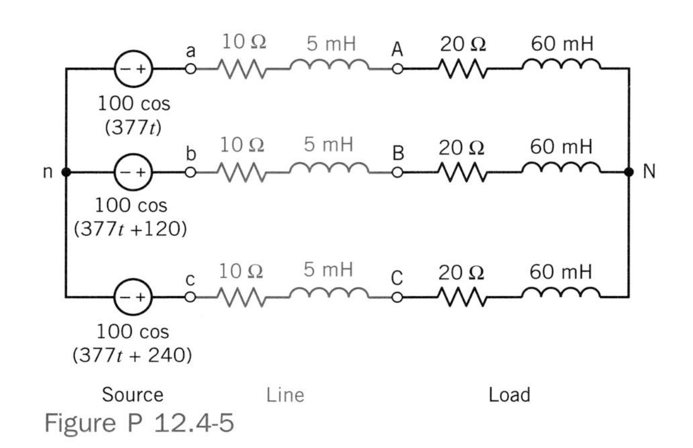

Figure P 12.4-5

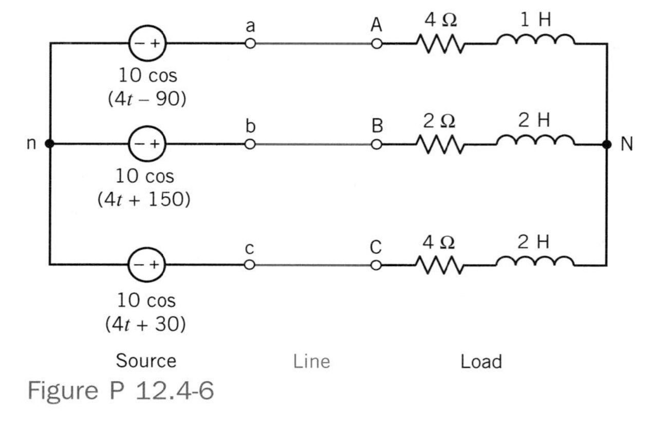

Figure P 12.4-6

P 12.4-7 A balanced Y-Y circuit is shown in Figure P 12.4-7. Find the average power delivered to the load.

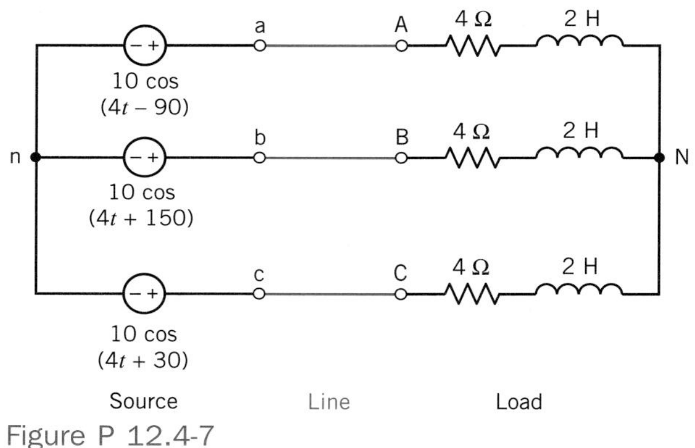

Figure P 12.4-7

Section 12.5 The Δ-Connected Source and Load

P 12.5-1 A balanced three-phase Δ-connected load has one line current

$$\mathbf{I}_B = 50\angle -40^\circ \text{ A rms}$$

Find the phase currents $\mathbf{I}_{BC}$, $\mathbf{I}_{AB}$, and $\mathbf{I}_{CA}$. Draw the phasor diagram showing the line and phase currents.

P 12.5-2 A three-phase circuit has two parallel balanced Δ loads, one of 5 Ω resistors and one of 20 Ω resistors. Find the magnitude of the total line current when the line-to-line voltage is 480 V rms.

Section 12.6 The Y-to-Δ Circuit

P 12.6-1 Consider a three-wire Y-to-Δ circuit. The voltages of the Y-connected source are $\mathbf{V}_a = \frac{208}{\sqrt{3}}\angle -30^\circ$ V rms, $\mathbf{V}_b = \frac{208}{\sqrt{3}}\angle -150^\circ$ V rms, and $\mathbf{V}_c = \frac{208}{\sqrt{3}}\angle 90^\circ$ V rms. The Δ-connected load is balanced. The impedance of each phase is $\mathbf{Z} = 12\angle 30^\circ$ Ω. Determine the line currents and calculate the power dissipated in the load.
Answer: $P = 9360$ W

P 12.6-2 A balanced Δ-connected load is connected by three wires, each with a 4 Ω resistance, to a Y source with $\mathbf{V}_a = \frac{480}{\sqrt{3}}\angle -30^\circ$ V rms, $\mathbf{V}_b = \frac{480}{\sqrt{3}}\angle -150^\circ$ V rms, and $\mathbf{V}_c = \frac{480}{\sqrt{3}}\angle 90^\circ$ V rms. Find the line current $\mathbf{I}_A$ when $\mathbf{Z}_\Delta = 39\angle -40^\circ$ Ω.
Answer: $\mathbf{I}_A = 17\angle 0.9^\circ$

P 12.6-3 The balanced circuit shown in Figure P 12.6-3 has $\mathbf{V}_{ab} = 380\angle 30^\circ$ V rms. Determine the phase currents in the load when $\mathbf{Z} = 3 + j4$ Ω. Sketch a phasor diagram.

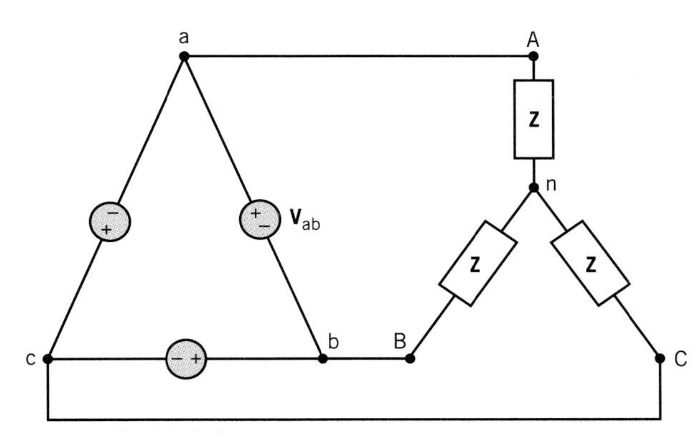

Figure P 12.6-3 A Δ-to-Y circuit.

P 12.6-4 The balanced circuit shown in Figure P 12.6-3 has $\mathbf{V}_{ab} = 380\angle 0°$ V rms. Determine the line and phase currents in the load when $\mathbf{Z} = 9 + j12\ \Omega$.

Section 12.7 Balanced Three-Phase Circuits

P 12.7-1 The English Channel Tunnel rail link is supplied at 25 kV rms from the United Kingdom and French grid systems. When there is a grid supply failure, each end is capable of supplying the whole tunnel but in a reduced operational mode.

The tunnel traction system is a conventional catenary (overhead wire) system similar to the surface main line electric railway system of the United Kingdom and France. What makes the tunnel traction system different and unique is the high density of traction load and the end-fed supply arrangement. The tunnel traction load is considerable. For each half tunnel the load is 180 MVA. (Barnes and Wong 1991).

Assume that each line-to-line voltage is 25 kV rms and the three-phase system is connected to the traction motor of an electric locomotive. The motor is a Y-connected load with $\mathbf{Z} = 150\angle 25°\ \Omega$. Find the line currents and the power delivered to the traction motor.

P 12.7-2 A three-phase source with a line voltage of 45 kV rms is connected to two balanced loads. The Y-connected load has $\mathbf{Z} = 10 + j20\ \Omega$, and the Δ load has a branch impedance of 50 Ω. The connecting lines have an impedance of 2 Ω. Determine the power delivered to the loads and the power lost in the wires. What percentage of power is lost in the wires?

P 12.7-3 A three-phase source has a Y-connected source with $v_a = 5\cos(2t + 30°)$ connected to a three-phase Y load. Each phase of the Y-connected load consists of a 4 Ω resistor and a 4-H inductor. Each connecting line has a resistance of 2 Ω. Determine the total average power delivered to the load.

Section 12.8 Power In a Balanced Load

P 12.8-1 Find the power absorbed by a balanced three-phase Y-connected load when

$$\mathbf{V}_{CB} = 208\angle 15° \text{ V rms} \quad \text{and} \quad \mathbf{I}_B = 3\angle 110° \text{ A rms}$$

Answer: $P = 620$ W

P 12.8-2 A three-phase motor delivers 20 hp operating from a 480-V rms line voltage. The motor operates at 85 percent efficiency with a power factor equal to 0.8. Find the magnitude and angle of the line current for phase A.

P 12.8-3 A three-phase balanced load is fed by a line-to-line voltage of 220 V rms. It absorbs 1500 W at 0.8 power factor lagging. Calculate the phase impedance if it is (a) Δ connected and (b) Y connected.

P 12.8-4 A 600-V rms three-phase circuit has two balanced Δ loads connected to the lines. The load impedances are $40\angle 30°\ \Omega$ and $50\angle -60°\ \Omega$, respectively. Determine the line current and the total average power.

P 12.8-5 A three-phase feeder simultaneously supplies power to two separate balanced three-phase loads. The first total load is Δ connected and requires 39 kVA at 0.7 lagging. The second total load is Y connected and requires 15 kW at 0.21 leading. Each line has an impedance $0.038 + j0.072\ \Omega$/phase. Calculate the line-to-line source voltage magnitude required so that the loads are supplied with 208 V rms line-to-line.

P 12.8-6 A building is supplied by a public utility at 4.16 kV rms. The building contains three balanced loads connected to the three-phase lines:

(a) Δ connected, 500 kVA at 0.85 lagging.
(b) Y connected, 75 kVA at 0.0 leading.
(c) Y connected; each phase has a 150 Ω resistor parallel to a 225 Ω inductive reactance.

The utility feeder is five miles long with an impedance per phase of $1.69 + j0.78\ \Omega$/mile. At what voltage must the utility supply its feeder so that the building is operating at 4.16 kV rms?

P 12.8-7 The diagram shown in Figure P 12.8-7 has two three-phase loads that form part of a manufacturing plant. They are connected in parallel and require 4.16 kV rms. Load 1 is

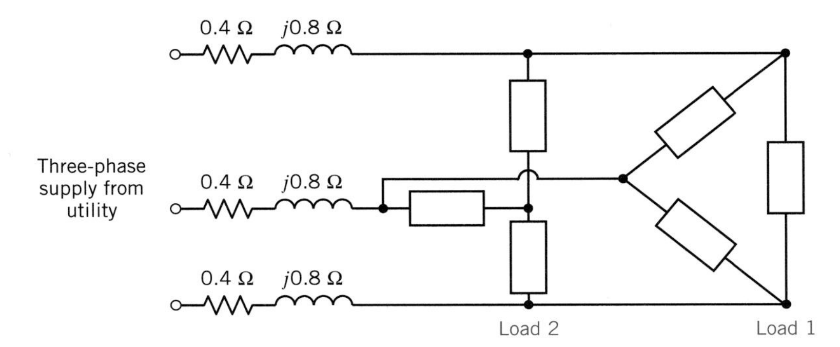

Figure P 12.8-7
A three-phase circuit with a Δ load and a Y load.

1.5 MVA, 0.75 lag pf, Δ connected. Load 2 is 2 MW, 0.8 lagging pf, Y connected. The feeder from the power utility's substation transformer has an impedance of $0.4 + j0.8\ \Omega$/phase. Determine

(a) The required magnitude of the line voltage at the supply.
(b) The real power drawn from the supply.
(c) The percentage of the real power drawn from the supply that is consumed by the loads.

P 12.8-8 The balanced three-phase load of a large commercial building requires 480 kW at a lagging power factor of 0.8. The load is supplied by a connecting line with an impedance of $5 + j25$ mΩ for each phase. Each phase of the load has a line-to-line voltage of 600 V rms. Determine the line current and the line voltage at the source. Also, determine the power factor at the source. Use the line-to-neutral voltage as the reference with an angle of 0°.

Section 12.9 Two-Wattmeter Power Measurement

P 12.9-1 The two-wattmeter method is used to determine the power drawn by a three-phase 440-V rms motor that is a Y-connected balanced load. The motor operates at 20 hp at 74.6 percent efficiency. The magnitude of the line current is 52.5 A rms. The wattmeters are connected in the A and C lines. Find the reading of each wattmeter. The motor has a lagging power factor.

P 12.9-2 A three-phase system has a line-to-line voltage of 4000 V rms and a balanced Δ-connected load with $\mathbf{Z} = 40 + j30\ \Omega$. The phase sequence is *abc*. Use the two wattmeters connected to lines *A* and *C*, with line *B* as the common line for the voltage measurement. Determine the total power measurement recorded by the wattmeters.

Answer: $P = 768$ kW

P 12.9-3 A three-phase system with a sequence *abc* and a line-to-line voltage of 200 V rms feeds a Y-connected load with $\mathbf{Z} = 70.7\angle 45°\ \Omega$. Find the line currents. Find the total power by using two wattmeters connected to lines *B* and *C*.

Answer: $P = 400$ W

P 12.9-4 A three-phase system with a line-to-line voltage of 208 V rms and phase sequence *abc* is connected to a Y-balanced load with impedance $10\angle -30°\ \Omega$ and a balanced Δ load with impedance $15\angle 30°\ \Omega$. Find the line currents and the total power using two wattmeters.

P 12.9-5 The two-wattmeter method is used. The wattmeter in line *A* reads 920 W, and the wattmeter in line *C* reads 460 W. Find the impedance of the balanced Δ-connected load. The circuit is a three-phase 120-V rms system with an *abc* sequence.

Answer: $\mathbf{Z}_\Delta = 27.1\angle -30°\ \Omega$

P 12.9-6 Using the two-wattmeter method, determine the power reading of each wattmeter and the total power for Problem 12.6-3 when $\mathbf{Z} = 0.868 + j4.924\ \Omega$. Place the current coils in the *A*-to-*a* and *C*-to-*c* lines.

PSPICE PROBLEMS

SP 12-1 Determine the line voltage and the phase current at the load for Problem 12.4-2.

SP 12-2 Determine the line current of Problem 12.6-2 using PSpice.

SP 12-3 A circuit has a Y-to-Y connection as shown in Figure 12.6-1 with $R = 2\ \Omega$ and $\mathbf{Z} = 8 + j9\ \Omega$. Determine the line current and the total power delivered to the load.

VERIFICATION PROBLEMS

VP 12-1 A Y-connected source is connected to a Y-connected load (Figure 12.4-1) with $\mathbf{Z} = 10 + j4\ \Omega$. The line voltage is $V_L = 416$ V rms. A student report states that the line current $\mathbf{I}_A = 38.63$ A rms and that the power delivered to the load is 16.1 kW. Verify these results.

VP 12-2 A Δ load with $Z = 40 + j30\ \Omega$ has a three-phase source with $V_L = 240$ V rms (Figure 12.4-2). A computer analysis program states that one-phase current is $4.8\angle -36.9°$A rms. Verify this result.

DESIGN PROBLEMS

DP 12-1 A balanced three-phase Y source has a line voltage of 208 V rms. The total power delivered to the balanced Δ load is 1200 W with a power factor of 0.94 lagging. Determine the required load impedance for each phase of the Δ load. Calculate the resulting line current. The source is a 208-V rms *ABC* sequence.

DP 12-2 A three-phase 240-V rms circuit has a balanced Y load impedance **Z**. Two wattmeters are connected with current coils in lines *A* and *C*. The wattmeter in line *A* reads 1440 W, and the wattmeter in line *C* reads zero. Determine the value of the impedance.

DP 12-3 A three-phase motor delivers 100 hp and operates at 80 percent efficiency with a 0.75 lagging power factor. Determine the required Δ-connected balanced set of three capacitors that will improve the power factor to 0.90 lagging. The motor operates from 480-V rms lines.

DP 12-4 A three-phase system has balanced conditions so that the per-phase circuit representation can be utilized as shown in Figure DP 12.4. Select the turns ratio of the step-up and step-down transformers so that the system operates with an efficiency greater than 99 percent. The load voltage is specified as 4 kV rms, and the load impedance is 4/3 Ω.

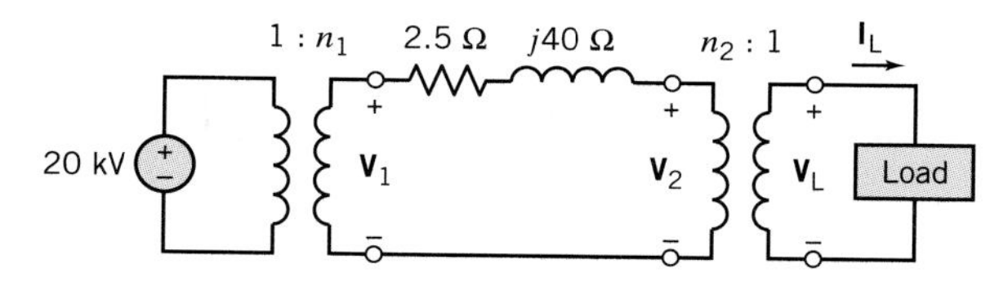

Figure DP 12.4 Per-phase diagram with two transformers.

CHAPTER 13

Frequency Response

Preview

Figure 13.0-1 illustrates a familiar situation. A linear, time-invariant circuit with a sinusoidal input has reached steady state. We know that the steady-state response of this circuit will be sinusoidal and will have the same frequency as the input. Suppose this circuit is disturbed by changing the frequency of the input sinusoid. After the transient part of the response dies out, the circuit will again be at steady state. We will notice that the steady-state response has changed. Of course, the frequency of the response has changed to the new frequency of the input, but more than that has changed. Both the magnitude, B, of the response and the phase angle, ϕ, have changed.

Repeating this experiment several times, we observe that the circuit acts differently at different frequencies. For example, it may well be the case that at some frequencies the output is about the same size as the input, while at other frequencies the output is much smaller than the input. Similarly, the output sinusoid may be delayed with respect to the input sinusoid. At some frequencies, this delay may be a significant fraction of the period of the sinusoid, while at other frequencies this delay may be almost negligible.

It is important that we be able to characterize the behavior of a circuit for a range of input frequencies. Gain and phase shift are properties of a linear circuit that describe the behavior of the circuit as the input frequency changes. Suppose an input voltage consists of several sinusoidal components. For example, human speech typically contains sinusoids in the range 20 Hz to 20,000 Hz. A circuit that is intended to amplify human speech must treat all sinusoids with frequencies in the range 20 Hz to 20,000 Hz "in the same way" or the amplified signal will be distorted. Gain and phase shift allow us to be precise about what we mean when we say "in the same way." The

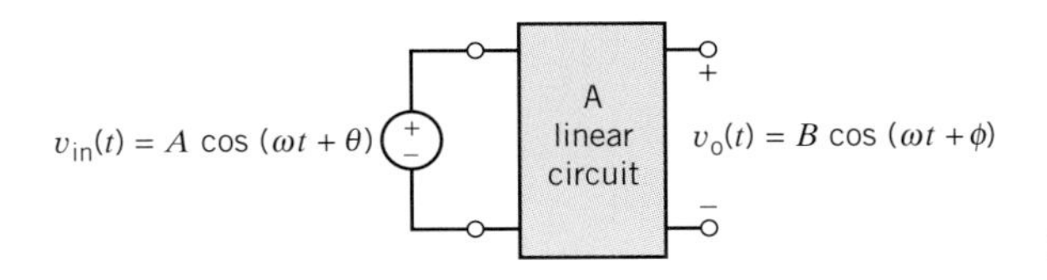

Figure 13.0-1 The steady-state response of a linear circuit with a sinusoidal input.

amplifier output will be an undistorted copy of the input voltage only if the gain of the circuit is constant over the entire frequency range and the phase shift is proportional to the input frequency.

In 1938 William Hewlett and David Packard built an audio oscillator, as shown in Figure 13.0-2. This oscillator provided a sinusoidal steady-state signal in the frequency range 20 Hz to 20 kHz. The audio oscillator made it possible to measure the gain and phase shift of a circuit, an exercise now routinely performed by undergraduate engineering students.

Figure 13.0-2 Hewlett-Packard's first product, the model 200A audio oscillator (preproduction version). William Hewlett and David Packard built an audio oscillator in 1938, from which the famous firm grew. Courtesy of Hewlett-Packard Company.

13.1 Design Challenge

RADIO TUNER

Three radio stations broadcast at three different frequencies, 700 kHz, 1000 kHz, and 1400 kHz. Figure 13.1-1 shows a simplified diagram of a radio receiver. The antenna receives signals from all three stations so that the input to the tuner will be a sum of these signals. Suppose this voltage is described by

$$v_i(t) = \sin(2\pi \cdot 7 \cdot 10^5 t + 135°) + \sin(2\pi \cdot 10^6 t) + \sin(2\pi \cdot 1.4 \cdot 10^6 t + 300°) \quad (13.1\text{-}1)$$

Consider the problem of tuning to the station that broadcasts at 1000 kHz. The tuner must eliminate the first and third terms of $v_i(t)$ to produce the output signal

$$v_o(t) = A \sin(2\pi \cdot 10^6 t + \theta)$$

The behavior of the tuner must be frequency dependent. That is, the effect of the tuner on each of the three terms of $v_i(t)$ must depend on the frequency of that term. The behavior of the frequency-dependent circuit is the subject of this chapter. We will return to the problem of designing the tuner at the end of this chapter.

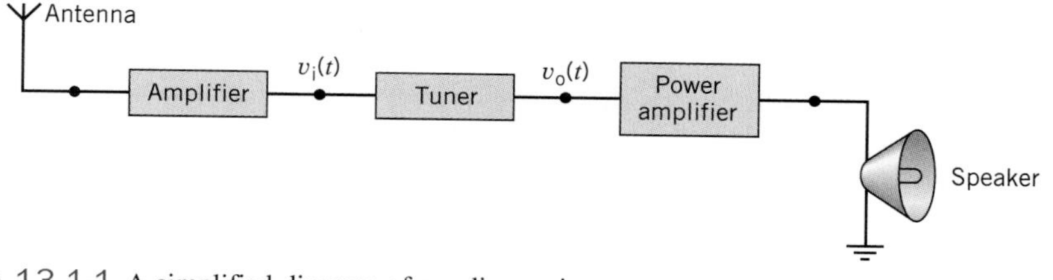

Figure 13.1-1 A simplified diagram of a radio receiver.

13.2 ELECTRONIC COMMUNICATION SYSTEMS

Our modern existence depends on communication systems such as television, radio, and the telephone. In this century, the telegraph emerged as an important communication system and then declined as competing systems came into existence. The rise and fall of the telegraph are shown in Figure 13.2-1.

The five communication systems that have done the most to change the twentieth century are, in order of occurrence, the telegraph, the telephone, radio, television, and the

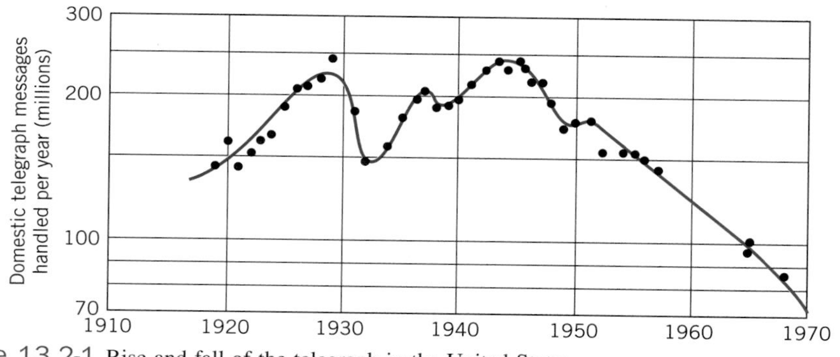

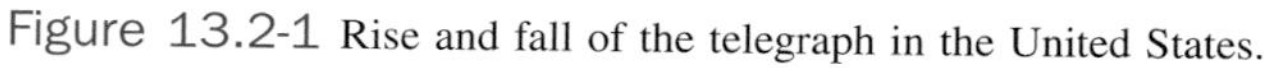

Figure 13.2-1 Rise and fall of the telegraph in the United States.

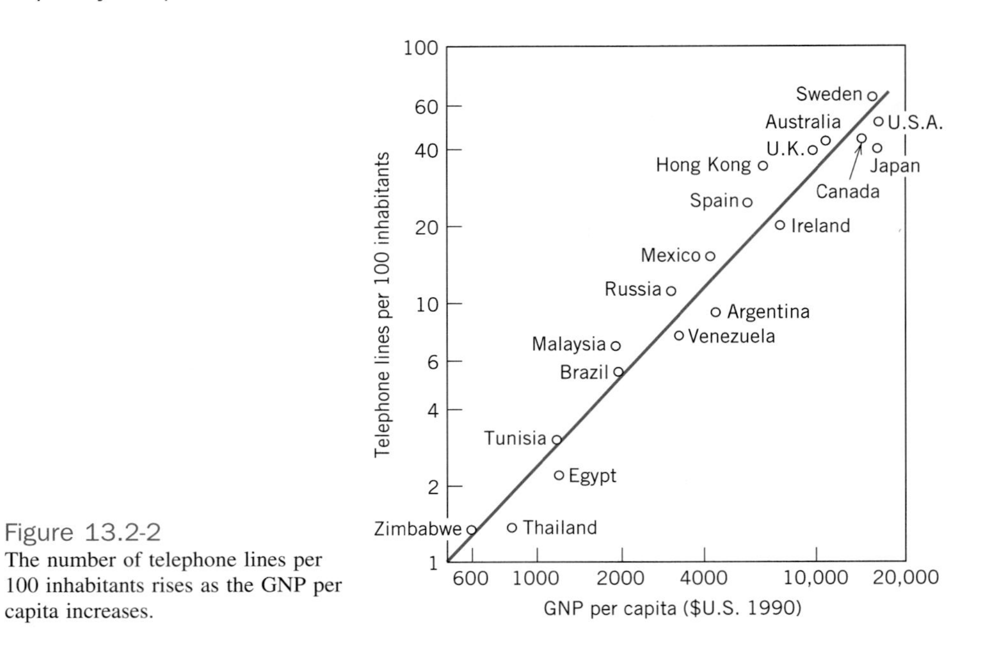

Figure 13.2-2
The number of telephone lines per 100 inhabitants rises as the GNP per capita increases.

space satellite. No sooner had Bell invented the telephone than people wanted it. By 1892, 240,000 telephones were in use in the United States. By 1956, America's 50 million telephones accounted for more than half of the world total. In America nearly 3.5 percent of the total gross national product is spent on telecommunication products and services. As shown in Figure 13.2-2, the number of telephone lines in a nation increases as the wealth of its inhabitants increases.

By 1950, 90 percent of U.S. households had a radio, as shown in Figure 13.2-3. With the introduction in 1972 of satellites for relaying telephone and television signals, the real cost of telecommunications continued the decline that has led to a massive use of telephones and wide distribution of television programs.

An international phone call from Europe to America will most probably be beamed 23,000 miles into space and bounced back to earth from one of 16 satellites, each of which is the size of a car. There are currently about 120,000 transatlantic satellite voice channels available, and cable provides another 10,000 channels.

As the use of radio and the telephone grew in the 1920s, many engineers designed circuits that operated at higher and higher frequencies. Coaxial cable was proposed as a telephone transmission line to carry signals with frequencies exceeding 10 MHz. By 1921 many radio engineers were transmitting signals at 3 MHz to 5 MHz. The late 1930s saw the advent of FM (frequency modulation) radio-transmitted signals above 30 MHz.

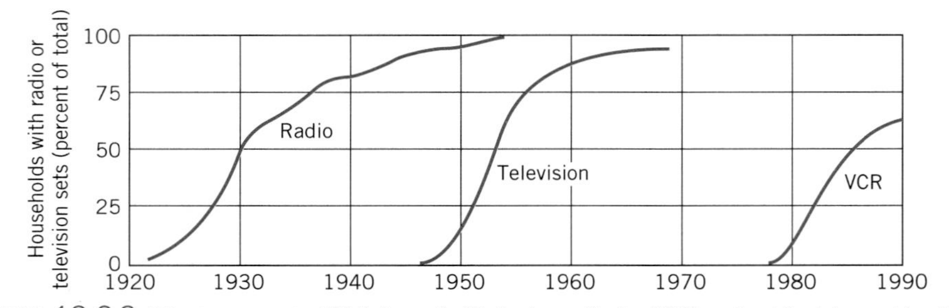

Figure 13.2-3 Ninety percent of U.S. households had a radio by 1950 and a television set by 1965. Sixty percent of U.S. households had a videocassette recorder by 1990.

13.3 GAIN, PHASE SHIFT, AND THE NETWORK FUNCTION

Gain, phase shift, and the network function are properties of linear circuits that are used to describe the effect that a circuit has on a sinusoidal input voltage or current. We expect that the behavior of circuits that contain reactive elements, that is, capacitors or inductors, will depend on the frequency of the input sinusoid. Thus we expect that the gain, phase shift, and network function will all be functions of frequency. Indeed, we will see that this is the case.

We begin by considering the circuit shown in Figure 13.3-1. The input to this circuit is the voltage of the voltage source, and the output, or response, of the circuit is the voltage across the 10 kΩ resistor. When the input is a sinusoidal voltage, then the steady-state response will also be sinusoidal and will have the same frequency as the input.

Suppose the voltages $v_{\text{in}}(t)$ and $v_{\text{out}}(t)$ are measured using an oscilloscope. Figure 13.3-2 shows the waveforms that would be displayed on the screen of the oscilloscope. Notice that the scales are shown but the axes are not. It is customary to take the angle of the input signal to be 0 deg, that is,

$$v_{\text{in}}(t) = A \cos \omega t$$

Then

$$v_{\text{out}} = B \cos(\omega t + \theta)$$

The **gain** of the circuit describes the relationship between the sizes of the input and output sinusoids. In particular, the gain is the ratio of the amplitude of the output sinusoid to the amplitude of the input sinusoid. That is

$$\text{gain} = \frac{B}{A}$$

The **phase shift** of the circuit describes the relationship between the phase angles of the input and output sinusoids. In particular, the phase shift is the difference between the phase angle of the output sinusoid and the phase angle of the input sinusoid. That is

$$\text{phase shift} = \theta - 0° = \theta$$

To be more specific, we need analytic representations of the sinusoids shown in Figure 13.3-2. The input voltage is the smaller of the two sinusoids and can be represented as

$$v_{\text{in}}(t) = 1 \cos 6283t \text{ V}$$

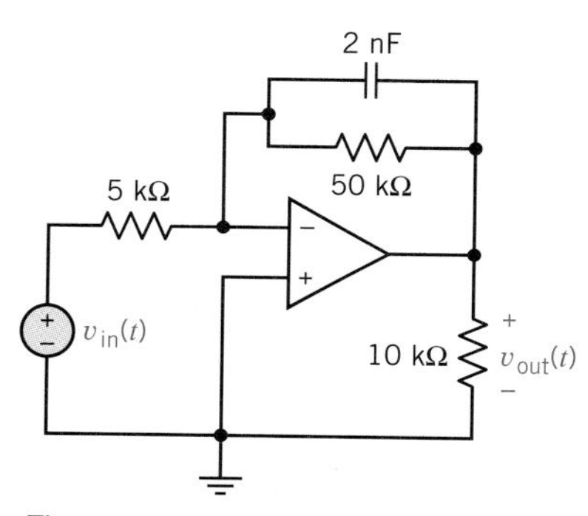

Figure 13.3-1 An op amp circuit.

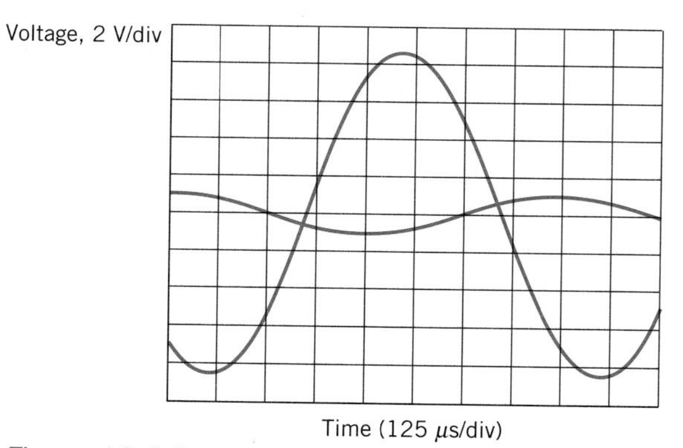

Figure 13.3-2 Input and output sinusoids for the op amp circuit of Figure 13.3-1.

The steady-state response is the larger of the two sinusoids and can be represented as

$$v_{out}(t) = 8.5 \cos(6283t + 148°) \text{ V}$$

The **gain** of this circuit at the frequency $\omega = 6283$ rad/s is

$$\text{gain} = \frac{\text{output amplitude}}{\text{input amplitude}} = \frac{8.5}{1} = 8.5$$

This gain is unitless because both amplitudes have units of volts. Since the gain is greater than 1, the output sinusoid is larger than the input sinusoid. This circuit is said to *amplify* its input. When the gain of a circuit is less than 1, the output sinusoid is smaller than the input sinusoid. This circuit is said to *attenuate* its input.

The *phase shift* of this circuit at the frequency $\omega = 6283$ rad/s is

$$\text{phase shift} = \text{output phase angle} - \text{input phase angle} = 148° - 0° = 148°$$

The phase shift determines the amount of time the output is advanced or delayed with respect to the input. Notice that

$$B \cos(\omega t + \theta) = B \cos\left(\omega\left(t + \frac{\theta}{\omega}\right)\right) = B \cos\left(\omega(t + t_0)\right)$$

where θ is the phase angle in radians and $t_0 = \theta/\omega$. The positive peaks of $B \cos(\omega t + \theta)$ occur when

$$\omega t + \theta = n(2\pi)$$

and solving for t we have

$$t = \frac{n(2\pi)}{\omega} - t_0 = nT - t_0$$

where n is any integer and T is the period of the sinusoid.

The positive peaks of $A \cos \omega t$ occur at $t = \dfrac{n(2\pi)}{\omega}$ and the positive peaks of $B \cos(\omega t + \theta)$ occur at $t = \dfrac{n(2\pi)}{\omega} - t_0$. A phase shift of θ rad is seen to shift the output sinusoid by t_0 seconds. When the frequency is 6283 rad/s, a phase shift of 148 deg or 2.58 rad causes a shift in time equal to

$$t_0 = \frac{\theta}{\omega} = \frac{2.58 \text{ rad}}{6283 \text{ rad/s}} = 410 \ \mu\text{s}$$

In Figure 13.3-2, the positive peaks of the input sinusoid occur at 0 ms, 1 ms, 2 ms, 3 ms, ... Positive peaks of the output sinusoid occur at 0.59 ms, 1.59 ms, 2.59 ms, 3.59 ms, ... Peaks of the output sinusoid occur 410 μs before the next peak of the input sinusoid. The output is *advanced* by 410 μs with respect to the input.

Notice that

$$v_{out}(t) = 8.5 \cos(6283t + 148°) = 8.5 \cos(6283t - 212°)$$

since a phase shift of 360 deg does not change the sinusoid. A phase shift of -212 deg or -3.70 rad causes a shift in time of

$$t_0 = \frac{-3.70 \text{ rad}}{6283 \text{ rad/s}} = -590 \ \mu\text{s}$$

Peaks of the output sinusoid occur 590 μs *after* the next peak of the input sinusoid. The output is *delayed* by 590 μs with respect to the input.

A phase shift that advances the output is called a **phase lead.** A phase shift that delays the output is called a **phase lag.**

At the frequency $\omega = 6283$ rad/s this circuit amplifies its input by a factor of 8.5 and advances it by 410 μs or, equivalently, delays it by 590 μs. The circuit of Figure 13.3-1 has a phase lead of 148 deg or, equivalently, a phase lag of 212 deg.

Now let us consider this circuit when the frequency of the input is changed. When the input is

$$v_{\text{in}}(t) = 1 \cos 3141.6t \text{ V}$$

the steady-state response of this circuit can be found to be

$$v_{\text{out}}(t) = 9.54 \cos(3141.6t + 163°) \text{ V}$$

The gain and phase shift of this circuit at the frequency $\omega = 3141.6$ rad/s are

$$\text{gain} = \frac{\text{output amplitude}}{\text{input amplitude}} = \frac{9.54}{1} = 9.54$$

and $\quad \text{phase shift} = \text{output phase angle} - \text{input phase angle} = 163° - 0° = 163°$

Changing the frequency of the input has changed the gain and phase shift of this circuit. Apparently, the gain and the phase shift of this circuit are functions of the frequency of the input. Table 13.3-1 shows the values of the gain and phase shift corresponding to several choices of the input frequency. As expected, the gain and phase shift changed when the input frequency changed. The **network function** describes the way the behavior of the circuit depends on the frequency of the input. The network function is defined in the frequency domain. It is the ratio of the phasor corresponding to the response sinusoid to the phasor corresponding to the input. Let $\mathbf{X}(\omega)$ be the phasor corresponding to the input to circuit and $\mathbf{Y}(\omega)$ be the phasor corresponding to the steady-state response of the network. Then

$$\mathbf{H}(\omega) = \frac{\mathbf{Y}(\omega)}{\mathbf{X}(\omega)} \tag{13.3-1}$$

is the network function. Notice that both $\mathbf{X}(\omega)$ and $\mathbf{Y}(\omega)$ could correspond to either a current or a voltage. Both the gain and the phase shift can be expressed in terms of the network function. The gain is

$$\text{gain} = |\mathbf{H}(\omega)| = \frac{|\mathbf{Y}(\omega)|}{|\mathbf{X}(\omega)|} \tag{13.3-2}$$

and the phase shift is

$$\text{phase shift} = \angle\mathbf{H}(\omega) = \angle\mathbf{Y}(\omega) - \angle\mathbf{X}(\omega) \tag{13.3-3}$$

Table 13.3-1 **Frequency Response Data for a Circuit**

f(Hz)	ω(rad/s)	Gain	Phase Shift
100	628.3	9.98	176°
500	3,141.6	9.54	163°
1,000	6,283	8.50	148°
5,000	31,416	3.03	108°
10,000	62,830	1.57	99°

Let's find the network function for the circuit shown in Figure 13.3-1. The first step is to represent this circuit in the frequency domain using impedances and phasors. Figure 13.3-3 shows the frequency domain circuit corresponding to the circuit in Figure 13.3-1. In this example, the phasor corresponding to the input is $\mathbf{V}_{in}(\omega)$, and the phasor corresponding to the output is $\mathbf{V}_{out}(\omega)$. We seek to find the network function $\mathbf{H}(\omega) = \mathbf{V}_{out}/\mathbf{V}_{in}$. Write the node equation at the inverting input node of the op amp and assume an ideal op amp. Then we have

$$\frac{\mathbf{V}_{in}(\omega)}{R_1} + \frac{\mathbf{V}_{out}(\omega)}{R_2} + j\omega C\,\mathbf{V}_{out}(\omega) = 0$$

This implies

$$\mathbf{H}(\omega) = \frac{\mathbf{V}_{out}(\omega)}{\mathbf{V}_{in}(\omega)} = \frac{-R_2}{R_1 + j\omega C R_1 R_2}$$

The gain of this circuit is

$$\text{gain} = |\mathbf{H}(\omega)| = H = \frac{R_2/R_1}{\sqrt{1 + \omega^2 C^2 R_2^2}}$$

The phase shift of this circuit is

$$\text{phase shift} = \angle \mathbf{H}(\omega) = 180^\circ - \tan^{-1}(\omega C R_2)$$

When $R_1 = 5\ \text{k}\Omega$, $R_2 = 50\ \text{k}\Omega$, and $C = 2$ nF,

$$\mathbf{H}(\omega) = \frac{-10}{1 + (j\omega/10{,}000)}$$

$$\text{gain} = |\mathbf{H}(\omega)| = \frac{10}{\sqrt{1 + (\omega^2/10^8)}}$$

$$\text{phase shift} = \angle \mathbf{H}(\omega) = 180^\circ - \tan^{-1}(\omega/10{,}000)$$

Notice that the frequency of the input has been represented by a variable, ω, rather than by any particular value. As a result, the network function, gain, and phase shift describe the way in which the behavior of the circuit depends on the input frequency. Earlier we considered the case when $\omega = 6283$ radians/second. Substituting this frequency into the equations for the gain and phase shift gives

$$\text{gain} = \frac{10}{\sqrt{1 + \dfrac{6283^2}{10^8}}} = 8.5$$

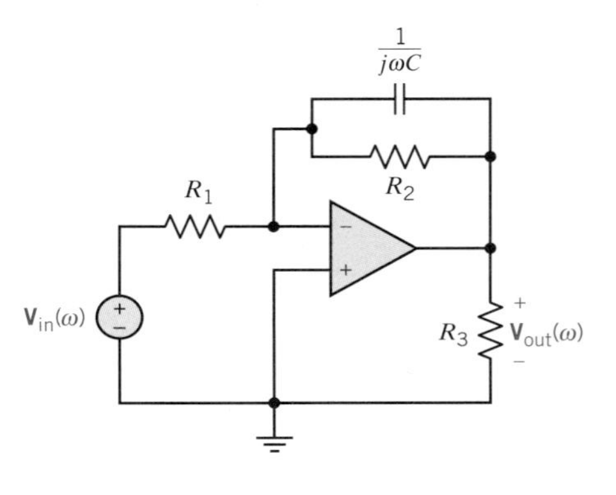

Figure 13.3-3
The frequency domain representation of the op amp circuit of Figure 13.3-1.

and $$\text{phase shift} = 180° - \tan^{-1}(6283/10000) = 148°$$

These are the same results as were obtained earlier by examining the oscilloscope traces in Figure 13.3-2. Similarly, each line of Table 13.3-1 can be obtained by substituting the appropriate frequency into the equations for the gain and phase shift.

Equations that represent the gain and phase shift as functions of frequency are called the **frequency response** of the circuit. The same information can be represented by a table or by graphs instead of equations. These tables or graphs are also called the frequency response of the circuit.

To see that the network function really does represent the behavior of the circuit, suppose that

$$v_{\text{in}}(t) = 0.4\cos(5000t + 45°)\ \text{V}$$

The frequency of the input sinusoid is $\omega = 5000$ radians/second. Substituting this frequency into the network function gives

$$\mathbf{H}(\omega) = \frac{-10}{1 + (j5000/10000)} = 8.94\angle 153°$$

Next $$\mathbf{V}_{\text{out}}(\omega) = \mathbf{H}(\omega)\,\mathbf{V}_{\text{in}}(\omega) = (8.94\angle 153°)(0.4\angle 45°) = 3.58\angle 198°$$

Back in the time domain, the steady-state response is

$$v_{\text{out}}(t) = 3.58\cos(5000t + 198°)\ \text{V}$$

Notice that the network function contained enough information to enable us to calculate the steady-state response from the input sinusoid. The network function does indeed describe the behavior of the circuit.

The circuit shown in Figure 13.3-1 is an example of a circuit called a first-order low-pass filter. First-order low-pass filters have network functions of the form

$$\mathbf{H}(\omega) = \frac{H_o}{1 + j\dfrac{\omega}{\omega_o}} \tag{13.3-4}$$

The gain and phase shift of the first-order low-pass filter are

$$\text{gain} = \frac{|H_o|}{\sqrt{1 + \dfrac{\omega^2}{\omega_o^2}}} \tag{13.3-5}$$

and $$\text{phase shift} = \angle H_o - \tan^{-1}(\omega/\omega_o) \tag{13.3-6}$$

The network function of the first-order low-pass filter has two parameters, H_o and ω_o. At low frequencies, that is, $\omega \ll \omega_o$, the gain is $|H_o|$ and so $|H_o|$ is called the dc gain. (When $\omega = 0$, $A\cos\omega t = A$, a constant or dc voltage.)

The other parameter of the network function, ω_o, is called the half-power frequency. To explain this terminology, suppose that the input to the first-order filter is

$$\text{input} = A\cos\omega_o t$$

In general, this input could be either a voltage or a current. It is important to notice that the input frequency has been set equal to ω_o. If the input was the voltage or current of a 1-ohm resistor, then the power dissipated in this resistor would be $A^2/2$ watts. Suppose, for convenience, that $H_o = 1$. Then the gain at $\omega = \omega_o$ is $\dfrac{1}{\sqrt{2}}$ and the phase shift is $-45°$ so

$$\text{output} = \frac{A}{\sqrt{2}}\cos(\omega_o t - 45°)$$

If this output was the voltage or current of a 1-ohm resistor, then the power dissipated in that resistor would be $A^2/4$ watts. When the input frequency is $\omega = \omega_o$, the power corresponding to the output is one-half the power corresponding to the input. For this reason, ω_o, is called the half-power frequency.

Next consider the problem of designing a first-order low-pass filter. Suppose we are given the following specifications:

$$\text{dc gain} = 2$$
$$\text{phase shift} = 120° \text{ when } \omega = 1000 \text{ radians/second}$$

Before designing a circuit to meet these specifications, we need to pay more attention to the phase shift. Consider Eq. 13.3-6. Both ω and ω_o will be positive, so $\tan^{-1}(\omega/\omega_o)$ will be between 0° and 90°. Also, $\angle H_o$ will be 0° when H_o is positive and 180° when H_o is negative. As a result, only phase shifts between −90° and 0° or between 90° and 180° can be achieved using a first-order low-pass filter. (Phase shifts that cannot be obtained using a first-order low-pass filter can be obtained using other types of circuit. That's a story for another day.) Table 13.3-2 shows two first-order low-pass filters, one for obtaining phase shifts between 90° and 180° and the other for obtaining phase shifts between −90° and 0°. Based on the phase shift, we select the circuit in the first row of Table 13.3-2. The specification on the dc gain gives

$$2 = |H_o| = \frac{R_2}{R_1}$$

Table 13.3-2 **First-Order Low-Pass Filter Circuits**

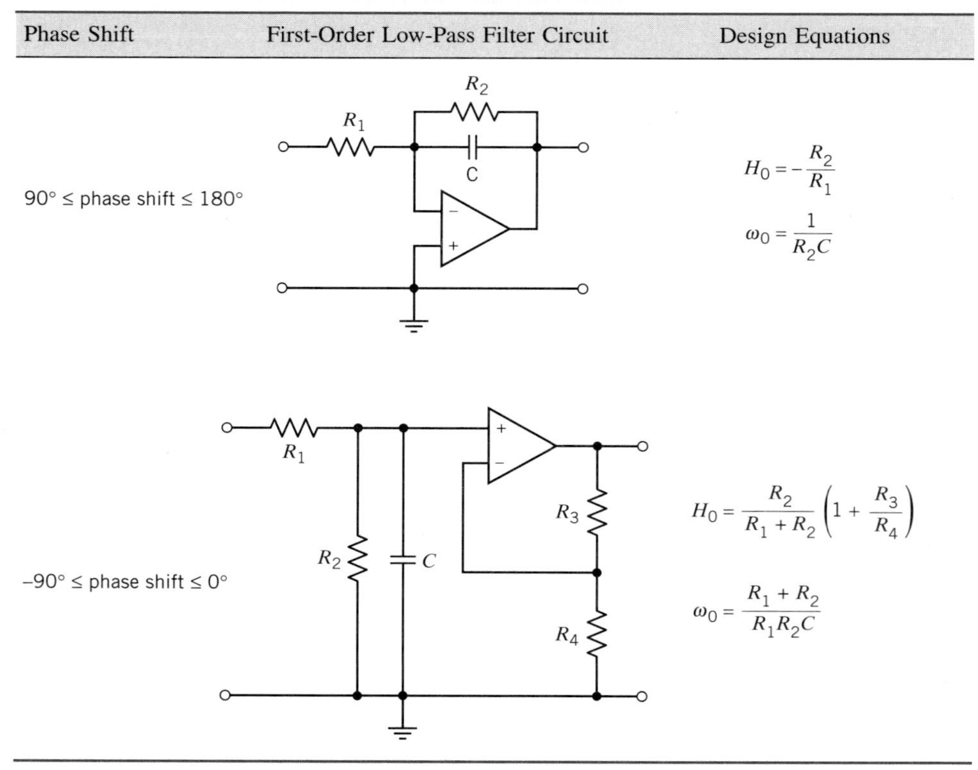

Phase Shift	First-Order Low-Pass Filter Circuit	Design Equations
90° ≤ phase shift ≤ 180°		$H_0 = -\frac{R_2}{R_1}$ $\omega_0 = \frac{1}{R_2 C}$
−90° ≤ phase shift ≤ 0°		$H_0 = \frac{R_2}{R_1 + R_2}\left(1 + \frac{R_3}{R_4}\right)$ $\omega_0 = \frac{R_1 + R_2}{R_1 R_2 C}$

The specification on phase shift gives

$$120° = 180° - \tan^{-1}(1000R_2C)$$

This is a set of two equations in the three unknowns R_1, R_2, and C. The solution is not unique. We will have to pick a value for one of the unknowns and then solve for values of the other two unknowns. Let's pick a convenient value for the capacitor, $C = 0.1\ \mu\text{F}$, and calculate the resistances.

$$R_2 = \frac{\tan(60°)}{1000 \times 0.1 \times 10^{-6}} = 17.32\ \text{k}\Omega$$

and

$$R_1 = \frac{R_2}{2} = 8.66\ \text{k}\Omega$$

We conclude that the circuit shown in the first row of Table 13.3-2 will have a dc gain $= 2$ and a phase shift $= 120°$ at $\omega = 1000$ rad/sec when $R_1 = 8.66\ \text{k}\Omega$, $R_2 = 17.32\ \text{k}\Omega$, and $C = 0.1\ \mu\text{F}$.

Exercise 13.3-1 The input to the circuit shown in Figure E 13.3-1 is the source voltage, v_s, and the response is the capacitor voltage, v_o. Suppose $R = 10\ \text{k}\Omega$ and $C = 1\ \mu\text{F}$. What are the values of the gain and phase shift when the input frequency is $\omega = 100$ rad/s?

Answer: 0.707, −45 deg

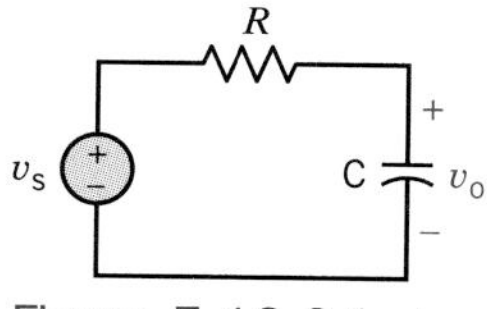

Figure E 13.3-1 An *RC* circuit.

Exercise 13.3-2 The input to the circuit shown in Figure E 13.3-2 is the source voltage, v_s, and the response is the resistor voltage, v_o. $R = 30\ \Omega$ and $L = 2\ H$. Suppose the input frequency is adjusted until the gain is equal to 0.6. What is the value of the frequency?

Answer: 20 rad/s

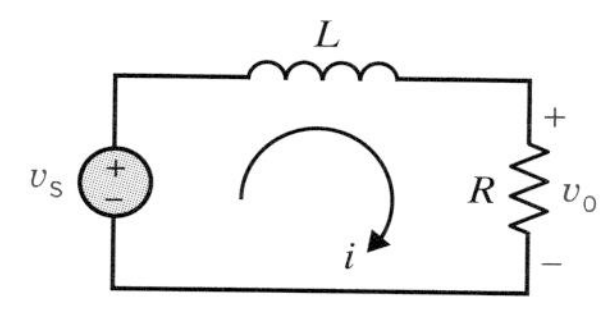

Figure E 13.3-2 The *RL* circuit.

Exercise 13.3-3 The input to the circuit shown in Figure E 13.3-2 is the source voltage, v_s, and the response is the mesh current, i. $R = 30\ \Omega$ and $\text{L} = 2$ H. What are the values of the gain and phase shift when the input frequency is $\omega = 20$ rad/s?

Answer: 0.02 A/V, −53.1 deg

Exercise 13.3-4 The input to the circuit shown in Figure E 13.3-1 is the source voltage, v_s, and the response is the capacitor voltage, v_o. Suppose $C = 1\ \mu\text{F}$. What value of R is required to cause a phase shift equal to −45 deg when the input frequency is $\omega = 20$ rad/s?

Answer: $R = 50\ k\Omega$

Exercise 13.3-5 The input to the circuit shown in Figure E 13.3-1 is the source voltage, v_s, and the response is the capacitor voltage, v_o. Suppose $C = 1\ \mu\text{F}$. What value of R is required to cause a gain equal to 1.5 when the input frequency is $\omega = 20$ rad/s?

Answer: No such value of R exists. The gain of this circuit will never be greater than 1.

13.4 BODE PLOTS

It is common to use logarithmic plots of the frequency response instead of linear plots. The logarithmic plots are called *Bode plots* in honor of H. W. Bode, who used them extensively in his work with amplifiers at Bell Telephone Laboratories in the 1930s and 1940s. A Bode plot is a plot of log-gain and phase angle values versus frequency, using a log-frequency horizontal axis. The use of logarithms expands the range of frequencies portrayed on the horizontal axis.

The network function **H** can be written as

$$\mathbf{H} = H\angle\phi = He^{j\phi} \tag{13.4-1}$$

The logarithm of the magnitude is normally expressed in terms of the logarithm to the base 10, so we use

$$\text{logarithmic gain} = 20\log_{10} H \tag{13.4-2}$$

and the unit is decibel (dB). The logarithmic gain is also called the "gain in dB." A decibel conversion table is given in Table 13.4-1.

The unit decibel is derived from the unit bel. Suppose P_1 and P_2 are two values of power. Both P_1/P_2 and $\log(P_1/P_2)$ are measures of the relative sizes of P_1 and P_2. The ra-

Table 13.4-1 **A Decibel Conversion Table**

Magnitude, H	20 log H (dB)
0.1	−20.00
0.2	−13.98
0.4	−7.96
0.6	−4.44
1.0	0.0
1.2	1.58
1.4	2.92
1.6	4.08
2.0	6.02
3.0	9.54
4.0	12.04
5.0	13.98
6.0	15.56
7.0	16.90
10.0	20.00
100.0	40.00

tio P_1/P_2 is unitless, while $\log(P_1/P_2)$ has the bel as its unit. The name bel honors Alexander Graham Bell, the inventor of the telephone.

The **Bode plot** is a chart of gain in decibels and phase in degrees versus the logarithm of frequency.

Let us obtain the Bode plots corresponding to the network function

$$\mathbf{H} = \frac{1}{(j\omega/\omega_0) + 1} = H\angle\phi \tag{13.4-3}$$

where

$$H = \frac{1}{\sqrt{1 + (\omega/\omega_0)^2}}$$

and

$$\phi = -\tan^{-1}\omega/\omega_o$$

The logarithmic gain is

$$20\log_{10} H = 20\log_{10}\frac{1}{\sqrt{1 + (\omega/\omega_o)^2}}$$

$$= 20\log_{10} 1 - 20\log_{10}\sqrt{1 + (\omega/\omega_o)^2} = -20\log_{10}\sqrt{1 + (\omega/\omega_o)^2}$$

For small frequencies, that is, $\omega \ll \omega_o$,

$$1 + (\omega/\omega_o)^2 \cong 1$$

so the logarithmic gain is approximately

$$20\log_{10} H = -20\log_{10}\sqrt{1} = 0\text{ dB}$$

This is the equation of a horizontal straight line. Since this straight line approximates the logarithmic gain for low frequencies, it is called the low-frequency asymptote of the Bode plot.

For large frequencies, that is, $\omega \gg \omega_o$,

$$1 + (\omega/\omega_o)^2 \cong (\omega/\omega_o)^2$$

so the logarithmic gain is approximately

$$20\log_{10} H = -20\log_{10}\sqrt{(\omega/\omega_o)^2}$$

$$= -20\log_{10}\omega/\omega_o = 20\log_{10}\omega_o - 20\log_{10}\omega$$

This equation shows one of the advantages of using logarithms. The plot of $20\log_{10} H$ versus $\log_{10}\omega$ is a straight line. This straight line is called the high-frequency asymptote of the Bode plot. Figure 13.4-1*a* and *b* compares the equation of the high-frequency asymptote to the more familiar standard form of the equation of a straight line, $y = mx + b$. The slope of the high-frequency asymptote can be calculated from two points on the straight line. This slope is given using units of dB/decade. In Figure 13.4-1*b* the gain in dB is plotted versus $\log\omega$, while in Figure 13.4-1*c* the gain in dB is plotted versus ω using a log scale. It is more convenient to label the frequency axis when a log scale is used for ω. The equation used to calculate the slope from two points on the line is the same in Figure 13.4-1*c* as it is in Figure 13.4-1*b*.

An interval between two frequencies with a ratio equal to 10 is called a decade, so the range of frequencies from ω_1 to ω_2 where $\omega_2 = 10\omega_1$ is called a decade. The difference between the logarithmic gains over a decade of frequency for $\omega \gg \omega_0$ using Eq.13.4-2 is

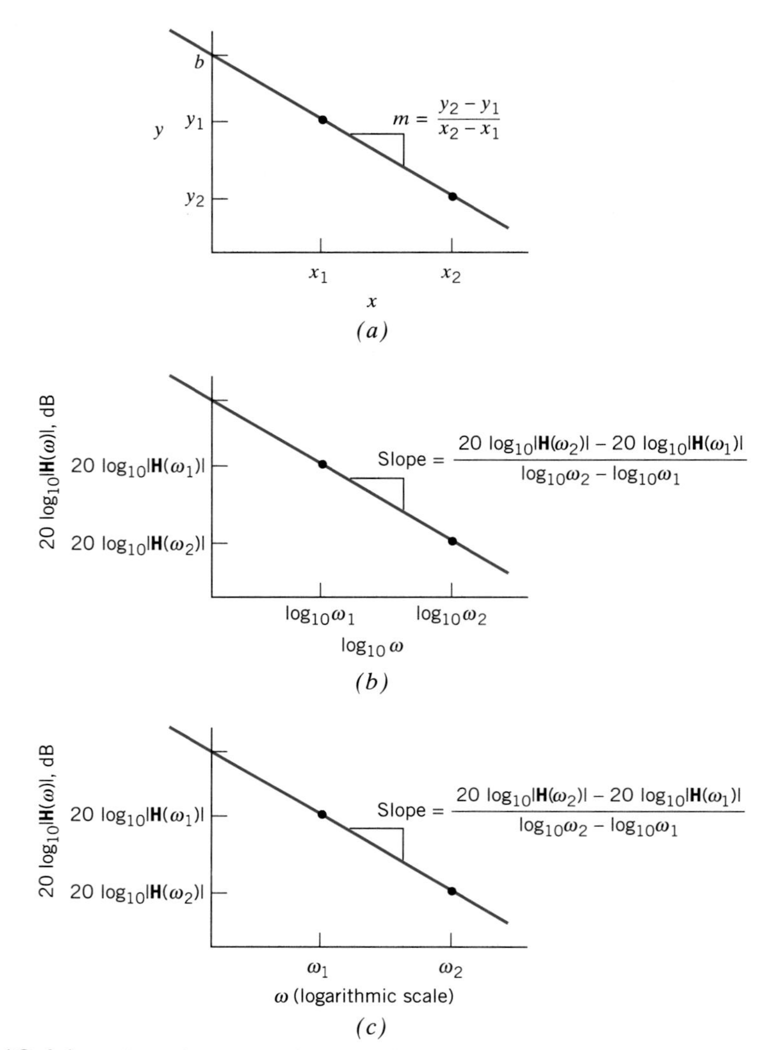

Figure 13.4-1 (*a*) Plot of y versus x for the straight line $y = mx + b$. (*b*) Plot of 20 log $|\mathbf{H}(\omega)|$ versus log ω for the straight line 20 log $|\mathbf{H}(\omega)| = 20 \log \omega_0 - 20 \log \omega$. (*c*) Plot of 20 log $|\mathbf{H}(\omega)|$ versus ω for the straight line 20 log $|\mathbf{H}(\omega)| = 20 \log \omega_0 - 20 \log \omega$.

$$\begin{aligned} 20 \log \mathbf{H}(\omega_1) - 20 \log \mathbf{H}(\omega_2) &= -20 \log \frac{\omega_1}{\omega_0} - \left(-20 \log \frac{\omega_2}{\omega_0}\right) \\ &= -20 \log \frac{\omega_1}{\omega_2} = -20 \log \left(\frac{1}{10}\right) = +20 \text{ dB} \end{aligned}$$

Thus, the slope of the high-frequency asymptote is -20 dB/decade.

The intersection of the low-frequency asymptote with the high-frequency asymptote occurs when

$$0 = 20 \log_{10} \omega - 20 \log_{10} \omega_0$$

that is, when

$$\omega = \omega_0$$

The low- and high-frequency asymptotes form a corner where they intersect. Since the asymptotes intersect at the frequency $\omega = \omega_0$, ω_0 is sometimes called the **corner frequency.**

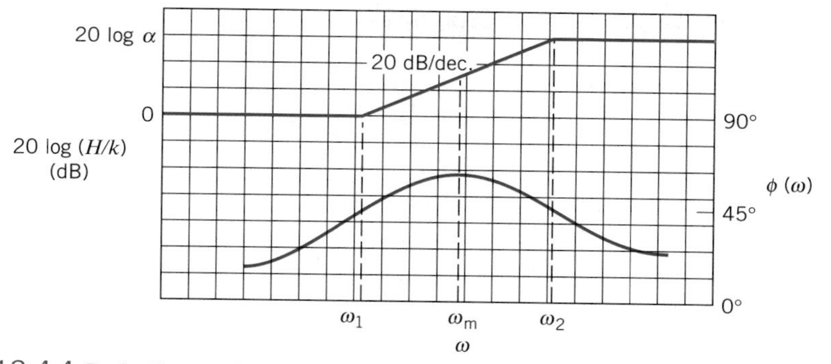

Figure 13.4-4 Bode diagram for the circuit of Figure 13.4-3.

$\omega_1 < \omega < \omega_2$, the equation of the asymptote does include a term involving $\log_{10}\omega$. The coefficient of $\log_{10}\omega$ is 20, indicating a slope of 20 dB/decade.

The effect of the dc gain k is limited to the term $20\log_{10} k$ which appears in the equation of each of the three asymptotes. Changing the value of k will shift the Bode plot up (increasing k) or down (decreasing k) but will not change the shape of the Bode plot. For this reason, we sometimes normalize the network function by dividing by the dc gain. The asymptotes of the Bode plot of the normalized network function are given by

$$20\log_{10}\left(\frac{H}{k}\right) \cong \begin{cases} 0 & \omega < \omega_1 \\ 20\log_{10}\omega - 20\log_{10}\omega_1 & \omega_1 < \omega < \omega_2 \\ 20\log_{10}\omega_2 - 20\log_{10}\omega_1 & \omega_2 < \omega \end{cases}$$

The phase angle of **H** is

$$\begin{aligned}\phi &= \angle k + \angle\left(1 + j\frac{\omega}{\omega_1}\right) - \angle\left(1 + j\frac{\omega}{\omega_2}\right) \\ &= 0 + \tan^{-1}\left(\frac{\omega}{\omega_1}\right) - \tan^{-1}\left(\frac{\omega}{\omega_2}\right)\end{aligned}$$

The phase Bode plot and the asymptotic magnitude Bode plot are shown in Figure 13.4-4. Notice that the slope of the asymptotic magnitude Bode plot changes as the frequency increases past ω_1 and changes again as the frequency increases past ω_2. Zeros, like ω_1, cause the slope to increase by 20 dB/decade. Poles, like ω_2, cause the slope to decrease by 20 dB/decade. The slope of every asymptote will be an integer multiple of 20 dB/decade.

The phase Bode plot shows that the phase shift is largest at the geometric mean of the corner frequencies, $\omega_m = \sqrt{\omega_1\,\omega_2}$.

Bode plots are frequently drawn using a computer and appropriate software. For example, Figure 13.4-5 shows how to use MathCAD to draw both Bode plots and asymptotic Bode plots.

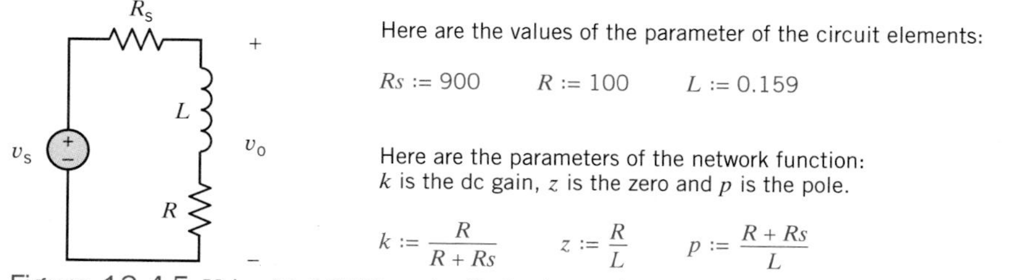

Using MathCAD to plot both the Bode plot and the asymptotic Bode plot.
Rs
L
R
vs
vo
+
−
Here are the values of the parameter of the circuit elements:
Rs := 900   R := 100   L := 0.159
Here are the parameters of the network function:
k is the dc gain, z is the zero and p is the pole.
k := R / (R + Rs)   z := R / L   p := (R + Rs) / L

Figure 13.4-5 Using MathCAD to plot Bode plots and asymptotic Bode plots.

We need N frequencies in the range ω min to ω max that are selected so that the values of log ω are equally spaced. Here is a procedure for generating those frequencies.

$\omega\text{min} := 100$ $\omega\text{max} := 100000$ $N := 100$ $n := 1..N$

$m := \ln\left(\frac{\omega\text{max}}{\omega\text{min}}\right)$ $\omega_n := \omega\text{min} \cdot e^{\frac{n}{N} \cdot m}$

Here is the network function:

$$H_n := k \cdot \frac{1 + j \cdot \frac{\omega_n}{z}}{1 + \frac{j \cdot \omega_n}{p}}$$

Here is the gain, in dB:

$G_n := 20 \cdot \log(|Hn|)$

Here is the magnitude Bode plot:

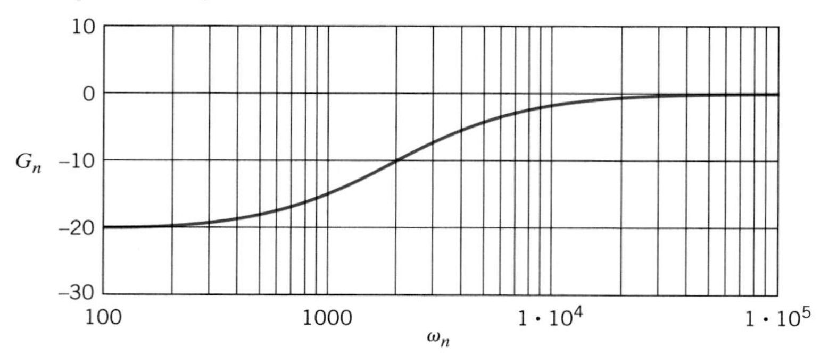

The asymptotic Bode plot is obtained by approximating factors $(1 + j\omega/\omega o)$ by 1 when $\omega < \omega o$ and by $\omega/\omega o$ when $\omega > \omega o$.

$H1_n := \text{if}\left(\omega_n < p,\ k \cdot \frac{\omega_n}{z},\ k \cdot \frac{p}{z}\right)$ This if statement separates the case when $\omega < p$ from the case when $\omega > p$.

$H2_n := \text{if}\left(\omega_n < z,\ k,\ H1_n\right)$ This if statement separates the case when $\omega < z$ from the case when $\omega > z$.

$H2$ is the approximate network function. The magnitude of $H2$ approximates the gain.

$AG_n := 20 \cdot \log(|H2_n|)$

After converting to dB, we are ready to plot the asymptotic Bode plot.

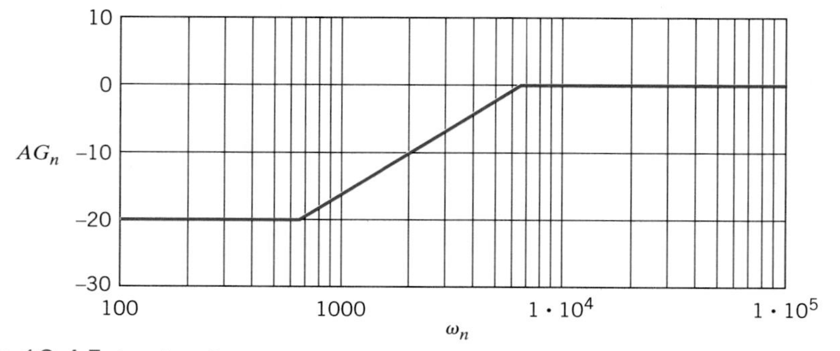

Figure 13.4-5 (continued)

Example 13.4-1

Find the asymptotic magnitude Bode plot of

$$\mathbf{H}(\omega) = K\frac{j\omega}{1 + j\frac{\omega}{p}}$$

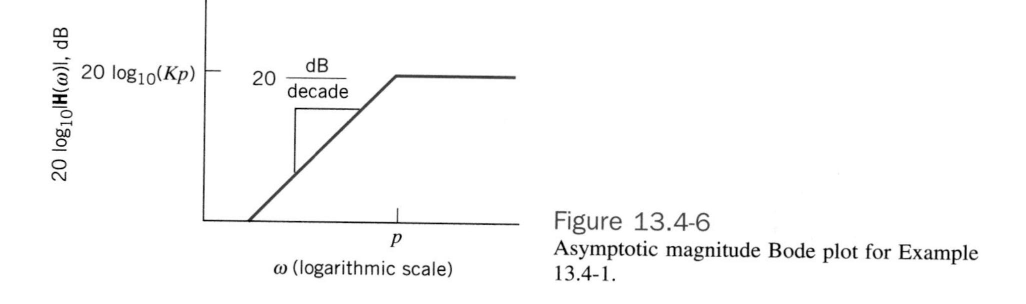

Figure 13.4-6
Asymptotic magnitude Bode plot for Example 13.4-1.

Solution

Approximate $\left(1 + j\frac{\omega}{p}\right)$ by 1 when $\omega < p$ and by $j\frac{\omega}{p}$ when $\omega > p$ to get

$$\mathbf{H}(\omega) \cong \begin{cases} K(j\omega) & \omega < p \\ Kp & \omega > p \end{cases}$$

The logarithmic gain is

$$20 \log_{10} |\mathbf{H}(\omega)| \cong \begin{cases} 20 \log_{10} K + 20 \log_{10} \omega & \omega < p \\ 20 \log_{10} Kp & \omega > p \end{cases}$$

The asymptotic magnitude Bode plot is shown in Figure 13.4-6. The $j\omega$ factor in the numerator of $\mathbf{H}(\omega)$ causes the low-frequency asymptote to have a slope of 20 dB/decade. The slope of the asymptotic magnitude Bode plot decreases by 20 dB/decade as the frequency increases past $\omega = p$.

Example 13.4-2

Find the asymptotic magnitude Bode plot of

$$\mathbf{H}(\omega) = \frac{\omega_0^2}{(j\omega)^2 + j2\zeta\omega_0\omega + \omega_0^2} = \frac{\omega_0^2}{\omega_0^2 - \omega^2 + j2\zeta\omega_0\omega}$$

Solution

The denominator of $\mathbf{H}(\omega)$ contains a new factor, one that involves ω^2. The asymptotic Bode plot is based on the approximation

$$(\omega_0^2 - \omega^2) + j2\zeta\omega_0\omega \cong \begin{cases} \omega_0^2 & \omega < \omega_0 \\ -\omega^2 & \omega > \omega_0 \end{cases}$$

Using this approximation, we can express $\mathbf{H}(\omega)$ as

$$\mathbf{H}(\omega) \cong \begin{cases} 1 & \omega < \omega_0 \\ -\dfrac{\omega_0^2}{\omega^2} & \omega > \omega_0 \end{cases}$$

The logarithmic gain is

$$20 \log_{10} |\mathbf{H}(\omega)| \cong \begin{cases} 0 & \omega < \omega_0 \\ 40 \log_{10} \omega_0 - 40 \log_{10} \omega & \omega > \omega_0 \end{cases}$$

The asymptotic magnitude Bode plot and the phase Bode plot are shown in Figure 13.4-7. The asymptotic Bode plot is a good approximation to the Bode plot when $\omega \ll \omega_o$ or $\omega \gg \omega_o$. Near $\omega = \omega_o$ the asymptotic Bode plot deviates from the exact Bode plot. At $\omega = \omega_o$ the value of the asymptotic Bode plot is 0 dB while the value of the exact Bode plot is

$$H(\omega_0) = \frac{1}{2\zeta}$$

As this equation and Figure 13.4-7 both show, the deviation between the actual and asymptotic Bode plot near $\omega = \omega_o$ depends on ζ. The frequency ω_o is called the *corner frequency.* The slope of the asymptotic Bode plot decreases by 40 dB/decade as the frequency increases past $\omega = \omega_o$. In terms of the asymptotic Bode plot, the denominator of this network function acts like two poles at $p = \omega_o$. If this factor were to appear in the numerator of a network function, it would act like two zeros at $z = \omega_o$. The slope of the asymptotic Bode plot would increase by 40 dB/decade as the frequency increased past $\omega = \omega_o$.

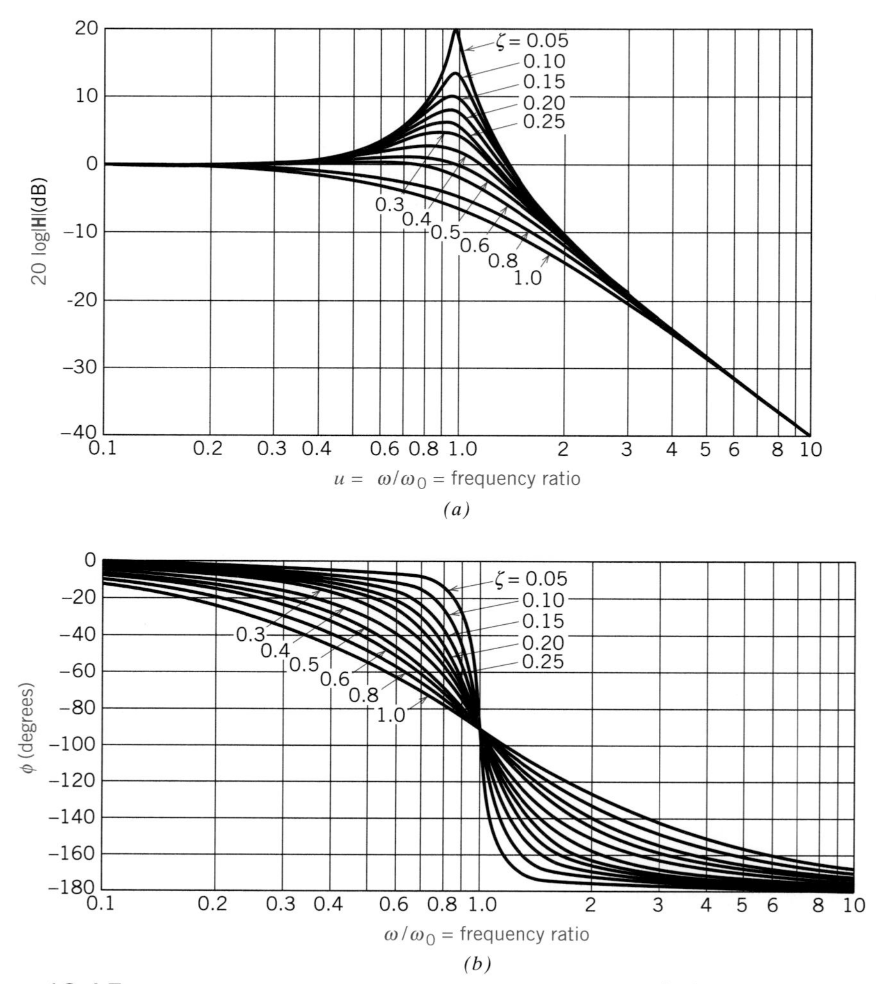

Figure 13.4-7 Bode diagram of $\mathbf{H}(j\omega) = [1 + (2\zeta/\omega_0)j\omega + (j\omega/\omega_0)^2]^{-1}$ for two decades of frequency.

Example 13.4-3

Find the asymptotic magnitude Bode plot of

$$\mathbf{H}(\omega) = \frac{5(1 + 0.1j\omega)}{j\omega(1 + 0.5j\omega)\left[1 + 0.6\left(\dfrac{j\omega}{50}\right) - \left(\dfrac{\omega}{50}\right)^2\right]}$$

Solution

The corner frequencies of $\mathbf{H}(\omega)$ are $z = 10, p = 2$ and $\omega_0 = 50$ rad/s. The smallest corner frequency is $p = 2$. When $\omega < 2$, then $\mathbf{H}(\omega)$ can be approximated as

$$\mathbf{H}(\omega) = \frac{5}{j\omega}$$

so the equation of the low-frequency asymptote is

$$20 \log_{10} |\mathbf{H}| = 20 \log_{10} 5 - 20 \log_{10} \omega$$

The slope of the low-frequency asymptote is -20 dB/decade. Let's find a point on the low-frequency asymptote. When $\omega = 1$

$$20 \log_{10} |\mathbf{H}| = 20 \log_{10} 5 - 20 \log_{10} 1 = 14 \text{ dB}$$

The low-frequency asymptote is a straight line with a slope of -20 dB/decade passing through the point $\omega = 1$ rad/s, $|\mathbf{H}| = 14$ dB.

The slope of the asymptotic Bode plot will change as ω increases past each corner frequency. The slope decreases by 20 dB/decade at $\omega = 2$, then increases by 20 dB/decade at $\omega = 10$, and finally decreases by 40 dB/decade at $\omega = 50$. The asymptotic magnitude Bode plot is shown in Figure 13.4-8.

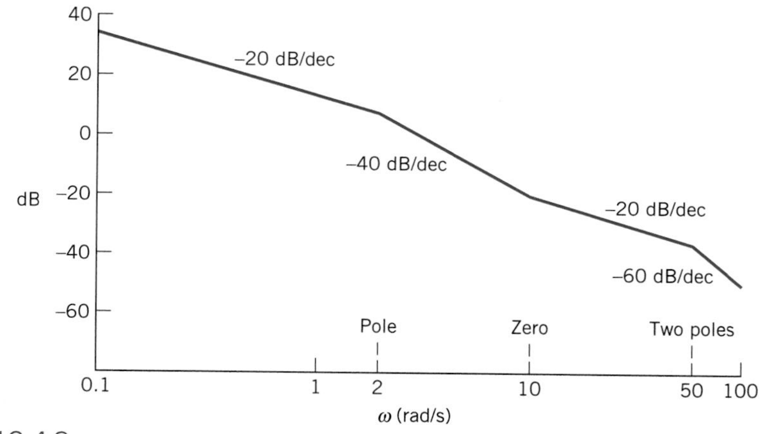

Figure 13.4-8 Asymptotic plot for Example 13.4-3.

Example 13.4-4

Let's design the circuit shown in Figure 13.4-3 to satisfy the following specifications.

1. The low-frequency gain is 0.1.
2. The high-frequency gain is 1.
3. The corner frequencies lie in the range 100 hertz to 2000 hertz.

We're confronted with two problems. First, can these specifications be satisfied using this circuit? Second, if they can, what values of R, R_s, and L are required?

Our earlier analysis of this circuit showed both that the low-frequency gain is less than 1 and that the high-frequency gain is equal to 1. This circuit can only be used to satisfy specifications that are consistent with these facts. Fortunately, the given specifications are consistent with these facts. The first specification requires

$$0.1 = \text{low-frequency gain} = k = \frac{R}{R + R_s}$$

Since this circuit has a high-frequency gain equal to 1, the second specification is satisfied.

Now let's turn our attention to the specifications on the corner frequencies. The specified frequency range is given using units of hertz, while the corner frequencies have units of radians/second. Since $\omega_1 > \omega_2$, the third specification requires that

$$(2\pi)100 < \frac{R}{L}$$

and

$$(2\pi)2000 > \frac{R + R_s}{L}$$

Our job is to find values of R, R_s, and L that satisfy these three requirements. We have no guarantee that appropriate values exist. If an appropriate set of values does exist, it may well not be unique. Let's try

$$R = 100\ \Omega$$

The specification on the low-frequency gain requires that

$$R_s = 9R = 900\ \Omega$$

The specification on the zero will be satisfied if

$$L = \frac{R}{(2\pi)100} = 0.159\ \text{H}$$

It remains to verify that these values of R, R_s, and L satisfy the specification on the pole frequency. Since

$$\frac{R + R_s}{L} = 6283 < 12566 = (2\pi)2000$$

The specification is satisfied.

In summary, when

$$R = 100\ \Omega, \quad R_s = 900\ \Omega, \quad \text{and} \quad L = 0.159\ \text{H}$$

the circuit shown in Figure 13.4-3 satisfies the specifications given above.

This solution is not unique. Indeed, when $R = 100\ \Omega$ and $R_s = 900\ \Omega$, then any inductance in the range $0.0796 < L < 0.159$ H can be used to satisfy these specifications.

Example 13.4-5

Design a circuit that has the asymptotic magnitude Bode plot shown in Figure 13.4-9*a*.

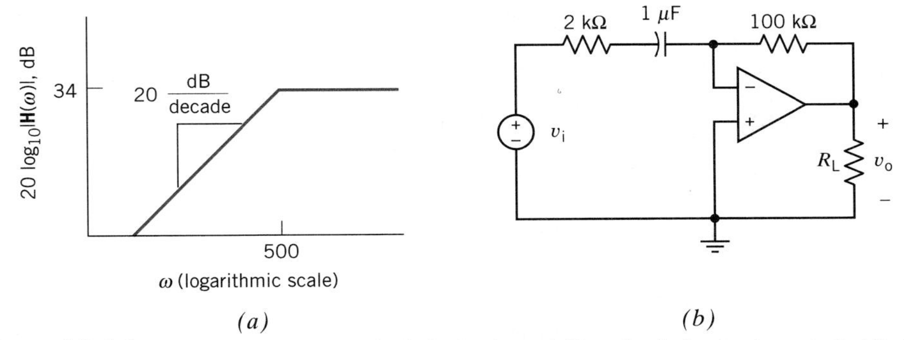

Figure 13.4-9 (*a*) An asymptotic magnitude Bode plot and (*b*) a circuit that implements that Bode plot.

Solution

The slope of this Bode plot is 20 dB/decade for low frequencies, that is, $\omega < 500$ rad/s, so $\mathbf{H}(\omega)$ must have a $j\omega$ factor in its numerator. The slope decreases by 20 dB/decade as ω increases past $\omega = 500$ rad/s, so $\mathbf{H}(\omega)$ must have a pole at $\omega = 500$ rad/s. Based on these observations

$$\mathbf{H}(\omega) = \pm k \frac{j\omega}{1 + j\dfrac{\omega}{500}}$$

The gain of the asymptotic Bode plot is 34 dB $= 50$ when $\omega > 500$ so

$$50 = \pm k \frac{j\omega}{j\dfrac{\omega}{500}} = \pm k \cdot 500$$

Thus $k = \pm 0.1$ and

$$\mathbf{H}(\omega) = \pm 0.1 \cdot \frac{j\omega}{1 + j\dfrac{\omega}{500}}$$

We need a circuit that has a network function of this form. Table 13.4-2 contains a collection of circuits and corresponding network functions. Row 4 of Table 13.4-2 contains the circuit that we can use. The design equations provided in row 4 of the table indicate that

$$0.1 = R_2 C$$

$$500 = \frac{1}{CR_1}$$

Since there are more unknowns than equations, the solution of these design equations is not unique. Pick $C = 1\ \mu\text{F}$. Then

$$R_2 = \frac{0.1}{10^{-6}} = 100\ \text{k}\Omega$$

$$R_1 = 500 \cdot 10^{-6} = 2\ \text{k}\Omega$$

The finished circuit is shown in Figure 13.4-9*b*.

Table 13.4-2 **A Collection of Circuits and Corresponding Network Functions**

Circuit	Network Function
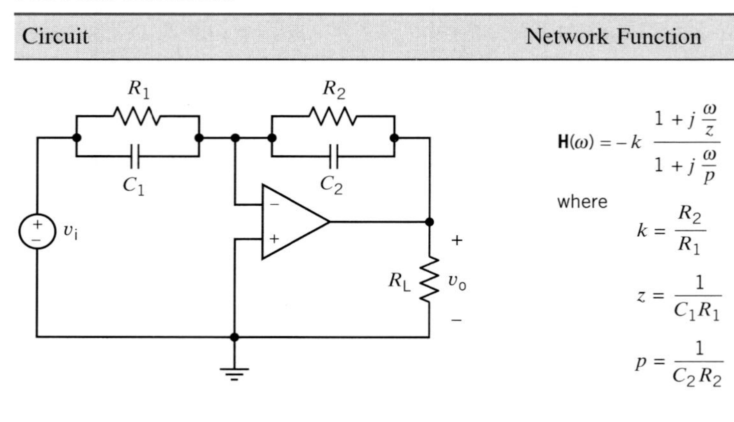	$\mathbf{H}(\omega) = -k \dfrac{1 + j\dfrac{\omega}{z}}{1 + j\dfrac{\omega}{p}}$ where $k = \dfrac{R_2}{R_1}$ $z = \dfrac{1}{C_1 R_1}$ $p = \dfrac{1}{C_2 R_2}$
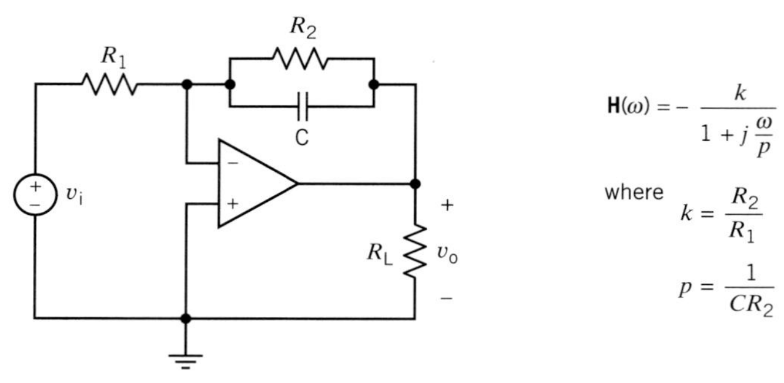	$\mathbf{H}(\omega) = -\dfrac{k}{1 + j\dfrac{\omega}{p}}$ where $k = \dfrac{R_2}{R_1}$ $p = \dfrac{1}{CR_2}$
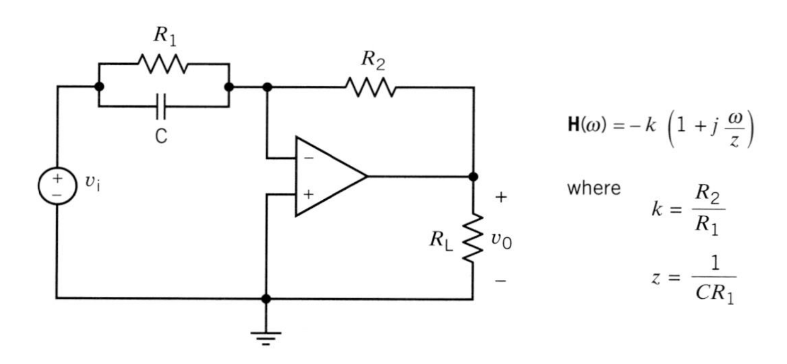	$\mathbf{H}(\omega) = -k\left(1 + j\dfrac{\omega}{z}\right)$ where $k = \dfrac{R_2}{R_1}$ $z = \dfrac{1}{CR_1}$
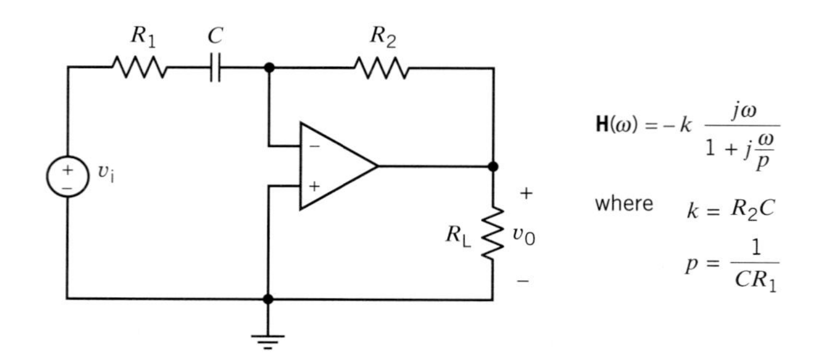	$\mathbf{H}(\omega) = -k \dfrac{j\omega}{1 + j\dfrac{\omega}{p}}$ where $k = R_2 C$ $p = \dfrac{1}{CR_1}$

Table 13.4-2 **A Collection of Circuits and Corresponding Network Functions** *(Continued)*

Circuit	Network Function
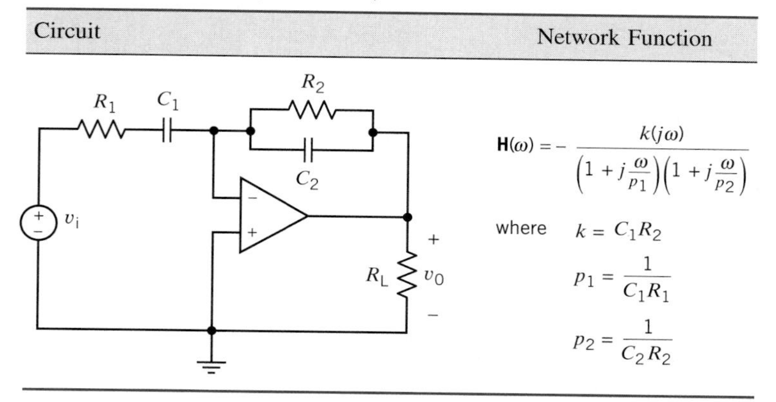	$\mathbf{H}(\omega) = -\dfrac{k(j\omega)}{\left(1 + j\dfrac{\omega}{p_1}\right)\left(1 + j\dfrac{\omega}{p_2}\right)}$ where $k = C_1R_2$, $p_1 = \dfrac{1}{C_1R_1}$, $p_2 = \dfrac{1}{C_2R_2}$

Exercise 13.4-1 (a) Convert the gain $|\mathbf{V}_o/\mathbf{V}_s| = 2$ to decibels. (b) Suppose $|\mathbf{V}_o/\mathbf{V}_s| = -6.02$ dB. What is the value of this gain "not in dB"?

Answer: (a) $+6.02$ dB (b) 0.5

Exercise 13.4-2 In a certain frequency range, the magnitude is $H = 1/\omega^2$. What is the slope of the Bode plot in this range, expressed in decibels per decade?

Answer: -40 dB/decade

Exercise 13.4-3 Consider the network function

$$\mathbf{H}(\omega) = \frac{j\omega A}{B + j\omega C}$$

Find (a) the corner frequency, (b) the slope of the asymptotic line for ω above the corner frequency in decibels per decade, (c) the slope of the Bode plot below the corner frequency, and (d) the gain for ω above the corner frequency in decibels.

Answer: (a) $\omega_0 = B/C$ (b) zero (c) 20 dB/decade (d) $20 \log_{10} \dfrac{A}{C}$

Exercise 13.4-4 A first-order circuit is shown in Figure E 13.4-4. Determine the ratio $\mathbf{V}_o/\mathbf{V}_s$ and sketch the Bode diagram when $RC = 0.1$ and $R_1/R_2 = 3$.

Answer: $\mathbf{H} = \left(1 + \dfrac{R_1}{R_2}\right)\dfrac{1}{1 + j\omega RC}$

Exercise 13.4-5 (a) Draw the Bode diagram of the network function $\mathbf{V}_o/\mathbf{V}_s$ for the circuit of Figure E 13.4-5.
(b) Determine $v_o(t)$ when $v_s = 10 \cos 20t$ V.

Answer: (b) $v_o = 4.18 \cos(20t - 24.3°)$ V

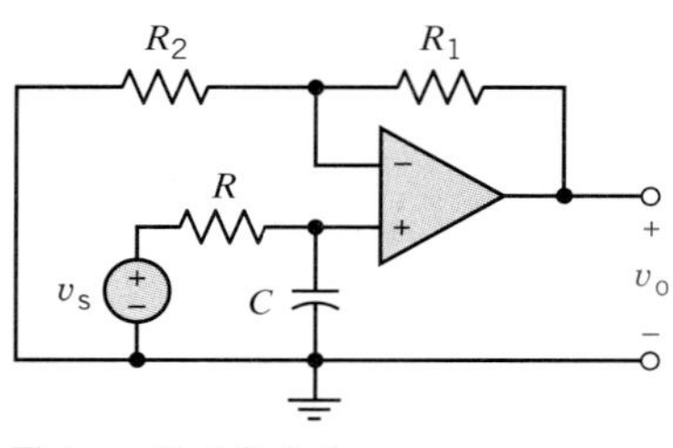

Figure E 13.4-4

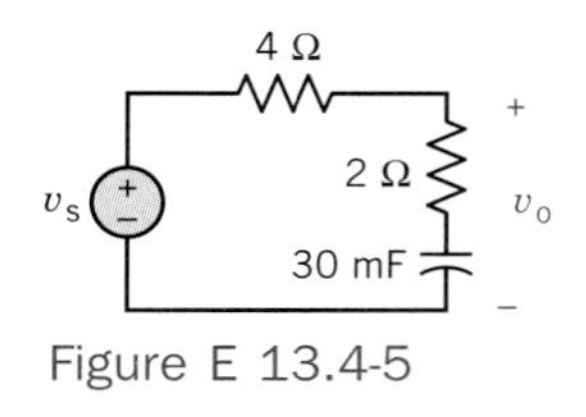

Figure E 13.4-5

Exercise 13.4-6 Draw the asymptotic magnitude Bode diagram for

$$\mathrm{H}(\omega) = \frac{10(1 + j\omega)}{j\omega(1 + j0.5\omega)(1 + j0.6(\omega/50) + (j\omega/50)^2)}$$

Hint: At $\omega = 0.1$ rad/s the value of the gain is 40 dB and the slope of the asymptotic Bode plot is -20 dB/decade. There is a zero at 1 rad/s, a pole at 2 rad/s, and a second-order pole at 50 rad/s. The slope of the asymptotic magnitude Bode diagram increases by 20 dB/decade as the frequency increases past the zero, decreases by 20 dB/decade as the frequency increases past the pole, and finally, decreases by 40 dB/decade as the frequency increases past the second-order pole.

13.5 RESONANT CIRCUITS

In this section we are going to study the behavior of some circuits that are called *resonant circuits.* We begin with an example.

Consider the situation shown in Figure 13.5-1*a*. The input to this circuit is the current of the current source, and the response is the voltage across the current source. Since the input to the circuit is sinusoidal, we can use phasors to analyze this circuit. We know that the network function of the circuit is the ratio of the response phasor to the input phasor. In this case, that network function will be an impedance

$$\mathbf{Z} = \frac{\mathbf{V}}{\mathbf{I}}$$

Figure 13.5-1*b* shows some data that were obtained by applying an input with an amplitude of 2 mA and a frequency that was varied. Row 1 of this table describes the performance of this circuit when $\omega = 200$ rad/s. At this frequency, the impedance of the circuit is

$$\mathbf{Z} = \frac{6.6\angle 48^\circ}{0.002} = 3300\angle 48^\circ\ \Omega$$

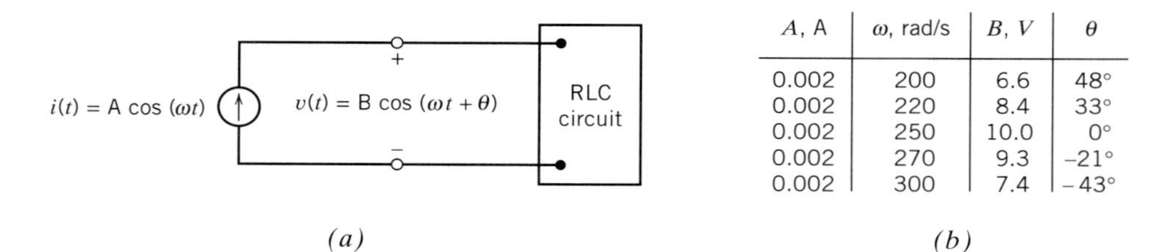

A, A	ω, rad/s	*B*, V	θ
0.002	200	6.6	48°
0.002	220	8.4	33°
0.002	250	10.0	0°
0.002	270	9.3	−21°
0.002	300	7.4	−43°

(*a*) (*b*)

Figure 13.5-1 (*a*) An *RLC* circuit with a sinusoidal input and (*b*) some frequency response data.

Let's convert this impedance from polar to rectangular form

$$\mathbf{Z} = 2208 + j2452\ \Omega$$

This looks like the equivalent impedance of a series resistor and inductor. The resistance would be 2208 Ω. Since the frequency is $\omega = 200$ rad/s, the inductance would be 12.26 H. Recall that in rectangular form impedances are represented as

$$\mathbf{Z} = R + jX$$

where R is called the resistance and X is called the reactance. When $\omega = 200$ rad/s, we say that the reactance of this circuit is inductive because the reactance could have been caused by an inductor.

The last row of the table describes the performance of this circuit when $\omega = 300$ rad/s. Now

$$\mathbf{Z} = \frac{7.4\angle -43°}{0.002} = 3700\angle -43° = 2706 - j2523$$

Since the reactance is negative, it couldn't have been caused by an inductor. This impedance looks like the equivalent impedance of a series resistor and capacitor.

$$R - j\frac{1}{\omega C} = 2706 - j2523$$

Equating the real parts shows that the resistance is 2706 Ω. Equating imaginary parts shows that the capacitance is 1.32 μF.

The reactance of this circuit is inductive at some frequencies and capacitive at other frequencies. We can tell when the reactance will be inductive and when it will be capacitive by looking at the last column of the table. When θ is positive, the reactance is inductive and when θ is negative, the reactance is capacitive. The frequency $\omega = 250$ rad/s is special. When the input frequency is less than 250 rad/s, the reactance is inductive, but when the input frequency is greater than 250 rad/s, the reactance is capacitive. This special frequency is called the **resonant frequency** and is denoted as ω_0. From the third row of the table we see that when $\omega = \omega_0 = 250$ rad/s

$$\mathbf{Z} = \frac{10\angle 0°}{0.002} = 5000\angle 0° = 5000 - j0$$

The reactance is zero. At the resonant frequency, the impedance is purely resistive. Indeed, this fact can be used to identify the resonant frequency.

Another observation can be made from Figure 13.5-1. The magnitude of the impedance is maximum when $\omega = \omega_0 = 250$ rad/s. When the frequency is reduced from ω_0, or increased from ω_0, the magnitude of the impedance is decreased.

Next, consider the circuit shown in Figure 13.5-2. This circuit is called the **parallel resonant circuit.** The equivalent impedance of the parallel resistor, inductor, and capacitor is

$$\mathbf{Z} = \frac{1}{\dfrac{1}{R} + j\omega C + \dfrac{1}{j\omega L}} = \frac{1}{\sqrt{\left(\dfrac{1}{R}\right)^2 + \left(\omega C - \dfrac{1}{\omega L}\right)^2}} \angle -\tan^{-1} R\left(\omega C - \frac{1}{\omega L}\right) \tag{13.5-1}$$

This circuit exhibits some familiar behavior. The reactance will be zero when

$$\omega C - \frac{1}{\omega L} = 0$$

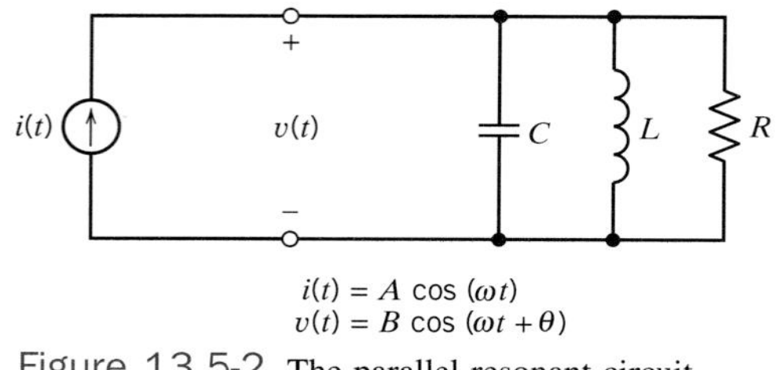

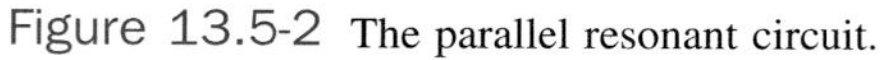

Figure 13.5-2 The parallel resonant circuit.

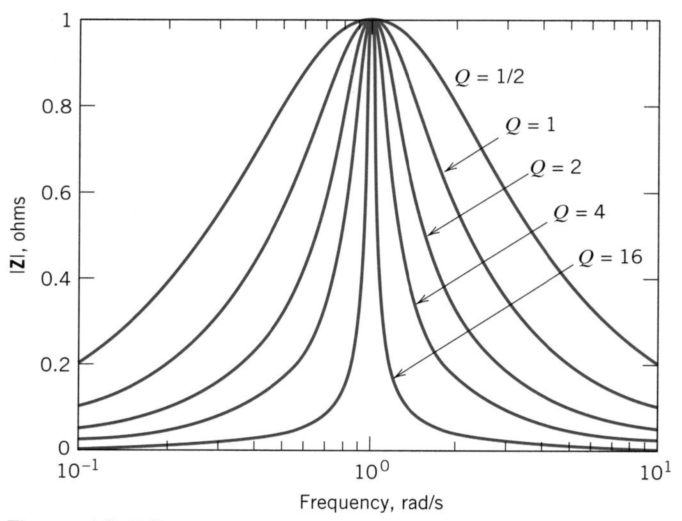

Figure 13.5-3 The effect of Q on the frequency response of a resonant circuit.

The frequency that satisfies this equation is the resonant frequency, ω_0. Solving this equation gives

$$\omega_0 = \frac{1}{\sqrt{LC}}$$

At $\omega = \omega_0$, $\mathbf{Z} = \mathrm{R}$. The magnitude of $\mathbf{Z}$ decreases as ω is either increased or decreased from ω_0. The angle of $\mathbf{Z}$ is positive when $\omega < \omega_0$ and negative when $\omega > \omega_0$, so the reactance is inductive when $\omega < \omega_0$ and capacitive when $\omega > \omega_0$.

The impedance can be put in the form

$$\mathbf{Z} = \frac{k}{1 + jQ\left(\dfrac{\omega}{\omega_0} - \dfrac{\omega_0}{\omega}\right)} \tag{13.5-2}$$

where

$$k = R, \quad Q = R\sqrt{\frac{C}{L}}, \quad \text{and} \quad \omega_0 = \frac{1}{\sqrt{LC}} \tag{13.5-3}$$

The parameters k, Q, and ω_0 characterize the resonant circuit. The resonant frequency, ω_0, is the frequency at which the reactance is zero and also where the magnitude of the impedance is maximum. The parameter k is the value of the impedance when $\omega = \omega_0$, so k is the maximum value of the impedance. Q is called the quality factor of the resonant circuit. The magnitude of the impedance will decrease as ω is reduced from ω_0 or increased from ω_0. The **quality factor** controls how rapidly $|\mathbf{Z}|$ decreases. Figure 13.5-3 illustrates the importance of Q. Both k and ω_0, have been set equal to 1 in Figure 13.5-3 to emphasize the relationship between Q and $|\mathbf{Z}|$.

Figure 13.5-3 shows that the larger the value of Q, the more sharply peaked is the frequency response plot. We can quantify this observation by introducing the bandwidth of the resonant circuit. To that end, let ω_1 and ω_2 denote the frequencies where

$$|\mathbf{Z}(\omega)| = \frac{1}{\sqrt{2}}\,|\mathbf{Z}(\omega_0)| = \frac{k}{\sqrt{2}}$$

There will be two such frequencies, one smaller than ω_0 and the other larger than ω_0. Let $\omega_1 < \omega_0$ and $\omega_2 > \omega_0$. The bandwidth, BW, of the resonant circuit is defined as

$$BW = \omega_2 - \omega_1$$

The frequencies ω_1 and ω_2 are solutions of the equation

$$\frac{k}{\sqrt{2}} = \frac{k}{\sqrt{1 + Q^2\left(\frac{\omega}{\omega_0} - \frac{\omega_0}{\omega}\right)^2}}$$

or

$$\sqrt{2} = \sqrt{1 + Q^2\left(\frac{\omega}{\omega_0} - \frac{\omega_0}{\omega}\right)^2}$$

Squaring both sides, we get

$$1 = Q^2\left(\frac{\omega}{\omega_0} - \frac{\omega_0}{\omega}\right)^2$$

Now taking the square root of both sides

$$\pm 1 = Q\left(\frac{\omega}{\omega_0} - \frac{\omega_0}{\omega}\right)$$

(The $\pm$ sign is required because $a^2 = b^2$ is satisfied if either $a = b$ or $-a = b$.) This equation can be rearranged to get the following quadratic equation

$$\omega^2 \mp \frac{\omega_0 \omega}{Q} - \omega_0^2 = 0$$

This equation has four solutions, but only two are positive. The positive solutions are

$$\omega_1 = -\frac{\omega_0}{2Q} + \sqrt{\left(\frac{\omega_0}{2Q}\right)^2 + \omega_0^2} \quad \text{and} \quad \omega_2 = \frac{\omega_0}{2Q} + \sqrt{\left(\frac{\omega_0}{2Q}\right)^2 + \omega_0^2}$$

Finally, we are ready to calculate the bandwidth

$$BW = \omega_2 - \omega_1 = \frac{\omega_0}{Q} \tag{13.5-4}$$

This equation says that the bandwidth is smaller; that is, the frequency response plot is more sharply peaked, when the value of Q is larger.

Example 13.5-1

Figure 13.5-4 shows a series resonant circuit. Determine the relationship between parameters k, Q, and ω_0 and the element values R, L, and C for the series resonant circuit.

Solution

The input to this circuit is the voltage of the voltage source, and the response is the current in the mesh. The network function is the ratio of the response phasor to the input phasor. In this case, the network function is the equivalent admittance of the series resistor, capacitor, and inductor.

$$\mathbf{Y} = \frac{1}{R + j\omega L + \frac{1}{j\omega C}} \tag{13.5-5}$$

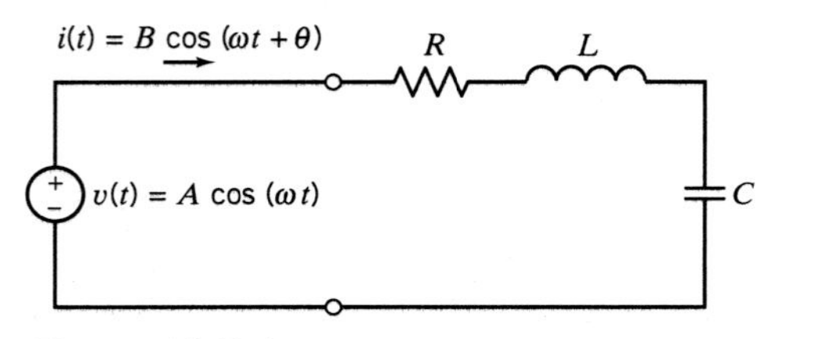

Figure 13.5-4 The series resonant circuit.

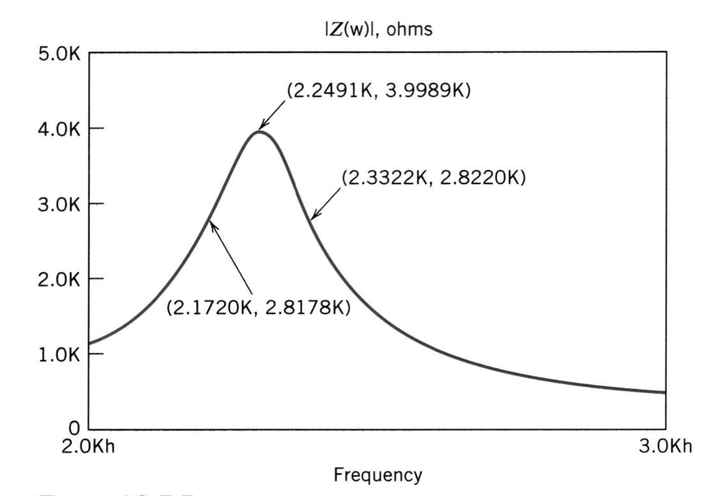

Figure 13.5-5 The magnitude frequency response of a resonant circuit.

In order to identify k, Q, and ω_0, this network function must be rearranged so that it is in the form

$$\mathbf{Y} = \frac{k}{1 + jQ\left(\dfrac{\omega}{\omega_0} - \dfrac{\omega_0}{\omega}\right)} \tag{13.5-6}$$

Rearranging Eq. 13.5-5

$$\begin{aligned}\mathbf{Y} &= \frac{1}{R + j\left(\omega L - \dfrac{1}{\omega C}\right)} \\ &= \frac{1}{R + j\sqrt{\dfrac{L}{C}}\left(\dfrac{\omega}{\dfrac{1}{\sqrt{LC}}} - \dfrac{\dfrac{1}{\sqrt{LC}}}{\omega}\right)} \\ &= \frac{\dfrac{1}{R}}{1 + j\dfrac{1}{R}\sqrt{\dfrac{L}{C}}\left(\dfrac{\omega}{\dfrac{1}{\sqrt{LC}}} - \dfrac{\dfrac{1}{\sqrt{LC}}}{\omega}\right)}\end{aligned}$$

Comparing this equation to Eq. 13.5-6 gives

$$k = \frac{1}{R},\ Q = \frac{1}{R}\sqrt{\frac{L}{C}},\quad \text{and} \quad \omega_0 = \frac{1}{\sqrt{LC}}$$

Example 13.5-2

Figure 13.5-5 shows the magnitude frequency response plot of a resonant circuit. What are the values of the parameters k, Q, and ω_0?

Solution

The first step is to find the peak of the frequency response and determine the values of the frequency and the impedance corresponding to that point. This frequency is the resonant frequency, ω_0, and the impedance at this frequency is k. This point on the frequency response is labeled in Figure 13.5-5. The frequency is

$$\omega_0 = (2\pi)2249 = 14130 \text{ rad/s}$$

The impedance is

$$k = 4000\ \Omega$$

Next, the frequencies ω_1 and ω_2 are identified by finding the points on the frequency response where the value of the impedance is $\frac{k}{\sqrt{2}} = 2828\ \Omega$. These points have been labeled in Figure 13.5-5. (The plot shown in Figure 13.5-5 was produced using PSpice and Probe. The cursor function in Probe was used to label points on the frequency response. Each label gives the frequency first, then the impedance. It was not possible to move the cursor to the points where the impedance was exactly 2828 Ω, so the points where the impedance was as close to 2828 Ω as possible were labeled.)

$$\omega_1 = (2\pi)2172 = 13647 \text{ rad/s} \quad \text{and} \quad \omega_2 = (2\pi)2332 = 14653 \text{ rad/s}$$

The quality factor, Q, is calculated as

$$Q = \frac{\omega_0}{BW} = \frac{\omega_0}{\omega_2 - \omega_1} = \frac{14130}{14653 - 13647} = 14$$

Now that the values of the parameters k, Q, and ω_0 are known, the network function can be expressed as

$$\mathbf{Z}(\omega) = \frac{4000}{1 + j14\left(\frac{\omega}{14130} - \frac{14130}{\omega}\right)}$$

Example 13.5-3

Design a parallel resonant circuit that has $k = 4000\ \Omega$, $Q = 14$, and $\omega_0 = 14130$ rad/s.

Solution

Table 13.5-1 summarizes the relationship between parameters k, Q, and ω_0 and the element values R, L, and C for the parallel resonant circuit. These relationships can be used to calculate R, L, and C from k, Q, and ω_0. First,

$$R = k = 4000\ \Omega$$

Next,

$$\frac{1}{\sqrt{LC}} = \omega_0 = 14130$$

Table 13.5-1 **Series and Parallel Resonant Circuits**

	Series Resonant Circuit	Parallel Resonant Circuit
Circuit		
Network function	$\mathbf{Y} = \dfrac{k}{1 + jQ\left(\dfrac{\omega}{\omega_0} - \dfrac{\omega_0}{\omega}\right)}$	$\mathbf{Z} = \dfrac{k}{1 + jQ\left(\dfrac{\omega}{\omega_0} - \dfrac{\omega_0}{\omega}\right)}$
Resonant frequency	$\omega_0 = \dfrac{1}{\sqrt{LC}}$	$\omega_0 = \dfrac{1}{\sqrt{LC}}$
Maximum value	$k = \dfrac{1}{R}$	$k = R$
Quality factor	$Q = \dfrac{1}{R}\sqrt{\dfrac{L}{C}}$	$Q = R\sqrt{\dfrac{C}{L}}$
Bandwidth	$BW = \dfrac{R}{L}$	$BW = \dfrac{1}{RC}$

and
$$R\sqrt{\frac{C}{L}} = Q = 14$$

Rearranging these last two equations gives
$$\frac{14\sqrt{L}}{4000} = \sqrt{C} = \frac{1}{14130\sqrt{L}}$$

So
$$L = \frac{4000}{14130\,(14)} = 20\text{ mH} \quad \text{and} \quad C = \frac{1}{14130^2(0.02)} = 0.25\ \mu\text{F}$$

Example 13.5-4

Figure 13.5-5 shows the magnitude frequency response plot of a resonant circuit. Design a circuit that has this frequency response.

Solution

We have already solved this problem. Three things must be done to design the required circuit. First, the parameters k, Q, and ω_0 must be determined from the frequency response. We did that in Example 13.5-2. Second, we notice that the given resonant frequency response is an impedance rather than an admittance, and we choose the parallel resonant circuit from Table 13.5-1. Third, the element values R, L, and C must be calculated from the values of k, Q, and ω_0. We did that in Example 13.5-3.

Exercise 13.5-1 For the *RLC* parallel resonant circuit when $R = 8\ \text{k}\Omega$, $L = 40$ mH, and $C = 0.25\ \mu$F, find (a) Q and (b) bandwidth.

Answer: (a) $Q = 20$ (b) $BW = 500$ rad/s

Exercise 13.5-2 A high-frequency *RLC* parallel resonant circuit is required to operate at $\omega_0 = 10$ Mrad/s with a bandwidth of 200 krad/s. Determine the required Q and L when $C = 10$ pF.

Answer: $Q = 50$, $L = 1$ mH

Exercise 13.5-3 A series resonant circuit has $L = 1$ mH and $C = 10\ \mu$F. Find the required Q and R when it is desired that the bandwidth be 15.9 Hz.

Answer: $Q = 100$, $R = 0.1\ \Omega$

Exercise 13.5-4 A series resonant circuit has an inductor $L = 10$ mH. (a) Select C and R so that $\omega_0 = 10^6$ rad/s and the bandwidth is $BW = 10^3$ rad/s. (b) Find the admittance $\mathbf{Y}$ of this circuit for a signal at $\omega = 1.05 \times 10^6$ rad/s.

Answers: (a) $C = 100$ pF, $R = 10\ \Omega$

(b) $\mathbf{Y} = \dfrac{10}{1 + j97.6}$

13.6 FREQUENCY RESPONSE OF OP AMP CIRCUITS

The gain of an op amp is not infinite; rather, it is finite and decreases with frequency. The gain $\mathbf{A}(\omega)$ of the operational amplifier is a function of ω given by

$$\mathbf{A}(\omega) = \frac{A_o}{1 + j\omega/\omega_1}$$

where A_o is the dc gain and ω_1 is the corner frequency. The dc gain is normally greater than 10^4 and ω_1 is less than 100. A circuit model of a frequency-dependent nonideal op amp is shown in Figure 13.6-1. This model is more accurate, but also more complicated, than the ideal op amp model.

Let us consider an example of an op amp circuit incorporating a frequency-dependent op amp.

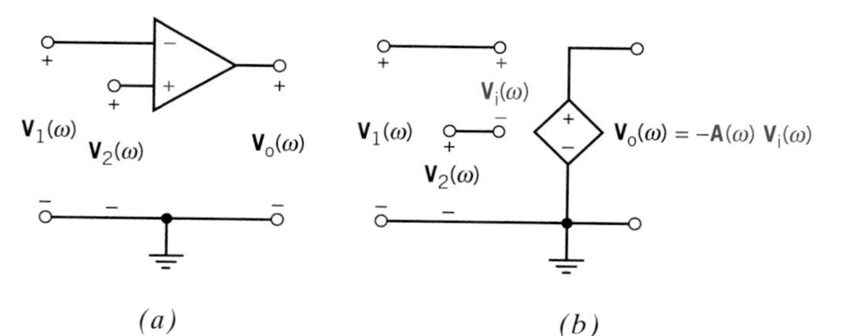

Figure 13.6-1 (*a*) An operational amplifier and (*b*) a frequency-dependent model of an operational amplifier.

Example 13.6-1

Consider the noninverting amplifier in Figure 13.6-2*a*. Replacing the op amp with a frequency-dependent op amp gives the circuit shown in Figure 13.6-2*b*. Suppose that $R_2 = 90\ \text{k}\Omega$ and $R_1 = 10\ \text{k}\Omega$ and that the parameters of the op amp are $A_o = 10^5$ and $\omega_1 = 10$ rad/s. Determine the magnitude Bode plot for both the gain of the op amp, $\mathbf{A}(\omega)$, and the network function on the noninverting amplifier, $\mathbf{V}_o/\mathbf{V}_s$.

Solution

The Bode plot of $20 \log|\mathbf{A}(\omega)|$ is shown in Figure 13.6-3. Note that the magnitude is equal to 1 (0 dB) at $\omega = 10^6$ rad/s.

Writing a node equation in Figure 13.6-2*b* gives

$$\frac{\mathbf{V}_i + \mathbf{V}_s}{R_1} + \frac{\mathbf{V}_i + \mathbf{V}_s + \mathbf{A}(\omega)\mathbf{V}_i}{R_2} = 0$$

The frequency-dependent model of the op amp is described by

$$\mathbf{V}_o = -\mathbf{A}(\omega)\mathbf{V}_i$$

Combining these equations gives

$$\frac{\mathbf{V}_o}{\mathbf{V}_s} = \frac{\mathbf{A}(\omega)}{1 + \dfrac{\mathbf{A}(\omega)}{k}}$$

where $k = (R_1 + R_2)/R_1$ is the gain of the noninverting amplifier when the op amp is modeled as an ideal op amp. Substituting for $\mathbf{A}(\omega)$, we get

$$\frac{\mathbf{V}_o}{\mathbf{V}_s} = \frac{A_o/(1 + j\omega/\omega_1)}{1 + (A_o/k)/(1 + j\omega/\omega_1)} = \frac{A_o}{1 + j\omega/\omega_1 + A_o/k} = \frac{A_C}{(1 + j\omega/A_2\omega_1)}$$

where A_c is the dc gain of the noninverting amplifier defined as $A_c = \dfrac{A_o}{1 + \dfrac{A_o}{k}}$ and $A_2 = 1 + \dfrac{A_o}{k}$. Usually, $1 \ll \dfrac{A_o}{k}$ so $A_c \cong k$ and $A_2 \cong \dfrac{A_o}{k}$. Then

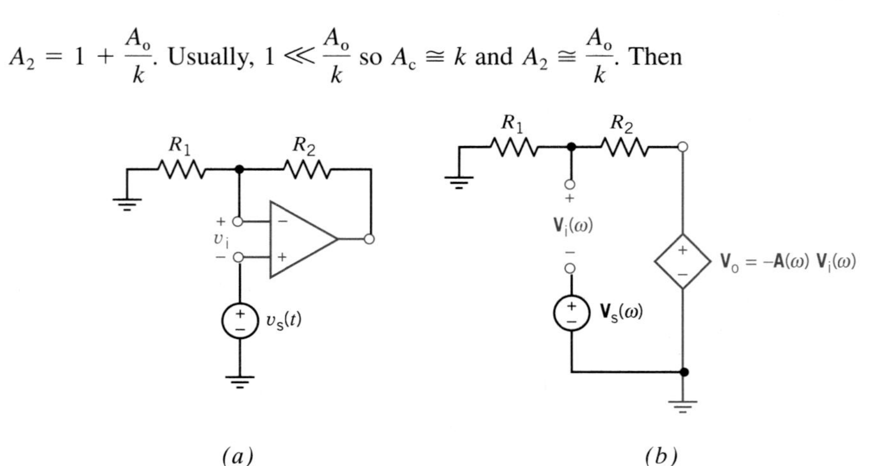

Figure 13.6-2 (*a*) a noninverting amplifier and (*b*) an equivalent circuit incorporating the frequency-dependent model of the operational amplifier.

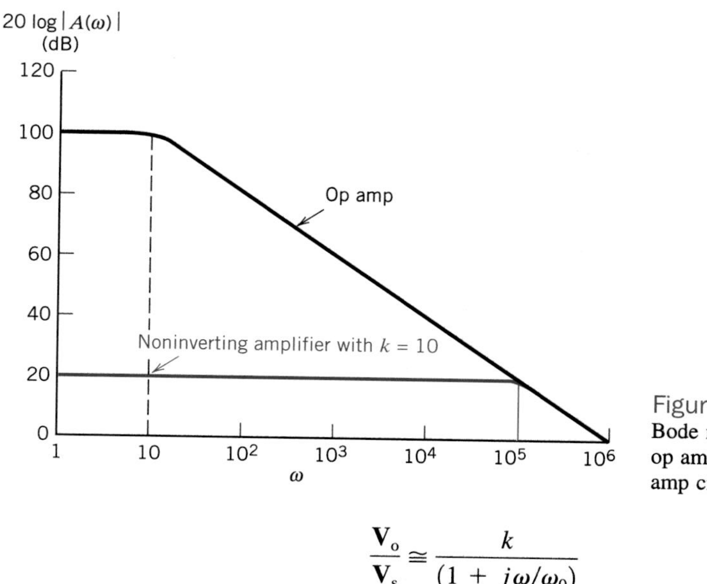

Figure 13.6-3
Bode magnitude diagram of the op amp and the noninverting op amp circuit (in color).

$$\frac{\mathbf{V}_o}{\mathbf{V}_s} \cong \frac{k}{(1 + j\omega/\omega_0)}$$

where $\omega_0 = \dfrac{A_o\omega_1}{k}$ is the corner frequency of the noninverting amplifier. Notice that the product of the dc gain and the corner frequency is

$$\omega_0 k = \frac{A_o\omega_1}{k}k = A_o\omega_1$$

This product is called the gain-bandwidth product. Notice it depends only on the op amp, not on R_1 and R_2.

For this example, $k = 10$ and $A_o = 100$ dB $= 10^5$, and thus we have $A_c = 10$, $A_2 = 10^4$, and $\omega_1 A_2 = 10^5$. Therefore,

$$\frac{\mathbf{V}_o}{\mathbf{V}_s} = \frac{10}{1 + j10^{-5}\omega}$$

This circuit has a magnitude curve as shown in color in Figure 13.6-3. Note that the noninverting op amp has a low-frequency gain of 20 dB and a break frequency of 10^5 rad/s. The gain-bandwidth product remains 10^6 rad/s.

13.7 FREQUENCY RESPONSE USING PSPICE

The frequency response of a circuit can be readily obtained using the .AC statement. The .AC statement is written as

```
.AC<sweep type><n><start freq><end freq>
```

The sweep type may be linear (LIN), octave (OCT), or decade (DEC). When LIN is selected, a sinusoidal steady-state analysis is performed at each of n frequencies linearly distributed over the range from "start freq" to "end freq." When OCT or DEC is selected, the analysis is performed at frequencies logarithmically distributed over the range from "start freq" to "end freq." When OCT or DEC is used, n indicates the number of points per octave or per decade.

The circuit shown in Figure 13.7-1a is a resonant circuit. Recalling that the resonant frequency is the frequency at which the imaginary part of $\mathbf{Z}(\omega)$ is zero, we find that

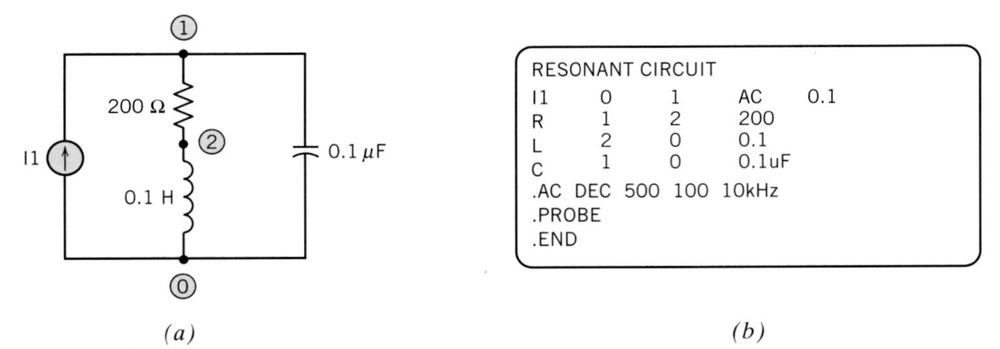

Figure 13.7-1 (*a*) An *RLC* resonant circuit and (*b*) the corresponding PSpice input file.

$$\omega_0^2 = \left[\frac{1}{LC} - \left(\frac{R}{L}\right)^2\right]$$

For the values of Figure 13.7-1*a*, we obtain $\omega_0 = 9798$ rad/s or $f_0 = 1559$ Hz. It is of interest to plot the frequency response over a range of frequencies that includes the resonant frequency, such as between 100 Hz and 10 kHz. We use the PSpice program shown in Figure 13.7-1*b* to analyze this circuit. The statement

```
I1 0 1 AC 0.1
```

indicates a sinusoidal current source directed from node 0 to node 1. The magnitude of the current is 0.1 A. No phase angle is given, so the phase angle is zero. The statement

```
.PROBE
```

causes PSpice to prepare a data file for Probe, the graphical post-processing program that is included with PSpice. Probe can be used to display the frequency response of the resonant circuit. The input to this circuit is the source current, and the response is the source voltage. Equivalently, the response is the node voltage at node 1. We use Probe to display VM(1)/I(I1), where VM(1) denotes the magnitude of the node voltage and node 1 and I(I1) denote the current source current. Figure 13.7-2 shows the frequency response. Several key points on the frequency response have been labeled using the cursor feature in

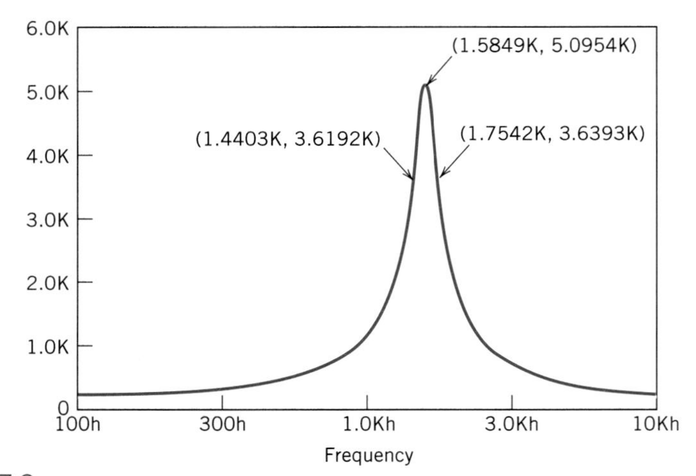

Figure 13.7-2 Frequency response of the resonant circuit. (Some points have been labeled using the cursor function provided by Probe. Each label gives the frequency in hertz followed by the magnitude of the impedance in Ω.)

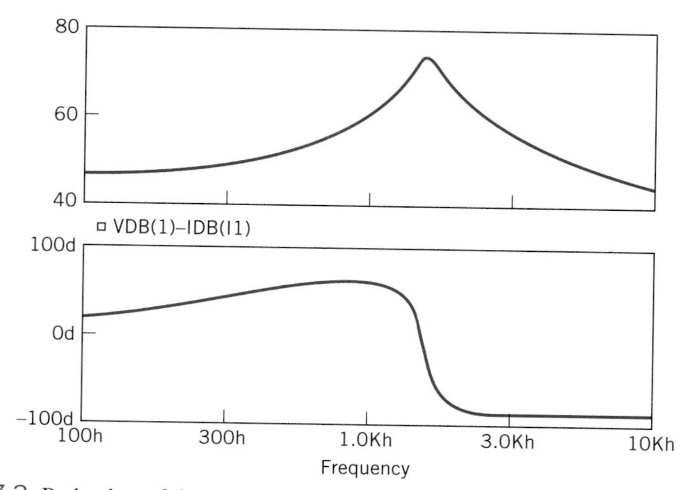

Figure 13.7-3 Bode plots of the resonant circuit of Figure 13.7-1*a*.

Probe. These points have been selected to provide the data required to calculate the resonant frequency and bandwidth of the circuit. The peak of the frequency response occurs at f_0 = 1,585 Hz and has a value equal to 5095. The two half-power frequencies are the frequencies where $|\mathbf{Z}| = 0.707(5095) = 3602$. These frequencies are 1440 Hz and 1754 Hz. Thus, the bandwidth is

$$BW = 1754 - 1440 = 314 \text{ Hz}$$

To obtain Bode plots of **Z**, use Probe to display VDB(1)-IDB(I1) and VP(1)-IP(I1). VDB(1) denotes the magnitude of the node voltage and node 1 in DB, and VP(1) denotes the phase of the node voltage and node 1. Figure 13.7-3 shows the Bode plots that describe $\mathbf{Z}(\omega)$.

13.8 PLOTTING BODE PLOTS USING MATLAB

MATLAB can be used to display the Bode plot or frequency response plot corresponding to a network function. As an example, consider the network function

$$\mathbf{H}(\omega) = \frac{K\left(1 + j\dfrac{\omega}{z}\right)}{\left(1 + j\dfrac{\omega}{p_1}\right)\left(1 + j\dfrac{\omega}{p_2}\right)}$$

Figure 13.8-1 shows a MATLAB input file that can be used to obtain the Bode plot corresponding to this network function. This MATLAB file consists of four parts.

In the first part, the MATLAB command `logspace` is used to specify the frequency range for the Bode plot. The command `logspace` also provides a list of frequencies that are evenly spaced (on a log scale) over this frequency range.

The given network has four parameters—the gain K, the zero z, and two pole p_1 and p_2. The second part of the MATLAB input file specifies values for these four parameters.

The third part of the MATLAB input file is a "for loop" that evaluates $\mathbf{H}(\omega)$, $|\mathbf{H}(\omega)|$, and $\angle\mathbf{H}(\omega)$ at each frequency in the list of frequencies produced by the command `logspace`.

The fourth part of the MATLAB input file does the plotting. The command

```
semilogx ( w/(2*pi), 20*log10(mag) )
```

```
% nf.m - plot the Bode plot of a network function

%---------------------------------------------------------------------------
%     Create a list of logarithmically spaced frequencies.
%---------------------------------------------------------------------------
wmin=10;                     % starting frequency, rad/s
wmax=100000;                 % ending frequency, rad/s
w = logspace(log10(wmin),log10(wmax));

%---------------------------------------------------------------------------
%              Enter values of the parameters that describe the
%                             network function.
%---------------------------------------------------------------------------
K= 10;               % constant
z= 1000;             % zero
p1=100;   p2=10000;   % poles

%---------------------------------------------------------------------------
% Calculate the value of the network function at each frequency.
%    Calculate the magnitude and angle of the network function.
%---------------------------------------------------------------------------
for k=1:length(w)
    H(k) = K*(1+j*w(k)/z) / ( (1+j*w(k)/p1) * (1+j*w(k)/p2) );
    mag(k) = abs(H(k));
    phase(k) = angle(H(k));
end

%---------------------------------------------------------------------------
%                             Plot the Bode plot.
%---------------------------------------------------------------------------
subplot(2,1,1), semilogx(w/(2*pi), 20*log10(mag))
xlabel('Frequency, Hz'), ylabel('Gain, dB')
title('Bode plot')
subplot(2,1,2), semilogx(w/(2*pi), phase)
xlabel('Frequency, Hz'), ylabel('Phase, deg')
```

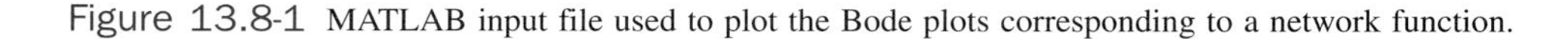

Figure 13.8-1 MATLAB input file used to plot the Bode plots corresponding to a network function.

does several things. The command `semilogx` indicates that the plot is to be made using a logarithmic scale for the first variable and a linear scale for the second variable. The first variable, frequency, is divided by 2π to convert to Hz. The second variable, $|\mathbf{H}(\omega)|$, is converted to dB.

The Bode plots produced using this MATLAB input file are shown in Figure 13.8-2.

The second and third parts of the MATLAB input file can be modified to plot the Bode plots for a different network function.

13.9 VERIFICATION EXAMPLES

Example 13.9-1

Figure 13.9-1*a* shows a laboratory setup for measuring the frequency response of a circuit. A sinusoidal input is connected to the input of a circuit having the network function $\mathbf{H}(\omega)$. An oscilloscope is used to measure the input and output sinusoids. The input voltage is used to trigger the oscilloscope so the phase angle of the input is zero. Frequency response data are collected by varying the input frequency and measuring the amplitude of the input voltage and the amplitude and phase of the output voltage.

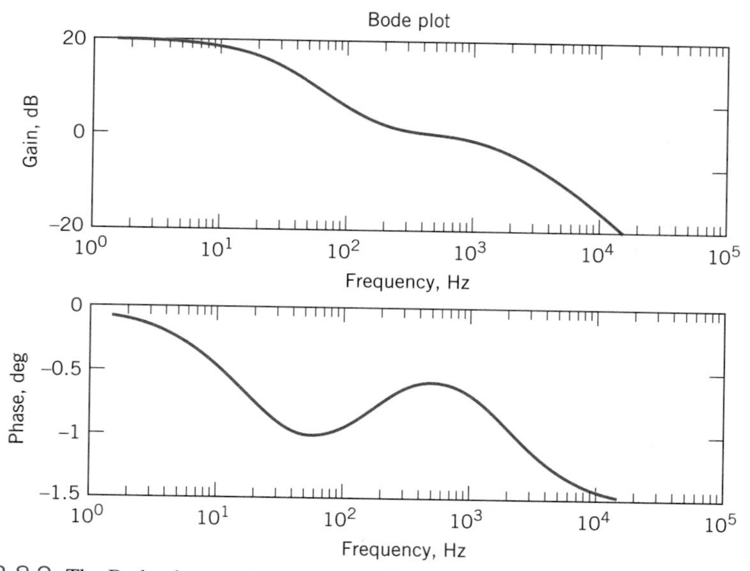

Figure 13.8-2 The Bode plots produced using the MATLAB input file given in Figure 13.8-1.

In this example, the desired frequency response is specified by the Bode plot shown in Figure 13.9-1*b*. Figure 13.9-1*c* shows frequency response data from laboratory measurements. In this example the amplitude, but not the phase angle, of the output voltage was measured. Does this laboratory data confirm that the circuit does indeed have the specified Bode plot?

Solution

The Bode plot has three features that we can look for in the frequency response data.

1. The dc gain is 14 dB.
2. The slope of the Bode plot is -20 dB/decade when $\omega \gg 200$ rad/s.
3. The corner frequency is 200 rad/s.

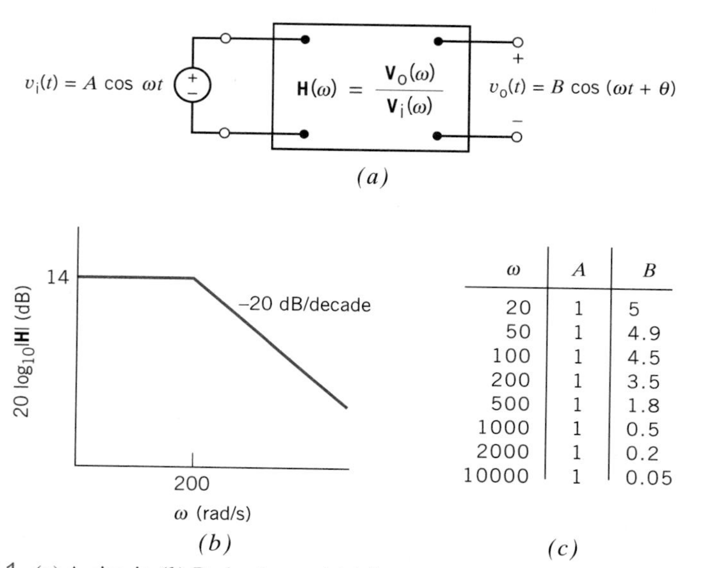

ω	A	B
20	1	5
50	1	4.9
100	1	4.5
200	1	3.5
500	1	1.8
1000	1	0.5
2000	1	0.2
10000	1	0.05

Figure 13.9-1 (*a*) A circuit, (*b*) Bode plot, and (*c*) frequency response data.

The lowest frequency at which frequency response data was taken is 20 rad/sec. At this frequency the gain was measured to be

$$|\mathbf{H}(20)| = \frac{B}{A} = \frac{5}{1} = 14\text{ dB}$$

which is equal to the dc gain specified by the Bode plot.

To identify the corner frequency from the frequency response data, we look for the frequency at which the gain is

$$\frac{\text{dc gain}}{\sqrt{2}} = \frac{5}{\sqrt{2}} = 3.536$$

The frequency response data indicate that the gain is 3.5 at a frequency of 200 rad/s. That agrees with the corner frequency of 200 rad/s of the specified Bode plot.

The slope of the frequency response at high frequencies is given by

$$\frac{20\log_{10}(0.05) - 20\log_{10}(0.5)}{\log_{10}(10000) - \log_{10}(1000)} = -20\text{ dB/decade}$$

which is the same as the slope of the Bode plot.

The frequency response data confirm that the circuit does indeed have the specified Bode plot.

Example 13.9-2

Your lab notes indicate that the circuit shown in Figure 13.9-2 was built using $R_1 = 10\text{ k}\Omega$, $R_2 = 50\text{ k}\Omega$, and $C = 10\text{ nF}$. The gain and phase shift of this circuit were measured to be 2.7 and 125° at 500 hertz. Is this information consistent?

Solution

The network function of this circuit is

$$\mathbf{H}(\omega) = -\frac{\frac{1}{j\omega C}\|R_2}{R_1} = -\frac{\frac{R_2}{R_1}}{1 + j\omega R_2 C}$$

$$= -\frac{\frac{50 \cdot 10^3}{10 \cdot 10^3}}{1 + j(2\pi \cdot 500)(50 \cdot 10^3)(10 \cdot 10^{-9})} = 2.685\angle 122.5°$$

The calculated gain and phase shift agree with the measured gain and phase shift. The lab notes are consistent.

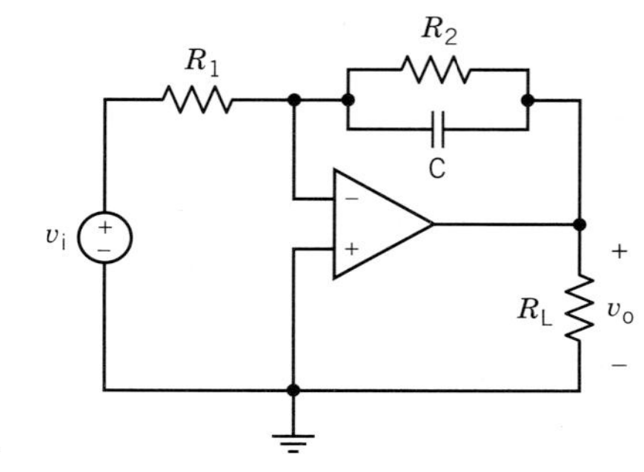

Figure 13.9-2
An op amp circuit.

Example 13.9-3

An old lab report from a couple of years ago includes the following data about a particular circuit:

1. The magnitude and phase frequency responses shown in Figure 13.9-3.
2. When the input to the circuit was

$$v_{\text{in}} = 4\cos(2\pi 1200)$$

the steady-state response was

$$v_{\text{out}} = 6.25\cos(2\pi 1200 + 110°).$$

Are these data consistent?

Solution

Three things need to be checked: the frequencies, the amplitudes, and the phase angles. The frequencies of both sinusoids are the same, which is good because the circuit must be linear if it is to be represented by a frequency response and the steady-state response of a linear circuit to a sinusoidal input is a sinusoid at the same frequency as the input. The frequency of the input and output sinusoids is

$$\omega = 2 \cdot \pi \cdot 1200 \text{ rad/s}$$

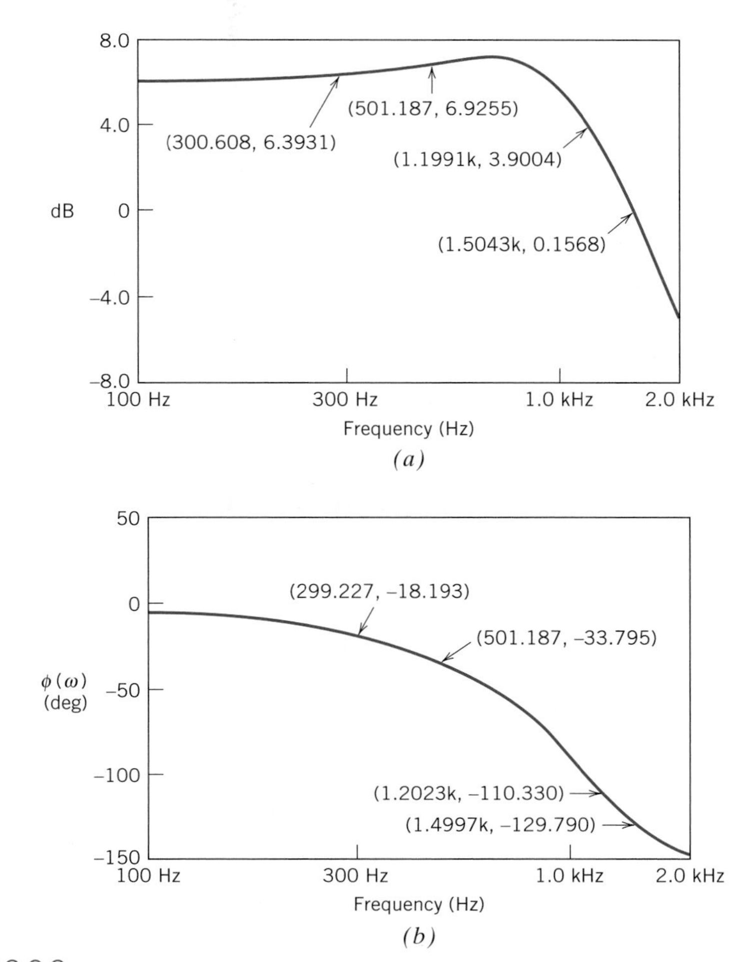

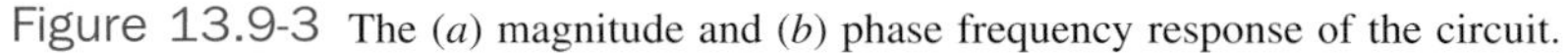
Figure 13.9-3 The (a) magnitude and (b) phase frequency response of the circuit.

or

$$f = 1200 \text{ Hz}$$

Fortunately, the gain and phase shift at 1200 Hz have been labeled on the frequency response plots shown in Figure 13.9-3. The gain at 1200 Hz is labeled as 3.9 dB, which means that

$$\frac{|\mathbf{V}_{\text{out}}|}{|\mathbf{V}_{\text{in}}|} = 3.9 \text{ dB} = 1.57$$

where $\mathbf{V}_{\text{in}}$ and $\mathbf{V}_{\text{out}}$ are the phasors corresponding to $v_{\text{in}}(t)$ and $v_{\text{out}}(t)$. Let us check this against the data about the input and output sinusoids. Since the magnitudes of the phasors are equal to the amplitudes of the corresponding sinusoids,

$$\frac{|\mathbf{V}_{\text{out}}|}{|\mathbf{V}_{\text{in}}|} = \frac{6.25}{4} = 1.56$$

This is very good agreement for experimental work.

Next consider the phase shift. The frequency response indicates that the phase shift at 1200 Hz is $-100°$, which means

$$\angle\mathbf{V}_{\text{out}} - \angle\mathbf{V}_{\text{in}} = -110°$$

Let us check this against the data about the input and output sinusoids. Since the angles of the phasors are equal to the phase angles of the corresponding sinusoids,

$$\angle\mathbf{V}_{\text{out}} - \angle\mathbf{V}_{\text{in}} = 110° - 0° = 110°$$

The signs of the phase angles do not match. At a frequency of 1200 Hz, a phase angle of 110° indicates that the peaks of the output sinusoid will follow the peaks of the input sinusoid by

$$t_0 = \frac{110°}{360°} \cdot \frac{1}{1200} = 0.255 \text{ ms}$$

while a phase angle of $-110°$ indicates that the peaks of the output sinusoid will precede the peaks of the input sinusoid by 0.255 ms. It is likely that the angle of the output sinusoid was entered incorrectly in the lab data.

We have found an error in the old lab report and proposed an explanation for the error.

13.10 Design Challenge Solution

RADIO TUNER

Three radio stations broadcast at three different frequencies, 700 kHz, 1000 kHz, and 1400 kHz. Figure 13.10-1 shows a simplified diagram of a radio receiver. The antenna receives signals from all three stations so the input to the tuner will be a sum of these signals. Suppose this voltage is described by

$$v_{\text{i}}(t) = \sin(2\pi \cdot 7 \cdot 10^5 t + 135°) + \sin(2\pi \cdot 10^6 t) + \sin(2\pi \cdot 1.4 \cdot 10^6 t + 300°) \quad (13.10\text{-}1)$$

Consider the problem of tuning to the station that broadcasts at 1000 kHz. The tuner must eliminate the first and third terms of $v_{\text{i}}(t)$ to produce the output signal

$$v_{\text{o}}(t) = A \sin(2\pi \cdot 10^6 t + \theta)$$

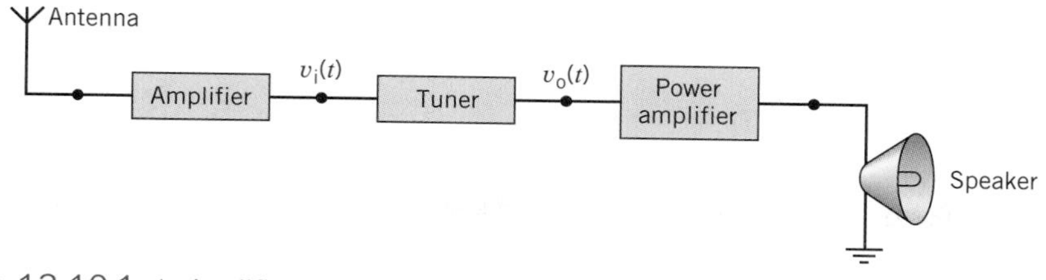

Figure 13.10-1 A simplified diagram of a radio receiver.

Describe the Situation and Assumptions

Let $\mathbf{H}(\omega)$ be the network function of the tuner. The tuner must have a gain approximately equal to one at 1000 kHz ($|\mathbf{H}(2\pi \cdot 10^6)| \cong 1$) and approximately equal to zero at 700 kHz and at 1400 kHz ($|\mathbf{H}(2\pi \cdot 7 \cdot 10^5)| \cong 0$ and $|\mathbf{H}(2\pi \cdot 1.4 \cdot 10^6)| \cong 0$). The tuner output will be

$$\begin{aligned} v_o(t) = &\ |\mathbf{H}(2\pi \cdot 7 \cdot 10^5)| \sin(2\pi \cdot 7 \cdot 10^5 t + 135 + \angle \mathbf{H}(2\pi \cdot 7 \cdot 10^5)) \\ &+ |\mathbf{H}(2\pi 10^6)| \sin(2\pi 10^6 t + \angle \mathbf{H}(2\pi 10^6)) \\ &+ |\mathbf{H}(2\pi \cdot 1.4 \cdot 10^6)| \sin(2\pi \cdot 1.4 \cdot 10^6 t + 300^\circ + \angle \mathbf{H}(2\pi \cdot 1.4 \cdot 10^6)) \end{aligned} \tag{13.10-2}$$

or

$$v_o(t) \cong \sin(2\pi \cdot 10^6 t + \theta)$$

where

$$\theta = \angle \mathbf{H}(2\pi \cdot 10^6)$$

State the Goal

The goal is to design a circuit consisting of resistors, capacitors, and op amps that has a gain equal to one at 1000 kHz and equal to zero at 700 and 1400 kHz.

Generate a Plan

The tuner will be based on a resonant circuit having $\omega_0 = 2\pi 10^6 = 6.283 \cdot 10^6$ rad/s and $Q = 15$. Figure 13.10-2 shows an op amp circuit called a simulated inductor. This circuit acts like a grounded inductor having an inductance equal to

$$L = \frac{C_2 R_1 R_3 R_5}{R_4} \tag{13.10-3}$$

Figure 13.10-3 shows how a parallel resonant circuit can be used to design the tuner. A parallel resonant circuit is shown in Figure 13.10-3*a*. The parallel resonant circuit must be modified if it is to be used for the tuner. The input to the tuner is a voltage, but the input to the parallel resonant circuit is a current. A source transformation is used to obtain a circuit that has a voltage input, shown in Figure 13.10-3*b*. Next, the inductor is replaced by the simulated inductor to produce the circuit show in Figure 13.10-3*c*. This is the circuit that will be used as the tuner.

The design will be completed in two steps. First, values of L, R, and C will be calculated so that the parallel resonant circuit has $\omega_0 = 6.283 \cdot 10^6$ rad/s and $Q = 15$. Next, the capacitor and resistors of the simulated inductor will be selected to satisfy Eq. 13.10-3.

Act on the Plan

First, design the resonant circuit to have $\omega_0 = 6.283 \cdot 10^6$ rad/s and $Q = 15$. Pick a convenient value for the capacitance, $C = 0.001\ \mu$F. Then

$$L = \frac{1}{\omega_0^2 C} = \frac{1}{(6.283 \cdot 10^6)^2 \cdot 10^{-9}} = 25.33\ \mu\text{H}$$

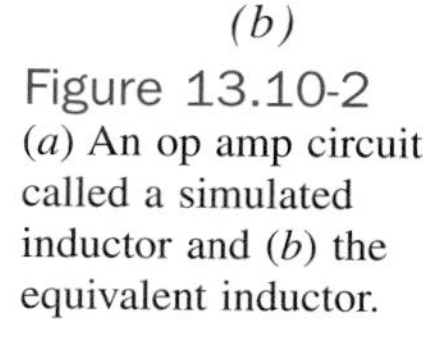

(a) (b)

Figure 13.10-2 (*a*) An op amp circuit called a simulated inductor and (*b*) the equivalent inductor.

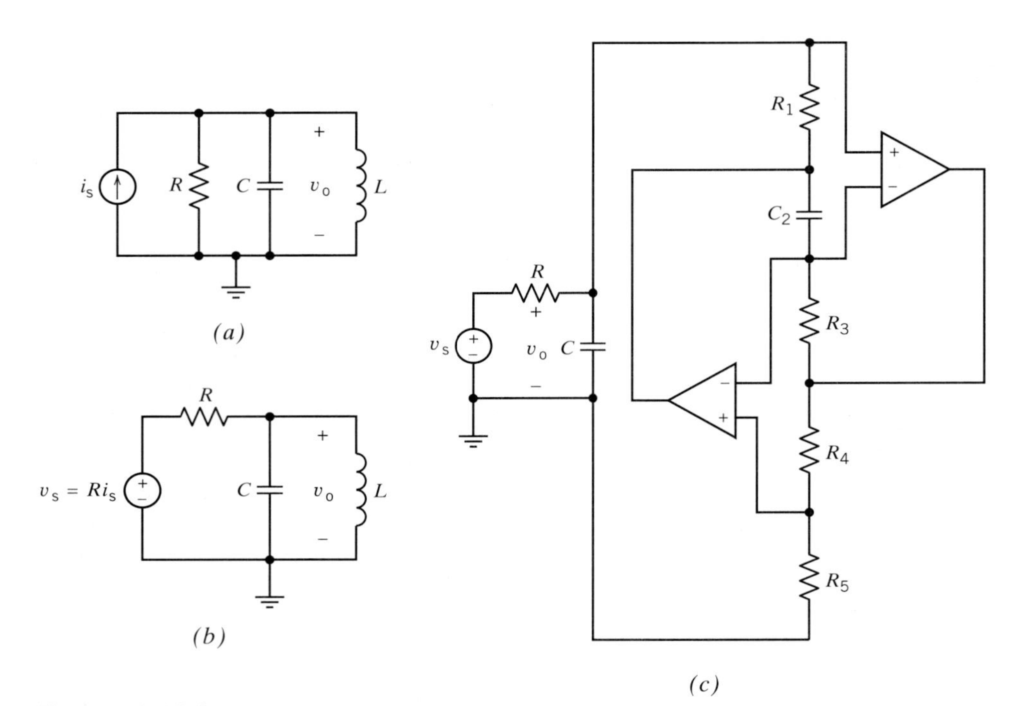

Figure 13.10-3 (*a*) A resonant circuit. (*b*) A band-pass filter. (*c*) An RC op amp band-pass filter.

and
$$R = Q\sqrt{\frac{L}{C}} = 15\sqrt{\frac{25.33 \cdot 10^{-6}}{10^{-9}}} = 2387\ \Omega$$

Next, design the simulated inductor to have an inductance of $L = 25.33$ mH. There are many ways to do this. Let's pick $C_2 = 0.001\ \mu$F, $R_1 = 1.5\ \text{k}\Omega$, $R_3 = 1.5\ \text{k}\Omega$, and $R_4 = 80\ \text{k}\Omega$. Then

$$R_5 = \frac{R_4 L}{C_2 R_1 R_3} = \frac{80 \cdot 10^3 \cdot 25.33 \cdot 10^{-6}}{10^{-9} \cdot 1.5 \cdot 10^3 \cdot 1.5 \cdot 10^3} = 900\ \Omega$$

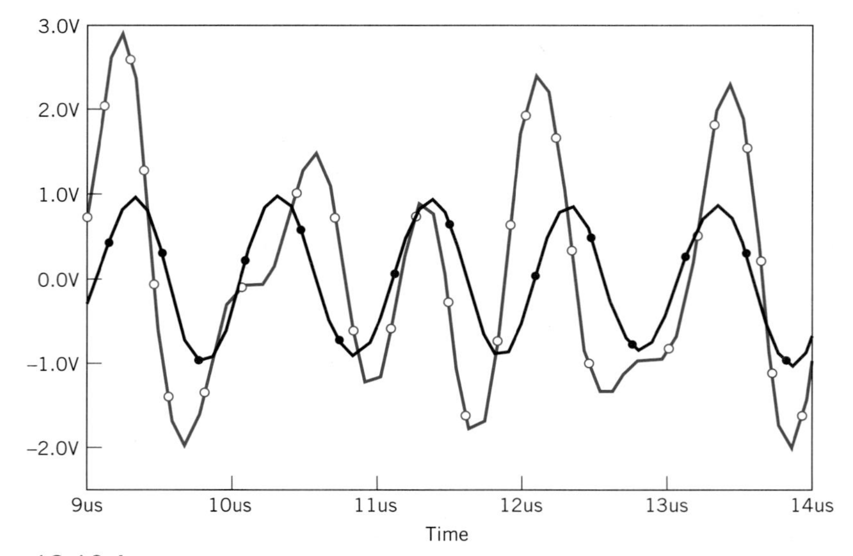

Figure 13.10-4 PSpice simulation of the radio tuner.

Verify the Proposed Solution

Figure 13.10-4 shows the results of a PSpice simulation of the tuner. The input to the circuit is $v_i(t)$ described by Eq. 13.10-1. This signal is not sinusoidal. The output of the filter is a sinusoid with an amplitude of approximately 1 and a frequency of 1000 kHz, as required by Eq. 13.10-2. Thus, the design specifications are satisfied.

13.11 SUMMARY

- Gain, phase shift, and the network function are properties of linear circuits that are used to describe the effect that a circuit has on a sinusoidal input voltage or current.
  - The gain of the circuit describes the relationship between the sizes of the input and output sinusoids. The gain is the ratio of the amplitude of the output sinusoid to the amplitude of the input sinusoid.
  - The phase shift of the circuit describes the relationship between the phase angles of the input and output sinusoids. The phase shift is the difference between the phase angle of the output sinusoid and the phase angle of the input sinusoid.
  - The network function describes the way the behavior of the circuit depends on the frequency of the input. The network function is defined in the frequency domain. It is the ratio of the phasor corresponding to the response sinusoid to the phasor corresponding to the input.
  - Table 13.4-2 tabulates the network functions of several common op amp circuits.
- The frequency response describes the way the gain and phase shift of a circuit depend on frequency. Equations, tables, or plots are each used to express the frequency response.
- Bode plots represent the frequency response as plots of the gain in decibels and the phase using a logarithmic scale for frequency. Asymptotic magnitude Bode plots are approximate Bode plots that are easy to draw. The terms *corner frequency* and *break frequency* are routinely used to describe linear circuits. These terms describe features of the asymptotic Bode plot.
- Some linear circuits exhibit a phenomenon called resonance. These circuits contain reactive elements but act as if they were purely resistive at a particular frequency, called the resonant frequency. Resonant circuits are described using the resonant frequency, quality factor, and bandwidth. Table 13.5-1 summarizes the properties of series and parallel resonant circuits.
- The gain of operational amplifiers depends on the frequency of the input. Using an op amp model that includes a frequency dependent gain makes our analysis more accurate but also more complicated. We use the more complicated model when we need the additional accuracy, and we use the simpler model when we don't.
- PSpice can be used to analyze a circuit and display its frequency response.
- MATLAB can be used to display the frequency response of a network function.

PROBLEMS

Section 13-3 Gain, Phase Shift, and the Network Function

P 13.3-1 Determine the network function $\mathbf{H} = \mathbf{V}_o/\mathbf{V}_s$ for the circuit of Figure P 13.3-1. Determine $|\mathbf{H}|$ and $\phi(\omega)$ and plot the magnitude and phase on a linear scale for $|\mathbf{H}|$ and ϕ versus a logarithmic scale for ω.

P 13.3-2 Consider the circuit shown in Figure P 13.3-2, where $\mathbf{V}_i$ and $\mathbf{V}_o$ are phasor voltages corresponding to v_i and v_o. (a) Find expressions for $\mathbf{V}_o/\mathbf{V}_i$ and $|\mathbf{V}_o/\mathbf{V}_i|$, both as a function of the radian frequency ω. (b) Sketch $|\mathbf{V}_o/\mathbf{V}_i|$ versus ω for $\omega > 0$. Does your sketch give the correct limiting results for $\omega \to 0$ and $\omega \to \infty$? Briefly justify your answers. (c) Sketch the phase of $\mathbf{V}_o/\mathbf{V}_i$ for $\omega \to 0$.

P 13.3-3 Determine the network function $\mathbf{H} = \mathbf{V}_o/\mathbf{V}_s$ for the circuit of Figure P 13.3-3. Sketch the gain, $|\mathbf{H}|$, and phase shift, $\phi(\omega)$, versus ω for $\omega > 0$.
Answer: $\mathbf{H} = 50{,}000/(j\omega + 70{,}000)$

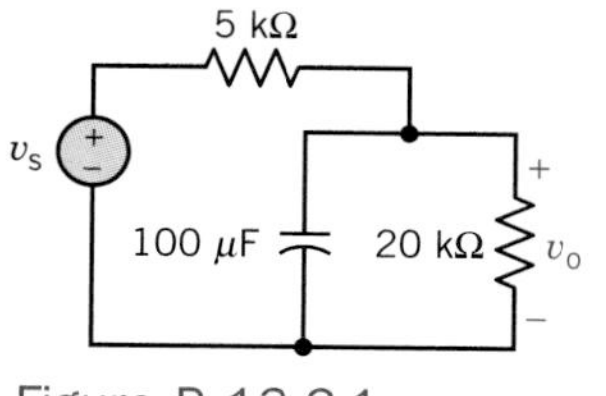

Figure P 13.3-1

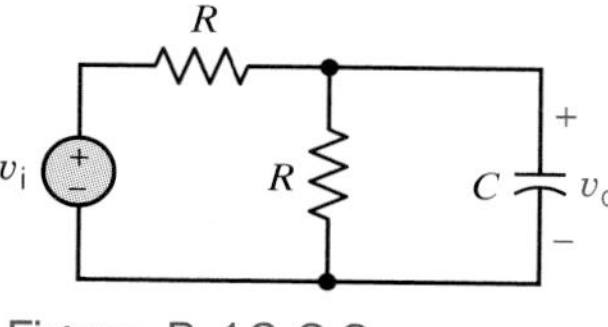

Figure P 13.3-2

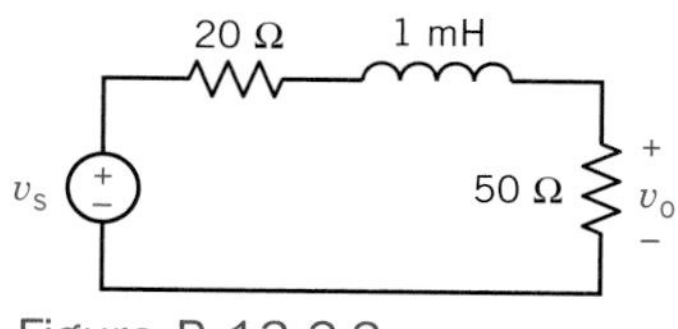

Figure P 13.3-3

P 13.3-4 A simple voltage divider with two resistors does not work as we expect at higher frequencies because of a parasitic capacitance shunting $\boldsymbol{R}_2$. To remedy the capacitance problem, another capacitance, C_1, is added in parallel with $\boldsymbol{R}_1$ as shown in Figure P 13.3-4. (a) Determine the relationship for $\boldsymbol{R}_1\boldsymbol{C}_1$ and $\boldsymbol{R}_2\boldsymbol{C}_2$ so that the output voltage is $\mathbf{V}_o = k\mathbf{V}_s$ and is not a function of frequency. (b) Plot the gain magnitude curve $\mathbf{V}_o/k\mathbf{V}_s$ (dB) for $\boldsymbol{R}_1 = \boldsymbol{R}_2 = 10\ \text{k}\Omega$, $\boldsymbol{C}_2 = 0.1\ \mu\text{F}$, and three values of $\boldsymbol{C}_1$: $1\ \mu\text{F}$, $0.1\ \mu\text{F}$, and $0.05\ \mu\text{F}$.
Answer: (a) $R_1C_1 = R_2C_2$

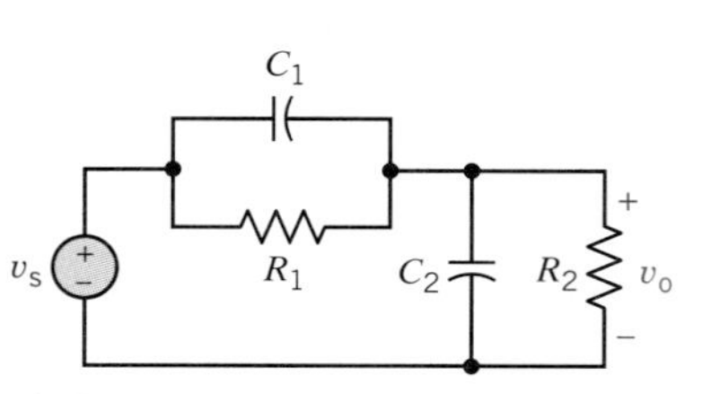

Figure P 13.3-4

P 13.3-5 The input to the circuit shown in Figure 13.3-5 is the source voltage $v_i(t)$, and the response is the voltage across R_L, $v_o(t)$. Find the network function.
Answer: $\mathbf{H}(\omega) = -5/(1 + j\omega/10)$

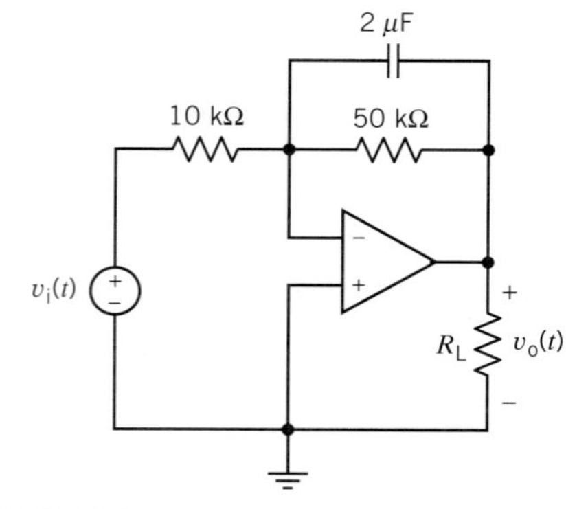

Figure P 13.3-5

P 13.3-6 The input to the circuit shown in Figure P 13.3-6 is the source voltage $v_i(t)$, and the response is the voltage across R_L, $v_o(t)$. Express the gain and phase shift as functions of the radian frequency ω.

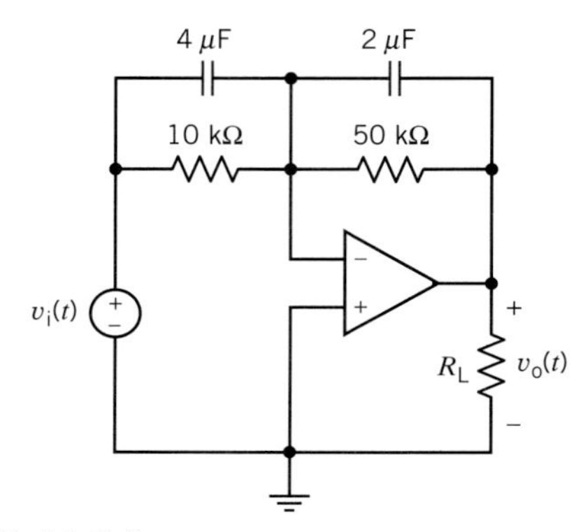

Figure P 13.3-6

P 13.3-7 The input to the circuit shown in Figure P 13.3-7 is the source voltage $v_i(t)$, and the response is the voltage across R_L, $v_o(t)$. The resistance R_1 is 10 kΩ. Design this circuit to satisfy the following two specifications:
a. The gain at low frequencies is 5.
b. The gain at high frequencies is 2.
Answer: $R_2 = 20\ \text{k}\Omega$, $R_3 = 30\ \text{k}\Omega$.

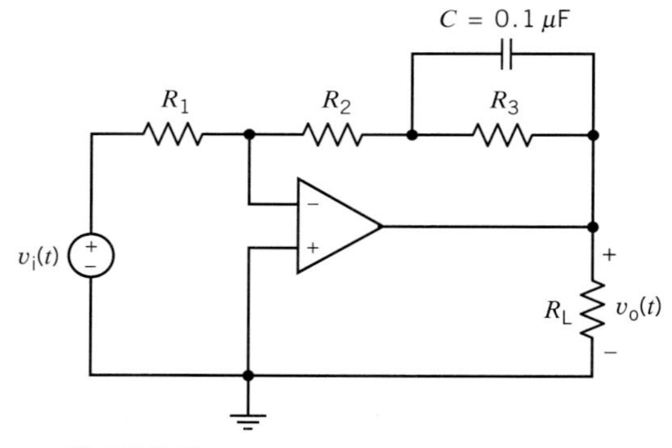

Figure P 13.3-7

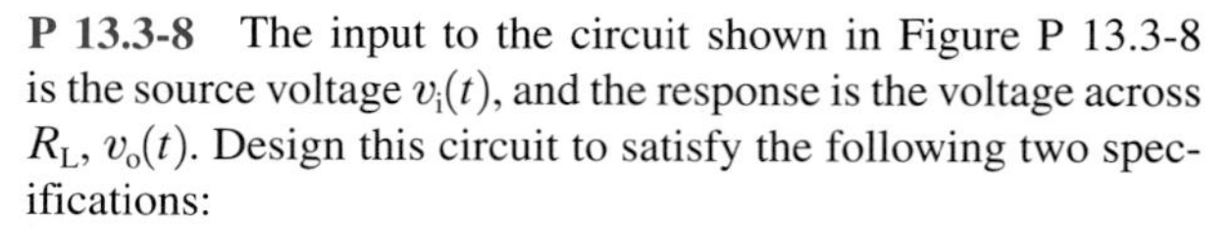

P 13.3-8 The input to the circuit shown in Figure P 13.3-8 is the source voltage $v_i(t)$, and the response is the voltage across R_L, $v_o(t)$. Design this circuit to satisfy the following two specifications:
a. The phase shift at $\omega = 1000$ rad/s is 135°.
b. The gain at high frequencies is 10.
Answer: $R_1 = 1\ \text{k}\Omega$, $R_2 = 10\ \text{k}\Omega$.

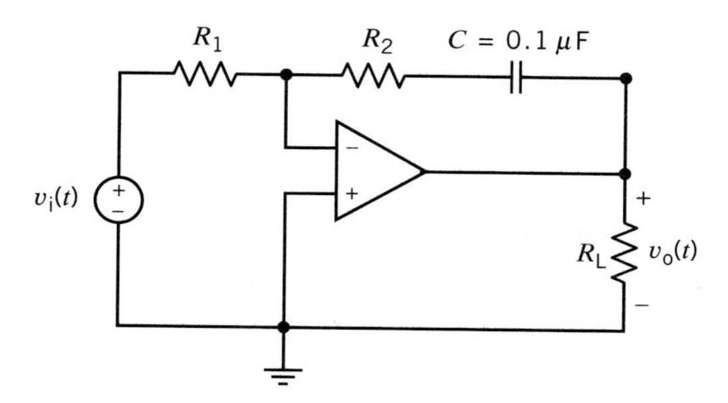

Figure P 13.3-8

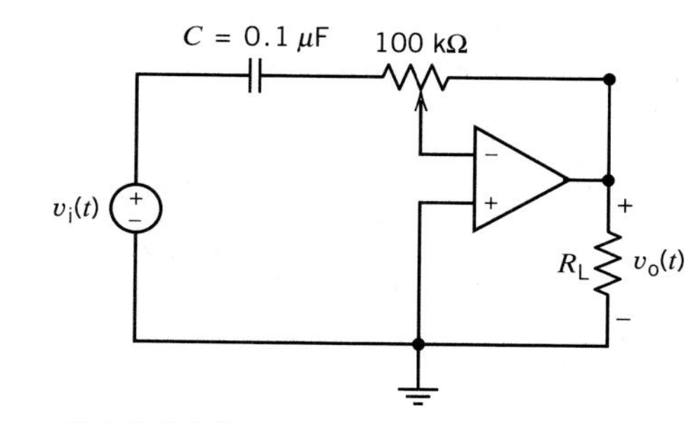

Figure P 13.3-10

P 13.3-9 The input to the circuit shown in Figure P 13.3-9 is the source voltage $v_i(t)$, and the response is the voltage across R_L, $v_o(t)$. Design this circuit to satisfy the following two specifications:

a. The phase shift at $\omega = 1000$ rad/s is 225°.
b. The gain at high frequencies is 10.

Answer: $R_1 = 10\ \text{k}\Omega$, $R_2 = 100\ \text{k}\Omega$.

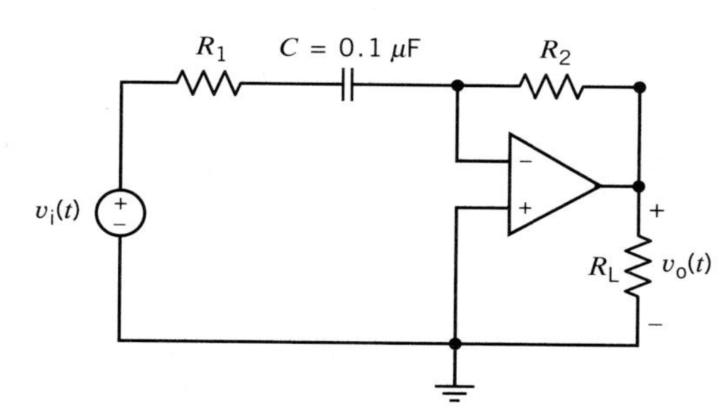

Figure P 13.3-9

P 13.3-10 The input to the circuit shown in Figure P 13.3-10 is the source voltage $v_i(t)$, and the response is the voltage across R_L, $v_o(t)$. Is it possible to adjust the potentiometer so that the high frequency gain of this circuit is equal to 4?

Answer: Yes.

P 13.3-11 The input to the circuit of Figure P 13.3-11 is

$$v_s = 50 + 30\cos(500t + 115°) - 20\cos(2500t + 30°)\ \text{mV}$$

Find the steady-state output voltage v_o for (a) $C = 0.1\ \mu\text{F}$ and (b) $C = 0.01\ \mu\text{F}$. Assume an ideal op amp.

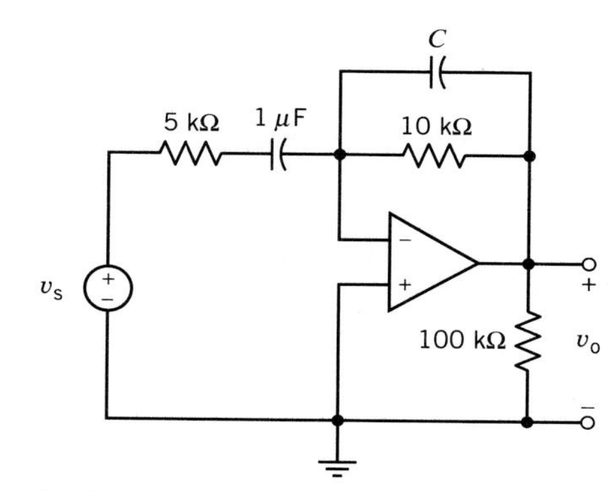

Figure P 13.3-11

P 13.3-12 The source voltage, v_s, shown in the circuit of Figure P 13.3-12*a* is a sinusoid having a frequency of 500 Hz and an amplitude of 8 V. The circuit is in steady state. The oscilloscope traces show the input and output waveforms as shown in Figure P 13.3-12*b*.

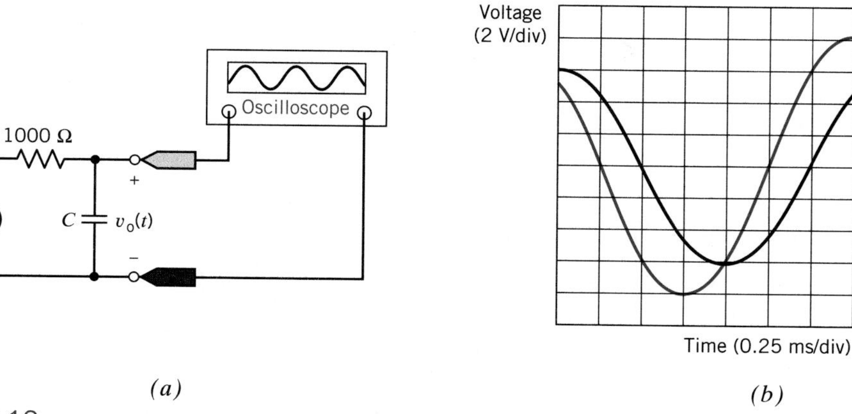

Figure P 13.3-12

(a) Determine the gain and phase shift of the circuit at 500 Hz.
(b) Determine the value of the capacitor.
(c) If the frequency of the input is changed, then the gain and phase shift of the circuit will change. What are the values of the gain and phase shift at the frequency 200 Hz? At 2000 Hz? At what frequency will the phase shift be $-45°$? At what frequency will the phase shift be $-135°$?
(d) What value of capacitance would be required to make the phase shift at 500 Hz be $-60°$? What value of capacitance would be required to make the phase shift at 500 Hz be $-300°$?
(e) Suppose the phase shift had been $-120°$ at 500 Hz. What would be the value of the capacitor?

Answers: (*b*) $C = 0.26\ \mu F$ (*e*) this circuit can't be designed to produce a phase shift $= -120°$

Section 13.4 Bode Plots

P 13.4-1 Sketch the magnitude Bode plot of $\mathbf{H}(\omega) = \dfrac{4(5 + j\omega)}{1 + j\dfrac{\omega}{50}}$.

P 13.4-2 Compare the magnitude Bode plots of $\mathbf{H}_1(\omega) = \dfrac{10(5 + j\omega)}{50 + j\omega}$ and $\mathbf{H}_2(\omega) = \dfrac{100(5 + j\omega)}{50 + j\omega}$.

P 13.4-3 The input to the circuit shown in Figure P 13.4-3 is the source voltage $v_{in}(t)$, and the response is the voltage across R_3, $v_{out}(t)$. The component values are $R_1 = 5\ k\Omega$, $R_2 = 10\ k\Omega$, $C_1 = 0.1\ \mu F$, and $C_2 = 0.1\ \mu F$. Sketch the asymptotic magnitude Bode plot for the network function.

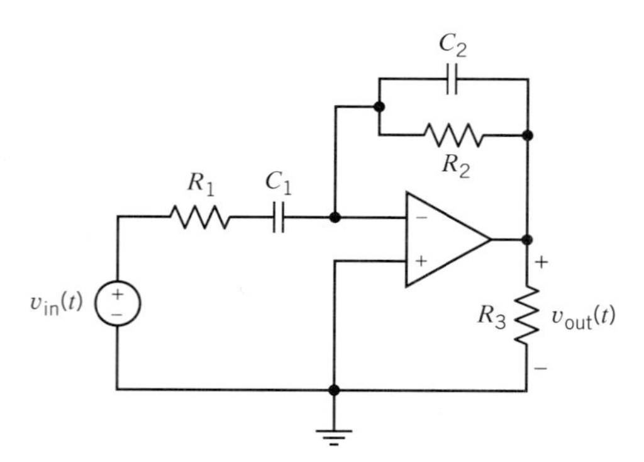

Figure P 13.4-3

P 13.4-4 The input to the circuit shown in Figure P 13.4-4 is the source voltage $v_s(t)$, and the response is the voltage across R_3, $v_o(t)$. Determine $\mathbf{H}(\omega)$ and sketch the Bode diagram.

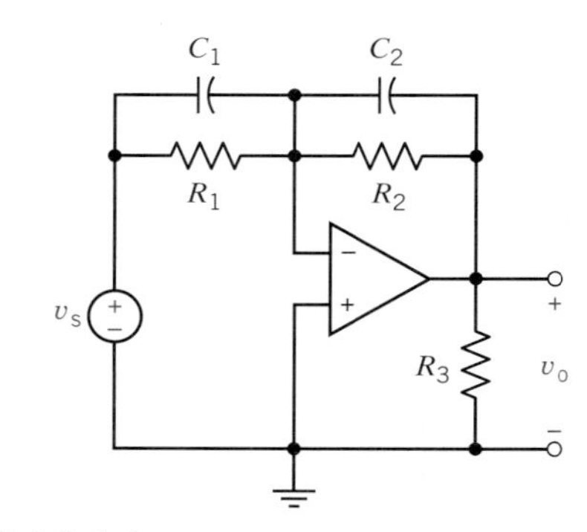

Figure P 13.4-4

P 13.4-5 The gain of a circuit is shown in Figure P 13.4-5 for $1 \le \omega \le 10^5$ rad/s. Find the network function, $\mathbf{H}(\omega)$, by estimating the asymptotic approximation for the Bode diagram of the circuit. The peak value of 20 log H is 43 dB at $\omega = 450$ rad/s. You may assume that this circuit's network function $\mathbf{H}(\omega)$ includes only first-order factors of the form $(1 + j\omega/\omega_0)$.

P 13.4-6 In the medical world, the stethoscope remains the most practical instrument available for listening for sounds arising within the heart and lungs. A recently introduced electronic stethoscope uses 10 band-pass amplifiers to let the physician concentrate on specific sounds (Bak 1986). The gain characteristic of one band-pass amplifier is represented by

$$\mathbf{H} = \frac{(j\omega/100 + 1)}{[1 + ju/10 + (ju)^2](j\omega/5000 + 1)}$$

where $u = \omega/\omega_0$ and $\omega_0 = 1000$. (a) Draw the exact Bode diagram of this filter for $100 \le \omega \le 10^4$.(b) Find the band-pass bandwidth BW. (c) Determine the Q of this circuit. (d) Find the resulting gain at ω_0.

P 13.4-7 Determine $\mathbf{H}(j\omega)$ from the asymptotic Bode diagram in Figure P 13.4-7.

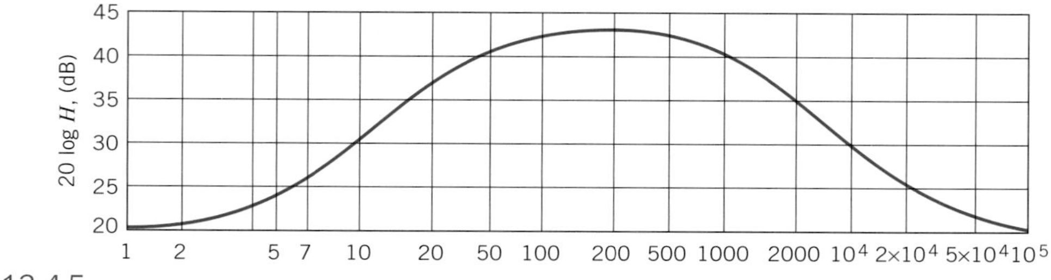

Figure P 13.4-5

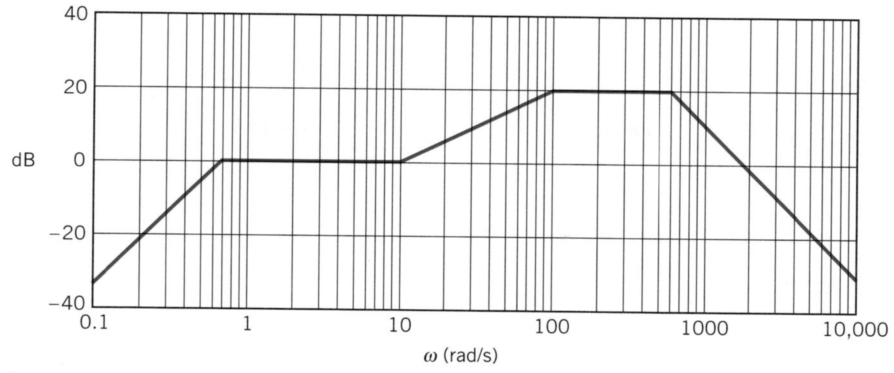

Figure P 13.4-7

P 13.4-8 The frequency response of an unknown circuit is measured and drawn with the asymptotes for 20 log H as shown in Figure P 13.4-8. The phase is also shown. Determine the equation for $\mathbf{H}(\omega)$, indicating the corner frequencies.

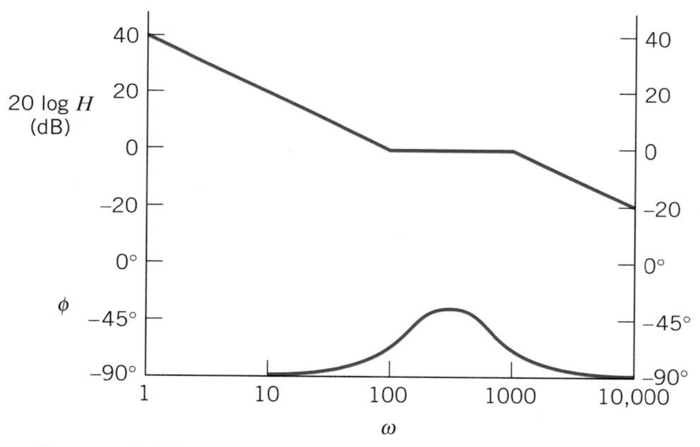

Figure P 13.4-8

P 13.4-9 A circuit has a voltage ratio

$$\mathbf{H}(\omega) = \frac{K(1 + j\omega/z)}{j\omega}$$

(a) Find the high- and low-frequency asymptotes of the magnitude Bode plot.

(b) The high- and low-frequency asymptotes comprise the magnitude Bode plot. Over what ranges of frequencies is the asymptotic magnitude Bode plot of $\mathbf{H}(\omega)$ within 1 percent of the actual value of $\mathbf{H}(\omega)$ in decibels?

P 13.4-10 Physicians use tissue electrodes to form the interface that conducts current to the target tissue of the human body. The electrode in tissue can be modeled by the RC circuit shown in Figure P 13.4-10. The value of each element depends on the electrode material and physical construction as well as the character of the tissue being probed. Find the Bode diagram for $\mathbf{V}_o/\mathbf{V}_s = \mathbf{H}(j\omega)$ when $R_1 = 1$ kΩ, $C = 1$ μF, and the tissue resistance is $R_t = 5$ kΩ.

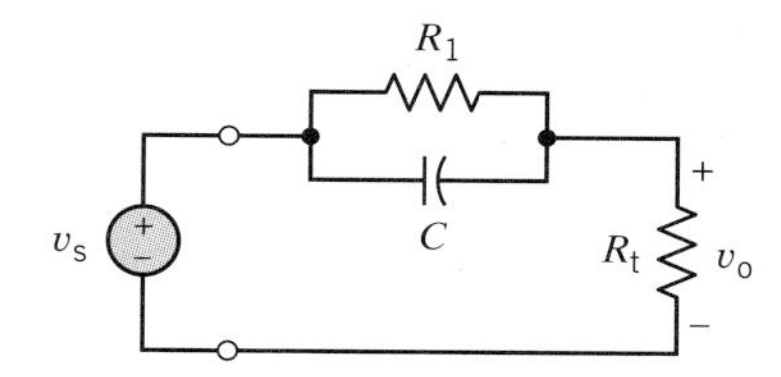

Figure P 13.4-10

P 13.4-11 Figure P 13.4-11 shows a circuit and corresponding asymptotic magnitude Bode plot. The input to this circuit shown is the source voltage $v_{in}(t)$, and the response is the voltage $v_o(t)$. The component values are $R_1 = 80$ Ω, $R_2 = 20$ Ω, $L_1 = 0.03$ H, $L_2 = 0.07$ H, and $M = 0.01$ H. Determine the values of K_1, K_2, p, and z.

Answer: $K_1 = 0.75, K_2 = 0.2, z = 333$ rad/s, $p = 1250$ rad/s

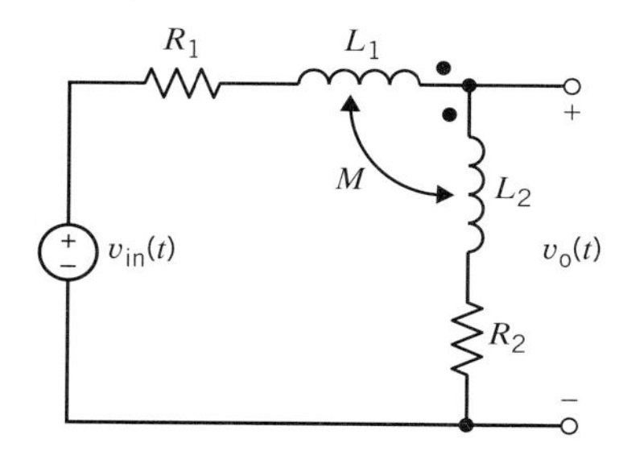

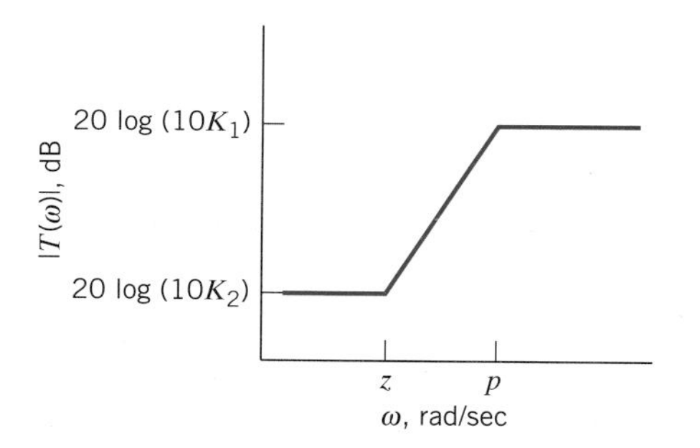

Figure P 13.4-11

P 13.4-12 The input to the circuit shown in Figure P 13.4-12 is the source voltage $v_{in}(t)$, and the response is the voltage across R_3, $v_{out}(t)$. The component values are $R_1 = 10\ \text{k}\Omega$, $C_1 = 0.025\ \mu\text{F}$, and $C_2 = 0.05\ \mu\text{F}$. Sketch the asymptotic magnitude Bode plot for the network function.

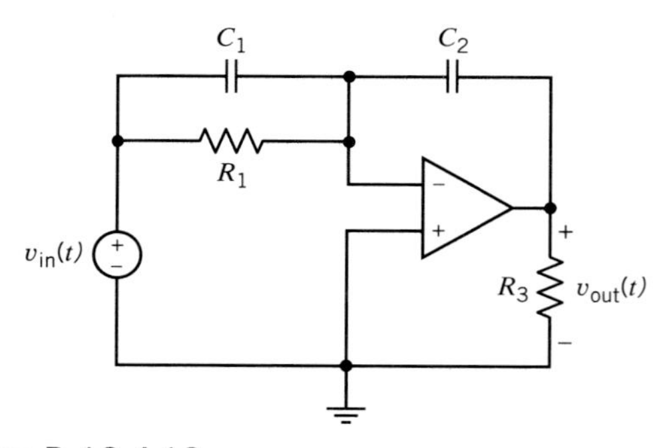

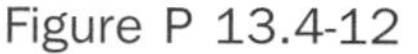
Figure P 13.4-12

P 13.4-13 Design a circuit that has the asymptotic magnitude Bode plot shown in Figure P 13.4-13.

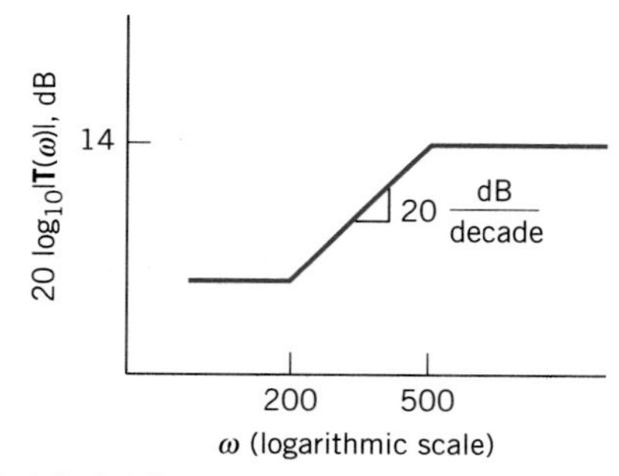

Figure P 13.4-13

P 13.4-14 Design a circuit that has the asymptotic magnitude Bode plot shown in Figure P 13.4-14.

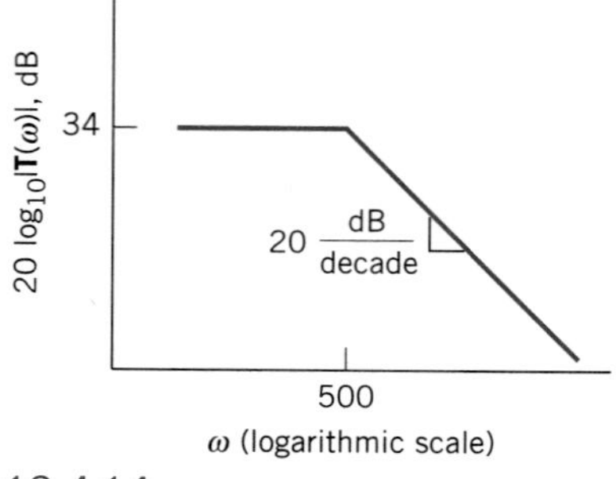

Figure P 13.4-14

P 13.4-15 Design a circuit that has the asymptotic magnitude Bode plot shown in Figure P 13.4-15.

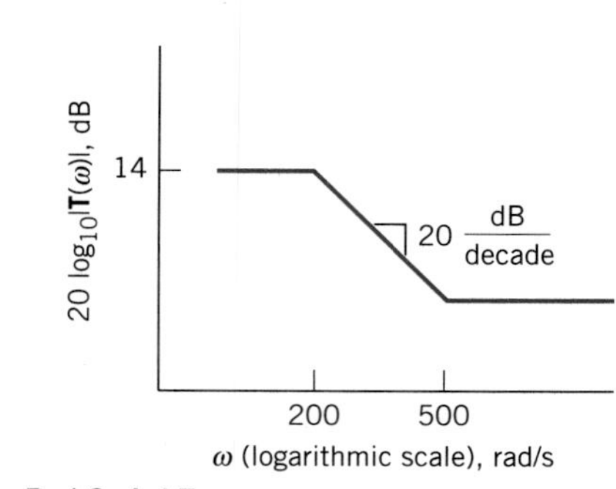

Figure P 13.4-15

P 13.4-16 Design a circuit that has the asymptotic magnitude Bode plot shown in Figure P 13.4-16.

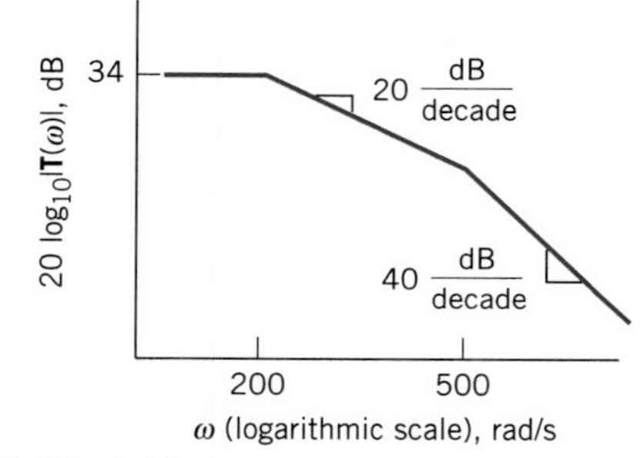

Figure P 13.4-16

P 13.4-17 The cochlear implant is intended for patients with deafness due to malfunction of the sensory cells of the cochlea in the inner ear (Loeb 1985). These devices use a microphone for picking up the sound and a processor for converting to electrical signals, and they transmit these signals to the nervous system. A cochlear implant relies on the fact that many of the auditory nerve fibers remain intact in patients with this form of hearing loss. The overall transmission from microphone to nerve cells is represented by the gain function

$$\mathbf{H}(j\omega) = \frac{10(j\omega/50 + 1)}{(j\omega/2 + 1)(j\omega/20 + 1)(j\omega/80 + 1)}$$

Plot the Bode diagram for $\mathbf{H}(j\omega)$ for $1 \le \omega \le 100$.

P 13.4-18 This low-pass amplifier with a network function $\mathbf{H} = \mathbf{V}_o/\mathbf{V}_s$ uses two identical operational amplifier circuits as shown in Figure P 13.4-18. It is desired to set the cutoff or break frequency at $\omega_c = 1000$ rad/s and the low-frequency gain at $|\mathbf{H}| = 0$ dB. (a) Find the required R_1, R_2, and C. (b) Find the attenuation in decibels for a signal at $\omega = 10{,}000$ rad/s.
Answer: (*a*) $R_1 = R_2 = 1\ \text{k}\Omega$, $C = 1\ \mu\text{F}$

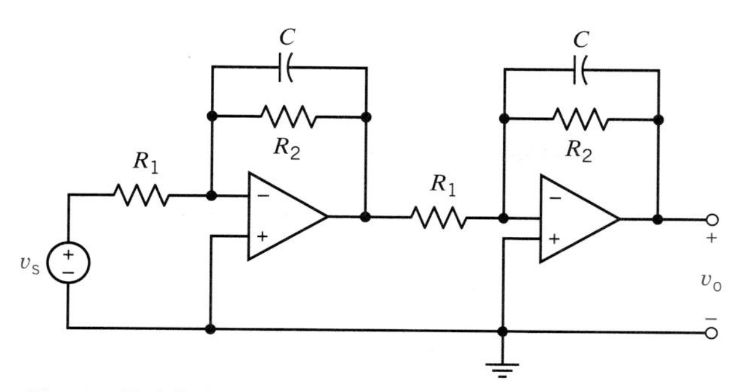

Figure P 13.4-18

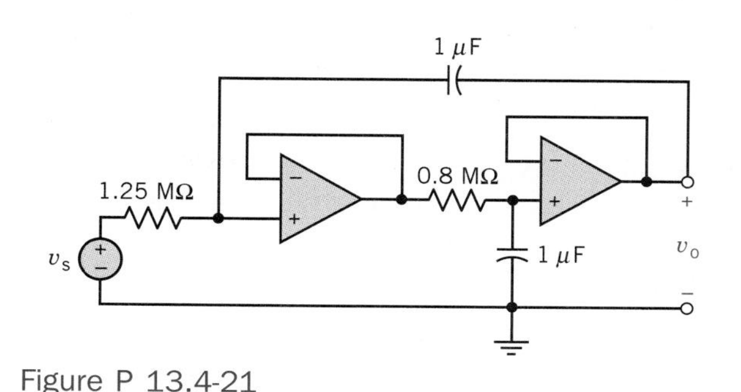

Figure P 13.4-21

P 13.4-19 An operational amplifier circuit is shown in Figure P 13.4-19, where $R_2 = 5$ kΩ and $C = 0.02$ μF. (a) Find the expression for the network function $\mathbf{H} = \mathbf{V}_o/\mathbf{V}_s$ and sketch the asymptotic Bode diagram. (b) What is the gain of the circuit, $|\mathbf{V}_o/\mathbf{V}_s|$, for $\omega = 0$? (c) At what frequency does $|\mathbf{V}_o/\mathbf{V}_s|$ fall to $1/\sqrt{2}$ of its low-frequency value?
Answer: (*b*) 20 dB (*c*) 10,000 rad/s.

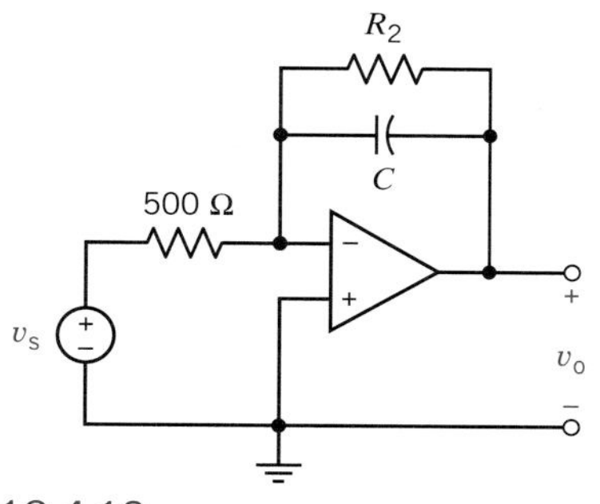

Figure P 13.4-19

P 13.4-20 A unity gain, low-pass filter is obtained from the operational amplifier circuit shown in Figure P 13.4-20. Determine the network function $\mathbf{H} = \mathbf{V}_o/\mathbf{V}_s$ and sketch the Bode diagram.

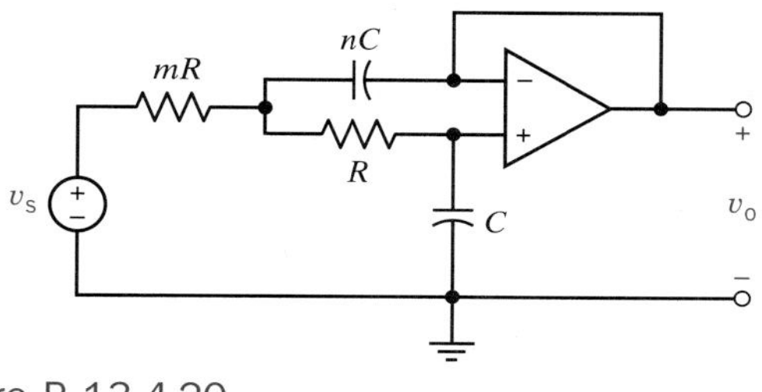

Figure P 13.4-20

P 13.4-21 Determine the network function $\mathbf{H}(\omega)$ for the op amp circuit shown in Figure P 13.4-21 and plot the Bode diagram. Assume ideal op amps.

Section 13.5 Resonant Circuits

P 13.5-1 For a parallel *RLC* circuit with $R = 10$ kΩ, $L = 1/120$ H, and $C = 1/30$ μF, find ω_0, Q, ω_1, ω_2, and the bandwidth *BW*.
Answer: $\omega_0 = 60$ krad/s, $Q = 20$, $\omega_1 = 58.519$ krad/s, $\omega_2 = 61.519$ krad/s, and $BW = 3$ krad/s

P 13.5-2 A parallel resonant RLC circuit is driven by a current source $i_s = 20 \cos \omega t$ mA and shows a maximum response of 8 V at $\omega = 1000$ rad/s and 4 V at 897.6 rad/s. Find *R*, *L* and *C*.
Answer: $R = 400$ Ω, $L = 50$ mH, and $C = 20$ μF

P 13.5-3 A series resonant RLC circuit has $L = 10$ mH, $C = 0.01$ μF, and $R = 100$ Ω. Determine ω_0, Q, BW.
Answer: $\omega_0 = 10^5$, $Q = 10$, and $BW = 10^4$

P 13.5-4 A quartz crystal exhibits the property that when mechanical stress is applied across its faces, a potential difference develops across opposite faces. When an alternating voltage is applied, mechanical vibrations occur and electromechanical resonance is exhibited. A crystal can be represented by a series *RLC* circuit. A specific crystal has a model with $L = 1$ mH, $C = 10$ μF, and $R = 1$ Ω. Find ω_0, Q, and the bandwidth.
Answer: $\omega_0 = 10^4$ rad/s, $Q = 10$, $BW = 10^3$ rad/s

P 13.5-5 Design a parallel resonant circuit to have $\omega_o = 2500$ rad/s, $\mathbf{Z}(\omega_o) = 100$ Ω, and $BW = 500$ rad/s.
Answer: $R = 100$ Ω, $L = 8$ mH,and $C = 20$ μF

P 13.5-6 Design a series resonant circuit to have $\omega_o = 2500$ rad/s, $\mathbf{Y}(\omega_o) = 1/100$ Ω, and $BW = 500$ rad/s.
Answer: $R = 100$ Ω, $L = 0.2$ mH, and $C = 0.8$ μF

P 13.5-7 The circuit shown in Figure P 13.5-7 represents a capacitor, coil, and resistor in parallel. Calculate the resonant frequency, bandwidth, and Q for the circuit.

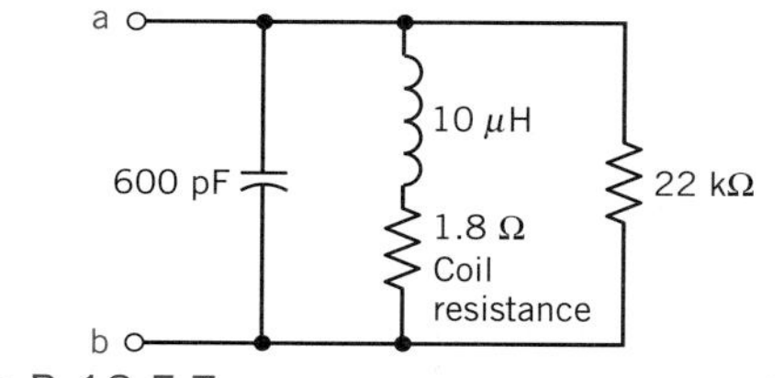

Figure P 13.5-7

P 13.5-8 Consider the simple model of an electric power system as shown in Figure P 13.5-8. The inductance, $L = 0.25$ H, represents the power line and transformer. The customer's load is $R_L = 100\ \Omega$, and the customer adds $C = 25\ \mu$F to increase the magnitude of $\mathbf{V}_0$. The source is $v_s = 1000 \cos 400t$ V, and it is desired that $|\mathbf{V}_o|$ also be 1000 V. (a) Find $|\mathbf{V}_o|$ for $R_L = 100\ \Omega$. (b) When the customer leaves for the night, he turns off much of his load, making $R_L = 1\ \text{k}\Omega$, at which point sparks and smoke begin to appear in the equipment still connected to the power line. The customer calls you in as a consultant. Why did the sparks appear when $R_L = 1\ \text{k}\Omega$?

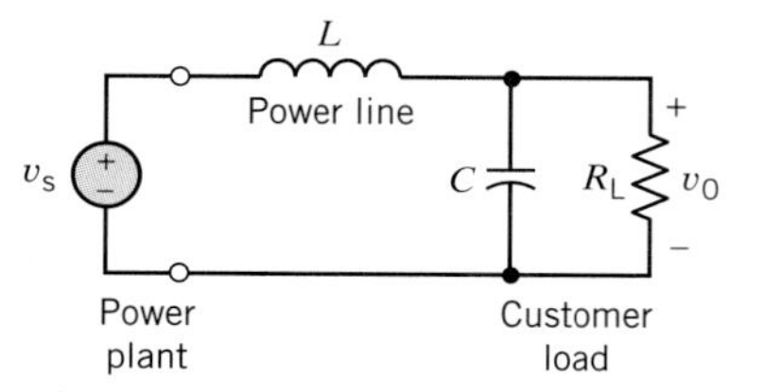

Figure P 13.5-8 Model of an electric power system.

P 13.5-9 Consider the circuit in Figure P 13.5-9. $R_1 = R_2 = 1\ \Omega$. Select C and L to obtain a resonant frequency of $\omega_0 = 100$ rad/s.

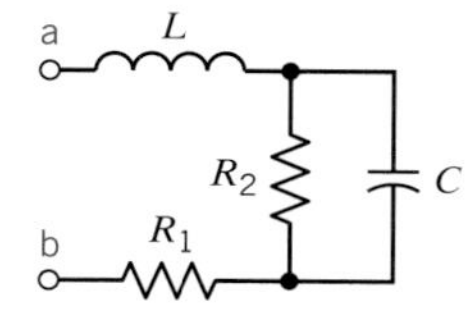

Figure P 13.5-9

P 13.5-10 For the circuit shown in Figure P 13.5-10, (a) derive an expression for the magnitude response $|\mathbf{Z}_{in}|$ versus ω, (b) sketch $|\mathbf{Z}_{in}|$ versus ω, and (c) find $|\mathbf{Z}_{in}|$ at $\omega = 1/\sqrt{LC}$.

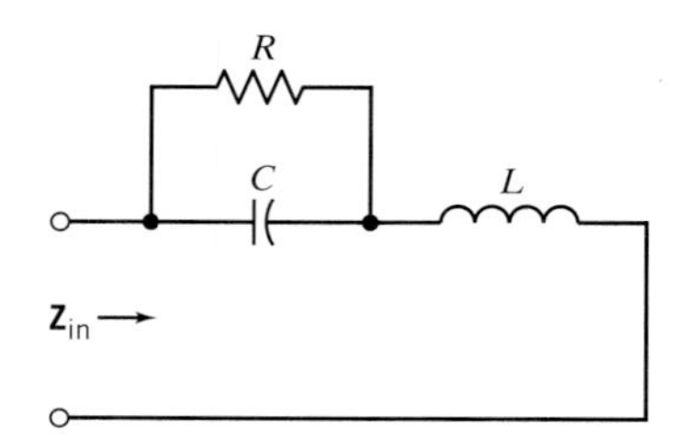

Figure P 13.5-10

P 13.5-11 The circuit shown in Figure P 13.5-11 shows an experimental setup that could be used to measure the parameters k, Q, and ω_0 of this series resonant circuit. These parameters can be determined from a magnitude frequency response plot for $\mathbf{Y} = \mathbf{I}/\mathbf{V}$. It is more convenient to measure node voltages than currents, so the node voltages $\mathbf{V}$ and $\mathbf{V}_2$ have been measured. Express $|\mathbf{Y}|$ as a function of $\mathbf{V}$ and $\mathbf{V}_2$.

Hint: Let $\mathbf{V} = \text{A}$ and $\mathbf{V}_2 = \text{B}\angle\theta$. Then $\mathbf{I} = \dfrac{(A - B\cos\theta) - jB\sin\theta}{R}$

Answer: $|\mathbf{Y}| = \dfrac{\sqrt{(A - B\cos\theta)^2 + (B\sin\theta)^2}}{AR}$

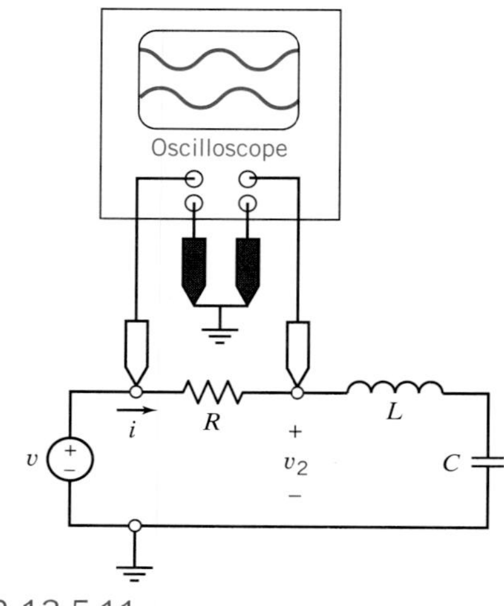

Figure P 13.5-11

PSPICE PROBLEMS

SP 13-1 Use PSpice to verify the answer of Exercise 13.3-1.

SP 13-2 Use PSpice to verify the answer of Exercise 13.3-2.

SP 13-3 Use PSpice to verify the answer of Example 13.4-4.

SP 13-4 Use PSpice to verify the answer of Example 13.4-5.

SP 13-5 Use PSpice to verify the answer of Example 13.5-2.

SP 13-6 Use PSpice to simulate the circuit shown in Figure P 13.3-11.

SP 13-7 Obtain the Bode diagram for $\mathbf{V}/\mathbf{I}$ of the circuit shown in Figure SP 13.7. Determine the bandwidth of the circuit.

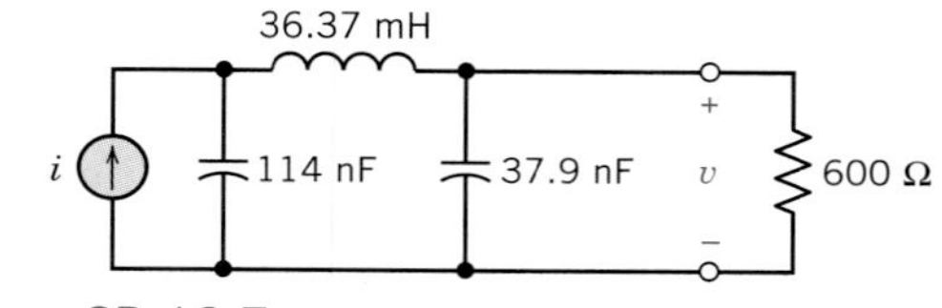

Figure SP 13.7

SP 13-8 The model of an amplifier for a phonograph stereo is shown in Figure SP 13.8. Plot the dB magnitude portion of the Bode diagram for $\mathbf{V}_o/\mathbf{V}_s$ for 20 Hz to 20 kHz.

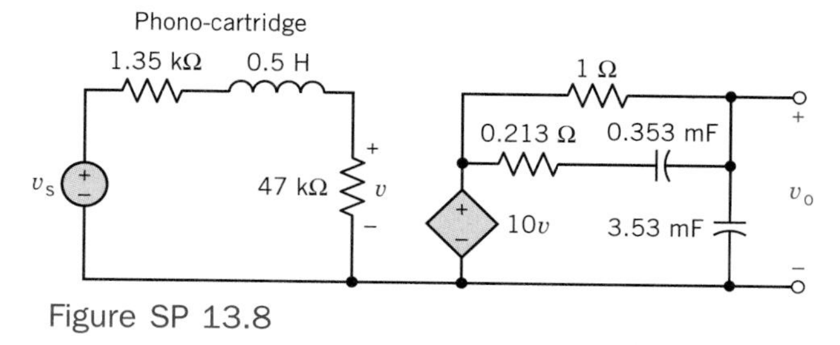

Figure SP 13.8

SP 13-9 A circuit with a transformer is shown in Figure SP 13.9. The input current i is a 1-mA sinusoidal signal. Determine and plot the magnitude of the output signal, v, for a frequency range between 0.94 MHz and 1.06 MHz. This circuit is designed to pass a signal between 0.98 MHz and 1.02 MHz and to severely reject signals at (1.00 ± 0.05) MHz. Discuss the performance of the circuit.

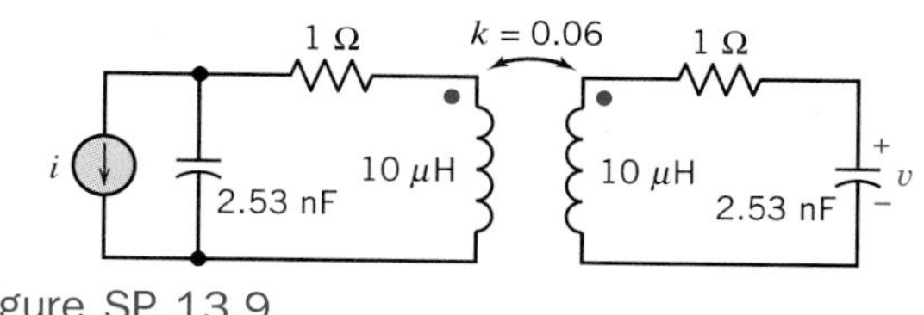

Figure SP 13.9

VERIFICATION PROBLEMS

VP 13-1 Circuit analysis contained in a lab report indicates that the network function of a circuit is

$$\mathbf{T}(\omega) = \frac{1 + j\dfrac{\omega}{630}}{10\left(1 + j\dfrac{\omega}{6300}\right)}$$

This lab report contains the following frequency response data from measurements made on the circuit. Do these data seem reasonable?

ω, rad/s	200	400	795	1585	3162
$\lvert\mathbf{T}(\omega)\rvert$	0.105	0.12	0.16	0.26	0.460

ω, rad/s	6310	12600	25100	50000	100000
$\lvert\mathbf{T}(\omega)\rvert$	0.71	1.0	1.0	1.0	1.0

VP 13-2 A parallel resonant circuit (see Figure 13.5-2) has $Q = 70$ and a resonant frequency $\omega_0 = 10{,}000$ rad/s. A report states that the bandwidth of this circuit is 71.43 rad/s. Verify this result.

VP 13-3 A series resonant circuit (see Figure 13.5-4) has $L = 1$ mH, $C = 10\ \mu$F, and $R = 0.5\ \Omega$. A software program report states that the resonant frequency is $f_0 = 1.59$ kHz and the bandwidth is $BW = 79.6$ Hz. Are these results correct?

VP 13-4 An old lab report contains the approximate Bode plot shown in Figure VP 13.4 and concludes that the network function is

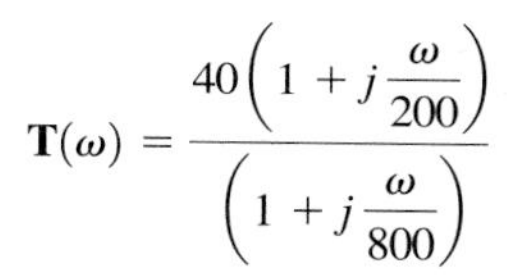

$$\mathbf{T}(\omega) = \frac{40\left(1 + j\dfrac{\omega}{200}\right)}{\left(1 + j\dfrac{\omega}{800}\right)}$$

Do you agree?

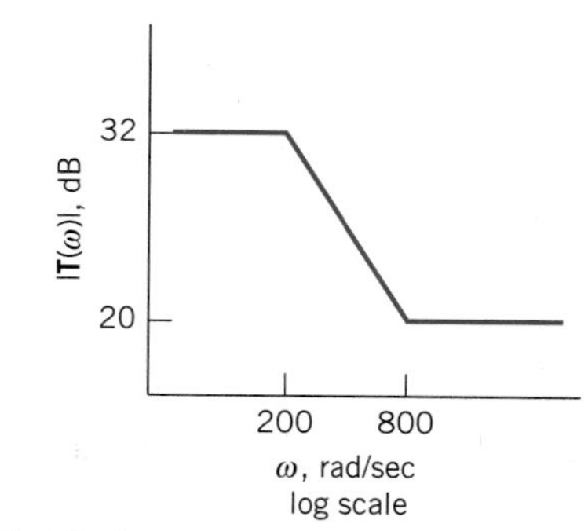

Figure VP 13.4

DESIGN PROBLEMS

DP 13-1 Design a circuit that has a low-frequency gain of 2, and a high-frequency gain of 5, and makes the transition of $H = 2$ to $H = 5$ between the frequencies of 1 kHz and 10 kHz.

DP 13-2 Determine L and C for the circuit of Figure DP 13.2 in order to obtain a low-pass filter with a gain of −3 dB at 100 kHz.

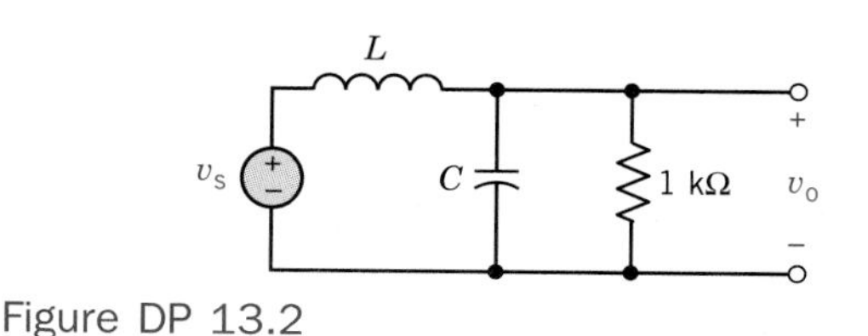

Figure DP 13.2

DP 13-3 British Rail has constructed an instrumented rail car that can be pulled over its tracks at speeds up to 180 km/hr and will measure the track-grade geometry. Using such a rail car, British Rail can monitor and track gradual degradation of the rail grade, especially the banking of curves, and permit preventive maintenance to be scheduled as needed well in advance of track-grade failure.

The instrumented rail car has numerous sensors, such as angular-rate sensors (devices that output a signal proportional to rate of rotation) and accelerometers (devices that output a signal proportional to acceleration), whose signals are filtered and combined in a fashion to create a composite sensor called a compensated accelerometer (Lewis 1988). A component of this composite sensor signal is obtained by integrating and high-pass filtering an accelerometer signal. A first-order low-pass filter will approximate an integrator at frequencies well above the break frequency. This can be seen by computing the phase shift of the filter-transfer function at various frequencies. At sufficiently high frequencies, the phase shift will approach 90°, the phase characteristic of an integrator.

A circuit has been proposed to filter the accelerometer signal, as shown in Figure DP 13.3. The circuit is comprised of three sections, labeled A, B, and C. For each section, find an expression for and name the function performed by that section. Then find an expression for the gain function of the entire circuit, $\mathbf{V}_o/\mathbf{V}_s$. For the component values, evaluate the magnitude and phase of the circuit response at 0.01, 0.02, 0.05, 0.1, 0.2, 0.5, 1.0, 2.0, 5.0, and 10.0 Hz. Draw a Bode diagram. At what frequency is the phase response approximately equal to 0°? What is the significance of this frequency?

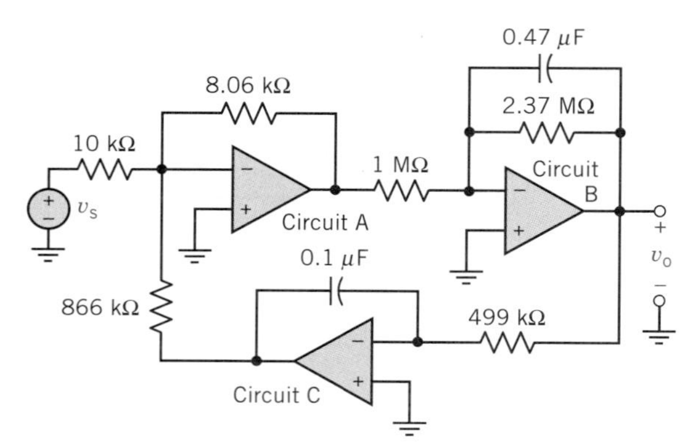

Figure DP 13.3

DP 13-4 Design a circuit that has the network function

$$\mathbf{H}(\omega) = 10\frac{j\omega}{\left(1 + j\dfrac{\omega}{200}\right)\left(1 + j\dfrac{\omega}{500}\right)}$$

Hint: Use two circuits from Table 13.4-2. Connect the circuits in cascade. That means that the output of one circuit is used as the input to the next circuit. $\mathbf{H}(\omega)$ will be the product of the network functions of the two circuits from Table 13.4-2.

DP 13-5 Strain-sensing instruments can be used to measure orientation and magnitude of strains running in more than one direction. The search for a way to predict earthquakes focuses on identifying precursors, or changes, in the ground that reliably warm of an impending event. Because so few earthquakes have occurred precisely at instrumented locations, it has been a slow and frustrating quest. Laboratory studies show that before rock actually ruptures—precipitating an earthquake—its rate of internal strain increases. The material starts to fail before it actually breaks. This prelude to outright fracture is called "tertiary creep" (Brown 1989).

The frequency of strain signals varies from 0.1 to 100 rad/s. A circuit called a band-pass filter is used to pass these frequencies. The network function of the band-pass filter is

$$\mathbf{H}(\omega) = \frac{Kj\omega}{\left(1 + j\dfrac{\omega}{\omega_1}\right)\left(1 + j\dfrac{\omega}{\omega_2}\right)}$$

Specify ω_1, ω_2, and K so that

1. The gain is at least 17 dB over the range 0.1 to 100 rad/s.
2. The gain is less than 17 dB outside the range 0.1 to 100 rad/s.
3. The maximum gain is 20 dB.

DP 13-6 Is it possible to design this circuit shown in Figure DP 13.6 to have a phase shift of –45 degrees and a gain of 2 both at a frequency of 1000 radians/second using a 0.1 microfarad capacitor and resistors from the range 1 k ohm to 200 k ohm?

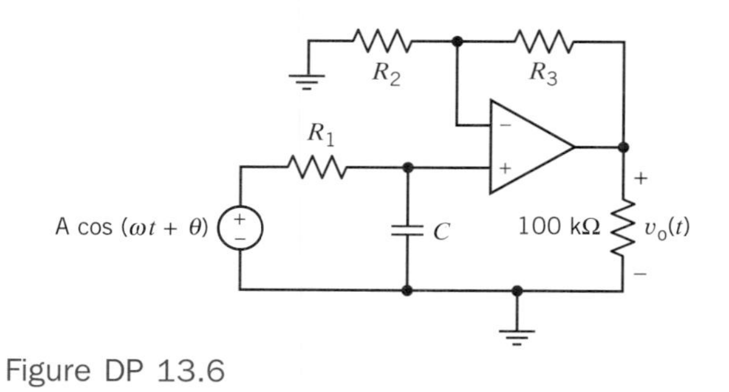

Figure DP 13.6

DP 13-7 Design the circuit shown in Figure DP 13.7*a* to have the asymptotic Bode plot shown in Figure DP 13.7*b*.

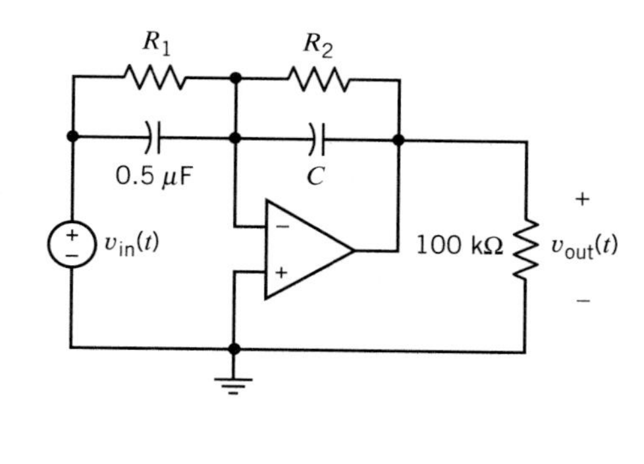

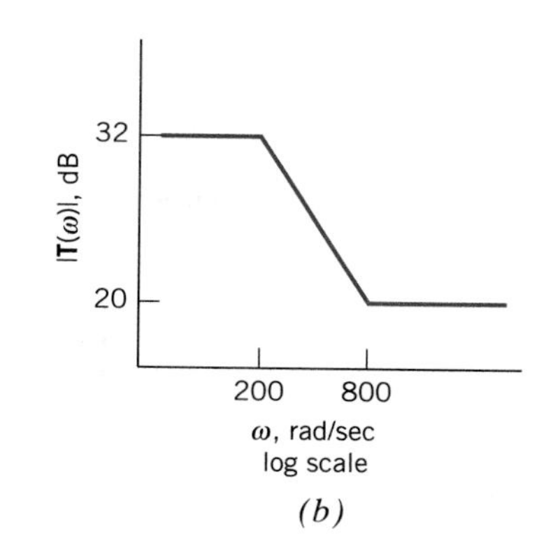

(a)

(b)

Figure DP 13.7

DP 13-8 For the circuit of Figure DP 13.8, select R_1 and R_2 so that the gain at high frequencies is 10 and the phase shift is 195° at $\omega = 1000$ rad/s. Determine the gain at $\omega = 10$ rad/s.

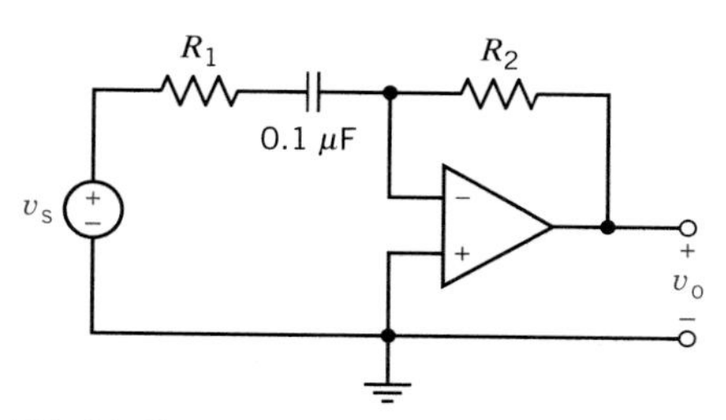

Figure DP 13.8

CHAPTER 14

The Laplace Transform

Preview

In this chapter we introduce a very powerful tool for the analysis of circuits. The Laplace transform enables the circuit analyst to convert the set of differential equations describing a circuit to the complex frequency domain, where they become a set of linear algebraic equations. Then, using straightforward algebraic manipulation, one may solve for the variables of interest. Finally, one uses the inverse transform to go back to the time domain and express the desired response in terms of time. This is a powerful tool, indeed!

We also introduce the concept of impedance in the complex frequency domain. Thus, we may again use the impedance method to analyze circuits using techniques such as Thévenin's theorem and source transformations as described in earlier chapters.

14.1 Design Challenge

SPACE SHUTTLE CARGO DOOR

The U.S. space shuttle docked with Russia's Mir space station several times. The electromagnet for opening a cargo door on the NASA space shuttle requires 0.1 A before activating. The electromagnetic coil is represented by L as shown in Figure 14.1-1. The activating current is designated $i_1(t)$. The time period required for i_1 to reach 0.1 A is specified as less than 3 seconds. Select a suitable value of L.

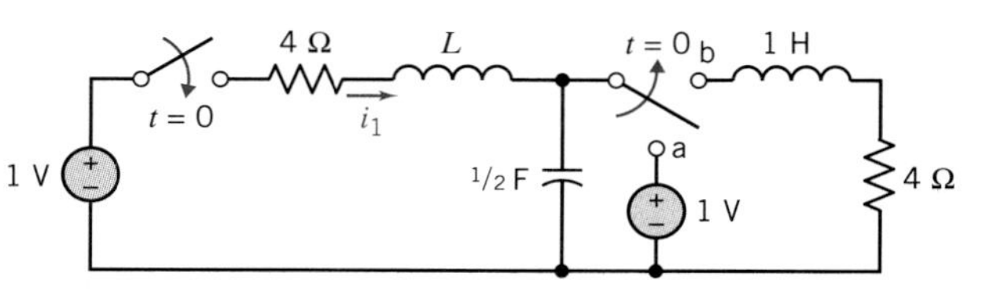

Figure 14.1-1
The control circuit for a cargo door on the NASA space shuttle.

Describe the Situation and the Assumptions

1. The two switches are thrown at $t = 0$ and the movement of the second switch from terminal a to terminal b occurs instantaneously.
2. The switches prior to $t = 0$ were in position for a long time.

State the Goal

Determine a value of L so that the time period for the current $i_1(t)$ to attain a value of 0.1 A is less than 3 seconds.

The circuit shown in Figure 14.1-1 is a third-order circuit because it contains three energy storage devices: a capacitor and two inductors. We will need to calculate the transient response of this third-order circuit in order to design the circuit. In this chapter we will see that the Laplace Transform can be used to calculate this transient response. We will return to this design problem at the end of the chapter.

14.2 COMMUNICATIONS AND AUTOMATION

The use of electric circuits for communications systems has grown over the past century. As society's ability to categorize, store, and transmit data increased, the concept of an information society became prevalent. An *information society* is one in which there is an abundance (some say an excess) in quantity and quality of information with all the necessary facilities for its distribution. Electrical communication systems facilitate the quick, efficient distribution and conversion of this information at a price that almost anyone can afford.

The balance of employment and products in the United States has been shifting toward information-related activities for more than 40 years. In 1988, according to some observers, about one-half of the U.S. worker population was employed in one form or another of information work, in contrast to 3 percent in agriculture and 22 percent in manufacturing.

Researchers have defined an *information ratio,* which is the percentage of household expenditures for various kinds of information-related activities, such as the time spent on reading, television, radio, and the like (Dordick 1986). The information ratio for selected countries is shown in Figure 14.2-1. Note how all nations are moving up the trend line as they increase their information ratio and their per capita income.

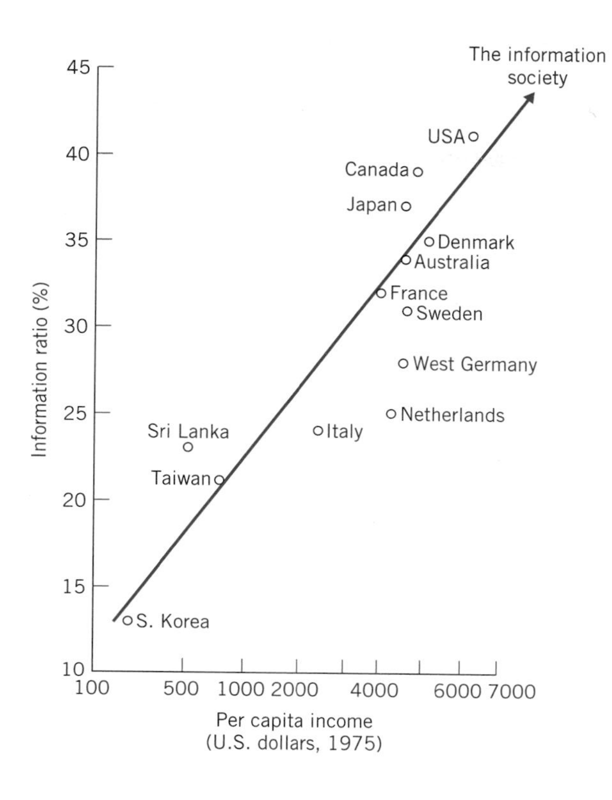

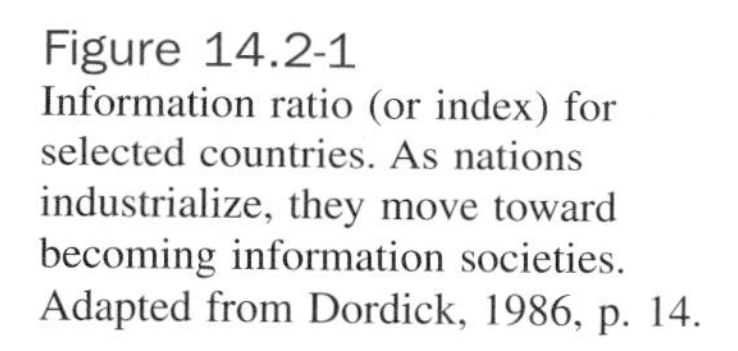

Figure 14.2-1
Information ratio (or index) for selected countries. As nations industrialize, they move toward becoming information societies. Adapted from Dordick, 1986, p. 14.

The information technologies, communications, and computers have reached the office, home, and factory. Industrialization is dependent on the use of information to control the production process in the factory. In 1908 Henry Ford developed his assembly line, which was to turn out millions of Model T's (Dorf 1974). The term *automation* first became popular in the automobile industry. **Automation** is the automatic operation and control of a process, device, or system. For example, a factory may use an automatically controlled machine to produce a product within specified tolerances.

A large impetus to the theory and practice of automation, also often called **automatic control,** occurred during World War II when it became necessary to design and construct automatic airplane pilots, gun control devices, and positioning systems for radar antennas. These new techniques moved to the factory after 1950.

It has been to the advantage of the producer to automate in order to reduce costs and increase quality. Furthermore, many conditions require remotely controlled devices. For example, the remotely controlled device shown in Figure 14.2-2 is used to enter and to move sensors about radiation hazard areas in nuclear power plants.

Figure 14.2-2 The Surveyor robot assumes a role in nuclear power plants where limitations or hazards exist for humans. The Surveyor can climb steps and carry sensors into radiation areas. Surveyor is an untethered, remote-controlled surveillance system. Courtesy of Electric Power Research Institute.

With the advent of the requirement to control complex devices such as the Surveyor vehicle or a robot, it became necessary to describe the characteristics and behavior of complex communication and control circuits and systems as they experience myriad excitation inputs.

In our preceding studies, we considered the analysis of circuits that experience one or more switch changes and have constant sources. Later, we considered the analysis of first- and second-order circuits described by differential equations. Then we considered the steady-state response of circuits with sinusoidal sources. In this chapter we consider a method of analysis that is particularly useful for the analysis of circuits incorporating independent sources that take on many forms, which may include a sinusoid, and exponential, or a step function.

14.3 LAPLACE TRANSFORM

As we have seen in earlier chapters, it is useful to *transform* the equations describing a circuit from the time domain into the frequency domain, then perform an analysis, and finally transform the problem's solution back to the time domain. You will recall that in Chapter 10 we defined the phasor as a mathematical transformation used to simplify the solution of the circuit response to a sinusoidal source at steady-state. The phasor transformation resulted in the problem being transformed to one of algebraic manipulation of complex numbers.

A **transform** is a change in the mathematical description of a physical variable to facilitate computation.

The transform method is summarized in Figure 14.3-1.

We are interested in using the transform that will take the differential equations describing a circuit into the complex frequency domain, where we may readily perform the required algebraic manipulations to determine the desired response. Then we may use an inverse transform to obtain the response in the time domain.

Pierre-Simon Laplace, who is shown in Figure 14.3-2, is credited with a transform that bears his name. The **Laplace transform** is defined as

$$\mathcal{L}[f(t)] = \int_0^\infty f(t)e^{-st}\,dt \tag{14.3-1}$$

and

$$\mathcal{L}[f(t)] = F(s) \tag{14.3-2}$$

so $\mathcal{L}$ implies the Laplace transform. Here the complex frequency is $s = \sigma + j\omega$. We only treat the one-sided (unilateral) transform for $t \geq 0$.

The notation implies that once the integral has been evaluated, $f(t)$, which is a time domain function, is transformed to $F(s)$, which is a frequency domain function.

We should stop and ask under what conditions the integral of Eq. 14.3-1 converges to a finite value. It can be shown that the integral converges when

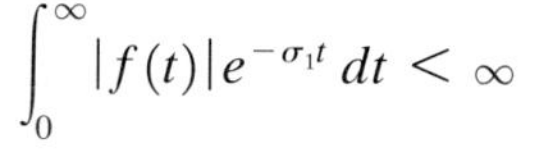

$$\int_0^\infty |f(t)|e^{-\sigma_1 t}\,dt < \infty$$

Figure 14.3-2
Pierre-Simon Laplace (1749–1827) is credited with the transform that bears his name. Courtesy of Burndy Library.

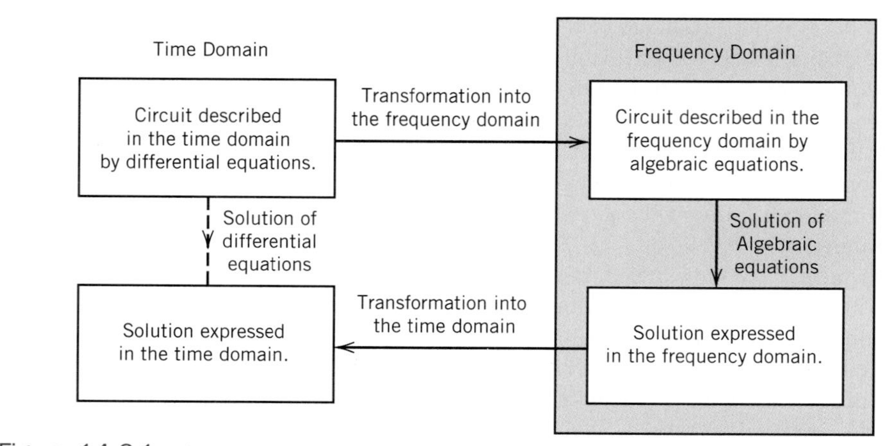

Figure 14.3-1 The transform method.

Table 14.3-1 **Important Laplace Transform Pairs**

$f(t), t \geq 0$	$F(s) = \mathcal{L}\{f(t)u(t)\}$
Step function, $u(t)$	$\frac{1}{s}$
e^{-at}	$\frac{1}{s+a}$
$\sin \omega t$	$\frac{\omega}{s^2+\omega^2}$
$\cos \omega t$	$\frac{s}{s^2+\omega^2}$
$e^{-at}f(t)$	$F(s+a)$
t^n	$\frac{n!}{s^{n+1}}$
$f^{(k)}(t) = \frac{d^kf(t)}{dt^k}$	$s^kF(s) - s^{k-1}f(0^+) - s^{k-2}f'(0^+) - \cdots - f^{(k-1)}(0^+)$

for some real positive σ_1. If the magnitude of $f(t)$ is $|f(t)| < Me^{\alpha t}$ for all positive t, the integral will converge for $\sigma_1 > \alpha$. The region of convergence is therefore given by $\infty > \sigma_1 > \alpha$, and σ_1 is known as the abscissa of absolute convergence. Functions of time, $f(t)$, that are physically possible always have a Laplace transform. Also, we will assume that $f(t) = 0$ for $t < 0$. We assume that the initial conditions for the circuit account for the circuit's behavior prior to $t = 0$. We will consider $f(t)$ to be continuous after $t = 0$. However, we can permit a discontinuity at $t = 0$ and assume the transform occurs for $t > 0$. Thus, we permit a discontinuity between $t = 0^-$ and $t = 0^+$.

If we have a transformation $\mathcal{L}[f(t)]$, then we must have an inverse transformation $\mathcal{L}^{-1}[F(s)] = f(t)$. If we develop a series of transformations using Eq. 14.3-1, we may obtain a table of **transform pairs** as listed in Table 14.3-1.

This table can be constructed by using Eq. 14.3-1. Again, note that we only consider the Laplace transform for signals which are zero for $t < 0$. The product $f(t)u(t) = 0$ for $t < 0$. For example, consider $f(t) = e^{-at}u(t)$, where $a > 0$ and $u(t)$ is the unit step function. Then we have

$$\mathcal{L}[e^{-at}u(t)] = F(s) = \int_0^\infty e^{-at}e^{-st}\,dt$$

$$= \left.\frac{-e^{-(s+a)t}}{s+a}\right|_0^\infty = \frac{1}{s+a} \tag{14.3-3}$$

since the value at the upper limit is zero because the convergence condition is satisfied by $\sigma > -a$.

Now let us find the transform of the familiar unit step function $u(t)$. We have

$$\mathcal{L}[u(t)] = F(s) = \int_0^\infty e^{-st}\,dt = \frac{1}{s}$$

where convergence is satisfied by $\sigma > 0$.

Let us obtain the transform of the first derivative of $f(t)$, denoted as $f'(t)$. We have

$$\mathcal{L}[f'(t)] = \mathcal{L}\left[\frac{df}{dt}\right] = \int_0^\infty f'(t)e^{-st}\,dt$$

Integrating by parts, we have

$$\mathcal{L}[f'(t)] = s\int_0^\infty f(t)e^{-st}\,dt + f(t)e^{-st}\Big|_0^\infty$$

Again, we assume that the integrated term vanishes at the upper limit, so that

$$\mathcal{L}[f'(t)] = sF(s) - f(0) \tag{14.3-4}$$

Thus, the Laplace transform of the derivative of a function is s times the Laplace transform of the function minus the initial condition.

For example, if $f(t) = e^{-t}$, we have

$$\mathcal{L}[f'(t)] = s\left(\frac{1}{s+1}\right) - 1$$

where $1/(s+1)$ is the Laplace transform of $f(t) = e^{-t}, t \geq 0$.

Another important property of the Laplace transform is linearity. That is, if $F_1(s)$ and $F_2(s)$ are the Laplace transforms of the time functions $f_1(t)$ and $f_2(t)$, respectively, then

$$\mathcal{L}[a_1 f_1(t) + a_2 f_2(t)] = a_1 F_1(s) + a_2 F_2(s) \tag{14.3-5}$$

for arbitrary constants a_1 and a_2.

Uniqueness is a fundamental property of the Laplace transform. Thus, if two time functions $f_1(t)$ and $f_2(t)$ have the same transform $F(s)$, then $f_1 = f_2$. Therefore, for $f(t)$ a unique function $F(s)$ exists. Conversely, given a Laplace transform $F(s)$, there is a unique time function $f(t)$. This property is written as

$$f(t) = \mathcal{L}^{-1}[F(s)]$$

meaning that $f(t)$ is the **inverse Laplace transform** of $F(s)$ and exists for $t \geq 0$.

The inverse Laplace transform is defined by the complex inversion integral

$$f(t) = \frac{1}{2\pi j}\int_{\alpha - j\infty}^{\alpha + j\infty} F(s)e^{st}ds \tag{14.3-6}$$

The uniqueness of the Laplace transform enables us to avoid this complex integration. Instead, we will build Laplace transform tables, such as Tables 14.3-1 and 14.4-1, by calculating the Laplace transform of several key functions. When we want to find the inverse Laplace transform, we will look in these tables. For example, suppose we want to find the inverse Laplace transform of

$$F(s) = \frac{2}{s}$$

Unfortunately, $2/s$ is not in Table 14.3-1 or 14.4-1. Linearity tells us that

$$\mathcal{L}^{-1}\left\{\frac{2}{s}\right\} = 2\,\mathcal{L}^{-1}\left\{\frac{1}{s}\right\}$$

Furthermore, $1/s$ is in Table 14.3-1. Thus, we find

$$f(t) = 2\,u(t)$$

Notice that we needed both a Laplace transform pair, $\mathcal{L}[u(t)] = 1/s$, and a Laplace transform property, linearity. The success of this method depends on having a supply of Laplace transform pairs (Tables 14.3-1 and 14.4-1) and Laplace transform properties (Table 14.3-2).

Table 14.3-2 **Laplace Transform Properties**

Property	Relationship
Linearity	$\mathcal{L}[a_1 f_1(t) + a_2 f_2(t)] = a_1 F_2(s) + a_2 F_2(s)$
Time delay	$\mathcal{L}[f(t - \tau)u(t - \tau)] = e^{-s\tau}F(s)$
Impulse	$\mathcal{L}[\delta(t)] = 1$
Frequency shift	$\mathcal{L}[e^{-at}f(t)] = F(s + a)$
Product of time and a function	$\mathcal{L}[tf(t)] = \frac{-dF(s)}{ds}$
	where $F(s) = \mathcal{L}[f(t)]$

Example 14.3-1

Find the Laplace transform of $\sin \omega t$.

Solution

Since

$$\sin \omega t = \frac{1}{2j}(e^{j\omega t} - e^{-j\omega t})$$

and since we know $\mathcal{L}(e^{at}) = 1/(s - a)$, we then have

$$\begin{aligned}\mathcal{L}(\sin \omega t) &= \frac{1}{2j}\left(\frac{1}{s - j\omega} - \frac{1}{s + j\omega}\right)\\ &= \frac{(s + j\omega) - (s - j\omega)}{2j(s - j\omega)(s + j\omega)}\\ &= \frac{\omega}{s^2 + \omega^2}\end{aligned}$$

Exercise 14.3-1 Find the Laplace transform of $f(t) = \cos \omega t$.

Answer: $F(s) = \dfrac{s}{s^2 + \omega^2}$

Exercise 14.3-2 Using the linearity property, find the Laplace transform of $f(t) = e^{-2t} + \sin t$.

Answer: $F(s) = \dfrac{s^2 + s + 3}{(s + 2)(s^2 + 1)}$

14.4 IMPULSE FUNCTION AND TIME SHIFT PROPERTY

Let us consider a special function called the **impulse function,** denoted as $\delta(t)$. First, we consider the pulse centered at $t = 0$ as shown in Figure 14.4-1. We have

$$\begin{aligned} f(t) &= \frac{1}{a} \qquad -\frac{a}{2} < t < \frac{a}{2}\\ &= 0 \qquad \text{all other } t \end{aligned} \tag{14.4-1}$$

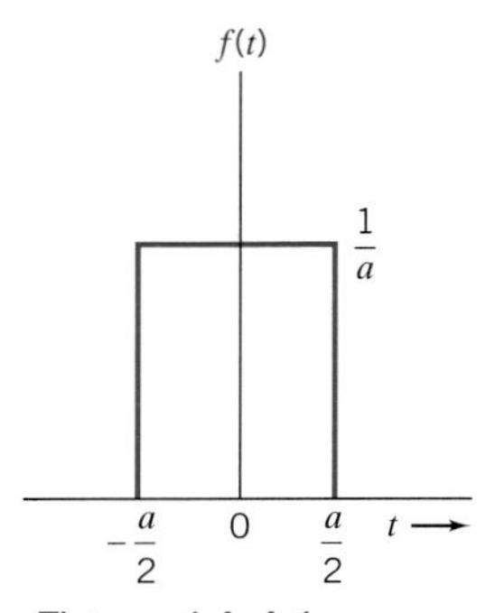

Figure 14.4-1 Pulse of width a and of height $1/a$ centered at $t = 0$.

where a is very small. Note that the area under the pulse remains equal to 1 regardless of the value of a. If a becomes progressively smaller, the area remains equal to 1, but we have a large pulse height with a very narrow width.

An **impulse** is a pulse of infinite amplitude for an infinitesimal time whose area $\int_{-\infty}^{\infty} f(t)\, dt$ is finite.

The **impulse function** is defined as

$$\delta(t) = 0 \quad \text{for } t \neq 0$$

and

$$\int_{-\infty}^{\infty} \delta(t)\, dt = 1 \tag{14.4-2}$$

Therefore, we have $f(t)\delta(t) = f(0)\delta(t)$ since $\delta(t) = 0$ for $t \neq 0$.

Let us determine the Laplace transform of the impulse function. By definition, we have

$$\mathcal{L}[\delta(t)] = \int_{0-}^{\infty} e^{-st}\delta(t)\, dt \tag{14.4-3}$$

where the lower limit is 0^- since we have an infinite discontinuity at $t = 0$. Since $\delta(t) = 0$ for $t \neq 0$, the integral of Eq. 14.4-3 is evaluated between 0^- and 0^+ to obtain

$$\mathcal{L}[\delta(t)] = e^{-st}\Big|_{\text{at } t=0} = 1$$

Now let us consider the transform of a time function shifted τ seconds in time. If we have $f(t)$ as shown in Figure 14.4-2*a* and we wish to shift (delay) it to τ seconds later, we may write it as

$$f(t - \tau)u(t - \tau)$$

where $u(t - \tau)$ is the unit step function, which is zero for $t < \tau$ and 1 for $t > \tau$. The delayed function is shown in Figure 14.4-2*b*.

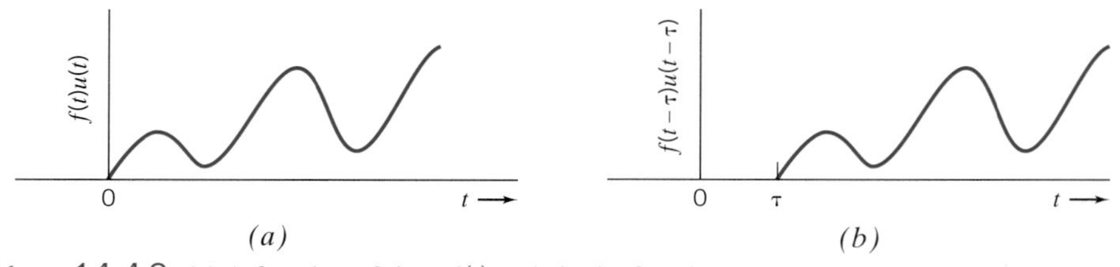

Figure 14.4-2 (*a*) A function of time $f(t)$ and (*b*) the function delayed by τ s.

To obtain the transform of the time-shifted function, we use the definition of the transform to obtain

$$\mathcal{L}[f(t - \tau)u(t - \tau)] = \int_{0}^{\infty} f(t - \tau)u(t - \tau)e^{-st}\, dt$$

$$= \int_{\tau}^{\infty} f(t - \tau)e^{-st}\, dt$$

Now let $t - \tau = x$ to obtain

$$\mathcal{L}[f(t-\tau)u(t-\tau)] = \int_0^{\infty} f(x)e^{-s(\tau+x)}\,dx$$

$$= e^{-s\tau}\int_0^{\infty} f(x)e^{-sx}\,dx$$

$$= e^{-s\tau}F(s) \tag{14.4-4}$$

Thus, for example, if we have a step function of amplitude A delayed by 2 seconds, we have the Laplace transform

$$\mathcal{L}[Au(t-2)] = e^{-2s}\left(\frac{A}{s}\right)$$

since $\mathcal{L}[Au(t)] = A/s$.

Another property can be obtained from the Laplace transform of $e^{-at}f(t)$ as follows:

$$\mathcal{L}[e^{-at}f(t)] = \int_0^{\infty} e^{-at}f(t)e^{-st}\,dt$$

$$= \int_0^{\infty} f(t)e^{-(s+a)}\,dt$$

$$= F(s+a) \tag{14.4-5}$$

This property is called the *frequency shift property.* The frequency shift property and the properties of the impulse, the time delay, and linearity are summarized in Table 14.3-2.

We may use the frequency shift property of Eq. 14.4-5 to obtain additional transform relationships. For example, consider

$$\mathcal{L}(e^{-at}\sin\omega t) = F(s+a)$$

Since

$$\mathcal{L}(\sin\omega t) = \frac{\omega}{s^2+\omega^2}$$

we have

$$\mathcal{L}(e^{-at}\sin\omega t) = \frac{\omega}{(s+a)^2+\omega^2} \tag{14.4-6}$$

Example 14.4-1

Find the Laplace transform of $g(t) = e^{-4t}u(t-3)$, which is shown in Figure 14.4-3.

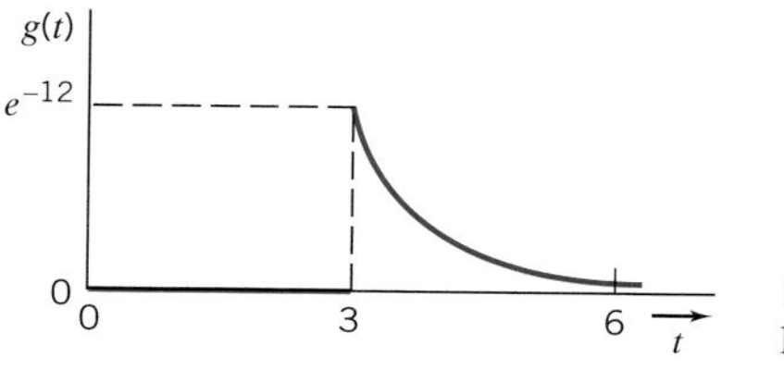

Figure 14.4-3
Delayed function $g(t) = e^{-4t}u(t-3)$.

Solution

First, find the transform of $u(t-3)$. Using the time delay property, we have

$$F(s) = \mathcal{L}[u(t-3)] = e^{-3s}\left(\frac{1}{s}\right)$$

Then, using the frequency shift property, we have

$$G(s) = \mathcal{L}[e^{-4t}u(t-3)] = \mathcal{L}[e^{-4t}f(t)] = F(s+4)$$

Therefore, we have

$$G(s) = e^{-3(s+4)}\frac{1}{s+4} = e^{-12}\frac{e^{-3s}}{s+4}$$

Another property of great utility arises for the product of time t and $f(t)$. Start with the derivative of $F(s)$ with respect to s, where $F(s) = \mathcal{L}[f(t)]$. Then we have

$$\frac{dF(s)}{ds} = \int_0^\infty \frac{d}{ds}[f(t)e^{-st}]dt = \int_0^\infty -tf(t)e^{-st}\,dt$$

Consequently,

$$\mathcal{L}[tf(t)] = -\frac{dF(s)}{ds} \tag{14.4-7}$$

where

$$F(s) = \mathcal{L}[f(t)]$$

This property is also included in Table 14.3-2.

For example, let us find the Laplace transform of $t\,u(t)$. From Table 14.3-2 we have

$$\mathcal{L}[t\,u(t)] = -\frac{dF(s)}{ds} = -\frac{d}{ds}\left(\frac{1}{s}\right)$$

where

$$F(s) = \mathcal{L}[u(t)] = 1/s$$

Completing the derivative,

$$\mathcal{L}[t\,u(t)] = \frac{1}{s^2}$$

These additional transform pairs are summarized in Table 14.4-1.

Table 14.4-1 **Additional Transform Pairs**

$f(t), t \geq 0$	$F(s)$
te^{-at}	$\dfrac{1}{(s+a)^2}$
t^n	$\dfrac{n!}{s^{n+1}}$
$e^{-at}\sin\omega t$	$\dfrac{\omega}{(s+a)^2+\omega^2}$
$e^{-at}\cos\omega t$	$\dfrac{(s+a)}{(s+a)^2+\omega^2}$
$e^{-bt}t^n$	$\dfrac{n!}{(s+b)^{n+1}}$

Exercise 14.4-1 Using the linearity property and the frequency shift property, find $\mathcal{L}[2u(t) + 3e^{-4t}u(t)] = P(s)$.

Answer: $P(s) = \dfrac{2}{s} + \dfrac{3}{s+4}$

Exercise 14.4-2 Using the time shift property, find $F(s) = \mathcal{L}[\sin(t-\tau)u(t-\tau)]$ where $\tau = 2$.

Answer: $F(s) = \dfrac{e^{-2s}}{s^2+1}$

Exercise 14.4-3 Find $\mathcal{L}(te^{-t}) = F(s)$.

Answer: $F(s) = \dfrac{1}{(s+1)^2}$

Exercise 14.4-4 Determine the Laplace transform of $f(t)$ shown in Figure E 14.4-4.

Hint: $f(t) = \left(5 - \dfrac{5}{3}t\right)u(t) - \dfrac{5}{3}\left(t - \dfrac{21}{5}\right)u\left(t - \dfrac{21}{5}\right)$

Answer: $F(s) = \dfrac{5e^{-4.2s} + 15s - 5}{3s^2}$

Exercise 14.4-5 Use the Laplace transform to obtain the transform of the signal $f(t)$ shown in Figure E 14.4-5.

Answer: $F(s) = \dfrac{3(1 - e^{-2s})}{s}$

Exercise 14.4-6 Determine the Laplace transform of $f(t)$ shown in Figure E 14.4-6.

Answer: $F(s) = \dfrac{5}{2s^2}(1 - e^{-2s} - 2se^{-2s})$

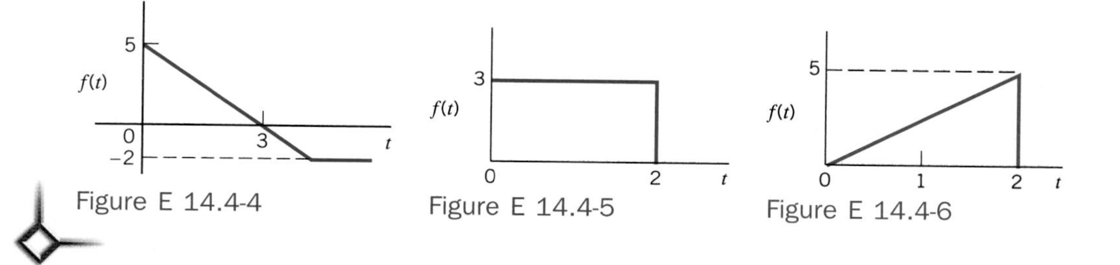

Figure E 14.4-4 Figure E 14.4-5 Figure E 14.4-6

14.5 INVERSE LAPLACE TRANSFORM

We have built up a set of transform pairs in Table 14.3-1 and Table 14.4-1. We can use these relationships between $F(s)$ and $f(t)$ to find

$$f(t) = \mathcal{L}^{-1}[F(s)]$$

where the symbol $\mathcal{L}^{-1}$ denotes the inverse transform.

Suppose we have

$$F(s) = \frac{2}{s} + \frac{3}{s+2} + \frac{9}{s^2+9}$$
$$= F_1(s) + F_2(s) + F_3(s) \quad (14.5\text{-}1)$$

We proceed to find the inverse transform for each term and obtain

$$f(t) = \mathcal{L}^{-1}[F(s)] = 2 + 3e^{-2t} + 3 \sin 3t \quad t \geq 0$$

Hence, if we wish to obtain the inverse transform of a general form of $F(s)$, we attempt to obtain a **partial fraction expansion** of $F(s)$ to get a series of terms similar to Eq. 14.5-1 so that we may find the inverse transform of each term.

For example, consider the case where we wish to find the expansion of $F(s)$ as follows:

$$F(s) = \frac{s+3}{(s+1)(s+2)} = \frac{A}{(s+1)} + \frac{B}{(s+2)}$$
$$= F_1(s) + F_2(s) \quad (14.5\text{-}2)$$

We may evaluate A by multiplying through by the denominator of $F_1(s)$ and setting s equal to -1. Following this procedure, we multiply by $s + 1$ to obtain

$$(s+1)\,F(s) = \frac{s+3}{s+2} = A + \frac{B(s+1)}{s+2}$$

Then we let $s = -1$ to obtain

$$\left.\frac{s+3}{s+2}\right|_{s=-1} = A + \left.\frac{B(s+1)}{s+2}\right|_{s=-1}$$

or

$$\frac{-1+3}{-1+2} = A + 0$$

Consequently, $A = 2$.

Following the same procedure for $F_2(s)$, we have

$$\left.\frac{s+3}{s+1}\right|_{s=-2} = B$$

or

$$B = -1$$

Therefore, substituting A and B into Eq. 14.5-2, we have

$$F(s) = \frac{2}{s+1} + \frac{-1}{s+2}$$

Taking the inverse Laplace transform of each term, we have

$$f(t) = 2e^{-t} - e^{-2t}, \quad t \geq 0$$

$F(s)$ of Eq. 14.5-2 has a numerator of order 1 and a denominator of order 2. In general, we write

$$F(s) = \frac{N(s)}{D(s)} = \frac{b_m s^m + b_{m-1}s^{m-1} + \cdots + b_1 s + b_0}{a_n s^n + a_{n-1}s^{n-1} + \cdots + a_1 s + a_0} \quad (14.5\text{-}3)$$

where the coefficients of the polynomials are real numbers. The function $F(s)$ is said to be a rational function of s since it is the ratio of two polynomials in s. The roots of the denominator polynomial $D(s)$ are the roots of the equation $D(s) = 0$ and are called the

poles of $F(s)$. The roots of the numerator polynomial are called the **zeros** of $F(s)$. Normally, we have the case $n > m$. For the case $n = m$, we first perform long division to obtain

$$F(s) = (b_m/a_n) + G(s)$$

Then we proceed to find $f(t)$ by inverse transformation of all the terms after obtaining the partial fraction of $G(s)$.

As an example of the special case $m = n$ for Eq. 14.5-3, consider

$$F(s) = \frac{s}{s+1}$$

Use long division to obtain

$$F(s) = 1 + \frac{-1}{s+1}$$

Then we obtain the inverse transform of each term to yield

$$f(t) = \delta(t) - 1e^{-t}, \quad t \geq 0$$

where $\delta(t)$ is the impulse function.

If we have repeated poles, we use a partial fraction expansion that includes all powers of the term $(s + p)$ up to the power of the repeated poles of $F(s)$. For example, if

$$F(s) = \frac{4}{(s+1)^2(s+2)}$$

we may arrange the partial fraction expansion as

$$F(s) = \frac{4}{(s+1)^2(s+2)} = \frac{A}{(s+1)^2} + \frac{B}{s+1} + \frac{C}{s+2} \tag{14.5-4}$$

First, we may evaluate C by multiplying through by $s + 2$ and letting $s = -2$. Then we have $C = 4$. To obtain A, we multiply through by $(s + 1)^2$ and let $s = -1$, obtaining

$$\left.\frac{4}{(s+2)}\right|_{s=-1} = A$$

or $A = 4$.

To find B, since we cannot use $s = -1$ again, we multiply both sides of Eq. 14.5-4 by the denominator polynomial, $D(s)$, to obtain

$$\begin{aligned} 4 &= A(s+2) + B(s+1)(s+2) + C(s+1)^2 \\ &= 4(s+2) + B(s+1)(s+2) + 4(s+1)^2 \end{aligned}$$

Equating coefficients of s^2 yields

$$B + 4 = 0$$

or $B = -4$. Then

$$F(s) = \frac{4}{(s+1)^2} - \frac{4}{s+1} + \frac{4}{s+2}$$

Recall from Table 14.4-1 that

$$\mathcal{L}^{-1}\left[\frac{a}{(s+b)^2}\right] = ate^{-bt}$$

Therefore, we have

$$f(t) = 4(te^{-t} - e^{-t} + e^{-2t}) \quad t \geq 0$$

Let us consider another approach. In the case of multiple roots, where a root r_i is repeated m times, the partial fraction expansion must include

$$\frac{b_1}{(s - r_i)} + \cdots + \frac{b_l}{(s - r_i)^l} + \cdots + \frac{b_m}{(s - r_i)^m}$$

It can be shown that the coefficient b_{m-k} is

$$b_{m-k} = \frac{1}{k!}\frac{d^k}{ds^k}[(s - r_i)^m F(s)]\bigg|_{s=r_i}$$

where $k = 0, 1, \ldots, m - 1$.

Therefore, to evaluate B of Eq. 14.5-4, we set $k = 1$ and evaluate

$$B = b_1 = \frac{d}{ds}(s - r_i)^2 F(s)\bigg|_{s=r_i} = \frac{d}{ds}(s + 1)^2 F(s)\bigg|_{s=-1}$$

$$= \frac{d}{ds}\frac{4}{s + 2}\bigg|_{s=-1} = \frac{-4}{(s + 2)^2}\bigg|_{s=-1} = -4$$

Often we will encounter an $F(s)$ that has two complex poles, as follows:

$$F(s) = \frac{N(s)}{(s + a + j\omega)(s + a - j\omega)} \tag{14.5-5}$$

where $N(s)$ is the unspecified numerator polynomial. The $F(s)$ may also be written as

$$F(s) = \frac{N(s)}{(s + a)^2 + \omega^2} = \frac{N(s)}{D(s)} \tag{14.5-6}$$

If $N(s) = s + a$, we know from Table 14.4-1 that

$$f(t) = e^{-at}\cos \omega t$$

If $N(s) = d$, we find from Table 14.4-1 that

$$f(t) = \frac{d}{\omega}e^{-at}\sin \omega t$$

Example 14.5-1

Find $f(t)$ when

$$F(s) = \frac{10}{(s + 2)(s^2 + 6s + 10)}$$

Solution

The roots of the quadratic $(s^2 + 6s + 10)$ are complex, and we may write $F(s)$ as

$$F(s) = \frac{10}{(s + 2)[(s + 3)^2 + 1]} = \frac{10}{(s + 2)(s + 3 - j)(s + 3 + j)}$$

Using a partial fraction expansion, we have

$$F(s) = \frac{10}{(s + 2)(s^2 + 6s + 10)} = \frac{A}{s + 2} + \frac{B}{s - r_1} + \frac{B^*}{s - r_1^*} \tag{14.5-7}$$

where $r_1 = -3 + j$ and r_1^* is the conjugate of r_1. It is easy to show that we obtain B^* for the partial factor $B^*/(s - r_1^*)$. Using the partial fraction method, we find

$$A = 5, \quad B = -\frac{5}{2} + j\frac{5}{2}, \quad B^* = -\frac{5}{2} - j\frac{5}{2}$$

Combining the two complex terms, we have

$$F(s) = \frac{5}{s+2} + \frac{B}{s - r_1} + \frac{B^*}{s - r_1^*}$$
$$= \frac{5}{s+2} + \frac{-5s - 20}{(s+3)^2 + 1}$$

Rearranging the second term into two terms for which the inverse will yield a cosine and a sine function, we have

$$F(s) = \frac{5}{s+2} - \frac{5(s+3)}{(s+3)^2 + 1} - \frac{5}{(s+3)^2 + 1}$$

Using Table 14.4-1, we obtain

$$f(t) = 5e^{-2t} - 5e^{-3t}\cos t - 5e^{-3t}\sin t$$
$$= 5e^{-2t} - 5e^{-3t}(\cos t + \sin t), \quad t \geq 0$$

Since the function $F(s) = \omega_n^2/(s^2 + 2\zeta\omega_n s + \omega_n^2)$ occurs frequently where the roots of the denominator are complex, we seek the inverse transform $f(t)$.

When we have

$$F(s) = \frac{\omega_n^2}{s^2 + 2\zeta\omega_n s + \omega_n^2}$$

we obtain

$$f(t) = \frac{\omega_n}{b} e^{-\zeta\omega_n t} \sin \omega_n b t$$

where $\zeta < 1$ and $b = \sqrt{1 - \zeta^2}$.

Another useful relationship, which you are invited to prove in Exercise 14.5-1, is for

$$F(s) = \frac{c + jd}{s + a - j\omega} + \frac{c - jd}{s + a + j\omega} = \frac{2[cs + (ca - \omega d)]}{(s+a)^2 + \omega^2} \tag{14.5-8}$$

Then

$$f(t) = 2me^{-at}\cos(\omega t + \theta) \quad t \geq 0$$

where

$$m = \sqrt{c^2 + d^2} \quad \text{and} \quad \theta = \tan^{-1}\left(\frac{d}{c}\right)$$

Four forms of $F(s)$ with complex poles are shown in Table 14.5-1 along with their inverse transforms.

A function $F(s)$ in the form of Eq. 14.5-8 has two complex poles that may be shown on the s-plane. Figure 14.5-1 shows a three-dimensional plot of $|F(s)|$ for the left-hand s-plane when $\omega_n = 5$ and $\zeta = 0.1$. Then, we have

$$F(s) = \frac{25}{s^2 + s + 25} = \frac{25}{(s - p_1)(s - p_1^*)}$$

where $p_1 = -0.5 + j4.97$.

Table 14.5-1 **Transforms of F(s) with Complex Poles**

$f(t),\ t \geq 0$	$F(s)$
1. $\frac{\omega_n}{b} e^{-\zeta\omega_n t} \sin \omega_n bt$ where $b = 1 - \zeta^2$	$\frac{\omega_n^2}{s^2 + 2\zeta\omega_n s + \omega_n^2}$
2. $e^{-at}(2c \cos \omega t - 2d \sin \omega t)$	$\frac{c + jd}{s + a - j\omega} + \frac{c - jd}{s + a + j\omega}$
3. $2me^{-at} \cos(\omega t + \theta)$ where $m = \sqrt{c^2 + d^2}$ and $\theta = \tan^{-1}(d/c)$	$\frac{me^{j\theta}}{s + a - j\omega} + \frac{me^{-j\theta}}{s + a + j\omega}$
4. $\frac{1}{\omega}(a^2 + \omega^2)^{1/2} e^{-at} \sin(\omega t + \phi)$ $\phi = \tan^{-1}(\omega/-a)$	$\frac{s}{(s + a)^2 + \omega^2}$

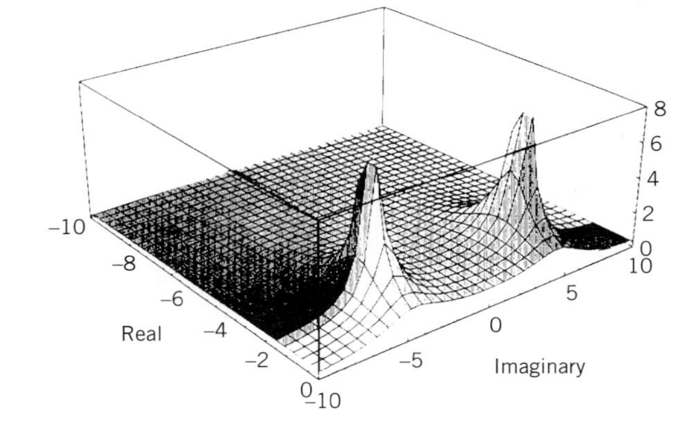

Figure 14.5-1
A three-dimensional plot of $|F(s)|$ on the left-hand s-plane. Note that the plot caps the magnitude at $s = p_1$ at 8 when it actually goes to infinity at that point. Courtesy of Mark A. Yoder of Rose–Hulman Institute.

Exercise 14.5-1 Prove that

$$\mathcal{L}^{-1}\left[\frac{cs + (ca - \omega d)}{(s + a)^2 + \omega^2}\right]$$

(for Eq. 14.5-8) is $f(t) = me^{-at}\cos(\omega t + \theta)$ where $m = \sqrt{c^2 + d^2}$ and $\theta = \tan^{-1}(d/c)$.

Exercise 14.5-2 Find the inverse transform of $F(s)$ expressing $f(t)$ in cosine and angle forms.

(a) $F(s) = \dfrac{8s - 3}{s^2 + 4s + 13}$ (b) $F(s) = \dfrac{3e^{-s}}{s^2 + 2s + 17}$

Answer: (a) $f(t) = 10.2e^{-2t}\cos(3t + 38.4°),\ t \geq 0$
(b) $f(t) = \frac{3}{4}e^{-(t-1)}\sin[4(t - 1)],\ t \geq 1$

Exercise 14.5-3 Find the inverse transform of $F(s)$.

(a) $F(s) = \dfrac{s^2 - 5}{s(s + 1)^2}$ (b) $F(s) = \dfrac{4s^2}{(s + 3)^3}$

Answer: (a) $f(t) = -5 + 6e^{-t} + 4te^{-t},\ t \geq 0$
(b) $f(t) = 4e^{-3t} - 24te^{-3t} + 18t^2e^{-3t},\ t \geq 0$

14.6 INITIAL AND FINAL VALUE THEOREMS

The initial value of a function $f(t)$ is the value at $t = 0$, provided that $f(t)$ is continuous at $t = 0$. If $f(t)$ is discontinuous at $t = 0$, then the initial value is the limit as $t \to 0^+$, where t approaches $t = 0$ from positive time.

A function's **initial value** may be found from

$$f(0) = \lim_{s \to \infty} sF(s) \tag{14.6-1}$$

To prove the relationship, substitute the definition of $F(s)$ on the right side of Eq. 14.6-1 to obtain

$$\lim_{s\to\infty} sF(s) = \lim_{s\to\infty} s\int_0^\infty f(t)e^{-st}\,dt = \lim_{s\to\infty}\int_0^\infty [f(t)se^{-st}]\,dt$$

where lim is used for the notation "limit." The integrand within the brackets is very small except near $t = 0$. Therefore,

$$\begin{aligned}\lim_{s\to\infty} sF(s) &\cong \lim_{s\to\infty} f(0)\int_0^\infty se^{-st}\,dt \\ &= f(0)\lim_{s\to\infty}\left[\frac{-se^{-st}}{s}\bigg|_0^\infty\right] \\ &= f(0)\end{aligned}$$

The initial value theorem is easy to apply for a specific $F(s)$. Consider the case where

$$F(s) = \frac{-2s^3 + 7s^2 + 2s + 9}{3s^4 + 3s^3 + 2s^2 + 6}$$

Hence

$$f(0) = \lim_{s\to\infty} sF(s) = \lim_{s\to\infty} s\left(\frac{-2s^3}{3s^4}\right) = \frac{-2}{3}$$

If we have

$$F(s) = \frac{\omega_0}{s^2 + as + \omega_0^2}$$

then

$$f(0) = \lim_{s\to\infty} sF(s) = \lim_{s\to\infty} s\frac{\omega_0}{s^2 + as + \omega_0^2} = 0$$

The *final value* of a function $f(t)$ is $\lim_{t\to\infty} f(t)$ where

$$\lim_{t\to\infty} f(t) = \lim_{s\to 0} sF(s)$$

To prove the relationship, first note that by the derivative property

$$\mathcal{L}\left(\frac{df}{dt}\right) = sF(s) - f(0) \tag{14.6-2}$$

and

$$\mathcal{L}\left(\frac{df}{dt}\right) = \int_0^\infty \left(\frac{df}{dt}\right)e^{-st}\,dt \tag{14.6-3}$$

Equating the left-hand sides of Eqs. 14.6-2 and 14.6-3, we have

$$\int_0^\infty \frac{df}{dt}e^{-st}\,dt = sF(s) - f(0) \tag{14.6-4}$$

and we take the limit as $s \to 0$ for both sides of Eq. 14.6-4 to obtain

$$\lim_{s \to 0} \int_0^\infty \frac{df}{dt} e^{-st}\, dt = \lim_{s \to 0} [sF(s) - f(0)] \tag{14.6-5}$$

We assume that $f(t)$ has a limit as $t \to \infty$ and use integration by parts for the integral of the left-hand side of Eq. 14.6-5 to obtain

$$\int_0^\infty \frac{df}{dt} e^{-st}\, dt = f(\infty) - f(0) \tag{14.6-6}$$

Therefore,

$$\lim_{s \to 0} \int_0^\infty \frac{df}{dt} e^{-st}\, dt = \lim_{s \to 0} [f(\infty) - f(0)] \tag{14.6-7}$$

Equating (14.6-6) and (14.6-7), we have

$$\lim_{s \to 0} [f(\infty) - f(0)] = \lim_{s \to 0} [sF(s) - f(0)]$$

Evaluating the limit, we obtain

$$f(\infty) = \lim_{s \to 0} sF(s) \tag{14.6-8}$$

and this relation is called the **final value theorem.** This theorem may be applied if, and only if, all the poles of $F(s)$ have a real part that is negative.

The final value theorem is a useful property since we may determine $f(\infty)$ directly from $F(s)$. However, one must be careful in using the final value theorem since the function may not have a final value as $t \to \infty$. For example, consider $f(t) = \sin \omega t$ where $F(s) = \omega/(s^2 + \omega^2)$. Now we know that $\lim_{t \to \infty} \sin \omega t$ does not exist. However, if we unwittingly used the final value theorem we would obtain

$$\lim_{s \to 0} sF(s) = \lim_{s \to 0} \frac{s\omega}{s^2 + \omega^2} = 0$$

while the actual function $f(t)$ does not have a limiting value as $t \to \infty$. We should not use the final value theorem in this case since the poles of $F(s)$ do not have a real part that is negative.

Consider the function

$$F(s) = \frac{N(s)}{D(s)} = \frac{N(s)}{(s - p_1)(s - p_2)\cdots(s - p_n)}$$

where p_i are the poles of $F(s)$. Since we take $\lim_{s \to 0} sF(s)$ for the final value, we require that the poles of $F(s)$ satisfy the requirement that $\mathrm{Re}\{p_i\} < 0$, except that one pole may be $p_j = 0$ since it cancels with the multiplicative factor s. Thus, we require that

$$\mathrm{Re}\{p_i\} < 0 \quad \text{for all } i$$

except that p_i may be zero for one value of i.

Look again at

$$F(s) = \frac{\omega}{s^2 + \omega^2}$$

when $f(t) = \sin \omega t$. Then we note that

$$F(s) = \frac{\omega}{(s + j\omega)(s - j\omega)}$$

and this function fails the requirement that $\text{Re}\{p_i\} < 0$ for all but one p_i. Thus, it does not have a final value.

Consider the function $F(s) = (s^2 + 4)/(s^3 + 3s^2 + 2s)$. We have

$$F(s) = \frac{s^2 + 4}{s(s + 1)(s + 2)}$$

Since two of the poles are $p_i < 0$ and one pole is at $p = 0$, we may proceed to find $\lim_{t\to\infty} f(t)$ as

$$f(\infty) = \lim_{s\to 0} sF(s) = \lim_{s\to 0} \frac{s^2 + 4}{(s + 1)(s + 2)} = \frac{4}{2} = 2$$

Exercise 14.6-1 Find the initial and final values of $f(t)$ when

(a) $F(s) = \dfrac{6s + 5}{s^2 + 2s + 1}$ (b) $F(s) = \dfrac{6}{s^2 - 2s + 1}$

Answer: (a) $f(0) = 6,\ f(\infty) = 0$

(b) $f(0) = 0,\ f(\infty)$, no final value

14.7 SOLUTION OF DIFFERENTIAL EQUATIONS DESCRIBING A CIRCUIT

We may solve a set of differential equations describing a circuit using the Laplace transform of the variable, $x(t)$, and its derivatives.

Consider the circuit of Figure 14.7-1 when it is known that the initial value of the inductor current is $i(0) = I_0$. We wish to determine $i(t)$, the response of the circuit excited by a source $v(t)$.

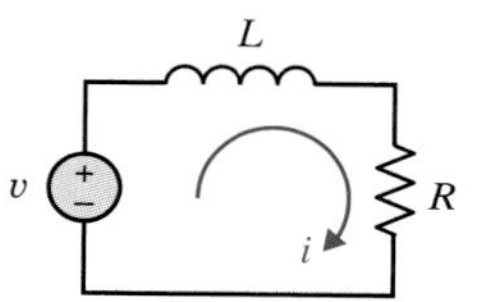

Figure 14.7-1 An *RL* circuit with an initial current $i(0) = I_0$.

The differential equation describing the circuit is

$$v = L\frac{di}{dt} + Ri \tag{14.7-1}$$

where $i(0) = I_0$. Recall from Table 14.3-1 that

$$\mathcal{L}\left(\frac{di}{dt}\right) = sI(s) - i(0)$$

where $I(s) = \mathcal{L}[i(t)]$. Also, we will use $V(s) = \mathcal{L}[v(t)]$. Then Eq. 14.7-1 becomes

$$V(s) = L[sI(s) - i(0)] + RI(s)$$

Let $i(0) = I_0$ and rearrange, obtaining

$$(R + Ls)\,I(s) = V(s) + LI_0$$

Then, since we wish to determine $i(t)$, we solve for $I(s)$ to obtain

$$I(s) = \frac{V(s) + LI_0}{R + Ls}$$

If, for example, $v(t) = M$ for $t \geq 0$, we have a step function and $V(s) = M/s$. Then $I(s)$ is

$$I(s) = \frac{M/s + LI_0}{R + Ls}$$

If $R = 1\ \Omega$, $L = \frac{1}{2}$ H, $I_0 = 1$ A, and $M = 2$ V, we have

$$I(s) = \frac{2/s + \frac{1}{2}}{1 + \frac{1}{2}s} = \frac{s + 4}{s(s + 2)} \tag{14.7-2}$$

To find $i(t)$ we perform the inverse transform by using the partial fraction expansion of Eq. 14.7-2 as follows:

$$I(s) = \frac{A}{s} + \frac{B}{s + 2}$$

Then we find that $A = 2$ and $B = -1$. Therefore,

$$i(t) = \mathcal{L}^{-1}\{I(s)\} = 2 - e^{-2t} \quad t \geq 0 \tag{14.7-3}$$

Let us use the initial and final value theorems to show how we may obtain $i(0)$ and $i(\infty)$ from $I(s)$. We find the initial value $i(0)$ from

$$i(0) = \lim_{s \to \infty} sI(s) = 1$$

which checks with the initial value given as $I_0 = 1$. The final value $i(\infty)$ is found from

$$i(\infty) = \lim_{s \to 0} sI(s) = 2$$

which checks with the final value evaluated from Eq. 14.7-3.

The general method of obtaining the solution of one or more differential equations describing a circuit is summarized in Table 14.7-1. It will, in a majority of cases, be most effective in identifying the variables of interest as the capacitor voltages and the inductor currents, since we will normally know the initial conditions of these variables.

Table 14.7-1 **Laplace Transform Method for Solving a Set of Differential Equations**

1. Identify the circuit variables such as the inductor currents and the capacitor voltages.
2. Write the differential equations describing the circuit and identify the initial conditions of the circuit variables.
3. Obtain the Laplace transform of all the terms of the differential equation.
4. Using Cramer's rule or a similar method, solve for one or more of the unknown variables, obtaining the solution in the s domain.
5. Obtain the inverse transform of the unknown variables and thus the solution in the time domain.

Example 14.7-1

Find $i(t)$ and $v_c(t)$ for the circuit shown in Figure 14.7-2 when $v_c(0) = 10$ V and $i(0) = 0$. The input source is $v_1 = 15u(t)$ V. Choose R so that the roots of the characteristic equation are real.

Solution

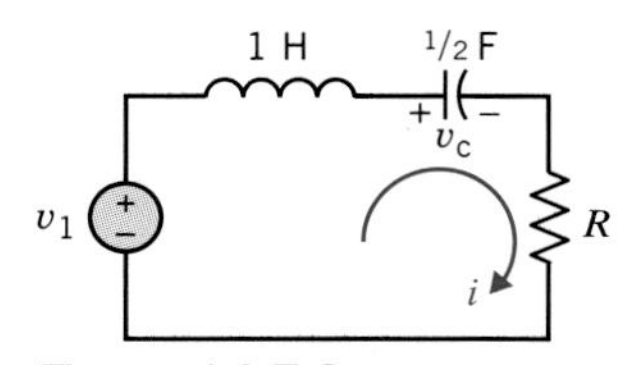

Figure 14.7-2 Circuit of Example 14.7-1.

We will identify the two variables of interest as $v_c(t)$ and $i(t)$. Then the KVL equation around the loop is

$$L\frac{di}{dt} + v_c + Ri = v_1 \qquad (14.7\text{-}4)$$

The differential equation describing the variable v_c is

$$C\frac{dv_c}{dt} = i \qquad (14.7\text{-}5)$$

These two differential equations are sufficient to solve for the two unknown variables. The Laplace transform of Eq. 14.7-4 is

$$L[sI(s) - i(0)] + V_c(s) + RI(s) = V_1(s) \qquad (14.7\text{-}6)$$

The Laplace transform of Eq. 14.7-5 is

$$C[sV_c(s) - v_c(0)] = I(s) \qquad (14.7\text{-}7)$$

Rearranging Eqs. 14.7-6 and 14.7-7 and noting that $i(0) = 0$, we have

$$(R + Ls)\,I(s) + V_c(s) = V_1(s)$$

and

$$-I(s) + CsV_c(s) = Cv_c(0)$$

Substituting the values for C, L, and $v_c(0)$, we obtain

$$(R + s)\,I(s) + V_c(s) = V_1(s) \qquad (14.7\text{-}8)$$

and

$$-I(s) + \tfrac{1}{2}sV_c(s) = 5 \qquad (14.7\text{-}9)$$

Since v_1 is a step of magnitude of 15 at $t = 0$, we have $V_1(s) = 15/s$. Substituting $V_1(s)$ into Eq. 14.7-8 and solving for $I(s)$ using Cramer's rule, we have

$$\begin{aligned} I(s) &= \frac{V_1(s)(\frac{1}{2}s) - 5}{(R + s)(\frac{1}{2}s) - (-1)} \\ &= \frac{(15/s)(s) - 5(2)}{(R + s)(s) + 2} \\ &= \frac{5}{s^2 + Rs + 2} \end{aligned} \qquad (14.7\text{-}10)$$

The inverse transform of $I(s)$ will depend on the value of R. The two roots of the denominator are equal when $R = 2\sqrt{2}$, and the roots will be real but unequal when $R > 2\sqrt{2}$. When $R < 2\sqrt{2}$ the roots are complex.

Assuming the value of $R = 3$ we have

$$\begin{aligned} I(s) &= \frac{5}{s^2 + 3s + 2} \\ &= \frac{5}{(s + 1)(s + 2)} \end{aligned}$$

Using a partial fraction expansion, we have

$$I(s) = \frac{A}{s + 1} + \frac{B}{s + 2}$$

where we find that $A = 5$ and $B = -5$. Therefore, using Table 14.3-1, we find that

$$i(t) = 5e^{-t} - 5e^{-2t}\ \text{A}, \quad t \geq 0$$

Note that $i(0) = 0$ and $i(\infty) = 0$ as required.

Solving Eqs. 14.7-8 and 14.7-9 for $V_c(s)$, we have

$$V_c(s) = \frac{(R + s)10 + 2V_1(s)}{(R + s)s + 2}$$
$$= \frac{10(R + s) + 2(15/s)}{s^2 + Rs + 2}$$
$$= \frac{10s^2 + 10Rs + 30}{s(s^2 + Rs + 2)}$$

Therefore, when $R = 3$ we have

$$V_c(s) = \frac{10(s^2 + 3s + 3)}{s(s + 1)(s + 2)}$$

Using partial fraction expansion, we have

$$V_c(s) = \frac{A}{s} + \frac{B}{s + 1} + \frac{C}{s + 2}$$

where $A = 15$, $B = -10$, and $C = 5$. Therefore, the capacitor voltage is

$$v_c(t) = 15 - 10e^{-t} + 5e^{-2t}\ \text{V}, \quad t \geq 0$$

Note that $v_c(0) = 10$ V and $v_c(\infty) = 15$ V as required.

Example 14.7-2

Find $v_c(t)$ for $t \geq 0$ for the circuit of Figure 14.7-3.

Solution

First, we identify the two variables as i and v_c. Determine the initial conditions by considering the circuit with the constant 12-V source connected for a long time. Then, replacing the inductor with a short circuit and the capacitor with an open circuit, as shown in Figure 14.7-4, it is clear that $i(0) = 60$ A and $v_c(0) = 12$ V.

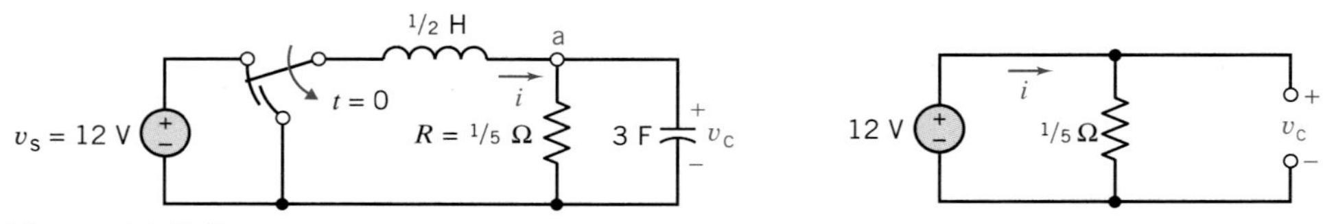

Figure 14.7-3 Circuit of Example 14.7-2.

Figure 14.7-4 Circuit of Figure 14.7-3 (Example 14.7-2) at $t = 0^-$.

We require two first-order differential equations in terms of v_c and i for $t \geq 0$. The KVL equation for the mesh with i of Figure 14.7-3 is

$$L\frac{di}{dt} + v_c = 0 \qquad (14.7\text{-}11)$$

The equation for the capacitor current i_c at node a is

$$i_c + \frac{v_c}{R} - i = 0$$

Since $i_c = C\,dv_c/dt$, we have

$$C\frac{dv_c}{dt} + \frac{v_c}{R} - i = 0 \tag{14.7-12}$$

Equations 14.7-11 and 14.7-12 describe the system completely in terms of the variables v_c and i. Taking the Laplace transform of Eqs. 14.7-11 and 14.7-12, we obtain

$$L[sI(s) - i(0)] + V_c(s) = 0$$

and

$$C[sV_c(s) - v_c(0)] + \frac{V_c(s)}{R} - I(s) = 0$$

Substituting the values for L, C, R, $i(0)$, and $v_c(0)$, we have

$$\tfrac{1}{2}[sI(s) - 60] + V_c(s) = 0$$

and

$$3[sV_c(s) - 12] + 5V_c(s) - I(s) = 0$$

Rearranging in a form more suitable for Cramer's rule, we obtain

$$sI(s) + 2V_c(s) = 60 \tag{14.7-13}$$

$$-I(s) + (3s + 5)V_c(s) = 36 \tag{14.7-14}$$

Solving for $V_c(s)$, we have

$$V_c(s) = \frac{36s + 60}{s(3s + 5) + 2} = \frac{36s + 60}{3s^2 + 5s + 2} = \frac{36s + 60}{3(s + \frac{2}{3})(s + 1)}$$

Using a partial fraction expansion, we have

$$V_c(s) = \frac{A}{s + \frac{2}{3}} + \frac{B}{s + 1}$$

where $A = 36$ and $B = -24$. Then obtaining the inverse transform, we have

$$v_c(t) = 36e^{-2t/3} - 24e^{-t}\ \text{V} \quad t \geq 0 \tag{14.7-15}$$

Exercise 14.7-1 Find $i(t)$ for the circuit of Figure E 14.7-1 when $i_1(t) = 7e^{-6t}$ A for $t \geq 0$ and $i(0) = 0$.

Answer: $i(t) = -\dfrac{35}{16}e^{-6t} + \dfrac{35}{16}e^{-2t}$ A, $t \geq 0$

Exercise 14.7-2 Find $v_2(t)$ for the circuit of Figure E 14.7-2 for $t \geq 0$.

Hint: Write the node equations at a and b in terms of v_1 and v_2. The initial conditions are $v_1(0) = 10$ V and $v_2(0) = 25$ V. The source is $v_s = 50 \cos 2t\, u(t)$ V.

Answer: $v_2(t) = \frac{23}{3}e^{-t} + \frac{16}{3}e^{-4t} + 12\cos 2t + 12 \sin 2t$ V, $t \geq 0$

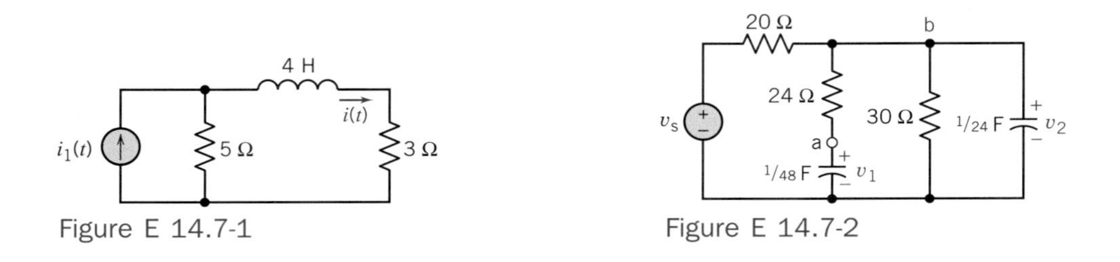

Figure E 14.7-1

Figure E 14.7-2

Exercise 14.7-3 Determine $f(t)$ for a differential equation

$$\frac{d^2f}{dt^2} + 5\frac{df}{dt} + 6f = 10e^{-3t} \quad t \geq 0$$

when $f(0) = 2$ and $\left.\dfrac{df}{dt}\right|_{t=0} = 0.$

Answer: $f(t) = -10te^{-3t} - 14e^{-3t} + 16e^{-2t}, t \geq 0$

14.8 CIRCUIT ANALYSIS USING IMPEDANCE AND INITIAL CONDITIONS

We have seen that we can represent a circuit in the time domain by differential equations and then use the Laplace transform to transform the differential equations into algebraic equations. In this section we will see that we can represent a circuit in the frequency domain using the Laplace transform and then analyze it using algebraic equations. This method will eliminate the need to write differential equations to represent the circuit.

The v-i relationship for the resistor is Ohm's law:

$$v(t) = i(t)\,R \tag{14.8-1}$$

Therefore, the Laplace transform relationship for a resistor R is

$$V(s) = I(s)\,R \tag{14.8-2}$$

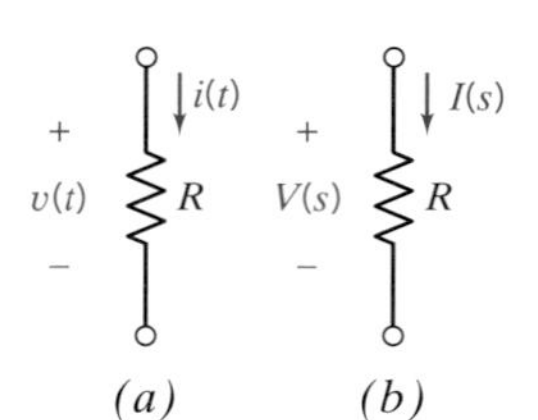

Figure 14.8-1 A resistor represented (a) in the time domain and (b) in the frequency domain using the Laplace transform.

Figure 14.8-1 shows the representation of the resistor in (a) the time domain and (b) the frequency domain using the Laplace transform. As the above equations suggest, the time and frequency domain representations of the resistor are very similar.

The impedance of an element is defined to be

$$Z(s) = \frac{V(s)}{I(s)} \tag{14.8-3}$$

provided all initial conditions are zero. Notice that the impedance is defined in the frequency domain, not the time domain.

In the case of the resistor, there is no initial condition to set to zero. Comparing Eqs. 14.8-1 and 14.8-2 shows that the impedance of the resistor is equal to the resistance.

A capacitor is represented by its time domain equation

$$v(t) = \frac{1}{C}\int_0^t i(\tau)\,d\tau + v(0) \tag{14.8-4}$$

The Laplace transform of Eq. 14.8-4 is

$$V(s) = \frac{1}{Cs} I(s) + \frac{v(0)}{s} \tag{14.8-5}$$

To determine the impedance of the capacitor, set the initial condition, $v(0)$, to zero. Then using Eq. 14.8-3 we obtain

$$Z_C(s) = \frac{1}{Cs}$$

as the impedance of the capacitor.

Eq. 14.8-5 is used to represent the capacitor in the frequency domain as shown in Figure 14.8-2*b*. The series connection of elements in Figure 14.8-2*b* corresponds to the sum of voltages in Eq. 14.8-5. The current through the impedance in Figure 14.8-2*b* produces the first voltage on the right side of Eq. 14.8-5 while the voltage source in Figure 14.8-2*b* supplies the second voltage on the right side of Eq. 14.8-5.

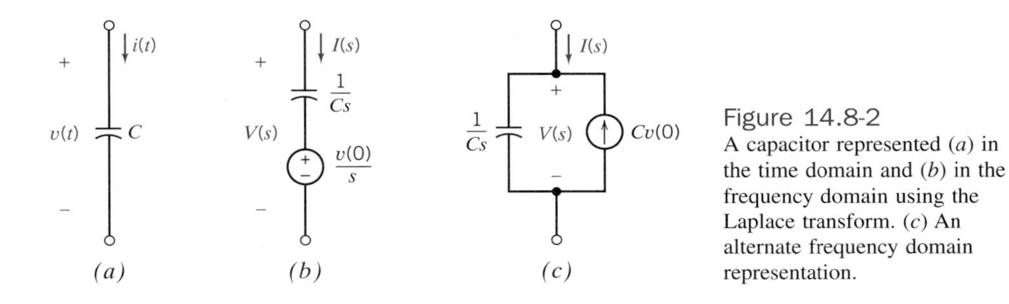

Figure 14.8-2
A capacitor represented (*a*) in the time domain and (*b*) in the frequency domain using the Laplace transform. (*c*) An alternate frequency domain representation.

Solving Eq. 14.8-5 for $I(s)$ gives

$$I(s) = CsV(s) - Cv(0) \tag{14.8-6}$$

Eq. 14.8-6 is used to represent the capacitor in the frequency domain as shown in Figure 14.8-2*c*. The parallel connection of elements in Figure 14.8-2*c* corresponds to the sum of currents in Eq. 14.8-6. The voltage across the impedance in Figure 14.8-2*b* produces the first current on the right side of Eq. 14.8-6 while the current source in Figure 14.8-2*b* supplies the current on the right side of Eq. 14.8-6. Notice that the reference direction for the current source in Figure 14.8-2*b* was chosen to correspond to the minus sign in Eq. 14.8-6.

An inductor is represented by its time domain equation

$$v(t) = L\frac{d}{dt}i(t) \tag{14.8-7}$$

The Laplace transform of Eq. 14.8-7 is

$$V(s) = LsI(s) - Li(0) \tag{14.8-8}$$

To determine the impedance of the inductor, set the initial condition, $i(0)$, to zero. Then using Eq. 14.8-3 we obtain

$$Z_L(s) = Ls$$

as the impedance of the inductor.

Eq. 14.8-8 is used to represent the inductor in the frequency domain as shown in Figure 14.8-3*b*. The series connection of elements in Figure 14.8-3*b* corresponds to the sum of voltages in Eq. 14.8-8.

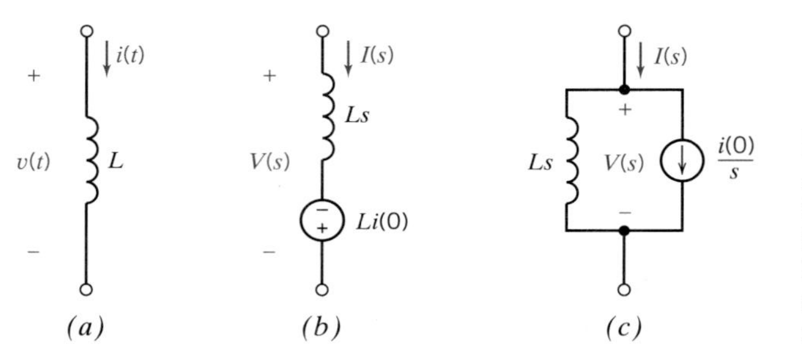

Figure 14.8-3
An inductor represented (*a*) in the time domain and (*b*) in the frequency domain using the Laplace transform. (*c*) An alternate frequency domain representation.

Solving Eq. 14.8-8 for $I(s)$ gives

$$I(s) = \frac{1}{Ls}V(s) + \frac{i(0)}{s} \tag{14.8-9}$$

Eq. 14.8-9 is used to represent the inductor in the frequency domain as shown in Figure 14.8-3*c*. The parallel connection of elements in Figure 14.8-3*c* corresponds to the sum of currents in Eq. 14.8-9.

Table 14.8-1 tabulates the time and frequency domain representation of circuit elements. In addition to resistors, capacitors, and inductors, Table 14.8-1 shows the frequency domain representations of independent and dependent sources and of op amps. Independent sources are specified by functions of time, $i(t)$ and $v(t)$, in the time domain and by the corresponding Laplace transforms, $I(s)$ and $V(s)$, in the frequency domain. Dependent sources and op amps operate the same way in the frequency domain as they do in the time domain.

To represent a circuit in the frequency domain, we replace the time domain representation of each circuit element by its frequency domain representation.

To find the complete response of a linear circuit, we first represent the circuit in the frequency domain using the Laplace transform. Next, we analyze the circuit, perhaps by writing mesh or node equations. Finally, we use the inverse Laplace transform to represent the response in the time domain.

Example 14.8-1

Determine the current in the inductor, i_2, for the circuit shown in Figure 14.8-4*a*.

Solution

Let's write and solve mesh equations. The series circuits that represent the capacitor and inductor in the frequency domain contain voltage sources rather than current sources. It's easier to account for voltage sources than current sources when writing mesh equations, so we choose the series representation for both the capacitor and inductor. The initial conditions are $v_c(0) = 8$ V and $i_L(0) = 4$ A. Figure 14.8-1*b* shows the frequency domain representation of the circuit.

The mesh current equations are

$$\left(1 + \frac{1}{s}\right)I_1(s) - \frac{1}{s}I_2(s) = \frac{12}{s} - \frac{8}{s}$$

Table 14.8-1 **Time Domain and Frequency Domain Representations of Circuit Elements**

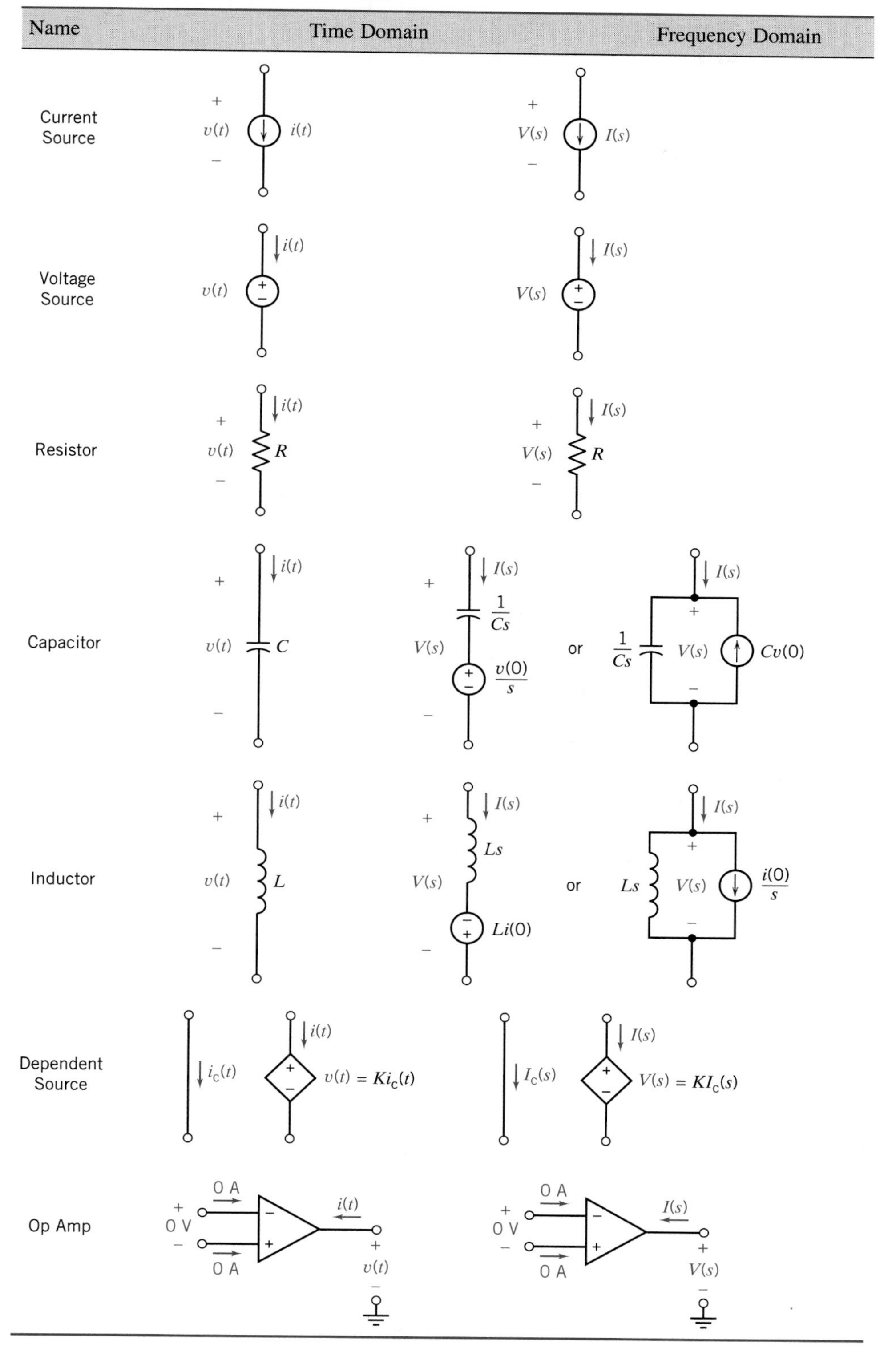

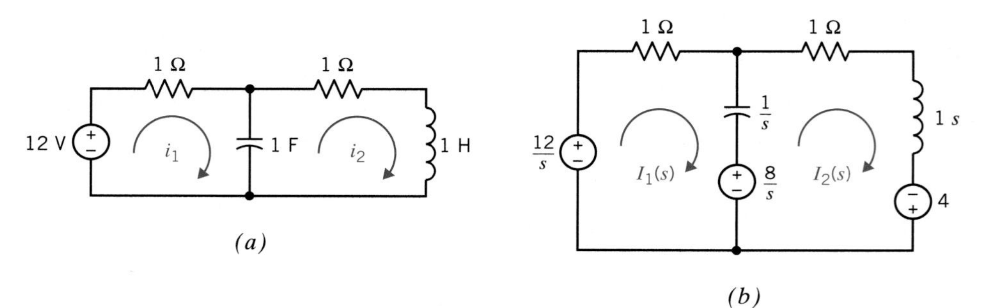

(b)

Figure 14.8-4 (*a*) Circuit with mesh currents. (*b*) Laplace transform model of circuit.

and
$$-\frac{1}{s}I_1(s) + \left(1 + s + \frac{1}{s}\right)I_2(s) = 4 + \frac{8}{s}$$

Solving for $I_2(s)$, we obtain

$$I_2(s) = \frac{4(s^2 + 3s + 3)}{s(s^2 + 2s + 2)}$$

The convenient partial fraction expansion is

$$\frac{I_2(s)}{4} = \frac{s^2 + 3s + 3}{s(s^2 + 2s + 2)} = \frac{A}{s} + \frac{Bs + D}{s^2 + 2s + 2}$$

Then, we determine that $A = 1.5$, $B = -0.5$, and $D = 0$. Then, we can state

$$\frac{I_2(s)}{4} = \frac{1.5}{s} + \frac{-0.5s}{(s + 1)^2 + 1}$$

Using the Laplace transform Tables 14.3-1 and 14.5-1, we obtain

$$i_2(t) = \{6 + 2\sqrt{2}e^{-t}\sin(t - 45°)\}\, u(t) \text{ A}$$

Checking the initial value of i_2, we get $i_2(0) = 4$ A, which verifies the correct value. The final value is $i_2(\infty) = 6$ A.

Example 14.8-2

The switch in the circuit shown in Figure 14.8-5*a* closes at time $t = 0$. Determine the voltage $v(t)$ after the switch closes.

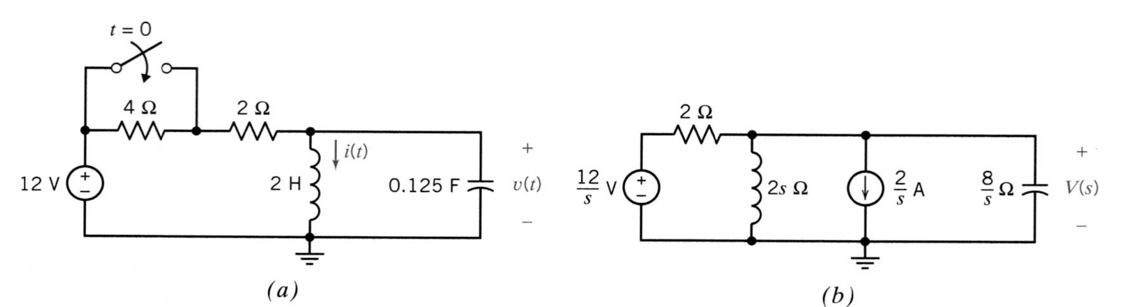

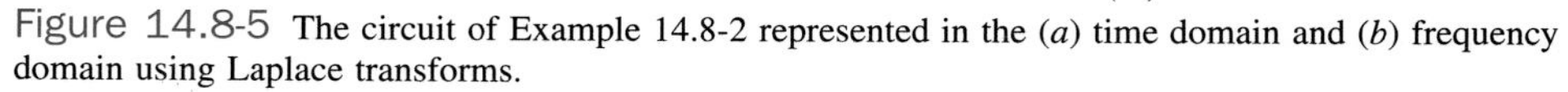
Figure 14.8-5 The circuit of Example 14.8-2 represented in the (*a*) time domain and (*b*) frequency domain using Laplace transforms.

Solution

Let's write and solve node equations. In the frequency domain, we will use the parallel model for the capacitor and inductor because the parallel models contain current sources rather than voltage sources. The initial conditions are $i(0) = 2$ A and $v(0) = 0$ V. Because $v(0) = 0$, the current of the current source in the frequency domain representation of the capacitor is zero. A zero current source is equivalent to an open circuit. Figure 14.8-5*b* shows the frequency domain representation of the circuit after the switch has closed.

Apply KCL at the top node of the inductor to get the node equation

$$\frac{V(s) - \frac{12}{s}}{2} + \frac{V(s)}{2s} + \frac{2}{s} + \frac{V(s)}{\frac{8}{s}} = 0$$

Solving for $V(s)$ gives

$$V(s) = \frac{32}{s^2 + 4s + 4} = \frac{32}{(s + 2)^2}$$

Finally, take the inverse Laplace transform to obtain $v(t)$

$$v(t) = \mathcal{L}^{-1}\left[\frac{32}{(s + 2)^2}\right] = 32te^{-2t}u(t) \text{ V}$$

Exercise 14.8-1 Determine the voltage $v_C(t)$ and the current $i_C(t)$ for $t \geq 0$ for the circuit of Figure E 14.8-1.

Hint: $v_C(0) = 4$ V.

Answer: $v_C(t) = (6 - 2e^{-0.67t})u(t)$ V and $i_C(t) = \frac{2}{3}e^{-0.67t}u(t)$ A

Exercise 14.8-2 Determine the voltage $v_o(t)$ for $t \geq 0$ for the circuit of Figure E 14.8-2.

Hint: $v_C(0) = 4$ V.

Answer: $v_o(t) = 24e^{0.75t}u(t)$ V. (This circuit is unstable.)

Exercise 14.8-3 Determine the current $i_L(t)$ for $t \geq 0$ for the circuit of Figure E 14.8-3.

Hint: $v_C(0) = 8$ V and $i_L(0) = 1$ A

Answer: $i(t) = \left(e^{-t}\cos 2t + \frac{1}{2}e^{-t}\sin 2t\right)u(t)$ A.

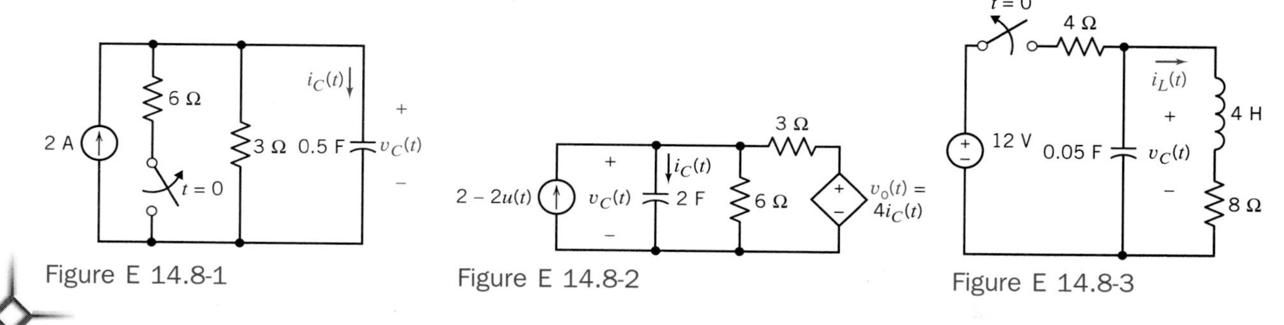

Figure E 14.8-1 Figure E 14.8-2 Figure E 14.8-3

14.9 TRANSFER FUNCTION AND IMPEDANCE

The **transfer function** of a circuit is defined as the ratio of the Laplace transform of the response of the circuit to the Laplace transform of the input to the circuit when the initial conditions are zero. For the circuit in Figure 14.9-1a, the input is the voltage source voltage, $v_1(t)$, and the response is the resistor voltage, $v_o(t)$. The transfer function of this circuit, denoted by $H(s)$, is then expressed as

$$H(s) = \frac{V_o(s)}{V_1(s)} \tag{14.9-1}$$

provided all initial conditions are equal to zero. In this case the only initial condition is the inductor current, so we require $i(0) = 0$.

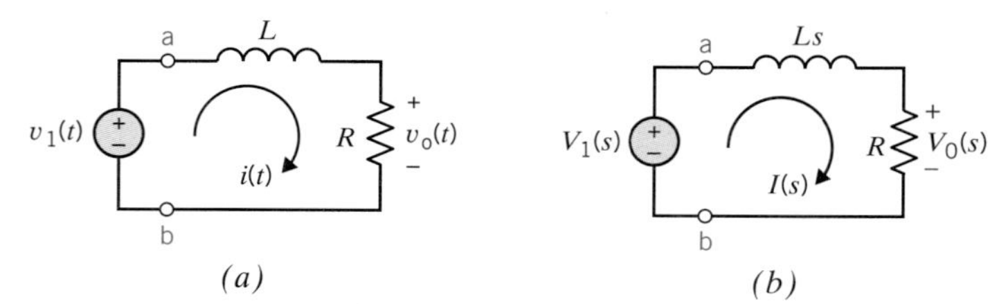

Figure 14.9-1 A circuit represented (a) in the time domain and (b) in the frequency domain using the Laplace transform.

We can write Eq. 14.9-1 as

$$V_o(s) = H(s)V_1(s) \tag{14.9-2}$$

which says that the Laplace transform of the response is equal the transfer function times the Laplace transform of the input, provided all initial conditions are equal to zero. We are going to get tired of saying "provided all initial conditions are equal to zero." A response subject to the requirement that all initial conditions are zero is called a zero-state response. With this terminology we can read Eq. 14.9-1 as "the transfer function is the ratio of the Laplace transform of the zero-state response to the Laplace transform of the input." Similarly, we can read Eq. 14.9-2 as "the Laplace transform of the zero-state response is the product of the transfer function and the Laplace transform of the input."

Two special cases are very significant. When the input is a step function then

$$V_1(s) = \mathcal{L}[u(t)] = \frac{1}{s}$$

and Equation 14.9-2 becomes

$$V_o(s) = \frac{H(s)}{s}$$

In this case the zero-state response is called the step response, that is

$$\textit{step response} = \mathcal{L}^{-1}\left[\frac{H(s)}{s}\right] \tag{14.9-3}$$

When the input is an impulse function then

$$V_1(s) = \mathcal{L}[\delta(t)] = 1$$

and Equation 14.9-2 becomes

$$V_o(s) = H(s)$$

In this case the zero-state response is called the impulse response, that is

$$\textit{impulse response} = \mathcal{L}^{-1}[H(s)] \tag{14.9-4}$$

It is important to notice that both the step response and the impulse response are zero-state responses, that is all initial conditions are set to zero.

Both the input to a circuit and the response of the circuit can be either a current or a voltage. When the input is a current and the response is a voltage, the transfer function is called an impedance. Similarly, when the input is a voltage and the response is a current, the transfer function is called an admittance. This terminology is consistent with our previous use of the term "impedance." For example, consider the row of Table 14.8-1 corresponding to the capacitor. Consider the frequency domain representation of the capacitor that contains a voltage source. The restriction that the initial condition is zero, $v(0) = 0$, causes the voltage source to be a zero voltage source, that is a short circuit. The frequency domain representation of the capacitor is reduced to a single element. When capacitor current is the input and the capacitor voltage is the response then the impedance of the capacitor is

$$Z_C(s) = \frac{V(s)}{I(s)} = \frac{1}{Cs} \tag{14.9-5}$$

Next, consider the frequency domain representation of the capacitor that contains a current source. The restriction that the initial condition is zero, $v(0) = 0$, causes the current source to be a zero current source, that is an open circuit. The frequency domain representation of the capacitor is again reduced to a single element. Once again, the impedance of the capacitor is given by Eq. 14.9-5.

A similar argument shows that setting the initial conditions to zero simplifies the frequency domain representation of the inductor to the single impedance

$$Z_L(s) = Ls \tag{14.9-6}$$

Example 14.9-1

For the circuit in Figure 14.9-1*a*, the input is the voltage source voltage, $v_1(t)$, and the response is the resistor voltage, $v_o(t)$. Find the transfer function of the circuit.

Solution

Figure 14.9-1*b* shows the frequency domain representation of the circuit, when all of the initial conditions are zero. In this case the only initial condition is the inductor current, so we require $i(0) = 0$. The requirement that $i(0) = 0$ reduces the frequency domain representation of the inductor to the impedance of the inductor.

Applying KVL to the mesh of the circuit in Figure 14.9-1*b* gives

$$V_1(s) = LsI(s) + RI(s)$$

Solving for $I(s)$ gives

$$I(s) = \frac{V_1(s)}{Ls + R}$$

The Laplace transform of the response is

$$V_o(s) = RI(s) = \frac{R}{Ls + R}V_1(s)$$

This result could have been obtained using voltage division. Finally, the transfer function is

$$H(s) = \frac{V_0(s)}{V_1(s)} = \frac{R}{Ls + R}$$

Example 14.9-2

Determine the step response of the circuit shown in Figure 14.9-2*a*.

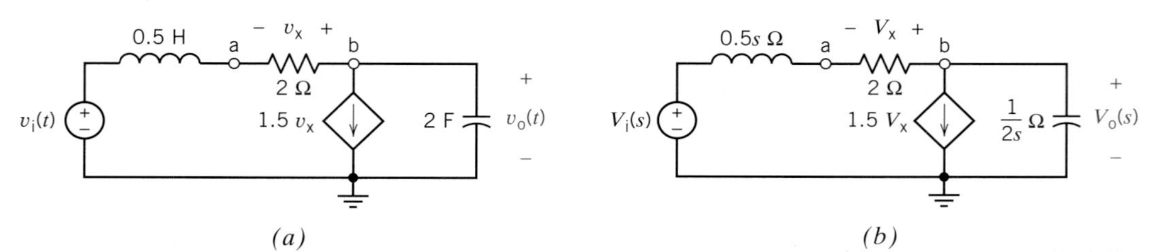

Figure 14.9-2 The circuit of Example 14.9-2 represented in the (*a*) time domain and (*b*) frequency domain using Laplace transforms.

Solution

Figure 14.9-2*b* shows the frequency domain representation of the circuit, when all of the initial conditions are zero.

Denote the node voltages at nodes a and b as V_a and V_b. The node equations are

$$\frac{V_a - V_i}{0.5\,s} - \left(\frac{V_b - V_a}{2}\right) = 0 \quad \Rightarrow \quad (4 + s)V_a - sV_b = 4V_i$$

and
$$\frac{V_b - V_a}{2} + 1.5(V_b - V_a) + 2sV_b = 0 \quad \Rightarrow \quad (1 + s)V_b = V_a$$

Solving for V_b gives

$$V_b = \frac{4}{(s + 2)^2}V_i$$

The response is $V_o = V_b$ so the transfer function is

$$H(s) = \frac{V_o(s)}{V_i(s)} = \frac{V_b(s)}{V_i(s)} = \frac{4}{(s + 2)^2}$$

The step response is

$$v_o(t) = \mathcal{L}^{-1}\left[\frac{H(s)}{s}\right] = \mathcal{L}^{-1}\left[\frac{4}{s(s + 2)^2}\right] = (1 - (1 + 2t)e^{-2t})u(t)$$

Example 14.9-3

Design the circuit of Figure 14.9-3*a* to have an impulse response equal to

$$h(t) = 2(e^{-t} - e^{-2t}) \quad t \geq 0$$

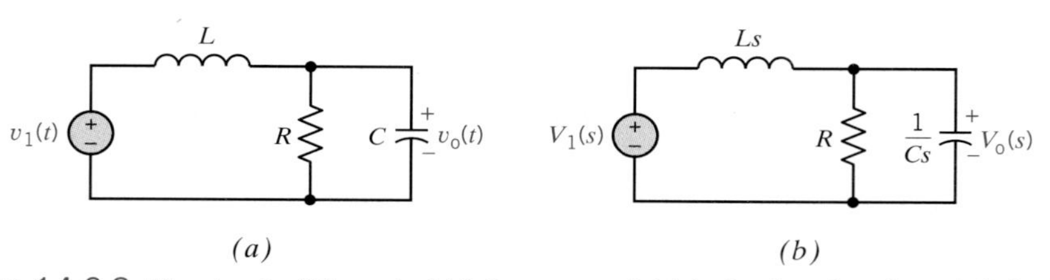

Figure 14.9-3 The circuit of Example 14.9-3 represented (*a*) in the time domain and (*b*) in the frequency domain using the Laplace transform.

Solution

From the given impulse response we have

$$H(s) = 2\left(\frac{1}{s+1} - \frac{1}{s+2}\right) = 2\frac{(s+2)-(s+1)}{(s+1)(s+2)} = \frac{2}{s^2+3s+2} \qquad (14.9\text{-}7)$$

Figure 14.9-3*b* shows the circuit represented in the frequency domain, using the Laplace transform. Using the voltage divider principle, we determine the transfer function of this circuit to be

$$H(s) = \frac{V_o(s)}{V_1(s)} = \frac{\dfrac{R\dfrac{1}{Cs}}{R+\dfrac{1}{Cs}}}{\dfrac{R\dfrac{1}{Cs}}{R+\dfrac{1}{Cs}} + Ls} = \frac{1/LC}{s^2 + (1/RC)s + 1/LC} \qquad (14.9\text{-}8)$$

Comparing Eqs. 14.9-7 and 14.9-8 gives $\frac{1}{LC} = 2$ and $\frac{1}{RC} = 3$. These equations don't have a single unique solution. To obtain one solution, choose $C = \frac{1}{12}$ F. Then $L = 6$ H and $R = 4\ \Omega$ are the required values. Other solutions can be obtained by changing the value of C and then recalculating L and R.

Exercise 14.9-1 The transfer function of a circuit is $H(s) = \frac{-5s}{s^2 + 15s + 50}$. Determine the impulse response and step response of this circuit.

Answer: (a) *impulse response* $= \mathcal{L}^{-1}\left[\frac{5}{s+5} - \frac{10}{s+10}\right] = (5e^{-5t} - 10e^{-10t})u(t)$

(b) *step response* $= \mathcal{L}^{-1}\left[\frac{1}{s+10} - \frac{1}{s+5}\right] = (e^{-10t} - e^{-5t})u(t)$

Exercise 14.9-2 The impulse response of a circuit is $h(t) = 5e^{-2t}\sin(4t)u(t)$. Determine the step response of this circuit.

Hint: $H(s) = \mathcal{L}[5e^{-2t}\sin(4t)u(t)] = \frac{5(4)}{(s+2)^2 + 4^2} = \frac{20}{s^2 + 4s + 20}$

Answer: *step response* $= \mathcal{L}^{-1}\left[\frac{H(s)}{s}\right] = \mathcal{L}^{-1}\left[\frac{1}{s} - \frac{s+4}{s^2+4s+20}\right]$

$$= \left(1 - e^{-2t}\left(\cos 4t + \frac{1}{2}\sin 4t\right)\right)u(t)$$

14.10 CONVOLUTION THEOREM

The *convolution* is defined for an output $y(t)$ as the integral

$$y(t) = \mathcal{L}^{-1}[H(s)F(s)] = \int_0^t h(\tau)f(t-\tau)\,d\tau \tag{14.10-1}$$

That is, the product of a transfer function $H(s)$ times a signal $F(s)$ in the s-domain is equivalent to the Laplace transform of an integral given on the right-hand side of Eq. 14.10-1.

To prove the convolution integral, consider the product $H(s)F(s)$, where

$$H(s) = \int_0^\infty h(\tau)e^{-s\tau}\,d\tau$$

The product is

$$H(s)F(s) = \int_0^\infty h(\tau)e^{-s\tau}F(s)\,d\tau \tag{14.10-2}$$

However, recall that the time shift property is

$$\mathcal{L}[f(t-\tau)] = e^{-s\tau}F(s) \tag{14.10-3}$$

Therefore, substituting Eq. 14.10-3 into Eq. 14.10-2, we have

$$H(s)F(s) = \int_0^\infty h(\tau)\mathcal{L}[f(t-\tau)]\,d\tau$$

$$= \int_0^\infty h(\tau)\left[\int_0^\infty f(t-\tau)e^{-st}\,dt\right]d\tau$$

Rearranging the order of the integrals, we have

$$H(s)F(s) = \int_0^\infty e^{-st}\left[\int_0^\infty h(\tau)\,f(t-\tau)\,d\tau\right]dt$$

$$= \mathcal{L}\left[\int_0^\infty h(\tau)\,f(t-\tau)\,d\tau\right] \tag{14.10-4}$$

Usually, the convolution is denoted by an asterisk so that when we seek the convolution of $h(t)$ with $f(t)$, we write

$$\mathcal{L}^{-1}[H(s)F(s)] = h(t) * f(t) = \int_0^\infty h(\tau)f(t-\tau)\,d\tau \tag{14.10-5}$$

The **convolution** of two functions of time, $f(t)$ and $h(t)$, is described as $f(t) * h(t)$ and is identical to the inverse transform of $F(s)H(s)$.

For example, find the convolution of $h(t) = e^{-t}$ and $f(t) = e^{-2t}$:

$$h(t) * f(t) = \mathcal{L}^{-1}[H(s)F(s)] = \mathcal{L}^{-1}\left[\left(\frac{1}{s+1}\right)\left(\frac{1}{s+2}\right)\right]$$

$$= \mathcal{L}^{-1}\left[\frac{1}{s+1} + \frac{-1}{s+2}\right] = e^{-t} - e^{-2t} \quad t \geq 0$$

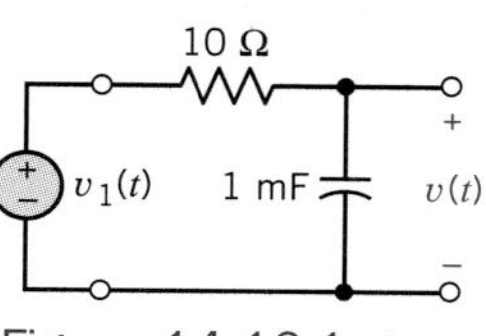

Figure 14.10-1 An *RC* circuit.

Example 14.10-1

Determine the output $v(t)$ for the circuit shown in Figure 14.10-1 using the convolution integral and the Laplace transform. The input voltage is $v_1(t) = 2e^{-100t}u(t)$ V.

Solution

Let us first determine the output using the convolution integral. The transfer function of the circuit is

$$H(s) = \frac{1}{RCs + 1} = \frac{100}{s + 100}$$

Therefore, the impulse response is

$$h(t) = 100e^{-100t}u(t)$$

Using Eq. 14.10-5, we have

$$\begin{aligned} v(t) = h(t) * v_1(t) &= 200\int_0^t e^{-100\tau}u(\tau)e^{-100(t-\tau)}u(t-\tau)\,d\tau \\ &= 200e^{-100t}\int_0^t d\tau \\ &= 200te^{-100t} \quad t > 0 \end{aligned}$$

We now proceed to determine $v(t)$ using $V_1(s)$ and $H(s)$. Using the Laplace transform, we find $V_1(s)$ as

$$V_1(s) = \mathcal{L}[2e^{-100t}u(t)] = \frac{2}{s + 100}$$

Then, the output is

$$V(s) = H(s)V_1(s) = \left(\frac{100}{s + 100}\right)\left(\frac{2}{s + 100}\right) = \frac{200}{(s + 100)^2}$$

Using the first entry in Table 14.4-1, we obtain

$$v(t) = 200te^{-100t} \quad t > 0$$

Exercise 14.10-1 (a) Find $H(s) = V_c(s)/V_1(s)$ for the circuit of Figure E 14.10-1. (b) Find the impulse response $h(t)$ of this circuit.

Answer: (a) $H(s) = \dfrac{1}{s + 1.25}$

(b) $h(t) = e^{-1.25t} \quad t > 0$

Figure E 14.10-1

Exercise 14.10-2 Find the convolution of $h(t) = e^{-2t}$ and $f(t) = u(t)$.

14.11 STABILITY

A circuit is said to be *stable* when the response to a bounded input signal is a bounded output signal. Consider a circuit represented by the transfer function $H(s)$. The impulse response is

$$h(t) = \mathcal{L}^{-1}\{H(s)\}$$

Thus, we require for stability that

$$\lim_{t \to \infty} |h(t)| = \text{finite}$$

If $H(s)$ is written in terms of its poles, we have

$$H(s) = \frac{N(s)}{(s - p_1)(s - p_2)\cdots(s - p_N)}$$

and, therefore,

$$h(t) = \sum_{i=1}^{N} A_i e^{p_i t} u(t)$$

Thus we require, for a stable circuit, that all the poles of $H(s)$ lie in the left-hand s-plane. A simple pole in the right-hand plane where $\sigma_1 > 0$ can be represented as

$$H(s) = \frac{K}{s - \sigma_1}$$

Then, $h(t) = Ke^{\sigma_1 t}$ and the response to an impulse is unbounded if $\sigma_1 > 0$.

Example 14.11-1

Determine $H(s) = V_3(s)/V(s)$ for the op amp circuit shown in Figure 14.11-1 and determine if the circuit is stable. Assume ideal op amps and the input signal is $v(t)$.

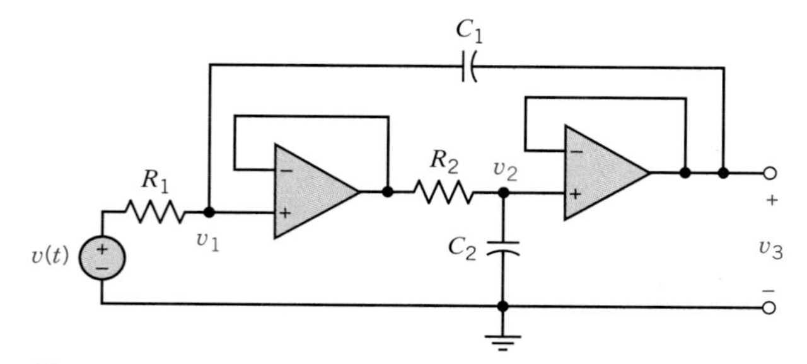

Figure 14.11-1 A two–op amp circuit with a feedback capacitor.

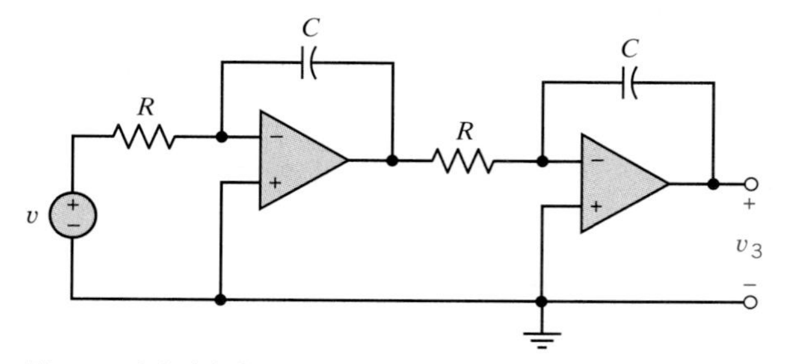

Figure 14.11-2 A two-stage op amp circuit.

Solution

The node equations at nodes 1 and 2, using the notation $V_n = V_n(s)$, are

1. $\dfrac{V_1 - V}{R_1} + (V_1 - V_3)s\,C_1 = 0$

2. $\dfrac{V_2 - V_1}{R_2} + V_2 s\,C_2 = 0$

Since, for an ideal op amp, $V_3 = V_2$ due to the buffer amplifier stage, we rewrite the two node equations as

1. $\left(\frac{1}{R_1} + sC_1\right)V_1 + (-sC_1)V_2 = \frac{V}{R_1}$

2. $\left(-\frac{1}{R_2}\right)V_1 + \left(\frac{1}{R_2} + sC_2\right)V_2 = 0$

Using Cramer's rule, we solve for V_2 in terms of V as

$$\frac{V_2(s)}{V(s)} = \frac{1}{R_1R_2C_1C_2s^2 + R_2C_2s + 1} = \frac{\omega_0^2}{s^2 + (1/R_1C_1)s + \omega_0^2}$$

and $\omega_0^2 = 1/R_1R_2C_1C_2$. This circuit is stable, since the two roots of the characteristic equation always lie in the left-hand s-plane for all positive values of R_1 and C_1.

Example 14.11-2

Determine $H(s) = V_3(s)/V(s)$ for the op amp circuit shown in Figure 14.11-2 and determine if the circuit is stable. Assume ideal op amps.

Solution

The circuit consists of two ideal integrators in cascade (see Section 7.10). Therefore, we have

$$\frac{V_3(s)}{V(s)} = \frac{1}{(RC)^2s^2}$$

The two poles of the circuit lie at the origin of the s-plane. If $V(s) = 1$, which represents an impulse input, we have

$$H(s) = V_3(s) = \frac{1}{(RC)^2s^2}$$

Using Table 14.3-1, we find the impulse response as

$$h(t) = \frac{t}{(RC)^2}$$

which is unbounded since $h(t) \to \infty$ as $t \to \infty$. Thus, this circuit is said to be unstable.

A stable circuit has a steady-state response while an unstable circuit does not. Thus, frequency response analysis and phasor analysis may only be applied to stable circuits.

The stability of a circuit can be determined from the circuit's frequency response examining both phase and magnitude. Stability cannot be determined from the magnitude of the frequency response alone. Indeed, both $H_1(s) = 1/(s + 1)$ and $H_2(s) = 1/(s - 1)$ have the same magnitude, but $H_1(s)$ stable and $H_2(s)$ is unstable.

The transfer function of a second-order circuit can be written as

$$H(s) = \frac{N(s)}{s^2 + a_1s + a_0}$$

$$= \frac{N(s)}{s^2 + 2\zeta\omega_0s + \omega_0^2} = \frac{N(s)}{s^2 + \omega_0s/Q + \omega_0^2}$$

When $\zeta = 0$ or $Q = \infty$, we have

$$H(s) = \frac{N(s)}{s^2 + \omega_0^2} = \frac{N(s)}{(s + j\omega_0)(s - j\omega_0)}$$

and this circuit is called an *oscillator* since the response is a sinusoidal signal. For example, the impulse response of

$$H(s) = \frac{100}{s^2 + 100}$$

is $h(t) = 10 \sin 10t$ for $t > 0$.

Exercise 14.11-1 The transfer function of a second-order circuit is

$$H(s) = \frac{25}{s^2 + (8 - k)s + 100}$$

Determine the required range of k for stable operation when k is $k \geq 0$.

Answer: $0 \leq k < 8$

Exercise 14.11-2 For the transfer function of Exercise 14.11-1, determine the required value of k so that the circuit is an oscillator.

Answer: $k = 8$

14.12 PARTIAL FRACTION EXPANSION USING MATLAB

MATLAB provides a function called *residue* that performs the partial fraction expansion of a transfer function. Consider a transfer function

$$H(s) = \frac{b_3 s^3 + b_2 s^2 + b_1 s^1 + b_0 s^0}{a_3 s^3 + a_2 s^2 + a_1 s^1 + a_0 s^0} \tag{14.12-1}$$

In Eq. 14.12-1, the transfer function is represented as a ratio of two polynomials in s. In MATLAB the transfer function given in Eq. 14.12-1 can be represented by two lists. One list specifies the coefficients of the numerator polynomial, and the other list specifies the coefficients of the denominator polynomial. For example,

$$\text{num} = [b_3 \;\; b_2 \;\; b_1 \;\; b_0]$$

and

$$\text{den} = [a_3 \;\; a_2 \;\; a_1 \;\; a_0]$$

(In this case both polynomials are third-order polynomials, but the orders of these polynomials could be changed.)

Partial fraction expansion can be used to represent $H(s)$ as

$$H(s) = \frac{R_1}{s - p_1} + \frac{R_2}{s - p_2} + \frac{R_3}{s - p_3} + k(s) \tag{14.12-2}$$

R_1, R_2, and R_3 are called residues, and p_1, p_2, and p_3 are the poles. In general, both the residues and poles can be complex numbers. The term $k(s)$ will, in general, be a polynomial in s. MATLAB represents this form of the transfer function by three lists:

$$R = [R_1 \;\; R_2 \;\; R_3]$$

is a list of the residues,

$$p = [p_1 \ \ p_2 \ \ p_3]$$

is a list of the poles, and

$$k = [c_2 \ \ c_1 \ \ c_0]$$

is a list of the coefficients of the polynomial $k(s)$.

The MATLAB command

```
[R, p, k] = residue(num, den)
```

performs the partial fraction expansion, calculating the poles and residues from the coefficients of the numerator and denominator polynomials. The MATLAB command

```
[n, d] = residue(R, p, k)
```

performs the reverse operation, calculating the coefficients of the numerator and denominator polynomials from the poles and residues.

Figure 14.12-1 shows a MATLAB screen illustrating this procedure. In this example

$$H(s) = \frac{s^3 + 2s^2 + 3s + 4}{s^3 + 6s^2 + 11s + 6}$$

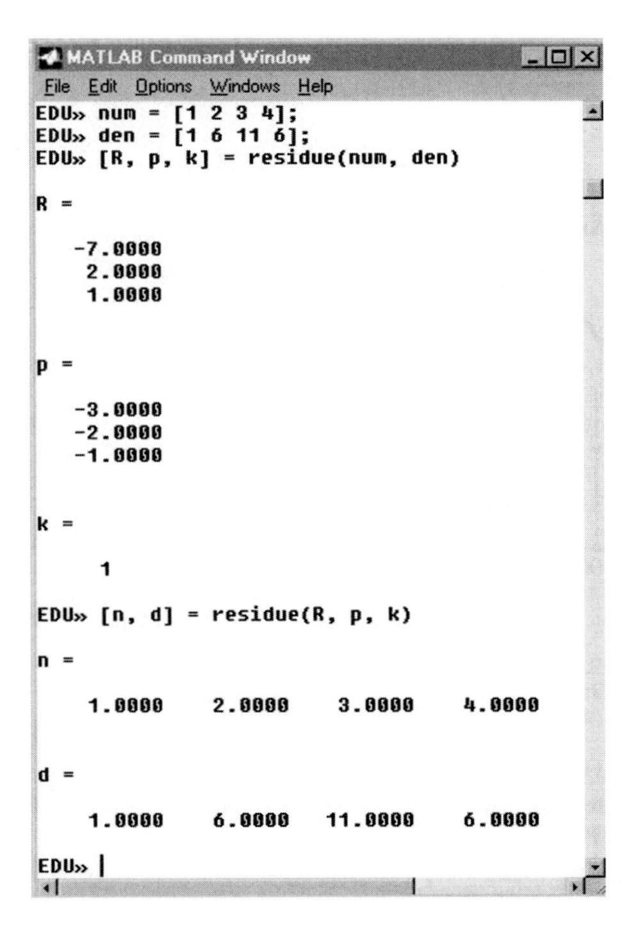

Figure 14.12-1
Using MATLAB to perform partial fraction expansion.

is represented as

$$H(s) = \frac{-7}{s+3} + \frac{2}{s+2} + \frac{1}{s+1} + 1$$

by performing the partial fraction expansion.

14.13 VERIFICATION EXAMPLES

Example 14.13-1

A circuit is specified to have a transfer function of

$$H(s) = \frac{V_0(s)}{V_1(s)} = \frac{25}{s^2 + 10s + 125} \tag{14.13-1}$$

and a step response of

$$v_0(t) = 0.1(2 - e^{-5t}(3\cos 10t + 2\sin 10t))u(t) \tag{14.13-2}$$

Are these specifications consistent? Let us verify their consistency.

Solution

If the specifications are consistent, then the unit step response and the transfer function will be related by

$$\mathscr{L}[v_0(t)] = H(s)\frac{1}{s} \tag{14.13-3}$$

where $V_1(s) = 1/s$.

This equation can be verified either by calculating the Laplace transform of $v_0(t)$ or by calculating the inverse Laplace transform of $H(s)/s$. Both of these calculations involve a bit of algebra. The final and initial value theorems provide a quicker, though less conclusive, check. (If either the final or initial value theorem is not satisfied, then we know that the step response is not consistent with the transfer function. The step response could be inconsistent with the transfer function even if both the final and initial value theorems are satisfied.) Let us see what the final and initial value theorems tell us.

The final value theorem requires that

$$v_0(\infty) = \lim_{s\to 0} s\left[H(s)\frac{1}{s}\right] \tag{14.13-4}$$

From Eq. 14.13-1, we substitute $H(s)$ obtaining

$$\lim_{s\to 0} s\left[\frac{25}{s^2 + 10s + 125}\cdot\frac{1}{s}\right] = \lim_{s\to 0}\left[\frac{25}{s^2 + 10s + 125}\right] = \frac{25}{125} = 0.2 \tag{14.13-5}$$

From Eq. 14.3-2, we evaluate at $t = \infty$ obtaining

$$v_0(\infty) = 0.1\,(2 - e^{-\infty}(2\cos\infty + \sin\infty)) = 0.1(2 - 0) = 0.2 \tag{14.13-6}$$

so the final value theorem is satisfied.

Next, the initial value theorem requires that

$$v_0(0) = \lim_{s\to\infty} s\left[H(s)\frac{1}{s}\right] \tag{14.13-7}$$

From Eq. 14.13-1, we substitute $H(s)$ obtaining

$$\lim_{s\to\infty} s\left[\frac{25}{s^2+10s+125}\cdot\frac{1}{s}\right] = \lim_{s\to\infty}\frac{25/s^2}{1+10/s+125/s^2} = \frac{0}{1} = 0 \quad (14.13\text{-}8)$$

From Eq. 14.13-1, we evaluate at $t = 0$ to obtain

$$\begin{aligned} v_0(0) &= 0.1(2 - e^{-0}(3\cos 0 + 2\sin 0)) \\ &= 0.1(2 - 1(3+0)) \\ &= -0.1 \end{aligned} \quad (14.13\text{-}9)$$

The initial value theorem is not satisfied, so the step response is not consistent with the transfer function.

Example 14.13-2

A circuit is specified to have a transfer function of

$$H(s) = \frac{V_0(s)}{V_1(s)} = \frac{25}{s^2+10s+125} \quad (14.13\text{-}10)$$

and a unit step response of

$$v_0(t) = 0.1(2 - e^{-5t}(2\cos 10t + 3\sin 10t))u(t) \quad (14.13\text{-}11)$$

Are these specifications consistent? (This step response is a slightly modified version of the step response considered in the previous example.)

Solution

The reader is invited to verify that both the final and initial value theorems are satisfied. This suggests, but does not guarantee, that the transfer function and step response are consistent. To guarantee consistency, it is necessary to verify that

$$\mathscr{L}[v_0(t)] = H(s)\frac{1}{s} \quad (14.13\text{-}12)$$

either by calculating the Laplace transform of $v_0(t)$ or by calculating the inverse Laplace transform of $H(s)/s$. Recall the input is a unit step, so $V_1(s) = 1/s$. We will calculate the Laplace transform of $v_0(t)$ as follows:

$$\begin{aligned} &\mathscr{L}[0.1(2 - e^{-5t}(2\cos 10t + 3\sin 10t))u(t)] \\ &\quad = 0.1\left[\frac{2}{s} - \frac{2(s+5)}{(s+5)^2+10^2} + 3\frac{10}{(s+5)^2+10^2}\right] \\ &\quad = 0.1\left[\frac{2}{s} + \frac{-2s+20}{s^2+10s+125}\right] \\ &\quad = \frac{2s+25}{s(s^2+10s+125)} \end{aligned}$$

Since this is not equal to $H(s)/s$, Eq. 14.13-12 is not satisfied. The step response is not consistent with the transfer function even though the initial and final values of $v_0(t)$ are consistent.

Exercise 14.13-1 A circuit is specified to have a transfer function of

$$H(s) = \frac{25}{s^2 + 10s + 125}$$

and a unit step response of

$$v_0(t) = 0.1(2 - e^{-5t}(2\cos 10t + \sin 10t))u(t)$$

Verify that these specifications are consistent.

14.14 Design Challenge Solution

SPACE SHUTTLE CARGO DOOR

Problem

The U.S. space shuttle docked with Russia's Mir Space station several times. The electromagnet for opening a cargo door on the NASA space shuttle requires 0.1 A before activating. The electromagnetic coil is represented by L as shown in Figure 14.14-1. The activating current is designated $i_1(t)$. The time period required for i_1 to reach 0.1 A is specified as less than 3 seconds. Select a suitable value of L.

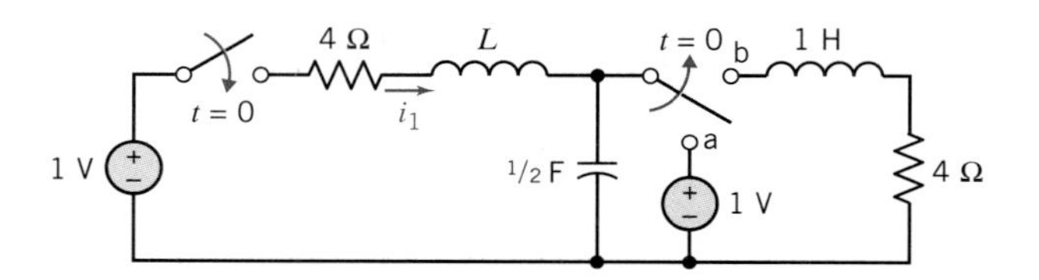

Figure 14.14-1 The control circuit for a cargo door on the NASA space shuttle.

Describe the Situation and the Assumptions

1. The two switches are thrown at $t = 0$ and the movement of the second switch from terminal a to terminal b occurs instantaneously.
2. The switches prior to $t = 0$ were in position for a long time.

State The Goal

Determine a value of L so that the time period for the current $i_1(t)$ to attain a value of 0.1 A is less than 3 seconds.

Generate a Plan

1. Determine the initial conditions for the two inductor currents and the capacitor voltage.
2. Designate two mesh currents and write the two mesh KVL equations using the Laplace transform of the variables and the impedance of each element.
3. Select a trial value of L and solve for $I_1(s)$.
4. Determine $i_1(t)$.
5. Sketch $i_1(t)$ and determine the time instant t_1 when $i_1(t_1) = 0.1$ A.
6. Check if $t_1 < 3$ seconds, and if not return to step 3 and select another value of L.

Goal	Equation	Need	Information
Determine the initial conditions at $t = 0$	$i(0) = i(0^-)$ $v_c(0) = v_c(0^-)$	Prepare a sketch of the circuit at $t = 0^-$. Find $i_1(0^-)$, $i_2(0^-)$, $v_c(0^-)$.	
Designate two mesh currents and write the mesh KVL equations.		$I_1(s)$, $I_2(s)$. The initial conditions $i_1(0)$, $i_2(0)$	
Solve for $I_1(s)$ and select L.			Cramer's rule
Determine $i_1(t)$.	$i_1(t) = \mathscr{L}^{-1}\{I_1(s)\}$		Use a partial fraction expansion.
Sketch $i_1(t)$ and find t_1.	$i_1(t_1) = 0.1$ A		

Act on the Plan

First, the circuit with the switches in position at $t = 0^-$ is shown in Figure 14.14-2. Clearly, the inductor currents are $i_1(0^-) = 0$ and $i_2(0^-) = 0$. Furthermore, we have

$$v_c(0) = 1 \text{ V}$$

Second, redraw the circuit for $t > 0$ as shown in Figure 14.14-3 and designate the two mesh currents i_1 and i_2 as shown.

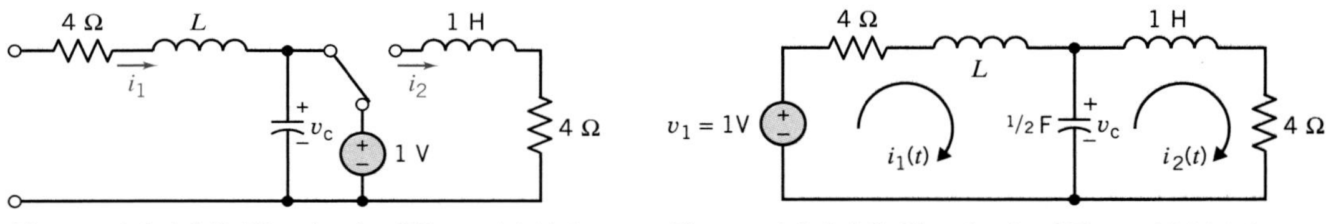

Figure 14.14-2 The circuit of Figure 14.14-1 at $t = 0^-$.

Figure 14.14-3 The circuit of Figure 14.14-1 for $t > 0$.

Recall that the impedance is Ls for an inductor and $1/Cs$ for a capacitor. We must account for the initial condition for the capacitor. Recall that the capacitor voltage may be written as

$$v_c(t) = v_c(0) + \frac{1}{C}\int_0^t i_c(\tau)\, d\tau$$

The Laplace transform of this equation is

$$V_c(s) = \frac{v_c(0)}{s} + \frac{1}{Cs} I_c(s)$$

where $I_c(s) = I_1(s) - I_2(s)$ in this case. We now may write the two KVL equations for the two meshes for $t \geq 0$ with $v_c(0) = 1$ V as

$$\text{mesh 1: } -V_1(s) + (4 + Ls)I_1(s) + V_c(s) = 0$$
$$\text{mesh 2: } (4 + 1s)I_2(s) - V_c(s) = 0$$

The Laplace transform of the input voltage is

$$V_1(s) = \frac{1}{s}$$

Also, note that for the capacitor, we have

$$V_c(s) = \frac{1}{s} + \frac{1}{Cs}(I_1(s) - I_2(s))$$

Substituting V_1 and V_c into the mesh equations, we have (when $C = 1/2$ F)

$$\left(4 + Ls + \frac{2}{s}\right)I_1(s) - \left(\frac{2}{s}\right)I_2(s) = 0$$

and

$$-\left(\frac{2}{s}\right)I_1(s) + \left(4 + s + \frac{2}{s}\right)I_2(s) = \frac{1}{s}$$

The third step requires the selection of the value of L and then solving for $I_1(s)$. Examine Figure 14.14-3; the two meshes are symmetric when $L = 1$ H. Then, trying this value and using Cramer's rule, we solve for $I_1(s)$, obtaining

$$I_1(s) = \frac{\left(\frac{2}{s}\right)\frac{1}{s}}{\left(4 + s + \frac{2}{s}\right)^2 - \left(\frac{2}{s}\right)^2} = \frac{2}{s(s^3 + 8s^2 + 20s + 16)}$$

Fourth, in order to determine $i_1(t)$, we will use a partial fraction expansion. Rearranging and factoring the denominator of $I_1(s)$, we determine that

$$I_1(s) = \frac{2}{s(s+4)(s+2)^2}$$

Hence, we have the partial fraction expansion

$$I_1(s) = \frac{A}{s} + \frac{B}{s+4} + \frac{C}{(s+2)^2} + \frac{D}{s+2}$$

Then, we readily determine that $A = 1/8$, $B = -1/8$, and $C = -1/2$. In order to find D, we use the differentiation method of Section 14.5 to obtain

$$D = \frac{1}{(2-1)!}\frac{d}{ds}\left[(s+2)^2 I_1(s)\right]_{s=-2}$$

$$= \left.\frac{-2(2s+4)}{s^4 + 8s^3 + 16s^2}\right|_{s=-2}$$

$$= 0$$

Therefore, using the inverse Laplace transform for each term, we obtain

$$i_1(t) = \frac{1}{8} - \frac{1}{8}e^{-4t} - \frac{1}{2}te^{-2t} \text{ A} \quad t \geq 0$$

Verify the Proposed Solution

The sketch of $i_1(t)$ is shown in Figure 14.14-4. It is clear that $i_1(t)$ has essentially reached a steady-state value of 0.125 A by $t = 4$ seconds.

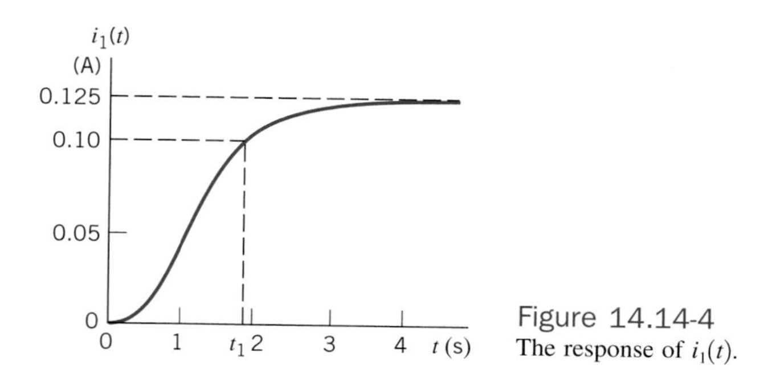

Figure 14.14-4
The response of $i_1(t)$.

In order to find t_1 when

$$i_1(t_1) = 0.1 \text{ A}$$

we estimate that t_1 is approximately 2 seconds. After evaluating $i_1(t)$ for a few selected values of t near 2 seconds, we find that $t_1 = 1.8$ seconds. Therefore, the design requirements are satisfied for $L = 1$ H. Of course, other suitable values of L can be determined which will satisfy the design requirements.

14.15 SUMMARY

- Pierre-Simon Laplace is credited with a transform that bears his name. The Laplace transform is defined as

$$\mathcal{L}\{f(t)\} = \int_0^\infty f(t)e^{-st}\,dt$$

- The Laplace transform transforms the differential equation describing a circuit in the time domain into an algebraic equation in the complex frequency domain. After solving the algebraic equation, we use the inverse Laplace transform to obtain the circuit response in the time domain. Figure 14.3-1 illustrates this process.
- Tables 14.3-1 and 14.4-1 tabulate frequently used Laplace transform pairs. Table 14.3-2 tabulates some properties of the Laplace transform.
- The inverse Laplace transform is obtained using partial fraction expansion.
- Table 14.8-1 shows that circuits can be represented in the frequency domain in a manner that accounts for the initial conditions of capacitors and inductors.
- To find the complete response of a linear circuit, we first represent the circuit in the frequency domain using the Laplace transform. Next, we analyze the circuit, perhaps by writing mesh or node equations. Finally, we use the inverse Laplace transform to represent the response in the time domain.
- The transfer function, $H(s)$, of a circuit is defined as the ratio of the response $Y(s)$ of the circuit to an excitation $X(s)$ expressed in the complex frequency domain.

$$H(s) = \frac{Y(s)}{X(s)}$$

This ratio is obtained assuming all initial conditions are equal to zero.
- The step response is the response of a circuit to a step input when all initial conditions are zero. The step response is related to the transfer function by

$$\text{Step response} = \mathcal{L}^{-1}\left[\frac{H(s)}{s}\right]$$

- The impulse response is the response of a circuit to a impulse input when all initial conditions are zero. The impulse response is related to the transfer function by

$$\text{Impulse response} = \mathcal{L}^{-1}[H(s)]$$

- A circuit is said to be *stable* when the response to a bounded input signal is a bounded output signal. All the poles of the transfer function of a stable circuit lie in the left-half s-plane.
- MATLAB performs partial fraction expansion.

PROBLEMS

Section 14.3 Laplace Transform

P 14.3-1 Find the Laplace transform, $F(s)$, when $f(t) = A\cos\omega t$, $t \geq 0$.

Answer: $F(s) = \dfrac{As}{s^2 + \omega^2}$

P 14.3-2 Find the Laplace transform, $F(s)$, when $f(t) = t$, $t \geq 0$.

P 14.3-3 Using the linearity property find the Laplace transform of $f(t) = e^{-3t} + t$, $t \geq 0$.

P 14.3-4 Using the linearity property find the Laplace transform of $f(t) = A(1 - e^{-bt})u(t)$.

Answer: $F(s) = \dfrac{Ab}{s(s + b)}$

Section 14.4 Impulse Function and Time Shift Property

P 14.4-1 Consider a pulse $f(t)$ defined by

$$\begin{aligned} f(t) &= A \quad 0 \leq t \leq T \\ &= 0 \quad \text{all other } t \end{aligned}$$

Find $F(s)$.

Answer: $F(s) = \dfrac{A(1 - e^{-sT})}{s}$

P 14.4-2 Consider the pulse shown in Figure P 14.4-2, where the time function follows e^{at} for $0 < t < T$. Find $F(s)$ for the pulse.

Answer: $F(s) = \dfrac{1 - e^{-(s-a)T}}{s - a}$

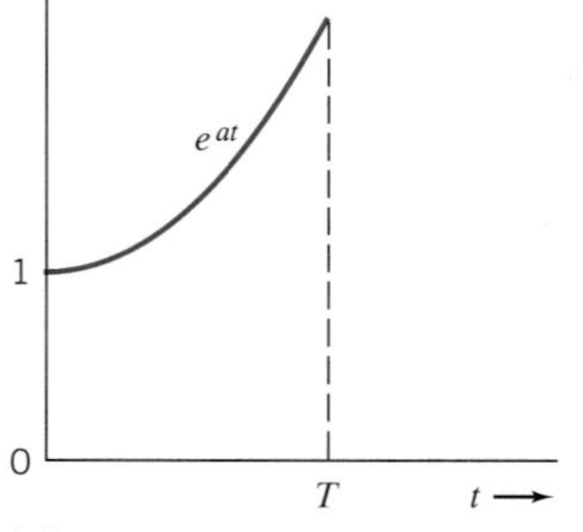

Figure P 14.4-2

P 14.4-3 Find the Laplace transform $F(s)$ for
(a) $f(t) = t^2e^{-3t}$, $t \geq 0$;
(b) $f(t) = \delta(t - T)$, $t \geq 0$;
(c) $f(t) = e^{-4t}\sin 5t$, $t \geq 0$.

P 14.4-4 Find the Laplace transform for $g(t) = e^{-t}u(t - 0.5)$.

P 14.4-5 Find the Laplace transform for $f(t) = \dfrac{-(t - T)}{T}u(t - T)$.

Answer: $F(s) = \dfrac{-1e^{-sT}}{Ts^2}$

Section 14.5 Inverse Laplace Transform

P 14.5-1 Find $f(t)$ when

$$F(s) = \frac{s + 3}{s^3 + 3s^2 + 6s + 4}$$

Answer: $f(t) = \frac{2}{3}e^{-t} - \frac{2}{3}e^{-t}\cos\sqrt{3}t + \dfrac{1}{\sqrt{3}}e^{-t} \times \sin\sqrt{3}t$, $t \geq 0$

P 14.5-2 Find $f(t)$ when

$$F(s) = \frac{s^2 - 2s + 1}{s^3 + 3s^2 + 4s + 2}$$

P 14.5-3 Find $f(t)$ when

$$F(s) = \frac{5s - 1}{s^3 - 3s - 2}$$

Answer: $f(t) = -e^{-t} + 2te^{-t} + e^{2t}$, $t \geq 0$

P 14.5-4 Find the inverse transform of

$$Y(s) = \frac{1}{s^3 + 3s^2 + 4s + 2}$$

Answer: $y(t) = e^{-t}(1 - \cos t)$, $t \geq 0$

P 14.5-5 Find the inverse transform of

$$F(s) = \frac{2s + 6}{(s + 1)(s^2 + 2s + 5)}$$

P 14.5-6 Find the inverse transform of

$$F(s) = \frac{2s + 6}{s(s^2 + 3s + 2)}$$

Answer: $f(t) = [3 - 4e^{-t} + e^{-2t}]\,u(t)$

Section 14.6 Initial and Final Value Theorems

P 14.6-1 A function of time is represented by

$$F(s) = \frac{2s^2 - 3s + 4}{s^3 + 3s^2 + 2s}$$

(a) Find the initial value of $f(t)$ at $t = 0$.
(b) Find the value of $f(t)$ as t approaches infinity.

P 14.6-2 Find the initial and final values of $v(t)$ when

$$V(s) = \frac{(s + 16)}{s^2 + 4s + 12}$$

Answer: $v(0) = 1$, $v(\infty) = 0$ V

P 14.6-3 Find the initial and final values of $v(t)$ when

$$V(s) = \frac{(s + 10)}{(3s^3 + 2s^2 + 1s)}$$

Answer: $v(0) = 0,\ v(\infty) = 10$ V

P 14.6-4 Find the initial and final values of $f(t)$ when

$$F(s) = \frac{-2(s + 7)}{s^2 - 2s + 10}$$

Answer: initial value $= -2$, final value does not exist.

Section 14.7 Solution of Differential Equations Describing a Circuit

P 14.7-1 Find $i(t)$ for the circuit of Figure P 14.7-1 when $i(0) = 1$ A, $v(0) = 8$ V, and $v_1 = 2e^{-at}u(t)$ where $a = 2 \times 10^4$.

Answer: $i(t) = \frac{1}{15}(-10e^{-bt} + 3e^{-2bt} + 22e^{-4bt})$ A, $t \geq 0,\ b = 10^4$.

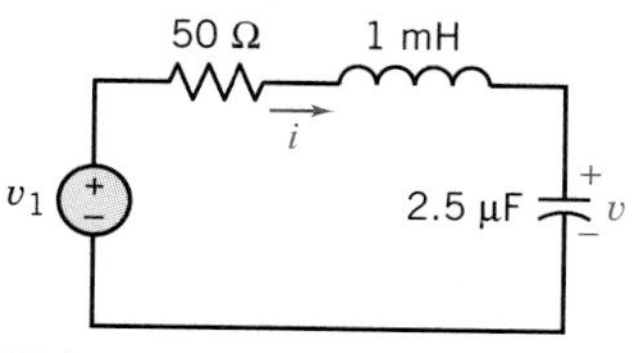

Figure P 14.7-1

P 14.7-2 All new homes are required to install a device called a ground fault circuit interrupter (GFCI) that will provide protection from shock. By monitoring the current going to and returning from a receptacle, a GFCI senses when normal flow is interrupted and switches off the power in 1/40 second. This is particularly important if you are holding an appliance shorted through your body to ground. A circuit model of the GFCI acting to interrupt a short is shown in Figure P 14.7-2. Find the current flowing through the person, $i(t)$, for $t \geq 0$ when the short is initiated at $t = 0$. Assume $v = 160 \cos 400t$ and the capacitor is initially uncharged.

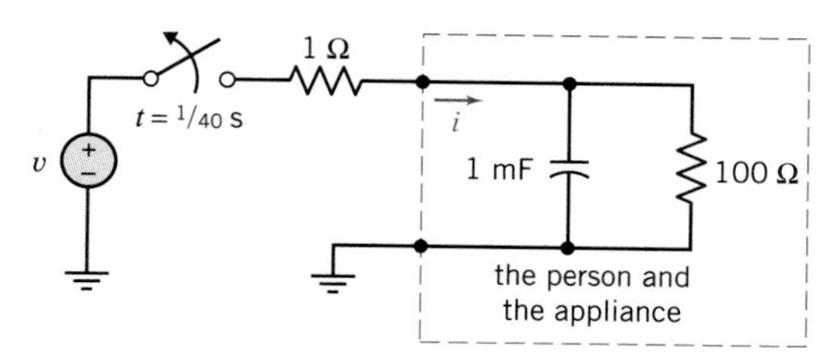

Figure P 14.7-2 Circuit model of person and appliance shorted to ground.

P 14.7-3 Using the Laplace transform, find $v_c(t)$ for $t > 0$ for the circuit shown in Figure P 14.7-3. The initial conditions are zero.

Hint: Use a source transformation to obtain a single mesh circuit.

Answer: $v_c = -5e^{-2t} + 5(\cos 2t + \sin 2t)$ V

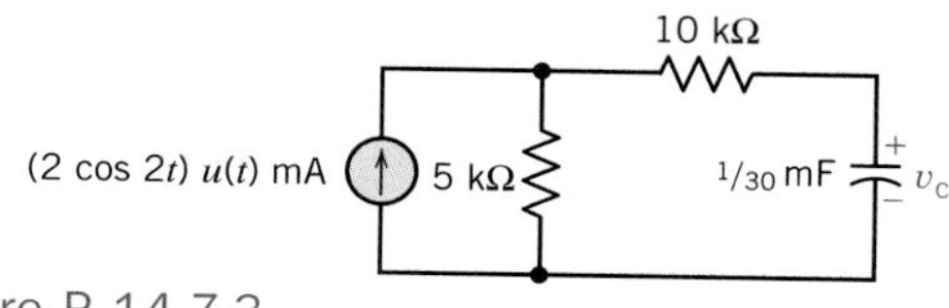

Figure P 14.7-3

P 14.7-4 Using Laplace transforms, find $v_c(t)$ for $t > 0$ for the circuit of Figure P 14.7-4 when (a) $C = 1/18$ F and (b) $C = 1/10$ F.

Answer: (a) $v_c(t) = -8 + 8e^{-3t} + 24te^{-3t}$ V

(b) $v_c(t) = -8 + 10e^{-t} - 2e^{-5t}$ V

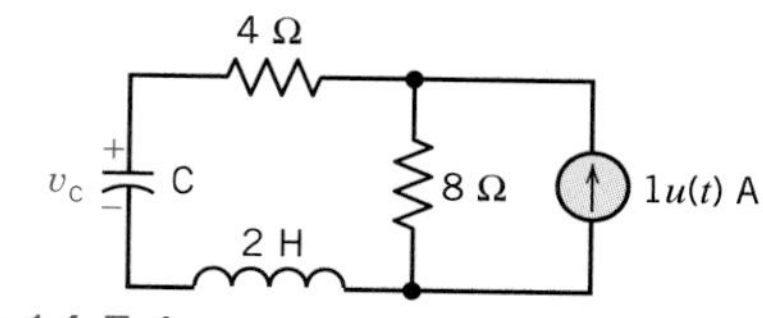

Figure P 14.7-4

P 14.7-5 Find $i(t)$ for the circuit of Figure P 14.7-5. Assume the switch has been open for a long time.

Answer: $i = -0.025e^{-200t} \sin 400t$ A, $t > 0$

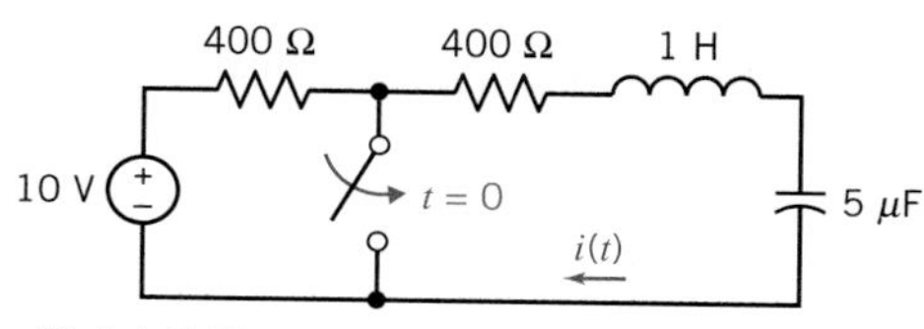

Figure P 14.7-5

Section 14.8 Circuit Analysis Using Impedance and Initial Conditions

P 14.8-1 Using Laplace transforms, find the response $i_L(t)$ for $t > 0$ for the circuit of Figure P 14.8-1.

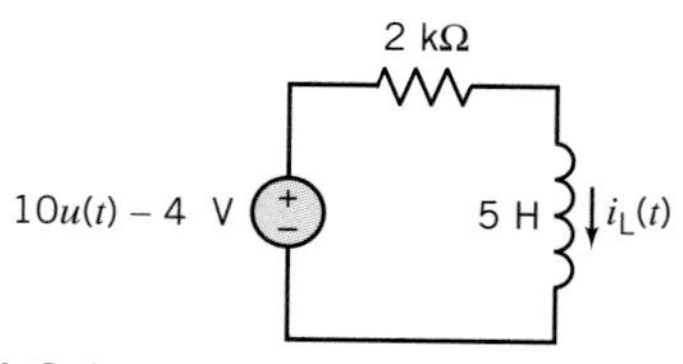

Figure P 14.8-1

P 14.8-2 Using Laplace transforms, find the response $i_L(t)$ for $t > 0$ for the circuit of Figure P 14.8-2.

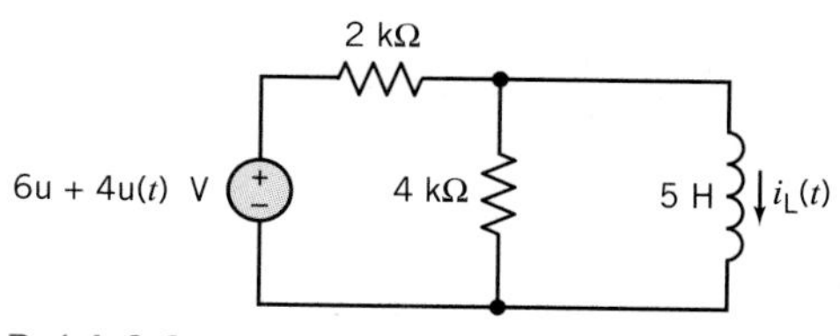

Figure P 14.8-2

P 14.8-3 Using Laplace transforms, find the response $v_c(t)$ for $t > 0$ for the circuit of Figure P 14.8-3.

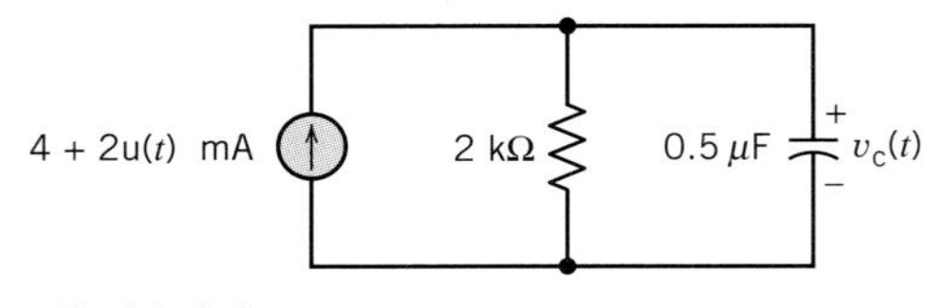

Figure P 14.8-3

P 14.8-4 Using Laplace transforms, find the response $v_c(t)$ for $t > 0$ for the circuit of Figure P 14.8-4.

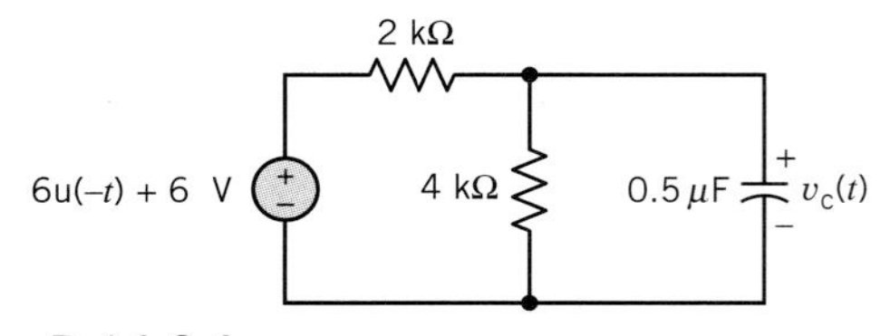

Figure P 14.8-4

P 14.8-5 Using Laplace transforms, find the response $v(t)$ for $t > 0$ for the circuit of Figure P 14.8-5 when $v_s = 6e^{-3t}\,u(t)$ V.
Answer: $v = \frac{44}{3}e^{-2t} + \frac{1}{3}e^{-5t} - 9e^{-3t}$ V

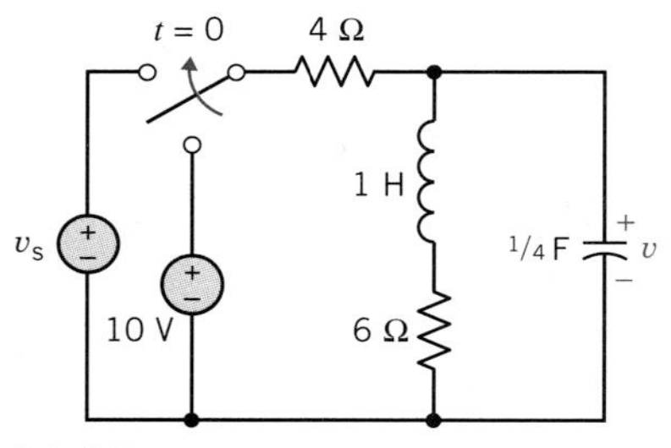

Figure P 14.8-5

P 14.8-6 Determine $v_0(t)$ when the capacitance has an initial voltage $v(0^-) = 5$ V, as shown in Figure P 14.8-6.

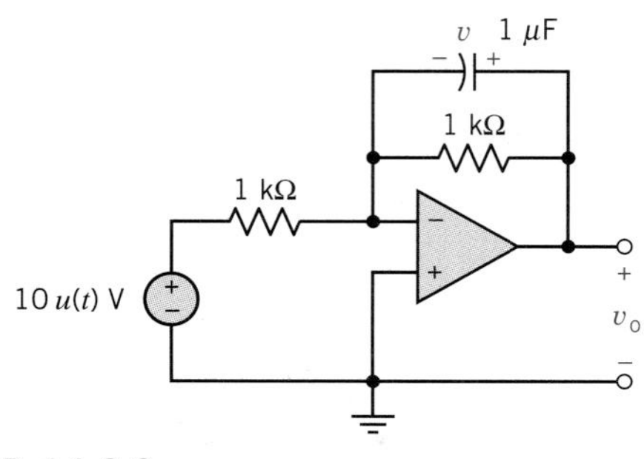

Figure P 14.8-6

P 14.8-7 The motor circuit for driving the snorkel shown in Figure P 14.8-7*a* is shown in Figure P 14.8-7*b*. Find the motor current $I_2(s)$ when the initial conditions are $i_1(0^-) = 2$ A and $i_2(0^-) = 3$ A. Determine $i_2(t)$ and sketch it for 10 seconds. Does the motor current smoothly drive the snorkel?

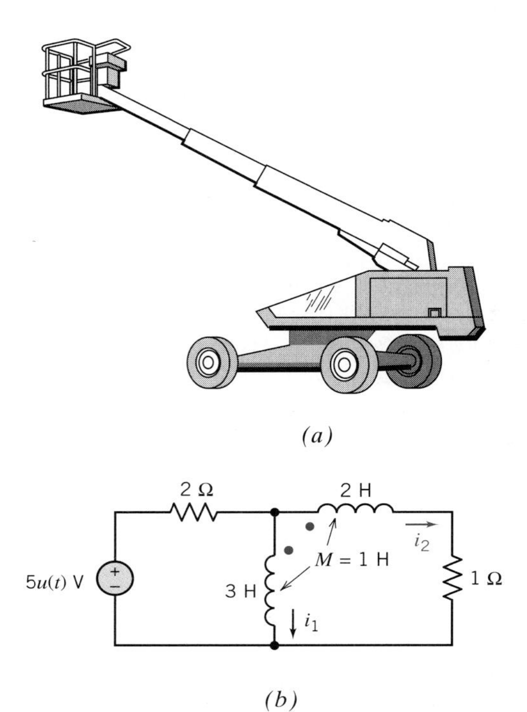

Figure P 14.8-7 Motor drive circuit for snorkel device.

Section 14.9 Transfer Function and Impedance

P 14.9-1 Consider the circuit of Figure P 14.9-1, where the combination of R_2 and C_2 represents the input of an oscilloscope. The combination of R_1 and C_1 is added to the probe of the oscilloscope to shape the response $v_0(t)$ so that it will equal $v_1(t)$ as closely as possible. Find the necessary relationship for the resistors and capacitors so that $v_0 = av_1$ where a is a constant.
Hint: Find the transfer function $V_0(s)/V_1(s)$. Choose R_1 and C_1 so that the transfer function does not depend on s.

P 14.9-2 Consider the circuit shown in Figure P 14.9-2. Show that by proper choice of L, the input impedance $Z = V_1(s)/I_1(s)$ can be made independent of s. What value of L satisfies this condition? What is the value of Z when it is independent of s?

P 14.9-3 A bridged-T circuit is often used as a filter and is shown in Figure P 14.9-3. Show that the transfer function of the circuit is

$$\frac{V_{out}(s)}{V_{in}(s)} = \frac{1 + (2R_1 + R_2)Cs + R_1R_2C^2s^2}{1 + 2R_1Cs + R_1R_2C^2s^2}$$

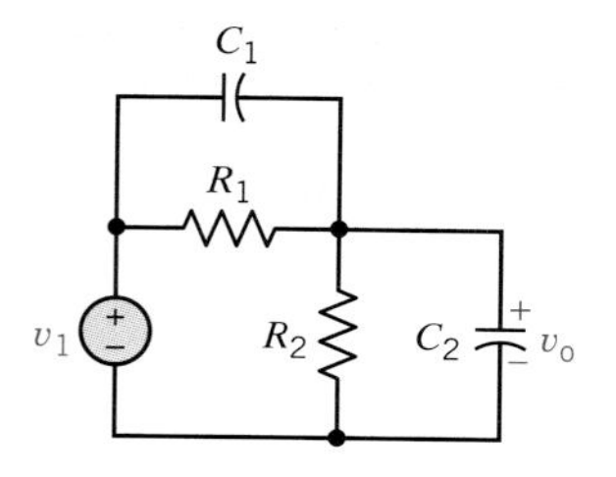

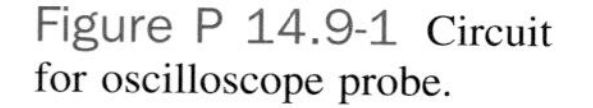

Figure P 14.9-1 Circuit for oscilloscope probe.

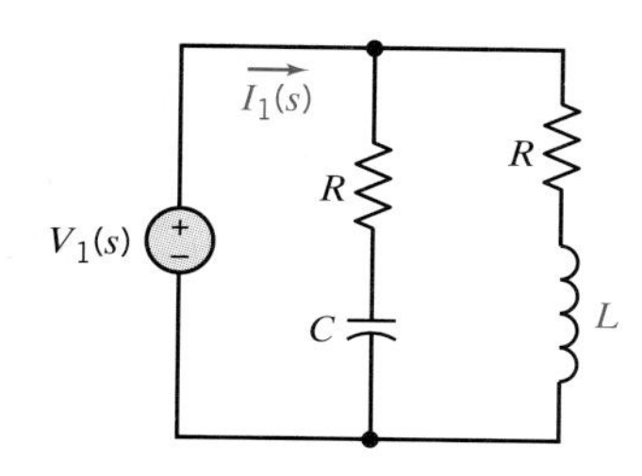

Figure P 14.9-2

Figure P 14.9-3 Bridged-T circuit.

P 14.9-4 (a) Find $H(s) = V_0(s)/V_1(s)$ for the circuit of Figure P 14.9-4. (b) Determine $v_0(t)$ when the initial current in the inductors is zero. Note that the voltage v_0 appears across the series combination of the 150-Ω resistor and the 2-mH inductor.

Answer: $3 - 20\, e^{-5\times 10^4 t}$ V, $t > 0$.

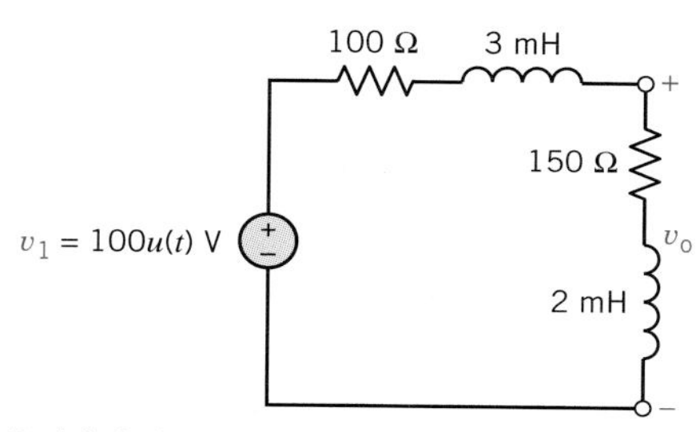

Figure P 14.9-4

P 14.9-5 An electric microphone and its associated circuit can be represented by the circuit shown in Figure P 14.9-5. Determine the transfer function $H(s) = V_0(s)/V(s)$.

Answer: $\dfrac{V_0(s)}{V(s)} = \dfrac{RCs}{(R_1Cs + 1)(2RCs + 1) - 1}$

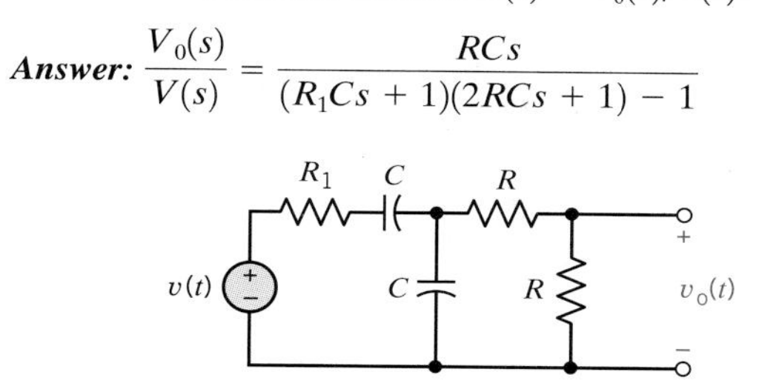

Figure P 14.9-5 Microphone circuit.

P 14.9-6 Determine $v(t)$ for the circuit of Figure P 14.9-6 when $v_s = 1\, u(t)$ V.

Hint: Replace parallel impedances with an equivalent impedance, then use voltage division.

Figure P 14.9-6

P 14.9-7 Modern electronic equipment generally needs one or more dc power supplies. Depending on the type of equipment, either a linear or a switch-mode stabilizer is used. Compared with the linear supply, the switch-mode power supply has some distinct advantages, including smaller size and higher efficiency for the same output power. They are used in television sets and computers. A circuit used to filter the high frequencies from the power supply output voltage is shown in Figure P 14.9-7 (Ruffell 1992). Determine the transfer function $V_0(s)/V(s)$.

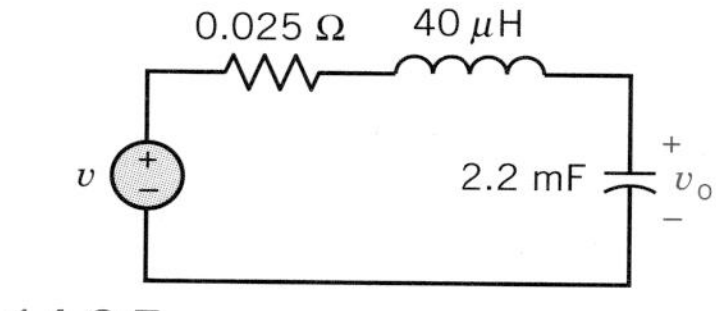

Figure P 14.9-7

P 14.9-8 Engineers had avoided inductance in long-distance circuits because it slows transmission. Heaviside proved that the addition of inductance to a circuit could enable it to transmit without distortion. George A. Campbell of the Bell Telephone Company designed the first practical inductance loading coils, in which the induced field of each winding of wire reinforced that of its neighbors so that the coil supplied proportionally more inductance than resistance. Each one of Campbell's 300 test coils added 0.11 H and 12 Ω at regular intervals along 35 miles of telephone wire (Nahin 1990). The loading coil balanced the effect of the leakage between the telephone wires represented by R and C in Figure P 14.9-8.
(a) Determine the transfer function $V_2(s)/V_1(s)$.
(b) When $C = 1$ mF and $R = (1/\sqrt{2})$ Ω determine $V_2(s)/V_1(s)$.

Answer: (a) $\dfrac{V_2(s)}{V_1(s)} = \dfrac{R}{RLCs^2 + (L + R_xRC)s + R_x + R}$

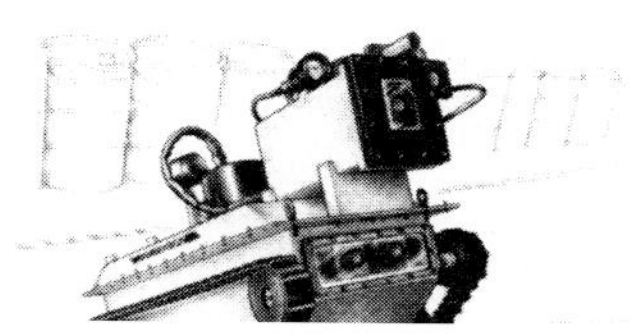

Figure P 14.9-8 Telephone and load coil circuit.

P 14.9-9 Obtain $H(s) = V_2/V_1$ for the circuit shown in Figure P 14.9-9. Assume an ideal op amp.

Answer: $$\frac{V_2(s)}{V_1(s)} = \frac{s - \dfrac{1}{RC}}{s + \dfrac{1}{RC}}$$

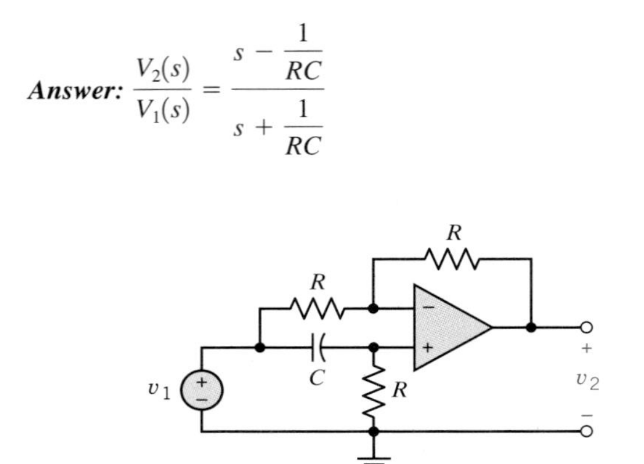

Figure P 14.9-9

P 14.9-10 An op amp circuit for a bandpass filter is shown in Figure P 14.9-10. Determine $V_0(s)/V(s)$. Assume ideal op amps.

Answer: $$\frac{V_o(s)}{V(s)} = \frac{-\dfrac{1}{R_2C_2}s}{s^2 + \dfrac{1}{R_1C_1}s + \dfrac{1}{R_1R_2C_1C_2}}$$

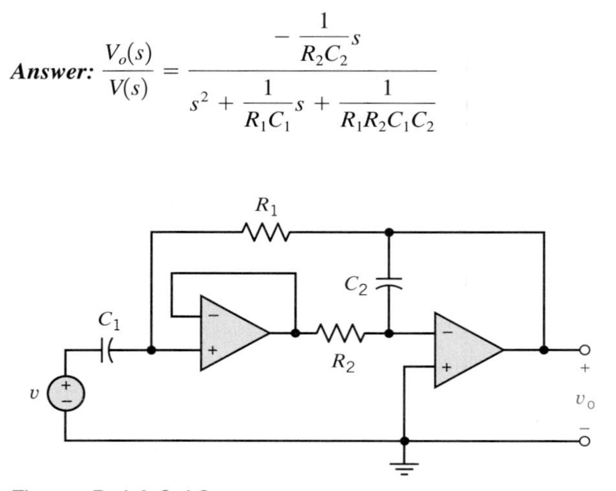

Figure P 14.9-10

P 14.9-11 A high-pass filter using op amps is shown in Figure P 14.9-11. Determine $V_0(s)/V(s)$. Assume ideal op amps.

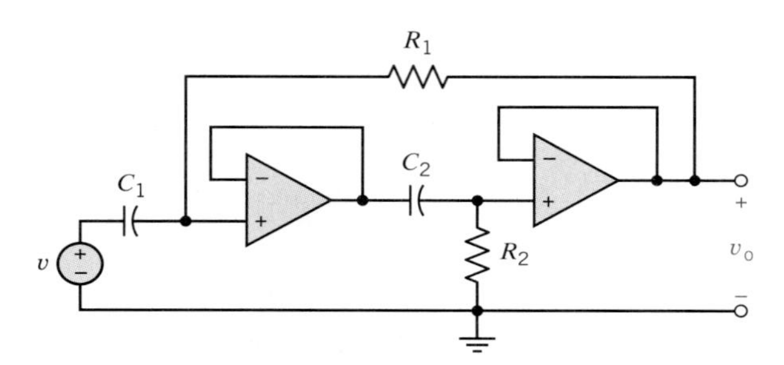

Figure P 14.9-11 High-pass filter.

P 14.9-12 A digital-to-analog converter (DAC) uses an op amp filter circuit shown in Figure P 14.9-12 (Garnett, 1992). The filter receives the pulse output from the DAC and produces the analog voltage, v_o. Determine the transfer function of the filter, $V_0(s)/V(s)$. Assume an ideal op amp.

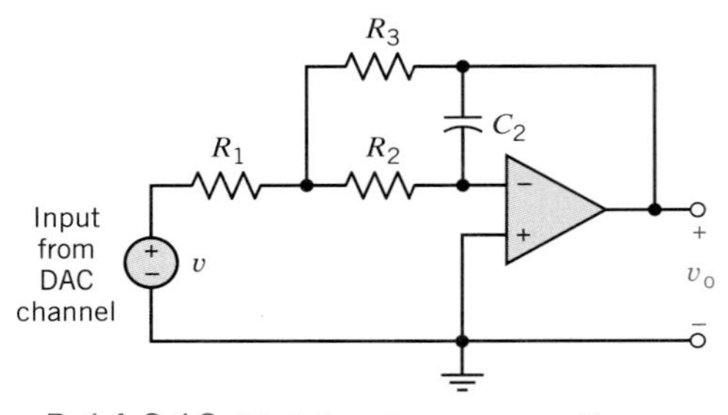

Figure P 14.9-12 Digital-analog converter filter.

P 14.9-13 The circuit shown in Figure P 14.9-13 can be used as a tone control for a stereo amplifier. Determine the transfer function $H(s)$. Assume an ideal op amp.

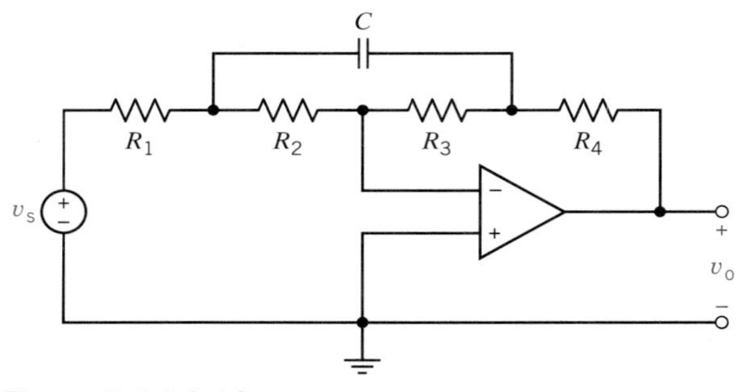

Figure P 14.9-13

P 14.9-14 A circuit has a transfer function

$$H(s) = \frac{2}{s(s+2)(s+1)^2}$$

Find the impulse response $h(t)$. Use the final and initial value theorems to verify the values of $h(t)$ at $t = 0$ and $t = \infty$.
Answer: $h(t) = (1 - 2te^{-t} - e^{-2t})\,u(t)$

P 14.9-15 A circuit has a transfer function

$$H(s) = \frac{400}{s^2 + 400s + 2 \times 10^5}$$

Find the impulse response.
Answer: $h(t) = (e^{-200t} \sin 400t)\,u(t)$

P 14.9-16 A circuit has a transfer function

$$H(s) = \frac{4(s+3)}{s^3 + 4s^2 + 4s}$$

Find the impulse response.

P 14.9-17 A series *RLC* circuit is shown in Figure P 14.9-17. Determine (a) the transfer function $H(s)$, (b) the impulse response, and (c) the step response for each set of parameter values given in the table below.

Table P 14.9-17

	L	C	R
a	2 H	0.025 F	18 Ω
b	2 H	0.025 F	8 Ω
c	1 H	0.391 F	4 Ω
d	2 H	0.125 F	8 Ω

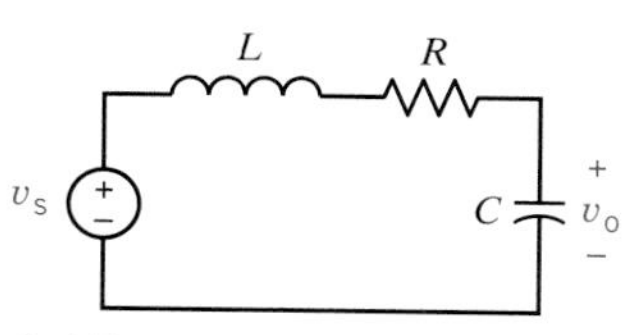

Figure P 14.9-17

P 14.9-18 The impulse response of a circuit is

$$h(t) = \sqrt{2}e^{-t/\sqrt{2}} \sin \frac{t}{\sqrt{2}}, \quad t \geq 0$$

Find $H(s)$.

Answer: $H(s) = \dfrac{1}{s^2 + \sqrt{2}s + 1}$

P 14.9-19 A circuit is described by the transfer function

$$\frac{V_0}{V_1} = H(s) = \frac{9s + 18}{3s^3 + 18s^2 + 39s}$$

Find the response $v_0(t)$ when the input v_1 is a unit impulse at $t = 0$.

P 14.9-20 A circuit is shown in Figure P 14.9-20*a*, which has a response to a unit step, $v_1 = 1\,u(t)$ V, as shown in Figure P 14.9-20*b*. (a) Obtain the transfer function $V_0(s)/V_1(s)$. (b) Using the transfer function and the step response, calculate *R*. (c) Find the value of *R* that would lead to critical damping of the step response.

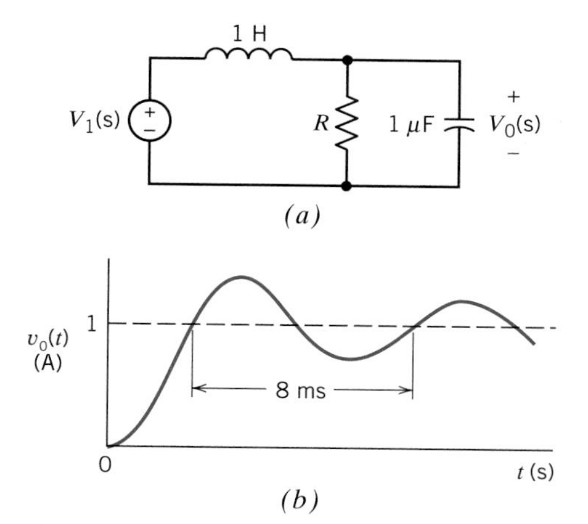

Figure P 14.9-20

Section 14.10 Convolution Theorem

P 14.10-1 Let $f(t)$ denote the 1-second pulse given by $f(t) = u(t) - u(t - 1)$. Determine the convolution $f(t) * f(t)$, which is the convolution of the pulse with itself. Draw $f * f$ versus time.

Answer: $f(t) * f(t) = tu(t) - 2(t - 1)u(t - 1) + (t - 2)u(t - 2)$

P 14.10-2 Consider a pulse of amplitude 2 and a duration of 2 seconds with its starting point at $t = 0$. (a) Find the convolution of this pulse with itself. (b) Draw the convolution $f(t) * f(t)$ versus time.

P 14.10-3 A circuit is shown in Figure P 14.10-3. Determine (a) the transfer function $V_2(s)/V_1(s)$ and (b) the response $v_2(t)$ when $v_1 = tu(t)$.

Answer: $v_2 = t - RC(1 - e^{-t/RC}), t \geq 0$

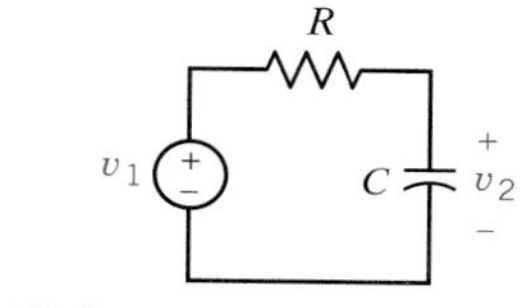

Figure P 14.10-3

P 14.10-4 Find the convolution of $h(t) = t$ and $f(t) = e^{-at}$ for $t > 0$ using the convolution integral and the inverse transform of $H(s)F(s)$.

Answer: $\dfrac{at - 1 + e^{-at}}{a^2}, t > 0$

Section 14.11 Stability

P 14.11-1 The transfer function of a circuit is

$$H(s) = \frac{k}{s + (3 - k)}$$

Determine the required range of k for stable operation.

P 14.11-2 The transfer function of a circuit is

$$H(s) = \frac{8}{s^2 + (6 - k)s + 8}$$

Determine the required range of k for stable operation.

P 14.11-3 The transfer function of a circuit is

$$H(s) = \frac{(s + 2)}{s^2 - 2s + 2}$$

Plot the poles and zeros on the s-plane and determine if the circuit is stable.

P 14.11-4 A circuit is shown in Figure P 14.11-4 with K unspecified. Determine the transfer function of the circuit and find the range of K so that the circuit is stable. Determine K so that the circuit acts as an oscillator.

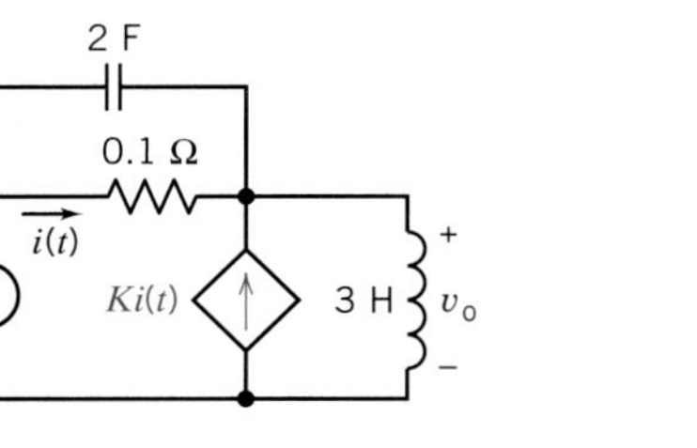

Figure P 14.11-4

P 14.11-5 Determine the transfer function $\frac{V_0(s)}{V_1(s)}$ of the circuit of Figure P 14.11-5. (a) Determine the value of K that causes the circuit to act as an oscillator. (b) Determine the impulse response of the circuit when $K = 1, R = 1\ \text{k}\Omega$, and $C = 0.5$ mF.

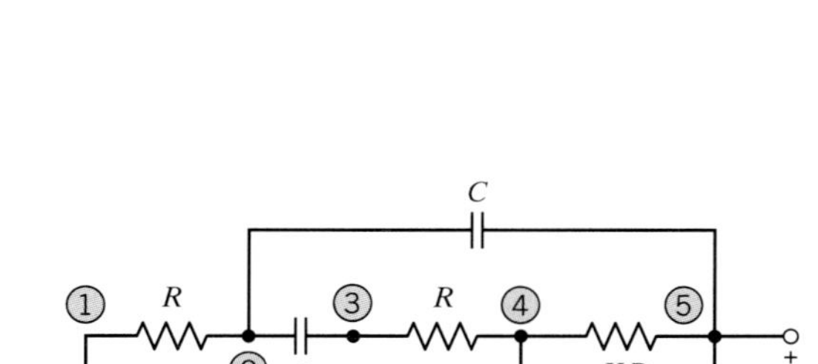

Figure P 14.11-5

PSPICE PROBLEMS

SP 14-1 Plot the response $v_c(t)$ for Example 14.7-2.

SP 14-2 The circuit of Figure SP 14.2*a* has no stored energy at $t = 0$ and the input signal is a pulse, as shown in Figure SP 14.2*b*. Determine $v(t)$ for a period of 10 seconds.

Figure SP 14.2

SP 14-3 Obtain $i(t)$ for the circuit of Figure 14.7-1 and verify the response obtained in Eq. 14.7-3 when $R = 1\ \Omega$, $L = 1/2$ H, $I_0 = 1$ A, and $M = 2$ V.

SP 14-4 Obtain the Bode diagram for the circuit of Problem 14.9-5 using PSpice.

SP 14-5 Plot $20 \log |\boldsymbol{H}(\omega)|$ for Problem 14.9-20 using PSpice and determine the maximum of $|\boldsymbol{H}(\omega)|$ and the frequency at which the maximum occurs.

SP 14-6 Determine and plot the unit step and unit impulse response of the circuit of Figure SP 14.6 for each set of parameter values listed in the following table.

Table SP 14.6

	R	C	K
a	50 kΩ	2 μF	1
b	50 kΩ	2 μF	1.25
c	50 kΩ	2 μF	5.25

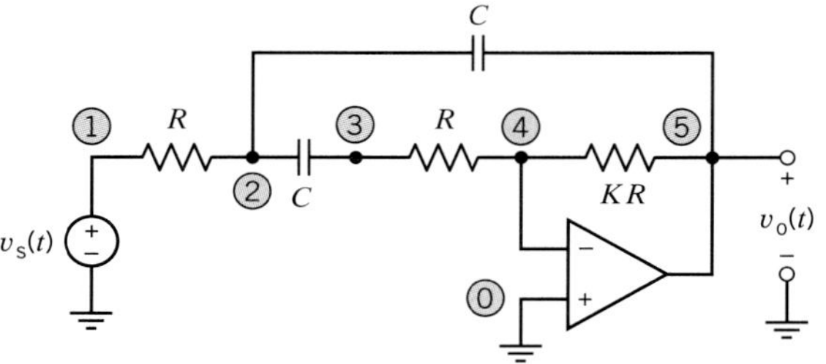

Figure SP 14.6

VERIFICATION PROBLEMS

VP 14-1 Computer analysis of the circuit of Figure VP 14.1 indicates that

$$v_C(t) = 6 + 3.3e^{-2.1t} + 2.7e^{-15.9t}\ \text{V}$$

and $$i_L(t) = 2 + 0.96e^{-2.1t} + 0.04e^{-15.9t}\ \text{A}$$

after the switch opens at time $t = 0$. Verify that this analysis is correct by checking that (a) KVL is satisfied for the mesh consisting of the voltage source, inductor, and 12 Ω resistor and (b) KCL is satisfied at node b.

Hint: Use the given expressions for $i_L(t)$ and $v_C(t)$ to determine expressions for $v_L(t)$, $i_C(t)$, $v_{R1}(t)$, $i_{R2}(t)$, and $i_{R3}(t)$.

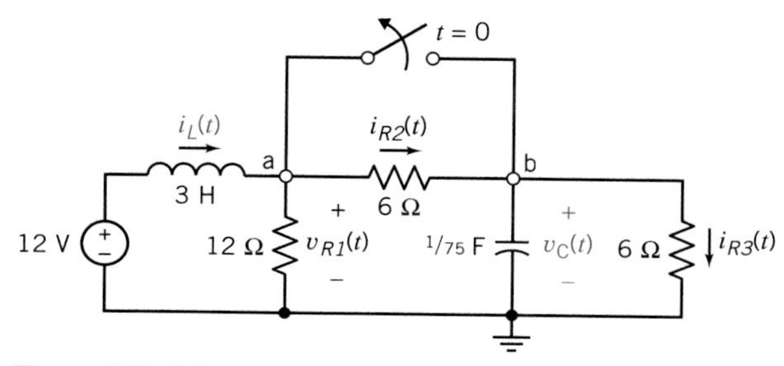

Figure VP 14.1

VP 14-2 Analysis of the circuit of Figure VP 14.2 when $v_C(0) = -12$ V indicates that

$$i_1(t) = 18e^{0.75t}\ \text{A} \quad \text{and} \quad i_2(t) = 20e^{0.75t}\ \text{A}$$

after $t = 0$. Verify that this analysis is correct by representing this circuit, including $i_1(t)$ and $i_2(t)$, in the frequency domain using Laplace transforms. Use $I_1(s)$ and $I_2(s)$ to calculate the element voltages and verify that these voltages satisfy KVL for both meshes.

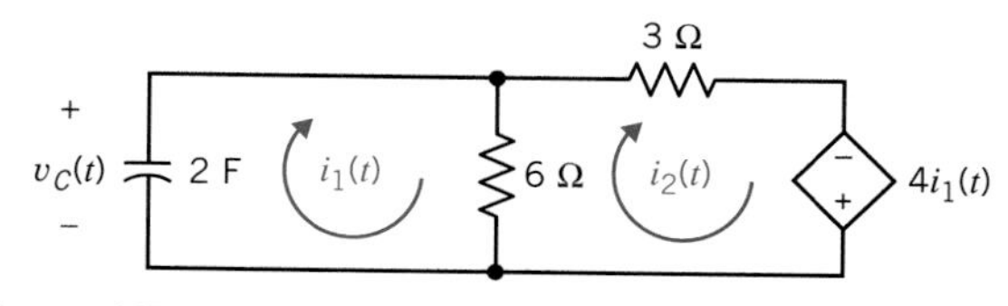

Figure VP 14.2

VP 14-3 Figure VP 14.3 shows a circuit represented in (*a*) the time domain and (*b*) the frequency domain using Laplace transforms. An incorrect analysis of this circuit indicates that

$$I_L(s) = \frac{s+2}{s^2+s+5} \quad \text{and} \quad V_C(s) = \frac{-20(s+2)}{s(s^2+s+5)}$$

(*a*) Use the initial and final value theorems to identify the error in the analysis. (*b*) Correct the error.

Hint: Apparently the error occurred as $V_C(s)$ was calculated from $I_L(s)$.

Answer: $V_C(s) = -\dfrac{20}{s}\left(\dfrac{s+2}{s^2+s+5}\right) + \dfrac{8}{s}$

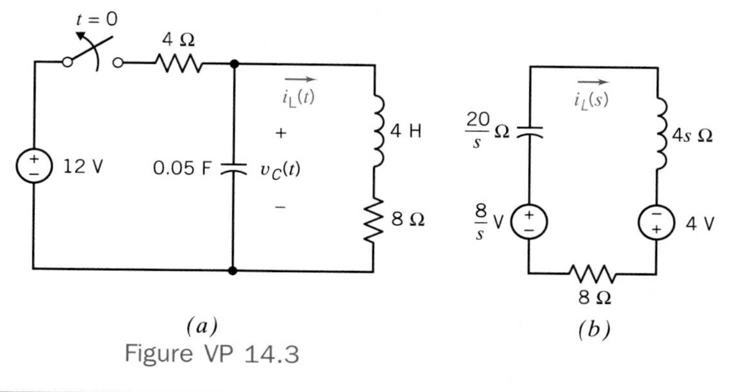

Figure VP 14.3

DESIGN PROBLEMS

DP 14-1 Design the circuit in Figure DP 14.1 to have a step response equal to

$$v_o = 5te^{-4t}u(t)\ \text{V}$$

Hint: Determine the transfer function of the circuit in Figure DP 14.1 in terms of k, R, C, and L. Then determine the Laplace transform of the step response of the circuit in Figure DP 14.1. Next, determine the Laplace transform of the given step response. Finally, determine values of k, R, C, and L that cause the two step responses to be equal.

Answer: Pick $L = 1$ H, then $k = 0.625$ V/V, $R = 8$ Ω, and $C = 0.0625$ F. (This answer is not unique.)

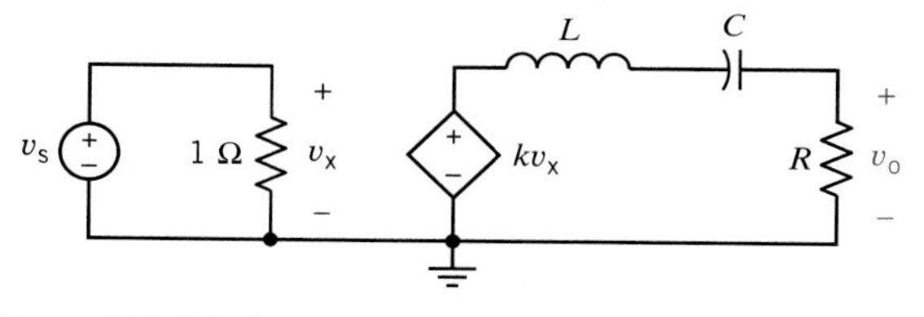

Figure DP 14.1

DP 14-2 Design the circuit in Figure DP 14.1 to have a step response equal to

$$v_o = 5e^{-4t}\sin(2t)\,u(t)\ \text{V}$$

Hint: Determine the transfer function of the circuit in Figure DP 14.1 in terms of k, R, C, and L. Then determine the Laplace transform of the step response of the circuit in Figure DP 14.1. Next, determine the Laplace transform of the given step response. Finally, determine values of k, R, C, and L that cause the two step responses to be equal.
Answer: Pick $L = 1$ H, then $k = 1.25$ V/V, $R = 8\ \Omega$, and $C = 0.05$ F. (This answer is not unique.)

DP 14-3 Design the circuit in Figure DP 14.1 to have a step response equal to

$$v_o = 5(e^{-2t} - e^{-4t})\,u(t)\ \text{V}$$

Hint: Determine the transfer function of the circuit in Figure DP 14.1 in terms of k, R, C, and L. Then determine the Laplace transform of the step response of the circuit in Figure DP 14.1. Next, determine the Laplace transform of the given step response. Finally, determine values of k, R, C, and L that cause the two step responses to be equal.
Answer: Pick $L = 1$ H, then $k = 1.667$ V/V, $R = 6\ \Omega$, and $C = 0.125$ F. (This answer is not unique.)

DP 14-4 Show that the circuit in Figure DP 14.1 cannot be designed to have a step response equal to

$$v_o = 5(e^{-2t} + e^{-4t})\,u(t)\ \text{V}$$

Hint: Determine the transfer function of the circuit in Figure DP 14.1 in terms of k, R, C, and L. Then determine the Laplace transform of the step response of the circuit in Figure DP 14.1. Next, determine the Laplace transform of the given step response. Notice that these two functions have different forms and so cannot be made equal by any choice of values of k, R, C, and L.

DP 14-5 The circuit shown in Figure DP 14.5 represents an oscilloscope probe connected to an oscilloscope. Components C_2 and R_2 represent the input circuitry of the oscilloscope, and C_1 and R_1 represent the probe. Determine the transfer function $H(s) = V_0(s)/V(s)$. Determine the required relationship so that the natural response of the probe is zero. Determine the required relationship so that the step response is equal to the step input to within a gain constant. Is this achievement physically possible?

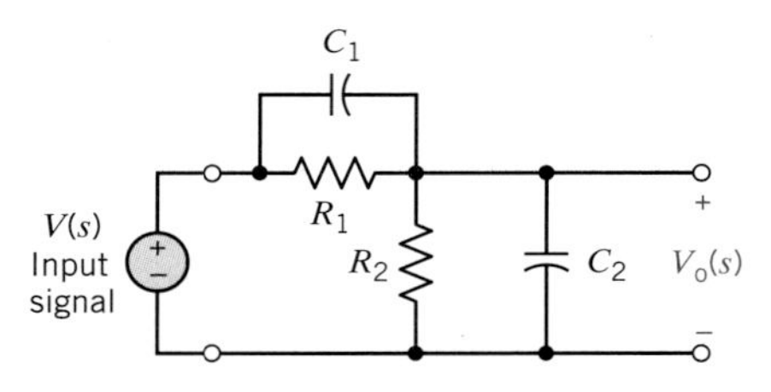

Figure DP 14.5 Oscilloscope probe circuit.

DP 14-6 A bicycle light is a useful accessory if you do a lot of riding at night. By lighting up the road and making you more visible to cars, it reduces the chances of an accident. While generator-powered incandescent lights are the most common type used on bikes, there are a number of reasons why fluorescent lights are more suitable. For one thing, fluorescent lights shine with a brighter light that fully covers the road, the rider, and the bike and really gets the attention of car drivers. Because they are shaped in narrow tubes that can mount alongside a bike's frame, fluorescent lights offer less wind resistance than a comparable headlight with a flat face. When used with a generator, a fluorescent light offers additional advantages over a conventional incandescent light, first, because it is more efficient—giving more light for the same pedaling effort—and, second, because it cannot be burned out by the overvoltage that a generator can produce when speeding down hills. Fluorescent lights also last longer than do incandescent bulbs, especially on a bicycle, where vibrations tend to weaken an incandescent bulb's filament.

A model of a fluorescent light for a bike is shown in Figure DP 14.6. Select L so that the bulb current rapidly rises to its steady-state value and only overshoots its final value by less than 10 percent.

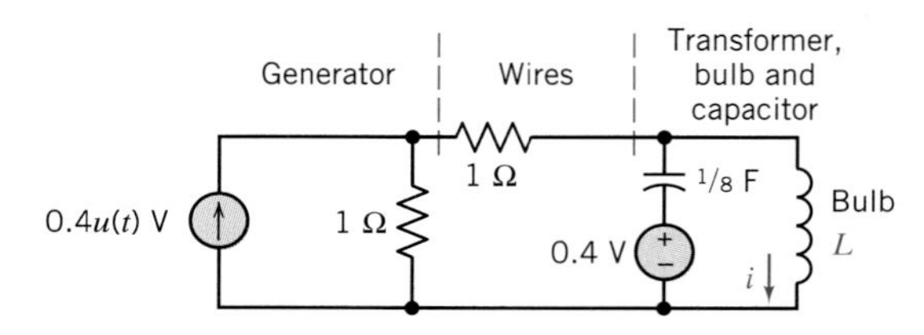

Figure DP 14.6 Fluorescent bicycle light circuit.

CHAPTER 15

Fourier Series and Fourier Transform

Preview

Periodic waveforms arise in many practical circuits. We have, heretofore, developed skills in the analysis of circuits that have sinusoidal waveforms as input sources. Thus, in this chapter we will endeavor to represent a nonsinusoidal periodic waveform by a series of sinusoidal waveforms. Then we can use superposition to find the response of the circuit to each sinusoid in turn and thus, in the aggregate, determine the response to the periodic waveform.

The idea of describing waveforms as a series of sinusoidal functions is useful to today's engineer. We daily experience speech synthesizers and music synthesizers that generate speech sounds or musical sounds as a result of generating an appropriate series of sinusoidal signals that activate a stereo speaker. In this chapter we study the representation of a waveform by a series of sinusoidal signals of integer multiple frequencies.

15.1 Design Challenge

DC POWER SUPPLY

A laboratory power supply uses a nonlinear circuit called a rectifier to convert a sinusoidal voltage input into a dc voltage. The sinusoidal input

$$v_{ac}(t) = A \sin \omega_0 t$$

comes from the wall plug. In this example, $A = 160$ V and $\omega_0 = 377$ rad/s ($f_0 = 60$ Hz). Figure 15.1-1 shows the structure of the power supply. The output of the rectifier is the absolute value of its input, that is,

$$v_s(t) = |A \sin \omega_0 t|$$

The purpose of the rectifier is to convert a signal that has an average value equal to zero into a signal that has an average value that is not zero. The average value of $v_s(t)$ will be used to produce the dc output voltage of the power supply.

The rectifier output is not a sinusoid, but it is a periodic signal. In this chapter we will see that periodic signals can be represented by Fourier series. The Fourier series of $v_s(t)$ will contain a constant, or dc, term and some sinusoidal terms. The purpose of the filter shown in Figure 15.1-1 is to pass the dc term and attenuate the sinusoidal terms. The output of the filter, $v_0(t)$, will be a periodic signal and can be represented by a Fourier series. Because we are designing a dc power supply, the sinusoidal terms in the Fourier series of $v_0(t)$ are undesirable. The sum of these undesirable terms is called the ripple of $v_0(t)$.

The challenge is to design a simple filter so that the dc term of $v_0(t)$ is at least 90 V and the size of the ripple is no larger than 5 percent of the size of the dc term.

We will return to this problem at the end of the chapter, after learning about Fourier series.

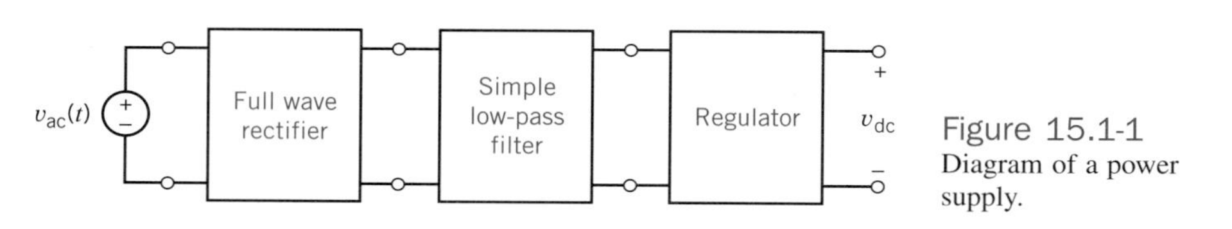

Figure 15.1-1
Diagram of a power supply.

15.2 CHANNELS OF COMMUNICATION

Accuracy is the basic requirement in the transmission of signals. Signals in their original form are transmitted through a channel. For example, Morse code dot-dash transmission over the original Atlantic cable was distorted by the *RC* nature of the cable and its insulation material. Thus, the transmission rate was limited, and the signal was distorted as it traveled over the cable.

The undersea cable transmits information slowly but uses only a narrow frequency channel. The telephone transmits information at an intermediate rate with moderate bandwidth requirements. Television transmits information at a high rate and requires a very wide frequency band. On long telephone lines, repeater amplifiers are spaced along the lines, regenerating the signal at points where it becomes weak. This allows transmission distance to be greatly extended.

Michael I. Pupin arrived in America in 1874 at age fifteen and spent the next five years preparing himself for admission to Columbia University. Pupin graduated with distinction in 1883 and after additional study at the University of Cambridge went to Berlin, where he worked under Helmholtz and G. Kirchhoff, receiving the doctorate in 1889. He then returned to Columbia to teach in the newly formed Department of Electrical Engineering.

Professor Pupin discovered that the periodic insertion of inductance coils in telephone lines would improve their performance by reducing attenuation and distortion. For telephone signals, the line capacitance again limits the high frequencies to about 3000 Hz and reduces accuracy in understanding speech over the telephone. Pupin's loading coils balanced out the capacitive effects, thus extending the frequency range of the lines and improving the accuracy of transmission. For a time such lines were called "Pupinized."

15.3 THE FOURIER SERIES

A number of eighteenth-century mathematicians, including Euler and D. Bernoulli, were aware that a waveform $f(t)$ could be approximately represented by a finite weighted sum of harmonically related sinusoids. Baron Jean-Baptiste-Joseph Fourier proposed in 1807 that a periodic waveform could be broken down into an infinite series of simple sinusoids which, when added together, would construct the exact form of the original waveform. There are certain periodic functions for which an infinite series will exactly represent the function. There are also other certain functions for which the infinite series provides a good, but inexact, representation.

Let us consider the periodic function

$$f(t) = f(t + nT) \quad n = \pm 1, \pm 2, \pm 3, \ldots$$

for every value of t, where T is the period. (The period is the smallest value of T that satisfies the equation above.)

The expression for a finite sum of harmonically related sinusoids called a *Fourier series* is

$$f(t) = a_0 + \sum_{n=1}^{N} a_n \cos n\omega_0 t + \sum_{n=1}^{N} b_n \sin n\omega_0 t \tag{15.3-1a}$$

or

$$f(t) = c_0 + \sum_{n=1}^{N} c_n \cos(n\omega_0 t + \theta_n) \tag{15.3-1b}$$

where $\omega_0 = 2\pi/T$ and a_0, a_n, and b_n (all real) are called the **Fourier trigonometric coefficients.**

In general, it is easier to calculate a_n and b_n than it is to calculate the coefficients c_n and θ_n. We will see in Section 15.4 that this is particularly true when $f(t)$ is symmetric. On the other hand, the Fourier series involving c_n is more convenient for calculating the steady-state response of a linear circuit to a periodic input.

Examining Eq. 15.3-1, we introduce some terminology. For $n = 1$, one cycle covers T seconds and the corresponding sinusoid $c_1 \cos(\omega_0 t + \theta_1)$ is called the *fundamental*. For $n = k$, k cycles fall within T seconds and $c_k \cos(k\omega_0 t + \theta_k)$ is called the kth **harmonic term.** Similarly, ω_0 is called the **fundamental frequency** and $k\omega_0$ is called the kth harmonic frequency. The function $f(t)$ can be represented to any degree of accuracy by increasing the number of terms in its Fourier series. Certain functions can also be represented exactly by an infinite number of terms.

A **Fourier series** is an accurate representation of a periodic signal and consists of the sum of sinusoids at the fundamental and harmonic frequencies.

The nature of the waveform $f(t)$ depends on the amplitude and phase of every possible harmonic component, and we shall find it possible to generate waveforms that have extremely nonsinusoidal characteristics by an appropriate combination of sinusoidal functions. The waveforms $f(t)$ that can be described by Eq. 15.3-1 satisfy the following mathematical properties:

1. $f(t)$ is a single-valued function except at possibly a finite number of points.
2. The integral $\int_{t_0}^{t_0+T} |f(t)|\, dt < \infty$ for any t_0.
3. $f(t)$ has a finite number of discontinuities within the period T.
4. $f(t)$ has a finite number of maxima and minima within the period T.

We shall consider $f(t)$ to represent a voltage or current waveform, and any voltage or current waveform that we can actually produce will certainly satisfy these conditions. We shall assume that the four conditions listed above are always satisfied.

Many electrical waveforms are periodic waveforms and can be described by a Fourier series. Once $f(t)$ is known, the Fourier coefficients may be calculated and the waveform is resolved into a dc term a_0 plus a sum of sinusoidal terms described by a_n and b_n. We will select N, the total number of terms, in order to achieve a faithful representation of $f(t)$ while accepting a certain degree of error.

Recall from calculus that sinusoids whose frequencies are integer multiples of some fundamental frequency $f_0 = 1/T$ form an orthogonal set of functions, that is,

$$\frac{2}{T}\int_0^T \sin\frac{2\pi nt}{T}\cos\frac{2\pi mt}{T}\, dt = 0 \quad \text{all } n, m \tag{15.3-2}$$

and

$$\frac{2}{T}\int_0^T \sin\frac{2\pi nt}{T}\sin\frac{2\pi mt}{T}\, dt = \frac{2}{T}\int_0^T \cos\frac{2\pi nt}{T}\cos\frac{2\pi mt}{T}\, dt$$
$$= \begin{cases} 0 & n \neq m \\ 1 & n = m \neq 0 \end{cases} \tag{15.3-3}$$

The coefficients of Eq. 15.3-1 can be obtained from

$$a_0 = \frac{1}{T}\int_{t_0}^{T+t_0} f(t)\, dt = \text{the average over one period} \tag{15.3-4}$$

$$a_n = \frac{2}{T}\int_{t_0}^{T+t_0} f(t)\cos n\omega_0 t\, dt \quad n > 0 \tag{15.3-5}$$

$$b_n = \frac{2}{T}\int_{t_0}^{T+t_0} f(t)\sin n\omega_0 t\, dt \quad n > 0 \tag{15.3-6}$$

We now demonstrate how Eqs. 15.3-4 through 15.3-6 are obtained using the orthogonal relationships of Eqs. 15.3-2 and 15.3-3. For example, to obtain a_k we multiply Eq. 15.3-1 by $\cos k\omega_0 t$ and integrate both sides over one period of T, obtaining

$$\int_0^T f(t)\cos k\omega_0 t\, dt = \int_0^T a_0 \cos k\omega_0 t\, dt$$
$$+ \sum_{n=1}^{N}\int_0^T (a_n \cos n\omega_0 t + b_n \sin n\omega_0 t)\cos k\omega_0 t\, dt \tag{15.3-7}$$

Evaluating the right-hand side of Eq. 15.3-7, we find that the only nonzero term is for $n = k$, so that

$$\int_0^T f(t) \cos k\omega_0 t \, dt = a_k(T/2) \qquad (15.3\text{-}8)$$

Therefore, we have derived Eq. 15.3-5 for all a_n. Following a similar approach and multiplying Eq. 15.3-1 by $\sin k\omega_0 t$, we can find Eq. 15.3-6. Equation 15.3-4 provides the dc term, which is the average value of $f(t)$ over the period T.

Example 15.3-1

Determine the Fourier series for the periodic rectangular wave shown in Figure 15.3-1. Plot the Fourier series approximation of the waveform when $N = 7$.

Solution

The fundamental radian frequency is $\omega_0 = 2\pi/T$. First, let us find a_0, the average value of $f(t)$ over the period T. Hence we obtain

$$a_0 = \frac{1}{T}\int_{-T/2}^{T/2} f(t)\, dt = \frac{1}{T}\int_{-T/4}^{T/4} 1 \, dt = \frac{1}{2}$$

We choose to integrate from $-T/2$ to $+T/2$ for convenience. Furthermore, the function is only nonzero from $-T/4$ to $+T/4$ for the selected period of integration. We call $f(t)$ of Figure 15.3-1 an *even function* since $f(t) = f(-t)$.

An **even function** $f(t)$ exhibits symmetry around the vertical axis at $t = 0$ so that $f(t) = f(-t)$.

Hence, Eq. 15.3-6 with $f(t) = 1$ for $-T/4 < t < T/4$ is

$$b_n = \frac{2}{T}\int_{-T/4}^{T/4} 1 \sin n\omega_0 t \, dt = 0$$

Therefore, all b_n from Eq. 15.3-6 will be zero. Thus, solving for a_n using Eq. 15.3-5, we obtain

$$\begin{aligned} a_n &= \frac{2}{T}\int_{-T/4}^{T/4} 1 \cos n\omega_0 t \, dt \\ &= \frac{2}{T\omega_0 n} \sin n\omega_0 t \Big|_{-T/4}^{T/4} \\ &= \frac{1}{\pi n}\left[\sin\left(\frac{\pi n}{2}\right) - \sin\left(\frac{-\pi n}{2}\right)\right] \end{aligned} \qquad (15.3\text{-}9)$$

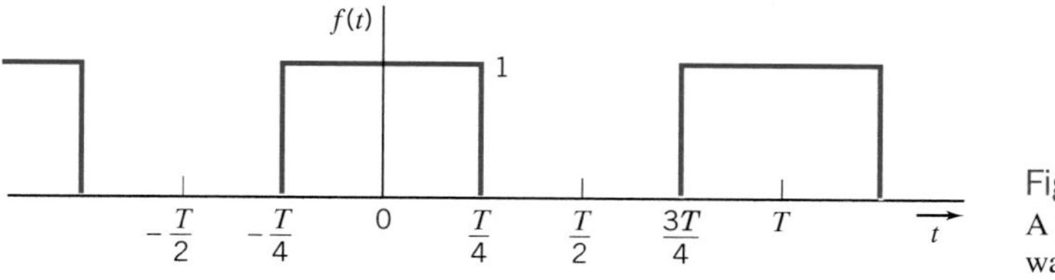

Figure 15.3-1 A periodic rectangular wave.

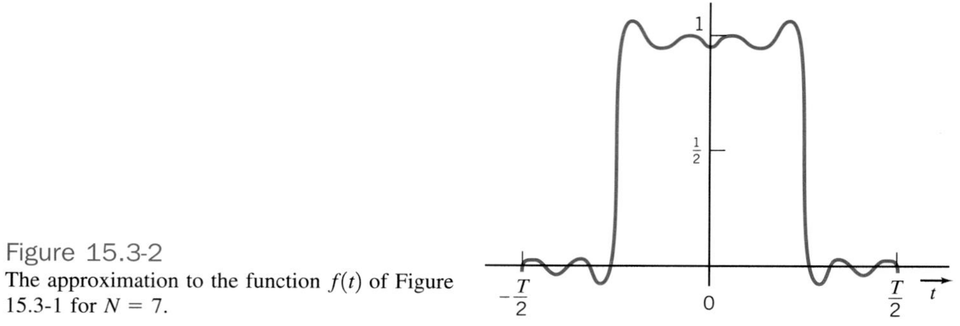

Figure 15.3-2
The approximation to the function $f(t)$ of Figure 15.3-1 for $N = 7$.

Equation 15.3-9 is equal to zero when $n = 2, 4, 6, \ldots$ and

$$a_n = \frac{2(-1)^q}{\pi n} \tag{15.3-10}$$

where $q = (n - 1)/2$ and $n = 1, 3, 5, \ldots$. Thus the Fourier series is

$$f(t) = \frac{1}{2} + \sum_{n=1,\text{ odd}}^{N} \frac{2(-1)^q}{\pi n} \cos n\omega_0 t$$

and $q = (n - 1)/2$. Then, a_1 and a_3 are

$$a_1 = \frac{2}{\pi} \quad \text{and} \quad a_3 = \frac{-2}{3\pi}$$

Similarly,

$$a_5 = \frac{2}{5\pi} \quad \text{and} \quad a_7 = \frac{-2}{7\pi}$$

Thus, the a_n coefficients decrease as $1/n$, and the coefficient of the seventh harmonic is one-seventh of the value of the fundamental. The approximation of the function $f(t)$ when $N = 7$ is shown in Figure 15.3-2. Note that the error is readily discernible. Nevertheless, the approximation may be satisfactory in many cases. This approximation neglects the terms for $N \geq 9$ where $a_9 = 2/9\pi = 0.0707$. In order to attain a good approximation we may seek to include all terms whose magnitude exceeds a specified percentage of the magnitude of the fundamental coefficient a_1.

Exercise 15.3-1 Determine the Fourier series when $f(t) = K$, a constant.
Answer: $a_0 = K, a_n = b_n = 0$ for $n \geq 1$

Exercise 15.3-2 Determine the Fourier series when $f(t) = A \cos \omega_0 t$.
Answer: $a_0 = 0, a_1 = A, a_n = 0$ for $n > 1, b_n = 0$

15.4 SYMMETRY OF THE FUNCTION $f(t)$

Four types of symmetry can be readily recognized and then utilized to simplify the task of calculating the Fourier coefficients. They are

1. Even-function symmetry,
2. Odd-function symmetry,

3. Half-wave symmetry,
4. Quarter-wave symmetry.

A function is *even* when $f(t) = f(-t)$, and a function is *odd* when $f(t) = -f(-t)$. The function shown in Figure 15.3-1 is an even function. For even functions all $b_n = 0$ and

$$a_n = \frac{4}{T}\int_0^{T/2} f(t) \cos n\omega_0 t \, dt$$

For odd functions all $a_n = 0$ and

$$b_n = \frac{4}{T}\int_0^{T/2} f(t) \sin n\omega_0 t \, dt$$

An example of an odd function is $\sin \omega_0 t$. Another odd function is shown in Figure 15.4-1.

Half-wave symmetry for a function $f(t)$ is obtained when

$$f(t) = -f\left(t - \frac{T}{2}\right) \tag{15.4-1}$$

These half-wave symmetric waveforms have the property that the second half of each period looks like the first half turned upside down. The function shown in Figure 15.4-1 has half-wave symmetry. If a function has half-wave symmetry, then both a_n and b_n are zero for even values of n. We see that $a_0 = 0$ for half-wave symmetry since the average value of the function over one period is zero.

Quarter-wave symmetry is a term used to describe a function that has half-wave symmetry and, in addition, has symmetry about the midpoint of the positive and negative half-cycles. An example of an odd function with quarter-wave symmetry is shown in Figure 15.4-2. If a function is odd and quarter-wave, then $a_0 = 0$, $a_n = 0$ for all n, $b_n = 0$ for even n. For odd n, b_n is given by

$$b_n = \frac{8}{T}\int_0^{T/4} f(t) \sin n\omega_0 t \, dt$$

If a function is even and quarter-wave, then $a_0 = 0$, $b_n = 0$ for all n, $a_n = 0$ for even n. For odd n, a_n is given by

$$a_n = \frac{8}{T}\int_0^{T/4} f(t) \cos n\omega_0 t \, dt$$

The calculation of the Fourier coefficients and the associated effects of symmetry of the waveform $f(t)$ are summarized in Table 15.4-1. Often the calculation of the Fourier series can be simplified by judicious selection of the origin ($t = 0$), since the analyst usually has the choice to arbitrarily select this point.

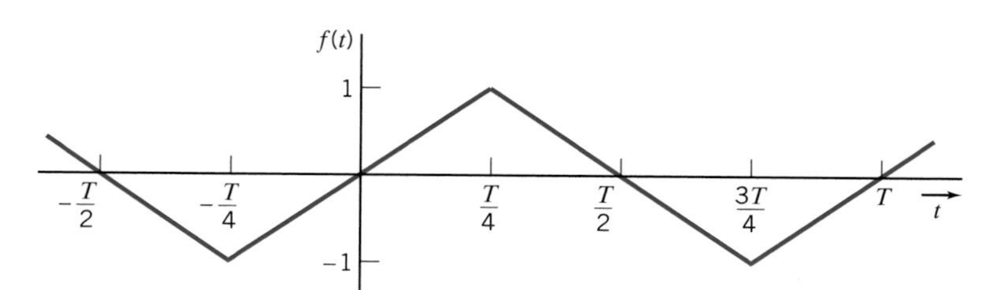

Figure 15.4-1 An example of an odd function with half-wave symmetry.

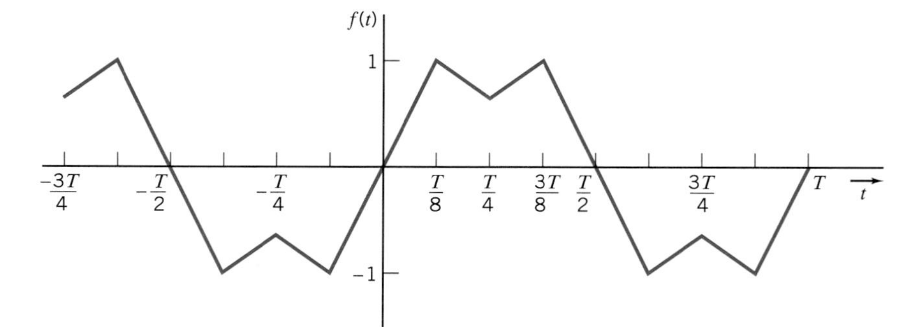

Figure 15.4-2 An odd function with quarter-wave symmetry.

Table 15.4-1 **Fourier Series and Symmetry**

Symmetry	Fourier Coefficients
1. Odd function $f(t) = -f(-t)$	$a_n = 0$ for all n $b_n = \frac{4}{T}\int_0^{T/2} f(t) \sin n\omega_0 t \, dt$
2. Even function $f(t) = f(-t)$	$b_n = 0$ for all n $a_n = \frac{4}{T}\int_0^{T/2} f(t) \cos n\omega_0 t \, dt$
3. Half-wave symmetry $f(t) = -f\left(t - \frac{T}{2}\right)$	$a_0 = 0$ $a_n = 0$ for even n $b_n = 0$ for even n $a_n = \frac{4}{T}\int_0^{T/2} f(t) \cos n\omega_0 t \, dt$ for odd n $b_n = \frac{4}{T}\int_0^{T/2} f(t) \sin n\omega_0 t \, dt$ for odd n
4. Quarter-wave symmetry Half-wave symmetric and symmetric about the midpoints of the positive and negative half-cycles	A. Odd function: $a_0 = 0, a_n = 0$ for all n $b_n = 0$ for even n $b_n = \frac{8}{T}\int_0^{T/4} f(t) \sin n\omega_0 t \, dt$ for odd n B. Even function: $a_0 = 0, b_n = 0$ for all n $a_n = 0$ for even n $a_n = \frac{8}{T}\int_0^{T/4} f(t) \cos n\omega_0 t \, dt$ for odd n

Example 15.4-1

Determine the Fourier series for the triangular waveform $f(t)$ shown in Figure 15.4-3. Each increment of time on the horizontal grid is $\pi/8$ s, and the maximum and minimum values of $f(t)$ are 4 and -4, respectively. Determine how many terms are required (find N) in order to retain all terms whose magnitude exceeds 2% of the magnitude of the fundamental term.

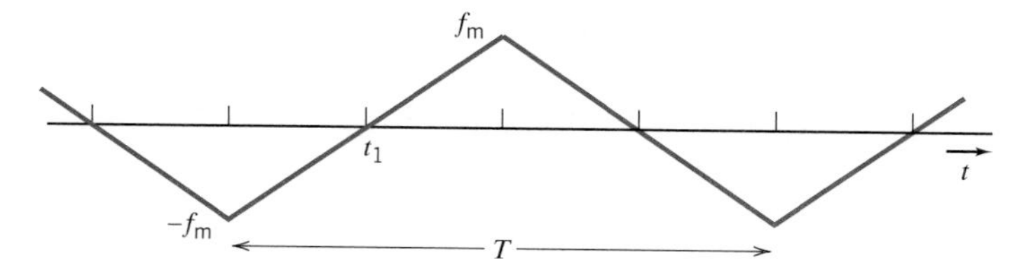

Figure 15.4-3 Waveform $f(t)$ for Example 15.4-1, where $f_m = 4$ and $T = \frac{\pi}{2}$ s.

Solution

We will attempt to obtain the most advantageous form of symmetry. If we choose t_1 as the origin, we obtain an odd function with half-wave symmetry. Since the function is then also symmetrical about the midpoint of the positive and negative half-cycles, we obtain quarter-wave symmetry.

We determine the fundamental frequency by noticing that $T = 4(\pi/8) = \pi/2$ s. Therefore,

$$\omega_0 = \frac{2\pi}{T} = 4 \text{ rad/s}$$

Using Table 15.4-1, we use entry 4A to obtain $a_0 = 0$, $a_n = 0$ for all n, $b_n = 0$ for even n. For odd n

$$b_n = \frac{8}{T}\int_0^{T/4} f(t) \sin n\omega_0 t \, dt$$

For the interval $0 \le t \le T/4$ we have

$$f(t) = \frac{f_m}{T/4}t = \frac{4f_m}{T}t$$

and since $T = \pi/2$ and f_m = maximum value = 4, we have

$$f(t) = \frac{4(4)}{\pi/2}t = \frac{32}{\pi}t \quad 0 \le t \le T/4$$

Then, we obtain

$$\begin{aligned} b_n &= \frac{8}{\pi/2}\left(\frac{32}{\pi}\right)\int_0^{T/4} t \sin n\omega_0 t \, dt \\ &= \frac{512}{\pi^2}\left[\frac{\sin n\omega_0 t}{n^2\omega_0^2} - \frac{t \cos n\omega_0 t}{n\omega_0}\right]_0^{T/4} \\ &= \frac{32}{\pi^2 n^2} \sin \frac{n\pi}{2} \quad \text{for odd } n \end{aligned}$$

Therefore, the Fourier series is

$$f(t) = 3.24 \sum_{n=1}^{N} \frac{1}{n^2} \sin \frac{n\pi}{2} \sin n\omega_0 t \quad \text{for odd } n$$

The first four terms of the Fourier series (up to and including $N = 7$) are

$$f(t) = 3.24(\sin 4t - \tfrac{1}{9}\sin 12t + \tfrac{1}{25}\sin 20t - \tfrac{1}{49}\sin 28t)$$

The next harmonic is for $n = 9$, and its magnitude is $3.24/81 = 0.04$, which is less than 2 percent of $a_1 = 3.24$. Therefore, we can attain the desired approximation by using the first four nonzero terms (including $N = 7$) of the series.

The Fourier series for selected functions $f(t)$ are provided in Table 15.4-2.

Table 15.4-2 **The Fourier Series for Selected Periodic Waveforms**

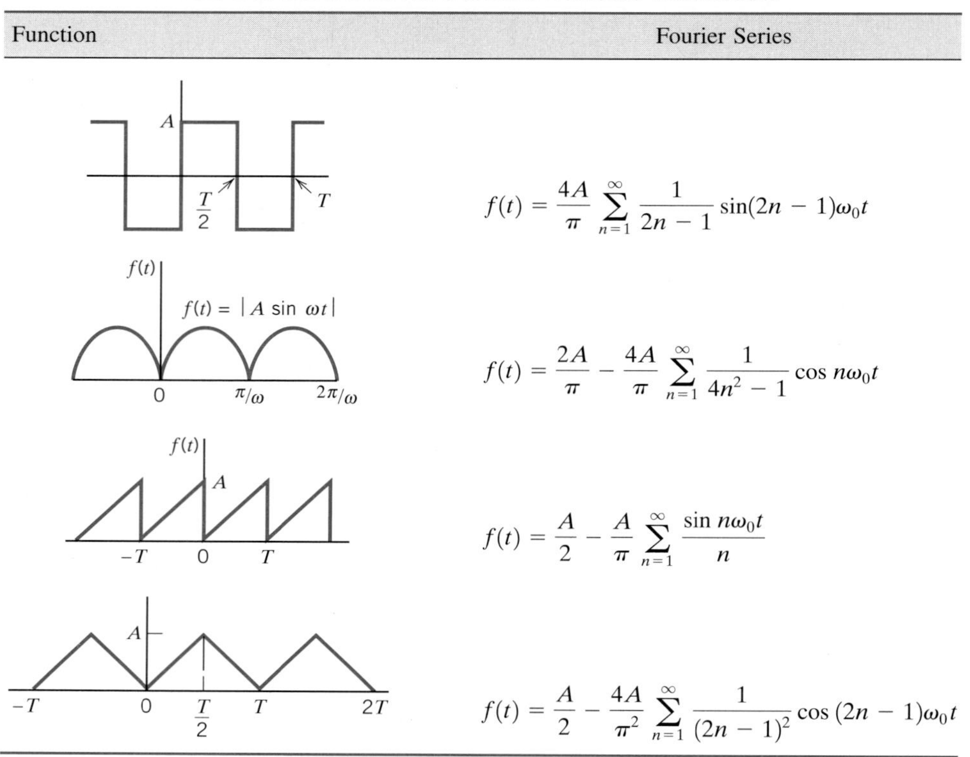

Function	Fourier Series
	$f(t) = \dfrac{4A}{\pi} \sum_{n=1}^{\infty} \dfrac{1}{2n-1} \sin(2n-1)\omega_0 t$
$f(t) = \lvert A \sin \omega t \rvert$	$f(t) = \dfrac{2A}{\pi} - \dfrac{4A}{\pi} \sum_{n=1}^{\infty} \dfrac{1}{4n^2 - 1} \cos n\omega_0 t$
	$f(t) = \dfrac{A}{2} - \dfrac{A}{\pi} \sum_{n=1}^{\infty} \dfrac{\sin n\omega_0 t}{n}$
	$f(t) = \dfrac{A}{2} - \dfrac{4A}{\pi^2} \sum_{n=1}^{\infty} \dfrac{1}{(2n-1)^2} \cos(2n-1)\omega_0 t$

Exercise 15.4-1 Determine the Fourier series for the waveform $f(t)$ shown in Figure E 15.4-1. Each increment of time on the horizontal axis is $\pi/8$ s, and the maximum and minimum are $+1$ and -1, respectively.

Answer: $f(t) = \dfrac{4}{\pi} \sum_{n=1}^{N} \dfrac{1}{n} \sin n\omega_0 t; \quad n \text{ odd}, \ \omega_0 = 4 \text{ rad/s}$

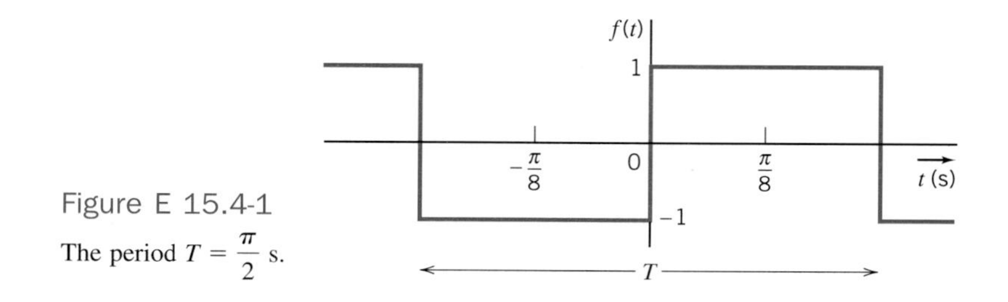

Figure E 15.4-1
The period $T = \dfrac{\pi}{2}$ s.

Exercise 15.4-2 Determine the Fourier series for the waveform $f(t)$ shown in Figure E 15.4-2. Each increment of time on the horizontal grid is $\pi/6$ s, and the maximum and minimum values of $f(t)$ are 2 and -2, respectively.

Answer: $f(t) = \dfrac{-24}{\pi^2} \sum_{n=1}^{N} \dfrac{1}{n^2} \sin(n\pi/3) \sin n\omega_0 t; \quad n \text{ odd}, \omega_0 = 2 \text{ rad/s}$

Exercise 15.4-3 For the periodic signal $f(t)$ shown in Figure E 15.4-3, determine if the Fourier series contains (a) sine and cosine terms and (b) even harmonics and (c) calculate the dc value.

Answer: (a) Yes, both sine and cosine terms; (b) no even harmonics; (c) $a_0 = 0$.

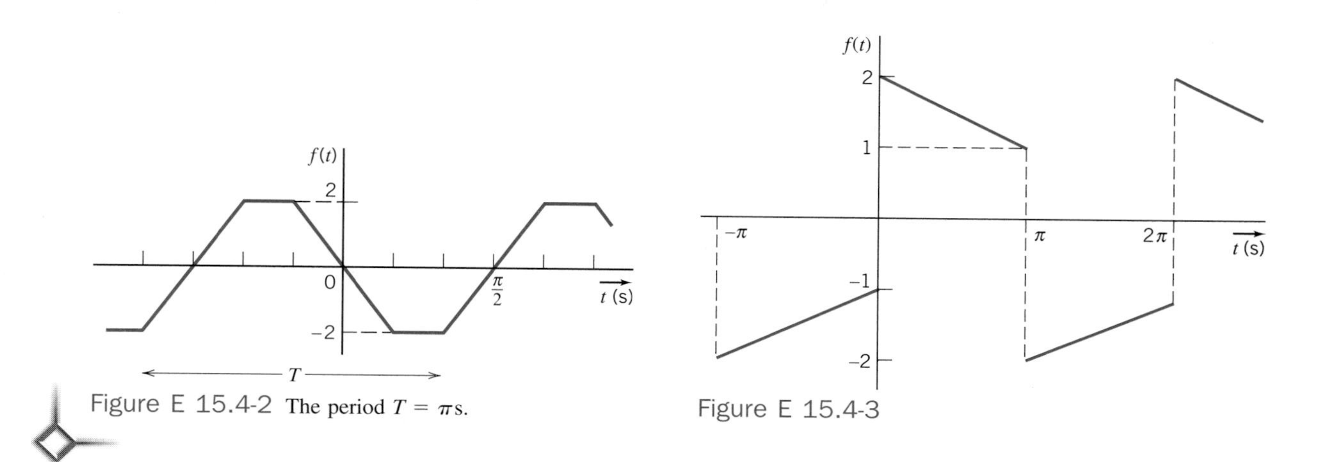

Figure E 15.4-2 The period $T = \pi$ s.

Figure E 15.4-3

15.5 EXPONENTIAL FORM OF THE FOURIER SERIES

The form of the Fourier series of Eq. 15.3-1a is often called the sine-cosine form or the rectangular form. An alternative is

$$f(t) = c_0 + \sum_{n=1}^{N} c_n \cos(n\omega_0 t + \theta_n) \tag{15.5-1}$$

where $c_0 = a_0$ = average value of $f(t)$ and $\mathbf{C}_n = (a_n - jb_n)/2$. Therefore

$$C_n = \sqrt{a_n^2 + b_n^2} \quad \text{and} \quad \theta_n = \begin{cases} -\tan^{-1}\left(\dfrac{b_n}{a_n}\right) & \text{if } a_n > 0 \\ 180^\circ - \tan^{-1}\left(\dfrac{b_n}{a_n}\right) & \text{if } a_n < 0 \end{cases}$$

or

$$a_n = 2\,C_n \cos\theta_n \quad \text{and} \quad b_n = 2\,C_n \sin\theta_n$$

Since the function $\cos(n\omega_0 t + \theta_n)$ may be written in *exponential form* using Euler's identity (see Appendix D), we have, with $N = \infty$,

$$f(t) = C_0 + \sum_{\substack{n=-\infty \\ n\neq 0}}^{\infty} \mathbf{C}_n e^{jn\omega_0 t} = \sum_{-\infty}^{\infty} \mathbf{C}_n e^{jn\omega_0 t} \tag{15.5-2}$$

where $\mathbf{C}_n$ are the **complex coefficients** defined by

$$\mathbf{C}_n = \frac{1}{T}\int_{t_0}^{t_0+T} f(t) e^{-jn\omega_0 t}\, dt \tag{15.5-3}$$

and where

$$\mathbf{C}_n = C_n e^{j\theta_n}$$

Furthermore, $\mathbf{C}_n = \mathbf{C}^*_{-n}$ so that the coefficients for negative n are the complex conjugates of $\mathbf{C}_n$ for positive n.

Example 15.5-1

Determine the complex Fourier series for the waveform shown in Figure 15.5-1.

Solution

The average value of $f(t)$ is zero, so $C_0 = 0$. As shown in Figure 15.5-1, the function is even. Then, using Eq. 15.5-3, we select $t_0 = -T/2$ and define $jn\omega_0 = m$ to obtain

$$\begin{aligned}
\mathbf{C}_n &= \frac{1}{T}\int_{-T/2}^{T/2} f(t)e^{-jn\omega_0 t}\,dt \\
&= \frac{1}{T}\int_{-T/2}^{-T/4} -Ae^{-mt}\,dt + \frac{1}{T}\int_{-T/4}^{T/4} Ae^{-mt}\,dt + \frac{1}{T}\int_{T/4}^{T/2} -Ae^{-mt}\,dt \\
&= \frac{A}{mT}\left(e^{-mt}\Big|_{-T/2}^{-T/4} - e^{-mt}\Big|_{-T/4}^{T/4} + e^{-mt}\Big|_{T/4}^{T/2} \right) \\
&= \frac{A}{jn\omega_0 T}(2e^{jn\pi/2} - 2e^{-jn\pi/2} + e^{-jn\pi} - e^{jn\pi}) \\
&= \frac{A}{2\pi n}\left(4\sin\frac{n\pi}{2} - 2\sin(n\pi)\right) \\
&= \begin{cases} 0 & \text{for even } n \\ \dfrac{2A}{n\pi}\sin n\dfrac{\pi}{2} & \text{for odd } n \end{cases} \\
&= A\frac{\sin x}{x}
\end{aligned}$$

where $x = n\pi/2$.

Since $f(t)$ is an even function, all $\mathbf{C}_n$ are real. We found $\mathbf{C}_n = 0$ for n even. Furthermore, we have for $n = 1$

$$\mathbf{C}_1 = \frac{A\sin\pi/2}{\pi/2} = \frac{2A}{\pi} = \mathbf{C}_{-1}$$

For $n = 2$ we obtain

$$\mathbf{C}_2 = \mathbf{C}_{-2} = A\frac{\sin\pi}{\pi} = 0$$

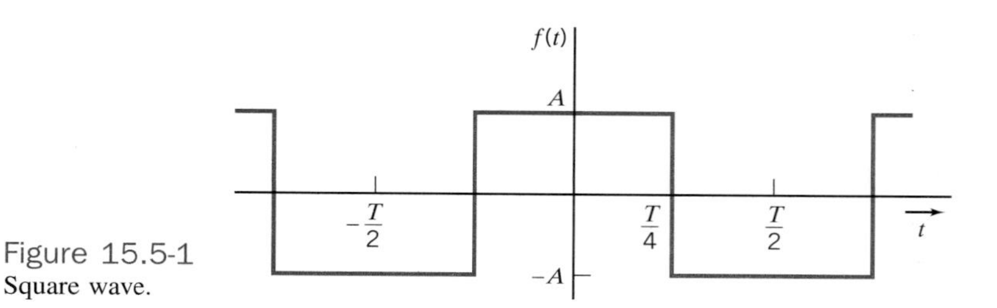

Figure 15.5-1
Square wave.

and for $n = 3$

$$\mathbf{C}_3 = \frac{A\sin(3\pi/2)}{3\pi/2} = \frac{-2A}{3\pi} = \mathbf{C}_{-3}$$

The complex Fourier series is

$$\begin{aligned} f(t) &= \cdots \frac{-2A}{3\pi} e^{-j3\omega_0 t} + \frac{2A}{\pi} e^{-j\omega_0 t} + \frac{2A}{\pi} e^{j\omega_0 t} + \frac{-2A}{3\pi} e^{j3\omega_0 t} + \cdots \\ &= \frac{2A}{\pi}(e^{j\omega_0 t} + e^{-j\omega_0 t}) + \frac{-2A}{3\pi}(e^{j3\omega_0 t} + e^{-j3\omega_0 t}) + \cdots \\ &= \frac{4A}{\pi}\cos\omega_0 t - \frac{4A}{3\pi}\cos 3\omega_0 t + \cdots \\ &= \frac{4A}{\pi}\sum_{\substack{n=1 \\ n=\text{odd}}}^{\infty} \frac{(-1)^q}{n}\cos n\omega_0 t \quad \text{where } q = \frac{n-1}{2} \end{aligned}$$

Note that the exponential series extends from $n = -\infty$ to $n = +\infty$. All voltages and currents of a circuit are real functions. Furthermore, for real $f(t)$ it follows that

$$|\mathbf{C}_n| = |\mathbf{C}_{-n}|$$

If it is known that $f(t)$ is real, only the positive-frequency part of the spectrum need be shown. If, in addition to being real, $f(t)$ is an odd function, then the coefficients $\mathbf{C}_n$ are all purely imaginary.

Example 15.5-2

Determine the complex Fourier coefficients for the square wave shown in Figure 15.5-2.

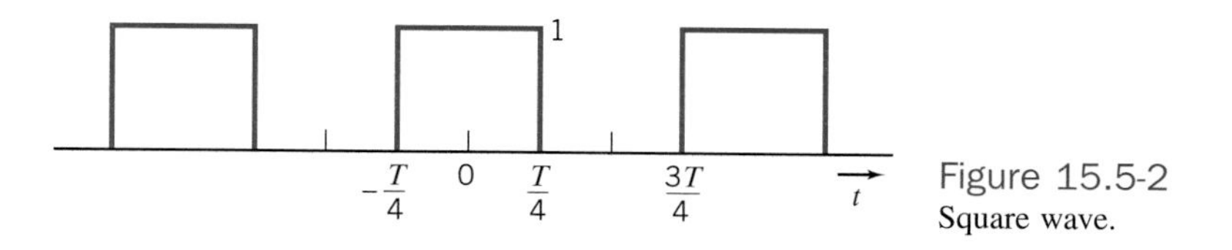

Figure 15.5-2
Square wave.

Solution

We note that the function is even, so we expect the coefficients to all be real. Using Eq. 15.5-3 and defining $m = jn\omega_0$, we have

$$\begin{aligned} \mathbf{C}_n &= \frac{1}{T}\int_{-T/4}^{T/4} 1e^{-mt}\,dt \\ &= \frac{1}{-mT} e^{-mt}\Big|_{-T/4}^{T/4} \\ &= \frac{1}{-mT}(e^{-mT/4} - e^{+mT/4}) \\ &= \frac{1}{-jn2\pi}(e^{-jn\pi/2} - e^{+jn\pi/2}) \\ &= 0 \quad n \text{ even}, n \neq 0 \\ &= (-1)^{(n-1)/2}\frac{1}{n\pi} \quad n \text{ odd} \end{aligned}$$

To find C_0 we determine the average value as

$$C_0 = \frac{1}{T}\int_0^T f(t)\,dt = \frac{1}{T}\int_{-T/4}^{T/4} 1\,dt = \frac{1}{2}$$

Finally, we note that the conditions of symmetry will imply that if a function is half-wave symmetric, $\mathbf{C}_n = 0$ for all even values of n and only odd harmonics exist (see Table 15.4-1). The Fourier coefficients for the complex Fourier series of selected waveforms are given in Table 15.5-1.

Exercise 15.5-1 Find the exponential Fourier coefficients for the function shown in Figure E 15.5-1.

Answer: $\mathbf{C}_n = 0$ for even n, $\mathbf{C}_n = \dfrac{2}{jn\pi}$ for odd n

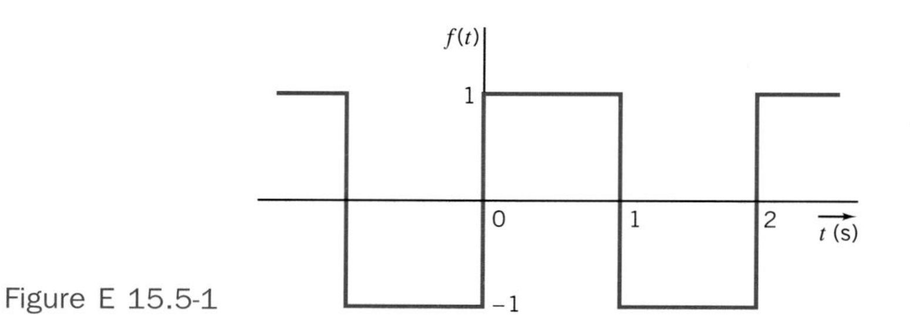

Figure E 15.5-1

Exercise 15.5-2 Determine the Fourier series in terms of sine and cosine terms for the waveform shown in Figure E 15.5-2.

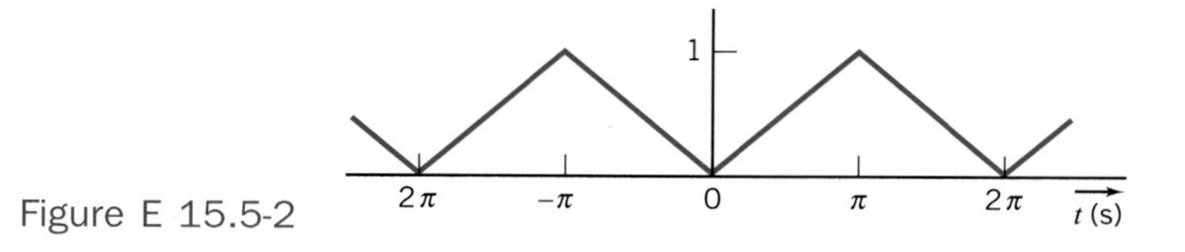

Figure E 15.5-2

Exercise 15.5-3 Determine the complex Fourier coefficients for the waveform shown in Figure E 15.5-3.

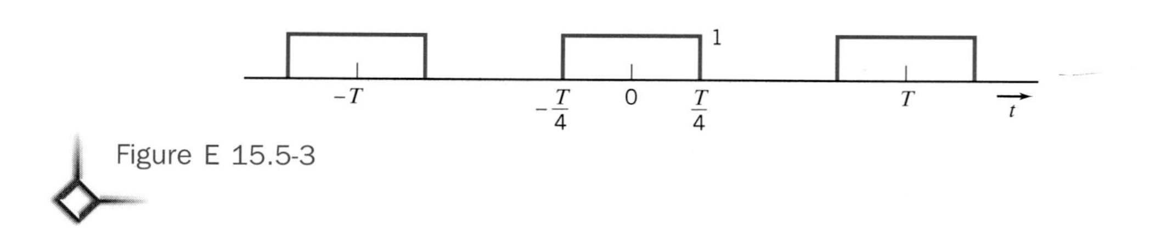

Figure E 15.5-3

Table 15.5-1 **Complex Fourier Coefficients for Selected Waveforms**

Waveform	Name of Waveform and Equation	Symmetry	$\mathbf{C}_n$
1.	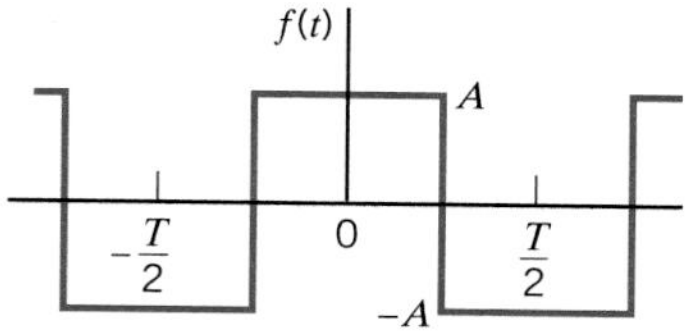 **Square wave** $f(t) = \begin{cases} A, & \frac{-T}{4} < t < \frac{T}{4} \\ -A, & \frac{T}{4} \le t < \frac{3T}{4} \end{cases}$	Even	$= A\frac{\sin n\pi/2}{n\pi/2}$, n odd; $= 0$, $n = 0$ and n even
2.	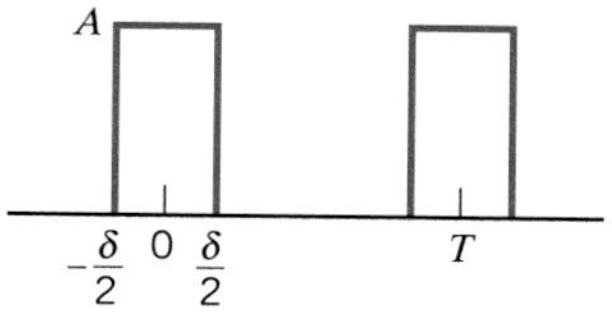 **Rectangular pulse** $f(t) = A, \quad \frac{-\delta}{2} < t < \frac{\delta}{2}$	Even	$= A\frac{\delta}{T}\frac{\sin(n\pi\delta/T)}{(n\pi\delta/T)}$
3.	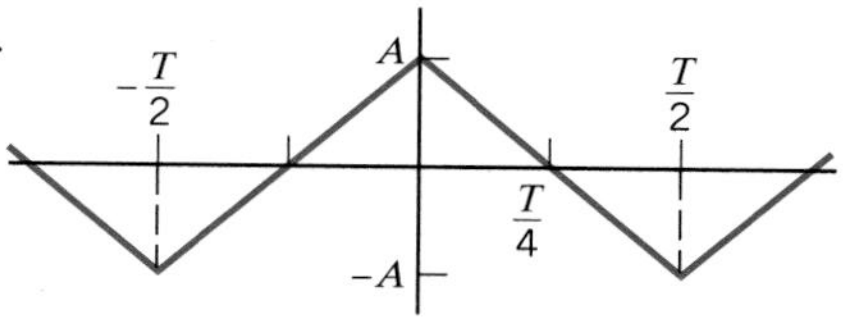 **Triangular wave**	Even	$= A\frac{\sin^2(n\pi/2)}{(n\pi/2)^2}$, $n \neq 0$; $= 0, \quad n = 0$
4.	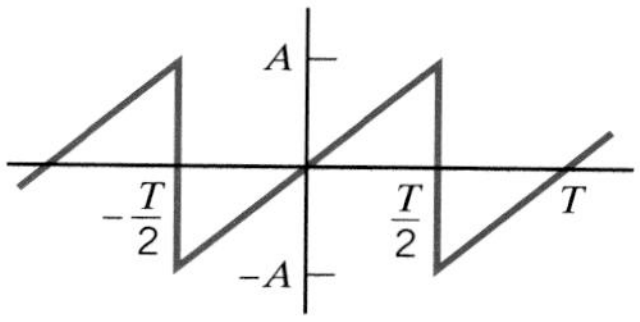 **Sawtooth wave** $f(t) = 2At/T \quad \frac{-T}{2} < t < \frac{T}{2}$	Odd	$= Aj(-1)^n/n\pi$, $n \neq 0$; $= 0, \quad n = 0$
5.	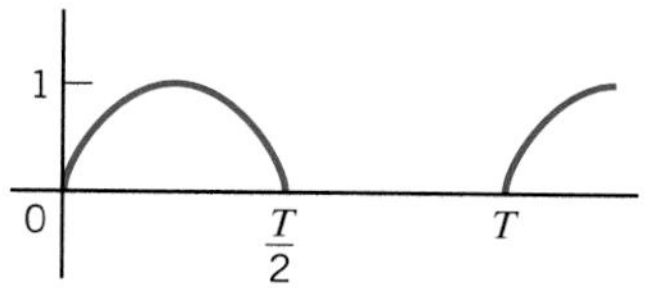 **Half-wave rectified sinusoid** $f(t) = \begin{cases} \sin \omega_0 t, & 0 \le t < T/2 \\ 0, & -T/2 \le t < 0 \end{cases}$	None	$= 1/\pi(1 - n^2)$, n even; $= -j/4, \quad n = \pm 1$; $= 0$, otherwise
6.	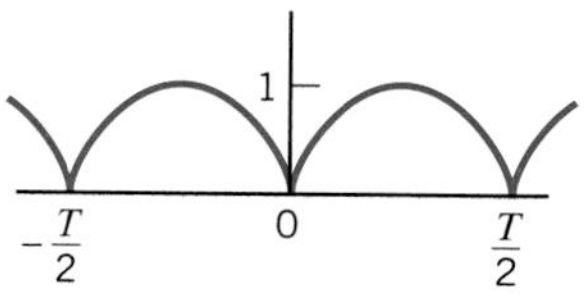 **Full-wave rectified sinusoid** $f(t) = \lvert\sin \omega_0 t\rvert$	Even	$= 2/\pi(1 - n^2)$, n even; $= 0$, otherwise

15.6 THE FOURIER SPECTRUM

If we plot the complex Fourier coefficients $\mathbf{C}_n$ as a function of angular frequency, we obtain a *Fourier spectrum.* Since $\mathbf{C}_n$ may be complex, we have

$$\mathbf{C}_n = |\mathbf{C}_n| \angle \theta_n \tag{15.6-1}$$

and we plot $|\mathbf{C}_n|$ and $\angle \theta_n$ as the *amplitude spectrum* and the *phase spectrum,* respectively. The Fourier spectrum exists only at the fundamental and harmonic frequencies and therefore is called a discrete or line spectrum. The amplitude spectrum appears on a graph as a series of equally spaced vertical lines with heights proportional to the amplitudes of the respective frequency components. Similarly, the phase spectrum appears as a series of equally spaced lines with heights proportional to the value of the phase at the appropriate frequency. The word "spectrum" was introduced into physics by Newton (1664) to describe the analysis of light by a prism into its different color components or frequency components.

Let us calculate the complex Fourier coefficients for the pulse of width δ repeated every T seconds, as shown in Figure 15.6-1. Then we will plot the *Fourier spectrum* for the function.

The **Fourier spectrum** is a graphical display of the amplitude and phase of the complex Fourier coefficients at the fundamental and harmonic frequencies.

The Fourier coefficients are

$$\mathbf{C}_n = \frac{1}{T}\int_{-T/2}^{T/2} f(t)e^{-jn\omega_0 t}\,dt \tag{15.6-2}$$

For $n \neq 0$, we have

$$\begin{aligned}\mathbf{C}_n &= \frac{A}{T}\int_{-\delta/2}^{\delta/2} e^{-jn\omega_0 t}\,dt \\ &= \frac{-A}{jn\omega_0 T}(e^{-jn\omega_0\delta/2} - e^{jn\omega_0\delta/2}) \\ &= \frac{2A}{n\omega_0 T}\sin\left(\frac{n\omega_0\delta}{2}\right) \\ &= \frac{A\delta}{T}\frac{\sin(n\omega_0\delta/2)}{(n\omega_0\delta/2)} \\ &= \frac{A\delta}{T}\frac{\sin x}{x}\end{aligned} \tag{15.6-3}$$

where $x = (n\omega_0\delta/2)$ and $n \neq 0$. When $n = 0$, we have

$$C_0 = \frac{1}{T}\int_{-\delta/2}^{\delta/2} A\,dt = \frac{A\delta}{T}$$

One may show that $(\sin x)/x = 1$ for $x = 0$ by using L'Hôpital's rule. Also, $(\sin x)/x$ is zero whenever x is an integer multiple of π, that is,

$$\frac{\sin(n\pi)}{n\pi} = 0 \quad n = 1, 2, 3, \ldots$$

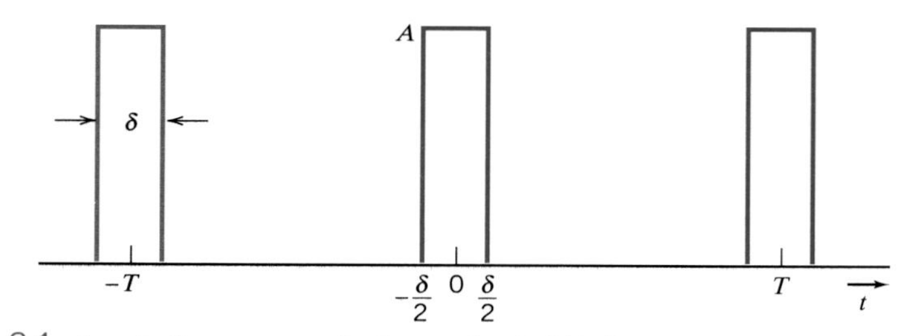

Figure 15.6-1 A periodic sequence of pulses each of width δ.

We plot $|\mathbf{C}_n|$ versus $\omega = n\omega_0$ in Figure 15.6-2*a* for n up to ± 15. Also, $|(\sin x)/x|$ is shown in Figure 15.6-2*a* in color. Note that $|\mathbf{C}_n| = 0$ when $n\omega_0\delta/2 = n\pi$, or

$$\frac{2\pi\delta}{2T} = \pi$$

This zero value occurs when $\delta/T = 1$ when $f(t) = A$ for all t. When $\delta = T/5$, as is the case for Figure 15.6-1, the zero value of $(\sin x)/x$ occurs at $\omega = 5\omega_0$, $\omega = 10\omega_0$, and so on. If δ/T is not an integer, the plot for $|\mathbf{C}_n|$ still follows $|(\sin x)/x|$ but the spectral lines are not equal to zero for any n.

The phase spectrum is plotted in Figure 15.6-2*b*. The plot of phase spectrum is a description of the fact that between ω_0 and $2\pi/\delta$ the angle is zero and between $6\omega_o = 12\pi/5\delta$ and $9\omega_o = 18\pi/5\delta$ the angle is π radians.

It is useful to consider the case where we keep δ fixed and vary the period T. As T increases, two effects are noticed: (1) the amplitude of the spectrum decreases in proportion to $1/T$ and (2) the spacing between lines decreases in proportion to $2\pi/T$. As the period T increases, the spacing between components, $\Delta\omega = 2\pi/T$, becomes smaller and there are more and more frequency components in a given range of frequency. The amplitudes of these components decrease as T is increased, but the shape of the spectrum does not change. We conclude that the shape, or envelope, of the spectrum is dependent only on the pulse shape and not on the repetition period T.

It is also instructive to keep T fixed and vary δ (with the restriction that $\delta < T/2$). As δ increases, (1) the amplitude of the spectrum increases in proportion to δ and (2) the frequency content of the signal is compressed within an increasingly narrow range of

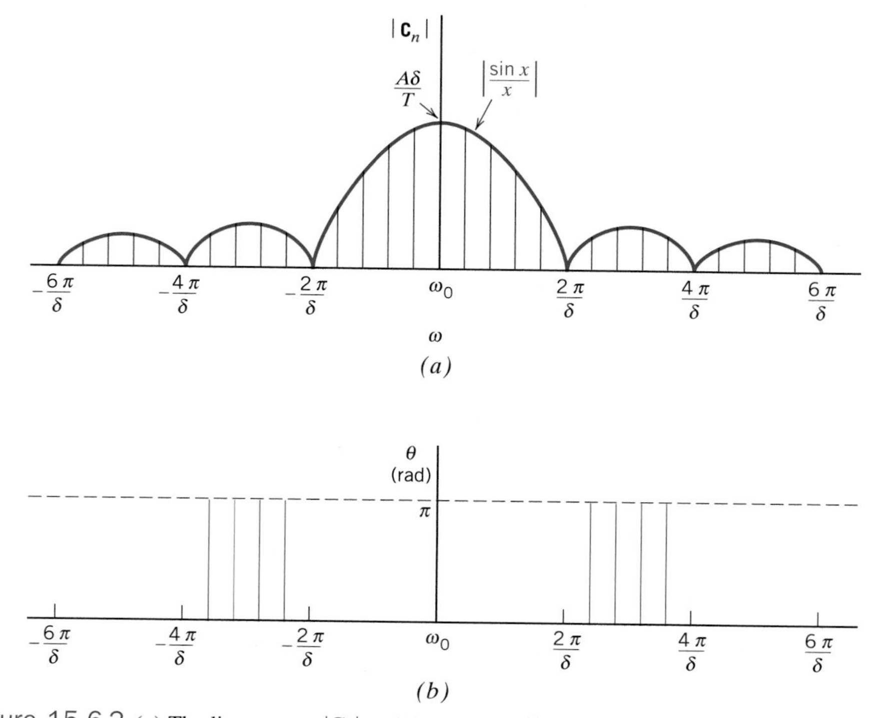

Figure 15.6-2 (*a*) The line spectra $|\mathbf{C}_n|$ and the function $|(\sin x)/x|$ in color for pulse width $\delta = T/5$ for each pulse. (*b*) Phase spectra for the pulse width $\delta = T/5$.

frequencies. Thus there is an inverse relationship between pulse width in time and the frequency spread of the spectrum. A convenient measure of the frequency spread is the distance to the first zero crossing of the $(\sin x)/x$ function.

Exercise 15.6-1 Plot $|\mathbf{C}_n|$ when $\delta = T/3$ and $\delta = T/8$ for the pulse of Figure 15.6-1, and compare the results with Figure 15.6-2.

15.7 THE TRUNCATED FOURIER SERIES

A practical calculation of the Fourier series requires that we truncate the series to a finite number of terms. Thus, the periodic function $f(t)$ is represented by a practical sum

$$f(t) \cong \sum_{n=-N}^{N} \mathbf{C}_n e^{jn\omega_0 t} = S_N(t) \tag{15.7-1}$$

Nevertheless, it can be shown that the Fourier coefficients emanate from the minimization of the mean-square error. The error for N terms is

$$\epsilon(t) = f(t) - S_N(t) \tag{15.7-2}$$

If we use the mean-square error (MSE) defined as

$$\text{MSE} = \frac{1}{T}\int_0^T \epsilon^2(t)\, dt \tag{15.7-3}$$

then the minimum of the MSE is obtained when the coefficients, $\mathbf{C}_n$, are the Fourier coefficients.

However, when we use a truncated series with $N < \infty$, we will usually experience an error. For example, if we use $N = 15$ terms to approximate a square wave, as shown in Figure 15.7-1, we find an oscillatory error. Furthermore, we observe an overshoot at each point of discontinuity where the function jumps from $-A$ to $+A$. The nineteenth-century American mathematician Josiah W. Gibbs showed that the overshoot remains at about 10 percent at the edge, regardless of the number of terms included in the expansion. This property is known as the **Gibbs phenomenon.**

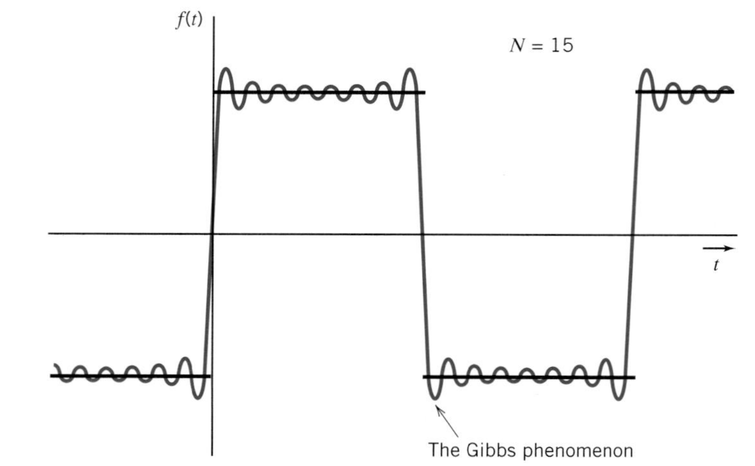

Figure 15.7-1 The Fourier series for $N = 15$ for a periodic square wave.

15.8 CIRCUITS AND FOURIER SERIES

It is often desired to determine the response of a circuit excited by a periodic input signal $v_s(t)$. We can represent $v_s(t)$ by a Fourier series and then find the response of the circuit to the fundamental and each harmonic. Assuming the circuit is linear and the principle of superposition holds, we can consider that the total response is the sum of the response to the dc term, the fundamental, and each harmonic.

Example 15.8-1

Find the steady-state response, $v_0(t)$, of the *RC* circuit shown in Figure 15.8-1*a*. The input, $v_s(t)$, is the square wave shown earlier in Figure 15.3-1. In this example, we will represent this square wave by the first four terms of its Fourier series. This Fourier series was calculated in Example 15.3-1. Assume that the parameters of the circuit are $R = 1\ \Omega$ and $C = 2$ F and that the period of the square wave is $T = \pi$ seconds.

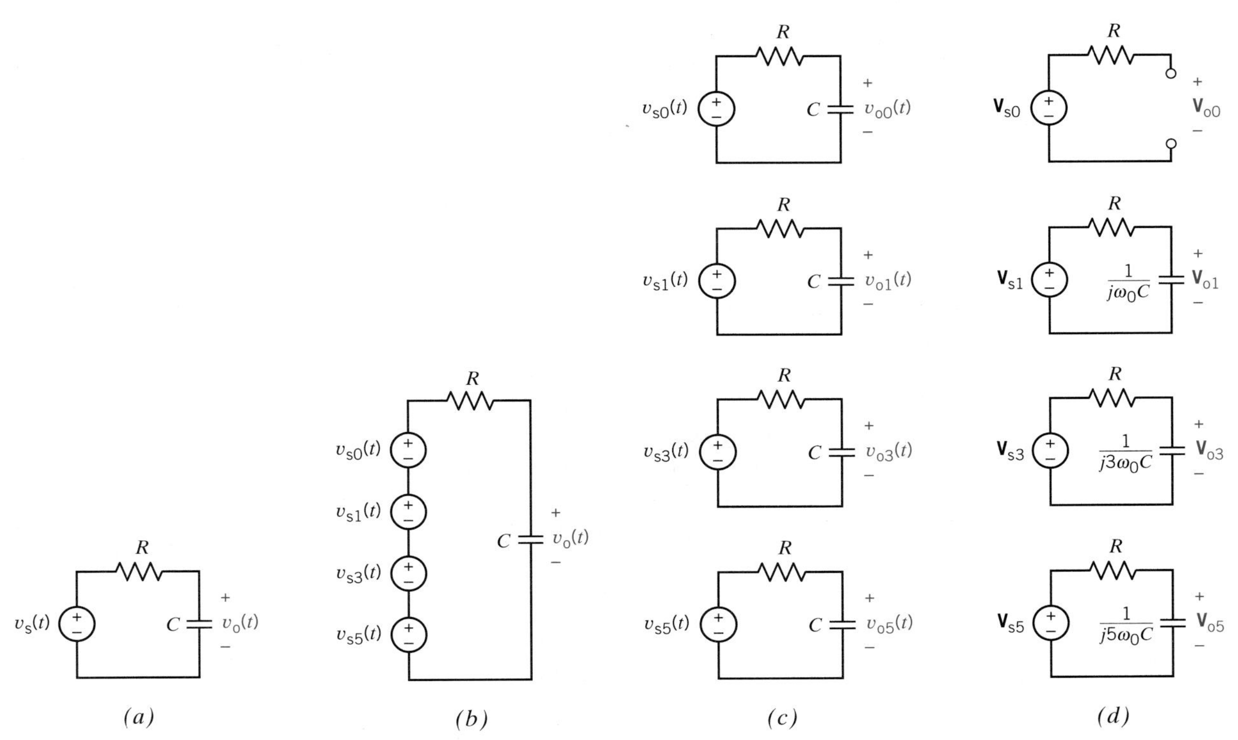

Figure 15.8-1 (*a*) An *RC* circuit excited by a periodic voltage $v_s(t)$. (*b*) An equivalent circuit. Each voltage source is a term of the Fourier series of $v_s(t)$. (*c*) Using superposition. Each input is a sinusoid. (*d*) Using phasors to find steady-state responses to the sinusoids.

Solution

In Example 15.3-1 we found that

$$v_s(t) \cong \frac{1}{2} + \sum_{\substack{n=1 \\ \text{odd}}}^{N} \frac{2(-1)^q}{n\pi} \cos n\omega_0 t$$

where $q = (n - 1)/2$ and $n = 1, 3, 5, \ldots$.

The input $v_s(t)$ will be represented by the first four terms of this series. Since $\omega_0 = 2$, we have,

$$v_s(t) = \frac{1}{2} + \frac{2}{\pi}\cos 2t - \frac{2}{3\pi}\cos 6t + \frac{2}{5\pi}\cos 10t$$

We will find the steady-state response, $v_0(t)$, using superposition. It is helpful to let $v_{sn}(t)$ denote the term of $v_s(t)$ corresponding to n. In this example, $v_s(t)$ has four terms, corresponding to $n = 0, 1, 3$, and 5. Then

$$v_s(t) = v_{s0}(t) + v_{s1}(t) + v_{s3}(t) + v_{s5}(t)$$

where
$$v_{s0}(t) = \frac{1}{2},\ v_{s1}(t) = \frac{2}{\pi}\cos 2t,$$

$$v_{s3}(t) = -\frac{2}{3\pi}\cos 6t \quad \text{and} \quad v_{s5}(t) = \frac{2}{5\pi}\cos 10t$$

Figure 15.8-1 illustrates the way that superposition is used in this example. First, since the series connection of the voltage sources with voltages $v_{s0}(t)$, $v_{s1}(t)$, $v_{s3}(t)$, and $v_{s5}(t)$ is equivalent to a single voltage source having voltage $v_s(t) = v_{s0}(t) + v_{s1}(t) + v_{s3}(t) + v_{s5}(t)$, the circuit shown in Figure 15.8-1*b* is equivalent to the circuit shown in Figure 15.8-1*a*.

Next, the principle of superposition is invoked to break the problem up into four simpler problems as shown in Figure 15.8-1*c*. Each circuit in Figure 15.8-1*c* is used to calculate the steady-state response to a single one of the voltage sources from Figure 15.8-1*b*. (When calculating the response to one voltage source, the other voltage sources are set to zero; that is, they are replaced by short circuits.) For example, the voltage $v_{o3}(t)$ is the steady-state response to $v_{s3}(t)$ alone. Superposition tells us that the response to all four voltage sources working together is the sum of the responses to the four voltage sources working separately, that is,

$$v_o(t) = v_{o0}(t) + v_{o1}(t) + v_{o3}(t) + v_{o5}(t)$$

The advantage of breaking the problem up into four simpler problems is that the input to each of the four circuits in Figure 15.8-1*c* is a sinusoid. The problem of finding the steady-state response to a periodic input has been reduced to the simpler problem of finding the steady-state response to a sinusoidal input. The steady-state response of a linear circuit to a sinusoidal input can be found using phasors. In Figure 15.8-1*d* the four circuits from Figure 15.8-1*c* have been redrawn using phasors and impedances. The impedance of the capacitor is

$$\mathbf{Z}_c = \frac{1}{jn\omega_0 C} \quad \text{for } n = 0, 1, 3, 5$$

Each of the four circuits corresponds to a different value of n, so the impedance of the capacitor is different in each of the circuits. (The frequency of the input sinusoid is $n\omega_0$, so each of the circuits corresponds to a different frequency.) Notice that when $n = 0$, $Z_c = \infty$ and therefore the capacitor acts like an open circuit. The four circuits shown in Figure 15.8-1*d* are very similar. In each case voltage division can be used to write

$$\mathbf{V}_{on} = \frac{1/jn\omega_0 C}{R + 1/jn\omega_0 C}\mathbf{V}_{sn} \quad \text{for } n = 0, 1, 3, 5$$

where $\mathbf{V}_{sn}$ is the phasor corresponding to $v_{sn}(t)$ and $\mathbf{V}_{on}$ is the phasor corresponding to $v_{on}(t)$. So

$$\mathbf{V}_{on} = \frac{\mathbf{V}_{sn}}{1 + jn\omega_0 CR} \quad \text{for } n = 0, 1, 3, 5$$

In this example, $\omega_0 CR = 4$ so

$$\mathbf{V}_{\text{on}} = \frac{\mathbf{V}_{\text{sn}}}{1 + j4n} \quad \text{for } n = 0, 1, 3, 5$$

Next, the steady-state response can be written as

$$\begin{aligned} v_{\text{on}}(t) &= |\mathbf{V}_{\text{on}}| \cos(n\omega_0 t + \angle V_{\text{on}}) \\ &= \frac{|\mathbf{V}_{\text{sn}}|}{\sqrt{1 + 16n^2}} \cos(n\omega_0 t + \angle \mathbf{V}_{\text{sn}} - \tan^{-1} 4n) \end{aligned}$$

In this example,

$$|\mathbf{V}_{\text{s0}}| = \frac{1}{2}$$

$$|\mathbf{V}_{\text{sn}}| = \frac{2}{n\pi} \quad \text{for } n = 1, 3, 5$$

$$\angle \mathbf{V}_{\text{sn}} = 0 \quad \text{for } n = 0, 1, 5 \text{ and } \angle \mathbf{V}_{\text{sn}} = 180° \text{ for } n = 3$$

Therefore, $v_{\text{o0}}(t) = \frac{1}{2}$

$$v_{\text{on}}(t) = \frac{2}{n\pi\sqrt{1 + 16n^2}} \cos(n2t + \angle \mathbf{V}_{\text{sn}} - \tan^{-1} 4n) \quad \text{for } n = 1, 3, 5$$

Doing the arithmetic yields

$$\begin{aligned} v_{\text{o0}}(t) &= \frac{1}{2} \\ v_{\text{o1}}(t) &= 0.154 \cos(2t - 76°) \\ v_{\text{o3}}(t) &= 0.018 \cos(6t + 95°) \\ v_{\text{o5}}(t) &= 0.006 \cos(10t - 87°) \end{aligned}$$

Finally, the steady-state response of the original circuit, $v_{\text{o}}(t)$, is found by adding up the partial responses

$$v_{\text{o}}(t) = \frac{1}{2} + 0.154 \cos(2t - 76°) + 0.018 \cos(6t + 95°) + 0.006 \cos(10t - 87°)$$

It is important to notice that superposition justifies adding the functions of time, $v_{\text{o0}}(t)$, $v_{\text{o1}}(t)$, $v_{\text{o3}}(t)$, and $v_{\text{o5}}(t)$ to get $v_{\text{o}}(t)$. The phasors $\mathbf{V}_{\text{o0}}$, $\mathbf{V}_{\text{o1}}$, $\mathbf{V}_{\text{o3}}$, and $\mathbf{V}_{\text{o5}}$ each correspond to a different frequency. A sum of these phasors has no meaning.

Exercise 15.8-1 Find the response of the circuit of Figure 15.8-1 when $R = 10\ \text{k}\Omega$, $C = 0.4$ mF, and v_s is the triangular wave considered in Example 15.4-1 (Figure 15.4-3). Include all terms that exceed 2 percent of the fundamental term.

Answer: $v_{\text{o}}(t) \approx 0.20 \sin(4t - 86°) - 0.008 \sin(12t - 89°)$ V

15.9 THE FOURIER TRANSFORM

The Fourier transform is closely related to the Fourier series and the Laplace transform. Recall that a periodic waveform $f(t)$ possesses a Fourier series. As we increase the period T the fundamental frequency ω_0 becomes smaller since

$$\omega_0 = \frac{2\pi}{T}$$

The difference between two consecutive harmonic frequencies is $\Delta\omega = (n + 1)\omega_0 - n\omega_0 = \omega_0 = 2\pi/T$. Therefore, as T approaches infinity, $\Delta\omega$ approaches $d\omega$, an infinitesimal frequency increment. Furthermore, the number of frequencies in any given frequency interval increases as $\Delta\omega$ decreases. Thus, in the limit $n\omega_0$ approaches the continuous variable ω.

Consider the exponential Fourier series

$$f(t) = \sum_{n=-\infty}^{n=\infty} C_n e^{jn\omega_0 t} \tag{15.9-1}$$

and

$$C_n = \frac{1}{T}\int_{-T/2}^{T/2} f(t)e^{-jn\omega_0 t}\, dt \tag{15.9-2}$$

Multiplying Eq. 15.9-2 by T and letting T approach infinity, we have

$$C_n T = \int_{-\infty}^{\infty} f(t)e^{-j\omega t}\, dt \tag{15.9-3}$$

Let $C_n T$ equal a new frequency function $F(j\omega)$ so that

$$F(j\omega) = \int_{-\infty}^{\infty} f(t)e^{-j\omega t}\, dt \tag{15.9-4}$$

where $F(j\omega)$ is the *Fourier transform* of $f(t)$. The inverse process is found from Eq. 15.9-1, where we let $C_n T = F(j\omega)$ so that

$$\begin{aligned} f(t) &= \lim_{T\to 0} \sum_{n=-\infty}^{\infty} C_n T e^{jn\omega_0 t} \frac{1}{T} \\ &= \lim_{T\to 0} \sum_{n=-\infty}^{\infty} F(j\omega) e^{jn\omega_0 t} \frac{\omega_0}{2\pi} \end{aligned}$$

since $1/T = \omega_0/2\pi$. As $T \to \infty$ the sum becomes an integral and the increment $\Delta\omega = \omega_0$ becomes $d\omega$. Then, we have

$$f(t) = \frac{1}{2\pi}\int_{-\infty}^{\infty} F(j\omega)e^{j\omega t}\, d\omega \tag{15.9-5}$$

Equation 15.9-5 is called the *inverse Fourier transform.* This pair of equations (Eqs. 15.9-4 and 15.9-5), called the Fourier transform pair, permits us to complete the Fourier transformation to the frequency domain and the inverse process to the time domain.

A given function of time $f(t)$ has a Fourier transform if

$$\int_{-\infty}^{\infty} |f(t)|\, dt < \infty$$

and if the number of discontinuities in $f(t)$ is finite. From a practical point of view, all pulses of finite duration in which we are interested have Fourier transforms.

The Fourier transform pair is summarized in Table 15.9-1.

Table 15.9-1 **The Fourier Transform Pair**

Equation	Name	Process
$F(j\omega) = \int_{-\infty}^{\infty} f(t)e^{-j\omega t}\,dt$	Transform	Time domain to frequency domain Conversion of $f(t)$ into $F(j\omega)$
$f(t) = \frac{1}{2\pi}\int_{-\infty}^{\infty} F(j\omega)e^{j\omega t}\,d\omega$	Inverse transform	Frequency domain to time domain Conversion of $F(j\omega)$ into $f(t)$

Example 15.9-1

Derive the Fourier transform of the aperiodic pulse shown in Figure 15.9-1.

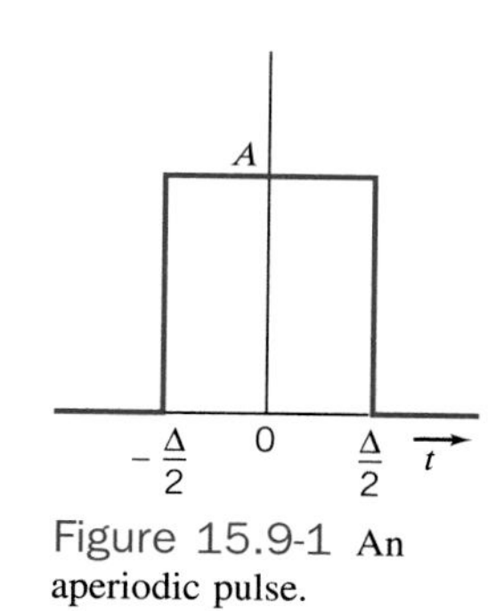

Figure 15.9-1 An aperiodic pulse.

Solution

Using the transform, we have

$$F(j\omega) = \int_{-\Delta/2}^{\Delta/2} Ae^{-j\omega t}\,dt = \frac{A}{-j\omega}e^{-j\omega t}\Big|_{-\Delta/2}^{\Delta/2}$$

$$= \frac{A}{-j\omega}(e^{-j\omega\Delta/2} - e^{j\omega\Delta/2}) = A\Delta\frac{\sin(\omega\Delta/2)}{\omega\Delta/2} \qquad (15.9\text{-}6)$$

Thus the Fourier transform is of the form $(\sin x)/x$, where $x = \omega\Delta/2$ as shown in Figure 15.9-2. Note that $(\sin x)/x = 0$ when $x = \omega\Delta/2 = n\pi$ or $\omega = 2n\pi/\Delta$ as shown in Figure 15.9-2. We will denote $(\sin x)/x = Sa(x)$.

Let us consider the shifted version of the rectangular pulse of Figure 15.9-1 where $A = 1/\Delta$ and the width of the pulse approaches zero, $\Delta \to 0$, while the area of the rectangle remains equal to 1. Then we have the *unit impulse* $\delta(t - t_0)$ so that

$$\int_a^b \delta(t - t_0)\,dt = \begin{cases} 1 & a \le t_0 \le b \\ 0 & \text{otherwise} \end{cases} \qquad (15.9\text{-}7)$$

We obtain the Fourier transform for a unit impulse at t_0 as

$$F(j\omega) = \int_{t_0-}^{t_0+} \delta(t - t_0)e^{-j\omega t}\,dt = e^{-j\omega t_0} \qquad (15.9\text{-}8)$$

When $t_0 = 0$ we have the special case

$$F(j\omega) = 1 \qquad (15.9\text{-}9)$$

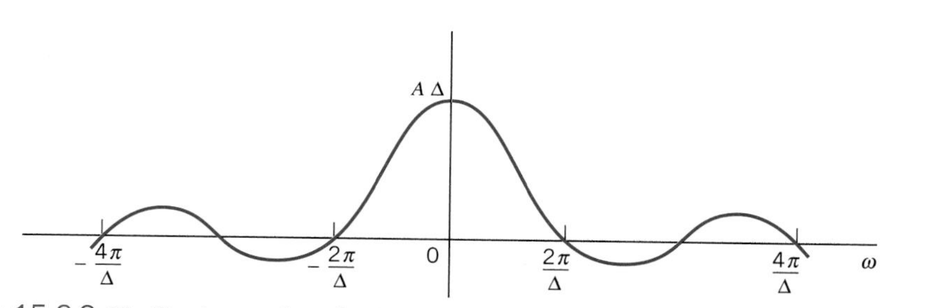

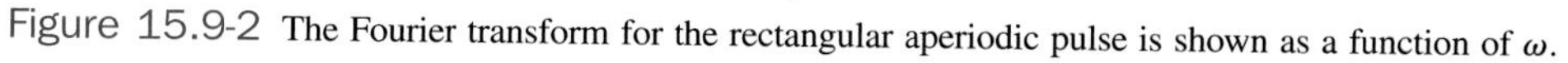

Figure 15.9-2 The Fourier transform for the rectangular aperiodic pulse is shown as a function of ω.

Thus, we note that $F(j\omega) = 1$ of a unit impulse located at the origin is constant and equal to 1 for all frequencies.

Exercise 15.9-1 Determine the Fourier transform of $f(t) = e^{-at}u(t)$, where $u(t)$ is the unit step function.

Answer: $F(j\omega) = \dfrac{1}{a + j\omega}$

15.10 FOURIER TRANSFORM PROPERTIES

We can derive some properties of the Fourier transform by writing $F(j\omega)$ in complex form as

$$F(j\omega) = X(\omega) + jY(\omega)$$

Alternatively, we have

$$F(j\omega) = |F(\omega)|\, e^{j\theta}$$

where $\theta = \tan^{-1}(Y/X)$. Note that we use $F(j\omega) = F(\omega)$ interchangeably. Furthermore,

$$F(-\omega) = F^*(\omega)$$

where $F^*(\omega)$ is the complex conjugate of $F(\omega)$.

If we have the Fourier transform of $f(t)$, we write

$$\mathcal{F}[f(t)] = F(\omega)$$

where the script $\mathcal{F}$ implies the Fourier transform. Then the inverse transform is written as

$$\mathcal{F}^{-1}[F(\omega)] = f(t)$$

Repeating the transformation equation, we have (Table 15.9-1)

$$F(\omega) = \int_{-\infty}^{\infty} f(t)e^{-j\omega t}\, dt \tag{15.10-1}$$

Then, if $\mathcal{F}[af_1(t)] = aF_1(\omega)$ and $\mathcal{F}[bf_2(t)] = bF_2(\omega)$, we have

$$\begin{aligned}\mathcal{F}[af_1 + bf_2] &= \int_{-\infty}^{\infty} [af_1 + bf_2]e^{-j\omega t}\, dt \\ &= \int_{-\infty}^{\infty} af_1e^{-j\omega t}\, dt + \int_{-\infty}^{\infty} bf_2e^{-j\omega t}\, dt \\ &= aF_1(\omega) + bF_2(\omega)\end{aligned}$$

This is known as the *linearity* property.

We now use the definition of the Fourier transform, Eq. 15.10-1, in the following examples to find several other properties.

Example 15.10-1

Find the Fourier transform of a time-shifted function $f(t-t_0)$.

Solution

$$\mathscr{F}[f(t-t_0)] = \int_{-\infty}^{\infty} f(t-t_0)e^{-j\omega t}\,dt$$

If we let $x = t - t_0$, we have

$$\mathscr{F}[f(t-t_0)] = \int_{-\infty}^{\infty} f(x)e^{-j\omega(x+t_0)}\,dx = e^{-j\omega t_0}F(\omega)$$

where $F(\omega) = \mathscr{F}[f(t)]$.

Selected properties of the Fourier transform are summarized in Table 15.10-1. We can use these properties to derive Fourier transform pairs.

Table 15.10-1 **Selected Properties of the Fourier Transform**

Name of Property	Function of Time	Fourier Transform
1. Definition	$f(t)$	$F(\omega)$
2. Multiplication by constant	$Af(t)$	$AF(\omega)$
3. Linearity	$af_1 + bf_2$	$aF_1(\omega) + bF_2(\omega)$
4. Time shift	$f(t-t_0)$	$e^{-j\omega t_0}F(\omega)$
5. Time scaling	$f(at),\ a > 0$	$\frac{1}{a}F\left(\frac{\omega}{a}\right)$
6. Modulation	$e^{j\omega_0 t}f(t)$	$F(\omega - \omega_0)$
7. Differentiation	$\frac{d^n f(t)}{dt^n}$	$(j\omega)^n F(\omega)$
8. Convolution	$\int_{-\infty}^{\infty} f_1(x)f_2(t-x)\,dx$	$F_1(\omega)F_2(\omega)$
9. Time multiplication	$t^n f(t)$	$(j)^n \frac{d^n F(\omega)}{d\omega^n}$
10. Time reversal	$f(-t)$	$F(-\omega)$
11. Integration	$\int_{-\infty}^{t} f(\tau)\,d\tau$	$\frac{F(\omega)}{j\omega} + \pi F(0)\delta(\omega)$

With the aid of the properties of the Fourier transform and the original defining equation, we can derive useful transform pairs and develop a table of these relationships. We have already derived the first three entries in Table 15.10-2, and we will add several more by using the properties of Table 15.10-1 and/or the original definition of the transformation.

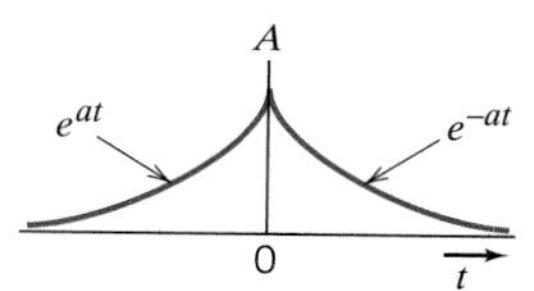

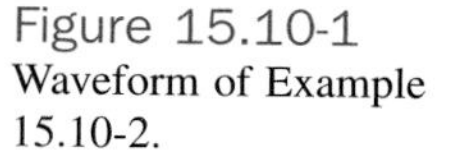
Figure 15.10-1 Waveform of Example 15.10-2.

Example 15.10-2

Find the Fourier transform of $f(t) = Ae^{-a|t|}$, which is shown in Figure 15.10-1.

Table 15.10-2 **Fourier Transform Pairs**

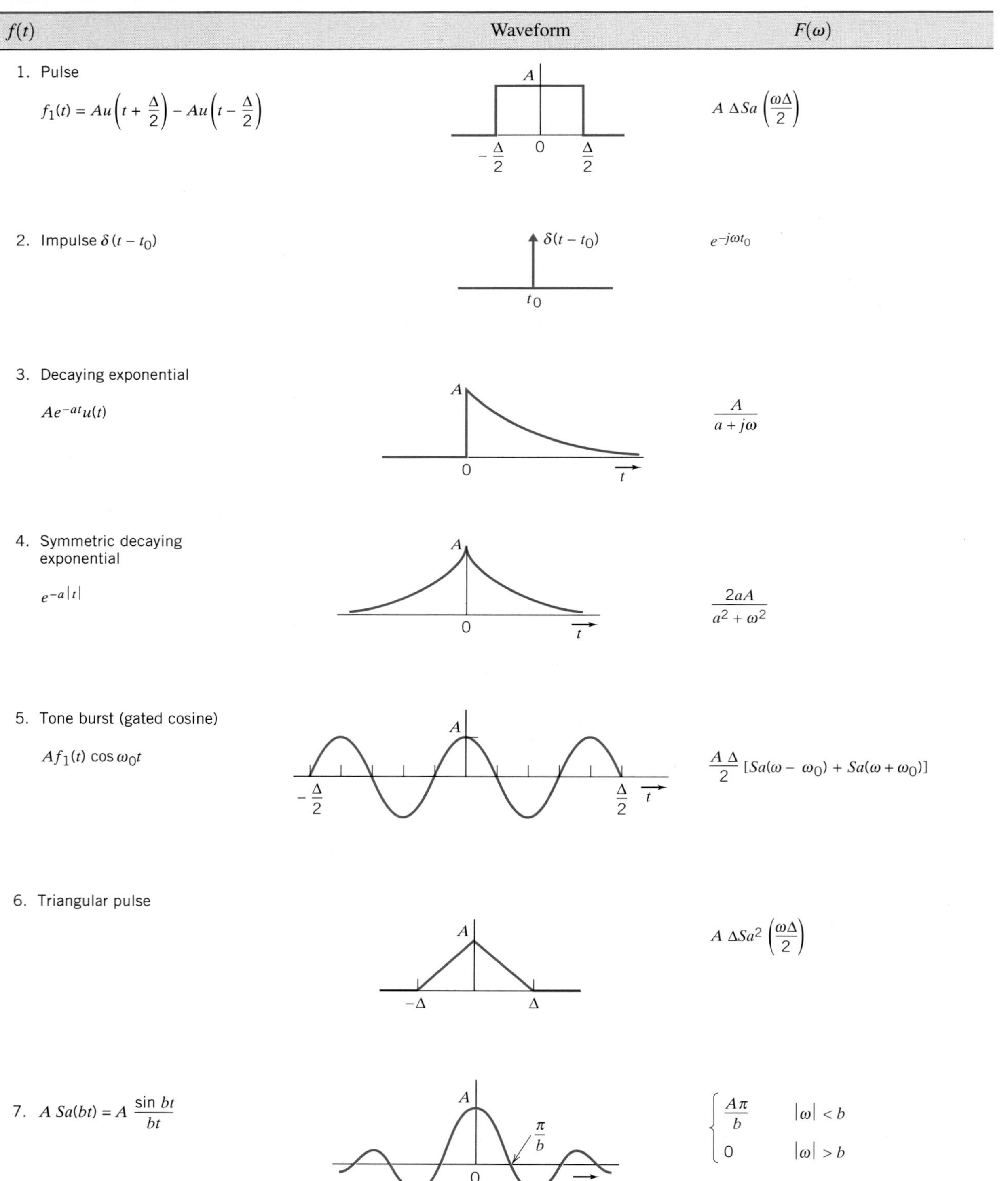

$f(t)$	Waveform	$F(\omega)$
1. Pulse $f_1(t) = Au\left(t + \frac{\Delta}{2}\right) - Au\left(t - \frac{\Delta}{2}\right)$		$A\,\Delta Sa\left(\frac{\omega\Delta}{2}\right)$
2. Impulse $\delta(t - t_0)$		$e^{-j\omega t_0}$
3. Decaying exponential $Ae^{-at}u(t)$		$\frac{A}{a + j\omega}$
4. Symmetric decaying exponential $e^{-a\|t\|}$		$\frac{2aA}{a^2 + \omega^2}$
5. Tone burst (gated cosine) $Af_1(t)\cos\omega_0 t$		$\frac{A\,\Delta}{2}[Sa(\omega - \omega_0) + Sa(\omega + \omega_0)]$
6. Triangular pulse		$A\,\Delta Sa^2\left(\frac{\omega\Delta}{2}\right)$
7. $A\,Sa(bt) = A\,\frac{\sin bt}{bt}$		$\begin{cases} \frac{A\pi}{b} & \|\omega\| < b \\ 0 & \|\omega\| > b \end{cases}$

(continued)

Table 15.10-2 **Fourier Transform Pairs** *(Continued)*

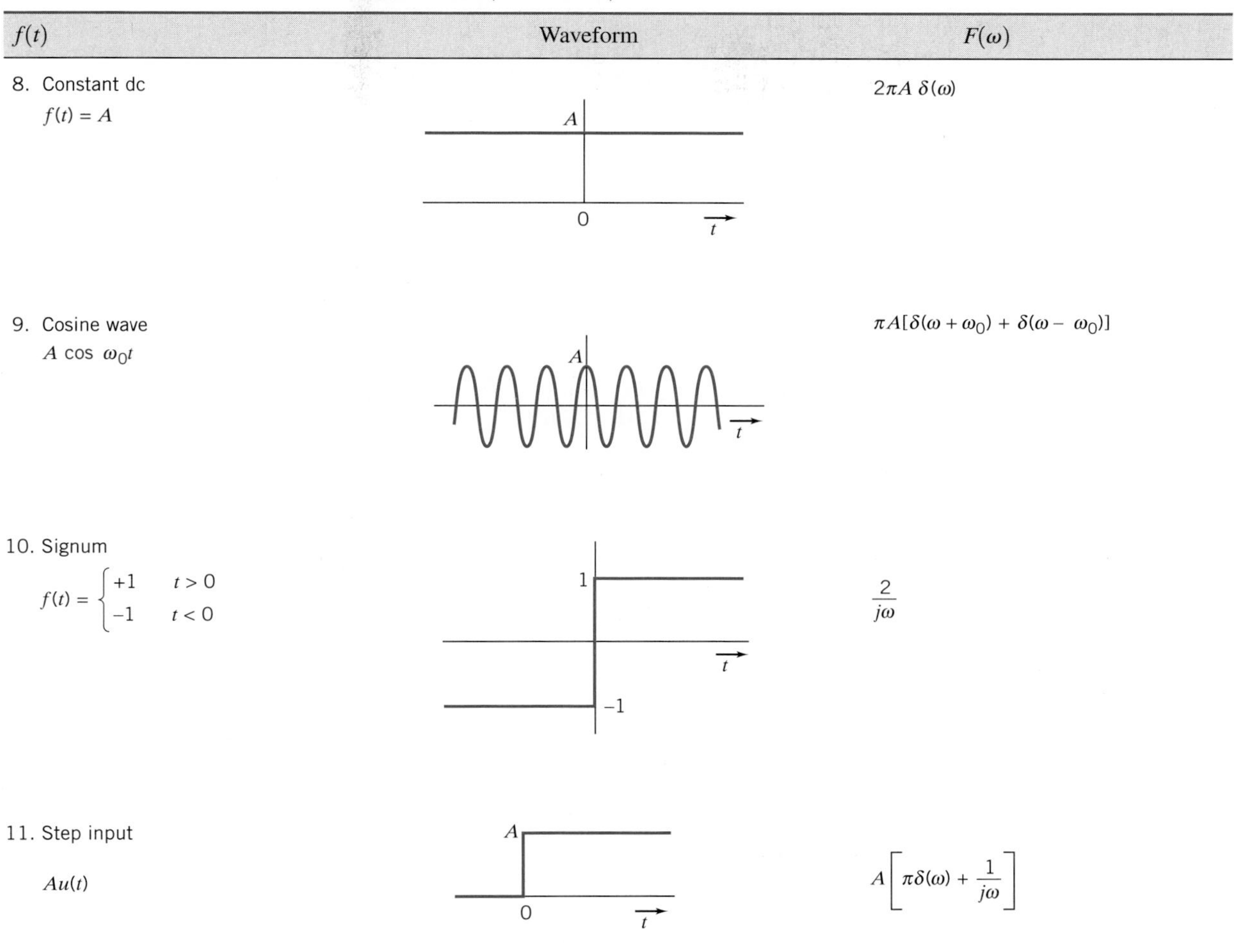

$f(t)$	Waveform	$F(\omega)$
8. Constant dc $f(t) = A$		$2\pi A\, \delta(\omega)$
9. Cosine wave $A \cos \omega_0 t$		$\pi A[\delta(\omega + \omega_0) + \delta(\omega - \omega_0)]$
10. Signum $f(t) = \begin{cases} +1 & t > 0 \\ -1 & t < 0 \end{cases}$		$\frac{2}{j\omega}$
11. Step input $Au(t)$		$A\left[\pi\delta(\omega) + \frac{1}{j\omega}\right]$

Note: $Sa(x) = (\sin x)/x$.

Solution

We will break the function into two symmetric waveforms and use the linearity property. Then,

$$f(t) = f_1(t) + f_2(t) = Ae^{-at}u(t) + Ae^{at}u(-t)$$

We have from entry 3 of Table 15.10-2

$$F_1(\omega) = \frac{A}{a + j\omega}$$

From property 10 of Table 15.10-1, we obtain

$$F_2(\omega) = F_1(-\omega) = \frac{A}{a - j\omega}$$

Using the linearity property, we have

$$F(\omega) = F_1(\omega) + F_2(\omega) = \frac{A}{a + j\omega} + \frac{A}{a - j\omega} = \frac{2Aa}{a^2 + \omega^2} \qquad (15.10\text{-}2)$$

This result is entry 4 in Table 15.10-2. Note that $F(\omega)$ is an even function.

Example 15.10-3

Find the Fourier transform of the gated cosine waveform $f(t) = f_1(t) \cos \omega_0 t$, where $f_1(t)$ is the rectangular pulse shown in Figure 15.9-1.

Solution

The Fourier transform of the rectangular pulse is entry 1 in Table 15.10-2 and is written as

$$F_1(\omega) = A\Delta(\sin x)/x$$

where $x = \omega\Delta/2$. The cosine function can be written as

$$\cos \omega_0 t = \frac{1}{2}(e^{j\omega_0 t} + e^{-j\omega_0 t})$$

Therefore,

$$f(t) = \frac{1}{2} f_1(t) e^{j\omega_0 t} + \frac{1}{2} f_1(t) e^{-j\omega_0 t}$$

Using the modulation property (entry 6) of Table 15.10-1, we obtain

$$F(\omega) = \tfrac{1}{2}F_1(\omega - \omega_0) + \tfrac{1}{2}F_1(\omega + \omega_0)$$

Therefore, using $F_1(\omega)$ from Eq. 15.9-6, we have

$$F(\omega) = \frac{A\Delta}{2} \frac{\sin[(\omega - \omega_0)\Delta/2]}{(\omega - \omega_0)\Delta/2} + \frac{A\Delta}{2} \frac{\sin[(\omega + \omega_0)\Delta/2]}{(\omega + \omega_0)\Delta/2}$$

or, using $Sa(x) = (\sin x)/x$, we have

$$F(\omega) = \frac{A\Delta}{2} Sa\left[(\omega - \omega_0)\frac{\Delta}{2}\right] + \frac{A\Delta}{2} Sa\left[(\omega + \omega_0)\frac{\Delta}{2}\right]$$

Exercise 15.10-1 Find the Fourier transform of $f(at)$ for $a > 0$ when $F(\omega) = \mathcal{F}[f(t)]$.

Answer: $\mathcal{F}[f(at)] = \dfrac{1}{a} F\left(\dfrac{\omega}{a}\right)$

Exercise 15.10-2 Show that the Fourier transform of a constant dc waveform $f(t) = A$ for $-\infty \le t \le \infty$ is $F(\omega) = 2\pi A \delta(\omega)$ by obtaining the inverse transform of $F(\omega)$.

15.11 THE SPECTRUM OF SIGNALS

The *spectrum,* also called the *spectral density,* of a signal $f(t)$ is its Fourier transform $F(\omega)$. We can plot $F(\omega)$ as a function of ω to show the spectrum. For example, for a rectangular pulse signal of Figure 15.9-1, we found that

$$F(\omega) = A\Delta Sa(\omega\Delta/2)$$

which is plotted in Figure 15.9-2. The spectrum of the rectangular pulse is real.

The Fourier transform of an impulse $\delta(t)$ is (item 2 of Table 15.10-2)

$$F(\omega) = 1$$

Thus, the spectrum of an impulse contains all frequencies, and a plot of the spectrum of the impulse is shown in Figure 15.11-1.

The Fourier transform of a constant dc signal of magnitude A is

$$F(\omega) = 2\pi A\ \delta(\omega)$$

which has a spectrum as shown in Figure 15.11-2. The integral of the impulse $\delta(\omega)$ has value unity at $\omega = 0$. The symbol for the impulse is a vertical line with an arrowhead.

For completeness, let us examine a function that has a Fourier transform that is complex. When $f(t) = Ae^{-at}u(t)$,

$$F(\omega) = \frac{A}{a + j\omega}$$

In order to plot the spectrum, we calculate the magnitude and phase of $F(\omega)$ as

$$|F(\omega)| = \frac{A}{(a^2 + \omega^2)^{1/2}}$$

and

$$\phi(\omega) = -\tan^{-1}\omega/a$$

The Fourier spectrum is shown in Figure 15.11-3.

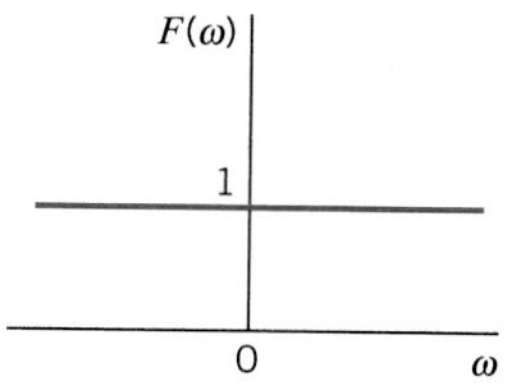

Figure 15.11-1 Spectrum of impulse $f(t) = \delta(t)$.

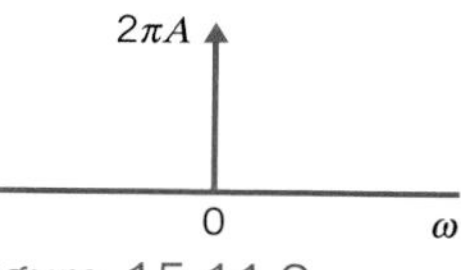

Figure 15.11-2 Spectrum of constant dc signal of magnitude A. The symbol for an impulse is a vertical line with an arrowhead.

The **Fourier spectrum** of a signal is a graph of the magnitude and phase of the Fourier transform of the signal.

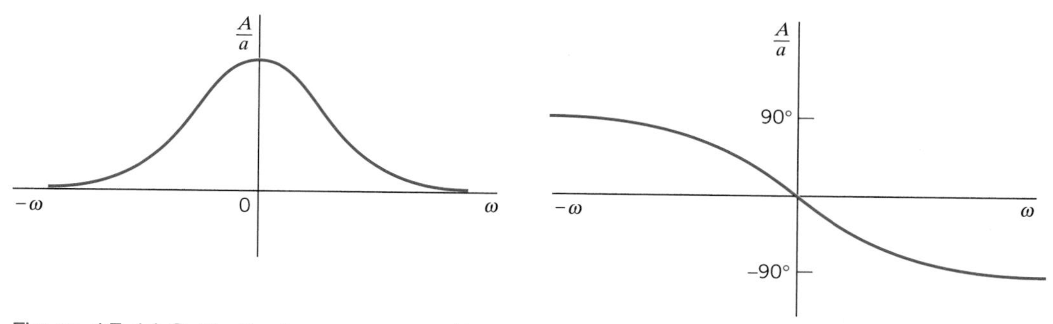

Figure 15.11-3 The Fourier spectrum for $f(t) = Ae^{-at}u(t)$.

Exercise 15.11-1 Calculate the Fourier transform and draw the Fourier spectrum for $f(t)$ shown in Figure E 15.11-1, where $f(t) = A \cos \omega_0 t$ for all t.

Answer: $F(\omega) = \pi A\delta(\omega + \omega_0) + \pi A\delta(\omega - \omega_0)$

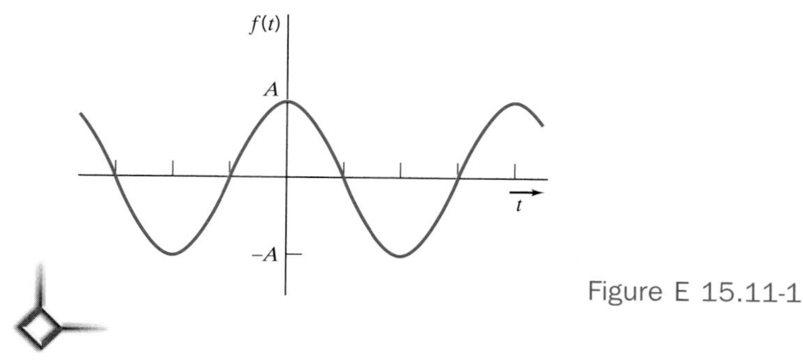

Figure E 15.11-1

15.12 CONVOLUTION AND CIRCUIT RESPONSE

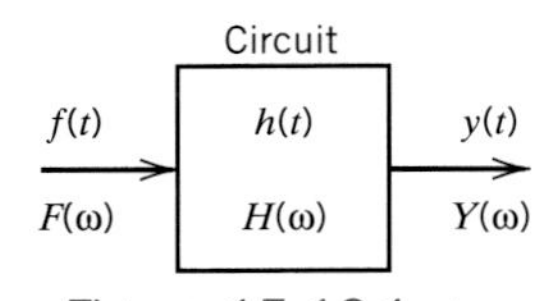

Figure 15.12-1 A linear circuit.

A circuit with an impulse response $h(t)$ and an input $f(t)$ has a response $y(t)$ that may be determined from the convolution integral. For the circuit shown in Figure 15.12-1, the convolution integral is

$$y(t) = \int_{-\infty}^{\infty} h(x)f(t-x)\,dx$$

If we use the Fourier transform of the convolution integral, we have

$$\mathcal{F}[y(t)] = \int_{-\infty}^{\infty}\int_{-\infty}^{\infty} h(x)f(t-x)\,dx\,e^{-j\omega t}\,dt$$

$$= \int_{-\infty}^{\infty} h(x)\int_{-\infty}^{\infty} f(t-x)e^{-j\omega t}\,dt\,dx$$

Let $u = t - x$ to obtain

$$\mathcal{F}[y(t)] = \int_{-\infty}^{\infty} h(x)\int_{-\infty}^{\infty} f(u)e^{-j\omega(u+x)}\,du\,dx$$

$$= \int_{-\infty}^{\infty} h(x)e^{-j\omega x}\,dx \int_{-\infty}^{\infty} f(u)e^{-j\omega u}\,du$$

or

$$Y(\omega) = H(\omega)F(\omega) \tag{15.12-1}$$

Thus, convolution in the time domain corresponds to multiplication in the frequency domain. When the input is an impulse, $f(t) = \delta(t)$, since $F(\omega) = 1$, we obtain the impulse response

$$Y(\omega) = H(\omega)$$

When the input is a sinusoid, the Fourier transform of the output is the steady-state response to that sinusoidal driving function.

Example 15.12-1

Find the response, $v_0(t)$, of the RL circuit shown in Figure 15.12-2 when $v(t) = 4e^{-2t}u(t)$ V. The initial condition is zero.

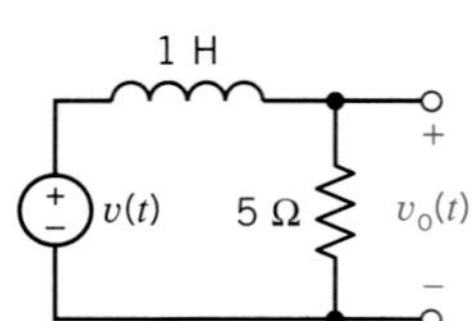

Figure 15.12-2 Circuit of Example 15.12-1.

Solution

Since $v(t) = 4e^{-2t}u(t)$, we obtain $V(\omega)$ as

$$V(\omega) = \frac{4}{2 + j\omega}$$

The circuit is represented by $H(\omega)$, and using the voltage divider principle, we have

$$H(\omega) = \frac{R}{R + j\omega L} = \frac{5}{5 + j\omega}$$

Then, we have

$$V_0(\omega) = H(\omega)V(\omega) = \frac{20}{(5 + j\omega)(2 + j\omega)}$$

Expand using partial fractions to obtain[1]

$$V_0(\omega) = \frac{-20/3}{5 + j\omega} + \frac{20/3}{2 + j\omega}$$

Using the inverse transform for each term (entry 3 of Table 15.10-2), we have

$$v_0(t) = \frac{20}{3}(e^{-2t} - e^{-5t})u(t) \text{ V}$$

The time domain responses obtained in this manner are responses of initially relaxed circuits. (No initial energy is stored.)

Example 15.12-2

Determine and plot the spectrum of the response $V_0(\omega)$ of the circuit of Figure 15.12-3 when $v = 10e^{-2t}u(t)$ V.

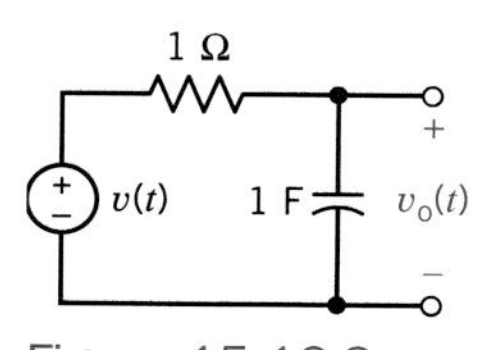

Figure 15.12-3 Circuit of Example 15.12-2.

Solution

The input signal $v(t)$ has a Fourier transform

$$V(\omega) = \frac{10}{2 + j\omega} = \frac{10}{(4 + \omega^2)^{1/2}} \angle -\tan^{-1}\omega/2$$

The circuit transfer function is

$$H(j\omega) = \frac{1/j\omega C}{R + 1/j\omega C} = \frac{1}{1 + j\omega} = \frac{1}{(1 + \omega^2)^{1/2}} \angle -\tan^{-1}\omega$$

Then the output is

$$V_0(\omega) = H(\omega)V(\omega) = \frac{10}{(2 + j\omega)(1 + j\omega)}$$

Therefore

$$|V_0| = \frac{10}{[(4 + \omega^2)(1 + \omega^2)]^{1/2}}$$

and

$$\phi(\omega) = \angle V_0(\omega) = -\tan^{-1}\frac{\omega}{2} - \tan^{-1}\omega$$

The calculated magnitude and phase for $V_0(\omega)$ are recorded in Table 15.12-1. For negative ω, $|V_0(\omega)| = |V_0(-\omega)|$ and

$$\phi(-\omega) = -\phi(\omega)$$

Therefore, the Fourier spectrum of $V_0(\omega)$ is represented by the plot shown in Figure 15.12-4.

Table 15.12-1 **Fourier Response for Example 15.12-2**

ω	0	1	2	3	5	∞
$\lvert V_0 \rvert$	5	3.16	1.58	0.88	0.36	0
$\phi(\omega)$	0°	−71.6°	−108.4°	−127.9°	−146.9°	−180°

[1]See Chapter 14, Section 5 for a review of partial fraction expansion.

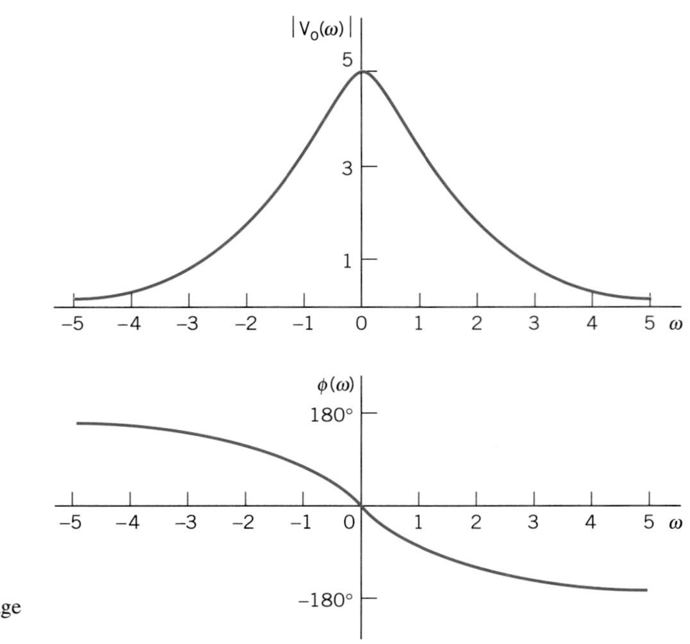

Figure 15.12-4
The amplitude and phase versus ω of the output voltage for Example 15.12-2.

Exercise 15.12-1 An ideal bandpass filter passes all frequencies between 24 rad/s and 48 rad/s without attenuation and completely rejects all frequencies outside this passband.

(a) Sketch $|V_0|^2$ for the filter output voltage when the input voltage is

$$v(t) = 120e^{-24t}u(t)\ \text{V}$$

(b) What percentage of the input signal energy is available in the signal at the output of the ideal filter?

Answer: (b) 20.5%

15.13 THE FOURIER TRANSFORM AND THE LAPLACE TRANSFORM

The tables of Laplace transforms, Table 14.3-1 and Table 14.4-1, developed in Chapter 14, can be used to obtain the Fourier transform of a function $f(t)$. Of course, the Fourier transform formally exists only when the Fourier integral, Eq. 15.9-4, converges. The Fourier integral will converge when all the poles of $F(s)$ lie in the left-hand s-plane and not on the $j\omega$ axis or at the origin.

If $f(t)$ is zero for $t \leq 0$ and $\int_0^{\infty} f(t)dt < \infty$, we can obtain the Fourier transform from the Laplace transform of $f(t)$ by replacing s by $j\omega$. Then

$$F(\omega) = F(s)|_{s=j\omega} \tag{15.13-1}$$

where

$$F(s) = \mathcal{L}\{f(t)\}$$

For example, if (entry 3 of Table 15.10-2)

$$f(t) = Ae^{-at}u(t)$$

then, from Table 14.3-1,

$$F(s) = \frac{A}{s + a}$$

Therefore, with $s = j\omega$ we obtain the Fourier transform

$$F(\omega) = \frac{A}{a + j\omega}$$

If $f(t)$ is a real function with a nonzero value for negative time only, then we can reflect $f(t)$ to positive time, find the Laplace transform, and then find $F(\omega)$ by setting $s = -j\omega$. Therefore, when $f(t) = 0$ for $t \geq 0$ and $f(t)$ exists only for negative time, we have

$$F(\omega) = \mathscr{L}\{f(-t)\}|_{s=-j\omega} \tag{15.13-2}$$

For example, consider the exponential function

$$\begin{aligned} f(t) &= 0 \qquad & t \geq 0 \\ &= e^{at} \qquad & t < 0 \end{aligned}$$

Then, reversing the time function, we have

$$f(-t) = e^{-at} \qquad t > 0$$

and, therefore,

$$F(s) = \frac{1}{s + a}$$

Hence, setting $s = -j\omega$, we obtain

$$F(\omega) = \frac{1}{a - j\omega}$$

Functions that are nonzero over all time can be divided into positive time and negative time functions. We then use Eqs. 15.13-1 and 15.13-2 to obtain the Fourier transform of each part. The Fourier transform of $f(t)$ is the sum of the Fourier transforms of the two parts.

For example, consider the function $f(t)$ with a nonzero value over all time where

$$f(t) = e^{-a|t|}$$

which is entry 4 in Table 15.10-2. The positive time portion of the function will be called $f^+(t)$, and the negative time portion will be called $f^-(t)$. Then,

$$f(t) = f^+(t) + f^-(t)$$

Hence

$$F(\omega) = \mathscr{L}\{f^+(t)\}_{s=j\omega} + \mathscr{L}\{f^-(-t)\}_{s=-j\omega}$$

In this case,

$$f^+(t) = e^{-at} \qquad t > 0$$

and

$$f^-(t) = e^{at} \qquad t < 0$$

Note that $f^-(-t) = e^{-at}$. Then,

$$F^+(s) = \frac{1}{s + a} \quad \text{and} \quad F^-(s) = \frac{1}{s + a}$$

Table 15.13-1 **Obtaining the Fourier Transform Using the Laplace Transform**

Case	Method
A. $f(t)$ nonzero for positive time only and $f(t) = 0, t < 0$	Step 1. $F(s) = \mathcal{L}\{f(t)\}$ 2. $F(\omega) = F(s)\|_{s=j\omega}$
B. $f(t)$ nonzero for negative time only and $f(t) = 0, t > 0$	Step 1. $F(s) = \mathcal{L}\{f(-t)\}$ 2. $F(\omega) = F(s)\|_{s=-j\omega}$
C. $f(t)$ nonzero over all time	Step 1. $f(t) = f^+(t) + f^-(t)$ 2. $F^+(s) = \mathcal{L}\{f^+(t)\}$ $F^-(s) = \mathcal{L}\{f^-(-t)\}$ 3. $F(\omega) = F^+(s)\|_{s=j\omega} + F^-(s)\|_{s=-j\omega}$

Note: The poles of F(s) must lie in the left-hand s-plane.

We obtain the total $F(\omega)$ as

$$F(\omega) = F^+(s)|_{s=j\omega} + F^-(s)|_{s=-j\omega} = \frac{1}{a + j\omega} + \frac{1}{a - j\omega} = \frac{2a}{\omega^2 + a^2}$$

The use of the Laplace transform to find the Fourier transform is summarized in Table 15.13-1. Remember that the method summarized cannot be used for sin ωt, cos ωt, or $u(t)$, since the poles of $F(s)$ lie on the $j\omega$ axis or at the origin.

Exercise 15.13-1 Derive the Fourier transform for

$$f(t) = \begin{cases} te^{-at} & t \geq 0 \\ te^{at} & t \leq 0 \end{cases}$$

Answer: $\dfrac{-j4a\omega}{(a^2 + \omega^2)^2}$

15.14 FOURIER SERIES OF A WAVEFORM USING PSPICE

PSpice can be used to calculate the coefficients of the Fourier series of a periodic voltage or current. There are two steps to the procedure.

1. Simulate the circuit containing the voltage or current to obtain a transient response.
2. Use the .FOUR command to tell PSpice to compute the coefficients and to put them in the PSpice output file.

The response that PSpice calls the transient response is the response that we called the complete response in Chapter 9. In other words, the PSpice transient response includes both transient and steady-state parts. The transient part must die out before the PSpice can calculate the coefficients of the Fourier series. Indeed, there must be at least one full period of the steady-state response at the end of the PSpice transient response if the coefficients are to be calculated accurately. (*Warning*: PSpice will calculate coefficients even if this condition is not satisfied. The coefficients will not be accurate. PSpice will not issue a warning.)

The syntax of the .FOUR command is

```
.FOUR <fundamental frequency> <variable name> . . .
```

Example 15.14-1

Consider the circuit shown in Figure 15.14-1*a*. The input to this circuit is the period voltage, $v_{in}(t)$, shown in Figure 15.14-1*b*. The output of this circuit, $v_{out}(t)$, will also be a periodic voltage. Both $v_{in}(t)$, and $v_{out}(t)$ will have the same fundamental frequency, 500 Hz. We will use PSpice to calculate the coefficients of the Fourier series of $v_{in}(t)$ and of $v_{out}(t)$. Figure 15.14-2 shows the PSpice input and output files. First, consider the input file. The command

.tran 10us 10ms uic

tells PSpice to run the simulation for 10 ms or 5 periods of $v_{in}(t)$. All transients will have died out before the final 2 ms. PSpice will calculate the Fourier series during the last full period of $v_{in}(t)$, i.e., during the last 2 ms of the simulation.

Notice that $v_{in}(t)$ is the node voltage at node 1 and that $v_{out}(t)$ is the node voltage at node 2 of the circuit. The command

.four 500Hz v(1) v(2)

tells PSpice that the fundamental frequency is 500 Hz and that we want the coefficients of the Fourier series of $v_{in}(t) = v(1)$ and $v_{out}(t) = v(2)$.

Next, consider the PSpice output file shown in Figure 15.14-2. The coefficients of the Fourier series of $v_{in}(t)$ are tabulated under the heading FOURIER COMPONENTS OF TRANSIENT RESPONSE V(1). The Fourier series of $v_{in}(t)$ can be constructed from the coefficients in the table shown in Figure 15.14-2 as

$$\begin{aligned} v_{in}(t) = 3 &+ 2.547\cos(\pi 1000t + 179^\circ) \\ &+ 0.8491\cos(3\pi 1000t + 177^\circ) \\ &+ 0.5098\cos(5\pi 1000t + 175^\circ) \\ &+ 0.3645\cos(7\pi 1000t + 174^\circ) \\ &+ 0.2839\cos(9\pi 1000t + 172^\circ) \end{aligned}$$

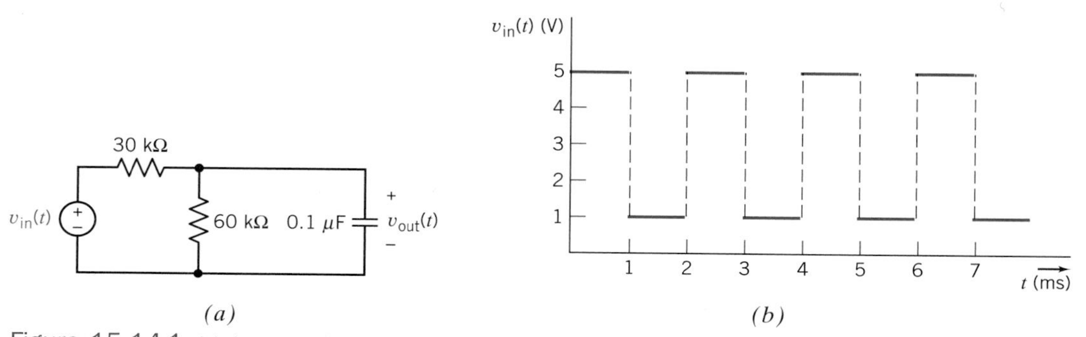

Figure 15.14-1 (*a*) An example circuit. (*b*) The periodic input voltage.

```
Illustration of the .FOUR command

Vin  1  0  pulse   ( 1  5  1ms  1us  1us  .999ms  2ms  )
R1   1  2  30k
R2   2  0  60k
C    2  0  0.10uF   ic=3

.tran  10us  10ms  uic
.four  500Hz  v(1)  v(2)

.end
```

(*a*)

```
FOURIER COMPONENTS OF TRANSIENT RESPONSE V(1)

DC COMPONENT = 3.000000E+00

HARMONIC FREQUENCY  FOURIER    NORMALIZED PHASE      NORMALIZED
   NO      (HZ)     COMPONENT  COMPONENT  (DEG)      PHASE (DEG)
   1     5.000E+02  2.547E+00  1.000E+00  1.791E+02   0.000E+00
   2     1.000E+03  4.831E-09  1.897E-09  9.740E+01  -8.170E+01
   3     1.500E+03  8.491E-01  3.334E-01  1.773E+02  -1.800E+00
   4     2.000E+03  4.741E-09  1.862E-09  1.048E+02  -7.435E+01
   5     2.500E+03  5.098E-01  2.002E-01  1.755E+02  -3.600E+00
   6     3.000E+03  4.594E-09  1.804E-09  1.120E+02  -6.710E+01
   7     3.500E+03  3.645E-01  1.431E-01  1.737E+02  -5.400E+00
   8     4.000E+03  4.394E-09  1.726E-09  1.191E+02  -6.001E+01
   9     4.500E+03  2.839E-01  1.115E-01  1.719E+02  -7.200E+00

   TOTAL HARMONIC DISTORTION = 4.291613E+01 PERCENT

FOURIER COMPONENTS OF TRANSIENT RESPONSE V(2)

DC COMPONENT = 2.007793E+00

HARMONIC FREQUENCY  FOURIER    NORMALIZED PHASE      NORMALIZED
   NO      (HZ)     COMPONENT  COMPONENT  (DEG)      PHASE (DEG)
   1     5.000E+02  2.668E-01  1.000E+00  9.818E+01   0.000E+00
   2     1.000E+03  1.236E-03  4.633E-03  6.229E+00  -9.195E+01
   3     1.500E+03  2.999E-02  1.124E-01  9.099E+01  -7.187E+00
   4     2.000E+03  6.190E-04  2.320E-03  5.789E+00  -9.239E+01
   5     2.500E+03  1.097E-02  4.111E-02  8.866E+01  -9.520E+00
   6     3.000E+03  4.135E-04  1.550E-03  6.811E+00  -9.137E+01
   7     3.500E+03  5.625E-03  2.119E-02  8.730E+01  -1.088E+01
   8     4.000E+03  3.102E-04  1.163E-03  8.159E+00  -9.002E+01
   9     4.500E+03  3.687E-03  1.382E-02  9.015E+01  -8.027E+00

   TOTAL HARMONIC DISTORTION = 1.224721E+01 PERCENT
```

(*b*)

Figure 15.14-2 PSpice (*a*) input and (*b*) output files for Example 15.14-1.

Similarly, the Fourier series of $v_{out}(t)$ can be constructed from the coefficients in the table labeled FOURIER COMPONENTS OF TRANSIENT RESPONSE V(2).

$$\begin{aligned} v_{out}(t) = 2 &+ 0.2668\cos(\pi 1000t + 98°) \\ &+ 0.02999\cos(3\pi 1000t + 91°) \\ &+ 0.01097\cos(5\pi 1000t + 89°) \\ &+ 0.005625\cos(7\pi 1000t + 87°) \\ &+ 0.0003687\cos(9\pi 1000t + 90°) \end{aligned}$$

15.15 VERIFICATION EXAMPLE

Figure 15.15-1 shows the transfer characteristic of the saturation nonlinearity. Suppose that the input to this nonlinearity is

$$v_{in}(t) = A\sin\omega t$$

where $A > a$. Verify that the output of the nonlinearity will be a periodic function that can be represented by the Fourier series

$$v_o(t) = b_1\sin\omega t + \sum_{\substack{n=3 \\ \text{odd}}}^{N} b_n\sin n\omega t \qquad (15.15\text{-}1)$$

where (Graham 1971)

$$B = \sin^{-1}\left(\frac{a}{A}\right)$$

$$b_1 = \frac{2}{\pi}A\left[B + \frac{a}{A}\sqrt{1 - \left(\frac{a}{A}\right)^2}\right]$$

and

$$b_n = \frac{4A}{\pi(1 - n^2)}\left[\frac{a}{A}\frac{\cos(nB)}{n} - \sqrt{1 - \left(\frac{a}{A}\right)^2}\sin(nB)\right]$$

Solution

The output voltage, $v_o(t)$, will be a clipped sinusoid. We need to verify that Eq. 15.15-1 does indeed represent a clipped sinusoid. A straightforward, but tedious, way to do this is to plot $v_o(t)$ versus t directly from Eq. 15.15-1. Several computer programs, such as spread-sheets and equation solvers, are available to reduce the work required to produce this plot. Mathcad is one of these programs. In Figure 15.15-2 Mathcad is used to plot $v_o(t)$ versus t. This plot verifies that the Fourier series in Eq. 15.15-1 does indeed represent a clipped sinusoid.

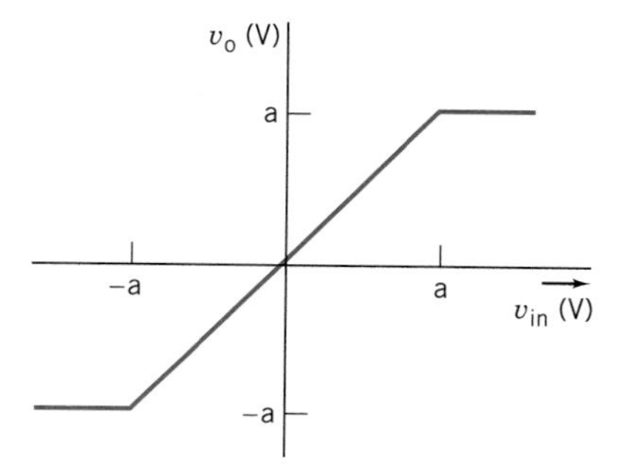

Figure 15.15-1 The saturation nonlinearity.

Plot a periodic signal from its coefficients.

Define n, the index for the summation:

$$N := 25 \quad n := 3,5,...,N$$

Define any parameters that are used to make it easier to enter the coefficient of the Fourier series:

$$A := 12.5 \quad a := 12 \quad B := \operatorname{asin}\left(\frac{a}{A}\right)$$

Enter the fundamental frequency:

$$\omega := 2 \cdot \pi \cdot 1000$$

Define an increment of time. Set up an index to run over two periods of the periodic signal:

$$T := \frac{2 \cdot \pi}{\omega} \quad dt := \frac{T}{200} \quad i := 1,2,...,400 \quad t_i := dt \cdot i$$

Enter the formulas for the coefficients of the Fourier series,

$$b_1 := \frac{2}{\pi} \cdot A \left[B + \frac{a}{A} \cdot \sqrt{1 - \left(\frac{a}{A}\right)^2} \right]$$

$$b_n := \frac{4 \cdot A}{\pi \cdot (1 - n^2)} \cdot \left[\frac{a}{A} \cdot \frac{\cos(n \cdot B)}{n} - \sqrt{1 - \left(\frac{a}{A}\right)^2} \cdot \sin(n \cdot B) \right]$$

Enter the Fourier series:

$$v(i) := b_1 \cdot \sin(\omega \cdot t_i) + \sum_{n=3}^{N} b_n \cdot \sin(n \cdot \omega \cdot t_i)$$

Plot the periodic signal:

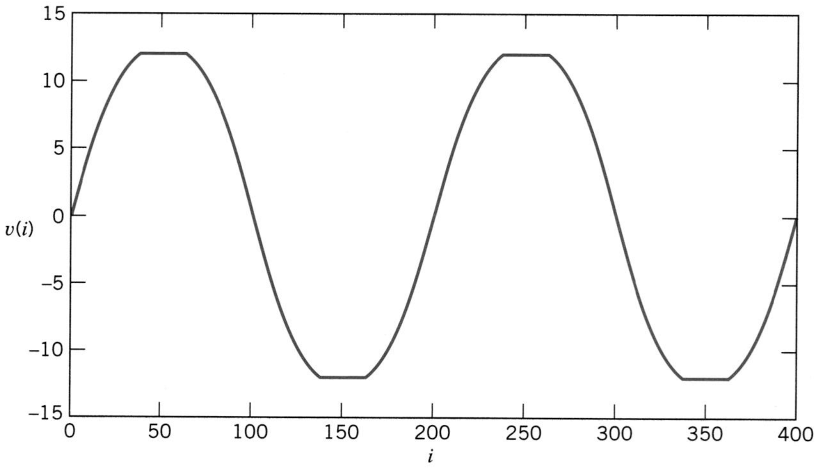

Figure 15.15-2 Using Mathcad to verify the Fourier series of a clipped sinusoid.

15.16 Design Challenge Solution

DC POWER SUPPLY

A laboratory power supply uses a nonlinear circuit called a rectifier to convert a sinusoidal voltage input into a dc voltage. The sinusoidal input

$$v_{ac}(t) = A \sin \omega_0 t$$

comes from the wall plug. In this example, $A = 160$ V and $\omega_0 = 377$ rad/s ($f_0 = 60$ Hz). Figure 15.16-1 shows the structure of the power supply. The output of the rectifier is the absolute value of its input, that is,

$$v_s(t) = |A \sin \omega_0 t|$$

The purpose of the rectifier is to convert a signal that has an average value equal to zero into a signal that has an average value that is not zero. The average value of $v_s(t)$ will be used to produce the dc output voltage of the power supply.

The rectifier output is not a sinusoid, but it is a periodic signal. Periodic signals can be represented by Fourier series. The Fourier series of $v_s(t)$ will contain a constant, or dc, term and some sinusoidal terms. The purpose of the filter shown in Figure 15.16-1 is to pass the dc term and attenuate the sinusoidal terms. The output of the filter, $v_o(t)$, will be a periodic signal and can be represented by a Fourier series. Because we are designing a dc power supply, the sinusoidal terms in the Fourier series of $v_o(t)$ are undesirable. The sum of these undersirable terms is called the ripple of $v_o(t)$.

The challenge is to design a simple filter so that the dc term of $v_o(t)$ is at least 90 V and the size of the ripple is no larger than 5 percent of the size of the dc term.

Define the Situation

1. From Table 15.4-2, the Fourier series of $v_s(t)$ is

$$v_s(t) = \frac{320}{\pi} - \sum_{n=1}^{N} \frac{640}{\pi(4n^2 - 1)} \cos(2 \cdot n \cdot 377 \cdot t)$$

Let $v_{sn}(t)$ denote the term of $v_s(t)$ corresponding to the integer n. Using this notation, the Fourier series of $v_s(t)$ can be written as

$$v_s(t) = v_{s0} + \sum_{n=1}^{N} v_{sn}(t)$$

2. Figure 15.16-2 shows a simple filter. The resistance R_s models the output resistance of the rectifier. We have assumed that the input resistance of the regulator is large enough to be ignored. (The input resistance of the regulator will be in parallel with R and will probably be much larger than R. In this case, the equivalent resistance of the parallel combination will be approximately equal to R.)

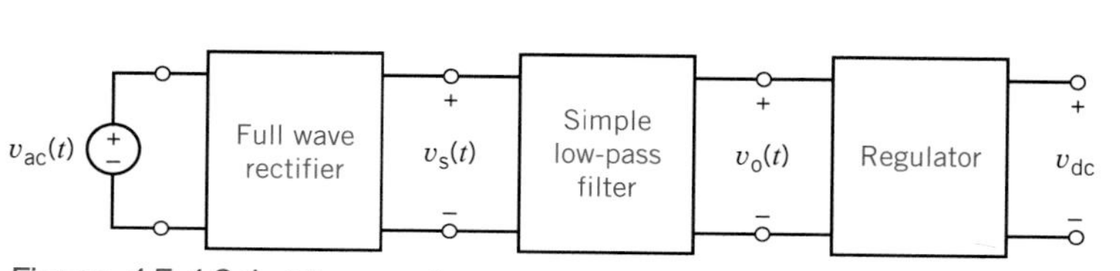

Figure 15.16-1 Diagram of a power supply.

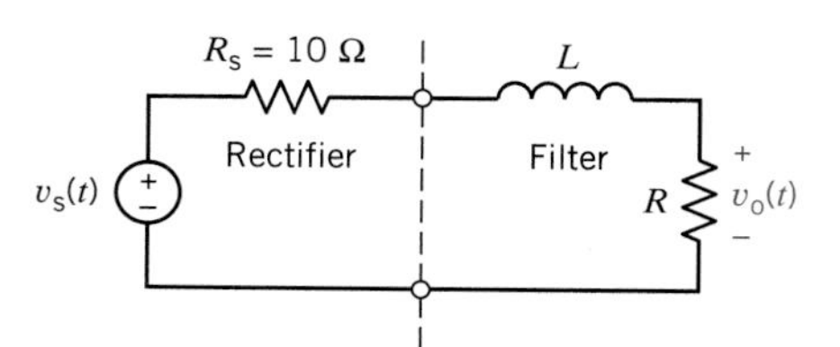

Figure 15.16-2 A simple *RL* low-pass filter connected to the rectifier.

3. The filter output, $v_o(t)$, will also be a periodic signal and will be represented by the Fourier series

$$v_o(t) = v_{o0} + \sum_{n=1}^{N} v_{on}(t)$$

4. Most of the ripple in $v_o(t)$ will be due to $v_{o1}(t)$, the fundamental term of the Fourier series. The specification regarding the allowable ripple can be stated as

$$\text{amplitude of the ripple} \leq 0.05 \cdot \text{dc output}$$

Equivalently, we can state that we require

$$\max\left(\sum_{n=1}^{N} v_{on}(t)\right) \leq 0.05 \cdot v_{o0} \qquad (15.16\text{-}1)$$

For ease of calculation, we replace Eq. 15.16-1 with the simpler condition

$$|v_{o1}(t)| \leq 0.04 \cdot v_{o0}$$

That is, the amplitude $v_{o1}(t)$ must be less than 4 percent of the dc term of the output (v_{o0} = dc term of the output).

State the Goal

Specify values of R and L so that

$$\text{dc output} = v_{o0} \geq 90$$

and

$$|v_{o1}(t)| \leq 0.04 \cdot v_{o0}$$

Generate a Plan

Use superposition to calculate the Fourier series of the filter output. First, the specification

$$\text{dc output} = v_{o0} \geq 90$$

can be used to determine the required value of R. Next, the specification

$$|v_{o1}(t)| \leq 0.04 \cdot v_{o0}$$

can be used to calculate L.

Act on the Plan

First, we will find the response to the dc term of $v_s(t)$. When the filter input is a constant and the circuit is at steady state, the inductor acts like a short circuit. Using voltage division

$$v_{o0} = \frac{R}{R + R_s} v_{s0} = \frac{R}{R + 10} \cdot \frac{320}{\pi}$$

The specification that $v_{o0} \geq 90$ requires

$$90 \leq \frac{R}{R + 10} \cdot \frac{320}{\pi}$$

or

$$R \geq 75.9$$

Let us select

$$R = 80\ \Omega$$

When $R = 80\ \Omega$,

$$v_{o0} = 90.54 \text{ V}$$

Next, we find the steady-state response to a sinusoidal term, $v_{sn}(t)$. Phasors and impedances can be used to find this response. By voltage division

$$\mathbf{V}_{on} = \frac{R}{R + R_s + j2n\omega_0 L}\mathbf{V}_{sn}$$

We are particularly interested in $\mathbf{V}_{o1}$:

$$\mathbf{V}_{o1} = \frac{R}{R + R_s + j2\omega_0 L}\mathbf{V}_{s1} = \frac{80}{90 + j754L} \cdot \frac{640}{\pi \cdot 3}$$

The amplitude of $v_{o1}(t)$ is equal to the magnitude of the phasor $\mathbf{V}_{o1}$. The specification on the amplitude of $v_{o1}(t)$ requires that

$$\frac{80}{\sqrt{90^2 + 754^2L^2}} \cdot \frac{640}{\pi \cdot 3} \leq 0.04\ \text{v}_{o1}$$

$$\leq 0.04 \cdot 90.54$$

That is,

$$L \geq 1.986 \text{ H}$$

Selecting

$$L = 2 \text{ H}$$

completes the design.

Verify the Proposed Solution

Figure 15.16-3*a* displays a plot of $v_s(t)$ and $v_o(t)$, the input and output voltages of the circuit in Figure 15.16-2. Figure 15.16-3*b* shows the details of the output voltage. This plot indicates that the average value of the output voltage is greater than 90 V and that the ripple is no greater than ± 4 V. Therefore, the specifications have been satisfied.

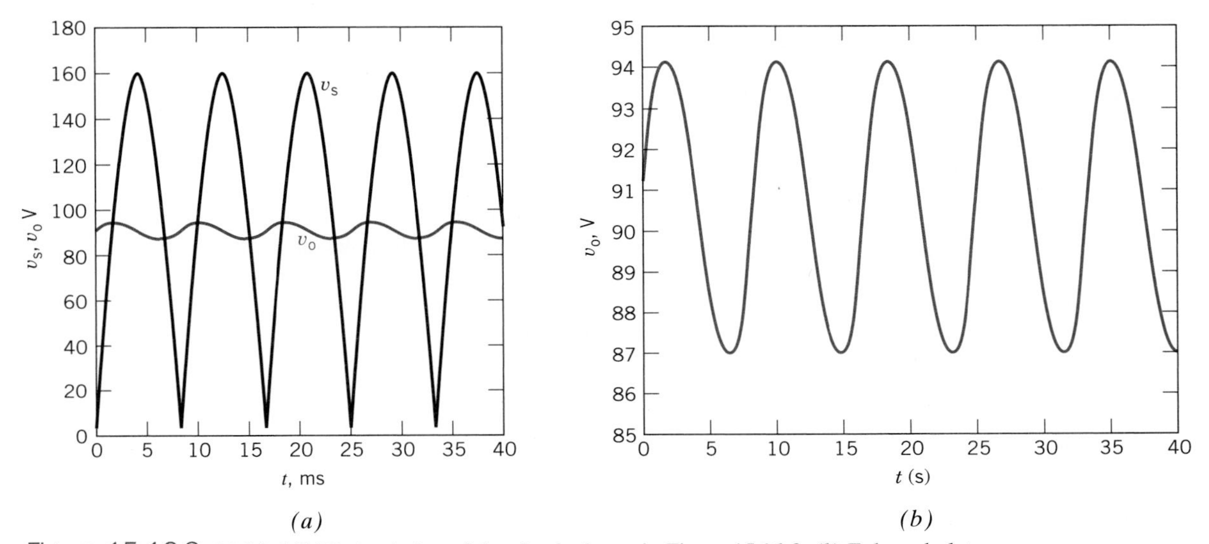

Figure 15.16-3 (*a*) MathCAD simulation of the circuit shown in Figure 15.16-2. (*b*) Enlarged plot of the output voltage.

15.17 SUMMARY

• Periodic waveforms arise in many circuits. For example, the form of the load current waveforms for selected loads is shown in Figure 15.17-1. While the load current for motors and incandescent lamps is of the same form as that of the source voltage, it is significantly altered for the power supplies, dimmers, and variable-speed drives as shown in Figure 15.17-1*b* and *c*. Electrical engineers have long been interested in developing the tools required to analyze circuits incorporating periodic waveforms.

• The brilliant mathematician-engineer Jean-Baptiste-Joseph Fourier proposed in 1807 that a periodic waveform could be represented by a series consisting of cosine and sine terms with the appropriate coefficients. The integer multiple frequencies of the fundamental are called the **harmonic frequencies** (or harmonics).

• The trigonometric form of the Fourier series is

$$f(t) = a_0 + \sum_{n=1}^{N} a_n \cos n\omega_0 t + \sum_{n=1}^{N} b_n \sin n\omega_0 t$$

The coefficients of the trigonometric Fourier series can be obtained from

$$a_0 = \frac{1}{T}\int_{t_0}^{T+t_0} f(t)\, dt$$

$$a_n = \frac{2}{T}\int_{t_0}^{T+t_0} f(t) \cos n\omega_0 t\, dt \qquad n > 0$$

$$b_n = \frac{2}{T}\int_{t_0}^{T+t_0} f(t) \sin n\omega_0 t\, dt \qquad n > 0$$

• An alternate form of the trigonometric form of the Fourier series is

$$f(t) = c_0 + \sum_{n=1}^{N} c_n \cos(n\omega_0 t + \theta_n)$$

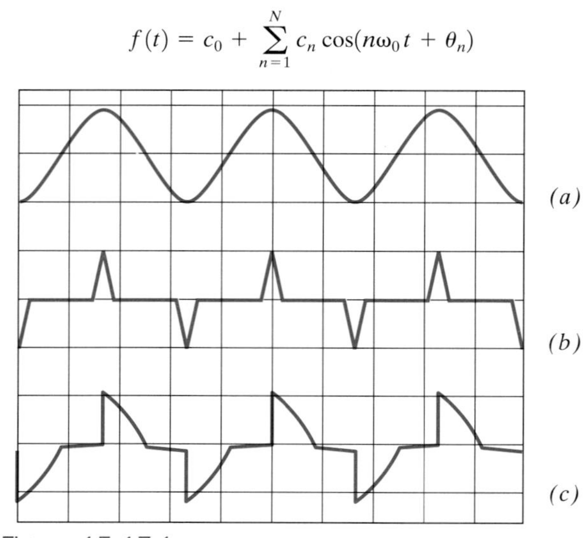

Figure 15.17-1 Load-current waveforms for (*a*) motors and incandescent lights, (*b*) switch-mode power supplies, and (*c*) dimmers and variable-speed drives. The vertical axis is current, and the horizontal axis is time. Source: *EPRI Journal*, August 1991.

where $c_0 = a_0$ = average value of $f(t)$ and

$c_n = \sqrt{a_n^2 + b_n^2}$ and

$$\theta_n = \begin{cases} -\tan^{-1}\left(\dfrac{b_n}{a_n}\right) & \text{if } a_n > 0 \\ 180^\circ - \tan^{-1}\left(\dfrac{b_n}{a_n}\right) & \text{if } a_n < 0 \end{cases}$$

• The Fourier coefficients of some common periodic signals are tabulated in Table 15.4-2.

• Symmetry can simplify the task of calculating the Fourier coefficients.

• The exponential form of the Fourier series is

$$f(t) = \sum_{-\infty}^{\infty} C_n e^{jn\omega_0 t}$$

where C_n are the complex coefficients defined by

$$C_n = \frac{1}{T}\int_{t_0}^{t_0+T} f(t)\, e^{-jn\omega_0 t} dt$$

• The line spectra consisting of the amplitude and phase of the complex coefficients of the Fourier series when plotted against frequency are useful for portraying the frequencies that represent a waveform.

• The practical representation of a periodic waveform consists of a finite number of sinusoidal terms of the Fourier series. The finite Fourier series exhibits the Gibbs phenomenon, that is, while convergence occurs as n grows large, there always remains an error at the points of discontinuity of the waveform.

• To determine the response of a circuit excited by a periodic input signal $v_s(t)$, we represent $v_s(t)$ by a Fourier series and then find the response of the circuit to the fundamental and each harmonic. Assuming the circuit is linear and the principle of superposition holds, we can consider that the total response is the sum of the response to the dc term, the fundamental, and each harmonic.

• The Fourier transform provides a frequency domain description of an aperiodic time domain function.

• A circuit with an impulse response $h(t)$ and an input $f(t)$ has a response $y(t)$ that may be determined from the convolution integral.

• The tables of Laplace transforms, Table 14.3-1 and Table 14.4-1, developed in Chapter 14, can be used to obtain the Fourier transform of a function $f(t)$.

• SPICE can be used to calculate the coefficients of the Fourier series of a periodic voltage or current. There are two steps to the procedure.

1. Simulate the circuit containing the voltage or current to obtain a transient response.
2. Use the .FOUR command to tell SPICE to compute the coefficients and to put them in the SPICE output file.

PROBLEMS

Section 15.3 The Fourier Series

P 15.3-1 Find the trigonometric Fourier series of Eq. 15.3-1 for a periodic function $f(t)$ that is equal to t^2 over the period from $t = 0$ to $t = 2$.

P 15.3-2 Find the trigonometric Fourier series of Eq. 15.3-1 for the function of Figure P 15.3-2. The function is the positive portion of a cosine wave.

P 15.3-3 A "staircase" periodic waveform is described by its first cycle as

$$f(t) = \begin{cases} 1 & 0 < t < 0.25 \\ 2 & 0.25 < t < 0.5 \\ 0 & 0.5 < t < 1 \end{cases}$$

Find the Fourier series for this function.

P 15.3-4 Determine the Fourier series for the sawtooth function shown in Figure P 15.3-4.

P 15.3-5 Find the Fourier series for

$$f(t) = |A \sin \omega t|$$

P 15.3-6 Find the Fourier series for the periodic function $f(t)$ that is equal to t over the period from $t = 0$ to $t = 2$ s.

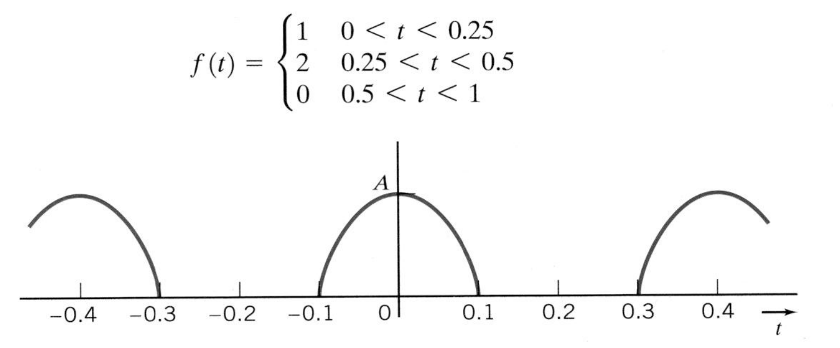

Figure P 15.3-2 Half-wave rectified cosine wave.

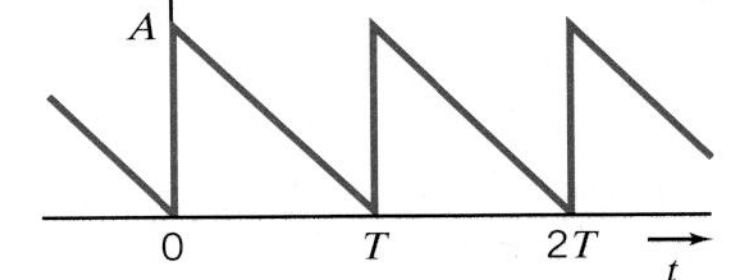

Figure P 15.3-4 Sawtooth wave.

Section 15.4 Symmetry of the Function $f(t)$

P 15.4-1 A sawtooth wave shown in Figure P 15.4-1 can be represented by a Fourier series.

(a) Find the trigonometric series.

(b) Plot the series consisting of the fundamental and two harmonics, and compare it to the original waveform for $-\pi \le t \le \pi$.

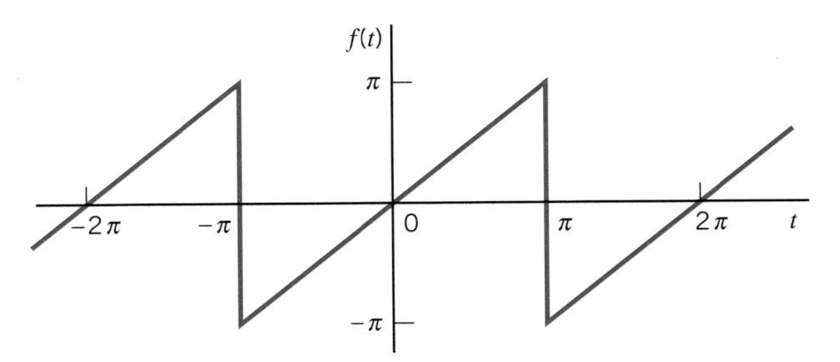

Figure P 15.4-1 Sawtooth wave.

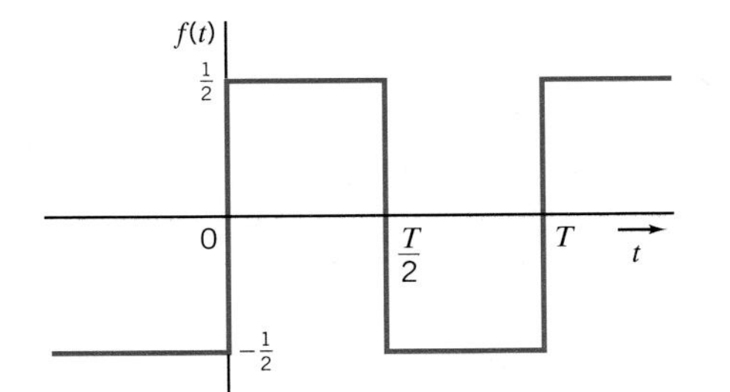

Figure P 15.4-2 Square waveform.

P 15.4-2 Find the sine-cosine form of the Fourier series for the waveform of Figure P 15.4-2.

P 15.4-3 Determine the Fourier series for the waveform shown in Figure P 15.4-3. Calculate a_0, a_1, a_2, and a_3.

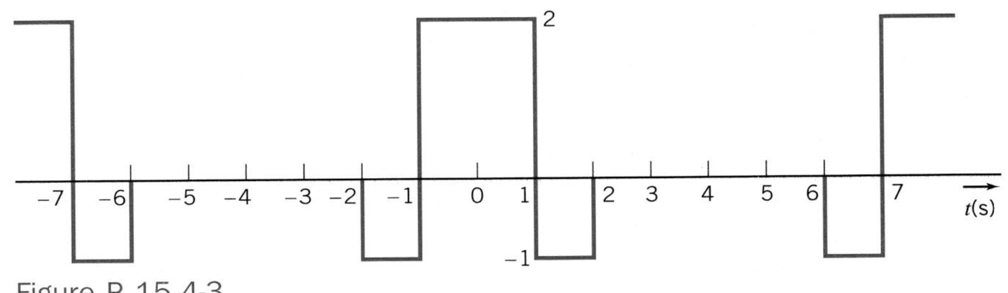

Figure P 15.4-3

P 15.4-4 Determine the Fourier series for

$$f(t) = |A \cos \omega t|$$

P 15.4-5 Determine the Fourier series for $f(t)$ shown in Figure P 15.4-5.

Answer: $a_n = a_0 = 0$; $b_n = 0$ for even n, $= 8/n^2\pi^2$ for $n = 1, 5, 9$, and $= -8/n^2\pi^2$ for $n = 3, 7, 11$

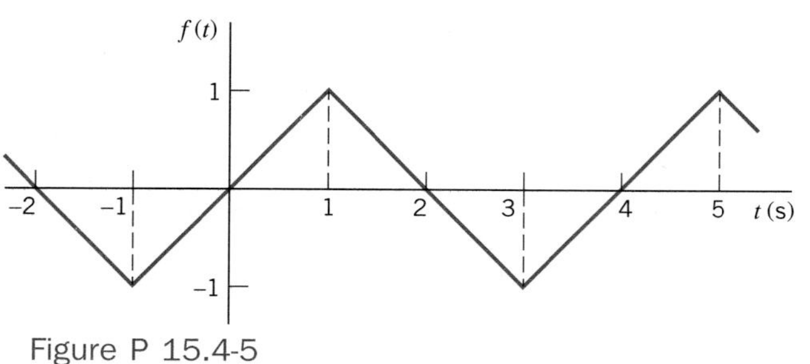

Figure P 15.4-5

P 15.4-6 Determine the Fourier series for the periodic signal shown in Figure P 15.4-6.

Answer: $f(t) = \frac{1}{2} + \frac{2}{\pi}\left(\sin t + \frac{1}{3}\sin 3t + \frac{1}{5}\sin 5t + \cdots\right)$

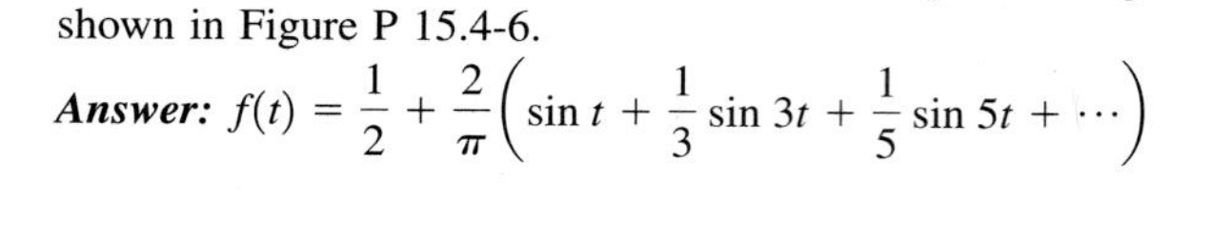

Figure P 15.4-6

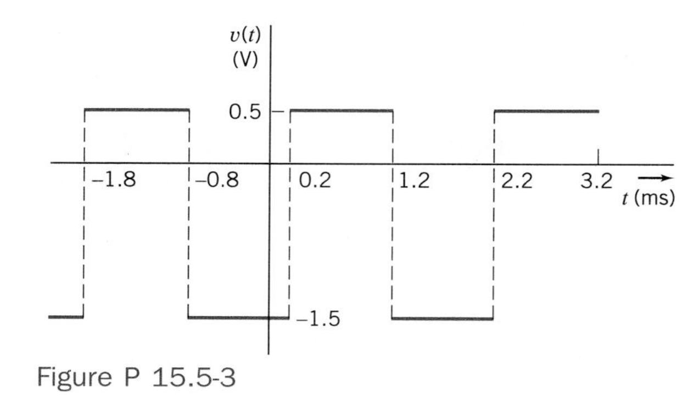

Figure P 15.5-3

Section 15.5 Exponential Form of the Fourier Series

P 15.5-1 Determine the exponential form of the Fourier series for the sawtooth wave shown in Figure P 15.4-1.

P 15.5-2 Determine the exponential form of the Fourier series for the waveform of Figure P 15.4-3.

P 15.5-3 Determine the exponential form of the Fourier series for the waveform of Figure P 15.5-3.

P 15.5-4 Determine the Fourier series for the waveform of Figure P 15.5-4.

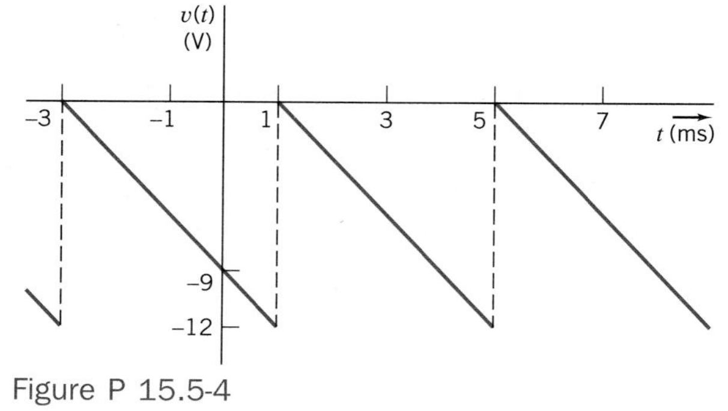

Figure P 15.5-4

P 15.5-5 Determine the exponential Fourier series of the periodic waveform shown in Figure P 15.5-5.

Answer: $C_n = \dfrac{1}{j2\pi n}$

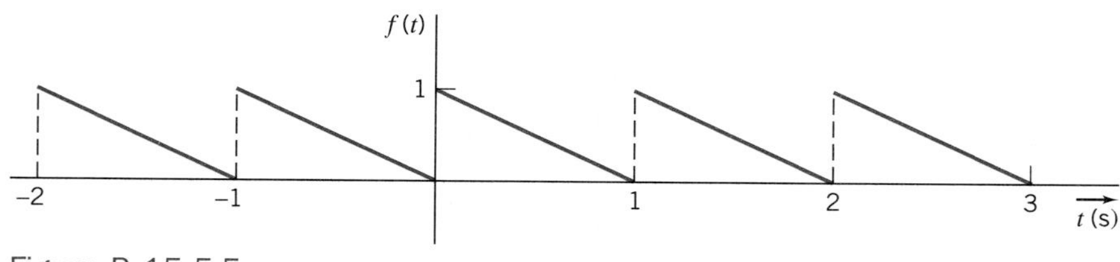

Figure P 15.5-5

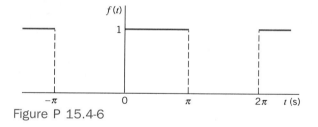

P 15.5-6 A periodic function consists of rising and decaying exponentials of time constants of 0.2 s each and durations of 1 s each as shown in Figure P 15.5-6. Determine the exponential Fourier series for this function.

Answer: $C_n = \dfrac{5}{(j\pi n)(5 + j\pi n)}, \; n = 1, 3, 5$

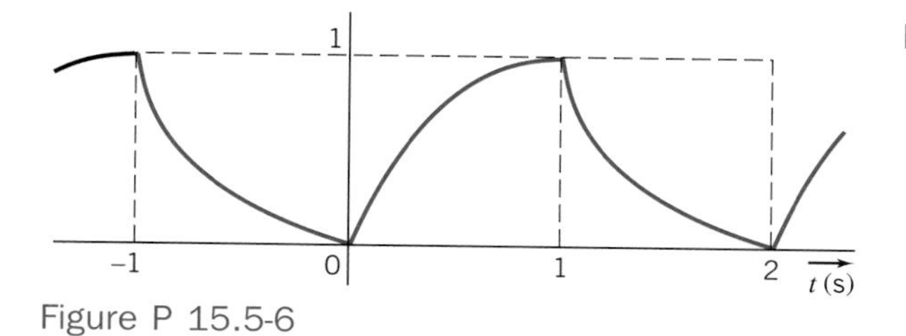

Figure P 15.5-6

Section 15.6 The Fourier Spectrum

P 15.6-1 Determine the cosine-sine Fourier series for the sawtooth waveform shown in Figure P 15.6-1. Draw the Fourier spectra for the first four terms including magnitude and phase.

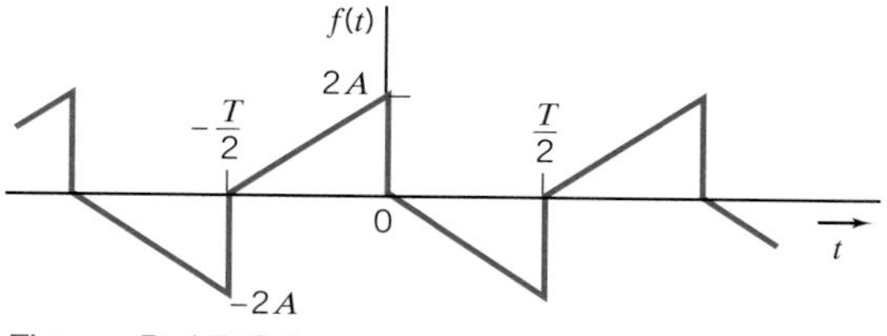

Figure P 15.6-1

P 15.6-2 The load current waveform of the variable-speed motor drive depicted in Figure 15.17-1c is shown in Figure P 15.6-2. The current waveform is a portion of $A \sin \omega_0 t$. Determine the Fourier series of this waveform, and draw the line spectra of $|\mathbf{C}_n|$ for the first 10 terms.

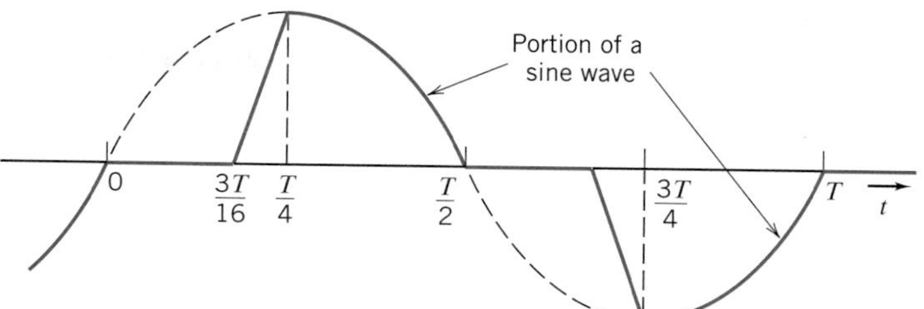

Figure P 15.6-2 The load current of a variable-speed drive.

P 15.6-3 Draw the Fourier spectra for the waveform shown in Figure P 15.5-3.

P 15.6-4 Draw the Fourier spectra for the waveform shown in Figure P 15.5-5.

P 15.6-5 Draw the Fourier spectra for the waveform shown in Figure P 15.5-6.

Section 15.8 Circuits and Fourier Series

P 15.8-1 A square wave, as shown in Figure P 15.8-1a, is applied to the RL circuit of Figure P 15.8-1b. Using the Fourier series representation, find the current $i(t)$.

P 15.8-2 In Section 15.2 we discussed the discovery by M. I. Pupin of the concept of telephone line balance. A length of cable can be represented by the RC circuit of Figure P 15.8-2, and the inductor L was inserted before being connected to another length of cable as shown. The voice sounds have significant harmonics up to 8000 Hz. Determine the required L when $R = 2\ \Omega$, $C = 4$ mF, and the resistance R_L of the receiver is 100 Ω. The goal is to pass all frequencies up to 8000 Hz and reject frequency content above 8000 Hz for each length of cable.

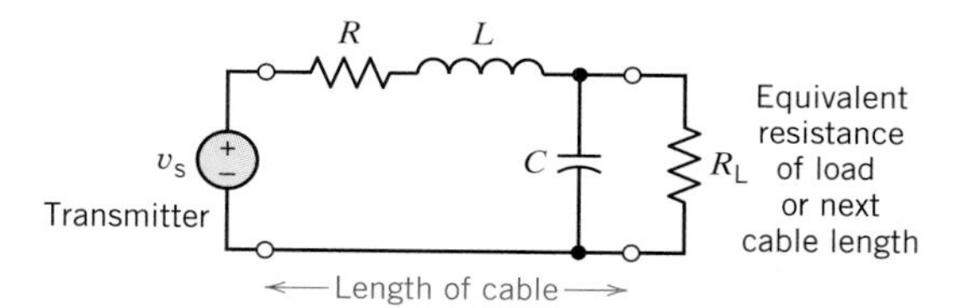

Figure P 15.8-2 A balanced cable according to Pupin.

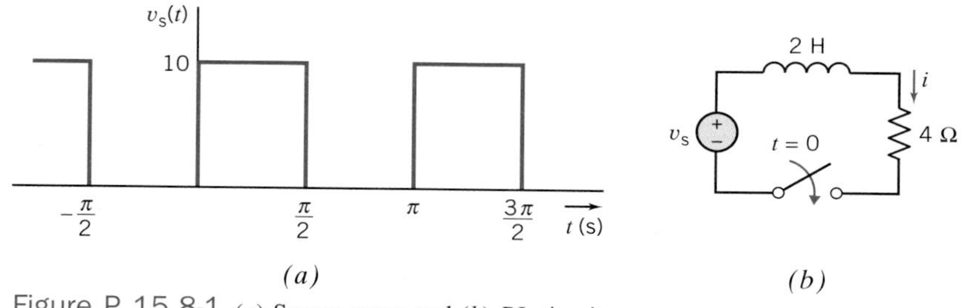

Figure P 15.8-1 (a) Square wave and (b) RL circuit.

P 15.8-3 Find the steady-state response for the output voltage, v_o, for the circuit of Figure P 15.8-3 when $v(t)$ is as described in P 15.5-4.

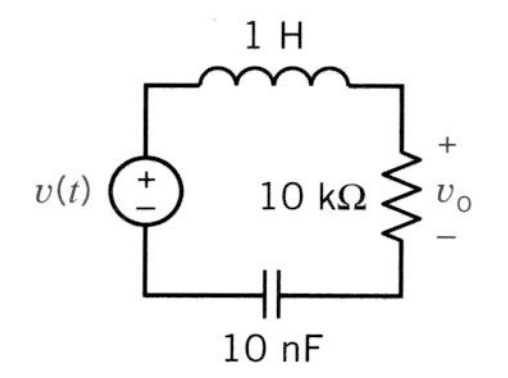

Figure P 15.8-3 A *RLC* circuit.

P 15.8-4 Determine the value of the voltage $v_o(t)$ at $t = 4$ ms when v_{in} is shown in Figure P 15.8-4*a* and the circuit is shown in Figure P 15.8-4*b*.

Section 15.9 The Fourier Transform

P 15.9-1 Find the Fourier transform of the function

$$f(t) = -u(-t) + u(t)$$

as shown in Figure P 15.9-1. This is called the signum function.

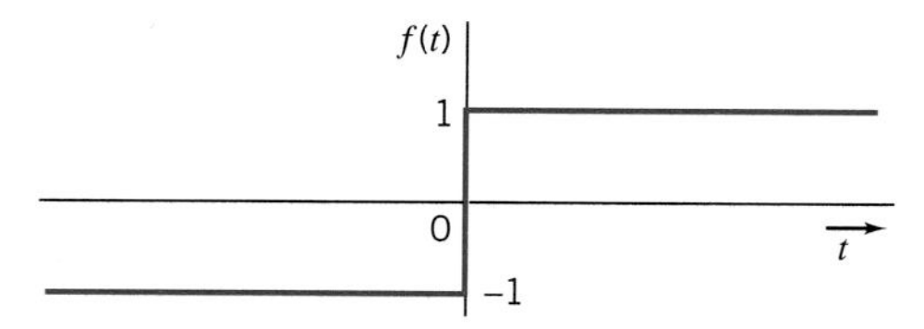

Figure P 15.9-1

P 15.9-2 Find the Fourier transform of $f(t) = Ae^{-at}u(t)$ when $a > 0$.

Answer: $F(\omega) = \dfrac{A}{a + j\omega}$

P 15.9-3 Find the Fourier transform of the waveform shown in Figure P 15.9-3.

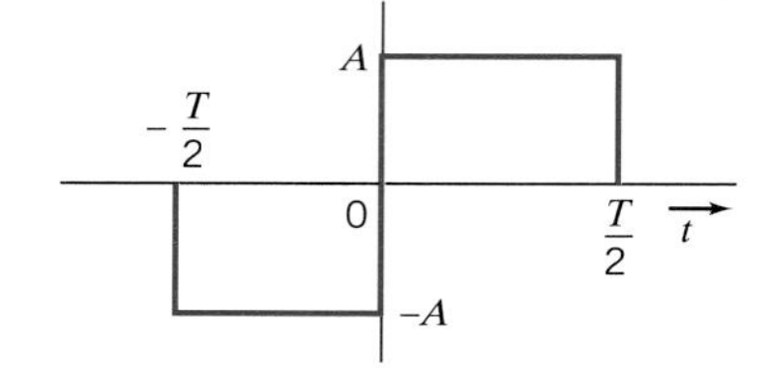

Figure P 15.9-3

P 15.9-4 Determine the Fourier transform of $f(t) = 10 \cos 50\, t$.

Answer: $F(\omega) = 10\pi\,\delta(\omega - 50) + 10\pi\,\delta(\omega + 50)$

P 15.9-5 Determine the Fourier transform of the pulse shown in Figure P 15.9-5.

Answer: $F(j\omega) = \dfrac{2}{\omega}(\sin\omega - \sin 2\omega) + \dfrac{j2}{\omega}(\cos\omega - \cos 2\omega)$

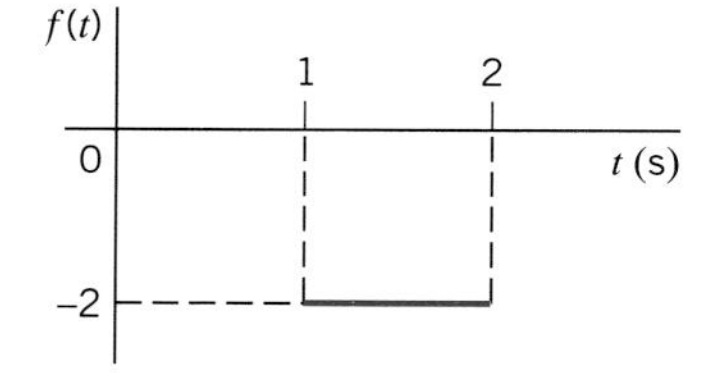

Figure P 15.9-5

P 15.9-6 Determine the Fourier transform of a signal with $f(t) = At/B$ between $t = 0$ and $t = B$ and $f(t) = 0$ elsewhere.

Answer: $F(j\omega) = \dfrac{A}{B}\left[\dfrac{-B}{j\omega}e^{-j\omega B} + \dfrac{1}{\omega^2}e^{-j\omega B} - \dfrac{1}{\omega^2}\right]$

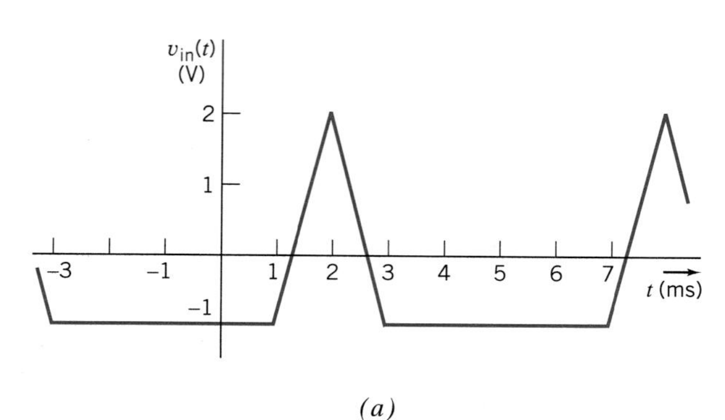

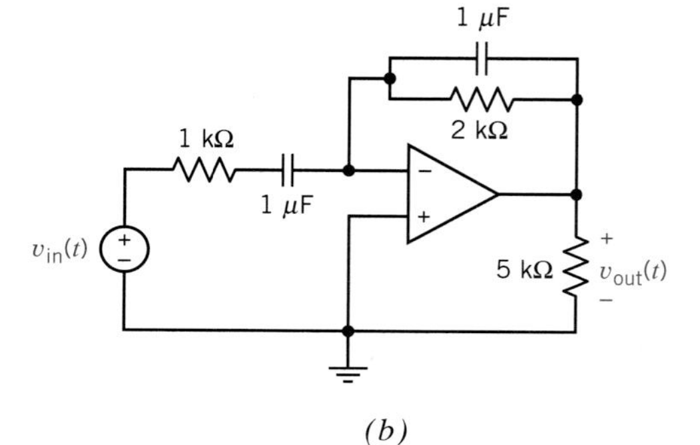

Figure P 15.8-4

15.16 Design Challenge Solution

DC POWER SUPPLY

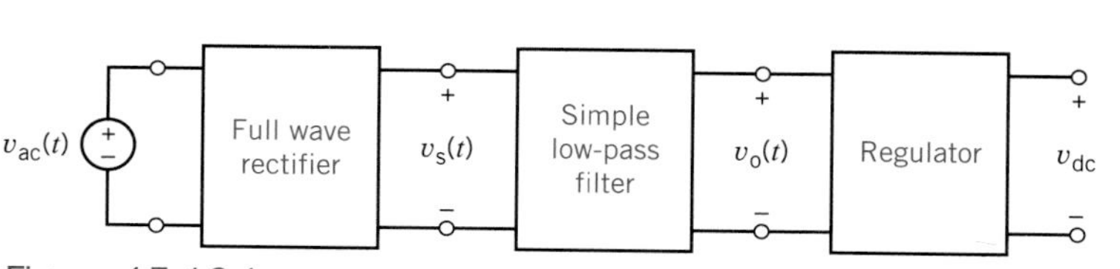

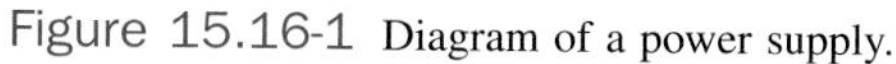

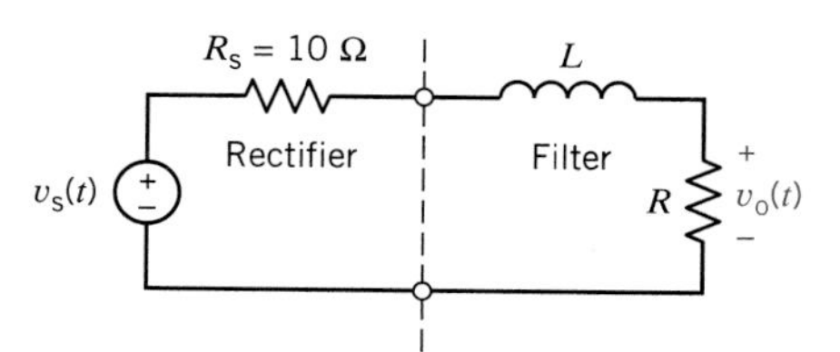

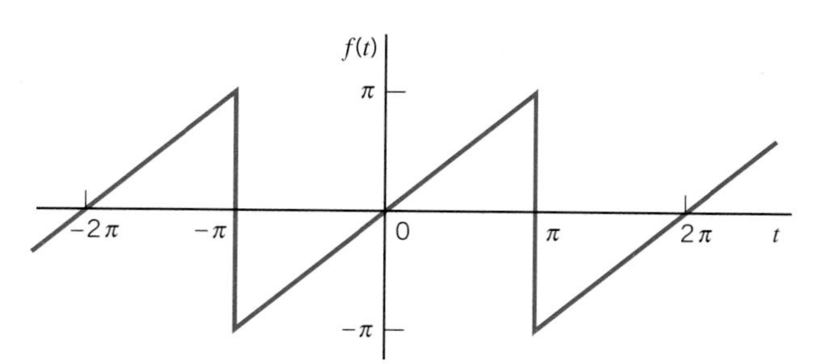

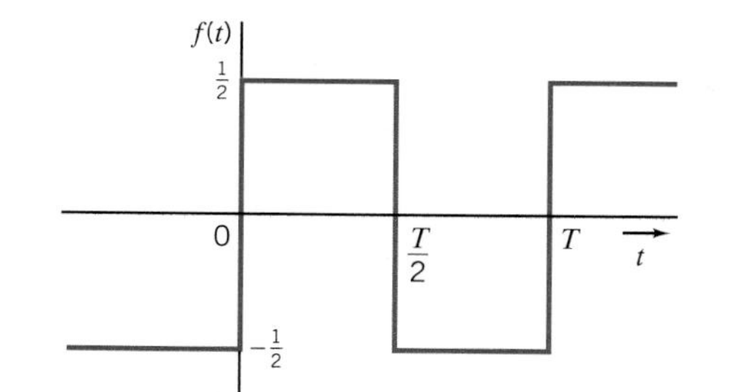

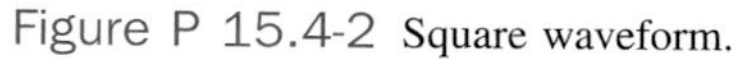

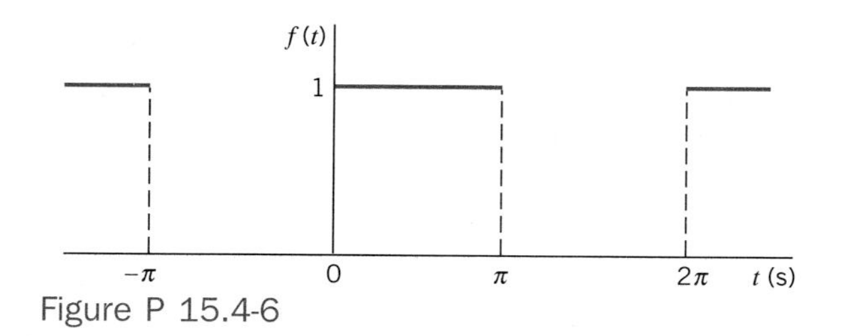

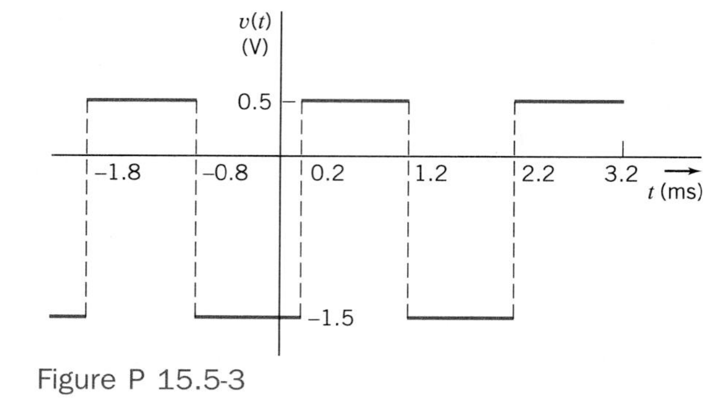

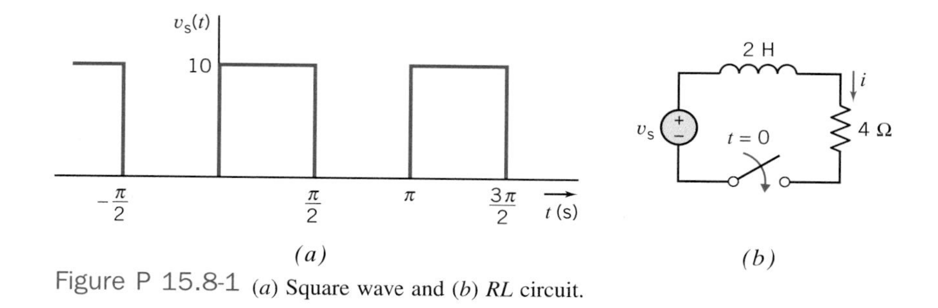

P 15.9-7 Determine the Fourier transform of the waveform $f(t)$ shown in Figure P 15.9-7.

Answer: $F(j\omega) = \dfrac{2}{\omega}(\sin 2\omega - \sin \omega)$

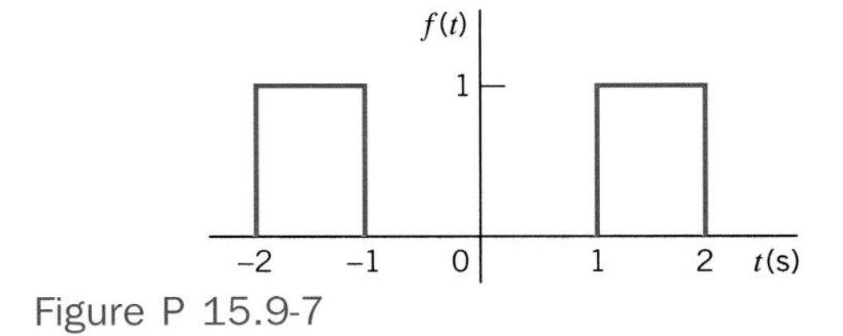

Figure P 15.9-7

Section 15.12 Convolution and Circuit Response

P 15.12-1 Find the current $i(t)$ in the circuit of Figure P 15.12-1 when $i_s(t)$ is the signum function, so that

$$i_s(t) = \begin{cases} +40\text{ A} & t > 0 \\ -40\text{ A} & t < 0 \end{cases}$$

Also, sketch $i(t)$.

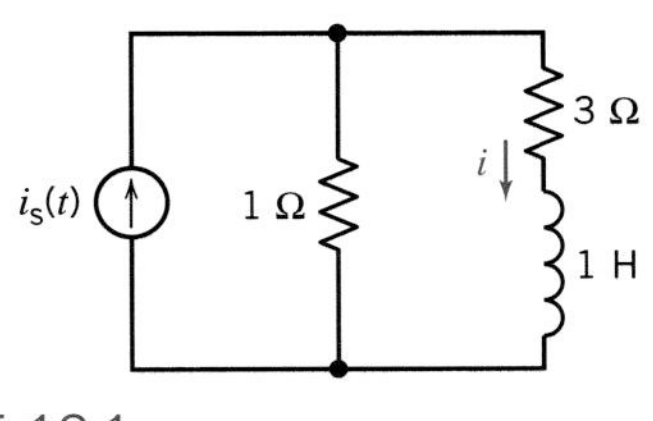

Figure P 15.12-1

P 15.12-2 Repeat Problem 15.12-1 when $i_s = 100 \cos 3t$ A.

P 15.12-3 The voltage source of Figure P 15.12-3 is $v(t) = 10 \cos 2t$ for all t. Calculate $i(t)$ using the Fourier transform.

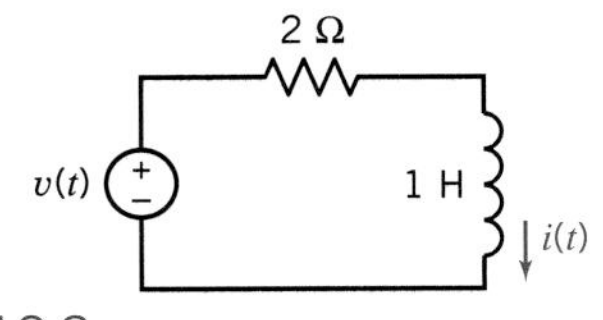

Figure P 15.12-3

P 15.12-4 Find the output voltage $v_0(t)$ using the Fourier transform for the circuit of Figure P 15.12-4 when $v(t) = e^t u(-t) + u(t)$ V.

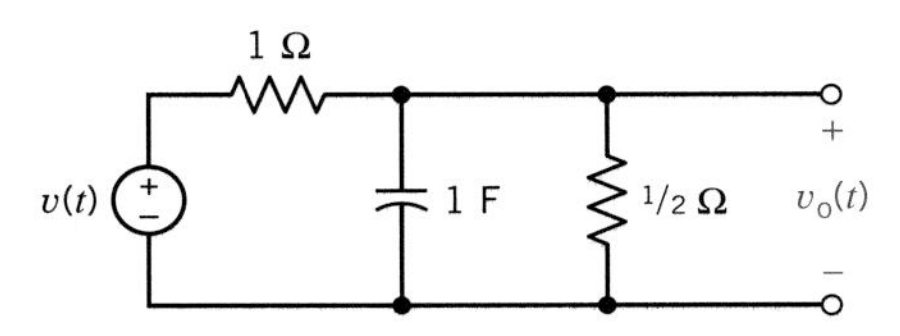

Figure P 15.12-4

P 15.12-5 The voltage source of the circuit of Figure P 15.12-5 is $v(t) = 15e^{-5t}$ V. Find the resistance R when it is known that the energy available in the output signal is two-thirds of the energy of the input signal.

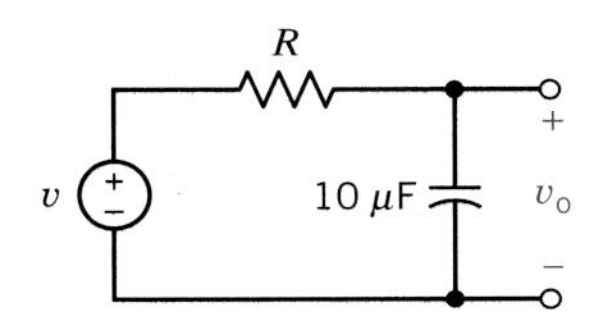

Figure P 15.12-5

P 15.12-6 The pulse signal shown in Figure P 15.12-6*a* is the source $v_s(t)$ for the circuit of Figure P 15.12-6*b*. Determine the output voltage, v_0, using the Fourier transform.

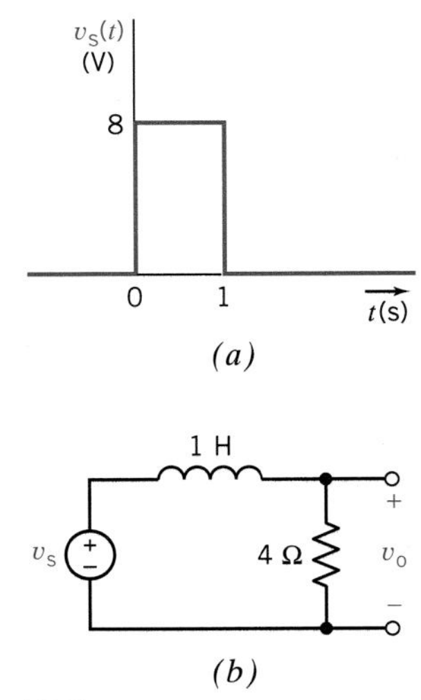

Figure P 15.12-6

PSPICE PROBLEMS

SP 15-1 Use PSpice to determine the Fourier coefficients and the Fourier series for P 15.4-1.

SP 15-2 Use PSpice to determine the Fourier coefficients for P 15.3-4.

SP 15-3 Use PSpice to determine the Fourier coefficients for a voltage square waveform as shown in Figure 15.5-1 when $T = 1$ ms and $A = 40$ V.

SP 15-4 Use PSpice to determine the Fourier coefficients for $v(t)$ shown in Figure SP 15.4.

SP 15-5 Use PSpice to determine the Fourier coefficients for $v(t)$ shown in Figure SP 15.5.

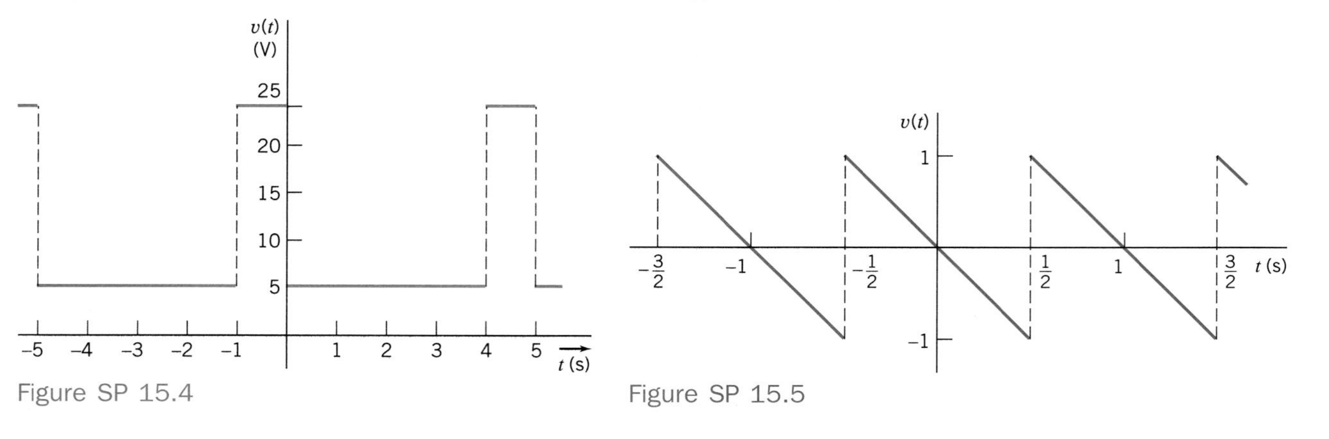

Figure SP 15.4

Figure SP 15.5

VERIFICATION PROBLEMS

VP 15-1 The Fourier series of $v_{in}(t)$ shown in Figure P 15.8-4 is given as

$$v_{in}(t) = \frac{1}{2} + \sum_{n=1}^{\infty} \frac{18}{n^2\pi}\left(1 - \cos\frac{n\pi}{3}\right)\cos\left(n\frac{\pi}{3}t - n\frac{2\pi}{3}\right) \text{ V}$$

Is this the correct Fourier series?
Hint: Check the average value and the fundamental frequency.
Answer: The given Fourier series is not correct.

VP 15-2 The Fourier series of $v(t)$ shown in Figure SP 15.3 is given as

$$v(t) = 9 + \sum_{n=1}^{\infty} \frac{40}{n\pi}\left(\sin\frac{n\pi}{5}\right)\cos\left(n\frac{\pi}{5}t - n\frac{\pi}{5}\right) \text{ V}$$

Is this the correct Fourier series?
Hint: Check the average value and the fundamental frequency.
Answer: The given Fourier series is not correct.

VP 15-3 The Fourier series of $v(t)$ shown in Figure SP 15.5 is given as

$$v(t) = 2\sum_{n=1}^{\infty} \frac{(-1)^n}{n\pi}\cos(n2\pi t) \text{ V}$$

Is this the correct Fourier series?
Hint: Check the average value and the fundamental frequency. Check for symmetry.
Answer: The given Fourier series is not correct.

DESIGN PROBLEMS

DP 15-1 A periodic waveform shown in Figure DP 15.1*a* is the input signal of the circuit shown in Figure DP 15.1*b*. Select the capacitance C so that the magnitude of the third harmonic of $v_2(t)$ is less than 1.4 V and greater than 1.3 V. Write the equation describing the third harmonic of $v_2(t)$, for the value of C selected.

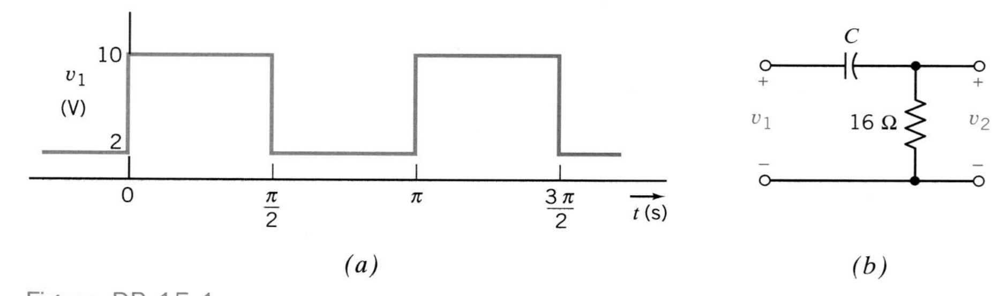

Figure DP 15.1

DP 15-2 A dc laboratory power supply uses a nonlinear circuit to convert a sinusoidal voltage obtained from the wall plug to a constant dc voltage. The wall plug voltage is $A \sin \omega_0 t$, where $f_0 = 60$ Hz and $A = 160$ V. The voltage is then rectified so that $v_s = |A \sin \omega_0 t|$. Using the filter circuit of Figure DP 15.2, determine the required inductance L so that the magnitude of each harmonic (ripple) is less than 4 percent of the dc component of the output voltage.

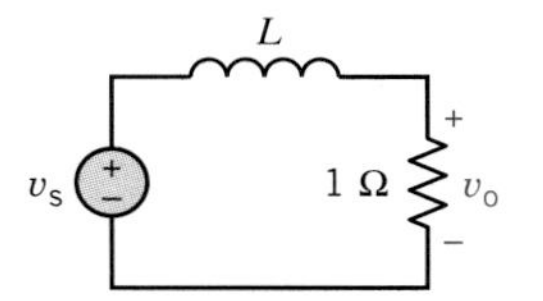

Figure DP 15.2 An *RL* circuit.

DP 15-3 A low-pass filter is shown in Figure DP 15.3. The input, v_s, is a half-wave rectified sinusoid with $\omega_0 = 800\pi$ (item 5 of Table 15.5-1). Select L and C so that the peak value of the first harmonic is $\frac{1}{20}$ of the dc component for the output, v_o.

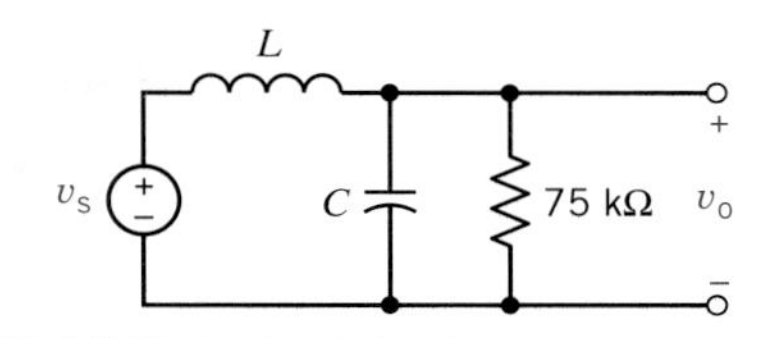

Figure DP 15.3 An *RLC* circuit.

CHAPTER 16

Filter Circuits

Preview

Transfer functions are used to characterize linear circuits. In a previous chapter, we learned how to analyze a circuit so that we could determine its transfer function. In this chapter, we learn how to design a circuit to have a specified transfer function. This design problem does not have a unique solution. There are many ways to obtain a circuit from a specified transfer function. A popular strategy is to design the circuit to be a cascade connection of second-order filter stages. This is the strategy we will use in this chapter.

The problem of designing a circuit that will have a specified transfer function is called filter design. In this chapter we will learn the vocabulary of filter design and describe second-order filter stages. Finally, we will learn how to connect these filter stages in order to obtain a circuit that has a specified transfer function.

16.1 Design Challenge

ANTI-ALIASING FILTER

Digital Signal Processing (DSP) frequently involves sampling a voltage, and converting the samples to digital signals. After the digital signals are processed, the output signal is converted back into an analog voltage. Unfortunately, a phenomenon called aliasing can cause errors to occur during digital signal processing. Aliasing is a possibility whenever the input voltage contains components at frequencies greater than one-half of the sampling frequency. Aliasing occurs when these components are mistakenly interpreted to be components at a lower frequency. Anti-aliasing filters are used to avoid these errors by eliminating those components of the input voltage that have frequencies greater than one-half of the sampling frequency.

An anti-aliasing filter is needed for a DSP application. The filter is specified to be a fourth-order Butterworth low-pass filter with a cutoff frequency of 500 hertz and a dc gain equal to one. This filter is to be implemented as an RC op amp circuit.

We will learn to design filter circuits in this chapter and will return to the problem of designing the anti-aliasing filter at the end of the chapter.

16.2 THE ELECTRIC FILTER

The concept of a filter was conceived early in human history. A paper filter was used to remove dirt and unwanted substances from water and wine. A porous material, such as paper, can serve as a mechanical filter. Mechanical filters are used to remove unwanted constituents, such as suspended particles, from a liquid. In a similar manner, an electric filter can be used to eliminate unwanted constituents, such as electrical noise, from an electrical signal.

The electrical filter was independently invented in 1915 by George Campbell in the United States and K. W. Wagner in Germany. With the rise of radio in the period 1910–1920, there emerged a need to reduce the effect of static noise at the radio receiver. As regular radio broadcasting emerged in the 1920s, Campbell and others developed the *RLC* filter using inductors, capacitors, and resistors. These filters are called *passive filters* because they consist of passive elements. The theory required to design passive filters was developed in the 1930s by S. Darlington, S. Butterworth, and E. A. Guillemin. The Butterworth low-pass filter was reported in *Wireless Engineering* in 1930 (Butterworth 1930).

When active devices, typically op amps, are incorporated into an electric filter, the filter is called an *active filter.* Since inductors are relatively large and heavy, active filters are usually constructed without inductors—using, for example, only op amps, resistors, and capacitors. The first practical active-*RC* filters were developed during World War II and were documented in a classic paper by R. P. Sallen and E. L. Key (Sallen and Key 1955).

16.3 FILTERS

We begin by considering an **ideal filter.** For convenience, suppose that both the input and output of this filter are voltages. This ideal filter separates its input voltage into two parts. One part is passed, unchanged to the output; the other part is eliminated. In other words, the output of an ideal filter is an exact copy of part of the filter input.

This is a familiar use of the word **filter.** For example, we expect an automotive oil filter to take a mixture of oil and dirt and separate it into two parts: oil and dirt. Ideally, the oil filter passes one part of its input, the oil, to its output without changing it in any way. The other part of the input, the dirt, should be completely eliminated. The oil filter stops the dirt from getting to the output.

To understand how an electric filter works, consider an input voltage

$$v_i(t) = \cos\omega_1 t + \cos\omega_2 t + \cos\omega_3 t$$

This input consists of a sum of sinusoids, each at a different frequency. (For example, periodic voltages can be represented in this way using the Fourier series.) The filter separates the input voltage into two parts, using frequency as the basis for separation. There are several ways of separating this input into two parts, and correspondingly, several types of ideal filter. Table 16.3-1 illustrates the common filter types. Consider the ideal **low-pass filter,** shown in row 1 of the table. The network function of the ideal low-pass filter is

$$\mathbf{H}(\omega) = \begin{cases} 1\angle 0^\circ & \omega < \omega_c \\ 0 & \omega > \omega_c \end{cases} \tag{16.3-1}$$

The frequency ω_c is called the **cutoff frequency.** The cutoff frequency separates the frequency range $\omega < \omega_c$, called the **pass-band,** from the frequency range $\omega > \omega_c$, called the **stop-band.** Those components of the input that have frequencies in the pass-band experience unity gain and zero phase shift. These terms are passed, unchanged, to the output of the filter. Components of the input that have frequencies in the stop-band experience a gain equal to zero. These terms are eliminated or stopped. An ideal filter separates its input into two parts: those terms that have frequencies in the pass-band and those terms that have frequencies in the stop-band. The output of the filter consists of those terms with frequencies in the pass-band.

Table 16.3-1 **Ideal Filters**

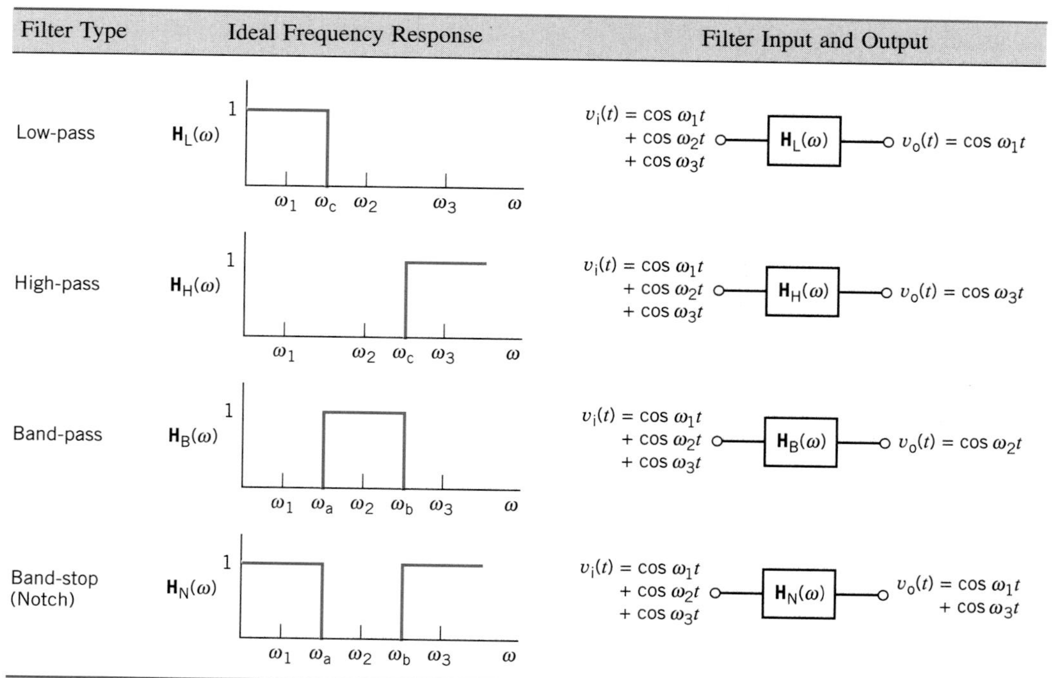

Unfortunately, ideal filters circuits don't exist. (This fact can be proved by calculating the impulse response of the ideal filter by taking the inverse Laplace Transform of the transfer function. The impulse response of an ideal filter would have to exist before the impulse itself. That is, the response would have to occur before the input that caused the response. Since that can't happen, ideal filter circuits don't exist.) Filters are circuits that approximate ideal filters. Filters divide their inputs into two parts, the terms in the pass-band and the terms in the stop-band. The terms in the pass-band experience a gain that is approximately one and also experience some phase shift. These terms are passed to the output, but they are changed a little. The terms in the stop-band experience a small gain that isn't quite zero. Because these terms aren't eliminated entirely, some small residue of these terms shows up in the filter output.

Butterworth transfer functions have magnitude frequency responses that approximate the frequency response of an ideal filter. Butterworth low-pass transfer functions are given by

$$H_L(s) = \frac{\pm 1}{D(s)} \tag{16.3-2}$$

We can choose either $+1$ or -1 for the numerator of $H_L(s)$. The polynomial $D(s)$ depends on the cutoff frequency and on the order of the filter. These polynomials, called Butterworth polynomials, are tabulated in Table 16.3-2 for $\omega_c = 1$ rad/s. There is a tradeoff involving the order of the filter. The higher the order, the more accurately the filter frequency response approximates the frequency response of an ideal filter; that's good. The higher the filter order, the more complicated the circuit required to build the filter; that's not good.

Example 16.3-1

We wish to design a low-pass filter that will approximate an ideal low-pass filter with $\omega_c = 1$ rad/s. Compare the fourth-order. Butterworth low-pass filter to the eighth order Butterworth low-pass filter.

Solution

The fourth row of Table 16.3-2 indicates that the transfer of the fourth-order Butterworth filter is

Table 16.3-2 **Denominators of Butterworth Low-Pass Filters with a Cutoff Frequency $\omega_C = 1$ rad/s**

Order	Denominator, D(s)
1	$s + 1$
2	$s^2 + 1.414s + 1$
3	$(s + 1)(s^2 + s + 1)$
4	$(s^2 + 0.765s + 1)(s^2 + 1.848s + 1)$
5	$(s + 1)(s^2 + 0.618s + 1)(s^2 + 1.618s + 1)$
6	$(s^2 + 0.518s + 1)(s^2 + 1.414s + 1)(s^2 + 1.932s + 1)$
7	$(s + 1)(s^2 + 0.445s + 1)(s^2 + 1.247s + 1)(s^2 + 1.802s + 1)$
8	$(s^2 + 0.390s + 1)(s^2 + 1.111s + 1)(s^2 + 1.663s + 1)(s^2 + 1.962s + 1)$
9	$(s + 1)(s^2 + 0.347s + 1)(s^2 + s + 1)(s^2 + 1.532s + 1)(s^2 + 1.879s + 1)$
10	$(s^2 + 0.313s + 1)(s^2 + 0.908s + 1)(s^2 + 1.414s + 1)(s^2 + 1.782s + 1)(s^2 + 1.975s + 1)$

$$H_4(s) = \frac{1}{(s^2 + 0.765s + 1)(s^2 + 1.848s + 1)} = \frac{1}{(s^2 + 0.765s + 1)} \times \frac{1}{(s^2 + 1.848s + 1)}$$

Similarly, the eighth row of Table 16.3-2 indicates that the transfer function of the eighth-order Butterworth filter is

$$H_8(s) = \frac{1}{(s^2 + 0.390s + 1)(s^2 + 1.111s + 1)(s^2 + 1.663s + 1)(s^2 + 1.962s + 1)}$$

$$= \frac{1}{(s^2 + 0.390s + 1)} \times \frac{1}{(s^2 + 1.111s + 1)} \times \frac{1}{(s^2 + 1.663s + 1)} \times \frac{1}{(s^2 + 1.962s + 1)}$$

Figure 16.3-1 shows the magnitude frequency response plots for these two filters. Both frequency responses show unity gain when $\omega \ll 1$ and a gain of zero when $\omega \gg 1$. Thus, both filters approximate an ideal low-pass filter with $\omega_c = 1$ rad/s. The eighth-order filter makes the transition from the pass-band to the stop-band more quickly, and so provides a better approximation to the ideal low-pass filter.

The transfer function of the fourth-order filter has been expressed as the product of two second-order transfer functions, while the transfer function of the eighth-order filter has been expressed as the product of four second-order transfer functions. Each of these second-order transfer functions will be implemented by a second-order circuit. Since all of these second-order circuits will be quite similar, it is reasonable to expect that the eighth-order circuit will be about twice as large as the fourth-order filter. That means twice as many parts, twice the power consumption, twice the assembly cost, twice the space, and so on.

The eighth-order filter performs better, but it costs more. In some applications the improved performance of the eighth-order filter justifies the additional cost, while in other applications it does not.

Example 16.3-2

Determine the transfer function of a third-order Butterworth low-pass filter having a cutoff frequency equal to 500 rad/s.

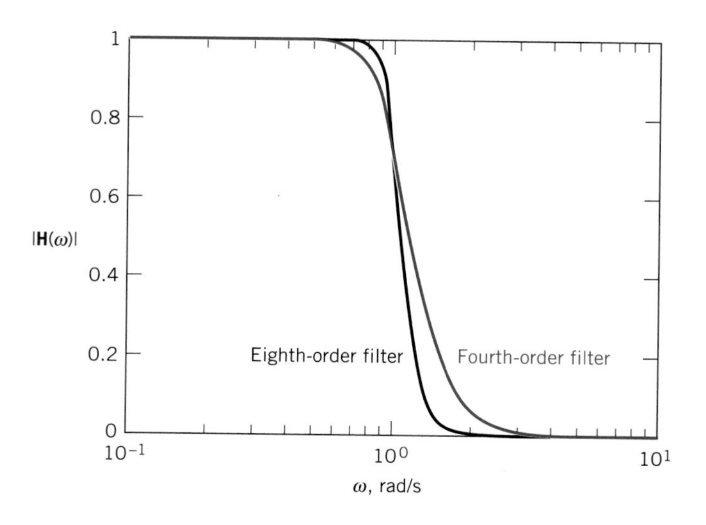

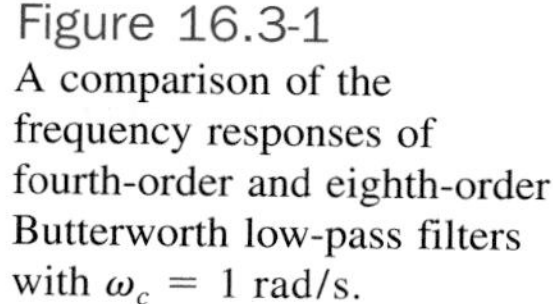

Figure 16.3-1
A comparison of the frequency responses of fourth-order and eighth-order Butterworth low-pass filters with $\omega_c = 1$ rad/s.

Solution

Equation 16.3-2 and Table 16.3-2 provide a third-order Butterworth low-pass filter with a cutoff frequency equal to 1 rad/s.

$$H_n(s) = \frac{1}{(s+1)(s^2+s+1)}$$

A technique called **frequency scaling** is used to adjust the cutoff frequency to $\omega_c = 500$ rad/s. Frequency scaling can be accomplished by replacing each s in $H_n(s)$ by $\frac{s}{\omega_c}$. That is

$$H(s) = \frac{1}{\left(\frac{s}{\omega_c}+1\right)\left(\left(\frac{s}{\omega_c}\right)^2+\frac{s}{\omega_c}+1\right)}$$

In this case $\omega_c = 500$ rad/s so

$$H(s) = \frac{1}{\left(\frac{s}{500}+1\right)\left(\left(\frac{s}{500}\right)^2+\frac{s}{500}+1\right)}$$

$$= \frac{500^3}{(s+500)\,(s^2+500s+500^2)}$$

$$= \frac{125000000}{(s+500)(s^2+500s+250000)}$$

$H(s)$ is the transfer function of a third-order Butterworth low-pass filter having a cutoff frequency equal to 500 rad/s.

Exercise 16.3-1 Find the transfer function of a first-order Butterworth low-pass filter having a cutoff frequency equal to 1250 rad/s.

Answer: $H(s) = \frac{1}{\frac{s}{1250}+1} = \frac{1250}{s+1250}$

16.4 SECOND-ORDER FILTERS

Second-order filters are important for two reasons. First, they provide an inexpensive approximation to ideal filters. Second, they are used as building blocks for more expensive filters that provide more accurate approximations to ideal filters.

The frequency response of second-order filters is characterized by three filter parameters: a gain k, the corner frequency ω_0, and the **quality factor** Q. Filter circuits are designed by choosing the values of the circuit elements in such a way as to obtain the required values of k, ω_0, and Q.

A second-order low-pass filter is a circuit that has a transfer function of the form

$$H_L(s) = \frac{k\omega_0^2}{s^2+\frac{\omega_0}{Q}s+\omega_0^2} \qquad (16.4\text{-}1)$$

This transfer function is characterized by three parameters: the dc gain k, the corner frequency ω_0, and the quality factor Q. When this circuit is stable, that is, when both $\omega_0 > 0$ and $Q > 0$, then the network function can be obtained by letting $s = j\omega$.

$$\mathbf{H}_L(\omega) = \frac{k\omega_0^2}{-\omega^2 + j\dfrac{\omega_0}{Q}\omega + \omega_0^2}$$

The gain of the filter is given by

$$|\mathbf{H}_L(\omega)| = \frac{k\omega_0^2}{\sqrt{(\omega_0^2 - \omega^2)^2 + \left(\dfrac{\omega_0}{Q}\omega\right)^2}}$$

$$\cong \begin{cases} k & \omega \ll \omega_0 \\ 0 & \omega \gg \omega_0 \end{cases}$$

When $k = 1$, this frequency response approximates the frequency response of an ideal low-pass filter with a cutoff frequency of $\omega_c = \omega_0$. When $k \neq 1$, the low-pass filter approximates an ideal low-pass filter together with an amplifier having gain equal to k. The quality factor, Q, controls the shape of the frequency response during the transition from pass-band to stop-band. Figure 16.4-1 shows the frequency response of a second-order low-pass filter ($k = 1$ and $\omega_c = \omega_0 = 1$) for several choices of Q. A Butterworth approximation to the ideal low-pass filter is obtained by choosing $Q = 0.707$.

Table 16.4-1 provides RLC circuits that can be used as second-order filters. Consider the low-pass filter shown in the first row of the table. The transfer function of this circuit is

$$H(s) = \frac{\dfrac{1}{LC}}{s^2 + \dfrac{R}{L}s + \dfrac{1}{LC}} \tag{16.4-2}$$

The relationship between the circuit parameters R, L, and C and the filter parameters k, ω_0, and Q is obtained by comparing Eq. 16.4-2 to Eq. 16.4-1. First, compare the constant terms in the denominators to see that the cutoff frequency of the filter is given by

$$\omega_0 = \frac{1}{\sqrt{LC}}$$

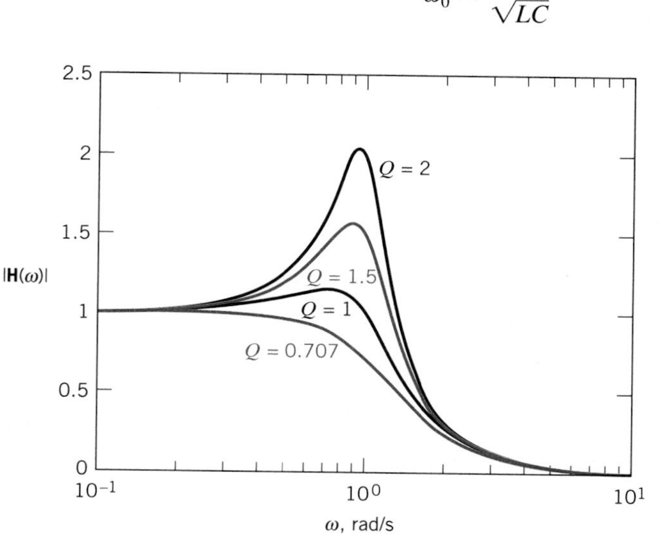

Figure 16.4-1 Frequency responses of second-order low-pass filters with four values of Q ($\omega_c = 1$ rad/s).

Table 16.4-1 **Second-Order *RLC* Filters**

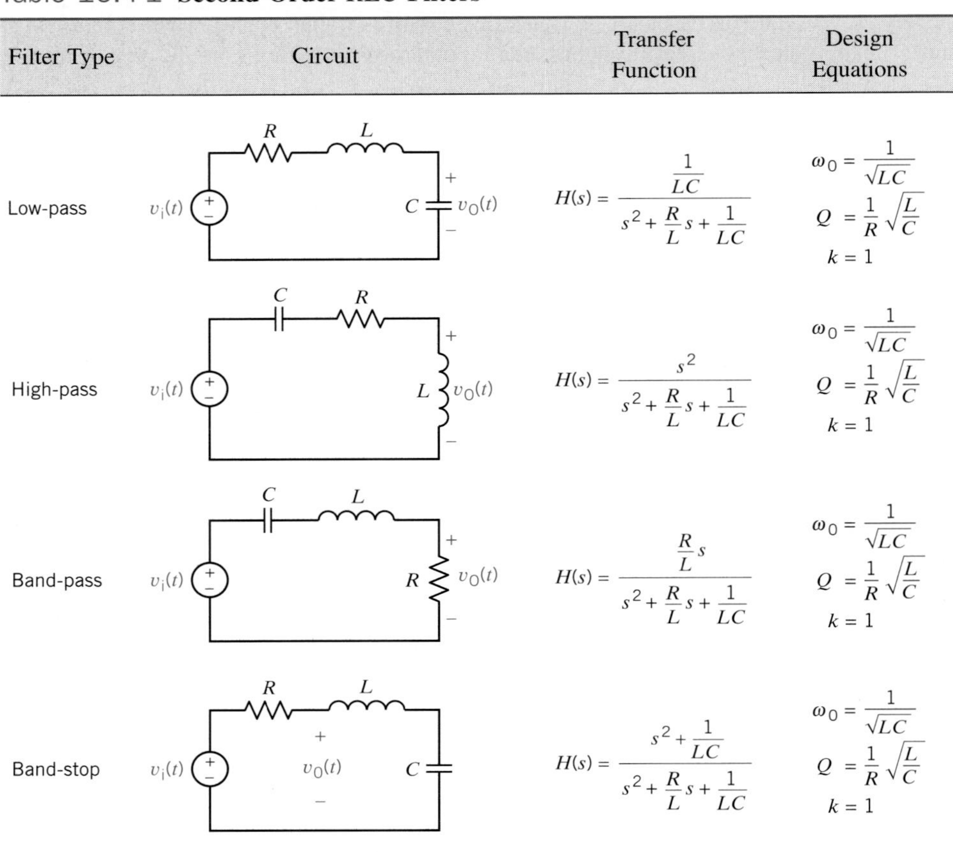

Filter Type	Circuit	Transfer Function	Design Equations
Low-pass		$H(s) = \dfrac{\frac{1}{LC}}{s^2 + \frac{R}{L}s + \frac{1}{LC}}$	$\omega_0 = \dfrac{1}{\sqrt{LC}}$, $Q = \dfrac{1}{R}\sqrt{\dfrac{L}{C}}$, $k = 1$
High-pass		$H(s) = \dfrac{s^2}{s^2 + \frac{R}{L}s + \frac{1}{LC}}$	$\omega_0 = \dfrac{1}{\sqrt{LC}}$, $Q = \dfrac{1}{R}\sqrt{\dfrac{L}{C}}$, $k = 1$
Band-pass		$H(s) = \dfrac{\frac{R}{L}s}{s^2 + \frac{R}{L}s + \frac{1}{LC}}$	$\omega_0 = \dfrac{1}{\sqrt{LC}}$, $Q = \dfrac{1}{R}\sqrt{\dfrac{L}{C}}$, $k = 1$
Band-stop		$H(s) = \dfrac{s^2 + \frac{1}{LC}}{s^2 + \frac{R}{L}s + \frac{1}{LC}}$	$\omega_0 = \dfrac{1}{\sqrt{LC}}$, $Q = \dfrac{1}{R}\sqrt{\dfrac{L}{C}}$, $k = 1$

Next, compare the coefficients of s in the denominators to see that

$$\frac{\omega_0}{Q} = \frac{R}{L}$$

Solving these two equations for Q gives

$$Q = \frac{1}{R}\sqrt{\frac{L}{C}}$$

Finally, comparing the numerators gives

$$k\omega_0^2 = \frac{1}{LC}$$

So the dc gain is

$$k = 1$$

Notice that ω_0 and Q are determined by the values of R, L, and C but that k is always one.

Many different circuits are used to build second-order filters. One of the popular filter circuits is called the **Sallen-Key filter.** Table 16.4-2 provides the information required to design Sallen-Key filters.

Table 16.4-2 **Sallen-Key Filters**

Filter Type	Circuit	Design Equations
Low-pass	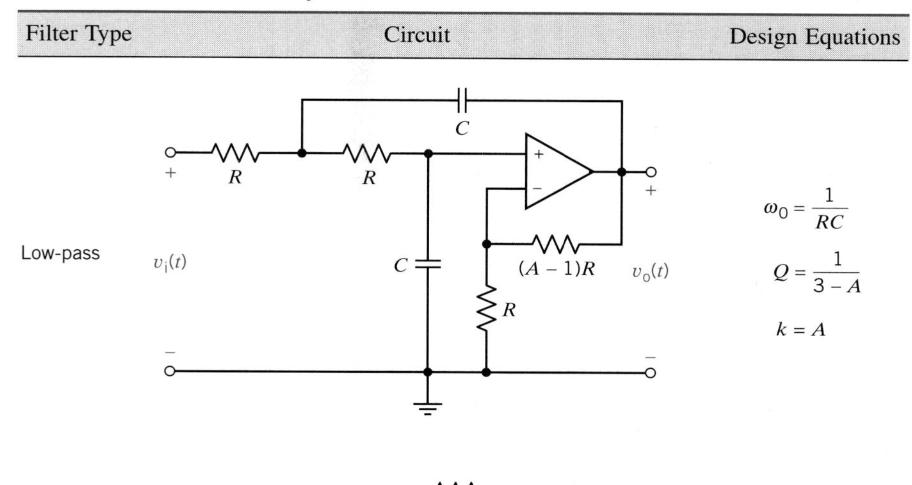	$\omega_0 = \frac{1}{RC}$ $Q = \frac{1}{3 - A}$ $k = A$
High-pass	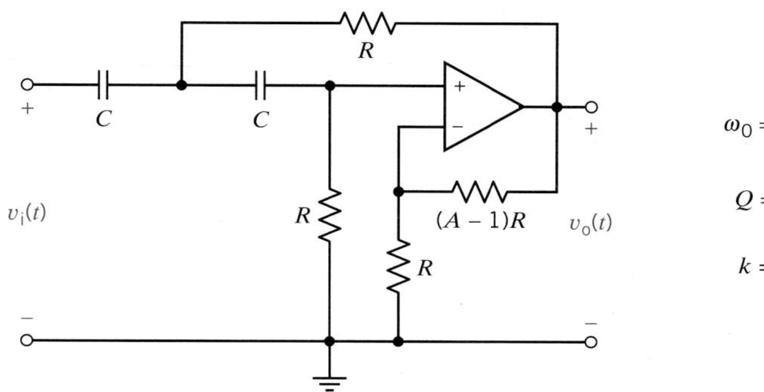	$\omega_0 = \frac{1}{RC}$ $Q = \frac{1}{3 - A}$ $k = A$
Band-pass	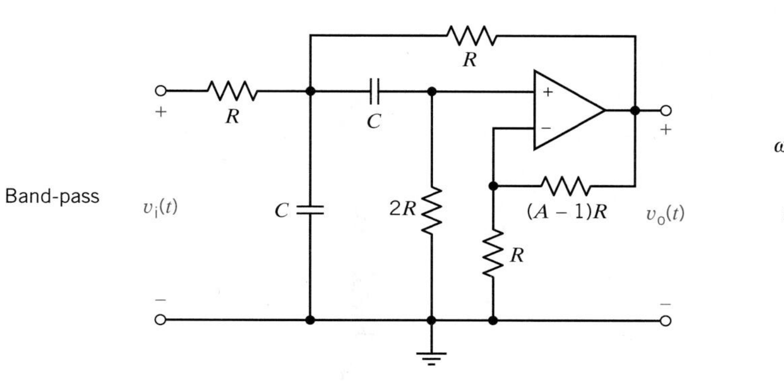 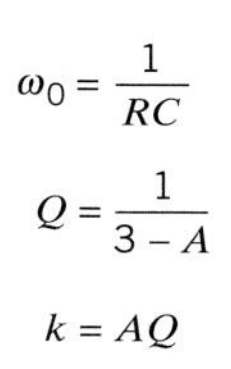	$\omega_0 = \frac{1}{RC}$ $Q = \frac{1}{3 - A}$ $k = AQ$
Band-stop	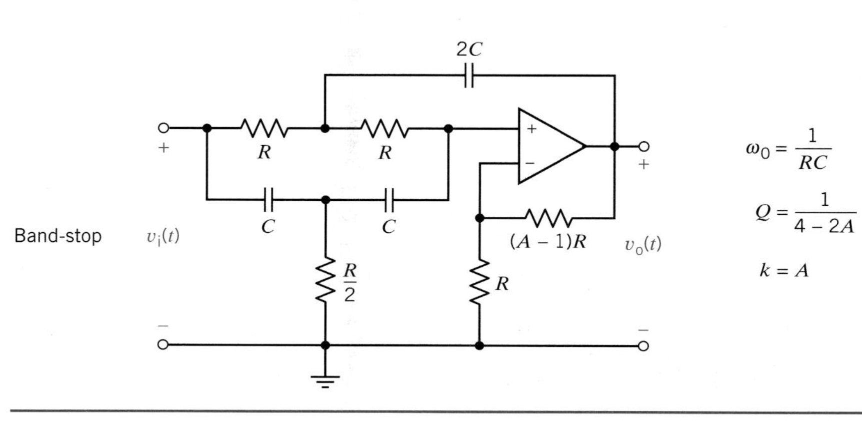	$\omega_0 = \frac{1}{RC}$ $Q = \frac{1}{4 - 2A}$ $k = A$

Example 16.4-1

Design a Butterworth second-order low-pass filter with a cutoff frequency of 1000 hertz.

Solution

Second-order Butterworth filters have $Q = \dfrac{1}{\sqrt{2}} = 0.707$. The corner frequency is equal to the cutoff frequency, that is

$$\omega_0 = \omega_c = 2\pi \cdot 1000 = 6283$$

The *RLC* circuit shown in the first row of Table 16.4-1 can be used to design the required low-pass filter. The design equations are

$$\frac{1}{\sqrt{LC}} = \omega_0 = 6283$$

and

$$\frac{1}{R}\sqrt{\frac{L}{C}} = Q = \frac{1}{\sqrt{2}}$$

The third design equation indicates that $k = 1$. This last design equation does not constrain the values of R, L, and C. Since we have two equations in three unknowns, the solution is not unique. One way to proceed is to choose a convenient value for one circuit element, say $C = 0.1\ \mu\text{F}$, and then calculate the resulting values of the other circuit elements

$$L = 1/(\omega_0^2 C) = 0.253\ \text{H}$$

and

$$R = \sqrt{\frac{2L}{C}} = 2251\ \Omega$$

If we are satisfied with this solution, the filter design is complete. Otherwise, we adjust our choice of the value of C and recalculate L and R. For example, if the inductance is too large, say $L = 1000$ H, or the resistance is too small, say $R = 0.03\ \Omega$, it will be hard to obtain the parts to build these circuits. Because there is no such problem in this example, we conclude that the circuit shown in the first row of Table 16.4-1 with $C = 0.1\ \mu\text{F}$, $L = 0.253$ H, and $R = 2251\ \Omega$ is the required low-pass filter.

Example 16.4-2

Design a second-order Sallen-Key band-pass filter with a center frequency of 500 hertz and a bandwidth of 100 hertz.

Solution

The transfer function of the second-order band-pass filter is

$$H(s) = \frac{k\dfrac{\omega_0}{Q}s}{s^2 + \dfrac{\omega_0}{Q}s + \omega_0^2}$$

The corresponding network function is

$$\mathbf{H}(\omega) = \frac{jk\dfrac{\omega_0}{Q}\omega}{\omega_0^2 - \omega^2 + j\dfrac{\omega_0}{Q}\omega}$$

Dividing numerator and denominator by $j\dfrac{\omega_0}{Q}\omega$ gives

$$\mathbf{H}(\omega) = \frac{k}{1 + jQ\left(\dfrac{\omega}{\omega_0} - \dfrac{\omega_0}{\omega}\right)}$$

We have seen network functions like this one earlier, when we discussed resonant circuits in Chapter 13. The gain, $|\mathbf{H}(\omega)|$, will be maximum at the corner frequency, ω_0. In the case of this band-pass transfer function, ω_0 is also called the center frequency and the resonant frequency. The gain at the center frequency will be

$$|\mathbf{H}(\omega_0)| = k$$

Two frequencies, ω_1 and ω_2, are identified by the property

$$|\mathbf{H}(\omega_1)| = |\mathbf{H}(\omega_2)| = \frac{k}{\sqrt{2}}$$

These frequencies are called the *half-power frequencies* or the 3 *dB frequencies*. The half-power frequencies are given by

$$\omega_1 = -\frac{\omega_0}{2Q} + \sqrt{\left(\frac{\omega_0}{2Q}\right)^2 + \omega_0^2} \quad \text{and} \quad \omega_2 = \frac{\omega_0}{2Q} + \sqrt{\left(\frac{\omega_0}{2Q}\right)^2 + \omega_0^2}$$

The bandwidth of the filter is calculated from the half-power frequencies

$$BW = \omega_2 - \omega_1 = \frac{\omega_0}{Q}$$

The Sallen-Key bandpass filter is shown in the third row of Table 16.4-2. Our specifications require that

$$\omega_0 = 2\pi \cdot 500 = 3142$$

and

$$Q = \frac{\omega_0}{BW} = 5$$

From Table 16.4-2, the design equations for the Sallen-Key band-pass filter are

$$\frac{1}{RC} = \omega_0 = 3142$$

and

$$A = 3 - \frac{1}{Q} = 2.8$$

Pick $C = 0.1\ \mu\text{F}$. Then

$$R = \frac{1}{C\omega_0} = 3183\ \Omega$$

Since $k = AQ$, the gain of this band-pass filter at the center frequency is 14. Also, one of the resistances is given by

$$(A - 1)R = 5729\ \Omega$$

The Sallen-Key band-pass filter is shown in Figure 16.4-2.

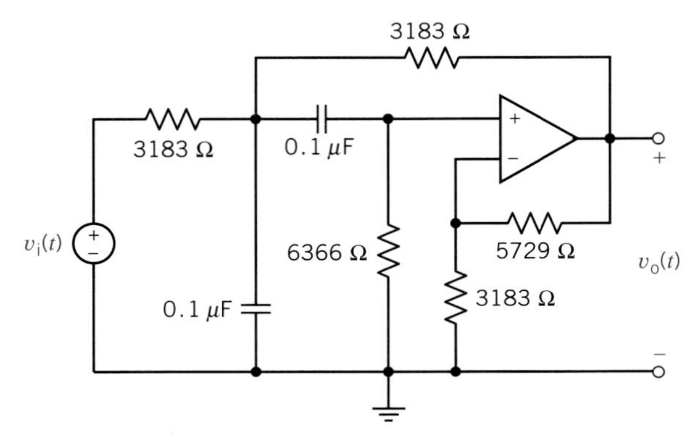

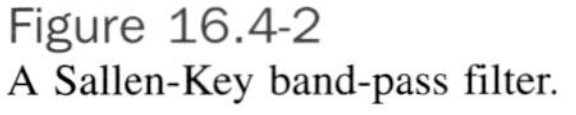

Figure 16.4-2
A Sallen-Key band-pass filter.

Example 16.4-3

Design a second-order band-stop filter with a center frequency of 1000 rad/s and a bandwidth of 100 rad/s.

Solution

The transfer function of the second-order band-stop filter is

$$H(s) = \frac{k(s^2 + \omega_0^2)}{s^2 + \dfrac{\omega_0}{Q}s + \omega_0^2}$$

Notice that the transfer functions of the second-order band-pass and band-stop filters are related by

$$\frac{k(s^2 + \omega_0^2)}{s^2 + \dfrac{\omega_0}{Q}s + \omega_0^2} = k - \frac{k\dfrac{\omega_0}{Q}s}{s^2 + \dfrac{\omega_0}{Q}s + \omega_0^2}$$

The network function of the band-stop filter is

$$\mathbf{H}(\omega) = \frac{k(\omega_0^2 - \omega^2)}{\omega_0^2 - \omega^2 + j\dfrac{\omega_0}{\mathrm{Q}}\omega}$$

When $\omega << \omega_0$ or $\omega >> \omega_0$, the gain is $|\mathbf{H}(\omega)| = k$. At $\omega = \omega_0$ the gain is zero. The half-power frequencies, ω_1 and ω_2, are identified by the property

$$|\mathbf{H}(\omega_1)| = |\mathbf{H}(\omega_2)| = \frac{k}{\sqrt{2}}$$

The bandwidth of the filter is given by

$$BW = \omega_2 - \omega_1 = \frac{\omega_0}{Q}$$

The Sallen-Key band-stop filter is shown in the last row of Table 16.4-2. Our specifications require that $\omega_0 = 1000$ rad/s and

$$Q = \frac{\omega_0}{BW} = 10$$

Table 16.4-2 indicates that the design equations for the Sallen-Key band-stop filter are

$$\frac{1}{RC} = \omega_0 = 1000$$

and

$$A = 2 - \frac{1}{2Q} = 1.95$$

Pick $C = 0.1\ \mu$F. Then

$$R = \frac{1}{C\omega_0} = 10\ k\Omega$$

The Sallen-Key band-stop filter is shown in Figure 16.4-3.

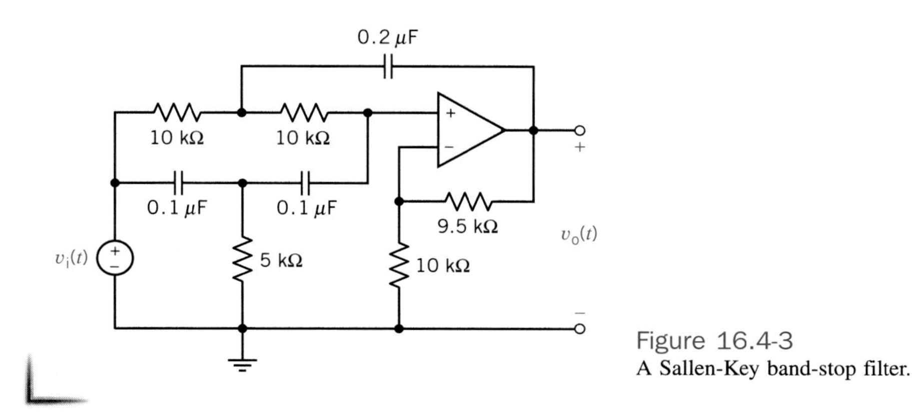

Figure 16.4-3
A Sallen-Key band-stop filter.

Example 16.4-4

Figure 16.4-4 shows another circuit that can be used to build a second-order filter. This circuit is called a *Tow-Thomas filter.* This filter can be used as either a band-pass or low-pass filter. When the output is the voltage $v_1(t)$, then the transfer function is

$$H_L(s) = \frac{-\dfrac{1}{R_k R C^2}}{s^2 + \dfrac{1}{R_Q C}s + \dfrac{1}{R^2 C^2}} \tag{16.4-3}$$

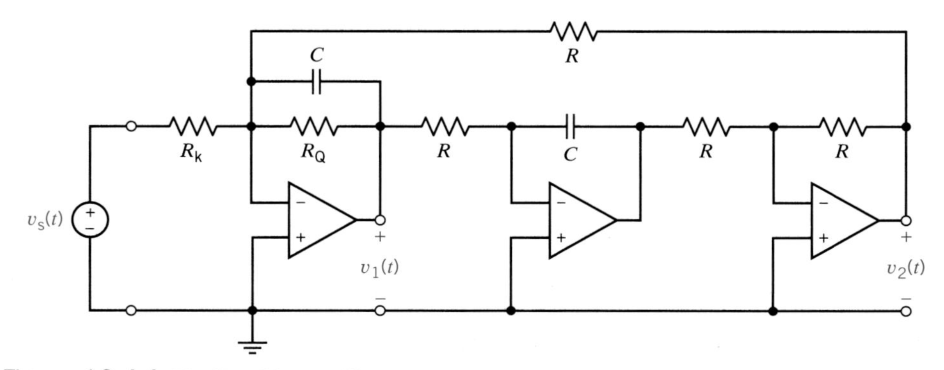

Figure 16.4-4 The Tow-Thomas filter.

and the filter is a low-pass filter. If, instead, the voltage $v_2(t)$ is used as the filter output, then the network function is

$$H_B(s) = \frac{-\frac{1}{R_k C} s}{s^2 + \frac{1}{R_Q C} s + \frac{1}{R^2 C^2}} \tag{16.4-4}$$

and the Tow-Thomas filter functions as a band-pass filter. Design a Butterworth Tow-Thomas low-pass filter with a dc gain of 5, and a cutoff frequency of 1250 hertz.

Solution

Since the Tow-Thomas filter will be used as a low-pass filter, the transfer function is given by Eq. 16.4-3. Design equations are obtained by comparing this transfer function to the standard form of the second order low-pass transfer function given in Eq. 16.4-1. First, compare the constant terms (i.e., the coefficients of s^0) in the denominators of these transfer functions to get

$$\omega_0 = \frac{1}{RC} \tag{16.4-5}$$

Next compare the coefficients of s^1 in the denominators of these transfer functions to get

$$Q = \frac{R_Q}{R} \tag{16.4-6}$$

Finally, compare the numerators to get

$$k = \frac{R}{R_k} \tag{16.4-7}$$

Designing the Tow-Thomas filter requires that values be obtained for R, C, R_Q, and R_k. Since there are four unknowns and only three design equations, we begin by choosing a convenient value for one of the unknowns, usually the capacitance. Let $C = 0.01\ \mu\text{F}$. Then

$$R = \frac{1}{\omega_0 C} = \frac{1}{(2\pi)(1250)(0.01)(10^{-6})} = 12732\ \Omega$$

A second-order Butterworth filter requires $Q = 0.707$, so

$$R_Q = QR = (0.707)(12732) = 9003\ \Omega$$

Finally

$$R_k = \frac{R}{k} = 2546\ \Omega$$

and the design is complete.

Example 16.4-5

Use the Tow-Thomas circuit to design a Butterworth high-pass filter with a high frequency gain of 5, and a cutoff frequency of 1250 hertz.

Solution

The Tow-Thomas circuit does not implement the high-pass filter, but it does implement the low-pass filter and the band-pass filter. The transfer functions of the second-order high-pass, band-pass, and low-pass filters are related by

$$H_H(s) = \frac{ks^2}{s^2 + \frac{1}{R_Q C}s + \frac{1}{R^2C^2}} = k + \frac{-\frac{1}{R_k C}s}{s^2 + \frac{1}{R_Q C}s + \frac{1}{R^2C^2}} + \frac{-\frac{1}{R_k R C^2}}{s^2 + \frac{1}{R_Q C}s + \frac{1}{R^2C^2}}$$
$$= k + H_B(s) + H_L(s) \tag{16.4-8}$$

A high-pass filter can be constructed using a Tow-Thomas filter and a summing amplifier. Both the band-pass and low-pass outputs of the Tow-Thomas filter are used. Equation 16.4-8 indicates that the band-pass and low-pass filters must have the same values of k, Q, and ω_0 as the high-pass filter. Thus, we require a Tow-Thomas filter having $k = 5$, $Q = 0.707$, and $\omega_0 = 7854$ rad/s. Such a filter was designed in Example 16.4-4. The high-pass filter is obtained by adding a summing amplifier as shown in Figure 16.4-5.

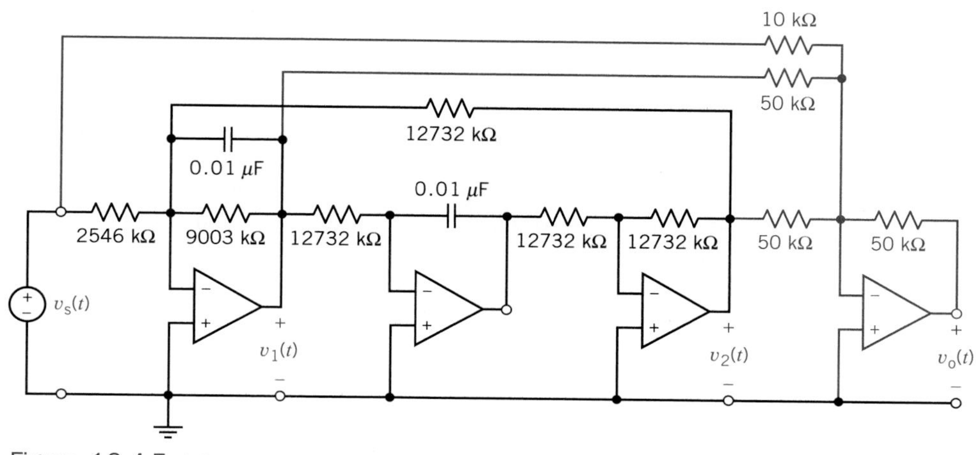

Figure 16.4-5 A Tow-Thomas high-pass filter.

16.5 HIGH-ORDER FILTERS

In this section we turn our attention to filters that have an order greater than two. These filters are called **high-order filters.** A popular strategy for designing high-order filters uses a cascade connection of second-order filters. The cascade connection is shown in Figure 16.5-1. In this figure, the transfer functions $H_1(s)$, $H_2(s)$, . . . , $H_n(s)$ represent second-order filters that are connected together to build a high-order filter. We refer to the second-order filter as "filter stages" to distinguish them from the high-order filter. That is, the high-order filter is a cascade connection of second-order filter stages. (When the order of the high-order filter is odd, a first-order filter stage is needed. Nonetheless, we talk about designing high-order filters as a cascade of second-order stages.)

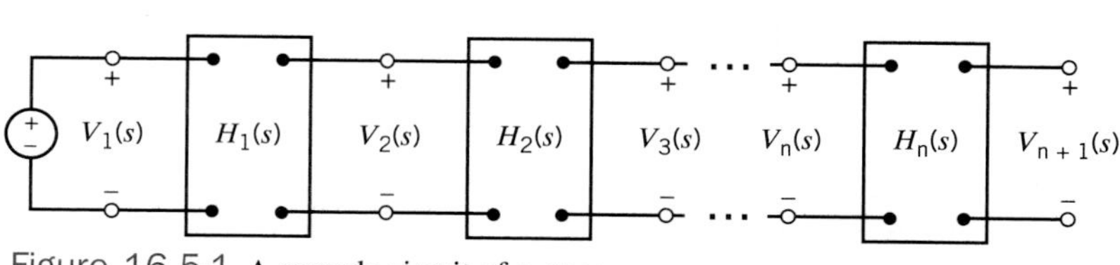

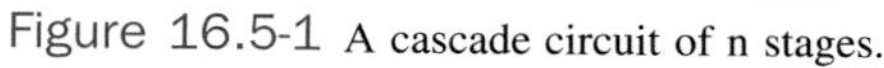
Figure 16.5-1 A cascade circuit of n stages.

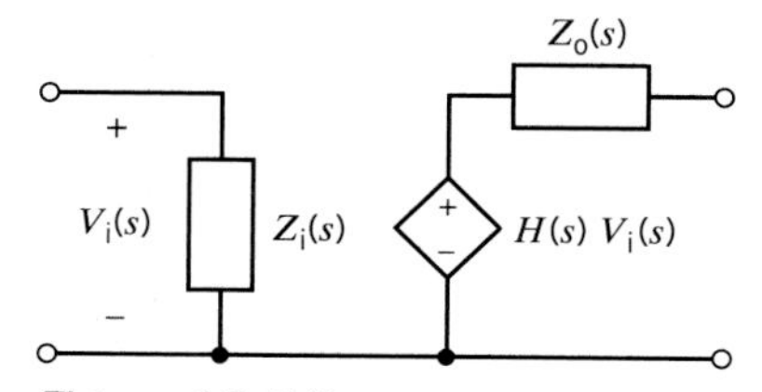

Figure 16.5-2 A model of one filter stage.

The cascade connection is characterized by the fact that the output of one filter stage is used as the input to the next stage. Unfortunately, the behavior of a stage will sometimes change when another stage is connected to it. We call this phenomenon **loading,** and we say that the second stage loaded the first. Generally, loading is undesirable and we try to avoid it. Figure 16.5-2 shows a model of a filter stage that is appropriate for investigating loading. This model includes the input and output impedance of the filter stage as well as the transfer function.

Figure 16.5-3 shows a high-order filter consisting of the cascade connection of two filter stages. Let's calculate the transfer function of the high-order filter. Starting at the output of the high-order filter, notice that there is no current in the output impedance, $Z_{o2}(s)$, of the second stage. Consequently, there is no voltage across $Z_{o2}(s)$, so

$$V_3(s) = H_2(s)V_2(s) \tag{16.5-1}$$

Next, we use voltage division to find $V_2(s)$.

$$V_2(s) = \frac{Z_{i2}}{Z_{o1} + Z_{i2}} H_1(s)V_1(s) \tag{16.5-2}$$

Connecting the second filter stage to the first stage has changed the output of the first stage. Without the second stage, there would be no current in $Z_{o1}(s)$. Consequently, there would be no voltage across $Z_{o1}(s)$, and the output of the first stage would be $V_2(s) = H_1(s)V_1(s)$. The second stage is said to load the first stage. This loading can be eliminated by making the input impedance of the second stage infinite, $Z_{i2}(s) = \infty$, or the output impedance of the first stage zero, $Z_{o1}(s) = 0$.

Combining Eqs. 16.5-1 and 16.5-2 gives

$$V_3(s) = H_2(s)\frac{Z_{i2}}{Z_{o1} + Z_{i2}} H_1(s)V_1(s)$$

Finally, the transfer function of the high-order filter is

$$H(s) = \frac{V_3(s)}{V_1(s)} = H_2(s)\frac{Z_{i2}}{Z_{o1} + Z_{i2}} H_1(s) \tag{16.5-3}$$

This equation simplifies to

$$H(s) = H_2(s)H_1(s) \tag{16.5-4}$$

when either the input impedance of the second stage is infinite, $Z_{i2}(s) = \infty$, or the output impedance of the first stage is zero, $Z_{o1}(s) = 0$. In other words, Eq. 16.5-4 can be used when the second stage does not load the first stage, but Eq. 16.5-3 must be used when the second stage does load the first stage. We will prove that the Sallen-Key filters have output impedances equal to zero. Therefore, there is no loading when Sallen-Key filter stages are cascaded. The transfer function of the high-order filter is the product of the transfer functions of the individual Sallen-Key filter stages. In contrast, the filters based on the series *RLC* circuit shown in Table 16.4-1 do not have output impedances that are equal to

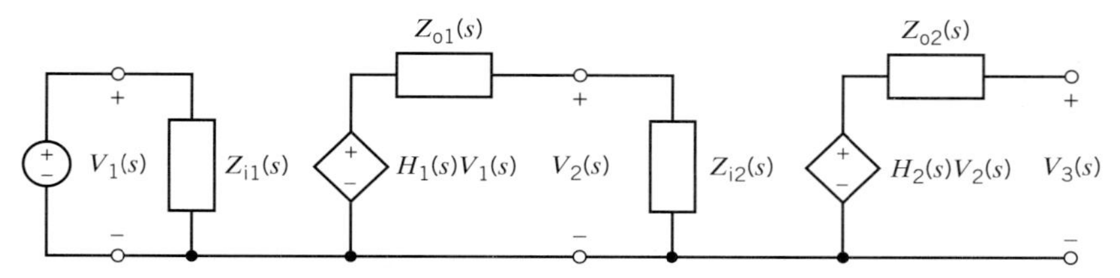

Figure 16.5-3 Cascade connection of two filter stages.

Table 16.5-1 **Measuring the Parameters of a Filter Stage**

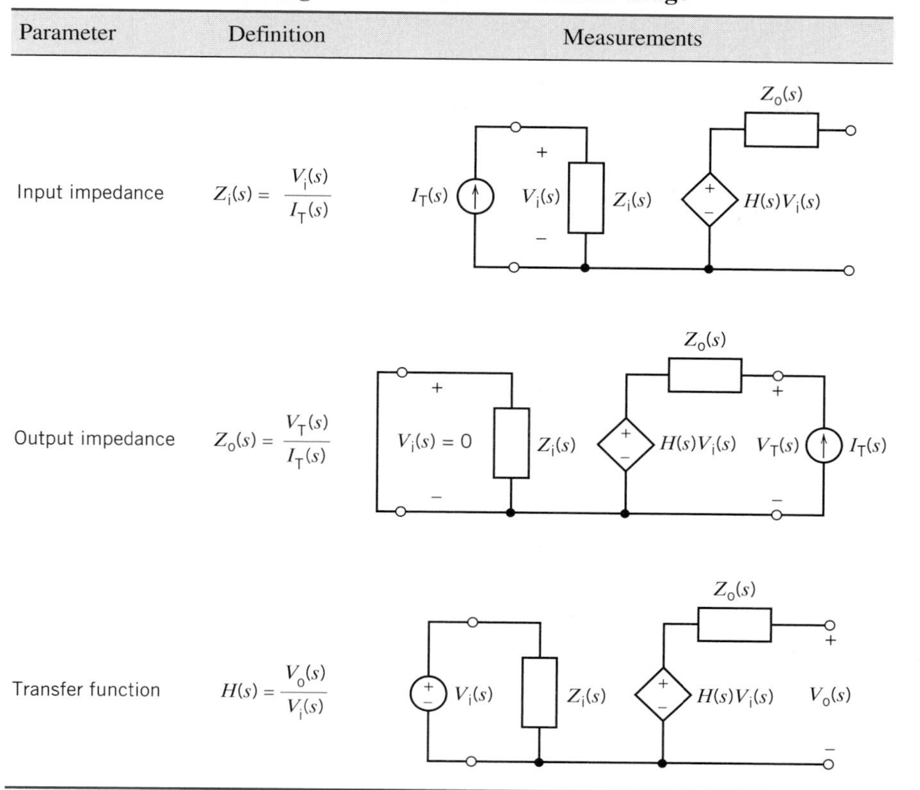

Parameter	Definition	Measurements
Input impedance	$Z_i(s) = \dfrac{V_i(s)}{I_T(s)}$	
Output impedance	$Z_o(s) = \dfrac{V_T(s)}{I_T(s)}$	
Transfer function	$H(s) = \dfrac{V_o(s)}{V_i(s)}$	

zero or input impedances that are infinite. If these filter stages were cascaded, the transfer function of the high-order filter would not be equal to the product of the transfer functions of the individual filter stages. Thus we can use cascaded Sallen-Key filter stages to design high-order filters without introducing loading.

Next, consider calculating the output impedance of a Sallen-Key band-pass filter. Table 16.5-1 shows how the parameters of the model of a filter stage can be calculated or measured. The second row of this table indicates that to calculate the output impedance, a short circuit should be connected to the filter input and a current source should be connected to the filter output. The voltage across the current source is calculated, and the ratio of this voltage to the current of the current source is the output impedance. Figure 16.5-4 shows a Sallen-Key filter with a short circuit across its input and a current source

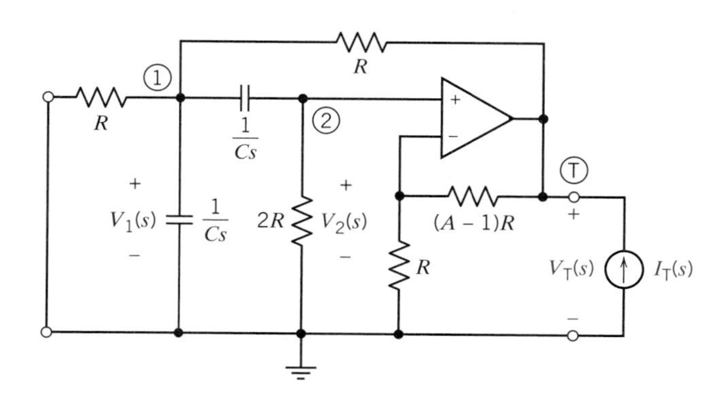

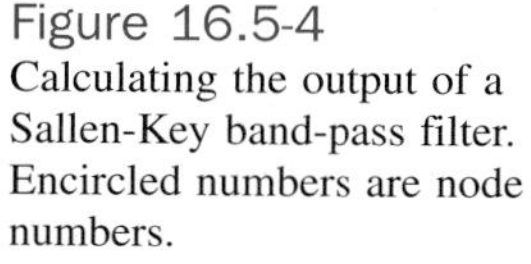

Figure 16.5-4 Calculating the output of a Sallen-Key band-pass filter. Encircled numbers are node numbers.

connected to its output. This circuit can be analyzed by writing node equations at nodes 1, 2, and T

$$\frac{V_1}{R} + CsV_1 + \frac{V_1 - V_T}{R} + (V_1 - V_2)Cs = 0$$

$$-(V_1 - V_2)Cs + \frac{V_2}{2R} = 0$$

$$\frac{V_2}{R} + \frac{V_2 - V_T}{(A - 1)R} = 0$$

Solving these node equations for V_T gives

$$[(RCs)^2 + (3 - A)\,RCs + 1]\,V_T = 0$$

Since the factor in brackets is not zero, this equation indicates that $V_T = 0$. The output impedance of the Sallen-Key band-pass filter is

$$Z_o = \frac{V_T}{I_T} = \frac{0}{I_T} = 0$$

Similarly, each of the Sallen-Key filters shown in Table 16.4-2 has an output impedance equal to zero.

High-order filters can be designed as a cascade connection of second-order filter stages. Filter stages that have an output impedance equal to zero are used so that the transfer function of the high-order filter will be the product of the transfer functions of the cascaded filter stages.

Example 16.5-1

Design a third-order Butterworth low-pass filter having a cutoff frequency of $\omega_c = 500$ rad/s and a dc gain equal to 1.

Solution

Equation 16.3-2 and Table 16.3-2 provide a third-order Butterworth low-pass filter having a cutoff frequency equal to 1 rad/s.

$$H_n(s) = \frac{1}{(s + 1)(s^2 + s + 1)}$$

Frequency scaling is used to adjust the cutoff frequency so that $\omega_c = 500$ rad/s.

$$H(s) = \frac{1}{\left(\frac{s}{500} + 1\right)\left(\left(\frac{s}{500}\right)^2 + \frac{s}{500} + 1\right)}$$
$$= \frac{500^3}{(s + 500)\,(s^2 + 500s + 500^2)}$$

$H(s)$ is the transfer function of a third-order Butterworth low-pass filter having a cutoff frequency equal to 500 rad/s. This transfer function can be expressed as

$$H(s) = \frac{-250000}{s^2 + 500s + 250000} \cdot \frac{-500}{s + 500} = H_1(s) \cdot H_2(s) \qquad (16.5\text{-}5)$$

A Sallen-Key low-pass filter can be designed to implement the second-order low-pass transfer function $H_1(s)$. Table 16.5-2 provides circuits and design equations for first-order

Table 16.5-2 **First-Order Filter Stages**

Filter Type	First-Order Circuit	Design Equations
Low-pass	C, R_1, R_2, $+$, $v_i(t)$, $-$, $+$, $v_o(t)$, $-$	$H(s) = \frac{-k}{s+p}$ where $p = \frac{1}{R_2 C}$ and $k = \frac{1}{R_1 C}$
High-pass	R_1, C, R_2, $+$, $v_i(t)$, $-$, $+$, $v_o(t)$, $-$	$H(s) = \frac{-ks}{s+p}$ where $p = \frac{1}{R_1 C}$ and $k = \frac{R_2}{R_1}$

filter stages. The circuit shown in the first row of this table can be used to implement $H_2(s)$. The first-order filter stages in Table 16.5-2 have output impedances equal to zero. Cascading these filter stages will not cause loading. Cascading the Sallen-Key filter with the first-order filter stage will produce a third-order filter with the transfer function $H(s)$.

First, let's design the Sallen-Key filter with transfer function

$$H_1(s) = \frac{-250000}{s^2 + 500s + 250000}$$

Values of the filter parameters k, ω_0, and Q are determined by comparing $H_1(s)$. with the standard form of the second-order low-pass transfer function given in Eq. 16.4-1. From the constant term in the denominator,

$$\omega_0^2 = 250000$$

Next, from the coefficient of s in the denominator

$$\frac{\omega_0}{Q} = 500$$

Finally, from the numerator,

$$k \cdot \omega_0^2 = 250000$$

So $\omega_0 = 500$ rad/s, $Q = 1$, and k $= 1$. The Sallen-Key low-pass filter is shown in row 1 of Table 16.4-2. Designing this filter requires finding values of R, C, and A. The design equations given in row 1 of the table indicate that

$$\omega_0 = \frac{1}{RC} \tag{16.5-6}$$

$$Q = \frac{1}{3 - A} \tag{16.5-7}$$

$$k = A \tag{16.5-8}$$

Equation 16.5-7 gives

$$A = 3 - \frac{1}{Q} = 3 - \frac{1}{1} = 2$$

but Eq. 16.5-8 gives

$$A = k = 1$$

Apparently, we can select A to get the correct value of Q, or we can select A to get the correct value of k, but not both. The dc gain is easy to adjust later, so we pick $A = 2$ to make $Q = 1$ and settle for $k = 2$. Equation 16.5-7 is satisfied by taking $C = 0.1\ \mu\text{F}$ and

$$R = \frac{1}{C\omega_0} = \frac{1}{(0.1 \times 10^{-6})(500)} = 20\ k\Omega$$

The Sallen-Key filter stage is shown in Figure 16.5-5a. The transfer function of this stage is

$$H_3(s) = \frac{-500000}{s^2 + 500s + 250000}$$

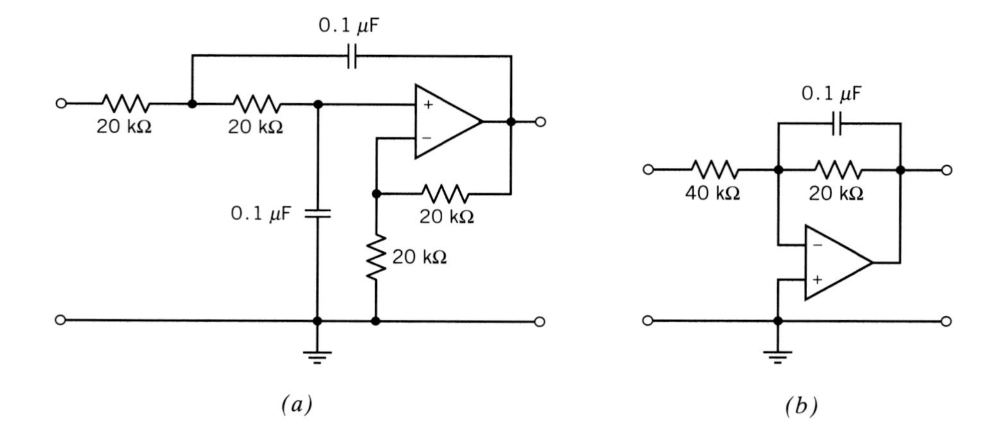

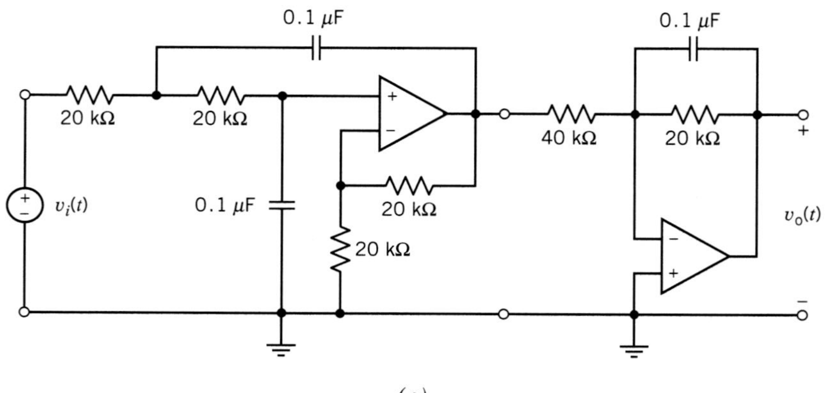

Figure 16.5-5 (a) A Sallen-Key filter stage, (b) a first-order filter stage, and (c) a third-order Butterworth filter.

The Sallen-Key filter stage achieved the desired values of ω_0 and $Q = 1$, but not the desired value of the dc gain. To compensate, we will adjust the dc gain of the first-order filter. The desired transfer function of the third-order filter can be expressed as

$$H(s) = \frac{-500000}{s^2 + 500s + 250000} \cdot H_4(s)$$

which requires

$$H_4(s) = \frac{-250}{s + 500}$$

The design equations in row 1 of Table 16.5-2 indicate that

$$500 = \frac{1}{R_2 C}$$

and

$$250 = \frac{1}{R_1 C}$$

Choose $C = 0.1\ \mu F$. Then

$$R_2 = \frac{1}{500 \cdot C} = \frac{1}{(500)(0.1 \times 10^{-6})} = 20\ k\Omega$$

and

$$R_1 = \frac{1}{250 \cdot C} = \frac{1}{(250)(0.1 \times 10^{-6})} = 40\ k\Omega$$

The first-order filter stage is shown in Figure 16.5-5*b*. Cascading the Sallen-Key stage and the first-order stage produces the third-order Butterworth filter shown in Figure 16.5-5*c*.

16.6 SIMULATING FILTER CIRCUITS USING PSPICE

PSpice provides a convenient way to verify that a filter circuit does indeed have the correct transfer function. Figure 16.6-1 illustrates a method of testing a filter design. The filter that is being tested here is a fourth-order notch filter consisting of two Sallen-Key notch filter stages and an inverting amplifier. This filter was designed to have the transfer function

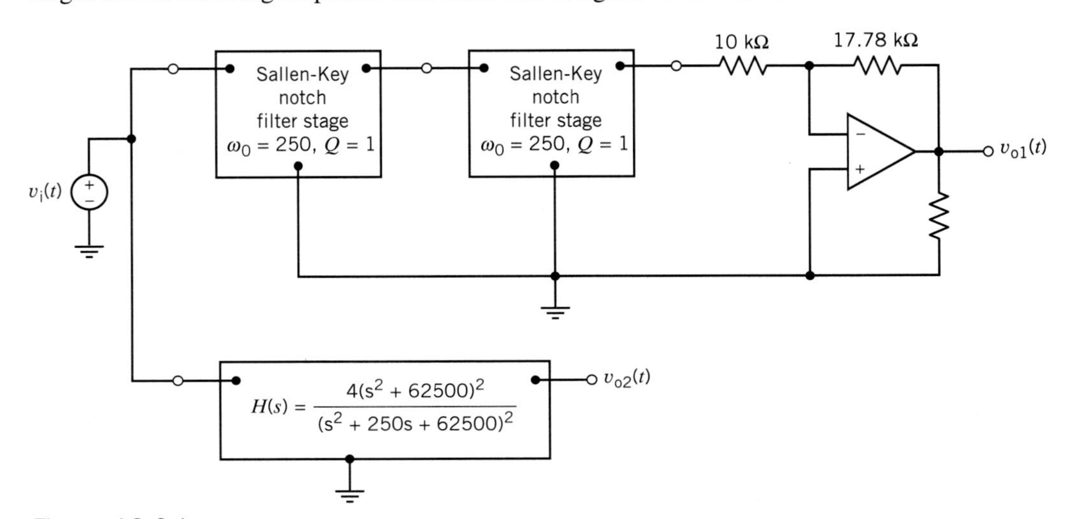

Figure 16.6-1 Verifying the transfer function of a fourth-order notch filter using PSpice.

$$H(s) = \frac{4(s^2 + 62500)^2}{(s^2 + 250s + 62500)^2}$$

The voltage source voltage, $v_i(t)$, is used as the input to two separate circuits. One of these circuits is the filter circuit consisting of the Sallen-Key stages and the inverting amplifier. The response of this circuit is the node voltage $v_{o1}(t)$. The other "circuit" implements $H(s)$ directly using a feature of PSpice. The response of this circuit is $v_{o2}(t)$. A single PSpice simulation produces the frequency responses corresponding to the transfer functions of both of these circuits, $V_{o1}(s)/V_i(s)$ and $V_{o2}(s)/V_i(s)$. Next, we use Probe, the graphical post processor included with PSpice, to display both frequency responses on the same axis. If these frequency responses are identical, we know that the filter circuit does indeed implement the transfer function $H(s)$.

Figure 16.6-2 shows the PSpice input file corresponding to Figure 16.6-1. Two aspects of this file require some explanation. First, notice that parameters are used in the subcircuit that represents the Sallen-Key filter stage. The line

```
.subckt sk_n in out params: C=.1uF w0=1krad/s Q=0.707
```

```
Testing a 4th order notch filter

Vin    1   0   ac   1
XSK1   1   2   sk_n   params:  C=.1uF w0=250 Q=1
XSK2   2   3   sk_n   params:  C=.1uF w0=250 Q=1
R1     3   4   10k
R2     4   5   17.78k
XOA    4   0   5   op_amp
RL     5   0   10G
XLP    1   6   4th_order_notch_filter

.subckt  sk_n  in out params: C=.1uF w0=1krad/s Q=0.707
R1   in    2      {1/C/w0}
R2   2     3      {1/C/w0}
C1   in    6      {C}
C2   6     3      {C}
C3   2     out    {2*C}
R3   6     0      {1/2/C/w0}
XOA  5     3      out         op_amp
R4   5     0      10kOhm
R5   out   5      {(1-1/Q/2)*10kOhm}
.ends   sk_n

.subckt op_amp inv non out
* an ideal op amp
E (out 0)   (non inv)   1G
.ends op_amp

.subckt  4th_order_notch_filter in out
R1    in  0  1G
R2   out  0  1G
E1   out  0  LAPLACE {V(in)} = { (4*(s*s+62500)*(s*s+62500)) /
+                     (s*s+250*s+62500)  *  (s*s+250*s+62500)  }
.ends 4th_order_notch_filter

.ac dec 100 1 1000
.probe V(1) V(5) V(6)
.end
```

Figure 16.6-2 PSpice input file used to test the fourth-order notch filter.

marks the beginning of the subcircuit named sk_n. (PSpice allows us to name, rather than number, nodes. The nodes "in" and "out" will connect this subcircuit to the rest of the circuit.) Three parameters are defined: *C*, w0, and *Q*. All are given default values, as required by PSpice. Expressions involving these parameters replace the values of some of the devices that comprise the subcircuit, for example, the line

```
R1 in 2 {1/C/w0}
```

indicates that resistor *R*1 is connected to nodes "in" and 2 and that the resistance of *R*1 is given by 1/C/w0. The values of parameters like *C* and w0 are given when the subcircuit is used. Consider the line

```
XSK2 2 3 sk_n params: C=.1uf w0=250 Q=1
```

which indicates that device XSK2 is a subcircuit sk_n. This line provides values for *C*, w0, and *Q*. These values will be used to calculate the resistance *R*1 that is used when sk_n implements XSK2. Different values of *C*, w0, and *Q* can be used each time the subcircuit sk_n is used to implement a different device. Table 16.6-1 provides PSpice subcircuits for the four Sallen-Key filter stages.

Next, consider the subcircuit

```
.subckt 4th_order_notch_filter in out
R1    in  0  1G
R2   out  0  1G
E1   out  0   LAPLACE {V(in)}  =  { 4*(s*s+62500)*  (s*s+62500)
/ +    (s*s+250*s+62500)(s*s+250*s+62500) }
.ends 4th_order_notch_filter
```

The key word LAPLACE indicates that controlled voltage of the VCVS is related to the controlling voltage using a transfer function. The controlling voltage of the VCVS is identified inside the first set of braces. The transfer function is given inside the second set of braces. The transfer function was too long to fit on the line describing the VCVS. The + sign at the beginning of the fourth line indicates that this line is a continuation of the previous line. Table 16.6-2 provides subcircuits describing second-order transfer functions.

Figure 16.6-3 shows the frequency responses produced using the PSpice input file shown in Figure 16.6-2. The frequency responses are identical and overlap exactly. The filter circuit does indeed implement the specified transfer function.

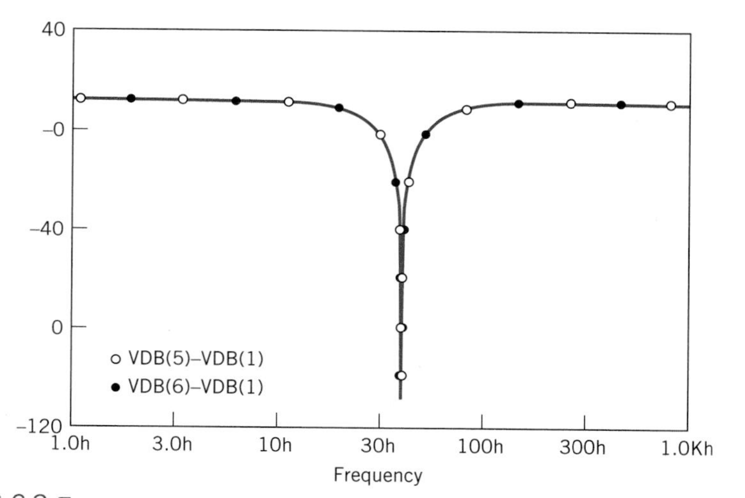

Figure 16.6-3 Frequency response plots used to verify the transfer function of the fourth-order notch filter.

Table 16.6-1 **PSpice Subcircuits for Sallen-Key Filter Stages**

Filter Stage	PSpice Subcircuit
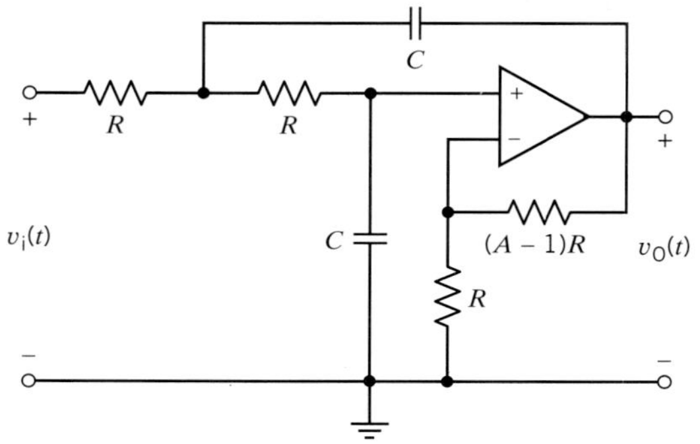	.subckt sk_lp in out params: C = .1uF w0 = 1krad/s Q = 0.707 R1 in 2 {1/C/w0} R2 2 3 {1/C/w0} C1 3 0 {C} C2 2 out {C} XOA 5 3 out op_amp R3 5 0 10kOhm R4 out 5 {(2 – 1/Q) *10kOhm} .ends sk_lp
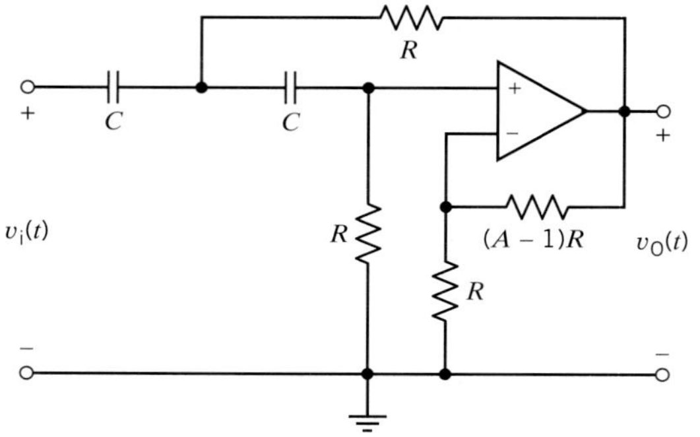	.subckt sk_hp in out params: C = .1uF w0 = 1krad/s Q = 0.707 R1 3 0 {1/C/w0} R2 2 out {1/C/w0} C1 in 2 {C} C2 2 3 {C} XOA 5 3 out op_amp R3 5 0 10kOhm R4 out 5 {(2 – 1/Q) *10kOhm} .ends sk_hp
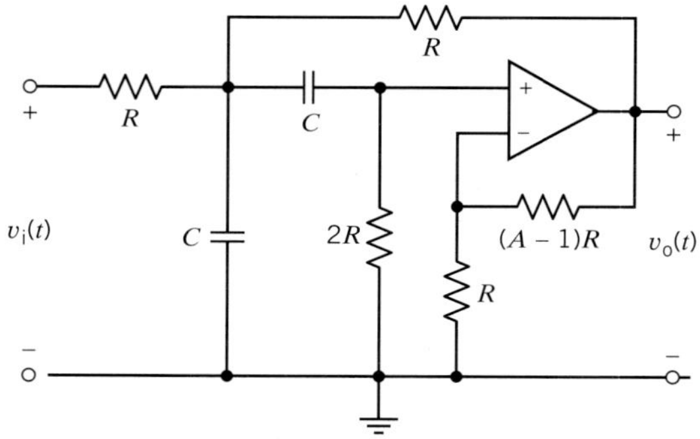	.subckt sk_bp in out params: C = .1uF w0 = 1krad/s Q = 0.707 R1 in 2 {1/C/w0} R2 2 out {1/C/w0} C1 2 3 {C} C2 2 0 {C} R3 3 0 {2/C/w0} XOA 5 3 out op_amp R4 5 0 10kOhm R5 out 5 {(2 – 1/Q)*10kOhm} .ends sk_bp
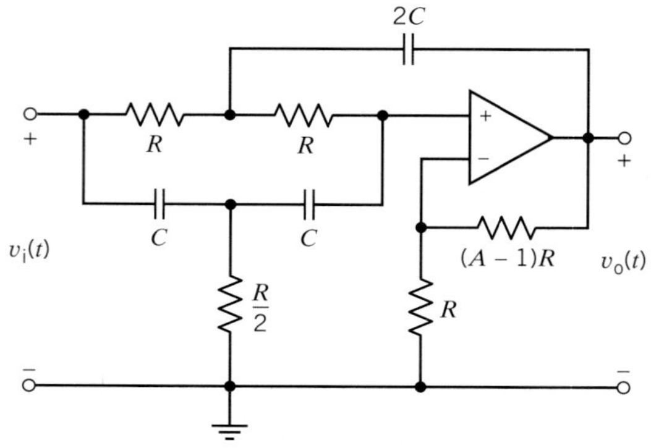	.subckt sk_n in out params: C = .1uF w0 = 1krad/s Q = 0.707 R1 in 2 {1/C/w0} R2 2 3 {1/C/w0} C1 in 6 {C} C2 6 3 {C} C3 2 out {2*C} R3 6 0 {1/2/C/w0} XOA 5 3 out op_amp R4 5 0 10kOhm R5 out 5 {(1 – 1/Q/2)*10kOhm} .ends sk_n

Table 16.6-2 **PSpice Subcircuits for Second-Order Transfer Functions**

Transfer Function	PSpice Subcircuit
Low-pass	.subckt lp_filter_stage in out params: w0 = 1krad/s Q = 0.707 k = 1 R1 in 0 1G R2 out 0 1G E out 0 LAPLACE {V(in)} = {(k*w0*w0)/(s*s + w0*s/Q + w0*w0)} .ends lp_filter_stage
High-pass	.subckt hp_filter_stage in out params: w0 = 1krad/s Q = 0.707 k = 1 R1 in 0 1G R2 out 0 1G E out 0 LAPLACE {V(in)} = {(k*s*s)/(s*s + w0*s/Q + w0*w0)} .ends hp_filter_stage
Band-pass	.subckt bp_filter_stage in out params: w0 = 1krad/s Q = 0.707 k = 1 R1 in 0 1G R2 out 0 1G E out 0 LAPLACE {V(in)} = {(k*w0*s/Q)/(s*s + w0*s/Q + w0*w0)} .ends bp_filter_stage
Band-stop (notch)	.subckt n_filter_stage in out params: w0 = 1krad/s Q = 0.707 k = 1 R1 in 0 1G R2 out 0 1G E out 0 LAPLACE {V(in)} = {(k*(s*s + w0*w0)/(s*s + w0*s/Q + w0*w0)} .ends n_filter_stage

16.7 VERIFICATION EXAMPLES

Example 16.7-1

Figure 16.7-1 shows the frequency response of a band-pass filter using PSpice. Such a filter can be represented by

$$\frac{\mathbf{V}_0(j\omega)}{\mathbf{V}_{\text{in}}(j\omega)} = \mathbf{H}(\omega) = \frac{H_0}{1 + jQ\left(\dfrac{\omega}{\omega_0} - \dfrac{\omega_0}{\omega}\right)}$$

where $\mathbf{V}_{\text{in}}(j\omega)$ and $\mathbf{V}_0(j\omega)$ are the input and output of the filter. This filter was designed to satisfy the specifications

$$\omega_0 = 2\pi 1000, \quad Q = 10, \quad H_0 = 10$$

Determine if the specifications are satisfied.

Solution

The frequency response was obtained by analyzing the filter using PSpice. The vertical axis of Figure 16.7-1 gives the magnitude of $\mathbf{H}(j\omega)$ in decibels. The horizontal axis gives the frequency in hertz. Three points on the frequency response have been labeled, giving

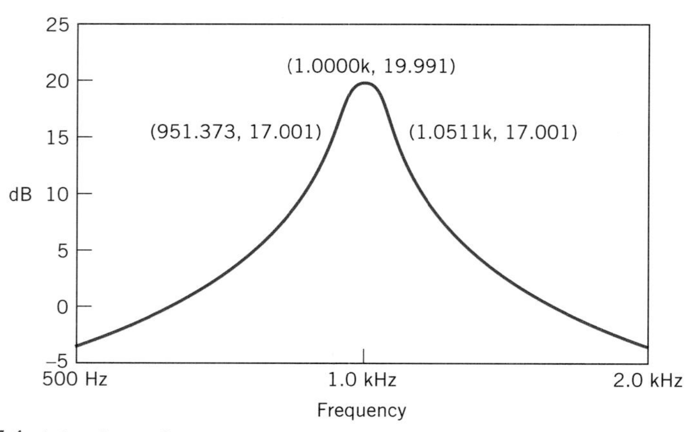

Figure 16.7-1 A band-pass frequency response.

the frequency and magnitude at each point. We want to use this information from the frequency response to check the filter to see if it has the correct values of ω_0, Q, and H_0.

The three labeled points on the frequency response have been carefully selected. One of these labels indicates that the magnitude of $\mathbf{H}(\omega)$ and frequency at the peak of the frequency response are 20 dB and 1000 Hz. This peak occurs at the resonant frequency, so

$$\omega_0 = 2\pi 1000$$

The magnitude at the resonant frequency is H_0, so

$$20 \log_{10} H_0 = 20$$

or

$$H_0 = 10$$

The other two labeled points were chosen so that the magnitudes are 3 dB less than the magnitude at the peak. The frequencies at these points are 951 Hz and 1051 Hz. The difference of these two frequencies is the bandwidth, BW, of the frequency response. Finally, Q is calculated from the resonant frequency, ω_0 and the bandwidth, BW

$$Q = \frac{\omega_0}{BW} = \frac{2\pi 1000}{2\pi(1051 - 951)} = 10$$

In this example, three points on the frequency response were used to verify that the band-pass filter satisfied the specifications for its resonant frequency, gain, and quality factor.

Example 16.7-2

ELab is a circuit analysis program that can be used to calculate the transfer function of a filter circuit (Svoboda 1997). (ELab can be obtained by downloading it from http://www.clarkson.edu/~svoboda/software.html on the internet.) Figure 16.7-2 shows the result of using ELab to analyze the Sallen-Key band-pass filter shown in Figure 16.4-2. This Sallen-Key filter was designed in Example 16.4-2 to have $\omega_0 = 3142$ rad/s, $Q = 5$, and $k = 14$. Use the data produced by analyzing the filter using ELab to verify that the filter does indeed have the required values of ω_0, Q, and k.

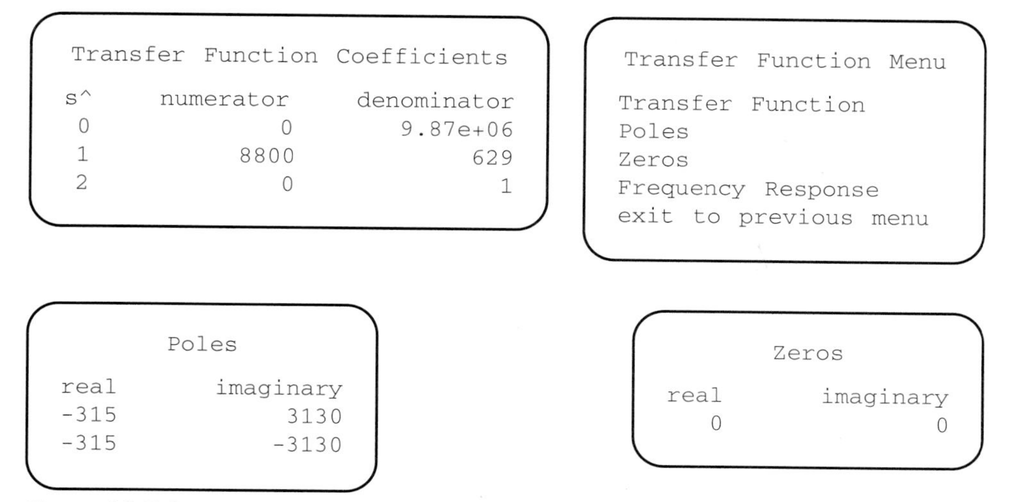

Figure 16.7-2 Using ELab to determine the transfer function of a band-pass filter.

Solution

The coefficients of the transfer function of the filter are given in the upper left-hand portion of Figure 16.7-2. The coefficients indicate that the transfer function of this filter is

$$H(s) = \frac{8800s}{s^2 + 629s + 9.87 \times 10^6} \tag{16.7-1}$$

The general form transfer function of the second-order band-pass filter is

$$H(s) = \frac{k\dfrac{\omega_0}{Q}s}{s^2 + \dfrac{\omega_0}{Q}s + \omega_0^2} \tag{16.7-2}$$

Notice that the coefficient of s^2 in the denominator polynomial is 1 in both of these transfer functions. Values of ω_0, Q, and k are determined by comparing the coefficients of the transfer functions in Equations 16.7-1 and 16.7-2.

The square root of the constant term of the denominator polynomial is equal to ω_0. Therefore

$$\omega_0 = \sqrt{9.87 \times 10^6} = 3142 \text{ rad/s}$$

Next, the coefficient of s in the denominator polynomial is equal to ω_0/Q. Therefore

$$Q = \frac{\omega_0}{629} = \frac{3142}{629} = 5$$

Finally, the ratio of the coefficient of s in the numerator polynomial to the coefficient of s in the denominator polynomial is equal to k. Therefore

$$k = \frac{8880}{629} = 14$$

The Sallen-Key band-pass filter shown in Figure 16.4-2 does indeed have the required values of ω_0, Q, and k.

16.8 Design Challenge Solution

ANTI-ALIASING FILTER

Digital Signal Processing (DSP) frequently involves sampling a voltage, and converting the samples to digital signals. After the digital signals are processed the output signal is converted back into an analog voltage. Unfortunately, a phenomenon called aliasing can cause errors to occur during digital signal processing. Aliasing is a possibility whenever the input voltage contains components at frequencies greater than one-half of the sampling frequency. Aliasing occurs when these components are mistakenly interpreted to be components at a lower frequency. Anti-aliasing filters are used to avoid these errors by eliminating those components of the input voltage that have frequencies greater than one-half of the sampling frequency.

An anti-aliasing filter is needed for a DSP application. The filter is specified to be a fourth-order Butterworth low-pass filter having a cutoff frequency of 500 hertz and a dc gain equal to one. This filter is to be implemented as an RC op amp circuit.

Describe the Situation and Assumptions

The anti-aliasing filter will be designed as a cascade circuit consisting of two Sallen-Key low-pass filters and perhaps an amplifier. The amplifier will be included if it is necessary to adjust the dc gain of the anti-aliasing filter.

The operational amplifiers in the Sallen-Key filter stages will be modeled as ideal operational amplifiers. Resistances will be restricted to the range 2 kΩ to 500 kΩ, and capacitances will be restricted to the range 1 nF to 10 μF.

State the Goal

The transfer function of a fourth-order Butterworth low-pass filter having a cutoff frequency of 500 hertz and a dc gain equal to one can be obtained in two steps. First, the transfer function of a fourth-order Butterworth low-pass filter is given by Eq. 16.3-2 and Table 16.3-2 to be

$$H_n(s) = \frac{1}{(s^2 + 0.765s + 1)(s^2 + 1.848s + 1)} \tag{16.8-1}$$

$H_n(s)$ is the transfer function of a filter having a cutoff frequency equal to 1 rad/s. Next, frequency scaling can be used to adjust the cutoff frequency 500 hertz = 3142 rad/s. Frequency scaling can be accomplished by replacing s by $\frac{s}{\omega_c} = \frac{s}{3142}$ in $H_n(s)$.

$$\begin{aligned} H(s) &= \frac{1}{\left(\left(\frac{s}{3142}\right)^2 + 0.765\left(\frac{s}{3142}\right) + 1\right)\left(\left(\frac{s}{3142}\right)^2 + 1.848\left(\frac{s}{3142}\right) + 1\right)} \\ &= \frac{3142^4}{(s^2 + 2403.6s + 3142^2)(s^2 + 5806.4s + 3142^2)} \end{aligned} \tag{16.8-2}$$

The goal is to design a filter circuit that has this transfer function.

Generate a Plan

We will express $H(s)$ as the product of two second-order low-pass transfer functions. For each of these second-order transfer functions we will

1. Determine the values of the filter parameters k, ω_0, and Q.
2. Design a Sallen-Key low-pass filter to have the required values of ω_0 and Q.

It's likely that the Sallen-Key filters won't have the desired values of the dc gain, so an amplifier will be required to adjust the dc gain. The anti-aliasing filter will consist of a cascade connection of the Sallen-Key filter stages and the amplifier.

Act on the Plan

Consider the first factor of the denominator of $H(s)$. From the constant term,

$$\omega_0^2 = 3142^2$$

So $\omega_0 = 3142$ rad/s. Next, from the coefficient of s in the denominator

$$\frac{\omega_0}{Q} = 2403.6$$

so

$$Q = \frac{3142}{2403.6} = 1.31$$

Next, design a Sallen-Key low-pass filter with $\omega_0 = 3142$ rad/s and $Q = 1.31$. The design equations given in row 1 of Table 16.4-2 indicate that

$$\omega_0 = \frac{1}{RC}$$

and

$$Q = \frac{1}{3 - A}$$

Pick $C = 0.1\ \mu$F. Then

$$R = \frac{1}{\omega_0 C} = \frac{1}{3142 \cdot 10^{-7}} = 3183$$

Also

$$A = 3 - \frac{1}{Q} = 3 - \frac{1}{1.31} = 2.24$$

The dc gain of this filter stage is $k = A = 2.24$, so the transfer function of this stage is

$$H_1(s) = \frac{2.24 \cdot 3142^2}{s^2 + 2403.6s + 3142^2}$$

Next, consider the second factor in the denominator of $H(s)$. Once again the constant term indicates that $\omega_0 = 3142$ rad/s. Now Q can be calculated from the coefficient of s to be

$$Q = \frac{3142}{5806.4} = 0.541$$

We require a Sallen-Key low-pass filter with $\omega_0 = 3142$ rad/s and $Q = 0.541$. Pick $C = 0.1\ \mu$F. Then

$$R = \frac{1}{\omega_0 C} = \frac{1}{3142 \cdot 10^{-7}} = 3183$$

and

$$A = 3 - \frac{1}{Q} = 3 - \frac{1}{0.541} = 1.15$$

The dc gain of this filter stage is $k = A = 1.15$, so the transfer function of this stage is

$$H_2(s) = \frac{1.15 \cdot 3142^2}{s^2 + 5806.4s + 3142^2}$$

The product of the gains of the filter stages is

$$H_1(s) \cdot H_2(s) = 2.576 \cdot H(s)$$

so

$$H(s) = 0.388 \cdot H_1(s) \cdot H_2(s)$$

The third stage of the anti-aliasing filter is an inverting amplifier having gain equal to 0.388. The anti-aliasing filter is shown in Figure 16.8-1.

Verify the Proposed Solution

Section 16.6 describes a procedure for verifying that a circuit has a specified transfer function. This procedure consists of using PSpice to plot the frequency response of both the circuit and the transfer function. These two frequency responses are compared. If they are the same, the transfer function of the circuit is indeed the specified transfer function.

Figure 16.8-2 shows the PSpice input file used to plot the frequency responses of both the circuit shown in Figure 16.8-1 and the transfer function given in Eq. 16.8-2. These frequency responses are shown in Figure 16.8-3. These frequency responses overlap exactly so that the two plots appear to be a single plot. Therefore, the filter does indeed have the required transfer function.

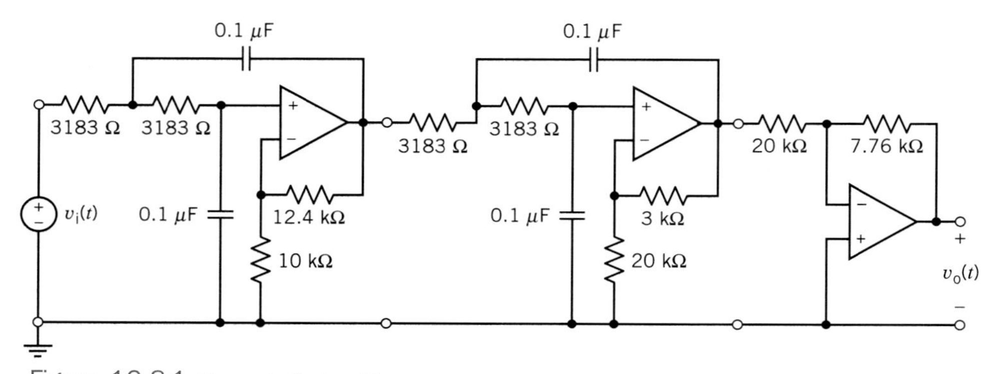

Figure 16.8-1 The anti-aliasing filter.

```
Verify the transfer function of the 4th order low-pass filter
Vin 1 0 ac 1
XSK1 1 2 sk_lp params: C={C} w0={w0} Q=1.31
XSK2 2 3 sk_lp params: C={C} w0={w0} Q=0.541
Ri 3 4 20000
Rf 4 5 7760
XOA 4 0 5 op_amp
X1 1 6 H1
X2 6 7 H2

.subckt sk_lp in out params: C=.1uF w0=1krad/s Q=0.707
R1 in 2 {1/C/w0}
R2 2 3 {1/C/w0}
C1 3 0 {C}
C2 2 out {C}
XOA 5 3 out op_amp
R3 5 0 10kOhm
R4 out 5 {(2-1/Q)*10kOhm}
.ends

.subckt op_amp inv non out
*an ideal op amp
E (out 0) (non inv) 1G
.ends op_amp

.subckt H1 in out
R1 in 0 1G
R2 out 0 1G
E out 0 LAPLACE {V(in)}={3142*3142/(s*s+2403.6*s+3142*3142)}
.ends H1

.subckt H2 in out
R1 in 0 1G
R2 out 0 1G
E out 0 LAPLACE {V(in)}={3142*3142/(s*s+5806.4*s+3142*3142)}
.ends H2

.ac dec 25 .01 5000
.probe V(7) V(5)
.param: C=0.1uF w0=3142 Q=2 k=2.5
.end
```

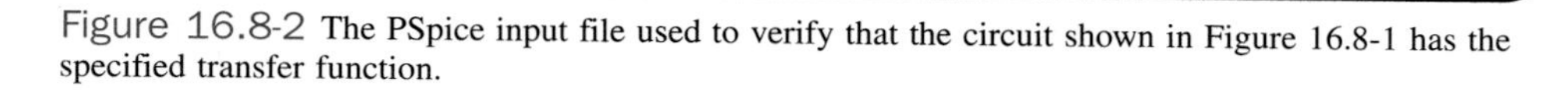

Figure 16.8-2 The PSpice input file used to verify that the circuit shown in Figure 16.8-1 has the specified transfer function.

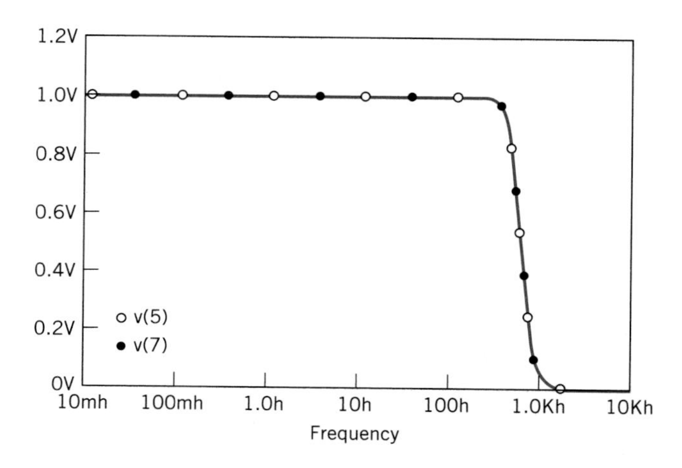

Figure 16.8-3 The frequency response of the circuit shown in Figure 16.8-1 and frequency response corresponding to the transfer function given in Eq. 16.8-2 are identical.

16.9 SUMMARY

- An ideal filter separates its input into two parts. One part is passed, unchanged, to the output; the other part is eliminated. In other words, the output of an ideal filter is an exact copy of part of the filter input.
- There are several ways of separating the filter input into two parts, and, correspondingly, several types of ideal filter. Table 16.3-1 illustrates the common filter types.
- Unfortunately, ideal filter circuits don't exist. Filters are circuits that approximate ideal filters.
- **Butterworth** transfer functions have magnitude frequency responses that approximate the frequency response of an ideal filter.
- The frequency response of second-order filters is characterized by three filter parameters: a gain k, the corner frequency ω_0, and the **quality factor** Q. Filter circuits are designed by choosing the values of the circuit elements in such a way as to obtain the required values of k, ω_0, and Q.
  - Table 16.4-1 provides the information required to design second-order *RLC* filters.
  - Table 16.4-2 provides the information required to design Sallen-Key filters.
- High-order filters are filters that have an order greater than two. A popular strategy for designing high-order filters uses a cascade connection of second-order filters.
- SPICE provides a convenient way to verify that a filter circuit does indeed have the correct transfer function.
- SPICE subcircuits reduce the complexity of simulations of high-order filters. Table 16.6-1 provides SPICE subcircuits for the four Sallen-Key filter stages.

PROBLEMS

Section 16.3 Filters

P 16.3-1 Obtain the transfer function of a third order Butterworth low-pass filter having a cutoff frequency equal to 100 hertz.

Answer: $H_L(s) = \dfrac{628^3}{(s + 628)(s^2 + 628s + 628^2)}$

P 16.3-2 A dc gain can be incorporated into Butterworth low-pass filters by defining the transfer function to be

$$H_L(s) = \frac{\pm k}{D(s)}$$

where $D(s)$ denotes the polynomials tabulated in Table 16.3-2 and k is the dc gain. The dc gain k is also called the pass-band gain. Obtain the transfer function of a third-order Butterworth low-pass filter having a cutoff frequency equal to 100 rad/s and a pass-band gain equal to 5.

P 16.3-3 High-pass Butterworth filters have transfer functions of the form

$$H_H(s) = \frac{\pm ks^n}{D_n(s)}$$

where n is the order of the filter, $D_n(s)$ denotes the nth order polynomial in Table 16.3-2, and k is the pass-band gain. Obtain the transfer function of a third-order Butterworth high-pass filter having a cutoff frequency equal to 100 rad/s and a pass-band gain equal to 5.

Answer: $H_H(s) = \dfrac{5 \cdot s^3}{(s + 100)(s^2 + 100s + 10000)}$

P 16.3-4 High-pass Butterworth filters have transfer functions of the form

$$H_H(s) = \frac{\pm ks^n}{D_n(s)}$$

where n is the order of the filter, $D_n(s)$ denotes the nth order polynomial in Table 16.3-2, and k is the pass-band gain. Obtain the transfer function of a fourth-order Butterworth high-pass filter having a cutoff frequency equal to 500 hertz and a pass-band gain equal to 5.

P 16.3-5 A band-pass filter has two cutoff frequencies, ω_a and ω_b. Suppose that ω_a is quite a bit smaller than ω_b, say $\omega_a < \dfrac{\omega_b}{10}$. Let $H_L(s)$ be a low-pass transfer function having a cutoff frequency equal to ω_b and $H_H(s)$ be a high-pass transfer function having a cutoff frequency equal to ω_a. A band-pass transfer function can be obtained as a product of low-pass and high-pass transfer functions, $H_B(s) = H_L(s) \cdot H_H(s)$. The order of the band-pass filter is equal to the sum of the orders of the low-pass and high-pass filters. We usually make the orders of the low-pass and high-pass filter equal, in which case the order of the band-pass is even. The pass-band gain of the band-pass filter is the product of pass-band gains of the low-pass and high-pass transfer functions. Obtain the transfer function of a fourth order band-pass filter having cutoff frequencies equal to 100 rad/s and 2000 rad/s and a pass-band gain equal to 4.

Answer:

$$H_B(s) = \frac{16000000 \cdot s^2}{(s^2 + 141.4s + 10000)(s^2 + 2828s + 4000000)}$$

P 16.3-6 In some applications band-pass filters are used to pass only those signals having a specified frequency ω_0. The cutoff frequencies of the band-pass filter are specified to satisfy $\sqrt{\omega_a \omega_b} = \omega_0$. The transfer function of the band-pass filter is given by

$$H_B(s) = k \left(\frac{\dfrac{\omega_0}{Q} s}{s^2 + \dfrac{\omega_0}{Q} s + \omega_0^2} \right)^m$$

The order of this band-pass transfer function is $n = 2m$. The pass-band gain is k. Transfer functions of the type are readily implemented as the cascade connection of identical second-order filter stages. Q is the quality factor of the second-order filter stage. The frequency ω_0 is called the center frequency of the band-pass filter. Obtain the transfer function of a fourth-order band-pass filter having a center frequency equal to 250 rad/s and a pass-band gain equal to 4. Use $Q = 1$.

Answer: $H_B(s) = \dfrac{250000s^2}{(s^2 + 250s + 62500)^2}$

P 16.3-7 A band-stop filter has two cutoff frequencies, ω_a and ω_b. Suppose that ω_a is quite a bit smaller than ω_b, say $\omega_a < \frac{\omega_b}{10}$. Let $H_L(s)$ be a low-pass transfer function having a cutoff frequency equal to ω_a and $H_H(s)$ be a high-pass transfer function having a cutoff frequency equal to ω_b. A band-stop transfer function can be obtained as a sum of low-pass and high-pass transfer functions, $H_N(s) = H_L(s) + H_H(s)$. The order of the band-pass filter is equal to the sum of the orders of the low-pass and high-pass filters. We usually make the orders of the low-pass and high-pass filter equal, in which case the order of the band-stop is even. The pass-band gains of both the low-pass and high-pass transfer functions are set equal to the pass-band gain of the band-stop filter. Obtain the transfer function of a fourth-order band-stop filter having cutoff frequencies equal to 100 rad/s and 2000 rad/s and a pass-band gain equal to 2.

Answer:

$$H_N(s) = \frac{2s^4 + 282.8s^3 + 40000s^2 + 56560000s + 8 \cdot 10^{10}}{(s^2 + 141.4s + 10000)(s^2 + 2828s + 4000000)}$$

P 16.3-8 In some applications band-stop filters are used to reject only those signals having a specified frequency ω_0. The cutoff frequencies of the band-stop filter are specified to satisfy $\sqrt{\omega_a\omega_b} = \omega_0$. The transfer function of the band-pass filter is given by

$$H_N(s) = k - H_B(s) = k - k\left(\frac{\frac{\omega_0}{Q}s}{s^2 + \frac{\omega_0}{Q}s + \omega_0^2}\right)^m$$

The order of this band-stop transfer function is $n = 2m$. The pass-band gain is k. Transfer functions of the type are readily implemented using a cascade connection of identical second-order filter stages. Q is the quality factor of the second-order filter stage. The frequency ω_0 is called the center frequency of the band-stop filter. Obtain the transfer function of a fourth-order band-stop filter having a center frequency equal to 250 rad/s and a pass-band gain equal to 4. Use $Q = 1$.

Answer: $H_N(s) = \dfrac{4(s^2 + 62500)^2}{(s^2 + 250s + 62500)^2}$

P 16.3-9 Transfer functions of the form

$$H_L(s) = k\left(\frac{\omega_0^2}{s^2 + \frac{\omega_0}{Q}s + \omega_0^2}\right)^m$$

are low-pass transfer functions. (This is not a Butterworth transfer function.) The order of this low-pass transfer function is $n = 2m$. The pass-band gain is k. Transfer functions of this type are readily implemented using a cascade connection of identical second-order filter stages. Q is the quality factor of the second-order filter stage. The frequency ω_0 is the cutoff frequency, ω_c, of the low-pass filter. Obtain the transfer function of a fourth-order low-pass filter having a cutoff frequency equal to 250 rad/s and a pass-band gain equal to 4. Use $Q = 1$.

P 16.3-10 Transfer functions of the form

$$H_H(s) = k\left(\frac{s^2}{s^2 + \frac{\omega_0}{Q}s + \omega_0^2}\right)^m$$

are high-pass transfer functions. (This is not a Butterworth transfer function.) The order of this high-pass transfer function is $n = 2m$. The pass-band gain is k. Transfer functions of the type are readily implemented using a cascade connection of identical second-order filter stages. Q is the quality factor of the second-order filter stage. The frequency ω_0 is the cut-off frequency, ω_c, of the high-pass filter. Obtain the transfer function of a fourth-order high-pass filter having a cutoff frequency equal to 250 rad/s and a pass-band gain equal to 4. Use $Q = 1$.

Section 16.4 Second-Order Filters

P 16.4-1 The circuit shown in Figure P 16.4-1 is a second-order band-pass filter. Design this filter to have $k = 1$, $\omega_0 = 1000$ rad/s and Q = 1.

P 16.4-2 The circuit shown in Figure P 16.4-2 is a second-order low-pass filter. Design this filter to have $k = 1$, $\omega_0 = 200$ rad/s, and $Q = 0.707$.

P 16.4-3 The circuit shown in Figure P 16.4-3 is a second-order low-pass filter. This filter circuit is called a Multiple-loop

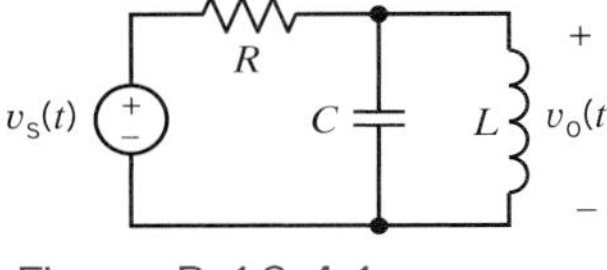

Figure P 16.4-1

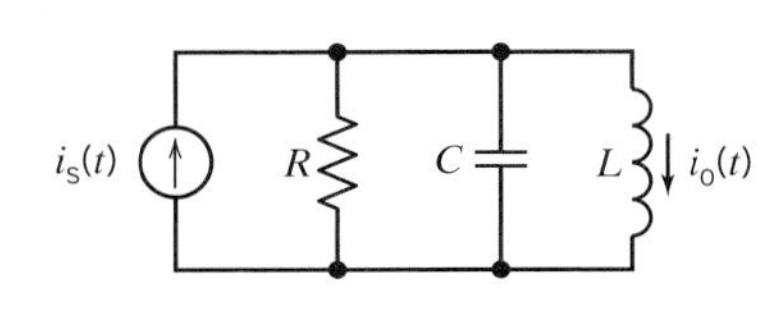

Figure P 16.4-2

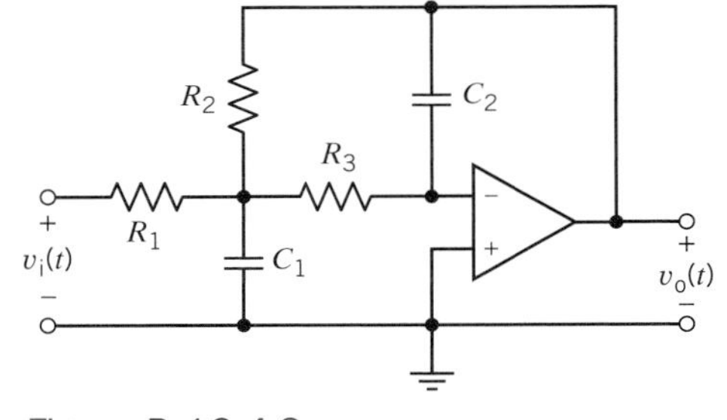

Figure P 16.4-3

Feedback Filter (MFF). The output impedance of this filter is zero, so the MFF low-pass filter is suitable for use as a filter stage in a cascade filter. The transfer function of the low-pass MFF filter is

$$H_L(s) = \frac{-\dfrac{1}{R_1R_3C_1C_2}}{s^2 + \left(\dfrac{1}{R_1C_1} + \dfrac{1}{R_2C_1} + \dfrac{1}{R_3C_1}\right)s + \dfrac{1}{R_2R_3C_1C_2}}$$

Design this filter to have $\omega_0 = 2000$ rad/s and $Q = 8$. What is the value of the dc gain?

Hint: Let $R_2 = R_3 = R$ and $C_1 = C_2 = C$. Pick a convenient value of C and calculate R to obtain $\omega_0 = 2000$ rad/s. Calculate R_1 to obtain $Q = 8$.

P 16.4-4 The circuit shown in Figure P 16.4-4 is a second-order band-pass filter. This filter circuit is called a Multiple-loop Feedback Filter (MFF). The output impedance of this filter is zero, so the MFF band-pass filter is suitable for use as a filter stage in a cascade filter. The transfer function of the band-pass MFF filter is

$$H_B(s) = \frac{-\dfrac{s}{R_1C_2}}{s^2 + \left(\dfrac{1}{R_2C_1} + \dfrac{1}{R_2C_2}\right)s + \dfrac{R_1 + R_3}{R_1R_2R_3C_1C_2}}$$

To design this filter, pick a convenient value of C and then use

$$R_1 = \frac{Q}{k\omega_0C}, \quad R_2 = \frac{2Q}{\omega_0C}, \quad \text{and} \quad R_3 = \frac{2Q}{\omega_0C(2Q^2 - k)}$$

Design this filter to have $k = 5$, $\omega_0 = 2000$ rad/s, and $Q = 8$.

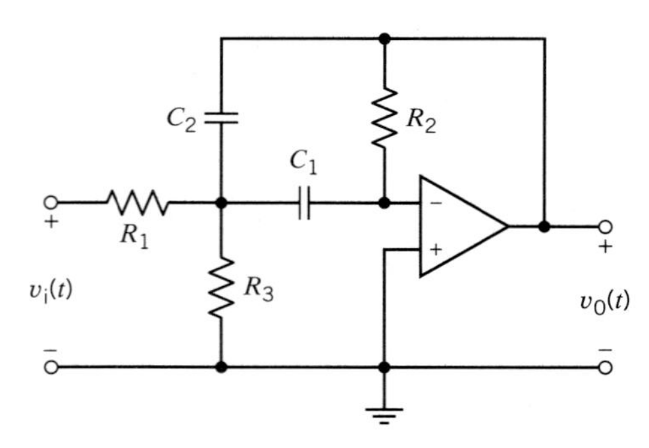

Figure P 16.4-4

P 16.4-5 The circuit shown in Figure P 16.4-5 is a low-pass filter. The transfer function of this filter is

$$H_L(s) = \frac{\dfrac{1}{R_1R_2C_1C_2}}{s^2 + \dfrac{1}{R_1C_1}s + \dfrac{1}{R_1R_2C_1C_2}}$$

Design this filter to have $k = 1$, $\omega_0 = 1000$ rad/s, and $Q = 1$.

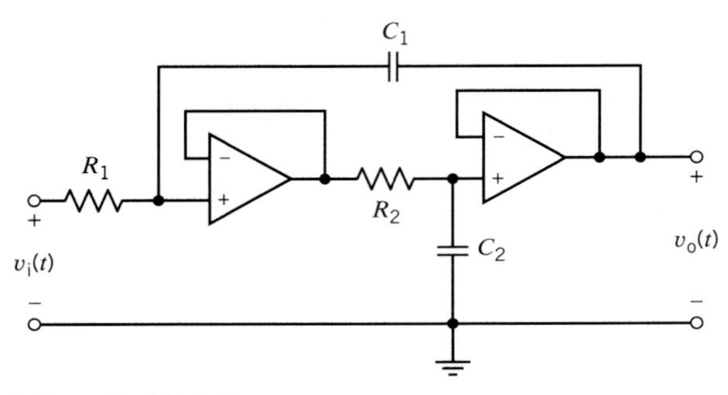

Figure P 16.4-5

P 16.4-6 The *CR:RC* transformation is used to transform low-pass filter circuits into high-pass filter circuits and vice versa. This transformation is applied to *RC* op amp filter circuits. Each capacitor is replaced by a resistor, while each resistor is replaced by a capacitor. Apply the *CR:RC* transformation to the low-pass filter circuit in Figure P 16.4-5 to obtain the high pass filter circuit shown in Figure P 16.4-6. Design a high-pass filter to have $k = 1$, $\omega_0 = 1000$ rad/s, and $Q = 1$.

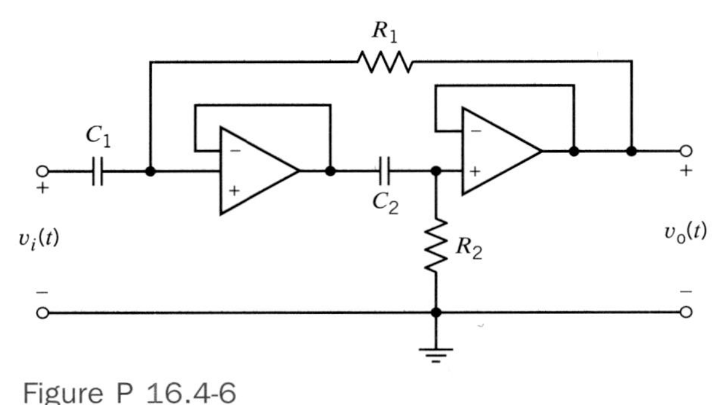

Figure P 16.4-6

P 16.4-7 We have seen that transfer functions can be frequency scaled by replacing s by s/k_f each time that it occurs. Alternately, circuits can also be frequency scaled by dividing each capacitance and each inductance by the frequency scaling factor k_f. Either way, the effect is the same. The frequency response is shifted to the left by k_f. In particular, all cutoff, corner, and resonant frequencies are multiplied by k_f. Suppose that we want to change the cutoff frequency of a filter circuit from ω_{old} to ω_{new}. We set the frequency scale factor to

$$k_f = \frac{\omega_{new}}{\omega_{old}}$$

and then divide each capacitance and each inductance by k_f. Use frequency scaling to change the cutoff frequency of the circuit in Figure P 16.4-7 to 250 rad/s.

Answer: $k_f = 0.05$.

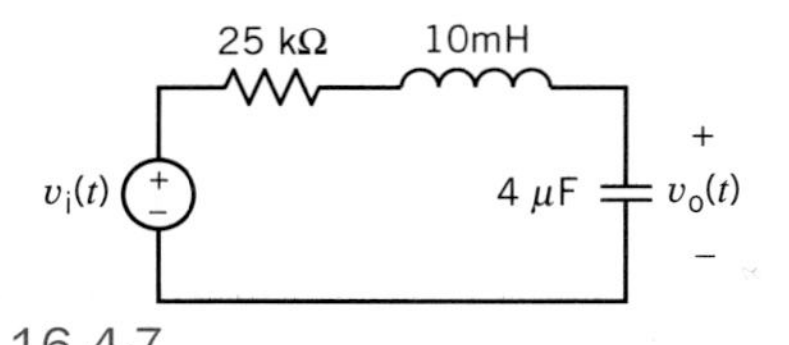

Figure P 16.4-7

P 16.4-8 Impedance scaling is used to adjust the impedances of a circuit. Let k_m denote the impedance scaling factor. Impedance scaling is accomplished by multiplying each impedance by k_m. That means that each resistance and each inductance is multiplied by k_m, but each capacitance is divided by k_m. Transfer functions of the form $H(s) = \frac{V_o(s)}{V_i(s)}$ or $H(s) = \frac{I_o(s)}{I_i(s)}$ are not changed at all by impedance scaling. Transfer functions of the form $H(s) = \frac{V_o(s)}{I_i(s)}$ are multiplied by k_m, while transfer functions of the form $H(s) = \frac{I_o(s)}{V_i(s)}$ are divided by k_m. Use impedance scaling to change the values of the capacitances in the filter shown in Figure P 16.4-8 so that the capacitances are in the range 0.01 μF to 1.0 μF. Calculate the transfer function before and after impedance scaling.

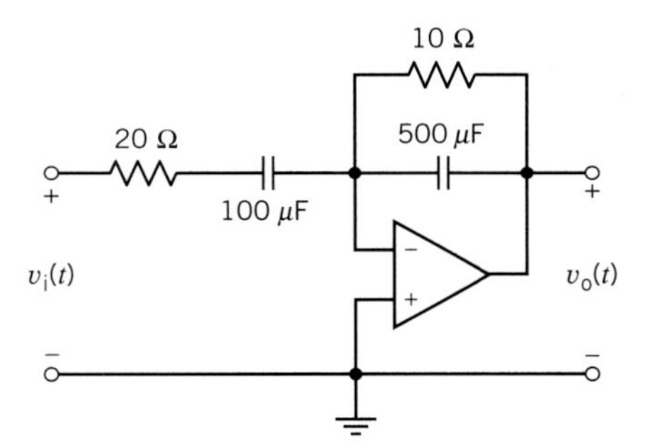

Figure P 16.4-8

P 16.4-9 A band-pass amplifier has the frequency response shown in Figure P 16.4-9. Find the transfer function, $H(s)$.
Hint: $\omega_0 = 2\pi(10 \text{ MHz})$, $k = 10 \text{ dB} = 3.16$, $BW = 0.2 \text{ MHz}$, $Q = 50$

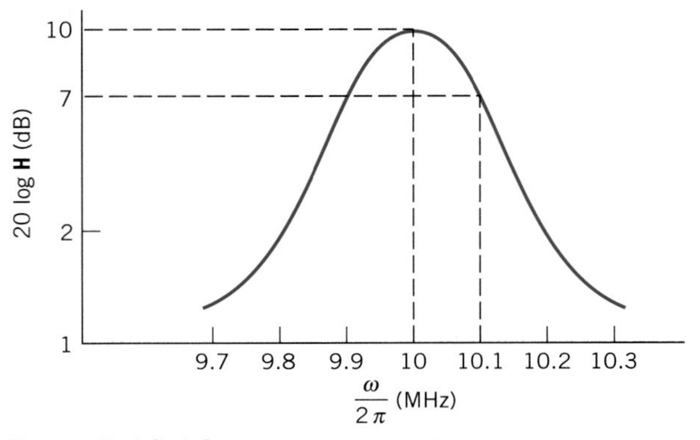

Figure P 16.4-9 A band-pass amplifier.

P 16.4-10 A band-pass filter can be achieved using the circuit of Figure P 16.4-10. Find (a) the magnitude of $\mathbf{H} = \mathbf{V}_o/\mathbf{V}_s$, (b) the low- and high-frequency cutoff frequencies ω_1 and ω_2, and (c) the pass-band gain when $\omega_1 \ll \omega \ll \omega_2$,

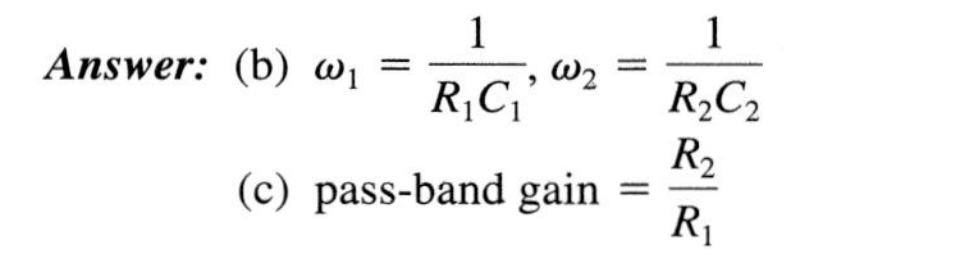

Answer: (b) $\omega_1 = \frac{1}{R_1C_1}$, $\omega_2 = \frac{1}{R_2C_2}$

(c) pass-band gain $= \frac{R_2}{R_1}$

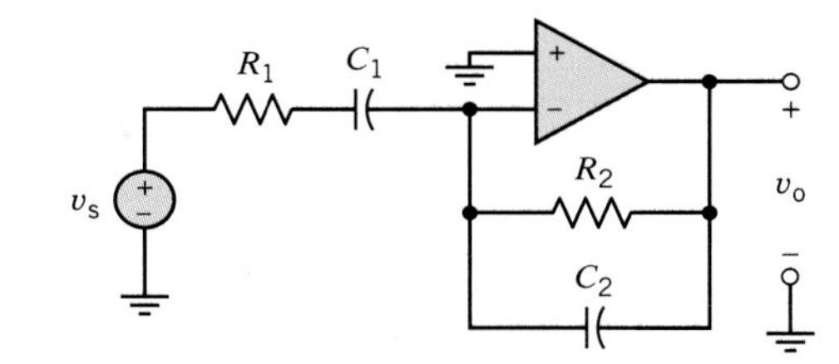

Figure P 16.4-10 A band-pass filter.

P 16.4-11 A unity gain, low-pass filter is obtained from the operational amplifier circuit shown in Figure P 16.4-11. Determine the network function $\mathbf{H}(\omega) = \mathbf{V}_o/\mathbf{V}_s$.

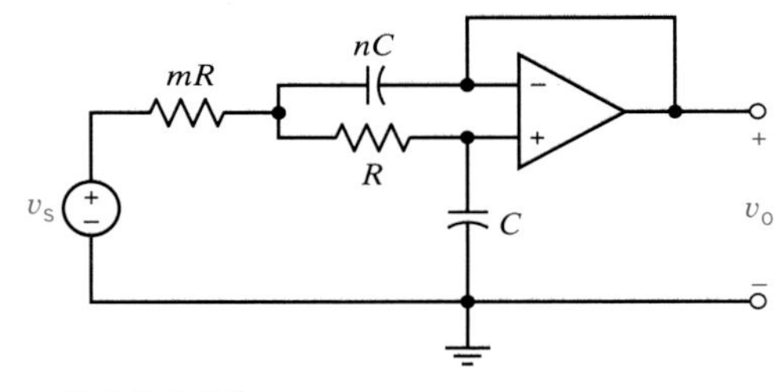

Figure P 16.4-11

P 16.4-12 A particular acoustic sensor produces a sinusoidal output having a frequency equal to 5 kHz. The signal from the sensor has been corrupted with noise. Figure P16.4-12 shows a band-pass filter that was designed to recover the sensor signal from the noise. The voltage v_s represents the noisy signal from the sensor. The filter output, v_o, should be a less noisy signal. Determine the center frequency and bandwidth of this band-pass filter. Assume that the op amp is ideal.

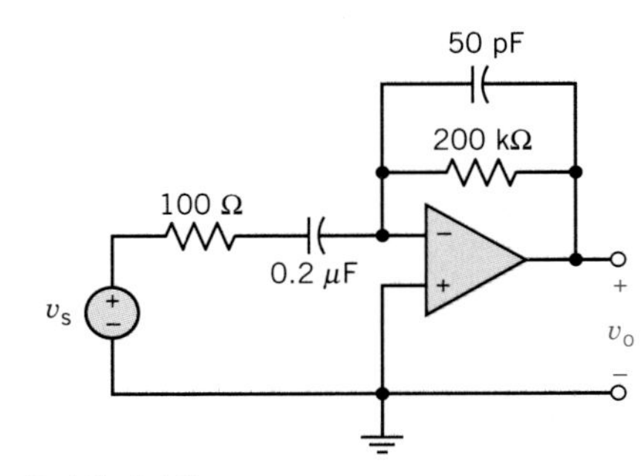

Figure P 16.4-12

P 16.4-13 A band-pass filter can be achieved by using two operational amplifiers in the circuit shown in Figure P 16.4-13 Assume the operational amplifiers are ideal and find the trar fer function $H = V_o/V_s$. Find ω_0, Q, and the bandwidth fo filter when $R_1 = 1\ \text{k}\Omega$, $R_2 = 100\ \Omega$, $C_1 = 1\ \mu\text{F}$, an 0.1 μF.

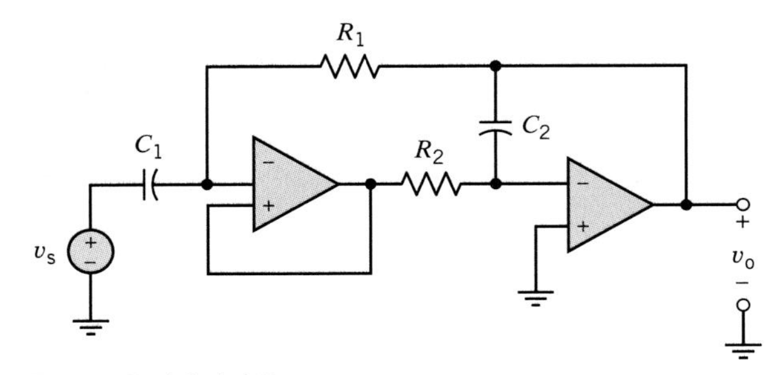

Figure P 16.4-13

P 16.4-14 An op amp circuit is shown in Figure P 16.4-14. Determine $H(s)$ when $a > 10$ and the op amp is ideal. Select a, R, and C so that the resonant frequency is 10^5 rad/s and the gain at the center frequency is 201.

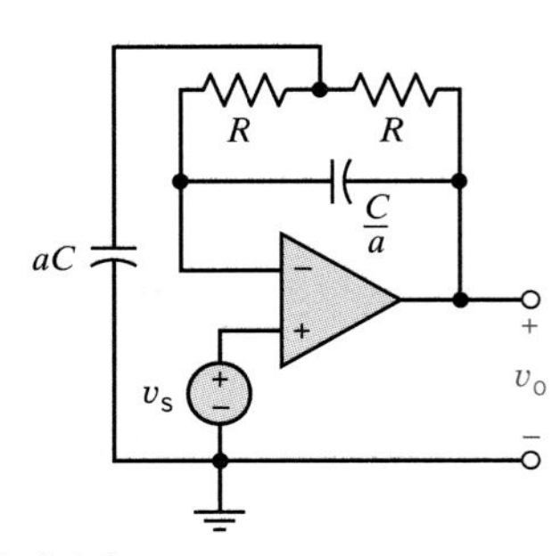

Figure P 16.4-14

P 16.4-15 A low-pass filter is shown in Figure P 16.4-15. Assume an ideal op amp and find the transfer function V_o/V_s. Find the cutoff frequency when $C = 20$ nF, $R = 1.2$ kΩ, and $R_1/R_2 = 1$.

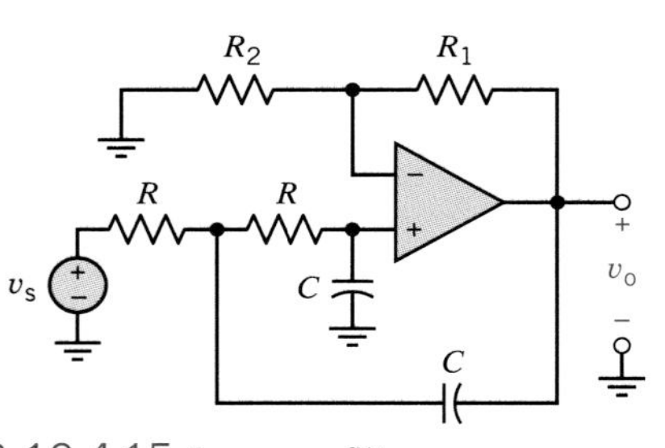

Figure P 16.4-15 Low-pass filter.

Section 16.5 High-Order Filters

P 16.5-1 Design a low-pass filter circuit that has the transfer function

$$H_L(s) = \frac{628^3}{(s + 628)(s^2 + 628s + 628^2)}.$$

Answer: See Figure SP 16-1.

P 16.5-2 Design a filter that has the transfer function

$$H_H(s) = \frac{5 \cdot s^3}{(s + 100)(s^2 + 100s + 10000)}.$$

Answer: See Figure SP 16-2.

P 16.5-3 Design a filter that has the transfer function

$$H_B(s) = \frac{16000000 \cdot s^2}{(s^2 + 141.4s + 10000)(s^2 + 2828s + 4000000)}.$$

Answer: See Figure SP 16-3.

P 16.5-4 Design a filter that has the transfer function

$$H_B(s) = \frac{250000s^2}{(s^2 + 250s + 62500)^2}.$$

Answer: See Figure SP 16-4.

P 16.5-5 Design a filter that has the transfer function

$$H_N(s) = \frac{2s^2}{s^2 + 2828s + 4000000} + \frac{20000}{s^2 + 141.4s + 10000}.$$

Answer: See Figure SP 16-5.

P. 16.5-6 Design a filter that has the transfer function

$$H_N(s) = \frac{4(s^2 + 62500)^2}{(s^2 + 250s + 62500)^2}.$$

Answer: See Figure SP 16-6.

P 16.5-7 (a) For the circuit of Figure P 16.5-7*a* derive an expression for the transfer function $H_a(s) = V_1/V_s$. (b) For the circuit of Figure P 16.5-7*b* derive an expression for the transfer function $H_b(s) = V_1/V_s$. (c) Each of the above filters is a first-order filter. The circuit of Figure P 16.5-7*c* is the cascade connection of the circuits in Figures P 16.5-7*a* and P 16.5-7*b*. Derive an expression for the transfer function $H_c(s) = V_1/V_s$ of the second-order circuit in Figure P 16.5-7*c*. (d) Why isn't $H_c(s) = H_a(s)\,H_b(s)$?

Hint: Consider loading.

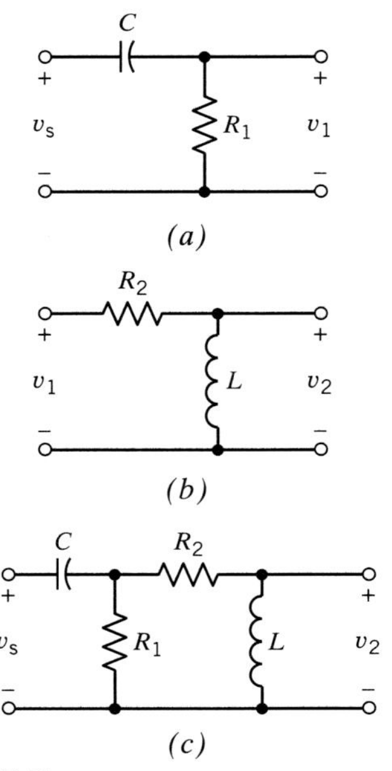

Figure P 16.5-7 (*a*) Circuit for $\mathbf{H}_1$. (*b*) Circuit for $\mathbf{H}_2$. (*c*) Circuit for $\mathbf{H}$.

P 16.5-8 Two filter stages are connected in cascade as shown in Figure P 16.5-8. The transfer function of each filter stage is of the form

$$H(s) = \frac{As}{(1 + s/\omega_L)(1 + s/\omega_h)}$$

Determine the transfer function of the fourth-order filter. (Assume that there is no loading.)

Amplifier 1: A = Gain = 100, –3 dB at f_l = 100 Hz, f_h = 10 kHz

Amplifier 2: A = Gain = 20, –3 dB at f_l = 10 Hz, f_h = 2 kHz

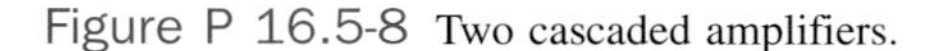
Figure P 16.5-8 Two cascaded amplifiers.

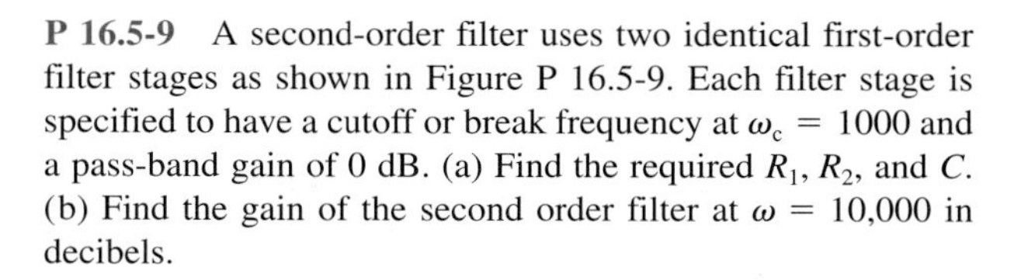
P 16.5-9 A second-order filter uses two identical first-order filter stages as shown in Figure P 16.5-9. Each filter stage is specified to have a cutoff or break frequency at ω_c = 1000 and a pass-band gain of 0 dB. (a) Find the required R_1, R_2, and C. (b) Find the gain of the second order filter at ω = 10,000 in decibels.

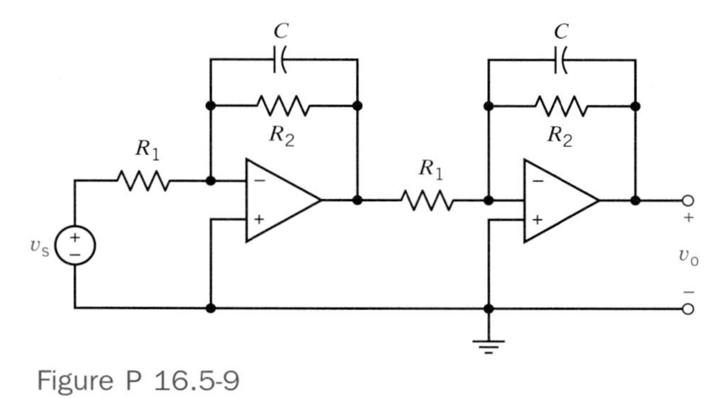

Figure P 16.5-9

PSPICE PROBLEMS

SP 16-1 The filter circuit shown in Figure SP 16.1 was designed to have the transfer function $H_L(s) = \frac{628^3}{(s + 628)(s^2 + 628s + 628^2)}$. Use PSpice to verify that the filter circuit does indeed implement this transfer function.

SP 16-2 The filter circuit shown in Figure SP 16.2 was designed to have the transfer function $H_H(s) = \frac{5 \cdot s^3}{(s + 100)(s^2 + 100s + 10000)}$. Use PSpice to verify that the filter circuit does indeed implement this transfer function.

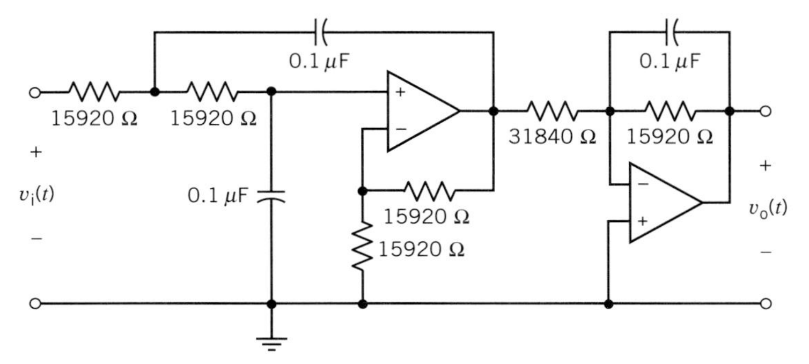

Figure SP 16.1

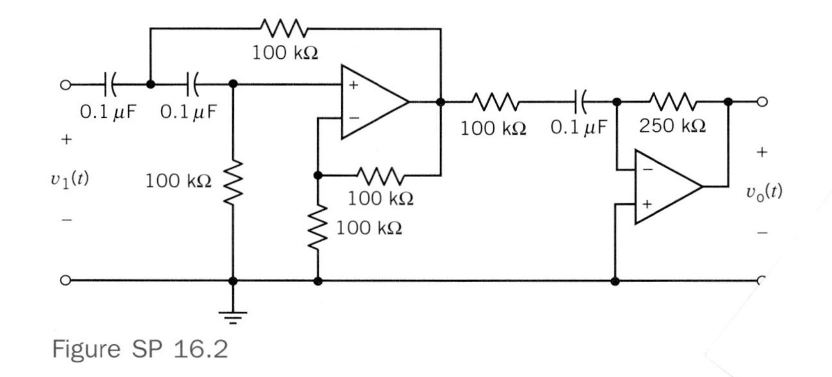

Figure SP 16.2

SP 16-3 The filter circuit shown in Figure SP 16.3 was designed to have the transfer function $H_B(s) = \dfrac{16000000 \cdot s^2}{(s^2 + 141.4s + 10000)(s^2 + 2828s + 4000000)}$. Use PSpice to verify that the filter circuit does indeed implement this transfer function.

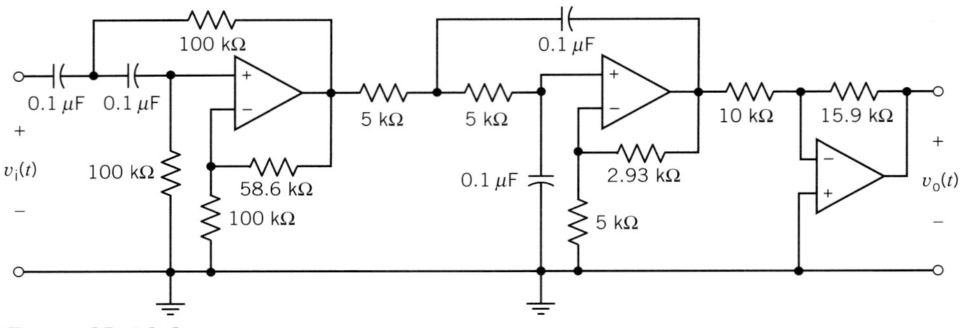

Figure SP 16.3

SP 16-4 The filter circuit shown in Figure SP 16.4 was designed to have the transfer function $H_B(s) = \dfrac{250000s^2}{(s^2 + 250s + 62500)^2}$. Use PSpice to verify that the filter circuit does indeed implement this transfer function.

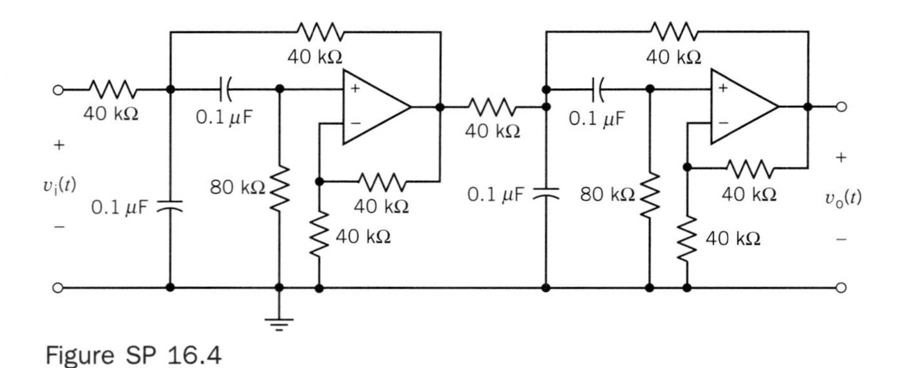

Figure SP 16.4

SP 16-5 The filter circuit shown in Figure SP 16.5 was designed to have the transfer function $H_N(s) = \dfrac{2s^2}{s^2 + 2828s + 4000000} + \dfrac{20000}{s^2 + 141.4s + 10000}$. Use PSpice to verify that the filter circuit does indeed implement this transfer function.

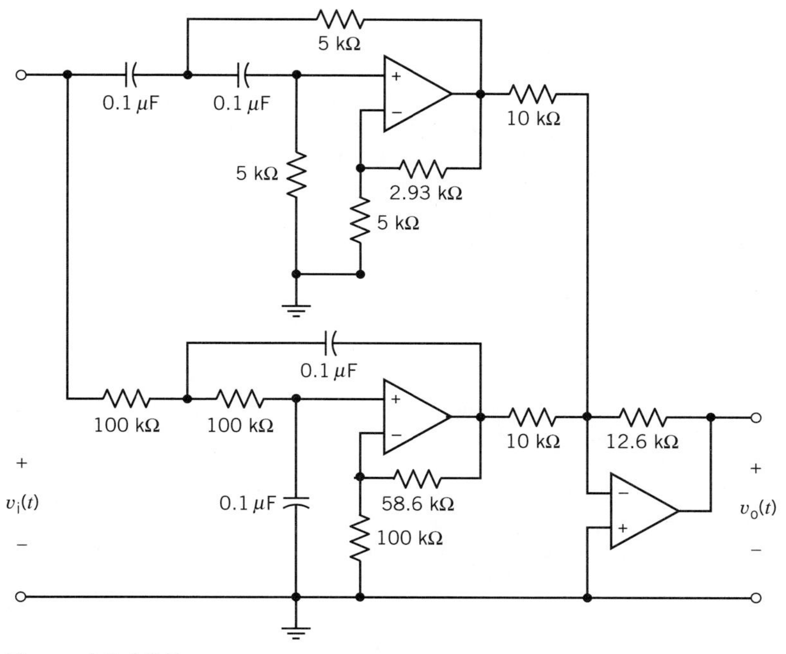

Figure SP 16.5

SP 16-6 The filter circuit shown in Figure SP 16.6 was designed to have the transfer function $H_N(s) = \dfrac{4(s^2 + 62500)^2}{(s^2 + 250s + 62500)^2}$. Use PSpice to verify that the filter circuit does indeed implement this transfer function.

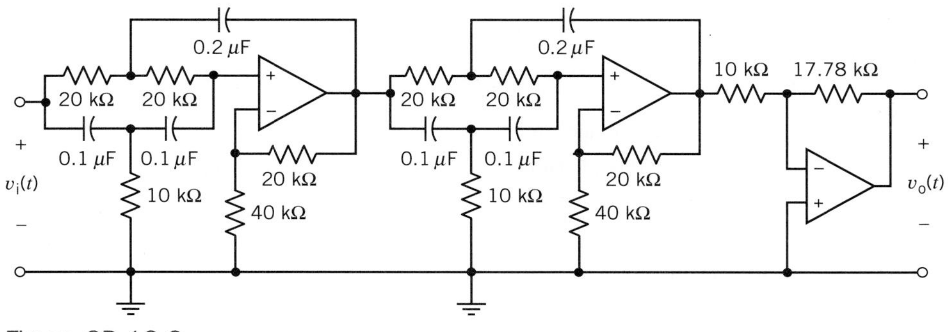

Figure SP 16.6

SP 16-7 A notch filter is shown in Figure SP 16.7. The output of a two-stage filter is v_1, and the output of a three-stage filter is v_2. Plot the Bode diagram of $\mathbf{V}_1/\mathbf{V}_s$ and $\mathbf{V}_2/\mathbf{V}_s$ and compare the results when L = 10 mH and C = 1 μF.

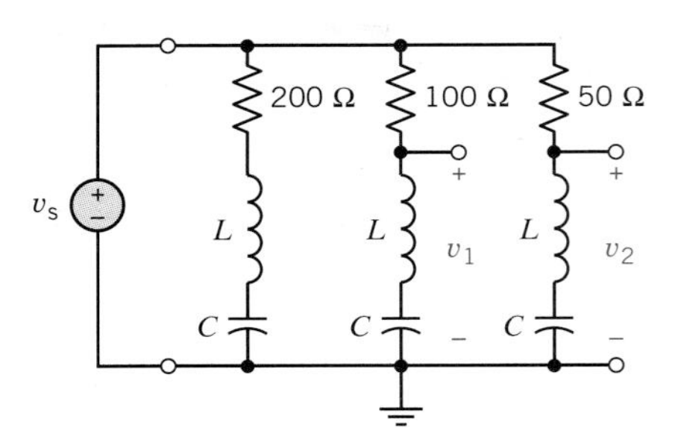

Figure SP 16.7

SP 16-8 An acoustic sensor operates in the range of 5 kHz to 25 kHz and is represented in Figure SP 16.8 by v_s. It is specified that the band-pass filter shown in the figure passes the signal in the frequency range within 3 dB of the center frequency gain. Determine the bandwidth and center frequency of the circuit when the op amp has $R_i = 500\ \text{k}\Omega$, $R_o = 1\ \text{k}\Omega$, and $A = 10^6$.

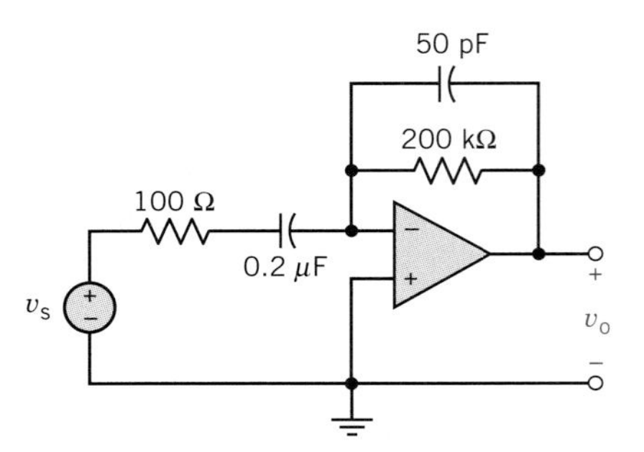

Figure SP 16.8

SP 16-9 Frequently, audio systems contain two or more loudspeakers that are intended to handle different parts of the audio-frequency spectrum. In a three-way setup, one speaker, called a woofer, handles low frequencies. A second, the tweeter, handles high frequencies, while a third, the midrange, handles the middle range of the audio spectrum.

A three-way filter, called a crossover network, is used to split the audio signal into the three bands of frequencies suitable for each speaker. There are many and varied designs. A simple one is based on series *LR*, *CR*, and resonant *RLC* circuits as shown in Figure SP 16.9. All speaker impedances are assumed resistive. The conditions are (1) woofer, at the crossover frequency: $X_{L1} = R_w$; (2) tweeter, at the crossover frequency $X_{C3} = R_T$; and (3) midrange, with components C_2, L_2, and R_{MR} forming a series resonant circuit with upper and lower cutoff frequencies f_u and f_L, respectively. The resonant frequency $= (f_u f_L)^{1/2}$.

When all the speaker resistances are 8 Ω, determine the frequency response and determine the cutoff frequencies. Plot the Bode diagram for the three speakers. Determine the bandwidth of the midrange speaker section.

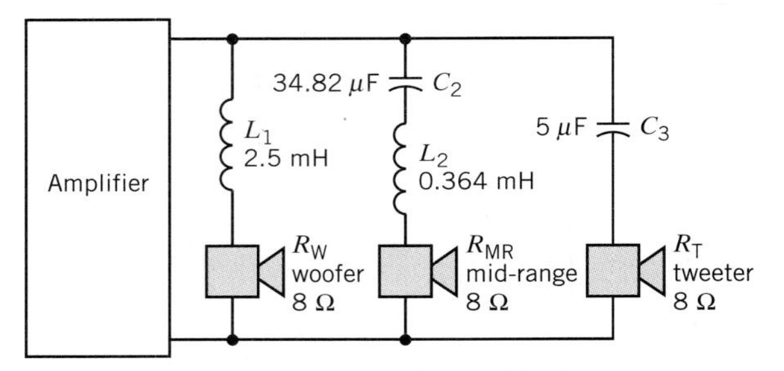

Figure SP 16.9 Three-way filter for a speaker system.

VERIFICATION PROBLEMS

VP 16-1 The specifications for a band-pass filter require that $\omega_0 = 100$ rad/s, $Q = 5$, and $k = 3$. The transfer function of a filter designed to satisfy these specifications is

$$H(s) = \frac{75s}{s^2 + 25s + 10000}$$

Does this filter satisfy the specifications?

VP 16-2 The specifications for a band-pass filter require that $\omega_0 = 100$ rad/s, $Q = 4$, and $k = 3$. The transfer function of a filter designed to satisfy these specifications is

$$H(s) = \frac{75s}{s^2 + 25s + 10000}$$

Does this filter satisfy the specifications?

VP 16-3 The specifications for a low-pass filter require that $\omega_0 = 20$ rad/s, $Q = .8$, and $k = 1.5$. The transfer function of a filter designed to satisfy these specifications is

$$H(s) = \frac{600}{s^2 + 25s + 400}$$

Does this filter satisfy the specifications?

VP 16-4 The specifications for a low-pass filter require that $\omega_0 = 25$ rad/s, $Q = .4$, and $k = 1.2$. The transfer function of a filter designed to satisfy these specifications is

$$H(s) = \frac{750}{s^2 + 62.5s + 625}$$

Does this filter satisfy the specifications?

VP 16-5 The specifications for a high-pass filter require that $\omega_0 = 12$ rad/s, $Q = 4$, and $k = 5$. The transfer function of a filter designed to satisfy these specifications is

$$H(s) = \frac{5s^2}{s^2 + 30s + 144}$$

Does this filter satisfy the specifications?

DESIGN PROBLEMS 

DP 16-1 Design a band-pass filter with a center frequency of 100 kHz and a bandwidth of 10 kHz using the circuit shown in Figure DP 16.1. Assume that $C = 100$ pF and find R and R_3. Use PSpice to verify the design.

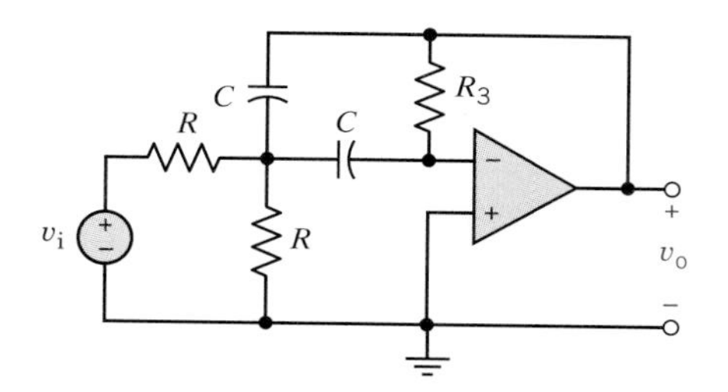

Figure DP 16.1

DP 16-2 A communication transmitter requires a band-pass filter to eliminate low-frequency noise from nearby traffic. Measurements indicate that the range of traffic rumble is $2 < \omega < 12$ rad/s. A designer proposes a filter as

$$H(s) = \frac{(1 + s/\omega_1)^2(1 + s/\omega_3)}{(1 + s/\omega_2)^3}$$

where $s = j\omega$.

It is desired that signals with $\omega > 100$ rad/s pass with less than 3-dB loss, while the traffic rumble be reduced by 46 dB or more. Select ω_1, ω_2, and ω_3 and plot the Bode diagram.

DP 16-3 A communication transmitter requires a band-stop filter to eliminate low-frequency noise from nearby auto traffic. Measurements indicate that the range of traffic rumble is 2 rad/s $< \omega <$ 12 rad/s. A designer proposes a filter as

$$H(s) = \frac{(1 + s/\omega_1)^2(1 + s/\omega_3)^2}{(1 + s/\omega_2)^2(1 + s/\omega_4)^2}$$

where $s = j\omega$. It is desired that signals above 130 rad/s pass with less than 4-dB loss, while the traffic rumble be reduced by 35 dB or more. Select ω_1, ω_2, ω_3, and ω_4 and plot the Bode diagram.

CHAPTER 17

Two-Port and Three-Port Networks

Preview

Many practical circuits have two ports of access called the input and output ports. One purpose of this chapter is to describe the parameters that can represent two-port networks. In addition, we will study the equations that describe two special three-port passive networks called the T and Π networks. Finally, we will discuss the relationships between the six sets of parameters that describe the two-port network and obtain the equations for a network consisting of two-port networks connected in series, parallel, and cascade.

17.1 Design Challenge

TRANSISTOR AMPLIFIER

Figure 17.1-1 shows the small signal equivalent circuit of a transistor amplifier. The data sheet for the transistor describes the transistor by specifying its h parameters to be

$$h_{ie} = 1250\ \Omega, \quad h_{oe} = 0, \quad h_{fe} = 100, \quad \text{and} \quad h_{re} = 0$$

The value of the resistance R_c must be between 300 Ω and 5000 Ω to ensure that the transistor will be biased correctly. The small signal gain is defined to be

$$A_v = \frac{v_o}{v_{in}}$$

The challenge is to design the amplifier so that

$$A_v = -20$$

(There is no guarantee that these specifications can be satisfied. Part of the problem is determining whether it is possible to design this amplifier so that $A_v = -20$.)

To solve this problem, we need to define h parameters that describe the transistor. We will return to this problem at the end of the chapter.

17.2 AMPLIFIERS AND FILTERS

Continued improvements in vacuum tubes and amplifier circuits made long-distance telephone lines possible, and in 1915 Alexander Graham Bell placed the first transcontinental telephone call to his famous assistant Thomas Watson. In 1921 Bell Telephone was experimenting with a 1000-mile telephone line with three voice channels, but the nonlinearity of even the best tubes introduced an intolerable amount of distortion.

Transmission of messages over long lines required the insertion of amplifiers at points along the line. These devices have two terminals for input and two terminals for output to the line. Harold S. Black, a 23-year-old engineer at Bell Laboratories, concluded that, in a rapidly growing country 4000 miles wide, a new approach would be required. First he tried to improve the amplifier tubes, but he decided that the necessary 1000-fold reduction in distortion could never be achieved that way. After hearing an inspiring talk by Charles Steinmetz, he clearly stated his problem: how to remove all the distortion from an imperfect amplifier. His first scheme was to compare the output (suitably reduced) to the input, amplify the difference (distortion) separately, and use it to cancel the distortion in the actual output. Within one day he had built a working *feed forward* amplifier.

But this 1923 invention required very precise balances and subtractions. For example, every hour on the hour somebody had to adjust the filament current to the tubes. For four

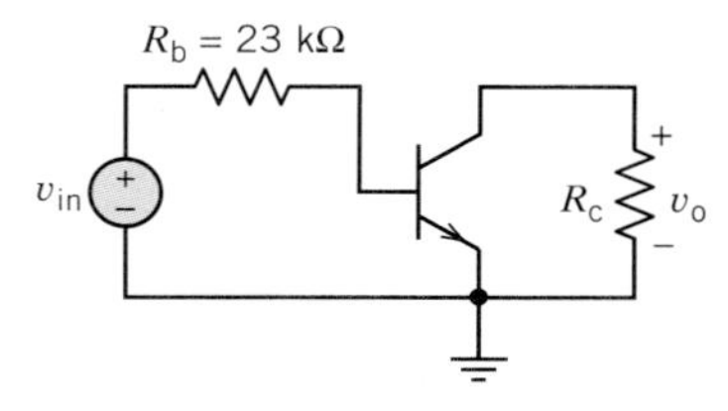

Figure 17.1-1
A transistor amplifier.

years, Black struggled, and failed, to turn his idea into a practical amplifier. As he related 50 years later in the December 1977 IEEE *Spectrum:*

Then came the morning of Tuesday, August 2, 1927, when the concept of the negative feedback amplifier came to me in a flash while I was crossing the Hudson River on the Lackawanna Ferry, on my way to work. I suddenly realized that if I fed the amplifier output back to the input, in reverse phase, I would have exactly what I wanted. . . . On a page of the *New York Times,* I sketched a simple diagram . . . and the equations for amplification with feedback.

Four months later, his goal was surpassed when a 100,000-to-1 reduction of distortion was realized in a practical one-stage amplifier. Now Black's negative feedback principle is applied in practically all amplifiers.

17.3 TWO-PORT NETWORKS

Many practical circuits have just two *ports* of access, that is, two places where signals may be input or output. For example, a coaxial cable between Boston and San Francisco has two ports, one at each of those cities. The object here is to analyze such networks in terms of their terminal characteristics without particular regard to the internal composition of the network. To this end, the network will be described by relationships between the port voltages and currents.

We study two-ports and the parameters that describe them for a number of reasons. Most circuits or systems have at least two ports. We may put an input signal into one port and obtain an output signal from the other. The parameters of the two-port network completely describe its behavior in terms of the voltage and current at each port. Thus, knowing the parameters of a two-port network permits us to describe its operation when it is connected into a larger network. Two-port networks are also important in modeling electronic devices and system components. For example, in electronics, two-port networks are employed to model transistors, op amps, transformers, and transmission lines.

A two-port network is represented by the network shown in Figure 17.3-1. A four-terminal network is called a *two-port network* when the current entering one terminal of a pair exits the other terminal in the pair. For example, I_1 enters terminal a and exits terminal b of the input terminal pair a–b. It will be assumed in our discussion that there are no independent sources or nonzero initial conditions within the linear two-port network. Two-port networks may or may not be purely resistive and can in general be formulated in terms of the s-variable or the $j\omega$-variable.

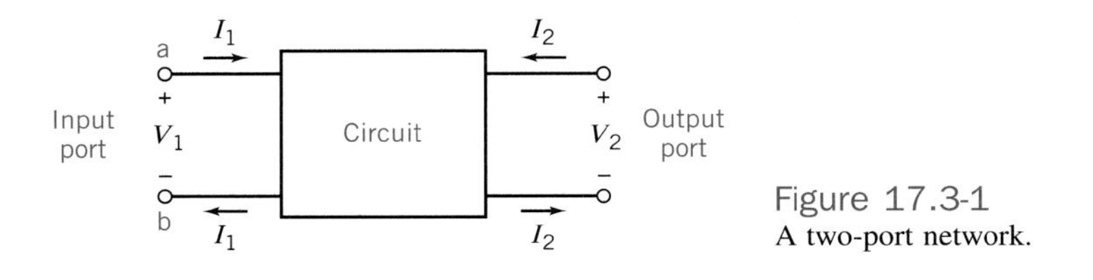

Figure 17.3-1
A two-port network.

A **two-port network** has two access points appearing as terminal pairs. The current entering one terminal of a pair exits the other terminal in the pair.

17.4 T-TO-Π TRANSFORMATION AND TWO-PORT THREE-TERMINAL NETWORKS

Two networks that occur frequently in circuit analysis are the T and Π networks as shown in Figure 17.4-1. When redrawn, they can appear as the Y or delta (Δ) networks of Figure 17.4-2.

If a network has mirror-image symmetry with respect to some centerline, that is, if a line can be found to divide the network into two symmetrical halves, the network is a *symmetrical network*. The T network is symmetrical when $Z_1 = Z_2$, and the Π network is symmetrical when $Z_A = Z_B$. Furthermore, if all the impedances in either the T or Π network are equal, then the T or Π network is *completely symmetrical*.

Note that the networks shown in Figure 17.4-1 and Figure 17.4-2 have two access ports and three terminals. For example, one port is obtained for the terminal pair a–c and the other port is b–c.

We can obtain equations for direct transformation or conversion from a T network to a Π network, or from a Π network to a T network, by considering that for equivalence the two networks must have the same impedance when measured between the same pair of terminals. For example, at port 1 (at a–c) for the two networks of Figure 17.4-2, we require

$$Z_1 + Z_3 = \frac{Z_A(Z_B + Z_C)}{Z_A + Z_B + Z_C}$$

To convert a Π network to a T network, relationships for Z_1, Z_2, and Z_3 must be obtained in terms of the impedances Z_A, Z_B, and Z_C. With some algebraic effort we can show that

$$Z_1 = \frac{Z_A Z_C}{Z_A + Z_B + Z_C} \tag{17.4-1}$$

$$Z_2 = \frac{Z_B Z_C}{Z_A + Z_B + Z_C} \tag{17.4-2}$$

$$Z_3 = \frac{Z_A Z_B}{Z_A + Z_B + Z_C} \tag{17.4-3}$$

Similarly, we can obtain the relationships for Z_A, Z_B, and Z_C as

$$Z_A = \frac{Z_1 Z_2 + Z_2 Z_3 + Z_3 Z_1}{Z_2} \tag{17.4-4}$$

$$Z_B = \frac{Z_1 Z_2 + Z_2 Z_3 + Z_3 Z_1}{Z_1} \tag{17.4-5}$$

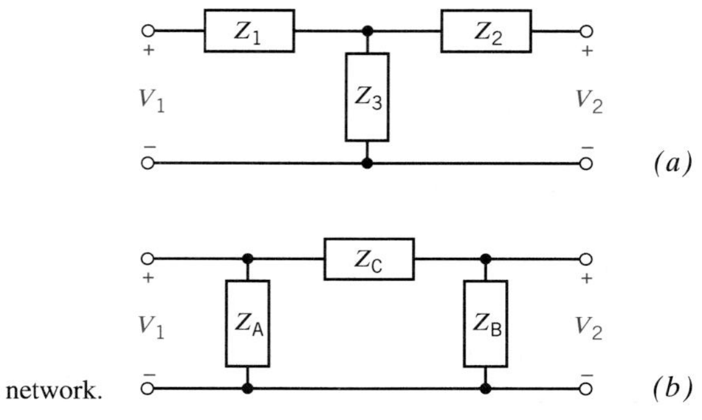

Figure 17.4-1
(*a*) T network and (*b*) Π network.

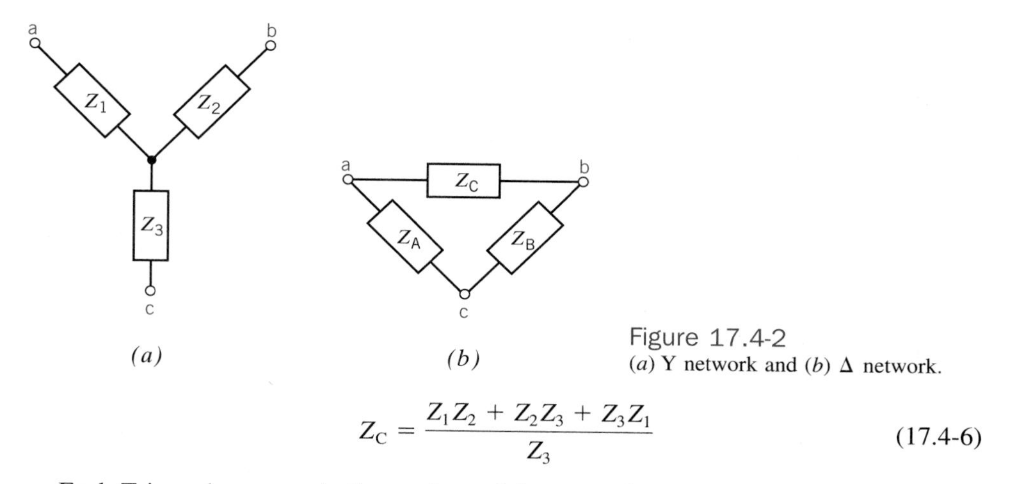

Figure 17.4-2
(*a*) Y network and (*b*) Δ network.

$$Z_C = \frac{Z_1 Z_2 + Z_2 Z_3 + Z_3 Z_1}{Z_3} \tag{17.4-6}$$

Each T impedance equals the product of the two adjacent legs of the Π network divided by the sum of the three legs of the Π network. On the other hand, each leg of the Π network equals the sum of the possible products of the T impedances divided by the opposite T impedance.

When a T or a Π network is completely symmetrical, the conversion equations reduce to

$$Z_T = \frac{Z_\Pi}{3} \tag{17.4-7}$$

and

$$Z_\Pi = 3Z_T \tag{17.4-8}$$

where Z_T is the impedance in each leg of the T network and Z_Π is the impedance in each leg of the Π network.

Example 17.4-1

Find the Π form of the T circuit given in Figure 17.4-3*a*.

Solution

The first impedance of the Π network, using Eq. 17.4-4, is

$$\begin{aligned} Z_A &= \frac{Z_1 Z_2 + Z_2 Z_3 + Z_3 Z_1}{Z_2} \\ &= \frac{j5(-j5) + (-j5)1 + 1(j5)}{-j5} \\ &= j5\ \Omega \end{aligned}$$

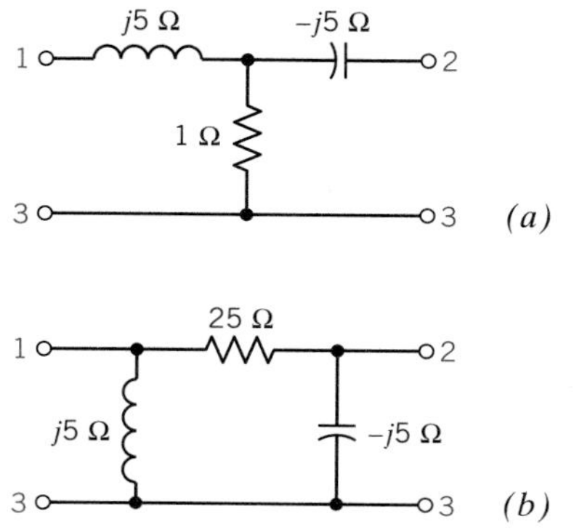

Figure 17.4-3
(*a*) T circuit of Example 17.4-1. (*b*) Π equivalent of T circuit.

Similarly, the second impedance using Eq. 17.4-5 is

$$Z_B = -j5\ \Omega$$

and the third impedance using Eq. 17.4-6 is

$$Z_C = 25\ \Omega$$

The Π equivalent circuit is shown in Figure 17.4-3*b*.

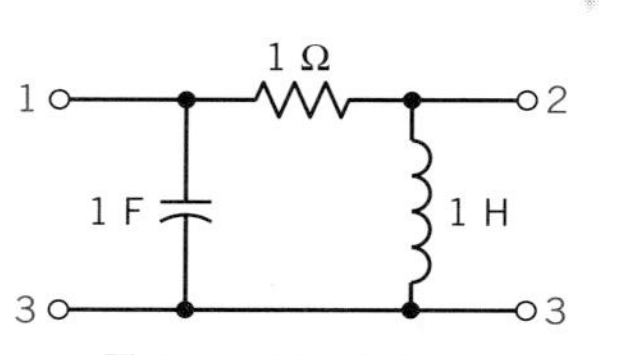

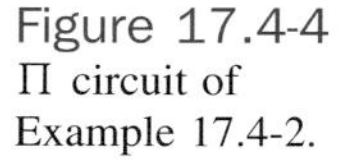

Figure 17.4-4
Π circuit of Example 17.4-2.

Example 17.4-2

Find the T network equivalent to the Π network shown in Figure 17.4-4 in the *s*-domain using the Laplace transform. Then, for $s = j1$, find the elements of the T network.

Solution

First, using Eq. 17.4-1, we have

$$Z_1 = \frac{(1)(1/s)}{s + 1 + 1/s} = \frac{1}{s^2 + s + 1}$$

Then, using Eq. 17.4-2, we have

$$Z_2 = \frac{1(s)}{s + 1 + 1/s} = \frac{s^2}{s^2 + s + 1}$$

Finally, the third impedance is (Eq. 17.4-3)

$$Z_3 = \frac{s(1/s)}{s + 1 + 1/s} = \frac{s}{s^2 + s + 1}$$

To find the elements of the T network at $s = j1$, we substitute $s = j1$ and determine each impedance. Then, we have

$$Z_1 = -j, \quad Z_2 = j, \quad Z_3 = 1$$

Therefore, the equivalent T network is as shown in Figure 17.4-5 for the value $s = j1$.

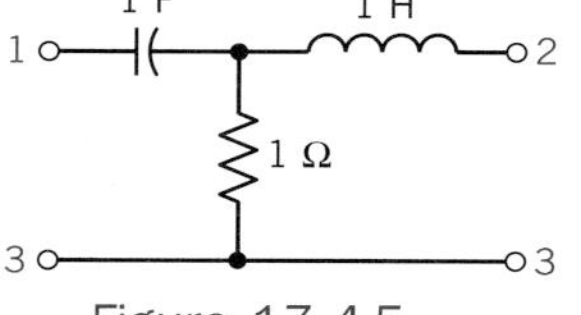

Figure 17.4-5
T circuit equivalent of the original Π circuit of Example 17.4-2 for $s = j1$.

Exercise 17.4-1 Find the T circuit equivalent to the Π circuit shown in Figure E 17.4-1.

Answers: $R_1 = 10\ \Omega$, $R_2 = 12.5\ \Omega$, $R_3 = 50\ \Omega$

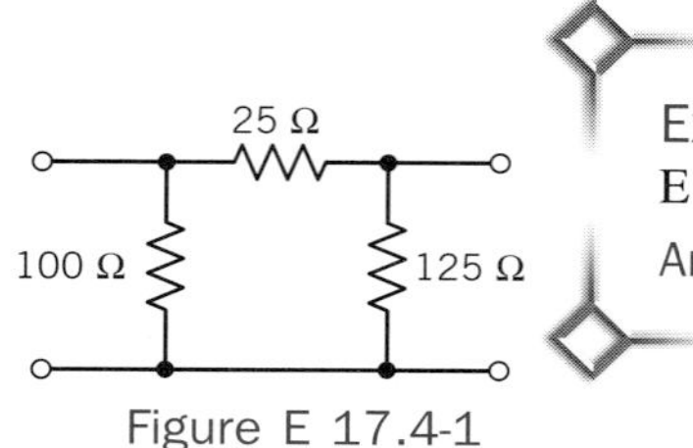

Figure E 17.4-1

17.5 EQUATIONS OF TWO-PORT NETWORKS

Let us consider the two-port network of Figure 17.3-1. By convention I_1 and I_2 are assumed to be flowing into the network as shown. The variables are V_1, V_2, I_1, and I_2. Within the two-port network two variables are independent and two are dependent, and we may select a set of two independent variables from the six possible sets: (V_1, V_2), (I_1, I_2), (V_1, I_2), (I_1, V_2), (V_1, I_1), and (V_2, I_2). We will also assume linear elements.

Table 17.5-1 **Six Circuit-Parameter Models**

Independent Variables (inputs)	Dependent Variables (outputs)	Circuit Parameters
I_1, I_2	V_1, V_2	Impedance Z
V_1, V_2	I_1, I_2	Admittance Y
V_1, I_2	I_1, V_2	Inverse hybrid g
I_1, V_2	V_1, I_2	Hybrid h
V_2, I_2	V_1, I_1	Transmission T
V_1, I_1	V_2, I_2	Inverse Transmission T'

The possibilities for independent (input) variables and the associated dependent variables are summarized in Table 17.5-1. The names of the associated six sets of circuit parameters are also identified in Table 17.5-1. For the case of phasor transforms or Laplace transforms with the circuit of Figure 17.3-1, we have the familiar impedance equations where the output variables are V_1 and V_2 as follows:

$$V_1 = Z_{11}I_1 + Z_{12}I_2 \tag{17.5-1}$$

$$V_2 = Z_{21}I_1 + Z_{22}I_2 \tag{17.5-2}$$

The equations for the admittances are

$$I_1 = Y_{11}V_1 + Y_{12}V_2 \tag{17.5-3}$$

$$I_2 = Y_{21}V_1 + Y_{22}V_2 \tag{17.5-4}$$

It is appropriate, if preferred, to use lowercase letters z and y for the coefficients of Eqs. 17.5-1 through 17.5-4. The equations for the six sets of two-port parameters are summarized in Table 17.5-2.

For linear elements and no dependent sources or op amps within the two-port network, we can show by the theorem of reciprocity that $Z_{12} = Z_{21}$ and $Y_{21} = Y_{12}$. One possible arrangement of a passive circuit as a T circuit is shown in Figure 17.5-1. Writing the two mesh equations for Figure 17.5-1, we can readily obtain Eqs. 17.5-1 and 17.5-2. Therefore, the circuit of Figure 17.5-1 can represent the impedance parameters. A possible arrangement of the admittance parameters as a Π circuit is shown in Figure 17.5-2.

Table 17.5-2 **Equations for the Six Sets of Two-Port Parameters**

Impedance Z	$\begin{cases} V_1 = Z_{11}I_1 + Z_{12}I_2 \\ V_2 = Z_{21}I_1 + Z_{22}I_2 \end{cases}$
Admittance Y	$\begin{cases} I_1 = Y_{11}V_1 + Y_{12}V_2 \\ I_2 = Y_{21}V_1 + Y_{22}V_2 \end{cases}$
Hybrid h	$\begin{cases} V_1 = h_{11}I_1 + h_{12}V_2 \\ I_2 = h_{21}I_1 + h_{22}V_2 \end{cases}$
Inverse hybrid g	$\begin{cases} I_1 = g_{11}V_1 + g_{12}I_2 \\ V_2 = g_{21}V_1 + g_{22}I_2 \end{cases}$
Transmission T	$\begin{cases} V_1 = AV_2 - BI_2 \\ I_1 = CV_2 - DI_2 \end{cases}$
Inverse transmission T'	$\begin{cases} V_2 = A'V_1 - B'I_1 \\ I_2 = C'V_1 - D'I_1 \end{cases}$

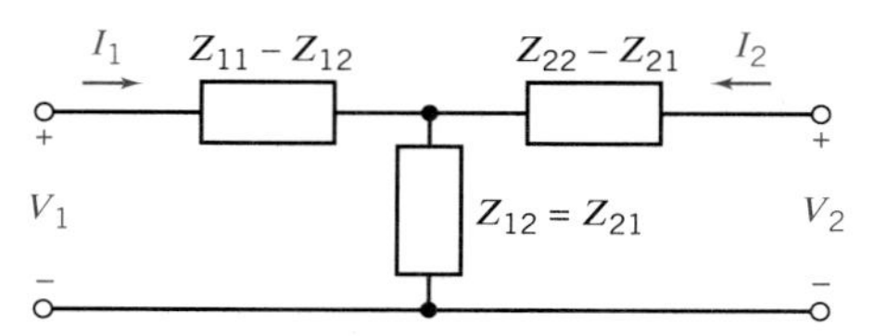

Figure 17.5-1 A T circuit representing the impedance parameters.

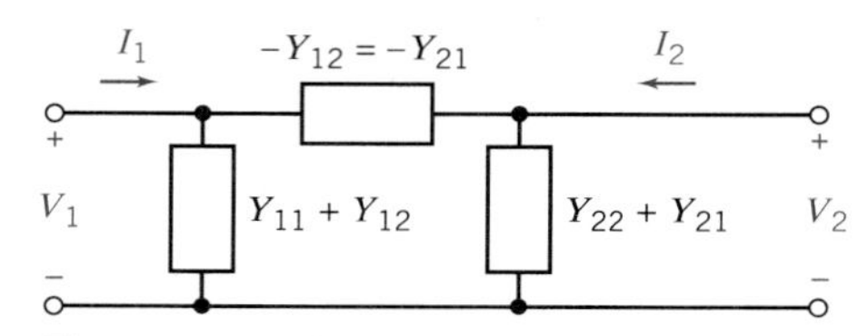

Figure 17.5-2 A Π circuit representing the admittance parameters.

Examining Eq. 17.5-1, we see that we can measure Z_{11} by obtaining

$$Z_{11} = \left.\frac{V_1}{I_1}\right|_{I_2=0}$$

Of course, $I_2 = 0$ implies that the output terminals are open-circuited. Thus, the Z parameters are often called *open-circuit impedances*.

The Y parameters can be measured by determining

$$Y_{12} = \left.\frac{I_1}{V_2}\right|_{V_1=0}$$

In general, the admittance parameters are called *short-circuit admittance parameters*.

Example 17.5-1

Determine the admittance and the impedance parameters of the T network shown in Figure 17.5-3.

Solution

The admittance parameters use the output terminals shorted and

$$Y_{11} = \left.\frac{I_1}{V_1}\right|_{V_2=0}$$

Then, the two 8-Ω resistors are in parallel and $V_1 = 28\, I_1$. Therefore, we have

$$Y_{11} = \frac{1}{28}\text{ S}$$

For Y_{12} we have

$$Y_{12} = \left.\frac{I_1}{V_2}\right|_{V_1=0}$$

so we short-circuit the input terminals. Then we have the circuit as shown in Figure 17.5-4.

Employing current division, we have

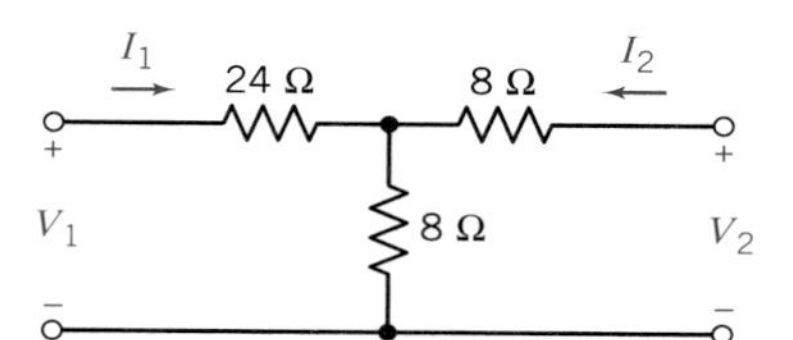

Figure 17.5-3 Circuit for Example 17.5-1.

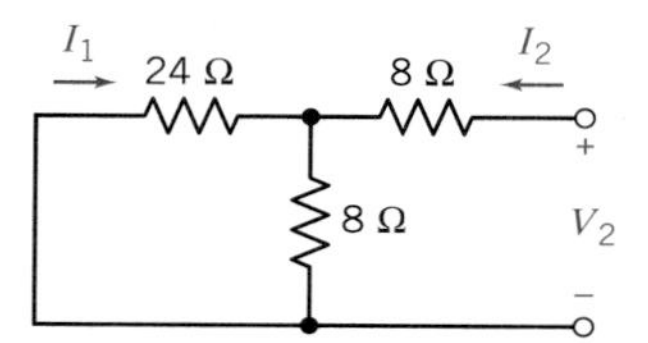

Figure 17.5-4 Circuit of Example 17.5-1 with the input terminals shorted.

$$-I_1 = I_2\left(\frac{8}{8 + 24}\right)$$

and
$$I_2 = \frac{V_2}{8 + [8(24)/(8 + 24)]} = \frac{V_2}{14}$$

Therefore
$$Y_{12} = \frac{I_1}{V_2} = \frac{-(V_2/14)(1/4)}{V_2} = -\frac{1}{56}\text{ S}$$

Furthermore,
$$Y_{21} = Y_{12} = -\frac{1}{56}\text{ S}$$

Finally, Y_{22} is obtained from Figure 17.5-4 as

$$Y_{22} = \left.\frac{I_2}{V_2}\right|_{V_1=0}$$

where
$$I_2 = \frac{V_2}{8 + [8(24)/(8 + 24)]} = \frac{V_2}{14}$$

Therefore,
$$Y_{22} = \frac{1}{14}\text{ S}$$

Thus, in matrix form we have $\mathbf{I} = \mathbf{YV}$ or

$$\begin{bmatrix} I_1 \\ I_2 \end{bmatrix} = \begin{bmatrix} \frac{1}{28} & -\frac{1}{56} \\ -\frac{1}{56} & \frac{1}{14} \end{bmatrix} \begin{bmatrix} V_1 \\ V_2 \end{bmatrix}$$

Now, let us find the impedance parameters. We have

$$Z_{11} = \left.\frac{V_1}{I_1}\right|_{I_2=0}$$

The output terminals are open circuited, so we have the circuit of Figure 17.5-3. Then

$$Z_{11} = 24 + 8 = 32\ \Omega$$

Similarly, $Z_{22} = 16\ \Omega$ and $Z_{12} = 8\ \Omega$. Then, in matrix form we have $\mathbf{V} = \mathbf{ZI}$ or

$$\begin{bmatrix} V_1 \\ V_2 \end{bmatrix} = \begin{bmatrix} 32 & 8 \\ 8 & 16 \end{bmatrix} \begin{bmatrix} I_1 \\ I_2 \end{bmatrix}$$

The general methods for finding the Z parameters and the Y parameters are summarized in Tables 17.5-3 and 17.5-4, respectively.

Table 17.5-3 **Method of Obtaining the Z Parameters of a Circuit**

Step IA	To determine Z_{11} and Z_{21}, connect a voltage source V_1 to the input terminals and open-circuit the output terminals.
Step IB	Find I_1 and V_2 and then $Z_{11} = V_1/I_1$ and $Z_{21} = V_2/I_1$.
Step IIA	To determine Z_{22} and Z_{12}, connect a voltage source V_2 to the output terminals and open-circuit the input terminals.
Step IIB	Find I_2 and V_1 and then $Z_{22} = V_2/I_2$ and $Z_{12} = V_1/I_2$.

Note: $Z_{12} = Z_{21}$ only when there are no dependent sources or op amps within the two-port network.

Table 17.5-4 **Method for Obtaining the Y Parameters of a Circuit**

Step IA	To determine Y_{11} and Y_{21}, connect a current source I_1 to the input terminals and short-circuit the output terminals.
Step IB	Find V_1 and I_2 and then $Y_{11} = I_1/V_1$ and $Y_{21} = I_2/V_1$.
Step IIA	To determine Y_{22} and Y_{12}, connect a current source I_2 to the output terminals and short-circuit the input terminals.
Step IIB	Find I_1 and V_2 and then $Y_{22} = I_2/V_2$ and $Y_{12} = I_1/V_2$.

Note: $Y_{12} = Y_{21}$ only when there are no dependent sources or op amps within the two-port network.

Exercise 17.5-1 Find the Z and Y parameters of the circuit of Figure E 17.5-1.

Answers: $\mathbf{Z} = \begin{bmatrix} 18 & 6 \\ 6 & 9 \end{bmatrix}$, $\quad \mathbf{Y} = \begin{bmatrix} \frac{1}{14} & -\frac{1}{21} \\ -\frac{1}{21} & \frac{1}{7} \end{bmatrix}$

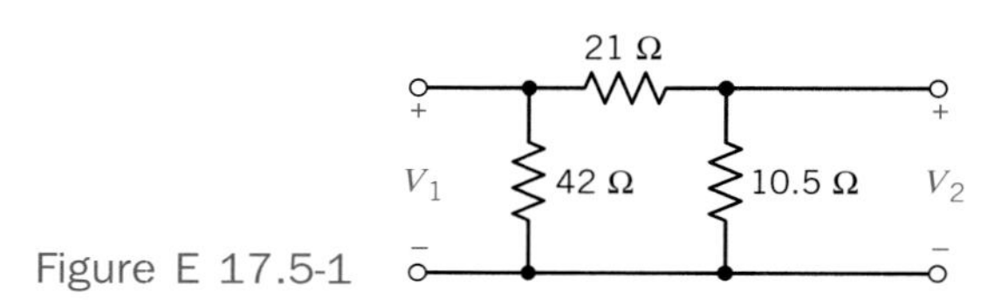

Figure E 17.5-1

17.6 Z AND Y PARAMETERS FOR A CIRCUIT WITH DEPENDENT SOURCES

When a circuit incorporates a dependent source, it is easy to use the methods of Table 17.5-3 or Table 17.5-4 to determine the Z or Y parameters. When a dependent source is within the circuit, $Z_{21} \neq Z_{12}$ and $Y_{12} \neq Y_{21}$.

Example 17.6-1

Determine the Z parameters of the circuit of Figure 17.6-1 when $m = \frac{2}{3}$.

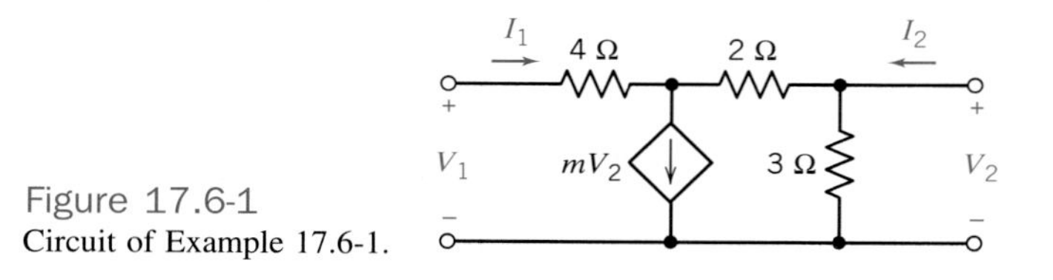

Figure 17.6-1
Circuit of Example 17.6-1.

Solution

We determine the Z parameters using the method of Table 17.5-3. Connect a voltage source V_1 and open-circuit the output terminals as shown in Figure 17.6-2a.

KCL at node a leads to

$$I_1 - mV_2 - I = 0 \tag{17.6-1}$$

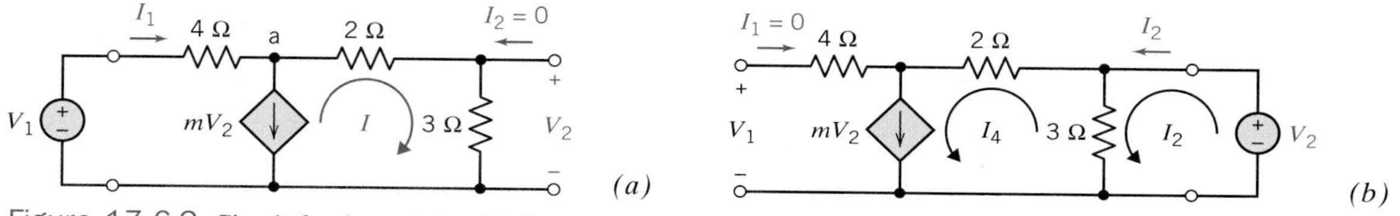

Figure 17.6-2 Circuit for determining (a) Z_{11} and Z_{21} and (b) Z_{22} and Z_{12}.

KVL around the outer loop is

$$V_1 = 4I_1 + 5I \tag{17.6-2}$$

Furthermore, $V_2 = 3I$, so $I = V_2/3$. Substituting $I = V_2/3$ into Eq. 17.6-1, we have

$$I_1 = mV_2 + \frac{V_2}{3} = (m + \tfrac{1}{3})V_2 \tag{17.6-3}$$

Therefore,

$$Z_{21} = \frac{V_2}{I_1} = 1\ \Omega$$

Substituting $I = V_2/3$ into Eq. 17.6-2, we obtain

$$V_1 = 4I_1 + \frac{5V_2}{3} = 4I_1 + \frac{5}{3}I_1 \tag{17.6-4}$$

Therefore,

$$Z_{11} = \frac{V_1}{I_1} = \frac{17}{3}\ \Omega$$

To obtain Z_{22} and Z_{12}, we connect a voltage source V_2 to the output terminals and open-circuit the input terminals, as shown in Figure 17.6-2. We can write two mesh equations for the assumed current directions shown as

$$V_1 + 5I_4 - 3I_2 = 0 \tag{17.6-5}$$

and

$$V_2 + 3I_4 - 3I_2 = 0 \tag{17.6-6}$$

Furthermore, $I_4 = mV_2$ so substituting into Eq. 17.6-6, we have

$$V_2 + 3mV_2 - 3I_2 = 0$$

or

$$V_2 = \frac{3}{3}I_2$$

Therefore,

$$Z_{22} = \frac{V_2}{I_2} = 1\ \Omega$$

Substituting $I_4 = mV_2$ into Eq. 17.6-5, we have

$$V_1 + 5mV_2 = 3I_2$$

or

$$V_1 + 5mI_2 = 3I_2$$

Therefore,

$$Z_{12} = \frac{V_1}{I_2} = (3 - 5m) = -\frac{1}{3}\ \Omega$$

Then, in summary, we have

$$\mathbf{Z} = \begin{bmatrix} \dfrac{17}{3} & -\dfrac{1}{3} \\ 1 & 1 \end{bmatrix}$$

Note that $Z_{21} \neq Z_{12}$, since a dependent source is present within the circuit.

Exercise 17.6-1 Determine the Y parameters of the circuit of Figure 17.6-1.

Answer: $Y = \begin{bmatrix} \frac{1}{6} & \frac{1}{18} \\ -\frac{1}{6} & \frac{17}{18} \end{bmatrix}$

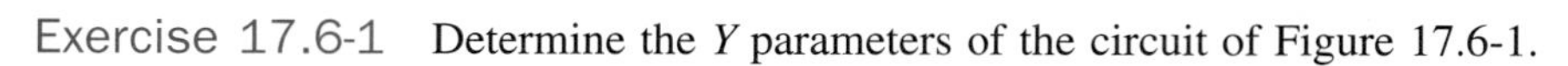

17.7 HYBRID AND TRANSMISSION PARAMETERS

The two-port hybrid parameter equations are based on V_1 and I_2 as the output variables, so that

$$V_1 = h_{11}I_1 + h_{12}V_2 \tag{17.7-1}$$

$$I_2 = h_{21}I_1 + h_{22}V_2 \tag{17.7-2}$$

or, in matrix form,

$$\begin{bmatrix} V_1 \\ I_2 \end{bmatrix} = \begin{bmatrix} h_{11} & h_{12} \\ h_{21} & h_{22} \end{bmatrix} \begin{bmatrix} I_1 \\ V_2 \end{bmatrix} = \mathbf{H} \begin{bmatrix} I_1 \\ V_2 \end{bmatrix} \tag{17.7-3}$$

These parameters are used widely in transistor circuit models. The hybrid circuit model is shown in Figure 17.7-1.

The inverse hybrid parameter equations are

$$I_1 = g_{11}V_1 + g_{12}I_2 \tag{17.7-4}$$

$$V_2 = g_{21}V_1 + g_{22}I_2 \tag{17.7-5}$$

or, in matrix form,

$$\begin{bmatrix} I_1 \\ V_2 \end{bmatrix} = \begin{bmatrix} g_{11} & g_{12} \\ g_{21} & g_{22} \end{bmatrix} \begin{bmatrix} V_1 \\ I_2 \end{bmatrix} = \mathbf{G} \begin{bmatrix} V_1 \\ I_2 \end{bmatrix} \tag{17.7-6}$$

The inverse hybrid circuit model is shown in Figure 17.7-2.

The hybrid and inverse hybrid parameters include both impedance and admittance parameters and are thus called *hybrid*. The parameters h_{11}, h_{12}, h_{21}, and h_{22} represent the short-circuit input impedance, the open-circuit reverse voltage gain, the short-circuit forward current gain, and the open-circuit output admittance, respectively. The parameters g_{11}, g_{12}, g_{21}, and g_{22} represent the short-circuit transfer admittance, the open-circuit forward current gain, the open-circuit forward voltage gain, and the short-circuit output impedance, respectively.

The transmission parameters are written as

$$V_1 = AV_2 - BI_2 \tag{17.7-7}$$

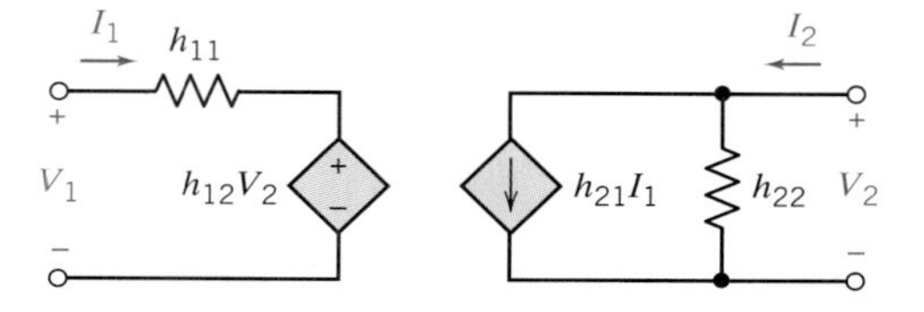

Figure 17.7-1
The h-parameter model of a two-port circuit.

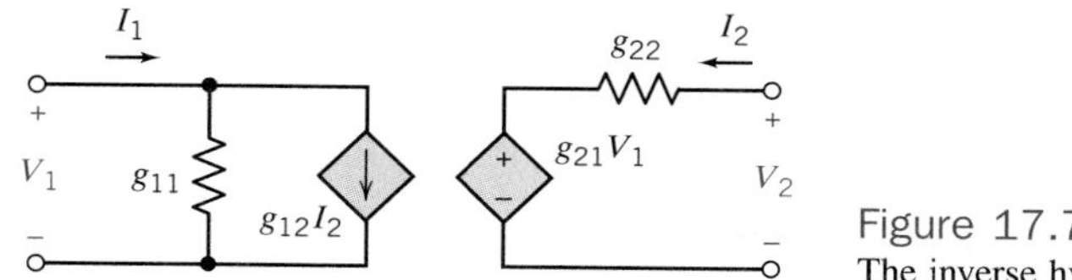

Figure 17.7-2
The inverse hybrid circuit (g-parameter) model.

$$I_1 = CV_2 - DI_2 \tag{17.7-8}$$

or in matrix form, as

$$\begin{bmatrix} V_1 \\ I_1 \end{bmatrix} = \begin{bmatrix} A & B \\ C & D \end{bmatrix} \begin{bmatrix} V_2 \\ -I_2 \end{bmatrix} = \mathbf{T} \begin{bmatrix} V_2 \\ -I_2 \end{bmatrix} \tag{17.7-9}$$

Transmission parameters are used to describe cable, fiber, and line transmission. The transmission parameters A, B, C, and D represent the open-circuit voltage ratio, the negative short-circuit transfer impedance, the open-circuit transfer admittance, and the negative short-circuit current ratio, respectively. The transmission parameters are often referred to as the $ABCD$ parameters. We are primarily interested in the hybrid and transmission parameters, since they are widely used.

Example 17.7-1

(a) Find the h parameters for the T circuit of Figure 17.7-3 in terms of R_1, R_2, and R_3.
(b) Evaluate the parameters when $R_1 = 1\ \Omega$, $R_2 = 4\ \Omega$, and $R_3 = 6\ \Omega$.

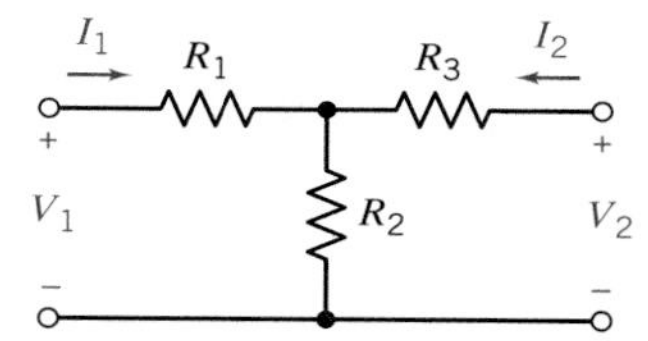

Figure 17.7-3 The T circuit of Example 17.7-1.

Solution

(a) First, we find h_{11} and h_{21} by short-circuiting the output terminals and connecting an input current source I_1, as shown in Figure 17.7-4a. Therefore,

$$h_{11} = \left.\frac{V_1}{I_1}\right|_{V_2=0} = R_1 + \frac{R_2R_3}{R_2 + R_3}$$

Then, using the current divider principle, we have

$$I_2 = \frac{-R_2}{R_2 + R_3} I_1$$

Therefore,

$$h_{21} = \left.\frac{I_2}{I_1}\right|_{V_2=0} = \frac{-R_2}{R_2 + R_3}$$

The next step is to redraw the circuit with $I_1 = 0$ and to connect the voltage source V_2 as shown in Figure 17.7-4b. Then we may determine h_{12} by using the voltage divider principle, as follows:

$$h_{12} = \left.\frac{V_1}{V_2}\right|_{I_1=0} = \frac{R_2}{R_2 + R_3}$$

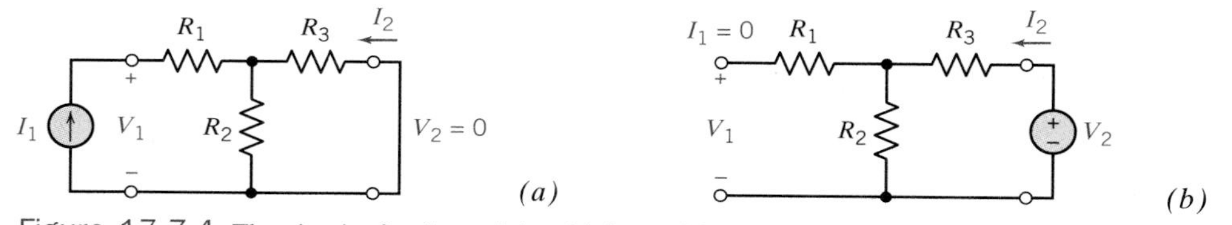

Figure 17.7-4 The circuits for determining (a) h_{11} and h_{21} and (b) h_{22} and h_{12}.

Finally, we determine h_{22} from Figure 17.7-4*b* as

$$h_{22} = \left.\frac{I_2}{V_2}\right|_{I_1=0} = \frac{1}{R_2 + R_3}$$

It is a property of a passive circuit (no op amps or dependent sources within the two-port) that $h_{12} = -h_{21}$. (b) When $R_1 = 1\ \Omega$, $R_2 = 4\ \Omega$, and $R_3 = 6\ \Omega$, we have

$$h_{11} = R_1 + \frac{R_2 R_3}{R_2 + R_3} = 3.4\ \Omega$$

$$h_{21} = \frac{-R_2}{R_2 + R_3} = -0.4$$

$$h_{12} = \frac{R_2}{R_2 + R_3} = 0.4$$

$$h_{22} = \frac{1}{R_2 + R_3} = 0.1\ \text{S}$$

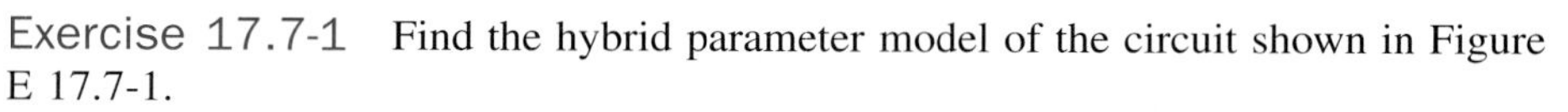

Exercise 17.7-1 Find the hybrid parameter model of the circuit shown in Figure E 17.7-1.

Answers: $h_{11} = 0.9\ \Omega$, $h_{12} = 0.1$, $h_{21} = 4.4$, $h_{22} = 0.6\ \text{S}$

Figure E 17.7-1

17.8 RELATIONSHIPS BETWEEN TWO-PORT PARAMETERS

If all the two-port parameters for a circuit exist, it is possible to relate one set of parameters to another, since the variables V_1, I_1, V_2, and I_2 are interrelated by the parameters. First, let us consider the relation between the Z parameters and the Y parameters. The matrix equation for the Z parameters is $\mathbf{V} = \mathbf{ZI}$ or

$$\begin{bmatrix} V_1 \\ V_2 \end{bmatrix} = \mathbf{Z} \begin{bmatrix} I_1 \\ I_2 \end{bmatrix} \tag{17.8-1}$$

Similarly, the equation for the Y parameters is $\mathbf{I} = \mathbf{YV}$ or

$$\begin{bmatrix} I_1 \\ I_2 \end{bmatrix} = \mathbf{Y} \begin{bmatrix} V_1 \\ V_2 \end{bmatrix} \tag{17.8-2}$$

Substituting for $\mathbf{I}$ from Eq. 17.8-2 into Eq. 17.8-1, we obtain

$$\mathbf{V} = \mathbf{ZYV}$$

or

$$\mathbf{Z} = \mathbf{Y}^{-1} \tag{17.8-3}$$

Thus, we can obtain the matrix $\mathbf{Z}$ by inverting the $\mathbf{Y}$ matrix. Of course, we can likewise obtain the $\mathbf{Y}$ matrix if we invert a known $\mathbf{Z}$ matrix. It is possible that a two-port network has a $\mathbf{Y}$ matrix or a $\mathbf{Z}$ matrix but not both. In other words, $\mathbf{Z}^{-1}$ or $\mathbf{Y}^{-1}$ may not exist for some networks.

If we have a known $\mathbf{Y}$ matrix, we obtain the $\mathbf{Z}$ matrix by finding the determinant of the $\mathbf{Y}$ matrix as ΔY and the adjoint of the $\mathbf{Y}$ matrix as

Table 17.8-1 **Parameter Relationships**

	Z		Y		h		g		T	
Z	Z_{11}	Z_{12}	$\frac{Y_{22}}{\Delta Y}$	$\frac{-Y_{12}}{\Delta Y}$	$\frac{\Delta h}{h_{22}}$	$\frac{h_{12}}{h_{22}}$	$\frac{1}{g_{11}}$	$\frac{-g_{12}}{g_{11}}$	$\frac{A}{C}$	$\frac{\Delta T}{C}$
	Z_{21}	Z_{22}	$\frac{-Y_{21}}{\Delta Y}$	$\frac{Y_{11}}{\Delta Y}$	$\frac{-h_{21}}{h_{22}}$	$\frac{1}{h_{22}}$	$\frac{g_{21}}{g_{11}}$	$\frac{\Delta g}{g_{11}}$	$\frac{1}{C}$	$\frac{D}{C}$
Y	$\frac{Z_{22}}{\Delta Z}$	$\frac{-Z_{12}}{\Delta Z}$	Y_{11}	Y_{12}	$\frac{1}{h_{11}}$	$\frac{-h_{12}}{h_{11}}$	$\frac{\Delta g}{g_{22}}$	$\frac{g_{12}}{g_{22}}$	$\frac{D}{B}$	$\frac{-\Delta T}{B}$
	$\frac{-Z_{21}}{\Delta Z}$	$\frac{Z_{11}}{\Delta Z}$	Y_{21}	Y_{22}	$\frac{h_{21}}{h_{11}}$	$\frac{\Delta h}{h_{11}}$	$\frac{-g_{21}}{g_{22}}$	$\frac{1}{g_{22}}$	$\frac{-1}{B}$	$\frac{A}{B}$
h	$\frac{\Delta Z}{Z_{22}}$	$\frac{Z_{12}}{Z_{22}}$	$\frac{1}{Y_{11}}$	$\frac{-Y_{12}}{Y_{11}}$	h_{11}	h_{12}	$\frac{g_{22}}{\Delta g}$	$\frac{g_{12}}{\Delta g}$	$\frac{B}{D}$	$\frac{\Delta T}{D}$
	$\frac{-Z_{21}}{Z_{22}}$	$\frac{1}{Z_{22}}$	$\frac{Y_{21}}{Y_{11}}$	$\frac{\Delta Y}{Y_{11}}$	h_{21}	h_{22}	$\frac{-g_{21}}{\Delta g}$	$\frac{g_{11}}{\Delta g}$	$\frac{-1}{D}$	$\frac{C}{D}$
g	$\frac{1}{Z_{11}}$	$\frac{-Z_{12}}{Z_{11}}$	$\frac{\Delta Y}{Y_{22}}$	$\frac{Y_{12}}{Y_{22}}$	$\frac{h_{22}}{\Delta h}$	$\frac{-h_{12}}{\Delta h}$	g_{11}	g_{12}	$\frac{C}{A}$	$\frac{-\Delta T}{A}$
	$\frac{Z_{21}}{Z_{11}}$	$\frac{\Delta Z}{Z_{11}}$	$\frac{-Y_{21}}{Y_{22}}$	$\frac{1}{Y_{22}}$	$\frac{-h_{21}}{\Delta h}$	$\frac{h_{11}}{\Delta h}$	g_{21}	g_{22}	$\frac{1}{A}$	$\frac{B}{A}$
T	$\frac{Z_{11}}{Z_{21}}$	$\frac{\Delta Z}{Z_{21}}$	$\frac{-Y_{22}}{Y_{21}}$	$\frac{-1}{Y_{21}}$	$\frac{-\Delta h}{h_{21}}$	$\frac{-h_{11}}{h_{21}}$	$\frac{1}{g_{21}}$	$\frac{g_{22}}{g_{21}}$	A	B
	$\frac{1}{Z_{21}}$	$\frac{Z_{22}}{Z_{21}}$	$\frac{-\Delta Y}{Y_{21}}$	$\frac{-Y_{11}}{Y_{21}}$	$\frac{-h_{22}}{h_{21}}$	$\frac{-1}{h_{21}}$	$\frac{g_{11}}{g_{21}}$	$\frac{\Delta g}{g_{21}}$	C	D

$\Delta Z = Z_{11}Z_{22} - Z_{12}Z_{21}$, $\Delta Y = Y_{11}Y_{22} - Y_{12}Y_{21}$, $\Delta g = g_{11}g_{22} - g_{12}g_{22}$, $\Delta h = h_{11}h_{22} - h_{12}h_{21}$, $\Delta T = AD - BC$

$$\text{adj } \mathbf{Y} = \begin{bmatrix} Y_{22} & -Y_{12} \\ -Y_{21} & Y_{11} \end{bmatrix}$$

Then

$$\mathbf{Z} = \mathbf{Y}^{-1} = \frac{\text{adj } \mathbf{Y}}{\Delta Y} \tag{17.8-4}$$

where $\Delta Y = Y_{11}Y_{22} - Y_{12}Y_{21}$.

The two-port parameter conversion relationships for the Z, Y, h, g, and T parameters are provided in Table 17.8-1.

Example 17.8-1

Determine the Y and h parameters if

$$\mathbf{Z} = \begin{bmatrix} 18 & 6 \\ 6 & 9 \end{bmatrix}$$

Solution

First, we will determine the Y parameters by calculating the determinant as

$$\Delta Z = Z_{11}Z_{22} - Z_{12}Z_{21} = 18(9) - 6(6) = 126$$

Then, using Table 17.8-1, we obtain

$$Y_{11} = \frac{Z_{22}}{\Delta Z} = \frac{9}{126} = \frac{1}{14}\ \text{S}$$

$$Y_{12} = Y_{21} = \frac{-Z_{12}}{\Delta Z} = \frac{-1}{21}\ \text{S}$$

$$Y_{22} = \frac{Z_{11}}{\Delta Z} = \frac{18}{126} = \frac{1}{7}\ \text{S}$$

$$h_{11} = \frac{\Delta Z}{Z_{22}} = \frac{126}{9} = 14\ \Omega$$

$$h_{12} = \frac{Z_{12}}{Z_{22}} = \frac{6}{9} = \frac{2}{3}$$

$$h_{21} = \frac{-Z_{21}}{Z_{22}} = \frac{-6}{9} = \frac{-2}{3}$$

$$h_{22} = \frac{1}{Z_{22}} = \frac{1}{9}\ \text{S}$$

Exercise 17.8-1 Determine the Z parameters if the Y parameters are

$$Y = \begin{bmatrix} \frac{2}{15} & \frac{-1}{5} \\ \frac{-1}{10} & \frac{2}{5} \end{bmatrix}$$

The units are siemens.

Answers: $Z_{11} = 12\ \Omega,\quad Z_{12} = 6\ \Omega,\quad Z_{21} = 3\ \Omega,\quad Z_{22} = 4\ \Omega$

Exercise 17.8-2 Determine the T parameters from the Y parameters of Exercise 17.8-1.

Answers: $A = 4,\quad B = 10\ \Omega,\quad C = \frac{1}{3}\ \text{S},\quad D = \frac{4}{3}$

17.9 INTERCONNECTION OF TWO-PORT NETWORKS

It is common in many circuits to have several two-port networks interconnected in parallel or in cascade. The *parallel* connection of two two-ports shown in Figure 17.9-1 requires that the V_1 of each two-port be equal.

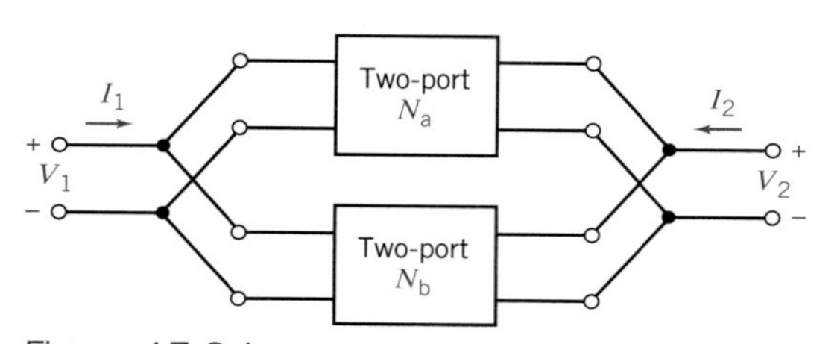

Figure 17.9-1 Parallel connection of two two-port networks.

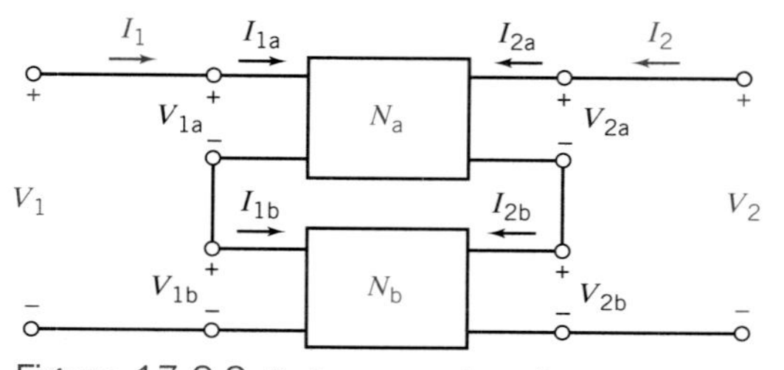

Figure 17.9-2 Series connection of two two-port networks.

Similarly, at the output port V_2 is the output voltage of both two-port networks. The defining matrix equation for network N_a is

$$\mathbf{I}_a = \mathbf{Y}_a\mathbf{V}_a \tag{17.9-1}$$

and for network N_b we have

$$\mathbf{I}_b = \mathbf{Y}_b\mathbf{V}_b \tag{17.9-2}$$

In addition, we have the total current $\mathbf{I}$ as

$$\mathbf{I} = \mathbf{I}_a + \mathbf{I}_b$$

Furthermore, since $\mathbf{V}_a = \mathbf{V}_b = \mathbf{V}$,

$$\mathbf{I} = \mathbf{Y}_a\mathbf{V} + \mathbf{Y}_b\mathbf{V} = (\mathbf{Y}_a + \mathbf{Y}_b)\mathbf{V} = \mathbf{Y}\mathbf{V}$$

Therefore, the Y parameters for the total network of two parallel two-ports are described by the matrix equation

$$\mathbf{Y} = \mathbf{Y}_a + \mathbf{Y}_b \tag{17.9-3}$$

For example,

$$Y_{11} = Y_{11a} + Y_{11b}$$

Hence, to determine the Y parameters for the total network, we add the Y parameters of each network. In general, the Y parameter matrix of the parallel connection is the sum of the Y parameter matrices of the individual two-ports.

The series interconnection of two two-port networks is shown in Figure 17.9-2. We will use the Z parameters to describe each two-port and the series combination. The two networks are described by the matrix equations

$$\mathbf{V}_a = \mathbf{Z}_a\mathbf{I}_a \tag{17.9-4}$$

and

$$\mathbf{V}_b = \mathbf{Z}_b\mathbf{I}_b \tag{17.9-5}$$

The terminal currents are

$$\mathbf{I} = \mathbf{I}_a = \mathbf{I}_b$$

Therefore, since $\mathbf{V} = \mathbf{V}_a + \mathbf{V}_b$, we have

$$\begin{aligned}\mathbf{V} &= \mathbf{Z}_a\mathbf{I}_a + \mathbf{Z}_b\mathbf{I}_b \\ &= (\mathbf{Z}_a + \mathbf{Z}_b)\mathbf{I} = \mathbf{Z}\mathbf{I}\end{aligned}$$

or

$$\mathbf{Z} = \mathbf{Z}_a + \mathbf{Z}_b \tag{17.9-6}$$

Therefore, the Z parameters for the total network are equal to the sum of the Z parameters for the networks.

When the output of one network is connected to the input port of the following network, as shown in Figure 17.9-3, the networks are said to be *cascaded*. Since the output variables of the first network become the input variables of the second network, the transmission parameters are utilized. The first two-port, N_a, is represented by the matrix equation

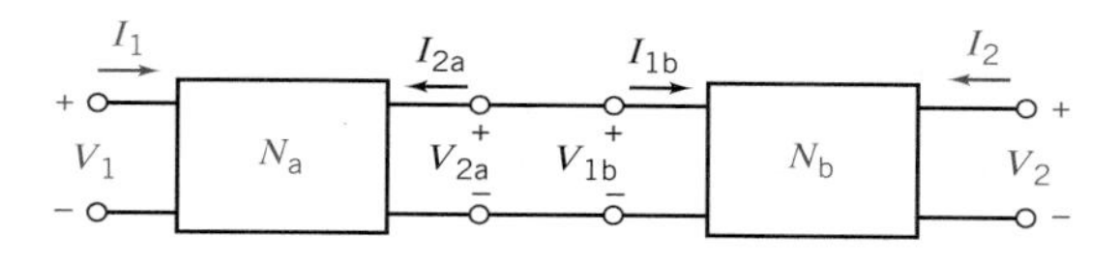

Figure 17.9-3
Cascade connection of two two-port networks.

$$\begin{bmatrix} V_{1a} \\ I_{1a} \end{bmatrix} = \mathbf{T}_a \begin{bmatrix} V_{2a} \\ -I_{2a} \end{bmatrix}$$

For N_b we have

$$\begin{bmatrix} V_{1b} \\ I_{1b} \end{bmatrix} = \mathbf{T}_b \begin{bmatrix} V_{2b} \\ -I_{2b} \end{bmatrix}$$

Furthermore, we note that at the input and output, we have

$$\begin{bmatrix} V_1 \\ I_1 \end{bmatrix} = \begin{bmatrix} V_{1a} \\ I_{1a} \end{bmatrix} \quad \text{and} \quad \begin{bmatrix} V_{2b} \\ -I_{2b} \end{bmatrix} = \begin{bmatrix} V_2 \\ -I_2 \end{bmatrix}$$

At the intermediate connection, we have

$$\begin{bmatrix} V_{2a} \\ -I_{2a} \end{bmatrix} = \begin{bmatrix} V_{1b} \\ I_{1b} \end{bmatrix}$$

Therefore,

$$\begin{bmatrix} V_1 \\ I_1 \end{bmatrix} = \mathbf{T}_a\mathbf{T}_b \begin{bmatrix} V_2 \\ -I_2 \end{bmatrix}$$

and

$$\mathbf{T} = \mathbf{T}_a\mathbf{T}_b \tag{17.9-7}$$

Hence, the transmission parameters for the overall network are derived by matrix multiplication, observing the proper order.

All of the preceding calculations for interconnected networks assume that the interconnection does not disturb the two-port nature of the individual sub-networks.

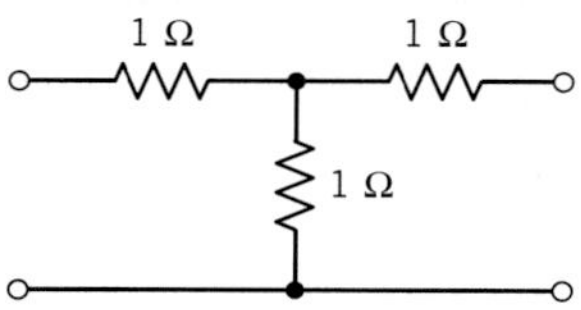

Example 17.9-1

For the T network of Figure 17.9-4, (a) find the Z, Y, and T parameters and (b) determine the resulting parameters after connecting two two-ports in parallel and in cascade. Both two-ports are identical as in Figure 17.9-4.

Figure 17.9-4
T network of Example 17.9-1.

Solution

First, we find the Z parameters of the T network. Examining the network, we have

$$Z_{12} = Z_{21} = 1\ \Omega$$
$$Z_{22} = Z_{11} = 2\ \Omega$$

Then, using the conversion factors of Table 17.8-1, we find

$$\mathbf{Y} = \begin{bmatrix} \frac{2}{3} & \frac{-1}{3} \\ \frac{-1}{3} & \frac{2}{3} \end{bmatrix}$$

and

$$\mathbf{T} = \begin{bmatrix} 2 & 3 \\ 1 & 2 \end{bmatrix}$$

Two identical networks connected in parallel will have a total $\mathbf{Y}$ matrix of

$$\mathbf{Y} = \mathbf{Y}_a + \mathbf{Y}_b$$

Since $\mathbf{Y}_a = \mathbf{Y}_b$, we have

$$\mathbf{Y} = 2Y_a = \begin{bmatrix} \frac{4}{3} & \frac{-2}{3} \\ \frac{-2}{3} & \frac{4}{3} \end{bmatrix}$$

Finally, when two identical networks are connected in cascade, we have a total **T** matrix of

$$\mathbf{T} = \mathbf{T}_a\mathbf{T}_b = \begin{bmatrix} 2 & 3 \\ 1 & 2 \end{bmatrix} \begin{bmatrix} 2 & 3 \\ 1 & 2 \end{bmatrix} = \begin{bmatrix} 7 & 12 \\ 4 & 7 \end{bmatrix}$$

Exercise 17.9-1 Determine the total transmission parameters of the cascade connection of three two-port networks shown in Figure E 17.9-1.

Answers: $A = 3,\ B = 21\ \Omega,\ C = \frac{1}{6}\ \text{S},\ D = \frac{3}{2}$

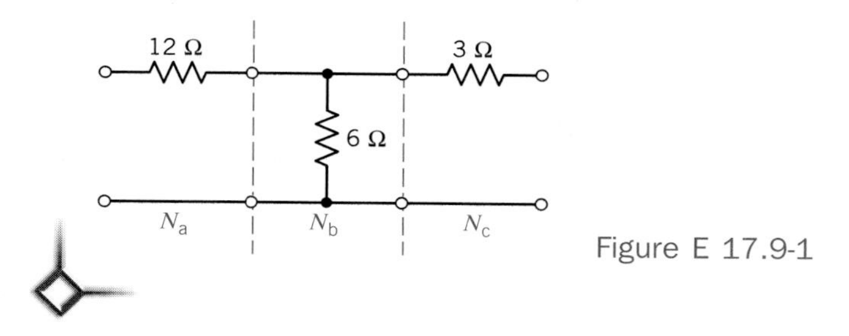

Figure E 17.9-1

17.10 USING PSPICE TO CALCULATE TWO-PORT PARAMETERS

PSpice can be used to calculate the two-port parameters. To do this, we use a 1-A or 1-V source while using open or short circuits to impose the necessary constraints. For example, let us determine the h parameters of the circuit shown in Figure 17.10-1. We recall that

$$h_{11} = \left.\frac{V_1}{I_1}\right|_{V_2=0}$$

and

$$h_{21} = \left.\frac{I_2}{I_1}\right|_{V_2=0}$$

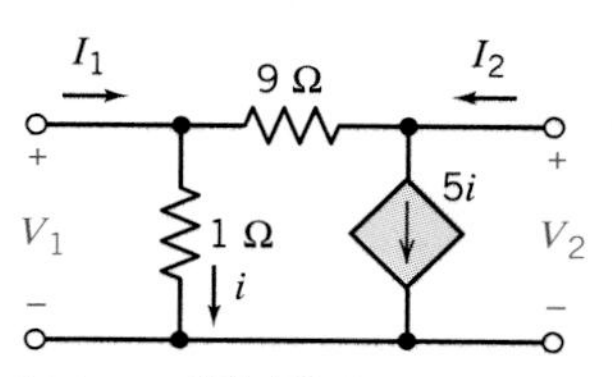

Figure 17.10-1 Two-port network.

Thus, we set $V_2 = 0$ by inserting a short circuit at the output and use $I_1 = 1$ A. These are the short-circuit parameters, and we use the .DC PSpice statement to find V_1 and I_2. Then $h_{11} = V_1$ and $h_{21} = I_2$.

Also, recall that

$$h_{12} = \left.\frac{V_1}{V_2}\right|_{I_1=0}$$

and

$$h_{22} = \left.\frac{I_2}{V_2}\right|_{I_1=0}$$

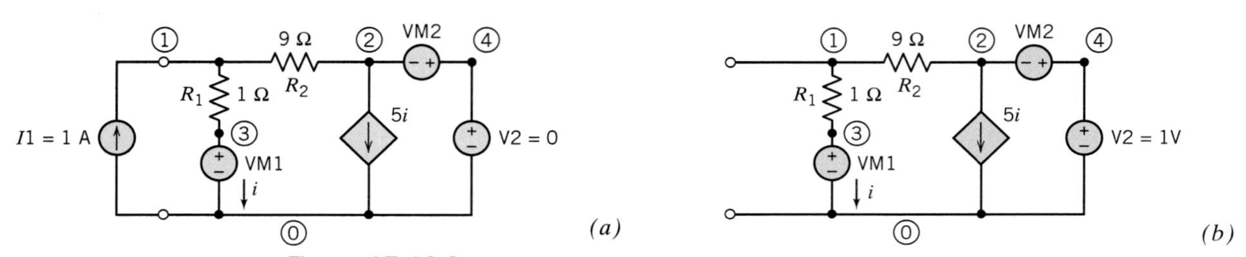

Figure 17.10-2 Two-port network redrawn to measure (*a*) h_{11} and h_{21} and (*b*) h_{12} and h_{22}.

We find these parameters by setting $I_1 = 0$ by open-circuiting the input terminals and inserting a source $V_2 = 1$ V at the output terminals. We then use another PSpice program to find V_1 and I_2. Then $h_{12} = V_1$ and $h_{22} = I_2$.

The circuit is redrawn to find h_{11} and h_{21} by short-circuiting the output as shown in Figure 17.10-2*a*. The dummy source VM2 is used to measure I_2, and the dummy source VM1 is used to measure i for the CCCS. The circuit to determine h_{12} and h_{22} is shown in Figure 17.10-2*b*, where $V_2 = 1$ V.

Since the two circuits redrawn to find the h parameters are relatively similar, we can readily write two PSpice programs from the circuits. The program to determine h_{11} and h_{21} is shown in Figure 17.10-3, and the program to determine h_{12} and h_{22} is shown in Figure 17.10-4.

After running the program of Figure 17.10-3, we find that $h_{11} = 0.9\ \Omega$ and $h_{21} = 4.4$. Using the program shown in Figure 17.10-4, we determine $h_{12} = 0.1$ S and $h_{22} = 0.6$.

17.11 VERIFICATION EXAMPLE

Example 17.11-1

The circuit shown in Figure 17.11-1*a* was designed to have a transfer function given by

$$\frac{V_o(s)}{V_{in}(s)} = \frac{2s - 10}{s^2 + 27s + 2}$$

Does the circuit satisfy this specification?

Solution

The h-parameter model from Figure 17.7-1 can be used to redraw the circuit as shown in Figure 17.11-1*b*. This circuit can be represented by node equations

```
CALCULATE H11 H21

I1     0    1    DC    1
R1     1    3    1
R2     1    2    9
V2     4    0    0  ;   SHORT CIRCUIT
E1     2    0    VM1  5
VM1     3    0    0    ;  MEASURE I
VM2     4    2    0    ;  MEASURE I2
.DC     I1    1    1    1
.PRINT     DC     V(1)    I(VM2)
.END
```

Figure 17.10-3 Program to determine h_{11} and h_{21}.

```
CALCULATE H12 H22

R1     1     3     1
R2     1     2     9
V2     4     0     DC    1
F1     2     0     VM1    5
VM1     3     0     0
VM2     4     2     0
.DC     V2     1     1     1
.PRINT     DC     V(1)     I(VM2)
.END
```

Figure 17.10-4 Program to determine h_{12} and h_{22}.

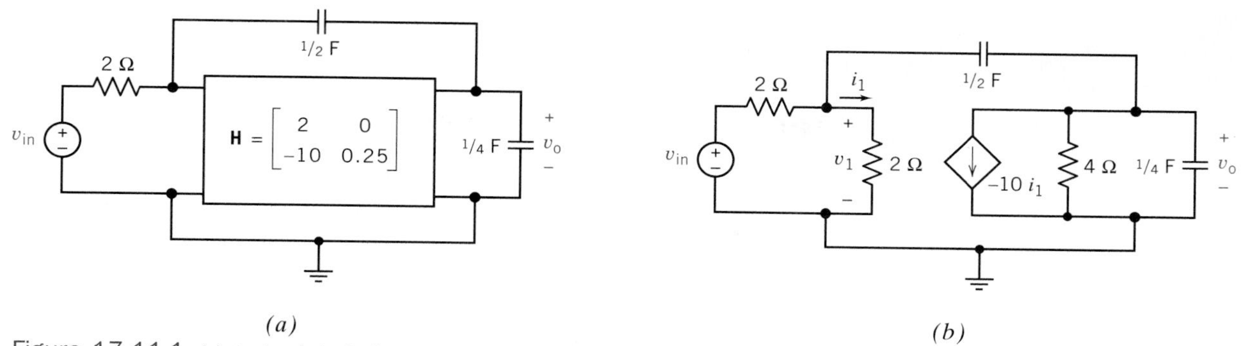

Figure 17.11-1 (*a*) A circuit including a two-port network. (*b*) Using the *h*-parameter model to represent the two-port network.

$$\begin{bmatrix} \left(1+\frac{s}{2}\right) & -\frac{s}{2} \\ \left(-5-\frac{s}{2}\right) & \left(\frac{3s}{4}+\frac{1}{4}\right) \end{bmatrix} \begin{bmatrix} V_1(s) \\ V_o(s) \end{bmatrix} = \begin{bmatrix} \frac{V_{in}(s)}{2} \\ 0 \end{bmatrix}$$

where $10I_1(s) = 5V_1(s)$ has been used to express the current of the dependent source in terms of the node voltages. Applying Cramer's rule gives

$$\frac{V_o(s)}{V_{in}(s)} = \frac{\frac{1}{2}\left(5+\frac{s}{2}\right)}{\left(1+\frac{s}{2}\right)\left(\frac{3s}{4}+\frac{1}{4}\right)-\frac{s}{2}\left(\frac{s}{2}+5\right)} = \frac{2s+20}{s^2-13s+2}$$

This is not the required transfer function, so the circuit does not satisfy the specification.

Exercise 17.11-1 Verify that the circuit shown in Figure E 17.11-1 does indeed have the transfer function

$$\frac{V_o(s)}{V_{in}(s)} = \frac{2s-10}{s^2+27s+2}$$

(The circuits in Figures 17.11-1*a* and E 17.11-1 differ only in the sign of h_{21}.)

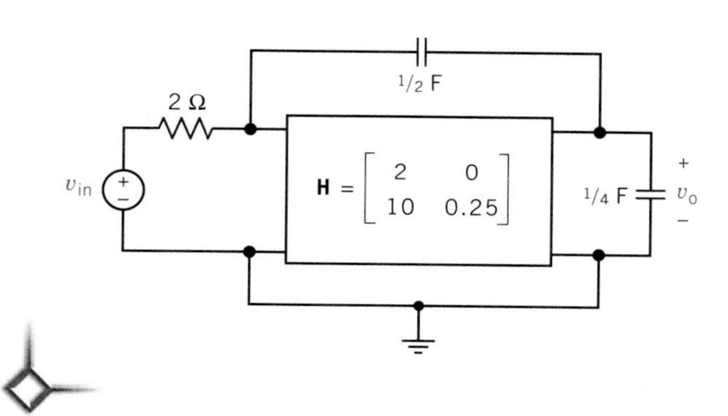

Figure E 17.11-1
A modified version of the circuit from Figure 17.11-1.

17.12 Design Challenge Solution

TRANSISTOR AMPLIFIER

Figure 17.12-1 shows the small signal equivalent circuit of a transistor amplifier. The data sheet for the transistor describes the transistor by specifying its h parameters to be

$$h_{ie} = 1250\ \Omega, \quad h_{oe} = 0, \quad h_{fe} = 100, \quad \text{and} \quad h_{re} = 0$$

The value of the resistance R_c must be between 300 Ω and 5000 Ω to ensure that the transistor will be biased correctly. The small signal gain is defined to be

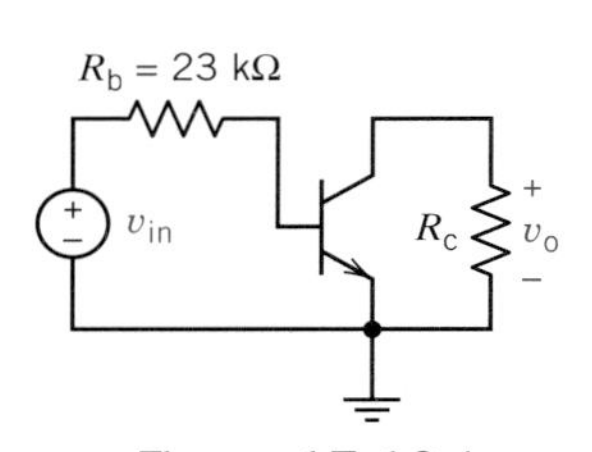

Figure 17.12-1 A transistor amplifier.

$$A_v = \frac{v_o}{v_{in}}$$

The challenge is to design the amplifier so that

$$A_v = -20$$

(There is no guarantee that these specifications can be satisfied. Part of the problem is to decide whether it is possible to design this amplifier so that $A_v = -20$.)

Describe the Situation and the Assumptions

1. R_c must be between 300 Ω and 5000 Ω.
2. The transistor is represented by h parameters. Figure 17.12-2a shows that the transistor can be configured to be a two-port network and represented by h parameters. Figure 17.12-2b shows an equivalent circuit for the transistor. This equivalent circuit is based on the h parameters. For this particular transistor, the values of the h parameters are

$$h_{ie} = 1000\ \Omega, \quad h_{oe} = 0, \quad h_{fe} = 100, \quad \text{and} \quad h_{re} = 0$$

Since

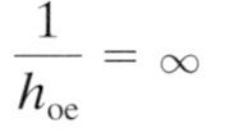

$$\frac{1}{h_{oe}} = \infty$$

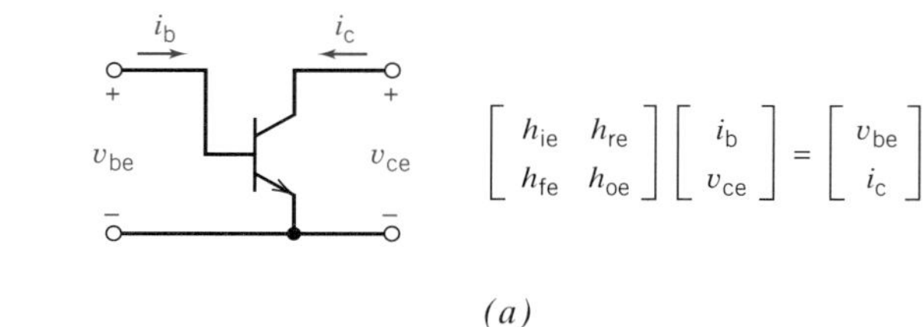

(a)

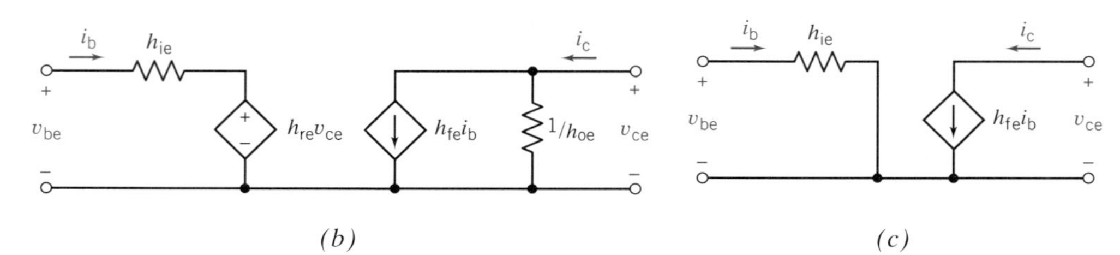

(b) (c)

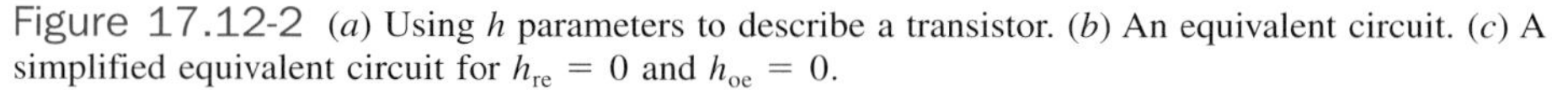

Figure 17.12-2 (a) Using h parameters to describe a transistor. (b) An equivalent circuit. (c) A simplified equivalent circuit for $h_{re} = 0$ and $h_{oe} = 0$.

the resistor at the left side of the equivalent circuit is an open circuit. Since

$$h_{re} = 0$$

the dependent voltage source is a short circuit. Figure 17.12-2*c* shows the equivalent circuit after these simplifications are made.

3. The voltage gain must be $A_v = -20$.

State the Goal

Select R_c so that $A_v = -20$.

Generate a Plan

Replace the transistor in Figure 17.12-1 by the equivalent circuit in Figure 17.12-2*c*. Analyze the resulting circuit to obtain a formula for the voltage gain, A_v. This formula will involve R_c. Determine the value of R_c that will make $A_v = -20$. If this value of R_c is between 300 Ω and 5000 Ω, then the amplifier design is complete. On the other hand, if this value of R_c is not between 300 Ω and 5000 Ω, then the specifications cannot be satisfied.

Act on the Plan

Figure 17.12-3 shows the amplifier after the transistor has been replaced by the equivalent circuit. Applying Ohm's law to R_c gives

$$v_o = -R_c 100 i_b$$

where the minus sign is due to reference directions. Next, apply KVL to the left mesh to get

$$v_{in} = 23{,}000 i_b + 1000 i_b$$

Then

$$A_v = \frac{v_o}{v_{in}} = \frac{-100\,R_c}{24{,}000}$$

Finally, set $A_v = -20$, obtaining

$$-20 = \frac{-100 R_c}{24{,}000}$$

Now solve for R_c to determine

$$R_c = 4800\ \Omega$$

Verify the Proposed Solution

First, the resistance $R_c = 4800\ \Omega$ is indeed between 300 Ω and 5000 Ω. Second, the gain of the circuit shown in Figure 17.12-3 is

$$\frac{v_o}{v_{in}} = \frac{-h_{fe} R_c}{R_b + h_{ie}} = -\frac{100 \times 4800}{23000 + 1000} = -20$$

Therefore, both specifications have been satisfied.

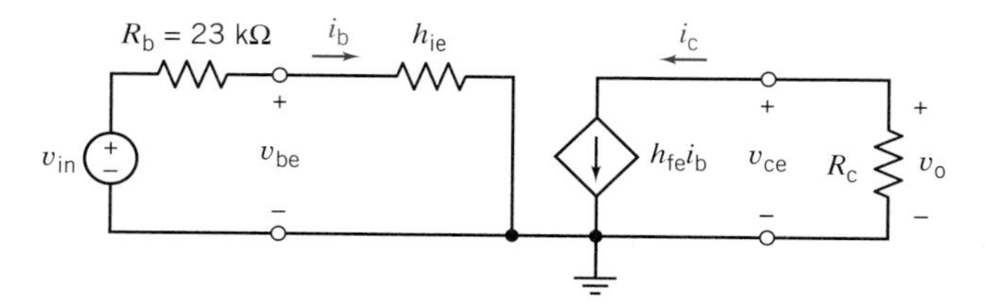

Figure 17.12-3
An equivalent circuit for the transistor amplifier.

17.13 SUMMARY

- A port is a pair of terminals together with the restriction that the current directed into one terminal is equal to the current directed out of the other terminal.
- Two-port models of circuits or devices are useful for describing the performance of the circuit or device in terms of the currents and voltages at its ports. The internal details of the circuit or device are not included in the two-port model, so the two-port model of a circuit may be considerably simpler than the circuit itself.
- The two-port model involves four signals—the current and voltage at each port. Two of these signals are treated as inputs and the other two are treated as outputs. There are six ways of separating the four signals into input and output signals, and so, six sets of two-port parameters. The six sets of two port parameters are called the impedance, admittance, hybrid, inverse hybrid, transmission, and inverse transmission parameters. Table 17.5-2 summarizes the six sets of two-port parameters.
- Table 17.8-1 summarized the equations used to convert one set of two-port parameters into another, e.g. to convert impedance parameters into hybrid parameters.
- We may use two-port parameters to describe the performance of the parallel, series, or cascade connection of two or more circuits.

PROBLEMS

Section 17.4 T-to-Π Transformation and Two-Port Three-Terminal Networks

P 17.4-1 Determine the equivalent resistance R_{ab} of the network of Figure P 17.4-1. Use the Π-to-T transformation as one step of the reduction.
Answer: $R_{ab} = 3.2\ \Omega$

P 17.4-2 Repeat Problem 17.4-1 when the 6-Ω resistance is changed to 4 Ω and the 10-Ω resistance is changed to 12 Ω.

P 17.4-3 The two-port network of Figure 17.3-1 has an input source V_s with a source resistance R_s connected to the input terminals so that $V_1 = V_s - I_1R_s$ and a load resistance connected to the output terminals so that $V_2 = -I_2R_L = I_LR_L$. Find $R_{in} = V_1/I_1$, $A_v = V_2/V_1$, $A_i = -I_2/I_1$, and $A_p = -V_2I_2/V_1I_1$ by using the Z-parameter model.

P 17.4-4 Using the Δ-to-Y transformation, determine the current I when $R_1 = 15\ \Omega$ and $R = 20\ \Omega$ for the circuit shown in Figure P 17.4-4.
Answer: $I = 385$ mA

P 17.4-5 Use the Y-to-Δ transformation to determine R_{in} of the circuit shown in Figure P 17.4-5.
Answer: $R_{in} = 673.85\ \Omega$

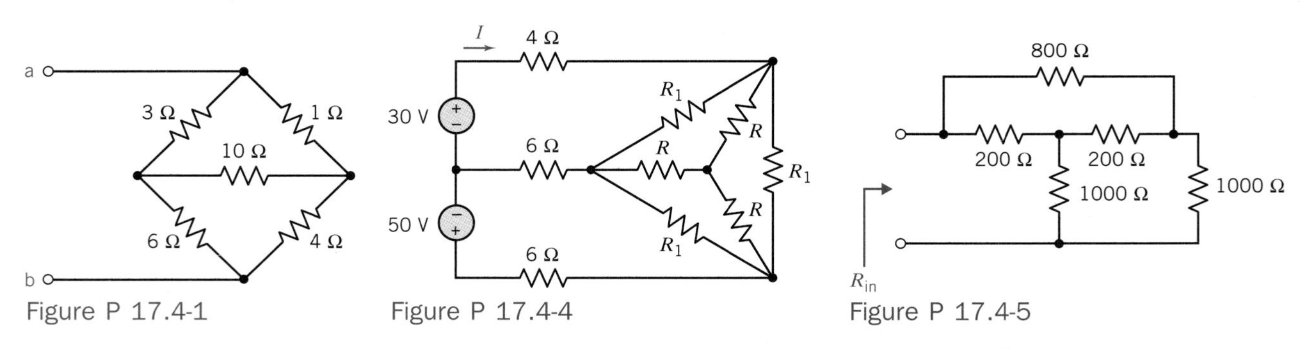

Figure P 17.4-1 Figure P 17.4-4 Figure P 17.4-5

Section 17.5 Equations of Two-Port Networks

P 17.5-1 Find the Y parameters and Z parameters for the two-port network of Figure P 17.5-1.

Figure P 17.5-1

P 17.5-2 Determine the Z parameters of the ac circuit shown in Figure P 17.5-2.
Answer: $Z_{11} = 2 - j4$, $Z_{12} = Z_{21} = -j4$, $Z_{22} = -j2$

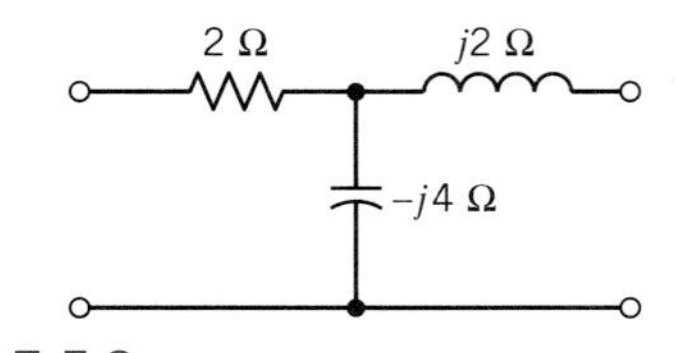

Figure P 17.5-2

P 17.5-3 Find the Y parameters of the circuit of Figure P 17.5-3 when $b = 4$, $G_1 = 2$ S, $G_2 = 1$ S, and $G_3 = 3$ S.

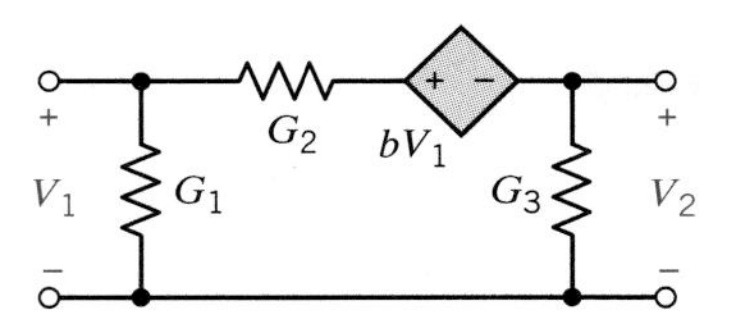

Figure P 17.5-3

P 17.5-4 Find the Y parameters for the circuit of Figure P 17.5-4.
Answer: $Y_{11} = 0.3$ S, $Y_{21} = Y_{12} = -0.1$ S, $Y_{22} = 0.15$ S

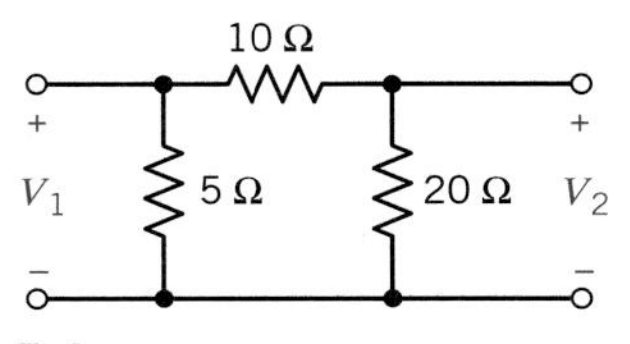

Figure P 17.5-4

P 17.5-5 Find the Y parameters of the circuit shown in Figure P 17.5-5.

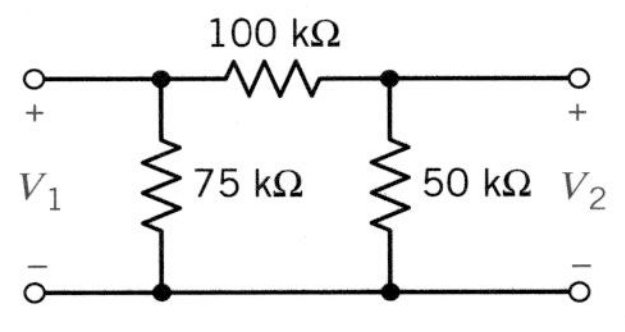

Figure P 17.5-5

P 17.5-6 Find the Z parameters for the circuit shown in Figure P 17.5-6 for sinusoidal steady-state response at $\omega = 3$ rad/s.
Answer: $Z_{11} = 3 + \mathrm{j}\ \Omega$, $Z_{12} = Z_{21} = -\mathrm{j}2\ \Omega$, $Z_{11} = -\mathrm{j}2\ \Omega$

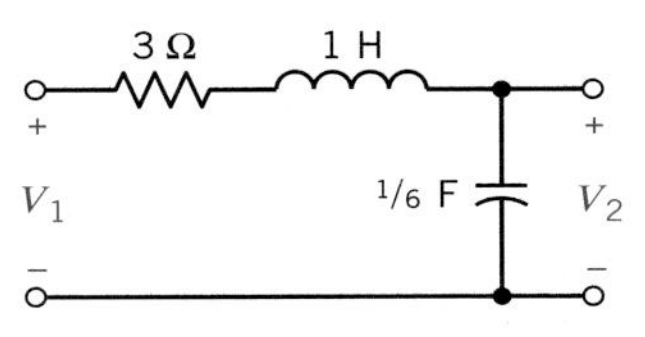

Figure P 17.5-6

P 17.5-7 Determine the impedance parameters in the s-domain (Laplace domain) for the circuit shown in Figure P 17.5-7.
Answer: $Z_{11} = (4s + 1)/s$, $Z_{12} = Z_{21} = 1/s$, $Z_{22} = (2s^2 + 1)/s$

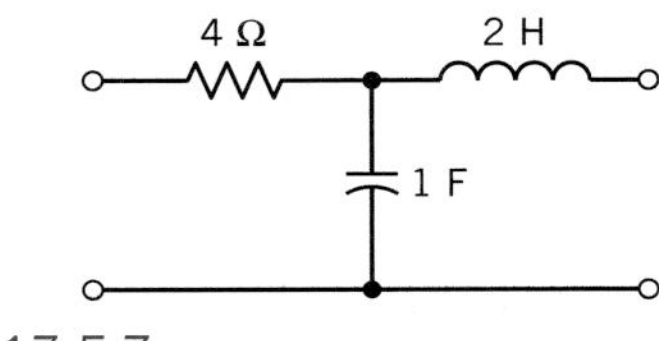

Figure P 17.5-7

P 17.5-8 Determine a two-port network that is represented by the Y parameters.

$$\mathbf{Y} = \begin{bmatrix} \dfrac{s+1}{s} & -1 \\ -1 & (s+1) \end{bmatrix}$$

P 17.5-9 Find a two-port network incorporating one inductor, one capacitor, and two resistors that will give the following impedance parameters:

$$\mathbf{Z} = \frac{1}{\Delta}\begin{bmatrix} (s^2 + 2s + 2) & 1 \\ 1 & (s^2 + 1) \end{bmatrix}$$

where $\Delta = s^2 + s + 1$.

P 17.5-10 An infinite two-port network is shown in Figure P 17.5-10. When the output terminals are connected to the circuit's characteristic resistance R_o, the resistance looking down the line from each section is the same. Calculate the necessary R_o.
Answer: $R_0 = (\sqrt{3} - 1)R$

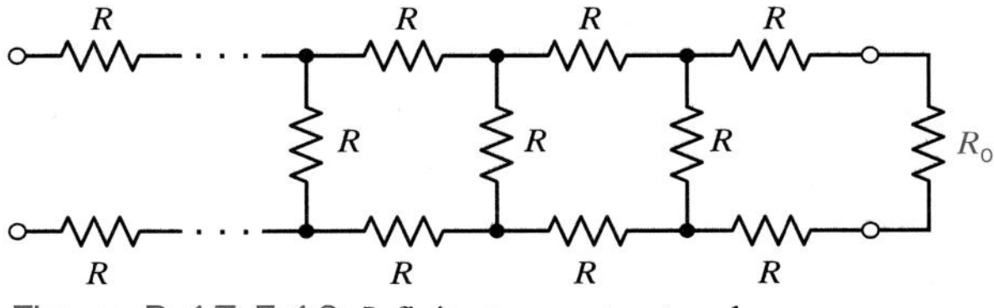

Figure P 17.5-10 Infinite two-port network.

Section 17.6 Z and Y Parameters for a Circuit with Dependent Sources

P 17.6-1 Determine the Y parameters of the circuit shown in Figure P 17.6-1.

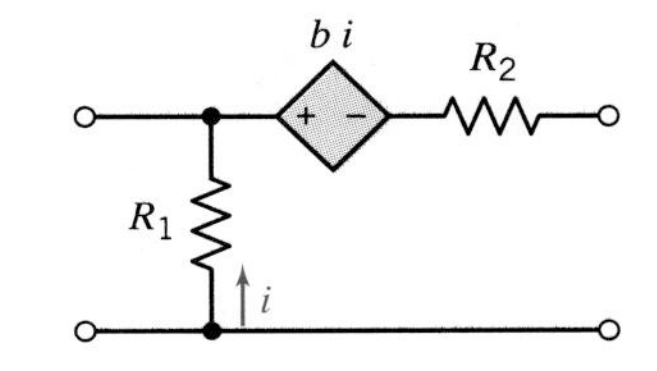

Figure P 17.6-1

P 17.6-2 An electronic amplifier has the circuit shown in Figure P 17.6-2. Determine the impedance parameters for the circuit.
Answer: $Z_{11} = 4$; $Z_{12} = 3(1 + \alpha)$; $Z_{21} = 3$; $Z_{22} = 5 + 3\alpha$

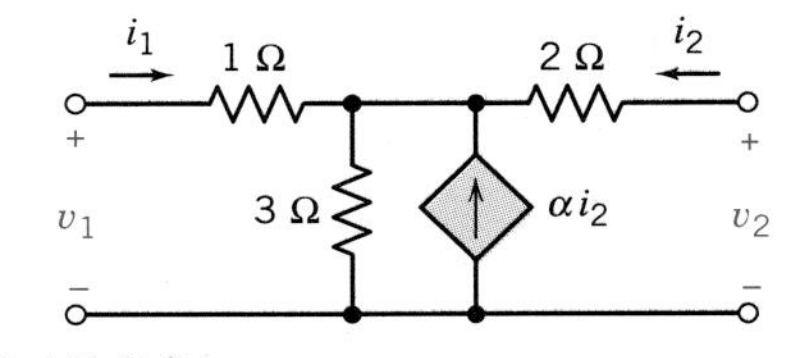

Figure P 17.6-2

P 17.6-3 (a) For the circuit shown in Figure P 17.6-3, determine the two-port Y model using impedances in the s-domain, (b) Determine the response $v_2(t)$ when a current source $i_1 = 1\,u(t)$ A is connected to the input terminals.

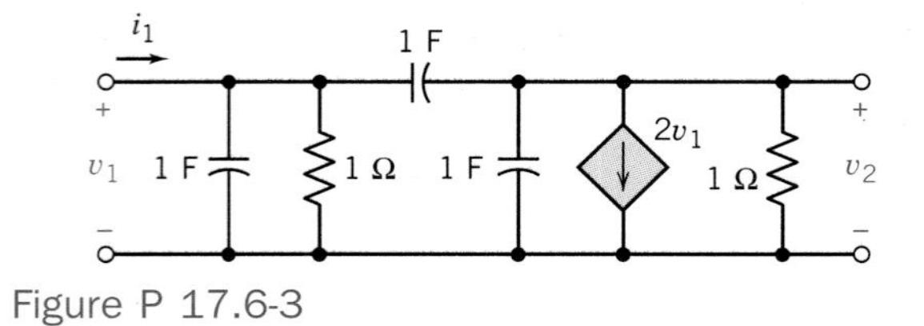

Figure P 17.6-3

P 17.6-4 One form of a heart-assist device is shown in Figure P 17.6-4*a*. The model of the electronic controller and pump/drive unit is shown in Figure P 17.6-4*b*. Determine the impedance parameters of the two-port model.

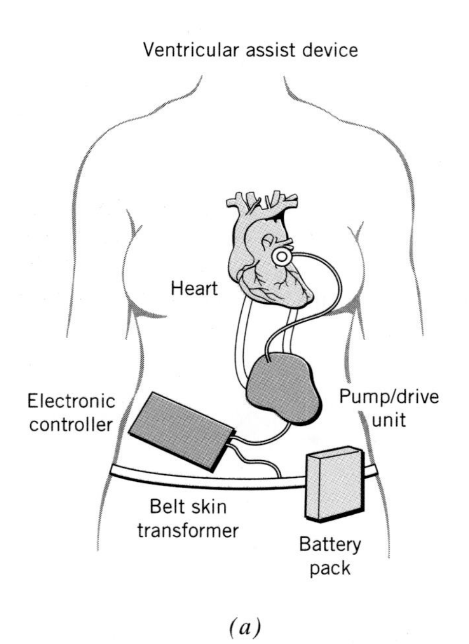

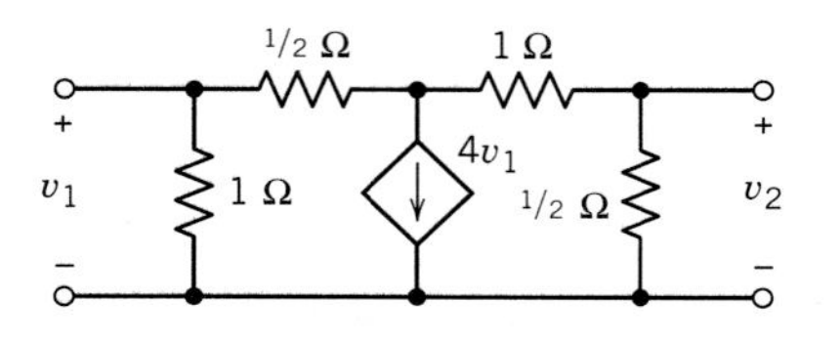

(*b*)

Figure P 17.6-4 (*a*) Heart-assist device and (*b*) model of controller and pump.

P 17.6-5 Determine the Y parameters for the circuit shown in Figure P 17.6-5.

Partial Answer: $Y_{12} = -\dfrac{1}{R_2},\; Y_{21} = \dfrac{-(1+b)}{R_2}$

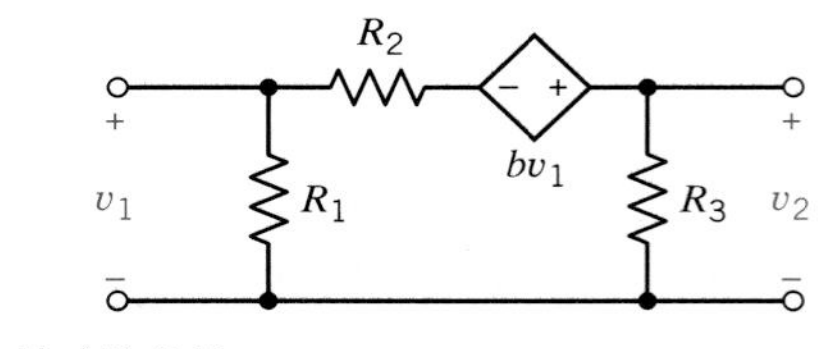

Figure P 17.6-5

Section 17.7 Hybrid and Transmission Parameters

P 17.7-1 Find the transmission parameters of the circuit of Figure P 17.7-1.

Answer: $A = 1.2$, $B = 6.8\ \Omega$, $C = 0.1$ S, $D = 1.4$

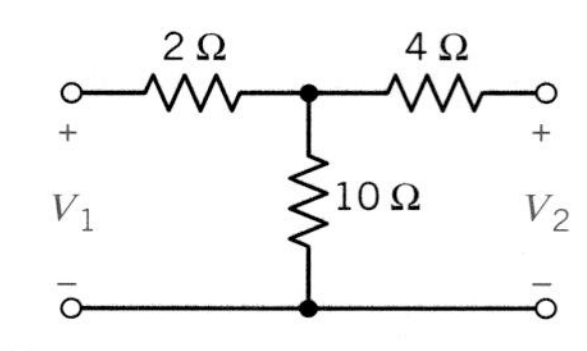

Figure P 17.7-1

P 17.7-2 An op amp circuit and its model are shown in Figure P 17.7-2. Determine the h-parameter model of the circuit and the **H** matrix when $R_i = 100\ \text{k}\Omega$, $R_1 = R_2 = 1\ \text{M}\Omega$, $R_o = 1\ \text{k}\Omega$, and $A = 10^4$.

Answer: $h_{11} = 600\ \text{k}\Omega$, $h_{12} = 1/2$, $h_{21} = -10^6$, $h_{22} = 10^{-3}$ S.

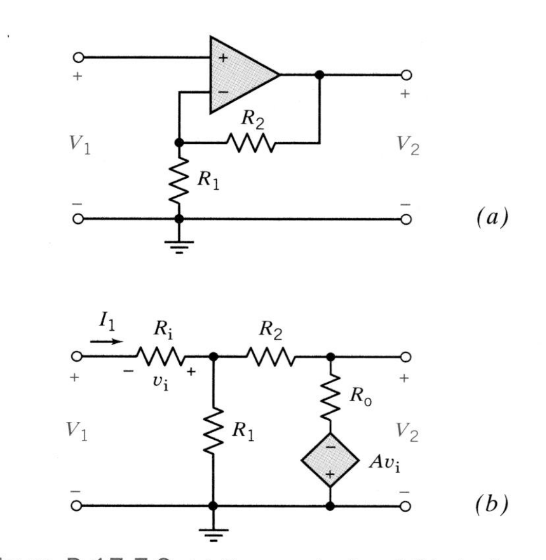

Figure P 17.7-2 (*a*) Op amp circuit and (*b*) circuit model.

P 17.7-3 Determine the h parameters for the ideal transformer of Section 11.9.

P 17.7-4 Determine the h parameters for the T circuit of Figure P 17.7-4.

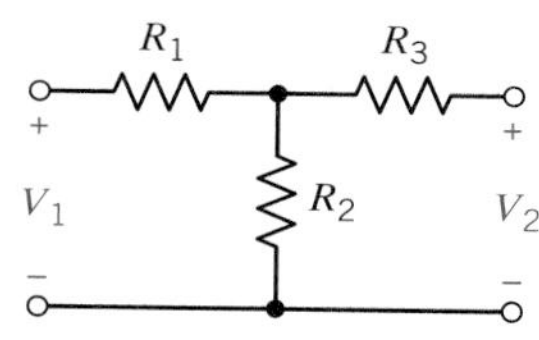

Figure P 17.7-4

P 17.7-5 A simplified model of a bipolar junction transistor is shown in Figure P 17.7-5. Determine the h parameters of this circuit.

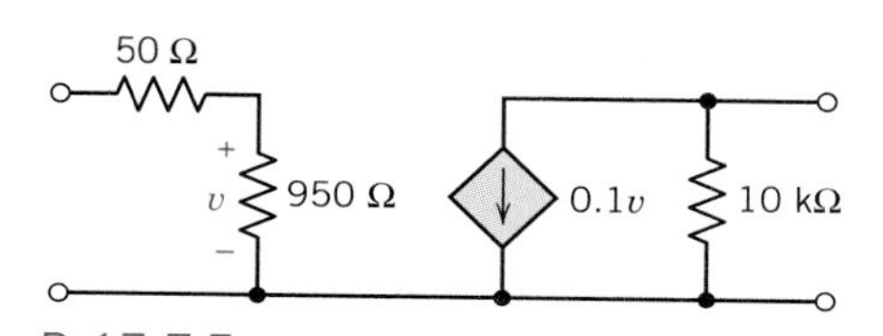

Figure P 17.7-5 Model of bipolar junction transistor.

P 17.7-6 A transformer can be represented by the inverse hybrid circuit model shown in Figure P 17.7-6. Open-circuit measurements indicate that $\mathbf{V}_1 = 1000$ V, $\mathbf{I}_2 = 0$, $V_2 = 500$ V, $\mathbf{I}_1 = 0.42$ A, and $P = 100$ W (secondary open circuit). Short-circuit measurements indicate that $\mathbf{I}_1 = 10$ A, $\mathbf{V}_1 = 126$ V, $P = 400$ W, and $\mathbf{V}_2 = 0$. Determine the inverse hybrid parameters.

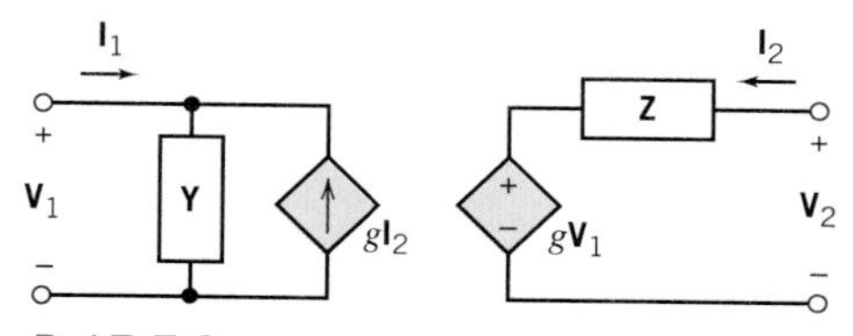

Figure P 17.7-6 Model of a transformer.

Section 17.8 Relationships Between Two-Port Parameters

P 17.8-1 Derive the relationships between the Y parameters and the h parameters by utilizing the defining equations for both parameter sets.

P 17.8-2 Determine the Y parameters if the Z parameters are (in ohms):

$$\mathbf{Z} = \begin{bmatrix} 3 & 2 \\ 2 & 6 \end{bmatrix}$$

P 17.8-3 Determine the h parameters when the Y parameters are (in siemens):

$$\mathbf{Y} = \begin{bmatrix} 0.1 & 0.1 \\ 0.4 & 0.5 \end{bmatrix}$$

P 17.8-4 A two-port has the following Y parameters: $Y_{12} = Y_{21} = -0.4$ S, $Y_{11} = 0.5$ S, and $Y_{22} = 0.6$ S. Determine the h parameters.

Answer: $h_{11} = 2\ \Omega$, $h_{21} = -0.8$, $h_{12} = 0.8$, $h_{22} = 0.28$ S

Section 17.9 Interconnection of Two-Port Networks

P 17.9-1 Connect in parallel the two circuits shown in Figure P 17.9-1 and find the Y parameters of the parallel combination.

Answer: $Y_{11} = 17/6$, $Y_{12} = Y_{21} = -4/3$, $Y_{22} = 5/3$

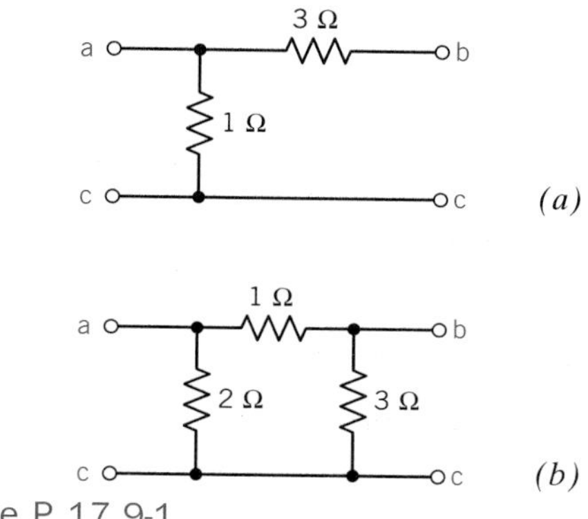

Figure P 17.9-1

P 17.9-2 For the T network of Figure P 17.9-2 find the Y and T parameters and determine the resulting parameters after the two two-ports are connected in (a) parallel and (b) cascade. Both two-ports are identical as defined in Figure P 17.9-2.

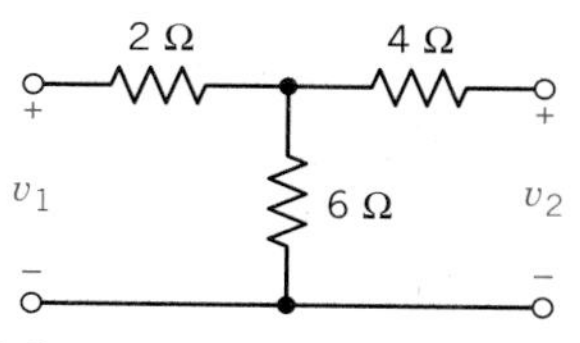

Figure P 17.9-2

P 17.9-3 Determine the Y parameters of the parallel combination of the circuits of Figure P 17.9-3*a* and *b*.

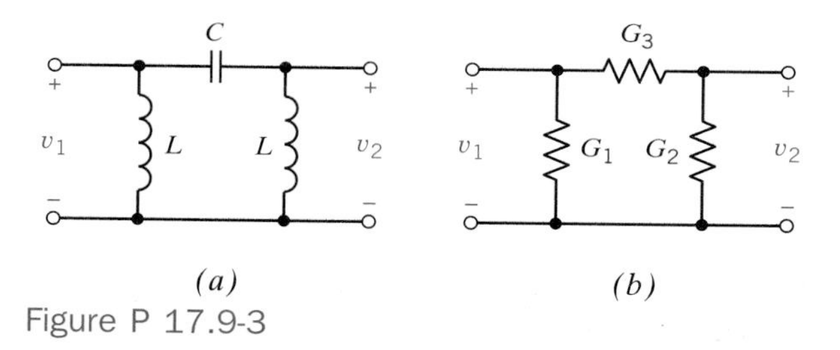

Figure P 17.9-3

PSPICE PROBLEMS

SP 17-1 Calculate the impedance parameters Z_{11} and Z_{21} of the circuit of Figure P 17.5-1.

SP 17-2 Calculate the admittance parameters Y_{11} and Y_{21} for the circuit of Figure SP 17.2.

SP 17-3 Determine the hybrid parameters of the circuit of Figure SP 17.3.

SP 17-4 Determine the h parameters h_{11} and h_{21} of the circuit shown in Figure SP 17.4.

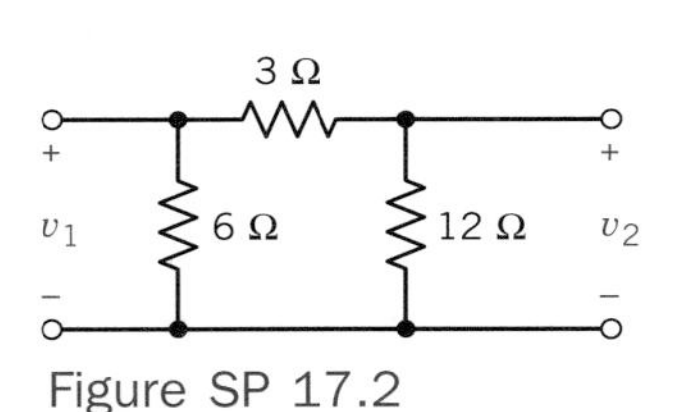

Figure SP 17.2

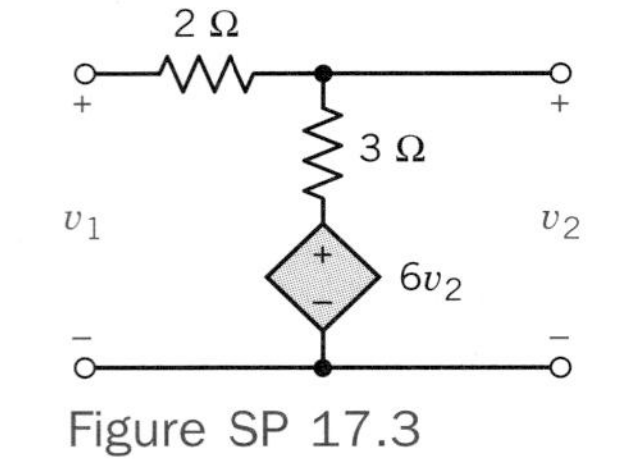

Figure SP 17.3

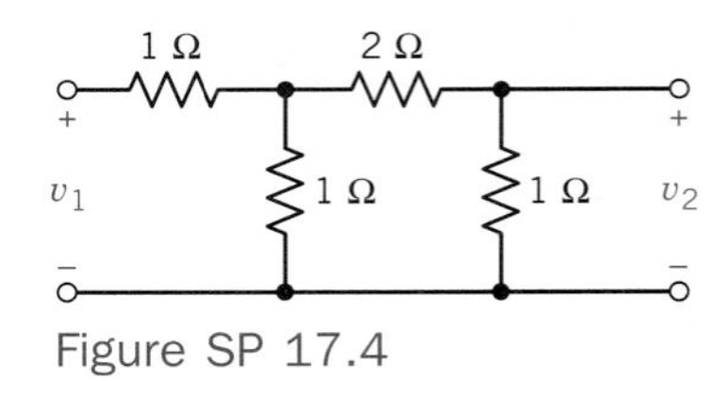

Figure SP 17.4

VERIFICATION PROBLEMS

VP 17-1 A laboratory report concerning the circuit of Figure VP 17.1 states that $Z_{12} = 15\ \Omega$ and $Y_{11} = 24$ mS. Verify these results.

VP 17-2 A student report concerning the circuit of Figure VP 17.2 has determined the transmission parameters as $A = \dfrac{2(s + 10)}{s}$, $D = -A$, $C = 0.1s$, and $B = \dfrac{-(3s^2 + 80s + 40)}{s^2}$. Verify these results when $M = 0.1$ H.

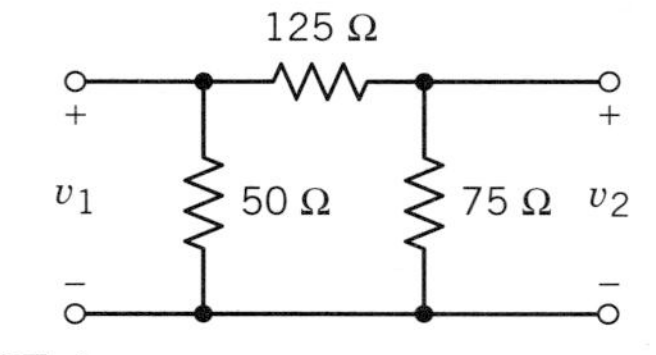

Figure VP 17.1

2 Ω, 2 Ω, M, v_1, 0.2 H, 0.2 H, v_2

Figure VP 17.2

DESIGN PROBLEMS

DP 17-1 Select R_1 and R so that $R_{in} = 16.6\ \Omega$ for the circuit of Figure DP 17.1. A design constraint requires that both R_1 and R be less than 10 Ω.

DP 17-2 The bridge circuit shown in Figure DP 17.2 is said to be balanced when $I = 0$. Determine the required relationship for the bridge resistances when balance is achieved.

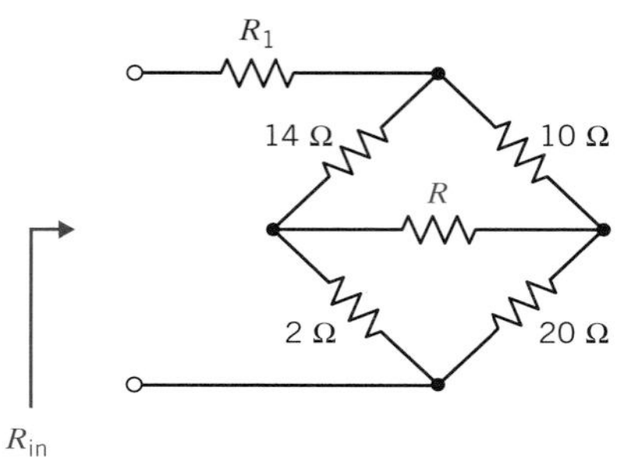

Figure DP 17.1

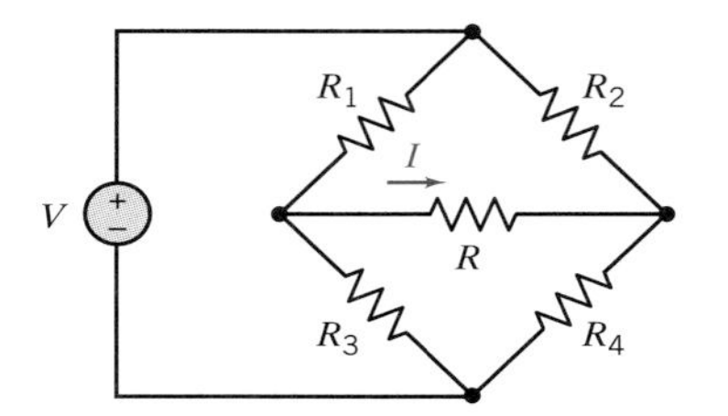

Figure DP 17.2 Bridge circuit.

DP 17-3 A hybrid model of a common-emitter transistor amplifier is shown in Figure DP 17.3. The transistor parameters are $h_{21} = 80$, $h_{11} = 45\ \Omega$, $h_{22} = 12.5\ \mu\text{S}$, and $h_{12} = 5 \times 10^{-4}$. Select R_L so that the current gain $i_2/i_1 = 79$ and the input resistance of the amplifier is less than 10 Ω.

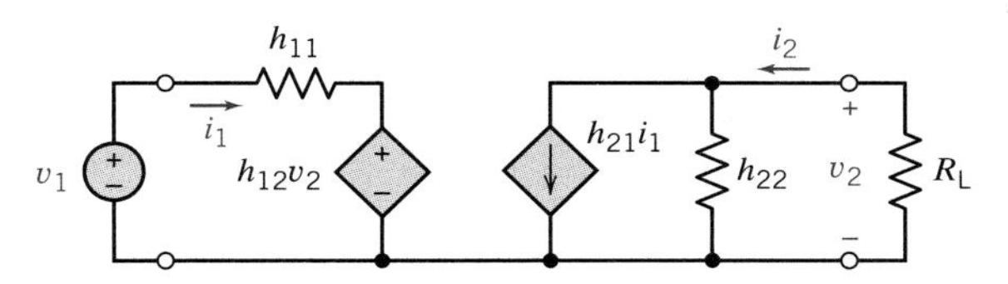

Figure DP 17.3 Model of transistor amplifier.

DP 17-4 A two-port network connected to a source v_s and a load resistance R_L is shown in Figure DP 17.4.

(a) Determine the impedance parameters of the two-port network.

(b) Select R_L so that maximum power is delivered to R_L.

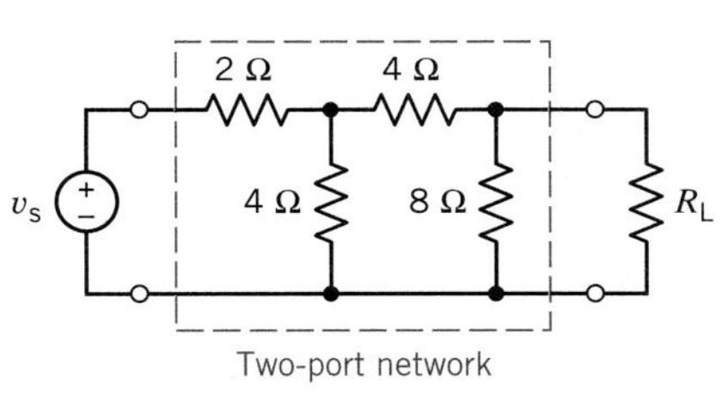

Figure DP 17.4

DP 17-5 (a) Determine the ABCD (transmission matrix) of the two-port networks shown in Figures DP 17.5*a* and DP 17.5*b*. (b) Using the results of (a), find the *s*-domain ABCD matrix of the network shown in Figure DP 17.5*c*. (c) Given $L_1 = (10/\pi)$mH, $L_2 = (2.5/\pi)$ mH, $C_1 = (0.78/\pi)$ μF, $C_2 = C_3 = (1/\pi)$ μF, and $R_L = 100\ \Omega$, find the open-circuit voltage gain V_2/V_1 and the short-circuit current gain I_2/I_1 under sinusoidal-state conditions at the following frequencies: 2.5 kHz, 5.0 kHz, 7.5 kHz, 10 kHz, and 12.5 kHz. (*Hint:* Use the appropriate entries of the ABCD matrix. Also note the resonant frequencies of the circuit.)

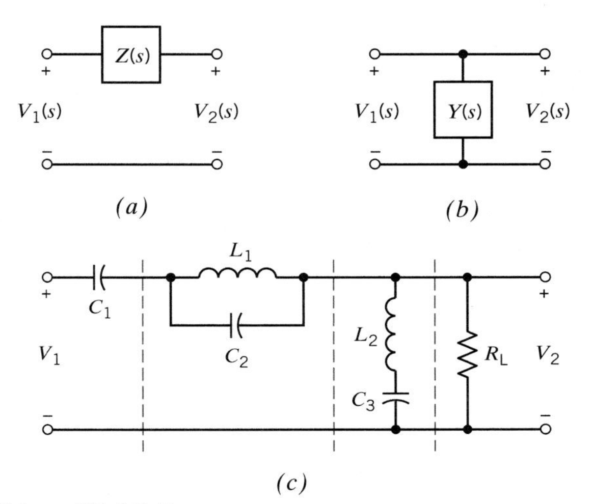

Figure DP 17.5

Matrices, Determinants, and Cramer's Rule

A.1 DEFINITION OF A MATRIX

There are many situations in circuit analysis in which we have to deal with rectangular arrays of numbers. The rectangular array of numbers

$$\mathbf{A} = \begin{bmatrix} a_{11} & a_{12} & \cdots & a_{1n} \\ a_{21} & a_{22} & \cdots & a_{2n} \\ \cdot & \cdot & & \cdot \\ \cdot & \cdot & & \cdot \\ \cdot & \cdot & & \cdot \\ a_{m1} & a_{m2} & \cdots & a_{mn} \end{bmatrix} \tag{A.1}$$

is known as a *matrix.* The numbers a_{ij} are called *elements* of the matrix, with the subscript i denoting the row and the subscript j denoting the column.

A matrix with m rows and n columns is said to be a matrix of *order* (m, n) or alternatively called an $m \times n$ (m by n) matrix. When the number of the columns equals the number of rows, $m = n$, the matrix is called a *square matrix* of order n. It is common to use boldface capital letters to denote an $m \times n$ matrix.

A matrix consisting of only one column, that is, an $m \times 1$ matrix, is known as a column matrix or, more commonly, a *column vector.* We represent a column vector with boldface lowercase letters as

$$\mathbf{v} = \begin{bmatrix} v_1 \\ v_2 \\ \cdot \\ \cdot \\ \cdot \\ v_m \end{bmatrix}$$

When the elements of a matrix have a special relationship so that $a_{ij} = a_{ji}$, it is called a *symmetrical* matrix. Thus, for example,

$$\mathbf{H} = \begin{bmatrix} 3 & -2 & 1 \\ -2 & 6 & 4 \\ 1 & 4 & 8 \end{bmatrix}$$

is a symmetrical matrix of order (3, 3).

A.2 ADDITION AND SUBTRACTION OF MATRICES

The addition of two matrices is possible for matrices of the same order. The sum of two matrices is obtained by adding the corresponding elements. Thus, if the elements of **A** are a_{ij} and the elements of **B** are b_{ij} and if

$$\mathbf{C} = \mathbf{A} + \mathbf{B}$$

then the elements of **C** that are c_{ij} are obtained as

$$c_{ij} = a_{ij} + b_{ij}$$

For example, the matrix addition for two 3×3 matrices is as follows:

$$\mathbf{C} = \begin{bmatrix} 2 & 1 & 0 \\ 1 & -1 & 3 \\ 0 & 6 & 2 \end{bmatrix} + \begin{bmatrix} 8 & 2 & 1 \\ 1 & 3 & 0 \\ 4 & 2 & 1 \end{bmatrix} = \begin{bmatrix} 10 & 3 & 1 \\ 2 & 2 & 3 \\ 4 & 8 & 3 \end{bmatrix}$$

From the operation used for performing addition, we note that the process is commutative, that is,

$$\mathbf{A} + \mathbf{B} = \mathbf{B} + \mathbf{A}$$

Also, we note that the addition operation is associative, so that

$$(\mathbf{A} + \mathbf{B}) + \mathbf{C} = \mathbf{A} + (\mathbf{B} + \mathbf{C})$$

To perform the operation of subtraction, we note that if a matrix **A** is multiplied by a constant α, then every element of the matrix is multiplied by this constant. Therefore, we can write

$$\alpha\mathbf{A} = \begin{bmatrix} \alpha a_{11} & \alpha a_{12} & \cdots & \alpha a_{1n} \\ \alpha a_{12} & \alpha a_{22} & \cdots & \alpha a_{2n} \\ \cdot & \cdot & & \cdot \\ \cdot & \cdot & & \cdot \\ \cdot & \cdot & & \cdot \\ \alpha a_{m1} & \alpha a_{m2} & \cdots & \alpha a_{mn} \end{bmatrix}$$

In order to carry out a subtraction operation $\mathbf{B} - \mathbf{A}$, we use $\alpha = -1$, and $-\mathbf{A}$ is obtained by multiplying each element of **A** by -1. For example,

$$\mathbf{C} = \mathbf{B} + \alpha\mathbf{A} = \mathbf{B} - \mathbf{A} = \begin{bmatrix} 2 & 1 \\ 4 & 2 \end{bmatrix} - \begin{bmatrix} 6 & 1 \\ 3 & 1 \end{bmatrix} = \begin{bmatrix} -4 & 0 \\ 1 & 1 \end{bmatrix}$$

A.3 DETERMINANTS

The *determinant* of a matrix is a number associated with a square matrix. We define the determinant of a square matrix **A** as Δ, where

$$\Delta = \begin{vmatrix} a_{11} & a_{12} & \cdots & a_{1n} \\ a_{21} & a_{22} & \cdots & a_{2n} \\ \cdot & \cdot & & \cdot \\ \cdot & \cdot & & \cdot \\ \cdot & \cdot & & \cdot \\ a_{n1} & a_{n2} & \cdots & a_{nn} \end{vmatrix}$$

For example, consider the determinant of a 2×2 matrix

$$\Delta = \begin{vmatrix} a_{11} & a_{12} \\ a_{21} & a_{22} \end{vmatrix}$$

In this case, the determinant Δ is defined to be

$$\Delta = a_{11}a_{22} - a_{12}a_{21} \tag{A.2}$$

In the second-order, or 2×2, case of Eq. A.1, the method of obtaining Δ in Eq. A.2 uses the diagonal rule, which is

$$\Delta = \begin{vmatrix} a_{11} & a_{12} \\ a_{21} & a_{22} \end{vmatrix}$$

Thus, Δ is the difference of the product of the elements down the diagonal to the right and the product of the elements down the diagonal to the left. The determinant of a 3×3 matrix is

$$\begin{aligned} \Delta &= \begin{vmatrix} a_{11} & a_{12} & a_{13} \\ a_{21} & a_{22} & a_{23} \\ a_{31} & a_{32} & a_{33} \end{vmatrix} \\ &= (a_{11}a_{22}a_{23} + a_{12}a_{23}a_{31} + a_{13}a_{32}a_{21}) - (a_{13}a_{22}a_{31} + a_{23}a_{32}a_{11} + a_{33}a_{21}a_{12}) \end{aligned}$$

In general, we are able to determine the determinant Δ in terms of cofactors and minors. The determinant of a submatrix of **A** obtained by deleting from **A** the ith row and the jth column is called the *minor* of the element a_{ij} and denoted as m_{ij}.

The cofactor c_{ij} is a minor with an associated sign, so that

$$c_{ij} = (-1)^{(i+j)}\, m_{ij}$$

As an example, consider the determinant

$$\Delta = \begin{vmatrix} 1 & -2 & 3 \\ 0 & 4 & -2 \\ 6 & -1 & -1 \end{vmatrix} \tag{A.3}$$

The minor of the element a_{11} is

$$m_{11} = \begin{vmatrix} 4 & -2 \\ -1 & -1 \end{vmatrix}$$

which is obtained by deleting the first row and the first column. The cofactor c_{11} is then

$$\begin{aligned} c_{11} &= (-1)^{(1+1)} m_{11} \\ &= m_{11} \end{aligned}$$

The rule for evaluating the determinant Δ of a $n \times n$ matrix is

$$\Delta = \sum_{j=1}^{n} a_{ij}\, c_{ij}$$

for a selected value of i. Alternatively, we can obtain Δ by using the jth column, and thus

$$\Delta = \sum_{i=1}^{n} a_{ij}\, c_{ij} \tag{A.4}$$

for a selected value of j.

As an example, let us evaluate the determinant in Eq. A.3. It is easiest to use the summation along the first column of the matrix of Eq. A.3 since $a_{21} = 0$. Then, using Eq. A.4, we have for $j = 1$

$$\Delta = \sum_{i=1}^{3} a_{i1}c_{i1} = a_{11}c_{11} + a_{21}c_{21} + a_{31}c_{31}$$

$$= 1(-1)^2\begin{vmatrix} 4 & -2 \\ -1 & -1 \end{vmatrix} + 6(-1)^4\begin{vmatrix} -2 & 3 \\ 4 & -2 \end{vmatrix} = -6 + 6(-8) = -54$$

A.4 CRAMER'S RULE

A set of simultaneous equations

$$\begin{array}{ccccccccc} a_{11}x_1 & + & a_{12}x_2 & + & \cdots & + & a_{1n}x_n & = & b_1 \\ a_{21}x_1 & + & a_{22}x_2 & + & \cdots & + & a_{2n}x_n & = & b_2 \\ \cdot & & \cdot & & & & \cdot & & \cdot \\ \cdot & & \cdot & & & & \cdot & & \cdot \\ \cdot & & \cdot & & & & \cdot & & \cdot \\ a_{n1}x_1 & + & a_{n2}x_2 & + & \cdots & + & a_{nn}x_n & = & b_n \end{array} \tag{A.5}$$

can be written in matrix form as

$$\mathbf{Ax} = \mathbf{b}$$

Cramer's rule states that the solution for the unknown, x_k, of the simultaneous equations of Eq. A.5 is

$$x_k = \frac{\Delta_k}{\Delta}$$

where Δ is the determinant of $\mathbf{A}$ and Δ_k is Δ with the kth column replaced by the column vector $\mathbf{b}$.

As an example, let us consider the simultaneous equations

$$\begin{aligned} x_1 - 2x_2 + 3x_3 &= 12 \\ 4x_2 - 2x_3 &= -1 \\ 6x_1 - x_2 - x_3 &= 0 \end{aligned}$$

In this case

$$\mathbf{A} = \begin{bmatrix} 1 & -2 & 3 \\ 0 & 4 & -2 \\ 6 & -1 & -1 \end{bmatrix}$$

and

$$\mathbf{b} = \begin{bmatrix} 12 \\ -1 \\ 0 \end{bmatrix}$$

The determinant of $\mathbf{A}$ was obtained in the preceding section as $\Delta = -54$. If we wish to obtain the unknown x_1, we have

$$x_1 = \frac{\Delta_1}{\Delta}$$

Then Δ_1 is Δ with the first column of **A** replaced by **b** so that

$$\Delta_1 = \begin{vmatrix} 12 & -2 & 3 \\ -1 & 4 & -2 \\ 0 & -1 & -1 \end{vmatrix} = 12(-6) - 1(-1)^3(5) = -67$$

Therefore, we have

$$x_1 = \frac{-67}{-54} = \frac{67}{54}$$

A.5 MULTIPLICATION OF MATRICES

Matrix multiplication is defined in such a way as to assist in the solution of simultaneous linear equations. The multiplication of two matrices **AB** requires that the number of columns of **A** is equal to the number of rows of **B**. Thus if **A** is of order $m \times n$ and **B** is of order $n \times q$, then the product is a matrix of order $m \times q$. The elements of a product

$$\mathbf{C} = \mathbf{AB}$$

are found by multiplying the ith row of **A** and the jth column of **B** and summing these products to give the element c_{ij}. That is,

$$c_{ij} = a_{i1}b_{1j} + a_{i2}b_{2j} + \cdots + a_{iq}b_{qj} = \sum_{k=1}^{q} a_{ik}b_{kj}$$

Thus we obtain c_{11}, the first element of **C**, by multiplying the first row of **A** by the first column of **B** and summing the products of the elements. We should note that, in general, matrix multiplication is not commutative, that is,

$$\mathbf{AB} \neq \mathbf{BA}$$

Also, we will note that the multiplication of a matrix of order $m \times n$ by a column vector (order $n \times 1$) results in a column vector of order $m \times 1$.

A specific example of multiplication of a column vector by a matrix is

$$\mathbf{x} = \mathbf{Ay} = \begin{bmatrix} a_{11} & a_{12} & a_{13} \\ a_{21} & a_{22} & a_{23} \end{bmatrix} \begin{bmatrix} y_1 \\ y_2 \\ y_3 \end{bmatrix}$$
$$= \begin{bmatrix} (a_{11}y_1 + a_{12}y_2 + a_{13}y_3) \\ (a_{21}y_1 + a_{22}y_2 + a_{23}y_3) \end{bmatrix}$$

Note that **A** is of order 2×3 and **y** is of order 3×1. Therefore, the resulting matrix **x** is of order 2×1, which is a column vector with two rows. There are two elements of **x**, and

$$x_1 = (a_{11}y_1 + a_{12}y_2 + a_{13}y_3)$$

is the first element obtained by multiplying the first row of **A** by the first (and only) column of **y**.

Another example, which the reader should verify, is

$$\mathbf{C} = \mathbf{AB} = \begin{bmatrix} 2 & -1 \\ -1 & 2 \end{bmatrix} \begin{bmatrix} 3 & 2 \\ -1 & -2 \end{bmatrix} = \begin{bmatrix} 7 & 6 \\ -5 & -6 \end{bmatrix}$$

For example, the element c_{22} is obtained as $c_{22} = -1(2) + 2(-2) = -6$.

APPENDIX B

Complex Numbers

B.1 A COMPLEX NUMBER

We all are familiar with the solution of the algebraic equation

$$x^2 - 1 = 0 \tag{B.1}$$

which is $x = 1$. However, we often encounter the equation

$$x^2 + 1 = 0 \tag{B.2}$$

A number that satisfies Eq. B.2 is not a real number. We note that Eq. B.2 may be written as

$$x^2 = -1 \tag{B.3}$$

and we denote the solution of Eq. B.3 by the use of an imaginary number $j1$ so that

$$j^2 = -1 \tag{B.4}$$

and

$$j = \sqrt{-1} \tag{B.5}$$

An *imaginary number* is defined as the product of the imaginary unit j with a real number. Thus we may, for example, write an imaginary number as jb. A complex number is the sum of a real number and an imaginary number, so that

$$c = a + jb \tag{B.6}$$

where a and b are real numbers. We designate a as the real part of the complex number and b as the imaginary part and use the notation

$$\text{Re}\{c\} = a$$

and

$$\text{Im}\{c\} = b$$

B.2 RECTANGULAR, EXPONENTIAL, AND POLAR FORMS

The complex number $a + jb$ may be represented on a rectangular coordinate place called a *complex plane.* The complex plane has a real axis and an imaginary axis, as shown in Figure B.1. The complex number c is the directed line identified as c with coordinates a, b. The *rectangular form* is expressed in Eq. B.6 and pictured in Figure B.1.

An alternative way to express the complex number c is to use the distance from the origin and the angle θ, as shown in Figure B.2. The *exponential form* is written as

$$c = re^{j\theta}$$

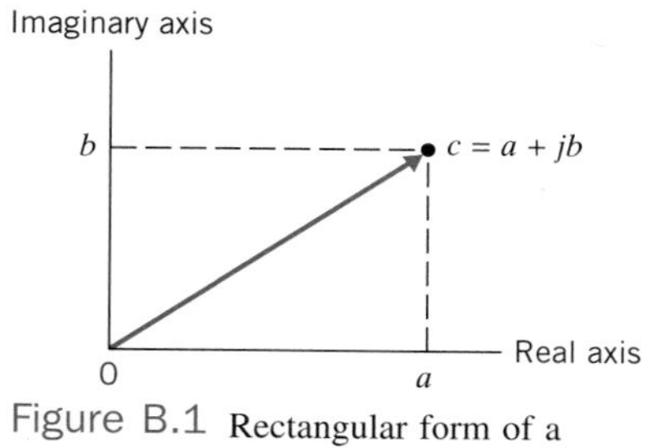

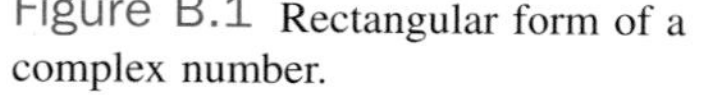
Figure B.1 Rectangular form of a complex number.

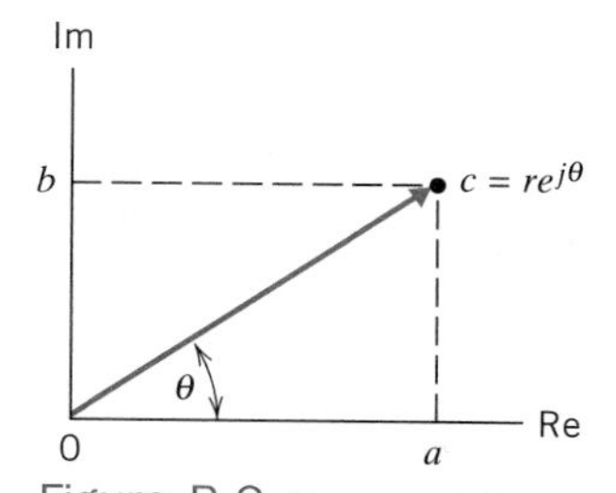

Figure B.2 Exponential form of a complex number.

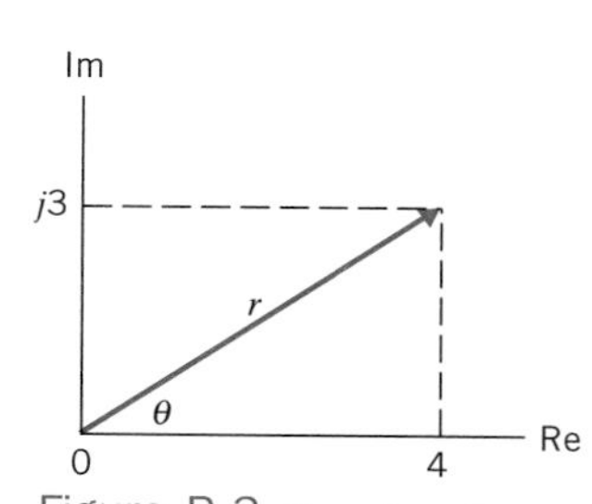

Figure B.3 Complex plane for Example B.1.

where

$$r = (a^2 + b^2)^{1/2}$$

and

$$\theta = \begin{cases} \tan^{-1}\left(\dfrac{b}{a}\right) & a > 0 \\ 180^\circ - \tan^{-1}\left(\dfrac{b}{-a}\right) & a < 0 \end{cases}$$

Note that $a = r\cos\theta$ and $b = r\sin\theta$. Also note that calculators give the principal value of the arc tangent and one must be sure that the final calculation is in the right quadrant.

The number r is also called the *magnitude* of c, denoted as $|c|$. The angle θ can also be denoted by the form $\underline{/\theta}$. Thus, we may represent the complex number in *polar form* as

$$c = |c| \underline{/\theta} = r\underline{/\theta}$$

Example B.1

Express $c = 4 + j3$ in exponential and polar forms.

Solution

First, draw the complex plane diagram as shown in Figure B.3. Then find r as

$$r = (4^2 + 3^2)^{1/2} = 5$$

and θ as

$$\theta = \tan^{-1}(3/4) = 36.9^\circ$$

The exponential form is then

$$c = 5e^{j36.9^\circ}$$

The polar form is

$$c = 5\,\underline{/36.9^\circ}$$

Note that if $c = 4 - j3$, then $\theta = \tan^{-1}(-3/4) = -36.9^\circ$. If $c = -4 + j3$ then $\theta = 180^\circ - \tan^{-1}\left(\dfrac{3}{4}\right) = 143.1^\circ$. One has to check continually to ensure the angle is in the right quadrant.

B.3 MATHEMATICAL OPERATIONS

The *conjugate* of the complex number $c = a + jb$ is denoted as c^* and is defined as

$$c^* = a - jb$$

In polar form, we have

$$c^* = r \angle -\theta$$

To add or subtract two complex numbers, we add (or subtract) their real parts and their imaginary parts. Therefore, if $c = a + jb$ and $d = f + jg$, then

$$\begin{aligned} c + d &= (a + jb) + (f + jg) \\ &= (a + f) + j(b + g) \end{aligned}$$

The multiplication of two complex numbers is obtained as follows (note $j^2 = -1$):

$$\begin{aligned} cd &= (a + jb)(f + jg) \\ &= af + jag + jbf + j^2bg \\ &= (af - bg) + j(ag + bf) \end{aligned}$$

Using the exponential form, we have

$$cd = r_1e^{j\theta_1} \times r_2e^{j\theta_2} = r_1r_2e^{j(\theta_1+\theta_2)}$$

Alternatively, we use the polar form to obtain

$$\begin{aligned} cd &= (r_1 \angle \theta_1)(r_2 \angle \theta_2) \\ &= r_1r_2 \angle \theta_1 + \theta_2 \end{aligned}$$

where $\qquad c = r_1 \angle \theta_1 \quad \text{and} \quad d = r_2 \angle \theta_2$

Division of one complex number by another complex number is easily obtained using the exponential form as follows:

$$\frac{c}{d} = \frac{r_1e^{j\theta_1}}{r_2e^{j\theta_2}} = (r_1/r_2)e^{j(\theta_1-\theta_2)}$$

Alternatively, we use the polar form as

$$\begin{aligned} \frac{c}{d} &= \frac{r_1 \angle \theta_1}{r_2 \angle \theta_2} \\ &= \frac{r_1}{r_2} \angle \theta_1 - \theta_2 \end{aligned}$$

It is easiest to add and subtract complex numbers in rectangular form and to multiply and divide them in polar form.

A few useful relations for complex numbers are summarized in Table B.1.

Table B.1 **Useful Relationships for Complex Numbers**

(1) $\frac{1}{j} = -j$
(2) $(-j)(j) = 1$
(3) $j^2 = -1$
(4) $1 \angle \pi/2 = j$
(5) $c^k = r^k \angle k\theta$

Example B.2

Find $c + d$, $c - d$, cd, and c/d when $c = 4 + j3$ and $d = 1 - j$.

Solution

First, we will express c and d in polar form as

$$c = 5\,\underline{/36.9^\circ}$$

and

$$d = \sqrt{2}\,\underline{/-45^\circ}$$

Then, for addition, we have

$$c + d = (4 + j3) + (1 - j) = 5 + j2$$

For subtraction, we have

$$c - d = (4 + j3) - (1 - j) = 3 + j4$$

For multiplication, we use the polar form to obtain

$$cd = (5\underline{/36.9^\circ})(\sqrt{2}\,\underline{/-45^\circ}) = 5\sqrt{2}\,\underline{/-8.1^\circ}$$

For division, we have

$$\frac{c}{d} = \frac{5\underline{/36.9^\circ}}{\sqrt{2}\underline{/-45^\circ}} = \frac{5}{\sqrt{2}}\underline{/81.9^\circ}$$

APPENDIX C

Trigonometric Formulas

$$\sin\alpha = \cos(90^\circ - \alpha) = -\cos(90^\circ + \alpha)$$
$$\sin\alpha = \sin(180^\circ - \alpha)$$
$$\tan\alpha = -\tan(180^\circ - \alpha)$$
$$\sin(\alpha \pm \beta) = \sin\alpha\cos\beta \pm \cos\alpha\sin\beta$$
$$\cos(\alpha \pm \beta) = \cos\alpha\cos\beta \mp \sin\alpha\sin\beta$$
$$\tan(\alpha \pm \beta) = (\tan\alpha \pm \tan\beta)/(1 \mp \tan\alpha\tan\beta)$$
$$\sin^2\alpha = (1 - \cos 2\alpha)/2$$
$$\cos^2\alpha = (1 + \cos 2\alpha)/2$$
$$\sin(-\alpha) = -\sin\alpha$$
$$\cos(-\alpha) = \cos\alpha$$
$$\sin\left(\omega t + \frac{\pi}{2}\right) = \cos\omega t$$
$$\cos\left(\omega t - \frac{\pi}{2}\right) = \sin\omega t$$
$$\sin 2\alpha = 2\sin\alpha\cos\alpha$$
$$\cos 2\alpha = \cos^2\alpha - \sin^2\alpha$$
$$\sin^2\alpha + \cos^2\alpha = 1$$

Euler's Formula

Euler's formula is

$$e^{j\theta} = \cos\theta + j\sin\theta \tag{D.1}$$

An alternative form of Euler's formula is

$$e^{-j\theta} = \cos\theta - j\sin\theta \tag{D.2}$$

To derive Euler's formula, let

$$f = \cos\theta + j\sin\theta$$

Differentiating, we obtain

$$\begin{aligned}\frac{df}{d\theta} &= -\sin\theta + j\cos\theta \\ &= j(\cos\theta + j\sin\theta) \\ &= jf\end{aligned}$$

When $f = e^{j\theta}$, we have

$$\frac{df}{d\theta} = jf$$

as required. Thus, we obtain the result, Eq. D.1.

Adding Eq. D.1 and Eq. D.2, we obtain

$$\cos\theta = \frac{1}{2}(e^{j\theta} + e^{-j\theta})$$

Similarly, subtracting Eq. D.2 from Eq. D.1, we obtain

$$\sin\theta = \frac{1}{2j}(e^{j\theta} - e^{-j\theta})$$

The equivalence of the polar and rectangular forms of a complex number is a consequence of Euler's formula. To verify this, consider a complex number $\mathbf{A} = a + jb$ where $a = r\cos\theta$ and $b = r\sin\theta$. Therefore,

$$\begin{aligned}\mathrm{A} &= r(\cos\theta + j\sin\theta) \\ &= re^{j\theta}\end{aligned}$$

by Euler's formula.

Appendix E

Standard Resistor Color Code

Low-power resistors have a standard set of values. Color-band codes indicate the resistance value as well as a tolerance. The most common types of resistors are the carbon composition and carbon film resistors.

The color code for the resistor value utilizes two digits and a multiplier digit in that order as shown in Figure E.1. A fourth band designates the tolerance. Standard values for the first two digits are listed in Table E.1.

The resistance of a resistor with the four bands of color may be written as

$$R = (a \times 10 + b)m \pm \text{tolerance}$$

where a and b are the values of the first and second bands, respectively, and m is a multiplier. These resistance values are for 2% and 5% tolerance resistors as listed in Table E.1. The color code is listed in Table E.2. The multiplier and tolerance color codes are listed in Tables E.3 and E.4, respectively. Consider a resistor with the four bands yellow, violet, orange, and gold. We write the resistance as

$$\begin{aligned} R &= (4 \times 10 + 7)\ \text{k}\Omega \pm 5\% \\ &= 47\ \text{k}\Omega \pm 5\% \end{aligned}$$

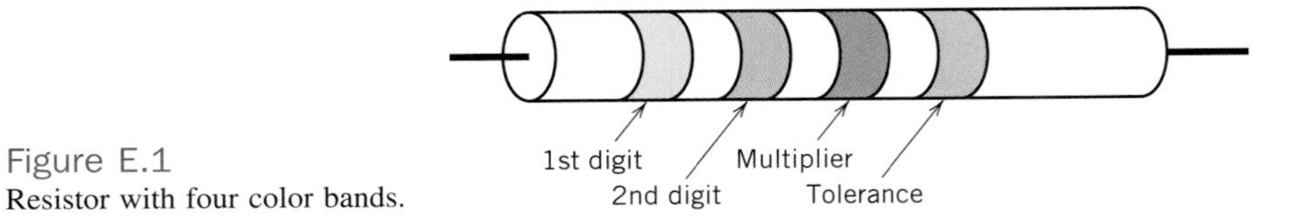

Figure E.1
Resistor with four color bands.

Table E.1 **Standard Values for First Two Digits for 2% and 5% Tolerance Resistors**

10	16	27	43	68
11	18	30	47	75
12	20	33	51	82
13	22	36	56	91
15	24	39	62	100

Table E.2 **Color Code**

0	black
1	brown
2	red
3	orange
4	yellow
5	green
6	blue
7	violet
8	gray
9	white

Table E.3 **Multiplier Color Code**

silver	0.01
gold	0.1
black	1
brown	10
red	100
orange	1 k
yellow	10 k
green	100 k
blue	1 M
violet	10 M
gray	100 M

Table E.4 **Tolerance Band Code**

red	2%
gold	5%
silver	10%
none	20%

Computer-Aided Analysis: SPICE

PREVIEW

This appendix is intended to serve as an introductory guide to SPICE, the circuit simulation program originally developed at the University of California at Berkeley. It is meant to provide the user with the means of learning to solve problems with this powerful computer tool. It contains sufficient detail and explanation to allow the user to advance rapidly to solving circuit analysis problems.

First, we provide an introduction to SPICE and its capabilities and describe the types of problems that SPICE can be used to solve. Section F.2 shows the user how to perform elementary circuit analysis. It gives the basic syntactical rules for the use of SPICE and shows how to solve simple dc circuits.

The next section deals with more sophisticated forms of dc circuit analysis including dc transfer curves and sensitivity analysis. The following sections explain how to perform time domain analysis and steady-state ac or frequency domain analysis. Finally, we show other analysis operations including temperature-dependent analysis, sensitivity analysis, analysis of circuits with mutual inductance, and Fourier analysis.

F.1 COMPUTER-AIDED ANALYSIS AND DESIGN

Engineers used mechanical calculators and slide rules for calculations in the analysis and design of circuits until the advent of digital computers in the 1950s. The world's first general-purpose electronic computer was built in 1946 by J. P. Eckert and J. W. Mauchly at the University of Pennsylvania. The nearly instantaneous response of electronic components made it possible for their ENIAC (Electronic Numerical Integrator and Computer) to multiply two 10-digit numerals in three-thousandths of a second. ENIAC employed 18,000 vacuum tubes and 6000 switches to perform 5000 additions per second. It was huge, taking up the walls of a 30-foot by 50-foot room, weighing 30 tons, and requiring eighty 12-hp fans to keep the components within temperature limits.

As computers became commercially available in the 1950s and 1960s, engineers worked to reduce the size of the components and the power required to operate the computer. The price per bit of memory dropped from 0.5 cent in 1973 to 0.0005 cent in 1991 (three orders of magnitude). Over the same time period, the number of gates per chip increased by three orders of magnitude and the power required to operate a computer declined by three orders of magnitude. With the reduction in size and power required, the personal computer and engineering workstations became available in the 1980s. Thus, by 1985 most engineers had sufficient computer facilities available to aid in the analysis and design of complex circuits.

In the 1970s, a mainframe computer program named SPICE was released. It was written at the University of California at Berkeley and put into the public domain.

SPICE is a general-purpose electric-circuit simulation program. The acronym SPICE stands for Simulation Program with Integrated Circuit Emphasis. The allowed components are resistors, capacitors, inductors, mutual inductances, independent dc and ac sources, dependent sources, transmission lines, diodes, and transistors.

In 1984, the first version of SPICE to run on a personal computer, named PSpice, became available. PSpice includes virtually all the features of general-purpose SPICE.

F.2 INTRODUCTION TO SPICE

Circuit analysis is an important part of the process of designing useful electric circuits. The use of computers to automate the process of circuit analysis has developed and evolved since the first digital computer became commercially available. One of the most complete circuit analysis tools is a software package known as SPICE. SPICE was developed to aid in the process of designing electronic circuits to be implemented as integrated circuits. This appendix will describe the use of SPICE in performing circuit analysis.

The process of designing useful circuits is generally an iterative one that involves several distinct steps that may be repeated several times over. First, the circuit requirements are described in detail. This step involves deciding what the circuit must be capable of when it is finally implemented. Next, the designer will synthesize a circuit that will likely satisfy the requirements. Synthesis is usually based on experience obtained from past synthesis and analysis operations. Once the synthesis is done, an analysis of the circuit must be done to validate the fact that the circuit will meet the requirements specified in the first step. If the original specifications are met, the design process is complete. If not, modifications must be made, and the process must be started again. These modifications are generally made in the synthesis step. The designer uses the information obtained from the analysis to change the configuration or component values so that the circuit behaves closer to the desired operation. In some cases it may also be necessary to modify the original specifications if the analysis shows them to be impractical. The number of times that the design loop must be traversed depends on the ability and experience of the designer.

Circuit analysis programs like SPICE are used to perform the analysis portion of the design cycle. These programs can save much time over performing the analysis using by-hand calculations. Further, once the problem has been set up, reanalysis can be done by simply editing the SPICE circuit description file and executing it again.

Finally, because of the way in which SPICE solves a circuit analysis, there will be some circuits that SPICE will not be able to solve. That is, SPICE's solution to the problem will not always be guaranteed to converge to a unique solution. In many cases, this situation can be corrected by modifying the circuit model. Of course this must be done carefully so that the modified model still represents the circuit accurately.

SPICE is capable of performing three main types of analysis. It can determine the dc behavior or selected output voltages with respect to changes in input voltages. This type of analysis is usually referred to as a dc analysis. The results of a dc analysis are sometimes called the transfer characteristics of the circuit. A single-point dc analysis also determines what is called the bias-point characteristics, that is, the behavior of the circuit when only a dc voltage is applied to the circuit. In most cases, this is the starting point to introduce and apply SPICE.

A second type of analysis that is usually required in order to determine fully a circuit's behavior is called a transient analysis. Transient analyses calculate circuit voltages and currents with respect to time. This assumes that there is a time-dependent stimulus that causes an effect on the rest of the circuit. To perform a time domain circuit analysis, SPICE first calculates the bias point and then calculates the circuit response to a time-dependent change in one or more voltage or current sources.

The third main type of analysis that SPICE can perform is called an ac analysis. This analysis type is also referred to as a sinusoidal steady-state analysis. Here, voltages and currents are calculated as a function of frequency. That is, output variable changes are calculated in response to changes in the amplitude, frequency, or phase of sinusoidal input voltage or current sources. Again, the starting point for an ac analysis is the calculation of the dc bias-point.

There are several subtypes of analysis that SPICE can perform. These subtypes are generally enhanced forms of one of the three main types of analysis. First, there is a Sensitivity Analysis that can be used to determine the dc response of a circuit to changes in the values assigned to circuit elements. For example, SPICE will calculate the change in a specified output voltage with respect to changes in selected circuit elements.

Sensitivity Analysis is a type of dc analysis that shows how variations in certain component values affect the overall dc behavior of a circuit. Temperature Analysis is used to determine changes in circuit response with respect to variations in component values due to changes in temperature. SPICE's Fourier Analysis allows the user to perform a spectral analysis of a circuit, calculating the Fourier coefficients for the sinusoidal components of voltages or currents in the circuit.

F.3 CIRCUIT DESCRIPTION WITH SPICE

SPICE can be used to perform many different types of circuit analysis. In order to familiarize the user with SPICE, this section will show how a very simple circuit can be analyzed with SPICE. The rules for creating a circuit description file are given, and the results of a SPICE analysis are shown.

For a circuit analysis program to perform its function, a circuit description that can be interpreted by the program must be generated. For SPICE, the topology of a circuit is described in terms of element models and the way in which they are interconnected. This information is contained in three parameters: element names, node numbers, and element values. Syntax for providing this information is standard for all versions of SPICE.

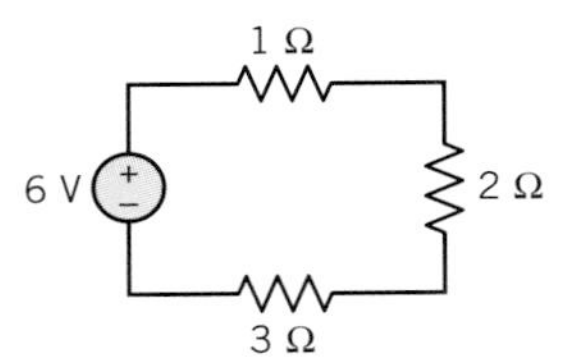

Figure F.1 Resistive circuit example with a dc source.

We will start by working through the example circuit shown in Figure F.1. This is a simple circuit consisting of three resistors and a dc voltage source. The objective is to determine the voltage across each resistor and the current flowing in the loop. The first step in creating the circuit description file is to name all of the elements. Next, each of the nodes is numbered. Figure F.2 is a modified version of Figure F.1 that contains node numbers and element names. Notice that the resistor names all begin with R and that the voltage source name begins with V. This is by convention and will be discussed later in this appendix. Further, nodes have been numbered starting at zero and continuing in consecutive order. Other node numbering (naming) schemes are available, and these will also be discussed later in this appendix.

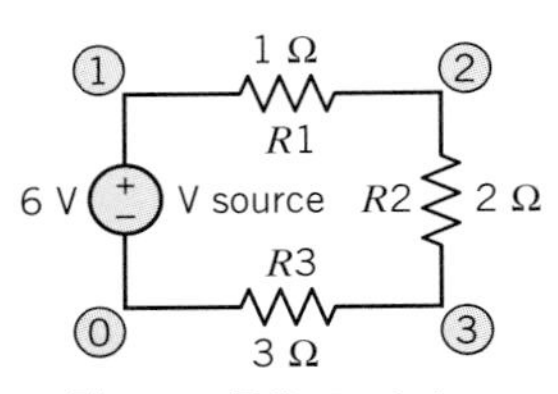

Figure F.2 Resistive circuit with SPICE element names and node numbers.

A circuit description file for SPICE consists of a set of lines or statements. Each statement either describes a circuit element, is used to convey information to SPICE about the type of analysis or output that is to be generated, or is a special control statement. The first statement in each circuit description file is an example of a special control statement: a title statement. SPICE assumes that the first statement in every circuit description file is a line that gives a title to the analysis being performed. This title will be printed across the top of each output sheet when a multiple-page output is generated. Figure F.3 contains a SPICE circuit description for the example circuit of Figure F.2 with a title of "3

```
            3 RESISTER CIRCUIT
****        CIRCUIT DESCRIPTION
************************************************
R1  1  2  1ohm
R1  2  3  2
R3  3  0  3
Vsource  1  0  6volts
.END
```

Figure F.3 Circuit description for the circuit of Figure F.2.

resistor circuit." Another important special control statement is seen at the end of the file. It is the .END statement that tells SPICE that it has read the entire circuit description file.

The general form of the statements used to describe individual elements is

```
<element name><+node><-node>...<element value>
```

The continuation dots in this general form are used to indicate that elements with more than two terminals will have more than two nodes and may have more than one value. The value given to an element is a number that indicates the size of the device in terms of a measurement unit corresponding to that device type. For instance, resistances are always given a value in ohms whereas voltage sources are given values in volts. The second line of Figure F.3 is an element description for the 1-Ω resistor. It indicates the resistor's name as **R1,** its node connections as between node 1 and node 2, and its element value as 1 Ω. Similarly, the other two circuit resistors are described in the two subsequent statements.

The voltage source uses the statement

```
V<name><+node><-node><voltage value>
```

The voltage source is named **Vsource** and is connected between node 1 and node 0. Voltage source descriptions must convey polarity information to SPICE. Thus, the convention that SPICE uses in a voltage element statement is that the first node is the positive node, followed by the negative node. The value of the dc voltage source is indicated to be 6 V.

SPICE uses a passive sign convention that always assumes the positive current flows into the first indicated node, which is assumed as the positive voltage node. The statement for a current source is

```
I<name><exit node><entry node><current value>
```

where all currents flow into the entry node.

The few lines in Figure F.3 complete the circuit description file necessary for SPICE to perform an analysis of the simple resistive circuit. If only a simple analysis is required, the next step is to submit this description to SPICE for processing.

One of the simplest analyses that SPICE is capable of performing is a single-point dc analysis. This is also referred to as an operating point analysis. An operating point analysis of the circuit in Figure F.1 may be obtained by submitting the circuit description file shown in Figure F.3 to SPICE. Since no type of circuit analysis is explicitly requested, an operating point analysis will be performed by default. An operating point analysis may be explicitly requested by including the statement .OP in the circuit description file.

SPICE is invoked by typing the program name, SPICE, and supplying the appropriate input and output filenames. The results of this analysis example are shown in Figure F.4.

```
     3 Resistor Circuit
****    SMALL SIGNAL BIAS SOLUTION    TEMPERATURE = 27.00 DEG C
*******************************************************************
NODE   VOLTAGE    NODE   VOLTAGE    NODE   VOLTAGE NODE   VOLTAGE
(  1)   6.0000    (  2)   5.0000    (  3)   3.0000
VOLTAGE SOURCE CURRENTS
NAME          CURRENT
Vsource       -1.000E+00
TOTAL POWER DISSIPATION 6.00E+00   WATTS
```

Figure F.4 Operating point analysis of the circuit of Figure F.2.

The first page printed, which repeats the input circuit description and includes any syntactical error messages or warnings, is not shown in this figure. SPICE labels the result of the operating point analysis as the SMALL SIGNAL BIAS SOLUTION on the top of the results page. The voltage calculated for each node is shown in tabular format. Following the node voltage table, SPICE prints the solution for the current flowing in each voltage source in the circuit and the total dc power dissipated by the resistors. Program performance characteristics are listed after all circuit analysis solutions are printed. The performance characteristics indicate whether or not the program terminated normally and how long it took to perform the circuit analysis.

On the top of each page containing results, SPICE prints the temperature at which the analysis is performed. SPICE may perform analyses at different temperatures. This capability is primarily useful when analyzing electronic circuits. However, the element that models a resistor's behavior in SPICE can have a temperature coefficient assigned to it. That is, a resistor's resistance value can be made to vary with temperature. A discussion of temperature-dependent analysis is beyond the scope of this appendix.

Notice that the current that SPICE calculates for the voltage source is -1 amp (see Figure F.4). SPICE uses the convention of reporting currents as the amount of current that flows into the positive node of an element, through the element, and out the negative node. Thus, the current flowing into the positive terminal of the voltage source is minus one ampere.

If a default analysis is permitted, the only solutions are nodal voltages. Print requests for specific currents or voltages will be ignored in the **.**PRINT statement. If specific branch currents or potential differences across circuit elements are required, then the more detailed **.**DC analysis must be invoked.

Many different element models are available to represent physical devices within a SPICE analysis. Table F.1 contains a list of the basic circuit elements available in SPICE. By convention, the key letter identified in the table is used to identify a particular element to SPICE. A key letter must be the first letter of an element name. The remaining letters in an element name may be alphabetic or numeric. For SPICE, an underscore or the dollar sign may also be used within an element name. Names can be longer than 80 characters, though names longer than 20 characters are rarely used.

Element names should be descriptive of the function of the element in the circuit if possible. Uppercase and lowercase letters may be used in names for most versions of SPICE. However, the program does not distinguish between the two types and hence, VCC is the same as Vcc, which is the same as vcc.

Nodes are the junction points where two or more circuit elements are joined. For SPICE, nodes must be numbered as positive integers. SPICE takes the node numbered 0 as the

Table F.1 **Basic Circuit Elements Available with SPICE**

C	Capacitor
E	Voltage-Controlled Voltage Source
F	Current-Controlled Current Source
G	Voltage-Controlled Current Source
H	Current-Controlled Voltage Source
I	Independent Current Source
K	Inductor Coupling (Transformer)
L	Inductor
R	Resistor
T	Transmission Line
V	Independent Voltage Source
X	Subcircuit Call

Table F.2 **Scale Factor Abbreviations**

Letter Suffix	Multiplying Factor	Name of Suffix
T	1E12	tera
G	1E9	giga
MEG	1E6	mega
K	1E3	kilo
MIL	25.4E-6	mil
M	1E-3	milli
U	1E-6	micro
N	1E-9	nano
P	1E-12	pico
F	1E-15	femto

ground node. Node numbers need not be chosen in consecutive order; i.e., a continuous sequence of numbers need not be used to identify all of the nodes. SPICE allows nodes to be identified by any character string, not just a string of numbers. Throughout this appendix, numbers will be used to identify nodes in order to maintain compatibility with most versions of SPICE.

The ground node, node 0, is the node to which other node voltages are referenced. SPICE requires that every node in the circuit have a dc path to ground. This is necessary because of the way in which SPICE solves for the unknown voltages and currents. SPICE issues a warning message if a dc path to ground does not exist from every node. Along with this requirement, SPICE requires that each node be connected to at least two elements. This precludes having a node or wire dangling, as you might have in a lab testing environment. A very large resistor, say on the order of 100 MΩ, can be used to provide the necessary dc path to ground and circumvent the restriction on dangling nodes. For most circuits, a 100-MΩ resistor will have little or no effect on the circuit's performance.

Component values may be represented in several different forms in SPICE. Standard decimal notation may be used. For example, numbers such as 10, 10., 10.0001, −10.0001 may be used whenever a value is needed. SPICE also accepts values in floating-point or scientific notation. In this notation, a number such as 1000 can be written as 1.0E3, which stands for 1×10^3. The letter E is used to separate the mantissa from the exponent of a base-10 floating-point number.

Because the range of values can be very large within a given SPICE circuit description, and to reduce the need for unnecessary typing and the errors that this typing might incur, the developers of SPICE included a set of abbreviations that may be used as suffixes to values. These suffixes are listed in Table F.2. For example, the value 0.004UF represents 0.000000004 F or 4E-9 F. Notice that this value could also have been written as 4NF.

It should be noted that in describing component values, any letter can follow the number, but if that letter is one of the designative modifiers it will be interpreted as a modifier. In other words, you could put 14a or 14 amperes or 14A to be more descriptive of the value. Care should be used to avoid typing 10F (for 10 farads) because F is the femto multiplier.

All well-written computer software is sufficiently documented so that someone unfamiliar with the code may readily use and modify it. This should also be true for circuit models written for use with SPICE. Two methods are available for specifying comments in a SPICE circuit model. Any line beginning with an asterisk (*) is taken to be a comment line. SPICE does not interpret any of the characters in a comment line, it simply prints the characters as is in the output listing. Comments may also be included at the end of lines that contain SPICE commands or element specifications. A semicolon is used to separate a comment from a command or element specification. Of course, the comments must always follow, not precede, the command or element specification.

Another important syntactical mechanism that you will eventually need to use is the line continuation operator. As your models become more complex, you will probably generate statements that are longer than a single line. To continue a line on the next line, the succeeding line should start with a plus sign character (+). Continuation lines may be carried on repeatedly, as long as succeeding continued lines begin with the plus character.

F.4 DC CHARACTERISTICS

Thus far, the simulations performed have calculated the behavior of a circuit when all of the circuit's sources are held at a single, individual fixed voltage or current. In many instances in the design process, it is necessary to obtain the analysis of a circuit for several different values of source voltage or current. This chapter discusses the use of a sweep analysis to determine the dc behavior of a circuit.

SPICE voltage or current sources may be swept over a range through the use of the .DC control statement. When the .DC statement is included in a simulation file, SPICE first performs the normal operating point analysis. Circuit unknowns are calculated with the single given source voltages or source currents. Next, SPICE performs the sweep analysis by stepping through a specified range of values for a particular input source. The result of using the .DC analysis type is the same as performing repeated .OP analyses where the value of a particular source is changed before each analysis. Obviously, it is much less work and much quicker to use the .DC statement rather than use multiple .OP analyses.

The syntax for the .DC statement is as follows:

```
.DC<source name><start value><stop value><increment>
```

where *source name* is the name of the voltage or current source that is to be swept through a range of values. This source must be defined elsewhere in the description file by a voltage or current source statement. *Start value* and *stop value* define the range of values over which the source will be swept. The *increment* value determines the step size between succeeding analysis points. *Start value, increment,* and *stop value* are interrelated to determine the number of points for which the circuit is analyzed.

It is easily seen that the number of analysis points is (*stop value* − *start value*)/*increment* + 1. Depending upon the version of SPICE being used, there may be a default limit for the number of points that can be included in a .DC analysis. SPICE does not limit the number of iteration points that it allows. Should the .DC parameters be incorrectly specified and cause SPICE to initiate an overly lengthy analysis, the program can be terminated by typing a control-C (hold down Ctrl key and press the C key at the same time).

As an example of the use of the .DC analysis capabilities, consider the circuit shown in Figure F.5. This circuit can be recognized as a model for Edison's parallel lamp scheme.

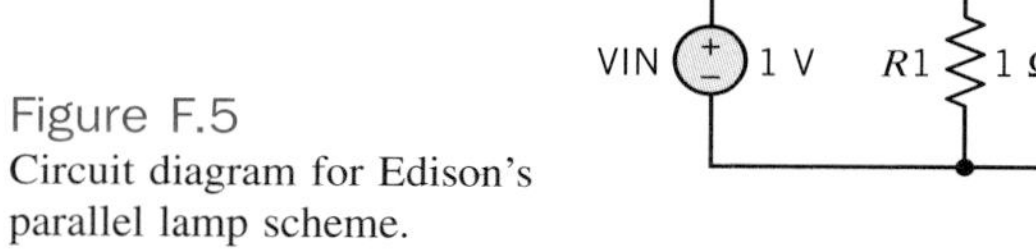
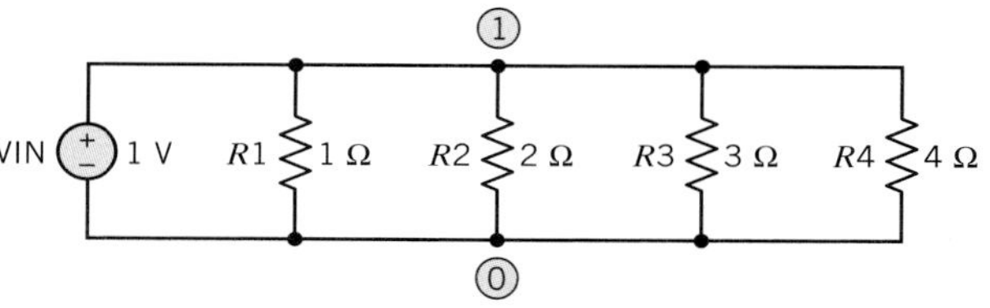

Figure F.5
Circuit diagram for Edison's parallel lamp scheme.

Each resistor represents an individual lamp. This scheme provides the same voltage to each lamp element, and as such, the current in individual lamps is independent of that in other lamps. The objective here is to determine the behavior of the circuit as the source voltage is increased from 0 V to 5 V. Of interest is the power dissipated in each lamp as the voltage supply increases in value. The power dissipated in a lamp can be used to determine the amount of light given off by the device. Although SPICE will not directly calculate the power supplied to an element, both the voltage across an element and the current through an element may be determined and provided as output by SPICE. The power dissipated in each element may then be calculated.

Figure F.6 contains the circuit description file for the circuit shown in Figure F.5. The .DC statement is used to sweep the source VIN over the range 0 to 5 V in steps of 0.2 V. Notice that the circuit description file is the same as that which would be used to calculate the operating point values, except for the addition of the .DC and .PRINT statements. The .PRINT statement is used to output the results of the .DC analysis.

SPICE presents the output of an operating point analysis automatically in a tabular form. This output displays the voltage value at each node of the circuit for the fixed driving source values. When an analysis is performed wherein the values are swept, such as the .DC analysis, it is convenient to provide a mechanism for choosing only the nodes of interest for output display. The .PRINT statement provides just such a mechanism. .PRINT will also be used with other analysis types that will be discussed in later sections. .PRINT allows you to choose which values from a particular analysis are to be displayed. By appropriately ordering the variables in the .PRINT statement, it is possible to order the resultant output tables for ease of value comparison.

The syntax for the .PRINT statement is

```
.PRINT<analysis type><output list>
```

where the *analysis type* specifies that the variables in the succeeding list are to correspond to values calculated for a particular analysis type. To obtain output generated by a .DC analysis, the *analysis type* DC should be specified. This will produce a listing for each output variable at each point in the .DC sweep analysis. The first column of the tabular output for a .DC analysis will contain values of the swept source over the full range specified in the .DC statement. Output values can be conveniently referred back to these source values.

The *output list* is a list of the node voltages, element voltages, or element currents whose values are to be displayed. Node voltage variables for the *output list* are of the

```
Edison's parallel lamp scheme
****    CIRCUIT DESCRIPTION
*****************************************************************
R1  1  0  1
R2  1  0  2
R3  1  0  3
R4  1  0  4
VIN  1  0  DC  1
.DC  VIN  0  5  .2
.PRINT  DC  I(VIN)  I(R1)  I(R2)  I(R3)  I(R4)
.PLOT  DC  I(VIN)  I(R1)  I(R2)  I(R3)  I(R4)
.PROBE
.END
```

Figure F.6 Circuit description for Edison's parallel lamp scheme.

form V(n1, n2), where the voltage drop from node n1 to n2 is specified. If only a single node, n1, is included in the list, the output voltage will be referenced to ground. That is, the n2 parameter will be assumed to be a 0. Output voltages may also be specified using the form V(element). Here, the voltage across the element will be output. *Element* may be any valid element name. For example, we print the voltage across a resistor R_2 by specifying V(R2).

Element currents are specified using the form I(element) where *element* is any valid element name, including the name of a source element. The printed values of current have the direction from n1 to n2 where n1 is the first node in the list and n2 is the second. For example, current flowing in the resistor described in the statement

```
R1 5 6 1Kohm
```

will have a positive value when current flows from node 5 to node 6. That is, when the voltage at node 5 is higher than the voltage at node 6, the current flowing in R1 will be reported as positive. This convention also applies to voltage sources. As with other elements, the printed values of current will be positive for currents flowing from n1 to n2, that is, from the positive node of the source to the negative node. This may seem to be the opposite of what is expected, but once recognized, this seeming discrepancy may be easily dealt with. For the voltage source described in the statement

```
VDD 7 8 6volts
```

the current in element VDD will be reported as positive when it flows from node 7 to node 8. Of course the voltage across the source will always have the same polarity no matter which way the current is flowing. To verify this, we refer again to the circuit in Figure F.2. Clearly, a 1-A current flows from node 0 to node 1 through the source. The input statement in Figure F.3 reads **Vsource 1 0 6volts** so that the .PRINT statement will yield the current from node 1 to 0, which is −1 A, as shown in Figure F.4.

Several voltages and currents may be printed in a single table. The actual number will depend on the version of SPICE being used and the default settings. Voltage and current values may be freely mixed in an output table.

The following is a set of valid .PRINT statements along with their meanings:

```
.PRINT V(5,2)                  ;print voltage between node 5 and node 2
.PRINT V(5)                    ;print voltage between node 5 and ground
.PRINT V(R1)                   ;print voltage across resistor R1
.PRINT 1(R1)                   ;print current through resistor R1
.PRINT V(1,2) V(2,3) I(VCC)    ;print voltage from node 1 to 2, the
                               ;voltage from node 2 to 3, and the current
                               ;through voltage source VCC in one table
```

Figure F.7 contains the output listing generated for the simulation of the circuit described in Figure F.6. This output shows the current through the source as well as through each of the resistive elements. Check to see that Kirchhoff's current law holds for each of the nodes of the circuit.

The number of digits printed in each column of the output table can be changed using the .OPTIONS statement. The syntax for this statement is

```
.OPTIONS[option name][option name = value]
```

where option name may be one of a list of keywords that may or may not require a *value* parameter. To change the number of digits printed, the keyword NUMDGTS followed by an integer value is used. The following statement changes the number of digits printed to six from the default of four:

```
.OPTIONS NUMDGTS = 6 WIDTH = 132
```

```
Edison's parallel lamp scheme

****   DC TRANSFER CURVES                    TEMPERATURE = 27.000 DEG C

**************************************************************

VIN          I(VIN)       I(R1)        I(R2)        I(R3)        I(R4)

0.000E+00    0.000E+00    0.000E+00    0.000E+00    0.000E+00    0.000E+00
2.000E-01   -4.167E-01    2.000E-01    1.000E-01    6.667E-02    5.000E-02
4.000E-01   -8.333E-01    4.000E-01    2.000E-01    1.333E-01    1.000E-01
6.000E-01   -1.250E+00    6.000E-01    3.000E-01    2.000E-01    1.500E-01
8.000E-01   -1.667E+00    8.000E-01    4.000E-01    2.667E-01    2.000E-01
1.000E+00   -2.083E+00    1.000E+00    5.000E-01    3.333E-01    2.500E-01
1.200E+00   -2.500E+00    1.200E+00    6.000E-01    4.000E-01    3.000E-01
1.400E+00   -2.917E+00    1.400E+00    7.000E-01    4.667E-01    3.500E-01
1.600E+00   -3.333E+00    1.600E+00    8.000E-01    5.333E-01    4.000E-01
1.800E+00   -3.750E+00    1.800E+00    9.000E-01    6.000E-01    4.500E-01
2.000E+00   -4.167E+00    2.000E+00    1.000E+00    6.667E-01    5.000E-01
2.200E+00   -4.583E+00    2.200E+00    1.100E+00    7.333E-01    5.500E-01
2.400E+00   -5.000E+00    2.400E+00    1.200E+00    8.000E-01    6.000E-01
2.600E+00   -5.417E+00    2.600E+00    1.300E+00    8.667E-01    6.500E-01
2.800E+00   -5.833E+00    2.800E+00    1.400E+00    9.333E-01    7.000E-01
3.000E+00   -6.250E+00    3.000E+00    1.500E+00    1.000E+00    7.500E-01
3.200E+00   -6.667E+00    3.200E+00    1.600E+00    1.067E+00    8.000E-01
3.400E+00   -7.083E+00    3.400E+00    1.700E+00    1.133E+00    8.500E-01
3.600E+00   -7.500E+00    3.600E+00    1.800E+00    1.200E+00    9.000E-01
3.800E+00   -7.917E+00    3.800E+00    1.900E+00    1.267E+00    9.500E-01
4.000E+00   -8.333E+00    4.000E+00    2.000E+00    1.333E+00    1.000E+00
4.200E+00   -8.750E+00    4.200E+00    2.100E+00    1.400E+00    1.050E+00
4.400E+00   -9.167E+00    4.400E+00    2.200E+00    1.467E+00    1.100E+00
4.600E+00   -9.583E+00    4.600E+00    2.300E+00    1.533E+00    1.150E+00
4.800E+00   -1.000E+01    4.800E+00    2.400E+00    1.600E+00    1.200E+00
5.000E+00   -1.042E+01    5.000E+00    2.500E+00    1.667E+00    1.250E+00
```

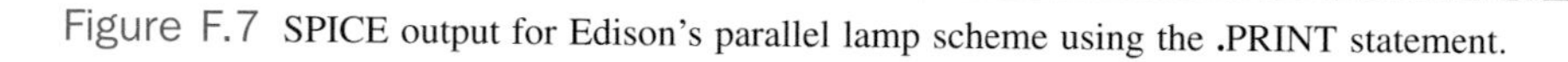

Figure F.7 SPICE output for Edison's parallel lamp scheme using the .PRINT statement.

This example also demonstrates the use of the WIDTH parameter to increase the width of the printed page from the default of 80 characters per line to 132 characters per line.

In many cases, it is advantageous to get a visual perspective of the data being presented as the output of a simulation. The .PLOT statement provides a means of obtaining such a perspective by having results plotted as a part of the output generated by SPICE. Unfortunately, this means of plotting produces only what is referred to as *character-printer* quality plots. This means that each plotted data point is represented by a character, with each step on the x-axis being the size of a line and each step on the y-axis being the size of a character column.

The syntax for using the .PLOT statement is very similar to that of the .PRINT statement. As with .PRINT, the type of analysis for which the plot is desired is specified first, followed by the variables that are to be plotted. The syntax is

```
.PLOT<analysis type><output list>[<min range>,<max range>]
```

where as with .PRINT, *analysis type* specifies one of several different possible types of analysis that SPICE can perform and *output list* specifies which of the circuit variables are to be plotted. When more than one output variable is specified in the *output list,* each variable in the list will be plotted on the same graph. Multiple .PLOT statements will produce plots on separate axes; one axis for each .PLOT statement.

PSpice is capable of producing standard line printer plots as described above. In addition, a program called PROBE has been written by MicroSim Corporation to supplement the plotting capabilities of PSpice. PROBE uses the graphics capabilities built into a PC to produce what is called *graphics printer quality plot.* The quality of the plot will depend on the resolution of your PC's display or printer.

To use PROBE, it is necessary to include a statement in the PSpice circuit description file so that data to be plotted by PROBE will be generated by PSpice. The data to be plotted will be put into a disk file named PROBE.DAT. When PROBE is executed, the software will check to determine that this file exists, and if not, PROBE will display an error message and terminate. The statement that must be included in the PSpice circuit description file is called the .PROBE statement. The syntax for the .PROBE statement is similar to that of the .PRINT and .PLOT statements:

```
.PROBE [output list]
```

The *output list* is optional and may contain a list of the node voltages or element currents that are to be put into the PROBE.DAT data file. If no parameters are contained in the *output list,* then all of the circuit voltages and currents are included in the PROBE.DAT file. This is how the .PROBE statement is typically used. All of the circuit variables are included in the data file, and as PROBE is run, individual variables are selected for plotting. In this way, it is unnecessary to rerun a PSpice analysis if, when a PROBE plot is done, it is suddenly recognized that an additional curve is desirable. Commands are entered via an on-screen menu, reducing the need to remember long lists of commands and their syntax. Multiple waveforms may be plotted on the same graph. Multiple graphs may be displayed on the same page. The command set provides for producing a hard copy of the graphs displayed on the screen.

Figure F.8 contains a graph generated by PROBE for the element currents in the circuit of Figure F.5.

Subcircuits are means for reducing the amount of typing necessary to describe large circuits that contain repeated blocks. A subcircuit for SPICE is very similar to a subroutine or a macro in a programming language. A set of circuit elements may be bundled into

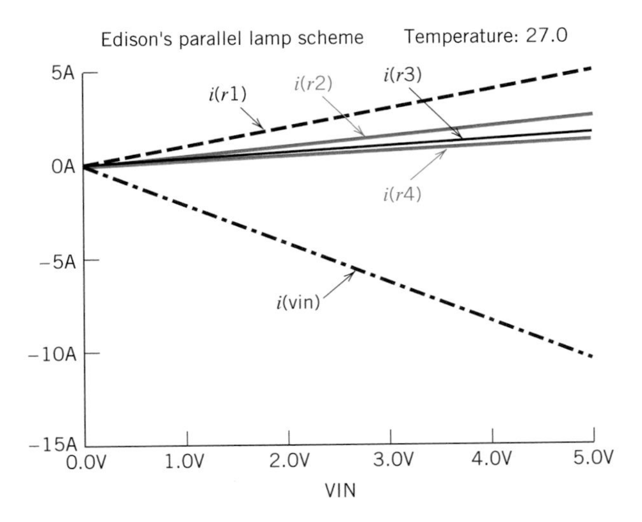

Figure F.8
Circuit data plotted with PROBE.

a block structure with a defined set of nodes at the interface boundary. This subcircuit may then be "called" and connected to other circuit elements as many times as it is needed in the overall circuit. More than one subcircuit may be defined within a circuit, and each subcircuit may be called more than one time.

A subcircuit is defined using the keyword **.SUBCKT** to start a subcircuit definition and **.ENDS** to signal the end of a subcircuit definition. The syntax for subcircuit definition is

```
.SUBCKT<subcircuit name><interface node list>
circuit element statements defining subcircuit
.ENDS
```

where the *subcircuit name* is a means by which to refer to a particular subcircuit. The *interface node list* is a list of nodes internal to the subcircuit to which external circuit elements will be attached. Node numbers contained in the subcircuit are unique to the subcircuit and may duplicate those in the external circuit. Only node 0 is an exception to this rule. Node 0 is the ground node and is common to all subcircuits and external circuits. As such, it is unnecessary to include node 0 in an interface node list.

SPICE has four basic types of dependent sources for use in circuit modeling: the *voltage-controlled voltage source* (VCVS), the *current-controlled current source* (CCVS), the *voltage-controlled current source* (VCCS), and the *current-controlled current source* (CCCS). Controlled sources may be linear functions of the controlling variable, or there may be a nonlinear relationship that defines how a source functions with regard to its control variable.

Linear dependent sources are syntactically specified in much the same way as independent sources. The difference is that in a dependent source, a controlling variable is specified, a multiplying or gain factor is specified, and the key letter that is used to begin the name of the element is different. The following is a list showing the syntax for each of the available dependent sources:

```
(VCVS) Ename <element nodes> <control nodes> <gain factor>
(CCVS) Hname <element nodes> <control current> <gain factor>
(VCCS) Gname <element nodes> <control nodes> <gain factor>
(CCCS) Fname <element nodes> <control current> <gain factor>
```

An example of a simple VCVS that is attached to nodes 1 and 2 and produces a voltage that is 10 times the voltage drop from node 5 to node 6 is

```
Esimple 1 2 5 6 10
```

In algebraic terms, this source would be described by the equation

$$v_1 - v_2 = 10 \times (v_5 - v_6)$$

Element nodes carry the same polarity conventions and current direction conventions as the independent sources. The first node specified is the positive node, and the positive current flows through the source from positive node to negative node. The control or sense nodes have the more positive node specified as the first node. The sense nodes have an infinite input impedance and therefore draw no current. That is, the controlled source senses the voltage across the controlling nodes but is not actually attached to those nodes. Control nodes may be the same as the element nodes.

Current-controlled sources will be a function of currents in specific branches of the circuit. The controlling branch current must be the current in an independent voltage source. This independent voltage source may have a zero or nonzero voltage value. If its value is zero, it is equivalent to a short circuit. Thus, if the controlling current is to be that which flows in a resistive branch, it is necessary to insert an independent voltage

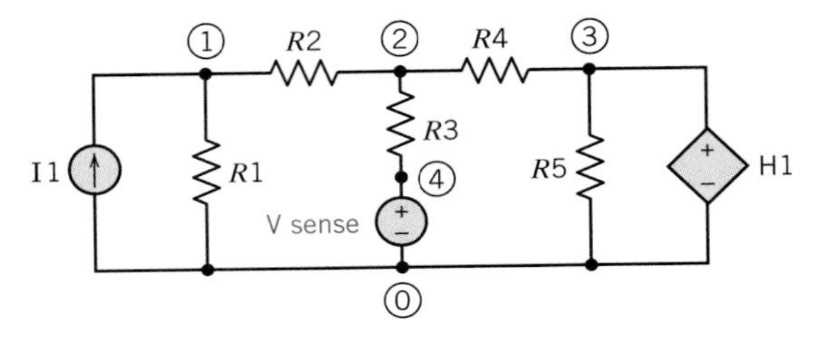

Figure F.9
Circuit with a CCVS H1.

source in series with the resistor. The value of the voltage source is set to zero so that the new source does not have any effect on the branch current.

An example of a current-controlled voltage source is

```
Hccvs 3 4 Vdummy 5.6
```

where the source name is *Hccvs* and it produces a voltage drop from node 3 to node 4. The controlling current flows through the independent voltage source *Vdummy* and the gain factor is 5.6 Ω. Notice that the gain factor here has the units of resistance and is called a transresistance. This is because to produce the units of volts, the output of the source, it is necessary to multiply the sensed current in amperes by a value in ohms. The term *transresistance* comes from the fact that a transfer is made from one set of nodes to another.

Rules for using the VCCS and CCCS are very similar to those for the VCVS and CCVS. Outputs of the dependent current sources are specified in the same way as those of the independent current sources. Controlling voltages and currents are specified in the same way as for the VCVS and CCVS. For the CCCS, a dimensionless gain factor is provided. A transconductance is supplied for the VCCS.

Figure F.9 is the schematic for a circuit containing a CCVS. The SPICE file for describing this circuit is shown in Figure F.10. Notice that the voltage source **Vsense** is in the circuit for the sole purpose of measuring the current through the resistor R_3. A single-point solution for the circuit using the **.OP** statement is shown in Figure F.11.

```
Current Controlled Voltage Source Example

****        CIRCUIT DESCRIPTION

*****************************************************************

I1  0  1  5amps
H1  3  0  Vsense  3.5
R1  1  0  10
R2  1  2  20
R3  2  4  30
R4  2  3  40
R5  3  0  50
Vsense  4  0  0volts
.OP
.END
```

Figure F.10 SPICE description of the CCVS circuit of Figure F.9.

```
Current Controlled Voltage Source Example
****     SMALL SIGNAL BIAS SOLUTION TEMPERATURE = 27.000 DEG C
*****************************************************************
NODE   VOLTAGE   NODE   VOLTAGE   NODE   VOLTAGE   NODE   VOLTAGE
(  1)   -39.207 (  2)   -17.621 (  3)    -2.056  (  4)     0.0000
VOLTAGE SOURCE CURRENTS
NAME          CURRENT
Vsense        -5.874E-01
TOTAL POWER DISSIPATION 0.00E+00 WATTS
```

Figure F.11 Solution for the CCVS circuit of Figure F.9.

F.5 TIME DOMAIN ANALYSIS

Electrical circuits all have some time-dependent characteristics that must be considered during the design cycle. Even the simple flashlight driven by dc batteries has a time dependent response. When the circuit's switch is closed, the dc source is applied to the circuit, causing a time-varying current response.

Thus far, we have discussed the use of SPICE in analyzing the performance of a circuit to which a dc source has been applied. In these analyses, it is assumed that none of the sources vary as a function of time. Further, capacitive elements were set to open circuits, and inductive elements were set to short circuits for the dc analyses. In this section, the behavior of a circuit in response to a time-varying stimulus will be discussed.

A typical circuit of interest is shown in Figure F.12. In this circuit, the switch is opened at arbitrary time $t = 0$ after being closed for a long time. If the switch has been closed for a long period of time before $t = 0$, it can be assumed that all transient effects have died out and that the circuit can be treated as a dc circuit. That is, the circuit can be analyzed with the capacitance set to an open circuit. A dc analysis provides that the voltage across the capacitor just before $t = 0$ is equal to the dc source voltage VINIT.

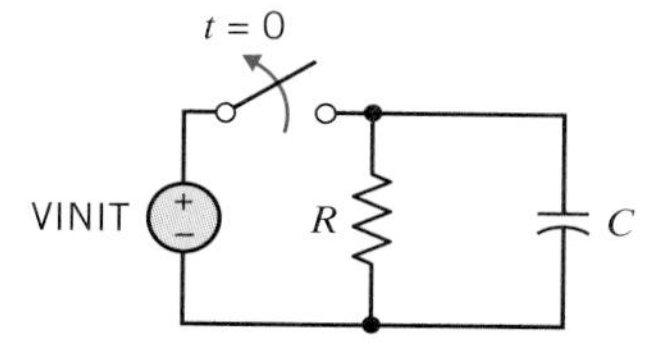

Figure F.12 *RC* circuit with dc source connected prior to $t = 0$.

Once the switch is opened at $t = 0$, the circuit becomes the simple connection of a resistor in parallel with a capacitor. The initial charge stored on the capacitor will cause a current to begin flowing through the resistor at $t = 0$. Current will flow until the initial voltage is reduced to zero by discharging the capacitor through the resistor. This is termed the natural response of the circuit. It is important to be able to calculate the voltage and current in this circuit as a function of time. The transient analysis capabilities built into SPICE can perform these calculations.

Unfortunately, SPICE does not have a built-in circuit element that directly models the behavior of a switch. A model for the effects of opening the switch must be built from existing element models. Figure F.13 contains the diagram of a circuit that will model this particular situation. Current source ISTEP is time-varying, changing as a step function when $t = 0$. Before $t = 0$, the current source produces sufficient current so that the appropriate initial voltage is developed across the terminals of the capacitor. At $t = 0$, the current source is set to zero, thus producing the effect of an open circuit.

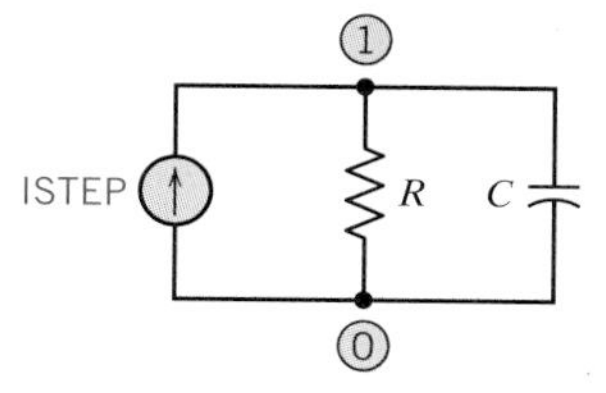

Figure F.13 Current source ISTEP supplies the initial conditions.

In order to determine how a circuit responds as a function of time, it is necessary to have models for sources that vary as a function of time. SPICE uses a modification of the previously used dc voltage or current source element statement to produce a time-varying source description. Several different functions of time are available. The general syntax for a time-varying source is

```
<source name><nodes><time function type><time parameters>
```

The *source name* and *nodes* use the same syntax as previously described for dc sources. The time function types include:

PULSE—pulse waveform that may be periodic
EXP—exponential waveform
SIN—sinusoidal waveform
PWL—piecewise linear waveform for constructing arbitrary functions of time

The PULSE input source waveform is shown in Figure F.14. The general form of the PULSE specification is

```
PULSE(<v1><v2><td><tr><tf><pw><per>)
```

where Table F.3 describes each of the parameters and their default values.

The PULSE source generates a waveform that is at $v1$ volts (or amps) at $t = 0$. After a delay time of *td*, the waveform makes a transition to $v2$ over a rise time interval *tr*. The waveform stays at $v2$ for a length of time equal to *pw*, then returns to $v1$, making the transition over the time interval *tf*. If *per* is specified, the waveform, starting from the beginning of *tr*, will repeat continuously with a period of *per*. The delay time is not included in the repeated waveform. Note that the value of $v1$ may be either greater or less than $v2$.

An example of a pulsed waveform current source is

```
ISPIKE 10 0 PULSE(0 100 10N 1P 1P 10P 20N)
```

This source produces a current flowing from node 10 to node 0. The current is initially at zero amps from $t = 0$ until $t = 10$ ns (1E-9 seconds). The current then rises linearly to 100 A over the next picosecond (1E-12 seconds) and stays there for 10 ps. Next, the current falls back to zero over a 1-ps interval. The current stays at zero for the next 19.988 ns, after which the waveform repeats its transition from 0 to 100 A.

As another example of a PULSE source, consider

```
VSTEP 5 6 PULSE(0 1 0 1P)
```

Only the first four PULSE parameters are specified, leaving the others to become the default values. This source produces a step waveform that changes from 0 to 1 V in 1 ps at time $t = 0$. The waveform stays at 1 V for the rest of the analysis since the pulse width parameter defaults to be equal to time TSTOP. This waveform approximates the ideal step function used in many different analysis problems.

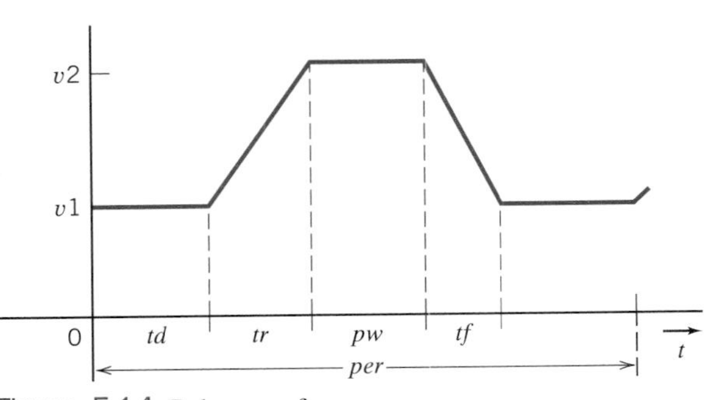

Figure F.14 Pulse waveform.

Table F.3 **PULSE Waveform Specifications**

Parameter	Default Value, Units
<*v1*> intial value	none, volts or amps
<*v2*> pulse value	none, volts or amps
<*td*> delay time	0, seconds
<*tr*> rise time	TSTEP, seconds
<*tf*> fall time	TSTEP, seconds
<*pw*> pulse width	TSTOP, seconds
<*per*> period	TSTOP, seconds

TSTEP and TSTOP are defined as part of the .TRAN statement in the next section. Even though a user might specify rise time and fall time equal to zero in the command line, a zero value will never be implemented.

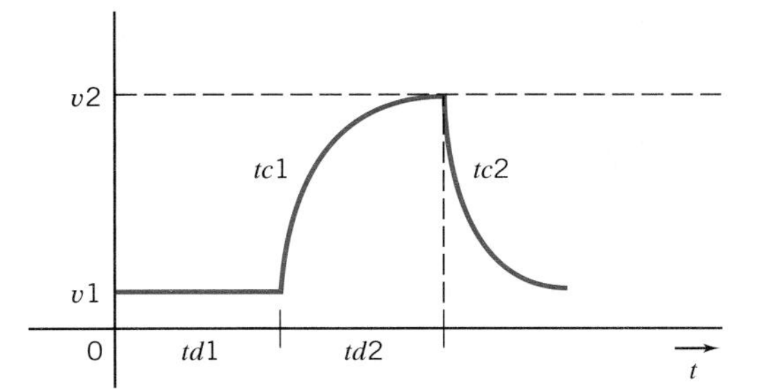

Figure F.15 Exponential waveform.

Table F.4 **EXP Waveform Specifications**

Parameter	Default Value, Units
<*v1*> initial value	none, volts or amps
<*v2*> peak value	none, volts or amps
<*td1*> rise delay time	0, seconds
<*tc1*> rise time constant	TSTEP, seconds
<*td2*> fall delay time	<*td1*> + TSTEP, seconds
<*tc2*> fall time constant	TSTEP, seconds

The EXP input source is similar to the PULSE source except for the fact that it produces a waveform that varies exponentially with time. The EXP input source waveform is pictured in Figure F.15. The general form of the EXP specification is

```
EXP (<v1><v2><td1><tc1><td2><tc2>)
```

where Table F.4 describes each of the parameters and their default values.

The EXP source generates a waveform that is at *v1* V (or amps) at $t = 0$. After a delay time of *td1,* the waveform makes a transition to *v2*. This transition takes the shape of an exponential waveform with a time constant of *tc1*. That is, the waveform is exponential. Fall delay *td2* determines the length of the interval over which the rise from *v1* to *v2* takes place. After interval *td2* expires, the waveform returns to *v1* exponentially, with the constant *tc2* specifying the exponential decay. Note that again the value of *v1* may be either greater than or less than *v2*.

An example of an exponential waveform voltage source is

```
VSAW 5 6 EXP(0 5 2M 500N 1M 1M)
```

This source produces a voltage drop from node 5 to node 6 that is initially 0 V until $t = 2$ ms. At 2 ms, the voltage exponentially rises to 5 V at a rate defined by the time constant 500 ns. The rising exponential lasts until $t = 3$ ms, after which the voltage drops back to zero exponentially with a time constant of 1 ms.

The SIN input source waveform is pictured in Figure F.16. The general form of the SIN specification is

```
SIN (<vo><va><freq><td><df><ph>)
```

where Table F.5 describes each of the parameters and their default values.

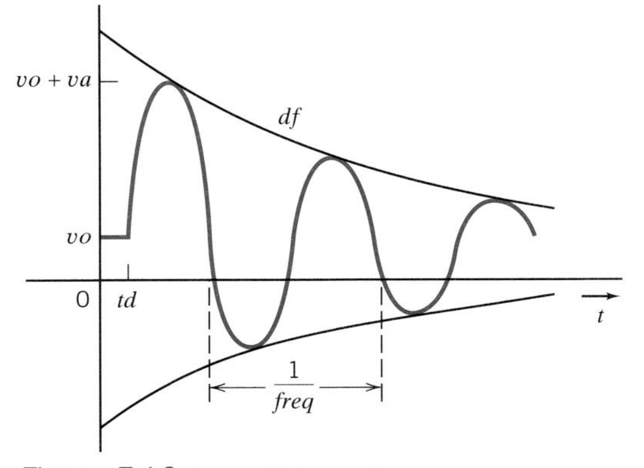

Figure F.16 Sinusoidal waveform.

Table F.5 **SIN Waveform Specifications**

Parameter	Default Value, Units
<*vo*> initial offset	none, volts or amps
<*va*> peak amplitude	none, volts or amps
<*freq*> frequency	1/TSTOP, hertz
<*td*> delay	0, seconds
<*df*> damping factor	0, 1/seconds
<*ph*> phase	0, degrees

The SIN source generates a waveform that is equal to vo V (or amps) at $t = 0$. After a delay time of td, the waveform begins to oscillate sinusoidally with an initial amplitude of va. The frequency of the sine wave is specified by *freq*. The amplitude of the waveform decays exponentially with the damping factor *df* specifying the decay rate. Parameter *ph* specifies the phase of the waveform in degrees with respect to $t = td$. An equation that describes the waveform mathematically is

$$v(t) = vo + va \times \exp\left(-(t - td)\,df\right) \times \sin(2\pi(freq(t - td) - ph/360))$$

A source specified with the time-varying SIN source is only for use in transient analyses. It cannot be used as a source in an ac analysis (ac analysis is explained in the next section).

An example of a sinusoidal time-varying waveform voltage source is

```
VRING 6 5 SIN(1 5 100K 50N 1M 90)
```

This source produces a voltage drop from node 6 to node 5 that is initially 1 V until $t = 50$ ns. At 50 ns, the source begins oscillating sinusoidally with a frequency of 100 KHz. The phase of the voltage is 90°, meaning that at $t = 50$ ns the voltage starts at its maximum amplitude (since sin 90° = 1) and begins declining toward zero. The magnitude of the waveform decays exponentially with an exponential damping factor of 1E-3.

The PWL (piecewise linear) input source waveform is shown in Figure F.17. The general form of the PWL specification is

```
PWL (<t1><v1><t2><v2>...<tn><vn>)
```

where Table F.6 describes each of the parameters and their default values.

The PWL source generates a waveform that is constructed of linear segments providing either a voltage or a current. Pairs of values *tn* and *vn* specify the position of the waveform at vertices. Between vertices the waveform is specified as a waveform in the shape of a straight line connecting respective vertices. This piecewise linear waveform can be used to model arbitrary-shaped waveforms.

An example of a piecewise linear time-varying waveform voltage source is

```
VTRIAN 16 15 PWL(0 0 1 1 2 0 3 1 4 0)
```

This source produces a voltage drop from node 16 to node 15 that is initially at 0 V at $t = 0$. It rises linearly to 1 V at $t = 1$ s. Next, the voltage drops linearly to 0 at $t = 2$ s. This pattern repeats until $t = 4$ s, after which the voltage remains at 0 V. Using the PWL source, it is possible to approximate the behavior of many waveforms.

The .TRAN statement is used to indicate to SPICE that a transient analysis is to be performed. Further, it specifies the interval over which the analysis is to take place, the interval over which output values should be printed, and optionally a maximum value for the analysis time step. The form of the .TRAN statement is

```
.TRAN[/OP]<t step><t stop> [<t start>] [<step ceiling>]
```

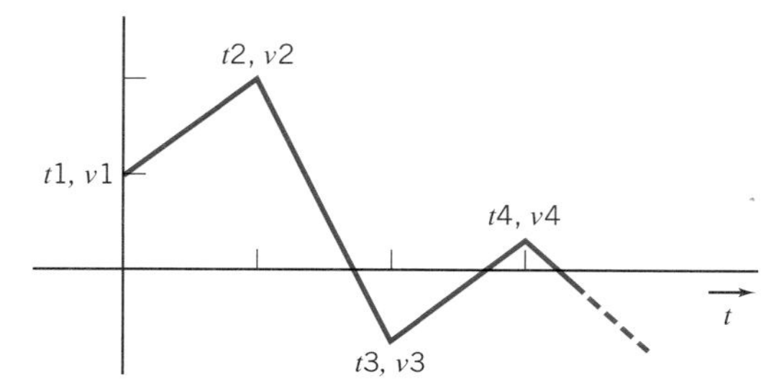

Figure F.17 Piecewise linear waveform.

Table F.6 **PWL Waveform Specifications**

Parameter	Default Value, Units
<*tn*> time at corner	none, seconds
<*vn*> value at corner	volts or amps

where the optional /OP parameter, when included, causes SPICE to produce an output listing containing the operating point analysis. Although this analysis is always done before a transient analysis, it is only printed when requested. Parameter *tstep* is the size of the interval between printed output points in either .PRINT tables or .PLOTs Parameter *tstop* specifies the time point at which the analysis (and the printing/plotting) should stop. Optional parameter *tstart* specifies the time point at which printing should start. All analysis begins at time zero, and by default, printing/plotting also begins at time zero. By specifying a value for *tstart* it is possible to avoid outputting unwanted portions of a table or plot. Optional parameter *step ceiling* specifies the maximum value that can be used as an internal time-step for the transient analysis. By default, most versions of SPICE set the maximum time step to be 1/50 of the duration of the analysis interval. In some cases, allowing the time step to become too large causes SPICE to skip over time intervals of interest, reporting little or no change in outputs over the interval. This problem can be circumvented by judicious use of the *step ceiling* parameter.

With the .TRAN statement *tstep* is the size of the interval between printed data points in either .PRINT tables or .PLOT plots. It should be noted that for any of the transient sources the value of the source at $t = 0$ is used by the SPICE program to calculate initial conditions at $t = 0$.

Output from a transient analysis can be generated in much the same way as in the .DC analysis. The .PRINT and .PLOT statements are used to generate the output of selected circuit values as a function of time. Data is output at intervals as specified in the .TRAN statement. The output statements take the following forms:

```
.PRINT TRAN <output values>
```

and

```
.PLOT TRAN <output values>
```

where the *output values* specify the node voltages or element currents in the same manner as described in the previous sections. The keyword TRAN is used to specify that SPICE should output the values calculated during the transient analysis for each of the output values contained in the subsequent list.

The .PROBE statement may also be used with PSpice to output the results of a transient analysis. As before, use of the simple .PROBE statement produces a data file containing the results of all of the circuit node voltages and element currents for use in later display operations. If a .TRAN analysis is the only analysis type specified, then only transient analysis outputs will be saved by the .PROBE statement. If any other types of analyses are included in the circuit description file, such as .DC or .AC data generated by these analyses will also be saved in the PROBE.DAT file.

Figure F.18 contains the SPICE circuit description for the source-free *RC* circuit shown schematically in Figure F.13. This circuit is used to model the behavior of the original source-free *RC* circuit, including providing the initial condition on the capacitor *C*. The current source used to supply the initial circuit condition is a time-varying source of type PULSE. The desired initial condition is to have 5 V across C1 at $t = 0$. With the resistor R1 equal to 1 kΩ in parallel with C1, it is necessary for ISTEP to provide 5 mA of current to R1. Examination of the element statement for ISTEP shows that the source will switch at $t = 0$ from providing 5 mA to 0 mA at $t = 0^+$. The .PLOT statement causes data generated by the transient analysis to be plotted.

Parameters for the .TRAN statement are chosen so that the events that take place in the circuit (i.e., the transients) can be easily viewed. If the *tstep* parameter is chosen to be too large, the transitions of interest will be seen compressed on the left side of the plot. However, if the *tstep* parameter is chosen to be too small, only a portion of the transition will be seen on the plot; most of the transition will be beyond the analysis interval. It may take one or two trial simulation runs to find the appropriate values for the .TRAN parameters. A reasonable rule of thumb is initially to choose the *tstep* parameter to be 1/10

```
RC Circuit with an Initial Voltage on C

****    CIRCUIT DESCRIPTION

**************************************************************

R1  1  0  1K
C1  1  0  1U
ISTEP 0  1  PULSE(5M  0  0  0  0); PROVIDES INITIAL CAPAC. VOLTAGE
.TRAN  .5M  10M
.PROBE
.PRINT TRAN V(1)
.PLOT TRAN V(1)
.END
```

Figure F.18 SPICE description of the source-free circuit of Figure F.13.

of the product of the smallest R and smallest C values in the circuit. Initially choose *tstop* to be 10 times the *tstep* value. For circuits containing resistors and inductors, choose *tstep* to be 1/10 of L/R, where L and R are the smallest inductor and the largest resistor values in the circuit. After viewing the output, adjustments can be made to these parameters so that the output will provide more or less detail as required.

The output for the RC circuit is shown in Figure F.19. As expected, the plot shows that the voltage of 5 V on the capacitor decays exponentially with a time constant equal to the product of R and C.

The RC circuit analysis example showed how initial conditions in a circuit could be set using standard circuit element models. SPICE contains built-in facilities to perform the same operations. There are three ways in which initial conditions may be set with these built-in facilities: using the .IC statement, using the .NODESET statement, or using the optional UIC parameter in the .TRAN and element statements.

The .IC statement, which stands for initial conditions, can be used to set specific nodes to initial voltages. The syntax for .IC is

```
.IC V(n1)=<node voltage>...
```

This list of node voltages will be used in the operating point analysis that precedes the transient analysis to set particular nodes to specified voltages. At the start of the transient analysis, the initial condition generator is removed from the circuit. Note that this statement can only be used with the transient analysis type and has no effect in a .DC or .AC analysis.

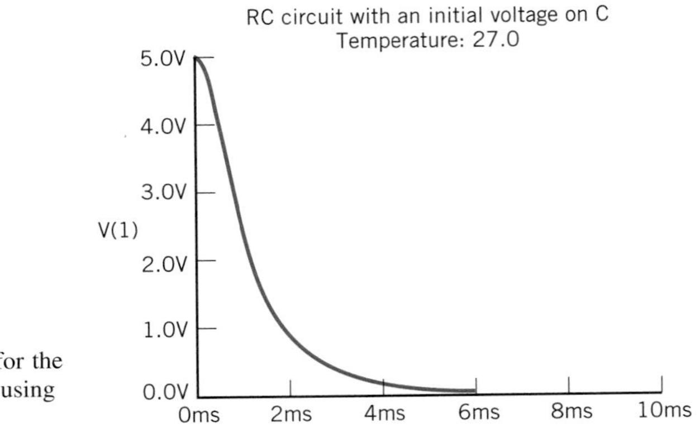

Figure F.19
Higher resolution output of V(1) for the source-free circuit of Figure F.13 using PROBE.

A different form of this statement is found in the .NODESET statement. The syntax is

```
.NODESET V(n1)=<node voltage>...
```

Statement .NODESET differs from .IC in that .NODESET only provides an initial node voltage guess for the simulator rather than forcing a node to a particular voltage. Effectively, .NODESET provides node voltages at the beginning of the operating point analysis. These values are used as the starting point in determining the actual operating point values. If no other forces in the circuit cause a change in these voltages, these become the starting point for the transient analysis. Node voltages specified in the .IC statement are effectively instantiated at the beginning of the operating point analysis and kept the same until the end of this analysis. Thus, conditions specified by the .IC statement will always be intact at the beginning of the transient analysis. This is not true when the .NODESET statement is used.

The standard form of the element statement for a capacitor or inductor is

```
<name><+ node><- node><element value><init. cond.>
```

For example, a capacitor C_1 of 1 mF can be represented as

```
C1 3 4 1M IC=1.5
```

where the initial condition is 1.5 V.

Individual elements may be set to some initial conditions using an optional parameter appended to the standard element statement. Two examples of initializing elements are given below for the C and L elements:

```
standard capacitor statement IC=<initial voltage>
standard inductor statement IC=<initial current>
```

where the units of voltage are volts and the units of current are amperes. In addition to specifying these initial conditions on the individual element statements, it is necessary to use the optional parameter UIC in the .TRAN statement. The following lines show the element statements for a capacitor and an inductor with initial conditions along with the .TRAN statement containing the necessary UIC parameter:

```
L1 10 11 2MH IC=1MA
C1 12 13 1U IC=2V
.TRAN[/OP]<tstep><tstop>[<tstart>][<step ceiling>]UIC
```

UIC causes SPICE to skip the operating point analysis and proceed to the transient analysis using the initial conditions specified on the element statements. We do not use the letter F following 1U in the line for C1, since F is the multiplier femto (Table F.2).

Example F.1

Determine and plot the capacitor voltage and current for the circuit shown in Figure F.20 when $R_1 = 1\ \text{k}\Omega$ and $C = 1\ \mu\text{F}$. The initial capacitor voltage is 2 V and $v_1 = 5$ V.

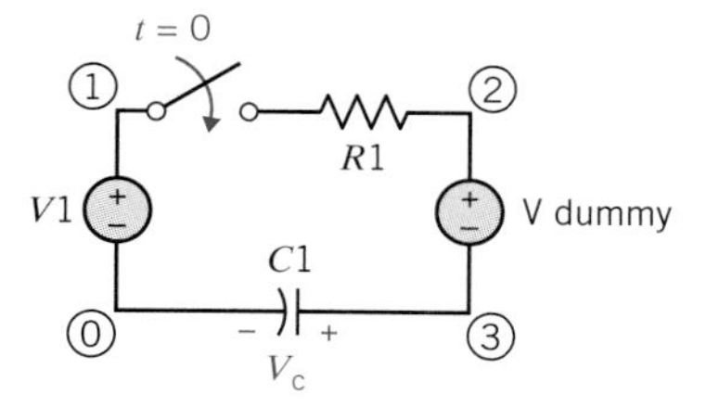

Figure F.20
Circuit diagram for a capacitor charging example.

Solution

Figure F.21 contains the circuit description file, and Figure F.22 and Figure F.23 contain PROBE plots of the voltage and current requested. Initial conditions are set using the .IC statement. The initial voltage can be seen as present at $t = 0$ with a corresponding initial current flow. After $t = 0$, the characteristic exponential increase in voltage with the corresponding current flow can be seen in the figures. We use Vdummy to obtain the capacitor current.

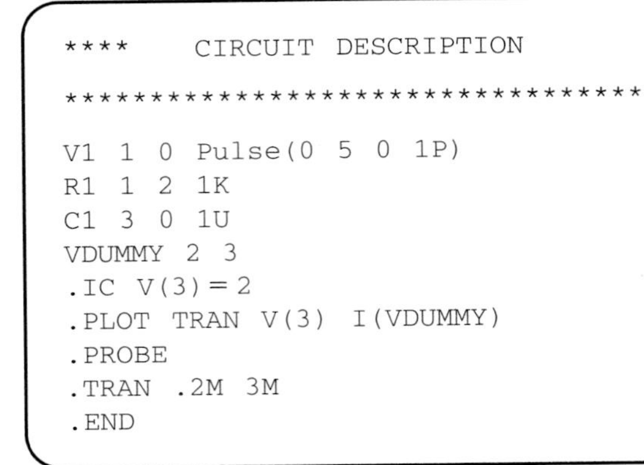

```
****     CIRCUIT DESCRIPTION

*********************************************************

V1  1  0  Pulse(0  5  0  1P)
R1  1  2  1K
C1  3  0  1U
VDUMMY  2  3
.IC  V(3)=2
.PLOT  TRAN  V(3)  I(VDUMMY)
.PROBE
.TRAN  .2M  3M
.END
```

Figure F.21 SPICE description for Example F.1.

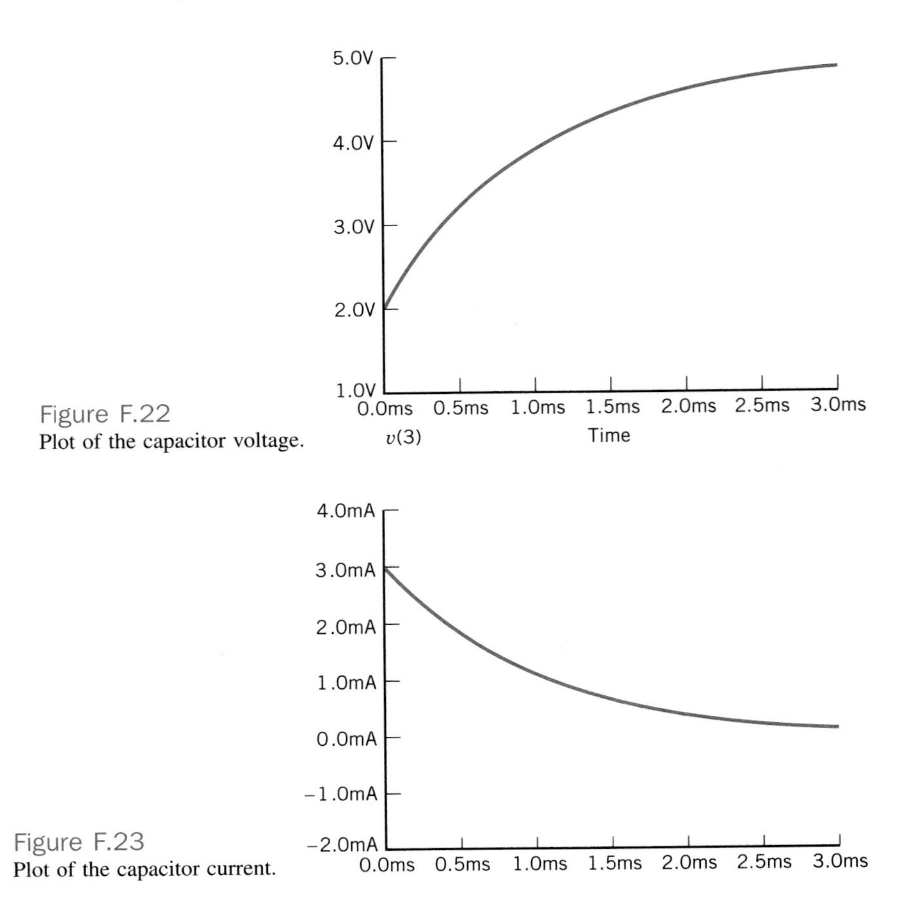

Figure F.22 Plot of the capacitor voltage.

Figure F.23 Plot of the capacitor current.

F.6 FREQUENCY DOMAIN ANALYSIS

Thus far we have covered two main types of analysis: dc analysis and time domain analysis. The third main type of analysis that is performed on a circuit is called frequency domain analysis. This type of analysis provides details on how circuits respond to stimuli of different frequencies. For SPICE, the term ac analysis is used to designate a frequency domain analysis. In an ac analysis, it is possible to observe the changes in magnitude and relative phase angle of node voltages and element currents. Unlike the transient analysis, which shows the circuit response as a function of time, the ac analysis contains no time-varying characteristics. In effect, the program assumes that each sinusoidal source has been applied for a very long time and thus calculates the steady-state response (amplitude and phase) for each frequency.

The **.**AC control statement specifies that the type of analysis SPICE is to perform is an ac analysis. The **.**AC statement further specifies the range of frequencies over which the analysis is to be performed and how the range of frequencies is to be swept. In the dc and transient analyses, the independent variables were swept linearly. That is, the increment between output points was a constant value. For the ac analysis, the independent variable frequency can be swept linearly, or it can also be swept logarithmically.

The general form of the **.**AC statement is

```
.AC <sweep type><n><start freq><end freq>
```

where n is related to the number of points to be output. Parameter *start freq* is the frequency at which the analysis is to begin while *end freq* specifies the last frequency point to be analyzed. The *sweep type* parameter may be chosen from the following list: LIN, OCT, DEC. Choosing LIN produces a linear range of frequencies as described above. For *sweep type* LIN, parameter n will be the total number of points output.

Using OCT, frequencies will be swept logarithmically from *start freq* to *end freq*. That is, the frequency will be increased in octaves. Parameter n indicates the number of points output per octave. An octave starts at *start freq* and ends at a frequency twice the starting frequency. The next octave starts where the last octave ends, and output continues until the *end freq* is reached. Sweep type DEC works in a similar manner. DEC specifies that the frequency will be increased in decades. The *start freq* is multiplied by ten to define the first decade. Parameter n indicates the number of output points per decade.

The **.**PRINT and **.**PLOT statements can be used to generate tabular and graphic output. For either output statement, the organization of the data on the x-axis will be determined by the sweep type specified on the **.**AC statement. Just as with the previous analysis types, an additional parameter in the **.**PRINT or **.**PLOT statement is used to indicate what output values for the ac analysis should be displayed. The form of the output statements is

```
.PRINT AC<ac output variables>
.PLOT AC<ac output variables>
```

Various formats for the data can be provided using the ac analysis. The particular value that is to be output is specified by a suffix placed on the output variables. A list of the possible suffixes and their meanings is contained in Table F.7. An example of the use of these output suffixes is

```
.PLOT AC VM(1)VM(2); plots the magnitude of the voltage at nodes 1 & 2
.PLOT AC IP(R3); plots the phase of the current in R3
```

As usual, the independent variable is printed or plotted in the first column of the output. Another thing to remember is that ac, dc, and transient analyses can be mixed in the

Table F.7 **AC Analysis Output Variable Suffixes**

Suffix	Value Output
no suffix	magnitude
M	magnitude
DB	magnitude in decibels (20 × log(value))
P	phase
R	real part
I	imaginary part

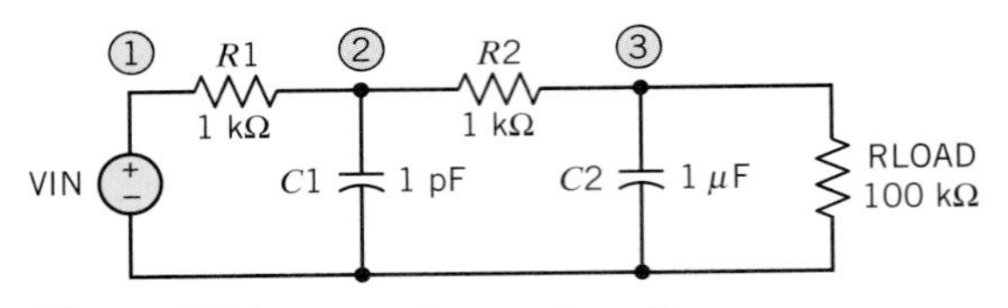

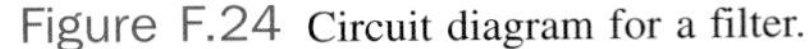
Figure F.24 Circuit diagram for a filter.

same simulation run. Outputs, of course, will be displayed in different sections of the listing.

Use of the .PROBE statement in PSpice is very much the same as for the previous two types of analysis. However, when an ac analysis is performed and the generic .PROBE (with no parameters) statement is used, even more data are saved in the PROBE.DAT file. This is because all of the different forms of value output are saved for possible use by the PROBE program.

Independent voltage sources for the ac analysis are specified in the same way as for the dc and transient analyses. Adding the keyword AC to an independent source statement indicates that the source is to be used in any subsequent .AC analysis. Two values may be specified as part of the ac source. The first value to follow the keyword AC is taken to be the magnitude of the source. If a second parameter is specified, it is taken to be the phase in degrees of the source. As discussed in the section on transient analysis sources, it is permissible to specify dc, transient, and ac values of a source all in the same statement.

An example circuit is shown in Figure F.24. This circuit acts as a low-pass filter. The SPICE description for the circuit is shown in Figure F.25. The magnitude of the source is set to 1 and the phase to zero. This is typical of single-source systems and allows the gain to be read directly from the output of the circuit. For more general circuits involving multiple sources, it may be necessary to specify different magnitudes and phases for the sources.

For the example, frequency is swept from 1 Hz to 100 kHz in octaves, with 2 output points per octave. Figure F.26 contains a .PROBE plot of the magnitude in decibels of the voltages at two nodes in the circuit. Examination of the figure shows that the output of

```
RC Filter
****    CIRCUIT DESCRIPTION
*********************************

VIN  1  0  AC  1  0
R1  1  2  1K
C1  2  0  1P
R2  2  3  1K
C2  3  0  1U
RLOAD  3  0  100K
.PROBE
.AC  OCT  12  1  100K
.PLOT  AC  V(2)  (0,1)  V(3)  (0,1)
.END
```

Figure F.25
SPICE description of the filter.

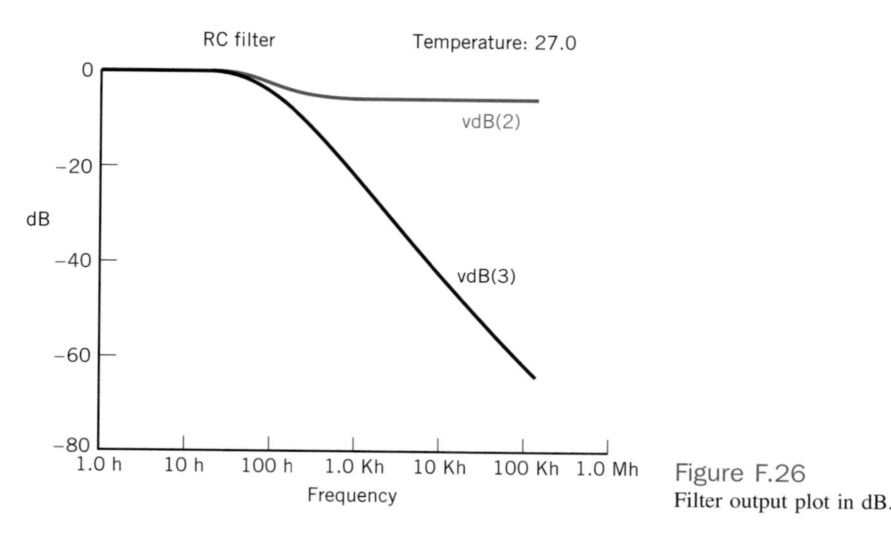

Figure F.26
Filter output plot in dB.

the first stage (i.e., the voltage at node 2) is reduced to 50% of the input value once the frequency reaches about 1 kHz. The output of the second stage (the voltage at node 3) is reduced by 40 dB at 10 kHz.

F.7 MUTUAL INDUCTANCE AND OP AMPS

SPICE has the built-in capability of modeling the behavior of a mutual inductance. In its simplest form, a mutual inductance model is given as a single statement similar to a simple inductance. Syntax for the simple form model mutual inductor statement is as follows:

```
K <name> L <name><L<name>><coupling value>
```

where the inductor *names* are those of the mutual inductors (there may be more than two), the *coupling value* is a constant k where $k = M/\sqrt{L_1 L_2}$, and M is the mutual inductance in henries. The value of k will always be $0 \leq k \leq 1$. A coupling value of $k = 0$ indicates that there is no coupling between inductors. The "dot" convention normally follows when specifying mutual inductances is also used with SPICE. A dot is assumed at the first node listed on each of the inductor statements specified on the mutual inductance statement. Thus, the current must flow into the dot of each inductor.

As an example of the use of the mutual inductance statement, consider the problem posed in Figure F.27. The problem is to determine the ratio of the phasor voltages $\mathbf{V}_2/\mathbf{V}_1$ when $\omega = 10$ rad/s.

To solve the problem, first the value of the coupling constant is calculated to be 0.707 and the frequency of interest is 1.592 Hz. These values are used in the circuit description as shown in Figure F.28. Notice that a large resistance RDUMMY is used to tie node 3 (and consequently node 4) to the rest of the circuit. SPICE necessitates that all elements have a dc path to ground and RDUMMY satisfies that need. Because of the size of the resistance, very little current flows in RDUMMY, and nodes 3 and 4 effectively are elec-

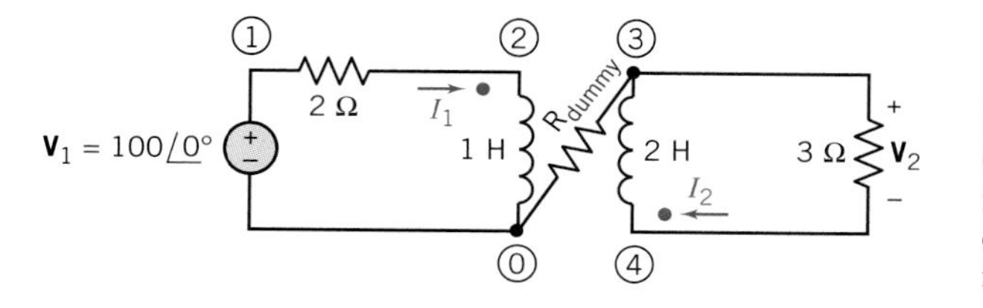

Figure F.27
Circuit diagram for the mutual inductance example. Note that the current flows into the dot of each inductor.

```
Simple Circuit with Mutual Inductance
****    CIRCUIT DESCRIPTION
**********************************************
v1  1  0  AC  100  0
R1  1  2  2
R2  3  4  3
L1  2  0  1
L2  4  3  2
KMUTUAL L1 L2 0.707
* COUPLING VALUE=M/(L1*L2)^   0.5
RDUMMY 3 0 100G
* RDUMMY keeps nodes 3 & 4 from floating
* no current in RDUMMY
.AC LIN 3 1.591 1.593
.PRINT AC VM(3,4) VP(3,4) I(RDUMMY)
.END
```

Figure F.28 Circuit description file for the mutual inductance example.

trically isolated from the rest of the circuit. The SPICE solution for the circuit at 1.592 Hz is shown in Figure F.29.

An ideal transformer can be modeled in a SPICE program by setting L_1 and L_2 large so that ωL_1 and ωL_2 are much greater than other impedances at the frequency of interest. Also, an ideal transformer requires $k = 1$. We will use $k = 0.999999$.

Circuits containing op amps can be readily analyzed using SPICE and the finite gain model of the op amp, which includes the input and output resistances and a VCVS with gain A.

We can compute the input resistance, output resistance, and the gain of an op amp circuit using the transfer function statement, which is

```
.TF <output variable><input variable>
```

For example, a finite gain model of an op amp is used in Figure F.30. We use the transfer function (.TF) statement as

```
.TF V(3) V(1)
```

to find the circuit gain V_3/V_1 and the circuit input resistance and output resistance. The circuit file contains the controlled source statement and calculates .TF at the .DC statement. When $R_1 = 10\ \text{k}\Omega$, $R_i = R_2 = 10\ \text{M}\Omega$, $R_3 = 50\ \text{k}\Omega$, $R_o = R_4 = 50\ \Omega$, and $A = 10^5$, the output printout is

```
Circuit with Mutual Inductance
****     AC ANALYSIS      TEMPERATURE=27.000 DEG C
*******************************************************
FREQ        VM(3,4)     VP(3,4)     I(RDUMMY)
1.591E+00   2.559E+01   1.267E+02   7.117E-15
1.592E+00   2.558E+01   1.267E+02   2.785E-15
1.593E+00   2.557E+01   1.266E+02   1.553E-15
```

Figure F.29 Solution for the mutual inductance example.

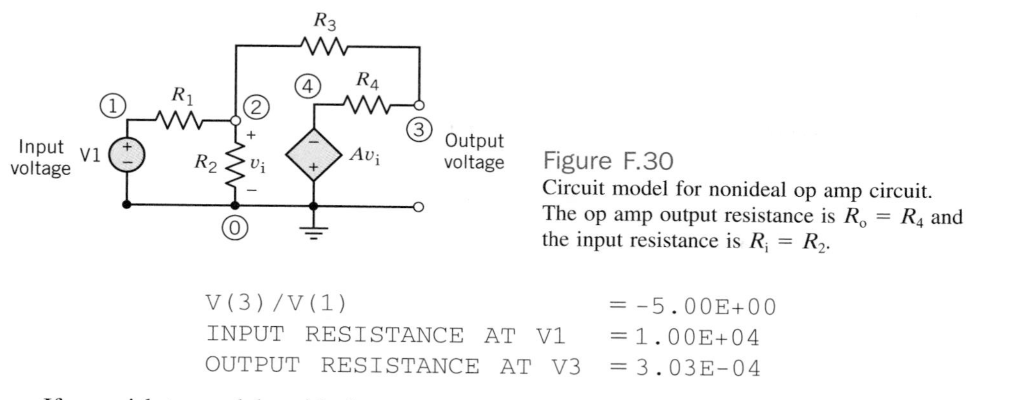

Figure F.30
Circuit model for nonideal op amp circuit. The op amp output resistance is $R_o = R_4$ and the input resistance is $R_i = R_2$.

```
V(3)/V(1)                    = -5.00E+00
INPUT RESISTANCE AT V1       =1.00E+04
OUTPUT RESISTANCE AT V3      =3.03E-04
```

If we wish to model an ideal op amp, we set R_i very large, $R_o = 0$, and A very large. For the preceding example, if we desire the ratio V(3)/V(1) for an ideal op amp, we set $R_4 = R_o = 0$, $R_i = R_2 = 100\ \text{M}\Omega$, and $A = 10^8$.

F.8 SUMMARY

This appendix explained and illustrated the utility of SPICE as an analytical and design tool. SPICE can be used to solve many complex circuits for their dc, ac, and transient responses.

A SPICE input file or *circuit file* contains (1) title and comment statements, (2) data statements, (3) solution control statements, (4) output control statements, and (5) an end statement. SPICE is capable of performing three main types of analysis. It can determine the behavior of selected output voltages with respect to changes in input voltages. This type of analysis is usually referred to as a dc analysis. A single-point dc analysis also determines what is called the bias-point characteristics, that is, the behavior of the circuit when only a dc voltage is applied to the circuit. In most cases, this is the starting point for either of two other types of analysis.

A second type of analysis that is usually required to determine fully a circuit's behavior is called a transient analysis. Transient analyses calculate circuit voltages and currents with respect to time. This assumes that there is a time-dependent stimulus that causes an effect on the rest of the circuit. To perform a time domain circuit analysis, SPICE first calculates the bias point and then calculates the circuit response to a time-dependent change in one or more voltage or current sources.

The third main type of analysis that SPICE can perform is called an ac analysis. This analysis type is also referred to as a sinusoidal steady-state analysis. Here, voltages and currents are calculated as functions of frequency. That is, output variable changes are calculated in response to changes in the frequency or phase of sinusoidal input voltage or current sources.

A summary to important statements used for SPICE analysis is provided in Table F.8.

Table F-8 **Summary of Important Statements**

Function	Statement
Element description	*<element name> <node> <node> <value>*
Operating point dc analysis	No statement required; occurs by default for a circuit list.
dc sweep of a voltage or current source	.DC ⟨*source name*⟩ ⟨*start value*⟩ ⟨*stop value*⟩ *<increment>*
Print output	.PRINT *<analysis type>* [*output list of variables*]
Plot output with low resolution and list numerical values	.PLOT⟨*analysis type*⟩ ⟨*output list*⟩
Higher resolution plot	.PROBE *<output list>*
Transient analysis for interval of time	.TRAN *<tstep> <tstop> <tstart>*
ac sweep of frequency; types are LIN, OCT, DEC	.AC⟨*sweep type*⟩ *<n>* ⟨*start freq*⟩ ⟨*end freq*⟩
n is related to number of points to be output	

APPENDIX G

Computer-Aided Anaylsis: DesignLab™

Several computer programs adapt SPICE for the personal computer. One such program is called DesignLab™. DesignLab™ includes Schematics™, PSpice®, and Probe®. Schematics™ is a graphical user interface, PSpice® implements SPICE, and Probe® is a graphical post processor. Schematics™ provides a convenient way to describe the simulation that PSpice® will perform. When appropriate, the results of a PSpice® simulation can be plotted using Probe®. PSpice® was developed by MicroSim Corporation. MicroSim Corporation merged with Orcad Corporation in 1998. An evaluation version of DesignLab™ can be downloaded at http://www.orcad.com/company/move.asp. Also, PSpice is included on the Orcad Demo CD. The Orcad Demo CD is available at http://www.orcad.com/Product/Schematic/evalrequest.asp.

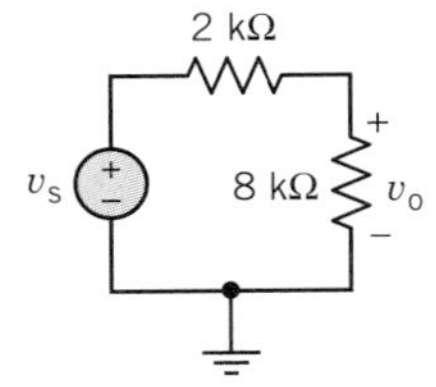

Figure G.1 A voltage divider.

Consider simulating the voltage divider shown in Figure G.1 using the evaluation version of DesignLab. First, Schematics is used to describe the simulation. This description has two parts. The circuit itself must be described, and the type of analysis must be specified.

Begin by starting Schematics. Figure G.2 shows the opening screen of Schematics. The top line of the screen shows the title of the program, Schematics, and a name of the simulation, "new1". A menu bar providing menus called File, Edit, Draw, . . . ,Window, Help is located under the title line. Two rows of buttons are located under the menu bar. A text

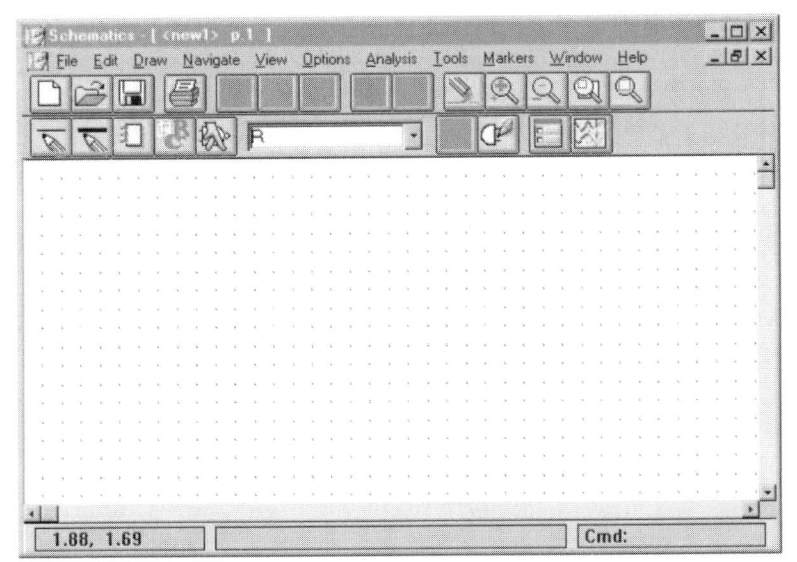

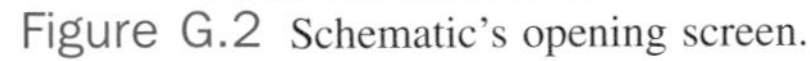

Figure G.2 Schematic's opening screen.

field is located in the lower row of buttons. That text field is used to specify the circuit elements that make up the circuit being simulated. A workspace is located beneath the lower row of buttons. The circuit being simulated is described by drawing it in this workspace. A line containing three message fields is located under the workspace. The center message field is of particular interest. This message field provides information about the Schematics screen. For example, move the mouse cursor to one of the buttons, say the right-most button in the bottom row. The center message field describes the function of the button. "Simulates the active schematic" is the function of the right-most button in the bottom row.

A circuit is described by drawing it in the workspace on the Schematic screen. Drawing the circuit shown in Figure G.1 requires that two resistors, a voltage source and some wires, be drawn. Also, the ground node must be labeled by drawing a ground symbol and connecting it to a node of the circuit. To draw a resistor, select **Draw/Place Part** from Schematics menus. (First, select Draw from the menu bar near the top of the Schematic screen. Next, select Place Part from among the options in the Draw menu.) Type R<Enter> to specify a resistor. (See Table G.1 for a list of the names that Schematics uses to identify various circuit elements.) A horizontal resistor will appear. Position this resistor using the mouse. Left click the mouse to fix the location of this resistor in the Schematics workspace. Another horizontal resistor will appear. Position this second resistor, left click the mouse to fix its location, then right click the mouse to indicate that no more resistors are needed. We want the second resistor to be vertical instead of horizontal. To rotate the second resistor, first select this resistor by left clicking the mouse on the resistor and then select **Edit/Rotate** from the Schematics menu bar. The Schematics screen is shown in Figure G.3.

The default value of a resistor is 1 kΩ. To change this value, select **Edit/Attributes.** The dialog box shown in Figure G.4 will pop up. Select the row labeled "Value" by left clicking the mouse on this row. Move the cursor to the field labeled "VALUE" and edit the resistor value.

Draw the voltage source by selecting **Draw/Place Parts** and then typing VDC<Enter>. Position and rotate the voltage source. Edit the value of the voltage.

Draw the wires to connect the circuit elements. Select **Draw/Wire.** Draw the wires using the mouse. Left click the mouse to mark the ends of the wire and to mark corners in the wire.

Label the ground node by selecting **Draw/Place Parts** and typing GND_ANALOG <Enter>. Position the ground symbol on the ground node.

Table G.1 **Circuit Element Names Used by MicroSim's Schemetics**

Name	Circuit Element
R	Resistor
L	Inductor
C	Capacitor
E	VCVS
G	VCCS
H	CCVS
F	CCCS
IDC, VDC	Constant current source or voltage source (DC analysis)
IAC, VAC	Sinusoidal current source or voltage source (AC analysis, frequency response, Bode plot)
IPULSE, VPULSE	Pulse current source or voltage source (Complete Response)
ISIN, VSIN	Sinusoidal current source or voltage source (Complete Response)
LM324	Op amp, including power supply terminals
GND_ANALOG	Ground

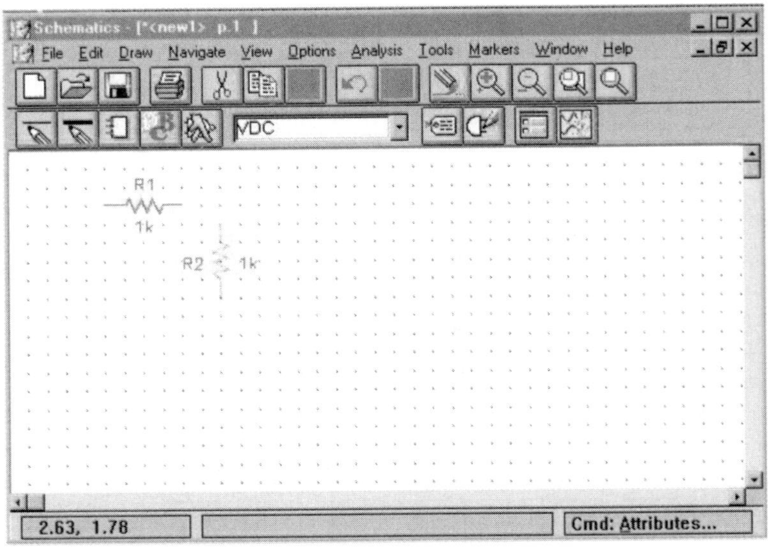

Figure G.3 Drawing the resistors.

In Figure G.1, the voltage across the 8 kΩ resistor is labeled as the output of the circuit. Since one node of the 8-kΩ resistor is the ground node, the voltage across the 8-kΩ resistor is equal to a node voltage. To label this node voltage in Schematics, select **Markers/ Mark Voltage.** A marker will appear. Position the marker using the mouse. Figure G.5 shows the Schematic screen after the circuit has been drawn and the output has labeled.

Now that the circuit has been drawn, the next step is to specify the type of analysis to be performed. Select **Analysis/Setup.** The Analysis Setup Dialog Box will pop up as shown in Figure G.6. Use the mouse to check boxes corresponding to the desired types of analysis. In Figure G.6 the box corresponding to DC Sweep has been checked. Next, left click the mouse on the button in the Analysis Setup Dialog Box corresponding to DC Sweep. The DC Sweep Dialog Box will pop up as shown in Figure G.7. Use the DC

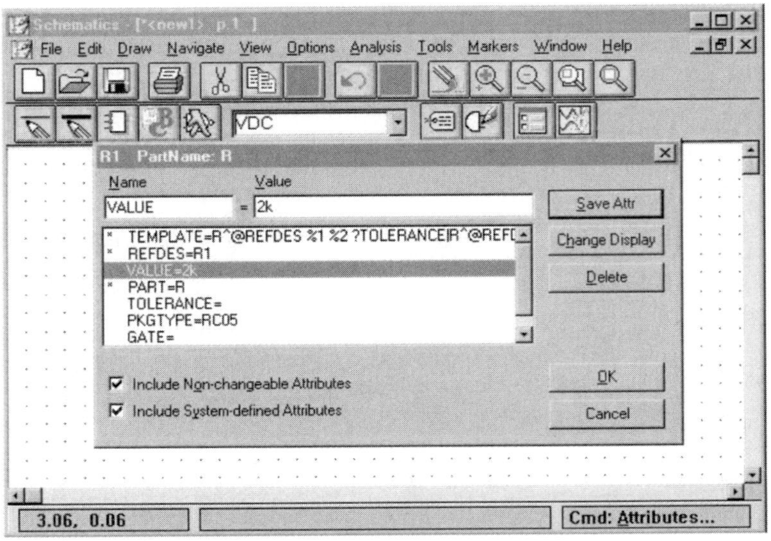

Figure G.4 Editing the properties of a circuit element.

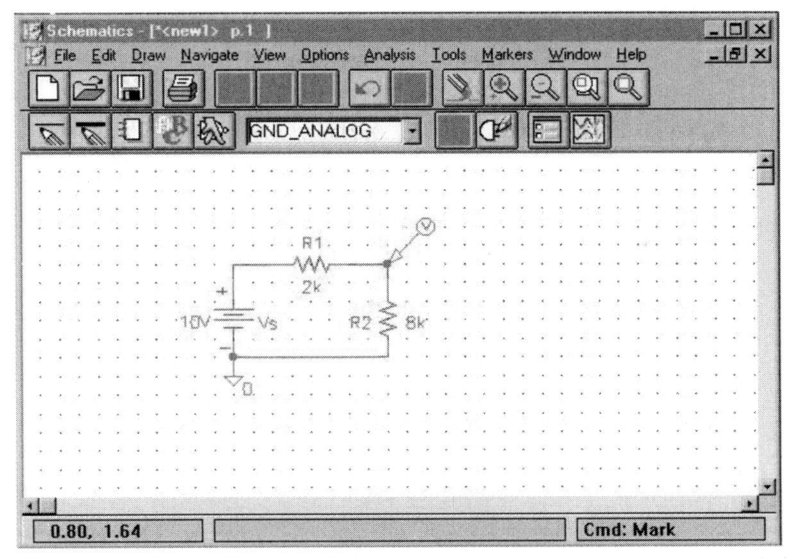

Figure G.5 A voltage divider. The top node of R2 is labeled as R2:2 and the bottom node is labeled as R2:1.

Sweep Dialog Box to describe the variation of a parameter of the circuit. In Figure G.7 the DC Sweep Dialog Box indicates that the voltage of the voltage source is to vary linearly from −10 V to 10 V in 1V increments. Close both the DC Sweep Dialog Box and the Analysis Setup Dialog Box.

Select **Analysis/Simulate** to execute PSpice. When PSpice finishes the simulation, it will automatically start Probe to display the simulation results. Figure G.8 shows that PSpice opened a window to report the progress of the simulation. After the simulation was finished, Probe opened a window to display the simulation results. The Probe screen in Figure G.8 shows a plot of V(R2:2) versus V_Vs. V_Vs denotes the voltage across the voltage source named Vs. V(R2:2) denotes the voltage at node 2 of the resistor named R2. (Schematics numbers the nodes of the resistor when it is placed in the schematics

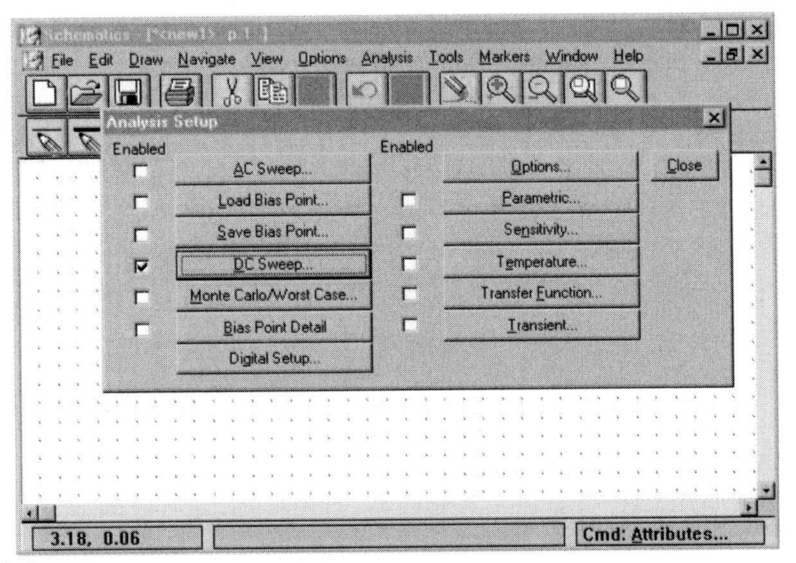

Figure G.6 Specifying the type of analysis.

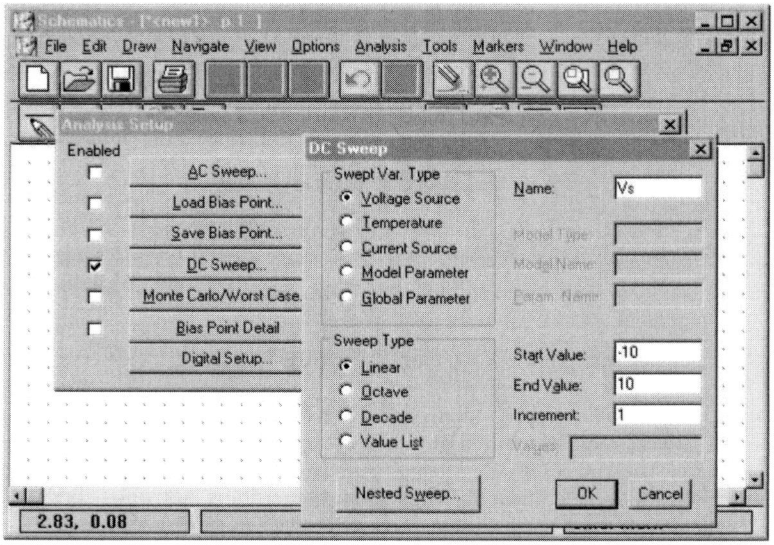

Figure G.7 The voltage source voltage varies from −10 V to 10 V.

workspace. In this case, the top node of the resistor is numbered R2:2 and the bottom node is numbered R2:1.)

Probe provides some tools that assist the user to read and display the values of points on the plot. Select **Tools/Cursor/Display** from the Probe menu bar to display Probe's cursors. Probe has two cursors. One cursor is positioned using the mouse or using the arrow keys. The other cursor is positioned by holding down the <Shift> key and using the arrow keys. The location of a cursor is marked by the intersection of a vertical dotted line and a horizontal dotted line. Probe also provides a table displaying the cursor positions. In Figure G.9, both Probe cursors have been positioned at points on the plot of V(R2:2) versus V_Vs. The table labeled "Probe Cursor" indicates that cursor A1 is located at the point V_Vs = −8.0000 V and V(R2:2) = −6.4000 V and cursor A2 is located at the point V_Vs = 8.0000 V and V(R2:2) = 6.4000 V.

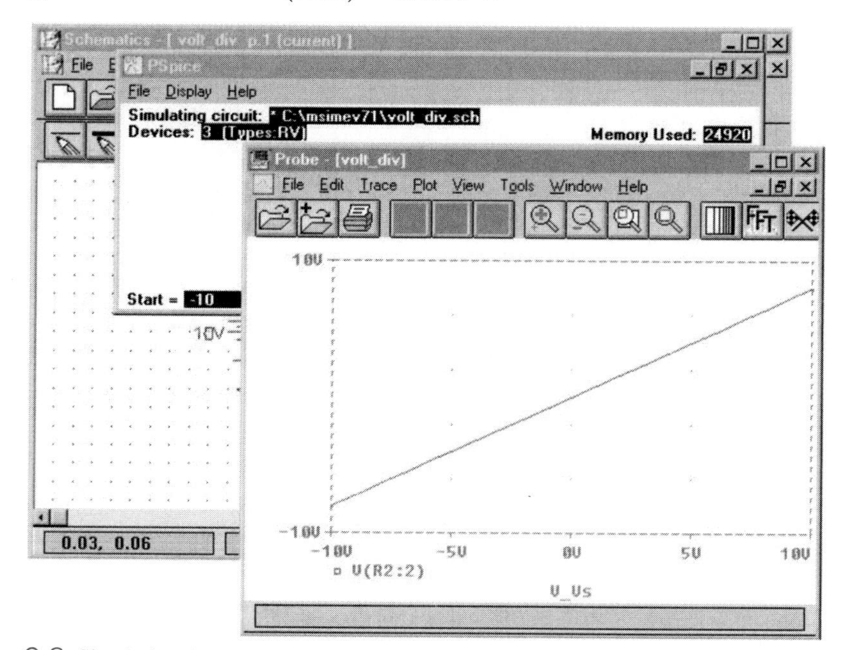

Figure G.8 Simulating the circuit using PSpice and displaying the results using Probe.

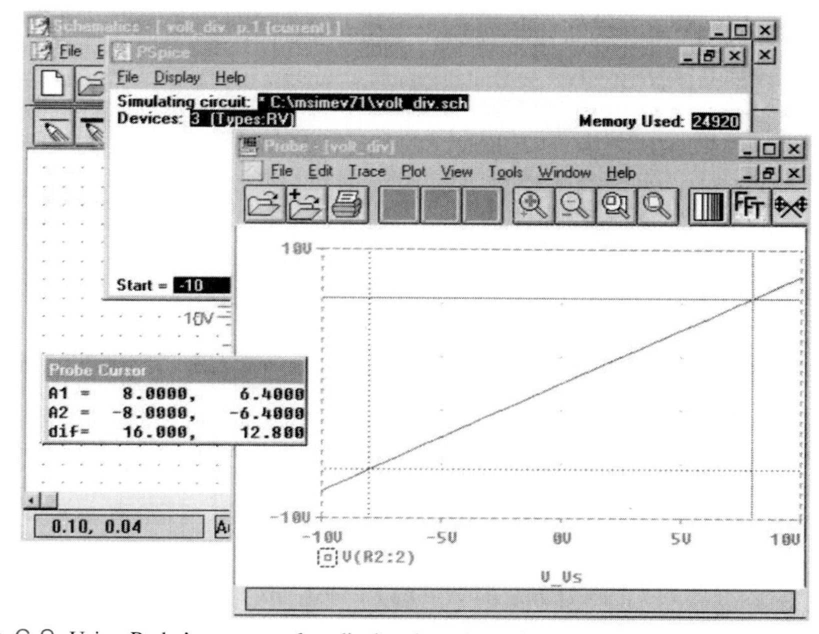

Figure G.9 Using Probe's cursor tool to display the values of points on the plot.

Let's change the simulation slightly. Close the Probe and PSpice windows to return the Schematics workspace. Suppose that we want to measure the current in R2 and the current in the voltage source instead of the voltage across R2. Left click the mouse on the voltage marker, then press <Delete> to remove this marker. Select **Markers/Mark Current into Pin** to obtain a marker for a current. Figure G.10 shows the voltage divider after current markers have been placed at a node of R2 and also at a node of Vs. A current marker can only be placed on a node of a device. That marker indicates the current that flows into that device from the rest of the circuit. In Figure G.10, one current marker indicates the current that is directed downward in the voltage source, while the other marker indicates the current that is directed downward in R2.

Select **Analysis/Simulate** to execute PSpice and then Probe. The Probe screen in Figure G.11 displays plots of $-I(R2)$ and $I(Vs)$ versus V_Vs. (By convention the current

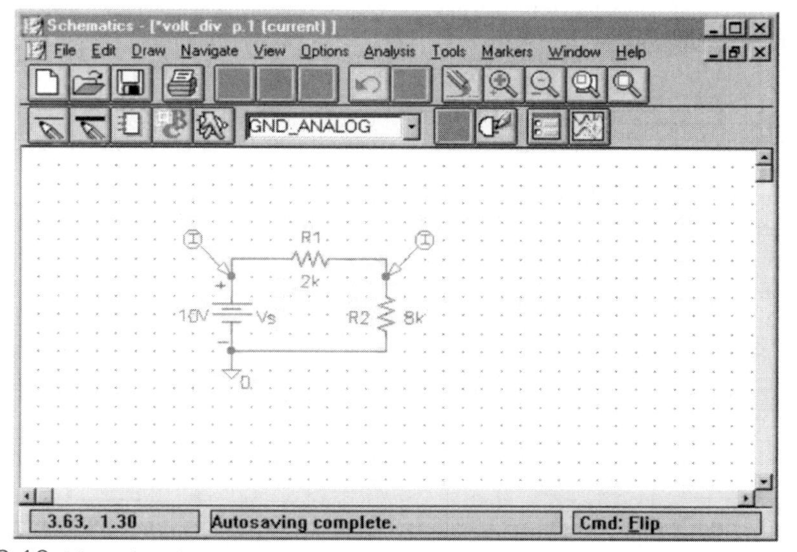

Figure G.10 Measuring element currents. Two current probes have been added to this circuit.

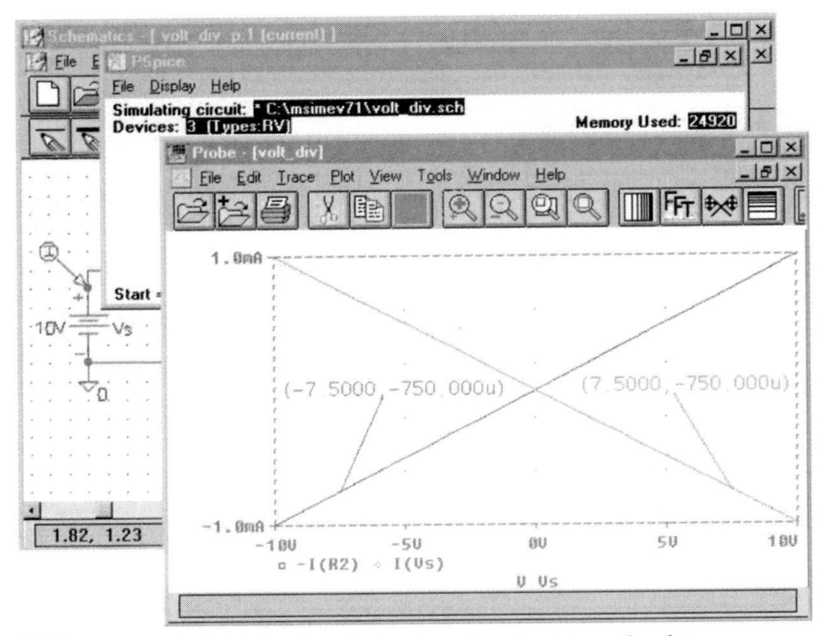

Figure G.11 Using Probe's label tool to display the value of points on the plot.

I(R2) is directed from node R2:1 of R2 toward node R2:2 of R2. Node R2:2 of R2 is the top node of R2, so I(R2) is directed upward in R2. The current indicated by the current marker at the top node of R2 is directed downward in R2, as is −I(R2). Ohm's law can be expressed in PSpice's notation as

$$\text{I(R2)} = \frac{\text{V(R2:1)} - \text{V(R2:2)}}{\text{R2}}$$

The current I(Vs) and the voltage Vs adhere to the passive convention.)

Figure G.11 illustrates Probe's Label tool. To display the value of a point on one of Probe's plots, first activate the cursor by selecting **Tools/Cursor/Display** and move the cursor to the point. (<Alt>→ moves the cursor from one plot to another.) Next, select **Tools/Label/Mark.** Finally, remove the cursor by selecting **Tools/Cursor/Display.** The label may be positioned using the mouse. The labels display (V_Vs, −I(R2)) and (V_Vs, I(Vs)). For example, −I(R2) = −750 μA when Vs = −7.5 V so

$$-750 \times 10^{-6} = -\text{I(R2)} = \frac{\text{Vs}}{\text{R1} + \text{R2}} = \frac{-7.5}{2 \times 10^3 + 8 \times 10^3} = -0.75 \times 10^{-3}$$

as expected, since −I(R2) is the current directed downward in R2.

Next, consider the first-order circuit shown in Figure G.12. Suppose we want to find the complete response of this circuit when the input voltage is a square wave. This circuit contains an op amp. The evaluation version of DesignLab™ includes a five-terminal model of the LM324 op amp. This model must be biased by adding power supplies. In Figure G.12, the dc voltage sources named VDC1 and VDC2 represent the power supplies that are used to bias the op amp.

The circuit in Figure G.12 requires two different types of voltage source. VDC+ and VDC− are dc sources, so the Schematic name is VDC (see Table G.1). In contrast, a pulse source is used to simulate a square wave. The Schematic name for a pulse source is VPULSE (Table G.1). Draw the voltage source by selecting **Draw/Place Parts** and then typing VPULSE<Enter>. Select **Edit/Attributes** to describe the square wave. The dialog box shown in Figure G.13 will pop up. In Figure G.13 this dialog box describes a square

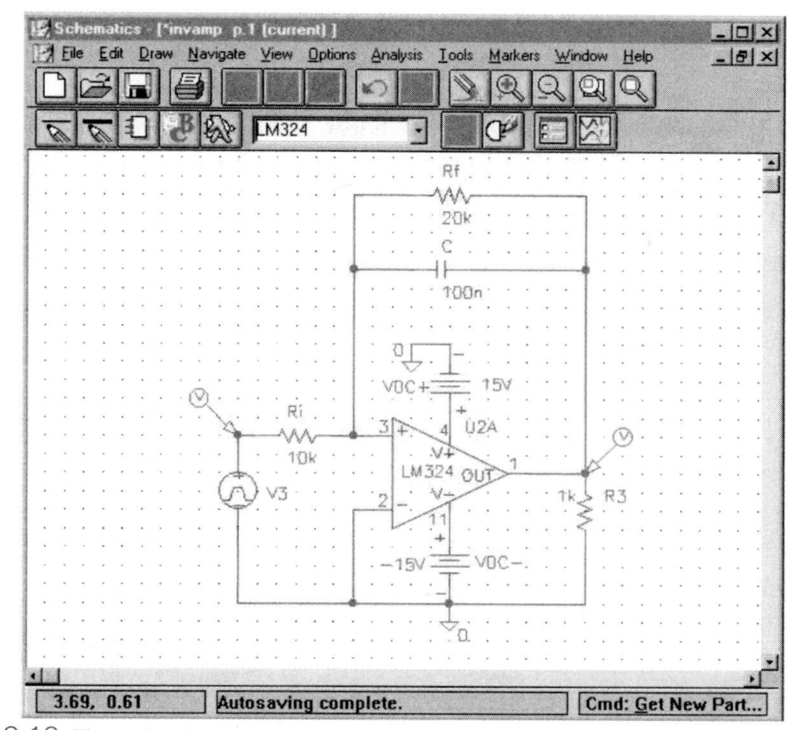

Figure G.12 First-order circuit that includes an LM324 op amp.

wave with a period equal to 20 ms and rise and fall times equal to 1 ms. The high and low voltage levels of the square wave are +5 V and −5 V, respectively.

Select **Analysis/Setup** to specify the type of analysis to be performed. The Analysis Setup Dialog Box will pop up as shown in Figure G.14. Select Transient Response using the check boxes. Left click the mouse on the Transient . . . button in the Analysis Setup Dialog Box. The Transient Dialog Box will pop up as shown in Figure G.14. Use the Transient Dialog Box to specify the duration of the simulation.

Select **Analysis/Simulate** to invoke PSpice and then Probe. The Probe screen in Figure G.15 displays plots of both V(R3:1), the output voltage, and V(V3:+), the input volt-

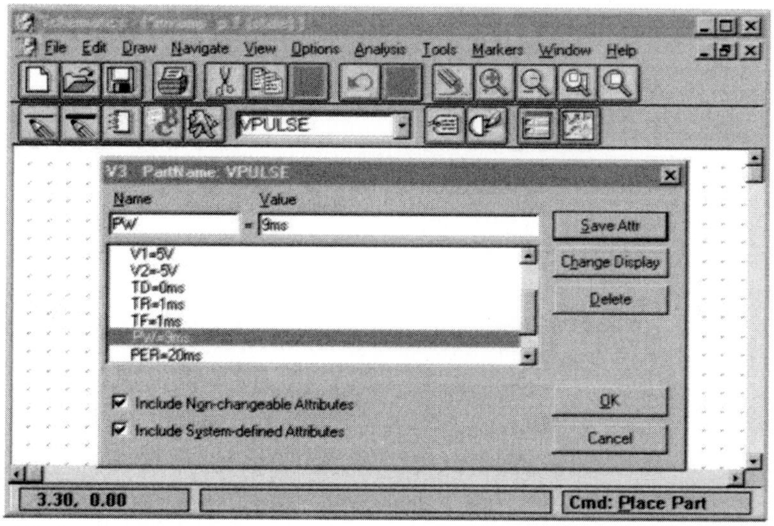

Figure G.13 Specifying a square wave input.

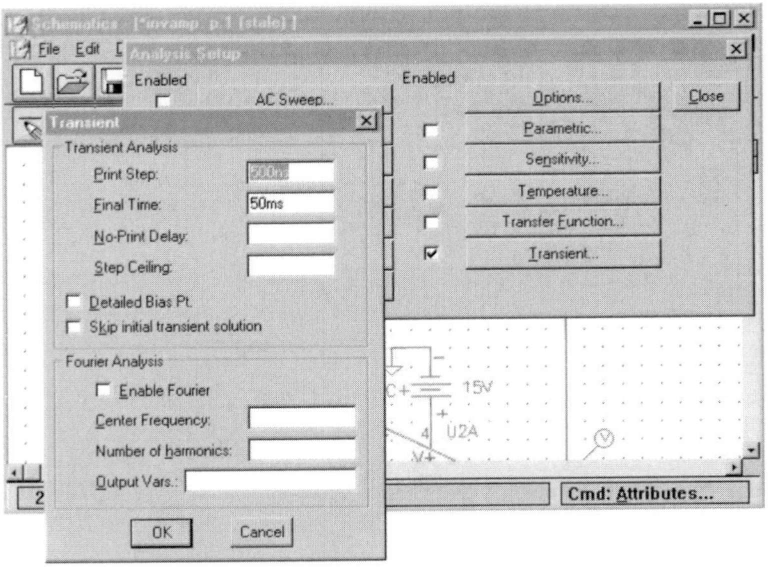

Figure G.14 Specifying a transient response.

age, versus time. A point on the plot of the output voltage has been labeled, showing that the value of the output voltage is 9.585 V when the time is 10.219 ms.

Suppose, instead, we want to obtain the magnitude Bode plot of the circuit shown in Figure G.12. The simulation must be changed in two ways. The voltage source must be changed, and the type of analysis must be changed. Let's change the source first. Remove the old voltage source by selecting and deleting it. Draw an ac voltage source by selecting **Draw/Place Parts** and then typing VAC<Enter> (see Table G.1). Select **Edit/Attributes** to describe the ac source. Specify a magnitude equal to 1 V and an angle equal to 0°. Using Schematic's notation, the network function is related to the input and output voltages by

$$20 \log_{10} \mathbf{H}(\omega) = 20 \log_{10}\left(\frac{\text{V(R3:1)}}{\text{V(V3:+)}}\right) = 20 \log_{10} \text{V(R3:1)} - 20 \log_{10} \text{V(V3:+)}$$

where, in this context, V(R3:1) denotes the phasor corresponding to the output voltage and V(V3:+1) denotes the phasor corresponding to the input voltage. We have specified

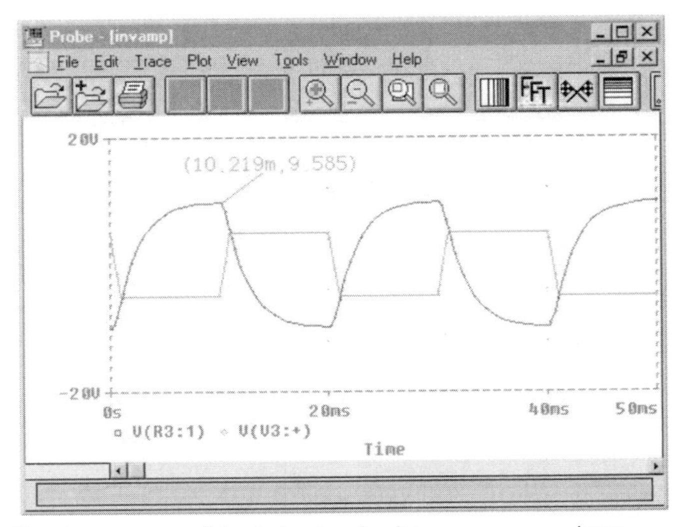

Figure G.15 Transient response of the first order circuit to a square wave input.

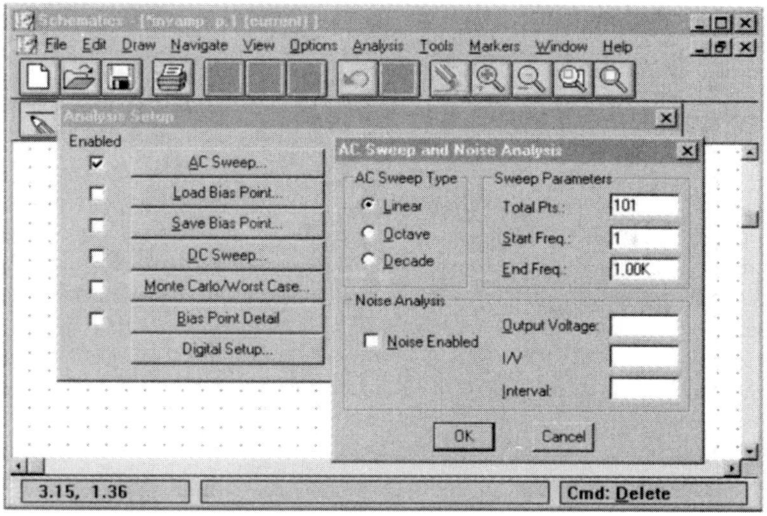

Figure G.16 Specifying a frequency response.

the input voltage to have a magnitude equal to 1 V and an angle equal to 0°, so V(V3:+) = $1\angle 0°$ and $20\log_{10}$ V(V3:+) = 0 dB. Thus

$$20\log_{10}\mathbf{H}(\omega) = 20\log_{10}\text{V(R3:1)} - 0 = \text{VdB(R3:1)}$$

where VdB(R3:1) is Probe's notation for $20\log_{10}$ V(R3:+). A voltage marker causes Probe to plot V(R3:1) rather than VdB(R3:1). Since that's not what we want, delete the voltage markers from the circuit drawn in the Schematics workspace.

Select **Analysis/Setup** to specify the type of analysis to be performed. The Analysis Setup Dialog Box will pop up as shown in Figure G.16. Select AC Sweep using the check boxes. Left click the mouse on the AC Sweep button in the Analysis Setup Dialog Box. The AC Sweep and Noise Analysis Dialog Box will pop up as shown in Figure G.16. Use this Dialog Box to specify the frequency range of the simulation.

Select **Analysis/Simulate** to invoke PSpice and then Probe. Select **Trace/Add** . . . from the Probe menus and type VdB(R3:1) to obtain the Bode plot shown in Figure G.17.

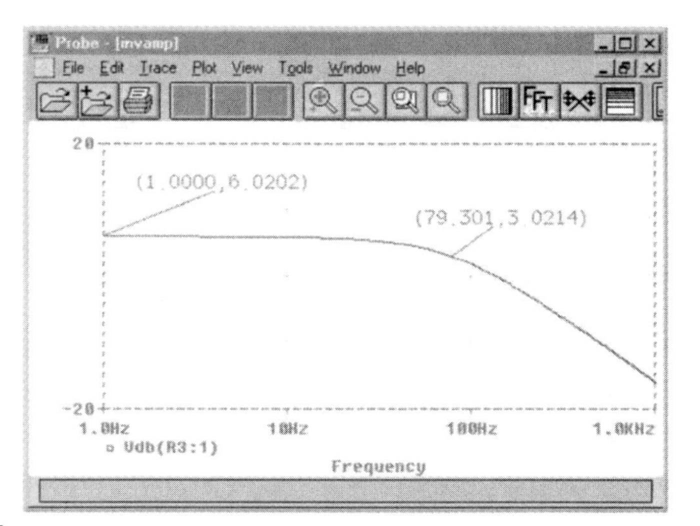

Figure G.17 Frequency response of the first order circuit.

References

1. Adler, Jerry, "Another Bright Idea," *Newsweek,* June 15, 1992, p. 67.
2. Agnew, J., "Simulating Audio Transducers with Pspice," *Electronic Design,* November 7, 1991, pp. 45–59.
3. Atherton, W. A., *From Compass to Computer,* San Francisco Press, San Francisco, 1984.
4. Bak, David J., "Stethoscope Allows Electronic Amplification," *Design News,* December 15, 1986, pp. 50–51.
5. Barnes, R., and Wong, K. T., "Unbalanced and Harmonic Studies for the Channel Tunnel Railway System," *IEE Proceedings,* March 1991, pp. 41–50.
6. Becker, J., "*RC* High-Pass Filter for Active Antennas," *Elektor Electronics,* February 1992, pp. 24–27.
7. Bernstein, Theodore, "Electrical Shock Hazards," *IEEE Transactions on Education,* August 1991, pp. 216–222.
8. Bracewell, Ronald N., "The Fourier Transform," *Scientific American,* June 1989, pp. 86–95.
9. Brown, S. F., "Predicting Earthquakes," *Popular Science,* June 1989, pp. 124–125.
10. Butterworth, S. "On the Theory of Filters," *Wireless World,* Vol. 7, October 1930, pp. 536–541.
11. Cantor, G., *Michael Faraday,* St. Martin's Press, New York, 1991.
12. Clark, W. A., *Edison: The Man Who Made the Future,* Putnam, New York, 1977.
13. Coltman, John W., "The Transformer," *Scientific American,* January 1988, pp. 86–95.
14. Doebelin, E. O., *Measurement Systems,* McGraw-Hill, New York, 1966.
15. Dordick, Herbert S., *Understanding Modern Telecommunications,* McGraw-Hill, New York, 1986.
16. Dorf, Richard, *The Electrical Engineering Handbook,* CRC Press, 1998.
17. Dorf, Richard C., *Modern Control Systems,* 7th Edition, Addison-Wesley, Reading, Mass., 1995.
18. Dorf, Richard C., *Technology, Society and Man,* Boyd and Fraser, San Francisco, 1974.
19. Dunsheath, P., *A History of Electrical Engineering,* Faber and Faber, London, 1962.
20. Edelson, Edward, "Solar Cell Update," *Popular Science,* June 1992, pp. 95–99.
21. Feldmann, Peter, and Rohrer, Ronald, "Proof of the Number of Independent Kirchhoff Equations in an Electrical Circuit," *IEEE Trans. Circuits,* July 1991, pp. 681–683.
22. Fourier, J. B. J., *Théorie Analytique de la Chaleur,* 1822, Dover Publishing, New York, 1985.
23. Frank, Peter H., "New Use of Electricity Seen in Medicine," *New York Times,* July 29, 1987, p. 28.
24. Gardner, Dana, "The Walking Piano," *Design News,* December 11, 1988, pp. 60–65.
25. Garnett, G. H., "A High-Resolution, Multichannel Digital-to-Analog Converter," *Hewlett-Packard Journal,* February 1992, pp. 48–52.
26. Graeme, Jerald, "Feedback Linearizes Current Source," *Electronic Design,* January 23, 1992, pp. 69–70.
27. Graham, Dunstan, *Analysis of Nonlinear Control Systems,* Dover Publishing, New York, 1971.
28. Jurgen, Ronald, "Electronic Handgun Trigger Proposed," *IEEE Institute,* February 1989, p. 5.
29. Kranzberg, M., and Pursell, C. W., *Technology in Western Civilization,* Oxford Univ. Press, London/New York, Vol. I, 1967.
30. Kullstam, Per A., "Heaviside's Operational Calculus," *IEEE Transactions on Education,* May 1991, pp. 155–166.
31. Lamarre, Leslie, "Problems with Power Quality," *EPRI Journal,* August 1991, pp. 14–23.
32. Lenz, James E., "A Review of Magnetic Sensors," *Proc. IEEE,* June 1990, pp. 973–989.
33. Lewis, Raymond, "A Compensated Accelerometer," *IEEE Transactions on Vehicular Technology,* August 1988, pp. 174–178.
34. Loeb, Gerald E., "The Functional Replacement of the Ear," *Scientific American,* February 1985, pp. 104–108.
35. Mackay, Lionel, "Rural Electrification in Nepal," *Power Engineering Journal,* September 1990, pp. 223–231.
36. *Mathcad User's Guide,* MathSoft Inc., Cambridge, MA, 1991.

37. McCarty, Lyle H., "Catheter Clears Coronary Arteries," *Design News,* September 23, 1991, pp. 88–92.
38. McMahon, A. M., *The Making of a Profession: A Century of Electrical Engineering in America,* IEEE Press, New York, 1984.
39. Metzger, T. L., "Electric Rockets," *Discover,* March 1989, pp. 18–22.
40. Meyer, H. W., *A History of Electricity and Magnetism,* Burndy Library, Norwalk, Conn., 1972.
41. Nahin, Paul J., "Behind the Laplace Transform," *IEEE Spectrum,* March 1991, pp. 60.
42. Nahin, Paul J., "Oliver Heaviside," *Scientific American,* June 1990, pp. 122–129.
43. Nye, David E., *Electrifying America,* MIT Press, Cambridge, MA, 1991.
44. Petroski, Henry, "Images of an Engineer," *American Scientist,* August 1991, pp. 300–303.
45. Pierce, John R., and Noll, A. M., "Signals: The Science of Telecommunications," *Scientific American Library,* W. H. Freeman, San Francisco, 1990.
46. Ruffell, J., "Switch-Mode Power Supply," *Elektor Electronics,* February 1992, pp. 62–66.
47. Sallen, R. P., and Key, E. L., "A Practical Method of Designing RC Active Filters," *IRE Transactions on Circuit Theory,* Vol. CT-2, March 1955, pp. 74–85.
48. Smith, E. D., "Electric Shark Barrier," *Power Engineering Journal,* July 1991, pp. 167–177.
49. Svoboda, J. A., "Elab, A Circuit Analysis Program for Engineering Education," *Computer Applications in Engineering Education,* Vol. 5, No. 2, 1997, pp. 135–149.
50. Svoboda, James A., "Using Spreadsheets in Introductory Electrical Engineering Courses," *IEEE Trans. Education,* November 1992, pp. 16–21.
51. Thomas, J. M., *Michael Faraday and the Royal Institution,* American Institute of Physics, New York, 1991.
52. Trotter, D. M., "Capacitors," *Scientific American,* Vol. 259, No. 1, 1988, pp. 86–90.
53. Williams, E. R., "The Electrification of Thunderstorms," *Scientific American,* November 1988, pp. 88–99.
54. Williams, L. P. "André-Marie Ampère," *Scientific American,* January 1989, pp. 90–97.
55. Wright, A., "Construction and Application of Electric Fuses," *Power Engineering Journal,* Vol. 4, No. 3, 1990, pp. 141–148.
56. Zorpette, Glenn, "Utilities Get Serious About Efficiency," *IEEE Spectrum,* May 1991, pp. 42–43.

Index

Historical Vignettes

Photograph of Thomas Edison from the historical vignette in Chapter 2.

Each chapter includes a section devoted to the history of electrical engineering. These sections introduce the pioneers of electrical engineering. Many of the names of the units for electrical quantities, e.g. Hertz, Ohms, Henries, and Amperes, honor these pioneers. The historical vignettes provide a context for the study of electric circuits by showing how the accomplishments of these pioneers contributed to current engineering practice.

Verification Examples and Problems

Engineers are frequently called upon to verify that a solution to a problem is indeed correct. For example, proposed solutions to design problems must be checked to confirm that all of the specifications have been satisfied. In addition, computer output must be reviewed to guard against data entry errors, and claims made by vendors must be examined critically.

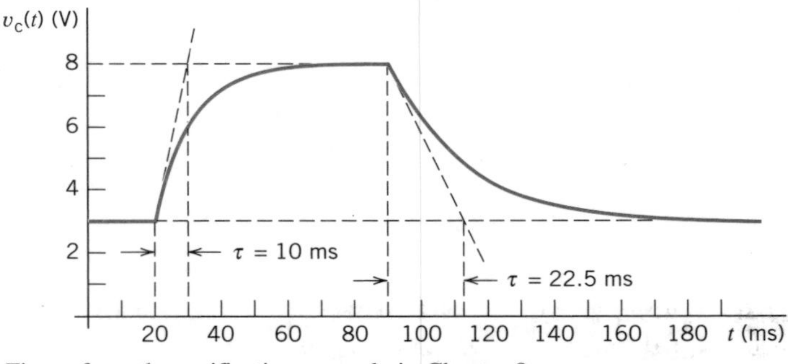

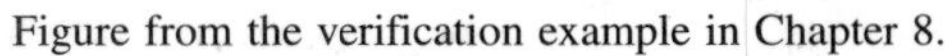

Figure from the verification example in Chapter 8.

Engineering students are also asked to verify the correctness of their work. For example, occasionally just a little time remains at the end of an exam. It is useful to be able to identify quickly those solutions that need more work.

This text includes some examples, called verification examples, that illustrate techniques useful for checking the solutions of the particular problems discussed in that chapter. At the end of each chapter are some problems, called verification problems, presented so that the reader will have an opportunity to practice these techniques.